Genetics Fundamentals Notes

Debasish Kar • Sagartirtha Sarkar
Editors

Genetics Fundamentals Notes

Editors
Debasish Kar
Department of Biotechnology
M S Ramaiah University of Applied Science
Bengaluru, Karnataka, India

Sagartirtha Sarkar
Department of Zoology
University of Calcutta
Kolkata, West Bengal, India

ISBN 978-981-16-7043-5 ISBN 978-981-16-7041-1 (eBook)
https://doi.org/10.1007/978-981-16-7041-1

This Springer imprint is published by the registered company Springer Nature Singapore Pte Ltd.
The registered company address is: 152 Beach Road, #21-01/04 Gateway East, Singapore 189721, Singapore

Contents

About the Editors

Debasish Kar is an Assistant Professor in the Department of Biotechnology at Ramaiah University of Applied Sciences, Bangalore. He pursued his MSc in Biotechnology from Utkal University, Bhubaneswar, India, followed by an MTech in Biotechnology from West Bengal University of Technology, Kolkata, India. Later, he completed his PhD in Biotechnology from IIT Kharagpur, India. Dr. Kar has several publications in international peer-reviewed journals. Dr. Kar has also published several books for competitive entrance examinations, viz. CSIR-NET, GATE, DBT-JRF, and ICMR-JRF, in the domain of life sciences and biotechnology.

Sagartirtha Sarkar is a Professor in the Genetics and Molecular Cardiology laboratory, Department of Zoology, Calcutta University, India. He completed his PhD from the University of Calcutta, India. He has worked at the Cleveland Clinic Foundation, Cleveland, Ohio, USA, as a postdoctoral fellow. Dr. Sarkar has also served as visiting scientist in the Department of Medicine, University of California, San Diego, CA, USA. He has more than 40 publications in peer-reviewed journals. He has supervised 18 doctoral and 5 MPhil students.

Part I

Classical Genetics

Fundamentals of Genetics

1

Shweta Panchal

Genetics is a dynamic and rapidly advancing field of biology that has found applications in all dimensions of human life. Breakthrough genetic discoveries are reported regularly in all major newspapers. Genetic tools are developing at a rate that has never been witnessed before in history. While understanding genetics and applying it is mostly driven for human and environmental betterment, it can also be misused and can raise ethical and moral concerns. For example, first half of the year 2019 witnessed two important discoveries that were reported by all major newspapers across the globe. One of the discoveries was genetic modification of a fungus, *Metarhizium*, to produce a spider toxin that kills malaria-causing mosquitoes in large numbers (Lovett et al. 2019). Malaria, being the biggest contributor of mortality worldwide, seeks rapid ways for disease prevention, especially in sub-Saharan countries. Studies like these help us to get a step closer to this goal clearly benefiting human society enormously. The other research news that got worldwide attention was the announcement of the birth of the world's first gene-edited babies by Chinese scientist He Jiankui (Cyranoski 2019). He was condemned by the scientific community for being irresponsible and reckless. Jiankui used a latest technology for genome editing called as CRISPR-Cas9 to edit specific genes in human embryos and allowed the babies to be born. The condemnation is mainly because scientists still don't know everything about this technology. CRISPR-Cas9 is based on a mechanism that some bacteria use to defend themselves against viruses by using an enzyme Cas9. This enzyme can be directed to make cuts in the DNA by providing a small RNA sequence for your site of interest in the DNA. This technology has revolutionized genetic manipulation in research which is exemplified by the 2020 Nobel Prize in Chemistry awarded to the discoverers of this powerful tool, Emmanuelle Charpentier and Jennifer Doudna. However, it has been shown that there are off-target effects too, meaning that the enzyme can cut into other sites in the

S. Panchal (✉)
Jawaharlal Nehru Centre for Advanced Scientific Research, Bangalore, India

D. Kar, S. Sarkar (eds.), *Genetics Fundamentals Notes*,
https://doi.org/10.1007/978-981-16-7041-1_1

DNA and can potentially inactivate a gene essential for proper functioning of the cell, like a tumour suppressor gene and thus lead to health problems. So, while most of the discoveries made in genetics have been extensively useful for humans and the environment, we are in an age where such scientific challenges need to be addressed by assembling an international community of experts while taking into consideration public opinions. It is indeed an interesting era for genetics.

While genetics has been used by civilizations for thousands of years for selective animal and plant breeding as well as for fermentations used in brewing and baking, the journey of understanding the underlying mechanisms in genetics begins in nineteenth century with prominent scientific giants in the field, Gregor Mendel and Charles Darwin. The understanding of inheritance, genetic changes and their role in evolution is critical to understanding all life.

Some fundamental concepts as a prelude to the understanding of genetics:

- The tree of life indicates that all organisms can be divided into three domains: Archaea, Bacteria and Eukarya. Bacteria and Archaea are prokaryotes, meaning that their cells lack a nuclear membrane and possess no membrane-bound organelles. All other organisms are eukaryotes belonging to Eukarya which have more complex cellular organization with membrane-bound organelles like mitochondria and chloroplasts, as well as membrane-bound nucleus.
- The gene is the basic unit of heredity. A gene is a unit of information in the DNA that encodes for a functional product and is involved in the expression of a trait or phenotype or characteristic.
- Genes occur in multiple forms that are called alleles. For example, a gene for the height of a pea plant can exist as an allele for tall plants or another allele for short plants.
- Genes confer phenotypes. Genes are inherited, and expression of these genes along with environmental effects determines the trait or the phenotype. The genetic information of an organism is called the genotype and the expressed trait is called the phenotype.
- The macromolecules of the cell that carry genetic information are deoxyribonucleic acid (DNA) or ribonucleic acid (RNA). Nucleic acids are polymers of repeating units called as nucleotides. A nucleotide is made of a sugar, a phosphate, and a nitrogen base. The nitrogen bases are of four types: adenine (A), cytosine (C), guanine (G) and thymine (T). In RNA, thymine is replaced with uracil (U). The sequence of these bases determines the DNA sequence or the genetic information of an organism. DNA is made of two complementary double strands of nucleotides.
- Genetic information is transmitted from DNA to mRNA to protein.
- Genes are located on the chromosomes. Large strings of DNA sequence are compacted in the form of a chromosome with the help of DNA-associated proteins. Each species possesses a specific number of chromosomes. For example, humans have 46 chromosomes, while bacteria possess only one chromosome.

- During mitosis and meiosis in the process of cell division, DNA replication takes place, the chromosomes separate, and the genomic content is divided equally into two daughter cells. Mitosis occurs in the somatic cells and meiosis occurs in the sex cells to form gametes.
- Mutations are permanent changes in the sequence of the DNA that can be passed on the future generations.
- Evolution is a process of genetic changes over a period of time in a population.

1.1 From Mendel Theory to Discovery of DNA

Genetics was being applied by our ancestors even before scientists started thinking about heredity. Human civilization was possible owing to application of genetics to domesticate plants and animals. The domesticated dog of today (*Canis domesticus*) is a result of years of artificial selection using controlled mating of parents of ancestral wolves (*Canis lupus*). Same is true for plants that have been selected for favourable traits from their wild ancestors. Thus, years of selective plant and animal breeding had indicated that useful traits can be selected by controlled mating. However, for centuries, people wondered how traits are transferred from one generation to the other and why some traits skip generations and appear later. However, it was not until 1865 that people began to understand the mechanism underlying this phenomenon, which is largely credited to the painstaking and extensive work of an Austrian monk Gregor Mendel (Fig. 1.1). Prior to Mendel, two misleading theories

Fig. 1.1 Photograph of Gregor Mendel (left) and his garden (left) as a part of the monastery's property in Brno where he conducted seminal experiments that laid the foundation of genetics

of inheritance were proposed. The first concept suggested that one parent contributes mostly to an offspring's inherited characteristics. Aristotle believed it to be a male by proposing a diagram of a fully formed homunculus inside the sperm. The second concept for inheritance was blending, suggesting mixing of parental characteristics in the progeny. This was more of an opinion rather than scientific concept, but it fit in the observations of human and animal features like skin colour, which is often a combination of both parents. However, blending was never observed in Mendel's experiments and both these theories were put to rest by his experiments. In contrast, by rigorous analysis, Mendel in 1865 discovered that individual traits are determined by discrete "factors" which we now know as genes, and are inherited from parents. His precise work on garden pea plants, which has now developed as a field of biology called as Mendelian genetics, was published in 1866 and went largely ignored, only to be rediscovered after his death in the 1900s by geneticists like Hugo de Vries and Thomas Hunt Morgan. Soon, Mendel's results were confirmed in a wide range of eukaryotic organisms suggesting that Mendelian principles of inheritance were a general theme in nature. In 1902, Sir Archibald Garrod became the first person to utilize Mendel's laws to explain the basis of inheritance in a human disease. He, along with a strong Mendel advocate, William Bateson, established that alkaptonuria was a recessive genetic disorder. This study paved the way for several future milestones with scientists providing evidences for molecular basis of inheritance in human diseases. The chromosome theory of inheritance by William Sutton and Theodor Boveri reaffirmed Mendel's laws of inheritance. In 1920, Thomas Morgan isolated the first genetic mutant of fruit flies and went on to use the fruit fly model to dissect and discover several details of classical genetics. Ronald A. Fisher, John B.S. Haldane and Sewall Wright established the field of population genetics in the 1930s by combining Mendelian genetics and evolutionary theory. Use of bacteria and viruses as simple genetic systems led to detailed study of the structure and function of genes. Elegant, breakthrough experiments by Frederick Griffith, Oswald Avery, Colin MacLeod, Maclyn McCarty, Alfred Hershey and Martha Chase proved that DNA is the molecule of heredity. Seminal papers published in the year 1953 by Rosalind Franklin, Maurice Wilkins, James Watson and Francis Crick established the three-dimensional double helical structure of DNA which ushered the field into the era of molecular genetics. Advances in techniques like recombinant DNA, gene cloning, polymerase chain reaction (PCR) and Sanger sequencing were a huge impetus to molecular genetics. This paved the way for development of gene therapy and launch of Human Genome Project leading to a new wave of genomics. Today, hundreds of genomes are being sequenced on a daily basis owing to immense improvement in sequencing technologies and analysis platforms. All this started with Mendel's detailed and robust experiments over a period of 8 years that paved the way for the golden era of research breakthroughs in life sciences and hence he is rightly known as the Father of Genetics.

1.1.1 Mendel's Work

Mendel served at an Augustinian monastery in the city of Brünn, Austria (now in Czech Republic), where natural sciences were taught. There, he embarked on a research project on plant hybridization in the monastery's own garden where he grew thousands of pea plants and provided the basis of genetics. Students should appreciate the meticulous scientific approach he used that resulted in robust data. He chose a plant which could be studied for the hypothesis he proposed. In addition, he designed the experiments carefully, and collected large amount of data, which was subjected to mathematical analysis and was highly reproducible. Mendel chose the humble pea plant, *Pisum sativum*, mainly for two reasons. First, peas had distinct observable traits that could be easily identified, and second, peas can both self-pollinate (pollen from the same flower) and be cross-pollinated (pollen from different flower from different plant), thus allowing the experimenter to generate specific pure lines as well as hybrid lines (Fig. 1.2). In addition, the plants were inexpensive and could produce large number of offspring. Using these pea plants, Mendel conducted experiments for about 8 years and consequently proposed the principles

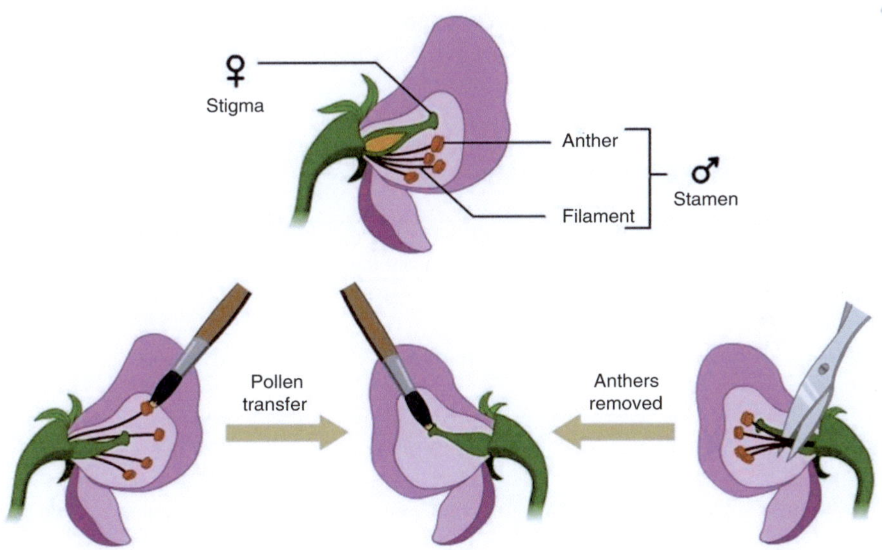

Fig. 1.2 Mendel's experimental model—the garden pea plant. Anatomy of the pea flower is shown (upper cartoon) indicating the male and female parts of the flower. The pollen is produced in the anthers, which land on the stigma, which contains the ovary, which later becomes the pea pod with the progeny (seeds). Lower cartoon: method for cross-pollinating pea plants is shown. To prevent self-fertilization, the anthers are removed from the female parent. Pollen from another plant is transferred to the stigma of this female plant using a paint brush. Each fertilized egg or ovule becomes an individual seed that can grow into a new pea plant. (Image credit: University of Waikato, www.biotechlearn.org.nz)

of inheritance. So whether it is pea plants or humans, genetic traits that follow these principles are known to be following Mendelian inheritance.

1.1.2 Mendelian Theory of Inheritance

Mendel studied seven different traits in the pea plant, each of which had two alternate forms. These traits were height (tall or short), seed colour (green or yellow) and seed shape (smooth or wrinkled), among others (Table 1.1). These were "either-or" traits, meaning there were no intermediate forms. This allowed the tracking of the traits in the subsequent generations by simple observations. Such traits are identified as discrete, as compared to continuous traits like skin colour in humans that show several intermediate forms. He collected lines with these discrete features and let them self-pollinate for several generations until he could confirm that they were pure breeding lines, meaning tall plants always produced tall progeny for several generations. These were the parental (P) generation. He then crossed these pure breeding lines of alternate forms and meticulously counted the hybrid plants to observe how the trait was inherited. These were monohybrid crosses, meaning mating was carried out between individuals that differ in only one trait. He carefully controlled the mating by making sure that no foreign pollen landed on the stigma of the flowers he chose. He found striking patterns of inheritance for all the traits

Table 1.1 Results of Mendel's garden pea hybridization experiments. Mendel's experiments with seven pea plant traits are shown here, with the results obtained for the F_1 and F_2 generations. [Adapted from "The results of Mendel's garden pea hybridizations" by OpenStax College, Biology (CC BY 3.0)]

Characteristic	Contrasting P_0 traits	F_1 offspring traits (dominant trait)	F_2 offspring traits	F_2 trait ratios
Flower colour	Violet vs. white	100% violet	705 violet	3.15:1
			224 white	
Flower position	Axial vs. terminal	100% axial	651 axial	3.14:1
			207 terminal	
Plant height	Tall vs. dwarf	100% tall	787 tall	2.84:1
			277 dwarf	
Seed texture	Round vs. wrinkled	100% round	5474 round	2.96:1
			1850 wrinkled	
Seed colour	Yellow vs. green	100% yellow	6022 yellow	3.01:1
			2001 green	
Pea pod texture	Inflated vs. constricted	100% inflated	882 inflated	2.95:1
			299 constricted	
Pea pod colour	Green vs. yellow	100% green	428 green	2.82:1
			152 yellow	

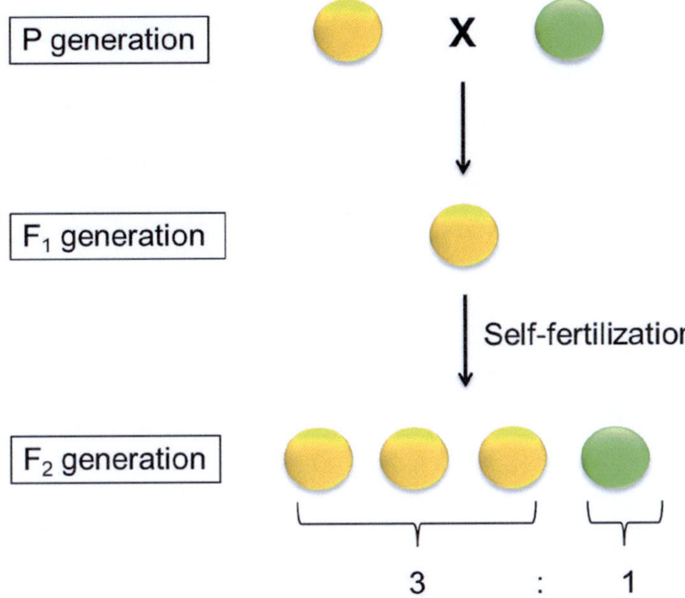

Fig. 1.3 Mendel's monohybrid cross. Cross-breeding of pure lines having yellow and green seeds (P generation) gives rise to plants with only yellow seeds (F1 generation). Self-pollination of F1 plants gives rise to F2 generation with individuals resembling the parent generation in the ratio of 3:1

individually studied. On crossing two pure bred lines of alternate traits, for example, plants with yellow versus green seeds, he observed that the first generation (F_1; first filial generation) always had plants with yellow seeds. He called this trait as the dominant trait and the trait with green seeds as recessive, which was "hidden" in this generation. This recessive trait reappeared in the F_2 generation when the F_1 plants were allowed to self-pollinate, although only in minority. However, the ratio of plants with yellow to green seeds observed in this generation was always 3:1 (Fig. 1.3).

Mendel's actual counts were 6022 yellow:2001 green seeds in this generation (3.01:1 ratio). This ratio was observed for all the traits that he studied (Table 1.1). These observations clearly refuted the theory of blending inheritance. Mendel was also able to perform reciprocal crosses, in which he studied whether a particular trait was transmitted via egg or sperm by reversing the traits of male and female parents. For example, he could use pollen from a plant with green seeds to fertilize the eggs of a plant with yellow seeds and vice versa. He observed that the progeny of these crosses was always similar in both cases, thus proving that the inherited factors are contributed by both parents equally.

He studied nearly 30,000 pea plants which provided robust statistical merit, and he communicated his results in 1865 to the scientific community. Thus, he concluded that traits like plant height and seed colour are controlled by a pair of

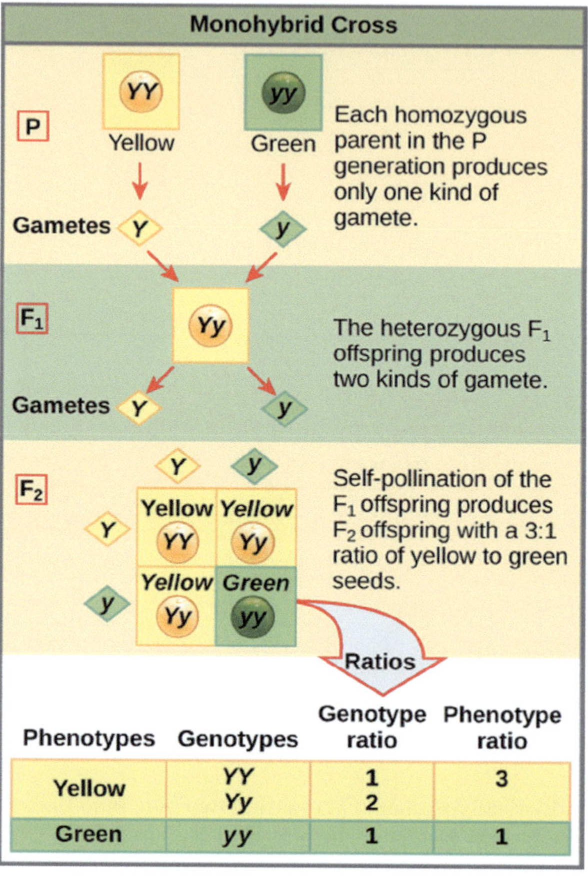

Fig. 1.4 Monohybrid cross ("Figure 12 02 02" By CNX OpenStax – (CC BY 4.0) via Commons Wikimedia). In the P generation, true-breeding pea plants (homozygous) for the dominant phenotype of yellow seeds and the recessive green seed phenotype are crossed. This cross produces F_1 heterozygote generation with all individuals having yellow seeds. Hence, yellow is the dominant trait here. On self-pollination, F_2 generation produces has a mix of yellow and green seed individuals, in the ratio of 3:1. Punnett square analysis can be used to predict the genotypes of the F_2 generation as shown in the box

heritable factors, which we now know as genes, received one from each parent. These genes have two alternate forms, known as alleles. For example, plant height has two alleles, tall and short alleles. The set of alleles carried by an organism is known as its genotype. The genotype determines the phenotype, which is the observable trait or feature, in this case, height of the plant. The individual is said to be homozygous when the two alleles for a trait are identical (AA or aa) and heterozygous if the alleles are different (Aa). Dominant and recessive alleles are designated by capital and small letters, respectively. In heterozygous situation, the dominant allele can mask the effect of the recessive allele. These two alleles segregate randomly during gamete formation such that each gamete (sperm or egg) randomly receives just one form. A Punnett square provides a simple method for tracing the gametes produced and determining the possible genotypes of the progeny (Fig. 1.4). Thus, according to the law of segregation, during meiosis, the two alleles for specific traits segregate randomly in individual gametes (egg or sperm) such that each gamete carries only one allele. During fertilization, these alleles unite and the genotype of the progeny is determined depending on the alleles received from the two parent gametes.

While monohybrid crosses established the law of segregation, Mendel had devised experiments involving dihybrid crosses to study the inheritance of two or more unrelated traits. Mendel questioned that if you have a dihybrid plant, which is heterozygous for two genes at the same time, how those alleles would segregate. He created a dihybrid by mating a pure breeding line of two alleles with a pure breeding line of the alternate forms of the alleles, for example, breeding plant with yellow round peas (YYRR) with green wrinkled peas (yyrr). The dihybrid of F_1 would have the genotype of YyRr displaying the dominant phenotype of yellow round peas. When Mendel allowed these dihybrids to self-fertilize, he found that plants of all combinations were produced, including parental type of yellow round and green wrinkled as well as new recombinant phenotypes of yellow wrinkled and green round. Mendel suggested that this was possible because the genes for both the traits assort independently. Thus, if a gamete has Y, it has equal probability of having R or r, and so a gamete can have four types of genotypes, viz. YR, yr, Yr or yR. Thus, on fertilization, any four kinds of eggs can be fertilized by any four kinds of sperms, thus leading to 16 possible outcomes for the zygote genotype. Four types of phenotypes can observed from these 16 combinations in a ratio of 9:3:3:1, as shown in Fig. 1.5. This is the basis for law of independent assortment, which states

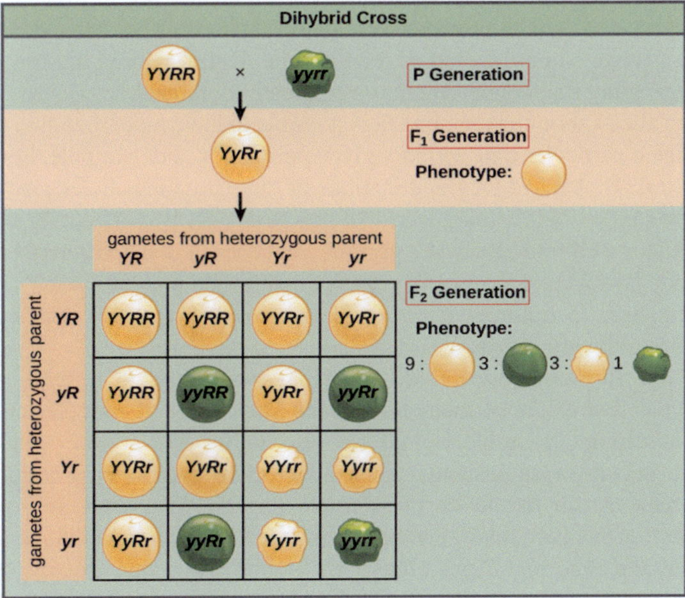

Fig. 1.5 Dihybrid cross ("Figure 12 03 02" By CNX OpenStax – (CC BY 4.0) via Commons Wikimedia). Pure breeding parent for two traits are crossed with pure breeding lines with recessive phenotype. F_1 generation has heterozygous plants with the dominant phenotype for both traits, and F_2 generation has plants with all combinations in the ratio of 9:3:3:1

that each pair of alleles segregates independently of other allele pairs during gamete production and subsequent fertilization from random union of gametes determines the observed phenotype. Hence, the inheritance of seed colour did not influence the inheritance of seed texture. This was held true for all the seven traits he studied, which gave the same phenotype ratios during dihybrid crosses.

1.1.3 Genetic Variation

Unlike the pea plant traits that Mendel examined, most traits, like skin colour, that we observe do not fall into the "either-or" category, meaning they have more than two forms. Some traits do not display clear-cut dominance or recessiveness, while some traits are multifactorial. Such traits do not follow the Mendelian ratios. These extensions to Mendelian inheritance can be thus divided into two categories, viz. single-gene inheritance and multigene inheritance.

Single-gene inheritance:

1. Incomplete dominance: In nature, sometimes a mixed or intermediate phenotype is observed in the presence of both the alleles, indicating that neither displays dominance to generate a phenotype. This was never encountered in Mendel's work because in a heterozygous situation, the dominant allele determined the phenotype. A typical example of incomplete dominance is the cross between a red snapdragon and a white snapdragon plant. According to Mendel's laws, the offspring flower should be either red or white. But in reality, the flowers are all pink suggesting that neither allele is completely dominant.
2. Multiple alleles: For a given gene, it is possible to have more than two alleles that determine a particular trait within a given population. For example, human blood types are A, B, AB and O. Combinations of three possible alleles give rise to the four blood types. While one individual can have only two alleles for the gene, in the population all four alleles are present which result in the different blood types observed. This results in genetic variation within the population.
3. Codominance: Two alleles may be simultaneously expressed with both alleles determining the phenotype.
4. Pleiotropy: In some cases, one gene may contribute to several observable phenotypes. For example, many men of the Maori tribe of New Zealand develop frequent respiratory problems and are also sterile. It was discovered that the problem occurs in an instance when a defective protein is produced by one gene necessary for the action of cilia and flagella. Men who are homozygous recessive for this particular gene have abnormally functional cilia for respiratory functions and abnormal flagella for sperm motion. Thus, one gene directly affects two separate functions in an individual.
5. Lethal alleles: In Mendel's experiments, all allelic combinations produced viable progeny. However, that is not the case for certain genes where homozygosity or heterozygosity affects the survival of the individual. Individuals homozygous for sickle cell anaemia allele die in childhood or early adulthood due to problems in the circulatory system.

Multigene inheritance: Several variations from Mendel's laws involve interactions between two and more genes to give rise to specific trait. In some cases, one gene's alleles can mask the effects of other gene's alleles, known as epistasis. In addition, several genes that lie physically close to each other on the chromosome and are genetically linked do not assort independently. Such genes are called linked genes and they do not follow Mendel's law of independent assortment and are expressed together as a cluster. In humans, the autosomal genes for red hair, freckles and a fair complexion are linked genes. Many characteristics like skin colour, human height and several diseases are controlled by multiple genes, several times in combination with environmental factors.

1.2 Model Organisms

Some organisms have been used as model organisms for several years to help scientists understand complex biological processes that cannot be easily studied in the organism of interest. Model organisms can be bacteria, fungi, rodents or plants. Use of humans in biomedical sciences is limited due to practical and ethical issues, and hence most of the understanding in cell biology and molecular biology is due to studies done in model organisms.

The important criteria that define model organisms include rapid growth and short generation times, high reproduction rates, ease of genetic manipulation, production of large number of offspring and ease of maintenance in standard laboratory conditions. Over the years, model organisms have been instrumental in helping scientists understand complex biological processes to discover the fundamental causes of a disease, production of better varieties of crops, understanding drug resistance in pathogens and so on. Since all organisms have a common point of origin in evolution, the basic molecular mechanisms and processes are the same, and hence studies in model organisms can be extrapolated to the organism of interest. For example, research in the single-celled eukaryote *Saccharomyces cerevisiae* has led to deep understanding of the cell cycle and has helped scientist develop drugs to target cell cycle of tumour cells. Studies with model organisms stand strong even in this era of next-generation sequencing where hundreds of genomes are sequenced daily. Complex processes of behaviour, disease and pathology require the need for a simplistic reductionist approach that is only possible in model organisms.

1.2.1 *Escherichia coli*

This model prokaryote is a Gram-negative, rod-shaped bacterium with size approximately 1 μm in length to 0.35 μm in width (Fig. 1.6), and sizes vary only slightly between strains. This bacterium was discovered by the German paediatrician and bacteriologist Theodor Escherich in the 1800s. He was investigating the causative agents for diarrhoea in babies which led to the identification of this bacterium in the lower intestine of humans, which he called *Bacillus communis* coli. It was later

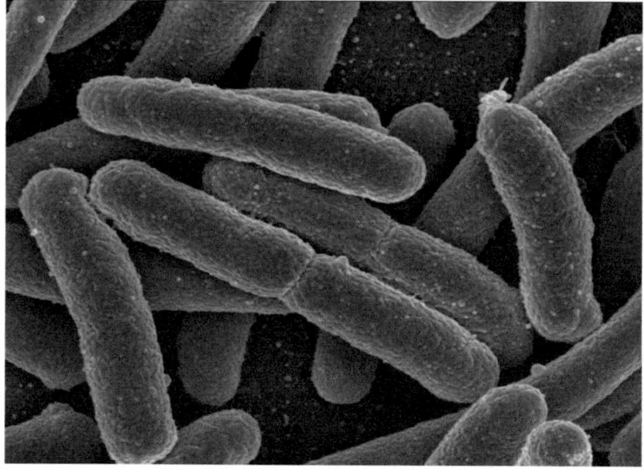

Fig. 1.6 Scanning electron micrograph of *E. coli*, grown in culture and adhered to a cover slip. Credit: Rocky Mountain Laboratories, NIAID, NIH

named *Escherichia coli* in his honour in 1958. Several strains of *E. coli* are normal inhabitants of the lower gastrointestinal tract of humans and other warm-blooded animals. While most strains are non-pathogenic, some strains like *E. coli* O157:H7 are infamous for causing food poisoning and bloody diarrhoea, which can be fatal. However, most strains of *E. coli* are classified as biosafety level 1, which makes this organism useful for teaching and demonstrations in undergraduate and high school levels. This bacterium belongs to a large family of Gram-negative bacteria called as *Enterobacteriaceae* which also includes some well-known pathogens like *Salmonella*, *Shigella*, *Yersinia* and *Klebsiella.*

E. coli has been considered as a molecular biologist's toolbox or a workhorse of molecular biology owing to its amazing properties. This bacterium can grow in minimal media with only one carbon source and in rich media where it can divide every 20 min. This incredible fast generation time makes it suitable to study rare genetic events in a relative short time. Fast generation time makes it possible to scale up the culture volume to industrial production making it extremely useful for biotechnological applications. *E. coli* is a facultative aerobe, which means it can grow in the presence as well as absence of oxygen. There is extensive knowledge about its genome, transcriptome, proteome as well as metabolome. The ~4.6 Mb genome of the bacterium is arranged in a closed circular double-stranded DNA. During bacterial cell division, the circular DNA undergoes replication, and then the cell undergoes binary fission such that each daughter cell receives one copy of the DNA. This entire process is highly regulated to ensure correct replication and segregation of the DNA. Since the haploid bacterial genome consists of one chromosomal DNA, a genetic mutation will directly lead to expression of a phenotype since a second, wild-type allele is absent. This makes understanding gene functions very straightforward. The phenotypes that can be observed for screening genetic mutations in *E. coli* are changes in colony morphology, resistance to antibacterials or bacteriophages, auxotrophic mutants and conditional temperature-sensitive (ts) mutations for essential genes. Hundreds of studies done using *E. coli* have led to amazing discoveries, several of which have been awarded the Nobel Prize. This includes work of Joshua Lederberg for his discoveries of genetic recombination and the organization of the genetic material of bacteria; Francois Jacob, André Lwoff and Jacques Monod for their discoveries on the genetic control of enzyme and virus synthesis; Max Delbruck, Alfred Hershey and Salvador Luria for their discoveries on the replication mechanism and the genetic structure of viruses; Arthur Kornberg for processes of DNA replication; and Jacob and Monod for gene regulation, among many others. This organism has been critical for development of recombinant DNA technology. Genetic engineering of *E. coli* plasmids was achieved to harbour and amplify transgenes or for transfer of the gene to another organism. Due to fast generation times, genetically engineered *E. coli* culture can be scaled up to large volumes and grown in bioreactors for production of useful compounds like insulin, growth hormones and drugs.

1.2.2 *Saccharomyces cerevisiae*

Yeasts have been exploited for thousands of years for the purpose or brewing and baking. Initially, the impetus to research on yeasts was mainly provided by its industrial applications. In modern science, budding yeast has been used extensively in laboratories and has provided understanding of several molecular and genetics questions in eukaryotic biology. *Saccharomyces cerevisiae* or baker's yeast is a single-celled eukaryote with short generation time and with a simple life cycle, which alternates between haploid and diploid phases. Like all eukaryotic cells, yeast cells also regulate cellular processes to determine the fate of the cell in a particular condition. The molecular basis of cellular processes in yeasts can be easily extended to multicellular eukaryotes. Yeast cells express and regulate genes, perform biological functions and differentiate using processes similar to the cells of multicellular organisms. However, unlike multicellular organisms, yeasts have experimental advantages like fast growth rate and ability to survive as haploids and diploid, thus making functional characterization of genes and pathways easier and faster. *S. cerevisiae* is indeed the first eukaryotic organism sequenced (1996), and genome analysis has indicated that it has more than 30% protein-coding genes homologous to humans. In addition, this budding yeast has also been utilized as a model to study drug resistance in pathogenic fungi, a concern that is growing at an alarming rate worldwide.

S. cerevisiae contains membrane-bound organelles like nucleus, mitochondria and endoplasmic reticulum. Yeast cells divide by budding in which the daughter bud pinches off the mother cell (Fig. 1.7). Yeast cells can divide once every 90 min under optimal laboratory conditions. Even though they are eukaryotes, yeast cells can be cultured like bacteria in agar plates and liquid growth media and can be stored for years at −80 °C by freezing in glycerol. They are inexpensive to grow and easy to maintain with hundreds of mutants that can be obtained for screening. One of the most widely used experimental strains of *S. cerevisiae* is S288C. This strain, isolated by Robert Mortimer, is now a standard laboratory strain that has been widely used as

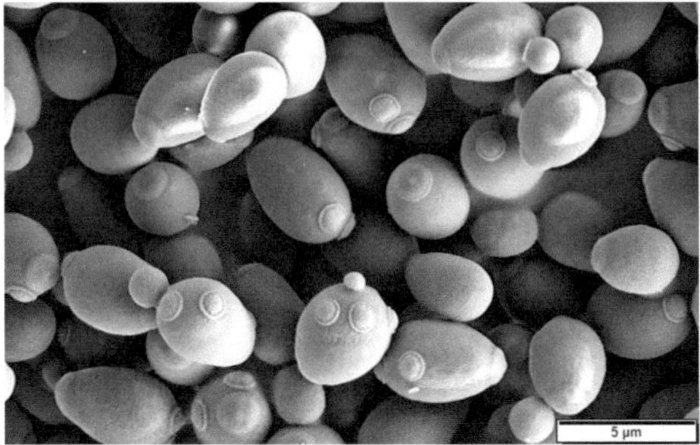

Fig. 1.7 Scanning electron micrograph of *S. cerevisiae* cells (Murtey and Ramasamy, 2016). Buds and bud scars can be seen here

the parental strain for the isolation of biochemical mutants, which continues even today in laboratories around the world.

The yeast life cycle is simple with alternating haploid (n) and diploid (2n) stages (Fig. 1.8). Haploid cells occur in two mating types, *a* and *α*. Both mating types can divide asexually by mitosis wherein with each round of cell cycle, a cell produces a daughter bud which pinches off from the mother bud at the end of cytokinesis, and thus maintain as stable haploids. However, opposite mating types can also engage in sexual reproduction, in which *a* and *α* cells communicate via pheromones that induce fusion of the cell and then the nucleus to give a diploid progeny. The diploid cell can divide mitotically and survive as a diploid. But in certain conditions like starvation, diploids undergo meiosis and sporulation to form an ascus with four haploid spores. The mating type is determined by mating type locus (MAT) present on chromosome III.

The cell cycle of *S. cerevisiae* has been a subject of intensive research over several decades and has provided valuable insight into human cell cycle and diseases related to it. It is easy to visualize different stages of yeast cell cycle by simple light microscopy by observing the yeast cell bud size. Unbudded cells are in G1 stage, cells with small buds are in S phase, and cells with large buds are in G2/M (Fig. 1.9). Scoring budding index has been a powerful tool in characterizing the changes in cell cycle due to changes in external or internal variables.

In 2001, three scientists, Leland Hartwell, Paul Nurse and Tim Hunt, shared the Nobel Prize for their independent studies on cell cycle regulation in yeast and humans. Specifically, Leland Hartwell uncovered the genetic basis of cell division in *S. cerevisiae* and contributed significantly on our understanding of the eukaryotic cell cycle with broad implications for human health and prevention and treatment of diseases like cancer. Several other discoveries using yeast genetics have been awarded prestigious prizes including the Nobel Prize. One such recent example is the 2016 Nobel Prize given to Yoshinori Ohsumi for his work on autophagy in budding yeast, which was used to identify several homologous autophagy genes in mammalian cells.

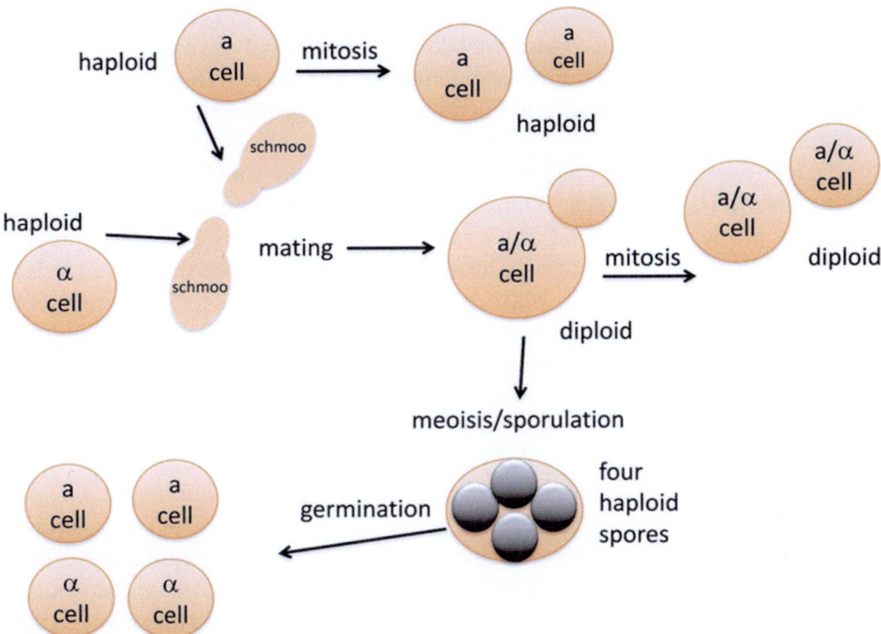

Fig. 1.8 Representative life cycle of *S. cerevisiae* (Duina et al. 2014). Haploid yeast cells (**a** cell or α cell) undergo mitotic cell division through budding to produce daughter cells. The two cell types (**a** cell and α cell) release pheromones, initiating the formation of schmoos and subsequent mating, which leads to the formation of a stable diploid (**a**/α cell). Diploid cells also divide mitotically by budding to produce genetically identical daughter cells. Under starvation conditions, diploids are induced to undergo meiosis, forming four haploid spores, which can germinate into two **a** cells and two α cells

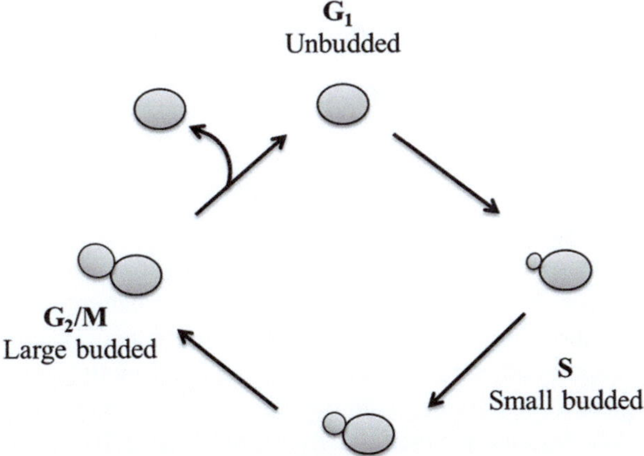

Fig. 1.9 Cell cycle of *S. cerevisiae.* (adapted from Hanson 2018) During asexual reproduction, cell cycle stage can be observed based on bud size as indicated here

Haploid *S. cerevisiae* genome has a size of around 12 Mb packed in 16 linear chromosomes in the nucleus which encode around 6000 polypeptides. The genome has relatively low number of intron-containing genes in *S. cerevisiae*. Due to this, the gene density of protein-coding genes is much higher than the gene density in human genome.

Several techniques have been developed over the years which make *S. cerevisiae* easy to genetically manipulate. Transformation of DNA in budding yeast is highly efficient. A plasmid constructed with a marker and a functional yeast gene of interest can be readily taken up by the cells from the external environment. Two kinds of plasmids can be transformed in yeast: Yeast integrative plasmids (Yips) are circular plasmids that are unable to replicate on their own and integrate readily at a region of homology by genetic recombination. Autonomously replicating plasmids can be yeast episomal plasmids (YEps) and yeast centromeric plasmids (YCps). These can exist independently in the cells and are useful as shuttle vectors to carry genes from other organisms and be replicated in yeast.

Yeast cells can be transformed with engineered DNA that will readily integrate into homologous region of interest for construction of deletion mutants, epitope tagging or any other genetic modification. Only a homology region of 40 bp is enough for efficient integration. This homologous recombination-based technology is one of the most useful tools that made *S. cerevisiae* an ideal model organism. Owing to the ease of genetic manipulation, entire libraries have been constructed and are available in depositories for research purposes. Examples are deletion library (collection of deletion mutants of non-essential genes), GFP clone library (collection of more than 4000 genes tagged with GFP for microscopy-based studies), TAP-tagged ORF library (more than 4000 genes tagged with TAP for protein purification studies) and several others. Important technological innovations in yeast have been widely useful for study of several fields of biology. Techniques like yeast two-hybrid screening for study of protein-protein interactions have led to significant discoveries before the era of next-generation sequencing.

1.2.3 *Drosophila melanogaster*

The common fruit fly, *Drosophila melanogaster*, has been used extensively as a model organism for over a century and continues to remain a useful system for studying genetics and cell biology. It has proved to be an ideal organism to study human genetic diseases, animal behaviour, development and neurobiology, evolution and pathogenesis. Rapid life cycle, small chromosome number and genome size, giant salivary chromosomes, ease of maintenance and genetic manipulation are the important experimental advantages that make it an ideal model organism.

Drosophila genetics mainly began in the lab of Thomas Hunt Morgan in the early 1900s, in what was famously called as "fly-room" (Fig. 1.10). Several crucial and historic experiments were performed in the fly-lab that laid the foundation of modern fly genetics. The sex-linked *white eye* mutation was discovered by Morgan in 1910. A PhD student of Morgan, Calvin Bridges, proved the chromosome theory of

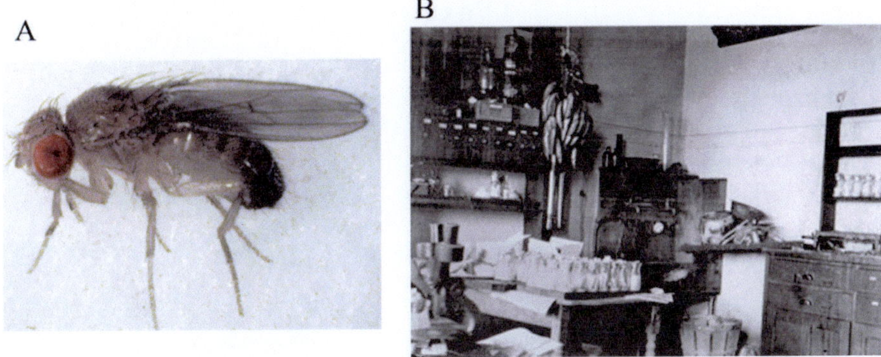

Fig. 1.10 Image of an adult *D. melanogaster* (**a**) (Jennings 2011) and Morgan's "fly-room" at Columbia University (Allocca et al. 2018) (**b**)

inheritance by showing that the nondisjunction of the sex-linked *white eye* gene correlates with the nondisjunction of the X chromosome. An undergraduate student in Morgan's lab, Alfred H. Sturtevant generated the first chromosome map by calculating the recombination frequencies. The mutagenicity of X-rays was demonstrated by Herman J. Muller. Several other discoveries continued to be made in the fly-room that added to the knowledge of transmission genetics. Indeed several Nobel prizes have been awarded to studies in *Drosophila* biology, with the most recent one awarded in 2017 to Jeffrey C. Hall, Micheal Rosbash and Michael W. Young for their discoveries of the molecular mechanisms controlling circadian rhythm.

A crucial factor for being a model organism is having similarity of cellular mechanisms with humans. *D. melanogaster* has homologues of approximately 75% genes involved in human diseases. Study of these genes can be extrapolated to humans, thus bypassing the ethical issues of biomedical research involving human subjects. *Drosophila* genome is about 137 Mb arranged in four pairs of chromosomes (one pair of sex chromosomes and three pairs of autosomes) with around 15,500 genes. Chromosome 1 is the X chromosome and chromosomes 2–4 are autosomes. The X chromosome is large and acrocentric, chromosomes 2 and 3 are large and metacentric, while chromosome 4 is a tiny acrocentric "dot" chromosome. A fly is a female if it has two X chromosomes, whereas an X and Y will designate a male. The small chromosome number and small genome size as compared to humans make it ideal for genetic studies by simplifying genetic manipulations. In addition, the fruit fly's salivary glands possess giant polytene chromosomes, features of which can be easily viewed through a light microscope by adding a chemical dye that gives unique banding patterns. These large chromosomes were the key tools for development of fly genetics and are still used today.

D. melanogaster is also very easy to grow and maintain and has a very short life cycle of 10 days. The fruit flies are regularly grown on a corn meal and sugar medium at room temperature. The reproduction rates are also very high, with several

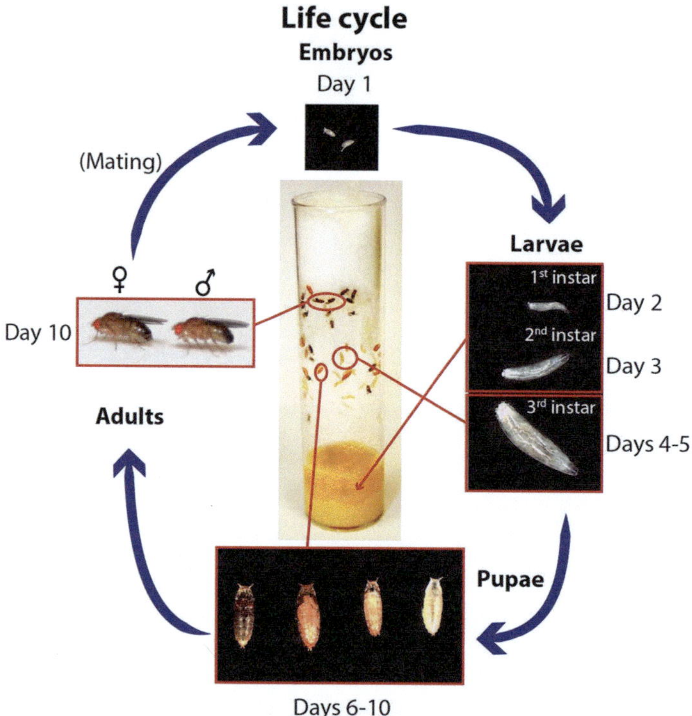

Fig. 1.11 Life cycle of *D. melanogaster* (Hales et al. 2015). *D. melanogaster* are cultured in glass vials closed using a cotton plug and containing food. The vial in the picture contains flies at different stages of growth. Depending on the growth stage, the organism can be in the food area or on the walls of the vial. The life cycle is completed within 9–10 days in laboratory conditions

hundred progeny produced on each mating. A female can produce around 3000 progeny in her lifetime and a male can sire over 10,000 offspring. The eggs laid by a female are about half millimetre long and can self-sustain embryonic development, which usually takes 24 h to complete. Embryogenesis ends with formation of first instar larva that grows in body size and moults to produce second and third instar larvae. After the third instar larva completes its growth, it crawls out of the food and pupates. It undergoes metamorphosis inside the protective pupal case. The metamorphosis involves radical change in the fly body plan where most adult structures like wings, eyes and genitalia develop. After the pupal development is complete, adult flies emerge out of the case in a process called as eclosion and become sexually mature in 8–12 h, thus repeating the life cycle (Fig. 1.11).

A range of genetic tools are available for genetic manipulations in *Drosophila*. A number of resources for the research community including stock centres and publicly available genome databases are available for *Drosophila* like FlyBase. In *Drosophila*, crossing-over occurs only on females making it possible to maintain gene linkage through male inheritance. This simplifies genetic manipulations and allows for a variety of genetic screens. Balancer chromosomes created by X-ray mutagenesis have multiple overlapping inversions, which are useful for stock-keeping of lethal alleles. Transgenes can be easily made in this organism using P element transposons, homologous recombination and more recently CRISPR/Cas9. Indeed, thanks to the abundance of genetic manipulation tools and publicly available fly stocks and databases, *Drosophila* will continue to be used as a key model organism for understanding the molecular basis of human diseases.

1.2.4 *Arabidopsis thaliana*

All of science, whether curiosity based or having direct application, is ultimately done for the benefit of human society and the environment we live in. Despite this understanding, model organisms appear out of obscurity leaving one wondering about its usefulness. One such model organism is the small stature plant, *Arabidopsis thaliana*, which surpasses all the agriculturally important plants like rice, wheat and beans in terms of the extensive research done. A small angiosperm, *A. thaliana*, is a weed and member of the Brassicaceae (mustard) family of plants which includes economically important species like cabbage and radish. Even though it has little direct usefulness to the human society, it has paved the way for breakthrough research in plant molecular biology and genetics. This plant was first described by Johannes Thal and he called it *Pilosella siliquata*. The name was later changed to *Arabidopsis thaliana* in his honour by Linnaeus. The credit for bringing this plant to genetic research limelight is given to Friedrich Laibach, who established its chromosome number during his PhD in 1907 and made seminal contribution to establish *Arabidopsis* as a fundamental model for plant genetics research by 1943.

A. thaliana, commonly known as thale cress, has a wide geographical distribution with different ecotypes growing in Europe, Asia and Africa. The *Arabidopsis* seed stock centres like Arabidopsis Biological Research Centre have been established to store seeds of different ecotypes (accessions) from natural populations to be made available for scientific research. The Columbia and Landsberg ecotypes are used widely as standards in genetic and molecular biology research in *Arabidopsis*.

The entire life cycle, starting from seed germination to maturation to produce seeds, is completed in around 6 weeks (Fig. 1.12). This is a significant advantage for using this plant when compared to crops and trees with much longer lifespan. The size of this plant is also small, with fully mature plant reaching about 15–20 cm in height and flowers which are about 2 mm in size. On seed germination, the seedling develops into a rosette plant which on maturation gives rise to an inflorescence stalk which produces flowers. The flowers can self-pollinate as the bud opens, but can also be cross-pollinated. The seeds from the flowers are produced in encases called as siliques. Often a single plant can produce more than 4000 seeds.

Due to this short and simple life cycle and high reproduction rates, genetic manipulation and isolation of large number of mutants is possible. Hundreds of small seeds can be cultivated in petri dishes for germination, and the small stature of the adult plant makes it easy to be grown in growth chambers or greenhouses in large

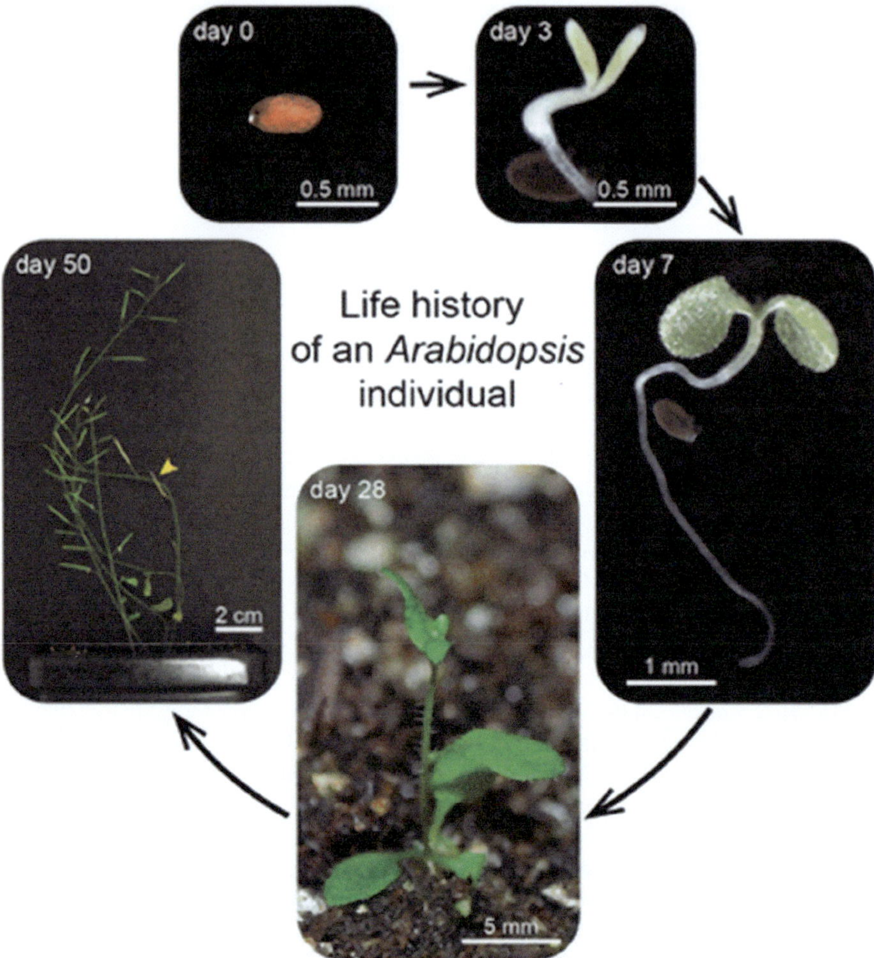

Fig. 1.12 Images of *Arabidopsis* life cycle (Woodward and Bartel, 2018). The small seed of the plant germinates with the radicle seen within 3 days to give rise to a seedling. The seed can be germinated till the seedling stage in petri dish with soft agar and growth medium. The seedling is then transferred to pots with soil and fertilizer. The seedling grows under fluorescent light to form a rosette plant with 25–28 days giving rise to stalks of seed pods called as siliques

numbers. Unlike several other plants, *A. thaliana* can grow indoors under weak fluorescent light and does not require co-culture with symbiotic organisms allowing growth in aseptic conditions with maximal control of biotic and abiotic factors. It is therefore not surprising to see hundreds of variables being tested individually and together to study the role of biotic and abiotic environment in plant physiology, plant defence and ecology. *A. thaliana* has a diploid genome of about 135 Mb with a haploid chromosome number of five. This is a very small genome as compared to other plant genomes like maize (2500 Mb) and rice (430 Mb) which are difficult to genetically manipulate. Whole-genome sequencing of *Arabidopsis* was completed in 2000 which was followed by availability of other resources like Affymetrix microarrays, identification of mutants through forward genetics and availability of precise insertion mutants of nearly all genes through reverse genetics. This led to a humungous rise in the numbers of papers published improving the pace of *Arabidopsis* research drastically. A useful approach in *Arabidopsis* research has been the use of reverse genetics to generate thousands of modified T-DNA insertional mutants. This is possible due development of simple method of transformation using genetically engineered *Agrobacterium tumefaciens* culture. The transferred DNA (T-DNA) is inserted into the plant genome by *Agrobacterium* naturally resulting in insertion of T-DNA in precise locations as designed. This method has added greatly to its use in molecular biology and led to amazing breakthroughs. Most importantly, the studies in *A. thaliana* can be extrapolated in other agriculturally and economically important plants like tomato, soybean, rice and so on. Research advancements in understanding plant hormone and defence pathways, circadian rhythm and environmental response have led to many applications in crops and other plants useful for humans. One important area of research which is highly relevant with increased need for food security is disease resistance. Plants are exposed to all kinds of pathogens, from bacteria, fungi, viruses and even nematodes. Developing plants that are genetically resistant to its pathogen is a major goal of most plant-breeding programmes. Research in *Arabidopsis* had led to identification of genes and genetic pathways important for plant immunity and disease resistance. These studies are useful for generating plants that have genetic resistance to pathogens, thus improving crop yield and output.

1.2.5 *Mus musculus*

When Mendel's laws were rediscovered in 1900, many in the scientific community questioned if using the pea plant and Mendel's laws of inheritance was enough to study inheritance of traits in humans. Soon the need to have a mammalian model for biomedical research became obvious. It was required that like other model organisms, the ideal mammal for genetic studies should be able to breed quickly producing large number of offspring and should display many easily scored, variable traits and that could be housed in large numbers in a laboratory space. These attributes were found in the common house mouse, *Mus musculus*. The most advantageous feature of this model organism for genetic analysis is the availability of hundreds of single-gene mutations. These mutations have arisen during its course of evolution for domestication during the "fancy-mouse" trade where mice were bred for different fur coats and other visible phenotypes. Early on researchers made use of these single-gene mutations to explain Mendel's laws and proved that they can be extrapolated to humans as well. One such scientist that used these trade mice was Clarence Cook Little, who became the father of modern lab mouse. He founded the Jackson Laboratory where he mated closely related mice for generations, creating the first inbred strains (Fig. 1.13).

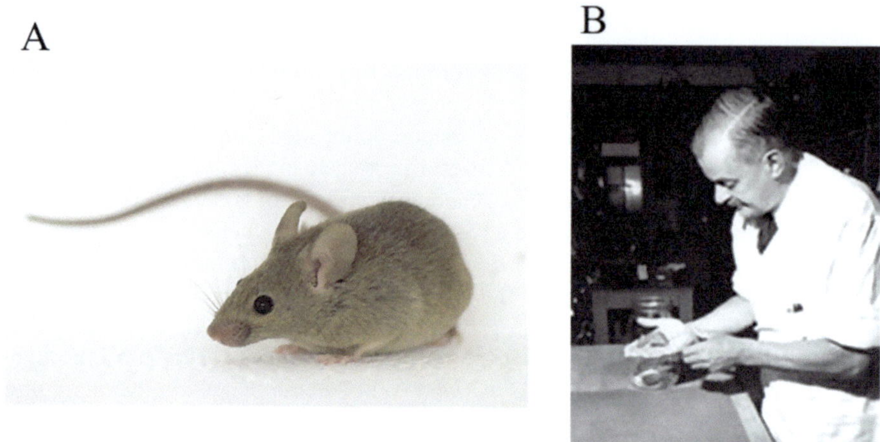

Fig. 1.13 The common house mouse, *Mus musculus*, is widely used as the mammalian model system in biology (**a**) (Phifer-Rixey and Nachman, 2015). Clarence Cook Little who created the first inbred mouse strain at Jackson Laboratory (**b**)(Clarke 2002)

Several other features make the mouse a mammal of choice of genetic studies. Mice have a short generation time of 8–9 weeks. They can breed in captivity, are docile and have a large little of right or more pups. They are small, are easy to handle and take up less laboratory space. While model organisms like fruit flies are used for several important genetic investigations, mice have an advantage of being a mammal and hence sharing complex traits with humans that are unavailable in fruit flies, worms and other model organisms. The mouse genome is around 3000 Mb arranged on 19 autosomes and 2 sex chromosomes (X and Y). This is similar to human genome of around 3000 Mb compacted in 22 autosomes and 2 sex chromosomes (X and Y). Almost every gene in the human genome has a homologue in mouse genome. Another important aspect is the conservation of synteny between the two genomes. This means that the genes that are closely linked on a locus in one species are also closely linked in the other species. Conserved synteny implies similar evolutionary trajectories of these genomes. Synteny also allows researchers to map homologous genes rapidly in the human genome.

The mouse's life cycle is similar to humans and other placental mammals only differing with respect to the timing for each step. The development stages both before and after birth are remarkably similar in all mammalian species. The male haploid sperm cell produces sperms via spermatogenesis, which it passes on during copulation. The females are born with all the haploid egg cells that they will have over their lifetime. After ovulation, the egg is fertilized by the sperm, the fusion of which activates the pathway of animal development (Fig. 1.14). From here on, mouse development is divided into two stages. First is the preimplantation stage where the zygote freely floats within the female body and mostly remains undifferentiated. This stage is beneficial for scientists because embryo at this stage can be removed from the female, cultured in a petri dish, manipulated genetically and then

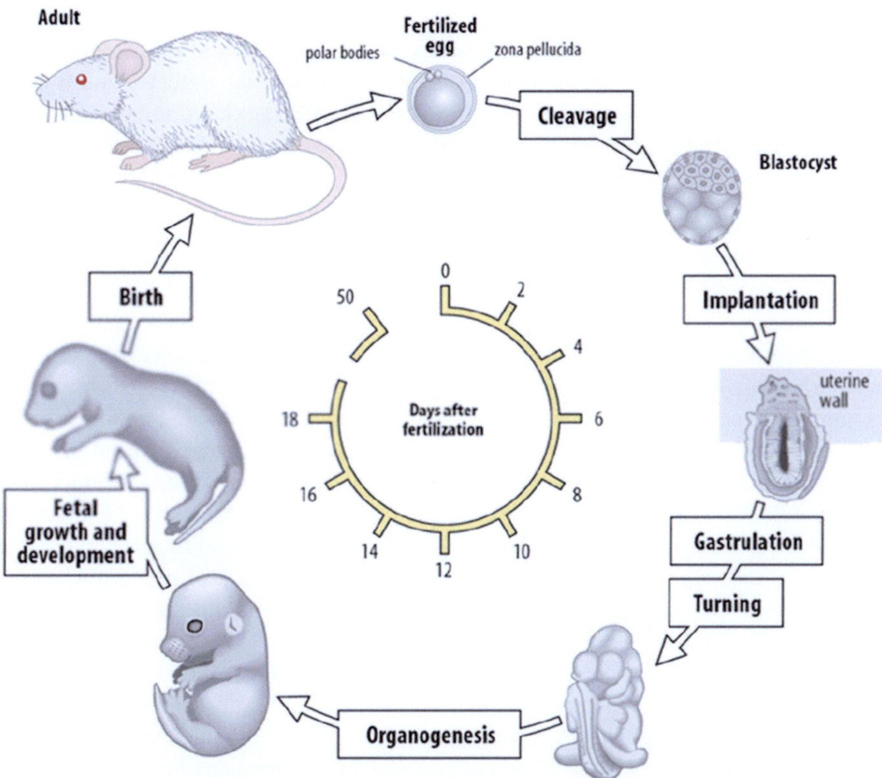

Fig. 1.14 Events in mouse development from fertilization to birth. The fertilized egg undergoes division and differentiation through various stages to develop into an embryo that undergoes organogenesis and development to form an adult mouse. The period from fertilization to birth lasts for about 21 days (Source: Dr. Brian E. Staveley, Department of Biology, Memorial University of Newfoundland)

placed back in the female body for further development. This cannot be done in the next stage, postimplantation, where the embryo grows and develops tissues and organs.

Advanced genetic techniques and tools have been developed which are helping scientists discover and characterize increasing number of human diseases. Powerful techniques for analysing the mouse genome include transgenic technology of addition of particular genes by nuclear injection into the germline in order to determine gene function and regulation. Creation of "knockout" mouse by homologous recombination and targeted mutagenesis advanced the field rapidly. New genetic techniques like transcription activator-like effector nucleases (TALENs) or the CRISPR/Cas9 system that use guided endonucleases allow precise genetic manipulations in the mouse genomes (Kaczmarczyk and Jackson, 2015). These and related techniques have led to invaluable mouse models of several number of human diseases. These mouse models include thousands of unique inbred strains and genetically engineered mutants that are available to the research community. There are mice strains prone to specific diseases like Lou Gehrig's and Huntington's disease, to different types of cancers, to lifestyle diseases like diabetes and obesity and even to behavioural and neurological disorders like anxiety, aggression, alcoholism and drug addiction. Immunodeficient mice are also available which are useful for research in AIDS and cancer.

1.3 Genetics and Evolution

From December 1831 to October 1836, Charles Darwin travelled across the globe as a naturalist in his HMS Beagle (Fig. 1.15). He studied and collected samples of hundreds of species that he encountered in varied environments that he visited. The famous finches were collected from different islands of Galapagos. About 23 years later, he published *On the Origin of Species* in 1859. In the book, Darwin makes remarkable observation about the specimens and fossils he collected. "The similar framework of bones in the hand of a man, wing of a bat, fin of a porpoise, and leg of the horse—the same number of vertebrae forming on the neck of the giraffe and of the elephant—and innumerable other such facts, at once explain themselves on the theory of descent with slow and successive modifications". He concluded that "all organic beings which have ever lived on this earth may be descended from some one primordial form". Darwin proposed that species undergo "descent with modification", which means that species evolve, and that all living organisms can trace their descent to a common ancestor. He suggested that the mechanism of evolution was by natural selection. Darwin's ideas had revolutionary impact on the way questions in biology were being asked. He proposed that there is variation of expression of a trait among the individuals of a population of a particular species. These trait variants can

Fig. 1.15 Charles Darwin (**a**) voyaged on the HMS Beagle (**b**) to study and collect specimens from around the globe to propose one of the biggest ideas in biology, which he published titled *On The Origin of Species by Means of Natural Selection* (**c**)

be passed down to future generations. In addition, he suggested that the some individuals are considered the fittest because of the possession of some of these variant traits and the variants are selected to be expressed more in the population. Darwin developed this idea not just by his own collection of species, but also from observations from artificial breeding of plants and animals. He noted: "The key is man's power of cumulative selection: nature gives successive variations; man adds them up in certain directions useful to himself". He applied this to selection by nature and wrote: "Can it, then, be thought improbable, seeing that variations useful to man have undoubtedly occurred, that other variations useful in some way to each being in the great and complex battle of life, should sometimes occur in the course of thousands of generations? ... This preservation of favourable variations and the rejection of injurious variations, I call Natural Selection". Even though these revolutionary insights gave a huge impetus to the scientific thought process, Darwin was unable to explain the source of this visible variation on which natural selection acts. Mendel's experiments on plant hybridizations were published in 1866, just 7 years after Darwin's publication of *On the Origin of Species*, and it is believed that Darwin received a copy of Mendel's paper, but he probably never read it. Several years later, scientists were able to merge genetics and the theory of evolution. After the discovery of DNA as the heritable material and deciphering the structure of DNA, scientists could finally understand that evolution is a process that begins at the molecular level, in the double helical molecule, DNA. A new field of evolutionary genetics emerged which dealt with study of genetic variation leading to evolutionary change. It is worth mentioning that Alfred Russel Wallace is jointly credited for coming up with the theory of evolution, at the same time as Darwin, wherein they co-published it in 1858, demonstrating their independent observations leading to the same conclusions. However, since the publication of *On the Origin of Species* by Darwin, Wallace was generally overshadowed.

1.3.1 Natural Selection

While on this globe-trotting voyage, Darwin made remarkable observations about birds called as finches on Galapagos Islands, a cluster of islands about 600 miles from mainland Ecuador. He observed that these finches closely resembled each other but had formed a graded series of beak sizes and shapes (Fig. 1.16). In 1860, he wrote, "seeing this gradation and diversity of structure in one small, intimately related group of birds, one might really fancy that from an original paucity of birds in this archipelago, one species had been taken and modified for different ends". With the help of his ornithologist friend, John Gould, Darwin realized that these variations in beaks could have evolved for specific functions. For example, the little warbler finch has a very fine needle-like beak that is perfect for picking insects. Another finch, called as woodpecker finch, has a harder robust beak that is used for beetle and termite larvae, while the cactus finches use long beaks for probing into cactus flowers. So basically, the variation of this trait occurred for the function required in the environment that the population was exposed to. Darwin called this

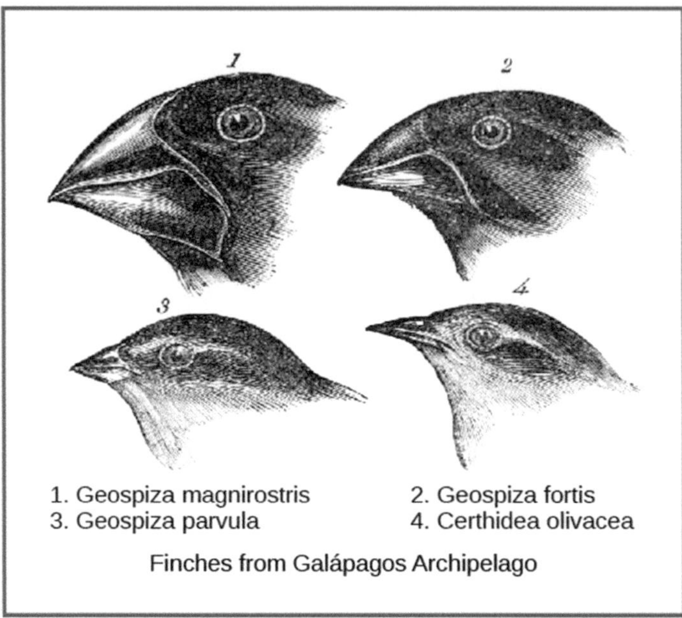

Fig. 1.16 Variation of beak shape in Darwin's finches. Darwin hypothesized that the beak of the ancestor species for these finches had adapted over time to the food source available, leading to formation of completely new species. This illustration shows the beak shapes for four species of ground finch: 1. *Geospiza magnirostris* (the large ground finch), 2. *G. fortis* (the medium ground finch), 3. *G. parvula* (the small tree finch) and 4. *Certhidea olivacea* (the green warbler-finch) (Source: The Galapagos Finches and Natural Selection. (2020, August 15). https://bio.libretexts.org/@go/page/13415)

mechanism as natural selection. Over time, the ancestral population of the finches had adapted to the food source available by acquiring changes in the beak and evolved. Different groups from the ancestral population would have become isolated from one another by geographical barriers or by other mechanisms. Once isolated, the groups would not be able to interbreed and were exposed to different environments. In each environment, natural selection acted as a drive to favour different traits. Over many generations, these changes in heritable traits accumulated in each isolated group such that the groups became a separate species. Hence, natural selection can also act as a mechanism driving speciation.

Natural selection is one of the core mechanisms of evolutionary change which leads to the evolution of adaptive traits. These selected traits are inherited and passed on to the next generation. The species that are better adapted have higher chances of survival in their environment. Evidence to support evolution and natural selection has accumulated over time, and now evolution is accepted as a robust scientific fact. Another example of natural selection was discovered among peppered moths near industrial cities in England. The moth population had varieties that varied in wing and body coloration. Most of the insects were pale in colour as compared to the dark coloured minority, so that they could easily camouflage on the birch trees to prevent being seen by their predators, the birds. In the nineteenth century, pollution of sooty smoke from the coal furnaces killed the lichen on the trees making the tree bark dark coloured. On account of this, the pale majority population of moths became visible when they landed on the blackened tree surfaces and were predated upon by the birds, while the dark coloured ones survived as they became camouflaged. The dark moths passed on the alleles for dark wing colour leading to offspring with the dark wing colour phenotype. Over several generations in continued environmental condition, the darker moths became more common and as many as 98% of the moth population became dark coloured. In today's world, one of the most relevant examples of natural selection or adaptive evolution is antimicrobial resistance, which is currently a global crisis. Pathogenic bacteria and fungi are able to evolve resistance to antimicrobials on account of repeated exposure to drugs such that the drugs are no longer able to control the infection.

While we understand now that evolution occurs over a long period of time as heritable traits change according to the environment, but what exactly changes? It had already been established that the unit of heredity is the DNA. Specifically, changes in the genes that affect the phenotype or the trait lead to evolution. Along with natural selection, there are other mechanisms that drive genetic variation and hence evolution. These are random mutations, gene flow and genetic drift. While mutation is actually the original source of any genetic variation, the mutation rate is generally less, except when driven by a strong selective force. Random and rare mutations that are desirable can be fixed in a population. Gene flow refers to movement of genes into or out of a population, mainly by movement of organisms to a different location. Genetic drift leads to changes in allele frequency due to chance events. This may lead to loss of some alleles completely.

1.3.2 Evolutionary Lineage

Biological evolution, as Darwin proposed, is descent with modification by genetic inheritance. This modification includes changes in genes or alleles as well as descent of different species from a common ancestor over many generations. The central idea thus is that all life on earth shares a common ancestor. Through a continuous process of descent with modification, the last universal common ancestor (LUCA) gave rise to the amazing diversity of species that we see today. The evolutionary or phylogenetic tree is a branching diagram depicting the relationships of species or other entities based on their similarities or differences of physical and genetic characteristics. The tree of life refers to such a phylogenetic tree of all the species tracing the relationship back to the common ancestor. Evolutionary lineage is a series of organisms connected by a continuous line of descent from ancestor to descendant over a period of time. Thus, evolutionary lineages are a part of the phylogenetic tree that allows us to study the evolutionary history of a species.

Figure 1.17 depicts a hypothetical phylogenetic tree showing relationships between species A, B, C, D and E. The study of the branches and from where they

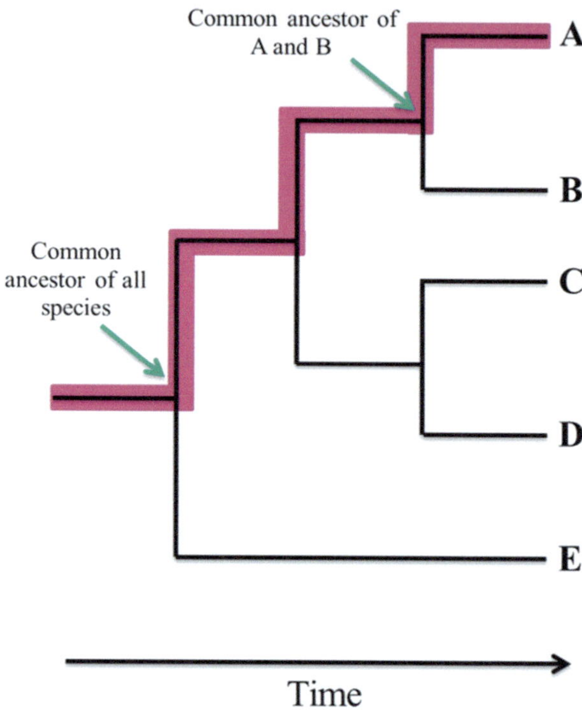

Fig. 1.17 A hypothetical phylogenetic tree displaying relationships between species A, B, C, D and E is shown. The common ancestors are indicated by arrow. The pink highlight indicates the evolutionary lineage of species A which can be traced to the last common ancestor

arise helps us to determine their evolutionary relationship. Each branch point or the internal node depicts a divergence event, where a single group of species split into two different species. Thus, the lineage of species A and B can be traced to the branch point from where they emerged, which is the common ancestor for A and B. This also indicates that A is more closely related to B than to any other species.

1.4 Genetics in Biological Research

The understanding of inheritance and its relation to evolutionary concept had resulted in development of a plethora of techniques which help scientists answer the most basic and important questions in biology. Development of tools for genetic manipulation has revolutionized the study of genetics and has been useful in development of recombinant products like insulin secreted by bacteria, pest-resistant crops and gene therapy to treat diseases in humans, among several other applications. Moreover, every day we are able to learn more about the mechanisms underlying cellular functions and organismal behaviour by applying genetic analysis in laboratories worldwide. Powerful techniques for isolating, recombining and analysing DNA techniques have allowed us to ask important questions in biological research. Most of the time to understand a biological pathway or a process, a scientist embarks on identifying and deducing the function of the gene(s) involved in that pathway. There are two main approaches for analysing gene function, viz. forward genetics and reverse genetics. Both these approaches are based on the traditional methodology of isolation of a mutation in a particular gene to identify the gene function, and both approaches are widely used by scientists today.

1.4.1 Forward Genetics

In forward genetics, genetic screens are developed by inducing mutations in the population using different means, described below. The aim is to induce mutation in every gene in the population by saturation mutagenesis. Then, the scientist begins studying an aberrant phenotype (a mutant in the population) to identify the gene that is responsible for producing this phenotype. For example, consider that a scientist is interested in identifying the gene responsible for cell division in budding yeast. First, the scientist will try to identify mutant individuals that are either slow growing or are arrested in their cell division cycle. On finding such mutant strains of budding yeast, the mutations can be mapped, thus allowing identification of the genes involved. The genes can be cloned and sequenced. The proteins formed by the genes can be isolated and their role in cell division can be biochemically studied. Hence, in forward genetics, the mutant phenotypes are known and are available before the corresponding genes are identified. One can find the names of many genes as per the phenotype that was used to identify them. One example is the *rosy* gene in *Drosophila* that encodes the enzymes xanthine dehydrogenase, mutation in which leads to lack of red pigmentation in the eye of the flies.

In early studies of genetics, scientists looked at naturally occurring variants or mutants in a population and studied them to identify and understand the gene function. However, discovery of mutagenic agents—chemicals that cause genetic mutations—allowed scientists to have large number of mutants with varied phenotypes. The earliest example of this is the use of X-rays by Hermann Muller in 1927 to induce mutations in *Drosophila melanogaster*. In addition to X-rays, there are different types of mutagen agents that are used today. X-rays cause breaks in double-stranded DNA, but may not be accurate for single genes and may lead to alterations in large pieces of the DNA. For finer mutations, scientists use chemicals like ethyl methanesulfonate (EMS) that cause point mutations and change at one nucleotide position. These mutations can be within a coding sequence of the gene or the regulatory region of the gene. However, such mutations are difficult to map. Another way for inducing mutations is by the use of transposons. Transposons, or transposable elements (TE), are DNA sequences that have the ability to move (or transpose) to different sites within the genome. If a transposon from a transcriptionally inactive site like heterochromatin moves into a genic region, it results in functional inactivation of the gene. This is called as insertional mutagenesis. Transposon mutagenesis leads to higher mutation frequency as compared to chemical mutagenesis, and availability of using a selectable marker for confirmation of transposition makes confirmation of mutants easy. The TE insertion can be easily mapped and the region can be cloned and sequenced. With this ability to generate mutations, scientists can develop genetic screens to understand gene function and dissect biological pathways. Forward genetic screens are hence unbiased with no prior knowledge of the gene product.

1.4.2 Reverse Genetics

Reverse genetics is an alternative to forward genetics, wherein the scientist begins with a genotype, alters its expression or sequence and then studies the effects of that alteration in the phenotype. A gene with an unknown function can be mutated, and its phenotype can be studied to understand its function in the cellular biology of an organism. The method of inducing mutations at specific locations in the genome is called site-directed mutagenesis. One strategy for site-directed mutagenesis is to use short oligonucleotides containing the desired mutated base. These oligonucleotides can anneal to single-stranded DNA template, and then are extended by DNA polymerase by polymerase chain reaction (PCR). The products of this amplified mutation are introduced into the cells and then mutants can be screened for the presence of the mutated gene.

Another method that is widely used in reverse genetics is transgenics. In this method, gene of interest is added to an organism that naturally lacks that gene, and the effect of the function of this gene in this organism is studied. This organism whose genetic makeup is now altered is called a transgenic organism and the foreign DNA added is called a transgene. Often, genes from the human genome are added to mice to obtain transgenic mice because of the ease of genetic manipulation of these genes in this model organism.

"Knockout" libraries of model organisms, made by using reverse genetics approach, are widely used in research today. Knockout mutant of a particular gene indicates complete inactivation of the gene resulting in loss of function. Knockout libraries are collections of mutants of a species in which almost all genes have been inactivated. In this library collection, each strain has a single mutation in a different gene. Such library can be generated by random mutation of a population using transposon insertions or T-DNA insertions (for plants). These libraries are then screened by PCR using a primer specific to the gene of interest and a primer specific to the inserted DNA. Such libraries are an invaluable resource for large-scale reverse genetics experiments where scientists aim to study the function of all genes in a model organism.

RNA interference (RNAi) is a gene silencing mechanism in different organisms. The 2006 Nobel Prize was awarded to Andrew Fire and Craig Mello for their work on RNAi which they used as a genetic approach to dissect biochemical pathways in *C. elegans*. RNAi is a naturally occurring cellular mechanism in which an organism's own gene is switched off by double-stranded RNA (dsRNA) in a process called as gene silencing. During RNAi, long dsRNA are cut into small fragments by an enzyme called as Dicer. These fragments called as small interfering RNAs (siRNA) bind to a family of proteins called as Argonaute proteins. On binding, one strand of the dsRNA is removed, and the single-stranded siRNA binds to the complementary mRNA (according to base pairing rules). At this stage, the Argonaute can cleave this target mRNA or regulate it in such a way that it is non-functional. The primary role of this process in the cell is to silence repetitive DNA of the cell or invading double-stranded RNA viruses. This natural process of the cell is exploited by scientists to design siRNAs complementary to the mRNA for the gene of their choice. The phenotypes obtained on RNAi silencing show partial loss-of-function phenotypes, called as knockdowns. The advantage of this approach is the ease and rapidity of the method and is especially useful for studying essential genes.

1.4.3 Manipulation of DNA

The above-mentioned approaches for genetic manipulation have been possible owing to the development of robust techniques for manipulation of DNA. Understanding gene function requires precise techniques to be able to isolate DNA, find the desired gene and isolate it or mutate it, insert the mutated gene in a cell and study the consequent effects. While the development of these techniques was not straightforward, they are now routinely used labs worldwide as basic molecular biology techniques.

The first step in any molecular genetics analysis is obtaining the DNA from a cell in isolation of other cellular components. Several methods are available for isolation of DNA and most of them rely on the same principle. The first step involved breakdown of the cell wall and membrane using lysis buffer which mostly contains a detergent. Macromolecules other than DNA (like proteins and RNA) are broken down using enzymes. The DNA is then precipitated with alcohol. DNA extracted from large number of cells can be seen as white stringy mass that can be spooled around a glass rod (Fig. 1.18).

A key development in molecular genetics was the discovery of restriction endonucleases, commonly known as restriction enzymes. These enzymes recognize specific short DNA sequences and make double-stranded cuts in the DNA at specific sites. These enzymes are naturally present in bacteria and act as an immune mechanism against bacterial viruses. Bacteria's own DNA is protected from its enzymes because it is modified by addition of methyl groups. Several types of restriction enzymes have been isolated, and the most commonly used in molecular biology are the type II restriction enzymes, most of which are commercially available. The restriction enzymes are named according to the species from where they were isolated. For example, *Eco*RI was isolated from *Escherichia coli*. Usually, restriction enzymes recognize 4–6 bp long DNA sequence. Some enzymes cut the DNA in such a way that the ends are staggered generating single-stranded overhangs. For

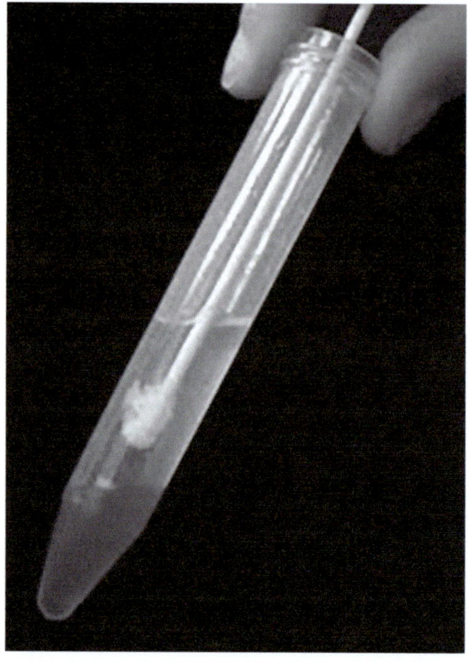

Fig. 1.18 White spool of strawberry DNA extracted from strawberry fruit (Teacher's manual by Carolina Biological Supply Company, USA). Total DNA was extracted by mashing strawberry fruit in an extraction buffer and precipitating the DNA with ethanol

example, the enzyme *Hin*dIII recognizes the following sequence and cuts the sugar-phosphate backbone of each strand at the point indicated by the arrow:

$$\begin{array}{c} \downarrow \\ 5' - A\,AGCTT - 3' \\ 3' - TTCGA\,A - 5' \\ \uparrow \end{array}$$

This generates fragments with the overhangs as shown below:

$$\begin{array}{ll} 5' - A & AGCTT - 3' \\ 3' - TTCGA & A - 5' \end{array}$$

Such staggered ends are called cohesive ends or sticky ends because they have sequence complementarity and can be easily paired or glued together. Hence, any two DNA fragments that are cut by this enzyme will give such complementary ends allowing us to join two different fragments together. This is called cutting and joining (ligating) DNA fragments. Some enzymes generate ends that are not sticky, but are blunt ends. *Pvu*II is an enzyme that cleaves in the following way:

5' – CAGCTG – 3' → 5' – CAG CTG – 3'
3' – GTCGAC – 5' → 3' – GTC GAC – 5'

A variety of recombinant DNA molecules can thus be generated by this process of cutting and joining at ambient conditions required for the enzyme action.

These changes made in the DNA have to be confirmed by visualization of the DNA. If the DNA fragment is cut by one enzyme that cuts at a site that occurs only once in your DNA fragment, then you should be able to observe two cut fragments with sizes that add up to the original fragment. To observe this, DNA fragments can be visualized by a technique called as agarose gel electrophoresis (Fig. 1.19). Agarose is a polysaccharide obtained from seaweed which is used to make a porous

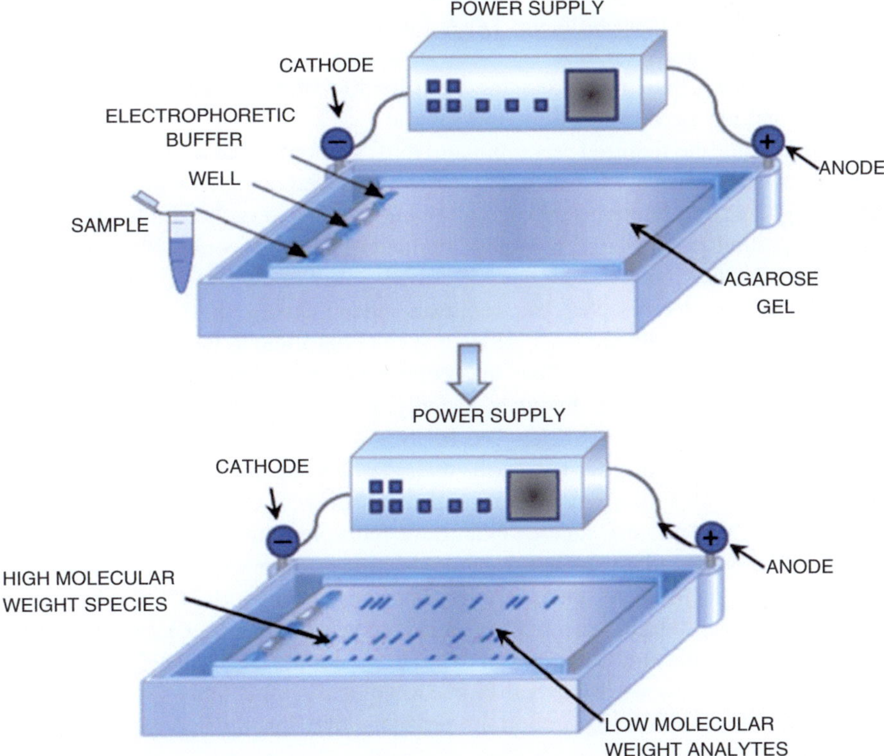

Fig. 1.19 Agarose gel electrophoresis (Drabik et al. 2016). A DNA sample is loaded into wells towards the cathode side in the agarose gel which is immersed in a buffer. The gel tank is connected to a power supply which passes electric current through the buffer. Due to this, DNA migrates from cathode towards anode with smaller DNA fragments migrating faster than larger fragments

gel matrix through which DNA molecules can move. DNA molecules are negatively charged ions at neutral or basic pH in an aqueous environment. In gel electrophoresis, DNA fragments get separated on the basis of their size, which is expressed in terms of number of base pairs present in that fragment. DNA samples are loaded into a well or a slot near the negative electrode of the gel matrix and drawn towards the positive electrode at the opposite end of the gel by applying electric current. Smaller molecules move through the pores in the gel faster than larger molecules, and this difference in the rate of migration separates the fragments on the basis of size. Standard DNA samples with known sizes are usually run alongside the molecules to provide a size comparison. DNA can be visualized by using fluorescent dyes like ethidium bromide that can intercalate between the DNA strands and fluoresce on exposure to UV. Distinct nucleic acid fragments appear as bands at different distances from the top of the gel depending on their size. DNA samples can also be probed using certain complementary short sequences. The short fragments called as probes are designed and labelled with radioactive or fluorescent dyes for detection. After running the DNA sample on agarose gel for separation, the DNA fragments are transferred onto a nylon membrane, and this process is called as blotting. This membrane with the DNA fragments can now be probed with the designed probes and visualized by X-ray or fluorescence. This technique is called Southern blotting, and it is used for confirmation of the DNA manipulation or mutation that has been induced in your sample.

On obtaining the required recombinant DNA, it is critical to amplify the product or have several copies of the DNA. This can be achieved by placing the recombinant DNA fragment in a bacterial cell and allowing the cell to replicate the DNA. This process is called as gene cloning as large number of identical copies or clones can be generated. Bacteria and yeasts have plasmids (also known as a vectors), small circular DNA molecules that can replicate independently of the cellular DNA. Plasmids occur naturally and have genes that can contribute to favourable properties to the organism carrying it, like antibiotic resistance. These plasmids can be engineered using restriction enzymes. Usually, plasmids used in molecular biology have multiple cloning site (MCS). The MCS is a short DNA sequence containing multiple sites that can be cut with different commonly available restriction endonucleases. This property makes plasmids suitable vectors for carrying the DNA sequence of interest. One cell can have multiple number of plasmids, which amplifies the clones that can be obtained. On introduction within a host cell, such plasmids replicate to make several copies, thus amplifying your DNA of interest as well (Fig. 1.20). A transgene like a human gene can thus be placed within the bacterial cell on a plasmid and allowed to amplify with the bacterial cell. Plasmids as vectors have been used for several biotechnological applications with large-scale production of economically important products like insulin and human growth hormone.

Any fragment of DNA can be amplified from the genome using a technique called as polymerase chain reaction (PCR). This technique was first developed by Kary Mullis and allows DNA fragments to be amplified billion times in just a few hours.

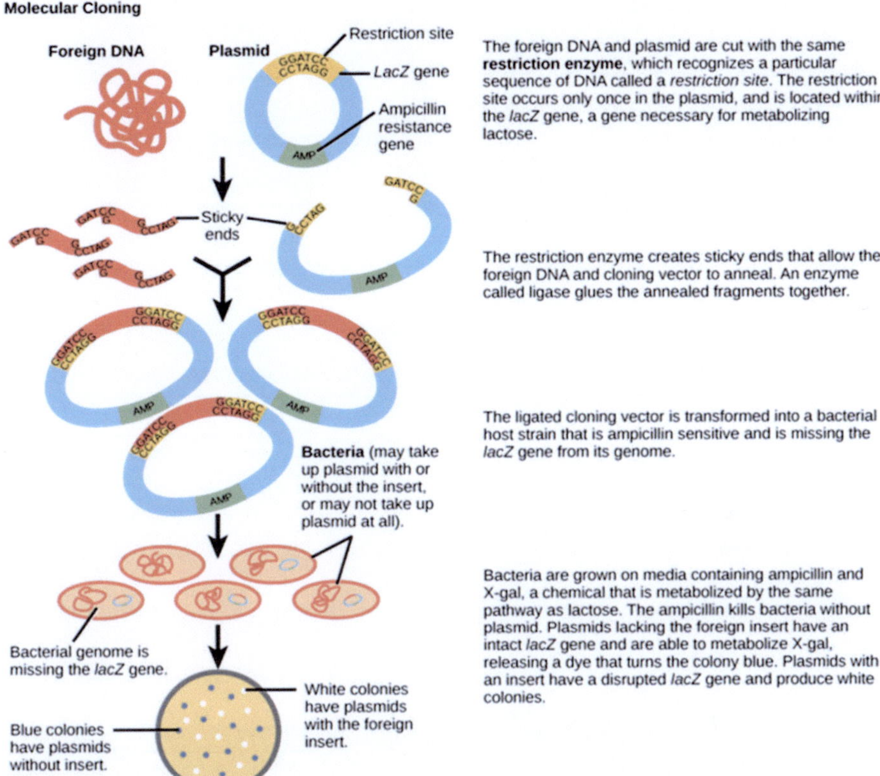

Fig. 1.20 Steps in molecular gene cloning (OpenStax College, Biotechnology. October 16, 2013. Provided by: OpenStax CNX. Located at: http://cnx.org/content/m44552/latest/Figure_17_01_06.png). This diagram shows the steps involved in molecular cloning of *lacZ* gene required for lactose metabolism. This gene is ligated into the plasmid using restriction enzymes. The bacteria that have the correct recombinant plasmid are screened by blue-white screening using the chemical X-gal

Even a single molecule of DNA can be used as a starting point to obtain several million copies by PCR. It is a robust and most widely used technique in molecular biology. The critical factor in a PCR reaction is the enzyme DNA polymerase. To replicate DNA, the parent or template DNA should be single-stranded. To achieve this, the temperature of the reaction is increased to 90–100 °C so that the hydrogen bonds between the two strands of the double-stranded DNA break. Primers or short complementary sequences are added to the reaction that binds to the single-stranded DNA at a particular temperature between 30 and 65 °C when the reaction is cooling from 90 to 100 °C. DNA polymerase is able to synthesize a complementary DNA strand starting from the site where the primer attaches. Thus, two new strands from two parent strands are produced. The whole cycle is then repeated several times and

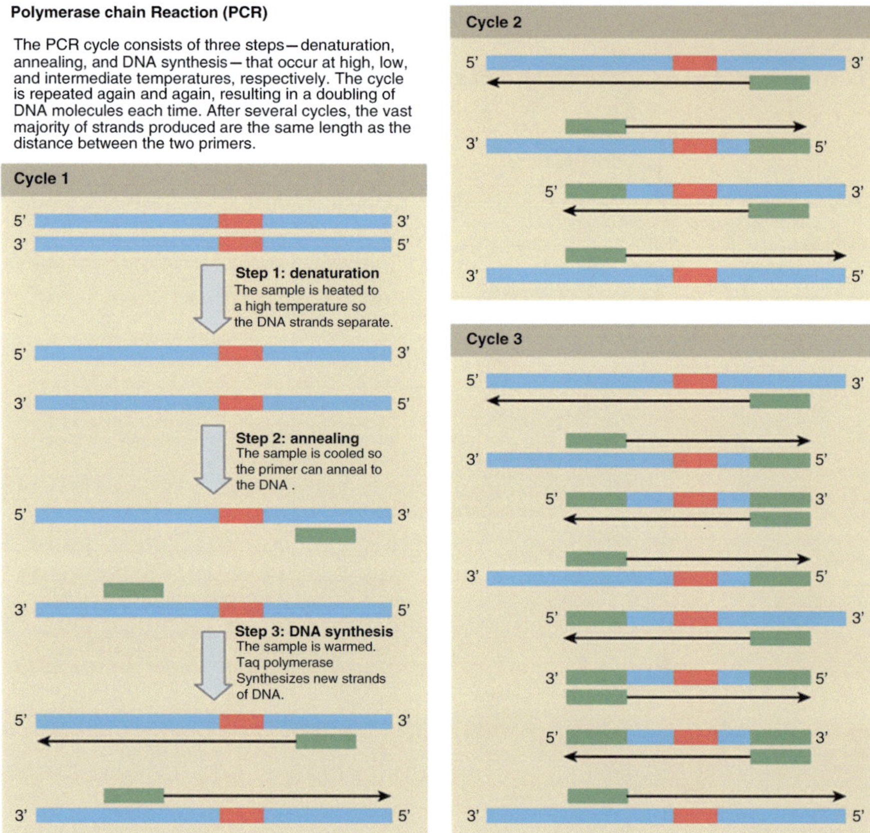

Fig. 1.21 PCR amplification (OpenStax College, Biotechnology. October 16, 2013. Provided by: OpenStax CNX. Located at: http://cnx.org/content/m44552/latest/Figure_17_01_04.jpg). PCR is used to amplify a specific sequence of DNA using thermostable DNA polymerase, primers and deoxynucleotides

the number of strands produced increases exponentially (Fig. 1.21). The critical discovery for this technique to work was the discovery of a DNA polymerase that was active such high initial temperatures at every cycle. This thermostable DNA polymerase was isolated from the bacterium *Thermus aquaticus* from the hot water springs of Yellowstone National Park, USA, and is known as *Taq* polymerase. In addition to amplification of DNA, PCR can also be used for amplifying sequences complementary to RNA. For this, the RNA is first converted to its complementary DNA (cDNA) using a viral enzyme called as reverse transcriptase. The cDNA is then subjected to regular cycles of PCR. This method is known as reverse-transcription PCR.

1.5 Pathway of Genetic Analysis

The study of genetics consists of three major sub-disciplines: classical genetics (also called transmission genetics), molecular genetics and population genetics (Fig. 1.22). Classical genetics includes the basic principles of heredity and inheritance of traits. The focus of study is an individual organism—how the organism inherits traits from the parents and then transmits traits to the next generation.

Molecular genetics deals with the nature of the actual genetic information underlying the inherited traits. This study includes the chemical nature of the gene and cellular processes that lead to the phenotype, including DNA replication, transcription, translation and gene regulation. The organization, structure and function of the gene are studied under molecular genetics.

Population genetics explores inheritance of traits and the underlying genetic mechanism in groups of individuals of the same species, which is called as a population. How the genetic composition and hence the trait changes spatially and

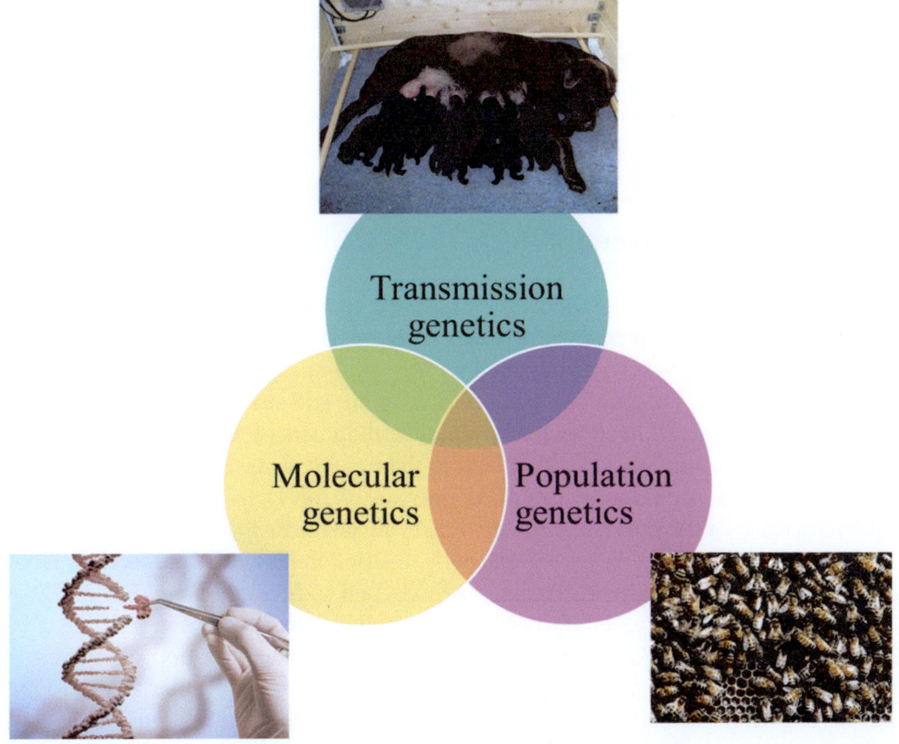

Fig. 1.22 Categories of genetics. The field of genetics can be subdivided into three different types—transmission genetics, molecular genetics and population genetics. Image source: top, @IngoDiBella via Flickr; bottom left—Livescience.com; bottom right—Time.com

temporally is studied. Hence, population genetics is fundamental to the study of evolution in which changes over time are studied.

1.5.1 Classical Genetics

When Gregor Mendel described the fundamental laws of inheritance through his rigorous experiments on garden pea plants, a new era of understanding in biology had commenced. His work and the work that followed gave rise to a field in biology that came to be known as Mendelian genetics, a synonym for classical or transmission genetics. The understanding for the mechanism for variation in traits within a population and inheritance of trait by next generations was mostly vague before Mendel's study. Mendel's hereditary experiments with pea plants led to the formation of law of segregation and law of independent assortment. The law of segregation describes how a pair of gene variants (alleles) is segregated in the reproductive cells (gametes). Mendel crossed two heterozygous plants (each with a different allele for a trait) and found that the trait in the offspring did not always match the trait of the parents, indicating that the alleles for the trait had segregated during the formation of gametes leading to different possible outcomes for the offspring's phenotype. Depending on the parental genotype, he could predict consistent ratios of phenotype in the offspring that can be produced. The law of independent assortment predicts the inheritance of two or more traits. Non-Mendelian inheritance patterns discovered later are also widespread in nature.

1.5.2 Molecular Genetics

Several lines of discoveries and inventions led to the formation of this field of genetic analysis. Identification of the unit of heritance, the gene, as a biochemical molecule, one gene-one enzyme theory, use of mutagens for making heritable changes in the genes and identification of the nature and structure of the inheritable molecule, the DNA, allowed for tremendous development of techniques to study the structure and function of genes that lead to the traits that were studied by classical genetics. Several techniques described previously for DNA manipulation using either forward or reverse genetics approach are used in the study of molecular genetics. Gene cloning employing plasmids, restriction enzymes and ligases, DNA amplification by PCR, hybridization methods and gel-based separation of DNA fragments are the basic tools of today that are used to answer questions in molecular genetics. Construction of whole-genome libraries and identification of mutations using PCR-based methods also helps to answer questions in molecular genetics. The advent of genome sequencing technologies in the last two decades has opened newer avenues and methods to answer basic questions in molecular genetics. Conventional sequencing methods are being constantly updated in order make genome sequencing rapid and easy. Today, we have technologies like Oxford Nanopore sequencing and Illumina sequencing, using which genome sequencing can be accomplished in a matter of few days. This is proven by hundreds of genomes being sequenced every day, and the technology being refined with newer and newer methods. Large-scale genome data (genomics) has opened up newer methods of computational analysis (bioinformatics). Different software and analysis platforms are continuously being updated for rapid and robust analysis of large quantities of "omics" data. Such data in collaboration with experimental data can answer major questions in biology that were not possible earlier. For example, if one wishes to know the binding regions of a protein on the genomic DNA of an organism, one can employ a technique called as "chromatin immunoprecipitation sequencing (ChIP-seq)" wherein the regions of the DNA that are bound by the protein are isolated and a library of the DNA fragments is made and then sequenced. Such whole-genome experiments are widely used by scientists and are accelerating the rate at which science is progressing.

1.5.3 Population Genetics

Population genetics is a branch of genetics that deals with the genetic composition and variation among individuals of a population within a species. In nature, often we observe individuals of a population displaying a variety of phenotype due to expression of different alleles of a gene, called as polymorphisms. This expression in a polymorphic population depends on the genetic structure as well as the environment. In population genetics, scientists try to understand the sources of such phenotypic variation in a population and predict how that population will evolve over time in the presence of different evolutionary factors. Organisms studied in population genetics are interbreeding, are sexually reproducing and have a common set of genes, known as the gene pool. Due to changes within the gene pool over time, the population evolves. The evolutionary forces that lead to these changes are also studied in population genetics. Genotypic and allelic frequencies are used to describe the genetic composition of a population. G.H. Hardy and Wilhelm Weinberg independently formulated a law, called as Hardy-Weinberg equation, that describes how reproduction and Mendelian principles affect these allelic and genotypic frequencies within a population. Allelic frequencies can be changed by several operational factors like mutations, migration, genetic drift and natural selection. Mutations can directly induce changes in the base composition of the DNA sequence. In natural selection, alleles that confer beneficial traits are selected and the ones that are deleterious are removed over time. Migration or large-scale movement of organisms from a population to another location leads to gene flow causing changes within the older as well and the new forming population. Genetic drift is a chance occurrence when some individuals have more offspring than others in the population, thus increasing the representation percentage of that allele. Usually, population geneticists develop mathematical models to study the patterns of genetic variation with a population. Modern population genetics however comprises of theoretical aspects, lab and field work.

1.6 Genetic Database

Decades of scientific research and the explosion of sequencing technologies have led to development of large number of databases which help connect scientific discoveries worldwide (Bianco et al. 2013). A biological database is an organized computer-based storage of information and data generated from scientific publications, experiments in research laboratories (in vitro and in vivo) and bioinformatics analysis (in silico). The information stored on a database is well-organized and easy to use. Databases are essential for continuous storing, sharing and updating data to keep the scientific community apprised of the latest research. Sharing and open data access to large-scale experimental projects has also led to large number of collaborative research projects. This is vital for rapid progress of scientific research which is ultimately beneficial to the human society. These databases are bioinformatics resources and tools that are open to public for information dissemination. Large number of sequenced genomes has led to development of not just an information storing database, but an interactive browser on which a user can view as well as analyse the sequence information deposited there. Such browsers are often linked to various bioinformatics tools for this purpose. In addition to genome sequence, other "omics" data like transcriptomics, proteomics and metabolomics collect experimental data which can be visualized and analysed on the databases.

Literature databases were one of the first scientific databases generated to collect and store all scientific publication in once place. Literature search is the first step of any scientific project that allows one to formulate a hypothesis based on the research already done in that particular field. The oldest scientific article database for biomedical research is PubMed, developed by National Center for Biotechnology Information (NCBI) which includes abstracts of the articles and links to the journal website. PubMed is the most widely used and updated site for bibliographic research in biomedical research.

NCBI also serves as an integrated and one of the largest and oldest platform for sharing and utilizing sequence-based resources in the scientific community. It provides an integrated data system for almost all existing genetic resources. It links the data to its original source as well as to a number of analytical tools that allow a researcher to obtain an in-depth valuable knowledge at the same location. Search on this website is extremely user-friendly with terms like gene symbol, gene name, marker name, text work or phrase related to the gene can be searched. The output displays the availability of your search term in different NCBI databases. It provides the options for a refined search depending on the user need. For example, search term "tnf" gives an interface as shown in Fig. 1.23. TNF (tumour necrosis factor) is a gene superfamily that regulates several cellular processes including immune response, cell proliferation and differentiation. As seen in the figure, the search gives links to several databases housed within NCBI, including genomes, proteins and genes that carry this search term. The user can easily navigate to the database of choice and visualize the data required. NCBI also houses several bioinformatics tools like Basic Local Alignment Search Tool (BLAST), conserved domain search tool, multiple sequence alignment tools and several others.

NCBI Databases

Results found in 33 databases for: **tnf** (1 errors)

Literature	
Bookshelf	4,181
MeSH	244
NLM Catalog	100
PubMed	164,186
PubMed Central	271,881

Genes	
Gene	20,668
GEO DataSets	9,422
GEO Profiles	871,786
HomoloGene	100
PopSet	53
UniGene	ERROR

Genetics	
ClinVar	2,140
dbGaP	0
dbSNP	1,216
dbVar	12,684
GTR	421
MedGen	115
OMIM	692

Proteins	
Conserved Domains	131
Identical Protein Groups	7,205
Protein	44,365
Protein Clusters	8
Sparcle	595
Structure	564

Genomes	
Assembly	0
BioCollections	0
BioProject	953
BioSample	1,794
Genome	34
Nucleotide	85,788
Probe	8,544
SRA	4,078
Taxonomy	5

Chemicals	
BioSystems	6,949
PubChem BioAssay	10,901
PubChem Compound	17
PubChem Substance	1,582

Fig. 1.23 Interface of NCBI website on search of the tern "tnf". The search term is found in several databases in the website. Depending on the question to be asked, the user can navigate to the databases and view and analyse the gene or protein of interest

Ensembl database (from the French word "ensemble" and "EMBL" European Molecular Biology Laboratory) database is a software system created by the EMBL-European Bioinformatics Institute (EBI) and the Wellcome Trust Sanger Institute (WTSI) to handle genome annotations of eukaryotic organisms. Like other databases, its aim is to provide genome sequence annotations as a free resource to the scientific community while integrating and linking other biological data. Several other databases exist and more are coming up as the bioinformatics data for biology keeps increasing at a rapid pace. Some of the important databases for model organisms are shown in Table 1.2.

Other important databases are listed here:

- National Center for Biotechnology Information (GenBank)—http://www.ncbi.nlm.nih.gov
- Ensembl—http://www.ensembl.org
- Human Gene Mutation Database (HGMD)—http://www.hgmd.cf.ac.uk/
- International Sequencing Consortium—http://www.intlgenome.org
- Kyoto Encyclopedia of Genes and Genomes (KEGG)—http://www.genome.ad.jp/kegg/

Table 1.2 Genetic databases for model organisms

Common name	Scientific name	Genetic database	URL
Baker's yeast	*Saccharomyces cerevisiae*	Saccharomyces Genome Database (SGD)	www.yeastgenome.org
Fruit fly	*Drosophila melanogaster*	FlyBase	www.flybase.org
Thale cress	*Arabidopsis thaliana*	The Arabidopsis Information Resource (TAIR)	www.arabidopsis.org
Common house mouse	*Mus musculus*	Mouse Genome Informatics (MGI)	www.informatics.jax.org
Worm/nematode	*Caenorhabditis elegans*	WormBase	www.wormbase.org
Zebrafish	*Danio rerio*	The Zebrafish Information Network (ZFIN)	www.zfin.org

- NCBI Entrez Web site—http://www.ncbi.nlm.nih.gov/Entrez/
- National Human Genome Research Institute (NHGRI)—http://www.genome.gov
- Online Mendelian Inheritance in Man (OMIM)—http://www.ncbi.nlm.nih.gov/omim
- Protein Data Bank (PDB)—http://www.pdb.org
- University of California at Santa Cruz (UCSC) Genome Browser—http://genome.ucsc.edu
- TIGR Gene Indices—http://www.tigr.org/tdb/tgi

1.7 Application of Genetics

Genetics has been exploited for thousands of years, even before the underlying mechanisms were unknown. Fermentation is used widely in the brewing and bakery industries since a very long time. Artificial selection of animals and plants by cross-breeding benefitted humans by providing improved products of dairy and increased yield of plant-based food products. In the recent past, understanding the inheritance mechanisms and development of genetic tools has opened myriad of opportunities for using genetic engineering in a variety of field that has benefitted humans directly as well as the environment. Some of the significant applications have been described below.

- Pharmaceutical products: Using recombinant DNA technology, easy to grow organisms like bacteria and yeast have been used as small "biofactories" for production of several compounds useful in medicine. Bacterial or yeast cells are transformed with a vector containing the gene for the product of choice derived from the human genome. These cells are grown in large bioreactors where they overexpress the transgene and produce large quantities of product that can be harvested and purified. One of the earliest examples of such recombinant technology was the production of human insulin. Insulin is essential for the control of blood sugar levels, and when the body cannot produce insulin, it leads to the disease diabetes mellitus. In such cases, patients have to take insulin from external sources to control blood sugar levels. Recombinant insulin is produced either in yeast or *E. coli*. The first genetically engineered, synthetic "human" insulin was produced using *E. coli* in 1978 by Arthur Riggs and Keiichi Itakura at the Beckman Research Institute in collaboration with Herbert Boyer at Genentech. Genentech in 1982 started selling the first commercially available biosynthetic human insulin under the brand name Humulin. Other medically useful products like growth hormones, vaccines, blood clotting factor, monoclonal antibodies as well as drugs are being produced using this technology. Recent advances have also led to use of plants for production of recombinant pharmaceutical products (Ma et al. 2003).

- Specialized microorganisms: Several microorganisms are also being used to recover oil from oil wells, break down toxic chemicals like oil spills and other pollutants and solubilize minerals from ores. Bioremediation is a field of biology where microorganisms like bacteria and fungi or their products are used to degrade or remove toxic compounds from the ecosystem.
- Agricultural products: Most of the crops that are agriculturally important today are quite different from their wild progenitors. Artificial breeding has caused several genetic modifications in these crops to select for desirable traits like high yield, disease and pest resistance, high nutritional value and so on. The Green Revolution led by Norman Borlaug relied heavily on genetic methods to develop high yielding crops. Norman Borlaug was awarded the Nobel Peace Prize because his revolution fed malnourished populations across the globe by introducing these techniques in the agricultural system of poorer countries. Such methods of conventional breeding also involve changes in the DNA sequence of the organism, just like the relatively recent genetic modification (GM) technology. However, genetic changes brought about by the GM technology are small in number, well defined, precise and targeted as compared to classical breeding methods where several genes of an organism may be involved. Today, a significant proportion of food products like corn and soybeans are genetically modified. However, the rules for modifying crops by genetic engineering are different across different countries with some concerns about their safety. One of the first transgenic plants produced by GM technology was Bt cotton. A gene from the bacteria *Bacillus thuringiensis* (Bt) that encodes for an insect toxin was cloned into the cotton genome, which made the plant resistant to the common pest bollworm. Bt cotton has gained tremendous success in countries like India as the cotton production boosted when transgenic plants were used.
- Genetic testing for diagnosis: Several human diseases have been found to have an underlying genetic and heritable component. Genetic disorders like sickle cell anaemia, Huntington disease and breast cancer are some of the examples. If there is a family history of a particular genetic disease, then identifying the genetic mutations and the genes involved in these diseases allows diagnosis of the disorder before it occurs and predicts the person's predisposition to that disease

so that prophylactic measures can be taken if necessary. For example, mutations in the human genes *BRCA1* and *BRCA2* notably increase the risk of breast and ovarian cancers in women at younger age. These two genes produce tumour suppressor proteins that help repair damaged DNA and thus maintain stability of the genome. Mutations that lead to non-functional proteins or proteins with altered function prevent this process of repair and hence lead to cancer. The deleterious *BRCA1* or *BRCA2* mutation can be inherited from mother or father such that each child with a parent carrying the mutation has 50% chance of inheriting the mutation. The effects of mutations in *BRCA1* and *BRCA2* are seen even when a person's second copy of the gene is normal. Identification of the mutation by sequencing the DNA can help patients in taking risk-reducing measures before the full-blown disease occurs.
During an infection, the causative pathogen can be identified using genetic tools. Several diagnostic kits are available for identification of specific species of the pathogen to allow for targeted treatment.

- Gene therapy: Gene therapy is the direct modification of the genes of a patient's cells to treat a disease. Gene therapy is mostly an experimental technique and is used only when no treatment options are available. However, it works well in certain kinds of diseases like haemophilia and severe combined immunodeficiency (SCID). The modifications in the genes can be carried out by introducing an engineered virus inside the cell or by collecting cells from the bone marrow or blood, editing them externally and using them to replace patient's cells. This method is currently at its infancy with no gene therapy products available on the market, but is promising and requires more in-depth research on the side effects and aspects of safety.

1.8 Promise of Genetics

The completion of the Human Genome Project paved the way for intense research on human genetic composition, identification of genes involved in diseases and study of cellular pathways that get altered in a disease. In the future, the area of functional genomics to study the function of genes in normal conditions will continue to be an active area of research as scientists will categorize more and more genes involved in human health. Gene therapy as a technique is likely to gain major advancements that will allow scientists to gain more success in treatment of diseases with mutations in a single gene. Medicinal genetics is headed towards comprehensive study of a disease by investigating multiple genes, pathways, systems, the effects of environment as well as genetic variations within a population. Newer methods will lead to study of complex diseases with such comprehensive outlook. Such studies are likely to utilize large sample sizes of thousands of patients whose genome data can be analysed. While robust, relatively cheap and easy to use sequencing platforms are already available today for analysis for large datasets like genomes, transcriptomes and proteomes, analysis tools, algorithms and software for handling such datasets are likely to be developed soon. These studies will allow doctors to predict predisposition of a patient to diseases like cancer, diabetes, cardiovascular diseases and so on. Another area of research that has benefitted greatly by advancements in genomics is the field of pharmacogenomics, the study of how genes are involved in a patient's response to drugs. While there have been only small studies so far, the future for these studies is promising. All these research areas will boost the arrival of personalized medicine. While the real-life implementation of personalized medicine still has a long way to go, there are good indications of major advancements in the steps leading to it. It can be predicted that precise genome editing techniques developed recently like meganucleases, zinc finger nucleases (ZFNs), transcription activator-like effector nucleases (TALENs) and clustered regularly interspaced palindromic repeats (CRISPR)/Cas will be improved and utilized in generation of genetically modified organisms like pathogen-resistant or drought-tolerant crops. While the promise of genetics in the future is likely to be fulfilled, major steps will be required for regulation of these methods to prevent misuse. The sad events of human experimentation during World War II exemplify how advancement of these technologies, especially genome editing, can be exploited. It will be the collective responsibility of the scientific community, policy makers and governments to ensure that does not happen and the science of genetics can be utilized for betterment of the human society.

Box 1.1 Scientific Concept: Forward and Reverse Genetic Approaches for the Analysis of Vertebrate Development in the Zebrafish—Nathan D. Lawson and Scot A. Wolfe

Forward and reverse genetics approaches have been widely used in bacterial, fungal and invertebrate models to study gene function and dissect cellular pathways. Study of developmental processes using these approaches has been successful in organisms like *C. elegans* and *D. melanogaster*, but poses major challenges for use in mammalian models like mouse because of complex internal development. As an alternative, utilizing zebrafish (*Danio rerio*) was proposed as a model system to study vertebrate development. This organism possesses several characteristics that make it suitable for such studies. For example, its clutch size per mating pair is large (>100), and it is externally fertilized with embryos developing rapidly and synchronously ex vivo. An important factor was the availability to perform detailed microscopic observations of the nearly transparent embryo at early stages of development. Zebrafish are small and easy to handle, and relatively less expensive to maintain as compared to other mammalian models. These factors have made it an established model organism to study vertebrate developmental pathways. As with other systems, forward genetics approach was initially utilized to identify gene functions in zebrafish. Several mutagens have been utilized to obtain genetic lesions with observable phenotypes. Initially, gamma rays were used which largely gave gross chromosomal aberrations which made accurate

(continued)

Box 1.1 (continued)

gene function prediction difficult. Later, *N*-ethyl-*N*-nitrosourea (ENU) was used which gave high mutagenic loads with phenotypes that can be linked to discrete genes. It is easy to expose individuals to the mutagen by simply adding the compound to the water, and ENU is currently the choice of mutagen for forward genetics in zebrafish. Other methods include using replication-deficient pseudotyped retroviruses or transposons as insertional mutagens. Methods to identify phenotype include serial microscopic observation of morphology and behaviour during embryo development, using cell- or tissue-specific molecular markers using whole mount in situ hybridization and immunostaining, and generating transgenic lines with tissue-specific fluorescent makers. Several studies employing these forward genetics techniques have resulted in identification of genes and pathways important for gastrulation and mesoderm induction, cilia formation, progression of infectious disease, regeneration and cardiovascular development. In many of these studies, pathways and genes conserved in diseases affecting humans have been revealed. However, given the large size of zebrafish genome, forward genetics approaches alone are unlikely to provide functions of all the genes. Additionally, availability of whole-genome sequence assembly and extensive expression analysis has provided the opportunity to develop tools to dissect gene function using reverse genetics approaches. These approaches include morpholino-mediated gene knockdown, Targeting Induced Local Lesions In Genome (TILLING), retroviral- and transposon-mediated mutagenesis and targeted gene inactivation via zinc finger nucleases. Of these, zinc finger nuclease (ZFN) is an exciting recent advance in genome editing in several organisms. ZFNs are a class of engineered DNA-binding proteins that can create double-strand breaks at specific locations leading to precise genome editing. Each ZFN consists of two functional domains, viz. a DNA-binding domain and a DNA-cleaving domain. The DNA-binding domain is designed to bind specific target loci, while the DNA-cleaving domain consisting of Fok1 nuclease creates a double-stranded break on dimerization. This fusion of the DNA-binding and DNA-cleaving domains creates a highly specific molecular scissors for precise gene editing. Use of ZFN in zebrafish is shown in Fig. 1.24. ZFN-mediated gene inactivation is sufficiently robust in zebrafish and causes minimal off-target damage in the genome. Despite several advances in the past few years, ZFN technology in zebrafish has limitations owing to the lack of knowledge of sequence-specific DNA recognition by ZFNs in the zebrafish genome. Technological advancements should allow use of this as well as other technologies in zebrafish rapidly in the coming years to be able to create several human disease models in this vertebrate model organism.

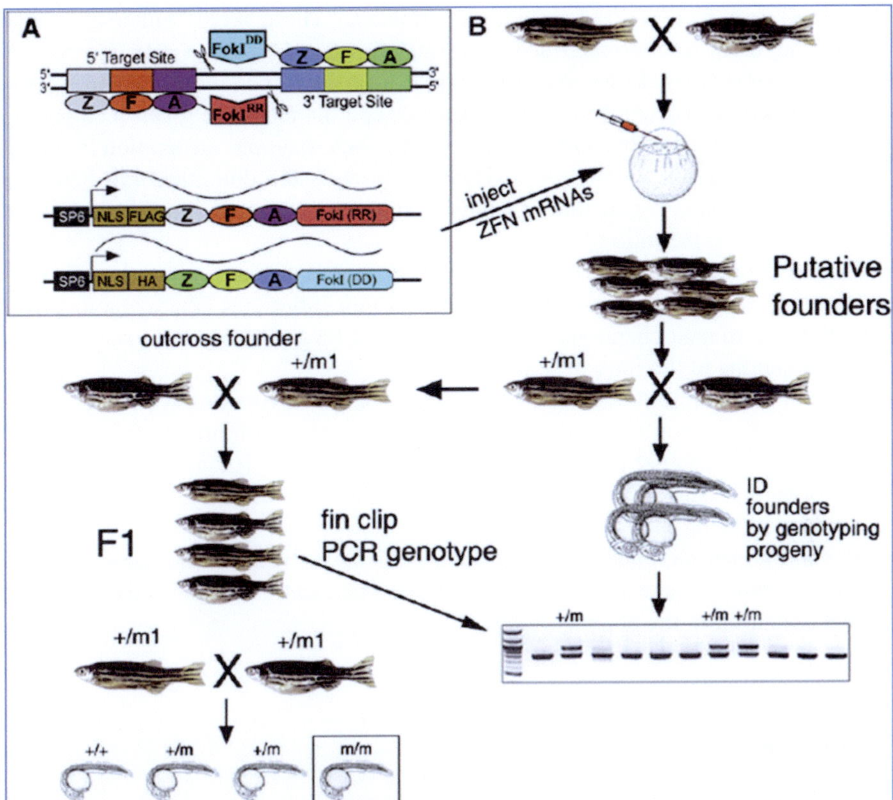

Fig. 1.24 Overview of ZFN-based gene inactivation. (**a**) A pair of ZFNs are designed to bind neighbouring sequences within the target gene of interest. DNA recognition is mediated by the ZFA, while the attached FokI nuclease domain generates a double-stranded break (DSB) upon dimerization. (**b**) mRNAs encoding each ZFN are prepared and then injected into one-cell embryos. Putative founders from these injections are raised to adulthood and out-crossed to identify carriers and the mutant alleles they transmit. Founders harbouring interesting alleles are out-crossed to generate an F1 population, and heterozygous F1 carriers are identified and then in-crossed to provide homozygous mutant embryos for phenotyping

1.9 Summary

- Genes confer phenotypes. Genes are inherited, and expression of these genes along with environmental effects determines the trait or the phenotype. The genetic information of an organism is called the genotype and the expressed trait is called the phenotype.
- Mendel studied seven different traits in the pea plant, each of which had two alternate forms. These traits were height (tall or short), seed colour (green or

yellow) and seed shape (smooth or wrinkled). These were "either-or" traits, meaning there were no intermediate forms.

- Monohybrid cross: In the P generation, true-breeding pea plants (homozygous) for the dominant phenotype of yellow seeds and the recessive green seed phenotype are crossed. This cross produces F1 heterozygote generation with all individuals having yellow seeds. Hence, yellow is the dominant trait here. On self-pollination, F2 generation produces has a mix of yellow and green seed individuals, in the ratio of 3:1.
- Dihybrid cross: Pure breeding parents for two traits are crossed with pure breeding lines with recessive phenotype. F1 generation has heterozygous plants with the dominant phenotype for both traits, and F2 generation has plants with all combinations in the ratio of 9:3:3:1.
- Natural selection is one of the core mechanisms of evolutionary change which leads to the evolution of adaptive traits. These selected traits are inherited and passed on to the next generation.
- In forward genetics, genetic screens are developed by inducing mutations in the population using different means.
- Reverse genetics is an alternative to forward genetics, wherein the scientist begins with a genotype, alters its expression or sequence and then studies the effects of that alteration in the phenotype.

References

Allocca M, Zola S, Bellosta P (2018) The fruit fly, *Drosophila melanogaster*: the making of a model (part I). In: *Drosophila melanogaster*—model for recent advances in genetics and therapeutics. https://doi.org/10.5772/intechopen.72832

Bianco AM, Marcuzzi A, Zanin V, Girardelli M, Vuch J, Crovella S (2013) Database tools in genetic diseases research. Genomics 101(2):75–85

Clarke T (2002) Mice make medical history. Nature. https://doi.org/10.1038/news021202-10

Cyranoski D (2019) The CRISPR-baby scandal: what's next for human gene-editing. Nature 566: 440–442. https://doi.org/10.1038/d41586-019-00673-1

Drabik A, Bodzoń-Kułakowska A, Silberring J (2016) Gel electrophoresis. In: Proteomic profiling and analytical chemistry, 2nd edn, pp 115–143. https://doi.org/10.1016/B978-0-444-63688-1.00007-0

Duina AA, Miller ME, Keeney JB (2014) Budding yeast for budding geneticists: a primer on the *Saccharomyces cerevisiae* model system. Genetics 197(1):33–48. https://doi.org/10.1534/genetics.114.163188

Hales KG, Korey CA, Larracuente AM, Roberts DM (2015) Genetics on the fly: a primer on the *Drosophila* model system. Genetics 201(3):815–842. https://doi.org/10.1534/genetics.115.183392

Hanson PK (2018) *Saccharomyces cerevisiae*: a unicellular model genetic organism of enduring importance. Curr Protoc Essent Lab Tech 16:e21. https://doi.org/10.1002/cpet.21

Jennings BH (2011) *Drosophila*—a versatile model in biology & medicine. Mater Today 14: 190–195

Kaczmarczyk L, Jackson WS (2015) Astonishing advances in mouse genetic tools for biomedical research. Swiss Med Wkly 145:w14186. https://doi.org/10.4414/smw.2015.14186

Lovett B, Bilgo E, Millogo SA, Ouattarra AK, Sare I, Gnambani EJ, Dabire RK, Diabate A, St. Leger RJ (2019) Transgenic *Metarhizium* rapidly kills mosquitoes in a malaria-endemic region of Burkina Faso. Science 364(6443):894–897. https://doi.org/10.1126/science.aaw8737

Ma J et al (2003) Genetic modification: the production of recombinant pharmaceutical proteins in plants. Nat Rev Genet 4:794–805. https://doi.org/10.1038/nrg1177

Murtey MD, Ramasamy P (2016) Sample preparations for scanning electron microscopy–life sciences. In: Modern electron microscopy in physical and life sciences. InTech, London

Phifer-Rixey M, Nachman MW (2015) The natural history of model organisms: insights into mammalian biology from the wild house mouse *Mus musculus*. eLife 4:e05959. https://doi.org/10.7554/eLife.05959

Woodward AW, Bartel B (2018) Biology in bloom: a primer on the *Arabidopsis thaliana* model system. Genetics 208(4):1337–1349. https://doi.org/10.1534/genetics.118.300755

Mendelian Principle of Inheritance 2

Dhruti Patwardhan

Genetics is the study of genes and their variation and heredity among organisms. Long before DNA was recognised as the genetic material, Gregor Mendel through his studies predicted the presence of such a factor responsible for heredity. Heredity had been observed in nature for centuries, but Mendel studied this phenomenon in a scientific manner, performed experiments and put forth his hypothesis that has withstood the test of time. Although some variations to principles of Mendelian inheritance have been observed, the basic framework of genetic inheritance initially proposed by him in essence remains true.

2.1 Mendelian's Monohybrid Cross

Gregor Mendel is widely regarded as the founder of genetics. He was a priest in an abbey in Brno where he conducted his experiments on pea plants. Mendel carried out his experiments on breeding of pea plants for about 7 years from 1856 to 1863 and presented his findings at the Brno Natural Science meetings in 1865. His paper which underlined the basic principles of inheritance was published in 1866. However, Mendel's work remained unnoticed until 1900 when other scientists like botanists Hugo de Vries, Erich von Tschermak and Carl Correns obtained similar results while working independently with plant breeding. They interpreted their results in the context of Mendel's theories and published their work supporting and drawing attention to Mendel's original work.

Mendel was successful in obtaining and interpreting his results due to a number of factors. One of the most important factors was his scientific approach and analytical reasoning. Others before him had crossed plants and described their results. Mendel, however, was able to formulate a hypothesis based on initial observations and design

D. Patwardhan (✉)
Indian Institute of Science, Bangalore, India

D. Kar, S. Sarkar (eds.), *Genetics Fundamentals Notes*,
https://doi.org/10.1007/978-981-16-7041-1_2

suitable experiments to test them. He also recorded the number of plants with different features across various crosses and tried to interpret these observations to fit a single framework of inheritance. One of the other reasons for his success was the choice of garden pea as his experimental model.

2.1.1 The Garden Pea

Pisum sativum, commonly known as the garden pea, was the ideal choice for genetic breeding experiments. Since Mendel worked in a monastery at Brno, he could easily access the monastery garden and greenhouse. Pea plants are easy to cultivate and can grow relatively rapidly with a life cycle of 1 year. He therefore invested years in following several generations of the plants. Pea plants are also able to produce numerous seeds which allowed Mendel to calculate mathematical ratios in the traits of offspring (seeds). Different varieties of peas are available and Mendel was able to choose those that differed in various traits and were purebred. He also chose to study features that were present in two easily distinguishable forms/traits like round seeds versus wrinkled seeds. He avoided those features which had a range of variable traits. He chose to study seven features which are shown in Fig. 2.1. Apart from these, some references also mention an eighth feature: seed coat colour which can be either green or white. Mendel noticed that a coloured seed coat always gives rise to plants bearing purple flowers, while the white seed coat gave rise to plants having white flowers. Seed coat colour is therefore sometimes mentioned instead of flower colour as the traits studied by Mendel.

With advances in molecular biology, studies have been performed to identify the genes and particular mutations responsible for the traits studied by Mendel. Genes can be classified into groups based on the structural or functional similarities of proteins they produce. They may also be grouped together if they all participate in a

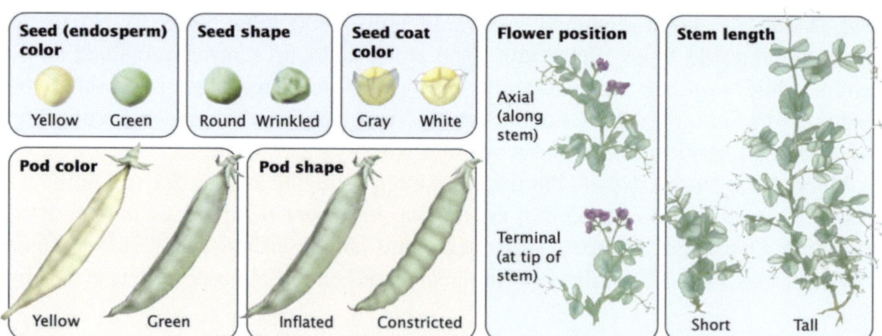

Fig. 2.1 The traits or plant characteristics studied by Mendel in his experiments on pea plants: Mendel studied seven different traits in pea plant for conducting his experiments. Traits being studied were present in either one of the two forms in the different varieties of pea plant (Pierce 2010)

Table 2.1 Group symbols and their functions for the traits studied by Mendel: Molecular studies to understand the traits studied by Mendel have resulted in identification of the genes responsible for the trait and their function. Of these R, LE, A and I have been cloned and well-studied. Less is known about the other genes

Trait	Group symbol	Gene function
Seed shape	R	Starch branching enzyme 1
Stem length	LE	GA3 oxidase 1
Flower colour	A	bHLH transcription factor
Pod colour	GP	Chloroplast structure in pod wall
Pod form	V	Sclerenchyma formation in pods
Position of flowers	FA	Meristem function
Seed colour	I	Stay-green gene

particular process. Gene group symbol and their functions for the traits studied by Mendel are given in Table 2.1.

The seed shape can be either round or wrinkled. The seeds differed in their starch, sugar and lipid content. Wrinkled seeds possess a higher amount of fructose, glucose and sucrose resulting in higher water retention due to osmotic pressure. A mutation in the *R* gene which codes for a starch branching enzyme 1 affected starch biosynthesis. This also further affected the protein and lipid biosynthesis in the seed ultimately changing its shape. Stem length was controlled by the *LE* gene. This gene codes for one of the GA3 oxidase genes which convert the gibberellin to an active form GA1. Gibberellin is a plant hormone which regulates the development of the plants including its length. Seed colour is influenced by the *I* gene which is important in chlorophyll degradation. Mutation in this gene leads to the appearance of green seeds. The flower colour is influenced by a gene which exhibits pleiotropic effects. This means that mutations in the gene can affect multiple traits. The gene codes for a basic helix-loop-helix (bHLH) transcription factor. It regulates multigene family-chalcone synthesis (CHS) genes which are responsible for flavonoid production. Flavonoids are secondary metabolites produced in plants and are responsible for pigmentation in plants, thus governing the flower colour. Other genes associated with the traits studied by Mendel have not been cloned and studied in molecular detail (Reid and Ross 2011).

To perform his experiments, Mendel crossed different varieties of pea plants. In order to understand how he achieved this, we need to know a little bit about the plant reproductive system. The male reproductive organ in a plant is called a stamen. It is composed of the filament and anther. The filament holds up the anther where pollen is produced. This pollen is carried by either wind, water or wildlife to the female reproductive organ. The female reproductive organ is called a pistil which consists of the stigma, style and ovary. The style holds the sticky stigma at the distal end, while the ovary is present at its proximal end. Stigma captures the pollen and allows it to germinate. Sperm carried in the pollen reaches the ovary through tubes formed during germination. Fertilisation occurs and an embryo is formed which is stored in the seed capsule. The seed remains dormant until favourable environmental conditions allow it to develop into a plant.

Pea plants often undergo self-pollination. This means that pollen from the flower will fall on the stigma of the same flower due to its close proximity. This happens even before the flower has opened. This type of pollination reduces genetic variability as the pollen and egg come from the same plant allowing them to maintain their characteristics. Plants which always pass on a specific trait to their offspring are called purebred varieties. Mendel grew pea plants for around 2 years in this manner to obtain purebred varieties for each trait. He also wanted to cross plants with different traits to see what traits were seen in the offspring. To achieve this, he opened the flowers and removed their anthers to prevent self-pollination. He then manually dusted pollen from the desired plant on the stigma of a flower from a different variety. This is called cross-pollination and resultant offspring are called hybrids. He obtained seeds from these cross-pollinated plants and observed their traits. He also grew these seeds through the next season to observe the traits of the hybrid plants (Fig. 2.2).

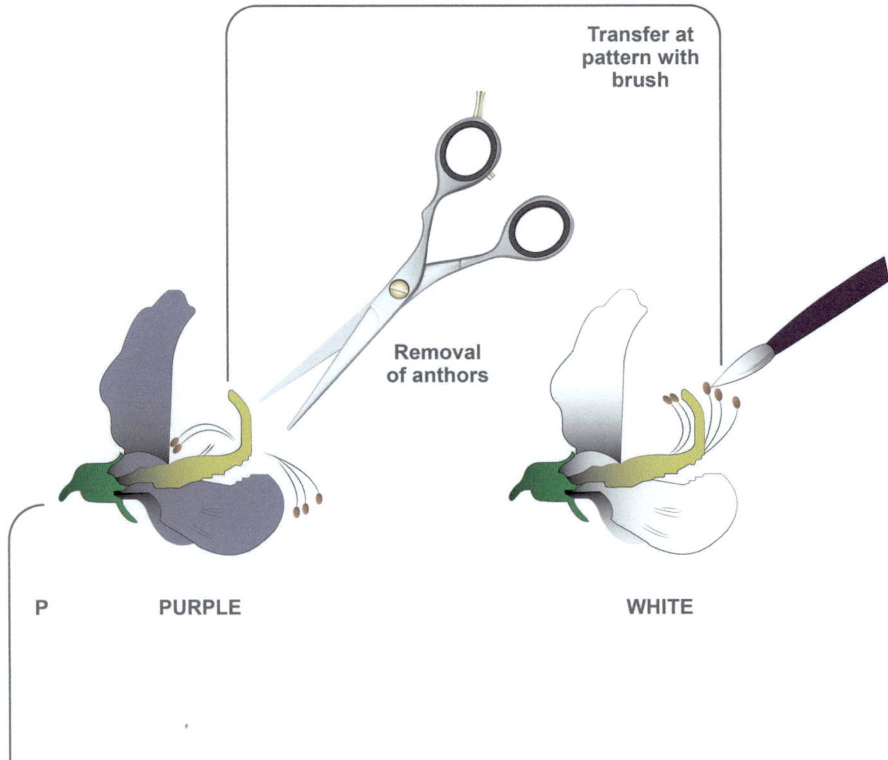

Fig. 2.2 Figure illustrating the male and female reproductive organs in a plant: To cross different varieties of plants, Mendel removed anthers from the flower to prevent self-pollination. He then dusted pollen from desired plant onto the stigma of this flower (Griffiths et al 2011)

2.1.2 Concept of Dominant and Recessive Traits

Mendel crossed different varieties of plants to study the traits inherited by the resultant offspring. He started by conducting monohybrid crosses, i.e. crosses between plants which differed by a single trait. Let us take the example of seed colour which can be either green or yellow. When Mendel crossed plants having green seed colour with those having yellow seed colour (referred to as the parental generation—P), he found that all the offspring called the first filial generation (F1) had yellow seed colour. He also carried out reciprocal crosses where instead of taking pollen from yellow seed plant and dusting it on the stigma of a green seeded plant, he took pollen from a green seeded plant and dusted it on the stigma of a yellow seed plant. In both cases, he found that plants in F1 generation had yellow seeds. Similarly, in crosses for the rest of the traits, he found that F1 generation always showed a single parental trait. It wasn't a mix of the parental traits, nor did the outcome change with various repetitions. This trait which was observed in the F1 generation was called the dominant trait and the trait which was lost was called the recessive trait. Mendel took this experiment one step further and allowed the plants from F1 generation to undergo self-pollination to create the F2 (second filial) generation. Most of the plants in the F2 generation had yellow seed colour, but surprisingly, there were a few plants in which seed colour was green. He counted the number of these plants and found that the number of plants having yellow seed colour was roughly thrice the number of plants having green seed colour.

Based on these results, Mendel made certain assumptions and put forth a hypothesis. Although the F1 generation always showed a single parental trait, the second parental trait reappeared in the F2 generation. This led him to assume that the F1 generation might have received genetic factors for both parental traits. Unless the F1 generation inherited genetic factors from both parents, it is impossible to explain the appearance of both parental traits in F2 generation. He hypothesised that offspring must inherit genetic factors from both parents and there must be two genetic factors in the plant for a single trait. The two genetic factors described here are what we now know as alleles. Alleles are, simply put, different forms of the same gene and are designated by a single letter. In this case, the allele for yellow seed colour is designated as Y and that for green seed colour is designated as y. Since the parental generation was purebred, the parental generation with yellow seed colour would have the alleles YY and the one with green seed colour would have the alleles yy. This composition of the alleles (YY or yy) is referred to as the genotype, and the trait physically expressed by the plant is called the phenotype (green or yellow seed colour).

He next assumed that the alleles separate when forming gametes and each allele gets segregated into one gamete. So the parental yellow seed coloured plants formed the gametes having allele Y and those from green coloured seed formed the gamete with allele y. In the F1 generation, these two gametes united and they had the genotype Yy. All F1 generation only had yellow coloured seeds. This trait was called the dominant trait. Although the allele for green coloured seed was present, it was masked and not expressed in the presence of Y. This was called the recessive trait. He concluded that, of the two parental traits, one trait is the dominant and the other is recessive. Only the dominant trait gets expressed even in the presence of the recessive trait.

2.1.3 Segregation of Alleles

He further hypothesised that the F1 generation having genotype Yy forms gametes having Y and y with equal probability. Therefore, half the gametes will have allele Y and the other half will have allele y and these will get paired randomly in the F2 generation. The resulting progeny might have the genotypes YY, Yy, yY and yy. Since yellow coloured seed is the dominant character, YY, Yy and yY will have yellow coloured seeds, and only yy will have green coloured seeds. This explains the 3:1 ratio of phenotypes observed by Mendel in his experiments. This ratio can only be obtained if we assume that the alleles get segregated with equal probability while forming gametes. When the genotype of a plant consists of the same alleles, they are called homozygous (YY, yy), and when the alleles differ they are called heterozygous (Yy) (Fig. 2.3).

2.1.4 Mendel's Analytic Approach

The findings from Mendel's monohybrid cross are formally stated in two laws known as law of segregation and law of dominance. Law of segregation states that during the formation of gametes, two alleles in an individual will separate such that each gamete will have one allele. Law of dominance states that hybrids of different alleles will express only one of the parental traits called the dominant trait. Mendel was able to draw meaningful insights from his work due to the analytic approach towards his experiments. The ratios Mendel obtained from his experiments were not perfect. Plants may die or wither before their characteristics can be noted. Some plants may fail to germinate. Therefore, the ratio of monohybrid cross that Mendel obtained was almost but not exactly 3:1. However, Mendel obtained numbers for multiple experiments and noted that the ratios were approximately 3:1 in all cases. He further went on to self-pollinate plants obtained from F2 generation to confirm his findings.

Let us take the example here of round vs wrinkled seeds. When plants having round seeds are crossed with those having wrinkled seeds, the resulting F1 generation has all round seeds. Here, round seeds is the dominant trait and represented by the allele R, while wrinkled seeds are represented by the allele r. We can therefore say that the parental generation had a genotype of RR and rr and the F1 generation has the genotype Rr. When these F1 plants are self-pollinated, 1/4 seeds are wrinkled (rr) and 3/4 are round (RR and Rr) in the resulting F2. On further self-pollinating the F2 generation, he observed that all the wrinkled seeds gave rise to wrinkled seeds on selfing. This can be explained by the fact that as the genotype of wrinkled seeds is rr, they will always form gametes carrying the allele r. Therefore, on self-pollination,

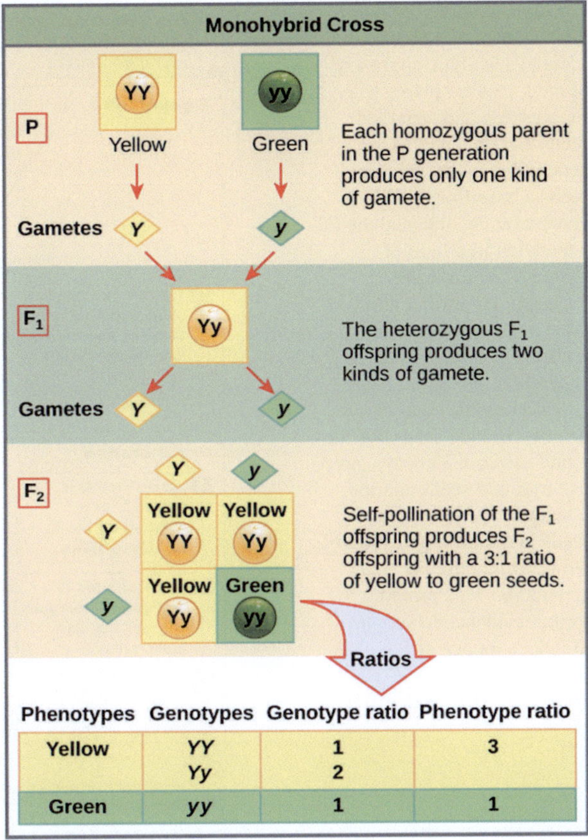

Fig. 2.3 Figure illustrating monohybrid cross between plants having yellow and green seed colour: The parental generation is homozygous and produces only one type of gametes. These get united in the F1 generation resulting in all plants having yellow coloured seeds. The F1 hybrids on selfing give rise to yellow and green seeds in a ratio of 3:1. The genotypic ratio is 1: 2:1 (The Punnett Square Approach for a Monohybrid Cross 2020)

they will always give rise to wrinkled seeds. Among the round seeds, 1/3 of the plants gave rise to only round seeds on self-pollination. The remaining 2/3 gave rise to a mix of round and wrinkled seeds in the ratio of 3:1. It follows that the seeds always giving rise to round seeds had the genotype RR. The remaining 2/3 seeds had the genotype Rr which, similar to the F1 generation, gives rise to the seeds in a ratio of 3:1. The results of these F3 generations give further evidence to support Mendel's hypothesis. These analytic approaches followed by Mendel were one of the strongest reasons why Mendel was able to come up with a reasonable hypothesis and support his claims through further experimentation (Fig. 2.4).

2.1.5 Test Cross: One Character

Homozygotes for the dominant allele as well as heterozygotes will display the dominant trait. To verify if the plant was homozygous or heterozygous for the dominant allele, Mendel crossed it with a plant showing the recessive trait and observed their progeny. This cross between plants showing dominant phenotype

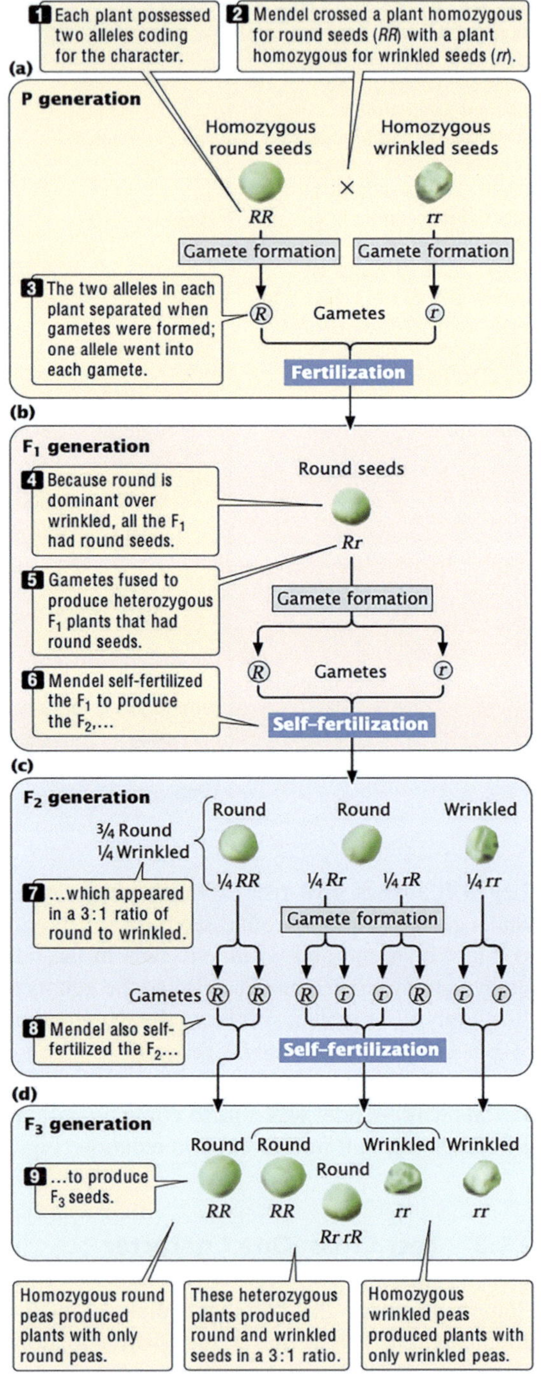

Fig. 2.4 Monohybrid cross till F3 generation shows Mendel's analytic approach: Mendel crossed plants with round seeds and wrinkled seeds and obtained F1 generation having all round seeds. On selfing the F1 generation, he obtained the F2 generation in a ratio of 3 round:1 wrinkled as expected. He further allowed the F2 to undergo self-pollination to obtain the F3 generation. He found that wrinkled seeds always gave rise to wrinkled seeds. Of the round seeds, 1/3 always gave rise to round seeds and the remaining 2/3 gave rise to seeds in the ratio of 3 round:1 wrinkled like the F2 generation. This provides further evidence to support Mendel's theory of segregation and dominance (Pierce 2010)

of unknown genotype with a plant of homozygous recessive genotype is called a test cross. Following Mendelian laws, if the plant being examined is homozygous for the dominant allele, all its progeny with the homozygous recessive parent will display the dominant phenotype. In the example below, if the purple flower is homozygous (having the genotype PP), it will produce gametes having P, and so all the F1 plants will show purple flowers. If, on the other hand, the purple flower is heterozygous (having the genotype Pp), it will produce two types of gametes having P or p. These, on crossing with p from the recessive parent, will give progeny with either purple or white flowers in a ratio of 1:1.

One of the recessive alleles will be provided by the parent showing the recessive phenotype as it will only produce gametes with the recessive allele. If a plant appears in the progeny displaying the recessive phenotype, it is clear that the other parent was carrying the recessive allele implying that the parent was heterozygous. If all the progeny show the dominant trait, the other parent has only provided gametes having the dominant allele implying that the parent was homozygous for the dominant allele. Test cross is a powerful method to examine unknown genotypes of an organism showing the dominant trait. This information is useful for breeders who would wish to choose homozygous plants for further breeding but cannot estimate the genotype based on observation of phenotype alone (Fig. 2.5).

2.2 Mendelian Dihybrid Cross

We have seen the results of crosses between plants differing in one trait. Mendel's next step was to study the pattern of inherited traits in crosses of plants differing in two traits. Let us take two traits in the pea plant: round vs wrinkled seeds and green vs yellow seeds. When Mendel crossed green round seeds with yellow wrinkled seeds, he observed that all the F1 progeny were yellow and round. In the monohybrid cross for each of the above traits, the F1 progeny expressed the dominant trait which was yellow colour and round shape. In the dihybrid cross too, the F1 hybrids expressed the two dominant traits. On selfing the F1 hybrids, he observed four phenotypes in the F2: yellow and round, green and round, yellow and wrinkled and green and wrinkled. On counting the number of plants in each category, he surmised that the plants were approximately in a ratio of 9:3:3:1 for the above combination of traits.

To make sense of the ratio that he obtained, Mendel made some logical deductions. He counted the number of yellow vs green and round vs wrinkled seeds and observed that they were in a ratio of 3:1 similar to the monohybrid cross. Mendel deduced that of all the F2 plants, 3/4 had yellow seeds and the remaining 1/4 had green seeds. Of the 3/4 yellow seeded plants, 3/4 had round seeds and 1/4 had wrinkled seeds. Similarly, 3/4 of the green seeds had round seeds and 1/4 had wrinkled seeds. This calculation gives the 9:3:3:1 ratio seen above. It appeared therefore that the dihybrid cross was a combination of 3:1 ratio for two traits. This will be easier to understand in the branched diagram in Fig. 2.6.

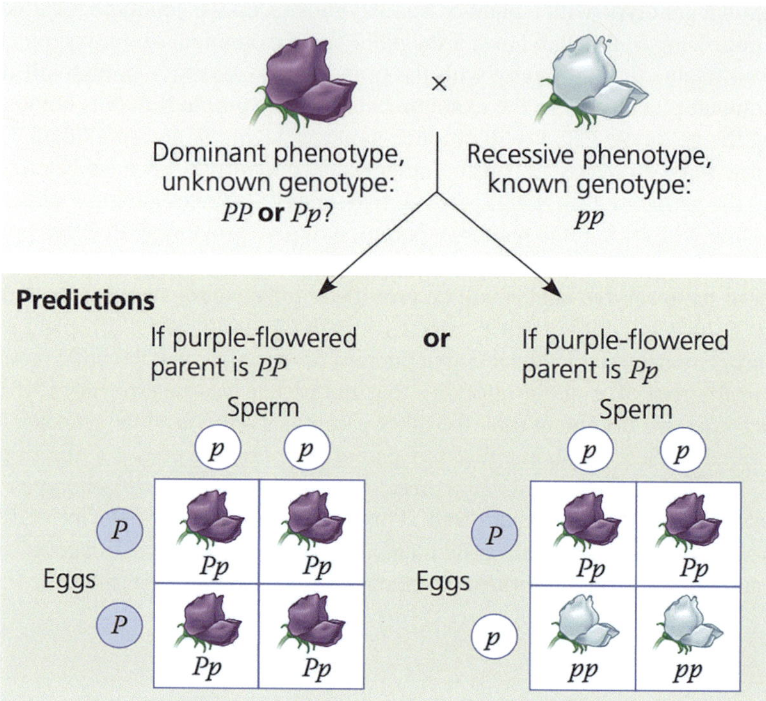

Fig. 2.5 Test cross involves crossing of a plant of unknown genotype with a plant showing the recessive trait: If the parent plant is homozygous, all its progeny will show the dominant trait. If the parent plant is heterozygous, half of its progeny will show the dominant trait and the other half will show the recessive trait. In the above illustration, test cross of a homozygous purple flower plant will result in all its progeny showing purple flowers. Test cross of a heterozygous purple flower plant will give flowers in the ratio of 1 purple:1 white (Reece et al 2011)

2.2.1 Independent Assortment

Mendel performed the dihybrid cross for a number of combinations of traits and always got a phenotypic ratio of 9:3:3:1. Let us now understand this ratio in biological terms. When a plant having yellow wrinkled seeds with the genotype YYrr is crossed with a plant having green round seeds with the genotype yyRR, hybrids with yellow round seeds having the genotype YyRr are produced. These hybrids can produce four gametes having four different combinations: YR, Yr, yR and yr. Each gamete carries one allele for each trait. On self-pollination, these gametes can merge in a variety of different combinations giving a phenotypic ratio of 9:3:3:1 as seen in the figure below (Fig. 2.7).

The fact that the dihybrid cross ratio is a combination of 3:1 ratio for each trait tells us that the gametes for each trait can assort independently. It means that allele Y has equal probability of pairing with either allele R or r to form a gamete. If one of the alleles in the cross above assorted preferentially with another allele, we would

¾ yellow seeds	¾ round seeds	¾ X 3/4 = 9/16 yellow round seeds
	¼ wrinkled seeds	¾ X ¼ = 3/16 yellow wrinkled seeds
¼ green seeds	¾ round seeds	¼ X ¾ = 3/16 green round seeds
	¼ wrinkled seeds	¼ X 1/4 = 1/16 green wrinkled seeds

Fig. 2.6 Phenotypic ratios obtained in a dihybrid cross: Each of the traits gives a 3:1 ratio. In the example above, we obtain seeds in a ratio of 3 yellow:1 green seeds. Each of these phenotypes also shows a ratio of 3 round:1 wrinkled seeds. This gives the overall 9:3:3:1 phenotypic ratio seen in a dihybrid cross

not obtain a phenotypic ratio of 9:3:3:1. This is called the law of independent assortment which states that different gene pairs can assort independently during gamete formation. However, genes which are close to each other on the same chromosome do not assort independently because they are held together on the same chromosome. In this case, alleles for different genes which are on the same chromosome always assort together during meiosis. The modified law of independent assortment can therefore be stated as 'Gene pairs present on different chromosomes assort independently of each other during formation of gametes'.

The tendency of genes which are close to each other to be inherited together is called linkage. Genes which get inherited together are classified into a single linkage group. Therefore, if any of the genes studied by Mendel belonged to same linkage group, their phenotypic ratios would have differed from the ones defined by Mendel. Mendel did not observe linkage between the genes that he studied and hence put forth the law of independent assortment. Mendel's work has been criticised on the basis that his data fits too well with his hypothesis and does not show as much variation due to chance as expected. Mendel's critics also cite the lack of evidence of linkage as one of the reasons to doubt Mendel's work. Recent work has, in fact, shown that the seven traits that Mendel studied belong to five different linkage groups of which only stem length and pod form show strong linkage. Mendel might have been lucky in his choice of traits for dihybrid cross. He might not have performed dihybrid crosses for this particular combination of stem length and pod form, or he would have been surprised by the resulting phenotypic ratios. Seed shape and pod colour show weak linkage, and all other traits are not linked allowing Mendel to obtain the same ratio for most of his dihybrid crosses. The debate about the validity of Mendel's work is discussed in detail in Box 2.1.

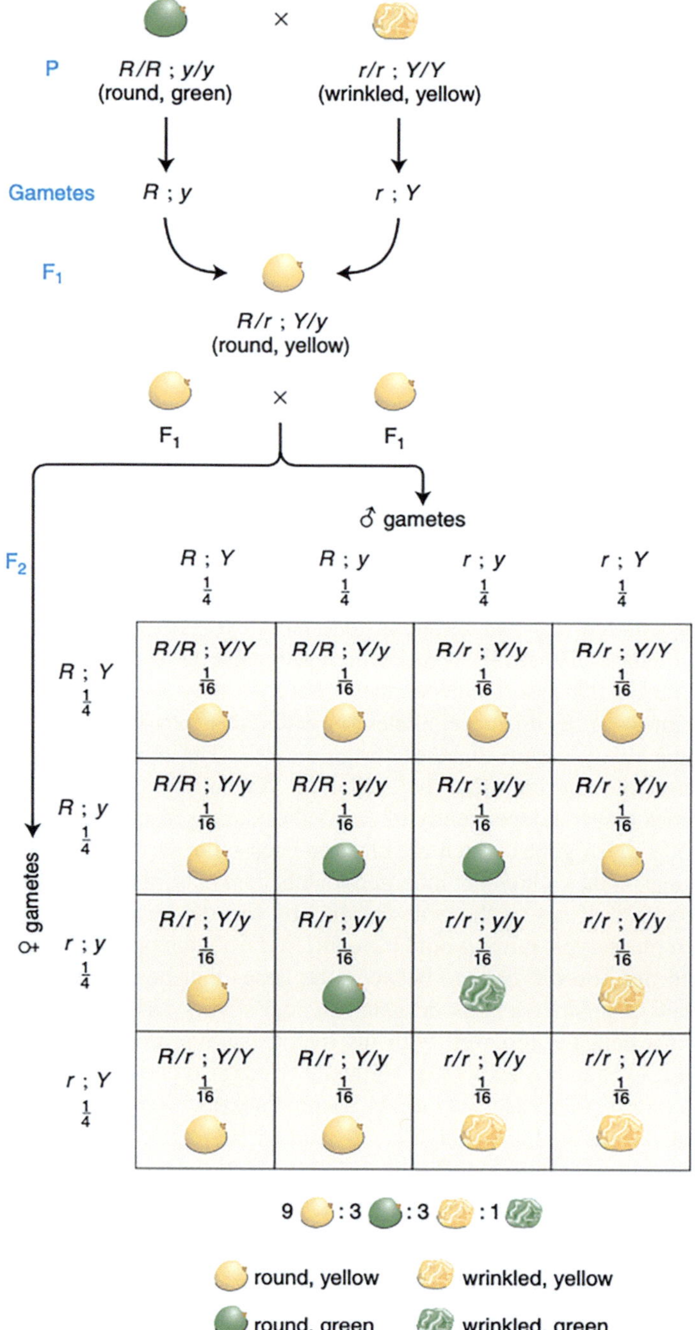

Fig. 2.7 Genotypes of progeny obtained from a dihybrid cross for wrinkled yellow seeds with round green seeds: The F1 progeny has yellow round seeds and can produce four types of gametes. These are shown on the top row and first left column of the square on the right. The various combinations of these gametes give a progeny with a phenotypic ratio of 9:3:3:1 (Griffiths et al. 2011)

2.2.2 Test Cross: Two Characters

Similar to the monohybrid cross, Mendel performed test cross for the dihybrid cross to verify his conclusions. As stated above, test cross involves crossing a plant of unknown genotype with a plant homozygous recessive for the traits under consideration. If we were to perform a test cross for the F1 produced from the dihybrid cross above, the tester (homozygous recessive individual) would be a plant having green and wrinkled seeds. In this case, we would expect the F1 to form gametes with all the combinations, RY, Ry, rY and ry, according to the law of independent assortment. This when fertilised with ry gametes from the tester would give the following different combinations: RrYy (round, yellow), Rryy (round, green), rrYy (wrinkled, yellow) and rryy (wrinkled, green). A phenotypic ratio of 1:1:1:1 would be expected. This is the result that Mendel obtained from his test cross for two characters providing evidence for his law of independent assortment.

The F1 hybrid obtained above is heterozygous for both traits. Let us see how the results would differ if the plant being tested was homozygous for either of the traits. If we take a plant having yellow, round seeds, its genotype could be either YyRr (discussed above), YYRr, YyRR or YYRR. If it is YYRR, it will produce only one type of gamete YR which when crossed with yr will produce all plants of YyRr genotype and single phenotype of yellow round seeds. If it is YYRr, two types of gametes will be produced: YR and Yr. This will give two phenotypes on test cross: yellow, round seeds (YyRr) and yellow, wrinkled seeds (Yyrr) in a ratio of 1:1. Similarly, if the genotype is YyRR, it will produce two phenotypes on performing test cross: yellow, round seeds (YyRr) and green, round seeds (yyRr) in a ratio of 1: 1. In this manner, we can detect the genotype of the individual based on the number and ratio of phenotypes produced (Fig. 2.8).

2.3 Mendelian Trihybrid Cross

Mendel's laws can be extended to obtain genotypic and phenotypic ratio of a trihybrid cross, i.e. a cross between plants differing in three traits. As the number of traits increase, the complexity of the ratio increases and several different methods can be used to predict the results. Let us take the example of three genes A, B and C having recessive alleles a, b and c. If a homozygous dominant plant with genotype AABBCC is crossed with a homozygous recessive plant aabbcc, the F1 hybrid will have a genotype of AaBbCc. When the F1 plant is further self-pollinated, it will produce eight different gametes as shown below (Fig. 2.9).

2.4 Application of Mendelian Principles

Mendel's principles of segregation, dominance and independent assortment can be applied to analyse inheritance of multiple traits. Various methods like Punnett square, forked line method or probability method can be utilised to obtain

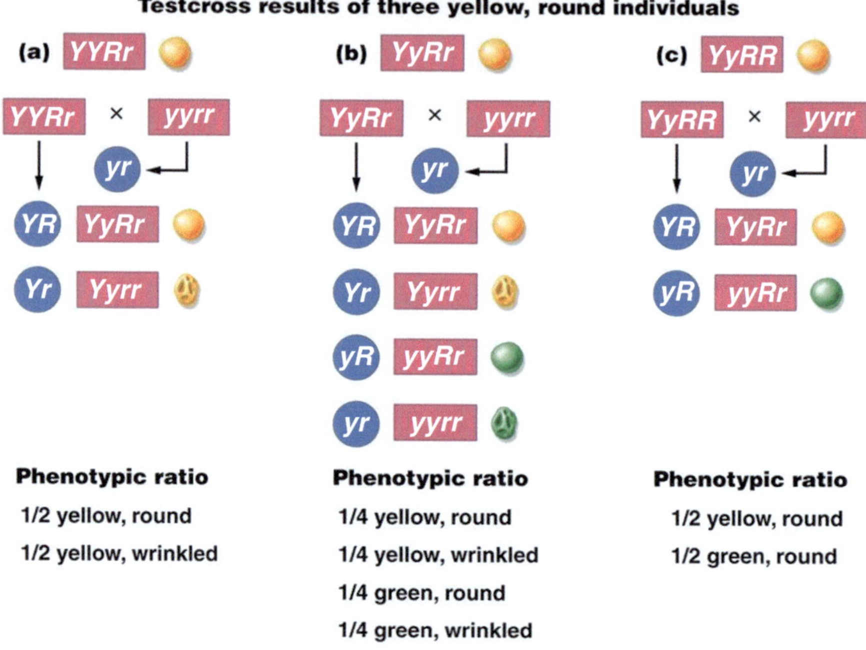

Fig. 2.8 Test cross for a dihybrid cross can reveal the parental genotype: Yellow round seeds with three different genotypes (YYRr, YyRr and YyRR) give rise to different phenotypes and phenotypic ratios in the progeny when subjected to a test cross. This allows us to estimate the genotype of the unknown individual. (Adapted from Klug et al. 2012)

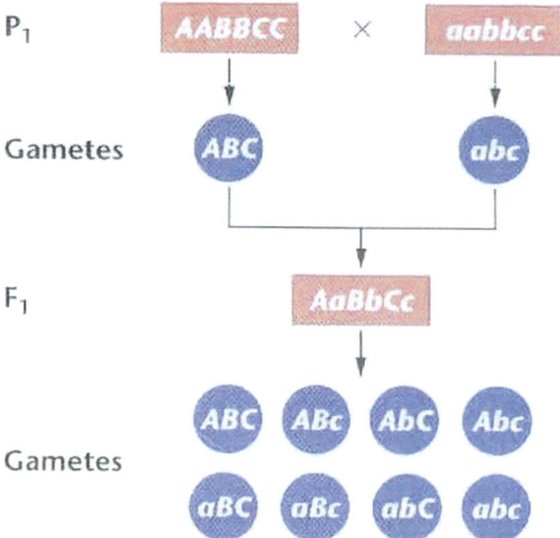

Fig. 2.9 Gametes produced during a trihybrid cross: Eight different gamete combinations of alleles and therefore eight different types of gametes may be produced by an individual heterozygous for three gene pairs (Klug et al. 2012)

phenotypic and genotypic ratios of cross between multiple traits. We will obtain the genotypic ratios for the trihybrid cross mentioned above using both Punnett square and forked line method.

2.4.1 The Punnett Square Method

In the Punnett square method, a grid is created with the gametes from one parent on the upper side and those from the other parent on the left side. Each cell or block within the grid contains a combination of alleles from both parents giving the genotype of the offspring resulting from the combination of the respective gametes. It is named after Reginald C Punnett who devised this method. It is a tabular representation of all possible combinations between the maternal and paternal alleles.

In the image below, gametes produced from F1 hybrid of a trihybrid cross are represented on the top row and left column of the grid. Since this is a self-pollination, the gametes on both sides are identical. In case the parents differ in genotype, gametes from one parent will be in the top row and will differ from gametes produced by the other parent which will be on the left column. The first box in the grid shows the genotype AABBCC which is produced if the gametes ABC and ABC (corresponding parental gametes in first row and first column) were to combine. In this manner, all possible genotypes of the progeny can be represented in the Punnett square in a simplified manner. The genotypes showing the same phenotypes are represented by the same colour of the cell in the grid. From this, we can infer that the phenotypic ratio of a trihybrid cross will be 27:9:9:9:3:3:3:1 (Fig. 2.10).

AaBbCc (top) / AaBbCc (left)	ABC	ABc	AbC	aBC	Abc	aBc	abC	abc
ABC	AABBCC	AABBCc	AABbCC	AaBBCC	AABbCc	AaBBCc	AaBbCC	AaBbCc
ABc	AABBCc	AABBcc	AABbCc	AaBBCc	AaBbcc	AaBBcc	AaBbCc	AaBbcc
AbC	AABbCC	AABbCc	AAbbCC	AaBbCC	AAbbCc	AaBbCc	AabbCC	AabbCc
aBC	AaBBCC	AaBBCc	AaBbCC	aaBBCC	AaBbCc	aaBBCc	aaBbCC	aaBbCc
Abc	AABbCc	AABbcc	AAbbCc	AaBbCc	AAbbcc	AaBbcc	AabbCc	Aabbcc
aBc	AaBBCc	AaBBcc	AaBbCc	aaBBCc	AaBbcc	aaBBcc	aaBbCc	aaBbcc
abC	AaBbCC	AaBbCc	AabbCC	aaBbCC	AabbCc	aaBbCc	aabbCC	aabbCc
abc	AaBbCc	AaBbcc	AabbCc	aaBbCc	Aabbcc	aaBbcc	aabbCc	aabbcc

Fig. 2.10 Punnett square showing genotypes produced from a trihybrid cross: The genotypic and phenotypic ratio for a cross between individuals of genotype AaBbCc and AaBbCc is shown in the Punnett square above. The gametes produced from each parent are shown in the top row and left column. Each grid represents a combination of the gametes in the respective row and column. The individuals showing the same phenotype are in the same colour. We can see that a phenotypic ratio of 27:9:9:9:3:3:3:1 is obtained

2.4.2 The Forked Line Method

Although useful, Punnett square can be too cumbersome to use when more than three traits are being analysed. In such instances, forked line method, also called the branch diagram, can be very useful. In this method, the genotypic or phenotypic outcome for one gene pair is first predicted. Then, outcome of the next gene pair is computed in conjunction with earlier gene pair. This method is followed for all the remaining gene pairs. In the figure below, phenotypic ratio of trihybrid cross is predicted using the forked line method. Here, three traits are considered: round vs wrinkled seeds, green vs yellow seeds and grey-brown vs white seed coat. We know the dominant phenotype for the other traits except seed coat where grey-brown is dominant over white. According to Mendel's law of segregation and random fertilisation, a monohybrid cross between round and wrinkled seeds will result in 3/4 seeds being round and the remaining 1/4 being wrinkled. This is the outcome for the first trait. Now, of the round seeds, 3/4 seeds would be yellow and 1/4 green, which is the outcome for the second trait. Similarly, 3/4 of the wrinkled seeds will be yellow and 1/4 green. Next, of the round and yellow seeds, 3/4 will have grey-brown coat and 1/4 will have a white coat. This gives us a total of 27 round, yellow and grey-white coat seeds (3/4 round $\times$ 3/4 yellow $\times$ 3/4 grey-brown coat $=$ 27/64). Similarly, proportion of round, yellow and white seed coat will be 3/4 $\times$ 3/4 $\times$ 1/4 $=$ 9/16. We can calculate the proportion for all phenotypes in this manner and obtain a phenotypic ratio of 27:9:9:9:3:3:3:1.

Calculations for genotypic ratio can be done in a similar manner. Although the phenotypic ratio for a monohybrid cross is 3:1, we need to bear in mind that the genotypic ratio is 1:2:1 (1 AA, 2 Aa and 1 aa). The genotypic calculation for a trihybrid cross using forked line method is illustrated below. There are a few rules of thumb that can be used to cross-check your calculations. First, count the number of heterozygous gene pairs in the cross. In a cross AaBbCc X AaBbCc, heterozygous gene pairs 3. 2^n will be the number of different gametes that can be produced from each parent. 3^n will be the number of different genotypes produced after fertilisation, and 2^n will be the number of different phenotypes produced in this cross. In the above example, therefore, $2^3 = 8$ different gametes are formed from each parent, $3^3 = 27$ different genotypes are produced in this cross and it gives rise to $2^3 = 8$ different phenotypes. These are the numbers that we get from our calculations with the forked line method as well as Punnett square method (Fig. 2.11).

2.4.3 The Probability Method

Probability is a mathematical tool that predicts the likelihood of occurrence of an event.

We can easily use probability to predict the outcome of a genetic cross. Probability can be calculated as the number of times an event occurs divided by the total number of events. For example, the probability of picking a queen of hearts from a deck of cards would be 1/52. This is because there is only one queen of hearts (one

Generation of F_2 trihybrid phenotypes

A or *a*	*B* or *b*	*C* or *c*	Combined proportion
3/4 *A*	3/4 *B*	3/4 *C*	(3/4)(3/4)(3/4) *ABC* = 27/64 *ABC*
		1/4 *c*	(3/4)(3/4)(1/4) *ABc* = 9/64 *ABc*
	1/4 *b*	3/4 *C*	(3/4)(1/4)(3/4) *AbC* = 9/64 *AbC*
		1/4 *c*	(3/4)(1/4)(1/4) *Abc* = 3/64 *Abc*
1/4 *a*	3/4 *B*	3/4 *C*	(1/4)(3/4)(3/4) *aBC* = 9/64 *aBC*
		1/4 *c*	(1/4)(3/4)(1/4) *aBc* = 3/64 *aBc*
	1/4 *b*	3/4 *C*	(1/4)(1/4)(3/4) *abC* = 3/64 *abC*
		1/4 *c*	(1/4)(1/4)(1/4) *abc* = 1/64 *abc*

Fig. 2.11 Forked line method for obtaining genotypic and phenotypic ratios in a trihybrid cross: For a monohybrid cross, genotypic ratios of 1:2:1 are obtained. Then for each of the genotypes, the ratio for the next set of genes is calculated and further on and the proportions multiplied at the end. Similarly, phenotypic ratios can be calculated by multiplying 3:1 ratios for each trait (Klug et al. 2012)

event) in a deck of 52 playing cards (all possible events). If, however, we were to calculate the probability that a card picked would be any queen, this probability would be 4/52 as there are four queen cards in a deck. There are two rules to be followed for the calculation of slightly more complicated probabilities.

1. **Multiplication rule**: The probabilities of co-occurrence of two or more events can be calculated by the product of their individual probabilities. This rule can only be used if the events are occurring independent of each other. If the occurrence of one event affects the probability of occurrence of the next event, then their combined occurrence cannot be a simple product of their independent probabilities. For example, the probability that two consecutive rolls of a dice are a three and a six in that order is $1/6 \times 1/6 = 1/36$. The individual probabilities of a three and a six in a roll of dice are 1/6 and 1/6, respectively. Their product will give the probabilities of their co-occurrence. We have already used the multiplication rule while calculating the dihybrid and trihybrid cross ratio in the branch diagram or forked line method. Multiplication rule is applied when the word *and* is used. In the above example, we wanted to find the probability of 'a three *and* a six in a roll of dice'.
2. **Addition rule**: We can calculate the probability of either of two or more mutually exclusive events occurring together by the addition rule. For example, if we just wanted to obtain the probability of occurrence of a three or a six in a roll of the dice, we would find the sum of individual probabilities. Thus, the probability is $1/6 + 1/6 = 2/6 = 1/3$.

The multiplication and addition rules can be used in predicting the outcome of genetic crosses instead of Punnett square or forked line method. Let us consider the cross between two plants having round seeds with genotypes Rr and Rr. The probability of wrinkled seeds can be calculated using multiplication rule. The probability of obtaining r allele from one parent is 1/2 and from the other parent is also 1/2. For a wrinkled seed, the genotype needs to be rr and its corresponding probability is 1/2 × 1/2 = 1/4. If we were to calculate probability of round seeds, both multiplication and addition rules need to be used. Round seeds can occur because of three genotypes: RR, Rr and rR. Their individual probabilities are as follows:

$$1/2\,\mathrm{R} \times 1/2\,\mathrm{R} = 1/4\,\mathrm{RR}$$

$$1/2\,\mathrm{R} \times 1/2\,\mathrm{r} = 1/4\,\mathrm{Rr}$$

$$1/2\,\mathrm{r} \times 1/2\,\mathrm{R} = 1/4\,\mathrm{rR}$$

Their combined probabilities would therefore be 1/4 + 1/4 + 1/4 = 3/4 round seeds. It is easier to use probability method for calculation of complex crosses with multiple traits as compared to Punnett square or forked line method.

2.4.3.1 Binomial Theorem

Binomial theory can be utilised when we want to calculate the occurrence of a specific set of outcomes among a large number of potential events. Let us consider the case of galactosemia. In this disorder, mutation in one of the galactose metabolising genes prevents the individual from converting galactose to glucose. The affected individual may show symptoms like lethargy, failure to gain weight and liver damage. This disorder can only occur if both copies of the gene are mutated. Thus, the parents of a child may each be carrying one mutated allele and not express the disorder. Only if the child inherits both mutated copies will he/she be affected. Let us assume that the gene affected here is a and its wild-type/normal allele is A. If we now want to find the probability that both kids of heterozygous parents are affected. The probability of one child being affected (aa) is 1/4 (Aa X Aa = 1/4 aa, 3/4 A-) and that of two children being affected will be 1/4 × 1/4 = 1/16.

Now, if we suppose the couple has three children and we want to find the probability that two children are unaffected and one child is affected. There are three scenarios in which this is possible:

$$1/4 \times 3/4 \times 3/4 = 9/64\ (\text{child 1 affected, other 2 unaffected})$$

$$3/4 \times 1/4 \times 3/4 = 9/64\ (\text{child 1 and 3 unaffected, child 2 affected})$$

$$3/4 \times 3/4 \times 1/4 = 9/64\ (\text{child 1 and 2 unaffected, child 3 affected})$$

$$\text{Total probability} = 9/64 + 9/64 + 9/64 = 27/64$$

This calculation becomes more complex for situations with more number of children and multiple different combinations. If we want to find the probability of this couple having five children, three of whom are affected and the remaining two are not, we can use the binomial expression. The binomial expression is of the form $(a + b)^n$ where a and b are probabilities of two alternate events and n is the number of times the event occurs. In the above case, we can define a as the probability that the child suffers from galactosemia (1/4), while b is the probability that the child remains unaffected (3/4). n here is the number of children which will be 5. The binomial can be expanded as follows:

$$(a+b)^5 = a^5 + 5a^4b + 10a^3b^2 + 10a^2b^3 + 5ab^4 + b^5$$

It follows the rule:

$$(a+b)^n = a^n + a^{n-1}b + a^{n-2}b^2 + a^{n-3}b^3 + \ldots .b^n$$

The expansion of $(a + b)^n$ consists of $n + 1$ terms. Each of these terms has a numerical coefficient. The coefficient of the first term is always 1. The second term has the coefficient same as the power to which the binomial is raised. So, in this case, it is 5. For the next coefficient, multiply the coefficient of the previous term with the exponent of a in that term. Divide this by the number of the term in the equation. Thus, to calculate the coefficient for a^3b^2 in the above expansion, $(5 \times 4)/2 = 10$ where coefficient of earlier term is 5 and exponent of a in that term is 4 and it is the second term in that equation. Similarly for the coefficient of a^2b^3, it can be calculated as $(10 \times 3)/3 = 10$. We can calculate the coefficients for the rest of the terms and expand the binomial.

Another method to calculate the coefficients in the equation is to use Pascal's triangle. We can determine the coefficients for each term in the binomial expression from the terms in front of the corresponding n. Notice that all terms other than 1 in Pascal's triangle are the sum of terms directly above them (Fig. 2.12).

Once we have the equation, we can obtain the probability of any combination of events by simply inserting the values of a and b. For example, to obtain the probability of three out of five children having galactosemia, the term we use is $10a^3b^2$:

$$10a^3b^2 = 10 \times (1/4)^3 \times (3/4)^2 = 90/1024 = 0.087$$

In this manner, we can easily calculate the probability of any combination of events. There is another method to do the above calculation. It uses the formula:

$$P = \frac{n!}{s!t!} \times a^s b^t$$

P is the overall probability of co-occurrence of two events X and Y. Event X has a probability of a occurring s times, while event Y has probability b occurring t times.

Pascal's Triangle

n	Numerical Coefficients
n	1
1	1 1
2	1 2 1
3	1 3 3 1
4	1 4 6 4 1
5	1 5 10 10 5 1
6	1 6 15 20 15 6 1
7	1 7 21 35 35 21 7 1
etc.	etc.

Fig. 2.12 Pascal's triangle: Pascal's triangle can be used to obtain coefficients for terms in the binomial expansion for any n. The terms other than 1 in Pascal's triangle are a sum of the terms directly above them

In the above case, X is the probability that the child is affected. Therefore, a is 1/4 and s is 3. Event Y is the probability that the child is unaffected. Here, b is 3/4 and t is 2. N is the total number of events which is 5 in this case. The symbol ! is for a factorial which is the product of all positive integers from 1 to n. For example, $5! = 5 \times 4 \times 3 \times 2 \times 1$.

The calculation therefore is:

$$\begin{aligned} P &= \frac{5!}{3!2!} \times (1/4)^3 (3/4)^2 \\ &= \frac{5 \times 4 \times 3 \times 2 \times 1}{(2 \times 1)(3 \times 2 \times 1)} \times (1/4)^3 (3/4)^2 \\ &= 0.087 \end{aligned}$$

This value is the same as that obtained from binomial theorem.

2.5 Test of Genetic Hypothesis

Crosses between two individuals of known genotypes yield a certain genotypic and phenotypic ratio. Based on Mendel's laws, we can predict a certain ratio. However, the experimental ratios may not match the expected values. Other than technical difficulties (like death of plants before the phenotype can be observed), chance plays a very important role in this deviation. This is easily illustrated with the example of a coin flip. We know that the probability of getting a heads or tails in a coin flip is 1:1.

If we do the coin toss for a large number of times, say 1000, we can expect that we will get a number close to 1:1. However, if we toss the coin only ten times, we might get seven heads and three tails or two heads and eight tails. This deviation from expected ratio is just a chance event.

Genetic ratios however can also be different, if there is some linkage between the traits being studied or if the gene is following some non-Mendelian pattern of inheritance. An experimenter needs to know if the deviation from expected ratios is just a matter of chance or it is of some biological significance. In such cases, we can make use of a chi-square test.

2.5.1 The Chi-Square Test

Chi-square test, also written as χ^2 test, is used to evaluate how well the observations support the null hypothesis. It is calculated from the sum of squared errors or sample variance. A chi-square test can only tell us if the resulting ratio of genetic crosses is deviating from the expected ratio merely due to chance. It cannot tell us if there is a mistake during crossing or during calculation of expected ratios or there are some complex inheritance patterns involved. In other words, it gives us a probability that the difference in observed and expected ratio can be due to chance alone.

Let us take an example to understand how to use the chi-square test. A monohybrid cross between two tall plants resulted in a progeny of 100 tall plants and 40 short plants. If we were to assume that the genes involved followed a Mendelian inheritance pattern, we would expect a ratio of 3 tall:1 dwarf plants. For a total of 140 plants, $3/4 \times 140 = 105$ plants should be tall and $1/4 \times 140 = 35$ plants should be short. We see that the observed ratio differs slightly from the expected ratio. Is this merely an effect of chance?

We start by establishing a null hypothesis (H_o). The null hypothesis is called so because it assumes that there is no real difference between our expected and observed outcomes and any deviation is a result of chance events. Through the probability derived from the chi-square test, we can then accept or reject the null hypothesis. Our null hypothesis for this example will be that the inheritance follows a ratio of 3:1. The formula for chi-square test is:

$$\chi^2 = \Sigma \frac{(O - E)^2}{E}$$

where E = expected value for that category

O = observed value for that category

Σ = sum of calculated values for all categories

Plugging in the above values in this equation:

$$\chi^2 = \frac{(100-105)^2}{105} + \frac{(40-35)^2}{35}$$
$$= 25/105 + 25/35$$
$$= 0.238 + 0.714 = 0.952$$

After this calculation, we have to determine the degrees of freedom (df) which is $n - 1$ where n is the number of different categories the value may fall into. Here, the plant can be tall or short. Thus, $n = 2$ and $\text{df} = 1$. Degrees of freedom are considered because more categories introduce more deviation in the results. We now have to interpret the χ^2 value in terms of its corresponding probability value. This calculation is very complex, and we instead make use of a standard table which provides probability values for different χ^2 values for each degree of freedom.

In the table below (Fig. 2.13), we can see that the calculated value of 0.952 for df 1 lies between p value of 0.5 and 0.1. We can interpret this as the probability that the observed deviation from expected value is due to chance is between 10 and 50%. Traditionally, scientists have accepted a p value cut-off of 0.05. That is to say that if the p value is above 0.05, we can accept the null hypothesis. If the p value is less than 0.05, it means that the probability that the deviation is due to chance is less than 5%. In this case, the null hypothesis is rejected. In our example, we can accept the null hypothesis and conclude that the variation seen in the ratios is a product of chance and that the inheritance indeed follows a 3:1 ratio.

Table 3.4 Critical values of the χ^2 distribution

					P				
df	**.995**	**.975**	**.9**	**.5**	**.1**	**.05**	**.025**	**.01**	**.005**
1	.000	.000	0.016	0.455	2.706	3.841	5.024	6.635	7.879
2	0.010	0.051	0.211	1.386	4.605	5.991	7.378	9.210	10.597
3	0.072	0.216	0.584	2.366	6.251	7.815	9.348	11.345	12.838
4	0.207	0.484	1.064	3.357	7.779	9.488	11.143	13.277	14.860
5	0.412	0.831	1.610	4.351	9.236	11.070	12.832	15.086	16.750
6	0.676	1.237	2.204	5.348	10.645	12.592	14.449	16.812	18.548
7	0.989	1.690	2.833	6.346	12.017	14.067	16.013	18.475	20.278
8	1.344	2.180	3.490	7.344	13.362	15.507	17.535	20.090	21.955
9	1.735	2.700	4.168	8.343	14.684	16.919	19.023	21.666	23.589
10	2.156	3.247	4.865	9.342	15.987	18.307	20.483	23.209	25.188
11	2.603	3.816	5.578	10.341	17.275	19.675	21.920	24.725	26.757
12	3.074	4.404	6.304	11.340	18.549	21.026	23.337	26.217	28.300
13	3.565	5.009	7.042	12.340	19.812	22.362	24.736	27.688	29.819
14	4.075	5.629	7.790	13.339	21.064	23.685	26.119	29.141	31.319
15	4.601	6.262	8.547	14.339	22.307	24.996	27.488	30.578	32.801

P, probability; df, degrees of freedom.

Fig. 2.13 Probability values for χ^2 distribution: Figure giving probability values for estimated χ^2 values at different degrees of freedom. The probability value keeps decreasing towards the right, while the χ^2 values keep increasing

Let us take another example. In a cross between plants having violet flowers and white flowers, violet flowers were observed in F1. On self-fertilisation, it was seen that 790 of the progeny had violet flowers and 210 had white flowers. Can we ascertain if this follows the Mendelian pattern of inheritance?

Null hypothesis—The pattern of inheritance follows Mendelian genetics and does not differ from a ratio of 3:1:

$$\chi^2 = \Sigma \frac{(O-E)^2}{E}$$

E: If the flower colour inheritance followed Mendelian genetics, we would see that 3/4 of the total flowers would be violet and 1/4 would be white since violet is the dominant character (because F1 flowers were violet). The expected numbers would therefore be:

Violet $3/4 \times 1000 = 750$, white $1/4 \times 1000 = 250$

$$\chi^2 = \frac{(790-750)^2}{750} + \frac{(210-250)^2}{250}$$
$$= 2.133 + 6.4 = 8.533$$

Degree of freedom here too is 1, as only two characters are being observed. *P* value for χ^2 8.533 at df 1 is less than 0.005

We can therefore reject the null hypothesis. The probability that deviation in ratio is purely due to chance is very less and the gene is probably following some non-Mendelian pattern of inheritance.

2.6 Application of Mendelian Principles in Human Genetics

Mendel's work has shed light on the inheritance of genes and traits. We can use this knowledge to analyse inheritance of various genetic diseases and traits in humans too. We can do this by obtaining information regarding occurrence of the trait being studied in the family of the affected individual.

2.6.1 Pedigree Analysis

Pedigree analysis is similar to a family tree for a specific trait. It is basically a chart which illustrates which family members have the traits being studied. This aids in understanding the method of inheritance of the trait. We can also predict the possible genotype of individuals for that trait which can help in predicting probability of inheritance of the trait in future generations.

A set of standardised symbols are used for illustrating a pedigree. Squares represent males, and circles represent females. A shaded box denotes individuals that expressed the phenotype being studied. A horizontal line between two individuals denotes mating. Their progeny are represented in the order of birth on a horizontal line connected to the parental mating line. Different generations are represented on descending levels. A double line connecting two individuals denotes consanguineous marriage. A marriage between second cousins or even more closely related individuals is referred to as a consanguineous marriage. Many studies have shown that consanguinity is one of the major contributors of birth defects and abnormalities. If an individual has a recessive gene, his progeny might inherit the gene but not express the phenotype. Thus, individuals belonging to the same family have a greater probability of carrying the recessive gene. A marriage between members of a family increases the probability of a child from this union inheriting two copies of a recessive gene and therefore suffering from a genetic disorder inherited in an autosomal recessive manner. Although consanguineous marriages have reduced over the years, they are still prevalent in the Middle East and parts of Asia and Africa. More symbols and their meanings are given in Fig. 2.14.

Let us examine the pedigree shown in Fig. 2.15. Individual 4 from generation III is the proband. This means that this individual was the first to be investigated for this

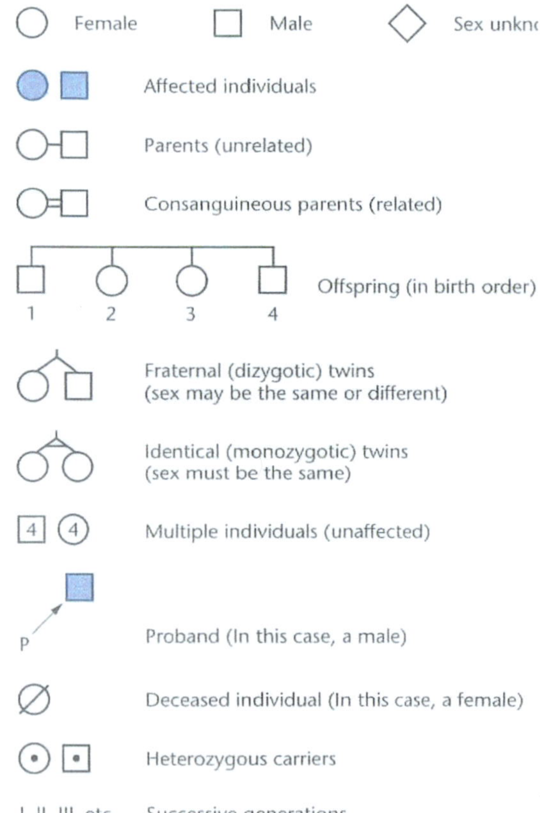

Fig. 2.14 Standard symbols used while drawing a pedigree chart: Proband denotes the first individual studied in the family for this trait. Consanguineous marriage means marriage between individuals from the same family, usually first cousins (Klug et al. 2012)

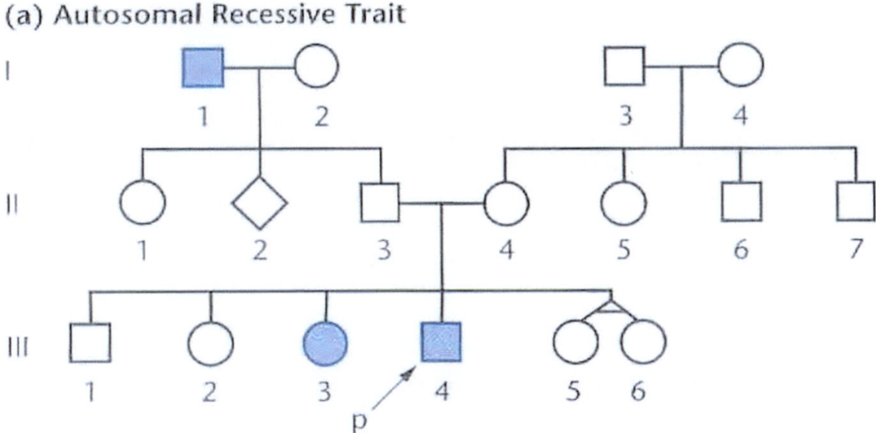

Fig. 2.15 Pedigree showing autosomal recessive mode of inheritance: The pedigree shown above shows that the character has skipped a generation. Fewer members of the family are affected and the trait is evenly distributed between males and females. Based on this, we can conclude that the mode of inheritance of the gene being studied is autosomal recessive (Klug et al. 2012)

phenotype and prompted construction of the pedigree. We can see that one of the siblings of individual 4 is also affected. None of the parental generation (Gen II) has any affected members. Among the grandparental generation (Gen I), individual 1 is affected. We can draw a few conclusions from this information. First is that the trait being studied is recessive. Based on Mendel's law of dominance, if the trait was dominant, at least one parent of the affected individual would have expressed the trait. Since none of the parents show the trait, they are most likely carriers of the recessive allele. This skipping of generation in expression of traits is a characteristic feature of recessive traits. Second, although there aren't many affected individuals, the trait seems to be passed equally between males and females (Gen III individuals). We can therefore assume that the recessive trait is on an autosome. Of the 23 pairs of chromosomes that are present in humans, 22 are autosomes and 1 pair is a sex chromosome (X and Y chromosomes). This means that the 22 pairs are inherited randomly between males and females. However, the sex chromosomes determine the sex of the individual. In humans, XX determines a female and XY determines males. Therefore, the inheritance of traits present on sex chromosomes will not follow Mendelian patterns and instead show different ratios for males and females. For example, genes on Y chromosomes will only be passed on to males and not to the females. The probability of inheritance of a mutated gene between males and females remains the same for an autosomal disorder irrespective of which parent carries the mutated gene. In case of recessive disorders, however, the probability of inheritance of mutated gene between the sons and daughters will differ based on whether the father or the mother is carrying the mutated gene. We will discuss this further in the next chapter. For the context of this discussion, it is enough to understand that any trait which seems to be passed equally between males and females is most likely present on an autosome.

2.6.2 Mendelian Segregation

We can also deduce from this pedigree that either individual I-3 or I-4 was heterozygous for the allele being studied. For individual III-4 to be affected, both his parents need to be heterozygous for the allele in question. Based on Mendel's law of segregation, for individual III-4 to be homozygous recessive, he has to inherit one recessive allele from each of his parents. Individual II-3 could have obtained the recessive trait from individual I-1 since he was affected. For individual II-4 to be a carrier, either individual I-3 or I-4 would have to be a carrier as they do not show the phenotype. We can determine the pattern of inheritance and composition of the genotype of an individual from a pedigree based on Mendel's laws of dominance and segregation.

Some examples of autosomal recessive disorders are cystic fibrosis, sickle cell anaemia and Tay-Sachs disease. Cystic fibrosis is caused by a defect in both copies of the cystic fibrosis transmembrane conductance regulator (*CFTR*) gene. The bodily fluids become thick and sticky. Due to this, the individuals suffer from respiratory and digestive problems. The abnormal mucous clogs airways and damages the pancreas. Tay-Sachs disease is a progressive neuronal disorder that affects the neurons in the brain and spinal cord. It is a rare disease in which infants start showing symptoms after 3–6 months. Their development slows and they develop muscle weakness. Progression of disease leads to loss of hearing, paralysis and seizures. The disease is caused due to two defective copies of the *HEXA* gene. This gene codes for the hexosaminidase A enzyme that plays a role in the breakdown of a fatty substance called GM2 ganglioside. Build of GM2 ganglioside is toxic for the neurons.

Let us now take the example of an autosomal dominant disorder. A typical pedigree is shown in Fig. 2.16 for inheritance of an autosomal dominant trait. We can immediately observe that this pedigree has at least one affected member in each generation. This is a typical characteristic of inheritance of a dominant allele. We can also see that the disorder has been passed on to both the males and the females. We can therefore infer that the allele is present on the autosomes. An example of autosomal dominant disorder is the Marfan syndrome. Marfan syndrome affects the connective tissue due to which a number of abnormalities in the heart, bones, joints, eyes and blood vessels can be observed. Marfan syndrome patients are tall and slender with long narrow faces. Their arm span exceeds their body height and they have elongated fingers and toes. Marfan syndrome is caused by a mutation in the *FBN1* gene which codes for the fibrillin 1 protein. Fibrillin 1 is instrumental in the formation of microfibrils. Microfibrils are threadlike filaments that provide strength and flexibility to the connective tissue. They also bind to growth factors

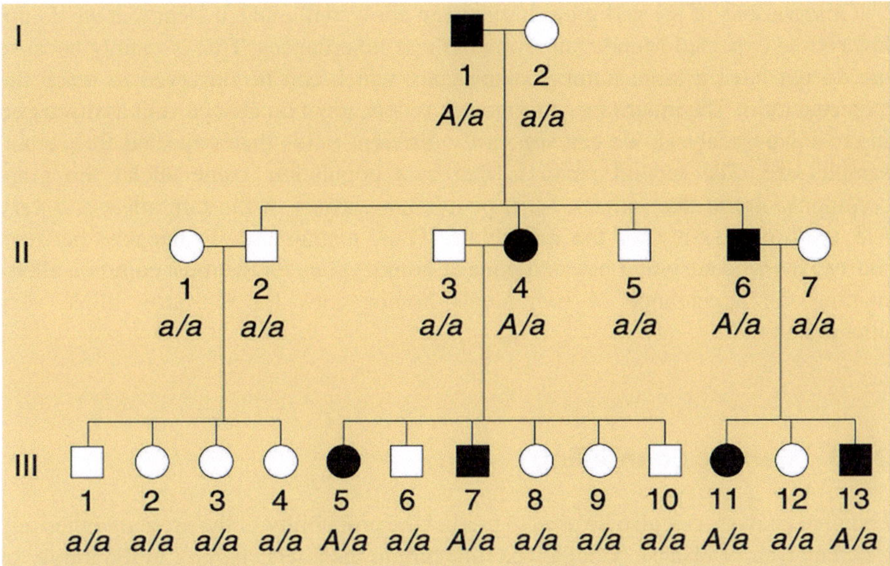

Fig. 2.16 Pedigree showing autosomal dominant mode of inheritance: The pedigree shown above has an affected member in each generation. A number of members of the family are affected and the trait is evenly distributed between males and females. Due to this, we can assume that the mode of inheritance of the trait being studied is autosomal dominant (Griffiths et al. 2011)

and control their release. Absence of functional fibrillin 1 reduces the amount of microfibrils leading to lack of control in the availability of growth factors. An excessive amount of available growth factors leads to overgrowth and abnormal tissue formation. Being an autosomal dominant disorder, the presence of even one mutated allele is sufficient for the manifestation of this disease.

Neurofibromatosis type 1 is also an autosomal dominant disorder. It is associated with a range of symptoms. Individuals suffering from the disease show a pigmentation change with appearance of dark patches of the skin. Benign tumours (non-cancerous) grow along the nerves in the brain and other parts of the body. In some cases, these tumours may turn cancerous. Additionally, these individuals may suffer from hypertension, macrocephaly, skeletal abnormalities and abnormal curvature of the spine. Some affected individuals may develop learning disabilities or attention deficit/hyperactivity disorder (ADHD). Neurofibromatosis type 1 is caused due to mutations in the *NF1* gene which codes for a protein called neurofibromin. This protein is produced in the neurons as well as glial cells like oligodendrocytes and Schwann cells. Neurofibromin acts as a brake for cell division and is known as a tumour suppressing gene. Non-functional neurofibromin leads to lifting of this brake and rampant and uncontrolled cell division leading to formation of tumours as seen in the disease.

Observations of the pedigree charts given above will make it clear that we do not always see expected Mendelian ratios in these inheritances. This is mainly because we do not have a large number of progeny which can be observed to reach the expected ratio. The inheritance of gametes is dependent on chance, and as discussed in chi-square analysis, we can see vastly different ratios than expected for a small sample size. The second factor is that in a population, some alleles are more commonly found than others. Most people are carriers of the rare allele and very few are homozygous for the rare allele. Thus, mating usually happens between individuals who are either heterozygous or homozygous for the most common allele, making the appearance of individuals homozygous for the rarer allele very uncommon.

2.6.3 Genetic Counselling

Pedigree analysis can also be used to predict the probability of the progeny inheriting a certain trait or disease. Couples having certain disorders running in the family or who themselves are affected may wish to know the probability of their children inheriting the disease. Couples with one of their children affected with a genetic disorder may seek to understand the possibility of their next child having the disease. Genetic counselling may help in such situations. Genetic counsellors will obtain information from the couples about affected family members and draw a pedigree. From this, they can deduce the mode of inheritance and further calculate the possibility of their unborn offspring inheriting the disorder. They can provide information and educate as well as address concerns of the family members regarding the disorders and provide support. They can also inform individuals about their genetic predisposition to certain diseases and lifestyle changes if any that can prevent or manage the disorder.

The Human Genome Project completed in 2003 was a 13-year-long study aimed at sequencing the entire human genome. This sequencing was carried out at multiple labs around the world and DNA was taken from a number of donors. The sequence is therefore a mosaic and not from any one individual. This prompted the 100,000 genome project in the UK which aims at sequencing 100,000 individuals comprising people with rare diseases, their families and cancer patients. With the mapping of these genomes, we can hope to understand more and more about our genes and the functions that they play in health and disease. We may be able to pinpoint the causes of a number of genetic diseases which remain unknown till now. Genomic sequences from patients will aid in developing diagnostics and therapeutics for individuals suffering from Mendelian disorders. It may allow us to get closer to personalised medicine where the analysis of an individual's genome may provide clues as to what treatment would be most effective for the individual.

Box 2.1 Scientific Concept: Gregor Mendel's Genetic Experiments: A Statistical Analysis After 150 Years (Jan Kalina)

Gregor Mendel is not only regarded as the founder of genetics but also as one of the pioneers of applying statistical principles in their experiments. Mendel's parents were peasants and bred fruit trees. After completing his secondary education, Mendel joined the St Thomas Augustinian abbey where he was ordained as a priest after some years. The abbey fostered an interest in science and encouraged Mendel to carry out his plant breeding experiments for which he was also given the use of one of the greenhouses in the abbey. In 1866, Mendel published his landmark paper which defined basic laws of genetics. He could not, however, enjoy the recognition of his work in his lifetime. Mendel passed away in 1884, aged 61 years, due to chronic nephritis. His work was rediscovered in 1900 independently by three scientists. In 1936, Robert Fisher reanalysed Mendel's data in light of the new statistical methods available. After a detailed analysis of the data, he concluded that Mendel's data seemed to be too much in agreement with his theoretical expectations. Given the variation caused by chance, his data seemed 'too good to be true'. He therefore concluded that Mendel must have falsified most, if not all, of his data to fit his theoretical assumptions (Kalina 2016).

Since then, a large number of papers have come out analysing both sides of the debate. In 2008, many scientists from different fields came together to write a review on the subject hoping to put an end to the argument. The authors were able to refute all arguments except the too good to be true claim for the data. There were few Mendel supporters like Pilgrim, who found fault with the way in which Fisher had performed his analysis (Pilgrim 1986). However, his arguments were later refuted by Edward (Edwards 1986). Then, there was a group of people who believed Fisher's analysis but found it too stringent and sought ways to analyse the data in an alternative manner. They too however found that Mendel might have adjusted the data to suit his hypothesis. Then there are those who believe that Fisher's analysis holds true given the assumptions. However, they try to provide suggestions for explaining the high p values other than deliberate malpractice. Some of the explanations provided are stopping experiments when the results seem to be good, error carried out by some anonymous assistant and discarding plants due to suspicion of some errors like pollen contamination and data selection for presentation (Novitski 2004).

In spite of the range of explanations provided, the bias towards expected values seen in Mendel's data has not been completely resolved. New statistical models to explain the bias are being proposed to provide a satisfactory explanation and end the controversy regarding falsification in Mendel's data. Based on the distribution of p values observed in Mendel's data and different simulation approaches, a plausible explanation of a two-stage model has been

(continued)

Box 2.1 (continued)
put forth. It states that Mendel repeated some of his experiments, presumably those which showed the largest deviation from expected values. He then reported only the best of two or a combination of results of the two which would result in values closer to those that were expected. This is called as a two-stage model where experiments were performed and results evaluated in second stage again for those experiments which did not meet the distribution. This speaks of an unconscious bias that Mendel introduced in his experimental approach but cannot be called as intentional scientific fraud (Pires and Branco 2010; Kalina 2016). Mendel's laws have stood the test of time and are still accepted today. His data might have been biased or adjusted but his conclusions about the nature of genes and their inheritance are still relevant today. We must therefore focus on Mendel's contribution to the field of genetics and quantitative biology and put an end to the debate as we seem to have some possible explanations.

2.7 Summary

- Gregor Mendel's painstaking decade long experiments and theories derived from them have laid the foundation of genetics, and he is known as the father of genetics. Mendel's three laws of genetics provide a framework for understanding the inheritance of genes. His work was rediscovered independently by three botanists in 1900.
- In monohybrid crosses, plants which differed in only one trait were crossed. The F1 hybrids carry the alleles for both parental traits but only express one of the parental traits which was termed as the dominant trait. The parental trait that was not expressed in the F1 hybrid was termed as the recessive trait. This was called as the law of dominance.
- The F1 hybrid produces gametes possessing the dominant and recessive alleles with equal probability. These can get paired randomly in F2 generation. This was called the law of segregation of alleles.
- Mendel also carried out dihybrid crosses where he crossed plants differing in two traits. He observed that the traits were inherited independently of each other. This was referred to as the law of independent assortment.
- He developed the method of test cross to determine the genotype of plants. Plants showing a dominant trait may either be heterozygous or homozygous for the dominant allele. Test cross involved crossing the plant in question with a plant showing the recessive trait. The ratio and phenotype of progeny from this cross could indicate heterozygosity or homozygosity of the plant being studied.
- Phenotypic and genotypic ratios for a cross with multiple traits can be predicted by using the methods of Punnett square, forked line method or probability

method. Statistical methods like chi-square goodness of fit test can be used to determine if the observed ratio from a cross vary from the expected ratios purely due to chance.

- Application of Mendelian principles can contribute to understanding and predicting genetic disorders in humans. Pedigrees can be constructed for a particular trait to understand mode of inheritance of a gene. Genetic counselling may help in counselling parents suffering from genetic disorders who want to understand the chances of their children inheriting the disease.

References

Edwards AWF (1986) Are Mendel's results really too close? Biol Rev 61:295–312

Griffiths AJ, Wessler SR, Lewontin RC, Gelbart WM, Suzuki DT, Miller JH (2011) An introduction to genetic analysis, 10th edn. Macmillan

Kalina J (2016) Gregor Mendel's genetic experiments a statistical analysis after 150 years. EJBI 12: 20–26

Klug WS, Cummings MR, Spencer CA, Palladino MA (2012) Concepts of genetics, 10th edn. Pearson Education, Inc.

Novitski E (2004) On Fisher's criticism of Mendel's results with the garden pea. Genetics 166: 1133–1136

Pierce BA (2010) Genetics: a conceptual approach. Macmillan

Pilgrim I (1986) A solution to the too-good-to-be-true paradox and Gregor Mendel. J Hered 77: 218–220

Pires AM, Branco JA (2010) A statistical model to explain the Mendel–Fisher controversy. Stat Sci 25:545–565

Reece JB, Meyers N, Urry LA, Cain ML, Wasserman SA, Minorsky PV, Jackson RB, Cooke BJ, Campbell NA (2011) Campbell biology. Pearson, Frenchs Forest, NSW

Reid JB, Ross JJ (2011) Mendel's genes: toward a full molecular characterization. Genetics 189: 3–10

The Punnett Square Approach for a Monohybrid Cross (2020) August 15. https://bio.libretexts.org/@go/page/13264

Extension of Mendelism

3

Rohini Keshava

3.1 Multiple Alleles

Mendelian principles of heredity dealt with the segregation of genes that had only two alternative forms (two alleles) at a gene locus. However, many genes exist in several alternative forms. In other words, any given gene can have several alternative variants/alleles, which occupy the same locus on the chromosome. When a gene exists in more than two alternative forms, it is called multiple allelism, and the allelic forms are called multiple alleles. Such alleles are said to constitute multiple allelic series. However, any given individual can possess only two of such alleles, on a pair of homologous chromosomes. Both homologous chromosomes can carry the same allele (homozygous) or carry different alleles (heterozygous).

3.1.1 ABO Blood Groups

Several genes in humans consist of multiple alleles. One of the best examples of the multiple allelic series in humans are those that determine the ABO blood grouping systems, i.e., multiple alleles of the ABO gene locus determine the ABO blood groups. Presently, as per the International Society of Blood Transfusion, there are about 33 blood group systems. However, of these 33 systems, one of the most significant is the ABO blood group system. It is of great clinical importance, particularly in transfusion medicine. The ABO system comprises of four different blood groups, viz., A, B, AB, and O. Worldwide distribution studies of these blood groups have shown that O group is the most common, and is followed by B, A and AB groups in the descending order of their abundance.

R. Keshava (✉)
Ramaiah University of Applied Sciences, Bangalore, India

D. Kar, S. Sarkar (eds.), *Genetics Fundamentals Notes*,
https://doi.org/10.1007/978-981-16-7041-1_3

3.1.2 The A and B Antigens

Characterization of the blood groups depends on the specific antigens located on the red blood cells (RBCs), i.e., erythrocytes. Currently known 33 blood group systems are represented by more than 300 antigens. The multiple allelic series of the ABO system determine the type of antigens on the RBCs, which in turn determines an individual's blood group. In an individual, the allelic pair occupying the ABO locus determines his/her blood group. Any given individual will belong to any one of the four blood groups. The two main antigens of the ABO blood system are called as "A" and "B" antigens. The presence or absence of these determines the type of blood group.

The multiple allelic series of the ABO blood group locus consists of three alleles, I^A, I^B, and i. Alleles I^A and I^B code for antigens A and B. Allele i does not code for any antigen. The dominance relationship among the ABO alleles can be represented as follows: $I^{\mathrm{A}} > i$, $I^{\mathrm{B}} > i$, $I^{\mathrm{A}} = I^{\mathrm{B}}$. This is an example of both dominance and codominance. When present in an allelic pair, both I^{A} and I^{B} are dominant, i.e., both phenotypes are equally expressed, and hence called codominant (refer to Sect. 3.2.4), whereas allele i is recessive to both these alleles. Hence, within this multiple allele series, the dominance relationship can be depicted as follows: $I^A = I^B > i$.

Presence of only I^{A} or I^{B}, in homozygous condition ($I^{\mathrm{A}}I^{\mathrm{A}}$ or $I^{\mathrm{B}}I^{\mathrm{B}}$), or in heterozygous condition ($I^{\mathrm{A}}i$ *or* $I^{\mathrm{B}}i$), results in the expression of A or B antigen, and hence the individual will be of either A or B blood group, respectively. If both I^{A} and I^{B} alleles are present in an individual, both A and B antigens are expressed and the individual is said to be of AB blood group. The genotype *ii* produces neither A nor B antigen on the RBCs and is said to be of O blood group. The possible genotypes and their corresponding blood types are shown in Table 3.1.

An antigen is a substance identified as foreign by the body and hence generates an immune response. In response to antigens, the immune system produces proteins called antibodies as a defense mechanism against the antigens. These antibodies possess the ability to specifically bind the antigens against which they are produced. The specific binding of the antibodies to the antigens results in large molecular aggregates that may precipitate. Such an antigen-antibody reaction is called as agglutination. However, an organism does not produce antibodies against its own antigens due to a mechanism called tolerance. For example, an individual whose blood group is A does not produce antibodies against antigen A, i.e., does not produce anti-A antibodies. Similarly, B group person will not produce anti-B antibodies and an AB group person will produce neither anti-A nor anti-B antibody. This also implies that an A group person will produce anti-B antibodies, an B group

Table 3.1 ABO blood typing system—genotypes and corresponding phenotypes

Genotype	Blood type
$I^{\mathrm{A}}I^{\mathrm{A}}$ or $I^{\mathrm{A}}i$,	A
$I^{\mathrm{B}}I^{\mathrm{B}}$ or $I^{\mathrm{B}}i$	B
$I^{\mathrm{A}}I^{\mathrm{B}}$	AB
ii	O

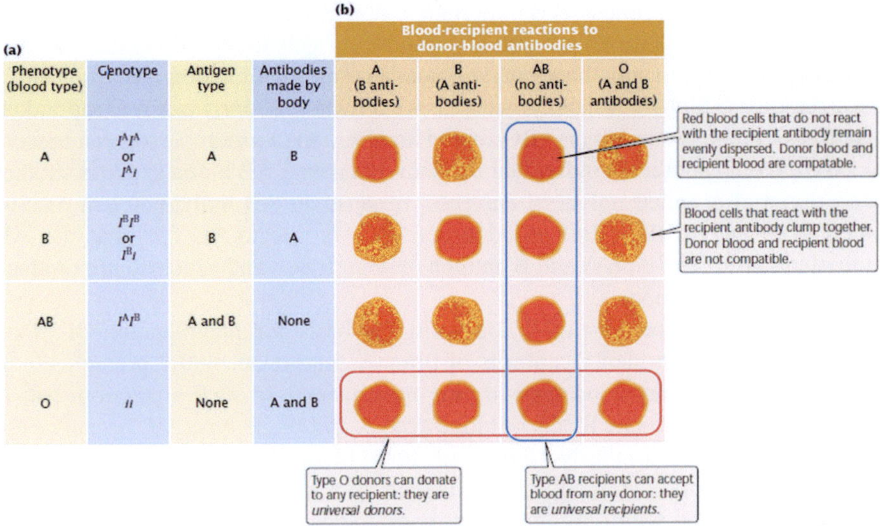

Fig. 3.1 ABO blood groups and transfusion compatibilities. (Left to right) Phenotype/blood group, and their corresponding genotypes, blood antigens, and serum antibodies of the respective blood groups, and their respective transfusion compatibilities are given

person will produce anti-A antibodies, and an O group person will produce both anti-A and anti-B antibodies.

Although antibodies generally are not produced without prior exposure to that particular antigen, the ABO system is an exception. In this case, the serum of individuals with A and B blood groups will possess anti-B and anti-A antibodies, respectively, even without prior exposure to these antigens. Likewise, both antibodies (anti-A and anti-B) will be present in the serum of an O group person. However, it has been suggested that the production of these antibodies in unexposed individuals is triggered by similar antigens present on the surfaces of commonly encountered bacteria. Therefore, if an individual is exposed to a non-self, blood antigen, then he/she is bound to generate an immune response, resulting in an agglutination or clumping of erythrocytes. For example, when an individual receives blood of incompatible type, i.e., whose antigens could stimulate an antibody production, the blood cells will agglutinate within the blood vessels and several blood vessels may become blocked. Recipient of such transfusion would go into a shock. Such a reaction may also be fatal to the recipient of such a transfusion. Therefore, matching the correct blood type is of great importance in blood transfusion, and hence the ABO blood grouping system is of high clinical significance (Fig. 3.1).

From Fig. 3.1, it can be seen that a person with blood group A can donate blood to A or AB group person; likewise, a person with B blood group can donate blood to a B or AB group person. However, as far as receiving the blood is concerned, A group can receive from both A and O; likewise, B group can receive from both B and O, because O group does not possess any antigens and hence does not cause agglutination in the recipient. It is important to note that although the blood serum of the O contains both anti-A and anti-B antibodies, it gets rapidly diluted during the process of transfusion and, therefore, will not cause an agglutination reaction. Due to these reasons, O blood group is considered as a universal donor, i.e., the O individual can donate blood to any other ABO type. However, an O group person can receive blood from only another O individual. An individual with AB blood group is another exception and is called universal acceptor, because an AB individual can receive blood from any of the ABO blood types. This is due to the fact that the AB individuals consist of both A and B antigen on their RBCs and hence do not produce anti-A or anti-B antibodies.

These agglutination reactions also form the basis for identification of blood groups in vitro. Such identification is of primary importance and a prerequisite to choose the compatible blood types during transfusions. In such a reaction, anti-A antibodies will agglutinate RBCs of blood type A and blood type AB, because both carry "A" antigen. Similarly, RBCs of both blood type B and AB will be agglutinated by anti-B antibody. Agglutination with only anti-A antibody indicates A blood group, and with only anti-B antibody indicates B blood group. However, agglutination with both indicates AB blood group. If there is no agglutination reaction with either antibodies, it indicates O blood group (Fig. 3.2).

3.1.2.1 Biochemistry of ABO Blood Antigens

Biochemically, ABO blood antigens are polysaccharide molecules, attached with proteins, forming glycoproteins. Antigens A and B, of the ABO blood grouping system, are both formed from a precursor substance called as the H substance or H antigen that is present in the RBC. The "H" substance is a mucopolysaccharide which is modified by enzymes encoded by I^A or I^B or both alleles, depending on whether the individual is of A, B, or AB blood type. The genes I^A and I^B produce enzymes called glycosyltransferases. Each of the glycosyltransferase catalyzes different modification of the H antigen (refer to Sect. 3.1.2.1), which forms the building block for the A and B antigens. The glycosyltransferases (encoded by I^A or I^B) attach either of the two sugar units (*N*-acetylgalactosamine or galactose) to the terminal sugars of the H substance, thus producing A or B antigen, respectively (Fig. 3.3). Hence, people with I^AI^A and I^Ai genotypes produce RBCs consisting of "A" polysaccharide, while those with genotypes I^BI^B and I^Bi produce RBCs with "B" polysaccharide. Likewise, individuals with genotype I^AI^B produce RBCs with both "A" and "B" polysaccharides, while those with *ii* genotype do not produce either of the polysaccharides on their RBC.

As explained earlier, the I^AI^B genotype shows codominance. This can be explained as follows: enzymes coded by both I^A and I^B alleles are about equally expressed resulting in the production of both A and B polysaccharides. The H substance will be modified by whichever enzyme it first interacts with. Once a particular enzyme modifies the H substance, it will no more be available for any modification with the other enzyme. This mechanism makes it possible that both A and B antigens are produced, in about equal proportions. The third allele, *i*, which is recessive to both I^A and I^B alleles, encodes a defective enzyme that is unable to

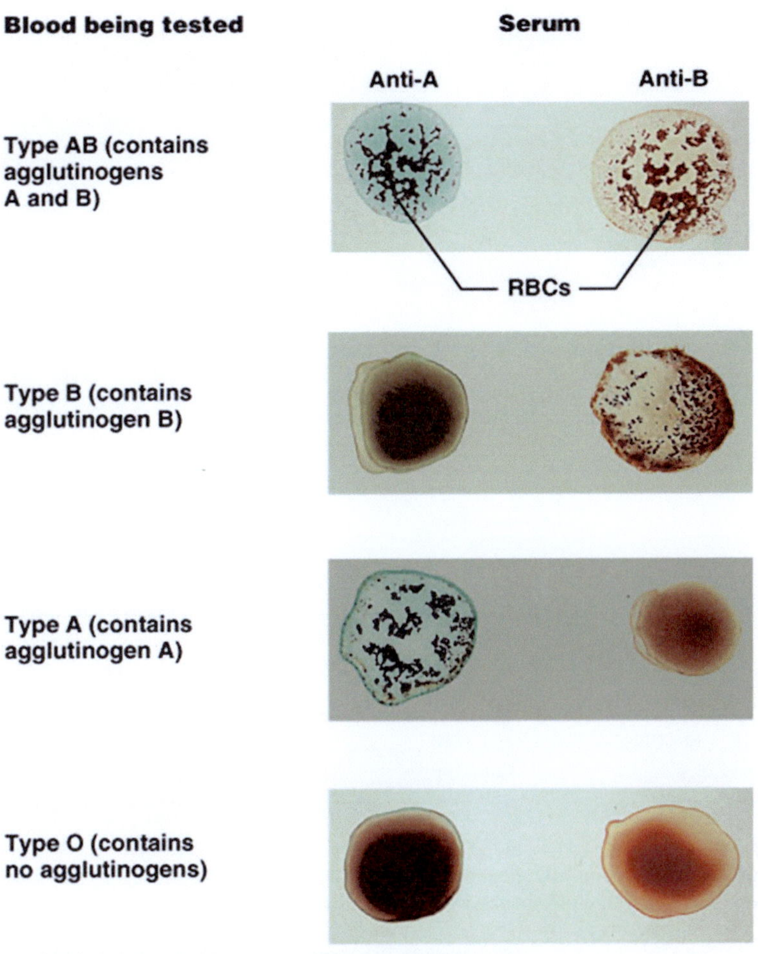

Fig. 3.2 Agglutination reaction in ABO blood groups. Blog group testing by addition of anti-A and anti-B antibodies is shown. Blood type AB shows agglutination with both antibodies. Blood type A and B show agglutination with anti-A and anti-B antibodies, respectively. Blood type O does not show agglutination with either of the antibodies

modify the H substance; hence, in homozygous (*ii*) condition, RBCs consist of only H antigen (Fig. 3.3).

Note: Currently, efforts are being made to enzymatically convert blood types A and B to O, thus increasing availability of universal donor blood. Blood transfusion is a vital part of health care and it requires cautious blood type matching in order to avoid adverse effects. Among the ABO blood types, only O is a universal donor. Therefore, for emergency situations, the O type is critical, but the resources or time involved in blood typing procedures is limited. Often there is a shortage of supply.

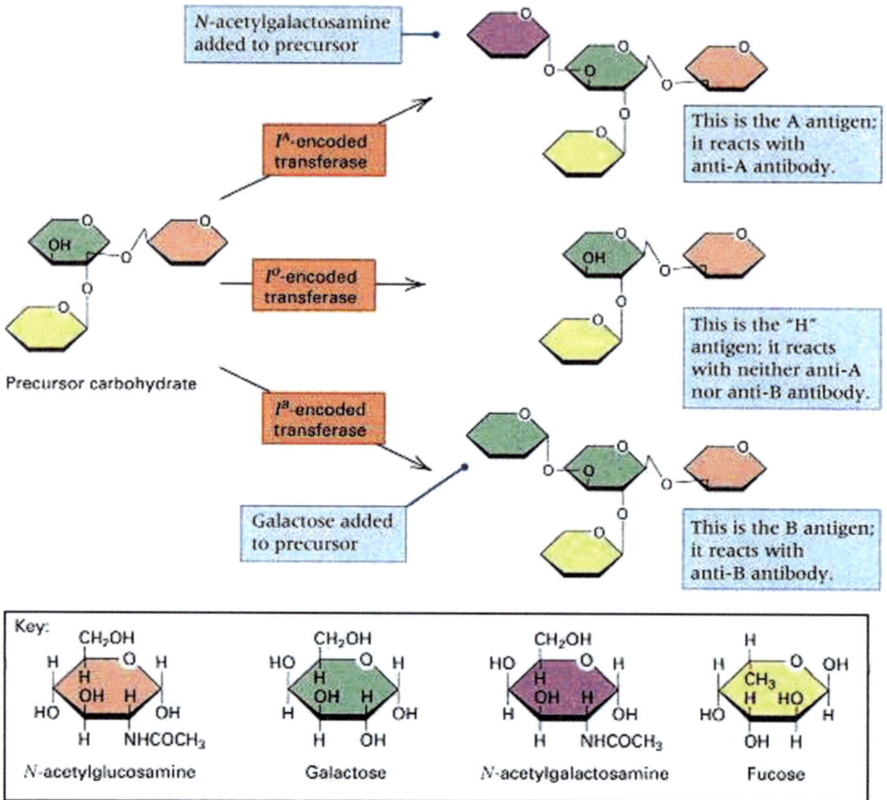

Fig. 3.3 Antigens of the ABO blood grouping system. The ABO antigens are located on the surface of RBCs and chemically these are carbohydrates. They are formed by glycosyltransferase-mediated modification of a precursor carbohydrate. Glycosyltransferases are encoded by alleles I^A and I^B. The enzyme coded by the *i* allele is inactive and the precursor carbohydrate remains unmodified, and is called the H substance. The enzyme coded by the I^A allele adds *N*-acetylgalactosamine and that of I^B allele adds galactose to the precursor H. The precursor H consists of fucose, galactose and *N*-acetylglucosamine

As already discussed above, the presence of an added sugar antigen differentiates A and B from O. A consists of GalNAc and B consists of Gal, additionally on the H-antigen characteristic of O type RBC. Therefore, removal of additional sugar antigen using an appropriate glycosidase has been attempted to convert A, B, or AB RBCs to O type and has met with significant success. The B-to-O was the first successful conversion and was demonstrated by Goldstein in 1982. Although enormous quantities of enzyme were required, proof of principle was successfully established and safe transfusions could be carried out in humans. By screening of bacterial libraries, novel α-galactosidases and α-*N*-acetylgalactosaminidases have been identified. This has allowed improvisations in B conversion and also led to successful A conversion. Screening of metagenomic libraries generated by feces analysis of an AB individual has led to detection of more efficient enzyme systems for the conversion of A.

3.1.3 The Bombay Phenotype

Bombay blood group and its variant para-Bombay are rare blood phenotype. In India, the frequency of both phenotypes combined is 1 in 10,000. It is slightly more common in Taiwan, affecting 1 out of 8000 people. A relatively large number of individuals with this phenotype were reported on a small French island 800 km east of Madagascar in the Indian Ocean, called Reunion Island. In Europe, one per million people possesses this phenotype.

The Bombay phenotype was named after the city Bombay, now known as Mumbai, in India, where it was first discovered in 1952 by Dr. Y. M. Bhende. An individual who had a peculiar blood type that reacted to other blood types in a way that was not observed before was discovered. When the serum from this individual was tested, it was found to contain antibodies that reacted with all RBCs of ABO phenotypes (i.e., blood groups O, A, B, and AB). The individual's RBCs apparently seemed to lack all antigens of the ABO blood group and an additional antigen that was earlier unknown. Based on these observations, a research paper was published in 1952, and the paper reported the presence of another antigen, related to the ABO blood grouping system, in addition to the known A and B antigens. This new blood antigen was called as the H substance or H antigen, and was found to be the building block for the A and B antigens of the ABO blood group system (refer to Sect. 3.1.2.1).

The "Bombay phenotype" refers to individuals whose RBCs lack the H antigen. Since the formation of A and B antigens is dependent on the presence of the precursor H substance, the RBCs of individuals with this phenotype also lack A and/or B antigens. Hence, these individuals produce anti-H, anti-A, and anti-B antibodies, and therefore during blood transfusion, they can receive blood only from another Bombay phenotype individual who also lacks the H, A, and B antigens.

The H antigen gene locus is represented by two alternate alleles, *H* and *h*. H antigen is coded by a dominant *H* allele, whereas the recessive form of this allele is an amorph, i.e., it does not code for any known product. The *HH*/*Hh* genotype codes for the presence of H antigen, which practically is found on all RBCs and is the building block for the production of the antigens within the ABO blood group. The deficiency of H antigen is known as the "Bombay phenotype." It is also known as *hh* blood group or Oh blood group. Being deficient in H by itself does not have any ill effects, but in case a blood transfusion is required for the individual with such a blood group, the donor also should be H deficient. If in case a transfusion is performed with even an O blood group donor, there can be a severe transfusion reaction. Since H antigen is required for the formation of ABO blood group antigens, its absence blocks the formation of the ABO blood group antigens. This can be misleading in paternity cases.

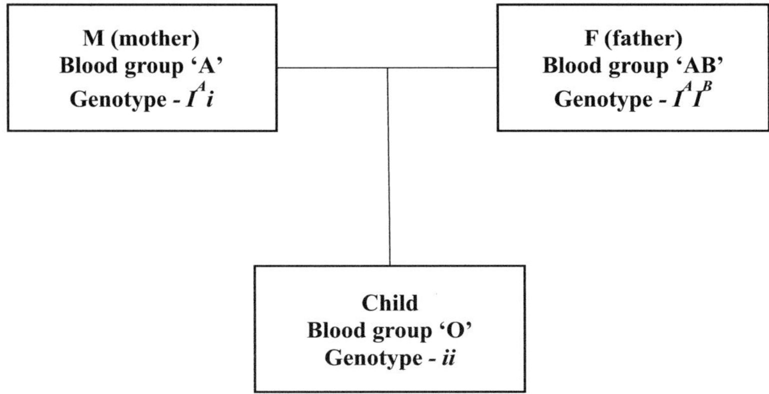

Fig. 3.4 Predicted ABO phenotypes and the genotypes of "M," "F" and child. F's ABO genotype does not consist of *i* allele. Paternity test would reveal that F is not the father of the child considering inheritance at the ABO locus

For example, consider the following hypothetical situation: "M" is the mother of a child, whose father "F" is in doubt of his paternity. In this case M's blood type was A (genotype I^Ai), her child's blood type was O (genotype *ii*), and F's blood type was AB (genotype I^AI^B). In this case, as the child's blood group is O, it must inherit an *i* allele from the father. Therefore, the possible genotypes of the father are I^Ai, I^Bi, or *ii*, i.e., the blood group of the child's father can be A or B or O. As per this situation, "F" cannot be the father of the child (Fig. 3.4). This could have been the final conclusion if only the ABO gene locus was considered. However, another gene locus also plays a role in the expression of the ABO blood phenotypes, i.e., the H antigen gene locus.

With respect to the H antigen gene locus, both "M" and "F" are carriers of incomplete *H* deficiency. In other words, both "M" and "F" are heterozygous, *Hh* at the H antigen gene locus. Therefore, their child is homozygous for the recessive *h* allele. Due to this, the child will not be expressing the H antigen and hence is unable to produce any ABO blood group antigens. Hence, despite inheriting the A or B allele from "F," the child's RBC lacks the A and B antigens similar to blood type O; therefore, it can be concluded that "F" is the father of the child (Fig. 3.5).

3.1.3.1 Biochemistry of H Antigen

As explained earlier (refer to Sect. 3.1.2.1), the biosynthesis of A and B antigens involves a series of enzymes, called the glycosyltransferases that transfer monosaccharides, resulting in antigens that are oligosaccharide chains. Similarly, the biosynthesis of the H antigen also involves such glycosyltransferases. Production of the H antigen involves a specific fucosyltransferase. Depending upon what alleles of the ABO system are present in an individual, the H antigen gets converted into either A, B, or both antigens (refer to Sect. 3.1.2.1). In a person of blood group O, the H antigen stays unmodified. Hence, the H antigen is characteristically present in individuals with O blood group.

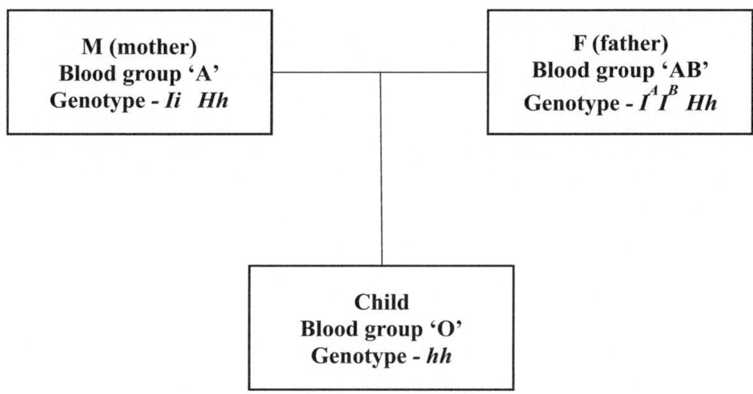

Fig. 3.5 Actual ABO and Hh genotypes of "M," "F," and child. Both "M" and "F" are heterozygous, *Hh*. Hence, their child is homozygous *hh* and lacks H antigen and is unable to produce any ABO antigens. Considering the H locus in addition to the ABO locus resolves the paternity issue and determines "F" is the father of the child

Table 3.2 Common and uncommon H phenotypes

Common H phenotypes	Uncommon H phenotypes
Secretor (common) • H antigen present on RBC • H antigen secreted in saliva • Anti-H is absent • Genotype: *H/H or H/h, Se/Se, or Se/se*	*Bombay phenotype* • H antigen absent in RBC • H antigen absent in saliva • Anti-H present in serum • Genotype: *h/h se/se*
Nonsecretor (common) • Presence of H antigen on RBC • H antigen not secreted in saliva • Anti-H not produced • Genotype: *H/H or H/h, se/se*	*Para-Bombay phenotype* • H antigen weakly expressed on RBC • Saliva may or may not contain H antigen • Anti-H is present in the saliva • Genotype: (*H*), *Se/Se or Se/se or se/se*

Two different loci of the genome encode two fucosyltransferases with similar substrate specificities, the *H* locus, consisting of FUT1 gene and the *Se* locus consisting of the FUT2 gene. The FUT1 gene is expressed in RBCs. At least one functional copy (i.e., at least one dominant *H* allele) of FUT1 gene is required for H antigen synthesis. The *HH* or *Hh* genotypes enable expression of active fucosyltransferase. The *hh* genotype makes inactive copies of FUT1 and results in Bombay phenotype. The *Se* locus consists of the FUT2 gene, and it is expressed in secretory glands. Individuals who possess at least one copy of the functional enzyme (*Se/Se* or *Se/se*) are "secretors." They produce soluble form of H antigen, and are found in bodily fluids such as saliva. "Nonsecretors" (*se/se*) cannot produce soluble H antigen. The FUT2 gene encoded enzyme is also involved in synthesis of Lewis blood group antigens. The two commonly found H phenotypes are "secretor" and "nonsecretor," whereas Bombay and the para-Bombay are less common. The comparison of the two is given in Table 3.2.

3.1.3.2 Clinical Significance of Anti-H Antibodies

3.1.3.2.1 Transfusion Reactions

If individuals with anti-H antibodies in their serum receive transfusions of blood that contains the H antigen (e.g., blood group O), they are at risk of an acute hemolytic transfusion reaction.

3.1.3.2.2 Hemolytic Disease of the Newborn (HDN)

Theoretically, the maternal production of anti-H antibodies during pregnancy could cause hemolytic disease in a fetus who has not inherited the Bombay phenotype from the mother. In practice, such cases of HDN have not been described, possibly due to the rarity of the Bombay phenotype.

3.1.4 *Drosophila* Eye Color

Drosophila consists of several multiple allelic genes. Genes that code for eye color of *Drosophila melanogaster* are one such example. A series called the "white-eye" is a good example. It is named so because it consists of alleles of the white locus (white-eye is one of the first discovered mutations of *Drosophila*). The series consists of several alleles, which in homozygous condition produce a series of eye colors of increasing intensity. The eye colors range from white through yellowish to red. Heterozygotes of these alleles produce eye colors that are intermediate to that of the parental homozygotes for those alleles. In this series, the wild type, i.e., the allele for red eye color is dominant to all the other alleles of the series, whereas the allele for white-eye color is recessive to all. Some of the alleles of this series and their symbols are given in Table 3.3. Note that the names of the alleles correspond to the eye color they produce. The dominance relationship/hierarchy of the series can be depicted as follows: $W > w^{co} > w^{bl} > w^{e} > w^{ch} > w^{a} > w^{h} > w^{bf} > w^{t} > w^{p} > w^{i} > w$.

Table 3.3 Allelic series of "white-eye" locus in *Drosophila melanogaster*

Allele	Symbol
White (recessive to all)	w
Ivory	w^{i}
Pearl	w^{p}
Tinged	w^{t}
Buff	w^{bf}
Honey	w^{h}
Apricot	w^{a}
Cherry	w^{ch}
Eosin	w^{e}
Blood	w^{bl}
Coral	w^{co}
Red (wild type—dominant to all)	W

3.2 Dominance Relationship

3.2.1 Dominance

The dominance concept states that among the pair of alleles in a genotype, only the dominant allele expresses itself in the phenotype and recessive allele gets suppressed or hidden. In other words, for a given pair of alleles, dominance is a condition in which a trait is expressed both in the homozygous and heterozygous conditions. Such an allele that expresses itself in the heterozygous condition is termed as dominant. The allele which is not expressed in the heterozygous condition is called recessive. A recessive allele can be expressed only in a homozygous state.

For example, in F_1 offspring of a monohybrid cross between round seed (R) and wrinkled seed (r) plants, plants inherit an *R* allele from round-seeded parent and an *r* allele from wrinkled-seeded parent (Fig. 3.6). But only the round trait encoded by *R* allele is *observed* in all the F_1 progeny, i.e., the heterozygote's phenotype is same as one of homozygous parent's phenotypes. Similar results were obtained in a cross between a tall (T) and short (t) plant (Fig. 3.7), where all the F_1 plants were tall. In these examples, the alleles for round (R) seed and tallness (T) are dominant, whereas the alleles for wrinkled seeds (r) and shortness (t) were recessive. The dominance concept is one of the most important conclusions derived by Mendel from the results of the monohybrid crosses.

3.2.2 Incomplete Dominance

All seven characters of pea plants that were chosen by Mendel clearly exhibited complete dominance, i.e., there was a clear dominance-recessive relationship among the pair of alleles. In other words, the dominant and the recessive traits of a character were clearly distinct, and there were no intermediate phenotypes. But as Mendel performed further studies, he observed that variations occurred in the dominance relationships of alleles of a gene. One such cross that he performed was related to the length of time taken by the pea plants for flowering. He performed a cross with two homozygous varieties with different flowering time. The flowering time between the two varieties differed by an average of 20 days. Interestingly, he observed that the F_1 offspring of this cross showed a flowering time that was intermediate of the two parents.

Such a phenomenon where the heterozygote has an intermediate phenotype compared to its homozygous parents is termed as incomplete dominance. The phenotype expressed by the heterozygous genotype will be in range between phenotypes expressed by homozygous genotypes. Incomplete dominance is also known by alternative terms such as semidominance and partial dominance.

3.2.2.1 Incomplete Dominance in Fruit Color of Eggplants

When a true breeding purple fruit (*PP*) producing plant is crossed with a true breeding white fruit (*pp*) producing plant, all F_1 (*Pp*) heterozygotes produce violet

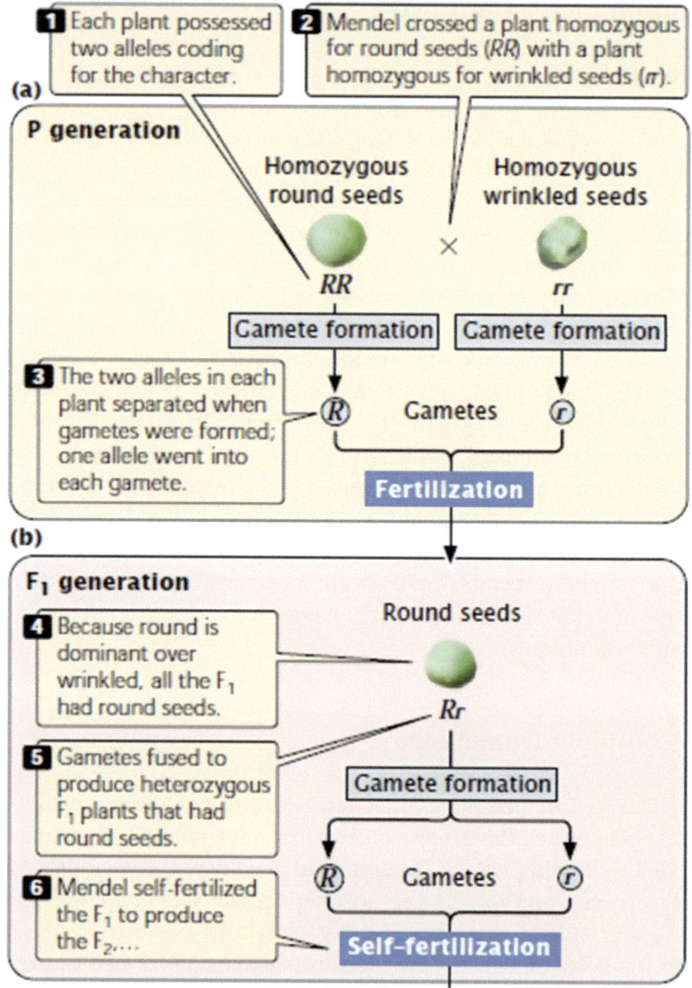

Fig. 3.6 Mendel's monohybrid cross depicting the law of segregation and dominance phenomenon. This is a monohybrid cross between purebred round seed (R) and wrinkled seed (r) pea plants. All plants of the F_1 generation expressed round seeds indicating that round shape of seeds is dominant over wrinkled shape. The F_2 generation produces both round and wrinkled shape plants in the ratio of 3:1, indicating the segregation of the two alleles

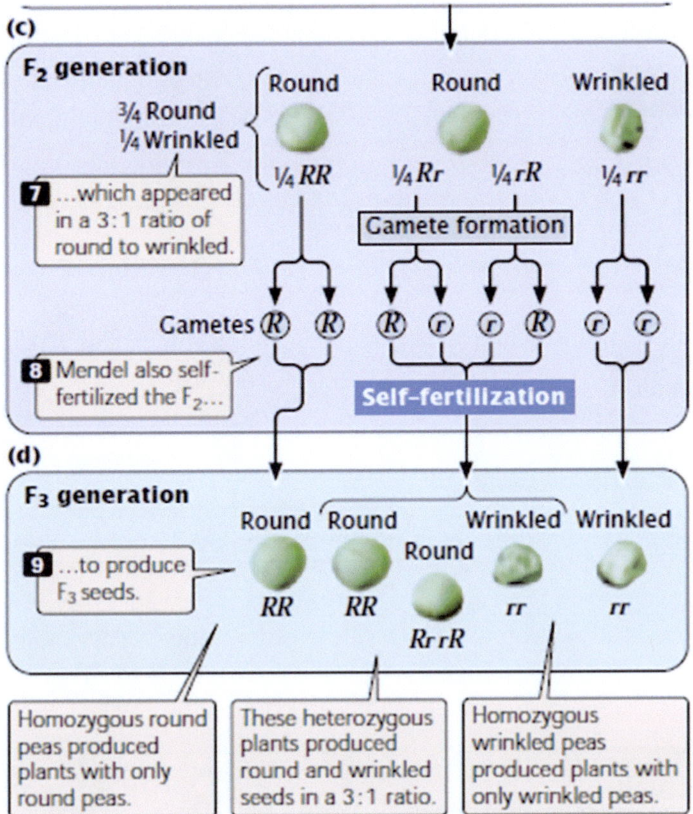

Fig. 3.6 (continued)

colored fruit (Fig. 3.8). After selfing F_1 plants, the F_2 obtained consists of 1/4 purple (*PP*), 1/2 violet (*Pp*), and 1/4 white (*pp*) (Fig. 3.8). This 1:2:1 phenotypic ratio is different than the usual 3:1 ratio obtained when alleles with complete dominance-recessive relationship are involved.

3.2.2.2 Incomplete Dominance in Feather Color in Chicken

Feather color in chicken provides another example of incomplete dominance. When a homozygous black and homozygous white chicken are crossed, they produce gray F_1 chickens. If these gray F_1 are further selfed, they produce F_2 in a phenotypic ratio equal to 1 black:2 gray:1 white.

3.2.2.3 Incomplete Dominance in Skin Pattern of Horses

Leopard spotting trait in horses exhibits incomplete dominance over unspotted trait. Horses with *LL* genotype (homozygous dominant) are white with numerous dark

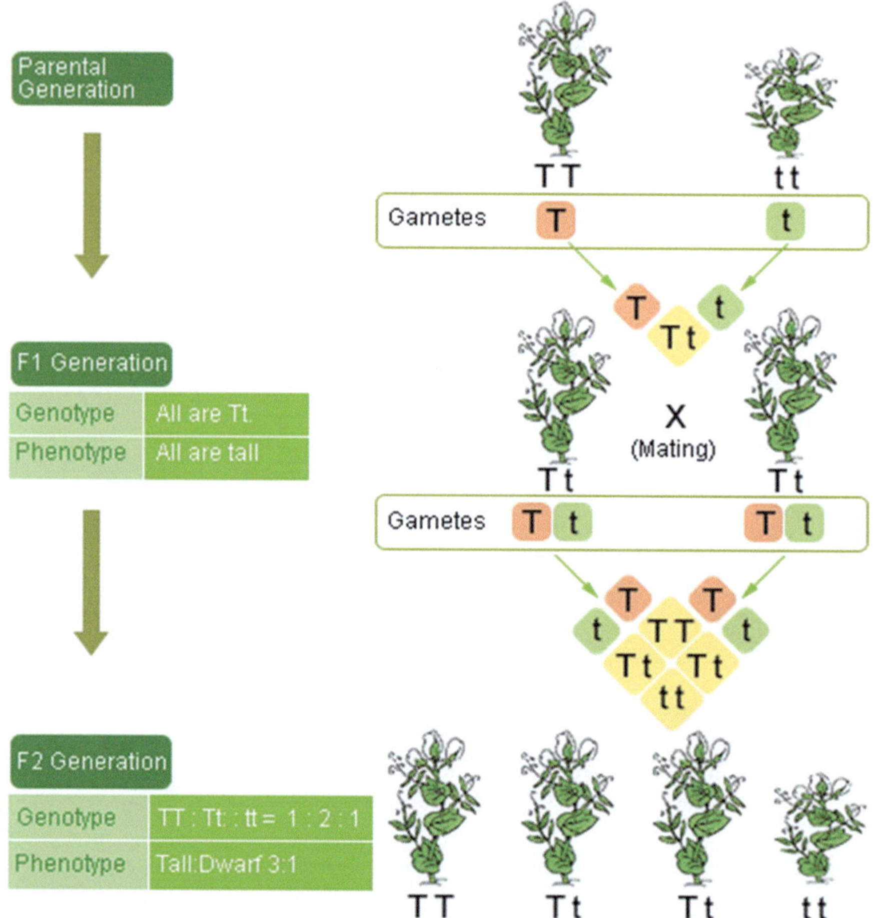

Fig. 3.7 Mendel's monohybrid cross depicting the law of segregation and dominance phenomenon. This is a monohybrid cross between purebred tall (T) and short (t) pea plants. All plants of the F_1 generation expressed tallness indicating that tall is dominant over short. The F_2 generation produces both tall and short plants in the ratio of 3:1, indicating the segregation of the two alleles

spots and those with *ll* genotype (homozygous recessive) are unspotted, whereas the heterozygote, with *Ll* genotype, has fewer spots (Fig. 3.9).

3.2.2.4 Mechanism of Incomplete Dominance

Most genes code for enzymes. For example, among a pair of dominant alleles coding for a particular enzyme, each allele often individually contributes to the total quantity of the enzyme in a cell or organism. A heterozygote will therefore produce an intermediate quantity of the enzyme, whereas a homozygote recessive will not produce the enzyme (Fig. 3.10). This example gives an insight into the mechanism

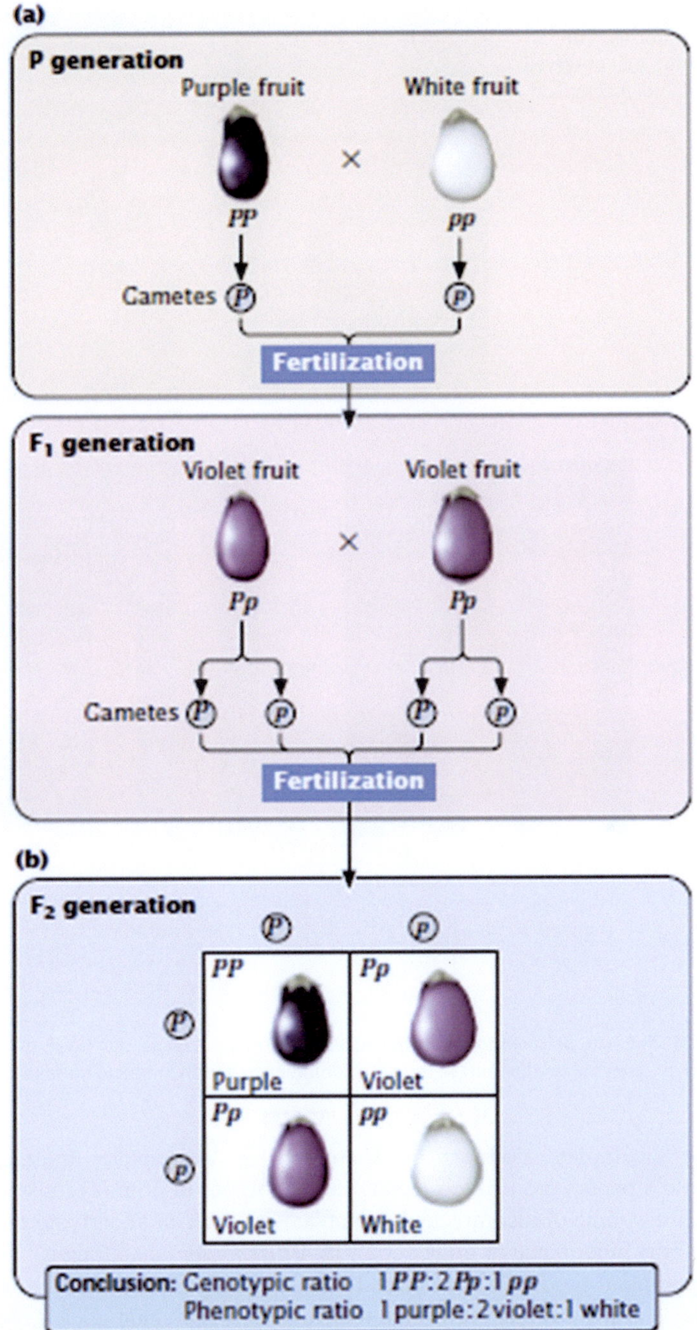

Fig. 3.8 Incomplete dominance in eggplant. The fruit color of the heterozygote is violet, which is intermediate to the purple and white fruit color of the parents

Fig. 3.9 Incomplete dominance in horses. The leopard spotting trait observed in horses exhibits incomplete dominance

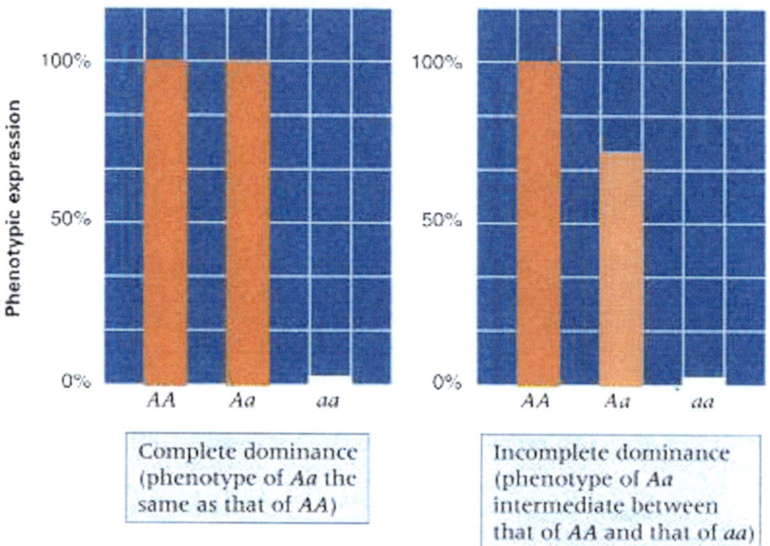

Fig. 3.10 Mechanism of incomplete dominance. The figure depicts the level of phenotypic expression (e.g., enzyme production) in a state of dominance and incomplete dominance

underlying incomplete dominance. Therefore, in incomplete dominance, the heterozygote's phenotype is intermediate to phenotypes of homozygous parents.

A typical example of incomplete dominance with such an underlying mechanism is related to the inheritance of flower color in *Antirrhinum* (snapdragon). Flowers of the wild-type plant produce a red colored anthocyanin pigment. The pigment is the end product of a sequence of enzymatic reactions. The dominant allele *I* encodes the wild-type enzyme, and is limiting to the rate of complete biosynthetic reaction. Therefore, the quantity of red pigment produced is proportional to the amount of enzyme expressed by *I* allele. An inactive enzyme is coded by the recessive *i* allele, and homozygous recessive (*ii*) flowers are ivory in color. The heterozygous (*Ii*) condition leads to a reduction in the quantity of the critical enzyme due to the presence of a single *I* allele, as compared with the amount of enzyme in the flowers of the wild-type homozygous for the dominant allele (*II*). Due to the reduction in the amount of the enzyme, the amount of red pigment is also reduced, compared to the wild-type. As a result, the flower color of the heterozygote is diluted and appears pink, which is intermediate of red and white (Fig. 3.11). When snapdragon plants with different flower color, such as a pure-breeding red-flowered variety and a pure-breeding ivory-flowered variety are crossed, F_1 plants produce pink flowers. The F_2 progeny produced red, pink, and white flowers in the ratio of 1:2:1, respectively. Further, among the F_2, the red-flowered plants produced only red-flowered plants, and similarly the ivory-flowered plants produced only ivory, whereas the pink-flowered plants produced again 1/4 red:1/2 pink:1/4 ivory.

Similar to snapdragons, *Mirabilis jalapa* (four o'clock plant) also shows another example of incomplete dominance. In this plant also, a cross between red flower producing and white flower producing plants results in pink-flowered offspring. If these pink-flowered F_1 are further selfed, an F_2 with the ratio of 1 red:2 pink:1 white is obtained. Similar to that observed in snapdragons, in this case also, pink flower is heterozygous and is intermediate of red and white homozygotes.

Incomplete dominance is frequently observed in quantitative phenotypes rather than in discrete phenotypes. A *quantitative* trait is a trait where it is possible to measure the phenotype on a continuous scale. In quantitative traits, the phenotype of heterozygous offspring is a measured value, usually falling in the range between phenotypes of homozygotes. Traits such as number of eggs laid, flowering time, amount of enzyme, height, weight, etc. are examples of quantitative traits. On the contrary, a *discrete* trait is either fully expressed or not expressed, and there is no intermediate phenotype. Round/wrinkled and yellow/green with respect to seeds in pea plants are examples of discrete traits.

3.2.2.5 Examples of Incomplete Dominance in Humans

A rare autosomal recessive genetic disorder, called Tay-Sachs disease, is characterized by a progressive neurodegeneration in the brain and spinal cord. Usually, infants get affected by the common form of this disease. In homozygous recessive condition, it causes death before the age of 3 in children. However, interestingly, the heterozygous individuals do not express the disease phenotype and appear normal.

The manifestation of the disease is due to a defective enzyme called hexosaminidase A, which is required for proper lipid metabolism. Enzyme activity is completely lacking in homozygotes for this recessive mutation, whereas about half the normal activity is observed in the heterozygotes. The individuals who are homozygous for the normal allele show full level of the enzyme. Since the heterozygote expresses an intermediate of the phenotypes of the homozygotes, this is a case of incomplete dominance.

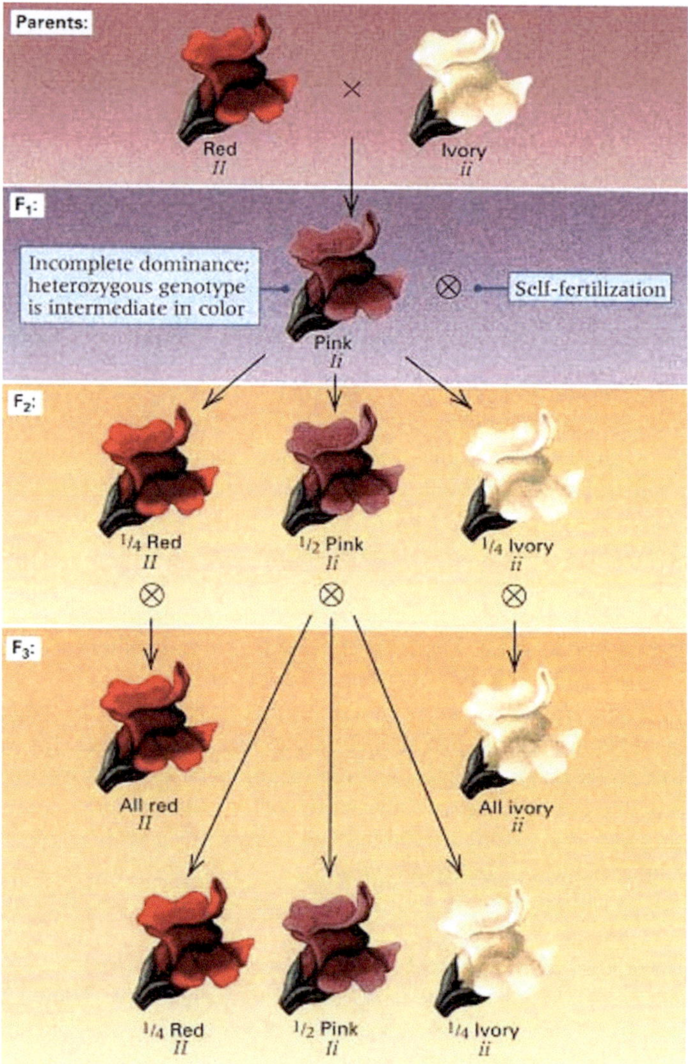

Fig. 3.11 Incomplete dominance in snapdragons. Flower color in snapdragons shows incomplete dominance, where the flower color of the F_1 heterozygote (pink) is intermediate to the red and white flower colors of homozygous parents

3.2.3 Complexities Involved in the Concept of Dominance

Although all traits described by Mendel had clear dominant-recessive patterns, examples of departures from strict dominance are also commonly observed, for example, even for a classical trait such as round versus wrinkled seeds in pea plants.

It is observed that to just consider round as dominant is an oversimplification of the phenomenon.

Owing to the fact that one gene controls several functions, which is often the case, the alleles that are responsible for the seed shape in peas not only determine the seed shape but also determine several other attributes. In the visible phenotype, round appears dominant over the wrinkled, and it only suggests that visually the effect of *WW* and *Ww* genotypes cannot be distinguished and they both appear round. But at the molecular level, they exhibit differences.

The shape of the wrinkled seeds is due to a genetic defect. The defect results in the deficiency of an active form of an enzyme called starch-branching enzyme I (SBEI). This enzyme is required for the synthesis of a branched-chain form of starch called as amylopectin. Compared with homozygous *WW* seeds, the heterozygous *Ww* seeds have only half the amount of SBEI. The seeds that are homozygous for the recessive allele, i.e., *ww*, have nearly no SBEI (Fig. 3.12).

Homozygous *WW* pea seeds contain large, well-rounded starch grains, and hence the seeds efficiently retain adequate water and shrink uniformly as they ripen, and do not become wrinkled. In homozygous *ww* seeds, the starch grains lack amylopectin and are irregular in shape. When these seeds ripen, they lose water very rapidly and shrink unevenly, resulting in the wrinkled phenotype (Fig. 3.12). The *w* allele also affects the shape of the starch grains in *Ww* heterozygotes. In heterozygotes, the seeds contain starch grains of intermediate shape (Fig. 3.12), and these seeds have high enough amylopectin content to enable uniform shrinking and thus prevent wrinkling (Fig. 3.12).

From these observations, it is clear that same pair of alleles can show complete dominance for one trait and show incomplete dominance for another trait. In this particular example, if only the overall shape of the seeds is considered, round is dominant over wrinkled and there are only two phenotypes. But if the shape of the starch grains as observed with a microscope is also considered, all three genotypes can be distinguished from each other, i.e., *WW*, with large, rounded starch grains; *Ww*, with large, irregular grains; and *ww*, with small, irregular grains. If we also consider the amount of the SBEI enzyme, the *Ww* genotype has an amount intermediate between the amounts in *WW* and *ww* genotypes (Fig. 3.12).

This particular example makes it clear that "dominance" is not a simple interaction between a particular pair of alleles irrespective of how the resulting phenotypes are observed. When a gene affects multiple traits, then a particular pair of alleles might show simple dominance for some traits but may differ in others. A given phenotype consists of many different physical and biochemical attributes, and dominance may be observed for only some of these attributes and not for others. Thus, dominance is a property of a pair of alleles in relation to a particular attribute of a given phenotype (Table 3.4).

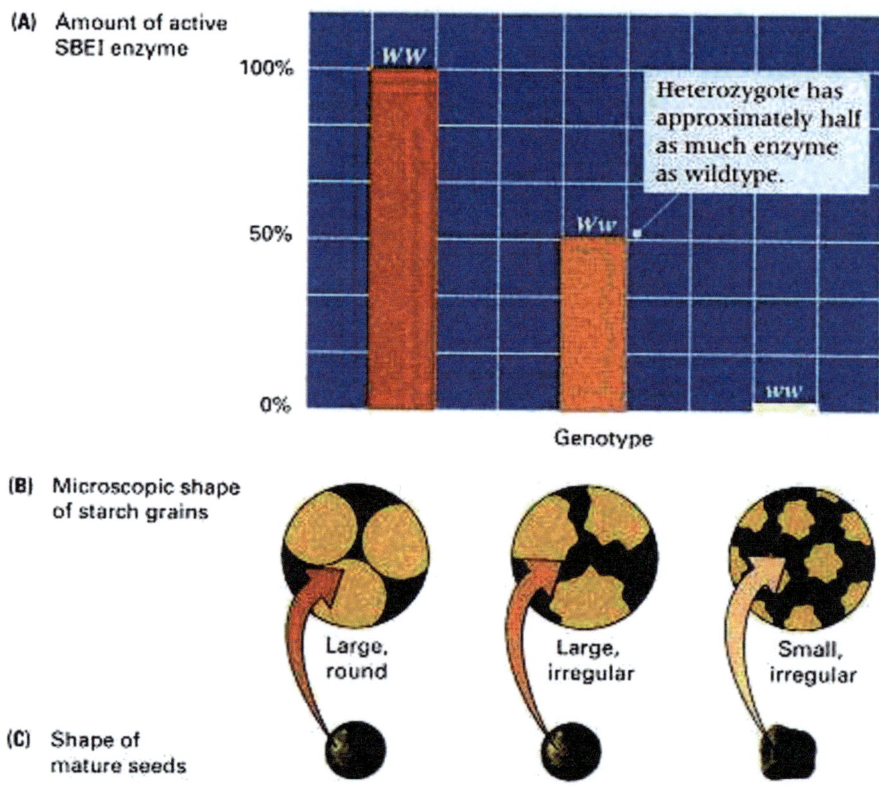

Fig. 3.12 Complexities involved in dominance where alleles affect multiple traits. Alleles W (round seed) and w (wrinkled seed) affect seed shape phenotype in pea plants. (**a**) Level of SBEI in the heterozygous plant is intermediate to that of homozygous parents—incomplete dominance. (**b**) Size and shape of the microscopic starch grains are intermediate in the heterozygote, compared to the homozygous parents—incomplete dominance. (**c**) The seed shape trait depicts complete dominance, with *W* coding for round and *w* coding for wrinkled seeds

Table 3.4 Comparison between the various forms of dominance relationships

Type of dominance	Definition
Dominance	Heterozygote's phenotype is same as one of homozygous parent's phenotypes
Incomplete dominance	Heterozygote's phenotype is intermediate to phenotypes of homozygous parents
Codominance	Heterozygote's phenotype expresses characteristics of both parental alleles

3.2.4 Codominance

Codominance refers to a condition in which the heterozygote's phenotype expresses characteristics of both parental alleles. The heterozygous genotype thus results in an equal mixture of the phenotypes of both of the corresponding homozygous genotypes. Unlike incomplete dominance here, the heterozygous phenotype is not intermediate between the homozygous genotypes (like pink colored flowers in snapdragons) but rather has equal representation of the characteristics of both homozygous genotypes. Some of the important examples of codominance are observed in the alleles coding for the antigens determining human blood groups.

3.2.4.1 Codominance in Human Blood Groups

The ABO human blood grouping system consists of multiple allelic series where alleles exhibit both simple dominance and codominance. I^A and I^B alleles of this multiple allelic series are codominant (refer to Sect. 3.1.2).

Another example of codominance in human blood grouping systems is observed in the MN blood group. The antigens M and N are determined by two alleles of a gene, one coding for M and the other coding for N. Homozygotes for M allele produce M antigen, and homozygotes for N allele produce N antigen, whereas both M and N antigens are produced in the heterozygote, because of their codominant nature. Accordingly, there are three possible blood types, i.e., M, N, and MN. The alleles for M and N antigen are represented as L^M and L^N. Hence, the genotypes for the M, N, and MN blood groups are L^ML^M, L^NL^N, and L^ML^N, respectively (Fig. 3.13). Here the letter L is used as a tribute to Karl Landsteiner, the discoverer of blood typing.

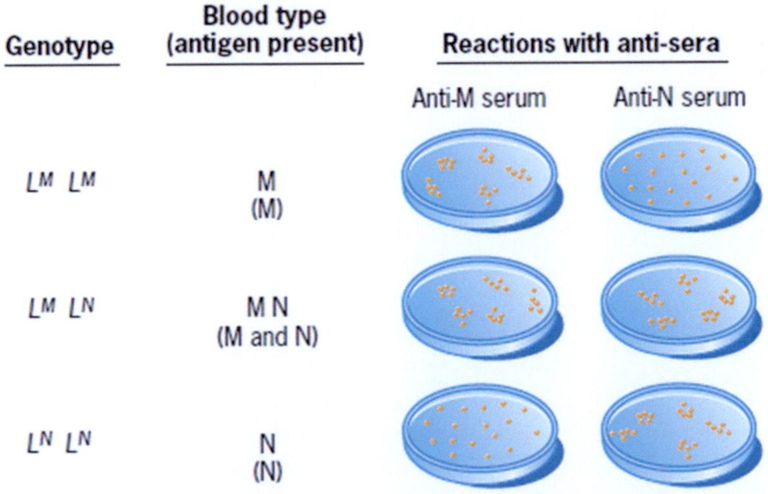

Fig. 3.13 Agglutination reactions to detect M and N antigens. The M and N antigens on the RBC are detected using specific anti-M and anti-N sera. The identification happens based on the agglutination caused by the antisera on binding to the specific antigen. Three blood types, M, N, and MN, can be identified

3.2.5 Lethal Alleles

An allele that usually causes death, at an early developmental stage, and often before birth is called as a lethal allele. Such an allele causes genotypes to be lost from the progeny of a particular cross. The ratio of progeny from a specific cross can be altered by lethal alleles. A peculiar pattern of inheritance was reported in mice in 1905, by Lucien Cuenot. A cross between two yellow mice yielded approximately 2/3 yellow and 1/3 nonyellow mice. A test cross of the yellow mice yielded all heterozygous yellow mice. Cuenot was unable to obtain true breeding yellow mice. Several discussions on these observations led to the realization that the lack of true breeding yellow mice in the progeny could be due to the lethality of the yellow allele when in homozygous condition (Fig. 3.14). In his experiment, Cuenot had originally crossed two heterozygous mice (*Yy* × *Yy*). As per Mendelian segregation, this cross would have resulted in 1/4 *YY*, 1/2 *Yy*, and 1/4 *yy* (Fig. 3.14). But in this particular mating, the homozygous *YY* mice were conceived but did not last until complete

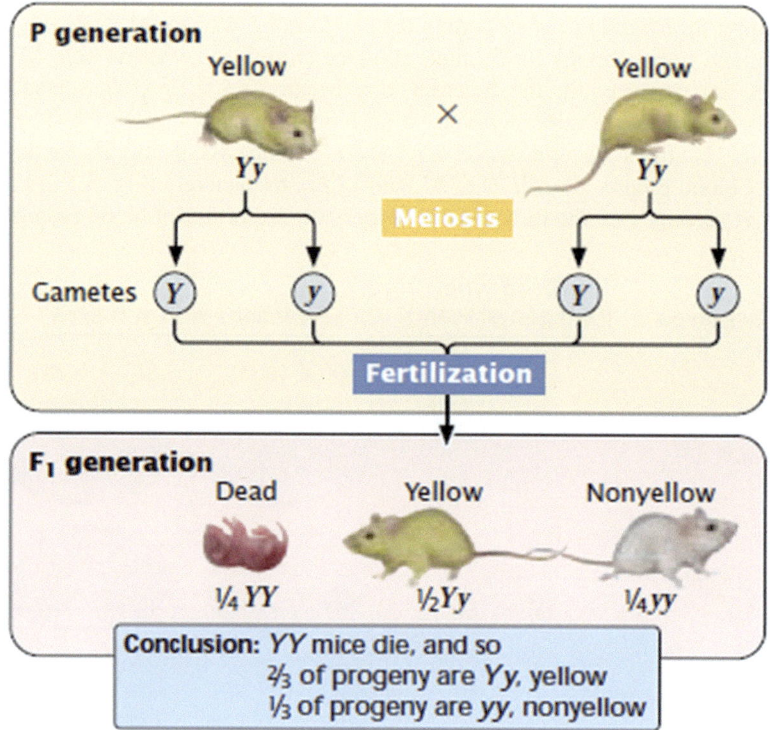

Fig. 3.14 Segregation of lethal alleles. Cross between two heterozygous yellow mice, resulting in approximately 2/3 yellow and 1/3 nonyellow offspring. Although the segregation is in the typical Mendelian pattern, presence of a lethal allele results in the given ratio of yellow and nonyellow mice. *Y* allele in homozygous condition causes lethality in the early development

development, resulting in progeny, with a *Yy* (yellow): *yy* (nonyellow), a ratio of 2:1 where all yellow mice were heterozygotes *(Yy)*.

Another case of a lethal allele is found in snapdragons. It was originally described in 1907 by Erwin Baur, in a strain of snapdragons called *aurea*, which consisted of yellow leaves. Two snapdragons with yellow leaves when crossed produced 2/3 progeny with yellow leaves and 1/3 progeny with green leaves. When two green-leaved plants were crossed, all progeny had green leaves. But when yellow was crossed with green, 1/2 progeny were green and 1/2 were yellow, indicating all yellow-leaved snapdragons to be heterozygous. As typically observed, a ratio of 2:1, among progeny of a cross between individuals of the same phenotype, indicated the presence of a recessive lethal allele.

In both cases, it was observed that one allele affected both survival and color. With respect to survival, the lethal alleles were recessive in nature, as they caused death in homozygous condition. But the same allele expressed dominance with respect to the color. However, lethal alleles can also be dominant in nature. In such a case, both homozygotes and heterozygotes do not survive. It is not possible for true dominant lethal alleles to be passed on from one generation to another unless they affect the individual after the onset of reproductive age, as observed in Huntington disease.

3.2.6 Sex-Limited Traits

Traits which are expressed particularly in one sex are called sex-limited traits. Although the genes coding for these traits are present in both the sexes, expression occurs in only one. In humans, formations of the breast and ovary are sex-limited traits in women. Likewise, facial hair and sperm production are sex-limited traits in men. In birds, the plumage patterns are sex-limited traits, where in many species, only the male possesses brightly colored plumage patterns. Horns in only males of certain sheep species and milk production only in mammalian females are other examples of sex-limited traits.

3.2.7 Sex-Influenced Traits

Sex-influenced traits are also called as sex-conditioned traits. They appear in both sexes but occur more frequently in one sex than the other. Premature pattern baldness, for example, is a sex-influenced trait in human beings. It is due to an allele that is differently expressed in the two sexes. In males, individuals develop bald patches, both in homozygous and heterozygous conditions for the allele. However, in females, only the homozygotes show a tendency for baldness, and that is usually limited to general thinning of the hair rather than balding. The difference has been suggested to be due to the requirement of the male hormone testosterone, for the full expression of the allele. Females produce much less of testosterone and are therefore rarely at risk of developing bald patches. The sex-influenced nature of this trait shows that intrinsic biological factors can play an important role in the regulation gene expression.

3.3 X-Linkage Feature Genes

In organisms with XY chromosomal sex determination, the sex chromosomes are heteromorphic unlike the autosomes which are homomorphic. Therefore, the patterns of inheritance of genes located on heteromorphic sex chromosomes are different when compared with autosomal inheritance. This happens because inheritance of alleles located on the sex chromosomes is linked with sex of the offspring. In XY sex determination, the sex of the offspring is dependent on the sperm that fertilizes the egg. The Y chromosome containing sperms determine male sex, while the X chromosome containing sperms determine the female sex. Therefore, alleles located on the X chromosome of a male parent are always inherited by his daughters but not his sons. Likewise, all sons receive the X chromosome from the mother. In such a crisscross inheritance, traits carried on the X chromosomes are transmitted to the sons by the mother and to the daughters by the father. For example, by the end of the eighteenth century, the blood-clotting disorder hemophilia was known to be associated with the X chromosome, because it was inherited mostly by men, and women could transmit to next generations without themselves being affected.

3.3.1 X-Linkage in *Drosophila*

X-linked pattern of inheritance was first demonstrated in *Drosophila* by T. H. Morgan in 1910. He observed white-eyed male in a culture of wild-type (red-eyed) flies (Fig. 3.15). He crossed a wild-type female with a white-eyed male. All offspring

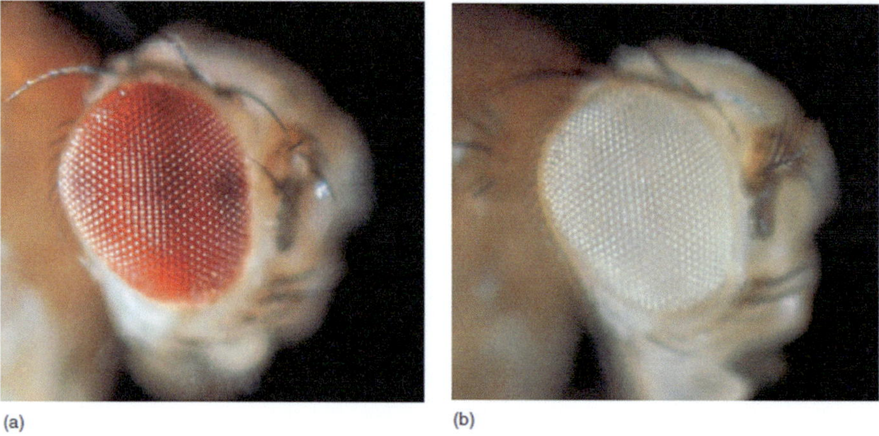

Fig. 3.15 *Drosophila* (fruit fly) eye color phenotype. (**a**) Wild-type red eye. (**b**) Mutant white-eye

Fig. 3.16 Inheritance pattern of the white-eye mutant trait in *Drosophila*

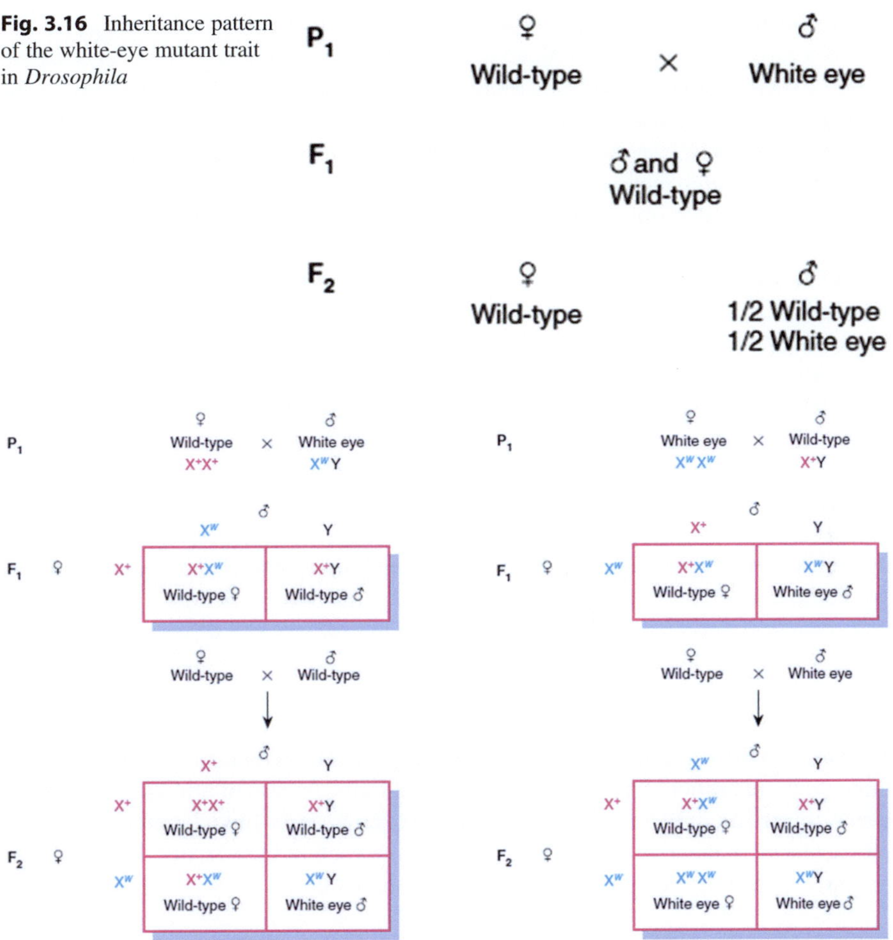

Fig. 3.17 X-linked inheritance of white-eye trait in *Drosophila*. The crosses depicted indicate nonreciprocity, an important feature of sex-linked inheritance

born out of that cross were wild-type. However, selfing of these F_1 flies produced offspring of two different types (Fig. 3.16). All the female offspring were wild-type, whereas one half of the males were white-eyed and the other half were wild-type. Morgan and other researchers interpreted these observations by hypothesizing that the gene for white-eye was positioned on the X chromosome (Fig. 3.17). X chromosomes carrying the white-eye allele and wild-type allele are denoted as X^w and X^+, respectively. The Y chromosome does not contain this gene locus and is denoted as Y.

Another characteristic with respect to X-linkage is as follows: since females possess two X chromosomes, they can either be homozygous or heterozygous for the given allele. But males have a single copy of X (Fig. 3.17) and a single copy of Y that are heteromorphic. Therefore, X-linked alleles can neither be homozygous nor heterozygous in males. An X-linked allele, not having a homolog on the Y, is said to be in a hemizygous condition. Hemizygosity causes a recessive allele to be expressed, even if present in single copy. Therefore, a male *Drosophila*, although with only one *w* allele, expresses white-eyed phenotype. Such a phenomenon is called pseudodominance because it resembles the way in which a single dominant autosomal allele determines the phenotype in a diploid heterozygous individual.

Nonreciprocity is another important feature of sex-linked inheritance. A cross is said to be reciprocal if for a given trait, the results of the cross are equivalent, with equal distribution of the trait in both sexes, irrespective of the sex of the parent that is possessing the trait. A cross is said to be nonreciprocal if parental sex alters the distribution of the trait in the following generations. Therefore, nonreciprocity indicates the difference in the outcome of such a cross based on the association of the trait with the sex of the organism. The inheritance pattern of sex linkage is not reciprocal and it is observed when a white-eyed female is crossed with wild-type male (Fig. 3.17). In this cross, the F_1 males are white-eyed, and F_1 females are wild-type. Further in the F_2, 50% of each sex are white-eyed. Such differences in the ratio of distribution of a trait among the two sexes and nonreciprocity of a cross indicate sex linkage, which are further confirmed by the crisscross pattern.

3.3.2 X-Linkage in Humans

Several X-linked traits are also observed in humans. The X-linked recessive traits are more easily identified than recessive autosomal traits. In case of humans also, a male needs to inherit only one recessive allele to show an X-linked trait; however, a female needs to inherit two such alleles, one from each of her parents. Therefore, the majority of people who show X-linked traits are male.

3.3.2.1 X-Linked Blood-Clotting Disorder: Hemophilia

Hemophilia is a disorder wherein the affected individuals lack a blood-clotting factor. In case of bleeding due to injuries, the hemophiliacs continue to bleed and, if it is not stopped by transfusion with clotting factor, it can cause death. In humans, the major type of hemophilia is caused by an X-linked recessive mutation, and therefore, almost all affected individuals are male. Males inherit the mutation from their carrier mothers (heterozygous). If they reproduce, they will further transmit the mutation to their daughters, who usually inherit a wild-type allele from their mother and hence do not develop hemophilia, but remain as carriers that will transmit the mutation to their sons. This results in more number of males than females being affected by the disease. On the contrary, other blood-clotting disorders which occur due to mutations in autosomal genes are equally distributed in both males and females.

The most famous case of X-linked hemophilia occurred in the early twentieth century, in the Russian imperial family (Fig. 3.18), where Czar Nicholas and Czarina Alexandra had four daughters and one son, and the son, Alexis, was afflicted with

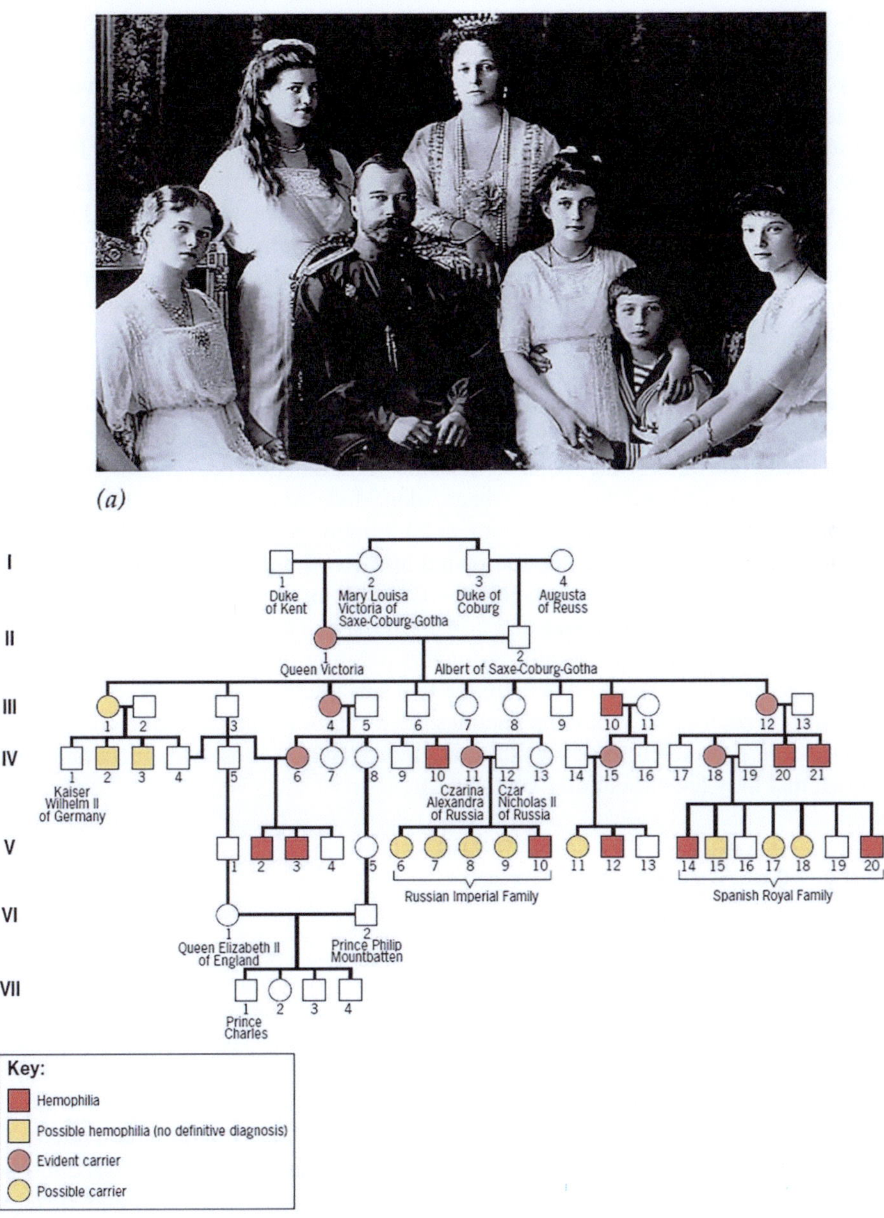

Fig. 3.18 Human X-linked disorder—hemophilia, the royal disease. (**a**) Russian imperial family of Czar Nicholas II. (**b**) Hemophilia in the royal families of Europe. Due to intermarriage, the hemophilia allele was transmitted from British royal family to Russian, German, and Spanish families

hemophilia. The X-linked mutation that caused the disorder in Alexis was transmitted to him by his mother, who herself was a heterozygous carrier. Czarina Alexandra happens to be the granddaughter of Queen Victoria of Great Britain, who was also a carrier. Through pedigree records, it is known that Queen Victoria transmitted the hemophilia allele to three of her nine children: Alice, Alexandra's mother; Beatrice, who had two sons with the disorder; and Leopold, who was also affected. The allele that was carried by Queen Victoria evidently arose as a new mutation in her germ cells, or her mother, father, or any other distant maternal ancestor. Throughout history, hemophilia has been a fatal disorder and most of the affected people deceased before the age of 20. Today, since more effective and relatively inexpensive treatments are available, hemophiliacs can live longer, healthy lives.

3.3.2.2 X-Linked Vision Disorder: Color Blindness

Color perception in humans is mediated by light-absorbing proteins present in the specialized cone cells of the retina. Three such light receptor proteins for absorption of blue, green, and red light have been identified. Abnormality in any of these receptor proteins can be the cause of color blindness. The classic type of color blindness involves faulty perception of red and green light. It follows an X-linked pattern of inheritance. Approximately 5–10% of human males are red-green color blind; however, a much lesser fraction of females, i.e., less than 1%, has this disability, suggesting that the alleles for this disorder are recessive. Studies at the molecular level have shown that there are two separate genes on the X chromosome for color perception. One of them encodes the receptor for green light, and the other encodes the receptor for red light. Detailed analyses of these two receptors have shown that they are structurally very similar.

3.4 Genotype to Phenotype

3.4.1 Penetrance

Phenotypic appearance of the genotypically determined traits is called as penetrance. Not all genotypes are able to "penetrate" the phenotype. When individuals possessing an appropriate genotype do not show the corresponding trait, such a trait is said to show incomplete penetrance. Polydactyly, the presence of extra fingers and toes in humans, is one such example (Fig. 3.19). A dominant mutant allele, *P*, causes this condition, but is not expressed in all of its carriers. In the pedigree shown in Fig. 3.19, the individual, symbolized III-2, must be a carrier even though he does not possess extra fingers or toes, because both his mother and his three children are polydactylous, which is an indication of the mutation being transmitted through III-2. Incomplete penetrance can be a grave problem in pedigree analysis as it can cause erroneous understanding of genotypes.

Another example of incomplete penetrance is as follows: a person could possess a genotype that can cause vitamin D-resistant rickets (a bone disease), but may not suffer from the disease. The allele which causes this disease is dominant and

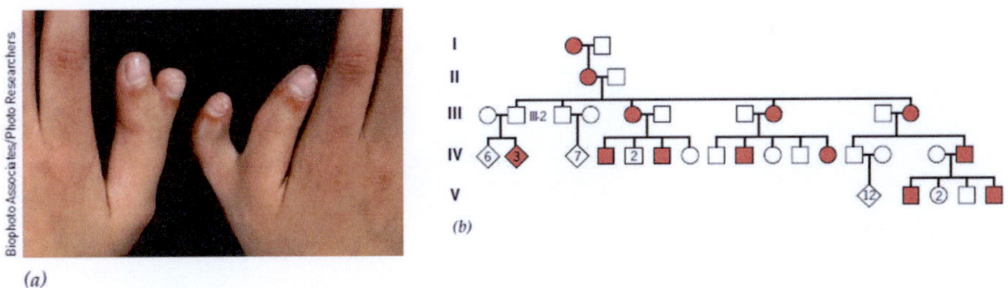

Fig. 3.19 Incomplete penetrance. (**a**) Polydactyly in humans is a phenotype showing extra fingers. (**b**) Pedigree showing inheritance pattern and the incomplete penetrance of the dominant trait polydactyly. It can be seen that a male offspring (III-2) in generation III is not expressing polydactyly although he is carrying the allele. The fact that one of his offspring expresses polydactyly confirms that III-2 carried the allele

sex-linked. Vitamin D-resistant rickets is different from vitamin D deficiency, because it cannot be treated by administering low levels of vitamin D. However, the disease condition can be treated with very high levels of vitamin D. Similar to polydactyly, in some family trees, unaffected parents give birth to children with vitamin D-resistant rickets. This is unexpected with respect to inheritance of dominant traits, because an affected offspring should have at least one affected parent. In this case, the parent however must be carrying the disease allele because the children are affected. Presence of low phosphorus levels in the parent's blood, which is a pleiotropic effect of the disease allele, proves the presence of the allele in the parent. Interestingly, the low-phosphorus attribute of the phenotype shows complete penetrance, whereas vitamin D-resistant rickets shows incomplete penetrance.

However, most genotypes show complete penetrance. For example, albinism (lack of melanin synthesis) is an autosomal recessive condition. Every individual who has the homozygous recessive genotype for the albinism allele will certainly express the disorder. There is no known example of a case where an individual with the genotype did not express the condition. Therefore, albinism is completely penetrant. However, certain genotypes, especially those that code for developmental traits, frequently exhibit incomplete penetrance.

3.4.2 Expressivity

The term expressivity is particularly used when a trait is not uniformly expressed among individuals that show a particular trait. A dominant mutation in *Drosophila*, called the Lobe eye mutation (Fig. 3.20), is one such example. The phenotypic expression of this mutation is extremely variable, in that it ranges from tiny compound eyes in some heterozygous flies to large, lobulated eyes in others. In between the two extremes, there exists an entire range of various phenotypes and hence the lobe mutation is said to show variable expressivity.

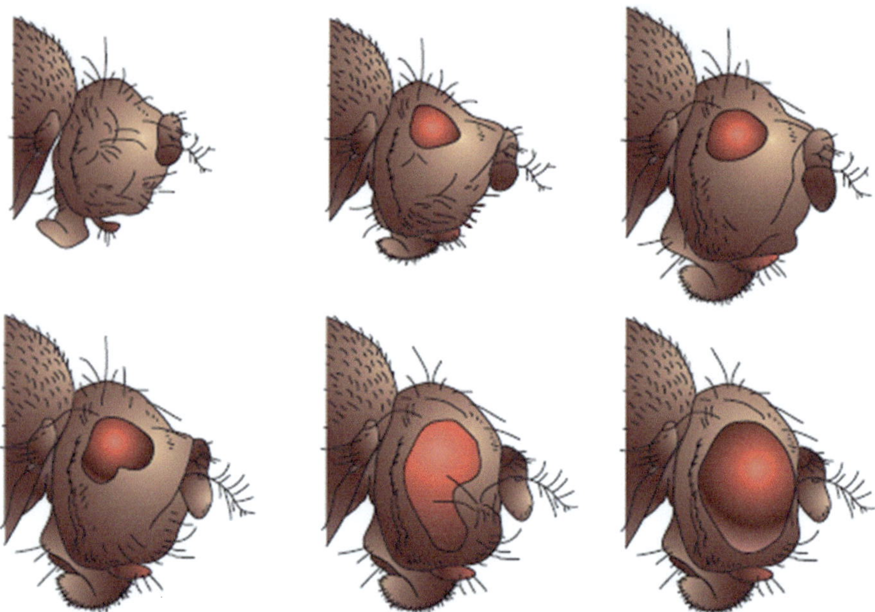

Fig. 3.20 Expressivity of Lobe eye mutation in *Drosophila*. The Lobe eye mutation of *Drosophila* shows variable expressivity. Although each of the flies is heterozygous for the mutation, the phenotypic expression varies from complete absence of an eye to nearly wild-type eye

Several developmental traits in addition to being incompletely penetrant also exhibit variable expressivity, ranging from mild to extreme. One such example is the cleft palate trait, which shows both variable penetrance and variable expressivity. Once penetrance is attained by the genotype, severity of the condition varies considerably, ranging from severe clefting of hard and soft palates to a mild external cleft. Failure to penetrate and variable expressivity are characteristics of the developmental traits in various organisms. Incomplete penetrance and variable expressivity both indicate that the pathway between a genotype and its expression as a phenotype goes through considerable modulation. Some of these modulations are suggested to be due to environmental factors, but some are also due to genetic factors. Breeding experiments have provided clear evidence that two or more genes can affect a particular trait.

3.4.3 Environmental Effects on Gene Expression

Phenotypes depend both on environmental and genetic factors, i.e., the genotype. The analyses of phenotypes have shown that genes do not act in isolation. Instead, they act in the background of an environment and in coordination with other genes. These analyses have also shown that a particular gene can influence various different traits.

A gene needs to function in the framework of both biological and physical environments. The physical environment factors are easier to study, because particular genotypes can be reared under controlled conditions in the laboratory, thus allowing an assessment of the effects of temperature, light, nutrition, and humidity.

3.4.3.1 Position Effect

The function and expression of a gene is very much dependent on its location or position in the genome/chromosome, to an extent that a change of location or position can alter its function. Such repositioning of genes usually occurs when chromosomal aberrations lead to rearrangement of chromosomal segments. When chromosomal rearrangements occur, genes that exist near the breakpoints become repositioned in the genome and they become flanked by novel neighboring genes and chromosomal segments. In several cases, such a repositioning of the gene affects its expression level or, in certain cases, it alters its ability to function; this is called position effect.

Examples of such effects have been extensively studied in organisms such as *Drosophila* and yeast. Position effect was first observed in the *Bar* eye mutation of *Drosophila*. The *Bar* eye mutation is a duplication of 16A region on the X chromosome of *Drosophila*, which reduces the eye to a bar shape, due to reduction in the number of facets. A *Bar* homozygote (*B*/*B*) and a *Doublebar*/wild-type heterozygote (*BB*/*B*_) both have four copies of the 16A region. Hence, it would be expected that both genotypes produce the same phenotype, i.e., with respect to the number of facets in the Bar eye. But interestingly, the number of facets in a *Bar* homozygote is 70 and in a heterozygote it is 45. Therefore, from this example, it is clear that in addition to the amount of genetic material, its arrangement or position where it is placed in the genome also plays a significant role in determining the extent of the phenotype.

The most common position effect observed in *Drosophila* results in a mottled or mosaic eye phenotype. The eye of such *Drosophila* is structured with interspersed patch cells expressing wild-type phenotype, and mutant cells, with an inactivated wild-type gene. This mosaic phenotype is called as a variegation, and hence this effect is called position-effect variegation (PEV).

Usually, PEV results from a chromosomal aberration that alters the position of a wild-type gene in euchromatin and relocates it to a new position near or within a heterochromatin (Fig. 3.21). In *Drosophila*, an inversion mutation in the X chromosome that changed the position of wild-type w^+ allele into heterochromatin was identified. This change of position resulted in the variegation phenotype, where patterns of red (wild-type) and white (mutant) facets appeared in the eyes of flies (Fig. 3.22).

Inactivation of the euchromatic region is known to occur through a length ranging from 1 to 50 bands in the polytene chromosome of *Drosophila*. The length depends on the particular chromosomal abnormality that has occurred. At the molecular level, this range approximately includes 20–1000 kb. Similar inactivation events are also reported to take place in cells of female mammals. It has been reported that when

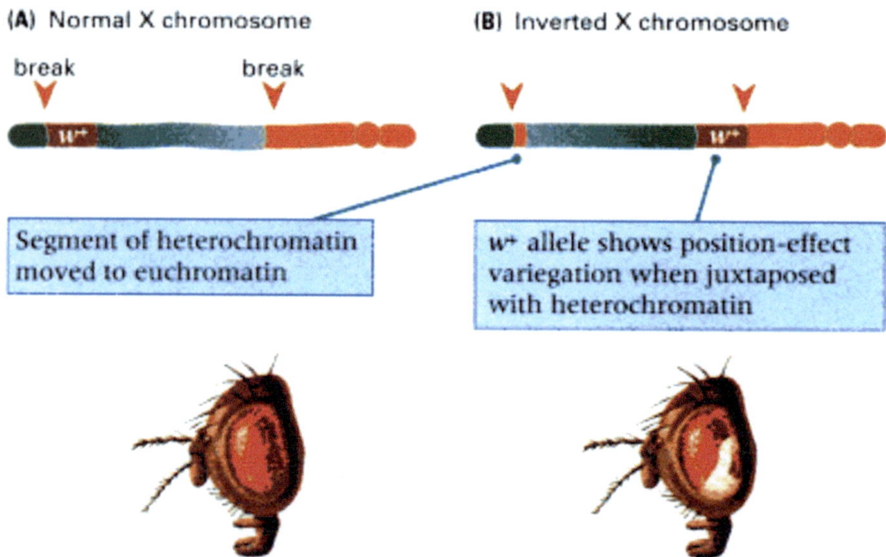

Fig. 3.21 Position-effect variegation (PEV). Commonly observed due to an inversion mutation, where chromosome rearrangement changes the location of a gene in euchromatin, to a location in or near heterochromatin. This is an inversion within X chromosome of *Drosophila melanogaster* involving wild-type allele of the white gene locus, causing its relocation near heterochromatin. The effect of PEV is observed as spotted red and white eyes. This phenomenon is called "position-effect variegation," because here the change occurs only in the gene position but not in the gene

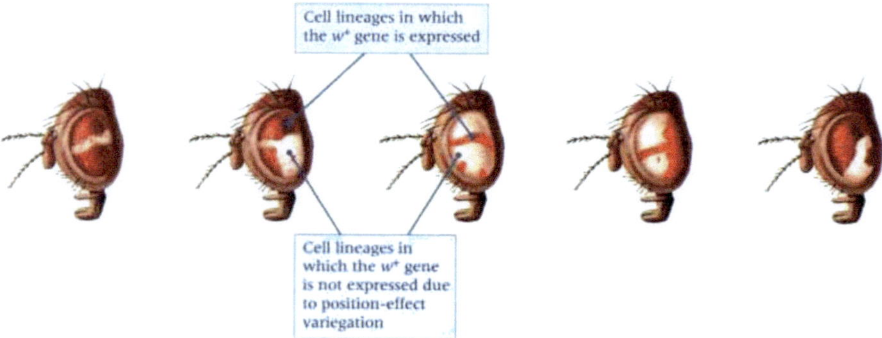

Fig. 3.22 Patterns of spotted eyes due to PEV in *Drosophila*. Each group of cells of the same color is a product of a single cell during the developmental process. Commonly, several small patches of red or a mixture of small and large patches is observed

euchromatic genes get translocated to the X chromosome, they become heterochromatic and inactive.

Note: PEV occurs when a gene usually located in euchromatin gets relocated near a heterochromatin due to transposition or chromosomal rearrangement. Intensive studies on PEV have been performed using white gene in *Drosophila*. Analysis of PEV in *Drosophila* at the biochemical, molecular, and genetic levels has provided useful insights into this process. The variegating phenotype shown by genes that abnormally get placed in heterochromatin is a result of gene silencing. A spread of the heterochromatin packaging across the heterochromatin/euchromatin border is suggested to cause the transcriptional silencing of the juxtaposed/repositioned gene. Several approaches have been used to identify the key contributors to this process. This has led to the characterization of several modifying enzymes and structural proteins that play important roles in establishment and maintenance of heterochromatin and the associated gene silencing. Heterochromatin formation is shown to critically depend on histone H3 methylation at lysine 9, with simultaneous interaction with other proteins and enzymes such as methyltransferases. It is shown that the spreading and maintenance of heterochromatin is dependent on multiple interactions between these proteins.

3.4.3.2 Temperature Effect

Environmental conditions such as temperature also affect gene expression. One such example is found in a *Drosophila* mutation called as *shibire*. *Shibire* is a Japanese word meaning “paralysis.” At 25 °C, the normal culturing temperature, *shibire* flies are viable and fertile. But these flies are highly sensitive to sudden shock, such that when a *shibire* culture vessel is shaken, the flies get temporarily paralyzed and fall to the bottom of the culture vessel. If the same culture of *shibire* flies is kept at a slightly higher temperature of 29 °C, all the flies fall to the bottom and die, even without exposure to a shock. Hence, the phenotypic expression of the *shibire* mutation is temperature-sensitive, i.e., it is dependent on the ambient temperature. It is a mutation that is viable at 25 °C, but is lethal at 29 °C. A probable explanation to this temperature sensitivity is that at 25 °C, the mutant gene produces a protein that is partially functional, whereas at 29 °C, the protein is completely nonfunctional.

3.4.3.3 Nutritional Effect

Phenylketonuria (PKU) is an example of a recessive disorder with faulty amino acid metabolism. Children homozygous for the mutant allele accumulate toxic metabolites in their brains which affect brain development and impair mental ability. The harmful aspects of PKU are correlated to a particular amino acid, phenylalanine, which is present in the diet of the individual suffering from this disorder. Although phenylalanine itself is not toxic, it gets metabolized into other toxic intermediates. Infants affected with PKU if fed with normal diets ingest enough amounts of phenylalanine that can bring about severe manifestations of the disorder. However, if such affected infants are fed with diets low in phenylalanine, they usually develop without serious mental impairment. Because PKU can be well diagnosed in newborns, the clinical impact of this disease can be significantly reduced by placing infants that are PKU homozygotes on a low-phenylalanine diet soon after birth. Thus, PKU clearly indicates the effect of nutrition on the gene expression. In this case, the presence of phenylalanine in the diet triggers the expression of the disease, and hence the same can be controlled by altering the diet of the individual.

3.4.3.4 Pleiotropy

It is now known that any given phenotype can be influenced by several genes. The converse of this is also true, i.e., one gene can influence many phenotypes. When a mutant gene affects many aspects of the phenotype, it is said to be pleiotropic (Greek for "to take many turns"). Such a phenomenon is called pleiotropy, and the effects caused are called pleiotropic effects.

PKU gene in humans is an example for pleiotropy. The primary effect of the recessive mutation in this gene is that it causes accumulation of toxic substances in the brain, leading to mental impairment. However, these mutations also interfere with the synthesis of the pigment melanin, leading to lightening hair color; and hence, individuals with PKU frequently possess light brown or blond hair. Biochemical tests also reveal the presence of rare compounds in the blood and urine of PKU patients, which is otherwise not found in normal individuals. This range of phenotypic effects is typical of most genes and results from interconnections between the biochemical and cellular pathways regulated by these genes.

Another example of pleiotropy is the mutation causing sickle cell anemia. Sickle cell mutation primarily affects the oxygen-carrying protein hemoglobin, present in the RBC. In addition, this mutation is also associated with several pleiotropic effects. In addition to its effect on RBC, it also affects several major organs and organ systems (Fig. 3.23). The sickle cell mutation is present in the β-globin gene that codes for a β polypeptide chain of hemoglobin. The molecular basis of the disease is illustrated in Fig. 3.24. The β-globin polypeptide chain is 146 amino acids long. The part of the β-globin gene coding for amino acids in positions 5 through 8 is shown in Fig. 3.24. The sickle cell mutation involves a base pair change indicated with an arrow. The mutation replaces the normal A-T base pair with a T-A base pair. Due to this mutation, the mRNA codon is replaced with a GUG instead of a GAG codon. GUG codes for valine (Val), while GAG codes for glutamic acid (Glu). Therefore, normal glutamic acid is replaced with valine in the β-globin polypeptide chain at position number 6 (Fig. 3.24). Thus, formed abnormal β polypeptide chain attributes long, needlelike polymer forming ability to hemoglobin. Due to the polymerization of hemoglobin, RBCs become deformed into crescent-like sickle-shaped cells. Some amount of these abnormal RBCs are immediately destroyed, reducing the blood oxygen-carrying capacity resulting in anemia. The remaining sickle-shaped RBC may clump and clog the capillaries, thus interrupting blood circulation.

The Glu → Val replacement results in profound pleiotropic effects. All of these effects are due to the breakdown of RBCs, or the reduced oxygen-carrying capacity of the blood, or the physiological alterations the body makes in an effort to compensate for the disease. Patients with sickle cell anemia suffer spells of severe pain. Anemia causes impaired growth, weakness, and jaundice. Affected people are also generally immunologically weakened, and they become susceptible to bacterial infections, which are the most frequent cause of death in children with the disease. Hence, sickle cell anemia is a severe genetic disease which can cause premature death.

In spite of its severity, it is quite prevalent in areas of Africa and Middle East. These regions also have an extensive incidence of malaria caused by *Plasmodium*

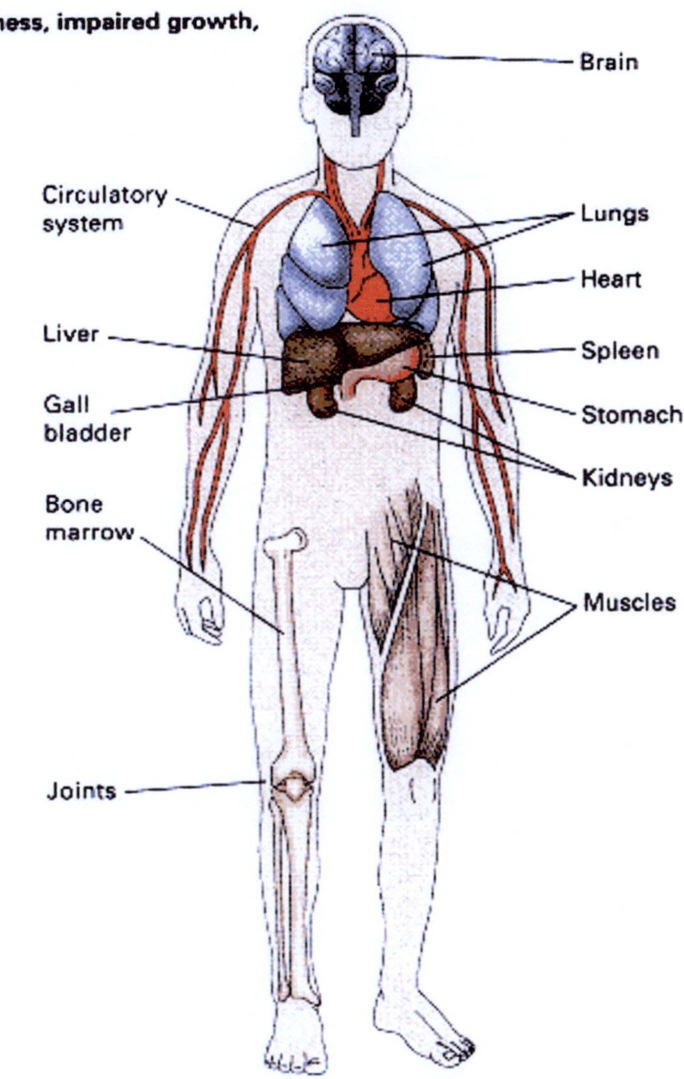

Fig. 3.23 Pleiotropy in sickle cell anemia. The sickle cell mutation affects several major organ systems in the human body, in addition to causing abnormal hemoglobin in RBC

falciparum. There is a significant association between the incidence of malaria and sickle cell anemia in these regions. The association is because of the mutant form of β-hemoglobin, which to a certain extent provides resistance against malarial infection. This is due to the fact that the malarial parasite requires the RBCs as part of its life cycle. The proliferation of the parasite in the RBCs is checked due to the sickling of the cells, and hence the severity of the malarial infection is reduced.

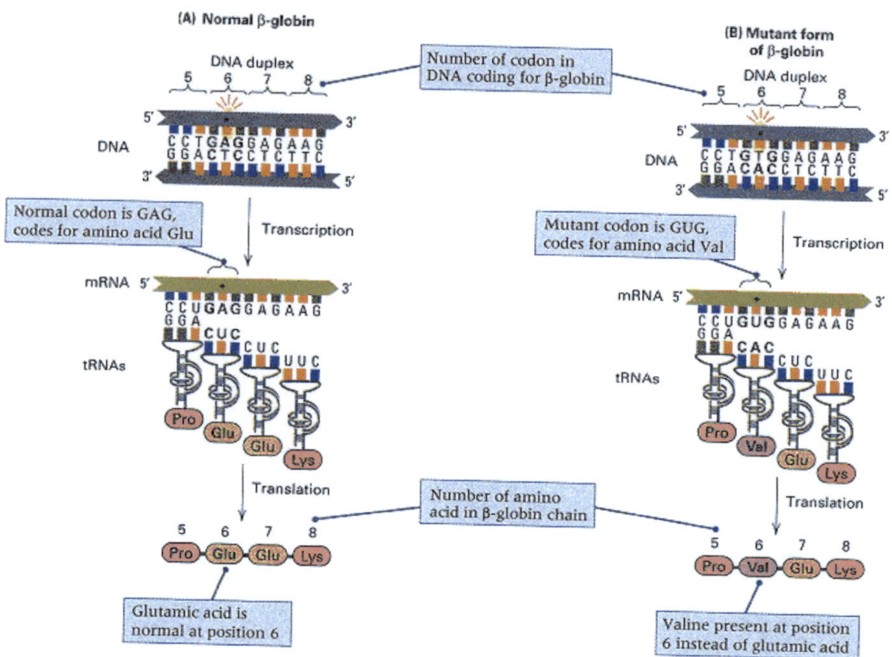

Fig. 3.24 Molecular basis of sickle cell anemia. (**a**) Shows a segment of the normal β-globin gene, and corresponding mRNA, coding for the amino acid sequence Pro-Glu-Glu-Lys. The GUG codon at the sixth position in the mRNA codes for glutamic acid (Glu). (**b**) Shows mutant form of the β-globin gene, showing an A-T transversion mutation. Likewise, the codon in mRNA alters from GUG to GAG, which in turn alters the amino acid at the sixth position from glutamic acid to valine (Val)

Another example of pleiotropy is the mutations that affect the formation of bristles in *Drosophila*. Wild-type flies possess smoothly curved, long bristles on the head and thorax. Flies which are homozygous for the *singed* bristle mutation have short, twisted bristles on the head and thorax. Thus, the wild-type form of the *singed* gene product is required for proper formation of bristles. In addition, it is also required for the production of healthy, fertile eggs. This fact is known because females homozygous for certain *singed* mutations are sterile; they lay fragile, ill-formed eggs that fail to hatch. However, these mutations do not affect male fertility. Thus, the *singed* gene pleiotropically regulates the formation of both bristles and eggs in females, whereas in males it regulated the formation of bristles.

3.4.4 Gene Interaction

In the dihybrid crosses performed by Mendel, each gene locus independently controlled the phenotype. In other words, one gene locus did not have any influence

Table 3.5 Proportions of F2 progeny in Mendel's dihybrid cross

9/16	*R_Y_*	Round, yellow
3/16	*R_yy*	Round, green
3/16	*rrY_*	Wrinkled, yellow
1/16	*rryy*	Wrinkled, green

on the functioning of the other gene locus. For example, when Mendel self-fertilized the F_1 plants obtained from a dihybrid cross between homozygous round yellow and wrinkled green plants, the phenotypic proportions in the F_2 progeny were as shown in Table 3.5.

This was a typical example of independent assortment. Here the independence observed is at two levels. Firstly, the genes at each locus were independently assorted during meiosis, and hence produced a 9:3:3:1 phenotypic ratio in the progeny. Secondly, each gene independently controlled particular phenotypes, i.e., *R* and *r* allele determined round and wrinkled seed shape, respectively. Similarly, the *Y* and *y* alleles determined yellow and green color, respectively. Alleles that controlled seed shape did not affect those that controlled color and vice versa.

Although genes often show independent assortment, they are not independent in their phenotypic expression. In several cases, a gene at one locus influences the expression of a gene at another locus. This kind of interaction among genes at different loci (nonallelic genes), which affects the phenotypic outcome, is termed gene interaction. Due to such gene interactions, products from different genes interact with one another to produce novel phenotypes, which cannot otherwise be formed by the effects of single loci. When there is involvement of two genes in the outcome of a single characteristic, the dihybrid phenotypic ratios observed in such crosses can be quite a deviation from the typical 9:3:3:1. Although gene interactions among three, four, or more loci are common, the examples discussed in this chapter are primarily focused on interaction of genes at two loci.

3.4.4.1 Production of Novel Phenotypes by Gene Interaction

Fruit color trait in a pepper, *Capsicum annuum*, is controlled by interactions between genes at two separate loci. These peppers come in four colors: red, brown, yellow, or green. In a plant homozygous for red peppers (*RRCC*) when crossed with a plant homozygous for green peppers (*rrcc*), all F_1 (*RrCc*) produce red peppers (Fig. 3.25). When the F_1 are self-pollinated (*RrCc* × *RrCc*), red, brown, yellow, and green peppers are obtained in the ratio of 9:3:3:1, respectively, in the F_2 (Fig. 3.25).

The red pigment in peppers is produced by a dominant allele *R* at one gene locus (locus 1); recessive allele *r* at the same locus does not produce red pigment. At another gene locus (locus 2), a dominant allele *C* causes green pigment chlorophyll to degrade, whereas the recessive allele *c* allows retention of chlorophyll. Therefore, genes at locus 1 and 2 interact with each other to produce the various colors observed in fruits of pepper *Capsicum annuum*. The genotypes and the phenotypes of the F_2 plants are given in Table 3.6.

This can be explained as follows: in case of *R_C_* genotype, dominant *R* produces red pigment, dominant *C* results in decomposition of chlorophyll, and together the

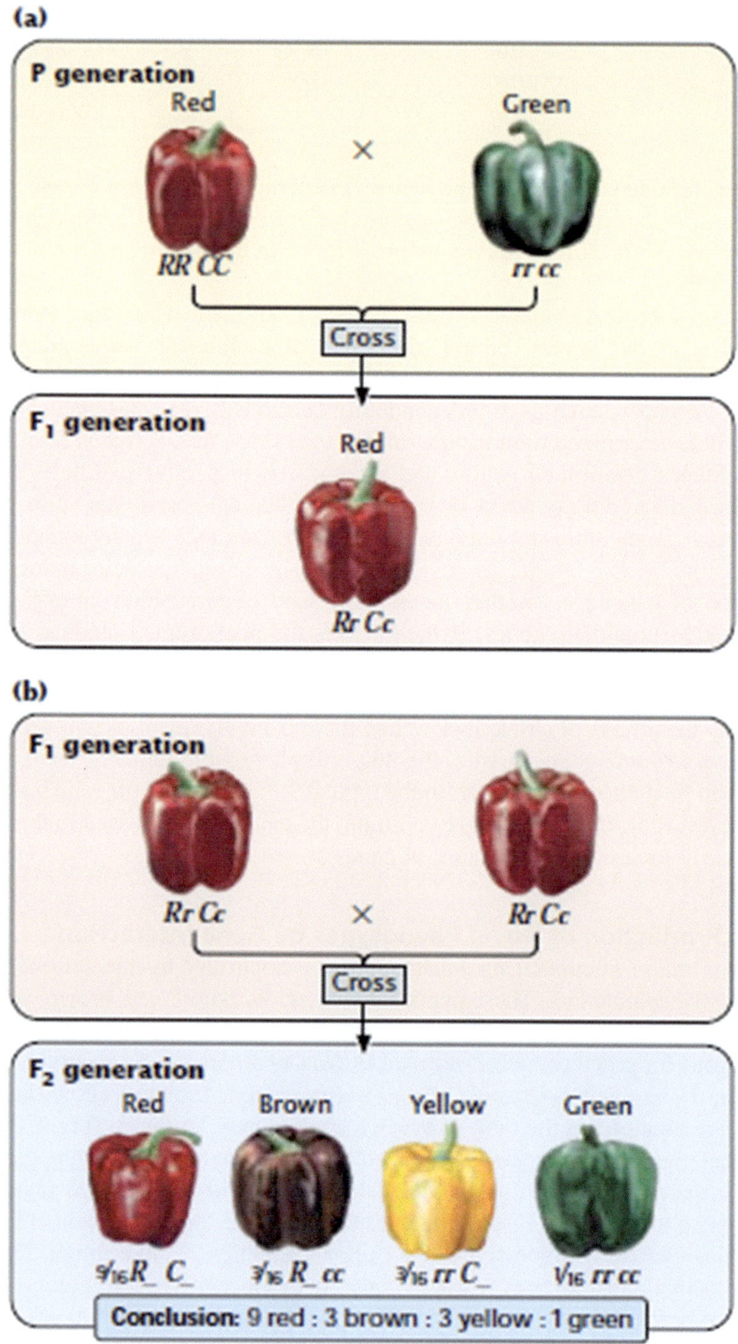

Fig. 3.25 Gene interaction in pepper *Capsicum annuum* resulting in novel phenotype. The fruit color in pepper is determined by the interaction between two gene loci, an example of a single character being influenced by two different genes

Table 3.6 Production of novel characters by gene interaction—F_2 genotypes and phenotypes for fruit color in *Capsicum annuum*

Genotype	Phenotype
*R_C_*	Red
R_cc	Brown
*rrC_*	Yellow
rrcc	Green

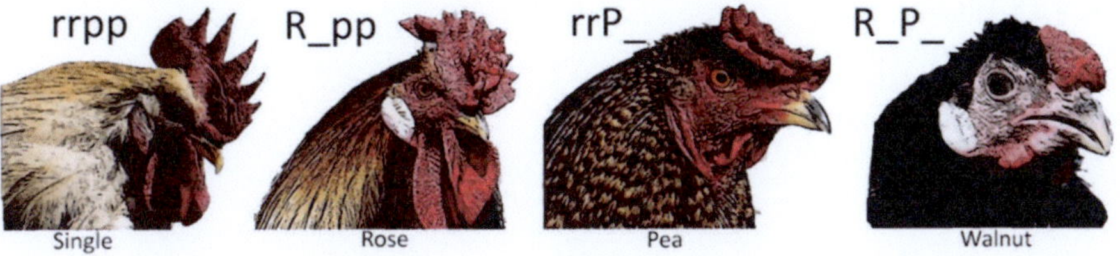

Fig. 3.26 Gene interaction resulting in chicken a novel phenotype. The combs of chicken are determined by the interaction between two gene loci, an example of a single character being influenced by two different genes

action of both nonallelic genes results in a red color. In the genotype *R_cc*, dominant *R* produces red pigment, recessive *c* retains chlorophyll, and the resultant is a brown color. In the genotype *rrC_*, recessive allele *r* does not produce red pigment, dominant allele *C* leads to the decomposition of chlorophyll, and as a result, the fruit color is yellow. In the genotype *rrcc*, recessive allele *r* does not produce red pigment and the recessive allele *c* retains chlorophyll and hence the fruit color is green.

Chicken combs provide another example of a novel phenotype produced due to gene interaction. A comb is a fleshy structure found on the heads of chicken. In this example, two loci (*R*, *r* and *P*, *p*) interact with each other and produce four types of combs (Fig. 3.26). The presence of at least one dominant *R* at locus 1, and at least one dominant *P* at locus 2 (genotype *R_P_*), produces a walnut comb in the chicken. Chicken with one dominant allele (*R*) at locus 1 when the alleles at the second locus are homozygous recessive (*pp*), i.e., a genotype *R_pp*, produces a rose comb. If the locus 1 is homozygous recessive (*rr*), and locus 2 consists of at least one dominant allele (*P_*), the chicken (*rrP_*) produces pea comb. When both loci are homozygous recessive (*rrpp*), the chicken produces a single comb.

3.4.4.2 Epistasis

Classically, epistasis is defined as a gene interaction where one gene overrides the influence of another gene at a different locus in such a way that its phenotype is suppressed. This is similar to phenomenon of dominance, but unlike dominance which involves alleles of the same gene, epistasis involves alleles of two different genes. The overriding gene is called epistatic, and the suppressed gene is called hypostatic. Genes that are epistatic genes may be either dominant or recessive.

Table 3.7 Recessive epistasis in coat color of Labrador retrievers—F_2 genotypes and phenotypes

Genotype	Phenotype
*B_E_*	Black
*bbE_*	Brown (usually called chocolate)
B_ee	Yellow
bbee	Yellow

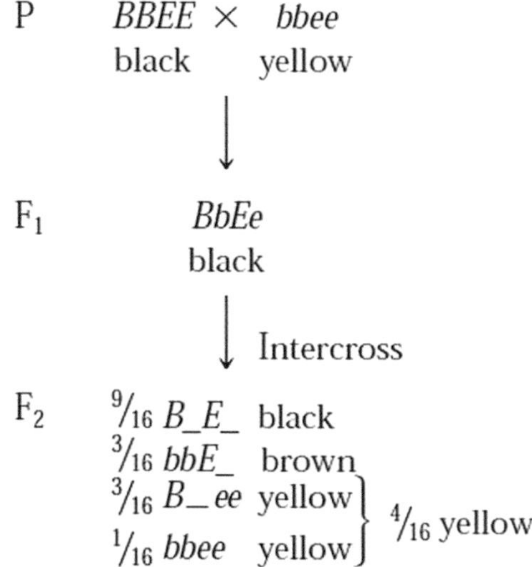

Fig. 3.27 Dihybrid cross depicting recessive epistasis in coat color of Labrador retrievers. The phenotypic ratio of recessive epistasis is 9:3:4

3.4.4.3 Recessive Epistasis

Coat color in Labrador retrievers is an example of recessive epistasis. These dogs have different coat colors such as black, brown, or yellow. Coat color is determined by two different loci, where one locus determines pigment type to be produced by the skin cells—dominant allele *B* codes for a black pigment; recessive allele *b* codes for a brown pigment. The second gene locus controls pigment deposition into the hair shaft—dominant allele *E* allows deposition of pigments (black or brown) into the hair shaft; recessive allele *e* prevents deposition of pigments (absence of pigments causes hair to appear yellow). If the second locus consists of genotype *ee*, the expression of the first locus (black or brown pigments) gets concealed. Therefore, genotypes with dominant alleles at both loci result in black coat color, homozygous recessive at locus 1, and at least one dominant allele at locus 2 results in brown coat color. But if the second locus is homozygous recessive, no pigments will be expressed irrespective of the genotype at locus 1. In the absence of any of the dark pigments, the coat color appears yellow. The possible genotypes and their corresponding phenotypes (coat color) are given in Table 3.7.

A cross between a homozygous dominant black Labrador and a homozygous recessive yellow Labrador produces black F_1 hybrids. An intercross among the F_2 results in the phenotypic ratio of 9 black:3 brown:4 yellow Labradors, as shown in Fig. 3.27.

From Fig. 3.27, it can be noted that genotypes *B_ee* and *bbee* are both yellow in color, although they possess alleles for black and brown coat color, respectively. Here the recessive allele *e*, at the second locus, is epistatic to both *B* and *b* alleles of the first locus, because *e* masks both their expression; likewise, *B* and *b* alleles are said to be hypostatic to the allele *e*. In this example, *e* is a recessive epistatic allele, because it can exert the epistatic effect only in the homozygous condition.

3.4.4.4 Dominant Epistasis

The phenomenon where a dominant allele of one gene locus masks the expression of alleles at another gene locus is called dominant epistasis. Dominant epistasis is observed in summer squash with respect to the inheritance of fruit color. Commonly, summer squashes come in yellow, white, or green color. In homozygous plant (*WWYY*) for white squash when crossed with a homozygous plant for green squash, all F_1 plants are white. When the F_1 are intercrossed, the following results are obtained in F_2 generation:

The proportions of phenotypes observed in the F_2 generation are 3/4 plants (12 out of 16) produced white squash, and 1/4 of the plants produced colored squash, i.e., 3 out of 16 produced yellow and 1 out of 16 produced green (3/16 + 1/16 = 4/16). These observations suggest that production of pigment is inhibited by a dominant allele (*W*) at one locus, resulting in white progeny, i.e., *W_* genotype inhibits pigment production and white squash is produced, and *ww* enables production of pigment producing colored squash. F_2 plants with homozygous recessive (*ww*) genotype at the W locus produced pigmented fruits. The fruits were either yellow or green in color, and were produced in the ratio of 3:1 as seen in Fig. 3.28. The type of pigment (green or yellow) is dependent on a second locus. The dominant *Y* allele (*Y_*) produces yellow pigment, and the recessive allele *(yy)* determines green. This second locus is expressed only in plants homozygous recessive (*ww*) for the first locus, i.e., in the absence of the dominant inhibitory *W* allele. Therefore, the plants producing yellow squash have the genotype *wwY_*, and those with genotype *wwyy*

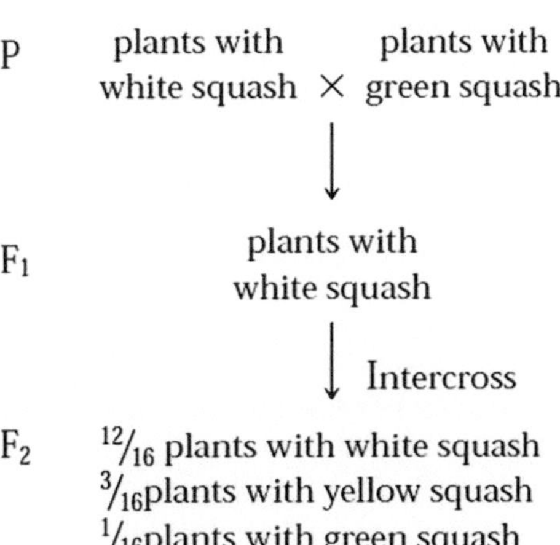

Fig. 3.28 Dihybrid cross depicting dominant epistasis in fruit color of summer squash. The phenotypic ratio of dominant epistasis is 12:3:1

Table 3.8 Dominant epistasis in summer squash—F_2 genotypes and phenotypes

Genotype	Phenotype
*W_Y_*	White squash
W_yy	White squash
*wwY_*	Yellow squash
wwyy	Green squash

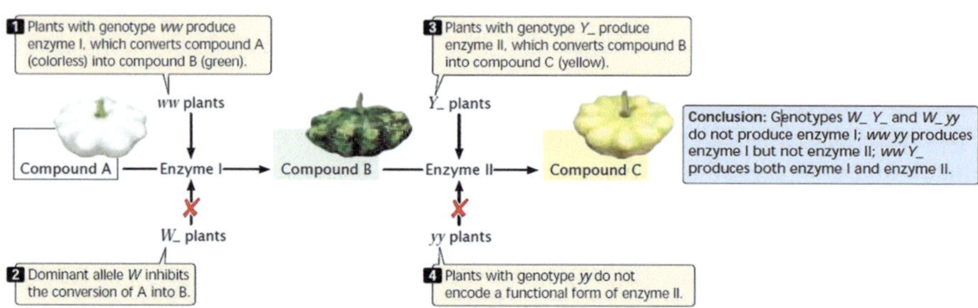

Fig. 3.29 Mechanism of dominant epistasis in summer squash. The yellow pigment in summer squash is synthesized by a two-step biosynthetic pathway. A colorless compound A is converted to a green compound B by enzyme I. Enzyme II converts compound B to a yellow colored compound C. The dominant allele *W* inhibits conversion from A to B. Plants with *Y* allele produce enzyme II which converts compound B to C. The *yy* genotype does not code for active enzyme II and hence compound B to C conversion does not occur

produce green squash. The possible genotypes and their respective phenotypes are given in Table 3.8.

Dominant allele *W* is epistatic and both *Y* and *y* are hypostatic, because presence of even one copy dominant *W* allele suppresses production of both pigments: yellow and green. This is in contrast to allele *e* in Labrador retriever where two copies of the recessive allele are required to bring about epistasis. Therefore, *W* is a dominant epistatic allele.

Summer squash is a good example to understand how epistasis occurs. This is a typical example of genes which take part in a series of reactions in a biochemical pathway. The mechanism of epistasis in summer squash can be explained as follows: production of yellow pigment is suggested to follow a two-step biochemical pathway (Fig. 3.29). The pathway starts with a colorless (white) precursor compound A which is converted to a green colored compound B, by enzyme I. In the second step, the green colored B is further converted by enzyme II to C which is yellow in color. Plants with genotype *ww* produce enzyme I, and the squash produced by these plants may be yellow or green, depending on presence or absence of enzyme II, respectively. Presence of allele *Y* at the second locus produces enzyme II and hence yellow fruits are obtained.

Presence of homozygous recessive *y* does not encode a functional enzyme II, and green squash are produced. Conversion of A to B is inhibited in the presence of dominant *W* at the first locus; plants with *W_* genotype cannot synthesize B and their fruits remain white, irrespective of the alleles at the second locus. Several examples of epistasis which involve such a mechanism, where a gene (such as *W*) is involved in an earlier step of a biochemical pathway, will be epistatic to genes (such as *Y* and *y*) that are involved in the later steps.

3.4.4.5 Duplicate Gene Interaction

Epistatic interaction between genes is not always in opposition to one another. Sometimes genes can have the same role in a particular process and substitute for one another. This kind of a mechanism is called duplicate gene interaction.

3.4.4.5.1 Duplicate Recessive Epistasis

When expression of dominant alleles is masked by recessive alleles at two gene loci, it is called duplicate recessive epistasis. Earlier this was called as complementary gene interaction because both genes (dominant alleles of both genes) are required to express the particular phenotype.

Flower Color in Peas

Bateson and Punnett, for the first time, explained epistasis based on their work in pea plants. They performed a cross with two pure-breeding varieties of pea bearing white flowers. All F_1 progeny had purple flowers. When these purple-flowered F_1 plants were selfed, the F_2 generation produced both purple and white flowers in the ratio 9.4:6.6, or approximately 9:7. Bateson and Punnett counted a total of 382 purple-flowered plants and 269 white-flowered plants in the F_2 generation. In a report published in 1909, Bateson first proposed an explanation and he called it the ability of one "allelomorphic pair" (pair of gene alleles) to mask the effect of alleles for another gene. Later he used the word "epistasis," (meaning, standing upon) to define the masking action of one gene by another.

The mechanism involved can be explained as follows: the alleles of genes controlling flower color in pea were designated *C* and *P*. The genes *C* and *P* are related to biosynthetic pathway involved in the synthesis of a pigment called anthocyanins. This biosynthetic pathway in peas happens in two steps. The first step requires gene *C* and the second step requires gene *P* (Fig. 3.30). If either of the gene is nonfunctional, the respective step fails and purple pigment is not produced. In such a case, the affected pea plants will bear only white flowers. Dominant alleles *C* and *P* code for functional enzymes involved in anthocyanin production, while the recessive alleles *c* and *p* code for nonfunctional enzymes. Therefore, if any one of the

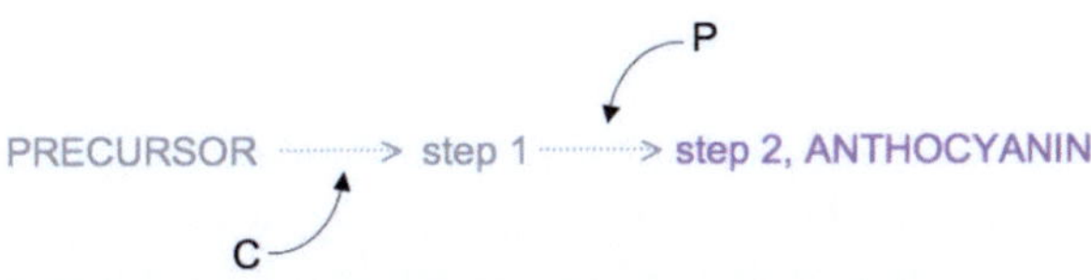

Fig. 3.30 Two-step anthocyanin biosynthesis process. Presence of genes *C* and P is required for anthocyanin synthesis. Anthocyanin pigment production involves biochemical conversion of a precursor compound. In the first step, precursor is converted into an intermediate by C. In the second step, intermediate is converted into anthocyanin by P

Table 3.9 Dihybrid cross between two purple-flowered (*CcPp*) pea plants

		Female Gametes			
		CP	***Cp***	***cP***	***cp***
	CP	*CCPP*	*CCPp*	*CcPP*	*CcPp*
Male Gametes	***Cp***	*CCPp*	*CCpp*	*CcPp*	*Ccpp*
	cP	*CcPP*	*CcPp*	*ccPP*	*ccPp*
	cp	*CcPp*	*Ccpp*	*ccPp*	*ccpp*

loci is homozygous recessive, anthocyanin pigment will not be synthesized (Table 3.9).

Albinism in Freshwater Snail *Physa heterostropha*

Albinism is a common genetic trait which is characterized by absence of a pigment, and is observed in both plants and animals. Production of pigments in most cases takes place through a multistep biosynthetic pathway, which is also applicable to albinism and hence may involve gene interaction. Researchers Robert T. Dillon and Amy R. Wethington studied albinism in *Physa heterostropha*, a common freshwater snail. They showed that the causal factor for albinism is the presence of homozygous recessive alleles, at either of two gene loci.

For their study, they collected inseminated snails from natural populations and placed them in cups of water. The snails laid their eggs in the cups. The progeny was studied after the eggs hatched. Some of the snails that hatched were albinos. When two such albino snails were mated, all of the F_1 were pigmented. The F2 obtained on intercrossing the F_1 progeny consisted of 9/16 pigmented snails and 7/16 albino snails, giving rise to a phenotypic ratio of 9:7, which is obtained when dominant alleles present at both loci (*A_B_*) are responsible for the production of pigment, and all other genotypes produce albinos as shown in Fig. 3.31.

The production of pigments in the snails is suggested to be due to a two-step biosynthetic pathway (Fig. 3.32), where pigment (compound C) is produced only after enzyme I converts compound A into compound B and enzyme II converts compound B into compound C. At least one dominant allele *A* is required at the first locus to produce enzyme I, and at least one dominant allele *B* is required at the second locus to produce enzyme II. Absence of compound C causes albinism, and it can happen in three ways. Firstly, production of enzyme I can be prevented due to recessive alleles in homozygous condition at the first locus (genotype *aaB_*), and thus compound B is never produced. Secondly, production of enzyme II may be prevented because of homozygous recessive alleles at the second locus (genotype *A_bb*), and thus compound B will not be converted into C. Thirdly, both loci may be occupied by two recessive alleles (*aabb*), resulting in the absence of enzymes I and

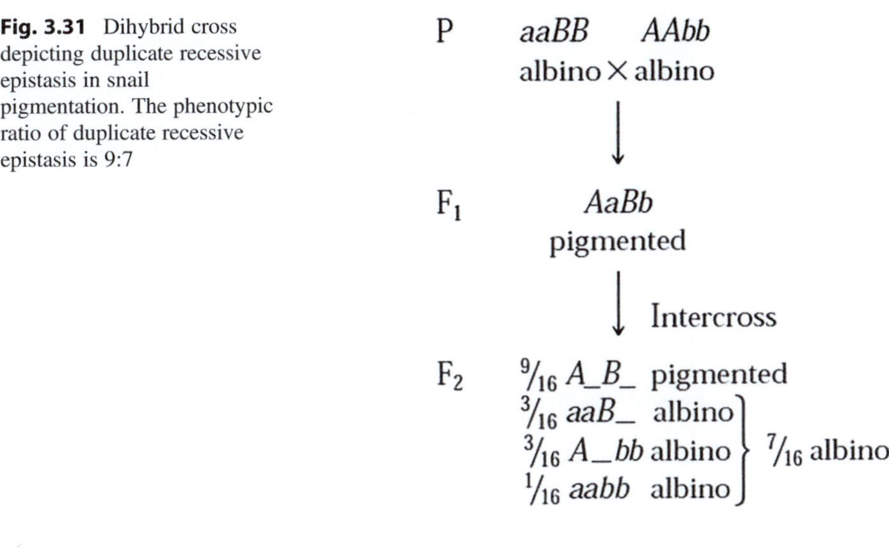

Fig. 3.31 Dihybrid cross depicting duplicate recessive epistasis in snail pigmentation. The phenotypic ratio of duplicate recessive epistasis is 9:7

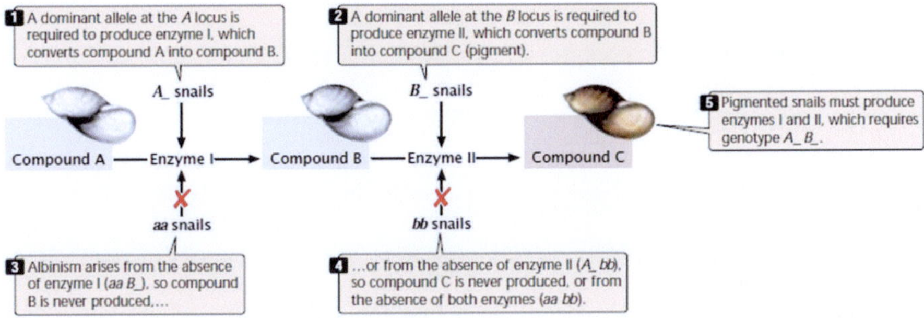

Fig. 3.32 Mechanism of duplicate recessive epistasis. Pigment synthesis in snails is by a two-step biosynthetic pathway. Compound A is converted to compound B by enzyme I. Enzyme II converts compound B to compound C (pigment). The dominant allele A is required to express enzyme I and dominant B is required for enzyme II expression. For pigmentation both enzymes are required (*A_B_*). Lack of either of them results in albinism

II. In this example, *a* is epistatic to *B*, and *b* is epistatic to *A*; *both* are recessive epistatic alleles because two copies of either allele *a* or *b* are needed to suppress pigment production. Epistasis in snails differs from the epistasis of coat color in Labrador retrievers in that suppression of pigment production in snails is caused by recessive alleles at either of the two loci, whereas in Labradors, recessive alleles at a single locus suppress pigment production.

3.4.4.5.2 Duplicate Dominant Epistasis

When expressions of recessive alleles present at two gene loci are masked by a dominant allele, it is known as duplicate gene action or duplicate dominant epistasis.

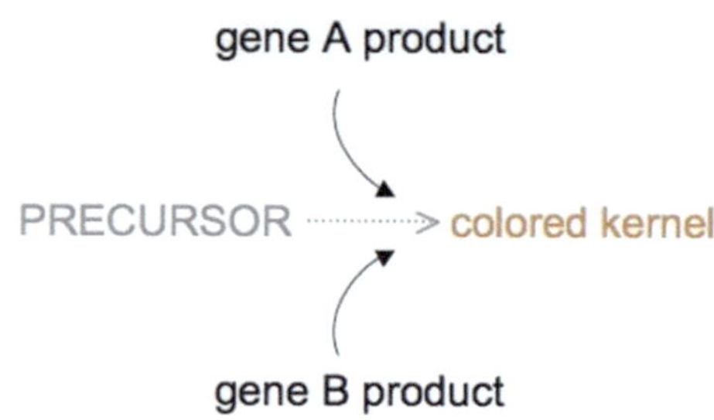

Fig. 3.33 Color production in wheat kernel. Kernel color in wheat is determined by two genes, A and B. Colored kernel will be produced if either gene is functional. Kernel will lack color (white) only if both genes are nonfunctional

Table 3.10 Dihybrid cross between wheat plants of genotype *AaBb*

		Female Gametes			
		AB	***Ab***	***aB***	***ab***
Male Gametes	***AB***	*AABB*	*AABb*	*AaBB*	*AaBb*
	Ab	*AABb*	*AAbb*	*AaBb*	*Aabb*
	aB	*AaBB*	*AaBb*	*aaBB*	*aaBb*
	ab	*AaBb*	*Aabb*	*aaBb*	*aabb*

Wheat Kernel Color

Kernel color in wheat is dependent on a biochemical reaction which converts a precursor molecule into a pigment. This biosynthetic reaction can be performed by either of the products of genes A or B (Fig. 3.33). Therefore, presence of either dominant *A* or *B* allele produces kernel color. But white kernel (lacking pigment) will be produced when neither dominant allele is present, i.e., when the genotype is double homozygous recessive. The outcome of such an interaction is a phenotypic ratio of 15:1 (15 color:1 noncolor) (Table 3.10).

Some of the possible gene interactions and their modified dihybrid ratios are given in Table 3.11.

The term "epistasis" was given by William Bateson to a particular type of genetic interaction, wherein one gene masks the effects of another gene. Earlier, different terms, such as complementary (duplicate recessive epistasis), supplementary (recessive epistasis), etc., were used to distinguish different forms of gene interaction. Ronald Fisher generalized the term epistasis to include several other forms of gene interactions. He included any multigene interactions that lead to a phenotype different than that of the expected phenotypes associated with combination of corresponding individual genes. Eventually, the term epistasis has evolved and presently it includes all gene interactions. It is now defined as any phenomenon

Table 3.11 Types of gene interactions

Ratio	Description	Name of interaction
9: 3:3: 1	**No gene interaction; each allele produces its own phenotype**	**Independent assortment without gene interaction**
9: 3:3:1	Both gene pairs show complete dominance; dominant alleles and also both homozygous recessives interact to produce new phenotypes	Not named, ratio similar to independent assortment, but phenotypes are formed by gene interactions
9:4:3	Both gene pairs show complete dominance; But, one gene when in homozygous recessive state, suppresses other gene's phenotype	Recessive epistasis
9:7	Both gene pairs show complete dominance; either of the genes can suppress the other gene when in homozygous recessive state	Duplicate recessive epistasis
12: 3:1	Both gene pairs show complete dominance; when one gene is dominant, the phenotype of the other gene is suppressed	Dominant epistasis
15:1	Both gene pairs show complete dominance; when either of the gene is dominant, it suppresses the other gene's effect	Duplicate dominant epistasis
13:3	Both gene pairs show complete dominance; but when either of the gene is dominant, it suppresses the other gene's effect	Dominant and recessive epistasis
9:6:1	Both gene pairs show complete dominance; when either of the gene is dominant, it suppresses the other gene's effect	Duplicate interaction
7:6:3	One gene pair shows complete dominance and the other shows partial dominance; when in homozygous recessive state, the first gene is epistatic to the second gene	No name
3: 6:3:4	One gene pair shows complete dominance and the other shows partial dominance; in homozygous recessive state, either gene hides other gene's effect; when both genes are in homozygous recessive state, the effects of the first gene are suppressed by the second	No name
11:5	Both gene pairs show complete dominance, only when both types of dominant alleles are present; if not, the recessive phenotype appears	No name

that involves different genes contributing to a single phenotype such that their effects are not just additive. Such genes are said to be epistatic. Therefore, epistasis is not restricted to the interactions of only two genes. Rather, epistasis can occur in all of the following scenarios: (1) whenever two or more loci interact to create novel phenotypes, (2) whenever an allele at one locus masks the effects of alleles at one or more other loci, and (3) whenever an allele at one locus modifies the effects of alleles at one or more other loci.

Box 3.1 Scientific Concept: Epistasis and Disease

When factors contributing to a disease phenotype are analyzed, epistasis forms a very significant factor along with gene mutations or environmental factors. Understanding epistatic gene interactions is believed to be crucial to fathom complex diseases, which include diabetes, cardiovascular disease, cancer, and Alzheimer's disease. Understanding the genetic basis and the causal factors of these diseases has been elusive when single genes are studied. But the growing focus on epistatic interactions is expected to contribute toward greater understanding of these complex human diseases. Researchers have attempted to understand epistatic interactions that could be involved in Alzheimer's disease. More than 100 probable epistatic interactions have been evaluated in sporadic cases of Alzheimer's disease. Sporadic cases are those cases of disease incidence which have no known hereditary background. Evaluation of epistatic interactions between associated pairs of genes involved measurement of magnitude and its statistical significance. Overall, 27 significant epistatic interactions have been shown to play a role in Alzheimer's disease process. They have been divided into five groups, which include (1) beta-amyloid production, (2) oxidative stress, (3) inflammation, (4) cholesterol metabolism, and (5) others. Certain epistatic interactions appeared to be synergistic, i.e., increased disease risk, while others appeared to be antagonistic, i.e., have a protective relationship.

Diabetes is another example of a complex disease wherein epistasis plays a role. In case of diabetes, both epistatic interactions and the environment play an important role. Interactions between numerous chromosomal loci have been discovered in type II diabetes patients. Experimental evidences have also shown involvement of epistasis in several other complex diseases such as autism, hypertension, cardiovascular disease, neurological disorders, and several types of cancer. Presently, it has become possible to detect relationships between epistatic interactions, genes, and networks at a systems level. With the currently available high-throughput experimental tools, it is possible to measure biochemical and molecular data. Examples of such high-throughput tools include DNA microarrays, bioinformatics, and computational methods. These tools can be applied to identify and understand epistatic relationships on a large scale, and the knowledge obtained can be applied for better diagnosis and treatment of complex diseases.

Box 3.2 Scientific Concept: Diseases with Complex Inheritance Patterns

Typically, monogenic disorders follow Mendelian patterns of inheritance. However, there are several disorders that exhibit complex patterns of inheritance. A common progressive neurodegenerative disorder such as Parkinson's disease (PD) is one such example. Earlier, PD was considered to be a sporadic disease without a genetic basis. However, research spanning over a decade have established the underlying role of genetic factors in PD pathogenesis. Interestingly, the causal factors include highly penetrant singular genes, unique variants with incomplete penetrance, and also the more common idiopathic forms of the disease. Therefore, it involves monogenic forms with Mendelian inheritance and also forms with non-Mendelian inheritance. However, the monogenic Mendelian forms are rare, but have provided insight into the genetic architecture underlying this disease. Monogenic mutations associated with both autosomal dominant and autosomal recessive PD have been characterized. In addition, a few gene variants exhibiting incomplete penetrance are shown to be strong risk factors for the disease incidence in some populations.

3.5 Extranuclear Inheritance

Extranuclear inheritance indicates transmission of characters through factors that reside outside of the nucleus. It is also called as non-Mendelian inheritance because Mendel's laws of heredity are not applicable to their inheritance patterns. Some of the other terminologies used include extrachromosomal, cytoplasmic, or nonchromosomal inheritance. Cytoplasmic inheritance involves several factors such as cellular organelles (e.g., chloroplast and mitochondria) containing their own DNA and parasitic or symbiotic particles (infective particles) that reside in the cytoplasm and possess their own genetic material. Organelles such as chloroplasts and mitochondria and infectious agents such as bacteria, viruses, and DNA molecules such as plasmids are all mediators of cytoplasmic inheritance. The genetic materials in these entities are also susceptible to mutation and their inheritance does not follow Mendelian rules and ratios.

Another type of extranuclear inheritance, called maternal effect, includes those traits and characters that are influenced by the mother's genotype or phenotype. When gametes are formed, the egg cell of the female gamete receives a greater volume of cytoplasm than the sperms or male gametes. This difference in the cytoplasmic content of the gametes is responsible for the maternal effect. During fertilization, the cytoplasmic contribution of the female gamete to the zygote is greater than that of the sperm. This results in an unequal contribution of the female parent toward the development of zygotes. Although chromosomal genes are contributed equally to the zygote by both parents, sex chromosomes being an exception, the sperm rarely contributes any additional material toward zygotic development. The role of the female parent in this regard is very significant. In addition to the chromosomal genes, the female parent often contributes to the initial cytoplasm and organelles to the zygote. Hence, zygotic development is facilitated by the maternal environment provided by the cytoplasm of the egg cell. Coiling in snails and moth pigments are examples of maternal effect.

Mitochondria and chloroplasts are cytoplasmic organelles with specialized functions. The unique feature of both these organelles is the presence of small circular organellar DNA, which make them semiautonomous. A specific subset of the total cellular genome is carried by these small circular organellar chromosomes. Genes concerned with energy production are located on the mitochondrial DNA, whereas genes important for photosynthesis are present in chloroplast DNA. However, neither of the organelle is completely independent, because they both depend on the nuclear genome to a certain extent for their functions and hence called semiautonomous. Another distinct feature of organellar genomes is their copy numbers, i.e., they are present in large number of copies in the cell. In addition, the organelles themselves are also present in multiple copies per cell. Collectively, any given cell consists of hundreds to thousands of organellar chromosomes. These organelle genes follow their own pattern of inheritance and are called uniparental inheritance, indicating that the progeny inherit organelle genes exclusively from one parent, which is in most cases the mother, and hence also called maternal inheritance.

Cytoplasmic inheritance can be deduced by performing crosses and studying the patterns of inheritance for several generations. It is specifically the results of the reciprocal crosses that reveal the cytoplasmic inheritance pattern. In any given cross, the variant phenotype will be transmitted to the progeny particularly by the female parent, but not the male parent. Hence, cytoplasmic inheritance pattern can be generally represented as follows:

Mutant ♀ × wild-type ♂ → all mutant progeny

Wild-type ♀ × mutant ♂ → all wild-type progeny

3.5.1 Extranuclear Genomes: Mitochondria

Several mitochondrial DNA (mtDNA) including that of humans have been sequenced. The genome/DNA of mitochondria (mtDNA) is compact, circular (Fig. 3.35), and double stranded. It consists of 16,569 base pairs. The genome includes 37 genes, of which 13 codes for polypeptides, 22 codes for transfer RNA (tRNA), and 2 codes for ribosomal RNA (rRNA) (Fig. 3.34). Majority of proteins required by the mitochondria for its functions are encoded by the nuclear genome. Each strand of the mtDNA duplex is transcribed into a single RNA which is further cleaved into smaller segments, releasing 22 tRNAs, along with 16S and 12S rRNA. The mitochondria are fairly self-sufficient with respect to availability of mRNAs required for protein synthesis. The oxidative phosphorylation process occurs within the mitochondrion, and it requires the participation of about 69 polypeptides. Human mtDNA carries 13 of these which include cytochrome b, two subunits of ATPase,

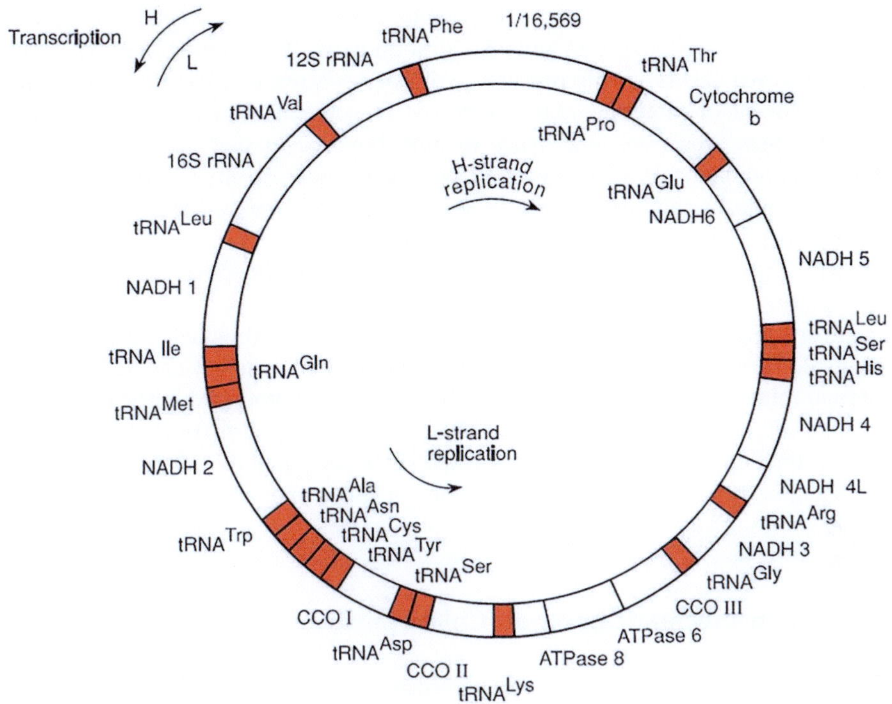

Fig. 3.34 Human mitochondrial DNA. The gene map of the human mitochondrial DNA shows the presence of a heavy (H) strand and light (L) strand. Except nine gene loci, all are located on the H strand (labeled on the outside). The remaining nine loci are located on the L strand (labeled on the inside). The origin of replication and the direction of transcription of the H and L strand are shown

three subunits of cytochrome-c oxidase, and seven subunits of NADH dehydrogenase. The synthesis of the rest of the polypeptides required for oxidative phosphorylation is regulated by the nuclear genes and takes place in the cytoplasm and then it gets transported into the mitochondrion.

The mitochondrial rRNA shares a great similarity to prokaryotic rRNA than to eukaryotic rRNA. This is demonstrated by sensitivity of mitochondrial ribosome to prokaryotic antibiotics such as chloramphenicol and streptomycin, which inhibit its function. The similarity observed between prokaryotes and mitochondria strongly supports the theory of symbiotic origin of mitochondria. This theory has been proposed by L. Margulis for both organelles, i.e., mitochondria and chloroplasts. The theory hypothesizes that initially cyanobacteria and free-living bacteria have been transformed into chloroplast and mitochondria, respectively. This is a theory that has been accepted. It is suggested that the prokaryotes invaded or were eaten up by primitive cells, which eventually evolved within these primitive cells to become the respective organelles.

Mitochondrial inheritance shows two general patterns in animals. Firstly, mitochondria are inherited in a maternal fashion where it is typically transmitted through the egg cell but not the male gamete. However, exceptions to the general rule of maternal inheritance of mitochondria do exist. A certain amount of "leakiness" however occurs in this process, as it has been recently shown in mice that nearly one out of thousand mitochondria is of paternal origin. Species, such as mussels, show biparental inheritance of mitochondria. In such species, the mitochondrial population of an offspring are obtained almost equally from both parents. Certain gymnospermous plants, e.g., coastal redwoods, show paternal inheritance of mitochondria, i.e., zygote receives only paternal mitochondria.

The second general pattern observed in mitochondrial inheritance is homoplasmy, which means existence of uniform populations of mitochondria within a cell or organism. In general, all mitochondria of an individual are genetically identical. However, phenomena such as biparental inheritance and leakiness of paternal mitochondria cause heteroplasmy, leading to mitochondrial heterogeneity within a cell or organism. When organelle populations are a mixture of two genetically distinct chromosomes, then they often show segregation of the two types into the daughter cells during cell division, a phenomenon called cytoplasmic segregation.

3.5.2 Extranuclear Genomes: Chloroplast

Chloroplasts are characterized by the presence of the photosynthetic pigment chlorophyll. The precursors of chloroplasts are called plastids, and they become chloroplasts after they develop chlorophyll. However, in certain conditions, plastids can fail to develop into chloroplasts, but still persist with reduced size and complexity. Such undeveloped plastids are referred to as proplastids, and each is approximately the size and shape of a mitochondrion. Similar to mitochondria, both DNA and ribosomes present in the chloroplast show prokaryotic affinities. The DNA of chloroplasts (cpDNA) is circular in shape and its size ranges from 85 kilobases (kb) in *Codium*, a green alga, to 2000 kilobases in *Acetabularia*, also a green alga. The size of cpDNA is nearly five times that of an animal mtDNA. Like mtDNA, the cpDNA also regulates synthesis of tRNA, ribosomal RNA, and certain proteins that constitute the organelle. Data obtained by sequencing several cpDNA has shown that about 100 genes are located in the chloroplast genome. Nearly 30 of these genes are known to code subunits of five protein complexes involved in photosynthesis: photosystem I, photosystem II, ribulose 1,5-bisphosphate carboxylase-oxygenase, cytochrome *b6-f* complex, and ATP synthase. Nearly 60 genes code for machinery involved in protein synthesis within the chloroplast (Fig. 3.36).

The prokaryotic affinities shown by the DNA and ribosomes of the chloroplast led to the hypothesis that chloroplast originated from symbiotic cyanobacteria (blue-green algae). These cyanobacteria have been shown to possess many similarities with the chloroplast, and the rRNAs of cyanobacteria have been shown to hybridize with the cpDNA. Similarities between mitochondria and chloroplasts have made it possible to predict the inheritance patterns of chloroplast mutations on the basis of knowledge derived from the study of mitochondrial genetics. However, in these

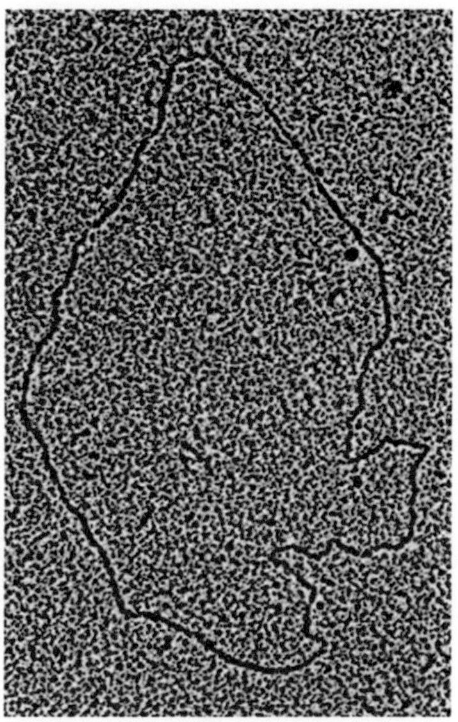

Fig. 3.35 Electron micrograph of circular mouse mitochondrial DNA. Magnification 48,000×

inheritances, it is important to discover and distinguish whether the mutation affecting chloroplast function has occurred in chromosomal DNA or cpDNA. If mutations undergo simple segregation, they will be chromosomal, whereas cpDNA mutations show cytoplasmic patterns of inheritance. Also, adding to the complexity, particularly in plants, is the presence of both mitochondria and chloroplasts, which makes it difficult to determine whether a given trait is due to a mutation in cpDNA or mtDNA. Similar to mitochondria, chloroplasts also exhibit homoplasmy and heteroplasmy. It is important to note that cytoplasmic inheritance in plants happens in two dimensions, i.e., via both chloroplast and mitochondria. In many cases, it may be hard to distinguish between the two. Note that heteroplasmic cells are also called cytohets or heteroplasmons.

3.5.3 Experimental Methods to Deduce Extranuclear Inheritance

Identification of extranuclear inheritance typically happened due to the observation of peculiar results from reciprocal crosses. The progeny of a reciprocal cross has to be followed for several generations; to deduct the patterns of extranuclear inheritance, any premature interpretation can become misleading. Nuclear transplantation has been utilized as powerful technique to identify extranuclear inheritance. The

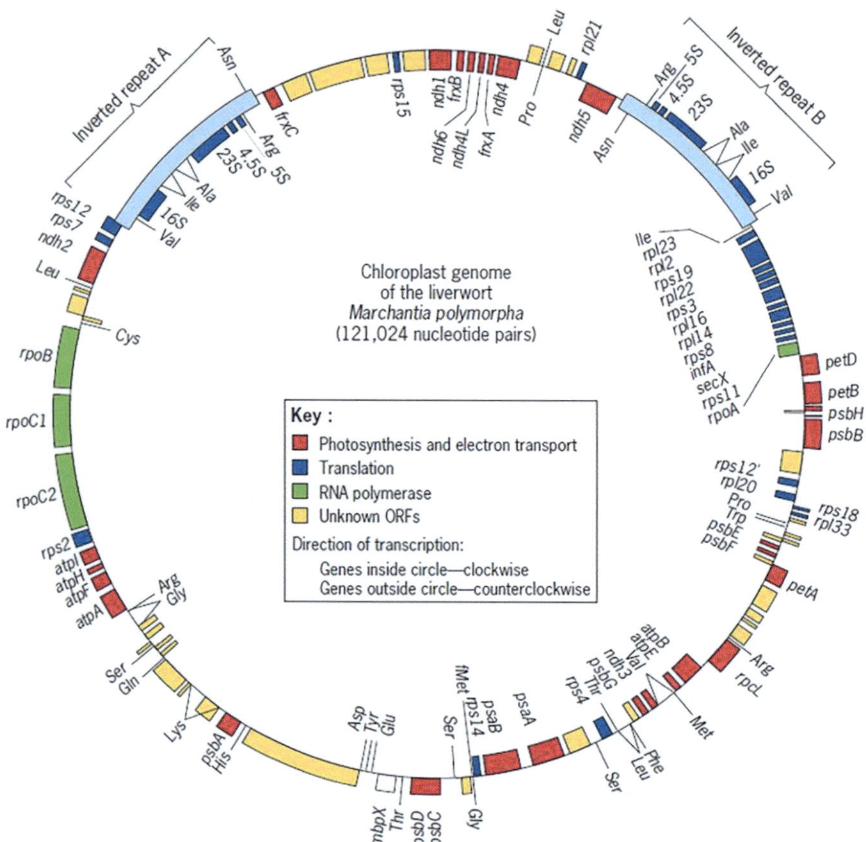

Fig. 3.36 Chloroplast DNA of *Marchantia polymorpha.* Circular DNA, 121,024 base pairs (bp) long, consisting of 128 genes. Majority of chloroplast proteins are encoded by nuclear genome and the rest are encoded by cpDNA. Consists of large inverted repeats A and B (IR_A and IR_B), small single-copy region (SSC), and a large single-copy region (LSC). The 128 possible genes include 4 rRNAs, 32 tRNAs, and 55 proteins

technique involves removal of a nucleus from a cell such as an amoeba or a frog egg by microsurgery or destroying using irradiation, and then substituted with nucleus from another source. A heterokaryon test is another experiment that can be performed with fungi such as *Neurospora* and *Aspergillus*, to determine extranuclear inheritance. In this technique, the ability of mycelia to fuse and form a heterokaryon, i.e., a cell containing nuclei from different strains, is utilized. The cytoplasm of a heterokaryon consists of nuclei of both strains, which subsequently produce spores (conidia) containing either of the two nuclei, and hence can be isolated. Isolated conidia can be cultured to form colonies, whose phenotype demonstrates whether the trait under study is determined by the cytoplasm or the nucleus.

It is also possible to isolate chromosomal genes in a particular cytoplasm by repeated backcrossing of offspring with the male parent. Each such cross reduces the quantity of the female chromosomal genes to half, but the cytoplasm remains similar to the female line. After several generations, female cytoplasm will consist of male genes, and the phenotypes resulting from this final cross will show whether the inheritance of a particular trait is chromosomal or extrachromosomal.

3.5.4 Extranuclear Inheritance: Examples

3.5.4.1 Maternal Effects

A phenomenon in which the phenotype of an offspring is influenced by maternal genotype or phenotype is called as maternal effect.

3.5.4.1.1 Snail Coiling

Shells of snails are coiled either dextrally (to the right) or sinistrally (to the left). The direction of coiling is observed by holding the snail and looking at the opening of the shell, top down from the apex. The snail is said to be coiled dextrally when the opening comes from the right-hand side. It is said to be coiled sinistrally if the opening comes from the left-hand side (Fig. 3.37).

The left panel of Fig. 3.37 shows fertilization between eggs of a dextral snail and sperms of a sinistral snail. All offspring are dextral, which could indicate that dextral

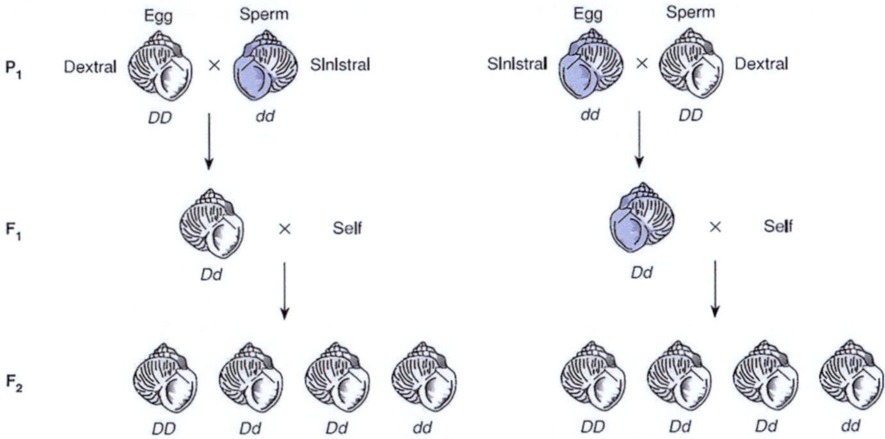

Fig. 3.37 Maternal effects—shell coiling in snails. Genotype of the mother and not phenotype determines coiling in snails. Figure depicts reciprocal crosses, where *D*, dominant, causes dextral and *d*, recessive, causes sinistral coiling. In both crosses shown, *DD* was crossed with *dd*. The F_1 in both crosses have *Dd* genotype, but express the mother's coiling phenotype. Offspring of *DD* mothers show dextral coiling, whereas offspring of *dd* mothers show sinistral coiling. F_2 in both crosses are identical because of identical genotypes (*Dd*) of F_1 mothers

coiling is dominant. When the F_1 snails were self-fertilized (snails are hermaphrodites), all offspring were dextrally coiled. If dextral coiling was a dominant trait, these results are unexpected. However, when F_2 were self-fertilized, one-fourth of offspring were sinistral, and three-fourths were dextral. If shell coiling is considered as a dominant trait, the continued self-fertilization should reveal the 3:1 phenotypic ratio obtained to be 1:2:1 in the following generations, but observations made did not correspond to it.

Results from a reciprocal cross (Fig. 3.37, *right*) showed that all F_1 have the same genotype similar to what was observed in the previous cross, but showed sinistral coiling, similar to the female parent. But further results were the same for both crosses. For both cases, it was observed that F_1 had a phenotype similar to the female parent although offspring of both crosses had the same genotype (*Dd*). The explanations for these observations are as follows: the genotype of the maternal parent determines phenotype of the offspring, with dextral dominant. Thus, the *DD* mother in Fig. 3.37 produces dextral F_1 with *Dd* genotype, whereas the *dd* mother produced sinistral progeny, which also had the same *Dd* genotype.

The spiral cleavage of zygotes of molluscs during development is involved in the determination of coiling phenotype. During such cleavage, the way the spindle is tipped with respect to the egg axis determines the coiling. Tipping of the spindle in one way produces sinistral and in another produces dextral coiling. It is the maternal cytoplasm that controls the direction of tipping, which is in turn under the regulation of the maternal genotype.

3.5.4.1.2 Moth Pigmentation

Pigmentation in the moth flour moth, *Ephestia kuehniella*, is another example of maternal effect. In flour moth, the cytoplasm of the egg cell influences the pigmentation phenotype of the offspring, under regulation of chromosomal genes. Kynurenine, which is a precursor molecule for pigment synthesis, accumulates in the eggs. When the recessive allele, *a*, is in homozygous condition (*aa*), it results in absence of precursor. The results of reciprocal crosses are different for larvae and adults. When a cross is performed with a nonpigmented female and a pigmented male, the results obtained adhere to Mendelian principles. But when a reciprocal cross is performed with a pigmented female (a^+a), all larvae produced are pigmented irrespective of their genotype, but will retain (a^+a) or lose (*aa*) pigments at adult stage based on their genotype (Fig. 3.38). The pigmentation observed in the larval stage is due to the residual kynurenine that was present in the egg's cytoplasm, which eventually got diluted out, and the adults showed pigmentation as per their own genotype.

3.5.4.2 Mitochondria-Mediated Inheritance

3.5.4.2.1 Poky Strain in *Neurospora*

In the haploid fungi *Neurospora*, studies on certain mutants can clearly establish maternal inheritance. For example, a mutant of *Neurospora* called poky (Fig. 3.39) shows a slow growth phenotype. Crosses in *Neurospora* can be performed in such

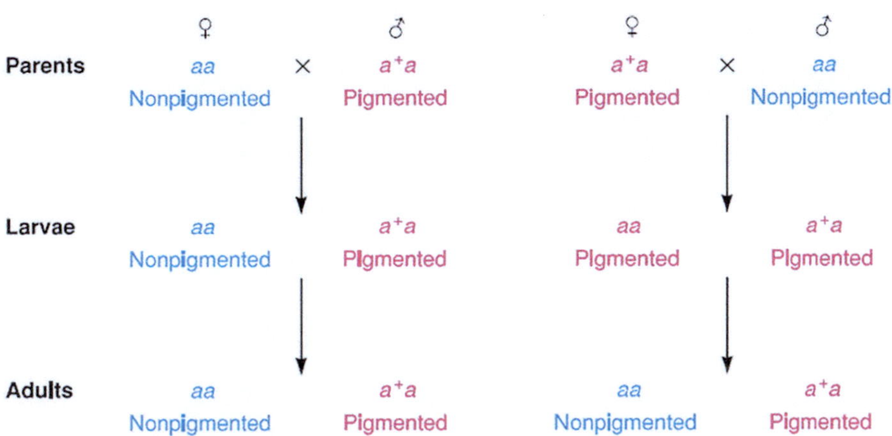

Fig. 3.38 Maternal effects—inheritance of pigmentation in larval and adult flour moth *Ephestia kuehniella*. Single locus controls presence (a^+) or absence (a) of kynurenine. Nonpigmented (aa) mother produces aa offspring that are also nonpigmented in both larval and adult stages (left). In a reciprocal cross (right), the pigmented mother with a^+a genotype produces aa offspring that are nonpigmented in the adult stage, but pigmented in the larval stage. This is due to the residual kynurenine present in the egg from a pigmented mother

way that cytoplasm is contributed by one of the parents and hence behaves as the maternal parent. Results of reciprocal crosses suggested that the mutation existed in the mitochondrial gene(s). From the reciprocal crosses, it was evident that the female parent or cytoplasmic parent's phenotype determines the phenotype of all offspring. The progeny of the reciprocal crosses are as follows:

Poky ♀ × wild-type ♂ → all progeny poky

Wild-type ♀ × poky ♂ → all progeny wild-type

Neurospora, being a fungus, does not possess chloroplasts, and hence the phenotype can be attributed to mitochondria based on its inheritance in reciprocal crosses. Hence, the poky mutation is now known to be in mitochondrial DNA.

3.5.4.2.2 Petites in *Yeast*

Yeast, a facultative anaerobe, can grow in both aerobic and anaerobic conditions. In aerobic conditions, the growth is characterized by distinctive colony morphology. In anaerobic conditions, the yeast forms smaller colonies, and also mitochondria are observed to be structurally reduced. Small anaerobic-like colonies can appear sometimes, even if the yeast is grown in aerobic conditions. However, in such small colonies, the mitochondrial structure appears perfectly normal. Such colonies are a result of mutations called as petite. A cross between wild-type and petites reveal three modes of inheritance (Fig. 3.40).

A petite caused by a chromosomal mutation is called a segregational petite, and follows Mendelian inheritance. Neutral petite is lost instantly when crossed with a wild-type. The suppressive petite shows variability in expression from one strain to another but is able to convert the wild-type mitochondria into the petite form. All

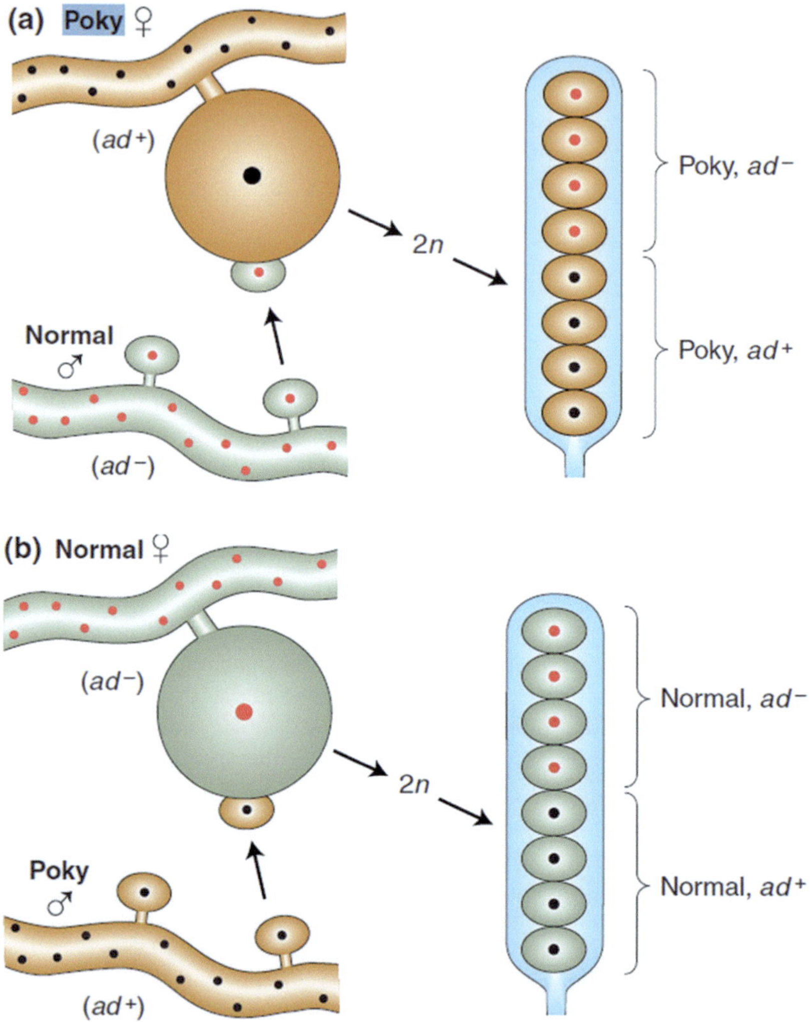

Fig. 3.39 Poky strain of *Neurospora*. Parent which contributes most of the cytoplasm to the progeny is called the female. Brown shading indicates mitochondria containing poky mutation. Green indicates normal mitochondrion. (**a**) All progeny is poky; (**b**) all progeny is normal. Both crosses indicate maternal inheritance pattern. The ad^+ (black) and ad^- (red) are representatives of nuclear genes shown to indicate 1:1 segregation

types of petites are formed due to a failure in mitochondrial function. Whether the regulation of the defective mitochondrial function resides in the mitochondria or within the cell's nucleus, they are usually deficient in one or the other cytochromes.

The composition of a DNA molecule can be measured by a technique called density gradient centrifugation. Here the term buoyant density is used to describe the equilibration position DNA attains when subjected to density gradient

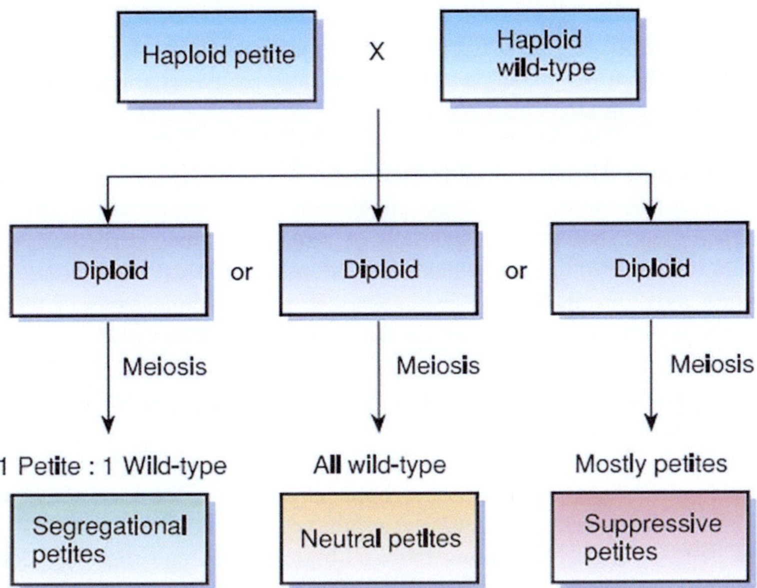

Fig. 3.40 Categorization of petite yeasts based on their segregation patterns. There are three categories of petites—segregational, neutral, and suppressive—depending on their meiotic segregation pattern of a cross between petite and wild-type diploids. In spores of segregational petite heterozygotes, a 1:1 ratio segregation is observed; heterozygous neutrals are lost; and suppressive petites behave as dominant under the same circumstances

centrifugation. Therefore, buoyant density is a measure used to indicate any changes in the composition of a DNA molecule when compared with the normal. When DNA from the petites is subjected to such an analysis, some petites do not show any change, while others show changes ranging from very small to complete absence of DNA. From these results, it can be inferred that petites can occur due to a point mutation, with no measurable change in the buoyant density, or can occur due to marked changes in DNA, or even total absence of DNA. Most petites are characterized by a mitochondrion, defective protein synthesis. Mitochondria that completely lack DNA are known to occur in neutral petites. When wild-type are mated with neutral petites, diploid cells are formed, in which normal mitochondria dominate. Almost every spore during meiosis obtains numerous normal mitochondria; therefore, all progeny is normal, whereas suppressive petites are known to influence normal mitochondria in the following ways: suppressive mitochondria may outnumber normal mitochondria by reproducing faster within a cell, or crossing over between the DNAs of wild-type and suppressive petite may affect normal DNA if the petite's DNA harbors severe damage.

Several crosses involving a wild-type and a suppressive petite, with mtDNA of known buoyant densities, have been performed. The mtDNA buoyant densities of petite offspring were variable. For example, if a normal strain with mtDNA buoyant density of 1.684 g/cm^3 is crossed with a suppressive petite of 1.677 g/cm^3 buoyant density, the mtDNA of the offspring had buoyant densities of 1.671, 1.674, and 1.683 g/cm^3. These results suggested that the suppressive character took over the colony by way of mtDNA recombination.

3.5.4.2.3 Human Mitochondrial Inheritance

In humans, root of certain diseases can be traced to mitochondrial pathologies. Human pedigrees showing transmission of rare phenotypes particularly through females but never through males have been identified. This pattern where a trait is transmitted only through females suggests cytoplasmic inheritance. Cytoplasmic inheritance in humans is mediated via the mtDNA and these rare phenotypes are caused by a mutation in mtDNA.

The first mitochondrial disease was reported in 1962 and was called as Luft disease, characterized by symptoms such as excessive sweating and general weakness. Douglas Wallace and colleagues in 1988 showed cytoplasmic mode of inheritance of another disease called Leber optic atrophy. As the name suggests, the disease affects the optical nerve. A large number of mitochondria are usually present in nerve cells, as their functioning requires high energy, and therefore defects in mitochondria are not tolerated by nerve cells. In addition to damaging the optic nerve, the disease also damages the heart. Maternal mode of inheritance of Leber optic atrophy has been shown in the pedigrees. Point mutations associated with the disease have been identified after sequencing mtDNA of individuals from affected families. The point mutations involve a change in nucleotide 11,778, in the NADH dehydrogenase subunit 4 gene, which is responsible for the disease phenotype. The point mutation replaces the nucleotide guanine with adenine at codon 340, which in turn substitutes amino acid arginine with histidine. Therefore, Leber optic atrophy is the first human disease to be specifically linked to a particular mutation in mtDNA.

Presently, more than 260 different pathogenic mutations involved in a spectrum of mitochondrial diseases have been reported, and the number is continuing to rise. Three types of mtDNA mutations are as follows: point mutations in genes coding for proteins and genes related to protein synthesis (rRNA or tRNA genes), and rearrangements in mtDNA, including deletions and insertions.

MERRF (myoclonic epilepsy with ragged red fiber) is another phenotype caused because of a point mutation mtDNA. The disease not only affects the muscle but also causes hearing disorders and affects the eye. Kearns-Sayre syndrome is another condition which exhibits an array of symptoms affecting the eyes, heart, muscles and brain, and is caused by a deletion in the mtDNA. Currently, it is known that large proportion of human diseases are associated with mitochondrial dysfunction. A spectrum of disorders such as neurodegenerative disorders, cardiovascular disorders, neurometabolic diseases, cancer, and obesity are associated with mtDNA mutations (Table 3.12). Inability of the mitochondria to produce adequate amounts of ATP is considered as one of the primary basis for most mitochondrial pathologies, and hence resulting in multisystemic disorders. Extremely severe clinical presentations are observed in high-energy demanding tissues, such as skeletal muscle, central nervous system, and heart muscles. However, it is also important

Table 3.12 A list of some mitochondrial diseases and their clinical features

Disorder	Primary features
CPEO (chronic progressive external ophthalmoplegia)	External ophthalmoplegia, bilateral ptosis
KSS (Kearns-Sayre syndrome)	Progressive external ophthalmoplegia, pigmentary retinopathy
Pearson syndrome	Sideroblastic anemia of childhood, pancytopenia, exocrine pancreatic failure
Leigh syndrome	Subacute relapsing encephalopathy
NARP (neurogenic weakness with ataxia and retinitis pigmentosa)	Late-childhood or adult-onset peripheral neuropathy, ataxia
MELAS (mitochondrial encephalomyopathy with lactic acidosis and stroke-like episodes)	Stroke-like episodes, seizures and/or dementia, ragged red fibers and/or lactic acidosis
MERRF (myoclonic epilepsy with ragged red fibers)	Myoclonus, seizures, cerebellar ataxia, myopathy
LHON (Leber hereditary optic neuropathy)	Subacute painless bilateral visual failure

to note that mitochondrial disorder can also occur due to mutations in the nuclear genome.

Some individuals with a mitochondrial disorder may be homoplasmic for mtDNA mutation, but heteroplasmic condition is more prevalent. In heteroplasmy, cytoplasmic segregation causes varying proportions of normal and mutant organelles to be transmitted to the progeny (Fig. 3.41). Heteroplasmic individuals have both mutant and wild-type mtDNA in various proportions. In such a case, a proportion of mutant to wild-type mtDNA determines disease expression. Therefore, higher level of mutant mtDNA is usually associated with increased severity of the clinical symptoms. This is called as the threshold effect. Only when the level of mtDNA mutation exceeds a critical threshold the associated defect can be detected. Even in an individual, the proportions of mutant and normal organelles vary temporally and based on the different tissue types. Accumulation of certain types of mitochondrial mutations over time has been considered as one of the possible causes of aging.

3.5.4.2.4 Mitochondrial Inheritance in Yeast: Antibiotic Influences

Correlating with the prokaryotic origin of the mitochondria, the protein synthesis machinery within the mitochondrion is prokaryotic in nature. Therefore, antibiotics such as erythromycin and chloramphenicol can inhibit protein synthesis in the mitochondrion. Exposure of yeast cells to these antibiotics causes a petite-type growth pattern. Yeast strains resistant to antibiotics can be obtained by growing them on a medium containing antibiotic. In such case, only mutant yeast cells, resistant to antibiotics, will grow in the medium. The resistance phenotype has been observed to be inherited through mtDNA, and not through cellular DNA. Crosses performed between resistant and sensitive (wild-type) yeast have shown a mitochondrial pattern of inheritance (Fig. 3.42). Diploid yeast colonies obtained from these crosses segregate into both resistant and sensitive cells. These results are

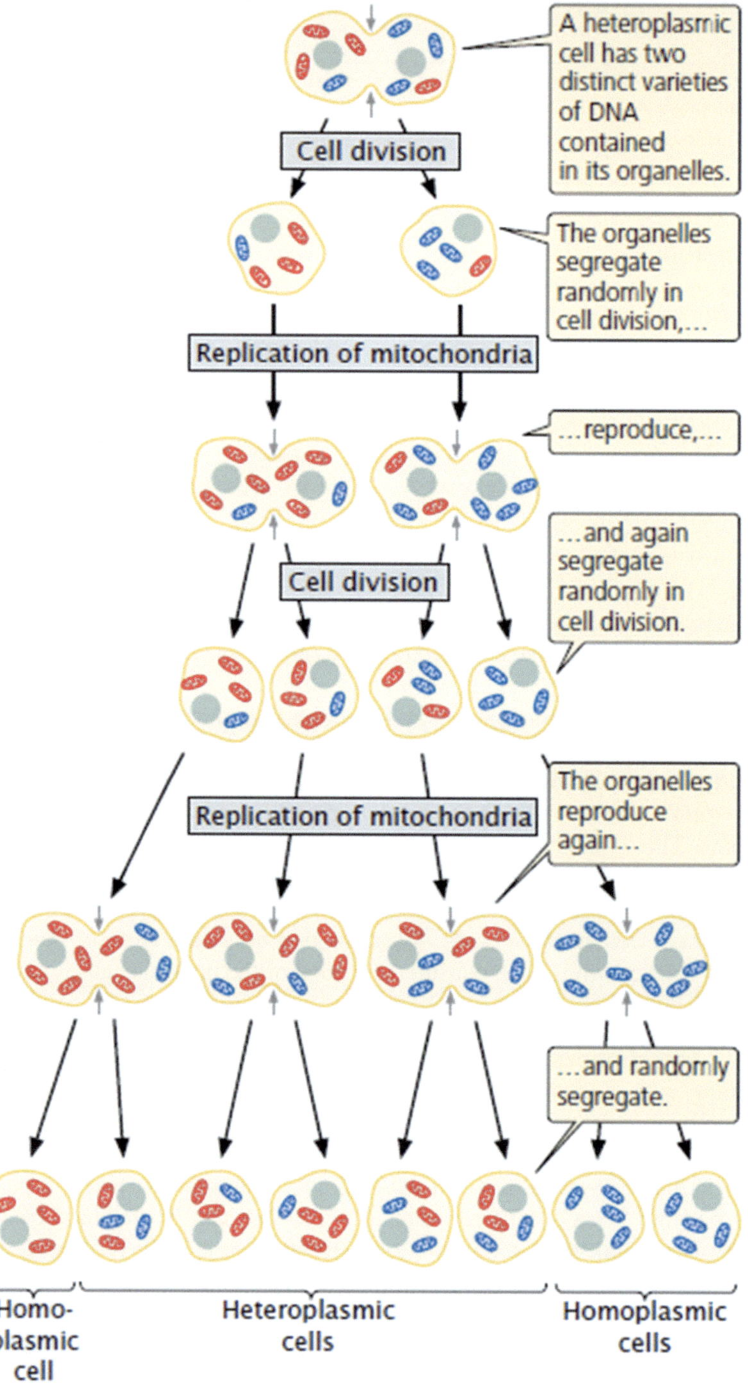

Fig. 3.41 Cytoplasmic segregation of mitochondria. In heteroplasmic cells, organelles randomly divide into the progeny. The diagram shows replicative segregation of mitochondria during mitosis. Similar process takes place in meiosis also

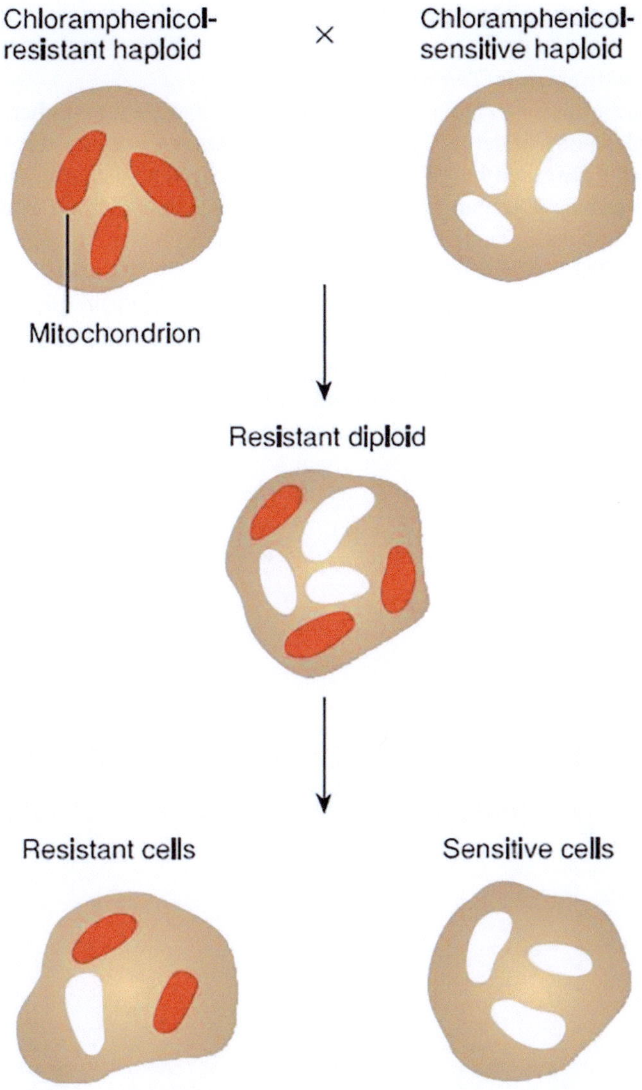

Fig. 3.42 Mitochondrial inheritance—chloramphenicol (antibiotic) resistance in yeast. Diploid cells formed from a cross between resistant and sensitive haploids produce both resistant and sensitive cells. Segregation of the resistance trait is non-Mendelian, and relies on random assortment of the mitochondria. Sensitive yeast does not possess resistant mitochondria. Resistant cells possess resistant mitochondria

unlike chromosomal gene inheritance, and random distribution of mitochondria during cell division such that a wild-type cell consisting of only sensitive mitochondria is formed is a possibility. Since only one to ten mitochondria are present per cell in some yeast, such random distribution can occur relatively at a high frequency resulting in yeast cells with only sensitive mitochondria.

3.5.4.3 Chloroplast-Mediated Inheritance

3.5.4.3.1 Variegation in *Zea mays*

As explained earlier in Sect. 3.5.1, cytoplasmic inheritance involves heteroplasmons. The heteroplasmons undergo cytoplasmic segregation in which the two types of organelles are distributed into different daughter cells. This is a process that mostly arises due to chance segregation at the time of cell division. A good example of such segregation can be seen in the variegated plants. Plants consisting of green and white patches are called variegated.

M. Rhoades studied variegation in corn (*Zea mays*) and the phenotype is controlled by a chromosomal locus called *iojap*. If the locus is homozygous for *iojap*, it inhibits development of proplastids into chloroplasts, thus causing variegation. Plastids affected by *iojap* lack ribosomes or ribosomal RNA and therefore are deficient in protein synthesis. Reciprocal crosses pertaining to this trait are illustrated in Fig. 3.43. The first cross between green female (*IjIj*) and variegated male (*ijij*) produces green F_1 (*Ijij*). Selfing of F_1 resulted in green (*IjIj* and *Ijij*) and variegated (*ijij*) with a genotypic ratio of 1:2:1, a typical Mendelian inheritance, with

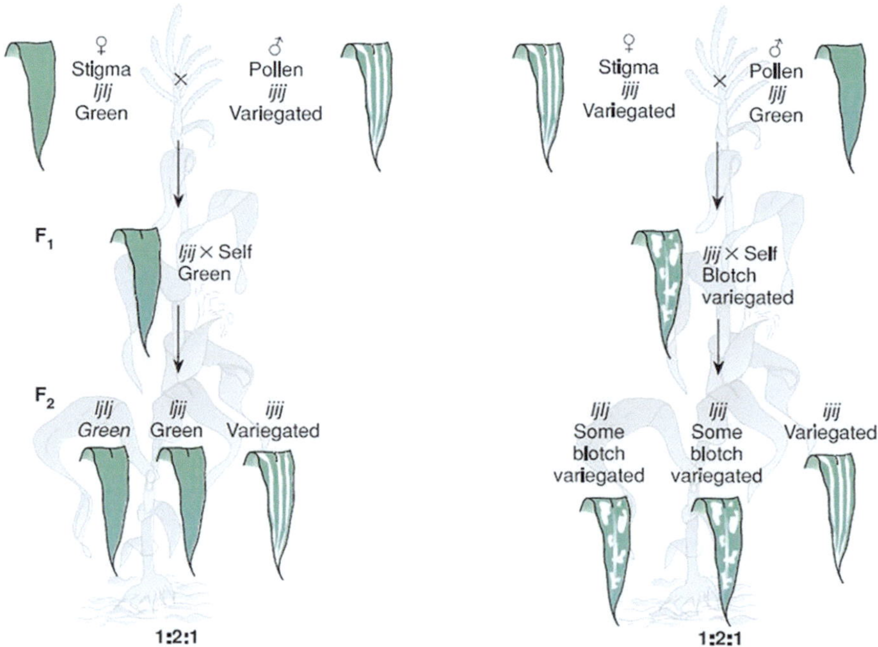

Fig. 3.43 Chloroplast inheritance—variegation in *Zea mays*. Reciprocal cross involving iojap gene for variegation in *Zea mays*. Variegation is induced by homozygous recessive, *ijij*. Blotch variegation shows irregular white areas instead of stripes seen in regular variegation. Heterozygous *Ijij* will be variegated if the mothers are variegated, because of transmittance of chloroplast from the mother. Pollen parents do not transmit chloroplast

variegation induced by the homozygous recessive genotype (*ijij*). However, a reciprocal cross between a variegated female (*ijij*) and green male (*IjIj*) resulted in blotch variegation in F_1 and F_2 plants carrying dominant *Ij* allele. These observations can be explained as follows: in the first cross, the ovule from the maternal parent carries normal chloroplast and pollen grains do not carry any chloroplasts; hence, normal chloroplasts are passed on to F_2 generation. However, in the F_2 generation, variegation is induced by the genotype *ijij*. In this case, chloroplasts belonging to the pollen parent are insignificant because they are not transmitted to the offspring. But in case of a reciprocal cross, where the female (ovule) parent is variegated, the ovule consists of abnormal proplastids in the cytoplasm. F_1 heterozygotes therefore inherit the cytoplasmic proplastids from the female parent, which remain unchanged even in the presence of normal dominant allele, *Ij*, leading to a blotchy variegation, with white spots consisting of colorless cells. Chloroplasts that get induced by the *ij* allele, to become proplastids, cannot revert back to normal type even if *Ij* allele is present. Evidences have suggested that *iojap* may bring about its effect by suppressing the chloroplast rather than causing a functional mutation.

3.5.4.3.2 Variegation in Four O'clocks

Maternal inheritance of variegation in *Mirabilis jalapa* (four o'clock plant) was discovered by Carl Correns in 1909. On the basis of the plant region on which the stigma parent (female parent) was located, he predicted phenotypic outcome with respect to color and variegation in the offspring. Flowers borne on white plant parts, irrespective of the pollen type that pollinated them, produced white colored plants. Similarly, flowers borne on green or variegated parts, irrespective of the pollen type that pollinated them, produced green or variegated plants, respectively. This clearly indicated maternal inheritance of variegation in *Mirabilis jalapa*. Figure 3.44 shows a frequently observed variegated leaf and branch phenotype in *Mirabilis jalapa*, Variegation observed in this plant is because of a mutant chloroplast gene. Mutation in cpDNA causes chloroplasts to be white in color, which in turn makes the branches and leaves white colored. Variegated branches display a mosaic of green and white patches. Therefore, these plants produce three types of branches—green, white, and variegated, and each branch produces flowers which consist of those respective compositions of chloroplasts. Whenever a cross is performed (Fig. 3.45), the maternal gamete determines leaf and branch color of the progeny. For example, an egg cell belonging to a flower on green branch consisting of normal chloroplast will produce all green progeny, regardless of the source of the pollen. Similarly, a white branch consisting of white chloroplasts produces all white progeny. But due to the lethality of the condition, descendants of white do not last beyond seedling stage.

The ova from variegated branches produce variegated zygotes consisting of both green and white chloroplasts (Fig. 3.45). Since these zygotes are heteroplasmic, they undergo cytoplasmic segregation. Division of such a zygote segregates green and white chloroplast into separate cells, which appear as distinct green and white sector characteristics of variegation. In this example, cytoplasmic segregation of an organelle population containing a mixture of genetically distinct chromosomes is clearly demonstrated.

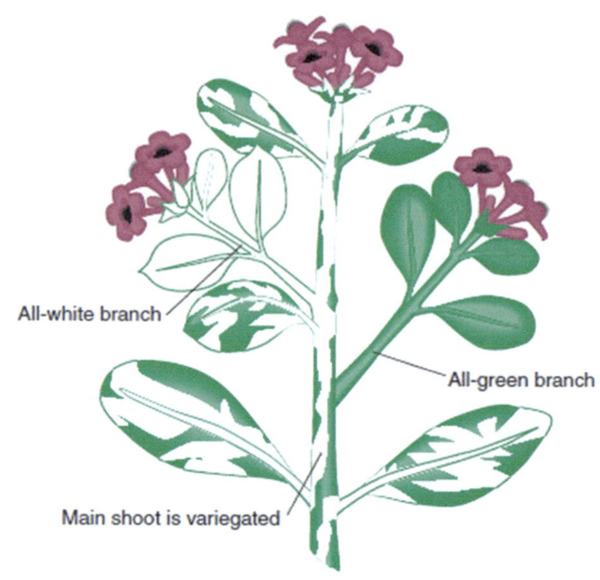

Fig. 3.44 Variegation in leaves of *Mirabilis jalapa*. Note that the plant consists of three types of branches, i.e., green, white, and variegated. Each of these branches gives rise to flowers that can be used for crosses

3.5.4.3.3 Antibiotic Resistance in *Chlamydomonas*

Chlamydomonas reinhardtii is a green alga, consisting of a single cell, and has been extensively studied to understand extrachromosomal inheritance. There are several advantages for choosing *Chlamydomonas* for these studies: (1) it consists of a single large chloroplast; (2) it can survive in culture techniques even if chloroplast is nonfunctional; and (3) it shows interesting non-Mendelian patterns of inheritance, associated with the mating type. Ruth Sager has extensively researched inheritance of streptomycin resistance in *Chlamydomonas*, and cells specific for antibiotic resistance can be selected by various means. Antibiotic sensitive normal cells get killed in a medium containing the antibiotic. But if these cells were initially grown in low antibiotic concentrations (100 g/mL), some cells exhibit resistance. Further, when these resistant cells are crossed with normal (sensitive) cells, 1:1 segregation of resistance phenotype is observed indicating the chromosomal inheritance of streptomycin resistance. If the experiment is repeated with higher concentrations of the antibiotic (500–1600 g/mL), resistant colonies will again be obtained, but a cross with wild-type will not result in a 1:1 segregation. *Chlamydomonas* cannot be distinguished into male or female sex, but are identified as mating types mt^+ and mt^-. Mating in *Chlamydomonas* occurs only between cells of opposite mating types. The mating type is encoded by a single gene locus consisting of a pair of alleles. A diploid zygote (product of fusion of two haploids of opposite mating types mt^+ and mt^-) undergoes meiosis to produce four haploid cells, of which two are mt^+ and two are mt^-. Resistance to higher concentration of antibiotics always segregates with mt^+ parent (Fig. 3.46), which is similar to maternal plastid inheritance in plants. Here it can be considered that mt^+ parent contributes cytoplasm to the zygote, while the mt^-

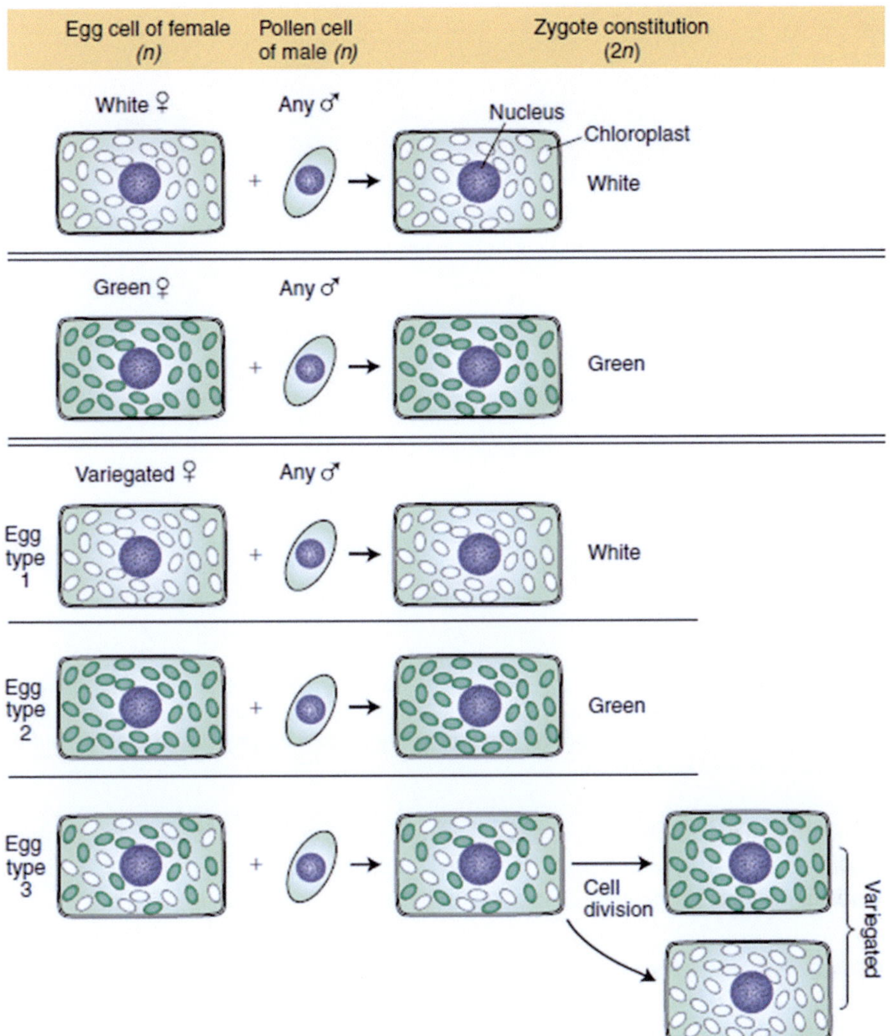

Fig. 3.45 Chloroplast inheritance in *Mirabilis jalapa*. Two different types of chloroplasts are shown in the figure, green and colorless/white. The chloroplast content of the female cells determines the type of offspring. Based on the chloroplast content and composition of the female branch, the offspring can be green, white, or variegated

parent is similar to the pollen parent. The mechanism underlying such extrachromosomal inheritance in *Chlamydomonas* involves preferential digestion of cpDNA of the mt^{-} parent.

Since Sager's discovery, several other mutations in *Chlamydomonas* have been discovered. All these have shown uniparental inheritance. All these mutations have linked with the chloroplast. It is important to note that in *Chlamydomonas* several

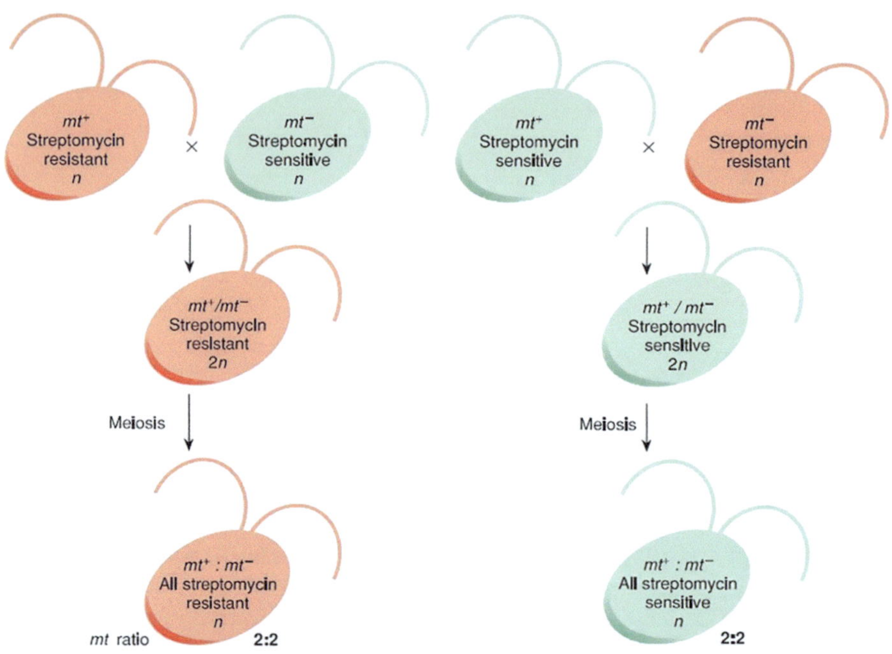

Fig. 3.46 Antibiotic resistance in *Chlamydomonas*. Streptomycin resistance is determined by the mt^+ parent. If the mt^+ parent is resistant (left), then all meiotic products and the heterozygous diploid will show resistance. If the mt^+ parent is sensitive (right), then all products of meiosis and the heterozygous diploid are sensitive

antibiotic resistance phenotypes are also transmitted via the mtDNA. But in these cases, it has been observed that it is always transmitted by the mt^- parent, which is opposite of what happens in chloroplast inheritance.

3.5.4.4 Infective Particles

3.5.4.4.1 Paramecium

Tracy Sonneborn discovered the cytoplasmic inheritance of a killer trait in *Paramecium*, a ciliated protozoan. *Paramecium* consists of two types of nuclei, a larger macronucleus and a smaller micronucleus. Micronuclei are two in number and are mainly involved in reproductive function. They consist of one macronucleus, which is polyploid in nature, and regulate the vegetative functions. *Paramecia* divide by binary fission, during which the micronuclei divide mitotically and macronuclei divide into halves by constriction.

Conjugation and autogamy are two processes wherein *Paramecia* undergo nuclear rearrangements of two types. The conjugation process involves coming together of two *Paramecia* belonging to different mating types, and the formation of a connecting bridge between them. The nuclear events that take place in each of the *Paramecia* after the formation of the connecting bridge are shown in Fig. 3.47.

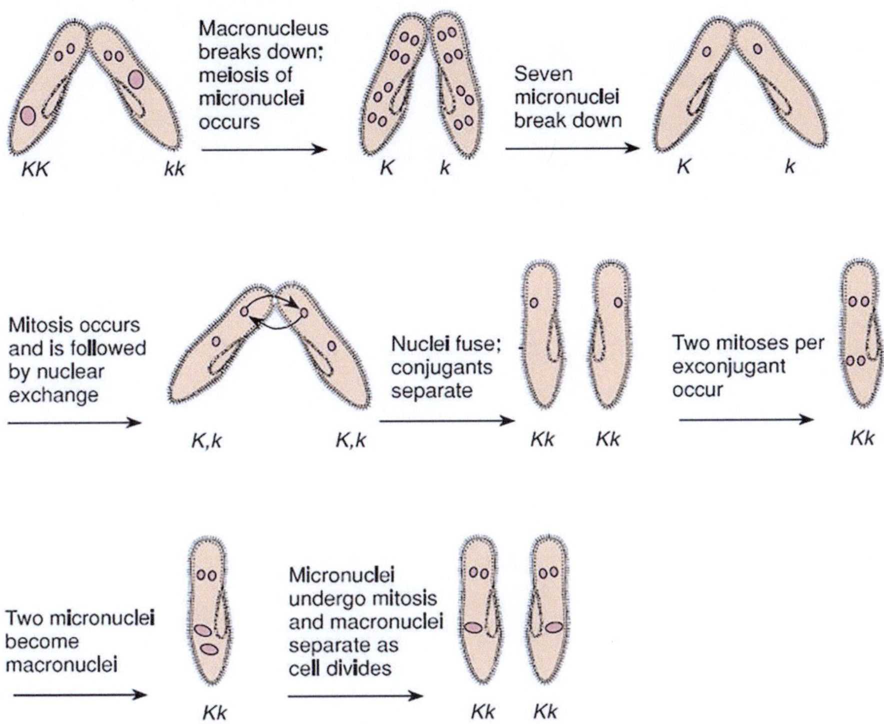

Fig. 3.47 Conjugation process in *Paramecium*. *K* and *k* are alleles of a gene present in micronuclei. Exconjugants of a conjugation between *KK* and *kk Paramecia* acquire *Kk* genotype

The process begins with temporary disintegration of the macronucleus in each cell, whereas micronuclei divide by meiosis to produce eight haploid micronuclei per cell. Out of the eight haploid micronuclei, seven disintegrate and the remaining one micronucleus divides by mitosis to form two haploid nuclei. At this stage, both the conjugant *Paramecia* consist of two haploid nuclei each. This is followed by a reciprocal exchange of nuclei between the two *Paramecia* that take place through the connecting bridge. After the reciprocal exchange, each *Paramecium* consists of one haploid nucleus of its own and one received from the other conjugant *Paramecium*. These two haploid nuclei undergo fusion to give rise to a diploid nucleus. After reciprocal exchange, the diploid nuclei of both conjugating cells are genetically identical. Two more mitoses occur in the diploid nuclei to give rise to four diploid nuclei in each cell. Two of these remain as micronuclei, whereas the other two become macronuclei. After the process is complete, the contact between the conjugant paramecia is released, and these separated cells are called exconjugants. These paramecia next undergo cell division, where the macronuclei separate, and the two micronuclei undergo one more mitosis before separation. At the completion of conjugation, each exconjugant gives rise to two daughter *Paramecia*. Each daughter *Paramecium* consists of one macronucleus and two micronuclei. Depending

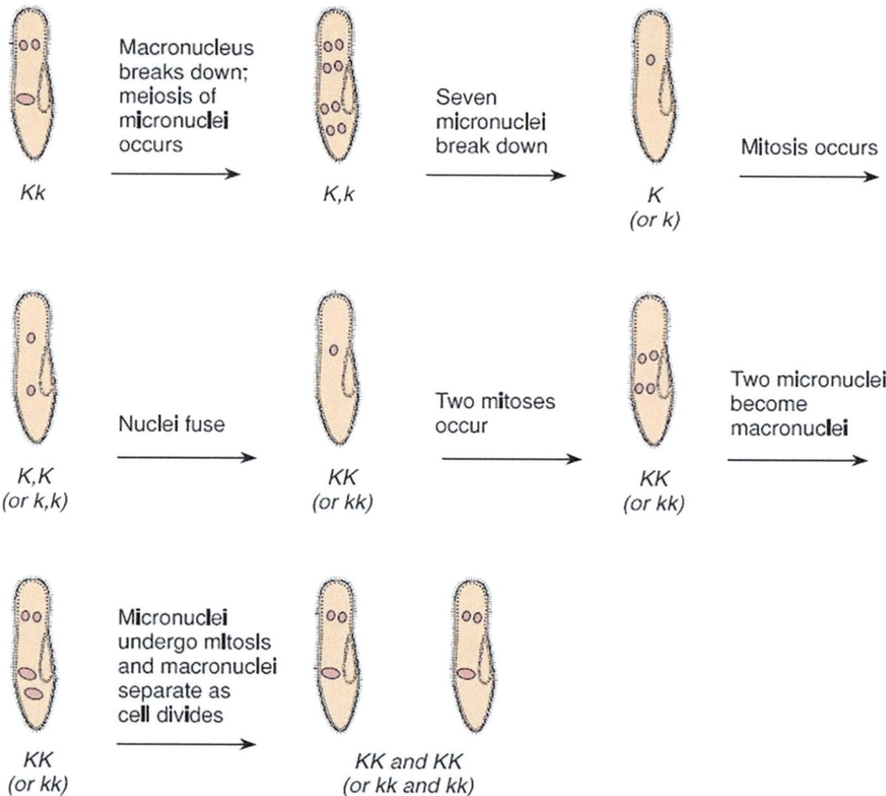

Fig. 3.48 Autogamy in *Paramecium*. *K* and *k* are alleles of a gene located in micronuclei. Autogamy in a heterozygote results in homozygosity for either *K* (*KK*) or *k* (*kk*)

particularly on the duration of contact (conjugation bridge) between the conjugating cells, cytoplasmic exchange may also occur in addition to exchange of nuclei. From Fig. 3.47, it can be seen that for a given pair of alleles (*K* and *k*), conjugation between two homozygous *Paramecia* (*KK* and *kk*) results in heterozygous (*Kk*) daughter *Paramecia*.

The process of autogamy involves a single *Paramecium* (Fig. 3.48). The nuclear events in autogamy are same as in conjugation, except for absence of nuclear reciprocal exchange, instead haploid nuclei of the same cell fuse to form the diploid product. If autogamy happens in a *Paramecium* heterozygous (*Kk*) for a given pair of alleles (*K* and *k*), after the completion of the process, the daughter cells become homozygous with respect to one of the alleles (*KK* or *kk*) with a 50% chance for each (Fig. 3.48).

3.5.4.4.2 Kappa Particles and Killer *Paramecium*

Sonneborn and his fellow researchers made some interesting observations when working with *Paramecium*. They observed that mixing of certain stocks of *Paramecium* resulted in killing of one of the stocks by the members of the other stock. Such *Paramecia*, which killed *Paramecia* of other stock, were called "killers," and those that got killed were called "sensitives." It was observed that sensitives could become temporarily resistant to killers during conjugation. If conjugation lasted long enough to allow cytoplasmic exchange, then both exconjugants became killers. This indicated that sensitives could acquire killer property if exchange of cytoplasm occurred. On the contrary, if such cytoplasmic exchange did not occur, then killers remained killers and sensitives remained sensitive. These observations indicated that transfer of certain cytoplasmic factors could be involved in conversion of sensitives to killers. Further studies by Sonneborn led to the discovery of cytoplasmic particles called kappa in the killer *Paramecia* (Fig. 3.49).

Initial studies on the killer trait could not establish any link between chromosomal genes and expression of killer phenotype in *Paramecium*. Later Sonneborn reported a case of conjugation in which the killer exconjugant of hybrid origin (originated from conjugation between killer and sensitive; see Fig. 3.47) underwent autogamy. Fifty percent of the *Paramecia* that were produced by autogamy had lost kappa particles and had become sensitive. Based on these results, he concluded that a particular nuclear gene (dominant allele *K*) is required to enable retention of kappa particles, and hence to maintain the killer status. This was subsequently verified by performing several crosses in *Paramecia*. Figure 3.50 shows the sequence of events that produce a killer *Paramecium* heterozygous for *K* allele (*Kk*) which, upon

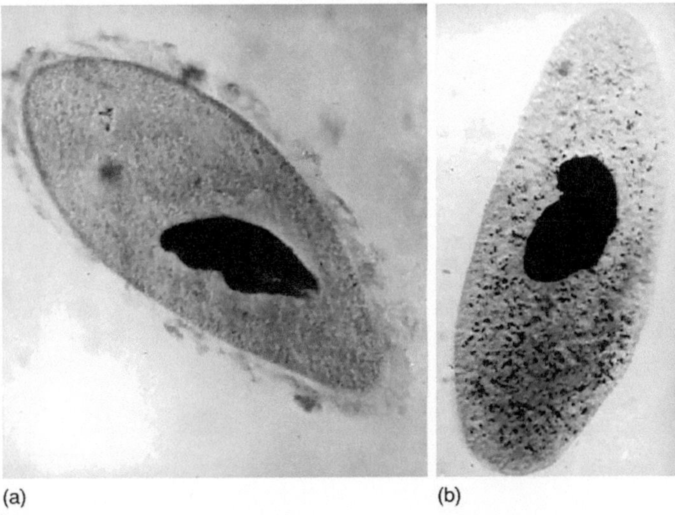

Fig. 3.49 (**a**) Normal (sensitive) *Paramecium* lacking kappa particles. (**b**) Killer *Paramecium*, containing kappa

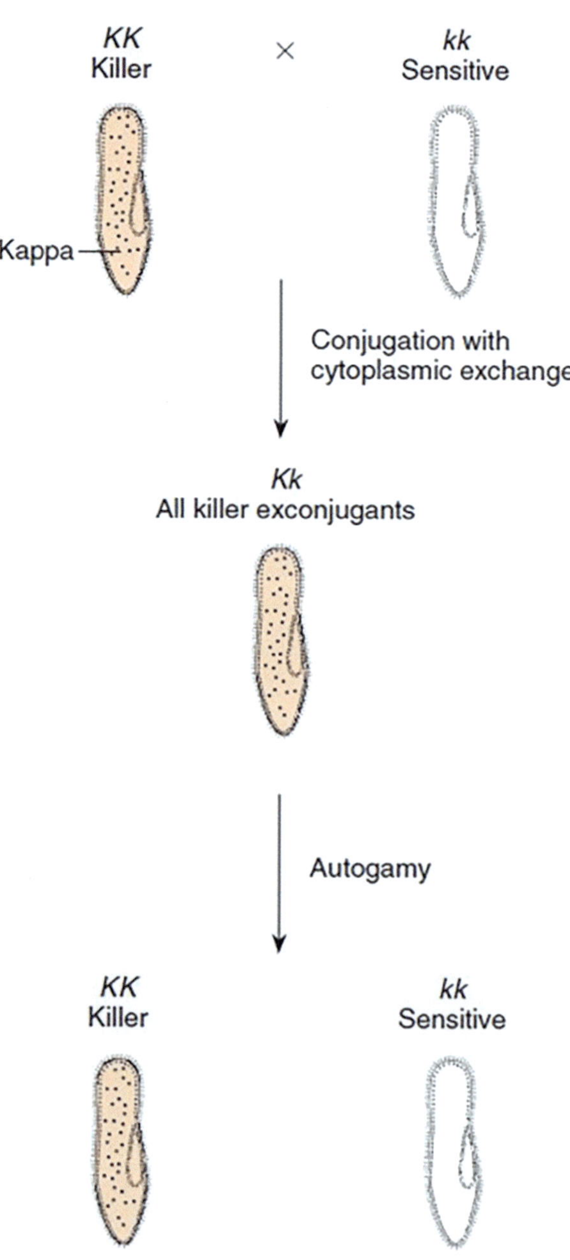

Fig. 3.50 Autogamy in heterozygous (Kk) killer *Paramecium*. There is a 50% chance of the heterozygote becoming either *KK* (killer) or *kk* (sensitive). The *Paramecium* with *kk* genotype becomes sensitive as it eventually loses the kappa particles

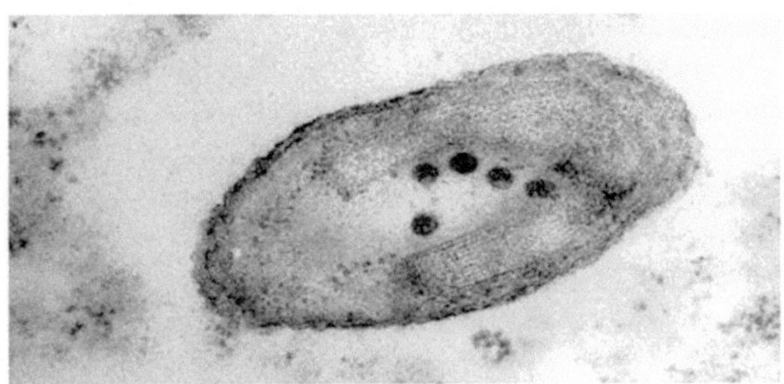

Fig. 3.51 A sectioned kappa particle (*Caedobacter taeniospiralis*) visualized under an electron microscope. The dark inclusions observed are kappa containing phage particles. Plane of the sectioning is through a rolled-up *R* body. Magnification 61,200×

autogamy, has a 50% chance to become either killer or sensitive. As seen in Fig. 3.50, *Paramecium* with genotype *kk* lose the kappa particles in the successive cell divisions and become sensitive. Kappa is presumed to be a bacterium as it possesses several prokaryotic features. Kappa was further studied by J. Preer and his colleagues, and was named *Caedobacter taeniospiralis*. Killer *Paramecia* execute killer effect by releasing a toxin called paramecin into its surrounding environment and the sensitives are killed.

Kappa are found in two visibly distinct forms, which can be distinguished when observed using bright phase contrast microscopy. There is a larger kappa called bright kappa, which consists of one or more discrete refractile regions. The other is a smaller kappa, called nonbright kappa, which does not contain refractile bodies (Fig. 3.51). These two forms are abbreviated as B kappa and N kappa, respectively. The refractile region is named R body. N kappa divide to produce more N kappa and are also highly infective. B kappa is associated with the killing activity. However, it is suggested that only a small portion of B kappa possess killing activity, and such B kappa are designated as killing or P kappa (P stands for paramecin).

3.5.4.4.3 Mate-Killer Infection and Mu Particles

Apart from kappa, *Paramecium* also harbors other infective agents. Another such infective agent is called the mate-killer infection. The killer cells in this case also possess visible, bacteria-like particles in the cytoplasm, and are called mu particles. Preer and colleagues named the bacteria *Caedobacter conjugatus*. The mode of action of the mu particles is different than that of the kappa, i.e., they do not produce any toxin, but instead kill their conjugating partners during conjugation. Like the *K*, allele for kappa, either *M*1 or *M*2 dominant alleles are shown to be required to maintain mu particles in the cytoplasm of *Paramecia*. Mate-killer become homozygous recessive (*m*1*m*1 or *m*2*m*2) by autogamy. In the homozygous recessive condition, all offspring eventually lose their mu particles, but are able to retain up to almost the eighth generation. By about the 15th generation, mu particles are present in only around 7% of the cells. The reduction in the quantity of mu over generations has been attributed to progressive dilution of a factor called metagon, which is known to be essential for the maintenance of mu in the cell. In homozygous recessive cells, metagon production is known to cease and hence results in loss of mu. These metagons appeared to be mRNA as they could be degraded by the enzyme RNase.

3.5.4.4.4 Sex-Ratio Phenotype in *Drosophila*

Another example of infective particles is found in *Drosophila*. The sex-ratio phenotype is found in *Drosophila* females, wherein such females mostly produce daughters and fewer sons. Several forms of sex-ratio phenotypes are known. There are forms that are inherited via chromosomal genes and also there are others which are nonchromosomal. Females with the nonchromosomal form also typically produce more number of daughters and fewer sons. An interesting feature of these nonchromosomal forms is that the sons do not transmit this trait to the next generation, whereas daughters transmit the sex-ratio trait to the next generations. This pattern is typical of cytoplasmic inheritance. About 50% of eggs from the sex-ratio females failed to develop, and cytoplasm from such eggs was collected and used to infect other females. A thorough examination of the cytoplasm of sex-ratio females has shown the presence of a spirochete (Fig. 3.52). Researchers could isolate the spirochete and cause sex-ratio phenotype in other female through infection with the isolated spirochetes. Therefore, it was established that the spirochete cytoplasmic inheritance caused sex-ratio phenotype in female *Drosophila*.

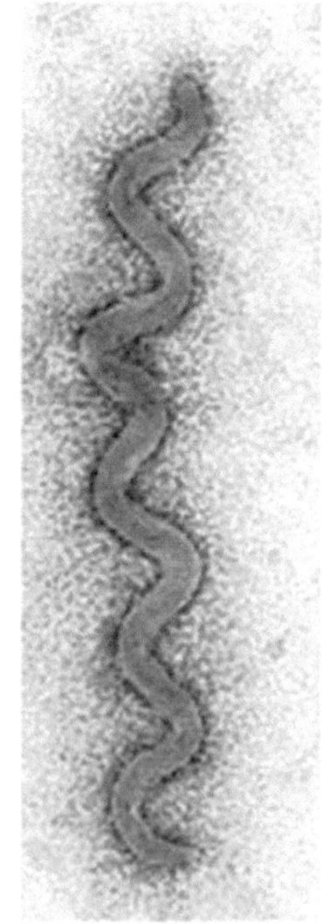

Fig. 3.52 Spirochete that transmits sex ratio, an extrachromosomal trait in *Drosophila*. Electron micrograph at magnification 22,700×

Box 3.3 Scientific Concept: Evolution of Approaches to Study Complex Diseases and Mapping of Gene Interactions (GIs)

GIs are now known to underlie various aspects of biology, including complex diseases, speciation, and evolution of sex. GIs are known to play a vital role in gene regulation, signal transduction, biochemical networks, and several other physiological and developmental pathways. These gene-gene interactions, put together, produce a phenotype. Also, it is known that some genes alter phenotype of other genes, the outcome of which is differences in penetrance and expressivity of those genes.

Several diseases in humans are a result of multiple genetic alterations and follow non-Mendelian inheritance patterns. In other words, these disease phenotypes occur due to a collection of mutations in several genes that individually control single phenotypic traits. Overall complex GIs make the disease phenotypes more than simply additive and hence raise the difficulty to point at the exact causative gene. In fact, GIs affect not only onset but also susceptibility of several complex diseases. These challenges have paved the way for new approaches to study these complex human diseases.

After completion of the Human Genome Project in 2003, a radical new idea of health called personalized medicine emerged. Personalized medicine seeks to assess the risk for a disease and determine appropriate treatment using one's genetic information obtained by sequencing. It was assumed that as more and more people got their genomes sequenced, discovery of disease-related genes would also happen alongside. Although this was true to an extent, it also increased the complexity of genetic data. As a result, even after 16 years of sequencing, where tens of thousands of genomes have been sequenced, gathering meaningful inference from the large volume of available genomic data is yet a major challenge.

The genome data also threw light on the vast genetic diversity prevalent in the human populations. It has been observed that these diversities bring about variability in the disease inheritance patterns of individuals. The situation is further complicated by environmental factors such as diet, lifestyle, etc., which also influence expression of a disease trait. Progressively, it has been realized that susceptibility to complex diseases such as heart disease, schizophrenia, etc. is an outcome of several combinations of subtle genetic changes scattered across the genome and not because of alterations in a single gene. There are also cases in which a single gene variant is potent enough to cause a disease, e.g., cystic fibrosis, hemophilia, etc. Even in such cases, two people affected by the same disease variant can experience vastly variable disease severity. Even more surprisingly, sequencing studies have identified people who carry damaging mutations, but are perfectly healthy. In such individuals, other unknown genetic variants within their genomes could be preventing disease expression. These observations clearly show that GIs play a crucial role in

(continued)

Box 3.3 (continued)
disease susceptibility and expression. Therefore, in addition to sequence data, it is also important to map the GIs so that the basis of a disease can be deciphered.

One of the primary challenges of modern genetics is to decipher the complexity involved in genotype to phenotype relationship, and to be able to apply the understanding to predict trait heritability. Traditional methods used to study heritable traits or diseases relied on pedigree construction, in which traits are mapped to the family members, revealing inheritance patterns. With the emergence of the molecular genetics, linkage mapping approaches were used to identify stretches of DNA in the human genome that were co-inherited with certain disease traits. This was done by using genomic landmarks such as single nucleotide polymorphisms (SNPs), as points of reference. Hence, linkage mapping has been a very useful strategy for studying genes involved in complex diseases such as type I diabetes. In spite of being useful, linkage mapping and other related approaches were limited with respect to their extent of use. These methods could not be used to study traits influenced by environmental factors, having low heritability, or affecting restricted groups of people. To overcome these limitations, newer strategies were evolved, which include whole-genome approaches, where whole-genome sequences of affected individuals are compared with normal control population. The differences between the two genomes are assumed to indicate possible "hot spots" for identifying genes causing those disease traits. Initially, sequencing approaches were not considered feasible due to excessive costs involved. But this scenario changed with advances in DNA sequencing technology and creation of comprehensive databases that recorded genome information. This resulted in application of whole-genome association studies for complex human disease traits. An interesting outcome of these studies was that scientists continually discovered new gene associations for complex disorders.

Currently, several tools are available to analyze the genotype-phenotype relationships, especially with respect to human diseases. Next-generation sequencing technologies have been applied to create a database of the millions of genetic variations in humans. This data, when used in combination with population-based registries and electronic health records, are producing extraordinary genomic and phenomic resources for such analysis. Genome-wide association studies (GWAS), scanning genomes of patient and healthy controls, have discovered thousands of genetic variations or mutations that are associated with diseases. As of 2018, the NHGRI (National Human Genome Research Institute) EBI (European Bioinformatics Institute) Catalog reported >50,000 associated loci for >3000 unique traits (https://www.ebi.ac.uk/gwas/). Most of these genetic variations are associated with common diseases that affect human populations globally.

(continued)

Box 3.3 (continued)

Although we currently understand the chief importance of GIs and also possess sequencing technologies that can comprehensively genotype entire genomes, mapping of vital GIs associated with naturally occurring variations within genomes of individuals remains extremely difficult. In this regard, many inbred model systems are available. These include organisms such as yeasts and worms. Also, fruit fly and mammal cell cultures are used as model systems. These systems provide an experimental setup for systematically mapping GIs. Understanding of gene interaction and their impact on the phenotype of the organism has been enabled by large-scale studies in such model organisms.

Mapping of GIs in the inbred animal model systems will enable an understanding of the fundamental principles of genetic networks, which in turn may make it possible to map GIs and networks in natural outbred populations taking into account the existing natural variations in individuals. The budding yeast, *Saccharomyces cerevisiae*, is the most well-studied model systems for deciphering GIs. The most widely applied method for analysis of yeast cell phenotypes involves independent gene mutations, and applies a mathematical tool for evaluation of GIs. In this method, the expected double mutant phenotype is considered to correspond to the product of those respective single mutant phenotypes. The scoring of GIs is done by distinguishing double mutants which deviate from the expected values in their phenotypes.

Studies in yeast cells emphasized the need to look beyond effect of individual genes in order to effectively understand their function. Yeast has been a suitable model due to its relatively smaller genome consisting of 6000 genes and already existing extensive database. In addition, many of the genes found in yeast cells are also found in humans. The first map showing the global genetic interaction network was created for a yeast cell. The model explains how thousands of genes synchronize with one another to perform life functions. Just as human societies are organized hierarchically from local communities up to countries, genes in cells form hierarchical networks to become organized into a cell. Instead of probing for single genes that underlie diseases, it would be more suitable to look for gene pairs. Studies to unearth the rules involved in combinatorial gene function have been performed in yeast cells. Using the yeast system, scientists have worked out how genes work in pairs. Studies for mapping GIs in yeast initially involved removing all possible gene pairs (18 million of them) and studying their effect. After completely studying gene pairs, the next step was to study trigenic combinations (36 billion possible combinations). It was found that similar to interactions between gene pairs, trigenic interactions also primarily occurred between functionally related genes. It was observed that such interacting genes coded for parts of the same molecular machinery or same part of a cell.

(continued)

Box 3.3 (continued)
Importantly with trigenic interactions, more unexpected partnerships between genes having unrelated functions and involved in different bioprocesses were observed.

If a similar study had to be performed with the human genome, it would involve examining approximately 200 million possible gene pairs for their association with a disease. But this has been relatively simplified, because of the know-how obtained from the yeast map. This has enabled researchers to start mapping genetic interactions in human cells. Recent advances in gene editing have made it possible to map relationships between disease genes by removing gene combinations from human cells.

Box 3.4 Scientific Concept: GI and Cancer
A characteristic feature of cancer is its heterogeneity. The disease consists of hundreds of distinct subtypes and varying genetic backgrounds. Identifying unique GIs specific to a particular cancer and its subtypes is of great significance. The Achilles heel of cancer therapy is systemic damage; hence, targeted therapy is the most pressing goal of current research. Genetic network analysis standardized in yeast has been now extended to human cells. It has been used to identify cancer cell line-specific GIs that can uncover unique functional relationships specific for those cancer types. Such unique GIs represent prospective therapeutic targets which can be inhibited either genetically or chemically to kill only cancer cells, without harming identical normal cells because they lack the specific alteration. For example, ovarian and breast cancer cells harboring BRCA1 or BRCA2 gene mutations are highly susceptible to PARP1 (poly (ADP)ribose polymerase 1) inhibition.

Using the cutting-edge CRISPR-Cas9 genome editing, researchers have developed a breakthrough technology, which allows faster creation of novel cancer treatments. This new technology was applied to identify a new prospective therapeutic target in a type of pancreatic cancer. This new treatment destroys cancer cells, by exploiting genetic faults inherent and unique to the type of pancreatic cancer. Function of every single gene expressed in pancreatic cancer was probed by the researchers to determine one of the receptors called Frizzled-5 is particularly required by mutant pancreatic cancer cells for their growth. Normally, Frizzled-5 activates a signaling pathway that controls cell division, cell differentiation, and cell death. However, it initiates tumor development when mutated or deregulated. After identifying the important role Frizzled-5 receptor plays in promoting growth of pancreatic cancer, the team of researchers quickly developed an antibody drug against the receptor to

(continued)

Box 3.4 (continued)
inhibit growth of these cancer cells. Thus, developed antibody has been shown to effectively kill cancer cells derived from patient tumor samples, and it was also shown to cause tumor shrinkage in mice. Most importantly, the antibody caused tumor shrinkage in mice but without harming the surrounding healthy cells. These studies hold promise of developing highly effective therapeutics that target only cancer cells, without causing systemic damage.

3.6 Summary

- When a gene exists in more than two alternative forms, it is called multiple allelism. One of the best examples of the multiple allelic series in humans are those that determine the ABO blood grouping system, and it is of great clinical importance, particularly in transfusion medicine. The multiple allelic series of the ABO system determine the type of antigens on the RBCs, which in turn determines an individual's blood group. The multiple allelic series of the ABO blood group locus consists of three alleles, I^A, I^B, and *i*. Bombay blood group and its variant para-Bombay are rare blood phenotypes.
- The dominance concept states that among the pair of alleles in a genotype, only the dominant allele expresses itself in the phenotype and recessive allele gets suppressed or hidden. A phenomenon where the heterozygote has an intermediate phenotype compared to its homozygous parents is termed as incomplete dominance. Codominance refers to a condition in which the heterozygote expresses the phenotype of both the alleles equally.
- An allele that usually causes death, at an early developmental stage, and often before birth is called as a lethal allele. Such an allele causes genotypes to be lost from the progeny of a particular cross.
- In organisms with XY chromosomal sex determination, the sex chromosomes are heteromorphic unlike the autosomes which are homomorphic. Therefore, the patterns of inheritance of genes located on heteromorphic sex chromosomes are different when compared with autosomal inheritance. X-linked pattern of inheritance was first demonstrated in *Drosophila* by T. H. Morgan in 1910. Hemizygosity causes a recessive allele to be expressed, even if present in single copy. Such a phenomenon is called pseudodominance. Nonreciprocity is another important feature of sex-linked inheritance.
- Phenotypic appearance of the genotypically determined traits is called as penetrance. Not all genotypes are able to "penetrate" the phenotype. However, most genotypes show complete penetrance. However, certain genotypes, especially those that code for developmental traits, frequently exhibit incomplete penetrance.

- The term expressivity is particularly used when a trait is not uniformly expressed among individuals that show a particular trait. Several developmental traits in addition to being incompletely penetrant also exhibit variable expressivity, ranging from mild to extreme.
- The function and expression of a gene is very much dependent on its location or position in the genome/chromosome, to an extent that a change of location or position can alter its function. In several cases, such a repositioning of the gene affects its expression level or, in certain cases, it alters its ability to function; this is called position effect.
- Any given phenotype can be influenced by several genes and conversely one gene can influence many phenotypes. When a mutant gene affects many aspects of the phenotype, it is said to be pleiotropic (Greek for "to take many turns"). Such a phenomenon is called pleiotropy, and the various effects caused are called pleiotropic effects.
- Although genes often show independent assortment, they are not independent in their phenotypic expression. In several cases, a gene at one locus influences the expression of a gene at another locus. This kind of interaction among genes at different loci (nonallelic genes), which affects the phenotypic outcome, is termed gene interaction. Classically, epistasis is defined as a gene interaction where one gene overrides the influence of another gene at a different locus in such a way that its phenotype is suppressed.
- Patterns of inheritance are typically of three types, and are categorized based on the gene location: *autosomal inheritance* is the inheritance of a gene located on the autosomes; *sex-linked inheritance* is the inheritance of a gene located on the sex chromosomes; and *cytoplasmic inheritance* is the inheritance of a gene located on organelle chromosomes such as chloroplast (cpDNA) and mitochondria (mtDNA).
- Extranuclear inheritance indicates transmission of characters through factors that reside outside of the nucleus. It is also called as non-Mendelian inheritance because Mendel's laws of heredity are not applicable to their inheritance patterns. Some of the other terminologies used include extrachromosomal, cytoplasmic, or nonchromosomal inheritance. Cytoplasmic inheritance involves several factors such as cellular organelles (e.g., chloroplast and mitochondria) containing their own DNA, and parasitic or symbiotic particles (infective particles) that reside in the cytoplasm and possess their own genetic material. Any given phenotype in cytoplasmic inheritance is transmitted to the progeny particularly by the female parent, but not the male parent.

Chromosome Mapping in Eukaryotes

4

Rohini Keshava

Construction of a series of descriptions indicating position and spacing of characteristics such as genes and other DNA sequence features along a chromosome is called chromosome mapping. The proposition of the chromosomal theory of inheritance by Sutton in 1903 led to the accumulation of evidence that genes are located on chromosomes. Morgan, through his *Drosophila* breeding experiments, determined location of white-eye gene locus on the X chromosome, and established chromosomes as carriers of genetic information. Also, it soon became obvious that the number of genes in any organism exceeded the number of chromosomes, suggesting location of several gene loci on an individual chromosome. Further studies on inheritances deviating from Mendelism, cell division (meiotic and mitotic) processes, and linear structure of eukaryotic chromosomes led to the concepts of genetic linkage, crossing over, and recombination.

All these together contributed toward the development of the chromosome mapping techniques. Mapping of positions of the genes on a chromosome was key to analyzing their functions. Although presently complete genomes are being sequenced, chromosome maps are still relevant. This is so because the functions of most genes identified through sequencing of genomes are unknown. Hence, it is required that the genes thus identified be correlated with information derived by their phenotypic analysis. In this regard, the chromosomal maps play a very vital role.

4.1 Linkage, Recombination, and Crossing over

The law of independent assortment was the second law of inheritance proposed by Mendel. This law stated that pairs of alleles segregate independently, i.e., if two hypothetical genes A and B are considered, inheritances of alleles of gene A and B

R. Keshava (✉)
Ramaiah University of Applied Sciences, Bangalore, India

D. Kar, S. Sarkar (eds.), *Genetics Fundamentals Notes*,
https://doi.org/10.1007/978-981-16-7041-1_4

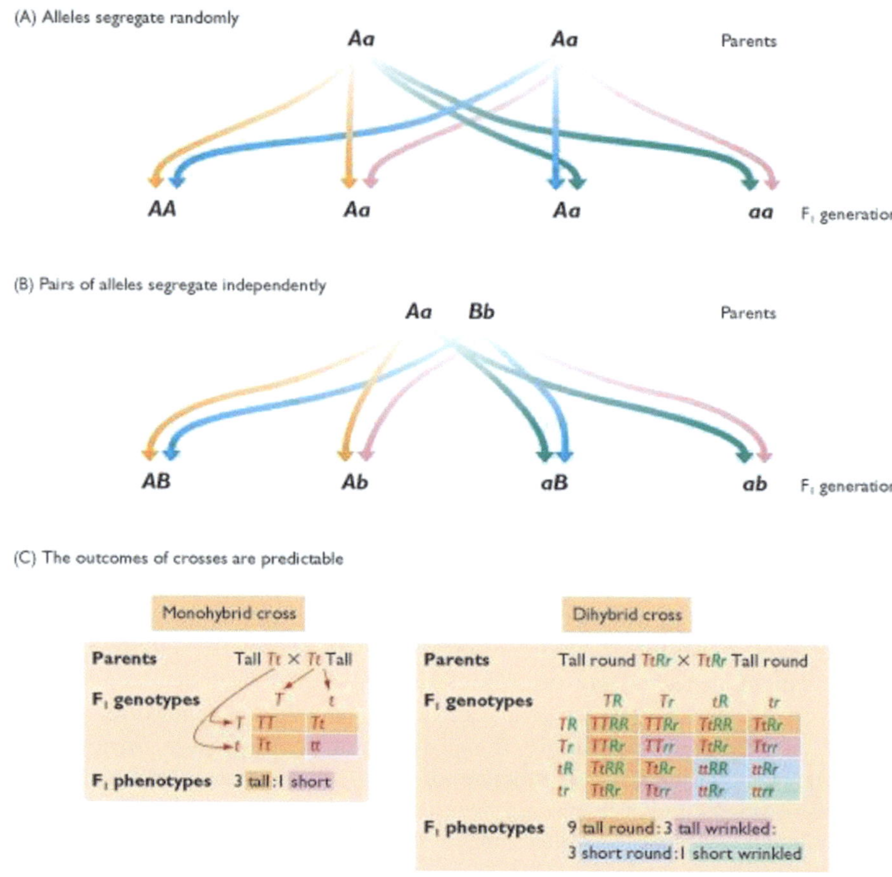

Fig. 4.1 Mendel's law of independent assortment. (Top) The second law of inheritance proposed by Mendel states that allelic pairs segregate independently. The figure depicts a parent heterozygous for a pair of genes A and B. It can be seen that among the F_1 generation, each member has equal chance of inheriting either allele of each of these genes from the parent. (Bottom) The independent segregation of the allelic pairs result in predictable outcome of the genetic crosses as depicted by the 9:3:3:1 phenotypic ratio of the dihybrid cross

are mutually exclusive and do not influence one another. Going by this law, it is possible to predict the outcomes of genetic crosses (Fig. 4.1).

Followed by the rediscovery of Mendel's work in 1900, it was soon observed that in meiotic cell division, pairing of homologous chromosomes takes place and that individual chromosomes of each pair segregate into separate daughter cells. These observations led to the assumption that homologous chromosomes behaved as whole units. Therefore, it was expected that all genes located on one chromosome are transmitted together (Fig. 4.2) without undergoing independent assortment. This concept was called as complete linkage.

However, the observations made by various scientists indicated otherwise. It was observed that pairs of genes either showed independent assortment, as it was

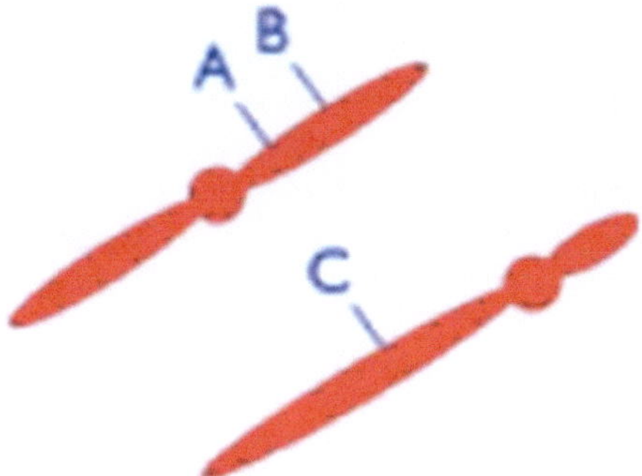

Fig. 4.2 Genes on the same chromosome were expected to show complete linkage. Since A and B are located on the same chromosome, they were expected to be always inherited together and hence show complete linkage. Therefore, the law of independent assortment was not applied to genes A and B, but was applicable to A and C or B and C as they were located on separate chromosomes

expected for genes present on different chromosomes, or showed partial or incomplete linkage (i.e., they were inherited together only sometimes, while at other times, they were apart) (Fig. 4.3).

Therefore, linkage or genetic linkage can be defined as the tendency of genes closely located on a chromosome to be frequently inherited together. The linkage phenomenon was first discovered in 1905, by William Bateson, Edith Rebecca Saunders, and Reginald C. Punnett (Fig. 4.3). Studies on the sweet pea plant indicated linkage between flower color and pollen shape. In 1911, studies conducted on inheritance patterns of fruit flies, *Drosophila melanogaster*, by Thomas Hunt Morgan, revealed that the eye color trait was linked with the sex of the fly. Morgan examined two genes on the X chromosome of *Drosophila.* One was a mutant gene for white-eye color, and the other was a mutant gene causing miniature wings. Morgan observed that the linkage was incomplete, i.e., he found that in female flies *white* and *miniature* alleles generally tended to remain together but however some recombinant X chromosomes with new combinations were produced.

These observations resulted in the emergence of the genetic linkage concept, which describes how two closely associated genes located on the same chromosome are often inherited together. Incomplete linkage is the general rule for genes located on the same chromosome. The reason for this incomplete linkage is the phenomenon of crossing over during meiosis. Crossing over leads to genetic recombination. Recombination during meiosis occurs due to the exchange of chromosomal fragments between non-sister chromatids.

4.1.1 Determination of Linkage

Bateson and Punnett's study in sweet pea (*Lathyrus odoratus*) included the gene influencing flower color (*P*, purple, and *p*, red) and the gene affecting the pollen grain shape (*L*, long, and *l*, round). Pure lines of the plants producing purple flowers

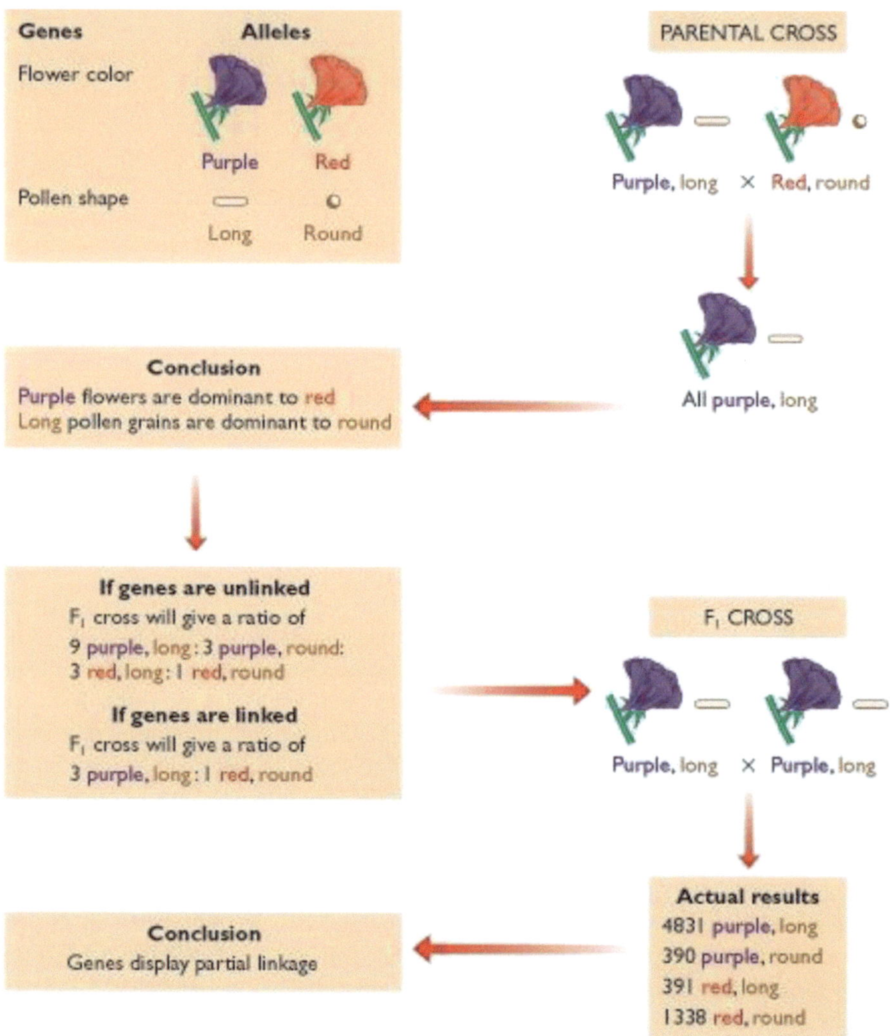

Fig. 4.3 Incomplete or partial linkage. Studies on inheritance conducted in the early twentieth century by Bateson, Punnett, and Saunders with sweet pea. The cross depicted in the figure shows that a result typical of a dihybrid cross is obtained in the parental cross, wherein all F_1 plants have the dominant phenotypes, viz., flowers with purple color and pollen grains that are long. But selfing of F_1 neither yielded an independent assortment ratio of 9:3:3:1 nor the complete linkage ratio of 1: 1. The unusual ratios obtained indicated partial linkage

and long pollen (*P*/*P*|·|*L*/*L*) were crossed with pure lines producing red flowers and round pollen grains (*p*/*p*|·|*l*/*l*). The F_1 heterozygotes (*P*/*p*|·|*L*/*l*) thus obtained were further selfed to obtain the F_2 plants. Figure 4.3 and Table 4.1 show the phenotypes and their proportions in the F_2 generation.

Although it was a dihybrid cross, the phenotypic ratio of the F_2 phenotypes deviated extremely from the expected 9:3:3:1 ratio as observed in the Mendelian

Table 4.1 Bateson and Punnett's breeding experiments in sweet pea by F_2 generation phenotypes

	Number of progeny	
Phenotype (and genotype)	Expected from independent assortment (9:3:3:1 ratio)	Observed
Purple, long ($P/–\|\|\cdot\|\|L/–$)	3911	4831
Purple, round ($/–\|\|P\cdot\|\|l/l$)	1303	390
Red, long ($p/p\|\|\cdot\|\|L/–$)	1303	393
Red, round ($p/p\|\|\cdot\|\|l/l$)	*435*	*1338*
	6952	6952

crosses. Also, the observed ratios could not be explained as modification of the Mendelian ratio. It was observed that the number of two phenotypic classes, i.e., the purple, long and the red, round, was larger than expected.

4.1.1.1 Coupling and Repulsion Hypothesis

In an attempt to explain the basis of the observations made in sweet pea, Bateson and Punnett hypothesized that the F_1 plants had produced more numbers of gametes with $P|\cdot|L$ and $p|\cdot|l$ combination than would be otherwise produced by independent assortment. Since these genotypes were the same as the gametes of the original pure lines, it was suggested that physical coupling of the dominant alleles *P* and *L* and the recessive alleles *p* and *l* could not have allowed their independent assortment in the F_1 generation. However, they could not explain the basis of this coupling phenomenon.

The proof and understanding of the coupling hypothesis put forth by Bateson and Punnett came with the application of *Drosophila* as a genetic tool. Thomas Hunt Morgan performed experiments with *Drosophila* and found similar deviation from Mendel's law of independent assortment. Morgan studied two autosomal genes, the genes affecting eye color (*pr*, purple, and pr^+, red) and wing length (*vg*, vestigial, and vg^+, normal) in *Drosophila.* Wild-type alleles of both genes are dominant. Morgan mated $pr/pr|\cdot|vg/vg$ flies with $pr^+/pr^+|\cdot|vg^+/vg^+$ and then performed a test cross of the F_1 double heterozygote females with double recessive male: $pr^+/pr|\cdot|vg^+/vg||♀||\times||pr/pr||\cdot||vg/vg||♂$.

Test cross is an extremely important tool used to reveal the gametic composition of a heterozygous offspring of the F_1 generation. This is so, because in such a cross, one parent called the tester (homozygous recessive for the concerned trait) contributes gametes consisting of only recessive alleles, and hence the phenotypes of the offspring reveal the gametic contribution of the other heterozygous parent. Therefore, for analysis only the meiosis in one parent (heterozygote) needs to be considered, thus simplifying the process. This is in contrast to analyzing progeny from an F_1 selfing experiment, where meiosis in both male and female parents has to be considered. The results obtained by Morgan are as follows, and it can be observed that the F_2 classes are specified by the alleles contributed by the F_1 female:

As seen in Table 4.2 the number of progeny deviated considerably from the Mendelian test cross ratio for independent assortment 1:1:1:1, thus indicating the coupling phenomenon among genes. The two largest groups were of $pr^+|\cdot|vg^+$ and

Table 4.2 Progeny from the test cross between a F_1 double heterozygote female ($pr^+/pr|\cdot|vg^+/vg|$) and a double recessive male ($pr/pr||\cdot||vg/vg$)

Genotype	Number of progeny
$pr^+||\cdot||vg^+$	1339
$pr||\cdot||vg$	1195
$pr^+||\cdot||vg$	151
$pr||\cdot||vg^+$	154
Total =	2839

Table 4.3 Progeny from the test cross between a F_1 double heterozygote female ($pr^+/pr|\cdot|vg^+/vg$) and a double recessive male ($pr/pr||\cdot||vg/vg$)

Genotype	Number of progeny
$pr^+||\cdot||vg^+$	157
$pr||\cdot||vg$	146
$pr^+||\cdot||vg$	965
$pr||\cdot||vg^+$	1067
	Total = 2335

$pr|\cdot|vg$. The test cross clearly indicates the combination of alleles within gametes of one of the parents in F_1. Thus, it enabled clear observation of the coupling phenomenon. The test cross also reveals a 1:1 ratio within the two parental and the two nonparental types.

The crossing experiment was repeated by changing the allelic combinations of the parents and hence the gametes. Further test cross of F_1 females was carried out:

$$\mathrm{P} \rightarrow pr^+/pr^+||\cdot||vg/vg||\times pr/pr||\cdot||vg^+/vg^+||$$

$$\downarrow$$

$$F_1 \rightarrow pr^+/pr|\cdot|vg^+/vg||$$

The progeny obtained as a result of the test cross were as follows:

The results of this test cross also clearly deviated from the Mendelian dihybrid test cross ratio of 1:1:1:1. In these results, the two largest groups were $pr^+||\cdot||vg$ and $pr||\cdot||vg^+$ (Table 4.3).

Similar to the earlier cross, in this too, the allelic combinations originally contributed to F_1 generation by parents form the most frequent groups in test cross progeny. The term repulsion was used by Bateson and Punnett to explain such phenomenon, as it appeared that the nonallelic dominant alleles "repelled" each other. This was in contrast to what was observed in case of coupling, where the dominant alleles appeared to stay together.

4.1.1.2 Coupling and Repulsion Explained

Morgan proposed that the genes regulating both phenotypes are situated *on the same homologous pair of chromosomes*. Hence, both *pr* and *vg* were suggested to be physically present on the same homologous chromosome and passed on from one parent; likewise, pr^+ and vg^+ were passed on from the other parent (Fig. 4.4). In case of repulsion, *pr* and vg^+ were suggested to be located on one parental chromosome and pr^+ and *vg* were located on the other. Therefore, repulsion could be visualized as another case of coupling, but the coupling was between a dominant allele of one

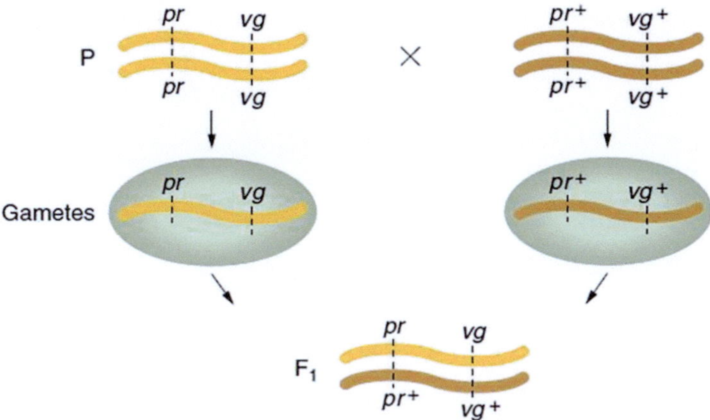

Fig. 4.4 Inheritance pattern of two allelic pairs located on the same homologous chromosome pair

gene and recessive of another. Hence, the occurrence of large number of parental allelic combinations in the progeny could be explained. However, this could not explain the occurrence of nonparental allelic combinations.

The terms coupling and repulsion are presently used to denote two types of linkage conformations in a double heterozygote. They are depicted as follows:

Coupling conformation $\dfrac{pr \quad vg}{pr^+ \quad vg^+}$

Repulsion conformation $\dfrac{pr \quad vg^+}{pr^+ \quad vg}$

The linkage between two dominant alleles or between two recessive alleles is termed as coupling. The term repulsion is used to indicate linkage between a dominant allele and a recessive allele. A test cross has to be performed or the genotypes of the parents have to be analyzed in order to determine the linkage conformations of a given double heterozygote. Alternatively, the coupling and repulsion phases are also termed respectively as the cis and trans (Fig. 4.5) arrangement of alleles on chromosomes.

4.1.1.3 Mechanism of Crossing over and Recombination

The physical process that underlies the genetic recombination is crossing over of non-sister chromatids of the homologous chromosomes. In an attempt to explain the physical basis of coupling and repulsion, Morgan hypothesized that pairing of homologous chromosomes occurred in meiosis, and occasionally chromosomes exchanged segments through an event of physical crossing over between them (Fig. 4.6). This phenomenon resulted in the new combinations, i.e., the nonparental combinations, which are also called as crossover products.

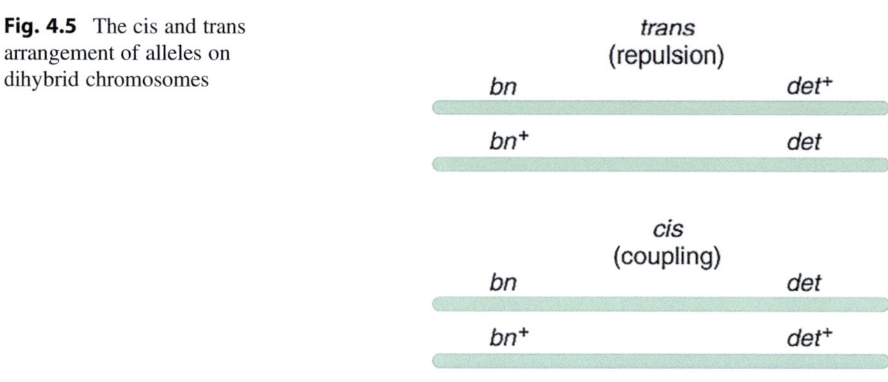

Fig. 4.5 The cis and trans arrangement of alleles on dihybrid chromosomes

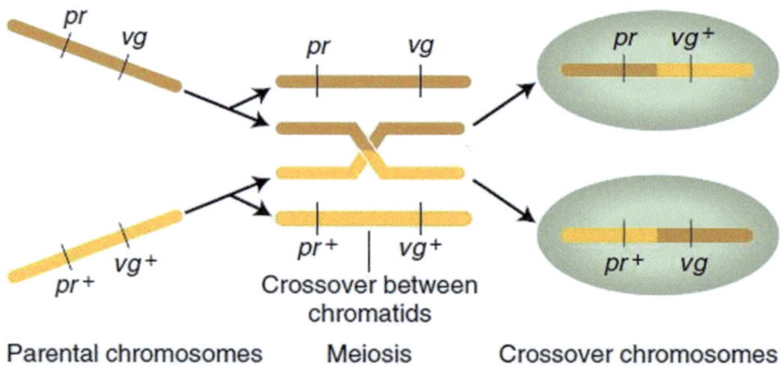

Fig. 4.6 Meiotic crossing over. One homolog from each parent is contributed to an individual offspring. The homologs undergo crossing-over and exchange parts of chromosomes, thereby producing gametic recombinant chromosomes whose allelic combinations are different than that of parental combinations

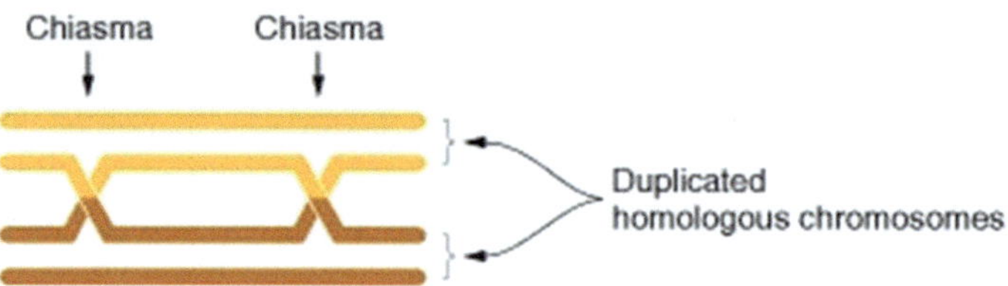

Fig. 4.7 Diagrammatic representation of crossing over and chiasma formation at meiosis. Chromatids of a pair of synapsed homologous chromosomes are represented as lines. Crossing over occurs between non-sister chromatids of homologous chromosomes

The physical phenomenon of crossing over was observed in meiosis, during which pairing of replicated homologous chromosomes occurs. Cross-shaped structures are often formed between the two non-sister chromatids of the paired homologous chromosomes (Fig. 4.7), and are called chiasma (singular) or chiasmata (plural). The discovery of the chiasmata corroborated with the concept of chromosomal crossing over proposed by Morgan. His success in linking this cytological phenomenon with results of his breeding experiments emphasized the significance of the chromosomal theory of inheritance. Such observations, of coupling and repulsion in F_1 selfing and test crosses, are commonly come across in genetics and are a clear deviation from independent assortment. In other words, it can be concluded that independent assortment does not occur when two genes are located close to each other on the same chromosome pair.

Crossing over occurs during prophase I of meiosis. There are two observable outcomes of crossing over and they are:

1. Chiasmata formation in the late prophase of meiosis I
2. Recombination of genes located on the opposite chromatids at the point of crossover

Although chiasmata can be visibly observed, the outcome of recombination can only be observed in the next generation, because it is in the offspring that the recombinant chromosomal genes are expressed.

The crossing over among homologous chromosomes results in production of recombinant gametes. This process is characterized by a physical exchange between the non-sister chromatids of homologous chromosomes (Fig. 4.8). As explained earlier, this event occurs during the prophase I of meiosis, where pairing of duplicated chromosomes occurs. At this stage, the paired chromosomes consist of four homologous chromatids, and they are said to form a tetrad. Although there are four homologous chromatids, only two of them take part in crossing over at any one point. This is followed by breakage of chromatids at the crossover site. Thus, detached fragments reattach with the opposite chromatid to produce recombinants. The other two chromatids do not undergo crossover and remain nonrecombinant at this site. Therefore, the outcome of each crossover event is two recombinant

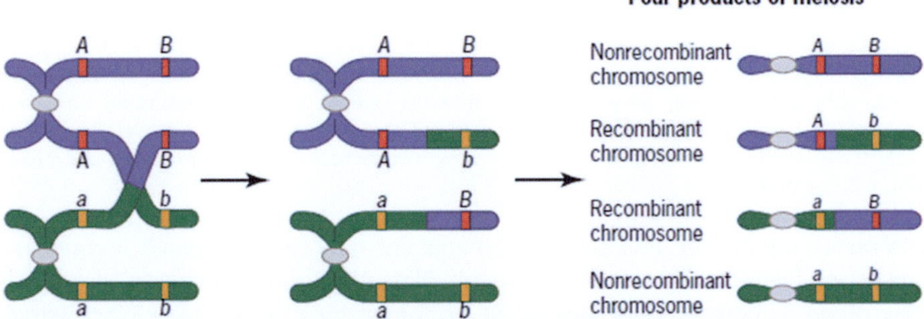

Fig. 4.8 Crossing over forms the basis of genetic recombination. The figure depicts the formation of recombinant chromosomes as a result of exchange of chromosomal segments between paired homologous chromosomes during meiosis. The homologs have been differentiated by using different colors. From each homolog, only one of the chromatids takes part in a specific event of recombination, thus producing two each of recombinant and nonrecombinant chromosomes

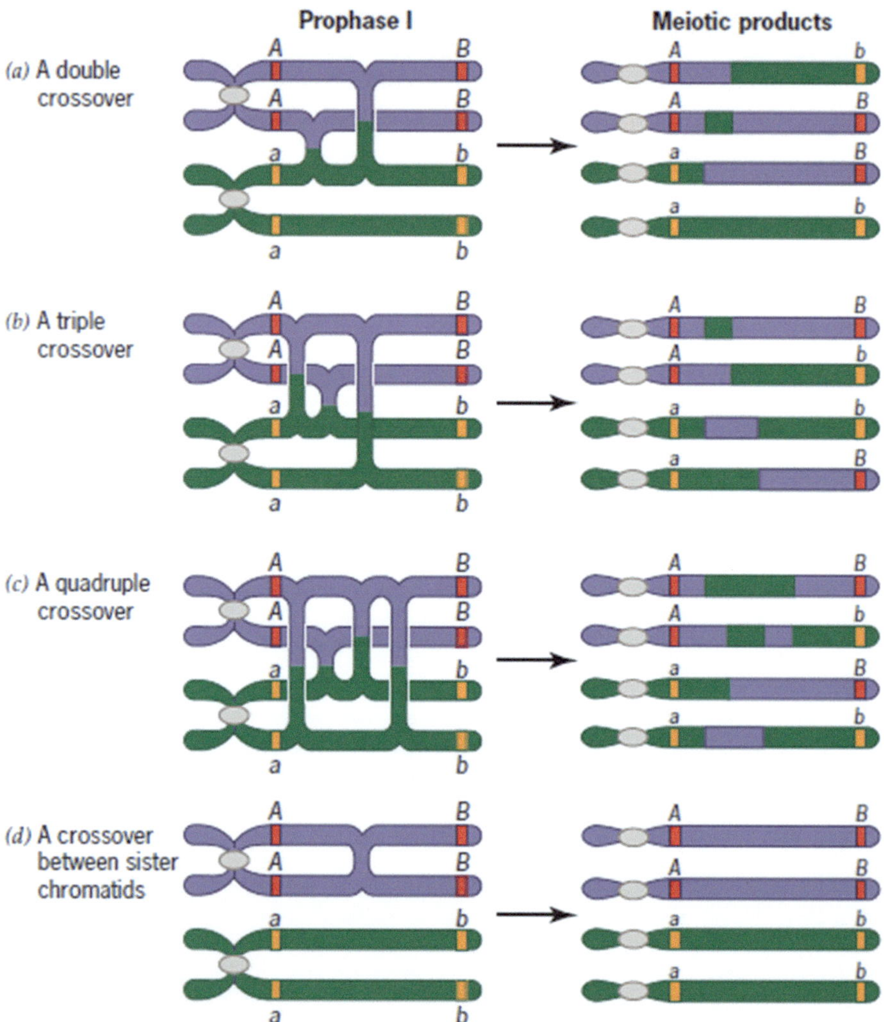

Fig. 4.9 Multiple meiotic crossovers. Multiple exchanges can take place between the chromatids of paired homologs during the prophase I of meiosis. Double, triple, and quadruple crossovers between non-sister chromatids are depicted in a, b, and c, respectively. Each crossover occurs at an exclusive location on the chromosome. If a crossover occurs between sister chromatids as depicted in d, it does not result in recombinants

chromatids out of four. It is important to note that only two chromatids involve in a crossover and exchange at any given point. Nevertheless, the other two chromatids may become involved in a crossover at a different point. Hence, in a given tetrad, there is a chance for multiple crossovers, resulting in multiple exchanges (Fig. 4.9), for example, two, three, or even four, which are called double, triple, or quadruple crossovers, respectively. However, it is important to note that if an exchange occurs between sister chromatids, genetic recombinants are not produced because they are identical. At the molecular level, the breaks in the chromatids at the chiasmata are caused by enzymes that act on the DNA constituting the chromatids. Further, enzymes also mediate the rejoining of the broken chromatid fragments to the opposite non-sister chromatid.

4.1.2 Proof That Crossing over Makes Recombination

The cytological evidence for the mechanism of crossing over was obtained in 1931 by two scientists: Harriet Creighton and Barbara McClintock. Through their experiment, they obtained evidence to prove that the genetic recombination event is associated with exchange of material between chromosomes. Morphologically distinguishable homologous chromosomes of maize were used by Creighton and McClintock for their study. The goal of the study was to determine if recombination between genes located on homologous chromosomes correlates with the physical exchange between these homologs. For this study, two different forms of chromosome 9 were selected. One of the chromosome 9 had a cytological aberration on both ends, whereas the other one was normal. On one of the ends, the abnormal chromosome 9 possessed a heterochromatic knob, and on the other end, a fragment of another chromosome was attached (Fig. 4.10).

These chromosomes were also genetically characterized, which enabled to detect recombination. One of the marker genes located on this chromosome coded for the kernel color (*C*, colored; *c*, colorless), and the other regulated texture of the kernel (*Wx*, starchy; *wx*, waxy). The following test cross (Fig. 4.11) was performed:

The progeny of this test cross was examined for recombination and the associated exchange of chromosomal segments between the two distinguished forms of

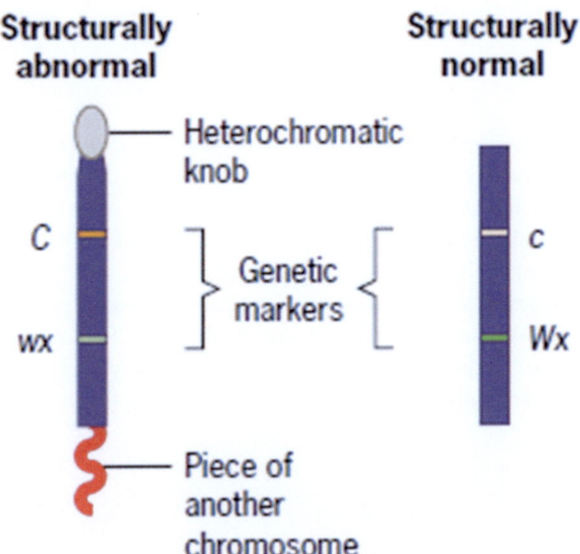

Fig. 4.10 Creighton and McClintock experiment—cytological evidence of crossing over. Two forms of chromosome 9 of maize which were genetically characterized were used in the experiment. Crossing over could be visually observed by the distinguishing morphology of the chromosome 9 and also the progeny could be scored for the characterized genetic markers

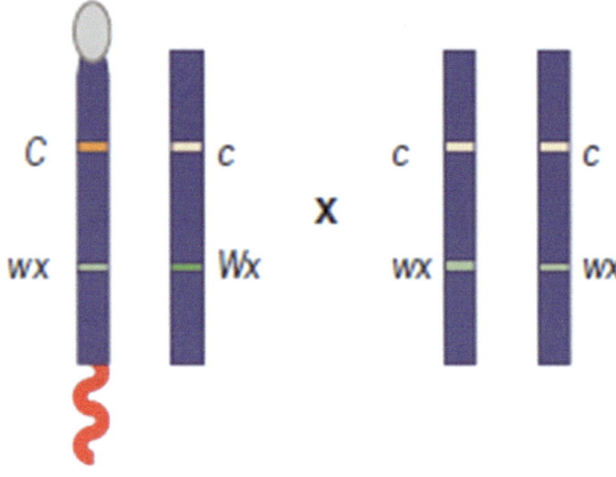

Fig. 4.11 Test cross between a F_1 heterozygote and homozygous recessive plant

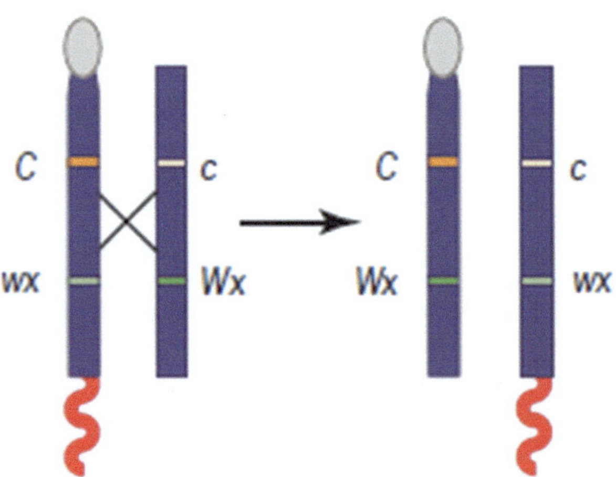

Fig. 4.12 Outcome of a test cross between a F_1 heterozygote and homozygous recessive plant. Recombinants *C Wx* and *c wx* consisted of a chromosome 9 with only one of the two cytological markers; exchange with the normal homolog of chromosome 9 via crossing over had resulted in a loss of the other marker

chromosome 9. The exchange of the chromosomal segments could be visualized due to the abnormal chromosome 9, and the genetic recombination could be identified by the expression of the marker genes. The results obtained were as follows: the recombinants *C Wx* and *c wx* consisted of a chromosome 9 which retained only one of the two cytological markers. The loss of the other marker could be clearly attributed to a chromosomal exchange event that occurred through crossing over with the normal chromosome 9 homolog in the previous generation (Fig. 4.12). These findings strongly indicated that physical exchange of chromatin material between paired homologous chromosomes formed the basis of recombination.

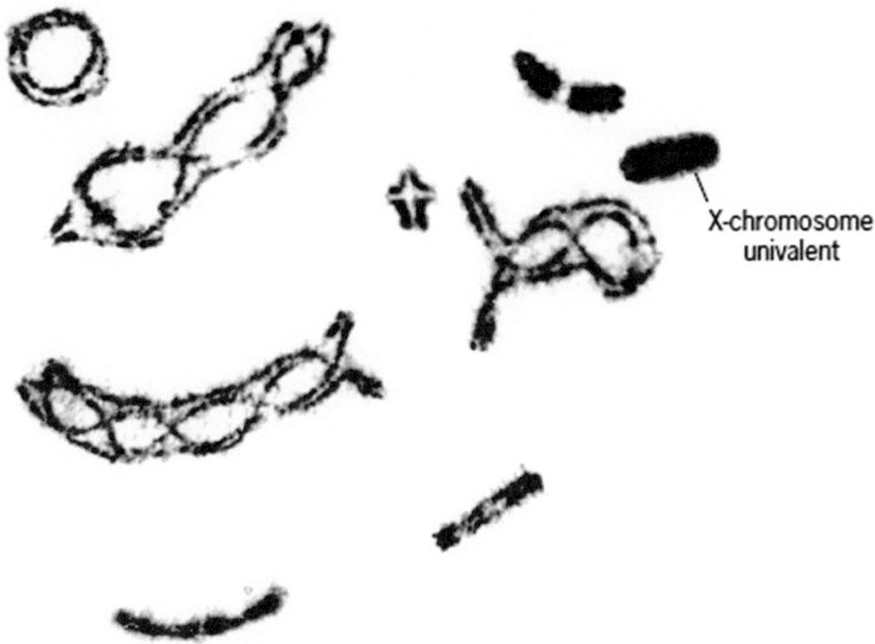

Fig. 4.13 Diplonema during meiosis in male grasshopper. Eight autosomal bivalents and X chromosome univalent can be seen. Among the autosomal bivalents, one chiasma each is seen in four of the smaller chromosomes, while the remaining have two to five chiasmata

4.1.3 Chiasmata and Crossing over

It is during the late prophase of meiosis I, chiasmata, the cytological evidence for crossing over can be clearly visualized. It is at this stage that paired homologous chromosomes slightly repel each other. Although the homologous chromosomes repel each other, they maintain a close contact at the centromere and also at each chiasma (Fig. 4.13). This slight repulsion between the paired homologs partly separates them, thus enabling accurate counting of chiasmata. It can be expected the large chromosomes form more chiasmata compared to small chromosomes. Therefore, the number of chiasmata can be considered to be proportional to the length of chromosomes. Although the chiasmata are visible in the late prophase of meiosis I, experimental evidence has suggested that crossing over occurs at an earlier stage. Heat shock experiments to alter the recombination frequency showed little effect when heat shocks were administered in late prophase, but earlier administration changed the recombination frequency.

Therefore, the crossing over events that lead to recombination occur rather early in the prophase of meiosis. Molecular studies on DNA synthesis revealed additional evidence in this regard. Although most of the DNA synthesis is completed during the interphase of meiosis, a small amount of synthesis is shown to occur in the prophase of meiosis I as well. This small amount of DNA synthesis has been attributed to the repair of broken chromatids possibly associated with the crossing over. Experimentally, it has been shown that this DNA synthesis occurred during early to mid-prophase, but not later. From this evidence, the time of crossing over was deduced to be during early to mid-prophase, i.e., much earlier than the appearance of chiasmata.

Most geneticists consider chiasmata as mere vestiges of the actual genetic exchange process. It is suggested that the chromatids, involved in the exchange process, remain entangled through most of the prophase. Eventually, the resolution of these entanglements occurs, resulting in separation of the chromatids to the opposite poles of a cell by the meiotic spindle apparatus. Hence, the chiasma is believed to represent an entanglement created by an earlier crossover event during the prophase. Several geneticists consider these entanglements resulting from crossing over as a means to hold the chromosomes of a bivalent together during prophase I of meiosis. In some organisms, prophase I is prolonged, i.e., it can extend up to 40 years in human females. If the crossovers do not exist, then paired homologous chromosomes may separate accidentally, particularly during extended periods. Also, the homologs thus separated may fail to disjoin properly during the subsequent anaphase. Defective chromosome disjunction during meiosis I ultimately yields gametes that are aneuploid. Hence, crossing over can be considered as a mechanism for holding paired homologous chromosomes together during cell division, thereby ensuring their appropriate segregation to each of the daughter cells. The possibility of nondisjunction can thus be minimized, thereby largely preventing the occurrence of aneuploid gametes.

4.1.4 Sturtevant and Mapping

Alfred Henry Sturtevant was one of Morgan's students. He worked in the "fly room," Morgan's research laboratory where the study of *Drosophila* genetics was born. As discussed, earlier Morgan contributed immensely to the concept of chromosomes as carriers of genes and linkage. He also proposed that closely linked genes are located close together on the same chromosome and rarely get shuffled by recombination, whereas loosely linked genes lie farther apart and hence are more frequently shuffled by recombination. Working along with Morgan on the recombination between genes, Sturtevant in 1911 proposed that variation in the strength of linkage can indicate their relative position along a chromosome, and hence can be used as a tool for mapping genes.

Sturtevant constructed a remarkably accurate and the first genetic map (Fig. 4.14) and established the basic gene mapping methodology which is used until today. Alfred Sturtevant furthered his research in gene mapping, *Drosophila* genetics, and other areas of biology such as evolution and became a leading geneticist.

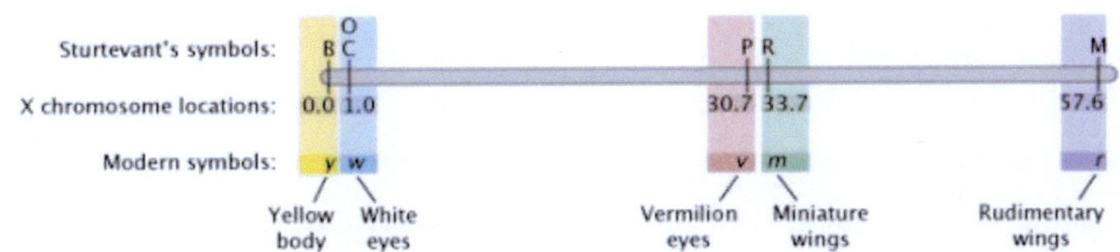

Fig. 4.14 Sturtevant's first genetic map. The first genetic map worked out by Sturtevant involved five genes on the *Drosophila* X chromosome. The genes were yellow body (y), white eyes (w), vermilion eyes (v), miniature wings, (m) and rudimentary wings (r). Marked above are the symbols initially used by Sturtevant; corresponding modern symbols along with their present X chromosome locations are marked below

4.1.5 Linkage Analysis

Linkage analysis forms the foundation of genetic mapping. Consider two genes A and B, each having two alleles *A,a* and *B,b*, respectively. Considering that these genes are located on the same chromosome, their behavior during meiosis can be analyzed as shown in Fig. 4.15. It is assumed that alleles *A* and *B* are located on one homolog and alleles *a* and *b* on the other. Two alternative outcomes are depicted in Fig. 4.15.

1. Meiosis without crossing over between genes A and B: Among the four products of meiosis, two of the gametes consist of *AB* genotype and the other two consist of *ab* genotype, i.e., 2 *AB* : 2 *ab*
2. Meiosis with crossing over between genes A and B: Results in four types of gametes with all possible genotypes: 1*AB* : 1*aB* : 1*Ab* : 1*ab*.

If the results of meiosis like the first example are scored in a hundred identical cells, wherein crossover never occurred, then the following genotypes will be present in the resulting gametes: 200 *AB* and 200 *ab*.

Here both genes A and B participate in meiosis as a single entity, thus showing complete linkage. But in the greater likelihood in at least some nuclei, crossovers will occur between A and B. In such a case, the allele pairs will not behave as a single entity and thus are not inherited together. Assuming crossovers occurred in 40 out of 100 meioses, the following outcome will be observed: 160 *AB*, 160 *ab*, 40 *Ab*, and 40 *aB*, thus displaying incomplete linkage. Here the gametes possess both the parental genotypes (*AB*, *ab*) and the recombinant genotypes (*Ab*, *aB*).

4.1.5.1 Linkage Analysis with Different Types of Organisms Can Be Categorized as Follows

- The first category includes linkage analysis in species wherein it is possible to perform planned breeding experiments, e.g., fruit flies and mice.
  This method is based on analyzing progeny derived from experimental crosses between parents of known genotypes. This method can be applied to all eukaryotes theoretically. However, this approach cannot be applied for human beings due to ethical considerations. Also, certain practical problems such as

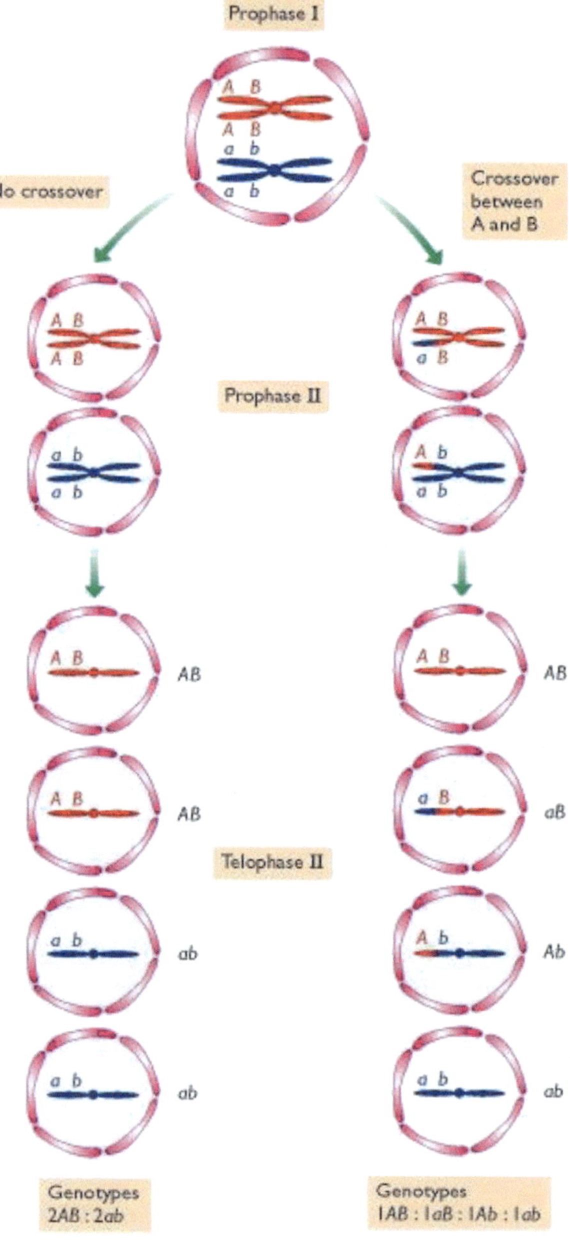

Fig. 4.15 Outcome of crossover between linked genes. The figure depicts a meiosis involving a homologous chromosome pair (the homologs are distinguished by red and blue color) consisting of linked genes A and B, with alleles *A*, *a*, *B*, and *b*. Two alternative scenarios during meiosis have been depicted. (Left) Meiosis without crossing over between the linked genes A and B. Among the four gametes formed in meiosis, two are *AB* genotype, and two are *ab* genotype; (Right) Crossover between A and B produces gametes of all four possible genotypes: *AB*, *aB*, *Ab*, and *ab*

prolonged gestation periods and time duration required by the newborn to mature and grow to a reproducible age limit the effectiveness of this approach to some animals and plants as well. As explained in Fig. 4.15, the key to gene mapping lies in the determination of genotypes of gametes resulting from meiosis.

In some cases, such as in *Saccharomyces cerevisiae* (yeast), a microbial eukaryote, gametes can be directly examined. It is possible to grow haploid colonies of these organisms and using biochemical tests, their genotypes can be determined. Gametes of higher eukaryotes can also be directly genotyped, by PCR-based analysis of DNA markers (such as RFLPs, SSLPs, and SNPs; refer to Sect. 4.4.2). Usually, spermatozoa are used for this purpose, but the typing of the sperms is a laborious process. Therefore, routine linkage analysis in higher eukaryotes does not involve examining the gametes directly. Rather, genotypes of the diploid progeny are determined. Diploid progeny is an outcome of gametic fusion, i.e., a genetic cross.

Nevertheless, a certain complication is inherent within a genetic cross, i.e., the resulting diploid progeny are a product of not one but two different meiosis (one in each parent), and in most organisms, there is an equal likelihood of crossover events during production of both male and female gametes. Therefore, it becomes very important that the crossover events occurring in each of the two meioses need to be deciphered. Hence, adequate care has to be taken while setting up a genetic cross. The standard procedure that is followed in these cases is to use a test cross. Figure 4.16 illustrates two possible test cross scenarios. The parental genotypes are a critical feature of the test cross.

In scenario I depicted in Fig. 4.16, one of the parents is a double heterozygote with a genotype *AB/ab* and consists of all four alleles. The genotypic notation *AB/ab* indicates that alleles *A* and *B* are present in one of the homologs of a homologous pair of chromosomes, and alleles *a* and *b* are located on the other homolog. Such double heterozygotes are obtained by crossing pure-breeding strains, e.g., *AB/AB* × *ab/ab*. A pure-breeding double homozygote is the second parent in this scenario (Fig. 4.16), i.e., in this parent, both homologs of a chromosome pair consist of *a* and *b* alleles, and the genotype is depicted as *ab/ab*.

The main objective of the analysis depicted in scenario I is to deduce genotypes of gametes contributed by the double heterozygous parent and to determine the recombinant fraction among them. It can be envisioned that all gametes contributed by the second double homozygous parent will consist of *ab* genotype, irrespective of whether they are parental or recombinant gametes.

Since both alleles *a* and *b* are recessive, in effect, meiosis in this parent will become invisible when genotypes of progeny are observed. This implies that genotypes of the diploid progeny can be represented clearly as the genotypes of gametes produced by the double heterozygous parent as shown in scenario I of Fig. 4.16. Hence, a test cross enables direct examination of a single meiosis. Further, this enables calculation of recombination frequency and map distance for the two genes being analyzed. Therefore, it can be concluded that for analysis of heterozygote genotypes (consisting of one dominant and one recessive alleles), the double homozygous parent has to be recessive for alleles of both phenotypes.

In scenario II of Fig. 4.16, DNA markers have been used for the analysis. In case of DNA markers, they can be codominant in nature, and therefore, the double homozygous parent can have a genotype of homozygous alleles in any combination

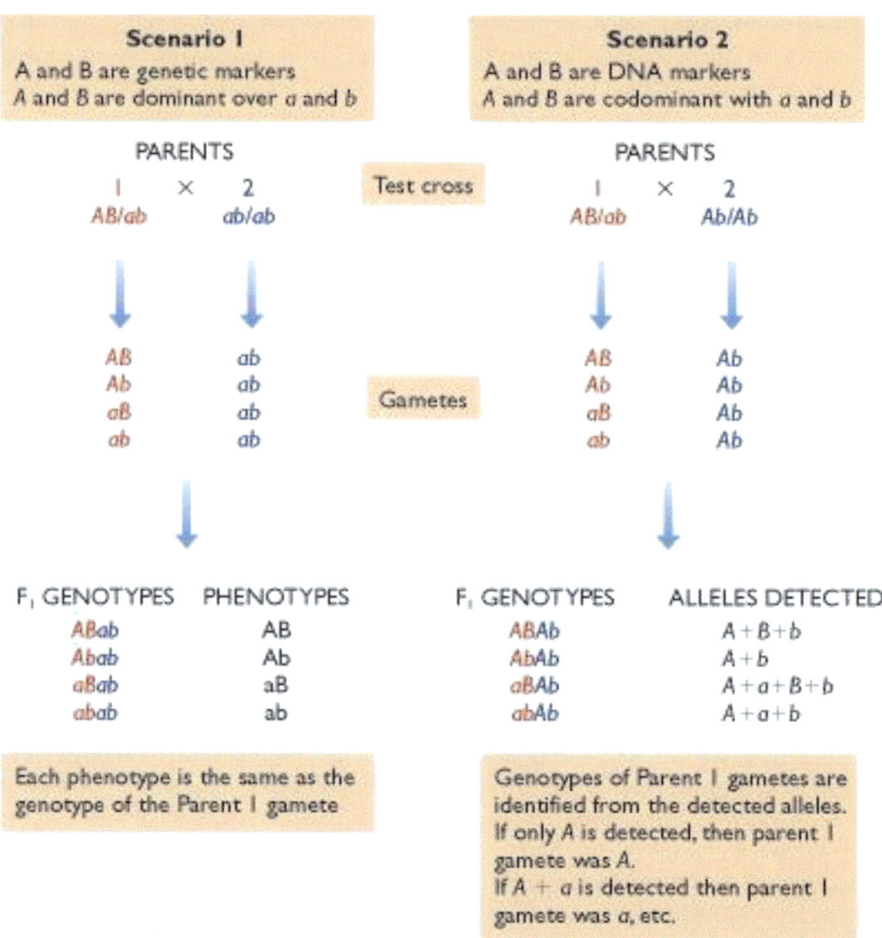

Fig. 4.16 Test cross for linkage analysis. (Left panel) Scenario I: genetic markers A and B, with alleles A, a, B, and b, are being analyzed. The progeny of test cross are scored by examining their phenotypes. Since the second parent is double homozygous recessive, no effective contribution is made to the progeny phenotypes. Hence, phenotype of F_1 individuals is same as the gametic genotypes of the first parent. (Right panel) Scenario II: DNA markers A and B whose allelic pairs are codominant are being analyzed. The figure shows the genotype of double homozygous parent to be *Ab*/*Ab*. Direct detection of the alleles present in each of the F1 progeny is made possible by PCR-based techniques, and hence genotypes of the gametes contributed by the first parent can be deduced in each individual offspring

as follows: *AB*/*AB*, *Ab*/*Ab*, *aB*/*aB*, or *ab*/*ab*. In this case, the markers on each allele are directly detected by PCR analysis and hence can be easily identified as depicted in scenario II of Fig. 4.16.

- The second category involves organisms where it is not possible to conduct planned breeding experiments, e.g., humans. In this case, family pedigrees will have to be made use of for linkage analysis (see Sect. 4.3).

4.2 Gene Mapping with Recombination Frequencies, Calculation of Map Distance, and Concept of Map Units

Morgan and Sturtevant's work had established a correlation between physical distances separating genes on chromosomes and their rates of recombination. According to Sturtevant (see Sect. 4.1.4), the occurrence of crossover events was random and was more likely to occur between two genes positioned far apart than those located close to each other on a chromosome. It was projected that recombination frequencies between genes could provide a convenient means of estimating their relative distances and positions (order of arrangement) on a chromosome. Hence, the practice of constructing genetic maps using recombination frequencies was introduced.

Distances on genetic maps are measured in map units, and are abbreviated as m. u.; 1% recombination is equal to one map unit. Map units are represented in centimorgans (cM), in honor of Thomas Hunt Morgan; one Morgan = 100 m.u.

A simple illustration of a genetic map is given below: consider three hypothetical genes *A*, *B*, and *C*, where distance between *AB* is 5 m.u., and *BC* is 10 m.u. Since genetic distances measured using recombination frequencies are nearly additive, distance between A and C is 15 m.u. According to this data, the location of B is in between A and C. Hence, on the basis of given map distances, the genetic map for A, B, and C can be constructed as follows:

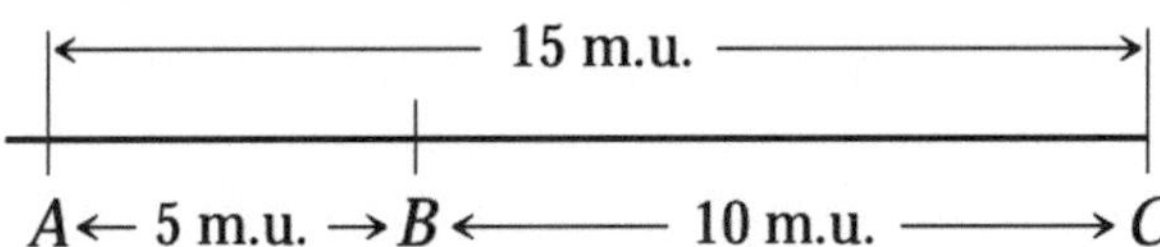

The map can also be redrawn by interchanging the positions of the markers C (left) and A (right):

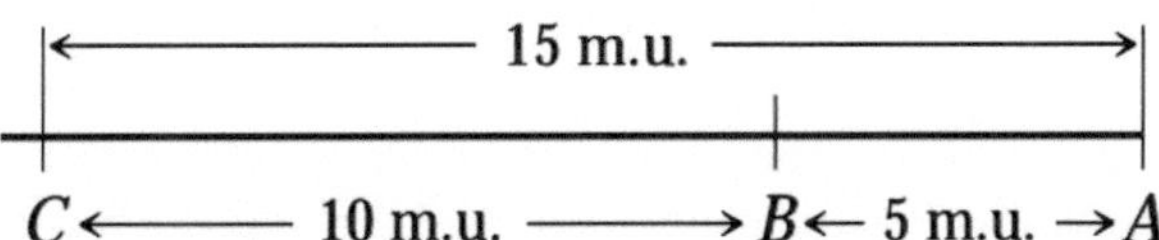

Both maps are equivalent and correct because relative positions of only three genes are known, and maximum information that could be deduced pertains to predicting which one of the three markers occupies the central position with respect

Table 4.4 Results of recombination between A, B, and C markers with D

Gene pair	Recombination frequency (%)
A and D	8
B and D	13
C and D	23

to their linear arrangement on the chromosome. If information about an additional gene is obtained, then the relative positions of the two markers, in this case A and C, can be deduced relative to the fourth marker. In this example, gene D is the fourth marker. Through genetic crosses, the following recombination frequencies between D and each of A, B, and C can be obtained:

From the data in Table 4.4, it is evident that C and D exhibit the highest percentage of recombination; hence, C and D must be the farthest, with A and B genes situated between them. With addition of data pertaining to D, the map can be redrawn as follows:

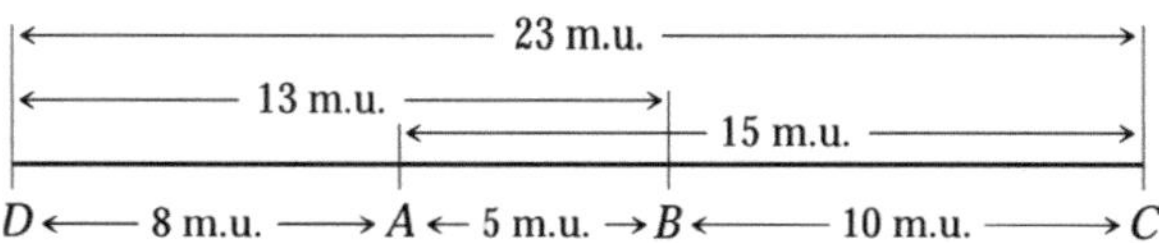

By performing a series of genetic crosses between various gene pairs, genetic maps can be constructed, and the linkage arrangements of several genes can be shown. Two points of significance need to be considered while constructing genetic maps based on recombination frequencies. Firstly, the recombination frequency between two genes located on a chromosome cannot exceed 50% as the rate of recombination for genes located on different chromosomes is also 50%. Therefore, it is not possible to distinguish between genes situated on different chromosomes and those located on the same chromosome but separated by larger distances. If 50% recombination is exhibited by a pair of genes, then the utmost conclusion one can derive is that they belong to different linkage groups, located either far apart on one chromosome or on separate chromosomes.

Secondly, test cross results for two genes located on the same chromosome, but relatively far apart, tend toward underestimation of the true recombination frequency, because the double crossovers that might take place between the two genes are not revealed (Fig. 4.17). Two separate crossover events occurring between the same two loci on a chromosome result are termed as a double crossover. In a single crossover, alleles on homologous chromosomes are swapped and recombinants are produced. The occurrence of a second crossover between the same two loci reverses effect of the first one and the original parental combination is restored (Fig. 4.17).

Since double crossovers result in only nonrecombinant gametes, it is not possible to distinguish between progeny produced without a crossing over and those that are produced by double crossovers. Also, overall leads to underestimation of

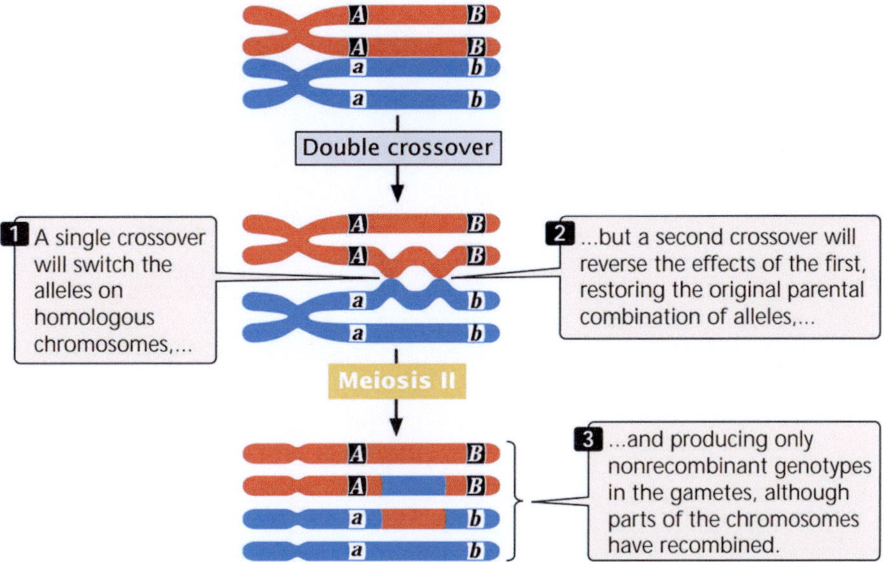

Fig. 4.17 Double crossovers result in nonrecombinant gametes. A single crossover swaps alleles on the homologous chromosomes and produces recombinants. But when a second crossover occurs between same two loci, it reverses the effect of the first crossover, i.e., the original parental combination is restored, leading to production of nonrecombinant gametes

recombination frequencies, because double crossovers go undetected. The frequency of double crossovers is higher between far apart genes, and hence genetic maps based on markers separated by short distances will always be more accurate than those based on markers located at larger distances.

4.2.1 Gene Mapping Through Two-Point Test Cross

A series of test crosses can be performed selecting gene pairs, and the recombination frequency obtained from them can be used for constructing genetic maps. When a test cross is performed involving two genes, it is named as a two-point test cross or in short called as two-point cross.

Consider the cross between wild-type females and males homozygous for *vestigial* (*vg*) wing and *black* (*b*) body mutations of *Drosophila* (Fig. 4.18).

The vestigial wing and the black body mutations are autosomal mutations that produce short wings and black body, instead of the wild-type long/normal wings (vg^+) and gray body (b^+). All F_1 offspring were gray bodied and long winged. This indicated the dominant nature of the wild-type alleles (vg^+ and b^+). A two-point test cross (Fig. 4.18) between the F_1 females and vestigial, black males produced F_2 progeny of four phenotypic classes, of which two were abundant and two were rare.

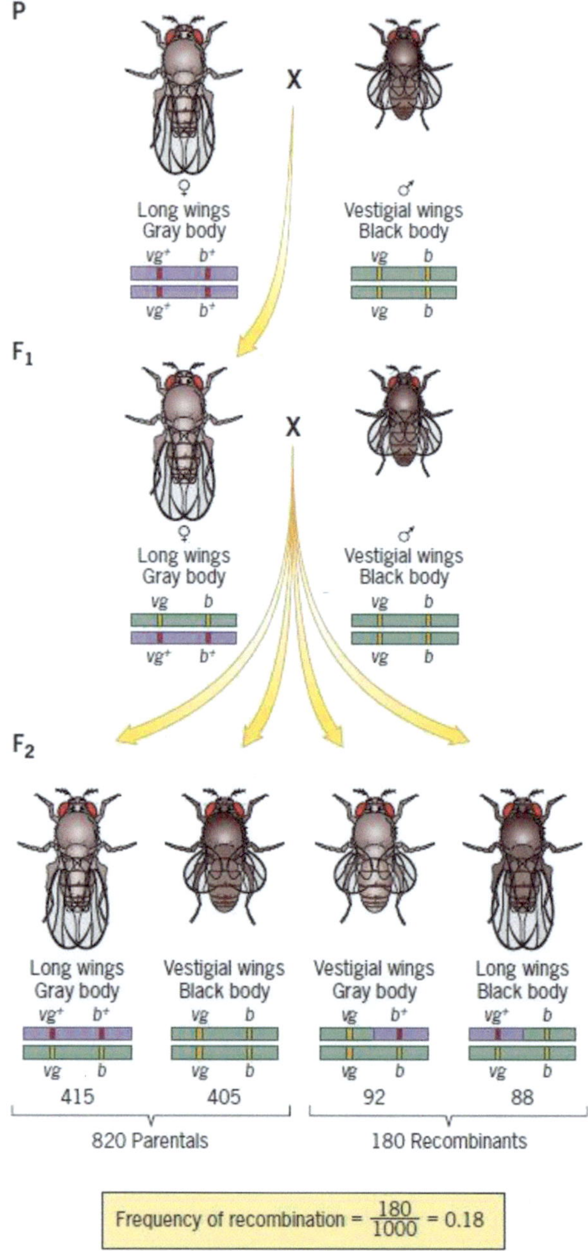

Fig. 4.18 A Two-point genetic cross involving two linked genes. The two linked genes analyzed in the cross are the genes coding for two mutant phenotypes vestigial wings (*vg*) and black body (*b*) in *Drosophila*

The phenotypes that were abundant were the same as parental phenotypes, and those that were rare were recombinants. As the number of recombinants was much lesser than 50% of the progeny, it clearly indicates the linkage relationship between *vestigial* and *black* genes. The observed linkage also suggested that these two genes are located on the same chromosome. The distance between these two genes can be determined by approximating the average number of crossovers between these genes in gametes of F_1 double heterozygous females, by calculation of recombination frequency in F_2 flies. It is observed that each such recombinant fly has inherited a chromosome that had crossed over once between *vg* and *b*. Therefore, the average number of crossovers in the sample of progeny is as follows:

Nonrecombinants		Recombinants		
(0) × 0.82	+	(1) × 0.18	=	0.18

In the calculation shown above, the number within the brackets represents the number of crossovers and the other number represents the frequency, for each class (recombinants and nonrecombinants) of flies. The inclusion of the nonrecombinant progeny in the calculation is for the purpose of calculating the average number of crossovers. Hence, all data has to be considered and not just that of the recombinants. The result of this analysis shows that, on an average, 18 chromosomes out of 100 had a crossover between *vg* and *b* during meiosis. This can be further interpreted to construct a genetic map wherein *vg* and *b* are 18 map units apart.

4.2.2 Gene Mapping Through Three-Point Test Cross

Data from test crosses including more than two genes can also be used for the purpose of recombination mapping. C. B. Bridges and T. M. Olbrycht designed an experiment in which they crossed wild-type male *Drosophila* with females homozygous for three X-linked recessive mutations—*scute* (*sc*) bristles, *echinus* (*ec*) eyes, and *crossveinless* (*cv*) wings. An intercross of the F_1 progeny was performed to produce F_2 flies. The F_2 flies obtained were further classified and counted. It is to be noted that F_1 females selected for the intercross were heterozygous for the three recessive mutations, i.e., one of their X chromosomes carried the recessive mutations, while the other X chromosome carried the respective wild-type alleles. Additionally, F_1 males possessed single X chromosome that carried the three recessive mutations. Consequently, this intercross becomes equivalent to a test cross where all the three genes in the F_1 female are present in a coupling configuration. F_2 flies derived from the intercross consisted of eight distinct phenotypic classes, two of which were parental and six were recombinant (Fig. 4.19). Among these classes, the flies belonging to the parental types were the most numerous. The recombinant classes were lesser in number and each represented a recombinant chromosome of different crossover events (Fig. 4.19).

In order to find out which crossovers gave rise to each type of recombinant, it is necessary firstly to determine the order of gene arrangement on the chromosome.

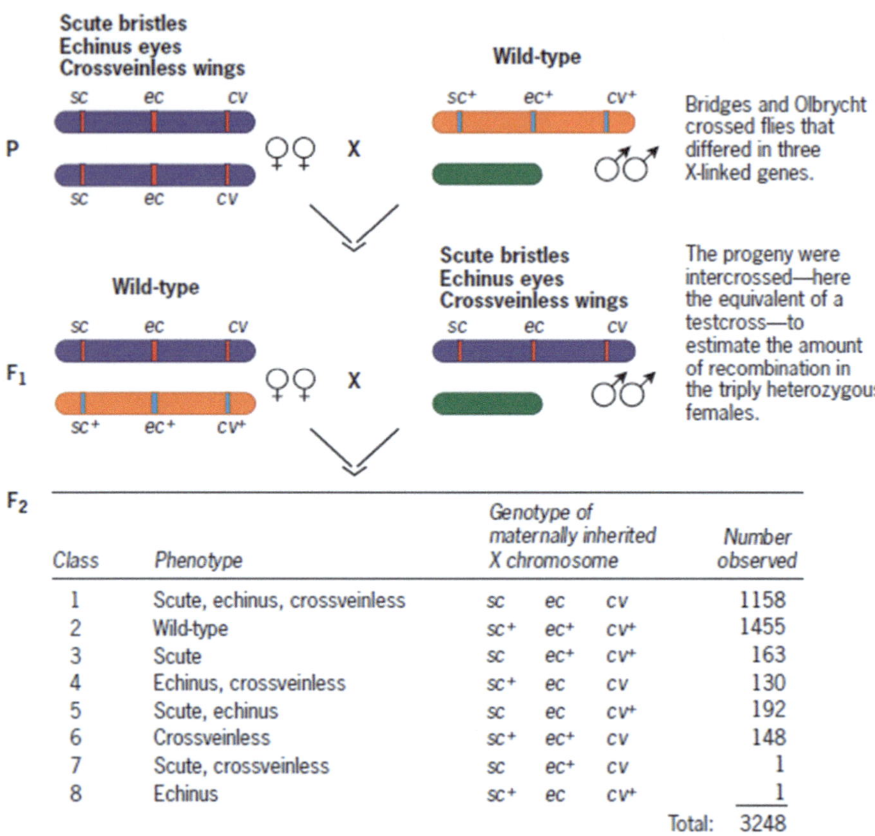

Class	Phenotype	Genotype of maternally inherited X chromosome			Number observed
1	Scute, echinus, crossveinless	sc	ec	cv	1158
2	Wild-type	sc+	ec+	cv+	1455
3	Scute	sc	ec+	cv+	163
4	Echinus, crossveinless	sc+	ec	cv	130
5	Scute, echinus	sc	ec	cv+	192
6	Crossveinless	sc+	ec+	cv	148
7	Scute, crossveinless	sc	ec+	cv	1
8	Echinus	sc+	ec	cv+	1
				Total:	3248

Fig. 4.19 Bridges and Olbrycht's three-point cross experiment. The three-point crosses were performed with X-linked genes *sc* (*scute bristles*), *ec* (*echinus eyes*), and *cv* (*crossveinless* wings) in *Drosophila*

4.2.2.1 Determining Gene Order

Considering the three X-linked genes, *scute* (*sc*) bristles, *echinus* (*ec*) eyes, and *crossveinless* (*cv*) wings, these can be arranged in three possible ways.

1. *sc—ec—cv*
2. *ec—sc—cv*
3. *ec—cv—sc*

It is required to determine which one of the three orders is correct. In order to determine this, the six recombinant classes have to be studied carefully. Four (classes 3, 4, 5, and 6) out of these six recombinants must have resulted from a single crossover in either one of the two regions as shown in Fig. 4.20 top left and right panel. The remaining two recombinants (class 7 and 8) must be products of a double crossing over, i.e., one exchange has taken place in each of the two regions as shown

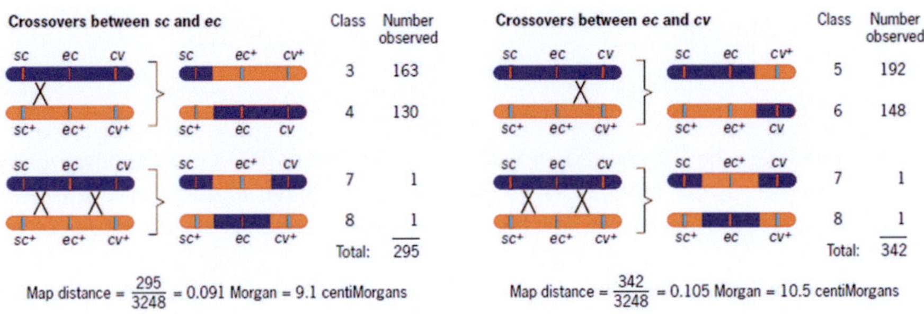

Fig. 4.20 Calculation of map distances between genes from three-point cross experiment of Bridges and Olbrycht. By estimating the average number of crossovers between each pair of genes, the distance between the genes is obtained

in Fig. 4.20 bottom left and right panel. As a double crossover swaps the position of the gene located at the center position relative to genes located on either side, it can be applied for determining the linear order of arrangement of these genes, i.e., their positions relative to one another can be elucidated. Naturally, the frequency of double crossovers is expected to be much lesser than the frequency of a single crossover. As a result, it can be observed that out of the six recombinant types, two are rare and hence must signify double crossover chromosomes. In the given example, classes 7 and 8, i.e., *sc* $ec^{+}cv$ and $sc^{+}ec$ cv^{+}, respectively, are the rare, double crossover types, consisting of one fly each (Figs. 4.19 and 4.20). Comparison with parental classes 1 and 2, i.e., *sc ec cv* and sc^{+} ec^{+} cv^{+}, respectively, shows that there has been a switch in the position of the *echinus* allele with respect to the *scute* and *crossveinless* alleles. These observations indicate that the *echinus* gene is located between the *scute* and *crossveinless*. Hence, it can be concluded that the correct gene order is *sc—ec—cv*.

4.2.2.2 Determining Distance Between Genes

After determination of the gene order, the next step is to determine the distances between these adjacent genes. The procedure involves computation of the average number of crossovers in the chromosomal regions between each pair of adjacent genes (Fig. 4.20).

Distances between pairs of genes can be derived by identifying recombinant classes that are a result of crossover between the concerned pair of genes. Therefore, the length of the region separating *sc* and *ec* can be calculated by considering the following four classes: 3 (*sc* $ec^{+}cv^{+}$), 4 (sc^{+} *ec cv*), 7 (*sc* ec^{+} *cv*), and 8 (sc^{+} *ec* cv^{+}). Classes 3 and 4 involve single crossovers between *sc* and *ec* and classes 7 and 8 involve two crossovers. One of the crossovers occurs between *sc* and *ec* and the other occurs between *ec* and *cv* alleles (Fig. 4.20). The average number of crossovers between *sc* and *ec* can be estimated by using the frequencies of these four classes:

Class 3		Class 4		Class 7		Class 8				
163	+	130	+	1	+	1				
			Total				=	295/3248	=	0.091

This result can be interpreted as follows: out of every 100 chromosomes going through meiosis in the females of F_1 generation, 9.1 have undergone a crossover between *sc* and *ec*. Hence, distance between *sc* and *ec* genes is 9.1 map units or 9.1 centimorgans.

Similarly, the distance between *ec* and *cv* can be obtained. Recombinant classes 5 (*sc ec cv*$^+$), 6 (*sc*$^+$ *ec*$^+$ *cv*), 7, and 8 involved a crossover between *ec* and *cv*. It is also required to consider the double recombinants since one of their two crossovers has occurred between *ec* and *cv*. Thus, by adding frequencies of all four classes, the following is obtained:

Class 5		Class 6		Class 7		Class 8				
192	+	148	+	1	+	1				
			Total				=	342/3248	=	0.105

From the above calculations, the distance between *ec* and *cv* is 10.5 map units. Thus, by combining data from both regions, a genetic map showing the relative positions and distances between these genes can be constructed.

sc—9.1—*ec*—10.5—*cv*

The map distances calculated in this way are additive. Therefore, the distance between *sc* and *cv* can be estimated by adding the map distance separating *sc* and *ec* to the distance separating *ec* and *cv*. The calculation is as follows:

$$9.1\ \text{cM} + 10.5\ \text{cM} = 19.6\ \text{cM}$$

The same can also be obtained by directly adding average number of crossovers between these genes:

Non-crossover classes		Single crossover classes		Double crossover classes		
1 and 2		3, 4, 5, and 6		7 and 8		
(0) × 0.805	+	(1) × 0.195	+	(2) × 0.0006	=	0.196

As observed in the above calculation, combined frequency of the different classes has been multiplied with the total number of crossovers (written in parentheses) that occurred within them. Hence, map distance is a product of the recombinant class frequency and number of crossovers that occurred within it.

Bridges and Olbrycht, in their recombination experiment, studied seven X-linked genes: *sc*, *ec*, *cv*, *ct* (*cut* wings), *v* (*vermilion* eyes), *g* (*garnet* eyes), and *f* (*forked* bristles). By calculating recombination frequencies between pairs of adjacent genes, they constructed genetic map of an X chromosome segment (Fig. 4.21). In their map, the extreme positions were occupied by *sc* and *f*. Each of the seven genes that

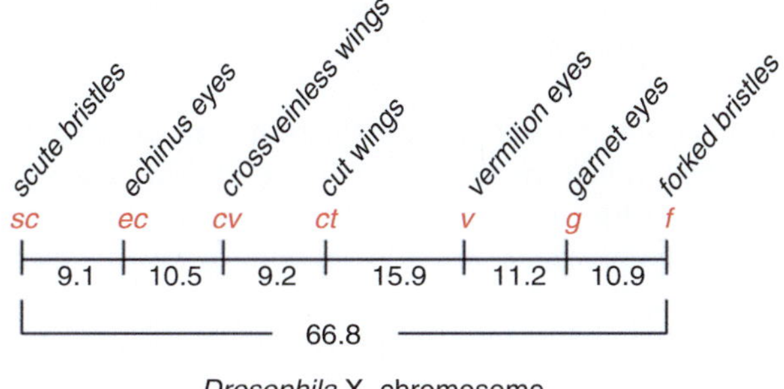

Fig. 4.21 Bridges and Olbrycht's map of seven X-linked genes in *Drosophila*. The map distances between these genes are given in centimorgans

Bridges and Olbrycht chose to study was *markers* representing a particular site on X chromosome. The total length of the mapped X chromosome segment was obtained by addition of map intervals between each pair of adjacent markers, and was estimated to be 66.8 cM. Hence, the average number of crossovers in this X chromosome segment was 0.668.

4.2.2.3 Interference and the Coefficient of Coincidence

Compared to two-point crosses, three-point crosses are more advantageous as they make enable detection of double crossovers. This makes it possible to determine if exchanges occurring in adjacent regions are mutually interdependent or independent. For example, in the given three-point cross (Fig. 4.20), we will be able to determine if occurrence of a crossover between *sc* and *ec* is independent of a crossover between *ec* and *cv* or whether occurrence of one has an inhibitory effect on the other. In general terms, it is possible to determine if crossover in one location on a chromosome inhibits formation of another in a nearby site.

This can be analyzed by calculating the expected frequency of double crossovers, by assuming that these events are independent and are done by multiplying the crossover frequencies of two chromosomal regions that are adjacent to one another. In the map constructed by Bridges and Olbrycht, the crossover frequency in the first region (*sc* and *ec*) was (163 + 130 + 1 + 1)/3248 = 0.091, and in the second region (*ec* and *cv*), it was (192 + 148 + 1 + 1)/3248 = 0.105. In case independence is assumed then, 0.091 × 0.105 = 0.0095 would be the expected double crossover frequency between *sc* and *cv*. When this is compared with observed frequency (2/3248 = 0.0006), it can be seen that it is much lesser than the expected frequency. Therefore, double crossovers between *sc* and *cv* are less frequent than would be expected in case of independence. This indicated that occurrence of one crossover inhibited occurrence of a second in the nearby vicinity, and this phenomenon is called interference. The coefficient of coincidence (*c*) indicates the extent of interference and can be defined as a ratio of the observed frequency to the expected frequency of double crossovers:

c	$=$	$\frac{\text{Observed frequency of double crossovers}}{\text{Expected frequency of double crossovers}}$		
	$=$	$\frac{0.0006}{0.0095}$	$=$	0.063

4.2.3 Mapping in *Neurospora* and Yeast: Haploid Mapping (Tetrad Analysis)

The *Ascomycetes* fungi retain the four haploid (tetrad) meiotic products in a saclike structure called an ascus and hence enable analysis of all products of meiosis, thereby making it possible to determine certain fundamental facts such as occurrence of DNA replication prior to crossing over and reciprocity of crossing over by employing several techniques.

Primarily, two fungi have served as models for these analyses. One is the *Saccharomyces cerevisiae* (common baker's yeast) and the other is *Neurospora crassa* (pink bread mold). Both these organisms retain the meiotic products as ascospores (Fig. 4.22).

4.2.3.1 Phenotypes of Fungi

The phenotypes of microorganisms generally can be classified into three broad types: colony morphology, nutrient requirements, and drug resistance. Several microorganisms are culturable in supporting medium such as agar in test tubes and petri plates (Fig. 4.22). Several substances can be added to the growth medium.

The wild-type *Neurospora* commonly grows in filamentous forms, while the yeast cells form colonies. Several mutations that alter colony morphology have been identified. One such alteration in the yeast (*ade* gene) results in red colored colonies. Similarly, in *Neurospora*, colonial (*col4*), tuft (*tu*), fluffy (*fl*), and dirty (*dir*) are all basic growth form mutants. Additionally, wild-type *Neurospora* shows sensitivity to sulfonamide, a sulfa drug, and also a mutant form of *Neurospora* (*Sfo*) requiring sulfonamide for its survival and growth also exists. Similarly, yeast also exhibits sensitivities to antifungal agents. The nutritional requirement phenotypes of these microorganisms give insights into biochemical metabolic pathways and also aid in genetic analysis. Growth of wild-type *Neurospora* can be observed in a minimal medium consisting of sugar and a nitrogen source along with some salts, organic acids, and biotin (a vitamin). *Neurospora* consists of several mutants, or strains, that are unable to grow on minimal medium, and require addition of some essential nutrients into the medium. For example, one such mutant strain of *Neurospora* grows in minimal medium only when it is supplemented with the amino acid arginine (Fig. 4.23).

It can be inferred that this mutant is deficient for one of the components (an enzyme) involved in the biosynthesis pathway of arginine (Fig. 4.24), which is functional in the wild-type *Neurospora*.

Several such nutritionally deficient strains with mutations in enzymes at various steps, particularly in long biochemical pathways, can be found. For example, several mutants at different loci for arginine biosynthesis are named as *arg*1, *arg*2, etc. Several such biosynthetic pathways in *Neurospora* and yeast harbor such mutations, and hence a variety of mutants exhibiting various nutritional requirements can be

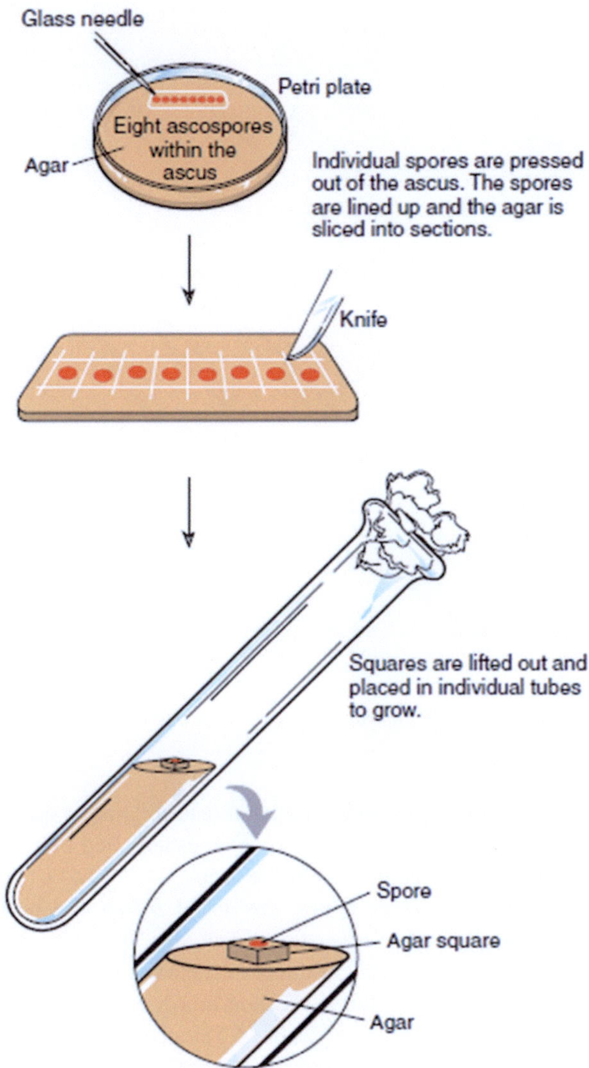

Fig. 4.22 Spore isolation technique in *Neurospora*

found. Mutations can also be induced by treatment with agents such as irradiation or chemicals. Such mutants are applied as tools for the analysis and mapping of chromosomes of microorganisms such as *Neurospora* and yeast.

4.2.3.2 Identification of Linkage and Mapping Genes in Yeast by Analysis of Unordered Spores

The budding or baker's yeast, *Saccharomyces cerevisiae*, occurs in both diploid and haploid forms (Fig. 4.25).

Haploid form of yeast generally forms in conditions of nutritional stress or starvation. With the return of optimal environmental conditions, haploid cells of a

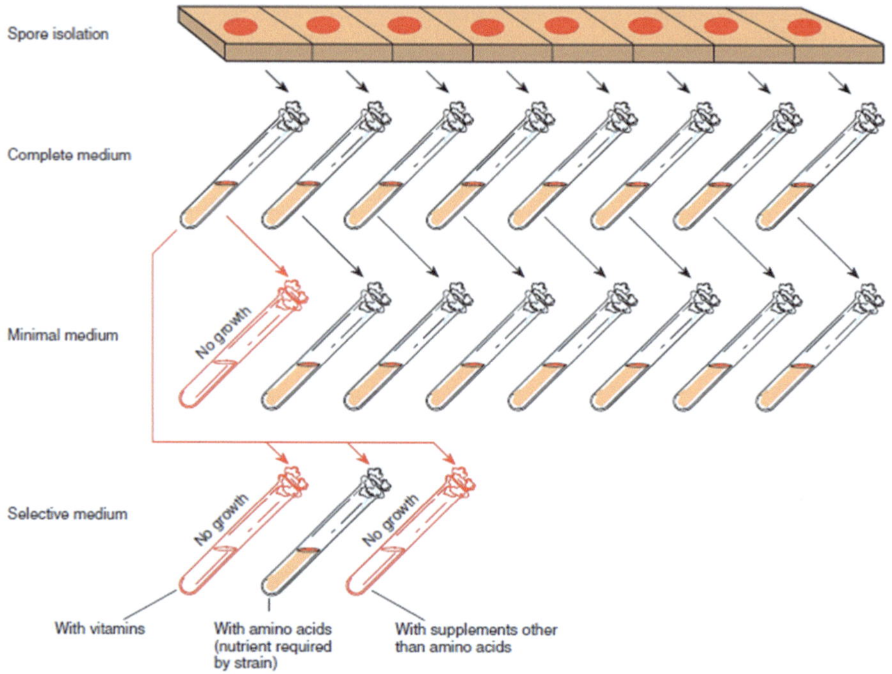

Fig. 4.23 Isolating nutritional requirement mutants in *Neurospora*

and α mating types undergo fusion resulting in the diploid yeast. Under conditions of starvation, the haploid stage is again established by meiosis. All products of meiosis in the yeast are contained in the ascus.

To understand mapping, the *a* and *b* loci can be taken as an example. When spores or gametes consisting *ab* and a^+b^+ fuse, the diploid formed undergoes meiosis. Such spores (products of meiosis) can be isolated and grown into haploid colonies. These colonies can then be observed for phenotypes encoded by the two loci. Only three patterns (Table 4.5) can occur in this case.

Therefore, there are three classes of spores. Class 1 spores are of two types, and are identical to parental haploid spores, and such an ascus is called parental ditype (PD). Class 2 consists of two types of recombinant spores and such an ascus is called nonparental ditype (NPD). Class 3 called as a tetratype (TT) comprises all four possible spore types.

Irrespective of the linkage status of the two loci, all three types of ascus can be generated. As it can be seen in Fig. 4.26, parental ditypes arise due to the lack of crossing over in case of linked loci, while nonparental ditypes arise due to double crossovers involving all four chromatids, also termed as four-strand double crossovers.

Therefore, it can be expected that the parental ditypes should be more in number than nonparental ditypes with respect to linked loci. However, in case of unlinked

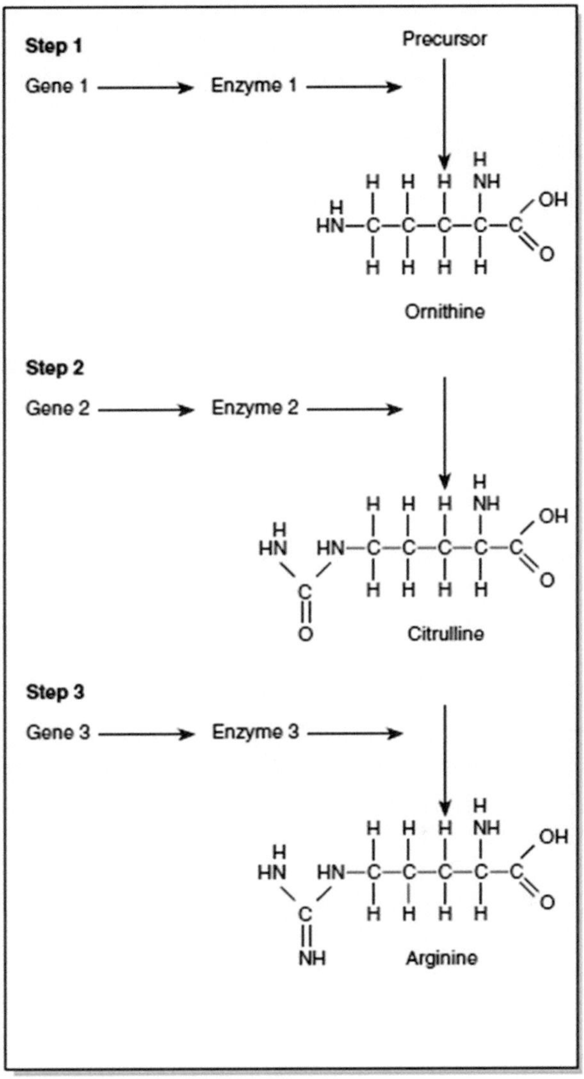

Fig. 4.24 Arginine biosynthetic pathway in *Neurospora*

Step 1

Gene 1 → Enzyme 1 → Precursor → Ornithine

Step 2

Gene 2 → Enzyme 2 → Citrulline

Step 3

Gene 3 → Enzyme 3 → Arginine

loci, parental and nonparental ditypes can arise through independent assortment, and in such a case, all types will be expected to arise in equal frequencies. Hence, by analyzing number of parental and nonparental ditypes, linkage between the loci can be determined. The parental ditypes will greatly exceed the number of nonparental ditypes if the genes are linked. The next step would be to calculate map distance between the given loci. From Fig. 4.26, it can be seen that all four chromatids in a nonparental ditype are recombinant, whereas only half the chromatids in a tetratype are recombinant. As mentioned earlier, 1% recombinant offspring is equal to 1 map

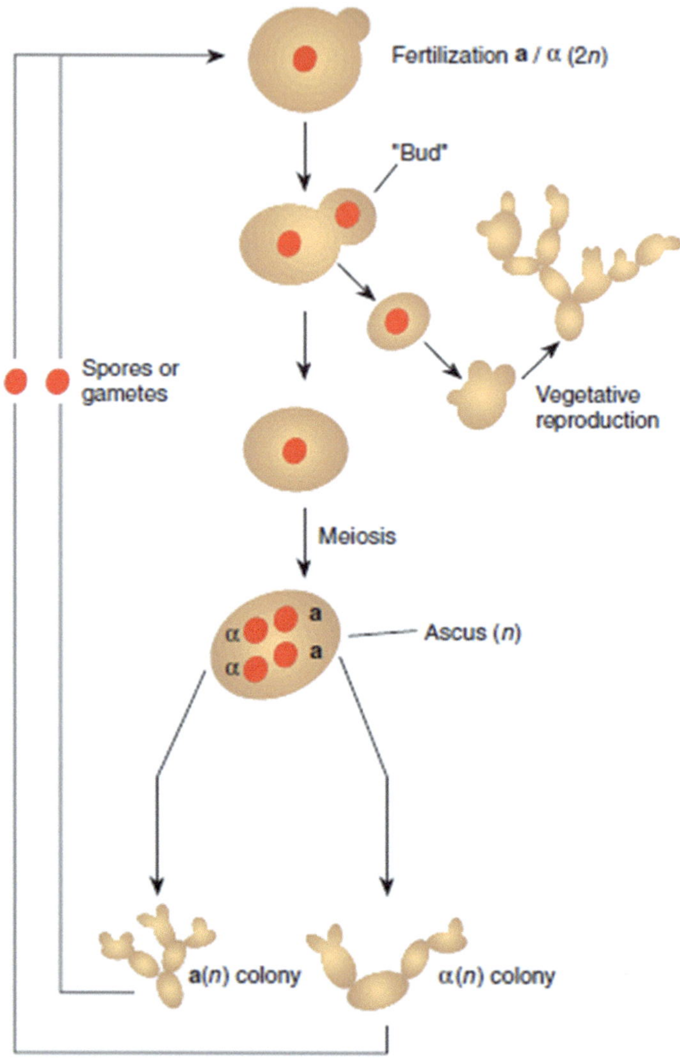

Fig. 4.25 Yeast life cycle. Mating types are represented by a and α. The haploid stage is denoted by n and the diploid stage is denoted by $2n$

Table 4.5 Meiosis in a dihybrid, aa^+bb^+, yeast and the resultant three ascus types

1 (PD)	2 (NPD)	3 (TT)
ab	ab^+	ab
ab	ab^+	ab^+
a^+b^+	a^+b	a^+b
a^+b^+	a^+b	a^+b^+
75	5	20

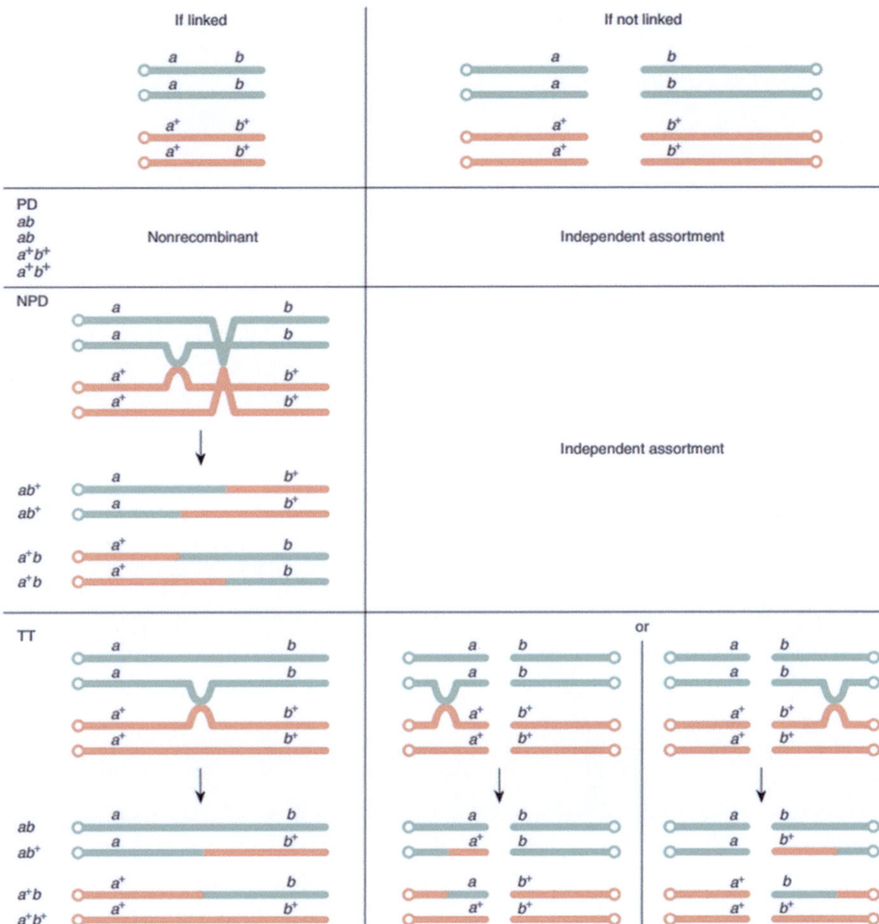

Fig. 4.26 Types of asci resulting from meiosis in yeast. PD (parental ditype), NPD (nonparental ditype), and TT (tetratype) asci formed in a dihybrid yeast by independent assortment or linkage at meiosis. Centromeres are depicted as open circles

unit, and using the following formula, the map distance between the loci can be calculated:

Map units	=	$\frac{(1/2)\text{ the number of TT asci}+\text{the number of NPD asci}}{\text{Total number of asci}}$	×	100						
Thus, for the data of Table 4.5										
Map units	=	$\frac{(1/2)20+5}{100}$	×	100	=	$\frac{10+5}{100}$	×	100	=	15

4.2.3.2.1 Mapping in *Neurospora*

Neurospora consists of ordered spores; life cycle of *Neurospora* is shown in Fig. 4.27.

Mating types in *Neurospora* are referred to as *A* and *a*. Mating between these two mating types results in fertilization and it takes place within an immature fruiting body. The diploid nucleus of the zygote undergoes meiosis immediately. Due to the narrow structure of the *Neurospora* ascus, the products of meiosis are arranged along

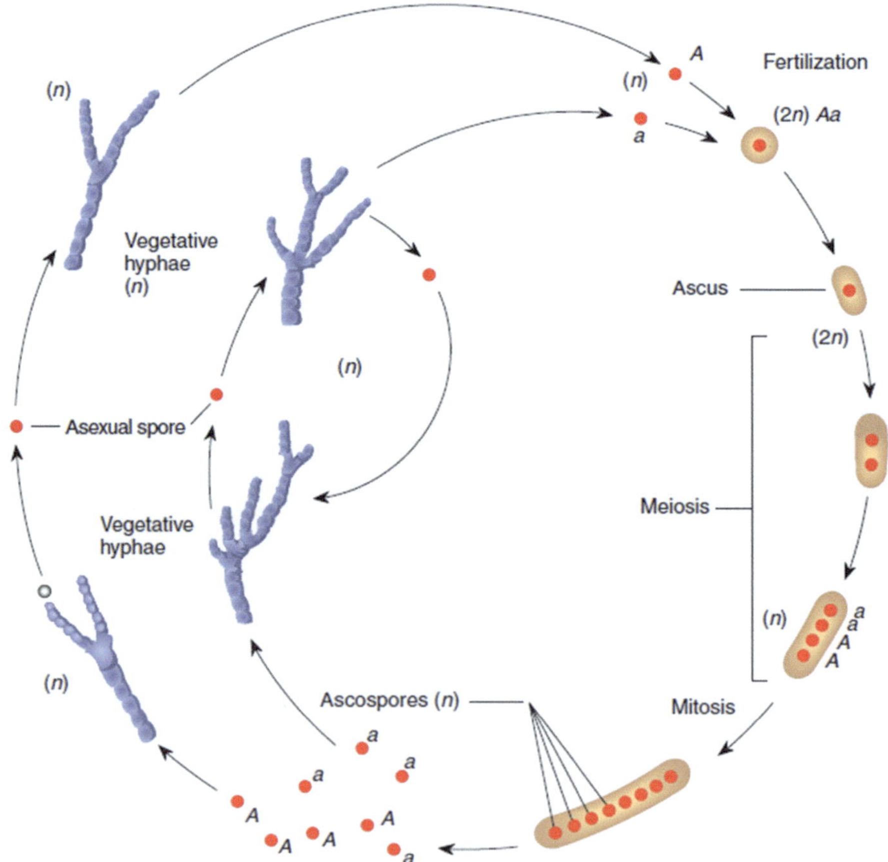

Fig. 4.27 *Neurospora* life cycle. The mating types are represented as *A* and *a*; *n* represents haploid stage; *2n* represents diploid stage

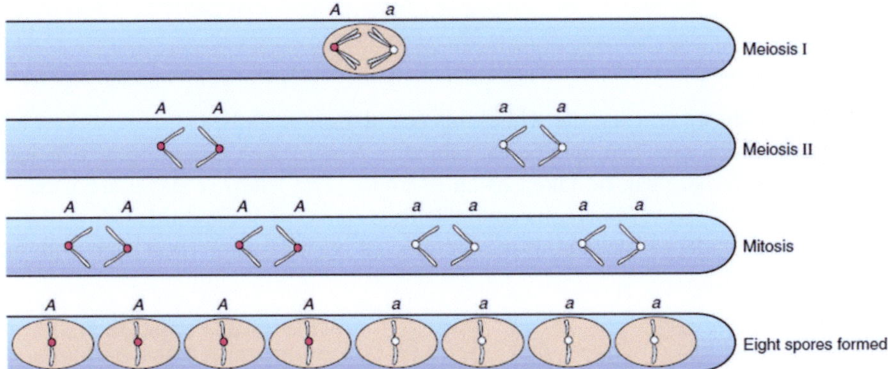

Fig. 4.28 Meiosis in *Neurospora*. The two mating types are denoted by *A* and *a*. Here *A* and *a* represent centromeres of the chromosomes of the two mating types, and the spores are labelled likewise. Note that, for convenience purpose only one pair of chromosomes is shown in the figure, although *Neurospora* consists of seven pairs of chromosomes

the ascus's longitudinal axis. Hence, the spores are said to be ordered (Fig. 4.28), i.e., if centromere of a chromosome of one mating type is labeled *A* and the other is labeled *a*, then at meiosis I, one each of *A* and *a* will be present in a tetrad. At the completion of meiosis, the four ascospores formed will be arranged in either one of these two orders, *a a A A* or *A A a a*.

In *Neurospora*, a mitosis takes place in each nucleus prior to the maturation of ascospores. This results in the formation of four pairs of spores instead of just four spores. The pairs are always identical provided they do not show any genetic phenomena such as a mutation or a gene conversion (Fig. 4.28). The ordered spores in *Neurospora* make it possible to map genetic loci relative to their centromeres.

4.2.3.2.2 First- and Second-Division Segregation

The centromeres in an ascus of *Neurospora* undergo a 4:4 segregation. There are two kinds of patterns that are observed among the loci on the chromosomes of spores and are dependent on incidence of crossover between the given locus and its centromere (Fig. 4.29). If a crossover did not take place between the given locus and its centromere, then the allelic pattern is same as the centromeric pattern, and is called first-division segregation (FDS), because the alleles separate from each other at meiosis I. Alternatively, if a crossover occurred, then different types of segregation patterns (2:4:2 or 2:2:2:2) called second-division segregation (SDS) can be observed. However, centromeres will always follow a FDS pattern because of the ordered nature of the spores in *Neurospora*, and hence will enable mapping distance between the given locus and its centromere. Considering simplest circumstances (Fig. 4.29), every SDS configuration will have four each of nonrecombinant and recombinant chromatids (spores). Hence, in an SDS ascus, half of the chromatids (spores) are recombinant, and the map distance can be calculated as follows, considering 1% recombinant chromatid is equal to 1 map unit:

Map distance	=	$\frac{(1/2)\text{ the number of SDS asci}}{\text{Total number of asci}}$	×	100

Table 4.6 shows an example where this calculation is applied.

Three-point crosses can also be examined in *Neurospora*. Mapping of two loci and their centromere can be taken as an example. The letters *a* and *b* are used to denote two loci. Fusion of mycelia results in the formation of dihybrids ($ab \times a^+b^+$), which further undergo meiosis. Data is collected by analyzing 1000 asci, without

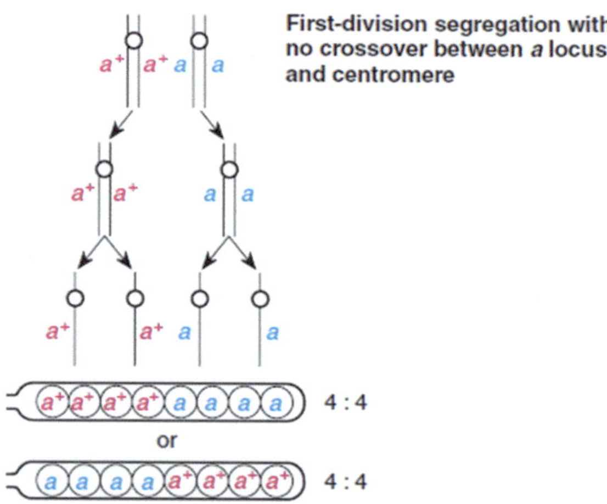

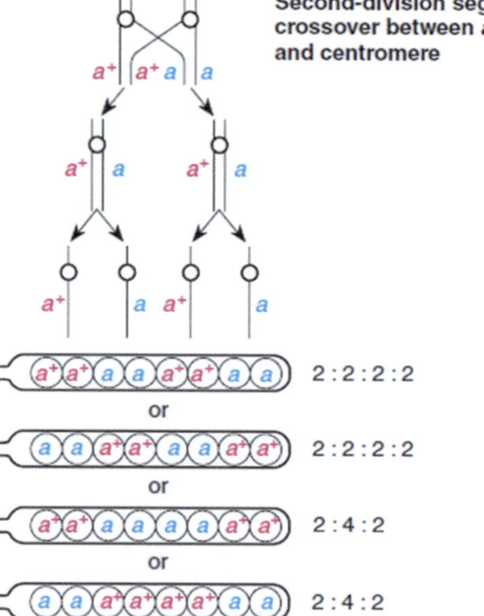

Fig. 4.29 *Neurospora* ascospore patterns. Six possible ascospore patterns with respect to one locus are depicted in the figure

Table 4.6 Meiosis in an $a^{+}a$ heterozygous *Neurospora* and the resulting genetic patterns (ten asci are examined)

	Ascus number									
Spore number	1	2	3	4	5	6	7	8	9	10
1	a	a	a^{+}	a	a	a^{+}	a	a^{+}	a^{+}	a^{+}
2	a	a	a^{+}	a	a	a^{+}	a	a^{+}	a^{+}	a^{+}
3	a	a	a^{+}	a^{+}	a^{+}	a^{+}	a	a	a	a^{+}
4	a	a	a^{+}	a^{+}	a^{+}	a^{+}	a	a	a	a^{+}
5	a^{+}	a^{+}	a	a^{+}	a	a	a^{+}	a	a^{+}	a
6	a^{+}	a^{+}	a	a^{+}	a	a	a^{+}	a	a^{+}	a
7	a^{+}	a^{+}	a	a	a^{+}	a	a^{+}	a^{+}	a	a
8	a^{+}	a^{+}	a	a	a^{+}	a	a^{+}	a^{+}	a	a
	FDS	FDS	FDS	SDS	SDS	FDS	FDS	SDS	SDS	FDS

Note:

$$\text{Map distance (alocus to centromere)} = (1/2)\%\text{SDS}$$
$$= (1/2)\ 40\%$$
$$= 20 \text{ map units}$$

changing the spore order. The data thus obtained are grouped as follows: since six different patterns can be shown by each locus (Fig. 4.29), 36 possible spore arrangements should be obtained when two loci are scored together (6×6).

Several of these 36 patterns are actually random variants of each other. This is so because at the first meiotic division, either of the centromeres of a tetrad can separate toward either of the poles (i.e., go left *or* right). Further, when splitting of centromeres takes place at meiosis II, movement of the spores within the progeny ascus is also random. Therefore, one single genetic event can yield up to eight "different" patterns. One such possible arrangement of the spore chromatids post crossing over is depicted in Fig. 4.30. Considering arrangements in which there is a crossover between *a* and *b* loci, all eight arrangements that produce ascus patterns as shown in Table 4.7 are equally possible. Then in that case, the 36 possible patterns reduce to only seven unique patterns as shown in Table 4.8.

4.2.3.3 Gene Order Determination

After having obtained the data shown in Table 4.8, the distance between each locus and the centromere and the linkage arrangement of two loci associated with the same centromere can be determined. It can be established by inspection that two loci are liked and also that they are linked to the same centromere. As shown in Table 4.8, examination of classes 1 (PD—parental ditype) and 2 (NPD—nonparental ditype) allows deduction of the linkage relationships among the loci and also that of the centromere. If the two loci under consideration are unlinked, both categories 1 and 2 will represent two equally probable alternative events when crossing over does not occur. Since in this case category 1 makes up almost 75% of all asci, it can be

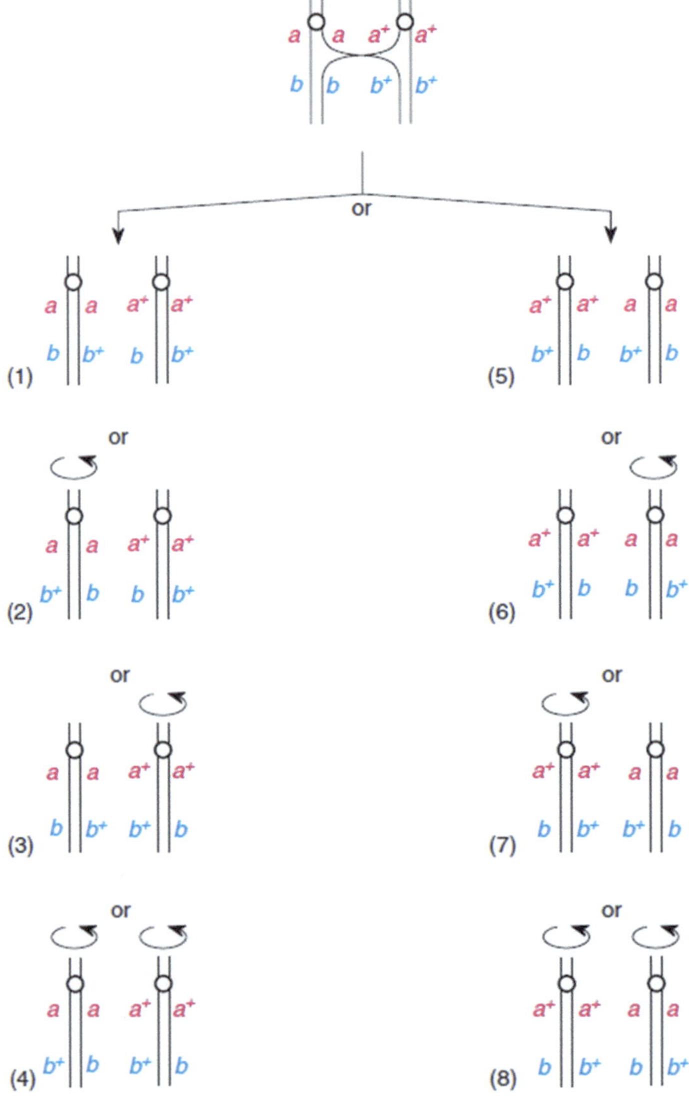

Fig. 4.30 Spore patterns in *Neurospora*. A single crossover between *a* and *b* loci can yield eight possible random arrangements. Circular arrows indicate rotation of a centromere from its position in the original configuration

deduced with certainty that the two loci are linked. In order to determine the distance between each of the locus and the centromere, it is required that for each locus 50% of SDS patterns are calculated. Classes 4, 5, 6, and 7 and classes 3, 5, 6, and 7 are the SDS patterns for loci *a* and *b*, respectively. Therefore, for each locus, the distance to the centromere, in map units, is calculated as follows:

Table 4.7 Eight out of 36 possible spore patterns in *Neurospora*, scored for loci, *a* and *b* (all are random variants of the same genetic event)

	Ascus number							
Spore number	1	2	3	4	5	6	7	8
1	ab	ab^+	ab	ab^+	a^+b^+	a^+b^+	a^+b	a^+b
2	ab	ab^+	ab	ab^+	a^+b^+	a^+b^+	a^+b	a^+b
3	ab^+	ab	ab^+	ab	a^+b	a^+b	a^+b^+	a^+b^+
4	ab^+	ab	ab^+	ab	a^+b	a^+b	a^+b^+	a^+b^+
5	a^+b	a^+b	a^+b^+	a^+b^+	ab^+	ab	ab^+	ab
6	a^+b	a^+b	a^+b^+	a^+b^+	ab^+	ab	ab^+	ab
7	a^+b^+	a^+b^+	a^+b	a^+b	ab	ab^+	ab	ab^+
8	a^+b^+	a^+b^+	a^+b	a^+b	ab	ab^+	ab	ab^+

Table 4.8 Meiosis in a dihybrid *Neurospora*, ab/a^+b^+, giving rise to seven unique classes of asci

	Ascus number						
Spore number	1	2	3	4	5	6	7
1	ab	ab^+	ab	ab	ab	ab^+	ab
2	ab	ab^+	ab	ab	ab	ab^+	ab
3	ab	ab^+	ab^+	a^+b	a^+b^+	a^+b	a^+b^+
4	ab	ab^+	ab^+	a^+b	a^+b^+	a^+b	a^+b^+
5	a^+b^+	a^+b	a^+b^+	a^+b^+	a^+b^+	a^+b	a^+b
6	a^+b^+	a^+b	a^+b^+	a^+b^+	a^+b^+	a^+b	a^+b
7	a^+b^+	a^+b	a^+b	ab^+	ab	ab^+	ab^+
8	a^+b^+	a^+b	a^+b	ab^+	ab	ab^+	ab^+
	729	2	101	9	150	1	8
SDS for *a* locus				9	150	1	8
SDS for *b* locus			101		150	1	8
Unordered	PD	NPD	TT	TT	PD	NPD	TT

For locus *a*:	(1/2)	$\frac{9+150+1+8}{1000}$	×	100
= 8.4 centimorgans				
For locus *b*:	(1/2)	$\frac{101+150+1+8}{1000}$	×	100
= 13.0 centimorgans				

The two distances obtained in the above calculation do not allow determination of gene order. As it can be seen in Fig. 4.31, there are two possibilities, i.e., the two loci can either be 21.4 or 4.6 map units apart. However, it can be determined as to which one of the two is correct. The solution for this is calculating distance between *a* and *b* using information on unordered spores. Therefore, the calculation for map distance is as follows:

$$\text{map units} = \frac{(1/2)\ \text{the number of TT asci} + \text{the number of NPD asci}}{\text{Total number of asci}} \times 100$$

$$= \frac{(1/2)118+3}{1000} \times 100 = 6.2$$

Since the above calculated map distance (6.2 map units) is closer to the distance expected if both loci were located on one side of the centromere, the second alternative depiction in Fig. 4.31 is accepted.

Another way to select between the alternatives shown in Fig. 4.31 is by finding what will be the status of *b* locus if a crossover occurs between *a* and its centromere. If arrangement 1 is correct, then crossover between *a* and its centromere should not alter *b*, and if arrangement 2 is correct, then most crossovers that move *a* relative to its centromere will also move *b*.

The asci classes 4, 5, 6, and 7 (Table 4.8) comprise all SDS patterns pertaining to *a* locus. In class 5 asci, 150 out of 168 have similar SDS patterns for *b* locus. Hence, 89% of the times, crossover between *a* and its centromere, also results in a crossover between *b* and its centromere, which provides clear evidence supporting arrangement 2 in Fig. 4.31.

The mapping procedure by tetrad analysis can be summarized as follows: for both unordered and ordered spores, indication of linkage comes from the occurrence of large numbers of parental ditypes compared with nonparental ditypes. With respect to unordered spores, like those observed in yeast, the distance between any two loci is equal to the sum of half the number of tetratypes and the number of nonparental ditypes, divided by the total number of asci, expressed as percentage. In case of

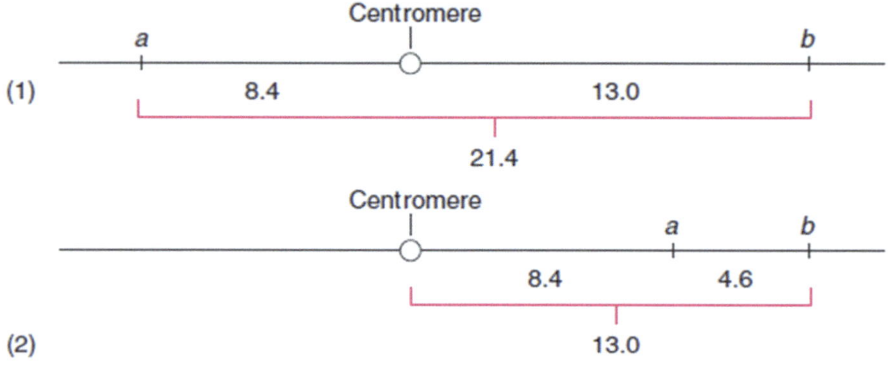

Fig. 4.31 Possible arrangements of *a* and *b* loci with respect to centromere. The distances are given in map units

ordered spores, like those in *Neurospora*, distance between a locus and its centromere is half the percentage of SDS. The method of mapping distance between two loci in ordered spores is similar to that applied for unordered spores.

4.2.3.4 Mapping of Centromere Using Linear Tetrads

In eukaryotic chromosomes, unlike other parts of the chromosome, mapping of special DNA sequences that form the centromeres cannot be done by utilizing recombination analysis as a tool. This is because they do not exhibit heterozygosity which can be used as markers. Nevertheless, centromeres *can* be mapped particularly in fungi such as *Neurospora* that produce tetrads that are linear. *Neurospora* is a haploid fungus, in which the meiotic divisions occur along the ascus's long axis, such that each product of meiosis produces eight linearly arranged ascospores, referred to as an octad. This octad comprises the tetrad (four products of meiosis) and a mitosis following the meiosis. Centromere mapping takes into consideration a gene locus and calculates the distance between this locus and the centromere. The method depends on the fact that meiotic crossover between a locus and a centromere within a tetrad results in different patterns of alleles, for example, an individual with *a* allele when crossed with another individual with allele *A* at the same locus ($a \times A$). According to the Mendelian law of segregation, the octad should comprise eight ascospores, including four each of genotype *a* and *A*. Hence, if a crossover did not occur between centromere and locus *a*/*A*, then the linear octad will consist of four ascospores of each allele placed adjacently in sets (Fig. 4.32). In case a crossover occurred between centromere and locus *a*/*A*, then the octad will consist one of four patterns, in which each of the alleles is arranged in sets of twos. Data from a particular cross of $A \times a$ *have been given* in Table 4.9.

In the given data in Table 4.9, the first two columns are products of meiosis without crossover between locus *A* and the centromere and are the FDS patterns (M_I). The remaining four patterns are the SDS patterns (M_{II}), which are all products of meiotic crossover between *A* and centromere (Fig. 4.33).

Hence, in this case, there has been no FDS, but the *A* and *a* alleles are segregated into separate nuclei during meiosis II. The formation of one of the M_{II} patterns is depicted in Fig. 4.33. The other patterns observed are also produced in a similar fashion but vary in the chromatids that move to the different poles in the second division of meiosis (Fig. 4.34).

The frequency of M_{II} pattern octads will be proportional to the distance between *a*/*A* region and centromere and hence can be used as a measure of distance between these points on the chromosome. In the given example, M_{II} frequency is equal to 14% (42/300). Using this frequency, the number of map units between the locus and the centromere can be calculated. The SDS frequency of 14% denotes the percentage of *meiosis*, whereas a map unit is the recombinant *chromatids* percentage produced during meiosis. Since meiotic crossover yields only 50% recombinants (4 out of 8; see Fig. 4.33), the M_{II} frequency of 14% must be divided by 2 to convert to map units (a frequency of recombinant *chromatids*). Therefore, in the given example, the distance between locus A and its centromere is 7 map units. These measurements can be further used to construct a chromosomal map.

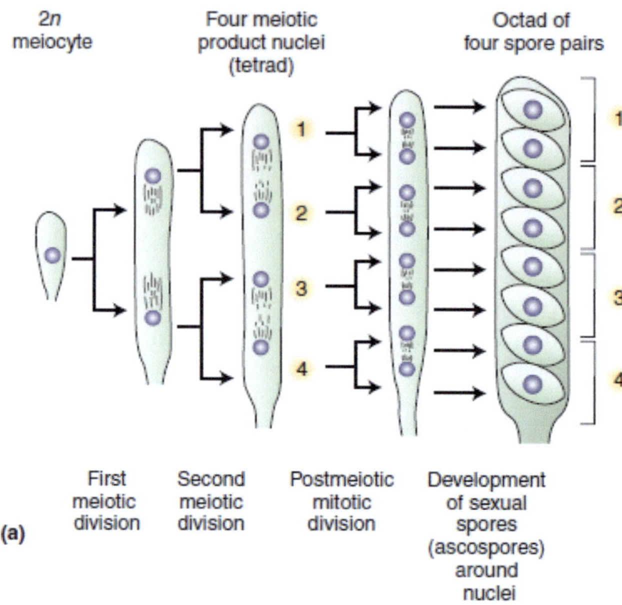

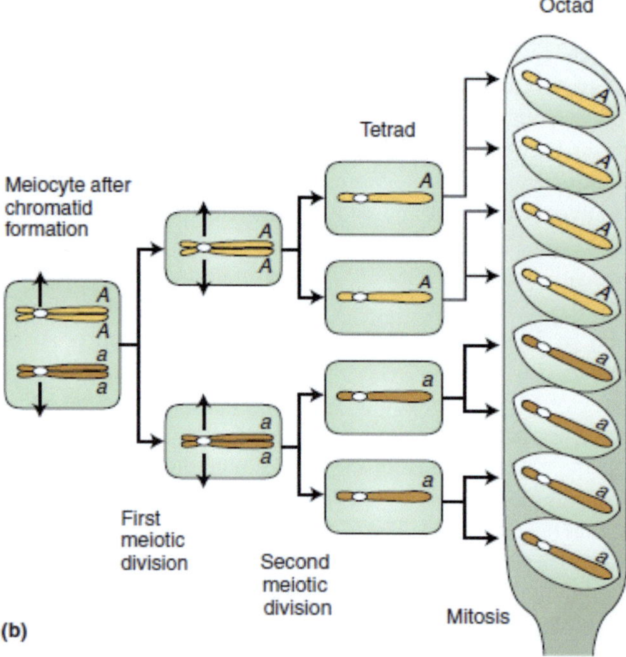

Fig. 4.32 Allelic segregation at meiosis in *Neurospora*. (**a**) Mitosis of the tetrad derived from meiosis produces octads which are present within the ascus. (**b**) A meiocyte with genotype *A/a* undergoes meiosis followed by mitosis, giving rise to equal number of *A* and *a* products, indicating law of segregation

Table 4.9 Data from a particular cross of $A \times a$

Octads					
A	*a*	*A*	*a*	*A*	*a*
A	*a*	*A*	*a*	*A*	*a*
A	*a*	*a*	*A*	*a*	*A*
A	*a*	*a*	*A*	*a*	*A*
a	*A*	*A*	*a*	*a*	*A*
a	*A*	*A*	*a*	*a*	*A*
a	*A*	*a*	*A*	*A*	*a*
a	*A*	*a*	*A*	*A*	*a*
126	132	9	11	10	12
Total = 300					

4.3 Construction of Genetic Maps of Human Genome

As explained in Sect. 4.1.5.1, linkage analysis in humans cannot rely on planned breeding experiments. In this case, family pedigrees will have to be made use of for linkage analysis. Primarily, disease alleles have been analyzed with the help of pedigrees.

4.3.1 Mapping by Pedigree Analysis: X-Linkage

The traits encoded by the X chromosome have unique inheritance patterns, and the gene loci on the X chromosome can be easily identified. Over 400 loci have been identified on the X chromosome. By using several different methods, it has been estimated that human chromosomes consist of between 50 and 100,000 loci. Initially, the X chromosome has been mapped using the pedigree analysis.

Once a gene is determined to be X-linked, it is necessary to determine the position of that gene on the X chromosome. It is also required to determine the distance (map units) of the gene and other loci on the X chromosome. Proper pedigrees can be used to determine position and map distances, if occurrence of crossing over can be ascertained. One such example is depicted in Fig. 4.35 and is called the "grandfather method." In the given example, the grandfather expresses one of the traits in consideration, which is color-blindness in this case. It can be observed that his grandson is glucose-6-phosphate dehydrogenase (G-6-PD) deficient, which indicates that the grandson's mother is a dihybrid for these two alleles. The alleles in the grandson's mother are in *trans* configuration, i.e., she has received the color-blindness allele from her father on one of her X chromosomes and an allele for G-6-PD-deficiency on her other X chromosome from her mother. Therefore, among the grandsons (Fig. 4.35), two are nonrecombinant (left), and two are recombinant (right). In theory, the map distance can be calculated by dividing the sum of the recombinant grandsons by the total number of grandsons. The same method could be

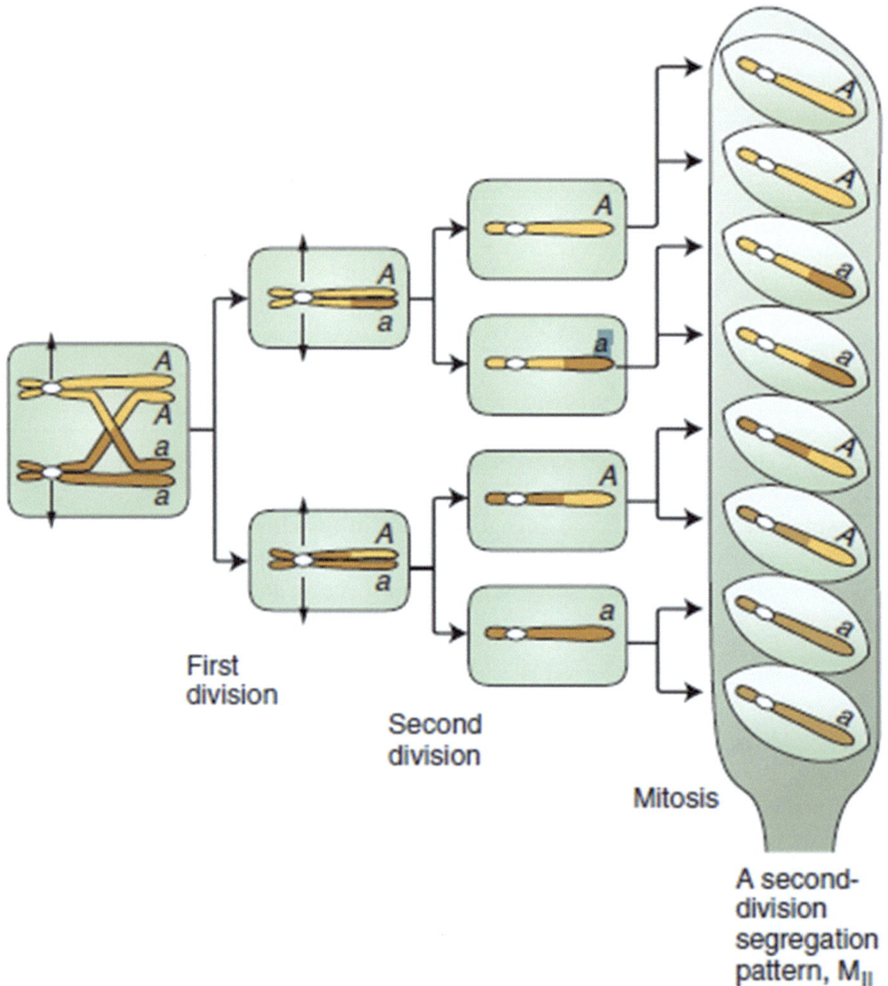

Fig. 4.33 Second-division segregation pattern in *Neurospora*. Segregation of *A* and *a* into separate nuclei takes place during the second meiotic division as a result of a crossover between the *A* locus and the centromere

applied in case the grandfather was both G-6-PD deficient and color-blind. In this case, the mother would be a dihybrid with *cis* configuration, and the sons would have a reverse arrangement. Here it is important to note that the grandfather's phenotype leads to the inference that the mother was a dihybrid, and reveals information on the *cis-trans* configuration of the alleles. Further, this allows the scoring of her sons as either nonrecombinant or recombinant.

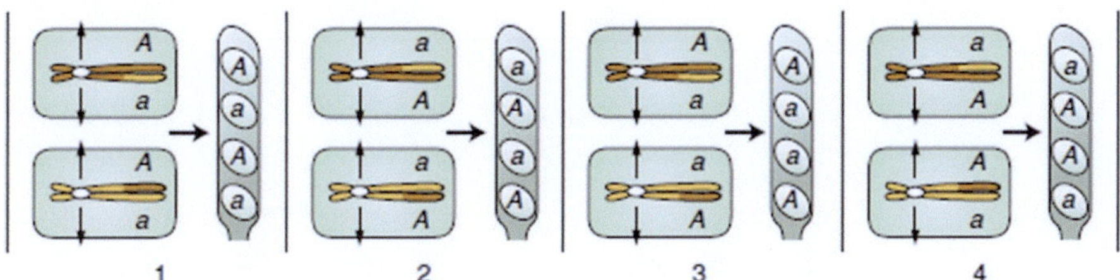

Fig. 4.34 Four SDS patterns in linear asci. Centromeres randomly attach to the spindle during the second meiotic division, resulting in the four arrangement patterns shown in the figure. The four arrangements occur with equal frequency

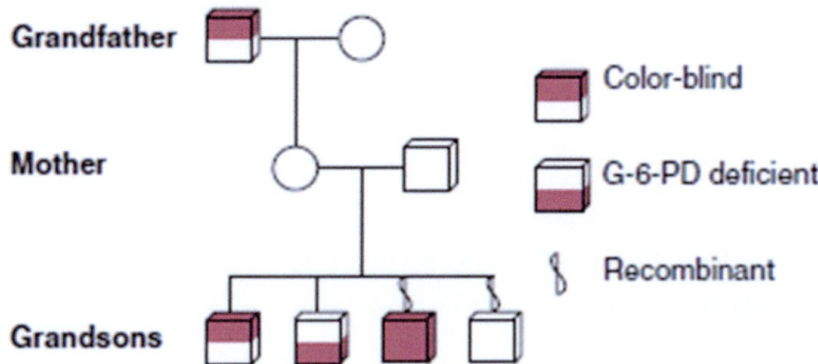

Fig. 4.35 Pedigree analysis for X-linkage. "Grandfather method" of crossover determination between genetic loci on human X chromosome. Color-blindness and G-6-PD alleles are considered for analysis

4.3.2 Mapping by Pedigree Analysis: Autosomal Linkage

Mapping of sex chromosomes through pedigree analysis is a relatively easier task. But when it comes to the autosomes, it is a much more complex process, because there are 22 autosomal linkage groups and it is practically impossible to use simple pedigrees to determine on which chromosome any two given loci are located on. With the help of pedigrees, it is possible to deduce if there is linkage between two gene loci, but the chromosomal location of the same cannot be deduced.

For example, nail-patella syndrome is a dominant trait whose symptoms include abnormal nail growth along with underdevelopment or absence of kneecaps, among other effects. In the given pedigree (Fig. 4.36) of this disorder, the male in generation II is a dihybrid, with the nail-patella allele (*NPS1*) associated with allele *A* of ABO blood group system and the normal nail-patella allele (*nps1*) associated with *B* allele of the ABO blood group system. Consequently, only one out of eight children (III-5) is a recombinant and the actual map distance is about 10%. Generally, because a greater number of crossovers occur in females, the map distances appear to be greater than in males.

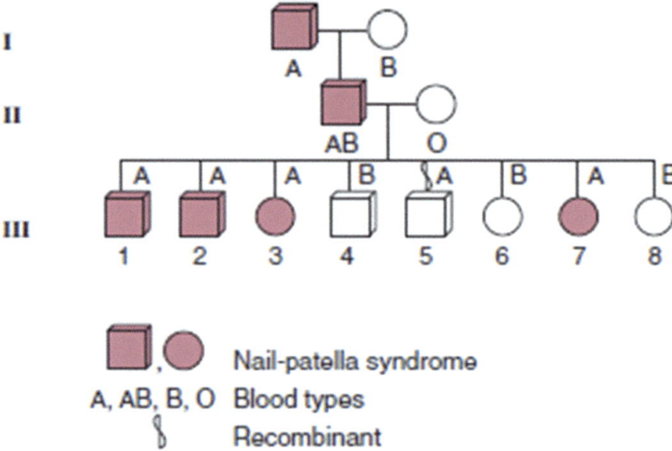

Fig. 4.36 Pedigree for autosomal linkage analysis. Analysis of the linkage between ABO blood group loci and the nail-patella syndrome

4.3.3 Using Lod Score Method to Assess Linkage in Human Pedigrees

In humans thousands of phenotypes are transmitted via the autosomes. Due to various reasons, the progress in mapping of the gene loci on human chromosomes was slow initially. The most important reason is that unlike in animals, performing controlled crosses in humans is impossible. Therefore, calculation of recombinant frequencies (RFs) had to be restricted to the occasional dihybrids that were produced by chance mating. Test cross equivalent mating is extremely rare. Another reason is that generally human matings produce smaller number of progeny. The smaller number of progeny limits availability data and hence making it difficult to derive statistically significant map distances. The third limitation is the vastness of the human genome and hence the distances between known genes, on an average, are very large.

Large sample sizes are essential in order to obtain reliable RF values. Nevertheless, even in cases of individual mating where the number of progeny is small, combining results of several such identical matings are essential to obtain a more reliable estimate. Calculation of Lod scores is a standard way of doing this. LOD is the acronym for "log of odds." To obtain a set of results in a given family, the LOD method employs calculation of two different probabilities. The first probability is calculated by assuming independent assortment, while the second is calculated by assuming a specific degree of linkage. This is followed by the calculation of the ratio (odds) between the two probabilities, and a logarithm of this ratio is called as the Lod value. Lod scores derived from different mating, but involving the same genetic markers, can be added as the logarithms are exponents. Therefore, a cumulative data set indicating presence or absence of a particular linkage value is obtained.

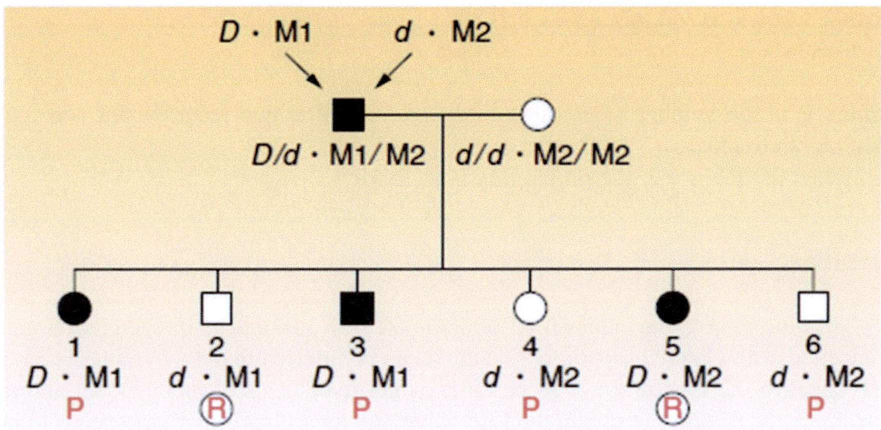

Fig. 4.37 Pedigree equivalent to a dihybrid test cross. D/d are alleles of a disease gene; M1 and M2 are alleles of a molecular marker. P indicates parental (nonrecombinant); R indicates recombinant

An example of how this calculation works is as follows: assume a hypothetical situation where a family is the product of a dihybrid test cross. Also, it needs to be assumed that for a dihybrid individual, the input gametes can be deduced, and hence the individual's gametes can be assessed for recombination events. The dihybrid in consideration is heterozygous for a dominant disease (*Dd*) allele and a molecular marker (M1M2). It is assumed that the dihybrid individual is a male, and is formed by union of gametes with genotypes (parental genotypes) *D*•M1 and *d*•M2. Consider this individual is married to a woman with genotype *d*/*d*•M2/M2. The six children of this marriage are shown in a pedigree depicted in Fig. 4.37. The pedigree also shows the parental and recombinant categories with respect to gamete contributed by the father.

Out of the six offspring, two are recombinants that would give a 33% RF. It can be considered that the genes are undergoing independent assortment and the offspring form a nonrandom sample. The probability of this outcome can be calculated under several hypotheses. The expected proportions of parental (P) and recombinant (R) genotypes are shown in Table 4.10 under independent assortment and under three RF values:

When assorted independently, in which case RF is equal to 50%, the probability of obtaining results is computed as follows:

Table 4.10 Expected proportions of parental (P) and recombinant (R) genotypes under three RF values and under independent assortment

	RF			
	0.5	0.4	0.3	0.2
P	0.25	0.3	0.35	0.4
P	0.25	0.3	0.35	0.4
R	0.25	0.2	0.15	0.1
R	0.25	0.2	0.15	0.1

$0.25 \times 0.25 \times 0.25 \times 0.25 \times 0.25 \times 0.25 \times B$
$= 0.00024 \times B$

where B is the number of possible birth orders for the two recombinant and four parental individuals.

When the RF is 0.2, the probability is as follows:

$0.4 \times 0.1 \times 0.4 \times 0.4 \times 0.1 \times 0.4 \times B$
$= 0.00026 \times B$

The ratio of both the values is 0.00026/0.00024 = 1.08 (it can be seen that B gets cancelled out). On the basis of these data, it can be inferred that the hypothesis of an RF equal to 0.2 is 1.08 times as possible as the hypothesis of independent assortment. Lastly, in order to obtain the Lod value, the logarithm of the ratio needs to be taken. Table 4.11 lists some ratios and their corresponding Lod scores:

These data indicate 30–40% RF because the largest Lod scores are generated by these hypotheses. However, these data alone are insufficient to credibly support any linkage model. In practice, for a specific RF value, a Lod score of at least 3 (obtained by adding scores from several matings) is considered as a convincing support. It can be noted that Lod score 3 represents RF that is 10^3 (1000) times as likely as the no linkage (independent assortment) hypothesis.

4.3.4 Assigning Genetic Loci to Chromosomes

The gene locus for the Duffy blood group, located on chromosome 1, was the first gene locus to be definitely assigned to a specific chromosome. Banding of chromosomes and somatic cell hybridization are the two techniques that have been very important for mapping of autosomes.

4.3.4.1 Chromosomal Banding

Around the 1970s, certain techniques to stain chromosomes were evolved. These techniques used certain histochemical stains that could stain the chromosomes, producing reproducible banding patterns on them. One such example was the Giemsa staining technique and the bands that formed were called G-bands. Before the advent of these staining techniques, mammalian chromosomes, including that of humans, were grouped into categories based on their general size, because

Table 4.11 Ratios and their corresponding Lod scores

	RF			
	0.5	0.4	0.3	0.2
Probability	0.00024	0.00032	0.00034	0.00026
Ratio	1.0	1.33	1.41	1.08
Lod	0	0.12	0.15	0.03

differentiating them was a difficult task. The banding techniques enabled identification of each human chromosome arranged in a karyotype.

4.3.4.1.1 ISCN Mapping System

A consistent numbering system is required for mapping genes on chromosomes and constructing relevant idiograms. The International System for Cytogenetic Nomenclature (ISCN) is a mapping that is used for this purpose. The numbering for a chromosome starts at the centromere according to the ISCN scheme. Based on centromere position, chromosomal short (p) and long (q) arms are identified; likewise, regions on the short arm will be designated p, and those on long arm with q. By convention, the p arm is always depicted on top in a karyotype.

Chromosomal arms are divided into regions, and starting from the centromere, these regions are assigned numbers in an increasing order according to the distance from the centromere. Specific morphological features such as a Giemsa band characterize these regions, and these are consistent with the chromosomes. On the short and long arm, the regions are labeled p1, p2, etc. and q1, q2, etc., respectively. Depending on resolution of staining procedures, within each region, additional bands may be detected, and these are further designated by adding another digit to the existing number of that region (Fig. 4.38), again in the increasing order of the distance from the centromere.

4.3.4.2 Somatic Cell Hybridization

Ability to distinguish the chromosomes allows application of somatic cell hybridization. It is a technique in which hybrid cells are formed by in vitro fusion of mouse/hamster cells with human cells.

Fusion of two cells initially results in a heterokaryon (a cell comprising nuclei from different origins). The different nuclei eventually fuse and a hybrid cell which tends to preferentially lose human chromosomes in the following generations is formed. After stabilization, it results in a cell consisting of the hamster or mouse chromosomes along with one or more human chromosomes. These chromosomes can be identified by applying banding techniques. The hybrid cells can be screened for human specific phenotypes, e.g., enzyme products, and these phenotypes can be assigned to particular human chromosomes present in those cells.

To map human genes, two particular tests are employed. One of the tests is called as the synteny test. Synteny means the same linkage group. The synteny test verifies the belongingness of two gene loci to the same linkage group. Two genes are considered syntenic if phenotypes encoded by them are always present or absent together in several hybrid cell lines. The other test, called the assignment test, determines chromosomal location of a particular gene locus. This is done by observing concordance between presence of a phenotype and a particular chromosome, or absence of a phenotype and a particular chromosome is in a hybrid cell line.

In 1970, the first synteny test for autosomes was performed, and it was demonstrated that the *B* locus of peptidase (*PEPB*) was linked to the *B* locus of lactate dehydrogenase (*LDHB*). Both enzymes required subunits for their formation,

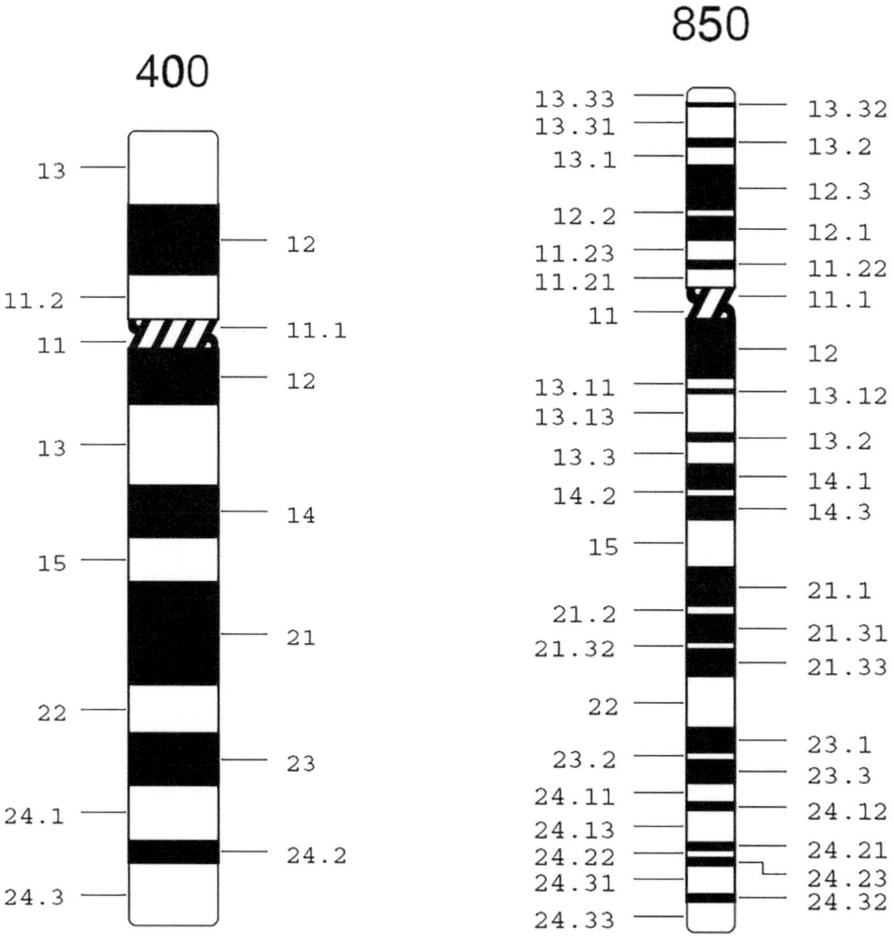

Fig. 4.38 Idiogram for chromosome 12. The chromosome is mapped as per the ISCN mapping system. (Left) A low-resolution map, obtained by staining of a metaphase chromosome with a stain like Giemsa. (Right) A high-resolution map, obtained by staining of a prometaphase chromosome. The number depicted on top of the ideogram represents the number of bands visualized

and the subunits were regulated by two loci, *A* and *B*, which were shown to be located on chromosome 12.

Another example of assignment test includes localization of tissue factor III, a blood-coagulating glycoprotein to chromosome 1. Results of the assignment test performed for the localization of the coagulating factor are depicted in Table 4.12. Twenty-nine human-mouse hybrid cell lines, or clones containing human chromosomes, and their tissue factor score, have been shown. Using the table location of the gene coding for tissue factor III can be clearly determined to be human chromosome 1. The concordance between the presence of human

Table 4.12 Use of human-mouse hybrid cell lines for assigning the gene for blood-coagulating factor III to chromosome 1 of humans

Hybrid Cell Line Designation	Tissue/ Factor Score	Human Chromosome Present																						
		1	2	3	4	5	6	7	8	9	10	11	12	13	14	15	16	17	18	19	20	21	22	X
WIL1	−	−	−	−	−	−	−	−	+	−	−	−	−	−	+	−	−	+	−	−	−	+	−	+
WIL6	−	−	+	−	+	+	+	+	+	−	+	+	−	−	+	−	−	+	−	+	+	+	−	+
WIL7	−	−	+	+	−	+	+	−	+	−	+	+	−	+	+	−	−	+	+	−	−	+	−	+
WIL14	+	+	−	+	−	−	−	+	+	−	+	−	+	−	+	+	−	+	−	−	−	−	−	+
SIR3	+	+	+	+	+	+	+	+	−	+	+	+	+	+	−	−	+	+	+	+	+	+	+	+
SIR8	+	+	+	+	+	+	−	+	+	+	+	+	+	+	+	+	+	+	+	−	−	+	+	+
SIR11	−	−	−	−	−	−	−	+	−	−	−	−	−	+	−	−	−	−	−	−	−	+	+	+
REW7	+	+	+	+	+	+	+	+	+	−	+	+	+	+	+	+	−	+	+	+	+	+	+	+
REW15	+	+	+	+	+	+	+	+	+	−	+	−	+	+	+	+	−	+	+	+	+	+	+	+
DUA1A	−	−	−	−	−	−	−	−	−	−	−	−	−	−	−	a	−	−	−	−	−	−	−	a
DUA1CsAzF	−	−	−	−	−	−	−	+	−	−	−	−	−	−	−	−	−	−	−	−	−	−	−	−
DUA1CsAzH	−	−	−	−	−	−	−	+	−	−	−	−	−	−	−	−	−	−	−	−	−	−	−	−
TSL1	−	−	−	+	+	−	−	−	−	−	+	+	−	+	+	−	+	+	+	−	+	−	−	−
TSL2	−	−	+	a	−	+	+	−	−	−	+	−	+	−	−	−	−	a	+	−	+	+	−	+
TSL2CsBF	−	−	−	−	−	+	−	−	−	−	−	−	−	−	−	−	−	−	−	−	−	−	−	−
XTR1	+	+	−	a	−	+	+	+	+	+	+	+	+	+	+	−	−	+	+	+	+	+	+	+
XTR2	−	−	−	a	−	+	−	−	+	−	+	−	+	+	−	−	−	−	+	−	+	+	−	a
XTR3BsAgE	+	+	−	a	−	+	+	+	+	+	+	−	−	+	+	−	−	+	+	+	−	+	−	a
XTR22	−	−	+	a	+	+	+	−	+	−	+	+	−	−	−	+	−	−	+	+	+	+	+	a
XER9	−	−	+	−	+	−	−	−	+	−	+	a	+	−	+	−	−	+	+	−	−	+	−	a
XER11	+	+	−	+	+	−	+	+	+	−	+	a	+	+	−	+	+	+	+	+	+	+	+	a
REX12	−	−	−	+	−	−	−	+	−	−	−	+	−	−	+	−	−	−	−	−	−	−	+	a
JSR29	+	+	+	+	+	+	+	a	+	a	+	+	+	+	+	+	+	+	+	+	+	+	+	+
JVR22	+	+	+	+	+	+	+	+	+	−	+	+	+	+	+	+	+	+	+	+	+	+	+	+
JWR22H	+	a	a	−	+	−	+	−	−	−	+	+	+	−	+	+	−	+	+	−	+	+	−	−
ALR2	+	+	+	+	+	+	+	+	−	+	+	+	+	+	+	+	+	+	−	+	+	+	+	+
ICL15	−	−	−	−	−	−	−	−	−	−	−	−	+	−	−	−	−	+	−	−	+	+	−	−
ICL15CsBF	−	−	−	−	−	−	−	−	−	−	−	−	+	−	−	−	−	−	−	−	+	+	−	−
MH21	−	−	−	−	−	−	−	−	−	−	−	−	−	−	−	−	−	−	−	−	−	+	−	−
% Discord[b]		0	32	17	24	31	21	21	31	21	24	30	21	21	28	14	24	21	28	17	34	41	21	27

Source: Reprinted with permission from S.D. Carson, et al., "Tissue Factor Gene Localized to Human Chromosome 1 (after 1p21)," *Science*, 229:229–291.

[a] A translocation in which only part of the chromosome is present

[b] Discord refers to cases in which the tissue factor score is plus, and the human chromosome is absent, or in which chromosome is present but the score is minus

chromosome 1 and the tissue factor III in the cell line is clearly established. Likewise, the absence of human chromosome 1 also corresponds to absence of tissue factor. Therefore, it can be said that that there is 100% concordance or zero discordance. None of the other chromosomes showed similar pattern with respect to the tissue factor III. Hence, it was established that the tissue factor III is located on chromosome I.

The task of determining the exact position of a particular locus on a chromosome is facilitated by particular cell lines which can be developed to incorporate broken chromosomes, which may lack certain parts or those parts may become incorporated in other chromosomes. These rearrangements reveal new linkage relationships that allow determination of the chromosomal region in which a particular locus is situated.

4.3.5 Chi-Square Test in Linkage Analysis

An objective statistical test called the chi-square (χ^2) test can be used to determine the presence or absence of linkage between two genes. If the recombination frequency (RF) is less than 50%, it can be inferred that the two genes are linked and are positioned on the same chromosome. It cannot be ascertained how much less than 50% RF signifies linkage; hence, it is not possible to directly test for linkage, as there is no precise linkage distance. Presence or absence of independent assortment is the only genetic criterion that can be precisely used to predict the presence or absence of linkage. Therefore, it becomes necessary to test the hypothesis for lack of linkage. If the observed results reject no linkage hypothesis, then it can be inferred as presence of linkage. Such a hypothesis is called null hypothesis. Since it allows precise experimental prediction that can be verified, it is generally of use in χ^2 analysis.

For example, the following specific data set can be tested for linkage using χ^2 analysis. Assume that a cross has been made between pure-breeding parents of genotypes A/A · B/B and a/a · b/b. A dihybrid A/a · B/b has been obtained, which is test crossed to a/a · b/b. A total of 500 progeny are obtained and classified as follows (depicted as gametes obtained from the dihybrid):

142	A · B	Parental
133	a · b	Parental
113	A · b	Recombinant
112	a · B	Recombinant
500		**Total**

The recombinant frequency calculated from these data is 225/500 = 45%. The RF value 45% is less than 50% as expected from independent assortment, and hence appears to be a case of linkage. However, the lower percentage of recombinant classes could also be a chance occurrence. Therefore, it is required that a χ^2 test be performed in order to calculate the likelihood that this result is based on chance. Calculation of the expectations (E) for each class is the first step in a χ^2 test. The hypothesis that needs to be tested here is that the two loci assort independently or in other words there is no linkage. Gametic E values are calculated by making simple predictions based on the first and second laws of Mendel as follows:

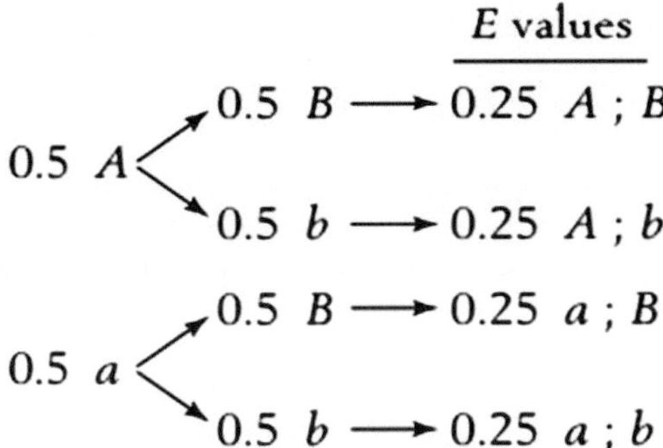

Hence, it might be asserted that if the dihybrid allele pairs are independently assorting, the gametic types should be in the ratio 1:1:1:1. Therefore, considering 1/4 of 500, i.e., 125, as the expected proportion of each gametic class seems rational. Nevertheless, it is important to note that 1:1:1:1 ratio can be expected only if all genotypes are equally viable. Often it is observed that genotypes are not equally viable due to the presence of certain alleles that affect the survival of individuals. Hence, allele ratios such as 0.6 *A*:0.4 *a* or 0.45 *B*:0.55 *b* may be observed rather than the 0.5:0.5 depicted above. These ratios can be used to predict independence. The observed genotypic classes from which the allelic proportions can be clearly observed are shown below:

Observed values	Segregation of *A* and *a*			
Segregation of *B* and *b*		**A**	***a***	**Total**
	B	142	112	254
	b	113	133	246
	Total	**255**	**245**	**500**

It can be seen that the allele proportions are 255/500 for *A*, 245/500 for *a*, 254/500 for *B*, and 246/500 for *b*. By multiplying allelic proportions, expected values under independent assortment can be calculated. For example, expected number of A B genotypes is obtained as follows:

Expected (*E*) value for A B
= (255/500) × (254/500) × 500 = 129.54

The entire grid of *E* values can be calculated using this approach, as follows:

Expected values	Segregation of *A* and *a*			
Segregation of *B* and *b*		**A**	***a***	**Total**
	B	129.54	124.46	254
	b	125.46	120.56	246
	Total	**255**	**245**	**500**

The value of χ^2 is calculated as follows:

Genotype	O	E	$(O - E)^2/E$
$A\ B$	142	129.54	1.19
$a\ b$	133	120.56	1.29
$A\ b$	113	125.46	1.24
$a\ B$	112	124.46	1.25
Total (equals the χ^2 value) = 4.97			

The χ^2 table is used further to find the corresponding probability (p) for the obtained χ^2 value (4.97). This requires calculation of the degrees of freedom. The number of nondependent values, in a statistical test, is generally referred to as the degrees of freedom. The product of number of classes denoted in rows minus one and number of classes denoted in columns minus one is equal to the number of degrees of freedom. In the given example, the degrees of freedom df = (2–1) × (2–1)- = 1. Hence, χ^2 value is located along the row corresponding to one degree of freedom in a χ^2 table. The value 5.021 is the closest to the χ^2 value of 4.97 in the χ^2 table and this corresponds to the probability 0.025, or 2.5%. Probability value p indicates the deviation from expectations. In the given example, probability is less than 5%, and hence the independent assortment, i.e., no linkage hypothesis, is rejected. Since the no linkage hypothesis is rejected, a probable linkage between the given loci is inferred.

4.4 Comparison of Genetic Map and Physical Map

Conventionally, genome mapping methods are grouped into two categories.

- Mapping that is based on genetic techniques is called genetic mapping. Maps are constructed using genetic techniques such as cross-breeding experiments or analysis of family histories, i.e., pedigrees in the case of humans. Position of genes and other genome sequence features are shown on these maps.
- Mapping wherein molecular biology techniques are used is called physical mapping. Using these techniques, DNA molecules are directly examined and maps are constructed. These maps show positions of genome sequence features, including genes.

4.4.1 Genetic Mapping

Genes were the first markers to be used in mapping of chromosomes. The first genetic maps were constructed for organisms such as the fruit fly, during the early twentieth century, and genes were used as the markers for this purpose (refer to Sect. 4.1.4). It is required that a heritable characteristic should exist in at least two alternative phenotypes, such that it can be used in genetic analysis, for example,

Table 4.13 Biochemical phenotypes used in mapping yeast chromosomes

Marker	Phenotype
ADE2	Requirement of adenine
CAN1	Canavanine resistance
CUP1	Resistant to copper
CYH1	Cycloheximide resistance
LEU2	Requirement of leucine
SUC2	Ability for sucrose fermentation
URA3	Requirement of uracil

the tall and dwarf pea plant phenotypes. Each variant phenotype is encoded by different alleles of a gene.

At the beginning, only genes that specified visually distinguishable phenotypes could be studied. For example, fruit fly chromosome maps that were initially constructed showed positions of genes that coded for phenotypes such as the body color, eye color, wing shape, etc. All of these phenotypes could be easily observed by visual examination of the flies either by the naked eye or by using a low-power microscope. Although this approach was useful in the early days, the number of visual phenotypes was the limitation, and also the analysis was complex in several cases as more than one gene can affect a single phenotype. In order to make gene maps more comprehensive, the biochemical characteristics of the phenotypes were used to distinguish the phenotypes. Biochemical phenotypes have been particularly significant in gene mapping of microbes and humans. The biochemical phenotypes used in mapping yeast chromosomes have been listed in Table 4.13.

Although visual characteristics have been used in humans, genetic variation studies have largely relied on biochemical phenotypes, identifiable via blood typing. In addition to the standard blood groups (such as ABO), serum and immunological protein variants, for example, the HLA (human leukocyte antigen) system, were also used for this purpose. These markers have an added advantage, because several of them have multiple alleles. Since gene mapping in humans is based on pedigree analysis, this becomes very relevant. Presence of multiple alleles increases the chance of marriage between individuals possessing allelic variants, and their inheritance patterns with the family lineages can be studied to derive useful information for the mapping purpose.

4.4.2 DNA Markers for Genetic Mapping

Even though the genes are very useful as markers, they are certainly not adequate. Particularly, in organisms such as the vertebrates and flowering plants which possess larger genomes, the maps cannot be based entirely on genes. This is because genes are widely spaced with large intergenic regions in most eukaryotic genomes. Therefore, using only genes for mapping will result in a map that is not very detailed. In addition, the fact that only a fraction of the total genes possess easily distinguishable allelic forms limits its use as a sole marker for mapping. Hence, very comprehensive gene maps cannot be constructed.

This clearly suggests that other types of markers are required. Markers other than genes that are utilized for mapping purposes are called DNA markers. Similar to gene markers, DNA markers should also comprise a minimum of two alleles to qualify for mapping purposes.

Three types of DNA sequence features, such as RFLPs (restriction fragment length polymorphisms), SSLPs (simple sequence length polymorphisms), and SNPs (single nucleotide polymorphisms), satisfy these requirements and hence are suitable for mapping purposes.

4.4.2.1 RFLPs (Restriction Fragment Length Polymorphisms)

The first DNA markers to be studied were the RFLPs. The digestion of a specific DNA molecule with a particular restriction enzyme will result in cleavage at specific sequence sites and produce same pattern of fragments reproducibly. However, this may not be the case with respect to genomic DNA, due to the polymorphic nature of the restriction sites. Some restriction sites will exist in two allelic forms, one consisting of the correct recognition sequence, whereas the other consisting of an altered restriction sequence, such that the enzyme can no longer recognize the restriction site. As a result, on digestion with that particular restriction enzyme, variable lengths of the restriction fragments will be generated, thus seen as a length polymorphism (Fig. 4.39).

It is estimated that the human genome consists of about 10^5 RFLPs. Each RFLP consists of only two alleles, i.e., either with presence or absence of the restriction site. Similar to mapping gene markers, inheritance pattern of RFLP alleles is analyzed and their position on the genome map is determined. Despite the number of RFLPs that can be available, their utility for gene mapping is limited due to

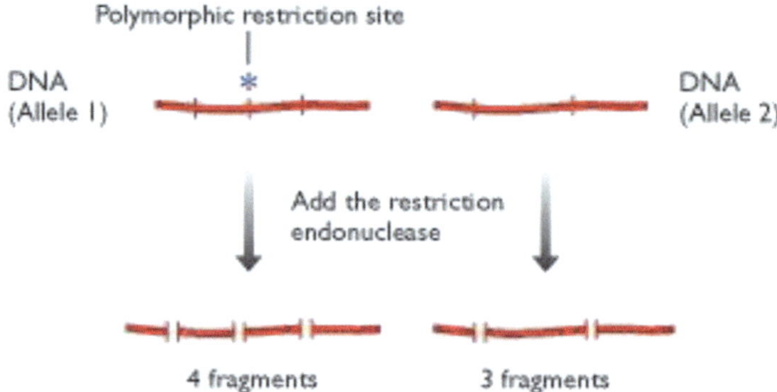

Fig. 4.39 RFLP marker. Allele 1 (left) consists of a polymorphic restriction site (indicated with an asterisk), while it is absent in Allele 2 (right). Restriction digestion with a specific restriction enzyme reveals this restriction fragment length polymorphism (RFLP) and is observed as variation in the pattern of fragments produced. In this example, Allele 1 (left) generates four fragments, whereas Allele 2 (right) produces three fragments

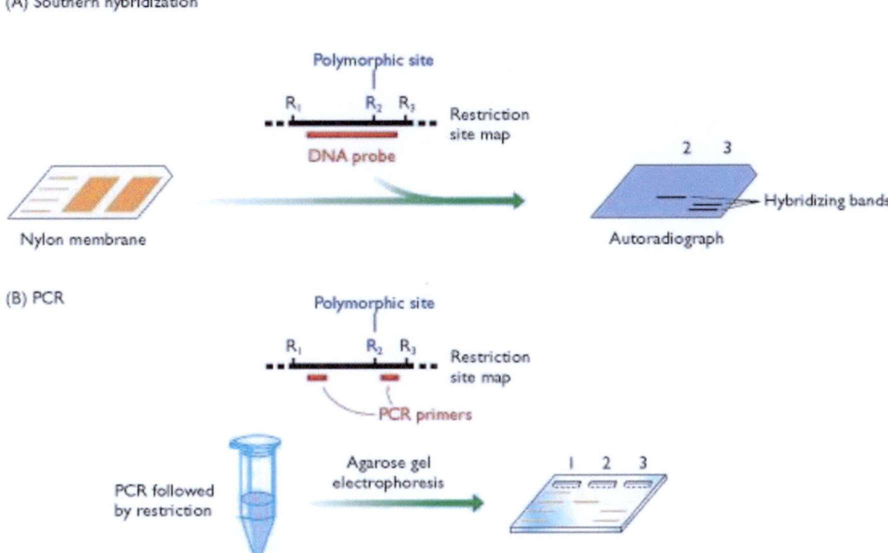

Fig. 4.40 RFLP scoring methods. (**a**) Scoring of RFLP by Southern hybridization. Involves restriction digestion of DNA with an appropriate restriction enzyme, separation of fragments on an agarose gel, transfer onto a nylon membrane, and analysis using a probe spanning the polymorphic restriction site. If the restriction site is absent, a single restriction fragment is detected (lane 2); presence of the restriction site yields two restriction fragments (lane 3). (**b**) Alternatively, PCR can also be used for RFLP analysis. Amplification is carried out with primers annealing on either side of the polymorphic restriction site, followed by restriction digestion of the PCR products, and agarose gel electrophoresis. Presence of the restriction site yields two bands, while its absence yields a single band

frequent absence of variation in a RFLP targeted for analysis among members of a family. Also, adding to the limitation is the tedious task of identifying one or two individual restriction fragments among several irrelevant fragments for a given restriction enzyme. For example, *Eco*RI, which recognizes 6-bp recognition sequence, is estimated to cut at about once every $4^6 = 4096$ bp and hence should produce nearly 800,000 fragments when human DNA is digested with it. Use of a probe that spans the polymorphic restriction site and analysis using Southern hybridization is one of the methods for scoring an RFLP (Fig. 4.40a). However, more recently, PCR-based methods are being used for scoring RFLPs. For this purpose, primers complementary to sequences on both ends of the polymorphic site are designed. RFLP analysis involves restriction digestion of amplified fragments followed by agarose gel electrophoresis (Fig. 4.40b).

4.4.2.2 SSLPs (Simple Sequence Length Polymorphisms)

SSLPs consist of an array of repeat sequences, with variable length. Alleles vary with respect to the number of repeat units (Fig. 4.41). In contrast to the RFLPs which consist of only two alleles, SSLPs can consist of multiple alleles wherein each allele

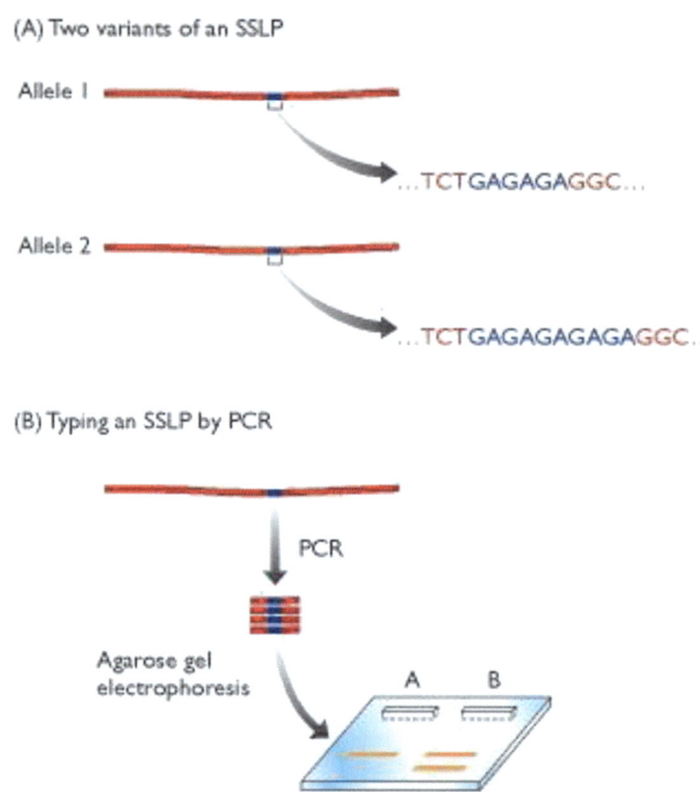

Fig. 4.41 Typing of SSLPs. (**a**) The figure represents two alleles of a SSLP. The motif "GA" is repeated three times in allele 1, whereas it is repeated five times in allele 2. (**b**) SSLP typing by PCR. Lane A consists of amplified products of the region surrounding SSLP. Lane B consists of DNA markers, representing size of the PCR amplified bands of the two alleles. It can be observed that the band in lane A is the same size as that of the larger bands in the marker lane, thus indicating that the tested DNA contained allele 2

can possess a different number of repeat units, thus can be of varying lengths. These kinds of repeat sequences are called microsatellites. It is estimated that the human genome consists of about 6.5×10^5 microsatellites.

4.4.2.3 SNP (Single Nucleotide Polymorphism)

SNPs are genomic positions which vary in a single nucleotide, for example, a G may be present in some individuals, while in others a C may be present (Fig. 4.42). Every genome consists of vast numbers of SNPs. Some of these SNPs also give rise to the RFLPs. About 1.42 million SNPs are estimated to be present in the human genome, and of these only 100,000 results in an RFLP.

Potentially, each SNP can exist as four alleles, due to the four possible nucleotides. But most of them exist in two forms, and hence pose the same limitation as the RFLPs with regard to their utility in human genetic mapping. Further, it is also

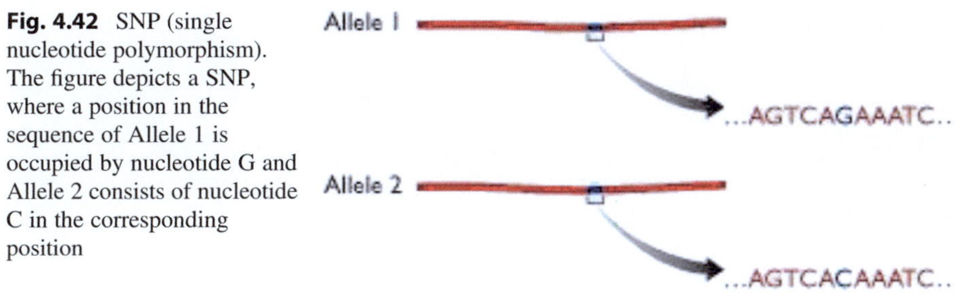

Fig. 4.42 SNP (single nucleotide polymorphism). The figure depicts a SNP, where a position in the sequence of Allele 1 is occupied by nucleotide G and Allele 2 consists of nucleotide C in the corresponding position

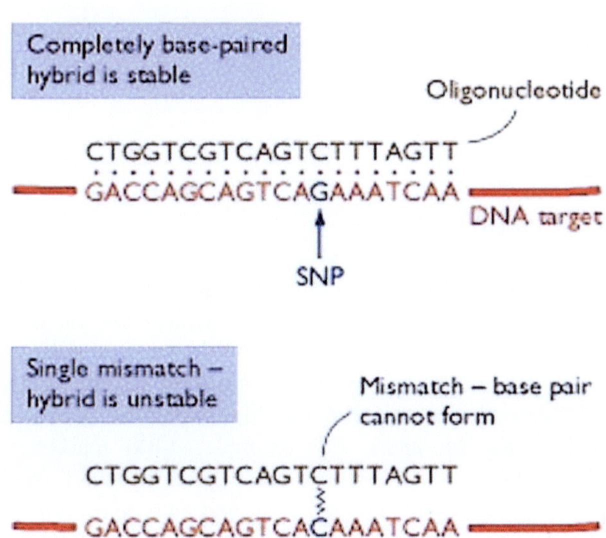

Fig. 4.43 Oligonucleotide hybridization for SNP typing. Highly stringent conditions are maintained during the assay. Due to the stringent condition, stable hybrids are formed only if the oligonucleotide completely base-pairs with the target site. Hybrids do not form even if there is a single mismatch. Stringent hybridization conditions are maintained by regulating incubation temperature and maintaining it a little below T_m (melting temperature) of the oligonucleotide; temperatures above T_m make even completely base-paired hybrids unstable, whereas temperatures more than 5 °C below the T_m may make even mismatched hybrids stable

possible that a particular SNP does not show any variability in a particular family chosen for the study. Due to their abundant numbers and availability of typing methods not requiring slow and labor-intensive gel electrophoresis make SNPs advantageous. These are rapidly detected by oligonucleotide hybridization analysis (Fig. 4.43), which is a very specific method enabling discrimination between two SNP alleles. DNA chip technology and solution hybridization (Fig. 4.44) are some of the screening strategies developed.

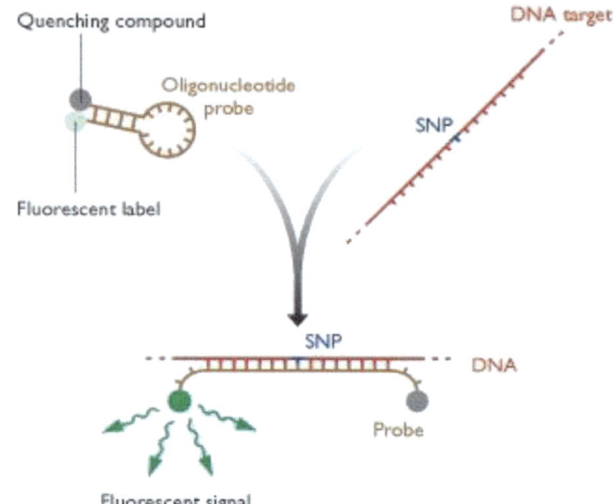

Fig. 4.44 Solution hybridization method of SNP detection. The oligonucleotide probe consists of two end-labels, one is a fluorescent dye, and the other is a quenching compound. Base pairing of the two ends of the oligonucleotide result in quenching of the fluorescent signal by the quencher. But when the oligonucleotide hybridizes with the target DNA, the fluorescence is emitted since the quencher is separated from the fluorescent dye

4.4.3 Physical Mapping

Genetic maps were insufficient for directing the genome sequencing projects, due to two reasons. The number of crossovers scored determines the resolution of the genetic map. With respect to microorganisms, this does not pose a problem because microorganisms can be cultured to obtain large numbers and thus several crossovers can be studied. This enables construction of a highly detailed genetic map, in which markers are just a few kb apart. For example, 1400 markers comprise the *Escherichia coli* genetic map at the rate of one per 3.3 kb on an average, and were adequate for directing the sequencing program, and an extensive physical mapping was not required. Similarly, for the *Saccharomyces cerevisiae* genome sequencing project, a fine-scale genetic map became available. It approximately comprised 1150 genetic markers, i.e., one per 10 kb on an average.

But with most other eukaryotes and humans, only few meioses can be studied because it is not possible to obtain large numbers of progeny. Hence, in these cases, linkage analysis will have limited resolving power. This implied that for such cases alternative mapping procedures will be required to supplement the genetic maps before large-scale DNA sequencing. To address this problem, several physical mapping techniques have been developed. The most important ones are as follows:

- Restriction mapping, which allows location of relative positions of restriction sites on a DNA molecule
- Fluorescence in situ hybridization (FISH), which allows marker localization to specific chromosomes using probes that hybridize to intact chromosomes
- Sequence-tagged site (STS) mapping, which involves mapping positions of short sequences with the help of PCR and/or hybridization analysis

4.4.3.1 Restriction Mapping

Using RFLPs as DNA markers for genetic mapping, the polymorphic restriction site positions can be located within a genome. Since very few restriction sites are polymorphic, several of them cannot be mapped by this technique (Fig. 4.45). Restriction mapping aims to increase the marker density on a genome map by mapping even non-polymorphic restriction sites using alternative methods.

4.4.3.1.1 Basic Methodology Involved in Restriction Mapping

In its simplest way, restriction maps are constructed by digesting DNA with two different restriction enzymes (each recognizes a different target sequence), and then comparing the sizes of the fragments obtained after digestion. For example, using restriction enzymes *Eco*RI and *Bam*HI (Fig. 4.46), DNA is first digested with each of the enzymes individually, and the size of the fragments obtained is characterized by agarose gel electrophoresis. Hence, the number of restriction sites pertaining to each of these enzymes present in the given DNA molecule can be worked out by these single digests. Further, the DNA can be digested by both these restriction enzymes together. This double restriction enzyme digestion enables determination of their relative positions in the given DNA molecule (Fig. 4.46).

4.4.3.1.2 Examining DNA Molecules Directly for Restriction Sites

Methods other than electrophoresis, such as optical mapping, can also be used for mapping restriction sites in DNA. In this the location of the restriction site on the given DNA, fragment is directly visualized with the help of a microscope. For this purpose, by using certain methods, such as gel stretching (Fig. 4.47) and molecular combing (Fig. 4.49), DNA is first attached to a glass slide in a stretched manner.

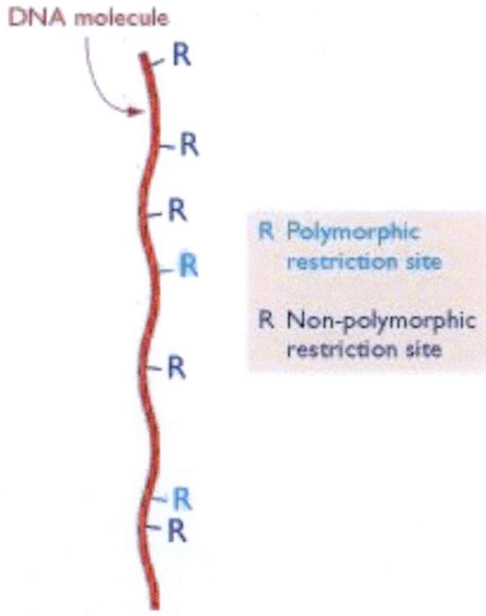

Fig. 4.45 Polymorphic restriction sites are important DNA markers used for mapping of genomes. All restriction sites are not polymorphic and techniques for mapping non-polymorphic sites have been developed. The above figure depicts the distribution of polymorphic and non-polymorphic restriction sites in a molecule of DNA

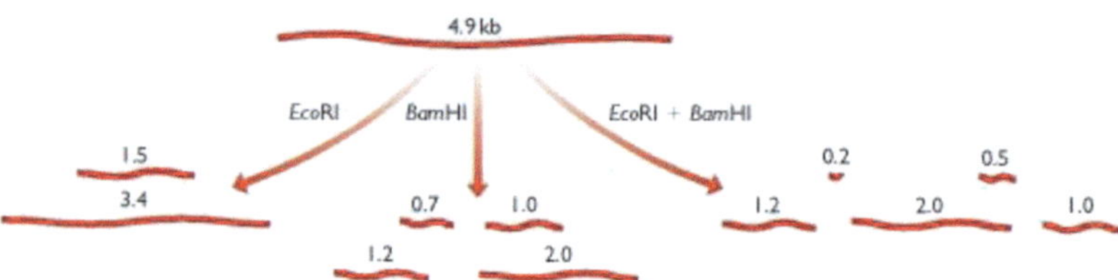

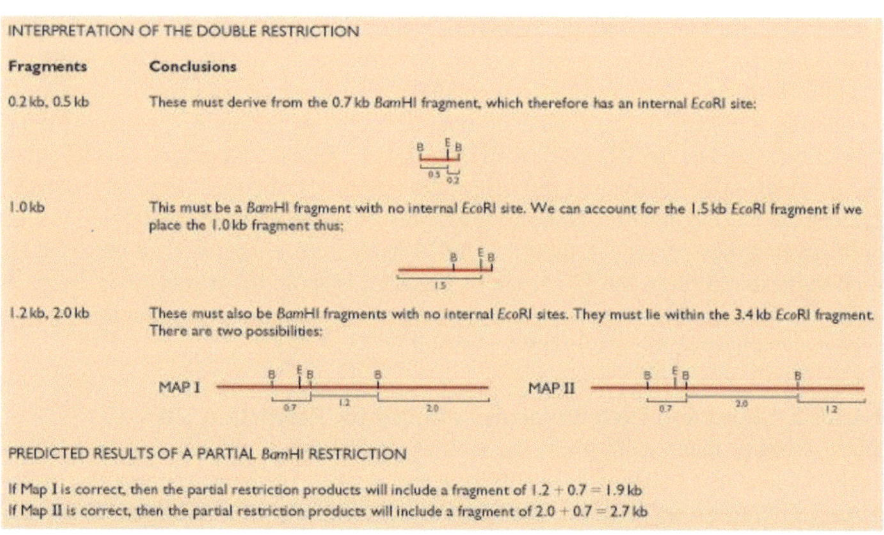

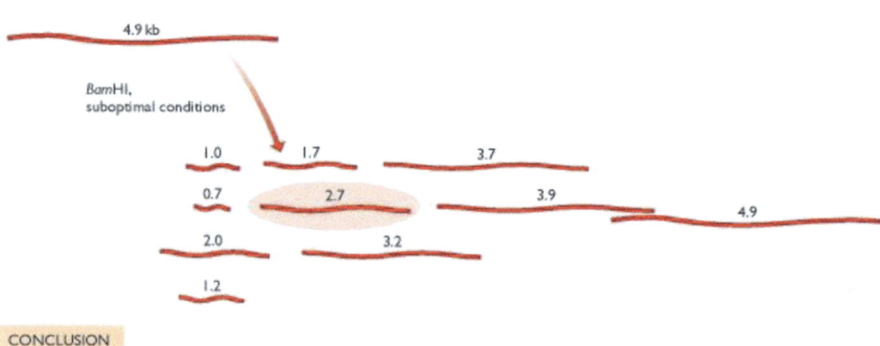

Fig. 4.46 Restriction mapping. The figure represents the basic restriction mapping procedure. In this example, the DNA molecule of size 4.9 has been digested using *Eco*RI and *Bam*HI represented as E and B, respectively. The fragments obtained by single and double restriction digestion are shown at the top. By comparing with single digests, the results of double digests can be interpreted to develop two alternative maps as described in the middle panel. Three restriction sites can be mapped using double restriction data. There are two alternative possibilities for the larger *Eco*RI fragment, as it consists of two *Bam*HI sites. This can be solved by a partial restriction of the original DNA molecule with only *Bam*HI by either using a suboptimal incubation temperature or incubating the reaction for a short duration of time. As shown in the bottom panel, the inclusion of 2.7 kb fragment in the products of the partial restriction shows MapII to be the correct one

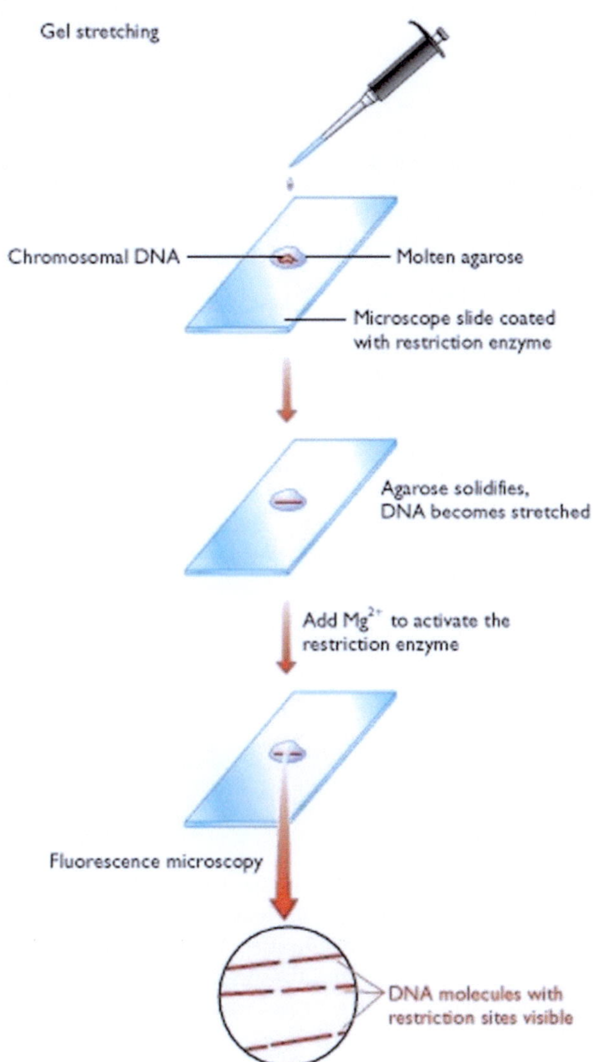

Fig. 4.47 Gel stretching and direct visualization of a restriction site. DNA molecules targeted for analysis are suspended in molten agarose and pipetted onto a slide coated with restriction enzyme. Solidification of the gel results in stretching of DNA molecules. At this stage, the enzyme is inactive due to the absence of magnesium ions, which is required for enzyme activity. The enzyme is activated by washing with a solution of magnesium chloride. The DNA molecules are cut by the enzyme, and the cut locations become visible as gaps as the DNA molecules gradually coil up. The visualization of DNA and hence the gaps is made possible by the addition of fluorescent dyes such as 4,6-diamino-2-phenylindole dihydrochloride (DAPI). The restriction site locations can be examined with a high-power fluorescence microscope

4.4.3.2 FISH (Fluorescence In Situ Hybridization)

FISH is the second type of physical mapping procedure. Similar to the optical mapping technique, FISH allows direct visualization of the position of a marker on an extended DNA molecule or a chromosome. In FISH, a DNA sequence marker is visualized after hybridization with a fluorescent probe (Fig. 4.48).

A fluorescent signal within a metaphase chromosome obtained by FISH is mapped by measuring its position relative to the end of chromosomal short arm (the FLpter value). The highly condensed nature of the metaphasic chromosome is a disadvantage and allows only a low-resolution (markers have to be at least 1 Mb apart) mapping. However, determination of the chromosome on which a new marker

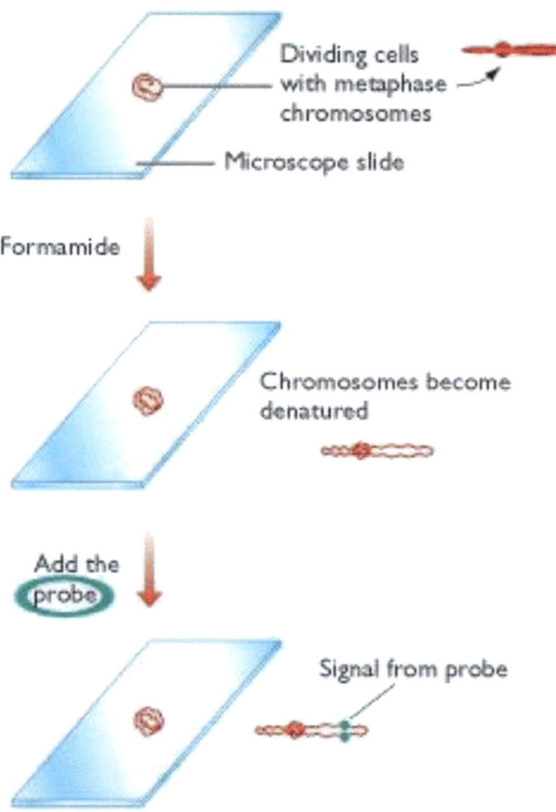

Fig. 4.48 FISH (fluorescence in situ hybridization). A sample of dividing cells is dried onto a slide, and treated with formamide to denature the chromosomes. The characteristic morphologies of the metaphase chromosomes are retained. The required fluorescently labeled probe is added to the chromosomes on the slide. The position on the chromosomal DNA to which the probe hybridizes is visualized by detection of the fluorescent signal that is emitted

is present has been one of the main applications of metaphase FISH. This provides a preliminary indication of its map position and can precede a fine scale mapping using various high-resolution FISH techniques. These techniques involve improvements in the chromosome preparation methods, leading to use of more extended chromosomes. There are two such methods:

- **Mechanically stretched chromosomes**—this involves inclusion of a centrifugation step in isolation of chromosomes, thus generating a shear force which causes chromosomes to stretch up to 20 times their normal length. The resolution is significantly improved (markers that are 200–300 kb apart can be distinguished).
- **Non-metaphase chromosomes**—chromosomes are highly condensed only during metaphase. Chromosomes are in an extended form at other stages of the cell cycle. Nuclei at prophase stage, consisting of adequately condensed and distinguishable chromosomes, have been used and have been found to be equivalent to mechanically stretched chromosomes without further advantage. Chromosomes at the interphase have been found to be more useful because chromosomes are most unpacked at this stage and a resolution down to 25 kb is possible. However, loss of chromosome morphology leads to lack of external reference points for mapping probe positions. Therefore, preliminary map information is obtained by

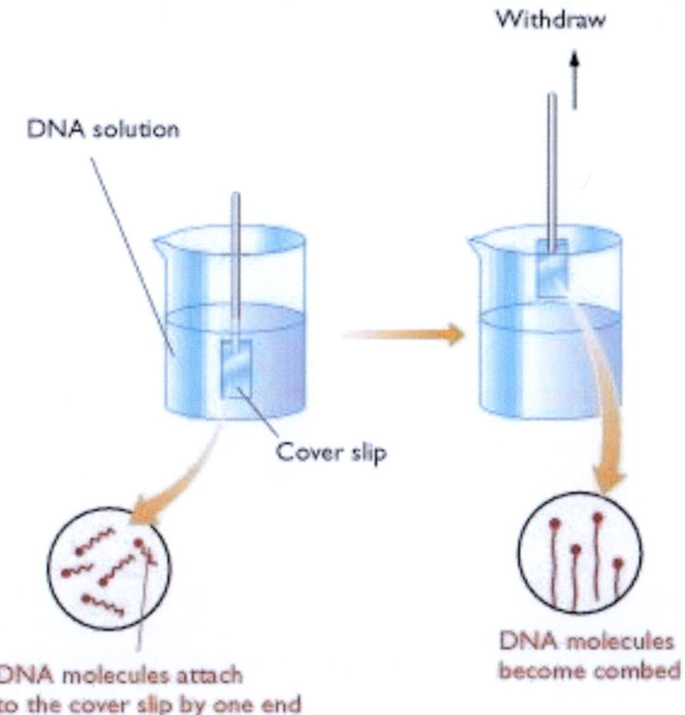

Fig. 4.49 Molecular combing. A solution containing the DNA molecules of interest is taken and a cover slip is dipped into it. The DNA molecules become attached by their ends to the cover slip. The cover slip is removed from the solution at a rate of 0.3 mm s^{-1} resulting in a "comb" consisting of parallel DNA molecules

using this technique, which is usually used to determine order of a series of markers in small chromosomal regions.

To increase FISH resolution to more than 25 kb, an approach called fiber-FISH, which utilizes purified DNA, is employed. For this purpose, DNA is prepared by gel stretching (Fig. 4.47) or molecular combing (Fig. 4.49) and enables a higher resolution, such that markers less than 10 kb apart can be distinguished.

4.4.3.3 STS (Sequence-Tagged Site) Mapping

Ideally, a rapid and technically undemanding high-resolution mapping procedure is required to generate a detailed physical map of a large genome. Neither restriction mapping nor FISH meet these requirements. In spite of the rapidness, ease, and

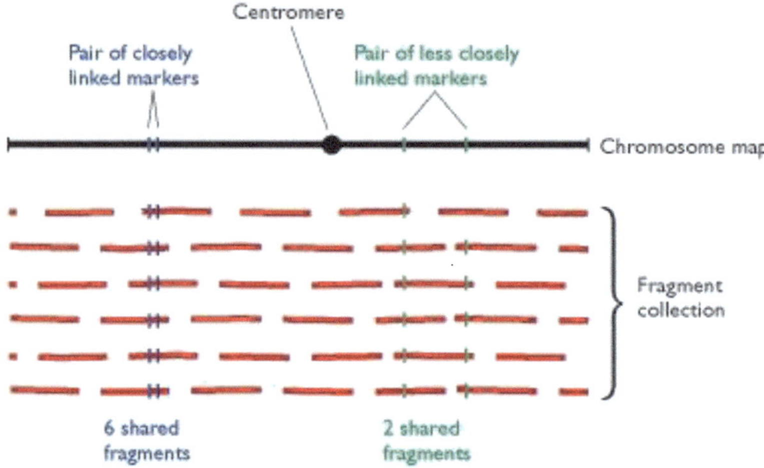

Fig. 4.50 Collection of fragments for STS mapping. The figure shows a set of DNA fragments that are appropriate for STS mapping. These are fragments spanning an entire chromosome, wherein each point on the chromosome is represented in about five fragments on an average. Markers shown in blue are closely located, and hence there is a high probability of finding them together on the same fragment. The two markers shown in green are further apart and hence there is less probability of finding them on the same fragment

usefulness of restriction mapping in obtaining detailed information, it is not suitable for large genomes. Although large genomes can be mapped using FISH and fiber-FISH, these are procedurally complex with slow rate of data accumulation because per experiment only three or four markers can be analyzed. Therefore, for constructing detailed physical maps, more powerful techniques are required. In this regard, the most powerful physical mapping technique is STS mapping. This has been used to generate high-resolution maps of large genomes.

An easily recognizable short (100–500 bp) sequence of DNA, occurring only once in a chromosome or genome, is called as a sequence-tagged site (STS). A set of STSs are mapped using a collection of overlapping DNA fragments (Fig. 4.50) derived from a single chromosome or whole genome. Data for construction of the map is obtained by determining fragments containing the STS, by hybridization analysis or PCR. PCR, however, is the preferred method as it is faster and amenable to automation.

How closely located two STSs are in a given genome determines the probability of these two STSs being located on the same fragment. If they are located close to one another, there is a higher probability that they will always be present on the same fragment. On the contrary, if they are located further apart, they may or may not be present on the same fragment with varying frequencies (Fig. 4.50). This data can be applied further for calculation of distance between markers, similar to mapping via linkage analysis (Sect. 4.1.5.1). As map distance is calculated using crossover frequency between two markers in linkage analysis, in STS mapping, map distance relies on *break* frequency between markers.

An STS is any unique DNA sequence. Two criteria must be satisfied by a DNA sequence to qualify as an STS. Firstly, it must be a known sequence, such that its

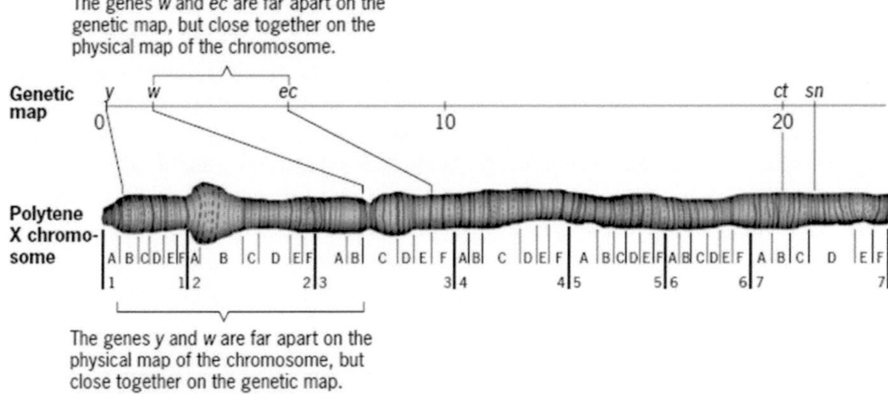

Fig. 4.51 Map of the polytene X chromosome. The figure represents the map of the left end of the *Drosophila* polytene X chromosome. It shows a comparison between the genetic map (above) and a physical map (below). The genes represented are as follows: *yellow* body (*y*), *white* eyes (*w*), *echinus* eyes (*ec*), *cut* wings (*ct*), and *singed* bristles (*sn*). It can be seen that *w* and *ec* are closer on the physical map than on the genetic map of the chromosome, whereas *y* and *w* are far apart on the physical map, but closer on the genetic map

presence or absence on various DNA fragments can be assayed by PCR. Secondly, an STS is required to occupy a unique chromosomal or genome location. If an STS sequence is present at more than one position, an ambiguity will be created in the mapping data. It has to be ascertained that the STSs should not be located within the repetitive DNA of a chromosome/genome.

The most common sources of STSs are the ESTs (expressed sequence tags), SSLPs, and random genomic sequences.

- *ESTs (expressed sequence tags)*: ESTs are short sequences representing genes. cDNA clones of protein-coding gene mRNAs are analyzed to obtain ESTs.
- *SSLPs*: In physical mapping, SSLPs can also be used as STSs. Polymorphic SSLPs, previously mapped using linkage analysis, are particularly valuable, as these establish a direct correlation between genetic and physical maps.
- *Random genomic sequences*: Sequences derived by sequencing random cloned genomic DNA fragments are called random genome sequences. Alternatively, sequences already deposited in the databases can be downloaded.

4.4.3.4 Genetic Distance vs Physical Distance

The occurrence of crossing over between homologous pairs of chromosomes forms the basis of all procedures involved in genetic distance measurement and construction of recombination maps. It can be expected that more frequent crossovers occur in long chromosomes than short ones. Hence, it can also be expected that this relationship will be shown in their genetic map lengths. Although this assumption is true for a large part, it has been observed that relatively more crossing over can occur in some regions of the chromosome than others. Therefore, distances represented on the genetic map do not show exact correspondence to the physical distance along the physical map of a chromosome (Fig. 4.51). Moreover, regions around the centromere and the ends of a chromosome are less likely to undergo crossing over; as a result, these regions appear condensed on a genetic map. Likewise, regions which undergo frequent crossovers are expanded in the genetic map. In spite of the lack of a uniform relationship between genetic and physical distances, there is a collinearity between genetic and physical maps of a chromosome, i.e., the loci on the chromosomes are present in the same order. Hence, mapping using the recombination frequency shows the exact order of the genes on a given chromosome. However, by using such a map, the actual physical distances between genes cannot be estimated.

Box 4.1 Scientific Concept: Estimation of Distances and Map Construction Using Radiation Hybrids: William Newell

Radiation hybrid (RH) mapping is a powerful technique applied for the purpose of mapping unique DNA sequences in the genomes. It is one of the preferred techniques for localization of new markers to specific regions of genomes. Markers such as the ESTs and STSs are used for analysis. The technique uses sampling of irradiation-induced chromosome fragments. This technique is originally developed by Goss and Harris (1975), and was further improvised for practical large-scale application by Cox et al. (1990) and Walter et al. (1994).

Goss and Harris (1975) irradiated human fibroblasts and the fragments were fused with recipient rodent cells. Scoring for the coretention of markers in a set of hybrid cells allowed the ordering of the markers on chromosomes 1 and X. This was based on the assumption that markers located far apart are more probable to become separated by irradiation than markers that are close, and hence are independently segregated in the component hybrid cells of the panel. Figure 4.52 depicts the marker distances for a given chromosome obtained by analysis of a panel of 100 hybrids.

Cox et al. (1990) extended this technique further by irradiating donor somatic cell hybrids, which consisted of a single copy of one human chromosome. The fragments obtained were fused with rodent cells. This particular technique was applied to map markers located on proximal and distal regions of chromosome 21. Subsequently, markers around disease loci, for example, BRCA1, neurofibromatosis type 2, Huntington's disease, and incontinentia pigmenti 1, in localized areas of the genome have been mapped using this technique.

A summary of chromosome-specific RH panels has been compiled. Recently, whole genome RH panels have been constructed and genome maps consisting of 100 s or 1000 s of markers have been developed. RH panels represent a sample of the donor genome. It specifically signifies a

(continued)

Box 4.1 (continued)

sample of radiation events that occurred in several donor chromosomes to give rise to the observed marker retention patterns in the RH panel.

Several statistical methods have been utilized for RH data analysis. Figure 4.53 shows the raw data showing the retention and breakage frequencies of the markers analyzed in a panel of hybrids. Presently, four main software packages, RHMAP, RHMAPPER, SAMAPPER, and MultiMap, have been used for the construction of complete maps from RH data. Retention of different chromosomal fragments in the different clones comprising an RH panel is modeled by each of these methods. The parameters in these models include breakage frequencies between all marker pairs, and retention frequencies of different fragments, which may depend on chromosomal location (Fig. 4.53). A chromosome map built on the distance data depicted in Fig. 4.52 is shown in Fig. 4.54. The raw data which forms the basis of the map is shown in Fig. 4.53.

This technique has proved to be powerful and has led to the construction of several good maps. However, the methods of analysis are being improved consistently. Particularly, none of the methods of analysis considered the probable irradiation induced physical rearrangements of chromosomes. It is suggested that several of the errors that have been encountered could be a result of such events. Studies in *Tradescantia* by Catcheside et al. (1946) have reported several different rearrangements in the chromosomes in addition to simple breaks. Similar studies involving human cells might improvise the models used for estimation of relative marker locations. For example, several conditions such as the temperature, stage in cell cycle, etc. are shown to influence the chromosome breakage efficiency and the types of alterations that occur (such as inversions, chromatid exchanges, isochromatid exchanges, etc.). It is also possible that some changes induced by the X-irradiation may prevent incorporation of human DNA into the rodent genome. It is required that these changes are quantified in various conditions and then incorporated into more accurate stochastic models.

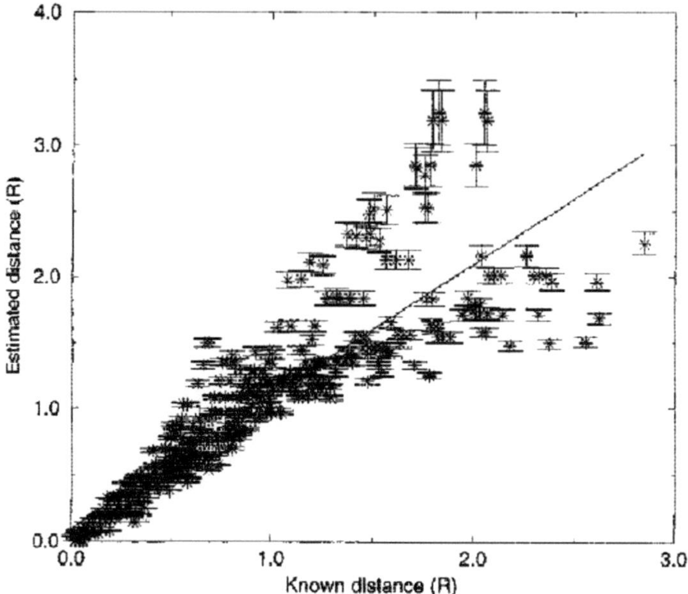

Fig. 4.52 Estimated vs. known distances for a panel of 100 hybrids, 100 markers, chromosome length 3 R. The distance between markers is measured in centirays (cR), where for each unit, there is a 1% probability of X-ray-induced breakage for a specific dosage in rads

4.5 Summary

- Construction of a series of descriptions depicting position and spacing of characteristic, identifiable biochemical markers, such as genes and other DNA sequence features comprising a chromosome, is called as chromosome mapping.
- Linkage or genetic linkage can be defined as the tendency of genes closely located on a chromosome to be frequently inherited.
- Test cross is an extremely important tool used to reveal the gametic composition of a heterozygous offspring of the F_1 generation.
- The crossing over among homologous chromosomes results in production of recombinant gametes.
- It is during the late prophase of meiosis I, chiasmata, the cytological evidence for crossing over can be clearly visualized.
- Distances on genetic maps are measured in map units, and are abbreviated as m. u.; one map unit = 1% recombination.
- Map units are denoted by centimorgans (cM); one Morgan = 100 m.u.
- When a test cross is performed involving two genes, it is named as a two-point test cross or in short called as two-point cross.
- Distance between gene pairs can be obtained by identifying recombinant classes resulting from crossover between concerned pair of genes.

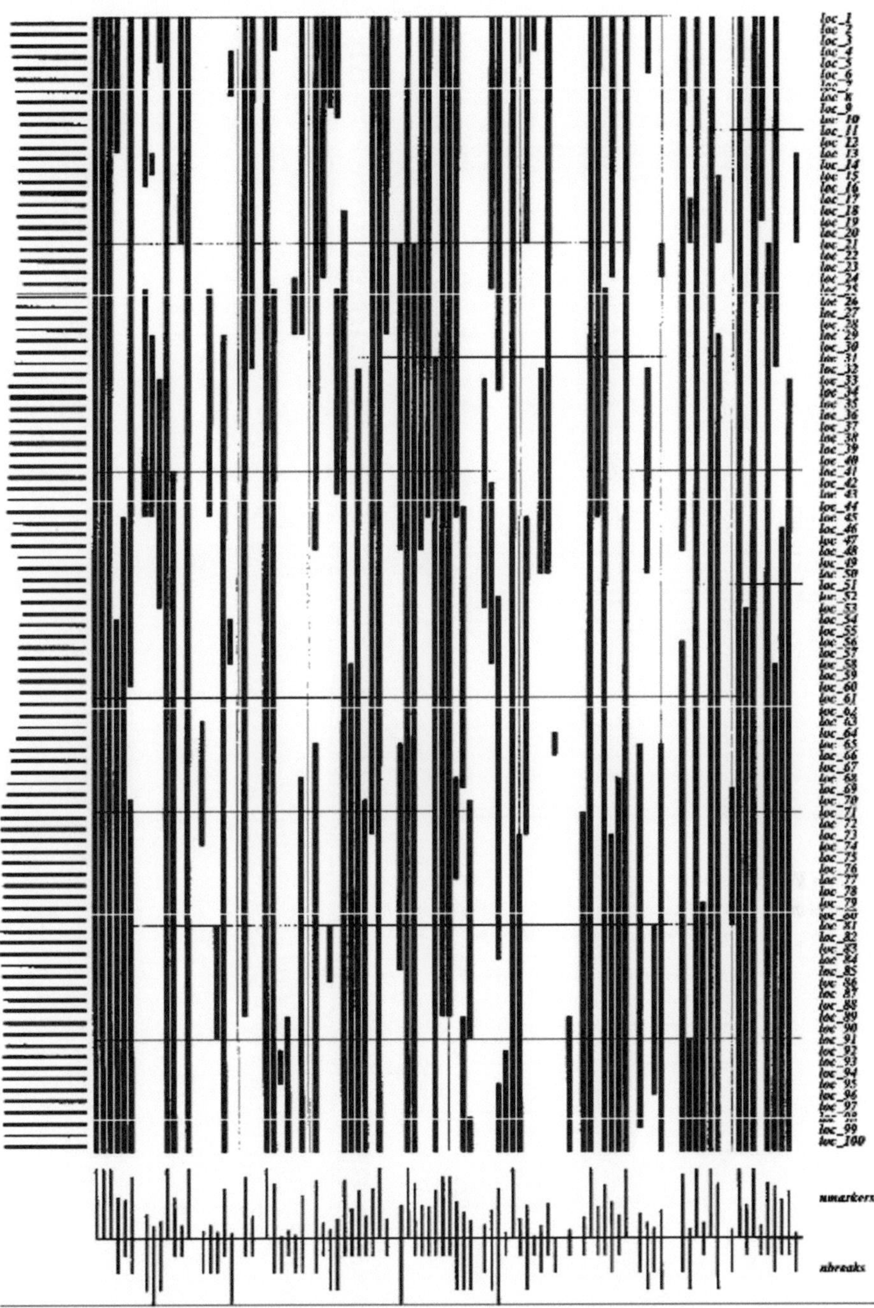

Fig. 4.53 Raw data used for the map (black) present and (white) absent. The retention frequencies are plotted to the left. (Bottom) Relative numbers of markers and breaks observed in each hybrid cell of the panel

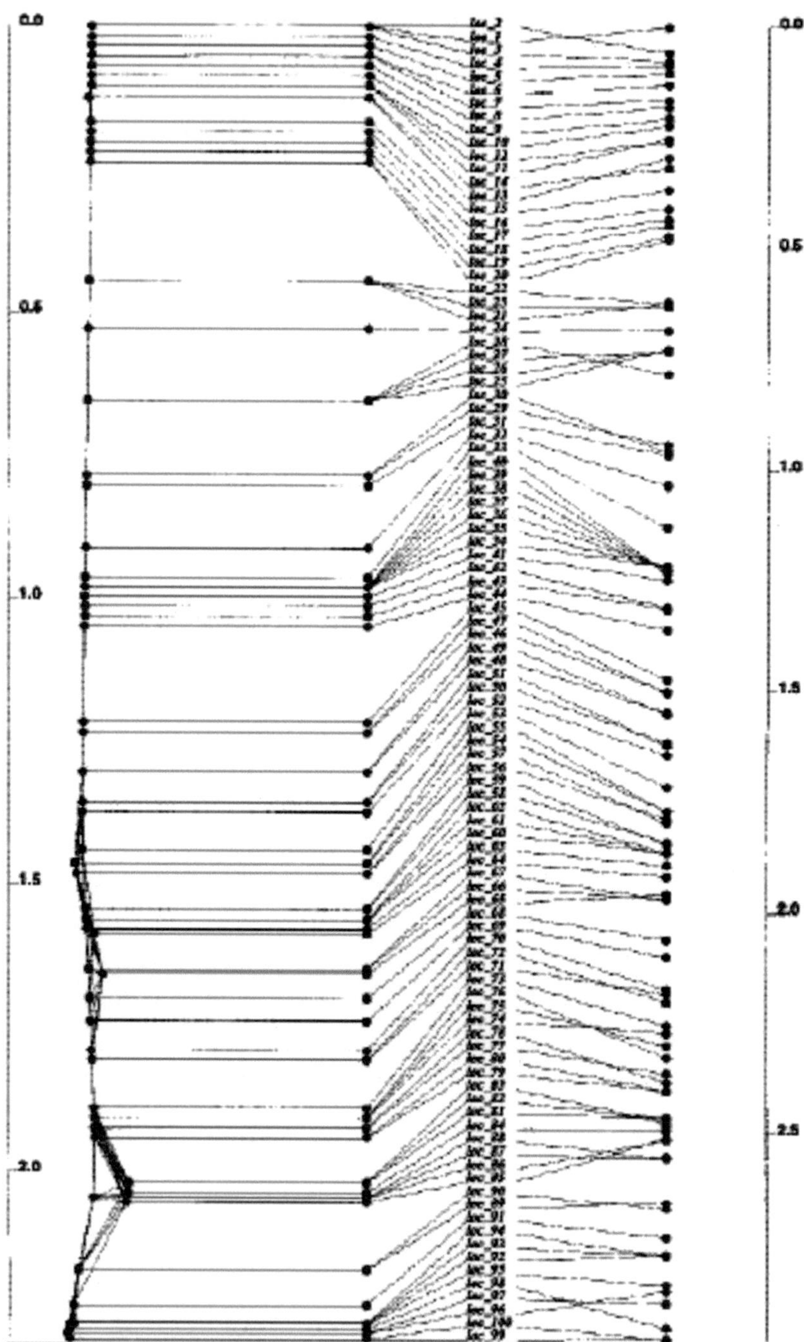

Fig. 4.54 Map constructed from the distances in Fig. 4.52. The optimal order is to the left of the names, the two-dimensional configuration at the extreme left, and the original (known) positions at the right

- Linkage analysis in humans cannot rely on planned breeding experiments. In this case, family pedigrees will have to be made use of for linkage analysis. Primarily, disease alleles have been analyzed with the help of pedigrees.
- Mapping that is based on genetic techniques is called genetic mapping. Maps are constructed using genetic techniques such as cross-breeding experiments or analysis of family histories, i.e., pedigrees in the case of humans. Position of genes and other genome sequence features are shown on these maps
- Mapping wherein molecular biology techniques are used is called physical mapping. Using these techniques, DNA molecules are directly examined and maps are constructed. These maps show positions of genome sequence features, including genes.
- Genes were the first markers to be used in mapping of chromosomes. The first genetic maps were constructed for organisms such as the fruit fly, during the early twentieth century, and genes were used as the markers for this purpose.
- Even though the genes are very useful as markers, they are certainly not adequate. Particularly, in organisms such as the vertebrates and flowering plants which possess larger genomes, the maps cannot be based entirely on genes. This is because genes are widely spaced with large intergenic regions in most eukaryotic genomes. Therefore, using only genes for mapping will result in a map that is not very detailed. In addition, the fact that only a fraction of the total genes possess easily distinguishable allelic forms makes it even more difficult to be used as a sole marker for mapping. Hence, gene maps cannot be made very comprehensive.
- RFLPs (restriction fragment length polymorphisms), SSLPs (simple sequence length polymorphisms), and SNPs (single nucleotide polymorphisms) are the DNA sequence features used for mapping.
- Genetic maps were insufficient for directing the genome sequencing projects as they have a low resolution. Several physical mapping techniques such as restriction mapping, fluorescence in situ hybridization (FISH), and sequence-tagged site (STS) mapping have been developed to obtain maps of higher resolution.

References

Catcheside DG, Lea DE, Thoday JM (1946) Types of chromosome structural change induced by the irradiation of Tradescantia microspores. J Genet 47:113–136

Cox DR, Burmeister M, Price ER, Kim S, Myers RM (1990) Radiation hybrid mapping: a somatic cell genetic method for constructing high-resolution maps of mammalian chromosomes. Science 250:245–250

Goss SJ, Harris H (1975) New method for mapping genes in human chromosomes. Nature 255: 680–684

Walter M, Spillett D, Thomas P, Weissenbach J, Goodfellow P (1994) A method for constructing radiation hybrid maps of whole genomes. Nat Genet 7:22–28

5 Study of Chromosome

Dhruti Patwardhan, S. A. Varshini, and Latha Galoth

5.1 Overview of Chromosome

All living things on earth have evolved from a single primordial ancestor which arose about 3.5–4 billion years ago. A consequence of this is that all organisms share a common system for storing and retrieving biological information. A genetic code consisting of deoxyribonucleic acids (DNA) or ribonucleic acids (RNA) is utilised. The same four nucleotides of either RNA or DNA are seen and the genetic code also remains constant in almost all organisms. Chromosomes are the way in which DNA is packaged and stored inside the cell. The method of packaging is different in prokaryotes and eukaryotes.

In eukaryotes, DNA is coiled around a set of proteins known as histones. This complex of DNA and histones is referred to as chromatin. The chromatin also coils around itself and creates a compact mass of DNA that can fit into the nucleus of a cell. This packaging does limit the accessibility of genes for expression and genes that need to be expressed are often unwound from this organisation. DNA in archaea is also organised around histones but they are different from the histones found in eukaryotes. Bacteria lack histones and their DNA is therefore not as compactly organised as those of eukaryotes. Gene expression, therefore, also becomes simpler in bacteria.

Chromosomes are very complex structures and have multiple levels of organisation. As stated above, the first level of organisation is the winding of DNA double helix around an octamer of histone proteins. This octamer consists of two copies each of H2A, H2B, H3, and H4 histones, also called the core protein. The

D. Patwardhan
Indian Institute of Science, Bangalore, India

S. A. Varshini (✉) · L. Galoth
Ramaiah University of Applied Sciences, Bangalore, India

D. Kar, S. Sarkar (eds.), *Genetics Fundamentals Notes*,
https://doi.org/10.1007/978-981-16-7041-1_5

DNA wraps twice around this core protein which consists of 145–147 bp of DNA. This arrangement of DNA coiled around an octamer of histone proteins is called a nucleosome. This is the fundamental unit of the chromosome and occurs repeatedly. Another histone known as the H1, present outside the nucleosome, clamps around the 20–22 bp of entry and exit points of the DNA and stabilises this structure. The nucleosome together with its associated H1 protein is called a chromatosome. Two chromatosomes are separated by 30–40 bp of linker DNA. The length of linker DNA may vary between different cell types. Together the nucleosomes and linker DNA appear as beads on string. The chromatosomes are then packed tightly together to form a fibre of 30 nm diameter. This fibre then forms loops of varying lengths averaging about 300 nm in length and anchored at its base to nuclear scaffold proteins. This 300 nm fibre is further compressed and folded to form 250 nm wide fibre. This is then supercoiled to form the chromatid which is 700 nm in width. During DNA replication, another copy of the chromatid is produced which is anchored together at the centromere. This structure is called the chromosome (Fig. 5.1).

In essence, each chromosome consists of a single strand of DNA which is tightly coiled and compactly packed in the cell with the help of histone proteins. Under normal circumstances, chromosomes are difficult to observe. However, during cell divisions, chromosomes condense further and become thick structures which can be observed. Chromosomes consist of three elements: centromere, telomeres and origins of replication.

Telomeres are the tips of the chromosomes. The ends of the chromosomes are vulnerable and may get degraded over time. Telomeres provide stability to the chromosomes and prevent useful information situated at its ends from getting degraded. Telomeres shorten in length with each cell division. They therefore play important roles in determining cell senescence, cancer and aging. Origins of replication are the points at which DNA replication is initiated. These cannot be easily seen under the microscope. Before mitosis can begin, each chromosome replicates itself to create an identical copy. This copy is called the sister chromatid and they are held together at the centromere (Fig. 5.2).

Centromere appears as a constricted structure which does not stain very well. This is the point at which protein assemblies called kinetochore form during cell division. Spindle microtubules attach to the kinetochore during cell division. These spindles pull the sister chromatids towards opposite ends of the cell. A single cell separates into two daughter cells, each of which inherits one of the sister chromatids. Thus, centromeres are very important to ensure correct separation of chromosomes into each cell during cell division. Chromosomes lacking centromeres are not attached by spindle fibres and are lost during cell division. Based on the position of centromere, chromosomes are divided into four types: metacentric, submetacentric, acrocentric and telocentric. In metacentric chromosomes, centromere is present at its centre. The chromosome arms on either side of a metacentric chromosome are roughly equal in length. In a submetacentric chromosome, the centromere is present slightly off centre, due to which one arm of the chromosome is slightly short (p) and one arm of the chromosome is slightly longer (q). When the centromere is present closer to its

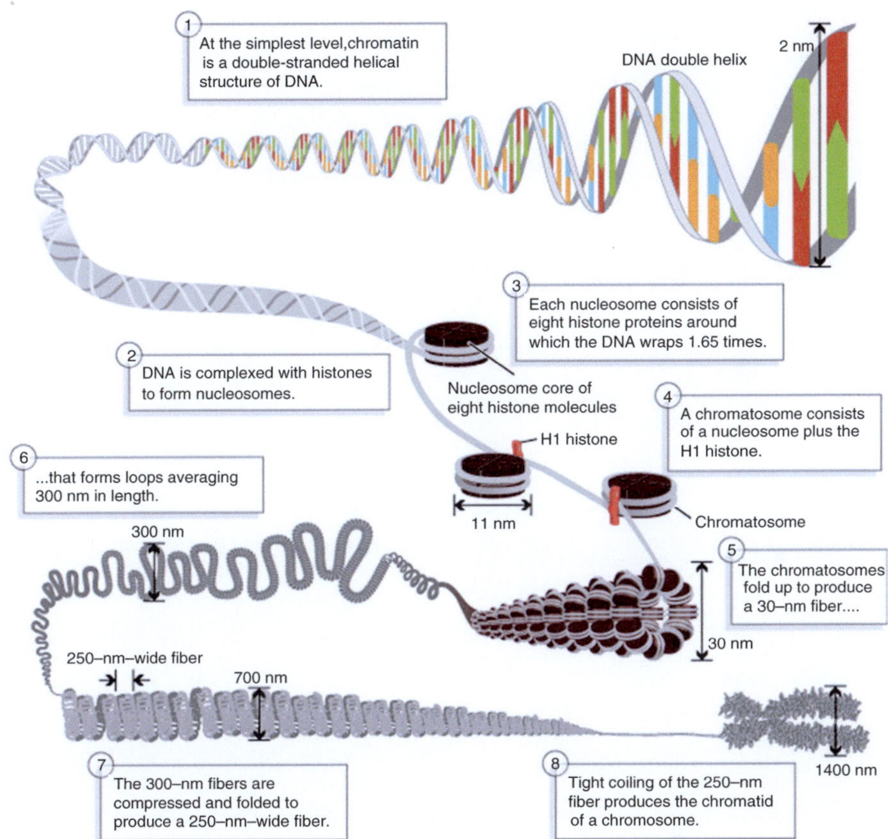

Fig. 5.1 Complex organisation of chromatin: The double helix of DNA is wound around an octamer of histone proteins (nucleosome) which, along with the H1 protein, forms a chromatosome. The chromatosome coils up to form a 30 nm fibre which loops up to produce a 300 nm fibre. This fibre is further folded up to form a 250 nm wide fibre which ultimately supercoils to form the chromosome (Annunziato 2008)

end than to its centre, the chromosomes are called acrocentric. The *p* arm of such chromosomes is very short and *q* is long. In a telocentric chromosome, the centromere is present on one end of the chromosome. This chromosome therefore has only one arm (Fig. 5.3).

Acrocentric human chromosomes like chromosomes 13, 14, 15, 21 and 22 have a secondary constriction other than the centromere in its short arm. These are called chromosomal satellites and are composed of repetitive DNA and tandem copies of ribosomal RNA genes. The position of these secondary constrictions remains constant, and they serve as markers for identification of these chromosomes. Due to the presence of large ribosomal RNA genes in this area, they form the nucleolar organising regions (NORs). Nucleoli are the site of ribosome biosynthesis. It has been seen that NORs are major hotspots for mutations in cancer which means that

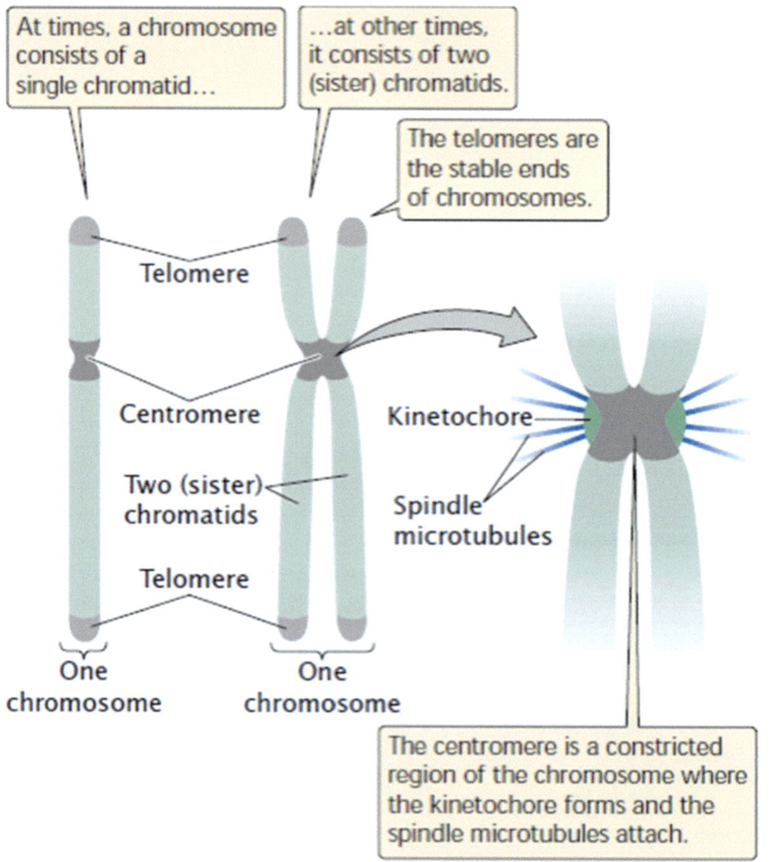

Fig. 5.2 Structure of an eukaryotic chromosome: Chromosomes replicate before cell division and the two sister chromatids are held together at its centromere. Kinetochore assembly takes place at the centromere and spindle microtubules attach to the kinetochore. Telomeres are present at chromosome terminals and help stabilise them (Pierce 2010)

the NOR region is prone to recombination in cancer. NORs have also been implicated in ageing and senescence.

5.1.1 Chromosome Number

Organisms vary in the number of chromosomes they possess. Bacteria typically possess a single chromosome, potatoes have 48 chromosomes and humans have 46 chromosomes. The complexity of organism has no relation with the number of chromosomes it possesses per cell. Eukaryotes possess chromosomes in pairs. This is because during sexual reproduction, organisms inherit one set of chromosomes from the male parent and another from the female. Thus, out of the 46 chromosomes

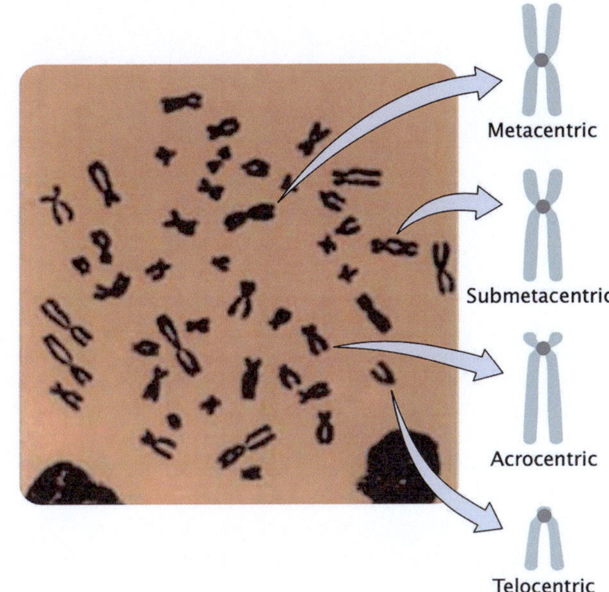

Fig. 5.3 Types of chromosomes: Chromosomes can be classified into four types based on the position of their centromeres. Telocentric chromosomes have their centromeres at the edges. Acrocentric chromosomes have centromere closer to one end of the chromosome. Metacentric chromosomes have centromere at its centre and submetacentric chromosomes have centromere closer to the centre than to its edges. Some acrocentric chromosomes have secondary constrictions besides the centromeres which are known as satellite chromosomes (Pierce 2010)

present in humans, 23 are inherited from the mother and 23 from the father. Each chromosome therefore has a pair which is referred to as the homologous chromosome. The pair of chromosomes are generally similar in shape, size and genes that they possess. If gene A is present on chromosome 1 at 1600 bp which codes for protein A, another gene also coding for protein A will be present on the same location in a homologous pair. These genes, however, may or may not be identical. The gene on homologous pair may contain some variation. These different forms of the same gene are called alleles. Genes may have multiple alleles. Such cells, possessing pairs of chromosomes, are called diploid. Most cells are diploid except cells which are involved in forming gametes, i.e. sperm or egg. The sperm and egg possess only one set of chromosomes and are called haploid. The fusion of single set of chromosome each from sperm and egg leads to the formation of a diploid organism. Chromosomes can be stained and visualised under a microscope. This is called a karyotype and it is useful in detecting the number and shape of chromosomes and chromosomal defects if there are any. Figure 5.4 shows a normal human karyotype and abnormal karyotype of a person having Down syndrome.

5.1.2 Autosome

Of the 23 pairs of chromosomes in humans, 22 pairs follow Mendelian pattern of inheritance and segregate equally between all offspring. These chromosomes are called autosomes. The remaining pair is called the sex chromosome because composition of this pair determines the sex of an individual. In human, females have two

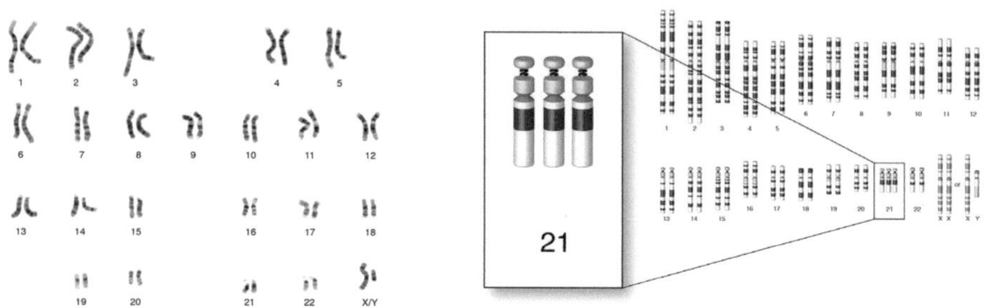

Fig. 5.4 Karyotype can give indication about chromosomal abnormalities: Karyotype is the complete set of chromosomes present in a eukaryotic cell. A. Humans normally have 23 pairs of chromosome—22 autosomes and 1 pair of sex chromosomes which can be either XX for females or XY for males. B. Karyotype for person having Down syndrome. The individual possesses an extra chromosome 21 shown by the red arrow. The total number of chromosomes in this individual is therefore 47 instead of the normal 46. Courtesy: National Human Genome Research Institute

X chromosomes and males have one X and one Y chromosome. Due to this, the inheritance of genes on these chromosomes may not follow Mendelian inheritance and may preferentially be inherited by daughters or sons.

5.1.3 Sex Chromosome

The term X chromosome comes from the term X body used by Hermann Henking while describing a structure that he observed in the nuclei of male insects. McClung recognised that this structure was a chromosome while studying grasshoppers and termed it as accessory chromosome, but it became generally known as the X chromosome. He stated that accessory chromosome determined sex of the individual based on the observation that female cells had one extra chromosome compared to males. In 1905, Nettie Stevens and Edmund Wilson, based on their observation in grasshoppers and other insects, showed that female cells have two X chromosomes, while male cells have only one X chromosome. In some insects, they also observed that both males and females had the same number of chromosomes, and males had another shorter chromosome instead of the second X chromosome. This was termed as the Y chromosome.

They also showed that females formed gametes containing only the X chromosome, while males formed gametes half of which had the X chromosome and the other half had Y chromosome. Thus, when a male sperm/gamete containing the X chromosome fused with the egg, it formed a female individual with XX chromosome. When the sperm containing Y chromosome fused with the egg, it led to the formation of a male individual with XY chromosome. This showed that sex of an organism is determined by the composition of its chromosomes. The X and Y chromosome are therefore called sex chromosomes and the nonsex chromosomes are called autosomes. Although X and Y chromosomes differ in their size and genes that they carry, they can pair during meiosis and get segregated. Pairing happens at the tips of X and Y chromosomes which are similar and carry the same genes. These regions are called pseudoautosomal regions.

Some organisms, like grasshoppers, lack the Y chromosome. In this case, sex determination occurs through the XX-XO system. That is, females have two X chromosomes and males have a single X chromosome. During gametogenesis, females produce gametes, all of which have the X chromosome. In males, half the gametes receive X chromosome, while the other half contains no sex chromosome. Since males in both XX-XO and XX-XY systems produce two types of gametes, they are referred to as the heterogametic sex. Females in both cases produce a single type of gamete and are referred to as the homogametic sex.

In some birds, moths and amphibians, the female is heterogametic and the male is homogametic. To prevent confusion with the XX-XY system, the chromosomes in these organisms are referred to as the Z and W chromosomes. Thus, the females have a ZW composition and on gametogenesis produce half of the gametes having Z chromosome and half of them having W chromosome. The males have a genetic composition of ZZ and produce sperms having the chromosome Z.

Some insects like bees, wasp and ants have no sex chromosome. Sex is determined from the number of sets of chromosomes it possesses. If the organism has a single set of chromosome (haploid), it is a male. If the organism has two sets of chromosome (diploid), it is female. Males can produce gametes through mitosis as they are already haploid. Females undergo meiosis to form gametes. A fusion of male and female gametes leads to a female offspring. An unfertilised female gamete can develop into a male offspring. Thus, genetically, males have a 50% chance of similarity since two types of gametes can be developed from a diploid organism. On the other hand, all females inherit the same male gametes and have a 50% chance of inheriting the same female gamete. In *Drosophila*, or fruit fly, the sex of the organism is determined by the ratio of sex chromosomes to autosomes. This is because the X chromosome contains female determinants and the autosomes contain the male determinants. Thus, if a fly has two sets of autosomes and one X chromosome (1X, 2A), the fly is male. In contrast, if the fly has two sets of autosomes and 2 X chromosomes (2X, 2A), it is female. In *Drosophila*, the Y chromosome is not involved in determination of sex. It contains genes essential for forming sperm in adults.

5.2 Chromosomal Basis of Heredity

Walther Flemming, while studying cell division, was able to stain the chromatin and published his results in 1878. Later, Heinrich Wilhelm Gottfried von Waldeyer-Hartz, in 1888, coined the term chromosome. However, the link between chromosomes or genes and heredity was not known. Independent observations by Walter Sutton and Theodor Boveri in 1902–1903, as well as rediscovery of Mendel's theories, led to the proposition of chromosomal theory of inheritance. Boveri while studying the development of sea urchins discovered that embryonic development was hampered if all chromosomes were not present. Sutton discovered that chromosomes are present in pairs in grasshopper and they separate during meiosis and stated that this may constitute the physical basis of Mendel's principles of heredity. The chromosomal theory of inheritance states that genes are present on chromosomes and they constitute the genetic material responsible for Mendelian inheritance.

5.2.1 Proof That Inheritance Linked to Chromosome

Thomas Hunt Morgan provided proof for the chromosomal basis of heredity through his experiments on fruit flies. Inspired by the rediscovery of Mendelian principles, Morgan started working on conducting genetic experiments initially on mice. He later moved on to work with *Drosophila melanogaster* or the fruit fly. His lab did not have very sophisticated instruments. The room was very small and untidy and contained just some bottles for rearing the flies and lens or microscopes to observe them. A lot of extremely important experiments in the field of genetics occurred in this primitive fly lab.

While conducting his experiments, Morgan came across a male *Drosophila* having white eye which stood out from the other flies which had red eyes. He conducted a series of genetic crosses to study the inheritance of this trait. Morgan started by crossing this white eyed male with a red eyed female. As expected, almost all the progeny had red eyes. Morgan actually found 3 white eyed males in a progeny of 1237 flies. He assumed that the white eyes had arisen due to spontaneous mutation. Thus, it seemed to be following Mendel's principles where red eye was the dominant character and white eye was the recessive character. He then crossed the F1 progeny with each other. Morgan expected to obtain a ratio of 3 red eyed progeny:1 white eyed progeny. Instead, he observed that all females had red eyes while half the males had red eyes and half had white eyes. Since the inheritance of the trait differed between males and females, Morgan hypothesised that the trait was linked to the X chromosome. Since females had two X chromosomes, the allele for white eyes would get masked and all female would have red eyes. In males, since only one X chromosome was present, they would automatically express the character dictated by the inherited allele on X chromosome.

If eye colour was indeed an X-linked trait, Morgan predicted that a reciprocal cross would yield different result. If a white eyed female was crossed with a red eyed male, all the F1 females would be red eyed and all the F1 males would be white eyed. The males would inherit X chromosome from the mother and Y chromosome from the father. Since the female is homozygous recessive, all males would inherit allele for white eyes from the mother. When the F1 generation is crossed with each other, half the F2 generation in both males and females would be red eyed and the other half would be white eyed. This becomes clearer in Fig. 5.5. Morgan obtained results that were in accordance with his expectations, and he therefore concluded that eye colour in *Drosophila* was an X-linked character.

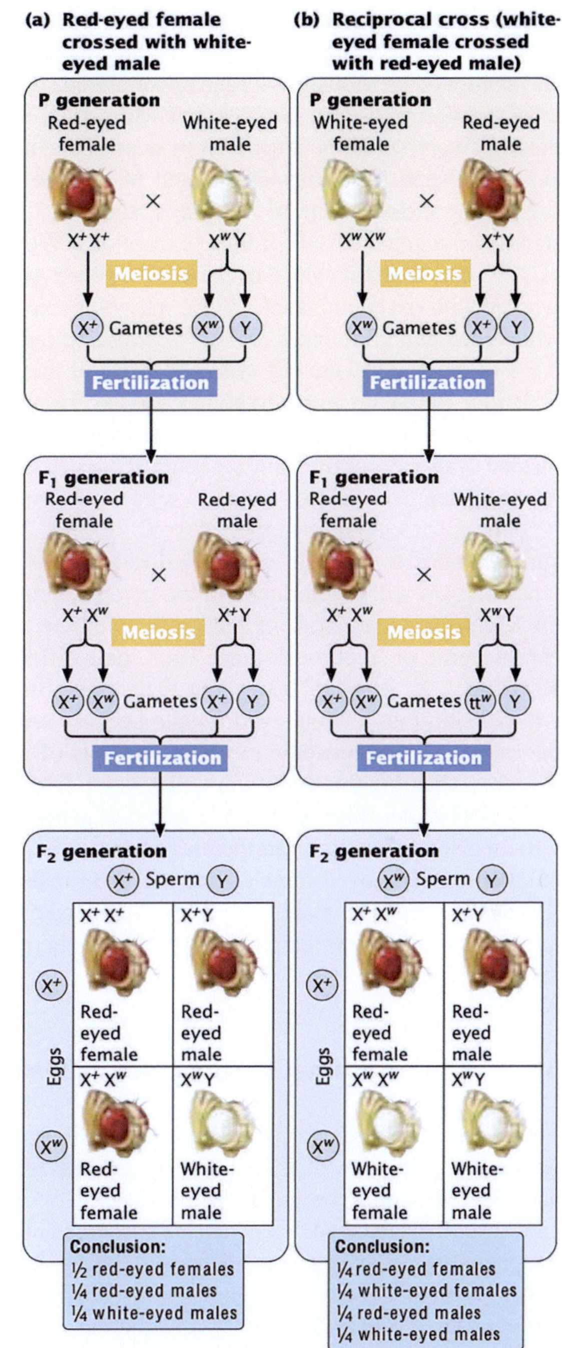

Fig. 5.5 Morgan's experiments on fruit flies showed that eye colour was an X-linked trait. When red eyed females were crossed with white eyed females, all F1 progeny had red eyes. On crossing them with each other, it was observed that all females in F2 had red eyes. Males in F2 had a ratio of 1 red eye:1 white eye. Reciprocal cross yielded different results. When white eyed females were crossed with red eyed males, all F1 females had red eyes and all F1 males had white eyes. The F2 progeny in both males and females in this cross had a ratio of 1 red eyes:1 white eye. w, allele for white eye; w+, wild type allele for red eye (Pierce 2010)

5.2.2 Nondisjunction: Proof of Chromosome Theory

As mentioned previously, Morgan encountered some white eyed flies during his first cross between red eyed females and white eyed males. Since the number of these flies was very small, he attributed their presence to development of new mutations. However, such flies with unexpected phenotype continued to appear even in the succeeding crosses. One of Morgan's students, Calvin Bridges, worked on understanding the genetic basis of these exceptions. We can denote the allele for red eyes as X^{w+} and that for white eyes as X^{w}. When white eyed females (X^wX^w) were crossed with red eyed male ($X^{w+}Y$), we would expect that all males (X^wY) would be white eyed and all females (X^wX^{w+}) would be red eyed. It was, however, seen that 2.5% of the males were red eyed and 2.5% of the females were white eyed.

Bridges came up with a hypothesis to explain this phenomenon. He assumed that this might be due to an error in the segregation of chromosomes during meiosis. Instead of two X chromosomes getting separated and forming two gametes with one X chromosome each, they failed to separate, forming gametes with one having two X chromosomes and another devoid of any X chromosome. The failure of chromosomes to separate was called nondisjunction. Due to nondisjunction, two types of eggs will be produced in the given example: one with X^wX^w and one with no X chromosome. Both of these can combine with a sperm bearing either X^{w+} chromosome or Y chromosome. Thus, four different combinations are possible— $X^wX^w\,X^{w+}$, X^wX^wY, $X^{w+}O$, or YO. O denotes absence of sex chromosome. X^wX^wY will develop into a white eyed female because sex in *Drosophila* is determined by the ratio of X chromosome to number of sets of autosomes. Since the ratio here is 1, embryo bearing this genotype will develop into a female. $X^{w+}O$ will develop into a red eyed male. Both $X^wX^w\,X^{w+}$ and YO genotypes are lethal and these embryos will die. Nondisjunction can therefore explain the unexpected appearance of a small number of white eyed females and red eyed males in a cross between white eyed females and red eyed male flies. Bridges also examined the chromosomes of the flies in question and found them to be the same as he had predicted (Fig. 5.6).

5.2.3 Chromosomal Basis of Mendelian Theory

The success of Bridge's experiment is not limited to the fact that he was able to predict the genotype of a fly showing unexpected phenotype in his culture. This actually meant that the phenotype expressed by an organism is dependent on the genotype. It established that genes are carried on chromosomes and these will dictate the phenotype of an organism. Thus, the 'units' defined by Mendel in his hypothesis which were later defined as genes were carried by chromosomes in the cell. The chromosomes contain hereditary information, and the experiments by Morgan and Bridges provided proof for chromosomal basis of heredity. They showed that the gene for eye colour was a sex linked because it was present on the X chromosome.

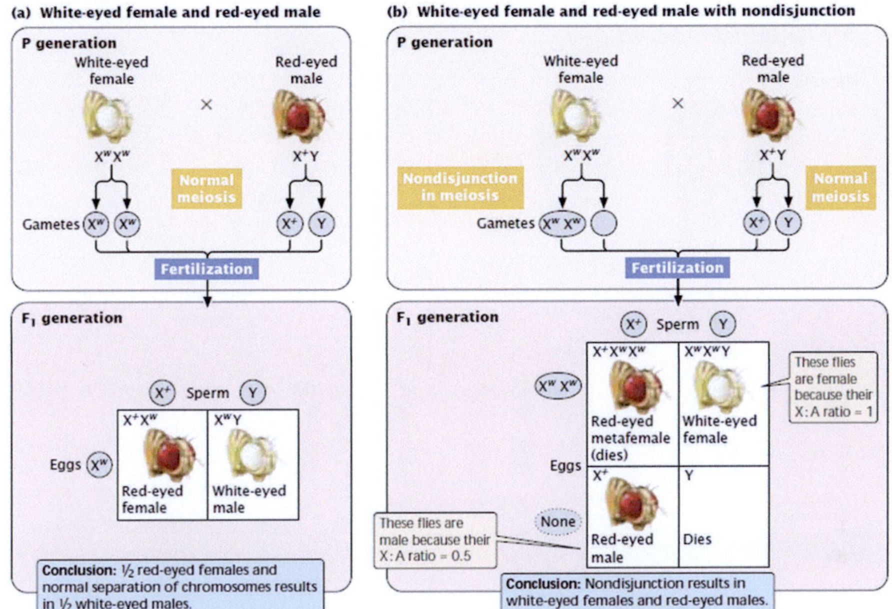

Fig. 5.6 Nondisjunction led to the appearance of unexpected phenotypes. (**a**) A cross between white eyed female and red eyed male yields red eyed females and white eyed males. (**b**) If there is nondisjunction during meiosis and the chromosomes fail to separate, one gamete has both X chromosomes and the other lacks a sex chromosome. In this case, the F1 generation will consist of white eyed females and red eyed males (Pierce 2010)

5.3 Sex-Linked Genes: Human

Sex-linked genes are also present in humans. Essentially, these are the genes which are either present on the X or Y chromosome or both. Their inheritance will vary among males and females. Disorders caused due to mutations of genes on X chromosomes are called X-linked disorders. Many more males than females are generally affected in an X-linked recessive disorder. Another feature of these disorders is that the affected father will never transmit the disorder to his son.

X-linked recessive disorders are caused due to a mutation in a gene present on the X chromosome, and they are inherited in a recessive manner. That means, that one defective copy is usually not enough to cause the disease as a copy on the other pair of the chromosome can make up for its loss. However, males possess a single copy of the X chromosome, and most of the genes present on this chromosome are absent on the Y chromosome. The defective copy of a gene on X chromosome remains uncompensated for in males, and they express the disorder in spite of carrying only one defective copy. This makes them more vulnerable to X-linked recessive disorder (Fig. 5.7). In the example below, if the father carries a mutation in the X chromosome, the gene is passed on to the daughters. The daughters do not express the

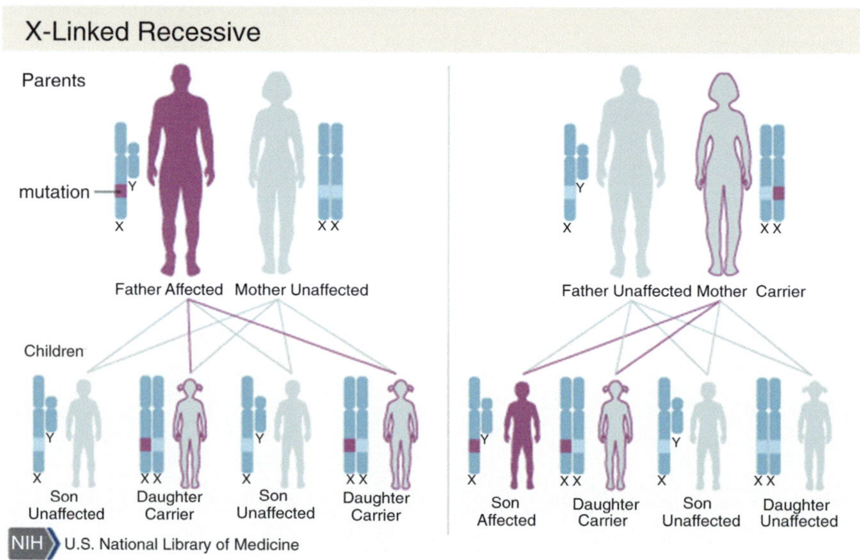

Fig. 5.7 Probability and pattern of inheritance of X-linked recessive disorders. In case of X-linked recessive disorders, if the father carries a mutation in the X chromosome, he will pass the chromosome to his daughters. Sons will not be affected since they will only inherit Y chromosome from their father. Since the daughters carry only one X chromosome with the affected gene, they will only be carriers and will not express the disorder. In case the mother carries a mutant gene on the X chromosome, she may pass it to both her sons and daughters. However, males being hemizygous will express the disorder if they inherit the mutant gene. Daughters will remain carriers (Courtesy of MedlinePlus from the National Library of Medicine)

disorder and are instead just carriers since they carry a single defective copy. The sons will remain unaffected as they will never inherit the X chromosome from their father. If the mother carries a mutation in the X chromosome, the gene may be passed equally between the sons and daughters. The daughter will be a carrier but the son will express the disorder. A female will only express an X-linked recessive disorder if she inherits a defective X chromosome from both the parents.

5.3.1 Haemophilia: X-Linked

Haemophilia is a disorder where clotting of blood is affected. Several proteins act in a cascade leading to clotting of blood. Mutations in any of these may cause haemophilia. One of the most common causes is mutation in a protein called factor VIII. Although there are medications available now, this disorder was thought to be potentially fatal earlier as the affected person would bleed profusely causing his or her death. Haemophilia is inherited in a recessive manner. Only one defective copy is not sufficient to cause the disease but both copies need to be mutated for the disease to manifest. Haemophilia is sometimes referred to as the royal disease because many members of European royalty suffered from this disease in the nineteenth and twentieth century. The mutation for haemophilia seems to have originated from Queen Victoria who passed on the defective gene to two of her daughters who were carriers. That is, they were not affected by the disorder because they had a normal copy of the gene on the other X chromosome. However, they did carry the genetic defect and passed it to their offspring. Queen Victoria's son also suffered from the disease. Since the male progeny inherits the X chromosome from the affected mother and Y chromosome from the father, he may inherit the allele with the genetic mutation. Since he also lacks another X chromosome to compensate for the mutation in the loci, he will suffer from the disorder.

5.3.2 Colour Blindness: X-Linked

Another example of X-linked recessive disorder is colour blindness. Colour blindness is a disorder in which individuals are unable to distinguish between some colours like blue and yellow or red and green. In humans, the retina in the eye consists of cone cells which aid in the perception of light. The cone cells consist of one of three different pigments which absorb light at different wavelengths. One type of pigment absorbs blue light, another absorbs red light, and a third absorbs green light. We are able to detect only three colours—red, blue and green, but processing of these signals by the brain allows us to perceive a host of various colours. The three pigments are coded by different genes. Gene for blue pigment is present on chromosome 7, while the gene for red and green pigment is present on the X chromosome. Colour blindness can occur if there are mutations in the loci coding for the red and green pigments. These mutations are generally recessive and inherited as an X-linked characteristic.

We can denote the normal allele as X^+ and the defective allele as X^-. If an affected male (X^-Y) were to mate with an unaffected female ($X^+ X^+$), their daughters would be carriers ($X^+ X^-$) but remain unaffected, while their sons would be normal (X^+Y). On the other hand, if an affected female ($X^- X^-$) mates with an unaffected male (X^+Y), their daughters would be carriers but remain unaffected ($X^+ X^-$), while all the males will be affected (X^-Y). Sons inherit the condition from their mothers due to the fact that they are hemizygous. Males are hemizygous for X-linked genes because they have a single X chromosome and therefore a single allele. Carrier daughters may pass on the condition to their sons. It may be noticed that the condition can pass from mothers to sons or it can be passed from fathers to grandsons through their daughters. X-linked recessive inheritance therefore shifts between sexes from one generation to another. For this reason, it also sometimes referred to as a criss-cross inheritance (Fig. 5.8).

5.3.2.1 X-Linked Dominant Disorders

Just as there are autosomal dominant disorders, there are also examples of X-linked dominant disorders. This means that the gene responsible for causing the disease is

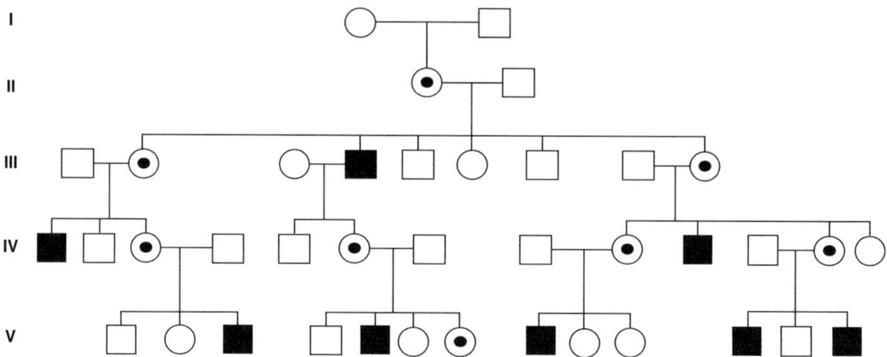

Fig. 5.8 Pedigree showing inheritance of an X-linked recessive disorder. Shaded boxes show affected individuals. More males exhibit this disorder than females. Females having a shaded centre are carriers and carry the mutated gene on one of the X chromosomes. This pedigree clearly exhibits criss-cross inheritance of disorder from carrier mother to affected son and from affected father to carrier daughter (Griffiths et al. 2002)

present on the X chromosome and is inherited in a dominant manner which means one copy of the mutant gene is sufficient to express the disorder. In these disorders, inheritance is not necessarily more common in males than in females unlike the X-linked recessive disorders. The pattern of inheritance depends on whether the father or mother carries the mutated gene. If the father carries the mutant gene, the trait will be exclusively passed on to his daughters. This is because sons will only inherit Y chromosome from the father and so the X chromosome is inherited by the daughters. In addition, being a dominant disorder, presence of one mutated copy is sufficient to cause a phenotype. In contrast, if the mother carries the defective X chromosome, the trait can be passed to both sons and daughters. It should be noted from this that a son having an X-linked dominant disorder will have necessarily inherited it from his mother (Fig. 5.9).

5.3.2.2 Rett Syndrome

Rett syndrome is an example of X-linked dominant disorder. It is caused due to the mutation in *MeCP2* gene. It is a developmental disorder and symptoms may start appearing after 6–18 months of age. They develop microcephaly (brain size is small) and breathing abnormalities. They may also have abnormal eye movements, irritability, seizures and scoliosis. They have language and communication problems and tend to make repeated movements. Rett syndrome is often seen in girls. This is because males having a mutation in *MeCP2* generally die in infancy. Rett syndrome belongs to a spectrum of disorders and it can range from mild to severe and atypical form of the disorder. The mutations in MeCP2 may affect synapses and communication between neurons. MeCP2 protein is believed to regulate activity of genes involved in brain function. It may lead to the disruption in neuronal functioning. It is however unclear how the MeCP2 mutations lead to symptoms seen in Rett syndrome.

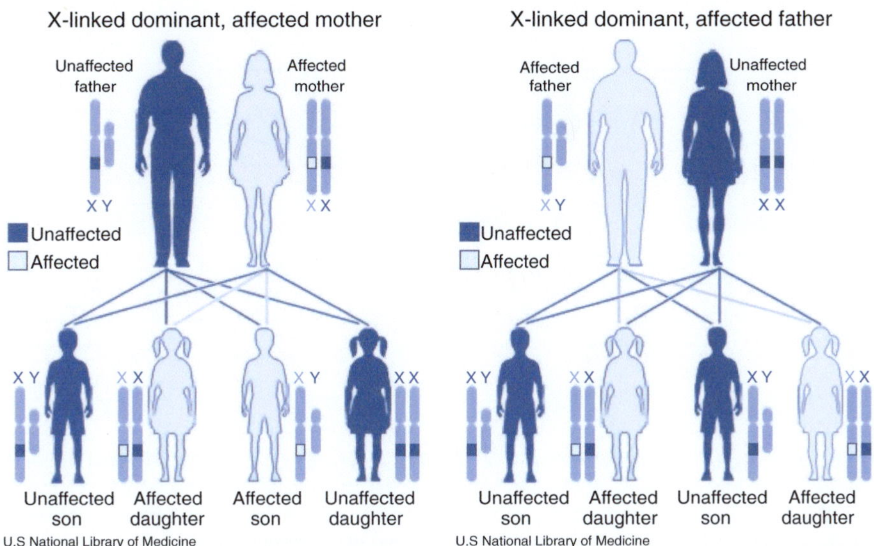

Fig. 5.9 Probability and pattern of inheritance of X-linked dominant disorder. If the mother carries an X chromosome with a mutated gene, both sons and daughters have an equal probability of inheriting the mutated gene and expressing the disorder. If the father carries an X chromosome with a mutated gene, his sons will remain unaffected. The daughters will inherit and express the disorder (Courtesy of MedlinePlus from the National Library of Medicine)

5.3.2.3 Fragile X Syndrome

Fragile X syndrome is another example of an X-linked dominant disorder. It causes severe developmental abnormalities including learning disabilities and cognitive impairment. Affected children may have delayed development and difficulty maintaining attention and focusing on a particular task. They may have difficulty in social interactions and communication. Typical features associated with fragile X syndrome is a long, narrow face and prominent jaw and forehead, large ears, flexible fingers and flat feet. It is more common in males than in females. Fragile X syndrome is caused by a mutation in the *FMR1* gene which codes for the FMRP protein. The *FMR1* gene has trinucleotide CGG repeats which can range from 5 to 40 in normal individuals. In affected individuals, there may be more than 200 repeats. This leads to a silencing of the gene, reducing the levels of FMRP protein. FMRP protein is thought to be involved in formation of synapses which helps neurons communicate (Fig. 5.10).

5.3.3 Genes on Y Chromosome

Genes on Y chromosome are passed exclusively from male to male. Especially genes that are present on the differential region of the Y chromosome, that is, the region on the Y chromosome which is not similar to the X chromosome, are present only in

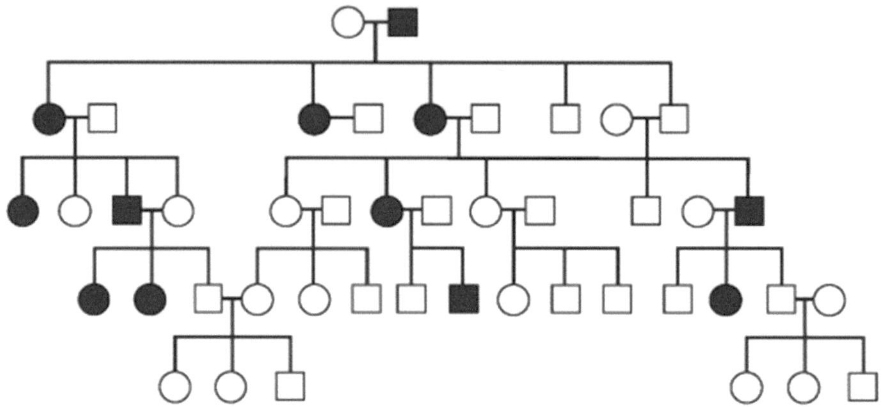

Fig. 5.10 Pedigree showing inheritance of X-linked dominant disorder. An X-linked dominant disorder can be inherited without any bias between males and females within a family. We can see that a father will pass the mutated gene on X chromosome to all his daughters. Sons may inherit the disorder from their mother carrying the mutated gene on the X chromosome

males. One of the primary genes that influence maleness is the sex-determining region Y (*SRY*) gene which is sometimes also referred to as the testis-determining factor. This gene is located on the differential region of the Y chromosome. The SRY protein is a transcription factor which regulates the expression of a number of different genes. It signals the foetus to develop male reproductive organs like testis and prevents the formation of female reproductive structures like uterus and fallopian tubes. Mutations on Y chromosome are associated with sterility. But this condition is not heritable and is caused by spontaneous mutations in the Y chromosome.

5.3.4 Genes Both on X and Y Chromosomes

Certain regions present on the ends of the X and Y chromosomes share common genes and can therefore pair up and split during meiosis. Since these genes are present on both chromosomes, they follow an autosomal form of inheritance rather than a sex-linked pattern of inheritance. For this reason, they are called pseudoautosomal regions or PAR. The PAR present on the short arms of X and Y region is called PAR1 and consists of about 2.6 Mb of DNA. PAR 2 present on the long arm of the chromosomes is shorter and spans about 320 kb. Pairing of PAR1 region is important for spermatogenesis. About 24 genes have been identified in the PAR1 region and they play a variety of roles. For example, *CD99* is involved in T cell adhesion process. *ASMT* is involved in the synthesis of melatonin. *CXYorf3* is a regulator of alternate splicing. There are a few genes for which function is not yet known.

Mutations in *SYBL1* located in PAR2 region are found to be associated with bipolar affective disorder. It codes for a member of synaptobrevins which play a role in cellular exocytosis. *IL9R* also located in PAR2 region has been linked to development of asthma. It belongs to the hematopoietin receptor subfamily. The *SHOX* (short stature homeobox-containing gene) is implicated in short stature of Turner syndrome. This gene is located in the PAR1 region.

5.4 Concept of Sex Determination

The sex of human embryos is not developed until the seventh week of gestation. Genes present on the Y chromosome signal for development of the testis at this point. In the absence of Y chromosome, the embryo develops into a female. The Y chromosome is therefore important for development of male characteristics.

5.4.1 Y Chromosome: Maleness

As stated before, regions on the tips of Y chromosomes called pseudoautosomal regions pair with the X chromosomes and segregate during gamete formation. The remaining region on the chromosome does not pair with the X chromosome and is referred to as the male-specific region of the Y (MSY). The MSY consists of both euchromatin and heterochromatin regions. Euchromatin near the PAR at the short end of the chromosome contains a gene called *SRY*. As stated in Sect. 5.3.3, this gene plays an important role in stimulating male gonad development in the embryo. It has been observed that there are males with two X chromosomes and no Y chromosomes. However, they do have the region of the Y chromosome containing the *SRY* gene attached to one of their X chromosomes. There are also cases where a female develops even when she has a genotype of XY. In such cases, the Y chromosome usually lacks the *SRY* region. These examples show that *SRY* gene is essentially responsible for male determination. It is not very clear how *SRY* triggers male development. It acts as a transcription factor which activates expression of a cascade of genes, some of which are present on the autosomes eg *Sox9* on chromosome 17 and *WT1* on chromosome 11. The action of these genes and their downstream partners promotes male gonad development and aids in blocking of female gonad development.

The Human Genome Project helped in getting a better picture of the genes present on Y chromosome. This information changed the perception of Y chromosome as the carrier of only a few genes. The MSY region is about 23 Mb long and consists of three regions—X transposed region, X degenerative region and ampliconic region. The X transposed region is derived from the X chromosome during the course of human evolution. It is similar to the Xq21 region of the modern X chromosome and has two genes. This region consists of 15% of the MSY region. The X degenerative region is more distantly related to the X chromosome. Many of the genes in this region are pseudogenes and are non-functional. There are 14 genes in this region that

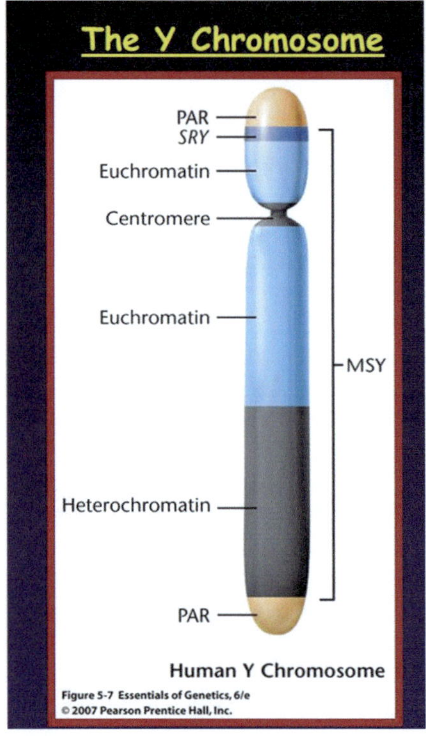

Fig. 5.11 The human Y chromosome. PAR is the pseudoautosomal region present on both the tips of the chromosome. MSY is called the male-specific region of the Y chromosome which does not pair with the X chromosome during meiosis. The *SRY* gene is located in the MSY near the PAR on the short arm of the chromosome. The MSY consists of both euchromatin and heterochromatin (Klug et al. 2007)

are capable of getting transcribed and *SRY* is one of them. This region forms 20% of the MSY. The ampliconic region has about 60 transcription units which are divided into nine gene families. Most of them have multiple copies. The members of each family are nearly identical to each other. Each repeat unit, called an amplicon, is present on the euchromatin region of the Y chromosome. Most of these genes are related to the development of the testis and its function. Mutations in these regions may lead to sterility in males. The ampliconic region forms about 30% of the MSY region (Fig. 5.11).

5.4.1.1 Klinefelter Syndrome

Individuals with Klinefelter syndrome have an underdeveloped testis which is unable to produce sperm. They have male genitalia and internal ducts. They are tall and have long limbs with larger hands and feet. There might be slight breast enlargement and presence of rounded hips. It was discovered later that these individuals had an abnormal karyotype. They had one Y chromosome and multiple X chromosomes ranging from 47, XXY to 49, XXXXY. The number here denotes the total number of chromosomes followed by the composition of the abnormal chromosomes. Here, 47, XXY means that the individual has the normal 44 autosomes and then has two X chromosomes and a Y chromosome instead of one X and one Y chromosome. More X chromosomes led to more severe symptoms.

This condition may arise due to nondisjunction of chromosomes during meiosis. The presence of Y chromosome allows the formation of male genitalia but it is abnormal and incomplete. This syndrome is present in 2 out of every 1000 male births.

5.4.1.2 Turner Syndrome

The affected female in Turner syndrome has underdeveloped ovaries but develops external female genitalia. These individuals are short, have underdeveloped breasts and have skin flaps on the back of the neck. They may also have a broad chest. They generally lack one X chromosome and are 45, X. Since these individuals lack a Y chromosome, they develop as females though this development is abnormal due to the lack of another X chromosome. Turner syndrome may also be a result of mosaics. In this case, the embryo is normal in the beginning. Mutations may occur in some cell during early development leading to a loss of one of the chromosomes. This mutated cell can divide and give rise to more abnormal cells. The individual thus has both the normal and abnormal cells and is a mosaic of these two cell types. The most common chromosome combinations are 45, X/46, XX and 45, X/46, XY. Turner syndrome affects 1 in every 2000 female births. This frequency is much lesser than the Klinefelter syndrome. One reason for his may be that many more of the 45, X foetuses get spontaneously aborted and die in utero.

5.5 Analysis of Sex-Linked Traits

5.5.1 X-Linked Recessive Inheritance

Expression of X-linked recessive inheritance is mostly seen in males as they have only one X chromosome and most of the alleles are hemizygous. Presence of defective allele on X chromosome is expressed in males, but it's not expressed in females as they have corresponding normal allele to mask the effect of defective allele.

The following are some criteria for understanding X-linked recessive inheritance (Adkison 2012):

1. Males who are not affected do not pass the disease on to their offspring.
2. Affected males' daughters are all heterozygous carriers.
3. Heterozygous women pass the mutant allele on to 50% of their sons (who are affected) and 50% of their daughters (who are heterozygous carriers).
4. Half of an affected male's sons would be affected if he marries a heterozygous woman, giving the false impression of male-to-male transmission (Fig. 5.12).

X-linked recessive inheritance is seen in case of transmission of haemophilia A. It is a blood disorder in which gene for clotting factor VIII is defective causing abnormally delayed clotting. It is mostly seen in males who receive defective genes from their unaffected mothers.

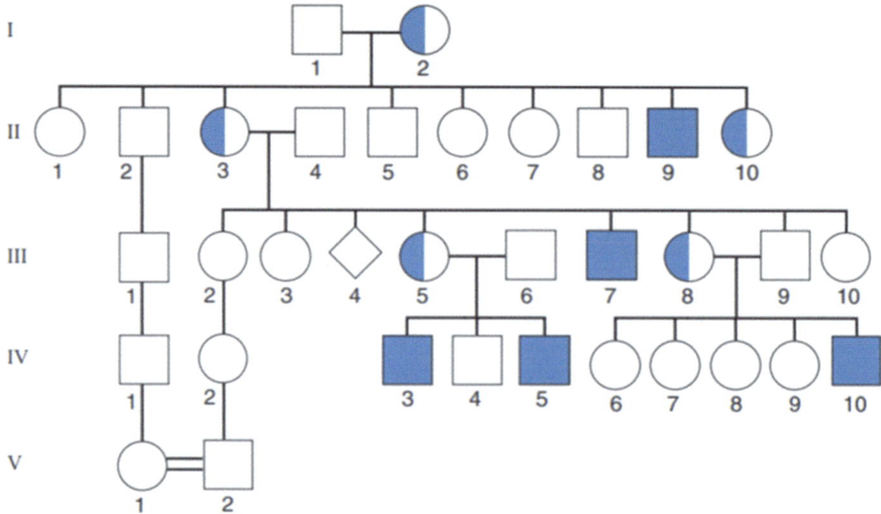

Fig. 5.12 X-linked inheritance of haemophilia A among descendants of Queen Victoria (1–2) of England

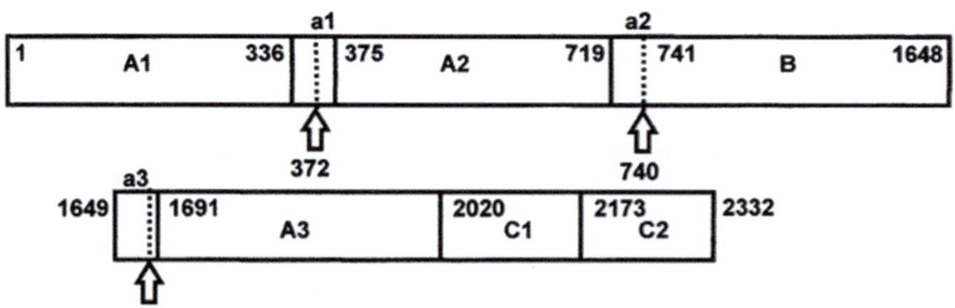

Fig. 5.13 Structure of the FVlll molecule linear structure

FVlll is a 300 kDa glycoprotein which is produced by the liver. FVlll gene is a dimeric protein containing a light chain of 80 kDa and heavy chain of 90–250 kDa. FVIII is an essential cofactor for the conversion by activated FIX (FIXa) of the zymogen factor X (FX) into active FXa (Fig. 5.13).

Mutations responsible for mild/moderate haemophilia A and resulting in reduced stability of the A2 domain are localised at or near the interface between the A domains.

A series of mutations in residues either only partially surface-exposed or located in the core of the FVlll molecule were predicted to result in impaired stability or folding of the FVlll molecule. Such alterations were predicted to result in accelerated intracellular degradation and poor secretion of a large number of FVlll variants.

5.5.2 X-Linked Dominant Inheritance

X-linked dominant inheritance occurs in both males and females, and only one mutant allele is involved. This type of mutation can be transmitted by a female to both male and female offspring; but males can transmit it only to females. Affected females transmit the affected allele to 50% of offspring, whereas males transmit mutant allele to 100% of females (Fig. 5.14).

In this case, heterozygous females are less affected than males as they have a normal, unmutated allele.

5.5.3 Y-Linked Inheritance

Y chromosome is smaller and carries lesser genes unlike X chromosome. It is transmitted from a father to a son. Y chromosome carries genes for:

- ASMTY (acetylserotonin methyltransferase)
- TSPY (testis-specific protein)
- IL3RAY (interleukin-3 receptor)
- SRY (sex-determining region)
- TDF (testis-determining factor)
- ZFY (zinc finger protein), PRKY (protein kinase, Y-linked)
- AMGL (amelogenin)
- CSF2RY (granulocyte-macrophage colony-stimulating factor receptor, alpha subunit on the Y chromosome)

It also inherits the genes for hypertrichosis of ears, webbed toes and porcupine man.

In hypertrichosis, a prominent amount of hairs of atypical length grow from the ears.

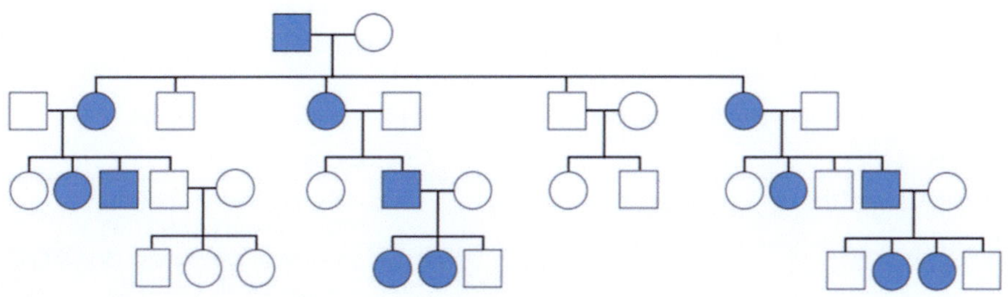

Fig. 5.14 Inheritance of X-linked dominant trait

5.6 Techniques for Studying Chromosome

Through the study of chromosomes, one can infer the number and morphology of chromosomes for the construction of karyotype and recombination studies through meiosis. To start with, the tissue sample is to be collected which can be anthers from bigger flowers or whole inflorescence from smaller flowers, and in the case of study of somatic chromosomes, root or shoot tips can be collected and fixed. To increase the number of metaphase cells, the somatic tissues may be pretreated before fixation. Furthermore, pretreatment increases chromosome condensation and makes morphological detection easier (Singh 2003). Followed by pretreatment and fixation, the cells are subjected to staining. For preparation of smear, the sample is placed over a slide and crushed in a drop of stain. Crushing of samples ensures spreading of cells and easy observation.

5.6.1 Cytological Fixatives

Cytological fixatives are chemical substances used to preserve or stabilise biological samples. A good fixative should have all of the following characteristics: rapid penetration strength, freezing the cell state without dissolving or distorting it, shrinking or swelling the cell, pathogen killing capacity and autolytic enzyme inactivation. One of the most important characteristics of a successful fixative is that it sensitises various cell constituents to particular stains so that they can be detected. Fixation helps to preserve cellular architecture and composition of cells in the tissue to allow them to withstand subsequent processing (Thavarajah et al. 2012). The process of fixation involves halting all the activities in a cell or a tissue in order to study cell activities across a time range. Data from different stages are united in order to study a particular process of cell. The fixative used is determined by the mechanism being studied, with the main focus being on the structure of the chromosome and its activity without interference from other cellular components.

On the basis of their oxidation potential, fixatives are classified into metallic and non-metallic compounds (Table 5.1). One of the most important characteristics of a good fixative is that it sensitises various cell constituents to particular stains, allowing them to be examined more easily. Since both of these properties cannot be present in a single fixative, a combination of one or more compatible chemicals must be combined to achieve successful results. Baker (1944) proposed that a

Table 5.1 Some common fixatives grouped on the basis of their oxidation potential

Reductants	Oxidants
Ethanol (C_2H_5OH)	'Osmium tetroxide (OsO_4)' 'Chromium trioxide (CrO_3)' 'Potassium dichromate ($K_2Cr_2O_7$)' 'Formaldehyde (HCOH)'
Mercuric chloride ($HgCl_2$)	'Acetic acid (CH_3COOH)'

Table 5.2 Some widely used fixatives are classified based on their reaction with albumin, with subgrouping based on their metallic appearance

Coagulants		Non-coagulants	
Metallic fixatives	Non-metallic fixatives	Metallic fixatives	Non-metallic fixatives
Mercuric chloride ($HgCl_2$)	Ethanol	Osmium tetroxide (OsO_4)	Formaldehyde (HCOH)
Chromium trioxide (CrO_3)		Potassium dichromate ($K_2Cr_2O_7$)	Acetic acid (CH_3COOH)

fixation mixture consisting of basic non-precipitant chemicals combined with a non-fixative salt would be the safest.

Coagulants and non-coagulants are the two types of fixatives depending on their reaction with albumin (Table 5.2). Non-coagulants form a meshwork, while coagulants react with albumin to form a coagulum. Non-additive fixatives add atoms to the protein backbone and interlink chains (Wolman 1955), while additive fixatives add atoms to the protein backbone and interlink chains (Baker 1966); some precipitate, while others do not.

5.6.1.1 Mercuric Chloride ($HgCl_2$)

Mercuric chloride is a compound that has been used for tissue fixation for a long time. It is a good protein precipitant since it reacts with amines, amides, amino acids and sulphydryl groups. It repairs nucleoproteins by reacting with the phosphate groups of nucleic acids. Mercuric chloride as a fixative has drawbacks in that it is corrosive in nature, and mercury, which is extremely poisonous, can be ingested through the skin and is a cumulative poison. Chromosomes pick up dyes well after fixation with mercuric chloride which is compatible with many fixatives, but one of the serious drawbacks is that it leaves behind metallic needle-like precipitates (Mayer 1918).

5.6.1.2 Potassium Dichromate ($k_2Cr_2O_7$)

At p^H greater than 4.6, potassium dichromate acts as a coagulant, interacting with chromic acid to preserve chromosome structure at lower p^H. It has a non-coagulant effect. It interacts with unsaturated lipids to render them insoluble, which is why it's used in mitochondrial research. The oxidation of proteins is combined with the interaction of reduced chromate ions to form cross links in the fixation reactions. Chromium ions react with the carboxyl and hydroxyl side chains of proteins, leaving the amino group free, allowing acidic dyes to stain the protein.

5.6.1.3 Chromic Acid (H_2CrO_4) or Chromium Trioxide (CrO_3)

Chromic acid is a strong oxidising agent and highly corrosive in nature. Reaction of chromic acid with proteins occurs in two steps. In the first step, coagulation and precipitation of protein occurs and in the second step, hardening due to $HCrO_4$— occurs. The nucleus and the chromosomes are well fixed, while the cytoplasm is coarsely coagulated. Since it reacts violently with organic reducers like ethanol and formalin, chromic acid should not be combined with them.

5.6.1.4 Osmium Tetroxide (OsO_4)

Osmium tetroxide is a highly radioactive and volatile crystalline solid found in nature. It must be treated with caution due to its high volatile property, as it can repair the conjunctiva of the eye and nasal mucosa. Intercalating between unsaturated bonds of lipids and phospholipids is part of the osmium tetroxide fixation reaction. Most proteins lack several double bonds, but certain amino acids, such as tryptophan and histidine, have double bonds in their rings, making them reactive with osmium tetroxide and forming a dark precipitate. Proteins that have been treated with osmium tetroxide are no longer reactive to acidic dyes. Osmium tetroxide is soluble in lipids and becomes black when exposed to ethanol. Due to the reaction with double bonds, it also blackens unsaturated lipids (Altmann 1894). Since it does not precipitate DNA, it cannot be used to analyse interface chromosomes. The disadvantages of using osmium tetroxide as a fixative include its high cost and the fact that it blackens tissues as it reacts with formaldehyde and ethanol to form reduced osmium dioxide, which is dark in colour.

5.6.1.5 Ethanol (C_2H_5OH)

70–100% ethanol is most commonly used as a chemical fixative. It acts as a dehydrating agent but does not react with macromolecules. It precipitates proteins and DNA by removing water molecules from them and making them insoluble in water. Its staining ability is unaffected because it does not react with macromolecules like DNA. Since it is a reductant that is oxidised to acetaldehyde and then to acetic acid, it is incompatible with oxidising fixatives including metallic oxides. The disadvantages of using ethanol as a fixative are that it shrinks and hardens the cells excessively. As a result, ethanol must be combined with other fixatives that reverse this effect.

5.6.1.6 Acetic Acid (CH_3COOH)

Acetic acid is a weak acid that works well as a solvent. It dissolves both polar and nonpolar compounds, making it ideal for dye dissolution. It's known as glacial acetic acid in its anhydrous form, and vinegar in its 4–8% form. It precipitates nucleic acid and dissolves histones, but it does not fix proteins, according to Pischinger (1937). It is understood that it causes excessive swelling of cellular elements, including chromosomes, but not nuclear protein objects. It prefers to fix pachytene chromosomes, but it can also fix metaphase and anaphase chromosomes. The stainability of cells is affected by acetic acid fixation. The cytoplasm is more sensitive to acidic dyes, while the metaphase and anaphase chromosomes are more sensitive to simple dyes. Since acetic acid has the ability to swell cells, it must be used in conjunction with a fixative that has the ability to shrink the cytoplasm.

5.6.1.7 Formaldehyde (HCHO)

Formaldehyde is often used to preserve specimens. It hydrates to form methylene glycol, a glycol (hydrated formaldehyde) in aqueous solution. Because of its ability to react with proteins, it is used as a fixative. It cross-links proteins by reacting with terminal and secondary amino groups, as well as immino groups. It forms bridges between the –CO and –NH groups at pH 4, but at pH 8, it reduces the sulphur bridges (-S-S-) to the –SH groups and then reacts to form the methylene bridge (-S-CH_2-S-) (Middlebrook and Phillips 1942) The property of bridge formation hardens the tissue, making tissue analysis easier. It isn't recommended for chromosome studies because it doesn't fix or precipitate DNA.

5.6.1.8 Propionic Acid (C_2H_5COOH)

It's for species with a limited chromosome count. It has a structure and solvent properties identical to acetic acid, but unlike acetic acid, it does not swell the chromosomes, so it is widely used as a fixative in chromosome research.

5.6.1.9 Chloroform ($CHCl_3$)

It is highly soluble in alcohol, ether or acetone and moderately soluble in water. Since it dissolves fat, it's mainly used on tissues with an oily or waxy coating. Since it dissolves the upper layers, it aids fixative penetration. It converts to carbonyl chloride in the presence of air and light.

5.6.1.10 Fixing Mixtures

None of the above-mentioned fixatives can be used alone as they have drawbacks. Hence, it is recommended to mix two or more fixatives together to overcome the drawbacks of each other and enhance the process of fixation.

5.6.1.11 Flemming's Weak Fluid (1882) and Flemming's Strong Fluid (1884) (Table 5.3)

Flemming's solution contains both coagulating agent chromium and additive fixative osmium tetroxide, hence considered a good fixative. The drawback of shrinking of cells due to osmic acid is overcome by the swelling effect of acetic acid. Osmium tetroxide aids in fixing the outer layer of cells due to its property of slow tissue penetration, whereas acetic acid penetrates the tissue well. Osmic acid causes blackening of tissues; hence, care should be taken not to keep the tissue in the fixative for a longer time.

Table 5.3 Composition of Flemming's weak fluid and Flemming's strong fluid

Flemming's weak fluid (1882)		Flemming's strong fluid (1884)	
Osmic acid	0.1%	Chromium trioxide, 5% aq	0–3 mL
Chromic acid aqueous	0.25%	Osmium tetroxide, 2% aq	0–4 mL
Glacial acetic acid	0.1%	Acetic acid, 20% aq	0–5 mL
		Distilled water	0–8 mL

5.6.1.12 Carnoy's Fluid (1886) or Carnoy's Fixative 1 (Table 5.4)

It is the most commonly used fixative and used for almost all plant samples. It is also mostly recommended for nuclear and mitochondrial organelles. The drawbacks of using ethanol alone as a fixative is that it fixes the cytoplasm alone and shrinks the tissue but does not fix the nucleus. This drawback is overcome by acetic acid which fixes nucleus and does not fix cytoplasm. This fixative aids in studying both squash and smear preparations. It is effective for both meiotic and mitotic studies of chromosomes.

5.6.1.13 Carnoy's Fluid II (1886) (Table 5.5)

It's mostly used to fix flower buds. Proportions of 1:3:4 and 1:1:3 are used in the Asteraceae and other families (Turner 1956). Since chloroform is present, the fixative is able to penetrate the waxy coating on the flower buds.

5.6.1.14 Navaschin's Fluid (Navaschin 1925) (Table 5.6)

Solutions A and B are prepared and stored separately as chromic anhydride present in solution A is an oxidiser and formaldehyde present in solution B is a reducer, hence, if mixed together, can cause oxidation reaction that can modify the fixation properties. It has a high penetration property. Hence, it requires lesser time for fixation, i.e. 3–4-h treatment is enough to act on a tissue. Since acetic acid can cause swelling up of cells, it must be used in lesser proportion for study of details of the chromosomes.

Table 5.4 Composition of Carnoy's fixative

Glacial acetic acid	1 part
Ethanol	3 parts

Table 5.5 Composition of Carnoy's fixative 2

Glacial acetic acid	1 part
Chloroform	3 parts
Absolute ethanol	6 parts

Table 5.6 Composition of Navaschin's fluid

Solution A	
Chromic anhydride	1.5 g
Glacial acetic acid	10 mL
Distilled water	90 mL
Solution B	
Formaldehyde aq. sol.	40 mL
Distilled water	60 mL

5.6.2 Cytological Stains

Live cells are transparent in nature; hence, it is difficult to observe them under the microscope. Stains and dyes are usually used to overcome this problem. According to Baker (1960), three factors influence the staining capacity of cellular components: (1) dye-component affinity, (2) component density, and (3) dye permeability within the same component. A chromophore is a group of atoms or electrons that causes an organic molecule to be coloured, and auxochromes are additional groups attached to chromophores that are known to increase the colour strength. Depending on the charge of the chromophore (basic or acidic), the dye is classified as basic or acidic. If the chromosome is positively charged, the chromophore is negative and vice versa. These can be cationic, with a positive charge that attracts the cell's basic components, or anionic, with a negative charge that attracts the acidic components. The ability of stain to stain is due to interatomic interactions such as double bond resonance. The majority of dyes are based on the delocalised electron structure of an aryl ring. Counterstaining is when one or more stains are used and each stain interacts differently with various cellular components.

5.6.2.1 Carmine

Carmine is a basic stain used to stain plant chromosomes. It is crimson red in colour having anthraquinone linked to glucose unit as colouring matter. It is obtained from an insect *Coccus cacti*. Carminic acid is the dye's main ingredient ($C_{22}H_{20}O_{13}$). It stains chromatin as a basic dye at acidic p^H, while it stains chromatin as an acidic dye at basic p^H. As a consequence, staining plant cells with carmine dissolved in acetic acid is favoured. Adding a few drops of ferric hydroxide as a mordant improves the stainability. Certain hard tissues should be stained with warm acetocarmine in HCl because it softens the tissue (Figs. 5.15 and 5.16).

5.6.2.2 Orcein

Orcein is derived from a variety of lichen types. The crude part, orcinol, is extracted and converted to orcein in the presence of air by ammonia. Orcein's chemical components were only discovered in the 1950s (Musso 1961 in Henwood 2003). It's made up more of alpha-amino orcein phenoxazone derivatives (Fig. 5.17). It is reddish brown in colour and readily soluble in ethanol, but less soluble in water. Dyer (1963) used a solution of 2 g natural orcein dissolved in 100 mL of a 1:1 mixture of lactic and propionic acids and diluted to 45% with water to analyse fresh pollen mother cells. It was also found to be ideal for preparing root tip chromosomes

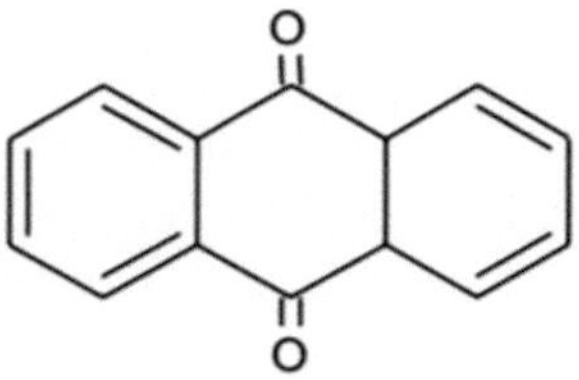

Fig. 5.15 Structure of anthraquinone

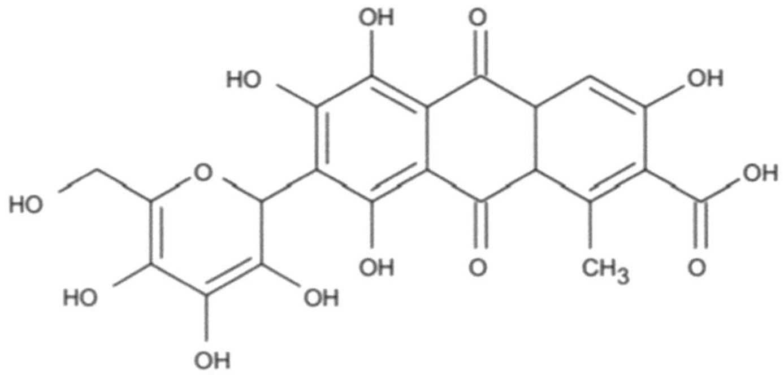

Fig. 5.16 Structure of carminic acid

Fig. 5.17 Structure of alpha-amino orcein

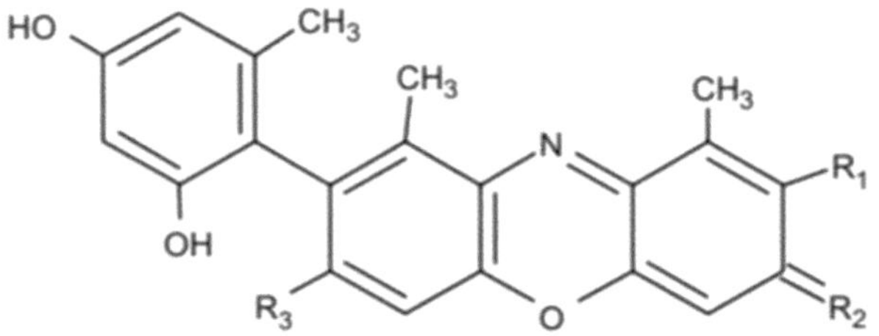

quickly for studying detailed morphology (Dyer 1963). Until mounting in acetic-orcein dye, acetic-orcein is combined with HCl for hydrolysis in root tip and shoot tip tests (Tijo and Levan 1950; Sharma and Sharma 1957).

5.6.2.3 Crystal Violet

It's a triarylmethane dye that's used for Gram's staining and histological staining. It's mostly used to describe bacteria. Alfred Kern was the first to synthesise it in 1883. The three triarylmethane groups attached to nitrogen (Fig. 5.18) serve as auxochromes, while the aryl rings function as chromophores. At neutral pH, it is bluish purple in hue, colourless at alkaline pH and yellow at acidic pH. It can't be used to research chromosomes because of its poor penetration.

5.6.2.4 Acetocarmine, Aceto-orcein (Table 5.7)

Procedure

- Dissolve 1 g carmine/orcein in 100 mL of 45% glacial acetic acid.
- Add aluminium granules and reflux for 24 h.
- Filter and store in dark bottle.
- Staining can be improved by adding 5 mL of 10% ferric chloride to 100 mL of solution.

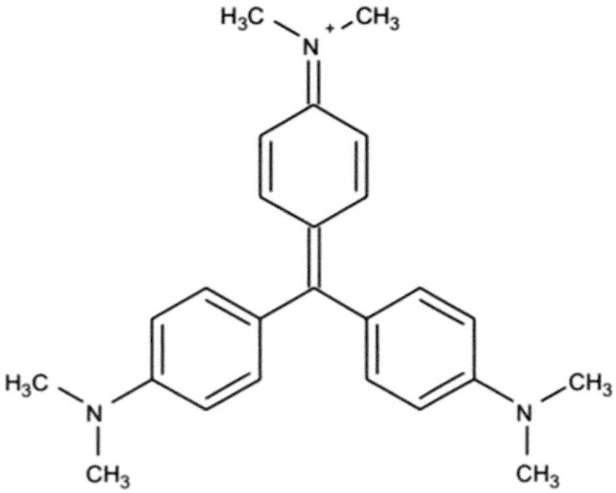

Fig. 5.18 Structure of crystal violet

Table 5.7 Composition of acetocarmine/aceto-orcein

Sl. no.	Constituents	
1.	Orcein, carmine	1 g
2.	Glacial acetic acid	45 mL
3.	Distilled water	55 mL

Table 5.8 Composition of Schiff's reagent

Constituents	
Basic fuchsine	0.5 g
Distilled water	100 mL
1 N HCl	10 mL
Potassium metabisulfite	0.5 g
Activated charcoal	0.5 g

5.6.2.5 Crystal Violet

With continuous stirring, 1 g of dye is added to 100 mL of boiling distilled water. Filter through Whatman filter paper and set aside for a week to mature.

5.6.2.6 Schiff's Reagent/Fuchsine Reagent (Table 5.8)

Procedure

- In 100 mL of boiled distilled water, 0.5 g of dye is added.
- Cool to 58 °C and strain into an amber bottle using Whatman filter paper.
- At 26 °C, add 10 mL of 1 N HCl and 0.5 g of potassium metabisulfite.
- Close the cover and set aside for 24 h; then wait until the solution becomes straw coloured or bleach it with a small amount of charcoal (0.25–0.5 g).
- Filter after a comprehensive shake.

5.6.3 Cytological Pretreating Agents

Prior to fixation, tissues are pretreated with agents to achieve well-spread, transparent and condensed chromosomes for studying their structure and behaviour. Pretreatment, according to La Cour (1935), is important for studying the spiral structure of chromosomes. The following modifications arise in mitotic cells as a result of tissue pretreatment:

- Increase the number of cells in metaphase.
- The chromosomes are constricted longitudinally.
- It elucidates the restraints.
- Increase the cytoplasm's viscosity.
- Allows for quick fixative penetration.

5.6.3.1 Colchicine

It is a water-soluble alkaloid having formula $C_{22}H_{25}NO_6$ (Fig. 5.19) and molecular weight 399.43 g/mol. Pelletier and Caventou were the first to isolate it in 1820. Geiger purified and named it in 1833, but O'Mara was the first to use it in 1939. Colchicine binds to microtubules and prevents tubulin polymerisation by combining with it (Molad 2002). The mitotic apparatus is made up of microtubules and is responsible for separating chromatids during division (Palevitz 1993). Colchicine pretreatment increases the number of metaphase cells, which increases the chances of seeing chromosomes in extremely condensed form.

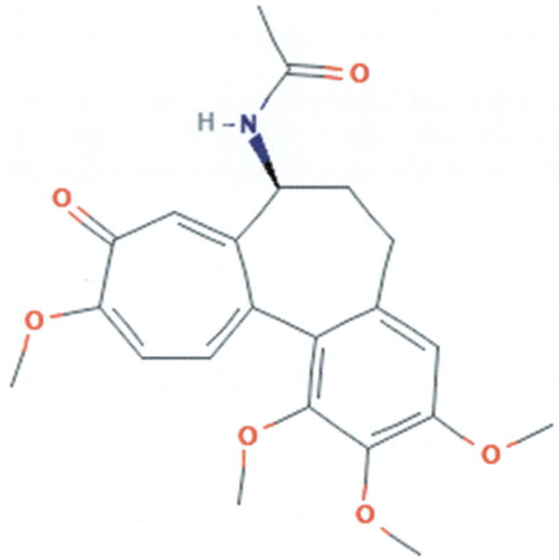

Fig. 5.19 Structure of colchicine

5.6.3.2 α-Bromonaphthalene

With a molecular weight of 207.07 g, it is a bromine derivative of naphthalene with the molecular formula C10 H7Br (Fig. 5.20). Similar to colchicine, it prevents the development of spindle fibres and stops dividing cells at metaphase.

5.6.3.3 8-Hydroxyquinoline

It has the molecular mass of 145.16 g/mol and the formula C9H7NO (Fig. 5.21). Owing to an increase in cytoplasm viscosity, it affects spindle formation and causes metaphase arrest, resulting in chromosome immobilisation, similar to other pretreating agents. It's best for plants with a limited number of chromosomes. For different plant species, different amounts of this chemical and different periods of time are suggested. For example, aloe vera (Vig 1968) received a 0.2% solution for 30 min, and orchid species received a 0.002 molar solution for 4–8 h (Pridgeon et al. 1999).

5.6.3.4 p-Dichlorobenzene (p-DCB)

It is a colourless solid with the molecular weight of 147 g/mol and the formula C6H4 Cl2 (Fig. 5.22). It induces chromatid separation, chromosome fragmentation and chromosome bridges at an early stage (1943, Carey and McDonough).

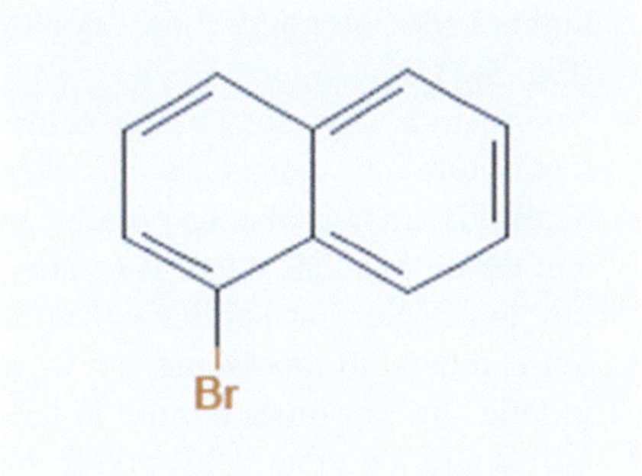

Fig. 5.20 Structure of α-bromonaphthalene

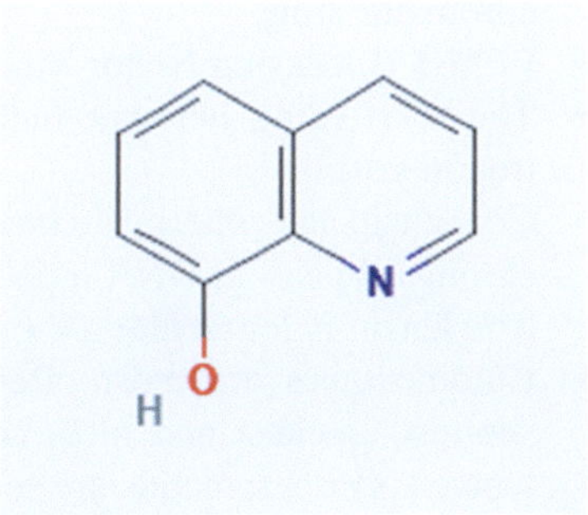

Fig. 5.21 Structure of 8-hydroxyquinoline

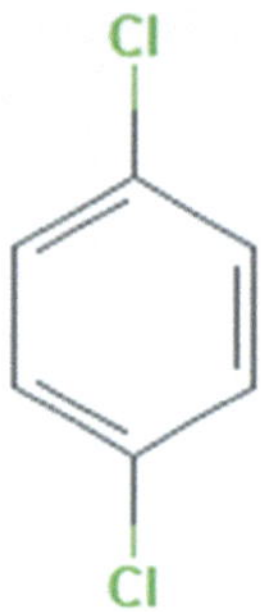

Fig. 5.22 Structure of p-dichlorobenzene

5.6.4 Analysis of Mitotic Chromosome

- The nucleosome fibre is the basic unit of chromatin organisation. It entails packing DNA into the nucleosome as a negative supercoil, resulting in a 10 nm diameter interface chromatin fibre. Histone octamer is wrapped twice with double-stranded DNA (H2a, H2b, H3, and H4).
- The nucleosome fibre folds into a solenoid (zigzag or crossed) to form the 30 nm chromatin fibre. The supercoiling is aided by histone H1.
- Based on protein-protein interaction or protein-DNA interaction between adjacent chromatin fibres, further folding or coiling of chromatin fibres to create higher-order domains that could fold into increasingly larger structures (Fig. 5.23).
- Non-histone chromosomal proteins form a scaffold that tightly packs the chromatin fibre into metaphase chromosomes.
- Scaffolds are non-histone proteins that vary depending on the organism's species and developmental stage. It creates a mechanism that keeps chromosome structure intact. Scl and Scll have been identified as two main scaffold proteins.
- Scl is related to topoisomerase II, which is present in the axial area of extended mitotic chromosomes but not in the loop domains.
- Scll is a protein that helps preserve the structural integrity of chromosomes. It is found in the middle of the long axis of mitotic metaphase chromosomes' chromatid arms.
- LFM-1 (Licensing Factor Model 1) is involved in cell cycle activities.
- The scaffolding network includes the kinetochore, actin-binding protein and tropomyosin.
- Condensin and cohesin are members of the SMC family and are required for sister chromatid pairing, DNA repair and replication and gene expression regulation.
- Condensin is found along the chromosome arms' central axis (Fig. 5.24).
- Chromosomes are most often studied during mitotic metaphase as they are shortest, thickest and most condensed. Mitosis is the phase of the cell cycle where sister chromatids are separated by spindle fibres.

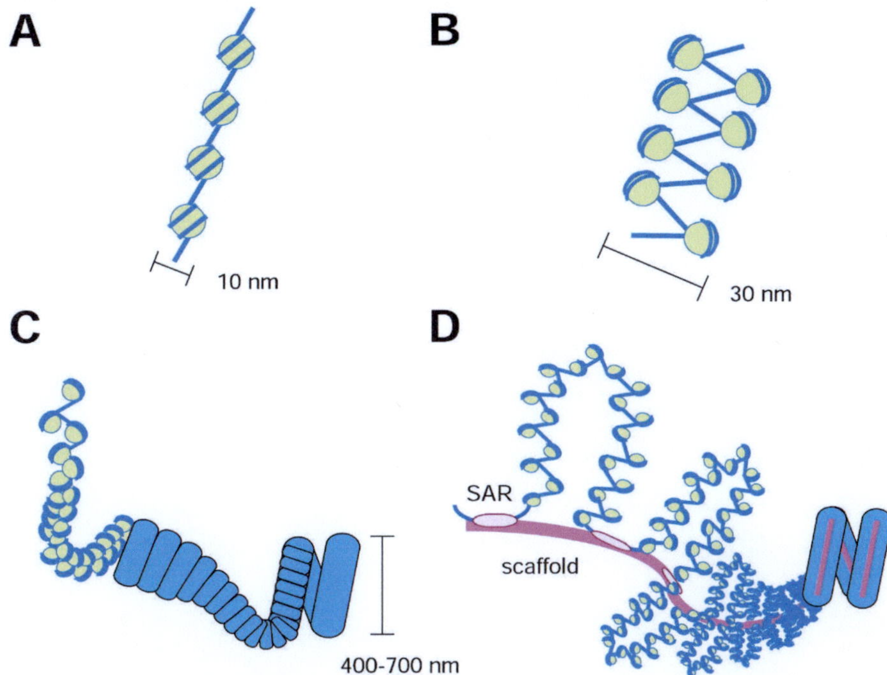

Fig. 5.23 Mitotic chromosome organisation. Diagram showing different levels of chromatin condensation, the 11 nm nucleosome is condensed into a 30 nm chromatin fibre which is then segregated into supercoiled domains or loops by attachment to chromosome scaffolds composed of non-histone proteins

- The cell copies its DNA during interphase, and chromatin is at its least condensed, that is, hundreds of thousands of times less condensed than during mitosis. This is why the enzyme complexes that copy DNA have the most access to chromosomal DNA, which is where the bulk of gene transcription takes place.

5.6.4.1 Prophase

During prophase, chromosomes begin to condense by enlisting the aid of condensin (reorganises chromosomes into their highly compact mitotic structure). Cohesin, a protein complex that binds replicated sister chromatids together, is largely removed from the arms of the sister chromatids, allowing them to separate. At the centromere, it is held. During prophase, the spindle starts to develop, and the two pairs of centrioles travel to opposite poles, while microtubules polymerise from the duplicated centrosomes.

5.6.4.2 Prometaphase

The nuclear envelope is broken up into tiny vesicles that are shared by future daughter cells. Since centrosomes are located outside the nucleus in animal cells, expanding cells' microtubules do not have access to them until the nuclear

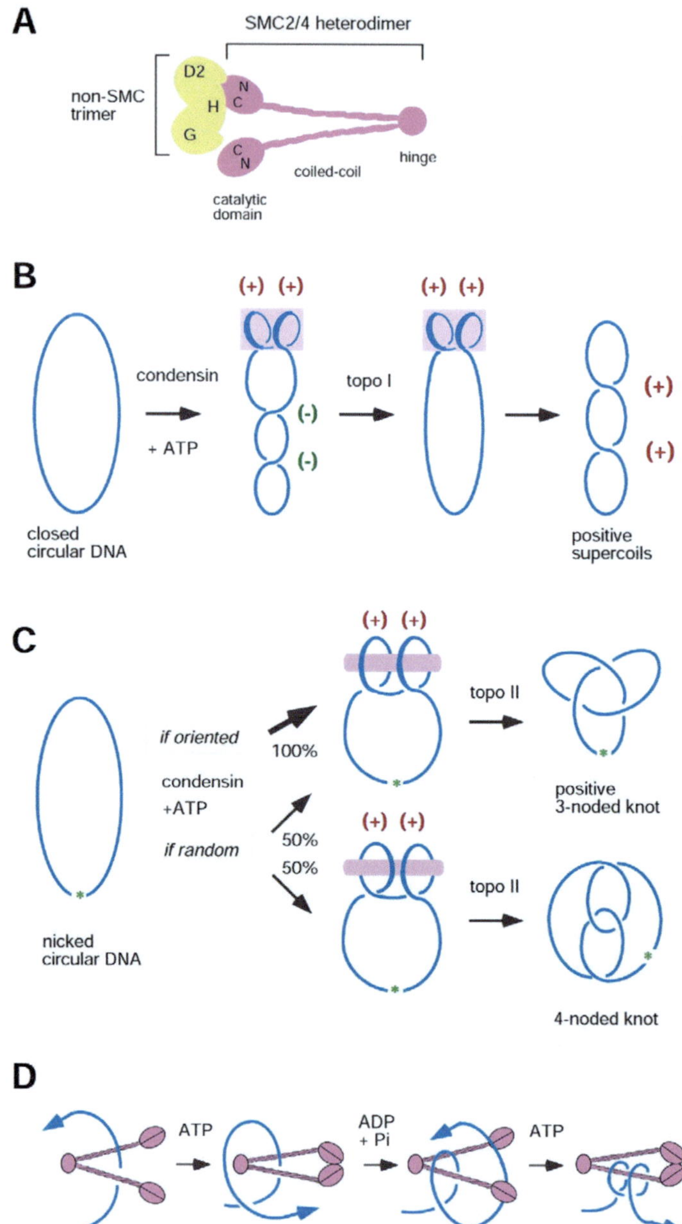

Fig. 5.24 Molecular architecture and actions of condensin. (**a**) Composition of condensin complex, (**b**) supercoiling assay, (**c**) knotting assay, (**d**) interaction of condensin with DNA

membrane tears apart. The chromosomes are bound to the mitotic spindle during mitosis. The kinetochore is a specialised chromosome area where both chromatids of each chromosome bind to the spindle.

5.6.4.3 Metaphase

During metaphase, chromosomes are in the most compacted state. The chromosomes align themselves across the equator of the mitotic spindle in the centre of the cell. The kinetochore microtubules pull the sister chromatids back and forth until they align along the cell's equator, known as the equatorial plane. The cell ensures that it is able to divide during mitosis by passing through the metaphase checkpoint. Anaphase can only be reached by cells with properly constructed spindles.

5.6.4.4 Anaphase

Owing to the degradation of cohesin molecules that were joining the sister chromatids by protease separase, the duplicated genetic material in the nucleus of the parent cell is divided into two identical daughter cells during anaphase. The mitotic spindle separates the chromosomes. During anaphase, there are two types of movements. The chromosomes travel towards the spindle poles as the kinetochore microtubules shorten, and the spindle poles separate as the non-kinetochore microtubules move past each other in the first section, and the spindle poles separate as the non-kinetochore microtubules move past each other in the second part.

5.6.4.5 Telophase and Cytokinesis

The chromosomes reach the poles during telophase. In their interface confirmations, the nuclear membrane reforms and chromosomes begin to decondense, which is accompanied by the cytoplasm being split into two daughter cells of the same genetic make-up.

5.6.5 Analysis of Human Karyotype

Karyotyping is the method of pairing and ordering all of an organism's chromosomes, resulting in a genome-wide snapshot of the chromosomes of an individual. They are used to detect genetic modifications such as chromosome number, deletion, replication, translocation and inversion using normal staining techniques.

Mitotic cells that have been arrested in metaphase or prometaphase are used to make karyotypes. Colchicine, which poisons the spindle formation, prevents cells from entering metaphase. The cells are then exposed to a hypertonic solution, which causes the cells' nuclei to swell and burst. After that, the nucleus is fixed, dropped on glass and stained with various stains to help in the analysis of the chromosome.

The majority of the karyotypes are stained with Giemsa staining, which provides a more consistent preparation and better individual band resolution and can be examined with a standard bright-field microscope. In G banding, AT-rich or

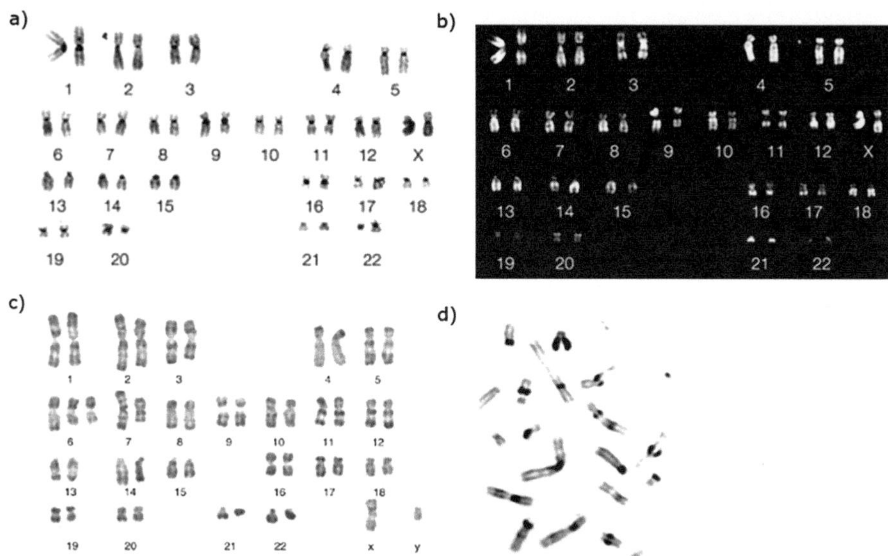

Fig. 5.25 Chromosome banding pattern by different staining techniques. (**a**) Giemsa banding, (**b**) Q-banding (**c**) R-banding and (**d**) C-banding are shown

Table 5.9 Classification of chromosomes based on size

Group.	Chromosomes	Description
A	1–3	Largest; 1 and 3 are metacentric. But 2 is submetacentric
B	4,5	Large; submetacentric with two arms very different in size
C	6–12, X	Medium size; submetacentric
D	13–15	Medium size; acrocentric with satellites
E	16–18	Small; 16 is metacentric, but 17 and 18 are submetacentric
F	19, 20	Small; metacentric
G	21, 22, Y″	Small; acrocentric with satellites on 21 and 22 but not on Y

heterochromatic regions appear to stain darker. GC-rich areas, on the other hand, tend to stain lightly (Fig. 5.25).

Human autosomes are numbered 1 to 22 and arranged in descending order by size in a karyogram. With the exception of chromosomes 21 and 22, all other chromosomes are present. Chromosome number 21 is the smallest. Sex chromosomes are usually found at the end of the chromosomes (Table 5.9).

Karyogram can be used to detect chromosomal abnormalities. Aneuploidy is caused by the absence or addition of a chromosome, e.g. trisomy 21 (Down syndrome).

5.6.6 Cytological Variation

Differences between members of the same species or between species are referred to as genetic variation. Allelic differences are caused by mutations in specific genes. Significant variations in chromosome structure are referred to as chromosomal aberrations. Usually, these have an effect on several genes. Typically, these have a multigene effect. They're also known as chromosome mutations.

Deletions, duplications, inversions and translocations are four types of chromosomal aberrations that can be observed under a microscope cytologically. However, certain cytological shifts are too subtle to identify. Missing chromosome fragments are referred to as deletions. Homozygous deletions may be lethal, while heterozygous deletions may be nonlethal or lethal, and can express recessive genes that were previously unknown. Duplications can trigger a genetic imbalance, resulting in phenotypic effects in the organism and a greater range of gene functions.

Inversions are caused by a 180-degree turn of a chromosome fragment. Inversion heterozygotes also have pairing problems during meiosis, resulting in the creation of inversion loops. When loops are crossed, the outcome is normally unviable. For pericentric and paracentric inversions, the crossover products would be distinct. A chromosomal segment is relocated to a different location in the genome during a translocation. Translocations generate duplication-deletion meiotic products in the heterozygous state, which may result in unbalanced zygotes and new gene linkages.

5.7 Monosomy: Cri-du-Chat Syndrome

Cri-du-chat syndrome, also known as 5p-syndrome or cat cry syndrome, is a rare genetic disorder caused by a deletion (missing piece) of genetic material on chromosome 5's small arm (the p arm).

A high-pitched catlike scream, mental retardation, slow growth, distinct facial features, small head size (microcephaly), widely spaced eyes (hypertelorism), low birth weight and poor muscle tone (hypotonia) in infancy are some of the signs. Over time, the catlike cry becomes less noticeable.

5.8 Trisomy

Trisomies are aneuploids; the addition of an extra chromosome ($2n + 1$) belongs to any one of the different chromosomes of a haploid complement. The addition of a chromosome defines that there are three homologous copies, so the total number of attainable trisomics is equal to the haploid set of chromosome number. This extra chromosome directs its type: normal (primary), isochromosome (secondary) and translocate (tertiary) or interchange forms. In the nomenclature system, a human chromosome complement (normal) is 46-XX (female) and 46-XY (male). Where a male with an extra X chromosome will be 47, XXY, and a female with trisomy 21 will be 47; XX, +21, such triplications are trisomies.

The presence of one extra chromosome results due to meiotic irregularities like an abnormal association during metaphase I and unequal segregation of chromosomes, bridge/laggard formation during anaphase I. This results in the occurrence of unbalanced gametes (i.e. $n + 1$) in trisomic individuals. Usually, autosomal aneuploids are miscarriage, except for aneuploids of some autosomes like chromosome 21. These chromosomes carry lesser genes and have a small size where the occurrence of extra copies is less liable than for bigger chromosomes. Apart from trisomy 21 results in human live births, Edwards syndrome (trisomy 18) and Patau syndrome (trisomy 13) are only human trisomies. The following are the description of viable human aneuploids.

5.8.1 Down Syndrome

Down syndrome or trisomy 21 is the most habitual genetic disorder that impacts foetal development, affecting 1 in 800 to 1 in 1000 every live-born infant. It was discovered by John Langdon Down in 1866. It is normally associated with trisomy in chromosome 21 in the G group of the acrocentric region of the smallest human autosome. The extra chromosome 21, i.e. there are 47 chromosomes, including two X chromosomes as well as the extra chromosome 21. The karyotype of the individual (Fig. 5.26a) is 47, XX, +21. Individuals with Down syndrome typically share similar features, and they show an obvious resemblance to one another (Fig. 5.26b). Due to an epicanthic fold (each eye); a flat/flush face with a round head; large tongue-swollen or protruding; underdeveloped and small ears; characteristically short; stubby fingers; Physical, psychomotor; while life expectancy is reduced to an average of about 50 years.

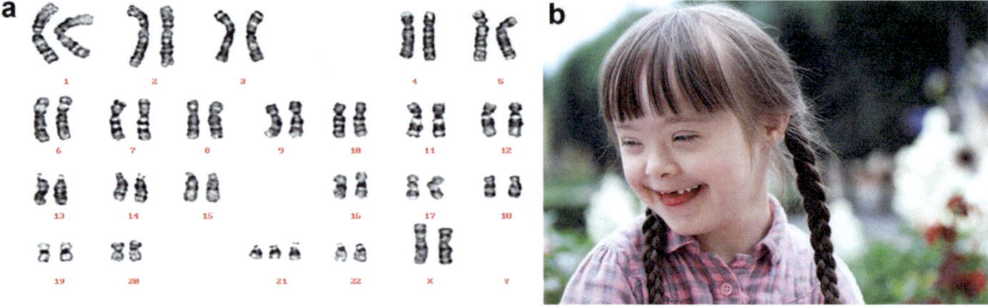

Fig. 5.26 (**a**) The karyotype; three representatives of the G group chromosome 21 are present. (**b**) A child with Down syndrome

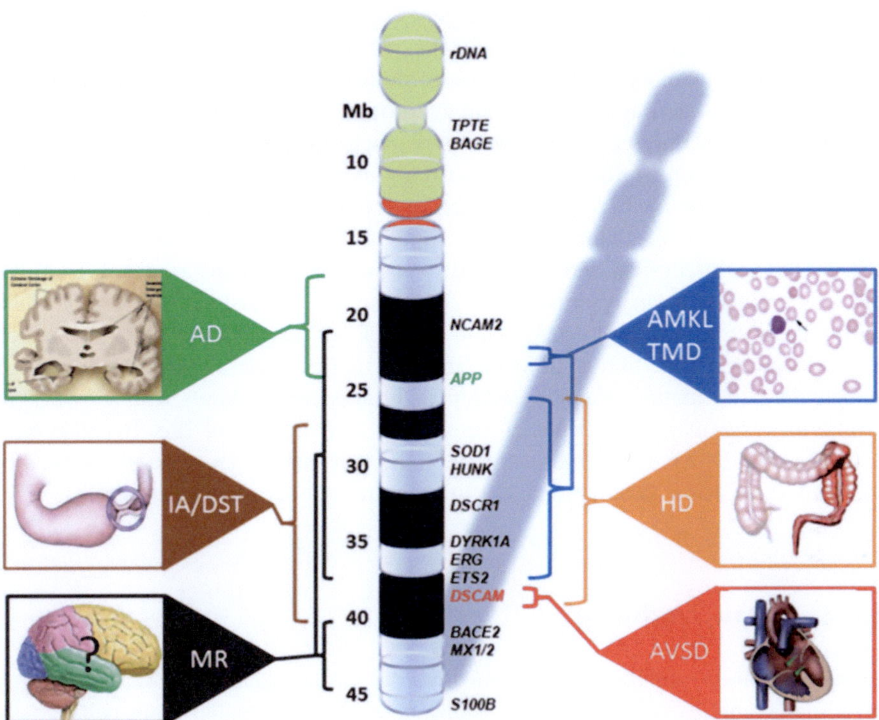

Fig. 5.27 HAS-21 regions linked to a specific Down syndrome phenotype. HSA21 short and long arm (yellow and blue), G-banding (light and dark region). The centromere is shown in red; numbers mean distance in Mb (megabases) from the HSA21 (distal end) short arm. Alzheimer disease (AD); acute megakaryoblastic leukaemia (AMKL)/transient myeloproliferative disorder (TMD); atrioventricular septal defect (AVSD); Down syndrome critical region; Hirschsprung disease (HD); imperforate anus (IA)/duodenal stenosis (DST); mental retardation (MR); recombinant DNA

Approximately 95% of cases of Down syndrome are because of the existence of one extra (third) copy of HAS-21 (human chromosome 21), while in some cases, trisomy of a part of HAS-21 directs to Down syndrome. In some smaller cases, mosaicism (state of being composed of cells of two genetically different types) of HAS-21 can lead to Down syndrome. Most often, the defect of paired homologues to disjoin in anaphase I or II, due to this chromosome composition, will be gametes with $n + 1$. Seventy-five per cent of these errors attributes to nondisjunction during maternal meiosis I leading to Down syndrome, and there is an evident maternal age effect for the birth of an individual with Down syndrome. There are high chances that an older mother will give birth to a baby with Down syndrome. Thus, the huge majority of Down syndrome individuals are conceived from older mothers.

Human segmental trisomies were analysed and have shown several dispersed HAS-21 sites linked to a certain Down syndrome phenotype (Fig. 5.27). A composite interaction of these causative genes (AVSD, AD, IA/DST, MR, HD) may exert a pathological effect at various stages of maturity or development. On the contrary, genes in charge for Down syndrome perhaps clustered into one chromosome region indicated as DSCR (Down syndrome critical region).

Genetic counselling early in certain pregnancies (certain women who get pregnant late during their reproductive years) is highly recommended. Diagnosis during pregnancy provides positive test outcome, or at high risk of having a newborn with Down syndrome. CVS (chorionic villus sampling) is a prenatal test where chronic villi (cervix; transcervical and abdominal wall; transabdominal) are taken from the placenta and used to analyse the foetal chromosome. The test is usually performed during the first trimester between 11 and 14 weeks of pregnancy. Another approach is amniocentesis (insertion of a needle into the mother's uterus to sample the fluid surrounding the foetus). Early detection of infants and children with Down syndrome makes a major difference in improving their quality of life. Because each newborn with Down syndrome is unique, treatment mainly depends on the need of the individual. Also, various stages of life may require numerous services.

5.8.2 Patau Syndrome

Patau syndrome is an infrequent genetic disorder for having an additional/extra copy of chromosome 13 (trisomy 13) in a few or entire body cells. Each cell usually involves 23 pairs of chromosomes, but an individual with Patau syndrome carries three copies of chromosome 13 (Fig. 5.28a, b), instead of two. Patau syndrome was first described by Klaus Patau in the year 1960. This affects 1 in every 5000 births and the risk of having a newborn with Patau syndrome is directly linked with the mother's age. More than nine out of ten children born with Patau syndrome die during the first year, whereas children with partial or mosaic trisomy 13 live for more

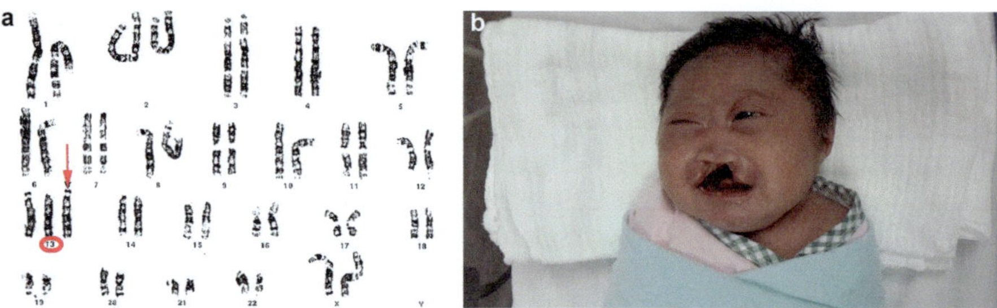

Fig. 5.28 (**a**) Karyotypic and phenotypic illustration of an individual with Patau syndrome. Three sets of chromosome 13 (D group) are present (43, 13+). (**b**) Infant with Patau syndrome

than a year. Individuals with Patau syndrome possess a wide range of health problems: holoprosencephaly (brain often divided into two half), cleft lip and cleft palate, microphthalmia (small eyes), anophthalmia (absence of one or both eyes), microcephaly (smaller in size than the normal head size); cutis aplasia (absence of skin from the scalp), ear malformation, deafness, capillary haemangiomas (red birthmarks), the presence of abnormal cysts in the kidney, reduced penis, enlarged clitoris in girls and polydactyly.

5.8.3 Edwards Syndrome

Edwards syndrome, called trisomy 18, is a rare but serious condition. The syndrome affects the survivability of an individual, and most often individuals will die before or shortly after being born. About 3 in 100 individuals born alive with Edwards syndrome will survive past their first birthday. It was first described by John H. Edwards and his colleagues. Every cell in our body carries 23 pairs of chromosomes, but an individual having Edwards syndrome carries three pairs of chromosome number 18 (instead of 2) (Fig. 5.29). Mosaic Edwards syndrome consists of extra chromosome 18 in some cells, whereas in some only a part of the

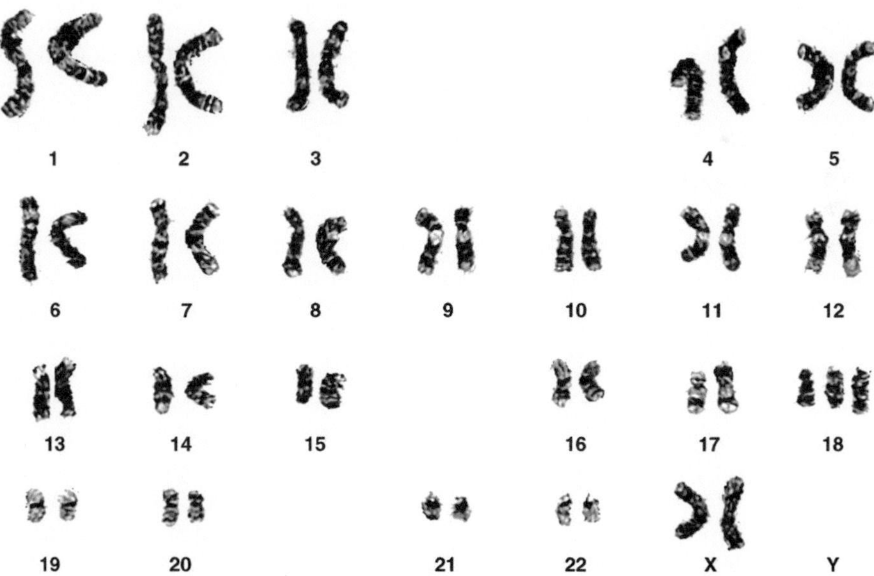

Fig. 5.29 Karyotype of infants having Edwards syndrome. Three members of chromosome 18 (trisomy 18), creating the 47, 18+ condition

extra chromosome 18 in the cells is present, called partial Edwards syndrome. It is characterised by multiple impairments, low sets of ears, small subsiding lower jaw, clamped fingers, cardiac malformations and deformation of the face, skull and feet.

5.9 Polyploidy

We are very familiar with haploids and diploids (n and $2n$). Often, the whole set of chromosomes fail to differentiate during meiosis or mitosis. Tetraploid offspring is occasionally produced by diploid parents. The chromosome doubling in the zygote results in the polyploid formation, or from cytologically non-reduced female and male gametes that will result in the development of functional tetraploid ($3n$) zygotes. Polyploidy involves triploids ($3n$), tetraploids ($4n$) and pentaploids ($5n$), including a higher number of chromosome sets. Polyploidy is most commonly present in plants like ferns and flowering plants, for example, *Hibiscus rosa-sinensis*, wheat and many tetraploids which are agriculturally most important plants (genus *Brassica*), and may occur during the division of cells, either during metaphase I in meiosis or mitosis. Also, polyploidy formation in human tissues is highly differentiated (heart muscle, liver, placenta and bone marrow) and occurs in a somatic cell of few animals: salmon, goldfish and salamander. Here we consider two major types of polyploids: auto-polyploidy, the multiple set of chromosomes, similar to the normal n complement of the related species; allopolyploidy, in which chromosome complements are from multiple species; and endoploidy, which is a repeated division of the chromosome in the absence of subsequent division of cell/nucleus.

5.9.1 Autopolyploidy

A type of polyploidy wherein the extra chromosome set is derived from an identical parental species or a parent. The cell or organism in autopolyploidy condition is called an autopolyploid. Common examples of autopolyploids in plants are *Tolmiea menziesii* (piggyback plant) and *Acipenser transmontanum* (white sturgeon). The nondisjunction of chromosomes during mitosis (which produces double ($4n$) the number of the chromosome) and meiosis (fusion of diploid gamete with normal haploid gamete to form triploid ($3n$) zygote) produces autopolyploid and is depicted in Fig. 5.30. In other words, the binding/fusion of diploid gametes ($2n$) results in offspring either showing triploid ($3n = n + 2n$) or tetraploid ($4n = 2n + 2n$) number. General results of autopolyploidy are seedlessness as in watermelon and bananas and inducing sterility in salmon and trout farming.

How polyploidy appears naturally is of considerable interest to the geneticist. Research on *A. thaliana* (*Arabidopsis thaliana*) explained that during the process, deviation from doubling of mRNA/set of RNA transcripts (transcriptome) in *A. thaliana* autotetraploid resulted due to increased size of the cell, wherein the size of the cell and associated phenotype effectively influenced the content of DNA

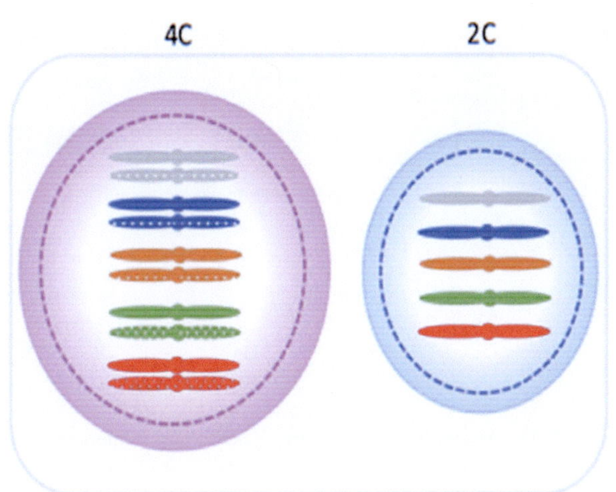

Fig. 5.30 Autotetraploid containing a double number of chromosome (2C), with the doubling content of DNA (4C). Diploid progenitors (2C) and autotetraploid (4C) resulting in larger cells and nuclei (shaded circles and nested circles with dotted lines)

rather than being completely determined by genotype. A separate chromosome state is maintained by the individual duplicated chromosome in the doubled nucleus. An increase in nuclear volume is directly related to the genome size, but doubling volume produced less genome doubling, which affects chromatin condensation and by altering the contacts between chromosomes (Fig. 5.31). As a result, autotetraploid showed increased inter-chromosomal interactions and reduced intra-chromosomal interaction. Overall, around 12% of the genome displayed position difference between structural domains such as loose vs condensed (LSD-CSD) that is between largely euchromatic and heterochromatic (A and B) regions commonly found in animals and plants.

5.9.2 Allopolyploidy

A type of polyploid strain or individual consisting of a chromosome set composed of more than two chromosome sets extracted more or less complete from variant species. Allopolyploid species that originated from a seed-setting shoot of a sterile F1 hybrid between *P. floribunda* Wall. and P. verticillata Forssk. represents a well-known case of new allopolyploid formation through somatic chromosome doubling. The production of unreduced gametes provides three possible ways: first, F1 hybrid individuals are formed through interspecies crossing; second, genome doubling takes place in a one-step process, that is, the union of unreduced gametes that are produced by the F1 hybrid; and this allopolyploid speciation undergoes the third pathway, the 'hybrid genome doubling' pathway (Fig. 5.32).

The new allopolyploids have a complete set of disomic chromosomes derived from the parental species. Karyotypic changes through recombination between homologous chromosomes may occur and lead to chromosomal diploidation. *Triticum* and *Aegilops* are wheat species which are the best systems for studying allopolyploid speciation. These species have two distinctive features: the first one is

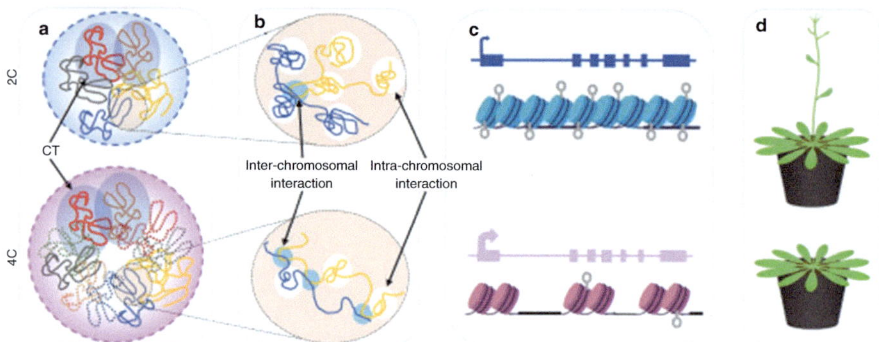

Fig. 5.31 Autopolyploid showing genetic and epigenetic effects in *A. thaliana* with its diploid progenitor (Col-0). (**a**–**c**) *A. thaliana* showing epigenetic response to autopolyploid. Within the diploid (2*n*) nucleus during interphase, every chromosome specifically occupies a space or a region inside the nucleus which is referred to as chromosome state or territory (CTs); in the autotetraploid, the nucleus of interphase wherein the chromosomes are of the doubled complement and the whole set appear to occupy the separate chromosomal state. The nucleus of *Arabidopsis* is also divided into transcriptionally oppressed and transcriptionally active structural domains, i.e. compact structural domain (repressed CSD-dark shades at the bottom of a nucleus) and active loose structural domain (LSD-unshaded interior part). Around 12% of autotetraploid shows chromatin switch states between these structural domains. Panels **a** and **b** shows the gametic number, 5-diploid and 10-tetraploid chromosomes. White circles represent the folding of chromatin structure producing intramolecular interactions. And blue circles represent intermolecular interactions. (**c**) Gene activity alters the chromatin reconstructions. FLC gene overexpression is correlated with the autotetraploid with increased accumulation of local chromatin loop along with decreased methylation of H3K-27me-3 represented in grey circles across the entire length of the gene. (**f**) FLC gene overexpression is related to late flowering, suggesting the possible link between the reconstruction of chromatin and its evolutionary aspects

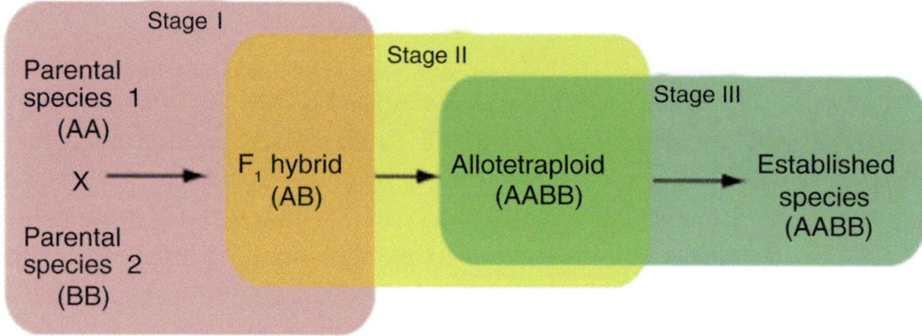

Fig. 5.32 Three major steps of allopolyploid speciation through the hybrid genome doubling pathway. Stage I, where F1 hybrid individuals are formed through interspecies crossing, forms the allopolyploid speciation. Stage II—actual doubling of genome; allopolyploidy individuals are formed through the union of unreduced gametes from F1 hybrids. Stage III—the new allopolyploid individuals formed via hybrid genome doubling start to propagate as reproductively isolated entities and finally become established as species

paternal and maternal lineages (known for *Triticum* and *Aegilops* polyploid species) and the second one is artificial interspecific crosses are easily made in those genera.

5.9.3 Endopolyploidy

Endopolyploidy refers to the augmentation of nuclear DNA within the cell. Endoreduplication is the most common process in plants. This involves the chromosomal DNA replication without the involvement of mitosis and no condensation/decondensation of chromatin structures which stays united either at the entire length (rarely) or at the centromeres. This kind of occurrence is most uneven in plants as shown in Fig. 5.33. Polyploidy may also occur in few tissues of diploid animals, like muscle tissues in humans. Prokaryotes lack nuclei as seen in *Epulopiscium fishelsoni* bacterium. And it is also most common in angiosperms such as flowering plants, where endopolyploidy is detected in few tissues at a higher level. Endopolyploidy mechanism involves internal and external factors, which can be observed as the key role of the developmental flexibility of plants.

5.10 Chromosome Shuffling

Change in the chromosome morphology refers to chromosome shuffling which results in structural change and the difference in gene sequence and number with or without altering gene structure or ploidy. Breaking of the chromosomal segment is known as chromosomal aberrations, that is, the chromosomal loss or gain or union

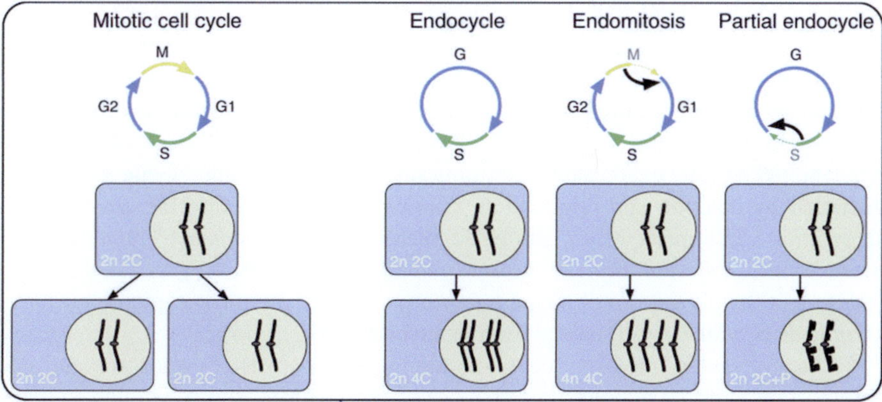

Fig. 5.33 During cell division, DNA replication at the S phase and chromosome segregation and cells divide in M phase producing two daughter cells. Endomitosis; partial M phase where cells enter the mitosis stage and perform segregation but exit before the division of the cell takes place. As a result, there will be the formation of 4C and 4*n* (tetraploid). Also, during the S phase, replication is present at a specific part of chromosomal DNA, which ends up showing the increase in DNA content (partially) $2C - 2n + P$

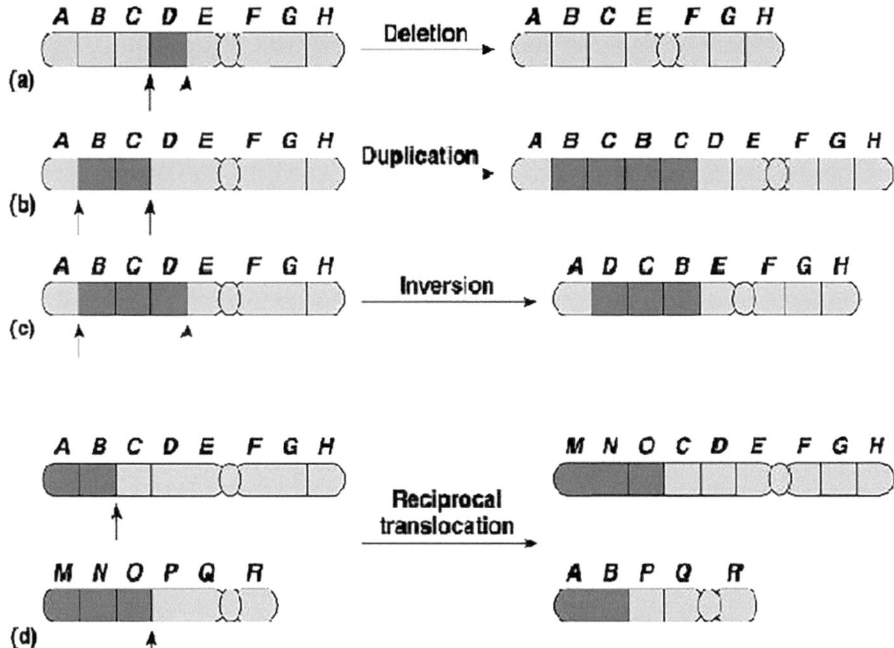

Fig. 5.34 Overview on the five different types of chromosomal aberration: gain, loss or rearrangement of chromosome segments

with different or same chromosomes. This aberration usually occurs spontaneously but physical and chemical factors are mainly influencing these structural changes.

Structural changes can exist within single or non-homologous or homologous chromosomes. This includes deletion (loss) or duplication (gain) of a chromosomal segment, or the translocation (exchange) of the chromosomal segment with different segment. In an inversion, segment of a chromosome gets deleted, turns 180 °C and is repositioned at the exact same position on the chromosome. Gene balance is disrupted by deletion and duplication, thus changing or altering the characters of an organism. In translocations, shifted segment involving genes gets movement and inserted into different linkage groups, whereas in inversion gene sequence is changed in the chromosome. Gene balance or altering the phenotypic character of organisms is neither by translocation nor inversion (Fig. 5.34).

In heterozygotes, meiosis occurs containing one normal chromosome and one chromosome with inversions. The chances of these chromosomes coming together are difficult, and crossing over between them will not take place or is uncommon. Aneuploidy formation is often occurring, and consequently, heterozygotes are comparatively less fertile than homozygotes due to which both pairs of chromosomes are either inverted or normal. Moreover, heterozygotic chromosomes also undergo translocation by producing the number of aneuploid gametes.

Chromosomal aberrations show direct link in evolution of organisms: increase in number of gene is mainly due to duplication. Inversions as well as translocations may result in formation of individuals (genetically isolated homozygous condition) that are more fertile than heterozygotes. In all chromosomal aberrations, the gene position effect is sometimes observed, in which a gene gets shifted to a new site on a chromosome exerting a different effect on the organism's phenotype.

5.10.1 Deletions

Deficiency occurs when deletion of part of a chromosome results in loss of one or more chromosomal genes. It was first observed by Bridges in 1917 in *Drosophila*. Deficiency arises from chromosomal breakage occurring at random position in either both or only one chromatid of a chromosome. The deletion caused by two breaks in the chromosome will result in an interstitial deletion (Fig. 5.35a). If there is only one break that occurs at the end of the chromosome, it is terminal deletion (Fig. 5.35b). Sometimes break occurs at both ends of a chromosome. Detection is based on the unpaired region of the normal chromosome that produces a loop during the pachytene stage (Fig. 5.35c). The deletion in wild-type allele of heterozygous organism may produce a mutant phenotype, while the same deletion in one allele of a homozygous wild-type organism would produce a normal phenotype. And also, deletion of the centromere results in an acentric chromosome that is lost, usually with serious or lethal consequences.

Deletion can be caused by the loss of bases during translocation, during unequal crossing over or during chromosomal crossovers within a chromosomal inversion. Deletions are responsible for an array of genetic disorders, and small deletions are less likely to be fatal; large deletions are often fatal depending on which gene is lost, whereas medium-sized deletion led to recognisable human disorders, e.g. Williams syndrome. Deletion of part of the short arm of chromosome 5 results in cri-du-chat syndrome. Deletion in the SMA-encoding gene causes spinal muscular atrophy. And some deletions are highly conserved (hCONDELs) which are responsible for evolutionary differences present among closely related species.

5.10.2 Duplications

Duplication is attachment of a chromosomal fragment resulting in the addition of one or more genes to a chromosome. This was first observed by Bridges in 1919. Whenever there is a duplication in a chromosome, there is a corresponding deletion in another chromosome. The origin of duplication is due to unequal crossing over during meiosis. This type of aberration also belongs to the category of unbalanced rearrangements as a segment of a chromosome is repeated. The segment can be repeated once or several times; the orientation of the repeat may be the same as the original segment (tandem repeat) or it may be in the opposite orientation (inverted duplication). The known human disorder is Charcot-Marie-Tooth disease type 1A

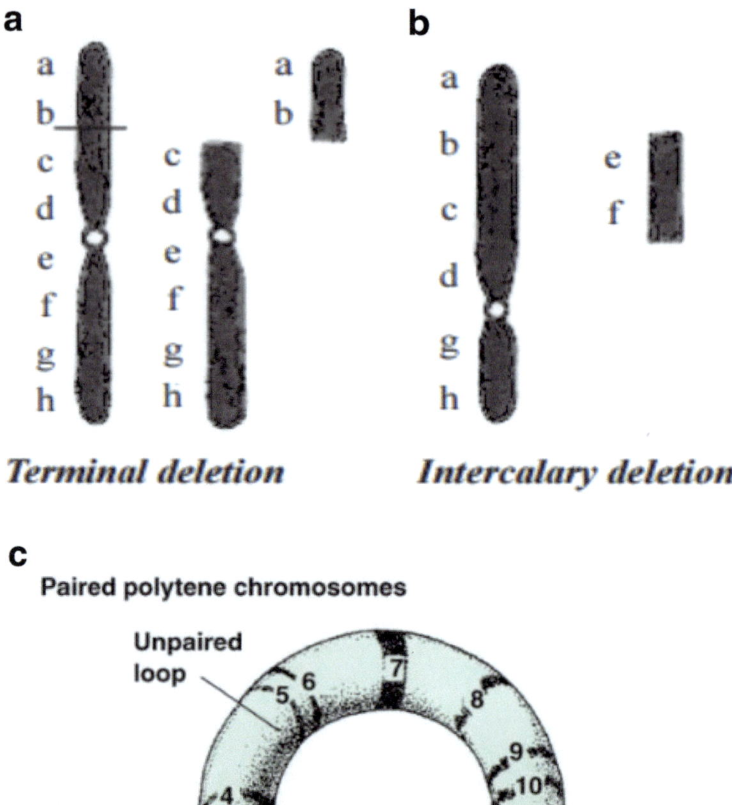

Fig. 5.35 Origins of (**a**) terminal deletion and (**b**) intercalary deletion. (**c**) For the formation of synapsis which occurs between a chromosome with a normal homologue and a large intercalary deficiency. The normal homologue having unpaired region must loop out of the linear structure into a compensation loop or deletion

which is caused by duplication of the gene encoding peripheral myelin protein 22 (PMP22) on chromosome 17.

The duplicated segment of a chromosome is present in another chromosome of the genome. Duplication (inter-chromosomal duplication) occurs in two forms: one

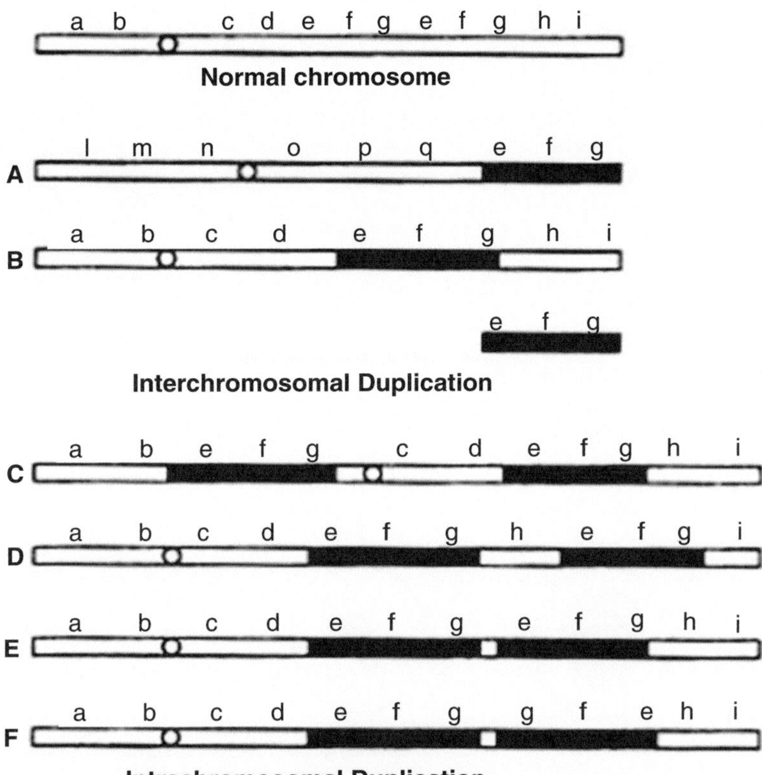

Fig. 5.36 Diagrammatic representation of some of the possible duplication types. (**a**) Duplicated segment is present in a non-homologous chromosome. (**b**) Duplicated segment is present separately as an acentric region. (**c**) Duplicated segment is present in the other arm of the same chromosome. (**d**) Duplicated segment is present in the same arm but is removed from the original segment. (**e**) Direct tandem duplication. (**f**) Reverse tandem duplication

is the segment of a chromosome which is duplicated is incorporated into a non-homologous chromosome, and other is the duplicated segment of a chromosome is present as a separate chromosome and the duplicated chromosome segment remains in the same chromosome (intra-chromosomal duplication) present either in the different arm or in the same arm but removed from the original segment (Fig. 5.36). If the segment remains in the same arm and next to the original segment, it is known as tandem duplication which results in direct tandem (gene order of the duplicated segments the same as that of the original segment) and reverse tandem (gene order of the duplicated segment is inverted).

The duplication was extensively studied in *Drosophila*, B locus (barred eye) of the X chromosome. This is the condition where the eye is narrower as compared to the normal eye shape. This is due to a part of the chromosome that is duplicated. The region of duplication is 16A (X chromosome).

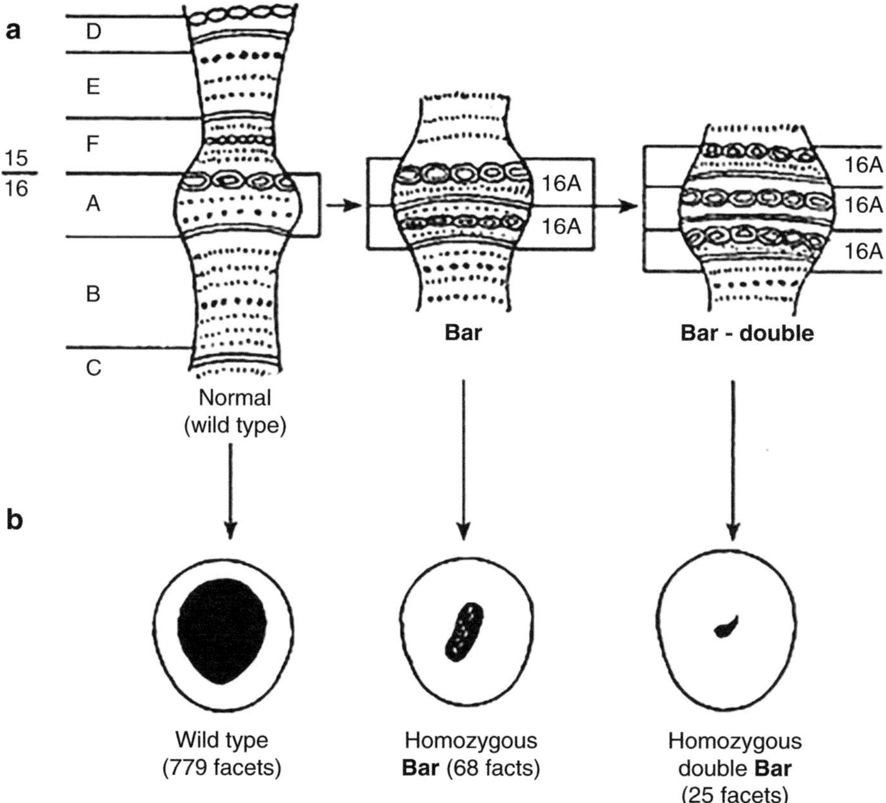

Fig. 5.37 (**a**) Ultra-bar formation from bar due to unequal crossover. (**b**) Position effect. (**c**) Duplication in the normal eye. Bar eye and a double bar eye (corresponding with chromosomal segment)

Individuals with barred eye develop ultra-bar (UB) (16A,16A (B)-16A,16A,16A) and normal wild form due to unequal crossing over (Fig. 5.37a, b). The different phenotypes in the homozygous bar and heterozygous UB individuals and each casing segment of 16A remain the same. And this is known as the position effect (Fig. 5.37c).

5.10.3 Inversion

Inversion shows the reverse gene order of a chromosome. Inversion is created when part of the chromosome turns 180°, detached or reinserted in such a manner that the genes are reversed in order. During meiotic prophase and chromosome break, entanglements of threads are formed which are presumable inversions. For instance, the segment is broken at two random places, and the two breaks can be nearby due to the formation of a loop in the chromosome. When they rejoin, the wrong ends get

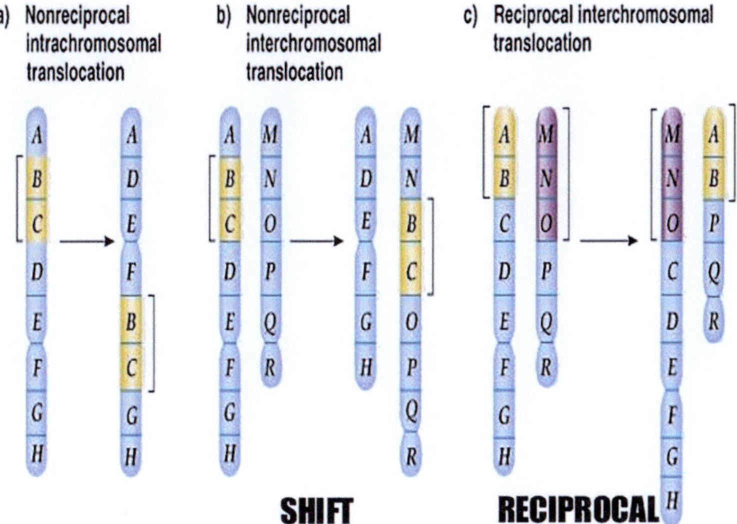

Fig. 5.38 Inversion

connected. One side of the loop connects with a different broken end from the one which is originally connected by joining the remaining two broken ends resulting in inverted or turned around loops.

During meiosis pairing between inverted chromosome and non-inverted (known as inversion heterozygous) chromosome results in the formation of inversion loop, if crossing over occurs within inversion loop, producing abnormal chromatids. If crossing over does not occur, homologues will segregate and form normal and two inverted chromatids. The mechanism is depicted in Fig. 5.38. Inversions are usually of two types: paracentric inversion (where inverted segments do not include centromere) and pericentric inversion (inverted segments include centromere). In a paracentric inversion, due to single crossover, the odd number of crossing over in inverted region forms dicentric (two centromeres) chromosome and acentric (no centromere) chromosome. This results in two chromatids: one carrying inversion and the other remaining normal. This formation can be observed during anaphase I in the form of a fragment and bridge (Fig. 5.39). Deficiency and duplication are due to crossing over between inverted segments. However, in pericentric inversion, the formation of pachytene is very similar to paracentric but differs during crossing over and configuration at the meiosis stage. Deficiency and duplication are present in two of the four chromatids. Due to this, gametes involving chromosomes do not function and result in gametic or zygotic lethality. (Fig. 5.40).

5.10.4 Translocation

A section of one chromosome is transferred to a non-homologous chromosome which is defined as translocation. If there is an exchange of a segment or part

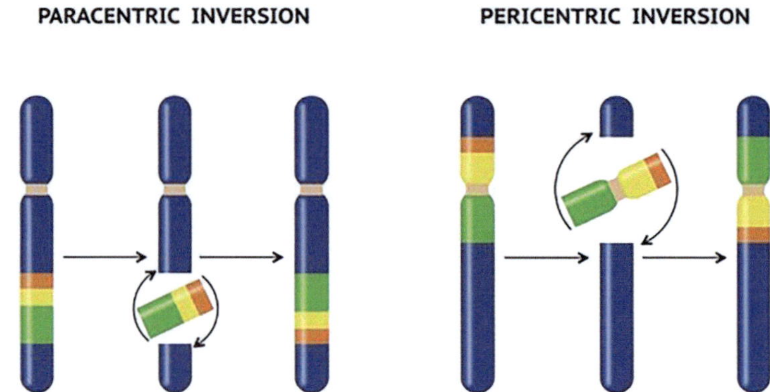

Fig. 5.39 Paracentric inversion and inversion during metaphase

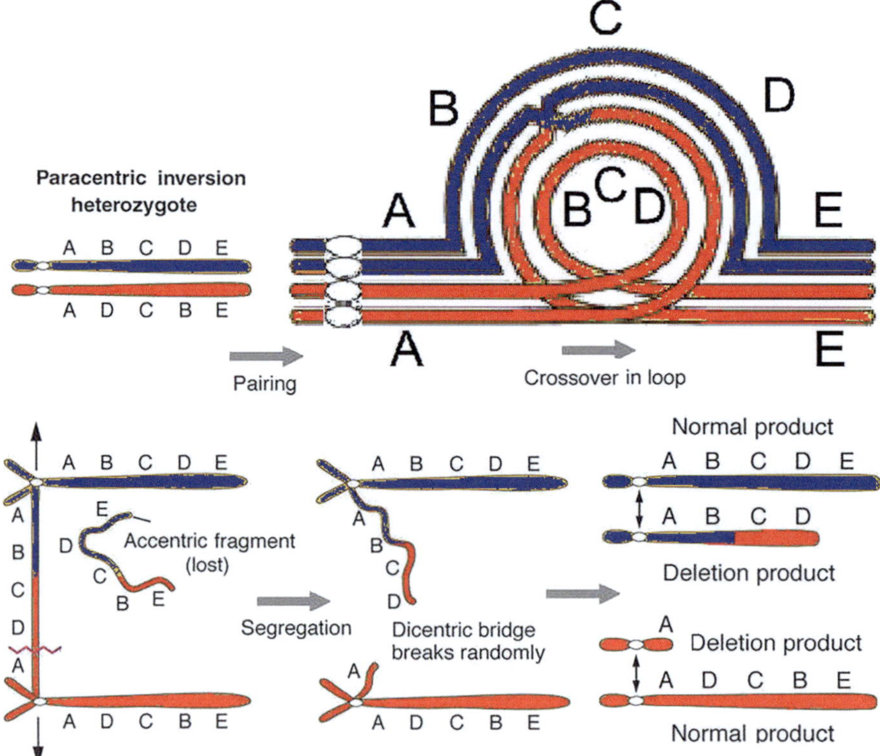

Fig. 5.40 Paracentric inversion, acentric fragment and dicentric bridge at anaphase I and result of double cross

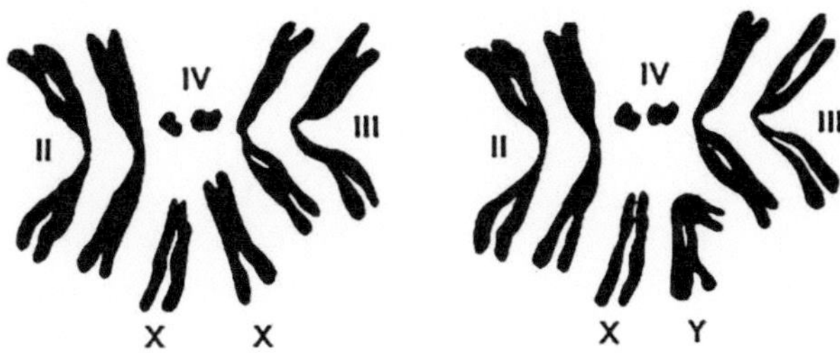

Fig. 5.41 *Drosophila* sex chromosome showing translocation

between non-homologous chromosomes, then it is known as reciprocal translocation. In this, there is no loss or addition of chromosomes. It was first observed in *Drosophila* having an unusual behaviour of a second chromosome gene (pale). The presence of a pale gene in the homozygous condition is lethal. Bridges studied that the presence of one gene on the third chromosome can suppress the pale gene which is also lethal in the homologous chromosome. Stern in the year 1926 studied translocation between some allele on the 'Y' chromosome to the 'X' (Fig. 5.41).

A chromosome having a single break and gets transferred onto the other chromosome end is known as simple translocation. Chromosome involves three breaks so that any two-break section of a chromosome can get inserted within the break developed in a chromosome that is non-homologous, and this shows shift or intercalary translocation. However, in reciprocal translocation, if there is a single break in homologous chromosome, an interchange of a chromosome segment between them develops (Fig. 5.42).

In homozygous condition, translocation forms an equal number of homologous pairs the same as the normal homozygous if centromere is not lost, but in heterozygous translocation, the result of pairing during meiosis is different (Fig. 5.43a). Reciprocal translocation results in four chromosome sets at each pachytene stage, and the formation of chiasmata between the chromosome forms quadrivalent gets disjoined in three different patterns of segregation in I meiotic division (Fig. 5.43b): alternate segregation, where non-homologous alternate chromosomes move to the same pole in zigzag/alternative pattern so that translocated and non-translocated chromosomes are present in different gametes with no duplication or deficiency; adjacent I segregation, where adjacent non-homologous chromosome moves to the same pole and each gamete contains non-translocated and translocated chromosome, and each gamete shows duplication and deficiency; and adjacent II segregation, which involves the movement of the adjacent centromeres to the same pole but these are now said to be homologous containing translocated and non-translocated chromosomes showing supplication and deficiency producing unbalanced gene components: unbalanced gamete formation in adjacent I and II segregation. The out-turn of these segregations inhibits the independent assortment between genes and non-homologous chromosomes due to the absence of deficiency and duplication (neither of single mutant phenotypes) in offspring.

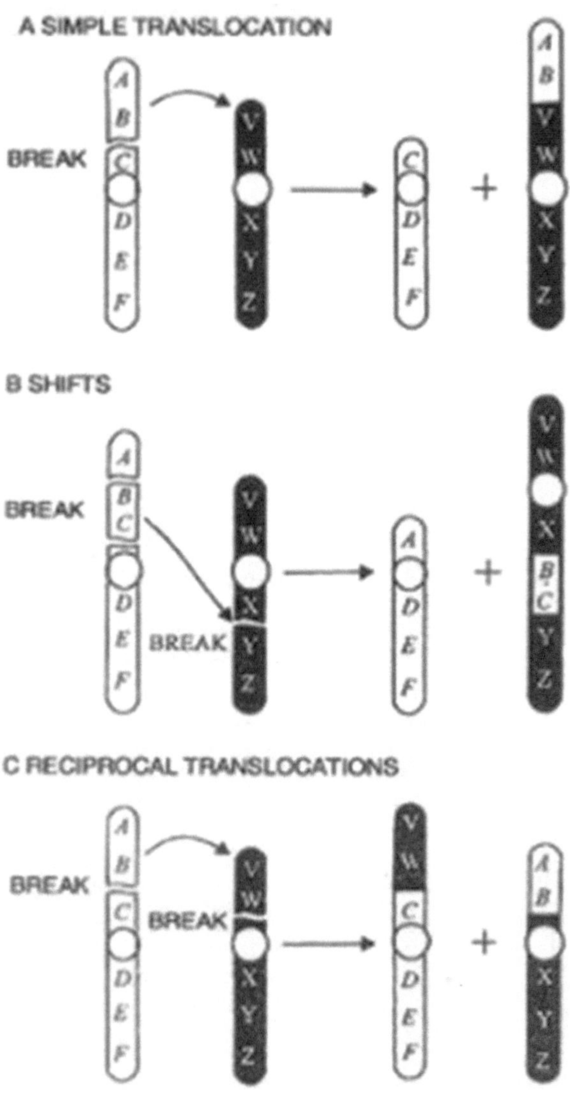

Fig. 5.42 Various forms of translocation showing simple, shift and reciprocal translocation

5.11 Fragile Site: Chromosomal Breakage

Fragile sites are constrictions, gaps or breaks on the chromosome in metaphase that develop when cells are subjected to perturbation of the deoxyribonucleic acid (DNA) replication stage (Fig. 5.44). The fragile site can be seen in all human chromosomes and named based on the band patterns (e.g. fra(X) (q27.3) was the first fragile site found on the X chromosome). Fragile sites are categorised as rare or common. Rare fragile site is present in very small portion of the total population having a maximal

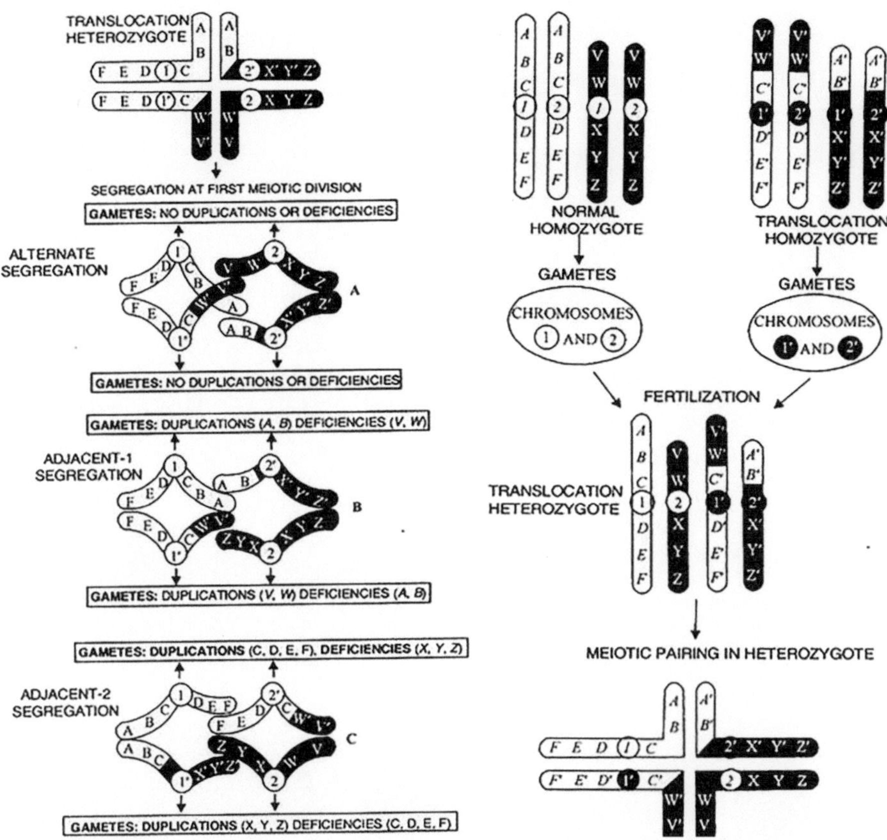

Fig. 5.43 (**a**) Translocation condition in homozygous and heterozygous forms. (**b**) Three different segregation patterns in translocation heterozygotes

frequency (1/20); the common fragile site is commonly present in every individual. There are around 89 common and 30 rare fragile sites that have been identified.

Rare fragile site FRAXA which is associated with fragile X syndrome is the common defect of hereditary mental disability. It is caused by the mutation of a single gene (FMR-1 fragile mental retardation), located on the X chromosome. Normally, gene consists of a sequence CGG which is repeated 30 times in a row. However, in fragile X syndrome, these three sequences are repeated around 300 times creating a microscopic gap on the length of the chromosome X. FMR-1 gene instructs the production of FMRP (fragile X mental retardation 1 protein) which is involved in the normal development of the brain (development of synapse, specialised connections between nerve cells). Mutation of the FMR-1 (gets switched off) results in blocking the synthesis of the protein by preventing normal brain development.

Fragile genes like FRA-3B, FHIT, WWO-X and FRA-16D are large tumour suppressor genes. Deletion at breakpoints within the fragile site (Fig. 5.45) is associated with many types of cancers (breast, lung and gastric cancers).

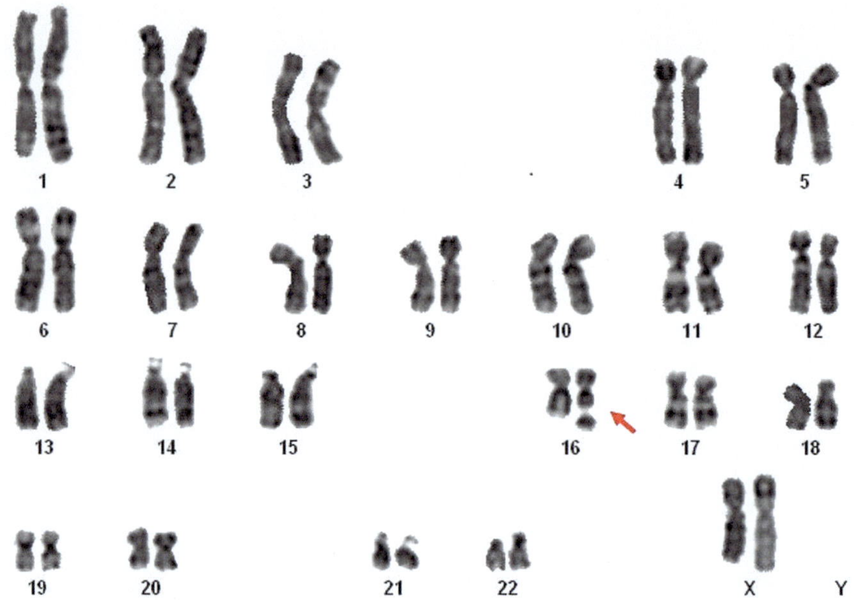

Fig. 5.44 Karyogram showing fragile site in chromosome 16q21

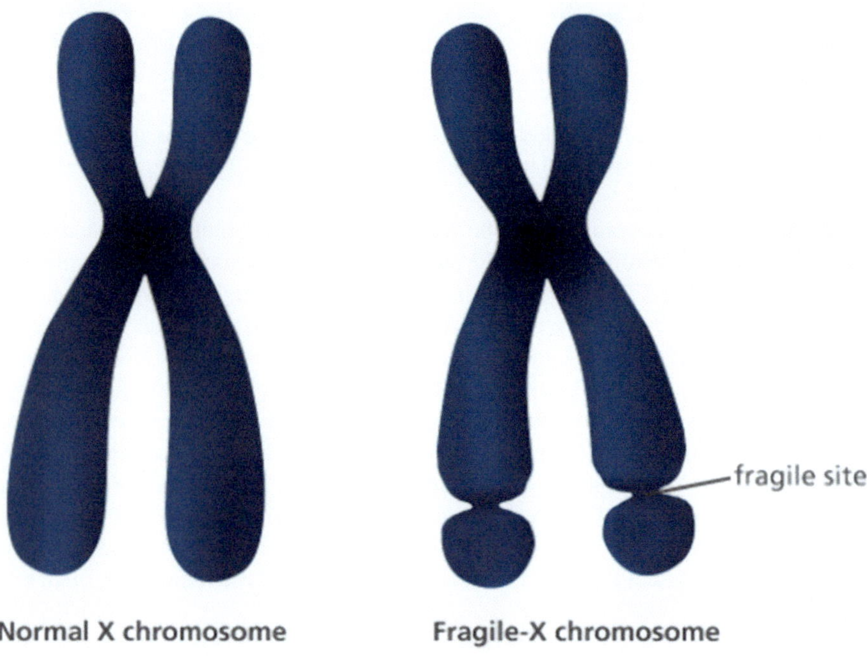

Fig. 5.45 Illustrative representation of normal and fragile X chromosomes

MicroRNA genes are commonly present at the fragile site which is involved in the alteration of the chromosome. Additionally, diseases like hepatitis B (HBV) and HPV-16 (human papillomavirus-16) are found to interact most likely around fragile sites which are described as the crucial significance for the development of cancer tumours. Various forms of syndromes, e.g. Jacobsen syndrome, breakage near or at FRA-11B (a part of the long arm of chromosome 11 is lost), and Seckel syndrome, a rare genetic disease due to presence of low levels of ATR, show a higher level of instability of chromosome at fragile site.

Box 5.1

Anaphase is the stage where condensation of the mitotic chromosome is important for the segregation of sister chromatids. Condensing and topoisomerase II (ATP hydrolysing enzymes) combination plays an important role in the reorganisation of chromatin involving the active assembling of chromosomes. Condensation of mitotic chromosomes involves a process of interphase chromatin conversion into rod-shaped structures by three different steps—condensation, individualisation and resolution. These are different from the condensation of heterochromatin and apoptotic condensation. The basic unit of chromatin folding is the nucleosomes as oligomers of histones on the DNA fibre that generate six- to sevenfold compaction by the coiling or folding of 30 nm fibre into a zigzag, solenoid or crossed fibre. To attain around 10,000–20,000 linear compactions of mitotic chromosome, 200–500 compactions of around 30 nm fibre during the mitotic condition are necessarily a must. Chromosome scaffold which is formed by DNA or protein (the backbone of the chromosome) determines the rod-shaped structure of the chromosome. In many models, two neighbouring elements bind to a scaffold even if they are separated by up to 100 kb of DNA forming a loop with the intervening DNA. The scaffold attachment region is a DNA sequence that functions as a cis-acting sequence in chromosome assembly. This DNA sequence is AT-rich which acts as a binding site of DNA topoisomerase II (a chromosome scaffold component).

Altering folding patterns of a mitotic chromosome is determined using variant dyes. SMA subunit and topoisomerase II condensing complex are the major biochemical compositions of the mitotic chromosome. These are known as topo II-IIα (CAP_B), and five subunit complexes are condensin (CAP, C, -D2, -E, G and H), chromokinesin (CAP-kip1/D) and chromatin remodelling ATPase (ISWI CAP-F). Mitotic chromosome fractions are rich in ATPases. Utilising the energy of ATP hydrolysis and energy utilised for microtubule-dependent movement of chromosomes induces the g, global and local conformational changes of chromosomes. Topo II (ATP-dependent) enzyme that crosses the strand of DNA by cleaving one strand of DNA via breaking passes the second strand of DNA and releases it.

(continued)

Box 5.1 (continued)

Invertebrate cells involve the process of removing catenated DNA between various chromosomes and convert each chromosome into a single mobile unit. Topo II association with the condensing complex is essential for the compaction of the chromosome. Cytologically identifiable pairs of sister chromatids are formed by the decatenation activity of topo II in prometaphase and are completely separated in anaphase. Throughout the cell cycle, the amount of topo II on chromosomes varies which is maximum in prometaphase and minimum after telophase. A large protein complex, condensin, is one of the profuse structural components of mitotic chromosomes.

SMC2 and SMC4 are the two core subunits of this complex, which by an antiparallel coiled-coil interaction gets folded intramolecularly, and at their hinge domains, they dimerise to form a V-shaped structure. An ATP binding catalytic domain is formed at each end of the SMC dimer by the folding of the SMC subunit which allows the association of N-terminal and C-terminal sequences. CAP-D2, -G and -H are the three non-SMC subunits that bind to one or both of the catalytic ends. Supercoiling assay and knotting assay are the two in vitro assays devised to show that condensin is capable of inducing super-helical tension in an ATP hydrolysis-dependent manner into DNA. The super-helical tension model, chiral looping model and protein-protein interactions that facilitate rod-shaped chromosome formation are the hypothetical models for the condensin contribution to the organisation of the chromosome. As observed in eukaryotic organisms so far, with the initiation of chromosome condensation, phosphorylation of histone H3 takes place at serine 10 in the prophase of mitosis.

Aurora B, a mitotic kinase, acts as a major physiological kinase that is responsible for phosphorylation of histone H3; this might contribute to the sister chromatids' resolution by possibly enhancing the chromatin fibres' structural flexibility or by increasing the electrostatic repulsion between the chromatids. Furthermore, work is necessary to understand the exact role of phosphorylation of histone H3 in mitotic chromosome dynamics. As we know, mitotic chromosome condensation is an active process that requires many ATP hydrolysing enzymes like topo II and the condensin complex; determination of how these two components contribute to the mitotic chromosome assembly by their combined action is an important challenge and requires further research and insights.

5.12 Summary

- Chromosomes are strands of DNA that are wound around histone proteins and are coiled and supercoiled to package the DNA in a compact form. Chromosomes have telomeres at their ends which stabilise the chromosomes. The chromosome also consists of a centromere where kinetochore complex forms and spindle fibres get attached during mitosis and meiosis. The sister strands formed after duplication of a chromosome are also held together at the centromere.
- Humans have 22 pairs of autosomes and 1 pair of sex chromosome. Sex of the individual depends on the composition of the sex chromosome. In humans, the Y chromosome contains genes which allow the development of male genitals and suppress the formation of female genitals during embryonic development. The presence of Y chromosome therefore determines maleness and absence of it leads to female development.
- Sutton and Boveri's independent observations in chromosomal inheritance and embryonic development led to the suggestion of chromosomal theory of inheritance around 1902. It stated that chromosomes are carrier of genetic material and are the units involved in Mendelian inheritance. Thomas Morgan and Calvin Bridge's work on fruit flies provided undeniable proof for the chromosomal basis of heredity.
- Mutations in genes present on sex chromosomes lead to sex-linked disorders. These are called sex linked because the inheritance of these disorders does not follow Mendelian inheritance and instead occurs preferentially in one of the sexes. X-linked recessive disorders are more prevalent in males than in females and show a criss-cross pattern of inheritance.

References

Adkison LR (2012) Mechanisms of inheritance. Elsevier's integrated review genetics. pp 28–50. https://doi.org/10.1016/b978-0-323-07448-3.00003-0

Altmann R (1894) Die Elementarorganismen und ihre Beziehungen zu den Zellen. Veit, Leipzig Armarego WL

Annunziato A (2008) DNA packaging: nucleosomes and chromatin. Nature Education 1(1):26

Baker JR (1944) Memoirs: the structure and chemical composition of the Golgi element. Q J Microsc Sci 2(337):1–71

Baker JR (1960) Cytological techniques: the principles underlying routine methods. Methuen, London

Baker JR (1966) Cytological technique. The principles underlying routine methods, 5th edn. Methuen, London

Chai CLL (2009) Chapter 4. Purification of organic chemicals purification of laboratory chemicals. Butterworth-Heinemann

Dyer AF (1963) The use of lacto-propionic orcein in rapid squash methods for chromosome preparations. Biotech Histochem 38:85–90

Griffiths AJ, Gelbart WM, Lewontin RC, Miller JH (2002) Modern genetic analysis, 2nd edn. Macmillan

Klug WS, Cummings MR, Spencer CA, Palladino MA (2007) Concepts of genetics, 6th edn. Pearson Education, Inc

La Cour LF (1935) Technic for studying chromosome structure. Stain Tech 10:57–60

Mayer P (1918) Ueber die Reinheit unserer Farbstoffe. Zeitsehr f wiss Mikr 34:305

Middlebrook WR, Phillips H (1942) The action of formaldehyde on the cystine disulphide linkages in wool: the subdivision of the combined cystine into two fractions differing in their reactivity towards formaldehyde. Biochem J 36(3–4):294

Molad Y (2002) Update on colchicine and its mechanism of action. Curr Rheumatol Rep 4(3):252–256

Navaschin M (1925) Morphologische Kernstudien der Crepis-Arten in bezug auf die Artbildung. Z Zellforsch 2:98–110

Palevitz BA (1993) Morphological plasticity of the mitotic apparatus in plants and its developmental consequences. Plant Cell 5(9):1001

Pierce BA (2010) Genetics: a conceptual approach. Macmillan

Pischinger A (1937) Untersuchungen über die Kernstruktur besonders über die Beziehungen der Struktur im Leben und nach der Fixier-ung. Zsch Zellf 26:249–280

Pridgeon AM, Cribb PJ, Chase JM, Rasmussen FN (1999) *Genera orchidacearum. General introduction, apostasioideae, cypripedioideae* (Reprinted edition). Oxford University Press.

Singh RJ (2003) Plant cytogenetics. CRC Press, Boca Raton, p 488

Sharma AK, Sharma A (1957) Permanent smears of leaf-tips for the study chromosomes. Stain Tech 32:167–169

Thavarajah R, Mudimbaimannar VK, Elizabeth J, Rao UK, Ranganathan K (2012) Chemical and physical basics of routine formaldehyde fixation. J Oral Maxillofac Pathol 16(3):400–405. https://doi.org/10.4103/0973-029X.102496

Tijo JH, Levan A (1950) The use of oxyquinoline in chromosome analysis. Anales de la Estacion Expt de Aula Dei 2:21–64

Turner B (1956) Chromosome numbers in the Leguminosae. Amer Jor Bot 43:577

Vig BK (1968) Spontaneous chromosome abnormalities in roots and pollen mother cells in aloeVera L. Bull Torrey Bot Club 95:254–261

Wolman M (1955) Problems of fixation in cytology, histology, and histochemistry. Int Rev Cytol 4:79–102

Genetic Study of Bacteria and Bacteriophage

6

Nidhi Sharma

6.1 Bacterial Genetics

Bacterial genomes are made up of a circular chromosome that possesses a single-stranded DNA (ssDNA) molecule of several million base pairs in total length. A classic example is *E. coli* genome that is widely organized in approximately 4.6 million base pairs of DNAs. No wonder that some bacteria contain multiple chromosomes, for example, *Vibrio cholerae* that causes cholera has two circular chromosomes and *Rhizobium meliloti* has three chromosomes. Apart from this composition, there are few bacteria that do not carry circular DNA and instead carry a linear chromosome. In addition to its own genome, the bacterial genome also possesses an additional structure known as "plasmid"—a small circular DNA molecule present in many numbers (usually known as copy number). Plasmids accommodate genes that are not essential at the functional point of view for bacteria but that might have been involved as an important factor for the life cycle and growth of their bacterial hosts. Some plasmids are involved in the mating process between bacteria (provide a channel for the exchange of genetic material during mating, which will be discussed in the next section). Some plasmid leads to an important role in the generation of antibiotic resistance among the population. Most plasmids are circular and of several thousand base pairs in length and, however, some of small size; about a hundred base pairs in the length have been found. Each plasmid structure carries at least an origin of replication (*ori*), a start point for DNA replication. The main function of *ori* is to let the plasmid replicate independently without involving a bacterial chromosome. Episomes are plasmids that are capable of freely replicating and integrating into the bacterial chromosomes. Episomes are categorized into several types based on importance and functionality. The F (fertility) factor of *E. coli* is one type of episome found robustly. F factor basically

N. Sharma (✉)
La Sapienza University of Rome, Rome, Italy

D. Kar, S. Sarkar (eds.), *Genetics Fundamentals Notes*,
https://doi.org/10.1007/978-981-16-7041-1_6

regulates the mating and gene exchange between *E. coli* cells (which will be discussed in the next section). In this part, we have focused on the bacterial mutant gene mutation process that contributes to genetic variation in the bacterial population. We begin with an overview of the chemical nature of the mutation in bacterial genetics and its effect at both molecular and organismal levels.

6.1.1 Bacterial Mutant Genetics

In microbiology, studying mutants is an important part. The study of mutants requires a vast knowledge of detection methods; geneticists should be able to find them out quickly even when they are rare and should be able to isolate them efficiently from wild type or least interested other mutants. What are the mutants? The bacterial population is the most important tool in genetics and biotechnology, and scientists have observed so many unusual new bacterial strains and are still observing to date. Every time new strain discovery is an outcome of occurrence of mutation ever so often. A slight variation in the genome of bacterial strains within a colony results in the rise of new features in the colonies and appears as different from its wild type and known as mutants.

Geneticist increases the likelihood of obtaining mutants by using mutagens to increase the rate of mutation from the usual one mutant per 10^7 to 10^{11} cells to about one per 10^3 to 10^6 cells. To study bacterial mutants, we should know the selection, isolation, and detection method that is described in the following section.

6.1.1.1 Mutant Detection

When it comes to detecting and collecting a mutant of a particular organism, one should know the wild-type characteristic of that organism to recognize the difference between a mutant and a wild type, and thus detection of mutants has been evolved along with the time as the necessity in the field of bacterial genetics. Mutation in prokaryotes is easy to detect even if it is a recessive mutation because it has been seen immediately and onward. A very simple example of albino mutants, which is a type of mutant of pigmented bacterium, can be visualized by observing the color differences of colonies. This mutation simply requires color detection. On the other hand, there are other detection methods available that are more complex. Replica plating method is one of them and is used for the detection of auxotrophic mutants, which has been explained in Fig. 6.1. This method distinguishes a mutant from the wild-type strain. Auxotrophic mutants are able to grow in the deficiency of particular biosynthetic end product that is necessary for an efficient growth of cells.

6.1.1.2 Mutant Selection

Specific incubation conditions can be a tool for the selection of specific mutants that can only grow under such conditions in which wild type will not be able to grow. Thus, this is an effective method for mutant selection. What exactly mutants do to a wild-type strain and what makes them grow in a specific condition? A selection method will either develop a reversion mutation (a mutation that restores the wild-

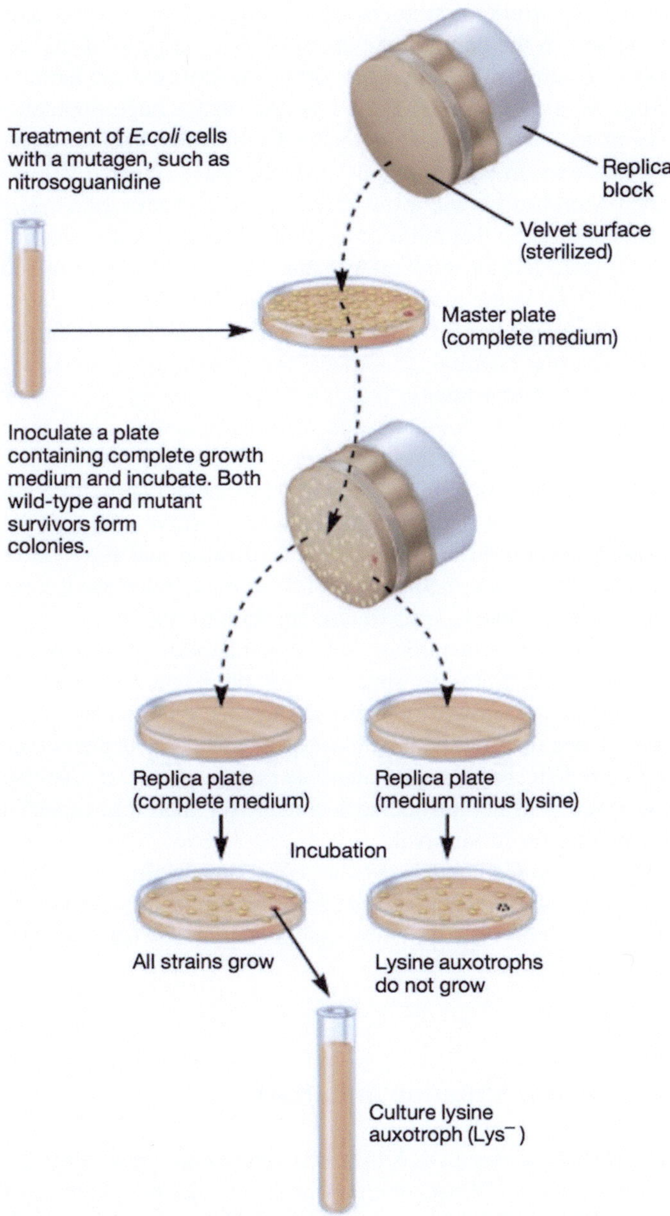

Fig. 6.1 Isolation method for bacterial mutants from the culture through replica plating method. Replica method for separation of bacterial mutants from the culture. Replica plating is popular for the isolation and detection of lysine auxotrophs. Auxotrophic mutants can be easily generated by using a mutagen. Both wild type and auxotroph are cultured together and plated on the complete medium. The culture containing wild type and auxotroph is plated on a complete medium. Immediate to colony formation, a soft velvet cloth is pressed on the culture plate, and bacterial colonies are picked by this piece of cloth. This traced colonies are transferred to another culture plate of minimal media (lacking Lysin) and in the same orientation as the master plate is Location of

type phenomenon if a wild-type strain is already mutated in the forward direction) or develop a resistance to the environmental stress. For example, if the purpose is to isolate a revertant (a strain that regains the former capability) from lac Z^{-} auxotroph, then the method would be simple. Let us assume that a huge population of lac Z auxotrophs is plated on the minimal media lacking the lactose, incubated, and examined for selection. In this case, only cells that have mutated which can regain the ability to grow on lactose will grow on the minimal media (Fig. 6.2). Out of this experiment, we can suggest that millions of strains can be plated, grown, and isolated on a single Petri dish, but those revertants that have the ability to manufacture the nutrient source lacking in the medium will grow selectively. Thus, this method can be a selective technique to test for mutation by a mere scanning of a single Petri dish. This method has successfully accounted for screening many environmental substances containing mutagenicity.

The selection method for mutants due to environmental stress is more likely similar to the previous approach that has been used for auxotrophs. It is known that wild-type strains may be susceptible to any environmental stress like virus (phage) attack or antibiotic treatment. Thus, plating a phage-sensitive wild-type bacterium on the minimal media produced colonies that are resistant and have become phage-mutant. This type of selection method is useful for many other environmental stress conditions, that is, virus attack, temperature, antibiotics, etc.

In addition to the abovementioned selection method, several bacterial strains utilize few specific substrates to survive. Not all bacteria use all the nutrient substrates present in the growth medium. Many bacteria just use a few primary carbon sources rather than the complete ones. In this case, bacteria can isolate by plating on the Petri dish containing an alternate carbon source. If colonies appear on the plate, this strain will be considered a mutant for that specific substrate being utilized in the culture for its survival.

The mutant selection method has eventually become useful for understanding the complexity of genes and biochemistry of a particular bacterial strain. Particularly, this technique is a significant tool for understanding the role of carcinogens in developing mutagenicity.

6.1.2 Spontaneous Mutation in Bacteria

Spontaneous mutation allows a DNA to be mutated without any exogenous factor or selective pressure such as UV rays or chemicals; instead, endogenous factors are the primary sources, that is, DNA damage, error in DNA replication machinery, less effective DNA repair machinery, and so on.

Fig. 6.1 (continued) auxotroph bacteria which must not grow on second plate and can compare to the master plate to spot the lysin mutant. This bacterium can be picked from the master plate and can further grow as a single culture of lysine auxotroph

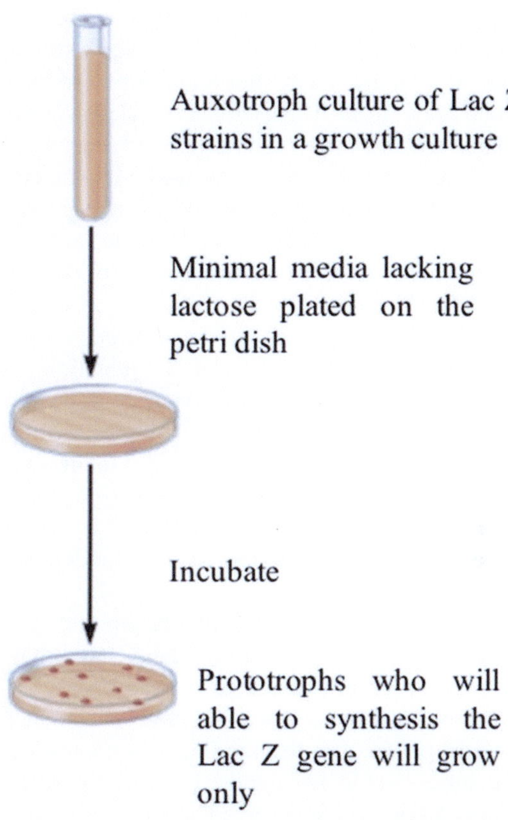

Fig. 6.2 Production and selection method for auxotrophic revertants (mutants). In this example, lac Z revertants are selected after plating a lac Z^- auxotroph because the agar contains a minimal medium that does not support auxotroph growth

Foster and her colleagues in 2012 have demonstrated the mutation rate in three different strains of very primitive and common bacteria *E. coli* by using the whole-genome sequencing (WGS) approach. They explained in their work that the mutation rate of 1–2 $\times 10^{-3}$ mutation per generation and per genome is natural for this bacterium which has not been induced by any external factors. Experimentally, they observed that at any given defined medium or growth condition, the mutation rate of any specific gene remains constant. Interestingly, if a small inoculum that may contain few mutants will be transferred to a culture medium, then the proportion of mutants in that growing culture is positively correlated with the progressively increased bacterial population. From the historical point of view, spontaneous mutation was first time spotted by “Salvador Luria and Max Delbrück” in 1943. In their experiment, they have found that when the *E. coli* was plated on the nutrient medium in the presence of T1 phage, it appears as a phage-resistant mutant before phage display on the plate. It means the *E. coli* observed on the plate were resistant to the phage attack, which might be the result of a spontaneous mutation. This discovery brought us the understanding of genetic diversity in the bacterial genome as a result of spontaneous mutation. Lately, this understanding has widened by including another piece of information based on spontaneous mutation which suggested that the occurrence of genetic changes due to spontaneous mutations is essential for developing antibiotic-resistant strains, host evasion, and acclimatization to the new environment, thus contributing to the bacterial evolution. Interestingly, a recent study from Tomasetti, C et al. has revealed that random mutagenesis in somatic cells of some tissue causes an error in DNA replication which transformed the normal cells into cancerous cells compared to other cells from different tissue at the same time. This study has underlined the role of spontaneous mutation in cancer risk; however, this study has been the reason for debate as well.

How can mutations in bacteria occur spontaneously? What is the method to detect them? The fluctuation test is one of the experimental techniques widely used for its detection. As we have discussed the Luria and Delbrück experiments where they have explained that if a mutation occurs spontaneously, then the incidence of the mutation at a different time in a different culture would be higher which led to the variation in resistant colonies throughout the cultures. This concept is known as the "fluctuation hypothesis." In this experiment, small inocula were grown in the culture medium, and the number of phage-resistant *E. coli* from this culture was measured and compared with other grown multiple *E. coli* colonies. According to the spontaneous hypothesis, we can understand two conditions:

1. If the phage resistance mutation occurs after exposure to the phage, then the number of phage-resistant mutants and mutation rate should be similar between the colonies of both sets of conditions.
2. On the other hand, if the mutation is spontaneous and occurs before the exposure to the phage, then the variability in the mutation for all the independently grown cultures would be highest because the difference in the size of the bacterial population which received the mutation at first will contribute to the overall observed variability till the end of generations. The data indicated that the mutations to phage resistance in *E. coli* occurred spontaneously with a constant probability per cell division.

Let's think about what could be those reasons governing spontaneous mutation? Spontaneous mutation does not need any exposure to external agents besides a mere error in DNA replication, or base substitution could bring the results as a spontaneous mutation. One of the reasons being considered is a malfunction in the machinery of DNA polymerase III (an enzyme associated with replication) during DNA synthesis. Adding, mispairing, or omitting a nucleotide on a parent DNA strand will appear as a mutated granddaughter DNA strand and thus lead to mutation in the continuity. In addition, some mobile genetic materials like transposons are also considered as the source for spontaneous mutation in the bacteria where these elements are present other than the nuclear chromosome.

Replication error appears when the nitrogenous base of a template nucleotide exists in a rare form such as tautomeric form. Tautomerization is a chemical process in which natural forms of nucleotide bases are (keto(C=O) and amino ($C=NH_2$)) converted into two rare structural isomers (imino (C=NH) and enol (C=OH)). These isomers of nucleotide base pairs make nonconventional hydrogen bonding and can

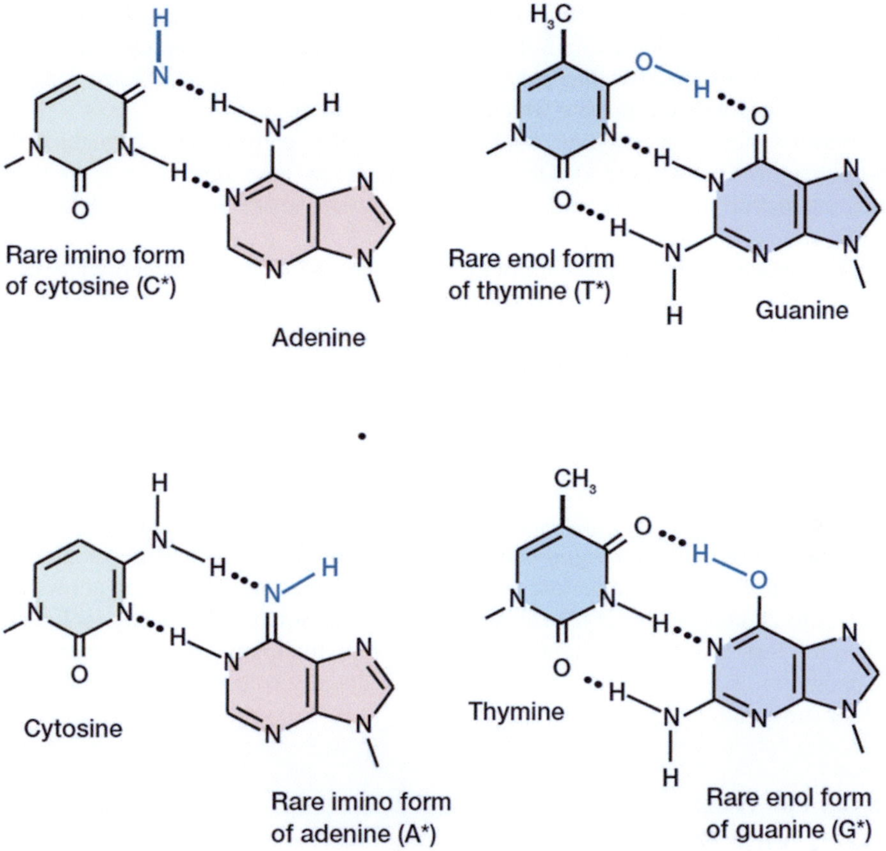

Fig. 6.3 Schematic diagram of base pairs undergoing tautomerization. Normally the keto form of the base formed a normal hydrogen bond-like A–T and C–G, but the enol tautomer instead produces A–C and G–T base pairs. The upper lane shows the normal pattern of A–T and C–G pairing, while the lower lane shows the rare bonding between (1) imino form of adenine and cytosine and (2) enol form of guanine and thymine

readily change to each other (Fig. 6.3). Thus, tautomeric shifts change the characteristics of the hydrogen bond among the four bases. In turn, this shift allows purine to bind with purine and pyrimidine to pyrimidine instead of pyrimidine–purine binding and eventually generates an alteration in the nucleotide sequence of the daughter strand after the first round of replication.

Such mutation based on tautomerization, known as **transition mutation**, is relatively common to be found. On the other hand, **transversion mutation** is another type of mutation in which purine may substitute to a pyrimidine and vice versa, but this mutation is not frequent due to stearic hindrance in the pairing of purine to purine and pyrimidine to pyrimidine.

Replication error also occurs when a purine and pyrimidine base leads to apurinic (loss of purine from the nucleotide sequence) or apyrimidinic (loss of pyrimidine base from the nucleotide sequence) sites in the sequence. Spontaneous loss of purine or pyrimidine base due to hydrolytic cleavage of n-glycosylic bonds with sugar moiety forms lesions, and thus polymerase enzyme is unable to synthesize complementary nucleotide on this site which leads to mutation.

Redox attack on guanine base results in the conversion of guanine to 8-hydroxydeoxyguanosine (8-OhdG). This modified base ultimately pairs with adenine instead of cytosine and produces G->T transversion in the end, during the replication.

Although most geneticists believe that spontaneous mutation is a random process without any induction through external agents, this theory lately in 1988 has been modified by John Cairns whose theory is an extended version of Luria and Delbrück's concept of spontaneous mutation. He stated that mutation does not happen only spontaneously in growing cells but preferably happens in nonproliferating cells by giving lethal conditions. In his experiments, he demonstrated that when mutant *E. coli* strain was subjected to grow on a complete medium, it was unable to utilize the lactose as a whole carbon source and interestingly if lactose were continuously added to the culture medium as an only carbon source, *E.coli* would more rigorously avoid the lactose to take as a carbon source. In this case, John found that lactose seemed to induce the mutation which allows the *E. coli* to choose the sugar over lactose for a carbon source. This study took us to interpret that this type of mutation is "adaptive mutation" where bacteria are supposed to choose the mutation that occurs frequently so that they can adapt to a better surrounding for their survival.

6.2 Viral Genetics

Every organism such as plants, animal, fungi, and bacteria are susceptible to viral infection. A virus is a simple replicating machinery that consists of a core nucleic acid and is protected by a surrounding protein coat usually known as a capsid. Virus can be categorized based on their shape and size and nucleic acid whether it is a double-stranded DNA or single-stranded DNA or sometimes a single-stranded RNA.

Virus has a particular class that infected their specific host, for example, a virus infected bacteria is called "bacteriophage," and this phage will not infect plants or animals directly but will be carried away with bacteria itself to further infect animals or plants. Phages have been involved in genetic research since the late 1940s. It has become an essential and advanced research tool in the area of genetic research since it contains a small and handy genome, reproduces rapidly, and produces a large number of progenies. The study of the phage genetic system has been employed long ago because this microorganism plays an important role in human society. In this section, we will focus on many unique aspects of phage genetics, that is, study of structure and life cycle of bacteriophage, detection methods for phage infection, application in genetic research, and so on.

The Discovery of phage included many efforts and stories that have been made by several scientists. First time, it has been spotted something vague in the river of India by Ernest Hanbury Hankin, a British bacteriologist in 1896. He simply reported that there is something in the water that had antibacterial properties and killed cholera, but he did not specify his finding. Later in 1915 another British bacteriologist Frederick Twort discovered that a very small molecule or an unknown thing kills bacteria in the bacterial culture; he had published his finding, but unfortunately, this work had been interrupted due to WWI. Later in 1917, Felix d'Herelle discovered a killing agent for bacteria at Pasteur Institute, France. He actually observed that when he added a filtrate that was collected from sewage to the bacterial culture of dysentery, in a few minutes, the culture colonies disappeared, and he named this filtrate as "invisible antimicrobial agent," and later this work was published. In the row of findings, a first-time phage study was developed in the Elvia institute, Tbilisi, Georgia, in 1923 to study this invisible thing and to develop phage therapy. In 1969, Max Delbrück, Alfred Hershey, and Salvador Luria discovered the replication of virus and virus genetics for which they were awarded the Nobel Prize in Physiology or Medicine.

6.2.1 Structure of Bacteriophage T4

T4 is one of the most extensively studied bacteriophages among all other phages such as T1–T7. T4 bacteriophage is specific to *E. coli* and has been demonstrated as a model for phage study by Delbruck and coworkers in 1944. In the modern time of genetic engineering, the study of phage included advanced tools and techniques particularly to understand its structure at the atomic level. Early discoveries on this bacteriophage have included the prediction of phage image using electron microscopy (EM) obtained by Brenner et al. in 1959. This work extended to an extensive study on phage head symmetry, tail, and baseplates through EM. For the first time, a complete T4 genome was sequenced in 2003. In the continuity of this work, high-resolution cryo-electron microscopy (cryo-EM) image was obtained which revealed a dome-shaped baseplate structure in the infectious virus. In subsequent years, star-shaped baseplate and prolate head structure of post-infection T4 were published. Thereafter, many other techniques like complementation assay (to study recombination in bacteriophage), cross-linking analysis (protein–protein interaction study), X-ray crystallography, and cryo-EM provide high-resolution atomic-level structure model for T4 phage. Studies from the structure model reveal the structural similarities among phage protein and bacterial protein which suggested the common evolutionary ancestry or coevolution with the bacterial host. Structural similarities among phages as well as of T4 components with bacterial proteins demonstrate common evolutionary ancestry or coevolution with bacterial hosts.

Bacteriophage T4 belongs to the family *Myoviridae* that infects *E. coli*. A basic T4 structure includes a head (capsid), tail, and baseplate. Phage has a rigid tail that is composed of many layers; the inner layer of the tail is surrounded by a contractile sheath that helps a phage during infection. Phage *Myoviridae* family like T4 contains a massive baseplate at the end of the tail with long attached fibers that guide the phage to find the receptor in the host cell and mediate the initial contact. A contractile tail is helpful in penetrating the bacterial outer membrane before the DNA delivery during the infection. Long attached fibers are made of six short tail fibers that are folded underneath the baseplate and unfold on the recognition of the host as a host sensor. The baseplate is a puncture device for phage located at the end of the long tail and the last element located at the end of the tail.

Capsid of T4 phage is assembled with three main components: (1) gp23 (48.7 kD) forms hexagonal capsid lattice, (2) gp24 forms pentamers of the vertices, and (3) gp20 forms a unique dodecameric entrance as a portal vertex that makes a gateway for DNA packaging and exit during the infection. Genetic material of T4 is a linear dsDNA of 168kbp which has 289 open reading frame (ORF).

6.2.1.1 DNA Packaging

DNA is translocated in the capsid through the central channel of portal protein on one vertex of the capsid. The portal complex at vertex works as a docking point for terminase; a viral ATPase complex that is bound to the end of the portal vortex provides an active packaging motor complex that assists the entry of viral DNA into the capsid. The packaging of DNA is evenly regulated till the threshold amount of DNA is entered into the capsid. Interestingly Smith et al. in 2001 have reported in their work that dense packaging of DNA after some point will raise the pressure inside the capsid approximately equal to 6Mpa. Termination of DNA packaging, after enough DNA has been densely packed into the capsid, is mediated by bolting of the active portal complex, which prevents the phage genome leakage. Portal complex on vertex alone is not enough in the packaging machinery; therefore, in addition, head completion proteins sometimes also called adaptor proteins are required to bind with the complex dodecameric portal and form the connecter at the end of the capsid.

If we take a closer look at the symmetry of capsid and tail, a mature or fully assembled capsid in a phage follows four to fivefold symmetry or so-called icosahedral symmetry. From the structure point of view, each multiple structural unit in the capsid has been displayed as a regular lattice with two, three, and five rotational symmetries. Portal protein shows dodecameric oligomerization pattern while tails show overall six- or threefold rotational symmetries, and protein repeats in this symmetry are shown in a helical pattern.

Brenner et al. in 1959 successfully illustrated phage image through electron microscopy (EM).

EM image of mature phage has been described in Fig. 6.4:

1. 1150 Å-long and 850 Å-wide icosahedron head encompassing genomic DNA
2. 925 Å-long and 240 Å-diameter contractile tail attached to the one end of the head through portal vertex
3. 270 Å-high and 520 Å-diameter hexagonal baseplate
4. 1450 Å-long six long tail fibers attached to baseplate

In the past decade, advanced techniques such as single-molecule tweezers and fluorescence studies revealed a few amazing facts about bacteriophage; T4 motor packages DNA at the rate of 2000 bp/s which is confirmed as the *fastest packaging motor* till date. Another fact that FRET-FCS studies on phage suggested that DNA in phage gets condensed during the translocation process.

6.2.2 Life Cycle of Bacteriophage T4

A phage life begins and ends during the infection period and has been characterized into two phases: lytic cycle and lysogenic cycle. Lytic cycle is a virulent phase since it infects a cell, destroys the cell, lyses the cell, and replicates and produces more phage particles. A lysogenic phage is a temperate phase because it infects the cell and incorporates dsDNA to the host cell, and no progeny will be produced. In lysogenic phase, foreign DNA such as viral DNA is incorporated into the bacterial chromosome and replicates multiple times as the host cells divide. However lysogenic phase also undergoes a lytic phase in some circumstances not capable of undergoing the lysogenic life cycle and therefore directly undergoes the lytic cycle.

Other than lytic and lysogenic, a phage also has been found in pseudolysogeny life. It is an unfavorable and unusual life cycle that occurs when a phage grows in unfavorable growth conditions that enable a phage to survive by preserving the phage genome until the host growth condition becomes advantageous again.

Lytic phase has been categorized as follows:

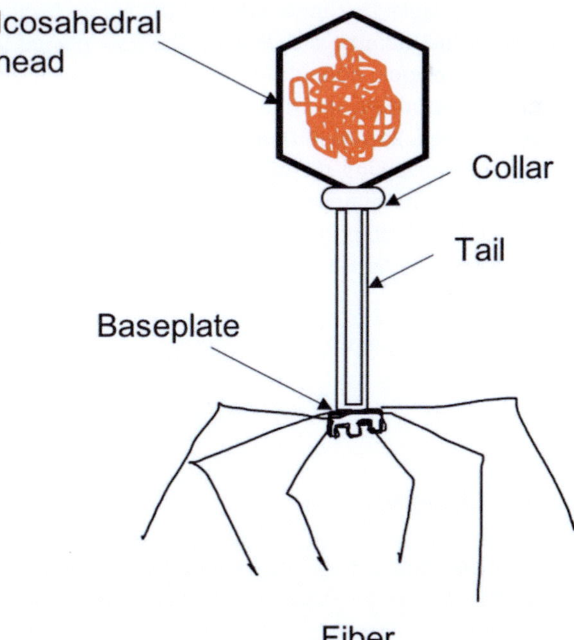

Fig. 6.4 Assembled mature bacteriophage T4. The assembly of T4 can be divided into three independent subassemblies: the head, the tail, and the long tail fibers. The tail binds to the head followed by the attachment of the fiber in the end. Six long tail fibers are then attached to form a viable T4 virion

1. Attachment to the bacterial cell and phage infection
2. Passage of DNA through the bacterial cell wall
3. Infected bacterium converted to phage factory
4. Production of phage DNA and protein
5. Assembly of phage
6. Release of phage particle

6.2.3 The Plaque Assay

Plaque assay has been developed to the necessity of phage detection in several medical conditions. Plaque assay has shown the ability of host cell to transform into plaque lawns if any bacterial colonies are being infected within the phage. Spot-on lawn assay has been identified as one of the simplified versions of the plaque assay which aims to identify potential viral plaques or virus infection in the growing bacterial culture. To this aim, approximately 1 μL aliquots of virus suspension would be enough to be applied to fresh lawns of diverse microbial host strains. For example, if the suspension of a phage (i.e., T4) is applied to the susceptible bacterial host (*E. coli*), then phage infects the bacterial cell, replicates independently, and promotes the lysis of the cell due to the occurrence of lytic phase and kills the bacteria in the end. This lysed bacterial cell indicates the formation of a clearing zone on the bacterial lawn known as plaques. In some circumstances where the lytic cycle is absent, bacterial colonies grow confluently. Any single plaque determines the number of phages that have infected the single bacteria and further keep infecting bacteria in the vicinity and develop the new plaque in the vicinity which together forms a big plaque, enough to be seen with naked eyes. Notably, plaque does not continue indefinitely, and the size of plaques is totally dependent on the type of phage, host cell, and condition of grown culture. Measuring the frequency of plaque formation or detecting the number of plaques formed on the bacterial lawn can be calculated by the dilution method. If phage suspension can be diluted in a serial dilution method, thus appropriate dilution factor can be used for calculation of plaque formation unit by using the following equation:

$$\textbf{Plaque forming unit}(\text{PFU}) = \frac{\text{Average Number of Plaques}}{\text{Dilution factor} \times \text{volume of diluted virus added}}$$

6.2.4 Lysogeny

Lysogeny is another life cycle or reproductive pathway of phage other than the lytic cycle. The very first evidence of lysogenic behavior of phage is confirmed by two subsequent experiments in 1920 and 1940 as follows:

1. In the 1920s, some remarkable results were obtained in the study of phage in the *E. coli* culture. Earlier microbial geneticists have performed an experiment in which they mixed two microbial strains: lysogenic-resistant and non-lysogenic-resistant strains. Lately, they have found that non-lysogenic-resistant strain was lysed in the culture. The geneticist explained that this phenomenon happened because resistant strain somehow causes lysis of nonresistant strain and thus resistant cell known as lysogeny or lysogens (a causative factor for cell to lysis). It is further important to understand that sometimes nonlysogenic bacteria might get infected with phage-derived lysogenic strain, and very few infected cells were not lysed but rather itself became lysogen.
2. However, in 1940, André Lwoff performed another experiment in which he studied the lysogenic bacteria *Bacillus megaterium* and he followed its behavior and cell divisions in the culture throughout. Once he established the culture, he separated each daughter progeny after each cell division. From this daughter progeny, he has put back one cell in a new culture while he followed the other cell throughout the cell division. In this experiment, he followed 19 cultures that represent ten consecutive generations. He also separated culture medium and found no phage existing in the free medium through which he confirmed that lysogenic behavior of the bacteria is followed by each cell division or each reproduction in the absence of any external phage in the medium.
   On the other hand, he spreads this separated phage-free medium, on the lawn of non-lysogenic bacteria, and astonishingly, he observed a spontaneous plaque formation in the culture. This observation he explained in his proposed hypothesis. Lysogenic behavior of phage passes from generation to generation, and this gives rise to pure noninfective strain but somehow this noninfective factor converted into infective phage although no free phages presence in the medium. This event is an exceptional case. In the term of microbiology, Lwoff named this factor as "prophage" which can change the noninfective factor to infective factor by chance or in an inductive manner.

Now we know why non-lysogenic-resistant cells have been found to be infected and lysed in the culture because prophage, which is a feature of lysogenic bacteria, eventually produced infective phage into the medium that causes the cells to lyse.

A common lysogenic cycle can be described in few steps (Fig. 6.5):

1. Linear phage DNA first injected to the host cell.
2. Initially phage mRNA will synthesize for short period, followed by synthesis of (a) repressor protein that inhibits the synthesis of the enzyme needed for lytic cycle and (b) site-specific recombination enzyme. Later, phage mRNA synthesis is turned off by repressor protein.
3. Phage DNA molecule incorporated into the bacterial chromosome.
4. As bacteria replicate and grow in the medium, phage DNA also multiplies and produces more phage progeny as part of the bacterial chromosome.

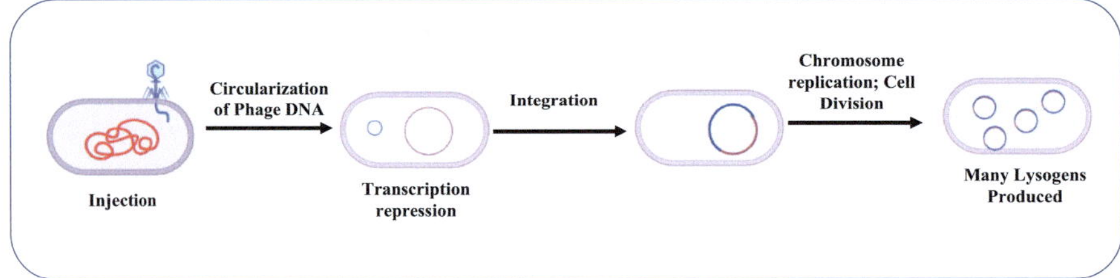

Fig. 6.5 A lysogeny phase after immediate to virus infection in the host cell. Infection of phage initiates circularization of phage DNA at first, followed by integration with bacterial chromosome and replication. Cell division reproduces more numbers of bacterial cells containing integrated phage DNA which accomplishes the lysogenic cycle of any phage

6.2.5 T4 Phage Modulates Bacterial Genetics

Other than its characteristic to invade bacterial hosts, T4 phage infection has been the subject of much scientific research to how phage has an impact on the bacterial genome. Such phenomenon can take place by either generalized transduction or introduction of phage-encoded protein whose expression results in changes of their host phenotype and activity. Phages have acquired these genes from their host and continued to evolve and change within its own genome. These extra genes can be named as "accessory genes" that can govern the biology of their bacterial host and find the tune in the way in which bacteria interact with their environments. Such observation has been possible by our ability to sequence phage genomes, and this information will serve as a start point for further study to determine how phage infection can contribute to their bacterial host's physiology endurance and evaluation. In many ways, phage affects the bacterial genome or phenotype, and few examples are given below:

6.2.5.1 Human Gut Microbiome Interaction

Human gut microbiome is another example of phage–bacteria interactions which consist of densely colonizing microorganisms including T4 phage and interacting with mammalian hosts. Metabolome data from several research on the role of phage in the gut microbiome suggested that T4 phage remains as a modulator of bacterial colonization in the human gut and supports the benefits out of this interaction to human health.

6.2.5.2 Host Communication

T4 phage further manipulates bacteria by accessing quorum sensing pathways which have effects on one or multiple behaviors of bacteria as shown in (Fig. 6.6). Genomic study of phage reveals that phage DNA sequence contains homologs of response regulator that is involved in quorum-sensing pathway. Also, it has been noticed that induction and release of temperate phages have been observed following the N-acyl homoserine lactone exposure (LuxS system of quorum-sensing pathway).

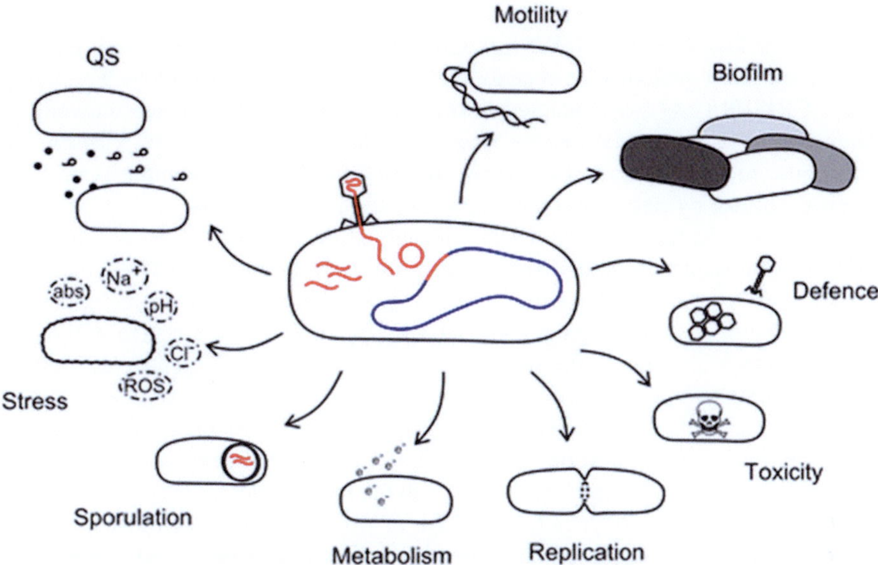

Fig. 6.6 T4 phage modulates bacterial genetics in several ways. After insertion of phage into a bacterial DNA, it further regulates several signaling in the bacterial cell which are essential for bacterial life cycle, stress responses, replication, metabolism, and so on

6.2.5.3 Host Replication

Phage uses bacterial machinery to manipulate its own replication and host replication system, and phages manipulating host replication can be seen in the dimorphic bacterium *Caulobacter crescentus*. Moreover, another example of phages influencing their host's replicative process is the phage-encoded homologs of MazG. MazG is a regulator of cell death in *E. coli*, and its expression influences bacterial replication in nutrient-limited environments. Interestingly, homologs of MazG have been found in phages infecting several diverse bacterial species, including several cyanophages, *Burkholderia cenocepacia* phages, and *Mycobacterium* phage L5.

6.2.5.4 Host Metabolism and Energy

Many phages that invade bacterial cells also control its metabolic regulation, such as carbon, nitrogen, and phosphate metabolism, and its utilization. Cyanophage is one example of such kind of modulation by phage in the host cell.

6.2.6 CRISPR/Cas9 Bacteria in Genetic Engineering

In the past decades, an endogenous machinery of bacteria/archaea has been used as a powerful tool in genome editing (called gene editing) technology which allows genetic material to be added, removed, or altered at particular locations in the genome. To this purpose, several techniques have been developed, and the very recent one is known as CRISPR/Cas9. CRISPR-Cas9 stands for clustered regularly interspaced short palindromic repeats and CRISPR-associated protein 9. Interestingly, CRISPR/Cas9 was adapted from a naturally occurring genome editing machinery in bacteria. Just like us, bacterial cells can be invaded by viruses, and in response to defense against the virus, the bacterial CRISPR immune system can thwart the attack by destroying the genome of the invading virus.

Interspersed between the short DNA repeats of bacterial CRISPRs are similarly short variable sequences called spacers that are derived from viral DNA that have attacked bacteria previously. However, this spacer helps bacteria to recognize the viral genome of its attacks again, and CRISPR (Fig. 6.7) defense system will cut up any viral DNA matching the spacer sequence. Thus, these spacers are termed "genetic memory."

Genetic manipulation such as gene insertion or gene overexpression is very well established; however, inhibition or abrogation of a particular gene is quite challenging until CRISPR/Cas9 came into existence in the last decade. CRISPR/Cas9 system has been categorized into types I, II, and III, of which type II is the most successful and widely used in genome editing since it requires one enzyme and one RNA to function as a DNA endonuclease. Moreover, the RNA components of the CRISPR/Cas9 system can be used separately by fusing the crRNA (mature CRISPR RNA) to

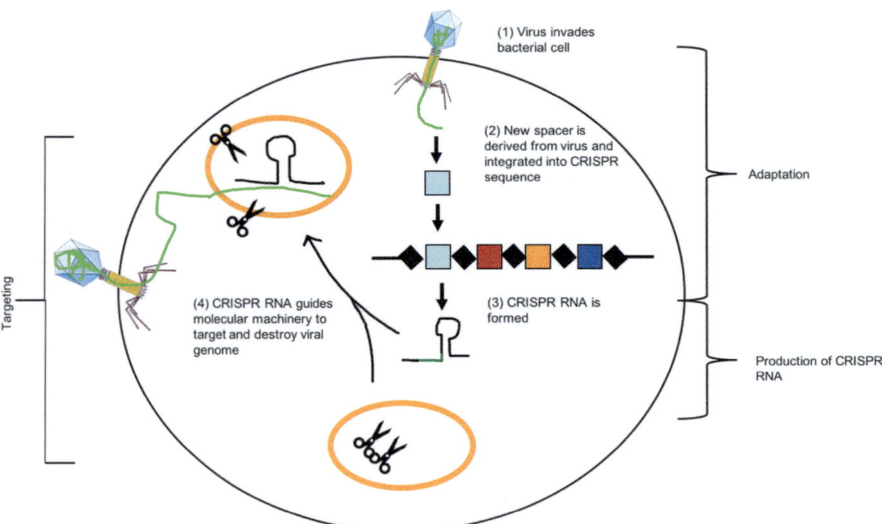

Fig. 6.7 CRISPR-mediated gene editing: CRISPR are regions in the bacterial genome that help participate in the defense against invading viral genome. These regions are composed of short DNA repeats (black diamond) and spacer (colored boxes). When a new virus infects a bacterium, a new spacer is generated by the viral genome and incorporated among existing spacers of CRISPR. CRISPR is transcribed and processed into short CRISSPR RNA molecule. This CRISPR RNA guides bacterial molecular machinery to a matching target sequence in the invading virus. The molecular machinery cuts and destroys the invading viral genome

the tracrRNA (trans-activating CRISPR RNA) generating a single guide RNA that recruits the Cas9 nuclease to specific genomic locations via standard Watson–Crick base pairing and facilitates double-strand break. The creation of site-specific double-strand breaks by the CRISPR/Cas9 complex then triggers genome editing through two different mechanisms: (1) repair through homologous recombination and (2) nonhomologous end joining. Notably, both pathways lead to functional inactivation of targeted genes with high efficiency, and thus CRISPR/Cas9 methodology has rapidly become the state-of-the-art technique for genetic manipulation of mammalian cells and genetically modified mice and has the potential to be used in a diverse range of gene therapy approaches in the future. Generation of the knockout mouse model for many disease studies has been possible by using adenovirus (AVV)-associated CRISPR/Cas9 system in recent years.

6.2.7 Application of CRISPR/Cas9

CRISPR/Cas9 is a simple and rapid tool that enables the efficient modification of endogenous genes in various species and cell types. A number of clinical trials using CRISPR/Cas9 system for genome editing are underway, and the first clinical trial involving CRISPR/Cas9-mediated gene modification has started in October 2016 at West China Hospital, Chengdu. CRISPR/Cas9 complex is nowadays an easy tool for many therapeutic approaches such as for immunotherapy in lung cancer, HIV, beta-thalassemia, Duchenne muscular dystrophy, hepatitis B virus (HBV) infection, and so on. CRISPR/Cas9 system is used in the current scenario and its applications are listed in Table 6.1.

6.3 Conjugation

6.3.1 Discovery of Conjugation

For many years it was thought that bacterial reproduction is only done by simple binary fission that splits a bacterial cell into two identical daughter cells excluding the exchange or recombination of genetic material. The very first evidence of exchanging the genetic material within the bacterial population was "conjugation"—a method of DNA transfer mediated by direct cell-to-cell contact. This result became part of the knowledge from a subsequent series of experiments conducted by Joshua Lederberg and Edward Tatum in 1946 (reported in Nature and the Journal of Bacteriology (JB) in 1946 and 1947). In this experiment, two auxotrophic strains were first selected and mixed which is further followed by incubation, culturing into a nutrient medium for many long hours, and plated on the minimal medium. Later, they observed a recombinant prototrophic colony on the minimal medium which has an incorporated recombination chromosome in each cell. Thus, this experiment suggested that the chromosome of two auxotrophs can associate with each other and undergo the recombination process (Fig. 6.8).

Table 6.1 CRISPR/Cas9 implementation and its applications in medical science

CRISPR/Cas9 use	Application
1. CRISPR/CAS9 delivery to edit the mutation in fumarylacetoacetate hydrolase (genetic disorder)	Gene editing in mouse type I tyrosinemia model in vivo
2. Gene correction by repairing the mutation in the CFTR gene for cystic fibrosis (genetic disorder)	Therapeutic
3. Duchenne muscular dystrophy (genetic disorder)	Therapeutic
4. Removing one or more exons from the mutated transcript by CRISPR/Cas9 system allowed for the production of truncated but still functional dystrophin protein in a mouse model of muscular dystrophy	Therapeutic
5. Generating autocatalytic mutations to generate homozygous loss-of-function mutations	Generating homozygous animal model
6. Gene deletion in mouse model to generate particular disease study model	Knock mouse model
7. Injection of CRISPR/Cas9 components into zygote/ early stage embryo to modify genetic structure permanently	Genome editing in human

However, Lederberg and Tatum failed to prove the concept of "physical contact between cells" which is the major requirement for gene transfer. But in the following years, 1950, Bernard Davis has demonstrated this gene transfer in "U tube" experiment. Bernard Davis constructed a U tube that contains two pieces of curved tube fused together at the base to form a "U" shape separated with a piece of fritted glass filter fixed between halves.

This filter does not allow the passage of bacteria; rather it allows the passage of the medium. During the incubation time, the medium was pumped back and forth through the filter to make sure that the medium is thoroughly switching between halves. After 4 h of incubation, bacteria were plated on the minimal media condition. And interestingly, Davis observed that when the auxotrophs were separated and cells were not in contact, the conjugation does not occur which means gene transfer needs direct contact (Fig. 6.9).

The next question rises to know what component or factor promotes conjugation. F factor (fertility factor)-associated gene transfer is the most common type of conjugation in bacteria that will be discussed in the next topic.

6.3.2 Discovery of Fertility Factor (F)

After being given the experimental evidence by Lederberg and Davis for conjugation, William Hayes in 1953 came up with the idea that genetic transfer occurred only in one direction in the abovementioned crosses. This is the reason that it has never been found that gene transfer in *E. coli* could be in a reciprocal manner. Thus, one cell must act as a donor, and the other cell must act as the recipient. This unidirectional gene transfer seemed to be compared with the original sexual difference between participants, according to which donor cell should be known as "male"

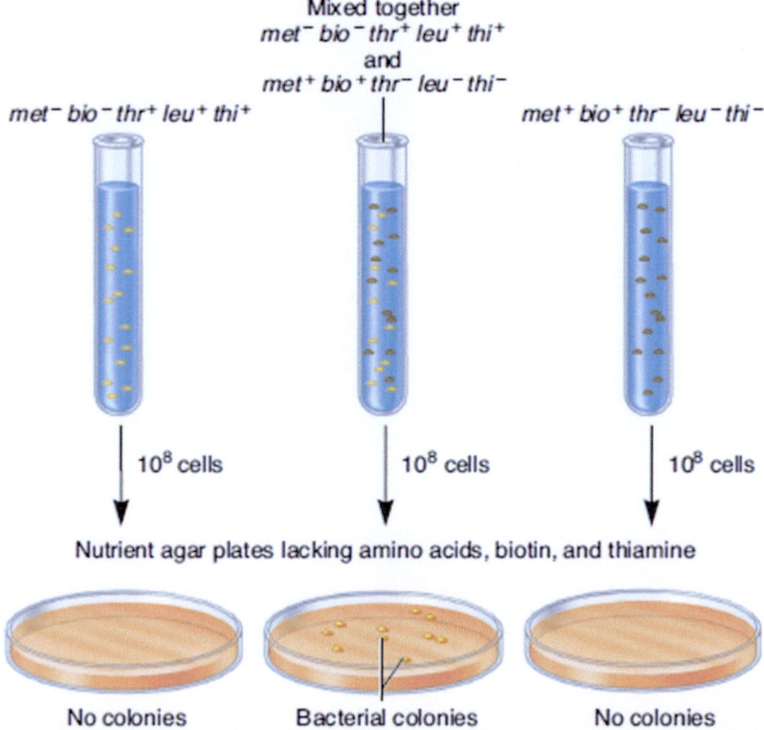

Fig. 6.8 Experimental setup by Lederberg and Tatum. Tube 1 contains a single auxotroph population that has met− bio− thr+ leu+ thi+, meaning this bacteria contains the functional gene only for thr, leu, and thi (amino acid) while methionine and biotin genes were absent and that this bacteria cannot grow on the minimal media which is lacking all the essential amino acid and biotin needed for any bacterial growth. In tube 3, another auxotroph population was present containing met+ bio+, thi−, leu−, thi− which is an opposite composition in tube 1. In tube 2, both populations were mixed and incubated for 4 h. Later when this population was plated on minimal media from these three tubes, the population from tube 1 and tube 3 was unable to grow on minimal media, while mixed population from tube 2 was successfully grown on the minimal media, which shows that transfer of gene between these two population has occurred somehow which has given the new mutant colonies containing met+ bio+, thr+, leu+, thi+ genes. This experimental evidence has proven the concept of gene transfer in the bacterial population

and recipient cell should be known as "female." Although such gene transfer is only possible in eukaryotic organisms but not in bacteria and hence conjugation is not a type of sexual reproduction at all. In bacterial gene transfer, one cell that has to transfer the gene behaves similar to a donor, and the other cell which is supposed to receive the donor's genetic material and change its own genetic makeup behaves similar to a recipient cell, while sexual reproduction has equally contributed to donor's and recipient's genetic information. Lately, it was discovered that gene transfer in *E. coli* through conjugation is eventually driven by one of the circular DNA plasmids known as fertility factor or "F factor" which is sometimes also called

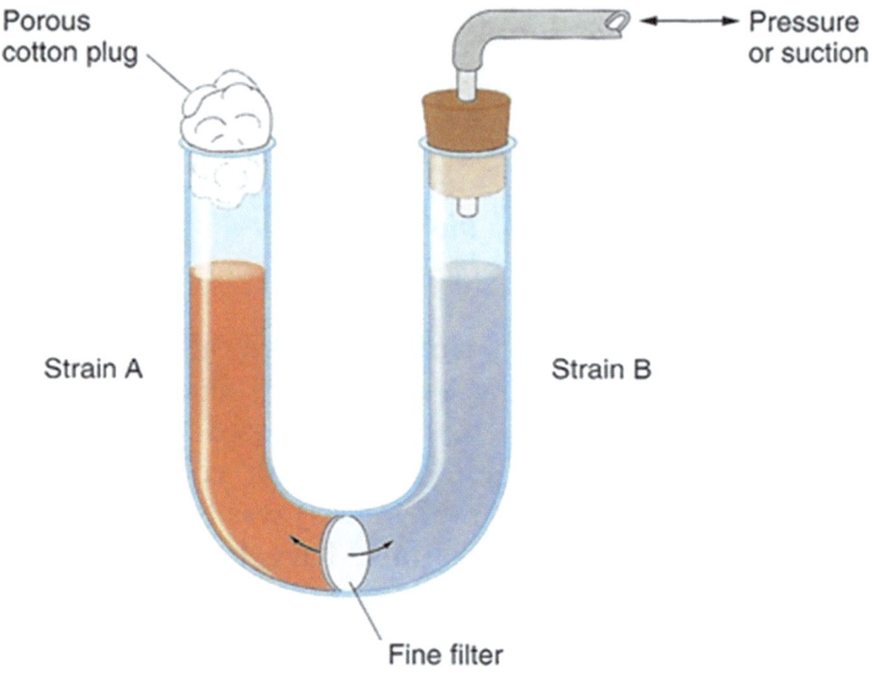

Fig. 6.9 Bernard Davis U tube experimental setup to prove that bacterial mating or physical contact is a must for gene transfer through conjugation. When auxotrophic strain A and auxotrophic strain B were plated on the minimal media and incubated for few hours, no growth was been observed on the media which confirms that no gene was transferred between the bacteria when they were separated through the filter paper

as sex factor. F factor is found in some species but not in all bacterial cells. First, we need to understand the characteristics of the F factor prior to following its role in conjugation.

The size of F factor varies from few kb to 100 kb in the form of duplex DNA keeping two distinct replication origin regions (Fig. 6.9). Among these two, the bigger one is denoted as *ori V* or vegetative replication region which is a point that supports the F factor to replicate autonomously in a particular situation when the plasmid is not being transferred such as cell division of F plasmid; this origin is bidirectional, whereas *ori T* is unidirectional and responsible for replication and transfer of F factor to the recipient cell. F factor shows the similar copy number of a plasmid as bacterial chromosome shows, and therefore one bacterium has one or two copies per bacterial chromosome.

The conjugation process is regulated by *sex pili* or F pili, a thin rod-like structure that appears as an extension of the cell wall. A protein subunit of pili is pilus coded by gene *tra* which polymerizes into pili. Bacteria carrying F plasmid (male or donor) attach to the recipient bacterial (female or recipient) cells for conjugative transfer. An F-positive bacterium has 23 pili on the surface and a *tra* operon encoding

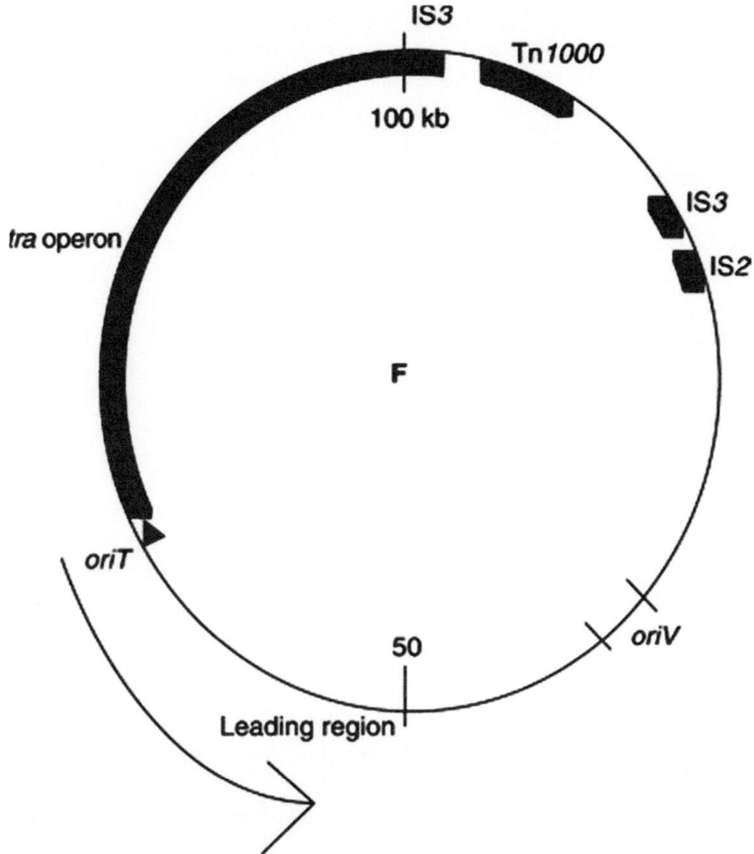

Fig. 6.10 A physical map of F plasmid (100 kb). This circular DNA is further divided into (1) *ori* V which is responsible for autonomous replication, (2) *ori* T which is the origin of replication and transfer of F plasmid, (3) *tra* operon which encodes the functional factor required for conjugation, and (4) IS3, IS2, and Tn 1000 which are transposable elements. The thin arrow indicates the direction of replication

30 functional genes that promote the transfer of the F plasmid (Fig. 6.10). Other than this, F plasmid has three *transposable elements* incorporated in the structure, in which two copies are of insertion sequence—IS2 and IS3—and one is transposons Tn1000 (sometimes known as γδ).

During the experiment, a variant of the F factor has been discovered accidentally by Hayes. He has observed that a variant from its original donor did not reproduce a recombinant after crossing with the recipient strain. This observation reflects that this donor cell apparently had lost the ability of gene transfer and had converted into recipient-like strain known as "sterile donor."

Through this analysis, Hayes realized that fertility characteristics (ability to donate) of *E. coli* could be silent and recover easily. Thus, he suggested that the donor ability of F factor shows a hereditary phenomenon.

6.3.3 F^+ and F^- Bacteria

As we have discussed in the previous section, gene transfer is done through the fertility factor (F factor) in a donor cell, designated as F^+ bacteria, whereas bacteria that is a recipient and lacking F factor is designated as F^- bacteria.

The F factor includes an origin of replication and genes required for conjugations as discussed previously (see Fig. 6.10). F^+ bacteria produce sex pili (singular known as pilus) that facilitate a physical contact between F^+ and F^- to pull them together (Fig. 6.11). The most important fact about conjugation is that this type of gene transfer can only take place cells that contain F and cells that lack the F factor. Detail mechanism is highlighted in Figs. 6.3 and 6.4.

6.3.4 Hfr Bacteria

Hfr (*high-frequency recombination*) bacteria simply became in existence immediately after $F^+ \times F^-$ mating. It has been suggested that this is the second type of F bacteria that maintains a higher frequency of conjugation than the F^+ bacteria. At first, Lederberg and Tatum have observed that conjugation between F^+ and F^- allows the transfer of genetic material in F plasmid but does not account for the transfer of chromosomal genes. Hfr-mediated conjugation transfers the donor chromosomal gene with a great efficiency into the recipient cell but does not change the

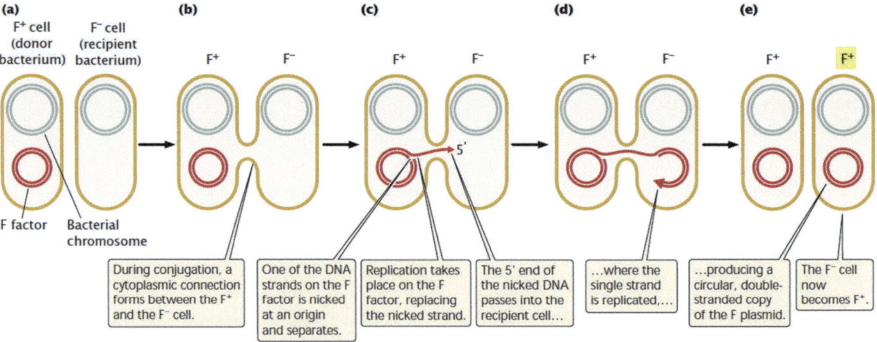

Fig. 6.11 Gene transfer mechanism in F+ and F^- cells. (**a**) F^+ is a donor cell that will transfer the F factor to the F^- (recipient cell). (**b**) A conjugation tube or a bridge begins to happen between these two cells. (**c**) Single-stranded DNA is generated by nick at the origin and separates it from the double-stranded circular DNA. (**d**) This 5′ nicked single-stranded DNA is transferred across the cells and enters in the recipient cell where single-stranded DNA begins to replicate and convert the F^- into F^+ cell. (Benjamin. A. Pierce., Genetics: A conceptual approach)

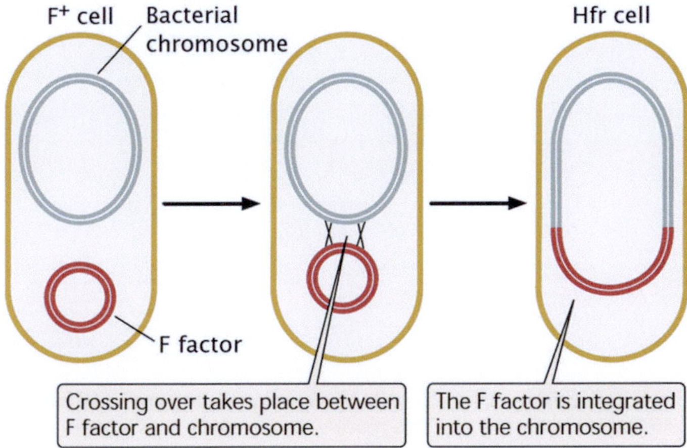

Fig. 6.12 Construction of Hfr cell from F^{+} cell. Integration of F factor to the bacterial chromosome in F^{+} cells which converted the F^{+} cell into Hfr strain containing both the features of plasmid and the bacterial chromosome. (Benjamin. A. Pierce., Genetics: A conceptual approach)

F^{-} into F^{+}, and this is because of a partial transfer of chromosome into F^{-} which does not change the cell into F^{+} unless the entire chromosome has been transferred. If the F factor is integrated into the chromosome and this chromosome has to be transferred, the chromosome will require 100 min in case of *E. coli*, but unfortunately, the conjugation breaks before the process is finished. At last, the F factor is not completely transferred to the recipient cell, and it remains F^{-}.

Thus, Hfr strains contain F integrated chromosome, and F^{+} cell further can form sex pili and conjugate with F^{-} cell (Fig. 6.12).

In the mating between Hfr and F^{-} cells (Fig. 6.13), integrated F first nicked at the one end on one strand. This nicked end moves toward F^{-} cell similar to the conjugation between F^{+} and F^{-} cells. Since the F factor is integrated with the bacterial chromosome, nick transfer further allows the transfer of chromosomal fraction into the recipient cell. The amount of transferred chromosomes depends on the duration of conjugation between the two cells and how long they are connected.

Once the nicked and single-stranded DNA is transferred to the recipient cell (F^{-}), it starts to replicate, and sooner the crossing/recombination between the donor and the recipient chromosome will take place. When the crossing over takes place in the recipient cell, degradation of the donor chromosome occurs instantly. The recombinant recipient chromosome remains intact in the cell and starts to replicate and pass to generations. It is already mentioned in the mating between Hfr and F^{-} cells that F^{-} will not become F^{+} or Hfr unless the entire F factor (F-integrated bacterial chromosome) will receive an F^{-}-recipient cell. This event seems to be a rare case or time-consuming because most of the conjugations last for only a short time and break cells apart any time before the chromosome could have been transferred.

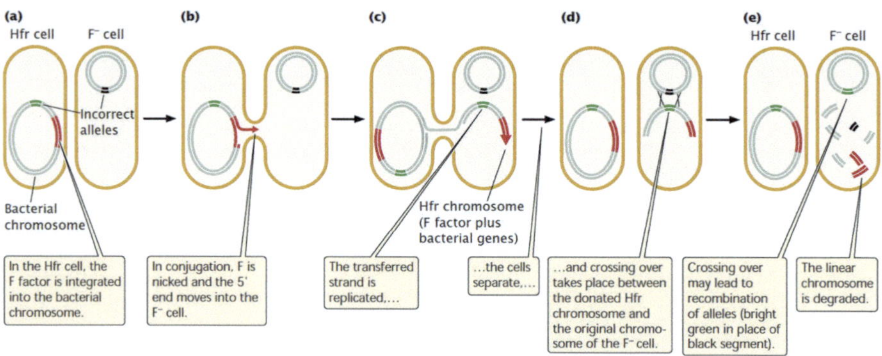

Fig. 6.13 Mating between Hfr cell and F$^-$ cell. Mating takes place between Hfr and F$^-$ cells which have been described in steps (**a**) to (**e**). This process is time taking so it occurs rarely in the nature

6.3.5 Mapping of Bacterial Chromosome

Before inventing genome sequencing, microbiologists had only one genetic approach to elucidate the organization of genes in bacterial chromosomes and that is linkage analysis. Genetic mapping using linkage analysis is a tedious task for them. To date, three methods for mapping have been identified: (1) interrupted conjugation, (2) transformation, and (3) transduction. All of them have important similarities and differences which are discussed in this section.

In bacterial genetics, to map the relative location of bacterial genes, Hfr conjugation method has been used more often. In Hfr conjugation, when Hfr and F$^-$ are mixed in one culture, Hfr strain begins the chromosome transfer immediately. Transfer of chromosome from donor to recipient is not synchronous but occurs at a constant rate and over a period of minutes. The mapping is done by using an interrupted conjugation. Within an interrupted conjugation, a bridge between donor and recipient is broken, and thus matting between Hfr and F$^-$ is interrupted at various intervals. This technique is organized after some time when the conjugation begins in a mixture and is interrupted by vigorous agitation in a blender (Fig. 6.9).

Through this method, the order of gene transfer and time intervals can be interpreted because timing is a direct reflection of the gene order on the bacterial chromosome (Fig. 6.14). In the given graph, X-axis of the curve demonstrates the time point at which the gene began to enter the recipient cell. The demonstrated curve is a circular chromosome map where the distance between every gene is equals the minutes lapsed until the gene is passed to the recipient cell. In the graph, the gene that is more distant from F factor (origin of replication) shows lower plateau compared to gene that is in the vicinity of F factor. In the given example of *E. coli*, trp is the distant gene from the F factor, and thus there is even a greater chance that the conjugation bridge will beak spontaneously before the trp gene is transferred to the cell. To generate the map, it is important to know that the transfer of genes always starts within the point of origin that is the F and thus the orientation and position of F on the chromosome determine the direction of the starting point of

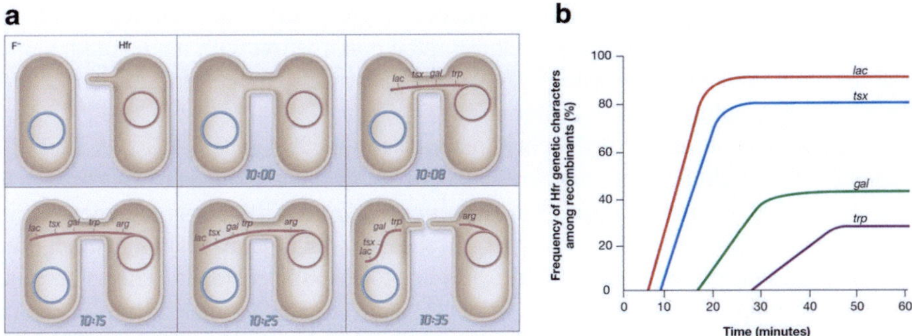

Fig. 6.14 Mapping of the bacterial chromosome by using Hfr and F^- mating system. Hfr and F^- interrupted conjugation experiment at time intervals. (**a**) Schematic diagram in the left panel shows that a linear transfer of genes has been paused and a discontinued conjugation bridge is taken into the consideration of the sequence of gene transfer from donor to the recipient cell. (**b**) In the right panel, the graph shows a relationship between time intervals at which a particular gene has been transferred into the recipient cell obtained by an interrupted conjugation experiment. From the graph, we can see that the gene order is lac–tsx–gal–trp

gene transfer. In the given example, the point of origin (F factor) is just immediately before the *lac* gene in the chromosome. Since the genome of *E. coli* is relatively larger, mapping is quite lengthy through Hfr strain. Therefore, the easy way to do so is to let the several Hfr strains integrate with F plasmid at different locations, and all the fractions of the map obtained through these different locations must be superimposed to create the entire map of *E. coli*. The overall map is adjusted to 100 min in the case of *E. coli*. In this sense, the term "minutes" not literally indicates the measurement of time but the distance between the genes on the map.

Gene mapping through transformation follows a few steps including the separation of DNA from the donor strain, fragmented and integrated to the recipient strain. Loci that are widely separated on the fragmented DNA from the donor chromosome and always carried by two different fragments, then the frequency of cotransformation is different from the single transformants of per10^3 recipient (a normal transformation occurs at the rate of one cell per 10^3 recipient). If the two loci are very close to each other and are carried by one fragment, the rate of cotransformation must be similar to a single transformation rate. Thus, the cotransformation will provide the information of the order of genes on the donor chromosome and will guide to map the genome.

Mapping through transduction is quite similar to transformation and also depends on the gene transfer but between two different bacterial traits. Gene transfer occurs through the bacteriophage. Similar to the transformation, small fragmented DNA will be cotransformed by phage from donor to the recipient strain. Rates of cotransformation in transduction will help to calculate the relative distance between genes and to create a genomic map.

In conclusion, all the three modes of gene transfer—interrupted conjugation, transformation, and transduction—are based on the same basic strategies that are used for mapping. The way of DNA transfer is slightly differing such as through the physical contact between bacteria in interrupted conjugation with interrupted conjugation, small naked fragmented DNA in transformation, and fragmented DNA through bacteriophage in transduction.

Using these techniques, researchers mapped about 2200 genes of *E. coli* K12 and compared this with the actual nucleotide sequence of the genome (i.e., physical map of the genome). Genome sequencing has revealed about 4300 possible genes. Thus, genetic analysis is defined over half of the potential genes. The genetic map approximates the physical map, but they do not correspond perfectly. This is because the genetic map is derived from genetic linkage frequencies that do not correlate exactly with the number of nucleotides that separate the two genes. Roughly speaking, 1 min of the *E. coli* genetic map corresponds to 40 kilobase of DNA sequence.

6.3.6 $F^+ \times F^-$ Mating

It was demonstrated by William Hayes in 1952 that the gene transfer observed by Lederberg and Tatum was one-directional and had performed between polar cells. Therefore, it is predictable that there must be cells having characteristics of a donor (F+ or fertile) and the recipient (F^- or infertile). This type of gene transfer is nonreversible.

An extra chromosome such as F factor in the F^+ strain encoded sex pili which is an essential need for plasmid transfer. Major role of sex pilus is to establish a physical cell–cell contact between the F^+ and F^- mating. Figure 6.11 shows the mechanism through which bacteria can transfer its plasmids such as F to the recipient cells. Once F^+ and F^- cells come into the vicinity, the F plasmid of F^+ strain directs the pili synthesis followed by its projection toward recipient cell to make contact and pull the recipient cell closer (Fig. 6.15). This protruding pilus makes a pore on the recipient cell, and thus F plasmid passes through this pore into the recipient cell. It is notable that during this transfer DNA does not transfer in a double-stranded form while it carries only one strand of F DNA, which initiates replication of the complementary strand in the conjugation tube (basically bridge like structure) which connects both the donor and the recipient cell. This replication concluded in two copies of F DNA, one remaining in donor and one appearing in the recipient cell as shown in (Fig. 6.15).

Replication of the F factor is accommodated by rolling mechanism and replication initiated by the help of a protein complex known as "relaxosome." This relaxosome first recognizes *ori T* (Fig. 6.10) site and nicks one strand from this point. Relax enzyme is a part of this relexosome and remain attached at the 5′ end of the nicked strand. During the replication of the F plasmid, replicated strand displaced and attached relax enzyme move along through the type IV secretion system to the recipient cell. Because pilus is embedded into secretion system, it has been suggested that the DNA moves through a lumen in the pilus.

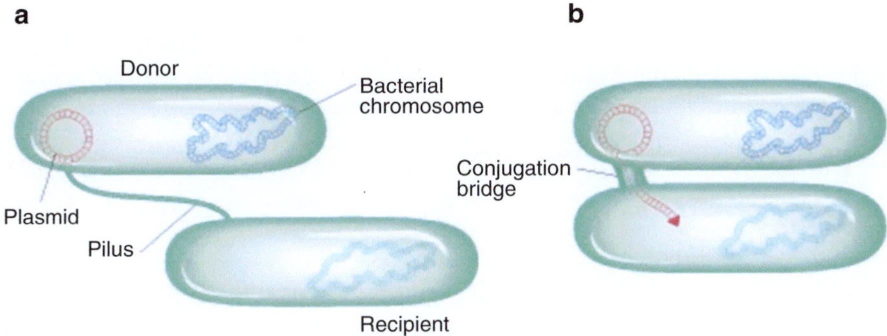

Fig. 6.15 F+ and F– mating system. Conjugation between F^+ and F^- initiates with pilus formation and pulls them together and during conjugation, shown in (**a**). This pilus formation and cell-to-cell contact are further followed by the formation of a bridge or a pore (essential passageway) between two cells. Single-stranded DNA passes into the recipient cell and becomes double strand by rolling replication mechanism as shown in (**b**)

6.3.7 F Plasmid

Plasmid has the capacity to replicate independently and to integrate into the bacterial chromosome. F plasmid is one of the important factors also known as fertility factor that integrates with bacterial chromosome to generate Hfr cells. Homologous recombination site present on either bacterial chromosome or F factor allows repairing and release of F plasmid. Occasionally integrated F plasmid exists from the bacterial chromosome by reverse recombination process. F factor is responsible for mating and gene exchange between bacteria and so the conjugation process is a totally F plasmid-dependent process. Most importantly, the F plasmid contains an origin site for replication and other genes required for conjugation and sex pili formation to make contact with the recipient cell (F^- cell). DNA is always transferred from F^+ to F^-. Thus, F plasmid is essential and needed for conjugation in bacteria. One more important thing about F plasmid is the integration of F plasmid into a bacterial mutant (known as F′) that is unable to replicate known as "integrative suppression." For its own replication, F plasmid uses many *E coli* replication proteins, but it does not use the *dnaA* protein usually required for bacterial chromosome replication. In case if dna A is mutated in bacteria and temperature is elevated to the 42 (on which dna A is inactivated), initiation of replication in bacterial chromosome will not be possible, but F plasmid will replicate since it has its own origin of replication. In *dnaA* mutant strain where the F plasmid is integrated into the chromosome, the replication of the chromosome is still possible and independent at high temperatures. However, replication does not occur at the origin of the chromosome and instead occurs at *oriV*, origin site on the F plasmid. In this way, integration of F plasmid suppresses the importance of *dnaA* as a phenotype by replacing it with F plasmid-derived replication.

Note that it is also tricky to select the F integrated positive strains, and therefore integrative suppression can be a possible way to select those that have integrated F plasmid. In a given example of mating between F′ *Lac*$^{+}$/*trp*^{s} and *Lac*$^{-}$ *(dnaA)/trp*^{r} would result in the colonies to be selected on high temperature and lactose/tryptophan selective medium. Colonies were successfully survived on the medium and high temperature should have contained an integrated F′ *Lac*$^{+}$.

6.3.8 R Plasmid

A bacterial chromosome often regulates antibiotic sensitivity or resistance. Alteration in the pathogenicity of bacteria is thought to occur due to another plasmid that allows the bacteria to grow in the host cell escaping the toxic effects of antibacterial drugs. Antibiotic resistance is a net result of a group of genes located on the R plasmid. Drug-resistant R plasmid has evolved in the past 60 years. The Discovery of antibiotic resistance came into existence after the incident in a Japanese hospital, where the first time an antibiotic resistance has been observed in *Shigella*; a bacteria causes dysentery, which later on with multiple treatments of antibiotics (penicillin, tetracycline, sulfanilamide, streptomycin, and chloramphenicol, etc.) has become resistant to drugs. Similar to the F plasmid, R plasmid contains circular small DNA that has divided into two fragments; one fragment contains a group of gene responsible for replication and copy number, transfer of gene, and sometimes resistance to the tetracycline, and another fragment includes the number of other resistance gene or factor (thus known as R factor or R plasmid). R plasmid serves several genes to act on antibiotic resistance and establish antibiotic resistance in the strains. R plasmid carrying the marker for resistance like chloramphenicol (Cam), streptomycin (Str). Stone, A. B. has mentioned in his study that R plasmid is a major factor that is responsible for the epidemic spread of multiple drug resistance throughout the bacterial population. For example, research shows that plasmid carrying genes for resistance to multiple antibiotics might have been transmitted from a cow udder infected with a hand towel. The practice of milking an infected cow followed by a farmer cleaning his hand with a towel was the reason for the antibiotic resistance passage from bovine to the human–microbe reservoir. This incident shows a severe problem in antibiotic therapy since plasmid transmission is not restricted to particular species or genera, and it may invade the unrelated species or population.

6.4 Transformation

Transformation is another method for gene transfer in bacteria that facilitates a DNA transfer without any donor bacteria or physical contact, rather a naked foreign DNA exogenously being uptakes by the recipient cell. This method suggested that bacterial genetic information is somehow transferable within the bacterial population where physical contact unlike conjugation is not needed. This discovery belongs to the time before when DNA structure was discovered.

6.4.1 Discovery of Transformation

Transformation is the second most abundant mode of gene transfer after conjugation. In the transformation, a naked DNA or partial DNA fragments are uptake by the recipient cell from the environment and incorporated into the recipient's chromosome as a result of recombination.

Transformation introduced a gene transfer method that allows the transfer of one genotype into another by the exchange of exogenous DNA. First-time transformation was confirmed in *Streptococcus pneumoniae* in 1928 by Frederick Griffith. Following this, in 1944, Oswald T. Avery, Colin M. MacLeod, and Maclyn McCarty explained that the "transforming principle" was none other than bacterial DNA in this method. This experiment was the first explanation that reveals the phenomenon like DNA can be transferrable.

Fredrik Griffith's experiment is shown in Fig. 6.16, where he used two distinguished *Streptococcus pneumoniae* strain; one is virulent and has lethal effect on most laboratory animals, and other one is nonvirulent and is not lethal for animals. Virulent strain designated as (S) and enclosed by polysaccharide cover (capsule like structure) and gives a smooth appearance when it grows on the medium so easily can be detectable on culture. Nonvirulent strain designated as (R) strain which is lacking any cover or capsule shows rough appearance when it grows on medium and thus can be detectable on medium. Griffith first boiled some virulent strains to kill them and injected these heat-killed strains into the mice, and he observed that these mice survived, and the cells do not show any lethal effect on the mice. In the next round, a mixture of heat-killed virulent cells and live nonvirulent cells were injected into the mice and mice did die. He isolated the live cells from the dead cells and grew them on the medium, gave the smooth colonies, and showed virulent characteristics on subsequent injection to the other mice. These results made him realize that heat-killed virulent (S) cells somehow converted the nonvirulent live cells (R) into live virulent cells (S) which was the reason for the death of mice (Fig. 6.16).

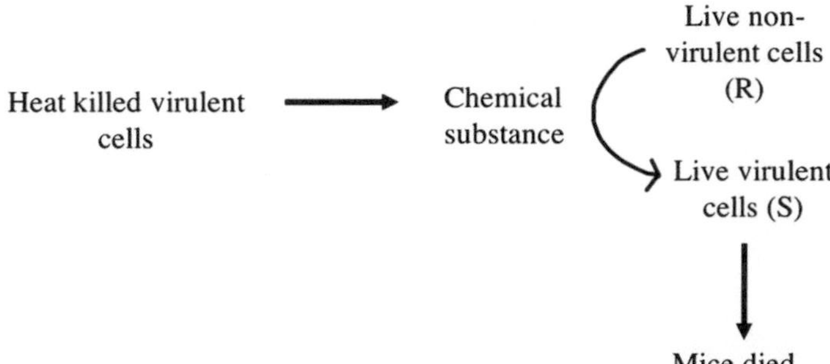

Fig. 6.16 Transformation of the nonvirulent live cell (R) into the live virulent cell (S). An experimental setup explains that chemical substances produced from heat-killed cells someway transferred into live nonvirulent cells and transformed it into live virulent cells (S). This substance can be any biomolecule whose competency is to be transformed is chemically unknown

But unfortunately, Griffith was not able to reason why this had happened and what made the nonvirulent live cells behave like virulent live cells?

Next to this achievement, a question was awaiting that what was the chemical composition of dead donor cells had caused this transformation? Was this protein or any other element? Since it was clear that this substance that had changed the genotype of the recipient cell must be something transmissible or heritable. A group of Oswald T. Avery, Colin M. MacLeod, and Maclyn McCarty solved this doubt when they proposed a study to destroy all the chemical substances of the dead cells and see the transforming capacity of this dead extract from the cell. They found that this mixture still could develop the ability of transformation. However, heat-killed virulent cell contains polysaccharide coat while live nonvirulent does not contain any coating; hence, it is assumed that the transforming agent could be this polysaccharide coat. Destroying polysaccharide coats does not diminish the transforming activity. Proteins, fats, and RNAs show similar results not to be transforming agents.

An interesting result obtained when the mixture was treated with DNase, it lost its transforming ability which strongly suggested that DNA was the genetic material that has transformed from one bacteria (dead virulent) to other bacteria (live nonvirulent), named transformation (Fig. 6.17).

It is noteworthy that natural transformation is a random process between two bacterium and DNA transfers from the donor bacterium to Hfr + bacterium. Any portion of the entire genome may be able to transfer.

6.4.2 Transformation Process

In the previous section, studies on transformation in *S. pneumoniae* have been intensively explained and now it has to be understood how this process takes place? An entire transformation involves (1) the uptake of foreign DNA from the surrounding and (2) its incorporation in the recipient bacterial chromosome or plasmid. Naturally occurring transformation takes DNA from the dead bacterium which might have been released to the environment after fragmentation of the DNA due to the death of the bacterium. Thus, the marine and soil microbiome utilizes this mode of transfer of genetic material at a massive level. A cell that uptakes the DNA is known as a competent cell that is the target for exogenous double-stranded DNA. This DNA was later cleaved and processed by endonuclease into double-stranded chunks of 5–15 kb size. The entire process of transformation is energy dependent. During the DNA uptake process, one strand of DNA is hydrolyzed by an envelope associated with exonuclease while the other strand is coupled with small proteins and move along with the plasma membrane. Thus, only one strand of DNA is available for interaction with recipient DNA. Fragmented single-stranded DNA aligns with the homologous region on the recipient genome and is integrated successfully.

However, transformation in *Hemophilus influenzae*, a gram-negative bacteria, unlike *S. pneumoniae* is different in many aspects of the transformation process. It

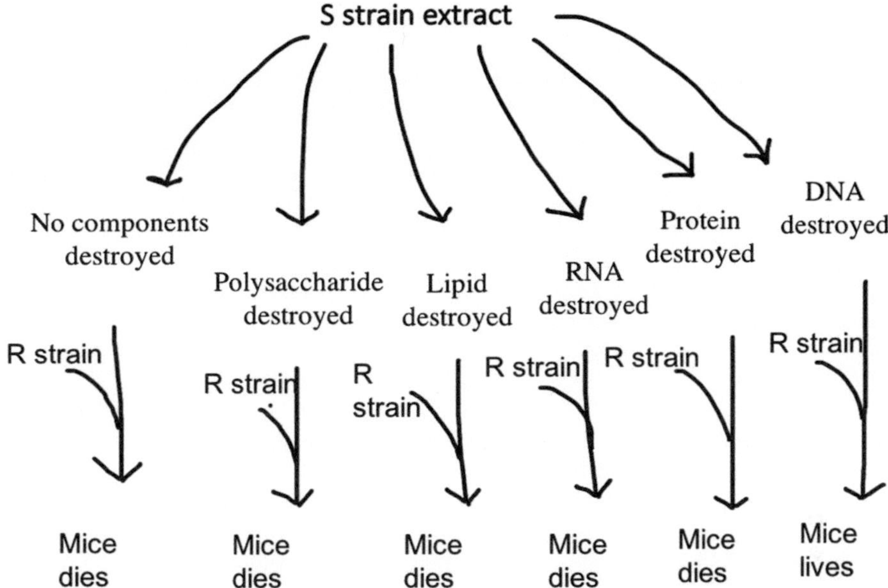

Fig. 6.17 Experimental evidence for the transformation of DNA in mice between different strains. S strain was recovered from the live cells, all the chemical components were destroyed step wise, and the extract mixture was reinjected with the R strain into the mice. As the result, (1) R strain converted into S strain when no component destroyed, (2) R strain converted to S stain when polysaccharide destroyed, (3) R strain converted to the S strain when lipid destroyed, (4) R strain converted into S strain when protein destroyed, and (5) R strain does not convert into S strain while DNA destroyed. Thus, these results confirm that when all the other components were destroyed but not DNA, R strain was not able to be transformed into virulent death causing lethal S strain, and destroying the DNA makes the R strain incapable to transformed into s strain and mice left alive

does not produce competence factor to make cells competent and it takes up DNA from closely related *S. pneumoniae*; thus, the case of *H. influenzae* transformation is a selective method while in the case of *S. pneumoniae* is less particular about the source of DNA.

It has been demonstrated that the specificity of *Hemophilus influenzae* transformation came from the 11 base pair sequence, 5′AAGTGCG-GTCA3′, which exists in a repeat over 1400 times in DNA, and this repeat must bind with competent cells. Nonetheless, transformation is not restricted to only a few selective bacterium DNA rather a DNA source for a competent call can be anything in an appropriate condition (Fig. 6.18).

However, recombination or crossover is common in bacteria but efficiently DNA uptake is limited. Even it has been seen that a species is capable of transformation, but a very small fragment of DNA is transformed in a growing population of bacteria. For a long time, bacterial geneticists were in the search of developing a novel technique to increase the transformation frequency that may enhance the DNA uptake into the cell. Hence, transformation supplemented with an artificial add-on

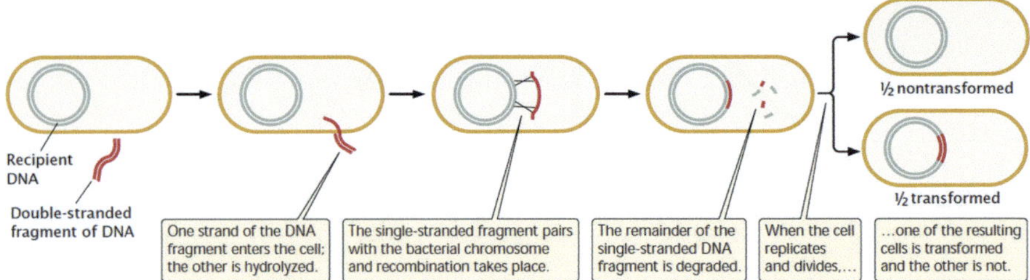

Fig. 6.18 Gene can be transferred between bacteria through transformation. Transformation in the bacterial cell initiates with DNA uptake and its integration to the bacterial chromosome that procures a recombinant DNA that contains both bacterial and exogenous DNA intact in the daughter bacterial cell

such as calcium chloride in the medium, heat shock, or electric field makes the cell membrane more porous and permeable to uptake DNA more efficiently. Increasing the DNA concentration will also work as an enhancer for transformation efficiency. These enriched techniques for the transformation of foreign DNA into any cell are vastly in practice for molecular biology studies in laboratories.

6.4.3 Transformation Linked Genes

Likewise, conjugation transformation has been used to map the bacterial gene, especially for those species that do not undergo conjugation or transduction. Mapping in such strains is only possible when they are entirely different genetic strains or traits that have to mate through the transformation. For example, a recipient strain might be auxotrophic for three nutrients $p^- \ q^- \ r^-$ (in the figure), mate with donor cell, and prototrophic for alleles $p^+ \ q^+ \ r^+$ (in the figure). DNA from the donor strain is treated and fragmented to increase competency. Fragmented donor DNA is added to the medium of the culture of recipient strain (competent cells). Eventually, fragmented donor DNA enters the recipient cell and immediately undergoes recombination. The recombination process must be followed by a homologous sequence on the recipient bacterial chromosome where the donor DNA is attached and intact throughout the procedure. Recipient cells that positively have received the genetic material from the donor cells through transformation are called "transformed."

How to understand that how many genes and what frequency they have transformed? To this end, we first need to observe the rate at which two or more genes are transferred together (usually termed as cotransformed). The recombination rate of these genes is the basic need for the measurement of transformation frequency. Gene can be mapped by observing the rate at which two or more genes are transferred together (cotransformed) in transformation. We assume that genes that are physically close to each other on the same DNA after fragmentation are more likely to be transformed contiguously into the competent cell. For example, in Fig. 6.19, gene p and q on the DNA of donor strain are physically linked so that

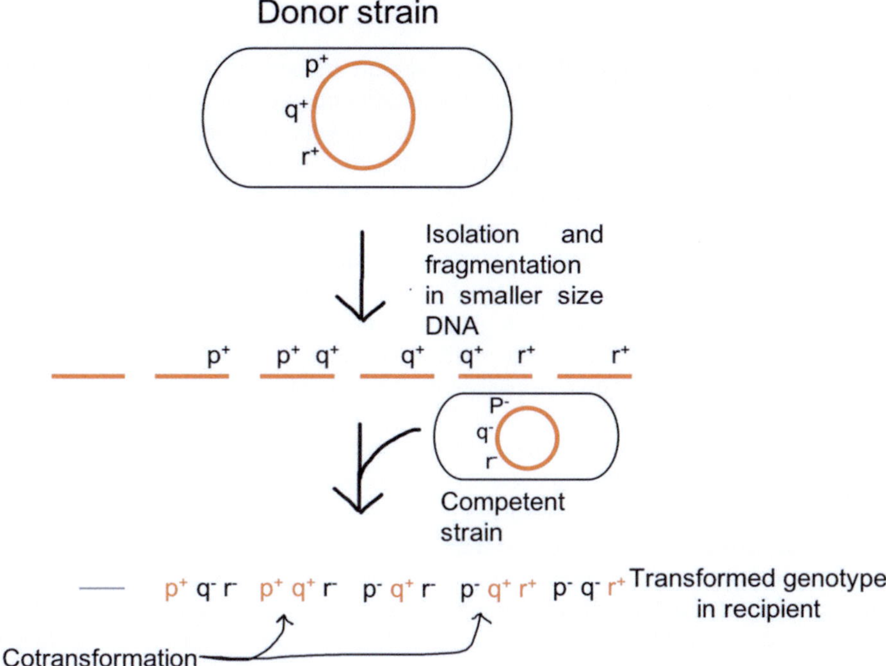

Fig. 6.19 Transformation and linkage for mapping the bacterial genome. Gene p and q are close enough to be transformed together, gene q and r are also close enough to transform together and therefore genotype observed as (1) $p^+ q^+ r^-$ and (2) $p^- q^+ r^+$ are cotransformed. Note that $p^+ q^+ r^+$ and $p^+ q^- r^+$ are rare genotypes because p^+ and r^+ are two distant genes transformed together, and thus, the rate of cotransformation is inversely proportional to the distance between the genes

they would preferably transform together. However, genes that are far apart are unlikely to be present on the same DNA fragment and rarely will be transferred together. In Fig. 6.19, we can observe that gene p and r are separated from each other and no fragments are produced containing $p^+ r^+$ on the same DNA and therefore we have not observed cotransformed for $p^+ q^- r^+$ which is the rarest event.

Thus, after performing the transformation, transformed colonies must have been obtained on selective media and performed the genotyping of each strain. Let us assume that if gene p and q are frequently cotransformed and gene q and r frequently cotransformed, then gene q must be in between p and r and the gene order on the DNA must be *p q r*.

6.5 Transduction

We have already spotted a light on bacterial gene transfer in the previous section. Transduction is the third type of gene transfer after conjugation and transformation. So, let us take a closer look at the third way of gene transfer in which gene is transferred between bacteria through the phage/bacteriophage. Transduction is a type of horizontal gene transfer that occurs naturally via phage infection.

Based on structure, the virus is simple, often composed of just a nucleic acid, and the genome is always protected by a protein coat known as a capsid. Phage does not replicate autonomously; instead, they first infect the cell, take control of host machinery, use it, and force the host cell to produce multiple copies of phage particles. Most phages initiate their replication immediately after infection. When a phage begins the replication, it reaches a certain number of copies or let us say a threshold number of copies that cause bursting of cells and produces many new phage progenies to further infect new bacterial host, such phages known as a virulent and the process is called "Lytic cycle." Few bacteriophages do not kill the bacteria immediately after infection and instead insert their genome into the bacterial genome without affecting it. The inserted phage called **prophage**. These bacteriophages passively replicated as the host cells genome does and thus this bacteriophage is known as **temperate bacteriophage** and the relationship between phage and its host cell is called **lysogeny** (Fig. 6.20). Temperate phages can remain inactive as an insert for many generations in their hosts. However, they are prone to some conditions like UV irradiation which can induce lytic cycle temperate phages.

Transduction is categorized into two: (1) generalized transduction in which any gene may transferred and (2) specialized transduction where only few genes are possibly transferred. How this transduction has been identified by researchers? It is discussed in the next section.

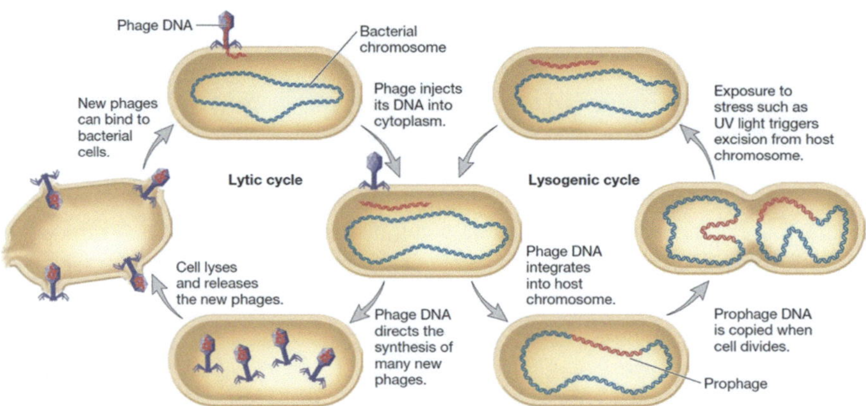

Fig. 6.20 Transduction comprised of lytic cycle and lysogenic cycle. A typical transduction process has both lytic and lysogenic cycles which either follows one each after or remains separated as shown in the image

6.5.1 Discovery of Transduction

Joshua Lederberg and Norton Zinder in 1951 were testing the recombination in *Salmonella typhimurim* by using the same techniques as Lederberg had found in *E. coli.* For their experiment, they used two distinguished strains of *Salmonella*; one was auxotroph for phe$^-$, trp$^-$, tyr$^-$ and the other was auxotroph for met$^-$, his$^-$. They mixed these two strains in one culture and plated on the minimal media, then similar to the E.coli experiment, they did not observe any wild-type strain When these two strains were plated on minimal medium, no wild-type cells were observed; however, at a low frequency of about $1/10^5$, recombination was observed. In the discovery of transduction, researchers referred to the U tube experiment with few modifications. They put the porous filter instead of the fritted filter paper in the conjugation experiment to prevent cell–cell contact. Later they observed that the agent responsible for the recombination is the size of the phage of P22 which is a known temperate phage of *Salmonella*. Furthermore, many studies together suggest that the vector for recombination is a P22 phage. But there was uncertainty among the researchers whether this filterable recombination agent is a phage or something else. Therefore, a comparison of the properties of this agent with phage, where it shows the similarity in the size, sensitivity to the antiserum, immunity to the hydrolytic enzyme, and so on has confirmed its virus alike characteristics.

As a result, Lederberg and Zinder have confirmed this new type of gene transfer through the virus and named it transduction instead of conjugation.

However, in the lytic cycle, sometimes phages interact with host DNA and integrate their own DNA that is then transferred to another host bacteria and insert its contents into bacterial DNA. Both temperate and virulent phages follow the transduction method to transfer the genes.

6.5.2 Generalized Transduction

Besides understanding the transduction further, it came with the next question that how transducing phages are reproduced after infection? To address this question, in 1965, K. Ikeda and J. Tomizawa had discussed the experiment on temperate phage P1 in *E. coli*. They highlighted in their experiment that when P1 infected the *E. coli* and lysed the donor cell where bacterial chromosome was broken up into small fragments; however, some of these pieces were captured mistakenly by phage particles and assembled in the head instead of phage DNA. Eventually, this has become the source of transducing phage.

During the infection, phage capsid (coat proteins) determines phage's ability to recognize and attack the host bacteria and transfer its content to the host cell. But now in the case of transducing phage, the transferrable material is the donor's chromosome which the phage had assembled during the transduction. Interestingly, transduction through the transducing phage could rise the situation of merodiploid (a partial diploid bacteria) since it transferred the donor's chromosome which recombined with the recipient chromosome and now the recipient cell will have two bacterial chromosomes which is a merodiploid situation (Fig. 6.21). This type of transduction allows the passage of any kind of host (bacterial) markers to other bacteria and is thus known as generalized transduction where any gene can possibly be transferred to another bacteria.

Phages P1 and P22 belong to the group that shows features of generalized transduction. Looking at their cycle, P1 phage is usually integrated to the host chromosome while P22 remains free in the cytoplasm.

6.5.3 Specialized Transduction

Unlike generalized transduction, specialized transduction will allow a few particular genes (few host markers) to be transduced as shown in Fig. 6.22.

λ phage is very well known for specialized transduction. λ prophage (shown as the red circle in the figure) integrated between gal and bio region of host chromosome (shown blue linear strand in figure) as shown in Fig. 6.23. Hence, λ can transduce only the gal and bio genes. Let us understand the mechanism of λ transduction.

Recombination between λ phage and host chromosome is catalyzed by a specific enzyme system. This enzyme guides λ phage to integrate at the same site as on chromosome which is denoted in the diagram with *att (*λ attachment site). This site also decides the lytic cycle and ensures that prophage excise at the correct point to produce a normal circular λ chromosome (see the figure, top left—a normal excision). However, errors in excision can occur as abnormal excision (see the figure top right abnormal excision). Under this type of excision, the prophage gene during the lytic cycle becomes circular but contains a nearby gene of the host chromosome and leaves behind some phage gene intact to the host chromosome (Fig. 6.23).

If we see the figure, we can assume that λ has a nearby gene gal on one side while bio on the other side. Resulting phage λ is now referred to as defective particle because it left gene behind on the host chromosome and carry either partial gal known as λdgal (λ-defective gal) or partial bio or λdbio. The resulting particles are defective due to the genes left behind and others transferred from the chromosome are referred to as λdgal (λ-defective gal) or λ dbio. This defective phage gene can be packaged into head and infect another host bacterium. In the second round of infection, λdgal gene will integrate to the λ attachment site on the second host chromosome (Fig. 6.23). This way, gal gene from the first host will transduce to the second host. It should be noticed that this type of transduction carries a few limited genes very near the original integrated prophage, so very few genes are able to transfer and are called specialized transduction.

6.5.4 Lambda (λ) Genetics

Lambda phage is the most extensively studied bacteriophage among all the other phages. Lambda phage is an important model system for latent infection of

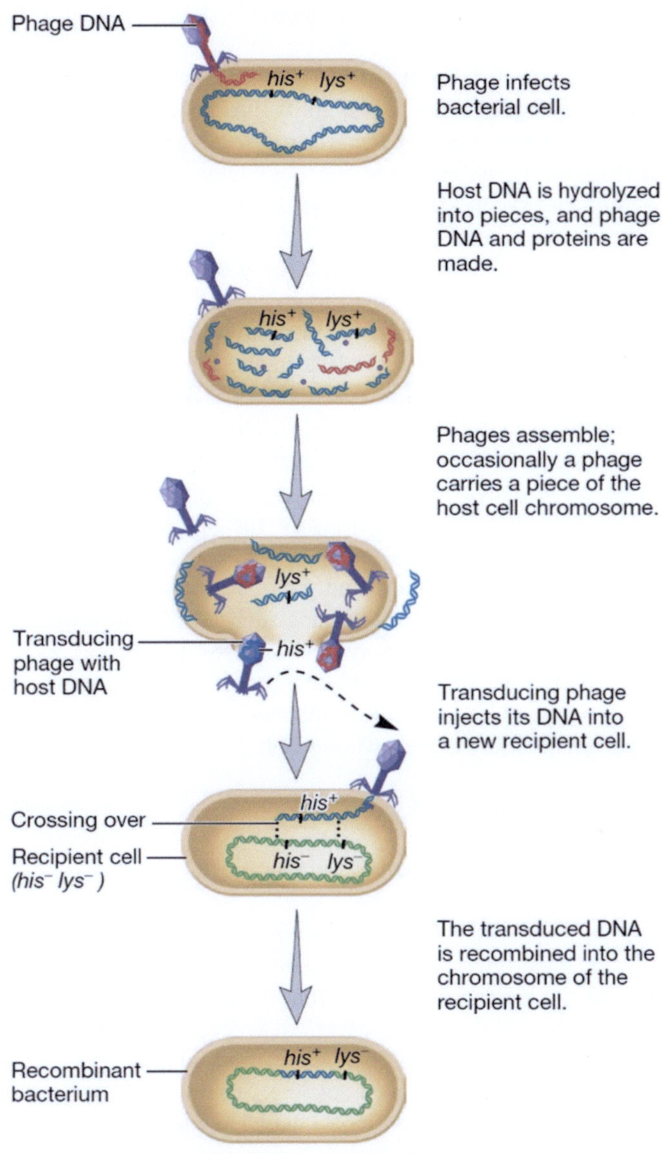

Fig. 6.21 Generalized transduction in bacteria. Generalized transduction is processed with infection of phage to the host cell and releases its genetic material into the cytoplasm. Once the phage infects the host cell, simultaneously host DNA hydrolyzed and synthesized partial phage DNA and proteins. Assemble phage contains a small fraction of the host bacterial chromosome. This phage further transduces and infects other host cells where the crossing over between phage DNA and bacterial chromosome takes place. In the above example, crossing over between bacterial

mammalian cells by a retrovirus, and this model system has been widely used for cloning purposes.

Lambda is the prototype of a group of phages that is a well-characterized virus with both lytic and lysogenic alternatives to its life cycle.

DNA inside the phage is linear but it circularizes on the infection to *E. Coli* chromosome (Fig. 6.24). At each end are complementary 12 bp long overhangs known as cos sequences (cohesive ends). Once inside the *E. coli* host cell, these pair up and the cohesive ends are ligated together by host enzymes forming the circular version of the lambda genome. Lambda can only be compatible for packaging of genome size of 37–52 Kb and also small fragments of extra DNA can be packaged into lambda genome without hindrance. Although accommodate longer insert, some of the lambda genomes must be removed. In the lambda genome, the left-hand region has essential genes for structural proteins while the right-hand region consists of genes responsible for replication and lysis. Cro has been believed to play an active role in switching lysogenic cells to the lytic state following induction. However such lambda replacement vectors cannot integrate to host genome and form lysogeny by themselves. The Middle region is necessary for integration and recombination (Fig. 6.22).

This lambda phage has made many undiscovered questions easy for a scientist to address and to develop advanced techniques such as how to sequence DNA and discovered essential enzyme for RNA synthesis. Also, studies on Lambda phage led to the discovery of (1) basic molecular biology principles of how gene transcription is halted with rho-dependent termination manner, (2) first transcription factor, and (3) gene regulation including "Operon" concept.

6.5.5 Nature of Transduction

Nature of transduction reflects here the question to understand the nature of prophage or prophage–host interaction. As sooner as the prophage is induced immediately after the infection, as more it will produce the prophage and the genome will be restored in the prophage. Before its discovery, it is necessary to understand whether a phage is a mere small invisible particle or a plasmid that lives in the bacterial cytoplasm or a part of the bacterial chromosome. From the past, it has been observed that a temperate phage lambda (λ) promotes the lysogenic cycle in its particular host bacterium *E.coli* used by Lederberg and Tatum (mentioned earlier). So far, the studies on lysogenic cycle of phage λ have introduced the λ phage as a first preference to refer for lysogenic and well-characterized known phage. If we look at the crosses between $F^+$and F^- cells obtain interesting results such as $F^- \times F^+ (\lambda)$ gives recombinant lysogenic recipient while $F^+ (\lambda) \times F^-$ results in nonlysogenic

---

Fig. 6.21 (continued) chromosome ($his^- Lys^-$) and phage DNA (his^+) results in $his^+ Lys^+$ positive recipient host cell. They again cross over

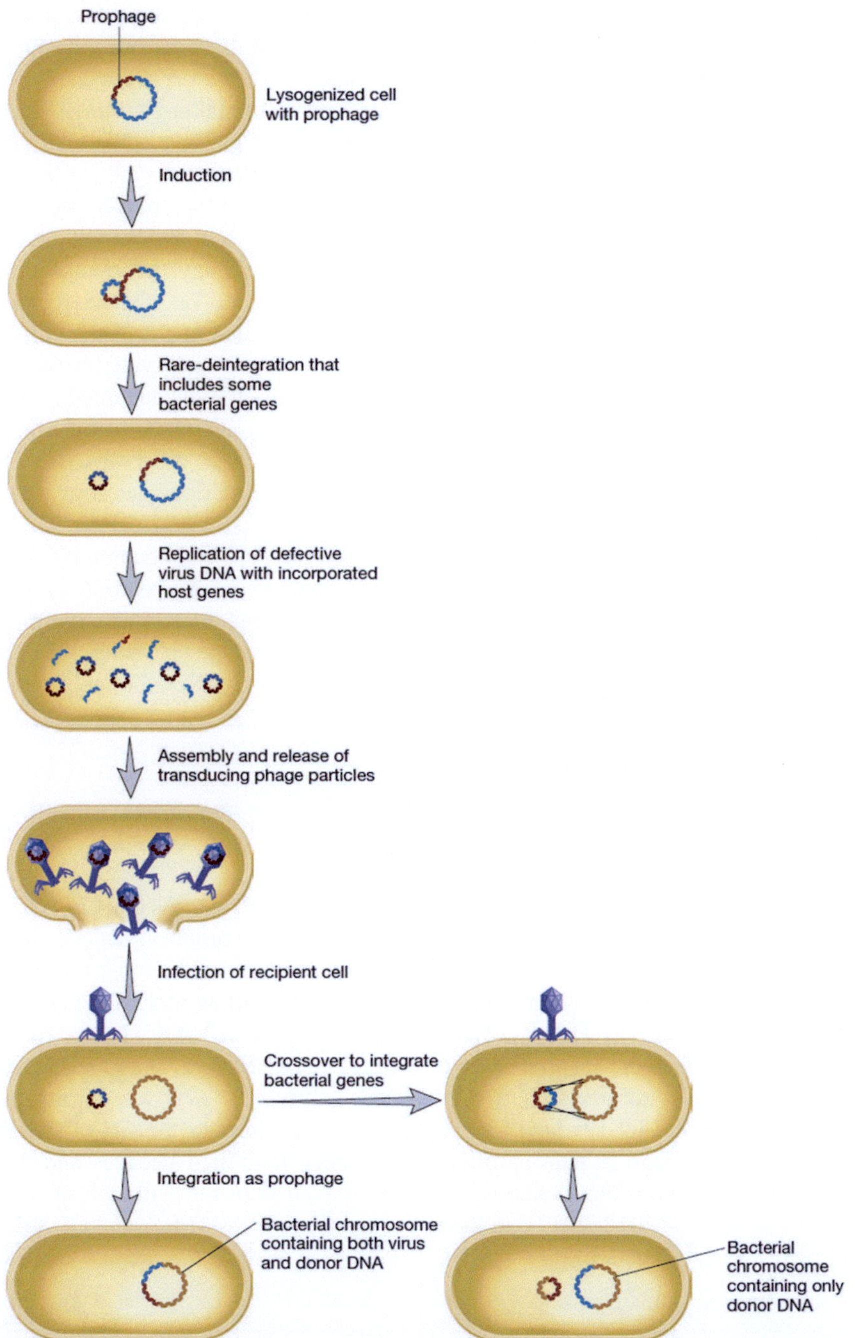

Fig. 6.22 Specialized transduction in host bacteria. In specialized transduction, prophage that contains some bacterial gene disintegrates on the specific induction. The disintegration of prophage produces a new circular chromosome in the same host cell. Replication in the host cells took place immediately and was followed by the assembly of new phage that released and infected other host

recombinants. These results were of more importance when Hfr strain had discovered and was used for crosses. If crosses happened between Hfr × F^- (λ), lysogenic F− exconjugants readily recovered with Hfr genes.

If a cross occurs between Hfr (lysogenic or lambda containing strain) with F^- (non-lysogenic or nonimmune recipient), entry of lambda prophage into non-lysogenic cell will immediately trigger the prophage into the lytic cycle. This process is known as Zygotic induction. On the other hand, if the cross between Hfr (λ) × F−(λ) occurs, the resultant recombinants are readily recovered, and no prophage lytic cycle occurs. From this observation we can say that cytoplasm of F− cell must have interchanged between two states (depends on whether recipient contains λ prophage). So when the recipient cell is a nonimmune cell, the entry of prophage will induce the lytic cycle. This cytoplasmic state is specifying the fact that prophage represses the multiplication of the virus and therefore when lambda infects the nonlysogenic cell, these prophage repressing factors are diluted immediately after the infection and thus virus will multiply and reproduce the progenies. But what if a virus specifies the repressing factor and why the virus does not shut off the replication of itself?

The answer is, it does because a fraction of infected cells can become lysogenic (prophage). But there is a race between lambda gene signal for reproduction and repressor specify signal to shut down the replication. In this way, a phage-directed cytoplasmic repressor model explains the immunity of lysogenic bacteria. In conclusion, a superinfected phage will immediately encounter the repressor and become inactivated.

6.6 Infectious (Bacterial/Viral) Disease

Coronavirus (CoVs) is a large family of single-stranded RNA virus that can infect a wide variety of animals including humans, causing respiratory, enteric fever, hepatic, and neurological disease. In human coronavirus mainly causes respiratory tract infection. Till date, six coronaviruses have been identified including (1) alpha-CoVs HCoV-NL63, (2) HCoV-229E, (3) beta-CoVs HCoV-OC43, (4) HCoV-HKU1, (5) severe acute respiratory syndrome-CoV (SARS-CoV), and (6) Middle East respiratory syndrome-CoV (MERS-CoV). However, a novel coronavirus, SARS-CoV-2 (COVID-19) lately in 2019, has been added to the list of existing coronavirus. Although human coronavirus has been identified decades ago, their clinical and epidemic importance was not recognized until the outbreak of SARS (2002) and MERS (2012–2017). In the next section, SARS, MERS, and COVID19 will be discussed in detail.

Fig. 6.22 (continued) cells. Multiple crossing overs between prophage and bacterial chromosome result in (1) bacterial chromosome containing only donor DNA and (2) bacterial chromosome containing both viral DNA and donor DNA

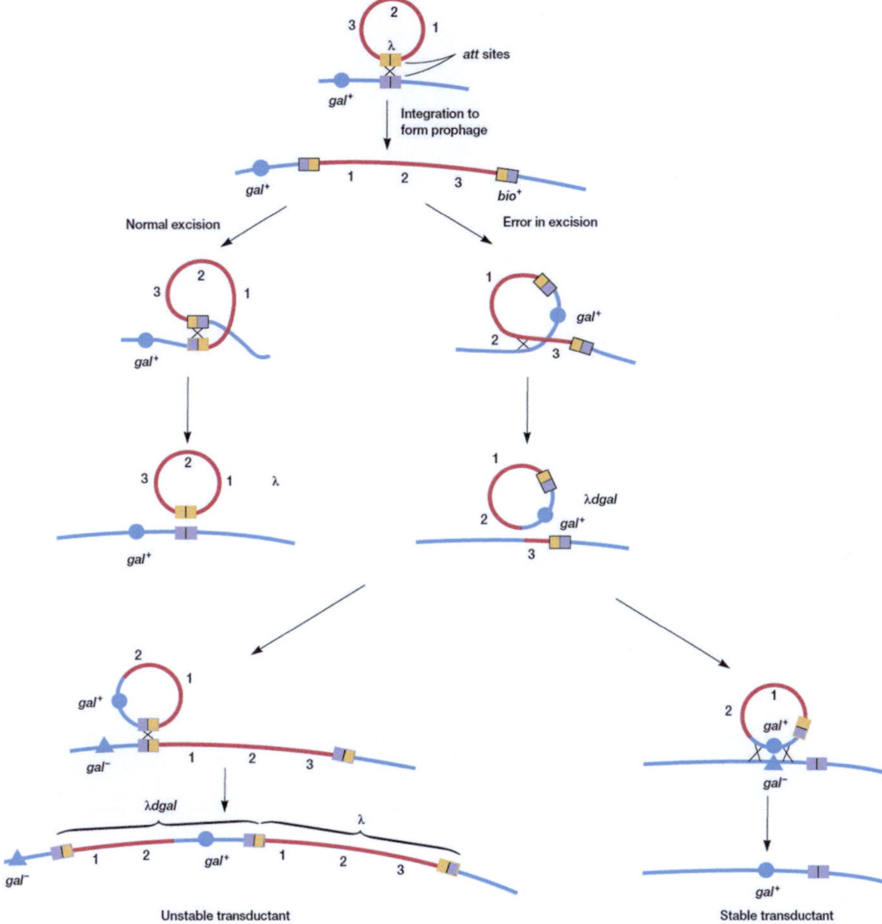

Fig. 6.23 The mechanism of transduction for phage lambda and *E. coli*. Integrated phage lies between *gal* and *bio* genes. When a normal excise occurs (top left) new phage is complete and does not contain any bacterial gene. While rare excise occurs (top right), either the *gal* or the *bio* genes are picked up by bacterial genes and some are lost. As a result, a defective lambda phage that contains a bacterial gene can transfer to the new host cell

6.6.1 SARS-CoV

SARS CoV virus was identified in 2003. SARS-CoV is a zoonotic disease and was thought to be an animal virus from an uncertain animal reservoir, like bats and civet cats, and found to be the first to infect humans in the Guangdong province of south China in 2002 (Fig. 6.25). However, these animals were only incidental hosts, as there was no evidence for the circulation of SARS-CoV-like viruses in palm civets in the wild or breeding facilities. Studies reported that bats are the reservoir of a wide variety of coronavirus including SARS-CoV-like and MERS-CoV-like virus.

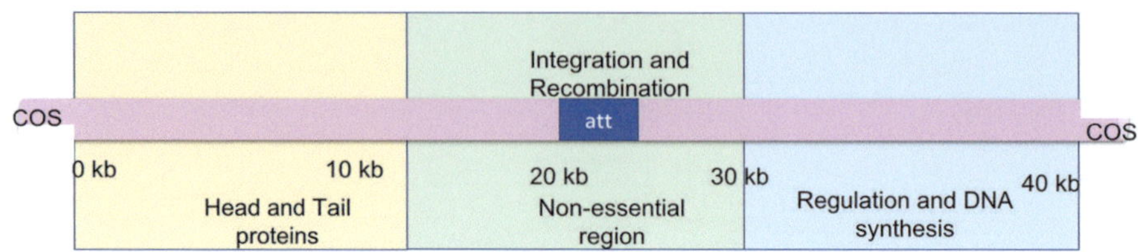

Fig. 6.24 Lambda replacement cloning vector. Lambda phage is easy to grow and therefore it has been modified to accept foreign DNA inserts. Both left and right ends are overhangs with cohesive regions known as COS and regulate the circularization of the DNA. The green region has genes that are nonessential for lambda growth and packaging but can be replaced by the foreign DNA insert (up to 23 kb) during the cloning. The yellow region codes for proteins essential for head and tail packaging

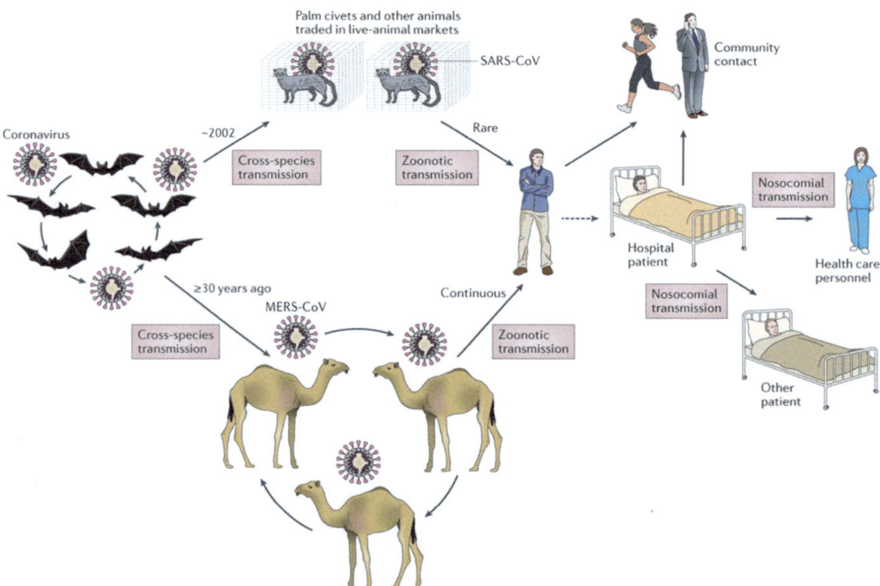

Fig. 6.25 Insights into SARS and MERS infection Cycle: SARS-CoV crossed the species barrier into masked palm civets and other animals in live animal market in China responsible for SARS-CoV infection occurred in late 2002. Later in 2012, the cross-infection in dromedary Camel was identified as MERS-CoV infection in the Middle East. SARS-CoV and MERS-CoV spread between humans mainly through nosocomial transmission, which results in the infection of healthcare workers and patients at a higher frequency than infection of their relatives

Transmission of this disease was primarily spread from person to person and appeared to occur in the second week of illness where excretion of virus in respiratory secretion and stool was on the peak. Lately, most of the cases of human-to-human transmission were due to negligence in the healthcare setting, absence of adequate infection control, and precautions. Symptoms of SARS include influenza-like fever, malaise, myalgia, headache, diarrhea, and shivering. Cough was initially dry, and shortness of breath and diarrhea are most prominent in first or second week of infection. Mostly severe cases developed rapidly, progressing to respiratory distress and required intensive care. Unlike COVID-19, SARS transmission counted as an epidemic since the geographic distribution was limited such as Toronto in Canada, Hong Kong, China, Chinese Taipei, Singapore, and Hanoi in Viet Nam, thus SARS infection.

6.6.2 MERS-CoV

Ten years after the emergence of SARS, in June 2012 a man in Saudi Arabia died with acute pneumonia and renal failure, which lately had identified infected with a novel coronavirus named Middle East respiratory syndrome coronavirus (MERS-CoV). MERS was also identified outside of Arabian Peninsula such as Jordan and United Kingdom, as a result of traveling; often, these imported MERS cases resulted in nosocomial transmission (Transmission usually occurs via healthcare workers, patients, hospital equipment, or interventional procedures). In the case of MERS, serological tests from dromedary camels from Oman and Qatar camel farms confirmed its transmission from camel to human first in Arab peninsula and later in the Middle East, Eastern Africa, and Northern Africa. Like SARS, MERSS causes acute respiratory syndrome as well which is associated with the upregulation of proinflammatory cytokines and chemokines. Immune response to SARS and MERS during the infection plays a key role in its spread since SARS-CoV and MERS-CoV use several strategies to avoid the innate immune response.

6.6.3 COVID-19

The 2019 novel coronavirus, also known as COVID-19 or SARS-CoV-2, is a novel human coronavirus that has emerged at the end of December 2019 in Wuhan, China. It is currently spreading all over the world in the form of a pandemic. Coronavirus disease (COVID 19) is a clinical syndrome associated with SARS-CoV-2 infection and is characterized by severe respiratory syndrome.

SARS-CoV-2 belongs to the same *Betacoronavirus genus* as the other coronaviruses that are responsible for severe acute respiratory syndrome (SARS-CoV) and the Middle East respiratory syndrome (MERS-CoV). Phylogenic analysis of novel coronavirus suggested that SARS-CoV-2 belongs together with SARS-CoV and Bat SARS-like coronavirus which is a different clade from MERS-CoV but more phylogenetically related to Bat SARS-like coronavirus (isolated from horseshoe Bats between 2015 and 2018) than SARS-CoV. Genomic comparison between SAARS-CoV-2 and SARS has shown that there are almost 380 amino acid substitutions and mainly contained in structural protein, and 27 mutations were found in viral spike protein (S) which is responsible for receptor binding and cell entry. It is assumed that due to this mutation, SARS-CoV-2 is less pathogenic than

Table 6.2 Overview of SARS-CoV-2, SARS-CoV, and MERS-CoV

Strain	Phylogenetic origin	Animal reservoir	Intermediate host	Receptor	Case fatality rate
SARS-CoV-2	Clade I, cluster IIa	Bats	Controversial	Angiotensin-converting enzyme 2 (ACE2)	2.3%
SARS-CoV	Clade I, cluster IIb	Bats	Palm civets	Angiotensin-converting enzyme 2 (ACE2)	9.5%
MERS-CoV	Clade II	Bats	Camels	Dipeptidyl peptidase 4 (DPP4)	34.4%

the SARS-CoV; however, further studies are still going on to understand its pathology and its contribution to other diseases like cancer (Table 6.2).

Similar to SARS-CoV, a mutation in the receptor-binding domain (RBD) of S protein in SARS-CoV-2, which directly interacts with human cell receptor—angiotensin converting enzyme 2 (ACE-2)—is thought to be the cause for its pathogenicity. Interestingly data on affinity analysis confirms that SARS-CoV-2 binds to ACE-2 more efficiently than SARS-CoV strain from 2003 although less efficiently than the 2002 strain.

ACE-2 is an ectoenzyme (an enzyme that has the catalytic site outside the plasma membrane and is mostly found in the endothelial cell) occurring in many tissues including, the lower respiratory tract, kidney, heart, and gastrointestinal tract. In vitro studies on SARS-CoV-2 show that inoculation of 2019nCoV (SARS-CoV-2) on the surface layer of human airway epithelial cells causes cytopathy effects and cessation of cilia movement. SARS-CoV induces downregulation of ACE-2 in lung epithelium, but SARS-CoV-2 shows higher affinity to ACE-2 and results in more severe lung infection than the SARS-CoV.

From the clinical aspects, virus loads are higher at the time of symptoms onset and are higher in the nose than the throat specimens which is why it is suggested to collect the patient specimens from the nose. In patients affected by COVID-19, viral loads progressively decrease within days following a different pattern from SARS, in which the highest shedding is recorded after 10 days of symptoms onset. Therefore, it has been suggested that SARS-CoV-2 can easily spread within the community than SARS even when mild or no symptoms (asymptomatic) are present. To date, there are no vaccines and drugs developed which can directly target COVID-19, but many clinical trials for vaccines are underway in different parts of the world, namely, Serum India technology (India), Oxford University (UK), Moscow's Gamaleya Institute (Russia), and AstraZeneca (USA).

Antiviral drugs including ribavirin, lopinavir, ritonavir, and remdesivir in combination with other drugs like chloroquine, cyclophilin chlorpromazine, loperamide, and cyclosporine A has been reported effective in some cases. In addition, antibody therapies and plasma therapies have been the leading proposed treatment in the case of MERS. Recently, plasma therapy has shown a potential advantage in COVID-19 treatment. Many countries such as India, United Kingdom, and United States are one of those who are successfully running a trial for plasma therapy and antibody therapies for COVID-19.

Box 6.1 Scientific Concept: Genetic Exchange Between Escherichia coli Strains in the Mouse Intestine—Jones R T et al.

Bacterial genome usually possesses the phenomenon of gene flow between species. Gene flow largely contributes to the evolution of bacterial species; however, the significance of gene flow is still unknown. In vitro conjugal transfer of genes in *E. coli* K-12 and some other species has been widely studied and also revealed that plasmids harbored by bacteria from the natural environment are continuously increasing. Such plasmids may introduce the determination of colicin production, multiple drug resistance, surface antigen, enterotoxin production, suppressor and mutagenic factors, hemolysin production, and so on. Some of these plasmids are enabled to conjugally transfer to the recipient cells and some of these has been identified as a fertility factor that supports the transfer of chromosomal gene at low frequency. Considering the abovementioned prospective gene flow among bacterial species in nature or in ecological niches, it has been observed that the study of gene transfer in some particular niche has become a difficult task due to the microbial complexity of the niche where these species are likely to occur. One of the common niches is the mammalian intestinal tract where enormous microorganism species live together and are associated within the tract facilitating the genetic transfer. In the given niche, *Enterobacteriaceae* is the common inhabitants where they possess the genetic transfer very often. So far, many studies have been focused to understand the in vivo transfer of genetic material between the *E. coli* K12 (Fig. 6.26) and the formation of recombinants. K12 strain is genetically compatible to investigate the in vivo gene transfer and to avoid the microbial contaminant, and germ-free mice have been used.

Recombination frequencies between three donor strains from in vitro liquid mating, which have been selected further for in vivo studies and the F^- recipient ×820 are shown in Table 6.3.Cross between $F^+ \times F^-$ gave the recombination frequencies of 5×10^{-5} for any maker while the cross between Hfr × F^- gives high frequencies for proximally transferred markers and decreased for markers situated at increasing distance from the origin of the chromosome transfer.

Recombination frequency for *pur E* marker (1.2×10^{-1}) is slightly lower than that for the more distantly transferred *pro C* (2.4×10^{-1}). It is because of integration of markers located next to the origin of chromosome transfer (i.e., *pur E* locus in Hfr strain OR74) is always lower than more distantly transferred markers.

Thus, proximally transferred markers showed high frequencies of inheritance, which further decreased as the distance from the origin of the

(continued)

Box 6.1 (continued)
chromosome transfer increased. Thus, these data show that the donor or recipient strain for in vivo experiments behaved in the predictable manner as they do in in vitro mating.

Therefore, gene transfer between microbial species is quite difficult to understand since the ecosystem is more complex. By creating a simplified ecosystem of microbial species through contaminating the intestine tract of mice with genetically well-characterized bacterial strains is one of the easy ways to study the gene transfer between two species. In this study, Hfr strain OR74 was chosen and maintained in the intestine of mice as its in vitro stable. Forty days after inoculation, F^- and F^+ strain/clone was isolated in vivo. The transition of Hfr $\Leftrightarrow$ F^+ $\Leftrightarrow$ F^- is unpredictable in in vivo and in vitro and will not remain the same during the gene transfer. The reason might be the influence of chemical, selection, and environmental factors. The study of recombinants in the natural environment based on selection and survival rate plays an important role to understand the relationship between its in vivo host and gene transfer pattern. It would be interesting to set up the experiments in vivo to compare the mechanism of gene transfer and formation of recombinants in animals similar to its natural habitats. The complexity of gene transfer among the microbial ecosystem in nature can be possible to study by performing in vivo experiments on well-characterized bacteria and produce new germ-free host animals.

Box 6.2 Scientific Concept: Role of Pili in Bacterial Conjugation—Ou, J. T et al.

Bacterial conjugation has been discovered in 1946 by Laderberg and Tatum where they have proved that cell–cell contact is required for the transfer of bacterial chromosomes from a donor to a recipient. Later, in 1958 Anderson et al. have provided the electron micrograph images of conjugation bridge formation between mating pairs which confirmed the hypothesis of Laderberg and Tatum. To understand the role of F-pili to F-factor and male-specific phages, Brinton proposed that F-pili not only serves in pair formation but is also important for tube formation through which bacterial chromosomes are transferred. Evidence from various studies and experiments suggested that F-pili connect the male and female and then bring them a close wall to wall contact by retracting the F-pili into male or female cells. To obtain clear evidence, microbiologists further studied the role of F-pili by isolating mating pairs by using a micromanipulator (Fig. 6.27). They compared the production of recombinants from loosely connected pairs with the ones produced by

(continued)

Box 6.2 (continued)

mates who were in intimate connect. Results from this study suggest that F-pili accelerates the transfer of bacterial chromosome, but close contact is one of the strong phenomena that facilitates an efficient transfer of bacterial chromosomes.

Micromanipulator is associated with the micropipette that separates microdroplets of medium containing single bacteria to be studied under high magnification. This microdroplet is larger enough to facilitate the bacterial cell division, and further this daughter bacterial cell can be taken from the drop and transferred to the other microdroplets.

During the experimental setup, the primary goal was to identify the mating pairs in the oil chamber of the micromanipulator where two mixed cultures were seeded. Morphological features observed under the microscope such as short and round denotes the male mate and long and thin denotes the female mate. Motility factor also supports the evidence to their identification such as male could be motile and could be seen pulling the nonmotile female through the medium. Mating pairs that have been chosen for this study have been described in Table 6.4.

From the given sources earlier, mating between F^+ (lac+, pro+ leu+ str^r $øII^r$, $MS2^s$) and F^-p678 (lac− pro+ leu− str^r $øII^s$, $MS2^r$) were done by mixing these two strains and placed them on an oil chamber and both loose and close contacted mating pairs were identified with the help of visualization under a microscope. The clones after transfer and cell division were tested whether they are resistant or sensitive to the phage øII or MS2. If they are resistant to øII and sensitive to mS2, this indicates that they were received the F factor. To analyze the close contact and loose contact between $F^+$and F^- p678, Table 6.3 further indicates that close mating pairs generate more viable F^+p678 clones with high frequency than the loose mating pairs. However, four viable F^+p678 clones from loose mate pairs suggest that F-pili might indeed aid as the means for transfer (Table 6.5).

During the experimental set up, it takes at least 6 min to isolate the first mating pairs after mixing male and female cells. Previous studies suggested that chromosomal bacterial transfer takes maximum 3–5 min and if we assume that the F factor takes a similar time like chromosome (3×10^4 nucleotide pairs/min) and it gives 7.5×10^4 Nucleotide pairs so it is predictable that time for transfer the F factor would be approximately 2 min. Therefore, one thing is clear that the F factor consumes a minimum 5 min to complete the transfer. However, it is difficult to exclude the fact that the F factor had already been transferred during unobserved close contacts before mating pair were isolated. To this end, experiments were further extended to mating between Hfr X F^- where bacterial chromosomal transfer could not take place at least 8 m in after the mixing the cultures.

(continued)

Box 6.2 (continued)

Evidence together suggest that initial contact between mating pair of cells appears to be a thin thread that is difficult to observe under light microscope and the simplest assumption is that this thin thread is F-pili that can be seen under electron microscope as a connecting bridge between male and female bacteria. It has also been proved that the mating pairs that were separated during mating from each other produce F-exconjugants that lately forms recombinant clones.

In the conclusion, such studies provide us with three important information;

F pili play a significant role in the mating between two *E. coli* bacteria by providing an external organelle for attachment of mating pairs.

F-pili is a medium for the transfer of bacterial chromosomes and F factor.

F-pili beings two mating pairs in proximity to create close contact between them and help to build the conjugation bridge more efficiently.

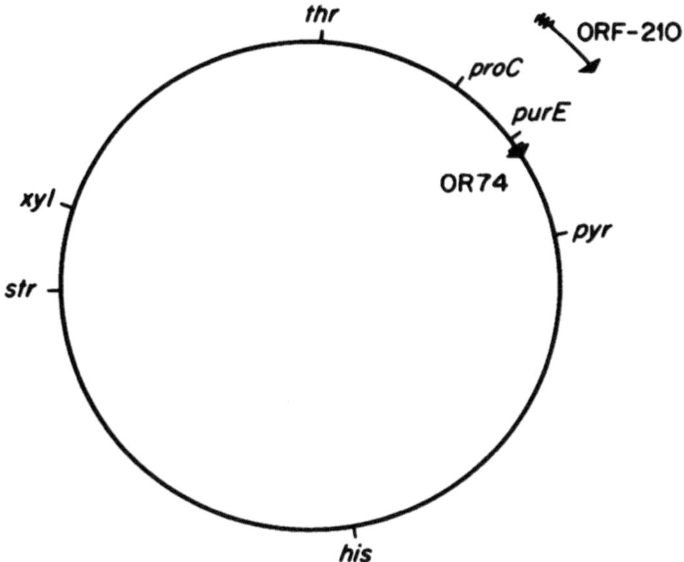

Fig. 6.26 Image shows Linkage map of E. coli K-12. The arrow indicates the origin, direction, and gradient of chromosome transfer of Hfr strain OR74. The F′ strain ORF-210 used in these studies had the same origin and direction of chromosome transfer as Hfr OR74 when F′ was integrated into the chromosome. The relevant fermentation and auxotrophic markers of F-recipient strain ×820 are shown (Jones and Curtiss (1970))

Table 6.3 Recombination frequencies obtained from in vitro liquid mating between the Hfr, F′ and F+ strains chosen for further in vivo studies and F⁻ recipient ×820[a] strain

Cross	Recombinant class selected	Recombination frequenices
χ209 F⁺ × χ820 F⁻	*purE*⁺, *str*ʳ	6.0 × 10^{-5}
	proC⁺, *str*ʳ	3.0 × 10^{-5}
	thr⁺, *str*ʳ	6.0 × 10^{-5}
	his⁺, *str*ʳ	3.5 × 10^{-5}
Hfr OR74 × χ820 F⁻	*purE*⁺, *str*ʳ	1.2 × 10^{-1}
	proC⁺, *str*ʳ	2.4 × 10^{-1}
	thr⁺, *str*ʳ	5.0 × 10^{-2}
	his⁺, *str*ʳ	3.0 × 10^{-5}
F′ ORF-210 × χ820 F⁻	*purE*⁺, *str*ʳ	3.5
	proC⁺, strʳ	8.0 × 10^{-2}
	thr⁺, *str*ʳ	2.4 × 10^{-2}
	his⁺, *str*ʳ	6.4 × 10^{-5}

Fig. 6.27 Image shows the arrangement of the equipment for micromanipulation. (a) Glove box, (b) hinged opening for glove box, (c) opening for arm insertion, (d) opening for microscope, (e) leveling platform, (f) fan, (g) heat strip, (h) thermoregulator, (i) 10-mL hypodermic syringe, (j) rubber tubing, (k) micropipette, (l) oil chamber, (m) micropipette receiver, (n) transformer,  (o) controlling level

6.7 Summary

- Bacteria and virus genomes show a potential scope for genetic studies. Bacterial and virus genome is small and haploid in nature. Bacterial genome is a simple circular ds DNA, while virus genome varies as ssDNA, dsDNA or ssRNA.
- Mutation is a very common feature in bacterial DNA and mutant selection; detection has been developed through various techniques including, replica

Table 6.4 *Escherichia coli* K-12 strains were used for recombination experiments

Strain	Mating type	Genetic character	Morphology	Source
Hayes	F^+	thi⁻, derived from HfrH, motile	Fat and round	Anderson
Hayes	Hfr	thi⁻, motile	Fat and round	
W3011 (Cavalli)	Hfr	met⁻, motile	Slim and long	E. Lederberg
P678	F^-	azi^r, lac⁻, gal⁻, mal⁻, xyl⁻, leu⁻, tsx^r, $tonA^r$, str^r, nonmotile	Slim and long	Anderson
W1-3	F^-	lac⁻, leu⁻, tsx^r, str^r, nonmotile	Fat and round	Tomizawa
W1-4	F^-	lac⁻, pro⁻, tsx^r, str^r, nonmotile	Fat and round	Tomizawa

Table 6.5 Transfer of F factor between loose and close pair mate in the cross F^+ X F^- p678

Types of contact between mating pairs	Close	Loose	P^a
No. of str^r exconjugants isolated	21	20	
No. of nonviable exconjugants	1	1	
No. of exconjugants that produced one or more clones of F^+ P678	16	4	0.01
Efficiency of formation of clones of F^+ P678	0.76	0.20	

[a] Value of P (χ^2 test) gives the probability that there is no difference between close- and loose-mating pairs in the production of recombinants

plate method, phage display, and plaque assay. The spontaneous mutation is one of the potential sources and have been reported as the first experimental evidence for microbial evolution which is a random process.

- Plasmids are extrachromosomal nongenetic DNA molecules that coexist with bacterial chromosomes. They rapidly and independently replicate in the bacterial cell. Episome is a plasmid that can either live freely or integrate into the bacterial chromosome.
- Plasmids are involved in DNA transfer in the bacterial population by conjugation, transformation, and transduction.
- Conjugation referred a physical interaction between two bacterial cells to exchange the DNA through the bridge, known as a conjugation tube. F factor (fertility factor) is responsible for DNA transfer. F plasmid containing cell only can transfer the DNA, known as F+cell while cells lacking F factor meant to

receive the DNA, know F− cell. Hfr cell contains F factor integrated with bacterial chromosome and shows the high frequency for DNA transfer.

- Mating between F+ and F− or Hfr and F− determines the rate of gene transfer from Hfr to F− in terms of the time unit. The rate of gene transfer represents the order of the gene on the chromosome.
- In transformation, bacteria uptake exogenous DNA from the environment without any physical contact between cells. Cotransformation is a rate at which linked genes transferred and the frequency of cotransformation defines the physical distance between a gene on the chromosome.
- The virus is auto replicating machinery with DNA or RNA presence in either circular or linear forms. Most common virus being studied in genetics is bacteriophage—a virus that infects the bacterial cell. Bacterial genome can be transferred through a phage known as transduction. Likewise, conjugation transformation, transduction also provides the information to map the gene order on the bacterial chromosome about the rate of cotransduction reveals the gene order on the bacterial chromosome.
- CRISPR/Cas9 is an advanced technique using strong bacterial adaptive immune machinery to adapt in genetic engineering to edit genes of interest in any kind of eukaryotic or prokaryotic cellular system.
- In generalized transduction, any gene can be transferred from one bacterium to another bacteria. In specialized transduction, a gene linked to the site of phage integration can only transfer from one bacterium to another bacteria.
- A phage life has been differentiated in the lysogeny phase and lytic phase. In the lytic phase, phage causes bacterial cell lysis over the infection and does not integrate with bacterial chromosomes. In lysogeny phase, phage DNA integrates with bacterial chromosome and remains dormant for generations.
- Coronavirus is single-stranded RNA virus that mainly causes acute respiratory syndrome in humans. Corona to date has been identified as SARS (2002), MERS (2012), and novel coronavirus (COVID-19) in 2019.

References

Brown TA (2016) Gene cloning and DNA analysis: an introduction. Wiley

Campbell NA, Reece J (2004) Biology, 7th edn. Benjamin Cummings/Pearson, Boston

Casjens SR, Hendrix RW (2015a) Bacteriophage lambda: early pioneer and still relevant. Virology 479–480:310–330. https://doi.org/10.1016/j.virol.2015.02.010

Casjens SR, Hendrix RW (2015b) Bacteriophage lambda: early pioneer and still relevant. Virology 479:310–330

De Wit E, Van Doremalen N, Falzarano D, Munster VJ (2016) SARS and MERS: recent insights into emerging coronaviruses. Nat Rev Microbiol 14(8):523

Firth N, Skurray R (1992) Characterization of the F plasmid bifunctional conjugation gene, traG. Mol Gen Genet MGG 232(1):145–153

Foster PL (2000) Adaptive mutation in Escherichia coli. Cold Spring Harb Symp Quant Biol 65:21–29

Griffiths AJF, Gelbart WM, Miller JH et al (1999) Modern genetic analysis. In: Bacterial transformation. W. H. Freeman, New York

Griffiths AJ, Miller JH, Suzuki DT, Lewontin RC, Gelbart WM (2000a) Bacterial conjugation. In: An introduction to genetic analysis, 7th edn. W. H. Freeman

Griffiths AJ, Miller JH, Suzuki DT, Lewontin RC, Gelbart WM (2000b) Mendel's experiments. In: An introduction to genetic analysis, 7th edn

Hochheiser K, Kueh AJ, Gebhardt T, Herold MJ (2018) CRISPR/Cas9: a tool for immunological research. Eur J Immunol 48(4):576–583

Holmes RK, Jobling MG (1999) Chapter 5: Genetics. In: Baron S (ed) Medical microbiology, 4th edn. University of Texas Medical Branch at Galveston, Galveston

Jones RT, Curtiss R (1970) Genetic exchange between Escherichia coli strains in the mouse intestine. J Bacteriol 103(1):71–80

Lederberg J, Tatum EL (1946) Gene recombination in Escherichia coli. Nature 158(4016):558–558

Lee H, Popodi E, Tang H, Foster PL (2012) Rate and molecular spectrum of spontaneous mutations in the bacterium Escherichia coli as determined by whole-genome sequencing. Proc Natl Acad Sci 109(41):E2774–E2783

Luria SE, Delbrück M (1943) Mutations of bacteria from virus sensitivity to virus resistance. Genetics 28(6):491

Mubarak A, Alturaiki W, Hemida MG (2019) Middle east respiratory syndrome coronavirus (MERS-CoV): infection, immunological response, and vaccine development. J Immunol Res 2019:6491738

Oppenheim AB, Adhya SL (2007) A new look at bacteriophage λ genetic networks. J Bacteriol 189(2):298–304

Redman M, King A, Watson C, King D (2016) What is CRISPR/Cas9? Arch Dis Child Educ Pract Ed 101(4):213–215. https://doi.org/10.1136/archdischild-2016-310459

Rossmann MG, Moras D, Olsen KW (1974) Chemical and biological evolution of a nucleotide-binding protein. Nature 250(5463):194

Stone AB (1975) R factors: plasmids conferring resistance to antibacterial agents. Sci Prog 62(245): 89–101

Sun S, Kondabagil K, Gentz PM, Rossmann MG, Rao VB (2007) The structure of the ATPase that powers DNA packaging into bacteriophage T4 procapsids. Mol Cell 25(6):943–949

Tatum EL, Lederberg J (1947) Gene recombination in the bacterium Escherichia coli. J Bacteriol 53(6):673

Tomasetti C, Li L, Vogelstein B (2017) Stem cell divisions, somatic mutations, cancer etiology, and cancer prevention. Science (New York, NY) 355(6331):1330–1334

Wikoff WR, Liljas L, Duda RL, Tsuruta H, Hendrix RW, Johnson JE (2000) Topologically linked protein rings in the bacteriophage HK97 capsid. Science 289(5487):2129–2133

Yap ML, Rossmann MG (2014) Structure and function of bacteriophage T4. Future Microbiol 9(12):1319–1327. https://doi.org/10.2217/fmb.14.91

Yin Y, Wunderink RG (2018) MERS, SARS and other coronaviruses as causes of pneumonia. Respirology 23(2):130–137

Part II

Molecular Genetics I: Analysis of Gene

Replication of DNA

7

Tanushree Banerjee

We observe around us a lot of variation amongst individuals, like height, skin colour, facial contour, eye colour, etc. All these variations are encoded in the language of four nucleotide bases present in the nucleic acid called deoxyribonucleic acid or DNA. The enormous information present in the nucleic acids also needs to be carried forwards from one generation to the next. That needs a highly precise and accurate method to replicate the genetic material.

Imagine when we had no idea about the existence of nucleic acids or the fact that DNA is the genetic material. It is interesting to think about how people ventured into this large unknown, step by step, to unravel the mystery of life. In this chapter we will learn about the classical studies which showed that DNA is the genetic material. We will also learn about how the genetic information is coded and organized and what are the precision mechanisms which have evolved to ensure error-free propagation of genetic information.

7.1 Classical Experiments: DNA as Genetic Material

Even before nucleic acids were discovered, people had figured out that there has to be some substance called genetic material which is causing the variation in individuals, and it is the same material which is being transferred to the progeny from the parent. Therefore, genetic material should have certain basic characteristics like the following:

T. Banerjee (✉)
Molecular Neuroscience Research Laboratory, Dr. D. Y. Patil Biotechnology and Bioinformatics Institute, Dr. D. Y. Patil Vidyapeeth, Pune, India
e-mail: tanushree.banerjee@dpu.edu.in

D. Kar, S. Sarkar (eds.), *Genetics Fundamentals Notes*,
https://doi.org/10.1007/978-981-16-7041-1_7

- **Variation:** It should be capable of shuffling the information for creating variation.
- **Stability:** It should be stable enough so as not to create intolerable variation.
- **Replication:** It should be able to replicate so that information can be passed on from the mother cell to the daughter cell.
- **Transmission:** It should be able to carry the information from one generation to the next.

During Crimean War in 1869, a Swiss biochemist Friedrich Miescher isolated a substance from the pus found in the bandages of war victims. Upon analyses he observed that the material contained carbon, hydrogen, oxygen, nitrogen and phosphorus. Upon searching he observed that the material is present in the nucleus of cells, so he called it **nuclein**. It was only in the 1900s that threadlike structures called chromosomes were discovered and over the next 40 years they were found to be made of protein and nucleic acids.

7.1.1 Transformation: Early Study

In 1928 Frederick Griffith, a British medical officer, was working with *Streptococcus pneumoniae*. He had two strains of the bacterium, smooth virulent strain and rough nonvirulent strain. Smooth strain had a polysaccharide coat which made it virulent and shiny. Rough strains could not produce the polysaccharide coat due to a mutation and due to which it was avirulent.

There are several types of S strains like IIS and IIIS. IIS mutates to form IIR, and IIIS mutates to form IIIR, but IIS never forms IIIR and likewise IIIS never forms IIR. R stains can also mutate back to S strains of the same type.

Griffith infected mice with various strains of *S. pneumoniae*. When infected with living IIR, mice lived, but when infected with living IIIS, mice died. IIIS bacteria could also be isolated from the dead mice. However, when mice were infected with heat-killed IIIS strains, mice lived. This showed that bacteria need to be alive and have polysaccharide coat to kill the mice.

Next, Griffith infected the mice with live IIR and heat-killed IIIS bacteria and the mice died. Griffith concluded that avirulent IIR was somehow transformed by virulent IIIS bacteria upon coming in contact with dead IIIS bacteria. He named this process **transformation** (Fig. 7.1). Griffith identified the material of the dead IIIS which transformed the live IIR as **transforming principle**. However, he could not establish the biochemical nature of the transforming principle and initially believed it to be a protein, but he was later proved wrong.

His experiments established that the transforming principle should be something which is stable enough to make the shiny capsule; it should be capable of replication so that it can be transmitted to the progeny cells. Therefore, Griffith's transforming principle had the properties of genetic material. The biochemical nature of this genetic material was later established by Avery, MacLeod and McCarty's experiment.

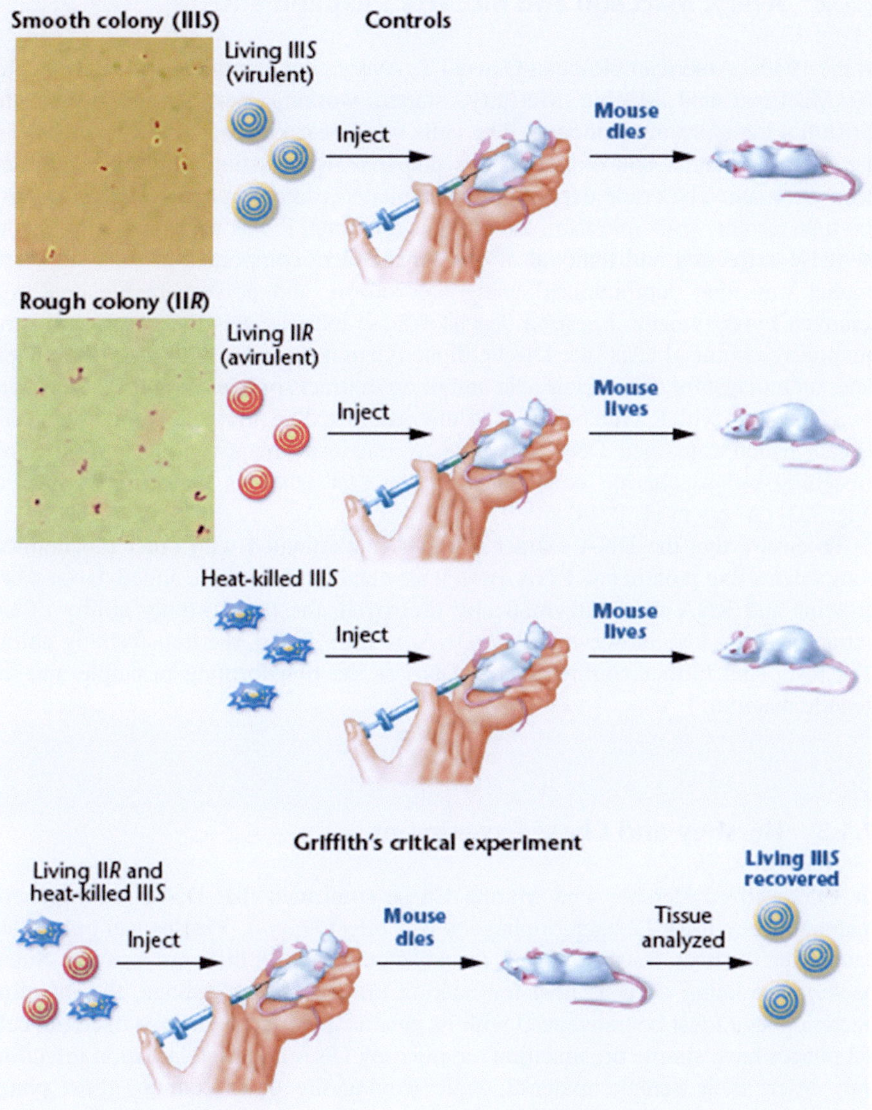

Fig. 7.1 Griffith's transformation experiment. Live bacterial colonies containing cells with capsules (type IIIS) when injected in mice, they died. When avirulent live bacterial cells without capsules (type IIR) were injected in mice, mice survived. When mice were infected with heat-killed IIIS bacteria, they survived. When live avirulent IIR and heat-killed virulent IIIS bacteria were injected together, the mice died (Adapted from: Klug et al. 2012)

7.1.2 Avery, MacLeod and McCarty's Experiment

In the 1930s, American biologist Oswald T. Avery, along with his colleagues Colin M. MacLeod and Maclyn McCarty, started working towards identifying the Griffith's transforming principle. The cells were heat killed at 65 °C. They lysed the S-type bacteria, and extracts were prepared using saline solution of sodium desoxycholate. The crude extract was precipitated using ethyl alcohol. To identify the transforming principle from the crude cell extract, it was necessary to carry out stepwise extraction and removal of one biochemical component at each step. The extract was then deproteinized using chloroform, and polysaccharide coat was removed by enzymatic digestion. Each of these extracts was then analysed using qualitative chemical tests like Dische diphenylamine reaction for deoxyribonucleic acid, orcinol test for ribonucleic acid and so on. Extracts from each of these steps was then incubated with R-type bacterial culture and plated on nutrient agar medium. The extract which contained DNA was able to transform the R-type bacteria to have glistening polysaccharide coat. This proved that DNA is the genetic material (Fig. 7.2) (Avery et al. 1944).

To ensure that the DNA extract was not contaminated with other biochemical components like protein and RNA, proteinases and RNAses were added. Even when proteins and RNA were enzymatically destroyed, the transforming ability of the extract was not lost. However, when DNAses were added, the transforming ability was lost. This further confirmed that DNA is the transforming principle and the genetic material.

7.1.3 Hershey and Chase Experiment

In 1952, Alfred Hershey and Martha Chase confirmed that DNA is the genetic material. They used T2 bacteriophage which infects *E.coli*. Viruses can only replicate inside the host. Bacteriophages need bacterial host for their propagation. Since, bacteria are being used as host for making progeny bacteriophage, the infecting bacteriophage must be transferred with its genetic material to the host bacterial cell. T2 phages have simple organization having only DNA and protein. Upon infection, they insert their genetic material while leaving the outer coat as ghost phage adsorbed on the surface of the bacterium.

Hershey and Chase used different radioactive isotopes for labelling the protein and DNA of the bacteriophage. S^{35} was used to label protein, and P^{32} was used to label DNA. T2 phages were grown in separate media containing either S^{35} or P^{32}. Bacteriophages were allowed to grow in those media for 4 h which was enough to incorporate the radioactive isotopes. Later, those two separately labelled phages were made to infect two separate cultures of *E. coli*. The phage coats were separated from the bacteria by agitation using a high-speed blender followed by centrifugation. The *E. coli* culture which was infected with S^{35}-labelled bacteriophages showed radioactivity in the culture medium and not inside the bacteria. The bacterial culture

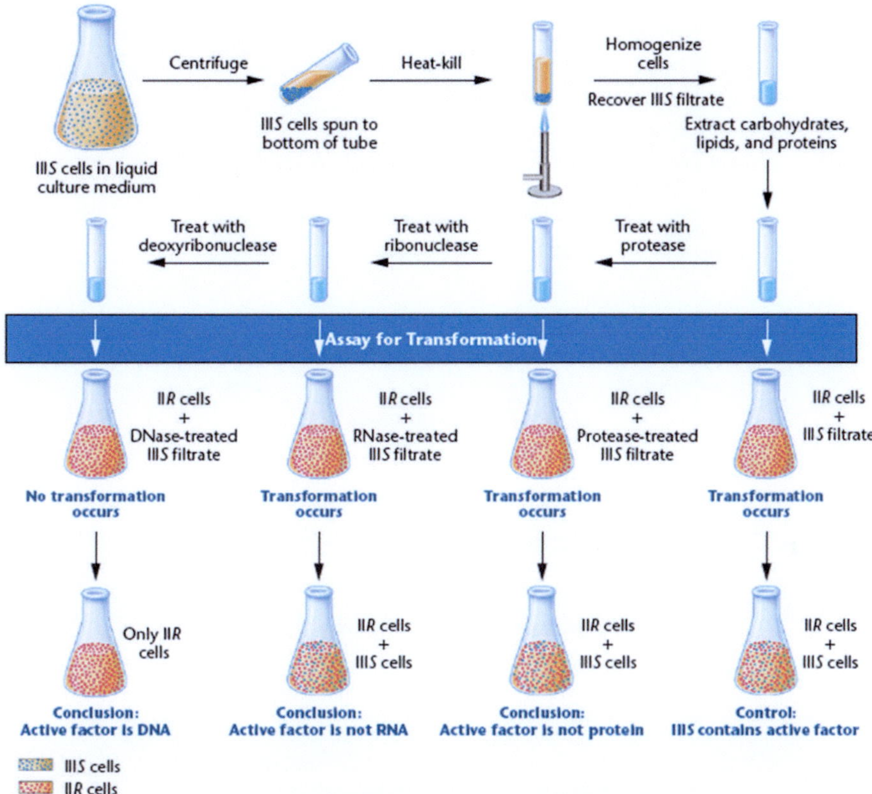

Fig. 7.2 Avery, MacLeod and McCarty's experiment showing DNA as the transforming principle. Virulent IIIS cells were heat killed, and carbohydrates, lipids and proteins were extracted from them. The extract was subsequently treated one by one in a stepwise manner with protease, ribonuclease and deoxyribonuclease. Avirulent live IIR bacterial cells were then incubated with these extracts which were treated with protease, ribonuclease and deoxyribonuclease. IIR cells were transformed to IIIS cells when treated with extracts incubated with protease and RNase. Transformation ability was lost when the IIR cells were incubated with deoxyribonuclease-treated extract (Adapted from: Klug et al. 2012)

which was infected by P^{32}-labelled phages showed activity in the bacterial cells (Fig. 7.3). This again proved that genetic material is DNA and not protein.

7.1.4 Transfection Study

Transfection is the process of introducing purified nucleic acid into eukaryotic cell. Although the pioneering experiments which proved DNA as the genetic material involved transformation of nucleic acid into bacterial cells and bacteriophages, currently transfection is widely used as a part of molecular cloning technique. It

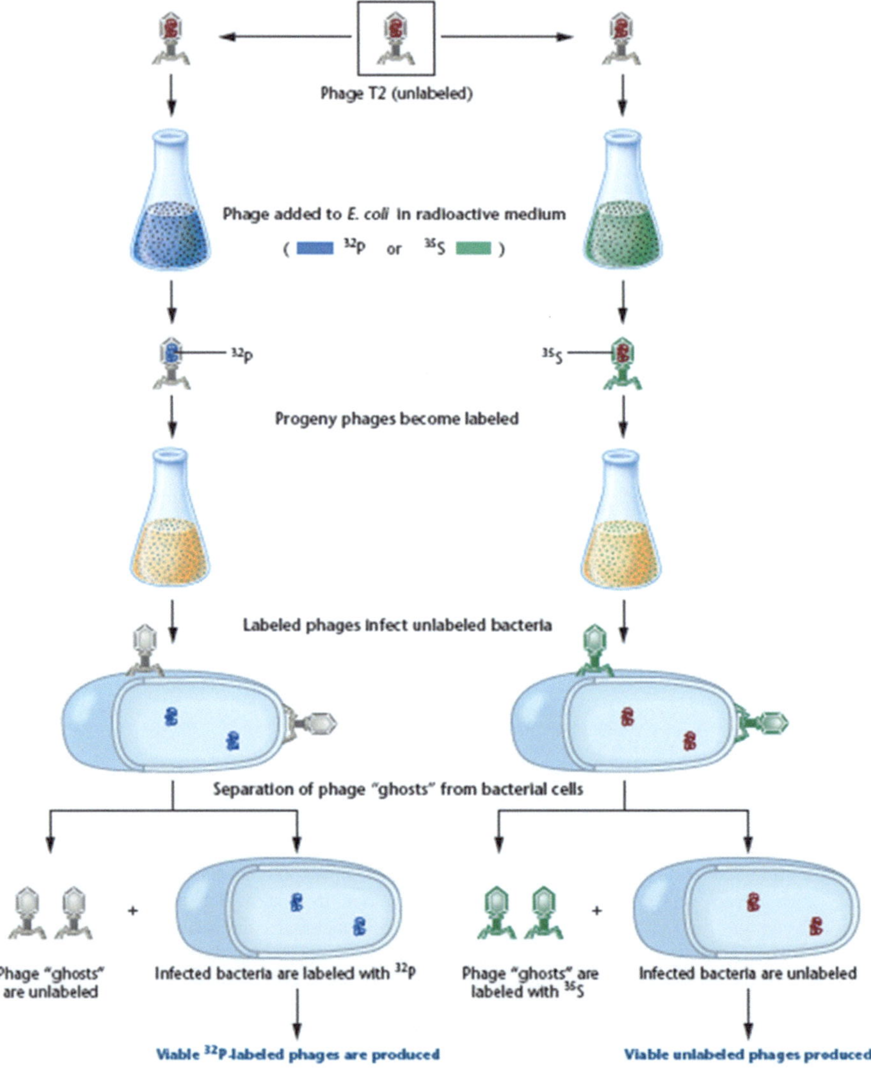

Fig. 7.3 Demonstration of DNA being the genetic material instead of protein by Hershey and Chase. T2 bacteriophages were added to media containing S^{35} and P^{32} radiolabel, in which *E. coli* was already growing. S^{35} was used to label protein, and P^{32} was used to label DNA. Progeny phages became labelled with S^{35} and P^{32}. Labelled phages were made to infect unlabelled bacteria. When phage ghosts were separated from the infected bacteria, P^{32}-labelled bacteria and unlabelled phage ghosts were obtained. However, S^{35}-labelled bacteria were not obtained (Adapted from: Klug et al. 2012)

means ‘infection by transformation of genetic material’. Transfection can be brought about by various techniques like electroporation, liposome mediated, nanoparticles, magnetofection, gene gun, etc. Viruses like lentivirus and adenovirus are also used as vectors for bringing about transfection.

7.2 Molecular Evidence: DNA as Genetic Material

When Friedrich Miescher isolated genetic material in 1869, he realized that this substance is present in the nucleus, and hence he called it nuclein. However, it took decades of further research to understand the distribution and organization of DNA.

7.2.1 Indirect Evidence: DNA Distribution

Around the 1880s, Walther Flemming observed threadlike structures present in the nucleus and named it as **chromatin or stainable material**. Using innovative microscopy, he observed threadlike bodies that formed during cell division and named it as 'mitosen'. He correctly deduced the movement of chromosomes during mitosis. Towards the end of the nineteenth century, advancements in microscopy techniques helped cytogenetics to progress further. Theodor Boveri took Flemming's observation to the next level by linking chromosomes as the hereditary material. He used roundworm *Ascaris megalocephala* embryo as a model organism for observing cell cleavage, chromosome distribution and reorganization during cell division. Walter Sutton in 1902 published the images of individual chromosomes undergoing various stages of meiosis. He identified 11 pairs of chromosomes, which had different sizes in the testes of *Brachystola magna*. However, it was in the early twentieth century that the work of Thomas Hunt Morgan could establish a correlation between inheritance of genetic traits to the behaviour of chromosomes, thereby proving 'chromosome theory of heredity'.

In twentieth century, chromosomes were stained with various dyes and were found to have a banding pattern. *Q*uinacrine *m*ustard (QM) was one such dye which was initially used to stain the chromosomes. It alkylates bases preferentially guanine and also intercalates between the DNA strands. The metaphase stage in human chromosomes from blood cultures, stained with QM, showed a distinct banding pattern of fluorescence. Chromosomes 3, 13–15 and Y were observed to be most intensely stained and were further analysed by ultrafluorimeter (Fig. 7.4).

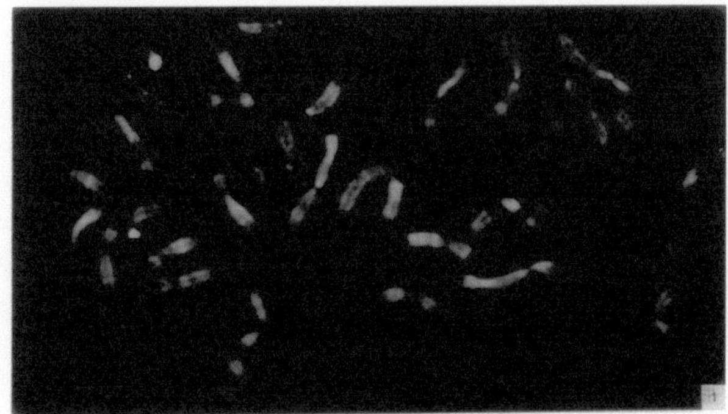

Fig. 7.4 Banding pattern of chromosomes in metaphase. Metaphase stage of human male chromosome from leukocyte culture stained with quinacrine mustard. Fluorescence image, × 2000. (Image adapted from: T Caspersson et al. Experimental Cell Research, 60, 1970, 315–319)

Later techniques like in situ hybridization methods were developed to analyse the chromosome distribution. The first generation of in situ hybridization methods used radioactive nucleic acid probes that produced autoradiographs when a small piece of photographic film was placed on top of a chromosome spread on a microscope slide. Decay of ^{32}P radioactive label in the probe exposed the photographic film, which was then developed in the same way as an autoradiograph. In chromosome autoradiographs, dark regions corresponded to the chromosome locations of a DNA sequence hybridized by the probe.

7.2.2 Indirect Evidence: DNA Mutagenesis

Further evidence that proved DNA to be the genetic material was the identification of variants amongst the wild-type organisms. In 1941, George Wells Beadle and Edward Lawrie Tatum's article 'Genetic Control of Biochemical Reactions in Neurospora' explained the mechanism by which genes control biochemical reactions and how those chemical reactions control the development of organism (Beadle and Tatum 1941). Initially, Beadle and Tatum observed variations in the eye colour of *Drosophila* and later switched to *Neurospora crassa*. It was easier to study *Neurospora* as it requires a simple growth medium, has short life cycle, and during reproduction produces ascospores which are easy to isolate for genetic and biochemical analysis. They exposed the fungus to X-rays and allowed random mutagenesis. Exposed fungus was then grown on minimal growth medium. Most of the mutants died on the minimal growth medium. Then they supplemented the growth medium with essential amino acids or vitamins to make complete medium. Mutants were able to grow on the complete medium. Those mutants were called as auxotrophs. Most of the time mutants required one essential nutrient for growth. This showed that during random mutagenesis only one metabolic pathway got affected. Each mutation affected the activity of a single enzyme. This led to 'one gene-one enzyme' hypothesis.

Another assay based on DNA mutagenesis was developed in the 1970s by Professor Bruce Ames. This assay tests the ability of a chemical compound or mixture to induce mutations in DNA. In this assay, mutant strains of the bacteria *Salmonella typhimurium (S. typhimurium)* were used. These haploid bacteria had some mutations in the gene encoding histidine synthesizing enzyme; their genotype is given as *his*−. Since histidine is essential for making proteins, the mutant bacteria lacking histidine synthesizing enzymes will die. If the growth media is supplemented with histidine, these mutants can survive. Sometimes the *his*− mutants revert to *his*+ state spontaneously by acquiring additional mutations. These new mutations are known as **secondary mutations**. Occurrence of secondary mutation causing reversal of phenotype is rare. Presence of certain chemicals can lead to an increase in the frequency of occurrence of secondary mutations and cause the *his*− mutants to become *his*+ again. Hence, obtaining revertant of *S. typhimurium* in media supplemented with trace amounts of histidine and chemicals proves the mutagenic ability of the added chemicals.

7.2.3 Direct Evidence: rDNA Study

In the 1970s, the advancement of science and advent of recombinant DNA technology provided direct evidences for DNA being the genetic material. In 1973, Herbert Boyer, University of California, and Stanley Cohen, Stanford University, used *E. coli* restriction enzyme to insert foreign DNA into bacterial plasmid. Restriction enzyme *Eco*RI, discovered by Boyer, cut DNA in a way that it creates staggered ends. Another DNA molecule cut by the same enzyme will have the same staggered ends. One DNA molecule cut by the enzyme *Eco*RI can latch onto another DNA molecule cut by the same enzyme as they possess the same staggered ends and are complementary to each other. Stanley Cohen discovered the process by which plasmid DNA could be isolated and inserted back into other bacteria. Plasmids are self-replicating DNA molecules present in bacteria apart from its chromosomal DNA. Plasmid pSC101, isolated by Cohen which was known to impart resistance to tetracycline, was cleaved by *Eco*RI. The linearized plasmid was mixed with another DNA molecule cut by the same enzyme. DNA from both sources joined together to form new loops. The bacteria *E. coli* was then transformed with recombinant DNA molecule and plated on tetracycline plates. Only the bacteria which carried the recombinant plasmid having tetracycline resistance gene could grow on the plate. Hence, the first study of interspecies molecular cloning provided concrete evidence that DNA imparted the tetracycline resistance phenotype and hence DNA is the genetic material.

7.2.4 RNA Serves as Genetic Material

Ribonucleic acid (RNA) is the other form of nucleic acid apart from DNA. If both these are nucleic acids capable of storing genetic information, then why has DNA evolved as the preferred genetic material over RNA? In the 1960s, it was shown that mRNA is capable of storing information and ribosomal RNA and transfer RNA are capable of translating the genetic material into proteins. Around two decades later, it was also shown that RNA has catalytic activity, and those RNAs are known as ribozymes. Considering the wide variety of functions, RNA seems to be a better genetic material than DNA. Based on these, theories were proposed which suggested that RNA was the first genetic material. Since it is single stranded unlike DNA, it is highly unstable and could be damaged easily. Therefore, DNA evolved as the preferred genetic material.

7.2.4.1 RNA as the Genetic Material in Viruses

Tobacco mosaic virus (TMV) was one of the early examples to have RNA as the genetic material. It just contains protein and RNA without any DNA. Gierer and Schramm, in 1956, showed that RNA alone is enough to infect tobacco plants. Hence, RNA was shown to be the genetic material.

In viruses, RNA could exist as a single-stranded or even as a double-stranded molecule. Based on the RNA, the viruses are classified as single-stranded RNA virus or double-stranded RNA viruses. Single-stranded RNA viruses could be further classified as positive- and negative-strand viruses. In positive-strand virus, the RNA directly codes for protein acting like mRNA. In negative-strand virus, the RNA is first transcribed into complementary positive strand by RNA-dependent RNA polymerase which then codes for the proteins.

There is one more class of viruses containing single-stranded RNA but have a DNA intermediate. These are known as retroviruses. An enzyme coded by the viral genetic material, reverse transcriptase, converts the viral RNA into complementary DNA which is then replicated into double-stranded DNA. This double-stranded DNA is then integrated into host DNA by the enzyme integrase to code for viral proteins along with the host proteins. An important member of this family is *h*uman *i*mmunodeficiency *v*irus (HIV).

Similar to retroviruses, there are mobile DNA elements, known as retrotransposons, which are found in eukaryotes. They also encode for reverse transcriptase enzyme. They are first transcribed into RNA and with the help of reverse transcriptase are converted back to DNA which then integrates with the chromosomal DNA at various regions. Since they can change their position by integrating at different regions, they are called as **jumping genes or mobile elements.**

7.3 Structure of DNA Helix

DNA is composed of four nucleotide bases which are linked to each other by phosphodiester bond into a polynucleotide chain. Bases of one polynucleotide chain can base pair with the bases of another polynucleotide chain to form double-stranded DNA duplex. The two polynucleotide chains are in opposite orientation with respect to each other.

7.3.1 Base Composition Study

Each nucleotide has three basic components: nitrogenous base, a cyclic five carbon sugar and a phosphate group attached to the 5′-carbon of the sugar. We will discuss about each component in detail:

Nitrogenous base: The nitrogenous bases could be of two types, single ringed called pyrimidines and double ringed called purines. Cytosine and thymine are pyrimidine bases, and adenine and guanine are purine bases. These nitrogen bases are attached to the 1′-carbon atom of the sugar by an *N*-glycosidic bond (Fig. 7.5). A base linked to the sugar is called nucleoside. In DNA, the bases are linked to deoxyribose sugar and hence forms deoxyribonucleoside, viz. adenine forms adenosine, guanine forms guanosine, cytosine forms cytidine and thymine forms thymidine (Fig. 7.6). In RNA, the bases are linked to ribose sugar and hence are ribonucleoside. Uracil is a base found in RNA instead of thymine. Upon being linked to ribose, it is called uridine.

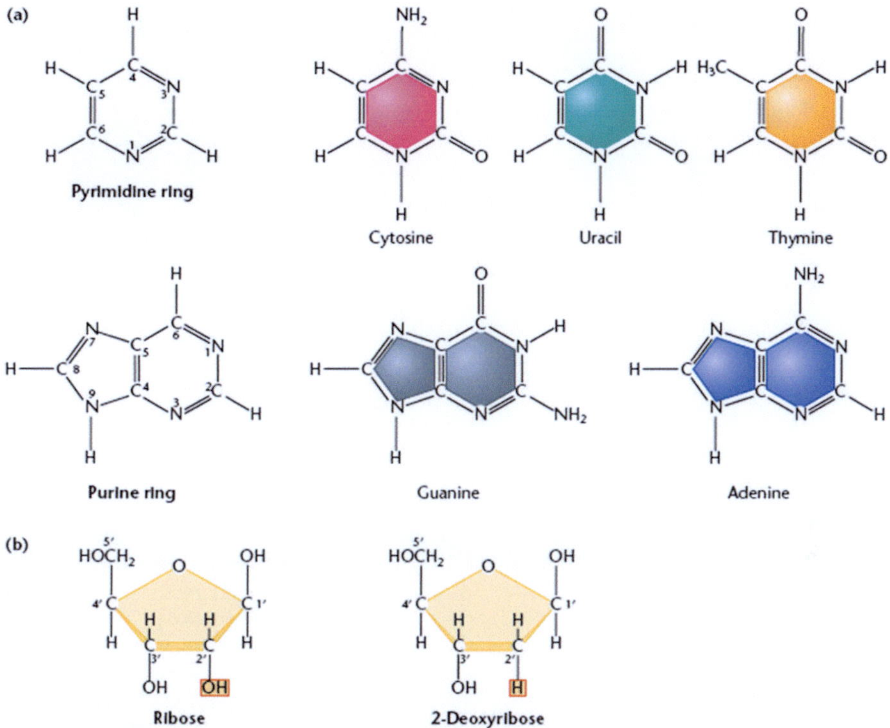

Fig. 7.5 Structure of bases. (**a**) Chemical structure of nitrogenous bases in RNA and DNA. (**b**) Chemical structure of pentose sugars in DNA and RNA (Adapted from: Klug et al. 2012)

Five-carbon sugar: The sugar present in nucleic acids have five carbons and have cyclic conformation. In RNA, hydroxyl groups are present in 2′ and 3′ position. In DNA the 2′ hydroxyl is absent, hence is called 2′ deoxyribose sugar.

Phosphate: The phosphate group is attached to the 5′-carbon of the sugar by a phosphoester linkage. When the phosphate group is attached to the nucleoside, it then becomes nucleotide. The nucleotides are covalently linked by a second phosphoester bond that joins the 5′-phosphate of one nucleotide to the 3′-OH group of the adjacent nucleotide. This bond between the phosphate group of 5-′-carbon and hydroxyl group of 3′ carbon is called **phosphodiester bond**. Nucleotides present in the polynucleotide chain have one phosphate and hence are called deoxynucleotide monophosphate (dNMP) where N is any nitrogenous base. However, nucleotides present freely have three phosphates and are called deoxynucleotide triphosphate (dNTP) (Fig. 7.7). Two of the three phosphates are removed during the formation of phosphodiester bond. The successive linkage of nucleotides by phosphodiester linkage results in a polynucleotide chain which has free 3′-OH at one end and free 5′-phosphate at the other end (Fig. 7.8).

Nucleoside

Uridine

Nucleotide

Deoxyadenylic acid

Ribonucleosides	Ribonucleotides
Adenosine Cytidine Guanosine Uridine	Adenylic acid Cytidylic acid Guanylic acid Uridylic acid
Deoxyribonucleosides	**Deoxyribonucleotides**
Deoxyadenosine Deoxycytidine Deoxyguanosine Deoxythymidine	Deoxyadenylic acid Deoxycytidylic acid Deoxyguanylic acid Deoxythymidylic acid

Fig. 7.6 Nomenclature of nucleosides and nucleotides. Structure and nomenclature of nucleosides and nucleotides of RNA and DNA (Adapted from: Klug et al. 2012)

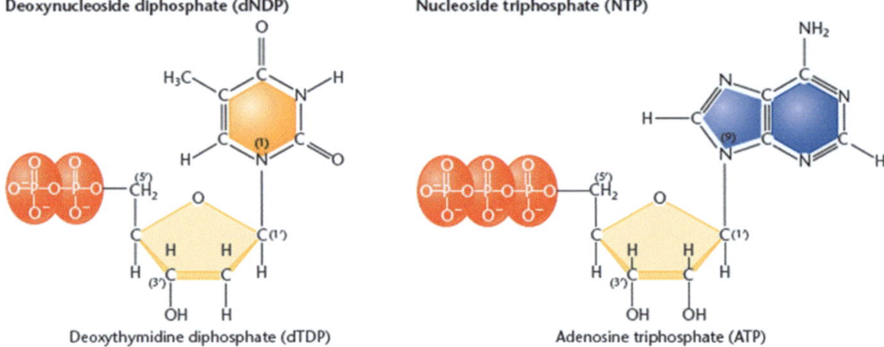

Fig. 7.7 Structure of nucleoside diphosphates and triphosphates. Structure of deoxythymidine diphosphate (dTDP) and adenosine triphosphate (ATP) (Adapted from: Klug et al. 2012)

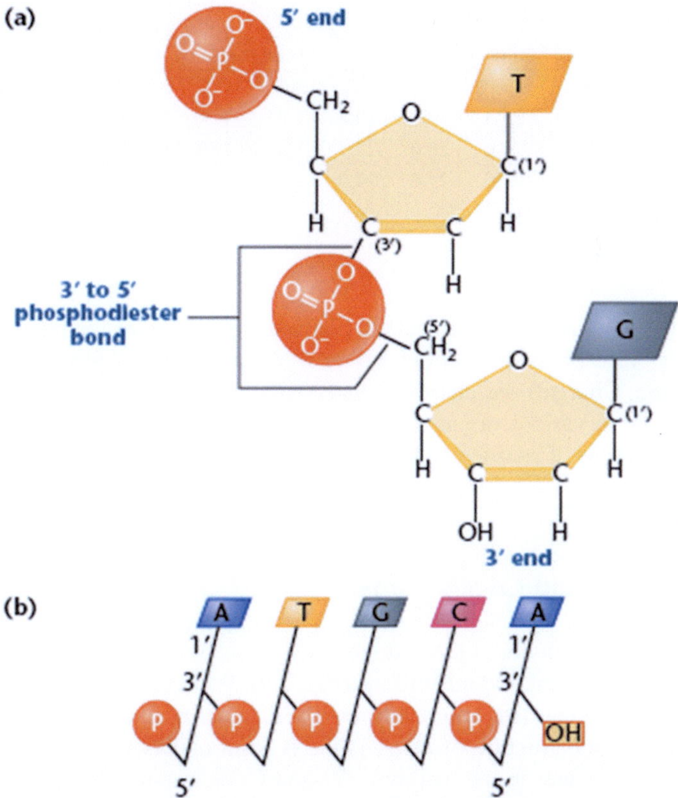

Fig. 7.8 Depiction of phosphodiester bond. (**a**) Phosphodiester bond between C-3′ and C5′ of the adjacent nucleotides. (**b**) Short hand notation for a polynucleotide chain (Adapted from: Klug et al. 2012)

7.3.2 X-Ray Diffraction Study

In 1953, Rosalind Franklin and Raymond Gosling solved the X-ray crystal structure of DNA.

From their crystallographic data, Franklin deduced that DNA existed in two forms, which are in equilibrium. They reasoned that the presence of two forms of DNA resulted in unclear diffraction pattern. These forms are A form, the dehydrated form of DNA, and B form, the hydrated form of DNA. Franklin could isolate the two forms of DNA by techniques like 'manipulation of the critical hydration of her specimens'. A and B forms were then separately crystallized, and those crystal structures became the basis for Watson and Crick's double helical DNA model. The diffraction pattern obtained by Franklin and Wilkins hinted of a two-stranded helical form. Patterns were observed to be consistent; hence they deciphered that helix diameter must be constant. The helical turn of DNA correlates to the horizontal lines in the picture which measures to 34 Å. They also calculated the gap between the

Table 7.1 Base composition of DNAs from various sources

	Base composition				Base ratio		Combined base ratio	
Organism	A	T	G	C	A/T	G/C	A + G/C + T	A + T/C + G
Human	30.9	29.4	19.9	19.8	1.05	1.00	1.04	1.52
E. coli	24.7	23.6	26.0	25.7	1.04	1.01	1.03	0.93
T7 bacteriophage	26.0	26.0	24.0	24.0	1.00	1.00	1.00	1.08

Source: Chargaff (1950)

base pairs to be 3.4 Å, which led them to conclude that there are ten nucleotide bases per turn. Franklins X-ray diffraction also showed that sugar phosphate backbone was on the outside. Chargaff did chemical analysis on the molar content of the bases in DNA molecule in various organisms and concluded [A] = [T] and [G] = [C], and total molar content of purines is equal to that of pyrimidines [A + G] = [C + T] (Table 7.1).

7.3.3 Watson and Crick's Model

Watson and Crick's model of DNA double helix was based on the X-ray diffraction pattern analysis of Franklin and Wilkins (Fig. 7.9). Watson and Crick combined the physical and chemical data and determined that the two strands are coiled around each other. The 3D structure of DNA is the B form of DNA. B form exists when there is plenty of water in the medium. Watson Crick's model based on Franklin's X-ray crystallography study has the following features:

1. DNA molecule consists of two polynucleotide chains which are wound around each other forming a right-handed (clockwise) helix.
2. The two chains are antiparallel, which means that the free 3′-OH group of one strand is opposite to the free 5′-phosphate group of the other strand and vice versa.
3. Sugar phosphate backbone is on the outer side of the helix with the bases occupying the inner side. The bases are perpendicular to the sugar phosphate backbone and are stacked on top of each other. The B form of DNA is narrow and elongated. Helical conformation creates major and minor groove along the axis.
4. The bases of one polynucleotide strand are hydrogen bonded to the opposite bases present in the other polynucleotide strand. Based on Chargaff's rule, they predicted that A is bonded with T by two hydrogen bonds and C is bonded with G by three hydrogen bonds. AT and GC are called complementary base pairing, and as per the space filling model, these are the only two permissible base pairings which can exist in the helix as per Chargaff's rule.
5. There are approximately 10 bp per 360° rotation of the helix. Each base pair is twisted 36° to the adjacent base. Each base pair is 0.34 nm apart, so ten base pairs in each turn encompasses 3.4 nm. The diameter of the helix is 2 nm.

(a)

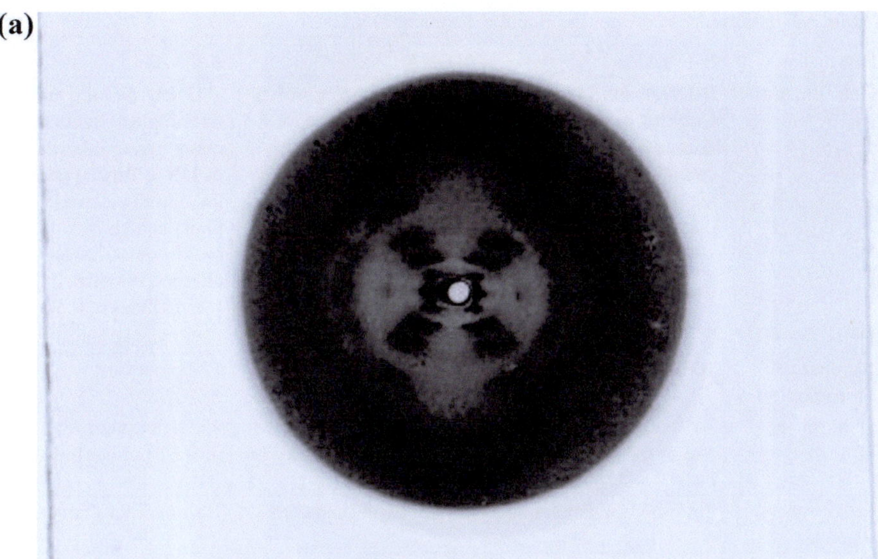

(b)

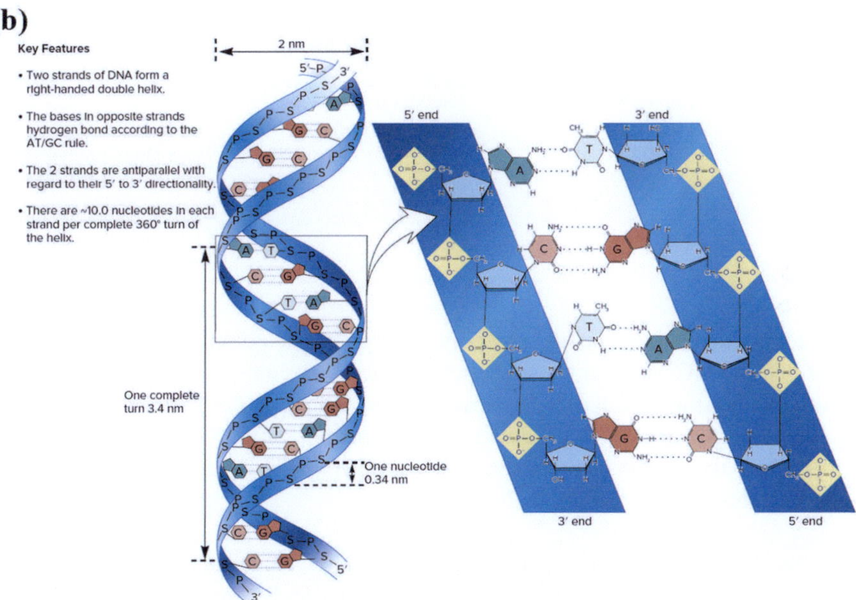

Fig. 7.9 X-ray diffraction pattern of DNA. (**a**) X-ray diffraction photograph obtained by Rosalind Franklin. The DNA fibres which were used for diffraction were of the B form. Diffraction pattern showed strong arcs on the periphery. These arcs represented the periodicity of nitrogenous bases, which are 3.4 Å apart. The inner cross pattern of spots indicated that DNA is helical in structure. (**b**) Watson and Crick DNA double helix model showing base pair interactions, major and minor groove, pitch and the diameter (Adapted from: Brooker 2018)

Table 7.2 Differences between A, B and Z forms of DNA

Parameters	B-DNA	A-DNA	Z-DNA
Conformation	B-DNA is the Watson-Crick form of the double helix. The two strands of the duplex are antiparallel and plectonemically coiled	Ribbon-like helix with a shorter, broader and more open core	Z-DNA is a long and thin duplex structure compared to other forms of DNA. Zigzag (hence the name) pattern in the phosphodiester backbone is due to alternating purine-pyrimidine sequence
Helix sense	Right-handed helix	Right-handed helix	Left-handed helix
Number of bases per turn	10 bp per turn	11 bp per turn	12 bp per turn
Helix pitch	34° helix pitch with rise of 0.34 nm per base	28° helix pitch with rise of 0.256 nm per base	45° helix pitch with rise of 0.37 nm per base
Helix diameter	2 nm helix diameter	2.3 nm helix diameter	1.8 nm helix diameter
Base pair orientation	Base pairs are almost centred over the helical axis	Base pairs are tilted and displaced laterally away from the central axis and closer to the major groove	Base pairs accommodate distortion
Sugar pucker	Sugars have C2′ endoconformation	Sugars have C3′ endoconformation	Guanine base has C3′ endoconformation sugar, and alternate cytosine has normal C2′ endoconformation
Glycosyl angle	Anti	Anti	Pyrimidine-anti Purine-syn

6. Keeping in account spatial constraint induced by the complementary base pairing of AT and GC, the two sugar phosphate backbones are not equidistant from each other along the helical axis. At certain point they are closer to the axis and at other points they drift apart from the axis. This unequal spacing results in formation of grooves; the wider grooves are called **major groove**, and narrower groove is called **minor groove** (Table 7.2).

If the water present in the medium is less, the DNA forms another right-handed helix known as A form of DNA. A form is shorter and wider than the B-DNA, and bases are tilted away from the main axis. A form of DNA is found in spores of some bacteria and some protein-DNA complexes (Fig. 7.10).

Another form of DNA found in nature is the Z-DNA. Unlike the A and B forms, it is a left-handed helix. Sugar phosphate backbone follows the zigzag path moving back and forth hence is called Z-DNA. Z-DNA has been shown to contain multiple stretches of alternating C and G nucleotides. Z-DNA-specific antibodies bind to regions of the DNA which are undergoing transcription. Hence, it is probable that

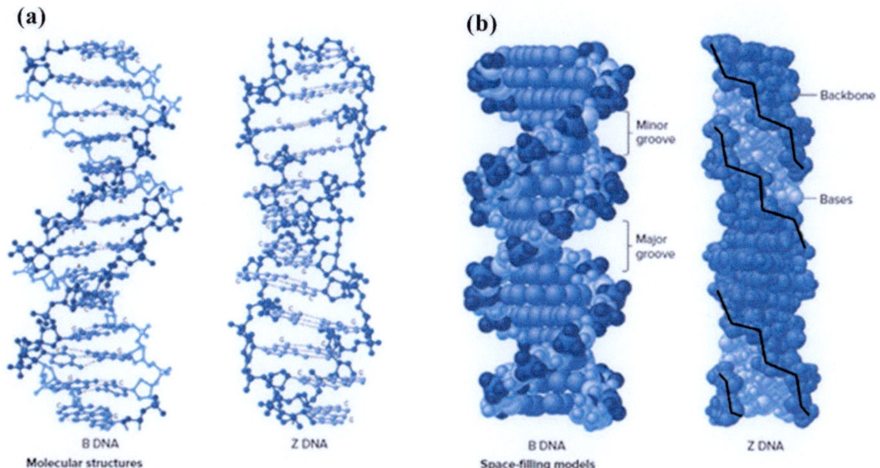

Fig. 7.10 Comparison of different forms of DNA: (**a**) Molecular structures deduced from X-ray crystallographic image of short fragments of B-DNA and Z-DNA. (**b**) Space filling model depicting that B-DNA has distinct minor and major grooves. In Z-DNA, due to the zigzag pattern, minor and major grooves are not well distinct. Black zigzag lines in the Z-DNA are connecting the phosphate groups of the DNA backbone (Adapted from: Brooker 2018)

Z-DNA may have role in gene expression. Other secondary structure of DNA like H-DNA also exists. In this form, a part of the double helix unwinds, and the single-stranded nucleotide chain then pairs with double helical region to form three stranded helices.

7.3.3.1 Triplex DNA

In 1957, Alexander Rich, David Davies and Gary Felsenfeld observed that DNA can attain triple helical forms in the in vitro conditions. Three synthetic strands of DNA were shown to form a triple helical structure. After 30 years, it was observed that naturally occurring DNA double helix can also base pair with a synthetic DNA especially at the major groove creating a DNA triple helix. The base pairing between the double helix and synthetic DNA followed the normal rules of base pairing as depicted in the Fig. 7.11. This observation was of particular interest to the scientific community as it gave a handle to regulate DNA transcription. Short synthetic DNA strands could be synthesized complementary to the DNA of interest and allowed to form DNA triple helix. Such triple helical structures were shown to inhibit the DNA transcription and hence emerged as an important tool for gene silencing.

7.3.3.2 Alternative Forms of DNA

Several alternative forms of DNA right-handed helix models have been shown to exist in the in vitro conditions. **C-DNA** exists at highly dehydrated conditions and has only 9.3 base pairs per turn. It has a helical diameter of 1.9 nm, and the base pairs are tilted relative to the axis. The sequences which lack guanine residues can exist in

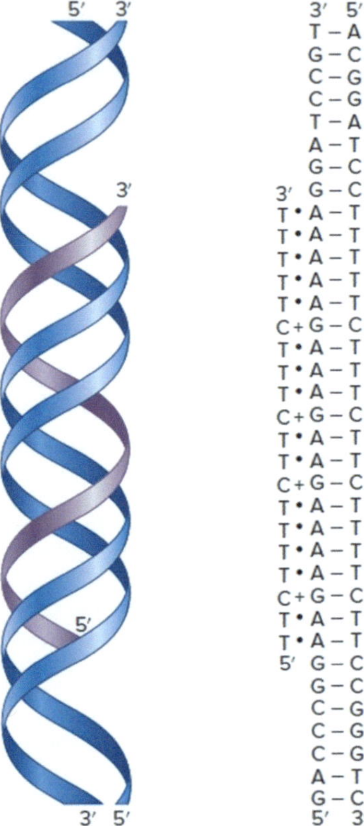

Fig. 7.11 DNA triplex: (**a**) In the ribbon model of DNA triple helix, it is shown that the major groove of the DNA double helix binds with a synthetic DNA strand. (**b**) The third strand base pairs with the complementary bases of the major groove following AT and GC rule. Cytosines of the third strand are protonated hence is positively charged (Adapted from: Brooker 2018)

two other forms in the in vitro conditions, **D** and **E**. D has eight bases per turn, and E has seven bases per turn. Recently discovered **P-DNA** (named after Linus Pauling) is narrower and longer with only 2.62 bases per turn. In this form of DNA, phosphate groups are towards the inside of molecules, and nitrogenous bases are towards the surface. Other secondary structures of nucleic acids are formed by guanine-rich sequences. These are helical structures formed by guanine tetrads. Guanine tetrads are formed by hydrogen bonding in a square plane. This type of hydrogen bonding is called as Hoogsteen base pairing. Such guanine tetrads can form from one, two or four strands of DNA molecules. These guanine tetrads can stack on top of each other to form **G-quadruplex**. Such quadruplexes have gene regulatory roles and are found at certain regions more frequently like in the telomeres (Fig. 7.12).

7.4 Analytical Study on DNA and RNA

7.4.1 Absorption of UV Light

DNA absorbs UV light. Its absorption maxima lie at 260 nm (Fig. 7.13).

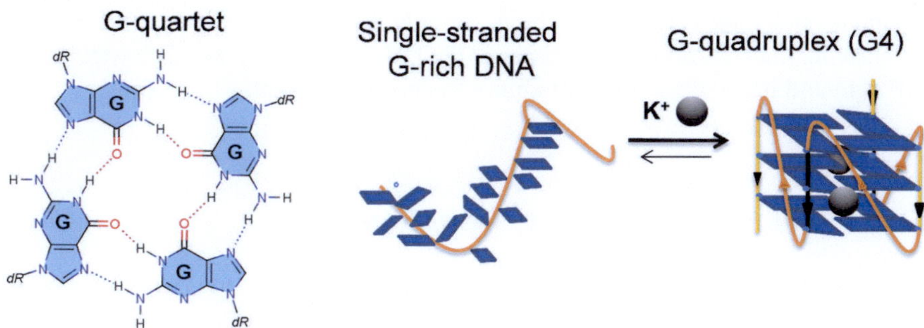

Fig. 7.12 G quadruplex: Guanine residues of intra or inter strands form Hoogsteen bond to form a planer structure which stack on top of each other to form G-quartet or G quadruplex (Adapted from: https://science.institut-curie.org/research/biology-chemistry-of-radiations-cell-signaling-and-cancer-axis/cmbc/team-teulade-fichou/probes-targeting-g-quadruplex-nucleic-acids)

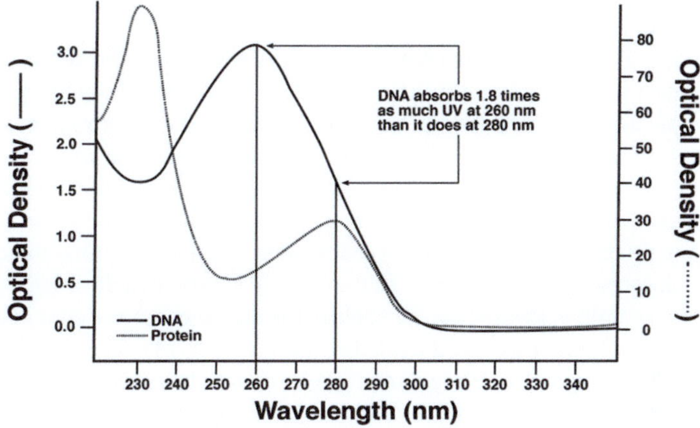

Fig. 7.13 Absorption spectra of DNA. Absorption of UV light by DNA vs protein at 260 nm. Absorption maxima lie at 260 nm for DNA. Another peak is observed at 280 nm

Table 7.3 Concentration of nucleic acid per A_{260} unit

Nucleic acid	Concentration (μg/mL) per A_{260} unit
dsDNA	50
ssDNA	33
ssRNA	40

As shown in the graph (this figure), DNA absorbs maximum UV light around 260 nm depicted by the broad peak. At 280 nm it absorbs only half of the maximum UV light (this figure). UV absorption is a property of the bases, and each base absorbs differently. Therefore, when the bases are not hydrogen bonded and are exposed to UV, they have higher ability to absorb the light. Different DNA preparations have slightly different absorption peak depending upon the DNA composition. DNA concentration can be determined by the equation: 1 OD_{260} unit = 50 μg/mL (Table 7.3).

Single nucleotides absorb even more strongly than single-stranded DNA because the stacking interactions between bases of the DNA strand decrease their exposure to UV light and hence decrease their ability to absorb UV light.

7.4.2 Sedimentation Velocity Centrifugation

The velocity with which a macromolecule moves is mainly a function of its molecular weight and its shape. The ratio of molecular velocity to centrifugal force is called the **sedimentation coefficient** designated as '*S*' or Svedberg unit. Therefore, $S =$ velocity/centrifugal force. The size of biological macromolecules is often referred to in terms of Svedberg units, like prokaryotic ribosome smaller subunits are called 30S subunit as its *S* value is 30. *S* value depends on molecular weight and shape. Therefore, changes in '*S*' value depicts changes in molecular shape (unfolding of biomolecules) or molecular weight (aggregation or dissociation) depending upon experimental conditions.

One common type of velocity centrifugation is density gradient centrifugation. In this technique a liquid medium is used whose density is varied along the length of the centrifugation tube. Sucrose medium is one of the earliest such media which was used to create density gradient. Various concentrations of sucrose solutions are made and layered one by one in the centrifugation tube. The highest concentration has the highest density and is layered at the bottom. Lowest density solution is layered on top. The density of the molecules to be sedimented is adjusted so that the molecules having lower density than the density of the sucrose solution are layered at the top of the tube. The sample is layered on the surface forming a band or a zone. Hence, this technique is also known as zonal centrifugation (Fig. 7.14). The density gradient stabilizes the liquid against mixing that might be caused by mechanical forces. After layering the sample, the tube is centrifuged by swinging bucket rotor. Once the centrifugation is complete, a hole is punched at the bottom, and successive layers are collected drop by drop. The presence of different macromolecules in the sample solution which was layered on top of the sucrose gradient could be analysed.

Equilibrium centrifugation is another velocity centrifugation technique based on density gradient. In this the macromolecules which are to be sedimented are suspended in a solution of CsCl whose concentration is approximately equal to that of the macromolecules. During centrifugation Cs^+ ions start sedimenting and diffusing throughout the tube. Continuous sedimentation and diffusion create a linear concentration gradient with maximum density at the bottom and decreasing towards the top. The macromolecules move in the density gradient and stops at the position where the density of CsCl solution is equal to that of the molecule. Hence, the macromolecules form a narrow band in the tube. If there are multiple types of macromolecules with different densities, they form multiple bands. The densest macromolecule will form the lowest band. This technique has very high resolution. The naturally occurring DNA molecule with N^{14} and density 1.708 g/cm^3 can be separated from N^{15} which has the density of 1.722 g/cm^3.

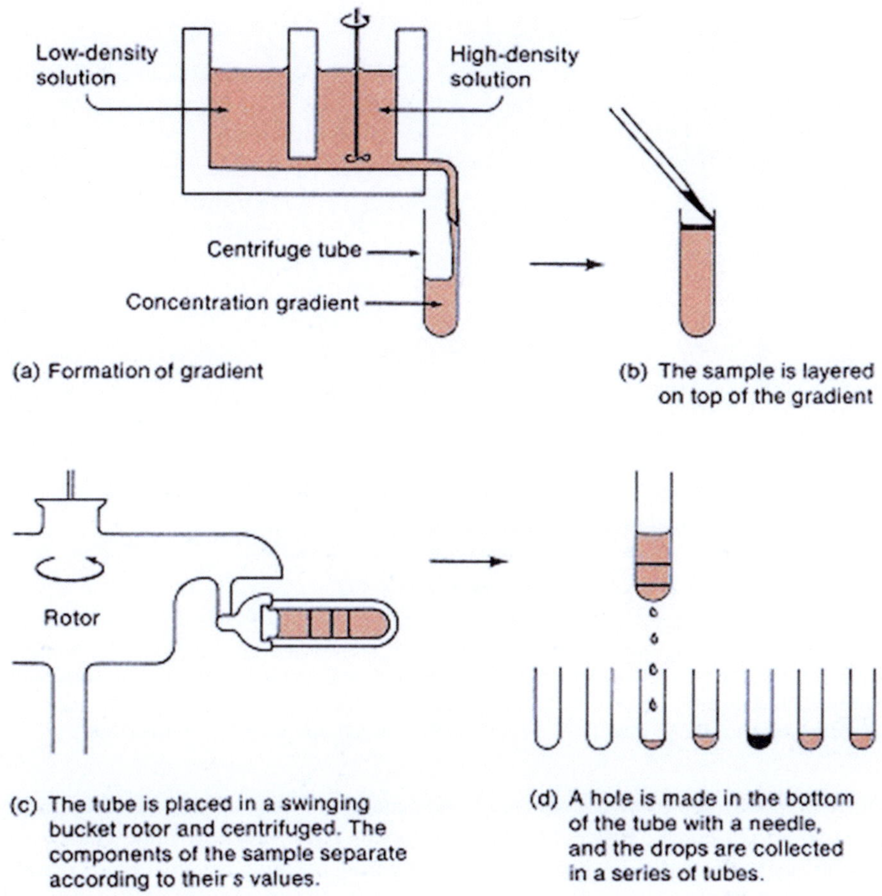

Fig. 7.14 Zonal centrifugation. (**a**) Formation of gradient. (**b**) Layering of sample on top of gradient. (**c**) Tube is placed in swinging bucket rotor for sample separation. (**d**) Collection of samples by punching hole at the bottom

Although both zonal and equilibrium centrifugation use a liquid medium of varying density for separation of macromolecules, these two methods are working on different principles. Therefore, on lengthening the centrifugation time, there might be mixing of different zones of the sucrose gradient with all the sucrose molecules settling at the bottom, but the CsCl gradient will always maintain the linear concentration gradient.

7.4.3 Denaturation and Renaturation of Nucleic Acids

DNA double helix has two polynucleotide chains. These are held together by the hydrogen bonding forces present between the complementary base pairs. Adenine is

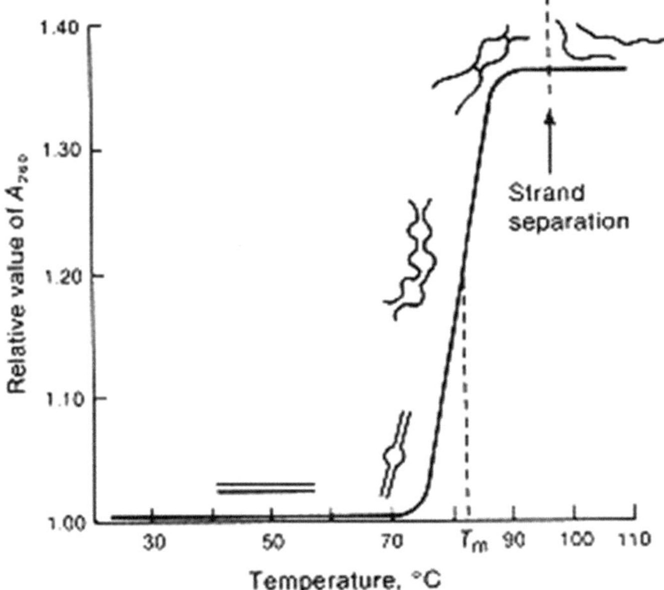

Fig. 7.15 DNA melting curve. Melting temperature, T_m, is depicted by a dotted line. DNA melting starts at around 70 °C. Between 70 and 90 °C, absorption at 260 nm constantly increases as the denaturation continues. Beyond 90 °C, curve reaches a plateau phase when the DNA strands are completely separated (Adapted from: https://members.tripod.com/arnold_dion/RecDNA/)

attached to thymine by two hydrogen bonds, and guanine is attached to cytosine by three hydrogen bonds. However, hydrogen bonds have very low bond energy and can be easily disrupted by heat. When the hydrogen bonds are disrupted, the two polynucleotide chains move apart from each other and lose the normal helical structure or the native state. This state is called as denatured state of DNA. The process of transition from the native state to the denatured state is called **denaturation**.

The physical property of DNA changes during the process of denaturation. Single-stranded DNA absorbs light at 260 nm more strongly than the double-stranded DNA. When the DNA solution is heated, it is observed that after 70 °C, the absorption at 260 nm increases linearly till 90 °C (Fig. 7.15). After that there is no more increase, but the absorption reaches a plateau phase. This process of denaturation where the DNA strands separate out from each other and become single stranded is melting of DNA, and the graph showing the linear increase in absorption as a function of temperature is called the **DNA melting curve**. Before the increase in absorption begins, the DNA is in the native double-stranded state. The linear increase in the absorption is due to the increase in the number of broken hydrogen bonds in the base pairs. When all the base pairs are separated, the initial part of the upper plateau is reached indicating complete denaturation and presence of only single-stranded DNA. The temperature at which the rise in A_{260} is half complete is called **melting temperature** and is designated as $\boldsymbol{T_m}$. AT has two hydrogen bonds, whereas GC has three hydrogen bonds. Therefore, more energy is required for breaking GC bonds than AT bonds. DNA having more GC content will have higher T_m than DNA with more of AT content. Hence, T_m varies with base composition.

DNA denaturation can also be brought about by increasing the pH. At high pH, the charge of several groups engaged in hydrogen bonding is changed, and hence they can no longer participate in hydrogen bonding interactions. At pH higher than 11.3, all the hydrogen bonds are disrupted, and the DNA is completely denatured. If the salt concentration is low, the strong negative charges present in the phosphate groups keep the DNA fully extended and single stranded. If the salt concentration is high enough to neutralize the negative charge of the phosphate groups, DNA folds back on itself forming intra-strand hydrogen bonds.

Very high temperature can also disrupt the phosphodiester linkage present in the sugar phosphate backbone leading to breakage of DNA strand. However, phosphodiester linkage is resistant to alkaline denaturation. Therefore, alkaline denaturation is considered to be a better method for denaturation.

When the denatured DNA returns back to its native form upon reducing the temperature or neutralizing the pH, it is called **renaturation or reannealing**. The reannealed DNA is called renatured DNA. Renaturation is a slower process than denaturation and hence requires optimal conditions like:

1. The concentration of salt must be high enough to neutralize the negative charge of the phosphate groups so as to remove electrostatic repulsion. Usually 0.15–0.5 molar NaCl is sufficient to attain renaturation.
2. The temperature should be high enough to remove non-specific, intra-strand hydrogen bonding, yet the temperature cannot be too high that inter-strand hydrogen bonds are not formed. The optimal temperature for renaturation is 20–25 °C below the T_m.

The rate-limiting step of renaturation is precise collision between the complementary strands so that the correct bases are coming opposite to each other. Since this is dependent on random motion, renaturation is also a function of concentration. Renaturation can be ascertained by decrease in absorption at 260 nm.

7.4.4 Hypochromic Effect

As discussed above, DNA absorbs UV light of wavelength 260 nm. The absorption increases as the DNA dissociates into single strands. The increase in absorption at 260 nm due to DNA denaturation is called **hyperchromic shift**. During DNA renaturation, the absorption at 260 nm decreases and hence is called the **hypochromic shift**. Therefore, it can be interpreted that hyperchromic shift is an indication of DNA denaturation and hypochromic shift depicts DNA renaturation. Hypochromic and hyperchromic shifts are indications of thermal stability of reassociated DNA duplexes in a solution (Fig. 7.16).

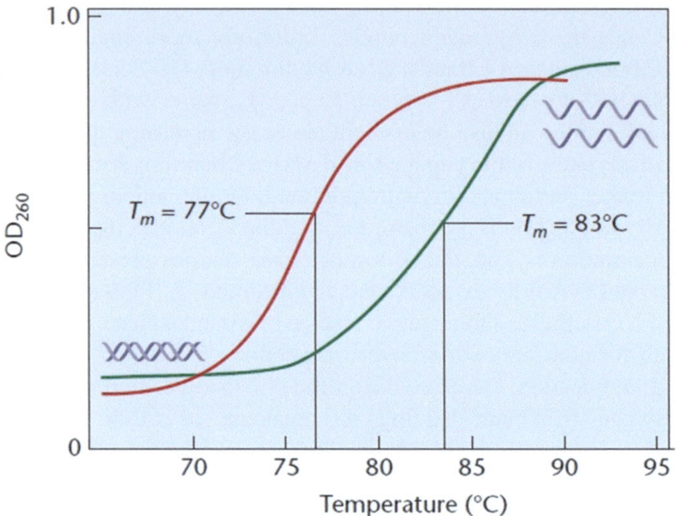

Fig. 7.16 A DNA melting curve depicting the increase in UV absorption versus temperature (hyperchromic effect) for two different DNA molecules having different GC content (Adapted from: Klug 2012). Molecule having higher GC content has higher T_m

7.4.5 FISH

*F*luorescence in situ *h*ybridization (FISH) is a molecular cytogenetic technique to identify certain DNA sequences. Fluorescent probes are used to hybridize with the complementary DNA sequence. Fluorescent microscopy is then used to localize the hybridized probe. FISH can also be used for identification and localization of certain RNA like mRNA, miRNA, etc. (Fig. 7.17). Cot hybridization is also used to isolate fractions of DNA for use in FISH experiments. The DNA repeats isolated by Cot fractionation can be used as blocking agents to reduce background signals during FISH techniques.

7.4.6 DNA Renaturation Kinetics

When DNA renaturation was proposed to be a function of length and complexity of DNA and solvent viscosity, following facts were already known:

1. The renaturation reaction is approximately second-order reaction.
2. The renaturation occurs at a maximum rate if the temperature is 20–30 °C below the melting temperature (T_m,) of the DNA.
3. A decrease in the molecular weight results in a reduction in the rate of renaturation.

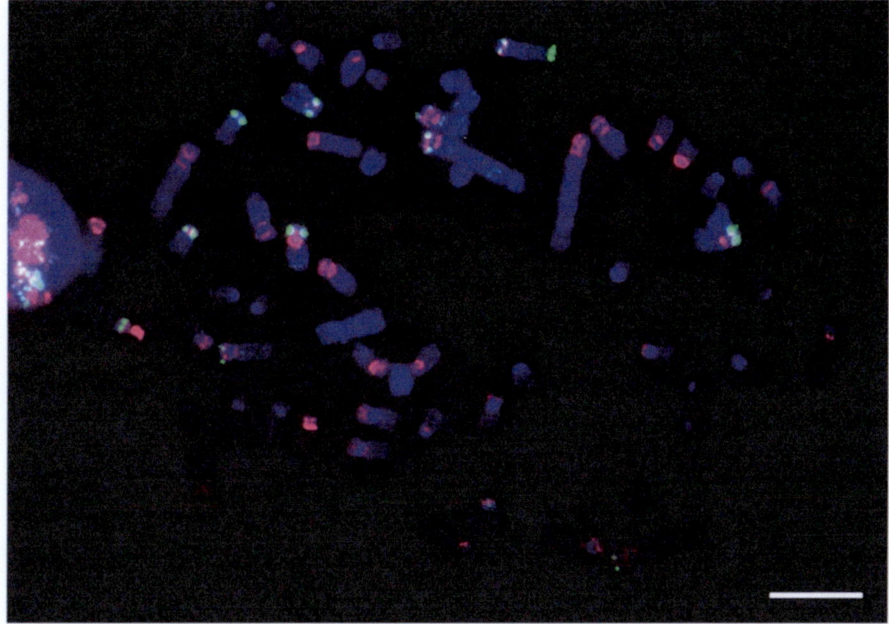

Fig. 7.17 Fluorescent in situ hybridization. Localization of telomeric DNA using DNA probe (red) and rDNA probe (green) on metaphase chromosomes from *C. porcellus*. (Adapted from Svetlana A. Romanenko et al. A First-Generation Comparative Chromosome Map between Guinea Pig (*Cavia porcellus*) and Humans, Plos One 2015)

4. The rate of renaturation of DNA of complex organisms, such as human, is much slower than the rate of renaturation of DNA of simpler organisms, such as bacteria.
5. Ionic strength of the electrolyte like NaCl acts as a factor for controlling rate of renaturation at concentration below 0.4 M.

The formation of first few base pairs is called the nucleation event. It is the rate-limiting event for renaturation, and the rest of the reannealing is the zippering up action which is the faster process. Since there are two rates of the renaturation reaction, it is a second-order rate kinetics. It was proposed that nucleation starts at certain preferred sites known as **nucleation sites**. Let's denote nucleation site by β. Let L be the average number of nucleotides per single strand of the denatured DNA preparation and P be the denatured phosphate concentration. Then $\beta P/2N$ is the average concentration of the nucleation site. k_n represents the rate constant for nucleation at any such site. Rate of nucleation is then $k_n(\beta P/2N)^2$. The rate of nucleation at all sites is $k_n(\beta P/2N)^2\beta N$. Rate of base pair formation is $v = k_n\beta^3(L/4N)P^2$.

Renaturation is dependent on temperature and concentration of DNA. When the temperature decreases below T_m, the reaction rate increases, with the maximum rate at 15–30 °C below T_m. If the temperature decreases further, then the renaturation rate decreases. Let N be the number of base pairs in non-repeating sequence of a DNA and L be the average number of nucleotides per single strand of the denatured DNA preparation; the rate constant is approximately $k_2 = 3 \times 105L_0.5N_1$. mole^{-1} s^{-1} at ($T_m$ − 25) °C and at [Na$^+$] = 1.0 mole in aqueous solution. This equation also depicts a second-order rate kinetics. If the GC content of the DNA is high, the reaction rate increases slightly. The reaction rate at maximum temperature is inversely proportional to the solvent viscosity. If the viscosity of the medium is changed by adding components like glycerol or $NaClO_4$, T_m can change significantly (Wetmur 1968).

7.4.7 Repetitive DNA

Repetitive DNA sequences renature faster than single copy sequences. Their reassociation rates and thermal stability of renatured duplexes are also significantly different. These differences could be studied by Cot analysis which yields valuable information about the types of DNA sequences that are present in genomes and their organization. Repetitive sequences are known to reassociate much faster than the single-copy sequences. Hence, depending upon the reassociation time, the amount of repetitive sequence present in the DNA can be revealed (Fig. 7.18) (Graham 2001).

There are various types of repetitive sequences present in eukaryotes. The fastest reassociating class is the 'foldback' sequence. These are sequences within the single DNA strand and can fold back upon itself. As there is no diffusion and random collision factors involved, the snapping back process is really rapid. A second class of repetitive DNA is satellite DNA sequences. They are tandemly repeated sequences. Tandem repeats range in length from a few base pairs to a few thousand base pairs. The third type of reassociating sequences are interspersed repeats. The number of copies of these sequences may vary from a few to hundreds of thousand

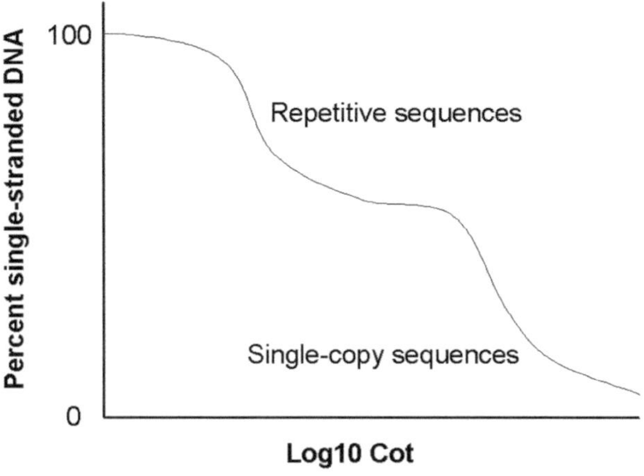

Fig. 7.18 C_0t curve of DNA. Log C_0t vs ssDNA represent the amount of repetitive and single-copy sequences (Adapted from: https://en.wikipedia.org/wiki/Cot_analysis)

copies. 'Alu' is one such repeat present in 500,000 copies. Fourth category is 'low copy long repeats'. Red-green colour vision gene belongs to this category. Genetic changes have occurred over time in all these groups, and they have acquired sequence divergence. Divergence amongst DNA sequence repeats can be measured by Cot analysis.

DNA concentration (C_o), the time allowed for reassociation (t) and the sequence organization of the DNA are the factors deciding the extent of reassociation. Since the extent of reassociation is directly proportional to both C_o and t, their arithmetic product (C_0t) denotes the extent of reassociation. A DNA solution whose reassociation is 50% complete at $C_0t = 100$ reassociates much more rapidly than the DNA solution whose reassociation is 50% complete at $C_0t = 10{,}000$. The DNA of the first solution is said to have much less sequence complexity. The sequence complexity of a genome is the total length of the different sequences it contains, measured in nucleotide pairs. For species with no repetitive DNA, the sequence complexity is equal to the genome size. Thermal stability of a renatured DNA can be assessed by C_0t analysis. If there is DNA mismatch, thermal stability is less, and hence lesser temperature will be required to separate the strands. The difference in the dissociation temperature when compared to the dissociation of native DNA can be used to calculate the percentage of DNA mismatch during reassociation. Greater is the difference; higher is the percentage of mismatch. The ability of RNA associating with DNA to form RNA/DNA duplex can be analysed by C_0t curve. In this case it is known as R_0t curve.

7.4.8 Electrophoresis of Nucleic Acids

DNA is negatively charged due to the presence of phosphate groups. In an electric field, the DNA moves from negative to positive electrode (Fig. 7.19). The rate of movement depends on the charge of the molecule. The mass of the molecule plays no direct role in electrophoretic migration; however, larger molecules face larger frictional forces which adversely affect their mobility. Hence, separation of molecules in the electrophoretic field is dependent on charge to mass ratio. The most common electrophoresis is gel electrophoresis. Agarose or polyacrylamide gels are prepared containing small slots for loading the DNA samples. Those slots are called wells. The speed of migration of DNA through the gel matrix will depend on the size of the DNA fragment (smaller fragment will run faster) and on its tertiary structure (supercoiled, relaxed or open circular or linear). Supercoiled DNA is very compact (less surface area and less frictional resistance) and hence moves fastest. For linear DNA, rate of migration is inversely proportional to molecular mass. The distance (D) moved is approximately $D = a - b \log M$ (molecular mass), in which a and b are constants, depending on buffer, gel concentration and temperature.

7.4.8.1 Pulse Field Gel Electrophoresis

Standard gel electrophoresis cannot separate DNA molecules larger than 20 kb. PFGE can separate as large as 200–3000 kb DNA molecules. During PFGE,

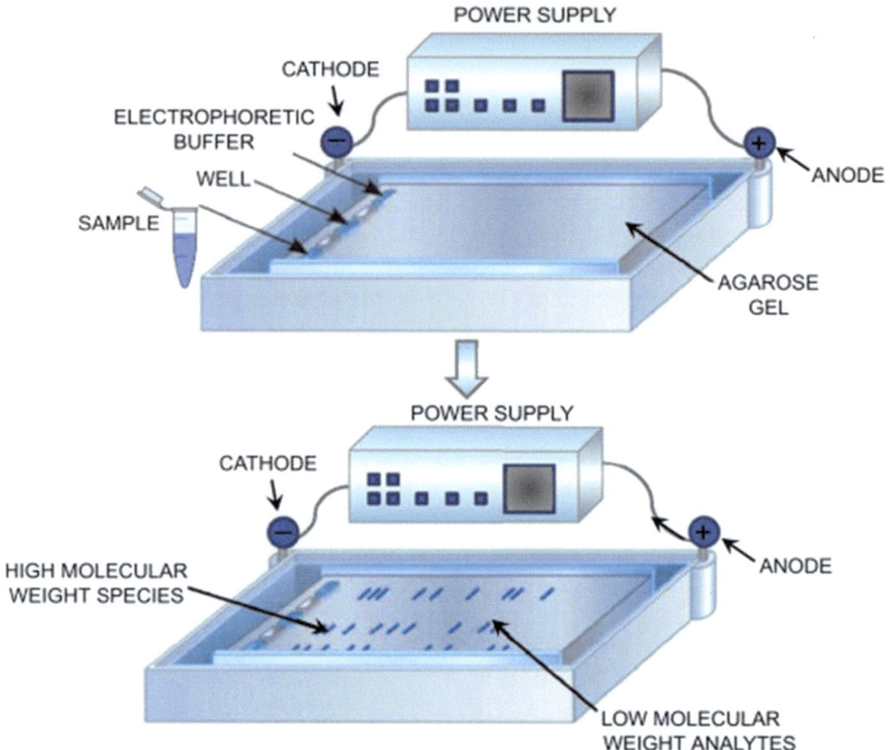

Fig. 7.19 Electrophoresis of nucleic acids. Unidirectional electric field is applied, and the DNA being negatively charged moves towards the positive electrode. Charge to mass ratio determines the mobility of proteins in the gel. The smallest fragment moves fastest, and the larger fragments stay near the well

orientation of electric field changes periodically (Fig. 7.20). This allows the DNA to bend like a snake through the gel matrix pores.

7.5 DNA Replication: Semi-conservative

DNA was proposed to be a double helix having antiparallel strands. That means, if one of the strands is running from 3′ to 5′, then the other strand is running 5′ to 3′. Therefore, 3′-OH group of one strand is opposite to the 5′-PO_3^- group of the other. Directionality is important for understanding the functions of enzymes involved in DNA replication.

In the 1950s, three different modes of replication were proposed. First model proposed **conservative** mode of replication. As per this hypothesis, the parental strands replicate to form daughter strands, and after replication, the parental strands stay together to form one double helix, and the new daughter strands form a new

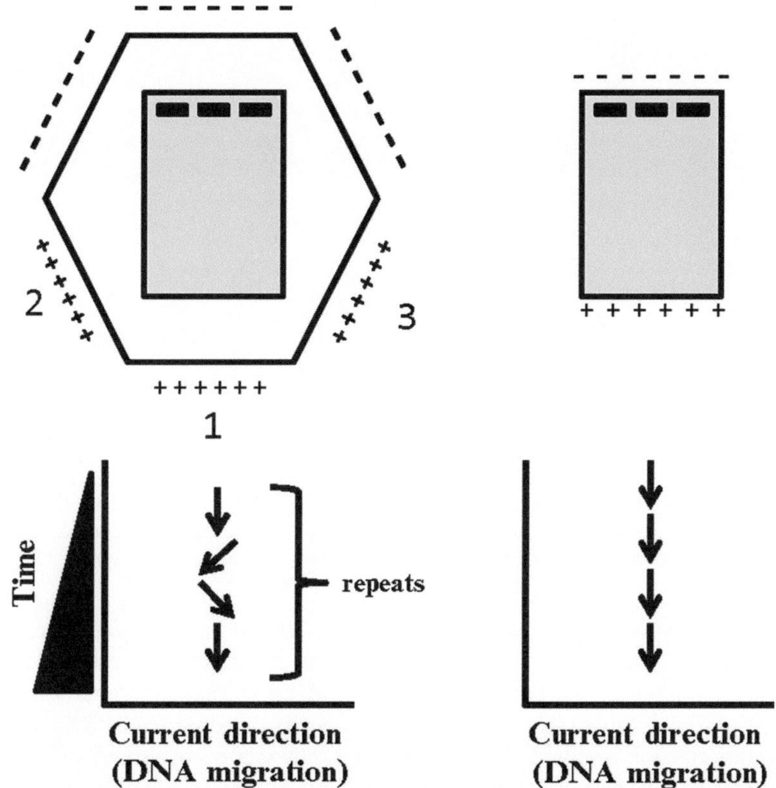

Fig. 7.20 Pulse field gel electrophoresis of DNA separation. The direction of electric field changes periodically, and depending on the direction of the electric field, direction of movement of DNA also changes. DNA moves through the gel like a snake (Adapted from: Sharma-Kuinkel et al. 2016)

double helix. Second model proposed **semi-conservative** mode of replication. As per this, the parental strands separate out, and each strand synthesizes the complementary daughter strand. The double helix is then formed between one parental and one daughter strand. **Dispersive** model proposed that the segment of parental strands is interspersed by newly synthesized daughter strands (Fig. 7.21).

In 1958, Matthew Meselson and Franklin Stahl showed that DNA replication actually follows semi-conservative model. They grew *E. coli* in a medium containing N^{15} for several generations (Meselson and Stahl 1958). The DNA of those bacteria was all labelled by heavy isotope of nitrogen having no N^{14} present in the DNA. They then grew the bacteria for different time periods in normal medium containing only N^{14} and no N^{15}. The rationale behind their methodology was that after every cycle of DNA replication N^{14} will be incorporated in the daughter strands. New strands incorporating N^{14} will be lighter than the parental strands containing N^{15} and hence will give separate bands in the CsCl density gradient. 10^8 *E. coli* were lysed after different time periods and subjected to CsCl density gradient centrifugation at

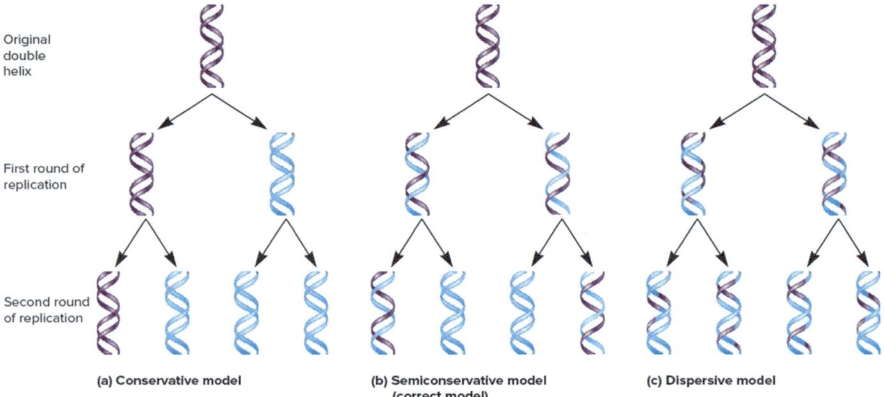

Fig. 7.21 Three possible models of DNA replication. The parent strands are shown in purple, and daughter strands are shown in blue. (**a**) In conservative model, the parental strands form a separate duplex, and daughter strands form a separate double strand. (**b**) In semi-conservative model, one of the parent strand base pairs with one of the daughter strands. (**c**) In dispersive model, both parent and daughter DNA are present in the same strand (Adapted from: Brooker 2018)

around 31,000 rpm. Since the sedimentation and diffusion are opposing forces, they produce a stable concentration gradient of the caesium chloride. These forces create a continuous increase of density along the direction of centrifugal force. The resulting gradient drives the DNA where its density is equal to the density of the CsCl solution. This is called equilibrium point. DNA stays as a band at this point. Lighter DNA will form wider band, and heavier DNA will form narrow band. The reason is lighter DNA is subjected to more diffusion force compared to the heavier DNA (Fig. 7.22).

Since the parental strands had N^{15} and with each replication cycle more and more N^{14} was getting incorporated, it was expected that there will be multiple DNA molecules with different densities. Density of the DNA molecules will be reducing after each cycle. Due to the centrifugal force, the densest band will be at the extreme end from the axis of centrifugation. As the density decreases, the bands will move closer to the axis. They aliquoted around 10^9 *E. coli*, lysed them, centrifuged at 45,000 rpm and allowed them to reach equilibrium with the CsCl gradient. The DNA in the tube was then observed by UV light. Important interpretations of the experiment are as follows (Fig. 7.23)

1. The bands are moving closer to the axis of rotation with time. So, the density of sample is reducing with incorporation of N^{14}.
2. Initially there was only one band, so all the DNA had same molecular weight. Then at 0.3 and 0.7 cycle, lighter bands appeared on the left. It indicated the presence of an intermediate density DNA species whose concentration was less than the initial band. The intensity of the band denoted the concentration of the DNA species.

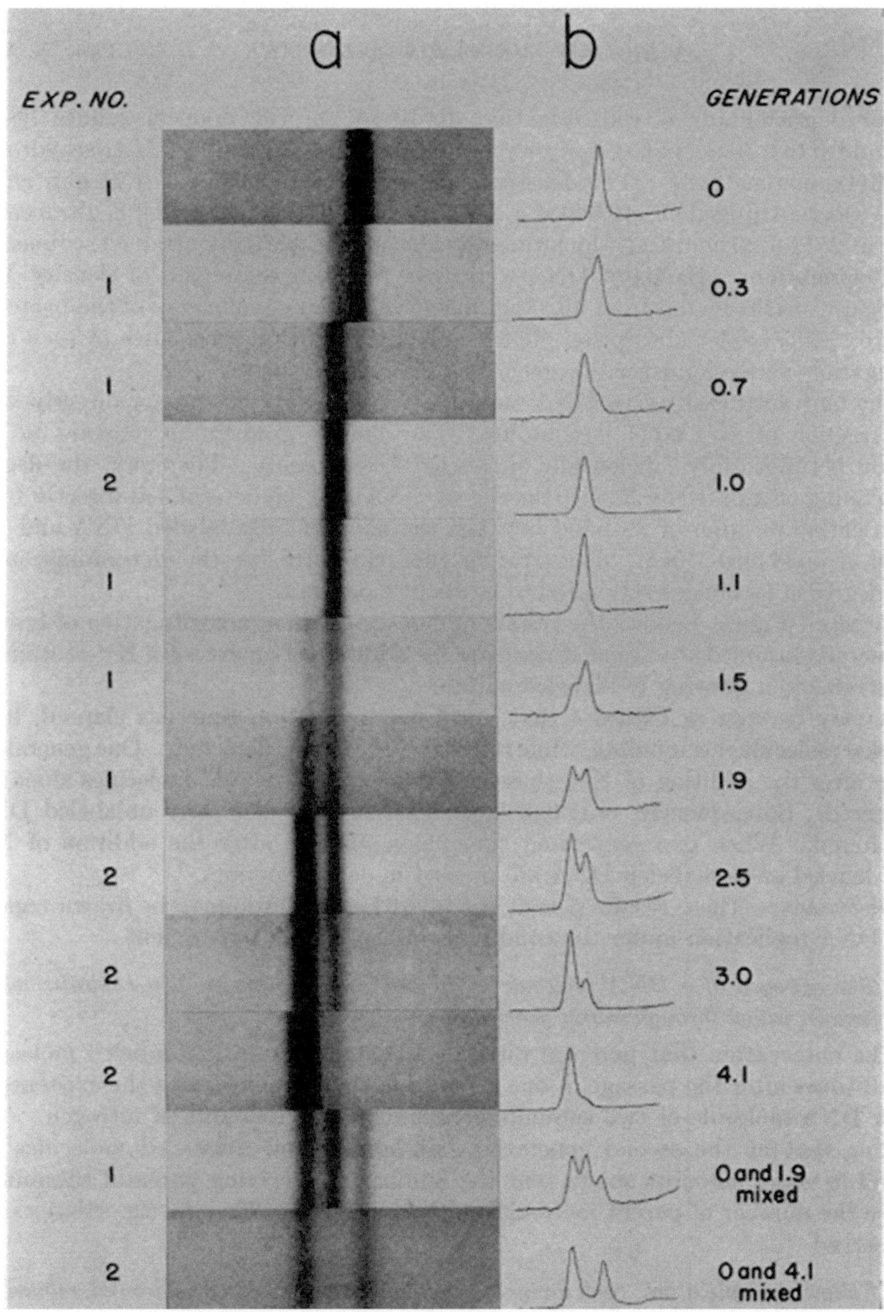

Fig. 7.22 Ultraviolet absorption photographs showing DNA bands resulting from density gradient centrifugation of bacterial lysates. Samples were collected at different time intervals after adding excess of N^{14} substrates to a growing N^{15}-labelled culture. Each photograph was taken after 20 h of centrifugation at 44,770 rpm. The time of sampling is measured from the time of the addition of N^{14} in units of the generation time. (Adapted from: Meselson M and Stahl F.W. 1958, 44, PNAS, 671–682)

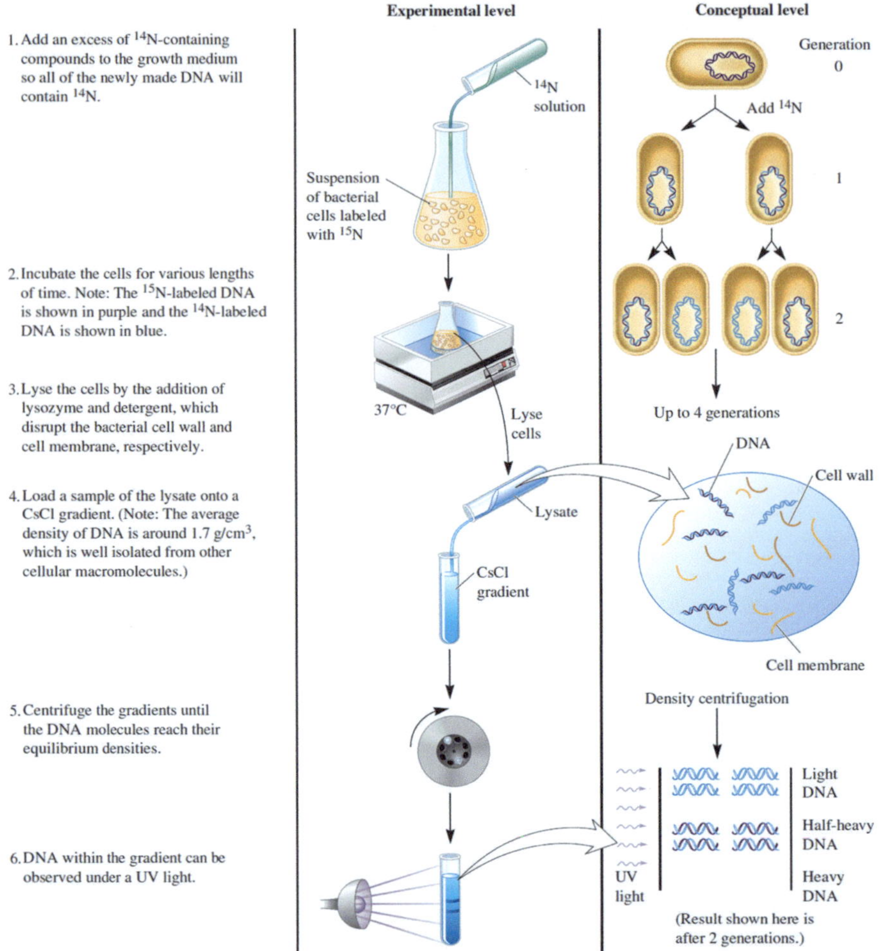

Fig. 7.23 The conceptual explanation of Meselson and Stahl experiment. Cells were initially grown in N^{15}-containing media so that the entire DNA is labelled with heavy isotope of nitrogen. Then N^{14} was added and incubated for various lengths of time and was then lysed. The lysate is then loaded on CsCl gradient and centrifuged. DNA in the gradient can be observed under UV light (Adapted from: Brooker 2018)

3. After one round of replication, a single band appeared which is half heavy as the initial band. It is consistent with both semi-conservative and dispersive model. In contrast, as per the conservative model, there should have been two bands, one heavy and one light. Clearly conservative model was incorrect.
4. To confirm whether it was semi-conservative or dispersive model, Meselson and Stahl checked further replication cycle. At 1.9 cycle, there were two bands, one for the mixture of heavy and light bands and the other for the all light DNA. As per dispersive model, there should have been only one band as each strand of

DNA would have carried 1/4 heavy DNA and 3/4 light DNA. Since single band was not obtained after two replication cycles, dispersive model was also discarded, and semi-conservative model was accepted.

7.6 Different Modes of DNA Replication

Till now we have learnt how DNA was shown to be the genetic material, what are the various techniques to analyse DNA, the structure of DNA and the semi-conservative mode of its replication. In this section we will learn that different organisms have different modes of replication. We will begin with the bacterial replication and then move on to eukaryotic replication.

Bacteria have single origin of replication: The Fig. 7.24 shows the autoradiogram of bacterial chromosomal DNA replication. The site at which the DNA replication begins is called **origin of replication**. At this site the DNA double strand separates from each other forming two single strands at that site. Cairns showed by autoradiography the replication in *E.coli*. The autoradiographic image appeared like Greek letter θ (Fig. 7.24). So, it was known as theta mode of replication. As the single strands are present for only a few base pairs and the rest of the DNA is still in the double-stranded form, it gives an appearance of two 'Y'-shaped fork facing each other. These are called **replication forks**. As the replication proceeds, the single-stranded region lengthens, and so the replication forks also increase in size. Hence, the replication proceeds in both the directions, opposite to each other. Therefore, it is called **bidirectional replication**. Eventually the replication forks extending on opposite directions meet each other to complete the replication. Since it is closed circular DNA, unwinding of DNA induces positive supercoiling. Most of the circular DNAs are negatively supercoiled; therefore, initial unwinding is not difficult; however as replication proceeds, unwinding becomes difficult due to the formation of positive supercoils. Those are removed by the action of topoisomerase

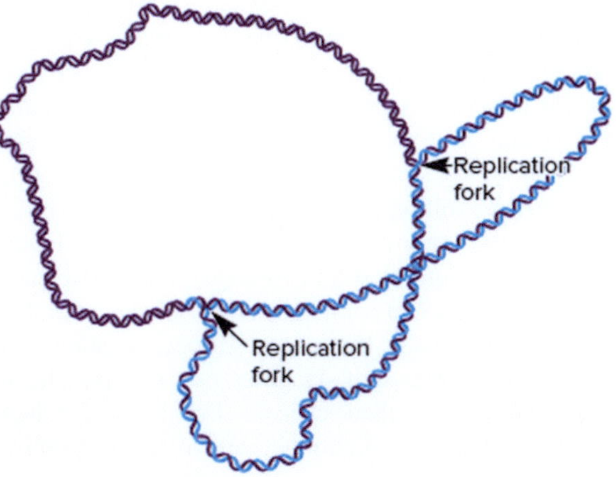

Fig. 7.24 θ-mode of replication in *E. coli.* In bacteria, as closed circular DNA proceeds for replication, the parent strands separate, and replication fork proceeds in opposite direction; it takes the form of Greek letter θ (Adapted from: Brooker 2018)

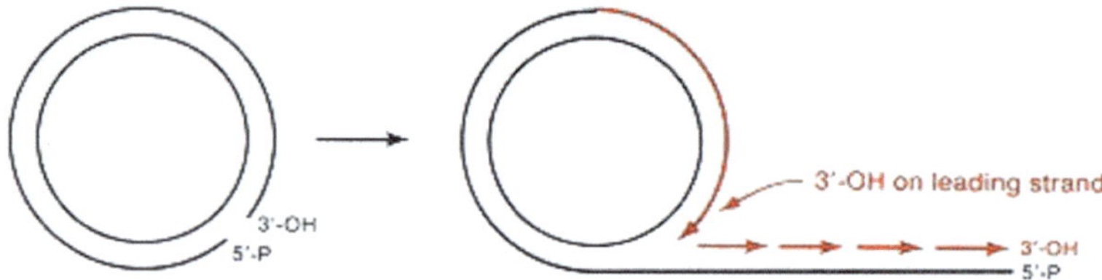

Fig. 7.25 Rolling circle mode of replication. The parent strand is shown in black and daughter strand in orange (Adapted from: Maloy et al. 1994)

II which creates a nick in one of the strands which swivels around the other strand to release the positive supercoil.

In some of the viruses, some bacterial plasmids and circular genomes like that of mitochondria replicate by **rolling circle replication**. In this mode of replication, one of the strands of closed circular double-stranded DNA is nicked. A free 3′-OH and free 5′ P are generated. The other strand remains intact and acts as template for the leading strand. DNA polymerase uses the free 3′-OH of the nicked strand and starts synthesizing the DNA complementary to the intact circular DNA (Fig. 7.25). Therefore, the newly synthesized DNA is covalently linked nicked strand. The new daughter molecules are present as linear DNA. Since the template DNA is present as a closed circle, the DNA polymerase synthesizes the complementary strand multiple times producing a linear concatemer of repeated sequences. These concatemeric DNAs are often essential in phage replication where these are packaged inside the viral particles. Even in bacterial mating, the donor DNA is transferred to the recipient by rolling circle replication. The parental strand for the lagging strand is displaced and replicates by usual discontinuous replication method.

Another rolling circle replication variant generates linear DNA from closed circular double-stranded DNA. It is called looped rolling circle replication. ϕ174 replicates by this mode. A phage protein (the A protein) nicks the viral strand replication origin and becomes covalently linked to the newly formed 5′ P. DNA pol III recognizes the newly formed 3′-OH group and starts synthesis displacing the nicked parental strand called the (+) strand. This strand becomes coated with SSB proteins and cannot act as template. Synthesis continues until the origin is reached. At this point the A protein binds to the 3′-OH group of the (+) strand and joins the 3′-OH and 5′-P groups of the (+) strand, dissociates and attaches to the newly synthesized (+) strand (Fig. 7.25) (Khan et al. 1997).

In looped rolling circle replication of phage ϕ174, the A protein nicks a supercoil terminus of a strand, known as the (+) strand. Rolling circle replication ensues to generate a daughter strand (orange) and a displaced (+) strand bound by single-strand binding (SSB) proteins. The displaced strand remains covalently linked to the A protein and hangs like a loop. Hence, is called **looped rolling circle replication**. When the synthesis of daughter (+) is complete, the entire parent (+) strand is displaced and is then cleaved from the daughter (+) strand. It is thereafter circularized by the joining activity of the A protein. The DNA molecule with the

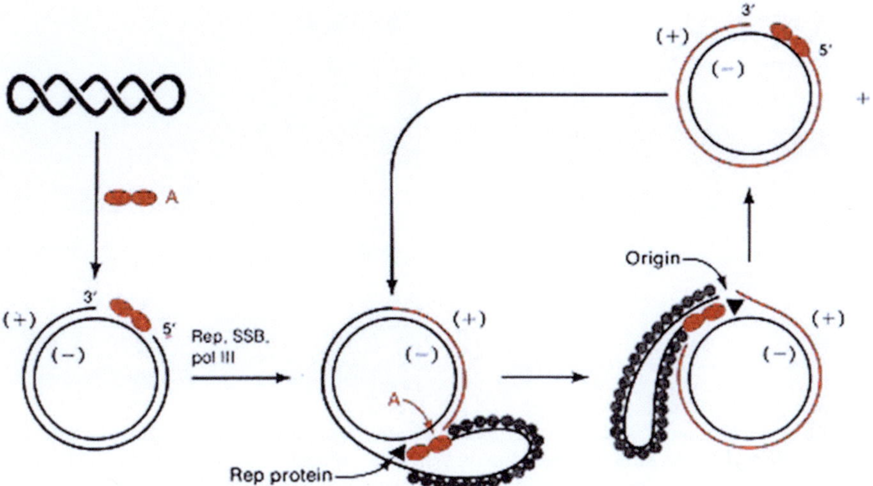

Fig. 7.26 Looped rolling circle replication of phage 174. Parent strands are shown in black and daughter strand in orange. As the two strands separate out, one of the strands is displaced and gets bound by SSB proteins. The displaced strand remains bound by A protein and hangs like a loop (Adapted from: Maloy et al. 1994)

new daughter strand is now ready for next round of replication. In this mode, the (−) strand is never cleaved (Fig. 7.26).

In eukaryotes, the chromosomes are longer and linear. Hence, to complete the process of replication faster, they have multiple origins of replication (Fig. 7.27). Evidence for multiple origins of replication was provided in 1968 by Joel Huberman and Arthur Riggs provided by adding a radiolabelled nucleoside (^{3}H-deoxythymidine) to a culture of actively proliferating cells. For a brief period of time, labelled thymidine was given, and then large amount of unlabelled thymidine was added. Chromosomes were then isolated and subjected to autoradiography. It was observed that labelled thymidine was incorporated at certain intervals. Therefore, it was proven that there are multiple origins of replication. The replication fork is formed similar to that of prokaryotes and proceeds bidirectionally. There are multiple replication forks which proceed simultaneously. Once the adjacent replication forks meet each other, they stop progressing, and replication eventually comes to an end when all the replications forks stop proceeding.

7.7 Mechanism of DNA Polymerase

DNA polymerase II has many subunits. Out of those, the β-subunit acts as a clamp (Table 7.5). The two β-subunits of the two DNA polymerase come together to form a ring. The hole of the ring is wide enough to accommodate one turn of DNA double strand. The clamp is loaded on the DNA with the help of other subunits which are

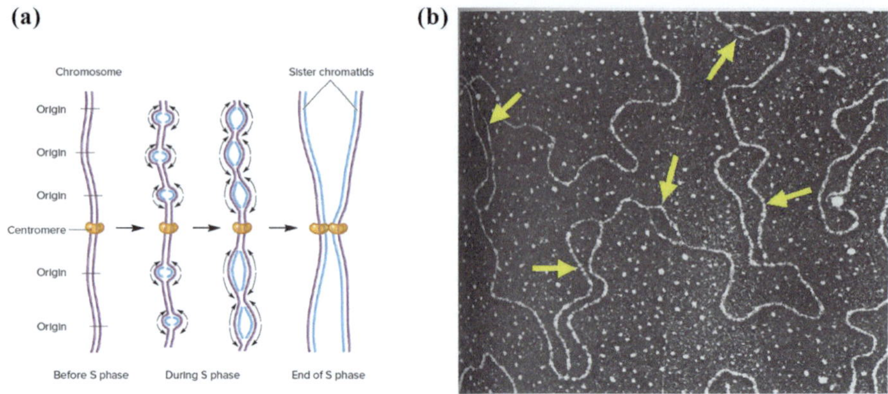

Fig. 7.27 Replication of eukaryotic chromosomes. (**a**) There are multiple origins of replication, and all replication forks proceed directionally. When the replication fork merges, the replication is completed forming two sister chromatids (Adapted from: Brooker 2018). (**b**) A microscopic image of multiple origins of replication in a eukaryotic chromosome. Each origin is observed as a replication bubble. Arrows are marked to identify the bubbles [Adapted from: A. Blumenthal, H. Kriegstein, and D. Hogness, 'The units of DNA replication in Drosophila melanogaster chromosomes', Cold Spring Harbor Symp. Auant. Biol. 38, 205–223 (1974)]

Table 7.5 Subunits of the DNA polymerase III holoenzyme

Subunit	Function	Groupings
α	5′–3′ polymerization	
ε	3′–5′ exonuclease	
θ	Core assembly	
γ, δ, δ' χ, ψ	Load enzyme on template (serves as clamp loader)	γ complex
β	Sliding clamp structure (processivity factor)	
τ	Dimerizes core complex	

together known as clamp loader. In the presence of β-subunit, the DNA polymerase has the synthesis rate of 750 nucleotides per second. In the absence of β-subunit, the synthesis rate is only 20 nucleotides per second as the enzyme falls off the template DNA after synthesis of only a few bonds. Therefore, the β-subunit is known to increase the **processivity** of the enzyme.

The α subunit is the catalytic subunit. In the catalytic site, the incoming nucleotide enters following the base pair rule. The incoming nucleotide has three phosphates which gets covalently linked to the last nucleotide added to the newly synthesized strand at its free 3′-OH group. The catalysis leads to removal of two terminal (β and γ) phosphates of the incoming nucleotide in the form of pyrophosphate. The pyrophosphate is later degraded into Pi. The ester bond is formed between the α-phosphate of the incoming nucleotide and 3′-OH of the last nucleotide added in the newly synthesized strand. Therefore, the direction of DNA synthesis is always 5′ to 3′.

Table 7.4 Properties of bacterial DNA polymerase I, II and III

Properties	I	II	III
Initiation of chain synthesis	–	–	–
5′–3′ polymerization	+	+	+
3′–5′ exonuclease activity	+	+	+
5′–3′ exonuclease activity	+	–	–
Molecules of polymerase/cell	400	?	15

DNA synthesis occurs with high amount of accuracy; only one mismatch occurs per 100 million bases. That is called as **fidelity** of the enzyme.

The structure of the active site of DNA polymerase allows only AT and GC pairing. Any incorrect nucleotide entering the active site leads to helix distortion Therefore, incorrect nucleotides can rarely occupy the active site. The correct nucleotide after entering the active site leads to conformational change in the active site and induced fit mechanism of catalysis ensues. AT and GC base pairs have very low potential energy and hence are extremely stable compared to incorrect base pairs. The stability of the bases is another factor leading to high accuracy of DNA synthesis.

DNA polymerase III has 3′ to 5′ exonucleolytic activity (Table 7.4). In case there is any mismatch, the enzyme has 3′ exonuclease site which can remove the wrongly incorporated base. This is called **proofreading** activity. Due to the proofreading ability of the enzyme, its fidelity increases further (Table 7.5) (Lamers et al. 2006).

7.8 Process of DNA Replication

Replication is initiated by DnaA protein at the origin: Origin of replication in *E. coli* is called *OriC*. Three types of DNA sequences are found in OriC: AT rich, DnaA box binding region and GATC methylation sites.

There are five DnaA boxes and three AT-rich regions. GATC methylation sites precede the AT-rich regions. DnaA protein in the ATP-bound state binds to the five DnaA boxes, and due to protein binding, DNA strands bend around the proteins (Fig. 7.28). Protein HU and IHF then bind and help to unwind the DNA at the AT-rich sites. Since AT has only two hydrogen bonds, it is relatively easy to be separated compared to GC-rich regions.

Once the DNA strands have separated, DnaA along with DnaC recruits a helicase to unwind the double helix. The separation of DNA strands to form single strands requires ATP. The DnaA protein remains bound to the single strand and moves along the replication fork as the directional replication proceeds (Fig. 7.29, Table 7.6).

GATC methylation sites are methylated at the adenine residues by the enzyme DNA adenine methyl transferase or Dam. All the methylation sites need to be methylated before the replication can begin. Since the newly made daughter strands are unmethylated, they cannot undergo replication again unless all the methylation sites are methylated.

Fig. 7.28 The sequences of OriC in *E. coli*. The AT-rich sequence is composed of three similar sequences that are 13 bp long and highlighted in blue and five DnaA boxes that are highlighted in orange. The GATC methylation sites are underlined (Adapted from: Brooker 2018)

7.8.1 DNA Replication at the Replication Fork

Several enzymes are involved in replication accomplishing different functions. In this section we will learn about those enzymes.

7.8.2 Unwinding of Double Helix

As we learnt in the earlier section, DNA helicases are required to unwind the DNA. It requires ATP to separate the two strands. As the strand separation proceeds, positive supercoils are induced in the region ahead of separated strands. DNA gyrase or topoisomerase II helps in releasing the positive supercoiling. To ensure that the separated strands stay in the single-stranded state, single-strand binding proteins bind to the single strand of DNA.

7.8.3 Synthesis of RNA Primer

The enzyme principally involved in replication is called DNA polymerase. However, this enzyme cannot start de novo synthesis of DNA strand. It can only extend

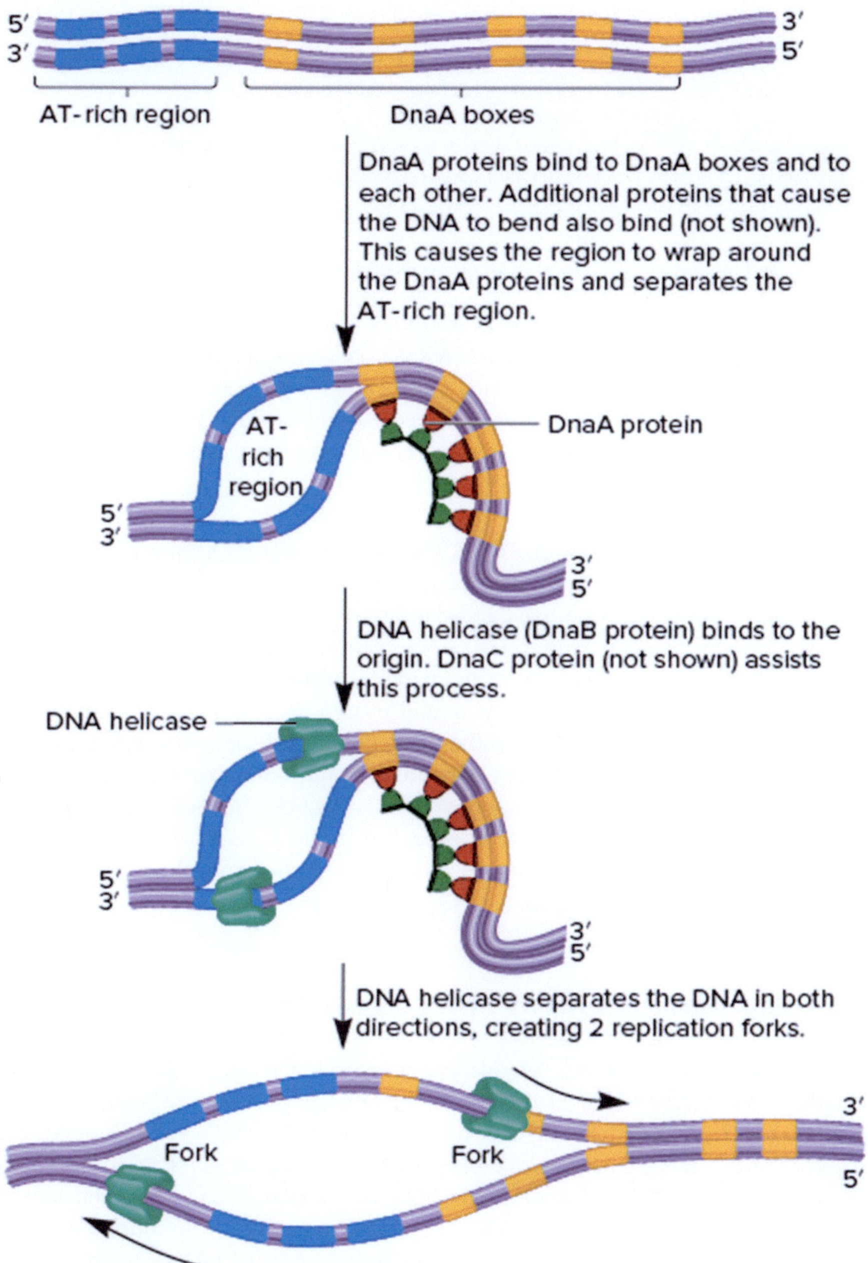

Fig. 7.29 Replication in *E. coli*. Proteins involved in replication at various stages have been indicated. AT-rich strands separate first, and DnaA protein binds to DnaA region. DNA helicase binds in the AT-rich region and proceeds in the opposite directions (Adapted from: Brooker 2018)

Table 7.6 Proteins involved in *E.coli* DNA replication

Common name	Function
DnaA proteins	Bind to DnaA box sequence within the origin to initiate DNA replication
DnaC proteins	Helps DnaA in the recruitment of DNA helicase
DNA helicase (DnaB)	Separate double-stranded DNA
Topoisomerase II (DNA gyrase)	Removes positive supercoiling in front of the replication fork
Single-strand binding proteins	Bind to single-stranded DNA and stabilizes in single-stranded form
Primase	Synthesizes short RNA primers
DNA polymerase III	Synthesizes DNA in the leading and lagging strand
DNA polymerase I	Removes RNA primers, fills in gap with DNA
DNA ligase	Covalently attaches adjacent Okazaki fragment
Tus	Binds to *ter* sequences and prevents the advancement of the replication fork

the polynucleotide chain to form DNA. Therefore, it requires a short RNA primer of 10–12 nucleotides, which can be extended by DNA polymerase to synthesize the new strand. Primase enzyme has the capability of synthesizing de novo short RNA primers. The direction of synthesis of polynucleotide chain synthesis is always from 5′ to 3′ direction. Therefore, one of the parental strands having 3′ to 5′ orientation acts as template for synthesis of new strand from 5′ to 3′ direction maintaining the antiparallel orientation. For the synthesis of this strand, only one RNA primer is required and is called the **leading strand**. For the other parental strand having orientation 5′ to 3′, the new strand synthesis requires several primers at short intervals, and those primers are then extended in the 5′ to 3′ direction, and those short DNA strands are later ligated together. This mode of synthesis is slower and hence called the **lagging strand**.

7.8.4 Synthesis of DNA by DNA Polymerase

In *E. coli* there are five DNA polymerase enzymes. DNA polymerases I and III are the major enzymes which are involved in DNA replication, whereas the DNA polymerases II, IV and V are involved in repair of damaged DNA. DNA polymerase III is the principal enzyme involved in replication. It has ten subunits. The α subunit carries out the function of synthesizing phosphodiester linkage between adjacent nucleotides (Table 7.5). The other nine subunits are involved in significant accessory roles in polynucleotide chain synthesis (Fig. 7.30, Table 7.5). DNA polymerase I is involved in removing the RNA primer and filling the gaps between the short polynucleotide chains (Fig. 7.31).

Different bacterial polymerases have different subunit compositions. However, the catalytic subunit in all these polymerases resembles the structure of the human fist (Fig. 7.32). The template traverse through the palm, thumb and fingers is

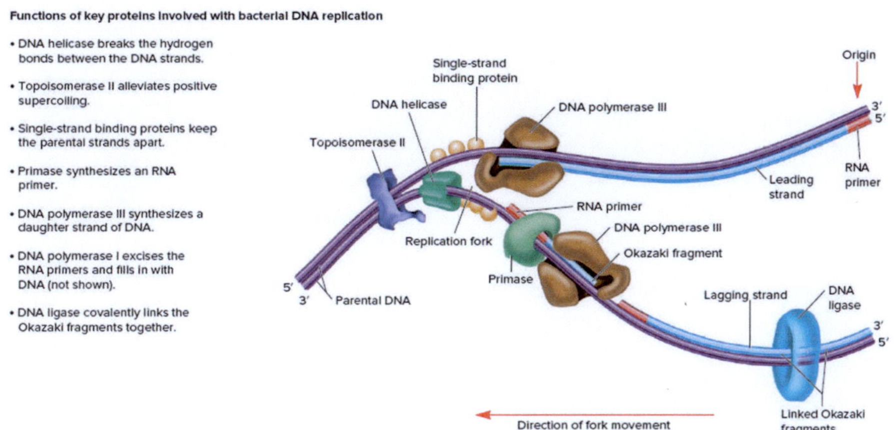

Fig. 7.30 Proteins involved in bacterial DNA replication. Proteins like topoisomerase II, DNA helicase, single-strand binding proteins, primase, DNA ligase and DNA pol III binding to the DNA strands at the replication fork (Adapted from: Brooker 2018)

wrapped around it. DNA polymerase allows attachment of incoming nucleotides only at the 3′ end. It also removes incorrect nucleotides though 3′ to 5′ exonuclease activity.

7.8.5 Synthesis of Leading and Lagging Strand

We learnt in the previous section that DNA polymerase lacks the ability of de novo polynucleotide chain synthesis and can extend polynucleotide only in 5′ to 3′ direction by adding new nucleotides at the 3′-OH end. Due to these limitations, both the DNA strands cannot be synthesized at the same rate. One of these becomes the leading strand, and the other becomes the lagging strand. In the leading strand, one RNA primer is made by primase, and DNA polymerase III extends it towards the opening of the replication fork in the 5′ to 3′ direction.

In the lagging strand, although the direction of synthesis is from 5′ to 3′, it occurs away from the opening of the replication fork. Several RNA primers are made by primase to facilitate synthesis of DNA fragments in the 5′ to 3′ direction. These fragments are 1000–2000 nucleotides long. In humans, these fragments are 100–200 nucleotides long (Fig. 7.33). All these fragments have short RNA primers which are later removed by DNA pol I. The shorter fragments of DNA are called Okazaki fragments after their discoverer, Reiji and Tsuneko Okazaki. DNA pol I has 5′ to 3′ exonuclease activity. It removes the RNA primer from 5′ to 3′ direction and then fills the vacant space by synthesizing DNA in the 5′ to 3′ direction. It uses the 3′ end of the adjacent Okazaki fragment to fill in the space. The adjacent Okazaki fragments are then ligated together by DNA ligase. DNA ligase in bacteria depends on NAD^+ for energy, and in eukaryotes and archaea, it depends on ATP.

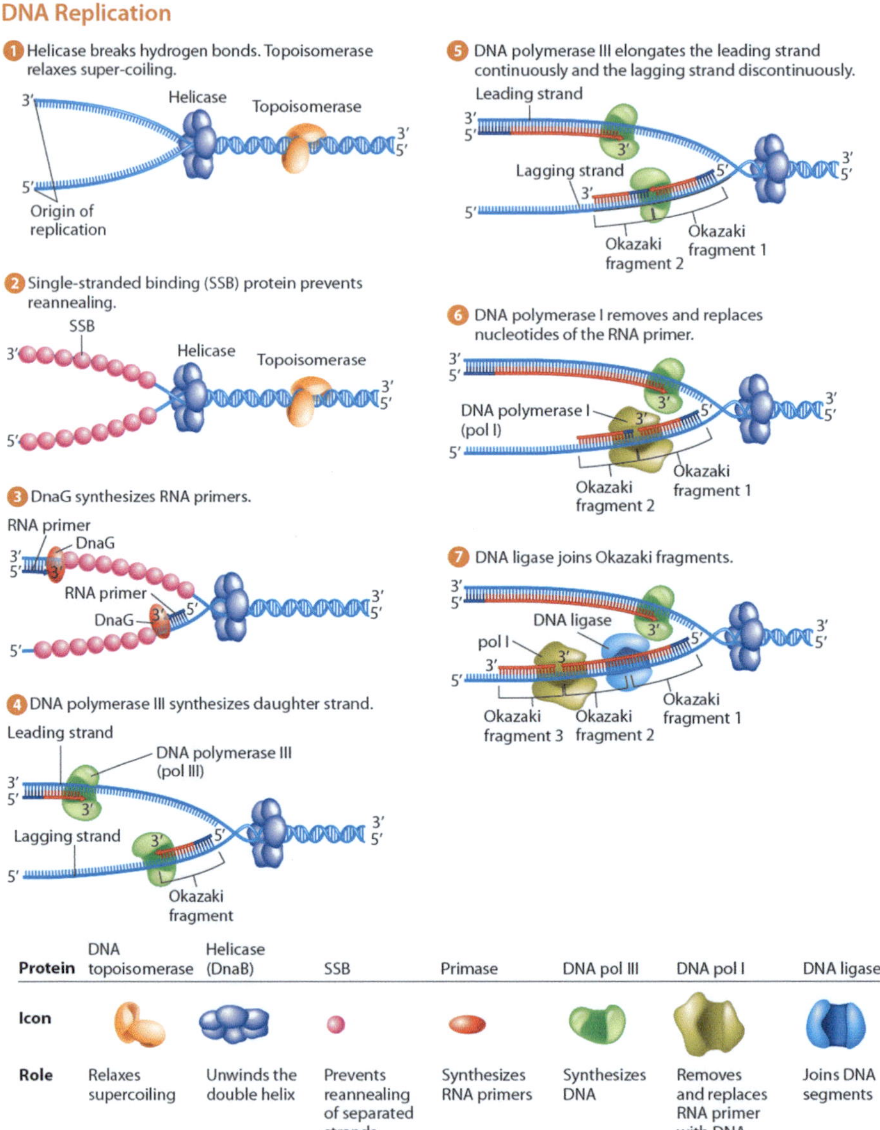

Fig. 7.31 DNA replication prokaryotes. (1) Helicase breaks hydrogen bonds and relaxes supercoiling. (2) SSB proteins stabilize single strand. (3) DnaG synthesizes RNA primers. (4) DNA pol III synthesizes daughter strands. (5) Leading and lagging strand synthesis. (6) DNA pol I removes RNA primers. (7) DNA ligase joins Okazaki fragments (Adapted from: Sanders and Bowman 2015)

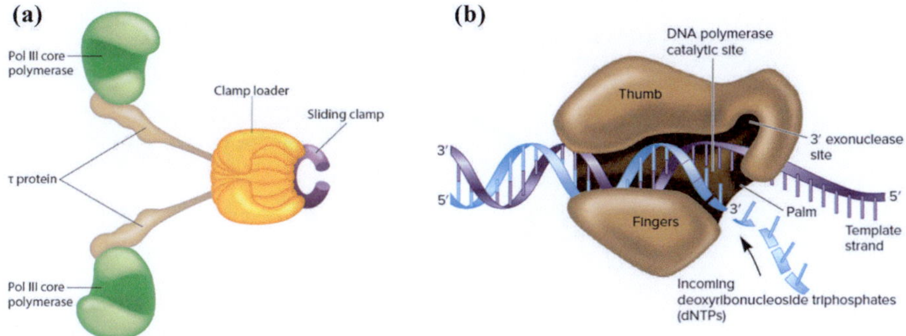

Fig. 7.32 DNA pol III holoenzyme. (**a**) Two core enzymes attached to tau arms, clamp loader and a sliding clamp (Adapted from: Sanders and Bowman 2015). (**b**) Model of pol III side view binding to DNA (Adapted from: Brooker 2018). The catalytic subunit of DNA polymerase resembles a hand that is wrapped around the template strand. Thus, the movement of DNA polymerase along the template strand is similar to a hand that is sliding along a rope (Lamers et al. 2006)

7.8.6 Formation of DNA Replication Complex

Separation of DNA strands and synthesis of RNA primers proceed along the complex of helicase and primase enzymes. These together make the **primosome** complex. This complex helps to lead the replication fork ahead. The primosome then physically associates with two DNA polymerase holoenzymes to form the **replisome** complex. Two DNA polymerases III act to replicate the leading and lagging strands (Fig. 7.34).

Two DNA polymerases move together as a unit. The lagging strand is looped out with respect to the DNA polymerase that synthesizes the lagging strand. This looping makes DNA accessible to the primase for RNA primer synthesis as well as to the polymerase so that it can synthesize DNA in 5′ to 3′ direction. Therefore, looping helps DNA polymerase dimer to move as a unit although there is a difference in the rate and mechanism of synthesis of the two strands. When the lagging strand polymerase reaches the end of Okazaki fragment, it gets released from the template and jumps to the RNA primer next to it. DNA clamp loader loads the DNA polymerase to the next RNA primer. The **proliferating cell nuclear antigen** (PCNA) protein functions as the sliding clamp in archaeal and eukaryotic replication, encircling the DNA template strand. Replication factor C (RFC) complex connects the DNA polymerases to the clamp loader and sliding clamp (Figs. 7.34 and 7.35).

Replication termination: In *E. coli*, opposite to the OriC sequence, there are termination sequences. Usually there are two termination sequences. One of the termination sequences, T1, inhibits the replication fork from left to right, and the other termination sequence, T2 prevents progression of replication fork from right to left (Fig. 7.36). Termination sequences are bound by proteins known as **termination utilization sequences** (tus). If one of the replication forks is halted by one of the termination sequences, the other replication fork progression stops when it meets the halted DNA polymerase. DNA ligase covalently links the two newly synthesized

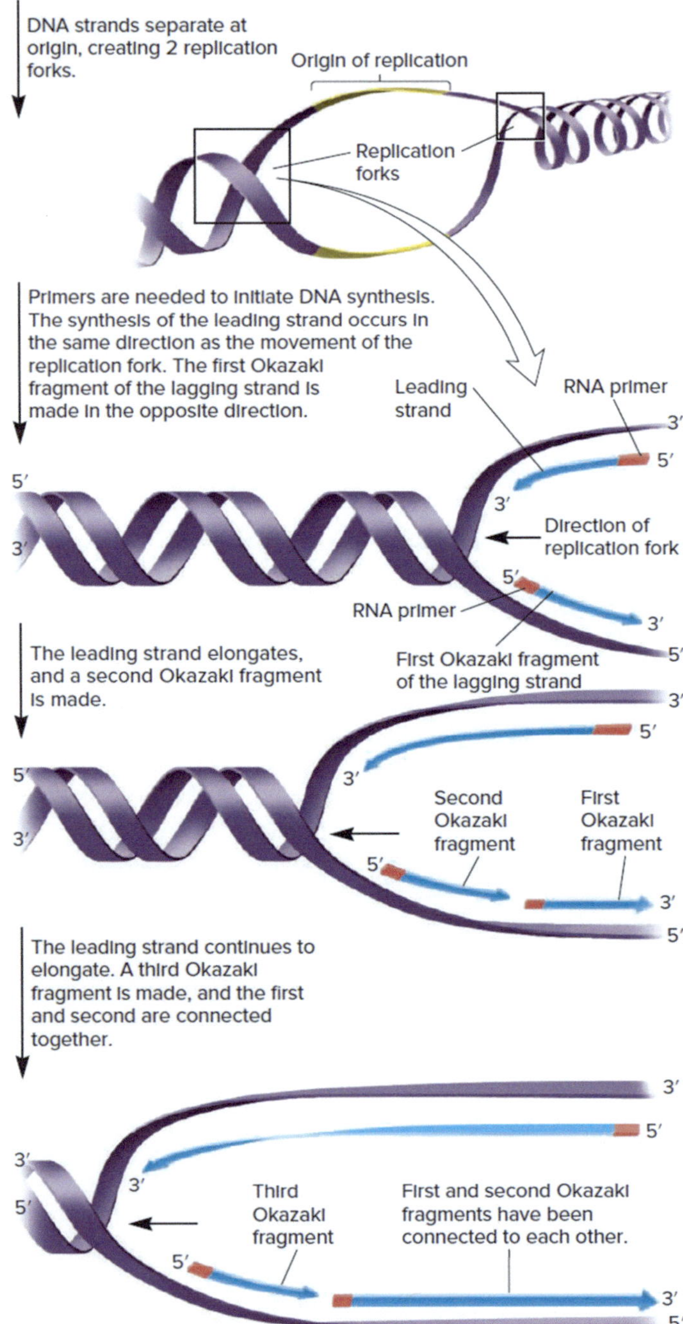

Fig. 7.33 Synthesis of leading and lagging strands. Parental strands are shown in purple, newly synthesized DNA in light blue and primers in light red (Adapted from: Brooker 2018)

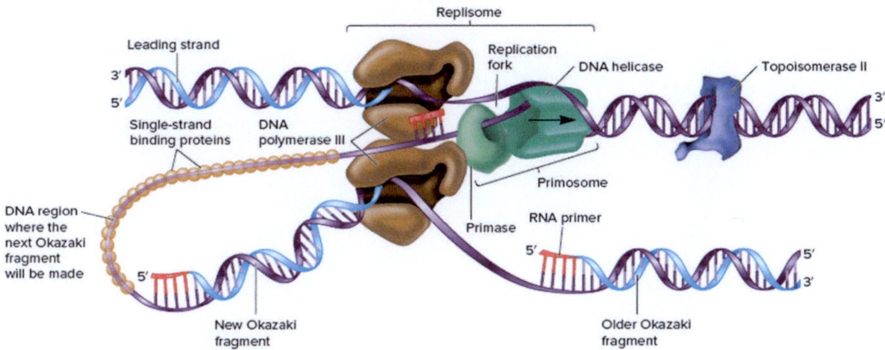

Fig. 7.34 A three-dimensional view of DNA replication. DNA helicase and primase associate together to form a primosome. The primosome associates with two DNA polymerase holoenzymes to form the replisome (Adapted from: Brooker 2018)

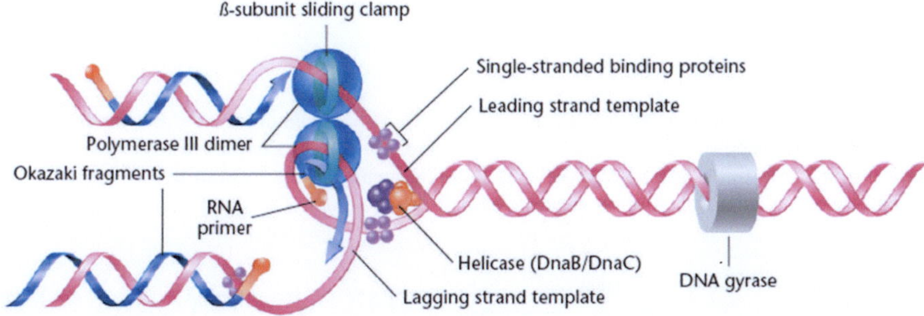

Fig. 7.35 DNA synthesis at a single replication fork. Enzymes and proteins involved in the process (Adapted from: Klug et al. 2012)

strands and forms two double-stranded closed circular DNA molecules. Sometimes, two double-stranded circular DNA molecules are intertwined after replication. Topoisomerase II then introduces a nick in one of the DNA molecules to release the intertwining.

7.9 Mechanism of DNA Ligase

DNA ligase is an enzyme that can join DNA chains to each other. The most well-studied ligases are those isolated from *E.coli* and T4 phage-infected *E. coli* known as T4 DNA ligase. Both the enzymes catalyse the synthesis of phosphodiester bonds between adjacent 3′-hydroxyl and 5′-phosphoryl termini in duplex DNA. Phosphodiester bond synthesis catalysed by the *E. coli* ligase is coupled to cleavage of the pyrophosphate bond of diphosphopyridine nucleotide (DPN). Hydrolysis of

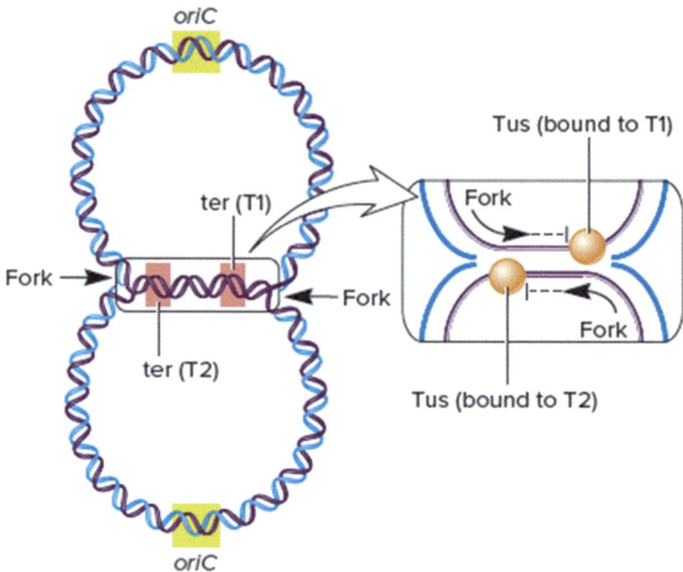

Fig. 7.36 Termination of DNA replication. Two sites in bacterial chromosome shown in rose-coloured rectangle and ter sequences designated T1 and T2. The T1 site prevents the advancement of fork from left to right, and T2 site prevents the advancement from right to left. Binding of Tus prevents the replication fork from proceeding past the ter sequences (Adapted from: Brooker 2018)

Fig. 7.37 Synthesis of a phosphodiester bond. It is formed between adjacent 3′-hydroxyl and 5′-phosphoryl group in duplex DNA by *E. coli* DPN and T4 (ATP) DNA ligases. (Adapted from: Lehman I R, Science, 186, 790–797)

the α-β-pyrophosphate bond of adenosine triphosphate (ATP) provides the energy for phosphodiester bond synthesis by the ligases.

In the case of the *E. coli* ligase, enzyme itself links with DPN to form ligase-adenylate. Next, the adenylyl group is transferred from the enzyme to the DNA. This creates a new pyrophosphate linkage between the adenosine monophosphate (AMP) and the 5′-phosphoryl terminus at the nick. Finally, the 5′-phosphate is attacked by the adjacent 3′-hydroxyl group to form a phosphodiester bond, and AMP is eliminated. The same sequence of reactions is catalysed by the T4 DNA ligase, except that in the first step the enzyme reacts with ATP rather than with DPN, and PPi is released (Fig. 7.37) (Lehman 1974).

7.10 Prokaryotic vs Eukaryotic DNA Replication

In eukaryote, the DNA is packaged more tightly with many proteins like histones to form linear DNA called chromosomes. The cell cycle is regulated much more tightly than prokaryotes. Therefore, eukaryotic replication is different from that of prokaryotes. In this section we will learn about the eukaryotic replication in detail highlighting its differences from prokaryotic replication.

DNA replication in both prokaryotes and eukaryotes require enzymes like primase, DNA polymerase, helicase, topoisomerase, single-strand binding proteins and DNA ligase. However, the molecular structure of these enzymes may be different in prokaryotes and eukaryotes.

As discussed earlier, eukaryotic chromosomes have multiple origins of replication. The molecular details of origin of replication in eukaryotes have been studied in *Saccharomyces cerevisiae*. The origins are known as *a*utonomously *r*eplicating *s*equences (ARS). These are about 50 bp in length and are rich in AT sequences. Certain consensus sequences like ATTTAT(A or G)TTTA are present in ARS. These are similar to bacteria having AT-rich regions called DnaA box.

In lower eukaryotes like *Saccharomyces cerevisiae*, origins are determined by DNA sequences. But in higher eukaryotes, origin may not be determined by DNA sequences. Chromatin packaging and histone modifications also play significant roles in deciding origins of replication.

In eukaryotes, a complex of proteins known as **origin recognition complex** (ORC) forms to initiate the formation of pre-replication complex (preRC) in G1 phase by promoting the binding of Cdc6, Cdt1 and a group of 6 MCM helicases. The binding of MCM is called DNA replication licensing. As S phase approaches, several protein kinases are recruited which removes the Cdc, Cdt1 proteins and ORC proteins. Once those proteins leave, other replication factors assemble at origin. MCM proteins moves in 3′ to 5′ direction and unwinds the double helix. As prokaryotes have clamp and a clamp loader, eukaryotes have PCDNA as sliding clamp. Replication factor C connects clamp to the polymerase in eukaryotes as Tau protein in prokaryotes.

7.10.1 Eukaryotes Have Many Different DNA Polymerases

Eukaryotes have many polymerases. Mammalian cells have over 12 different DNA polymerases. Four of these, alpha (α), epsilon (ε), delta (δ) and gamma (γ), have primary function of replicating DNA (Table 7.7). DNA polymerase γ replicates mitochondrial DNA, and α, ε and δ replicate DNA in nucleus. DNA polymerase α associates with primase. DNA polymerase ε and δ carry out the processive elongation of the DNA strands. DNA polymerase α and primase complex initiate the DNA synthesis at the replication fork and later are exchanged for ε and δ. ε plays a role in leading strand synthesis, and DNA polymerase δ is involved in lagging strand synthesis. Remaining DNA polymerases are involved in repair of damaged DNA, and some newly discovered DNA polymerases are called translesion replication

Table 7.7 Properties of eukaryotic DNA polymerases

DNA polymerase	Subunits	3′–5′ exonuclease	Function
α	4	No	RNA/DNA primers, initiation of DNA synthesis
δ	4	Yes	Lagging strand synthesis, DNA repair, proofreading
ε	4	Yes	Leading strand synthesis, DNA repair
γ	2	Yes	Mitochondrial DNA replication and repair
β	1	No	Base excision DNA repair
η	1	No	Translesion DNA synthesis
ζ	2	No	Translesion DNA synthesis
κ	1	No	Translesion DNA synthesis
ι	1	No	Translesion DNA synthesis
θ	1	No	DNA repair
λ	1	No	DNA repair
μ	1	No	DNA repair
ν	1	No	Unknown
Rev1	1	No	DNA repair

polymerases. Different types of translesion DNA polymerases are able to replicate over different kinds of DNA damage. For example, polymerase κ can replicate over DNA lesions caused by benzo[α]pyrene, an agent found in cigarette smoke, whereas polymerase η can replicate over thymine dimers, which are caused by UV light.

7.10.2 Removal of RNA Primers

In bacteria, RNA primers are removed by bacterial DNA polymerase I. None of the eukaryotic DNA polymerases can remove the RNA primer. Another enzyme called Flap endonuclease removes the RNA primer. DNA polymerase δ continues to extend the Okazaki fragment and reaches the short RNA primer of the next Okazaki fragment. This causes the RNA primer to form a short flap which is then removed by the FLAP endonuclease. If the RNA flap is long, then another enzyme called Dna2/helicase cuts the flap. It creates a shorter flap which is then removed.

7.10.3 End Replication Problem

The 3′ end of DNA cannot be replicated by DNA polymerase because RNA primer cannot be made upstream of this region. In this situation the DNA length would keep reducing in every replication cycle. Since bacterial DNA is circular, end replication problem does not exist in them. The loss of genetic information due to chromosome shortening is avoided by the presence of tandemly repeated sequences at the end of the chromosome, called telomeres. Enzyme telomerase synthesizes the tandem repeats of the telomeres. This enzyme was discovered by Carol Greider and Elizabeth Blackburn in 1984.

Action of Telomerase: Telomerase contains both RNA and protein component. Telomerase RNA component (TERC) contains a sequence complementary to the telomere repeat end. This allows telomere to bind to the 3′ overhang of the telomere. RNA sequence of the telomerase beyond the binding site acts as template for the synthesis of the end of the telomere DNA by adding six nucleotides (Fig. 7.38). Telomere lengthening is catalysed by the telomerase reverse transcriptase (TERT). TERT has two identical protein subunits that catalyse DNA synthesis having RNA as template. Following polymerization, it translocates to the new end of DNA, adding six nucleotides again.

7.10.4 Role of Telomeres in Cancer and Ageing

The length of telomere shortens with age. At birth the telomere length initially may be 8000 base pairs, but in an elderly person, it may be around 1500 bp. This decrease occurs as the activity of telomerase decreases with age. When telomeres become too short, cells undergo senescence and stop dividing (Fig. 7.39).

However, the cancer cells can undergo uncontrolled division which may not stop even if the telomere length is short. In them, the telomerase undergoes mutations which increase the activity of the enzyme.

7.11 Regulation of DNA Replication

We have learnt so far that DNA replication is semi-conservative, with continuous and discontinuous strand synthesis mostly occurring directionally. Some circular genomes replicate unidirectionally. Some of the viruses like M13 replicate by rolling circle replication.

In eukaryotes, the DNA replication is regulated by the cell cycle. In prokaryotes, DNA replication depends on cell size, availability of energy and precursor molecules. In presence of rich carbon sources and simple inorganic nutrients, doubling time is reduced. DNA synthesis is principally regulated at the initiation stage. In enriched medium the second round of initiation begins even before the first round of replication is complete. The rate of initiation is controlled relative to growth rate. Initiation of primer synthesis by the primase is the major rate controlling mechanism in prokaryotes. Rate of initiation is controlled by DnaA protein. Expression of DnaA protein is regulated by auto-feedback loop. There are more DNA copies per cell of those genes which are closer to *OriC* than of genes far away from *OriC*. It is expected that the genes whose products are needed in large amount are present close to the *OriC*. Rapidly growing cells can have multiple copies of the genome, while cells with a very low growth rate have only one copy.

The control of copy number of multi-copy number plasmid is mediated by the synthesis of anti-sense RNA of the replication initiator protein Rep. Rep anti-sense

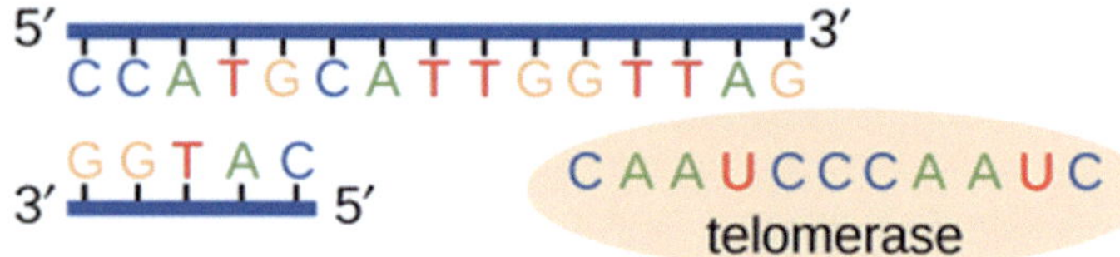

Telomerase has an associated RNA that complements the 3′ overhang at the end of the chromosome.

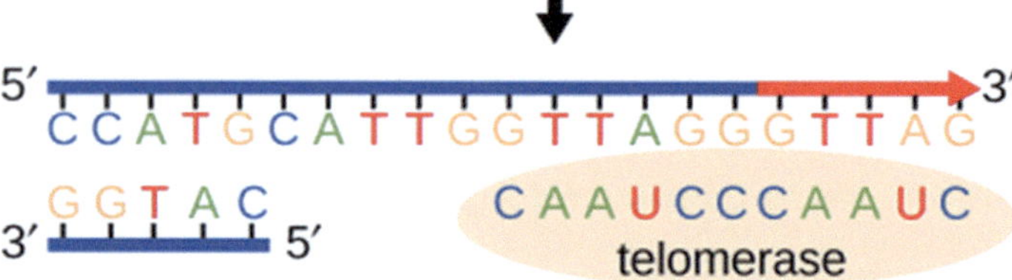

The RNA template is used to synthesize the complementay strand.

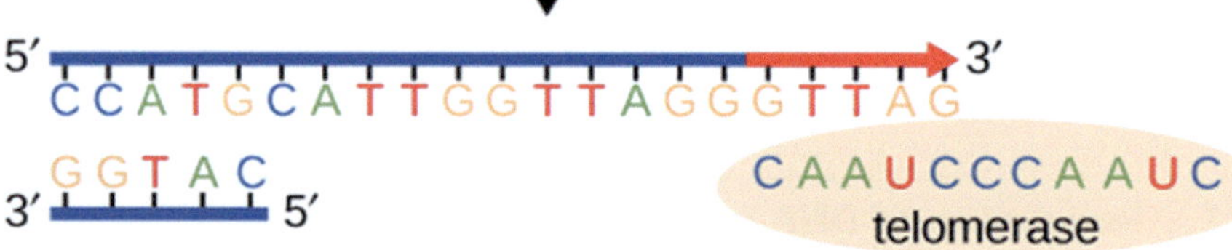

Telomerase shifts, and the process is repeated.

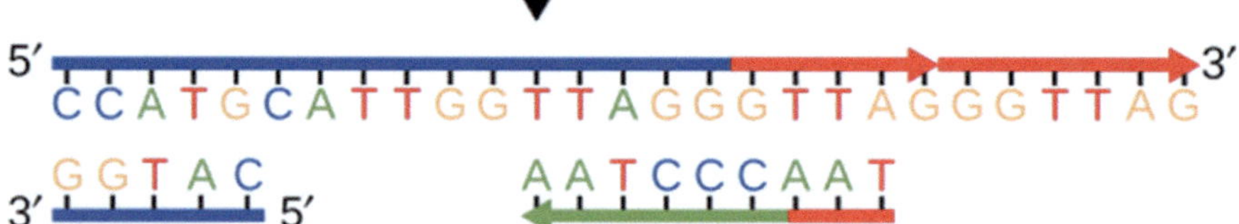

Primase and DNA polymerase syntesize the complementary strand.

Fig. 7.38 Replication of chromosome telomeres. Telomere lengthening is catalysed by the telomerase reverse transcriptase (TERT). TERT has two identical protein subunits that catalyse DNA synthesis having RNA as template. Following polymerization, it translocates to the new end of DNA, adding six nucleotides again (Adapted from: https://bio.libretexts.org/Courses/University_of_California_Davis/BIS_2A%3A_Introductory_Biology_(Britt)/02%3A_Face-2-Face/2.05%3A_DNA_Replication)

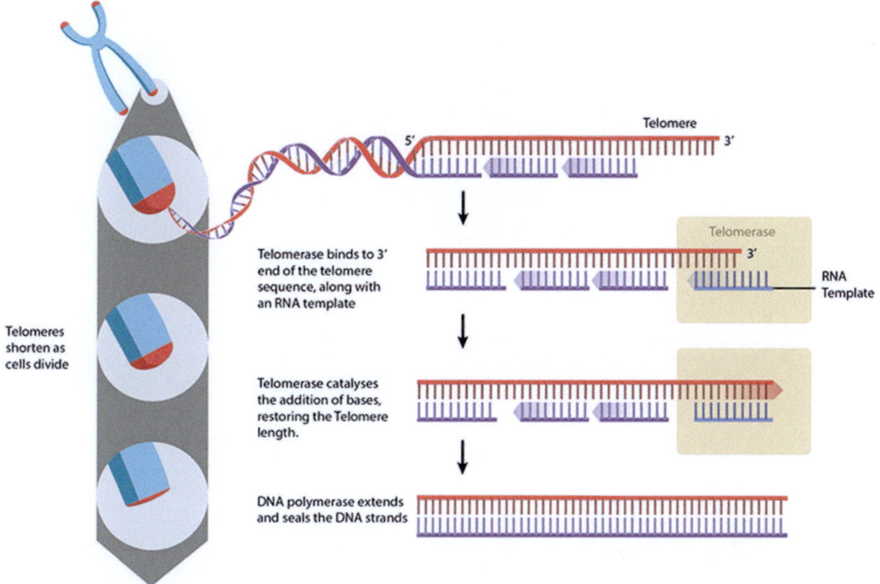

Fig. 7.39 Loss of DNA telomeres. Telomeres shorten with each cell division. Telomerase extends the telomere sequence along the RNA template (Adapted from: https://www.mechanobio.info/genome-regulation/what-are-telomeres/)

RNA is copied from the non-transcribed DNA strand and is therefore complementary to the normal RNA. Anti-sense RNA prevents synthesis of the Rep protein. Rep protein is required for initiation of DNA synthesis, and its concentration controls the frequency of initiation. Rep proteins encoded by plasmids bind to additional copies of binding sites called 'iterons', often present upstream of the *ori* sequences in the plasmids.

Regulation of eukaryotic genome replication is mainly brought about by cell cycle. It is very tightly regulated to ascertain that DNA replication occurs only once before division. In cell cycle, there are G0, G1 S, G2 and M phases. In the S phase, the DNA replication occurs, and at G1 checkpoint, regulatory proteins ensure that the cell is prepared for replication. At G2 checkpoint, DNA replication is cross-checked so that cell can proceed for division. DNA licensing is a process to ensure that chromatin is competent for DNA replication. A complex of proteins called origin recognition complex (ORC) bind to the *ori* sequences. Although ORC is present throughout the cell cycle, other proteins like MCM (minichromosome maintenance) are loaded stepwise. The loading of MCM proteins and organization of ORC is the important regulator for controlling the rate of initiation in eukaryotes.

Length of telomeres and activity of telomerase are also important for regulating DNA replication. If the length of telomeres decreases beyond a certain limit, the cell reaches senescence and stops replicating.

Fig. 7.40 Scheme of amino-labelled DNA probes for formaldehyde fixation. (**a**) Chemical reactions of formaldehyde fixation. (**b**) Scheme of formaldehyde-mediated cross-link between amino-labelled probe and cellular proteins in vicinity

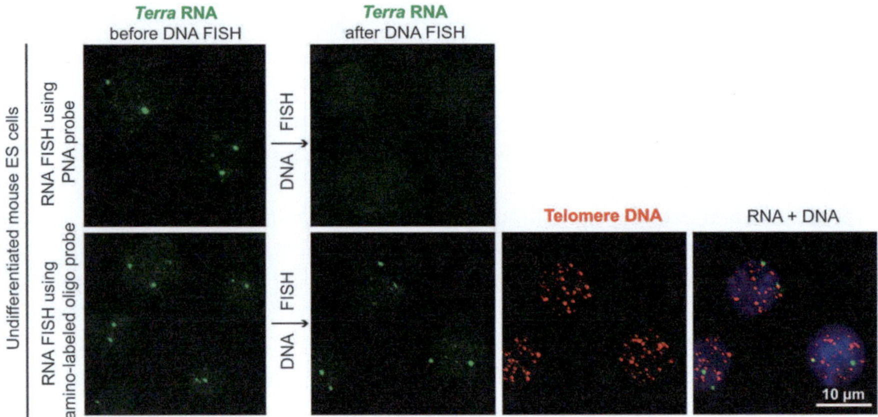

Fig. 7.41 RNA-DNA FISH of terra (RNA) and telomere (DNA) using an amino-labelled oligonucleotide probe. (**a**) Either a FITC-labelled PNA probe (green) or an amino-labelled oligonucleotide probe (fluorescently labelled with FAM, green) was used to detect terra in RNA FISH. Telomere DNA was detected by Cy3-labelled oligonucleotide probes (red) with the sequence of (GGGTTA). Nuclei were stained by DAPI (blue)

Box 7.1 Scientific Concept: Using Amino-Labelled Nucleotide Probes for Simultaneous Single Molecule RNA-DNA FISH, Reelina Basu et al.

In the earlier section, we have learnt about the importance of FISH (fluorescence in situ hybridization). This technique can be slightly modified to detect DNA and its corresponding RNA transcript in the cell by RNA-DNA FISH. The problem with this technique is that the RNA is too fragile to survive the harsh treatments of DNA FISH. Therefore, Reelina Basu et al. designed a method to amino label the oligonucleotide probes and simultaneously detect DNA and its RNA transcript (Basu et al. 2014). They selected a lnc (long noncoding) RNA *terra* (telomeric repeat-containing RNA) which is a repetitive sequence present in the DNA telomeres. Thirty-six nucleotide long oligo probes were synthesized by DNA synthesizer. An amino-modified thymine (amino-dT) was used to introduce an alkylamino group to the oligonucleotide probe with a density of at least one amino group every six nucleotides. The oligonucleotide with dual label of alkylamine and fluorophore was synthesized.

Terra RNA FISH was done using amino-labelled oligonucleotide probe and PNA probe (peptide nucleic acid). PNA probes are extremely stable and hence were chosen as control. RNA signals of terra were directly fixed right after RNA FISH. Terra signals detected by PNA probe were lost during the harsh treatments of DNA FISH, but amino-labelled probe signals could be easily detected even after DNA FISH. Authors could infer that not all telomere DNA is associated with terra RNA. The terra RNA transcripts are located in close proximity of telomeres, but some terra signals do not overlap with telomere DNA.

Synthetic amino-labelled oligo nucleotide probe was sufficient to detect signals of terra as it is a repetitive sequence and probe has multiple binding sites. For lnc RNA which exist in single copy, like HOTAIR, NEAT2 and xist, the amino-labelled probe needs to be created by nick translation so that there could be more incorporation of labelled amino group. More label incorporation would give more intense signal for detection of single copy lncRNA (Figs. 7.40 and 7.41).

Box 7.2 DNA Mutation Motifs in the Genes Associated with Inherited Diseases, Michal Ruzicka et al.

We learnt that DNA replication is a highly accurate mechanism. The error rate that is allowed is 1 in 10^9 or 10^{10}. Those errors are repaired by various mechanisms like nucleotide excision repair (NER), mismatch repair (MMR), homologous excision repair, post replication repair, etc. Despite all these

(continued)

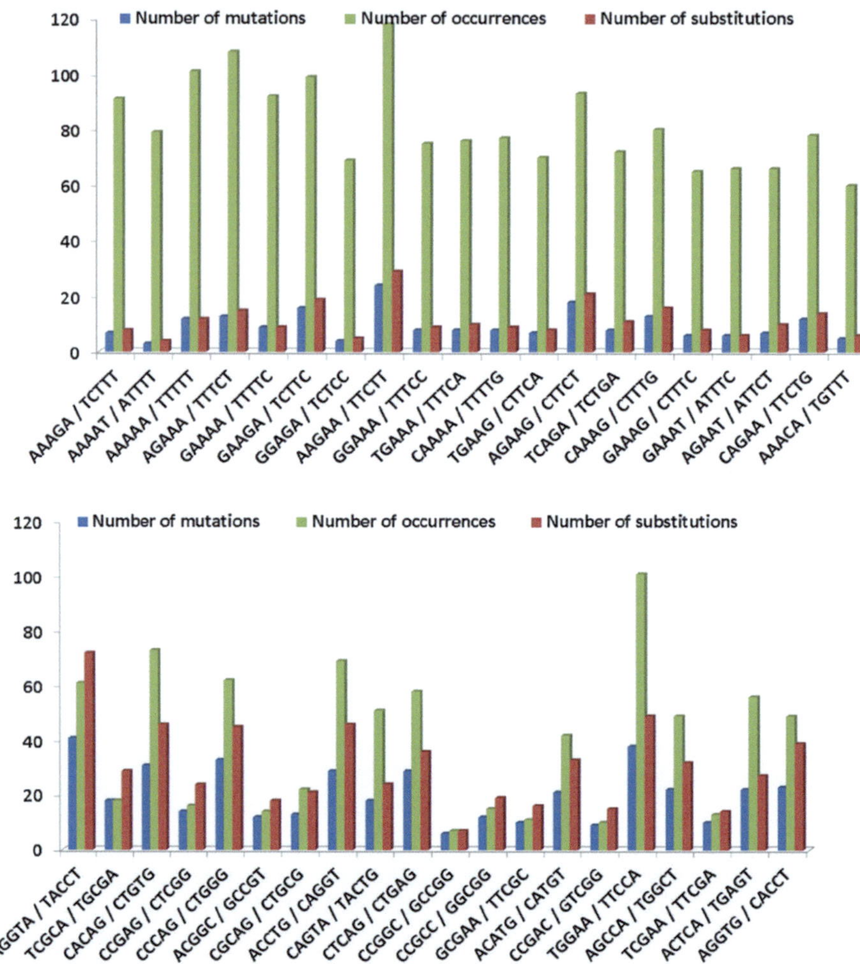

Fig. 7.42 Mutation hotspot and cold spot based on their occurrences. Occurrences of top 20 cold spots (top) and top 20 hotspots (bottom) in the five studied genes visualized with the number of detected mutations and substitutions in middle position (Adapted from: Ruzicka 2017)

Box 7.2 (continued)

repair mechanisms, mutations still occur. If the mutation is present in the germ cells, it gets easily inherited. There are certain sequence motifs which are more prone to mutation and are called mutation hotspots and other regions which are less prone to mutation than normal and are called cold spots. In this work the authors have analysed certain genes which are closely associated with genetic disorders like the PAH gene (associated with hyperphenylalaninemia), LDLR

(continued)

Box 7.2 (continued)
gene (associated with hypercholesterolemia), CFTR gene (associated with cystic fibrosis) and F8 and F9 genes (associated with hemophilia A and B, respectively) reported in Human Gene Mutation Database (HGMD). They analysed the sequences using advanced bioinformatics tools like molecular dynamics (MD) simulations implemented in the AMBER program package and free energy calculations using the adaptive biasing method (ABF) enhanced by the multiple walker approach (MWA).

They identified sequence motifs which could be divided into two groups: the hotspots and cold spots. They observed in one of the datasets that 18 out of 20 cold spots contain four or five purine tracts. The identified hotspots were observed to contain CpG dinucleotide sequence in the middle position of the motif. In the top 20 identified cold spots, CpG dinucleotide was not present.

Next, they analysed the reason why certain sequences are prevalent in hotspots and others in cold spots. In a cell, mismatches and small insertion/deletion are repaired by the MutSα protein in the MMR pathway. MutSα protein with bound ADP can diffuse freely on DNA and search for mismatches. Wherever it finds a mismatch, it binds to it and bends the DNA and repairs the sequence. The bending of DNA was quantified by the free energy change that is needed to bend a relaxed straight DNA to the bend conformation that is observed in the MutSα/DNA complex. The free energy profile showed that the hotspots are less flexible than the cold spots and hence are not repaired efficiently by MMR (Fig. 7.42) (Ruzicka, M., et al. Plos One. 12: e0182377, 2017).

7.12 Summary

- Molecular transformation factor was shown to be responsible for transformation of living R bacteria into S form by pioneering experiments of F Griffith. Avery, MacLeod and McCarty's transformation experiment proved that DNA is the genetic material. Hershey and Chase further demonstrated that DNA is passed to the progeny phages and hence carry the genetic information.
- Walther Flemming observed thread-like bodies forming during cell division and called it 'mitosen'. Later it was named as chromatin or stainable material present in the nucleus. Various dyes were later used for staining chromosomes like quinacrine mustard.
- DNA replication was shown to be semi-conservative by Meselson and Stahl using transient N^{15} labelling and then growing in excess of N^{14} followed by density centrifugation in CsCl gradient.
- Bacteria have closed circular DNA which replicates bidirectionally from single origin of replication. Around the origin of replication, the DNA becomes single

stranded for a few bases. Cairn showed by autoradiography that during DNA replication in *E.coli* it looks like Greek letter θ and hence is known as θ mode of replication.

- Viruses have rolling circle mode of replication forming concatemeric DNA. In this mode of replication, one of the strands of closed circular double-stranded DNA is nicked. A free 3′-OH and free 5′ P is generated, and the free 3′-OH is used for extending the new DNA stand. The other strand remains intact and acts as template for the leading strand.
- DNA contains deoxyribose sugar, phosphate group and nitrogenous base. RNA contains ribose sugar, phosphate group and nitrogenous base. RNA contains uracil unlike DNA.
- The bases are of two types: purines (adenine and guanine) and pyrimidines (cytosine and thymine). They bind to each other using hydrogen bonding. Adenine is linked to thymine by two hydrogen bonds, and cytosine is linked to guanine using three hydrogen bonds. Hydrogen bonding follows Chargaff's rule $[A + G] = [C + T]$.
- DNA has sugar phosphate backbone made of adjacent nucleotides linked by phosphodiester bond forming polynucleotide strand. Each polynucleotide strands has free 3′-hydroxyl and 5′-phosphate group.
- DNA double helix crystal structure was solved by Rosalind Franklin. Based on that crystal structure, Watson and Crick proposed DNA double helix model. Their model showed that DNA forms a double helical structure with sugar phosphate backbone lying on the outer edge of the helix and bases are stacked in the interior.
- There are 10 bp per 360° rotation of the helix. Each base pair is 0.34 nm apart, and therefore there are ten base pairs in each turn encompassing 3.4 nm. The diameter of B form of the helix is 2 nm. Diameter and pitch of the DNA vary in different forms of DNA like A, B and Z DNA.
- DNA denatures to get single stranded as temperature increases; it is called DNA melting. DNA absorbs UV light (Amax-260); the absorption increases when the DNA becomes single stranded. It is called hyperchromic shift. DNA concentration (Co), the time allowed for reassociation (t) and the sequence organization of the DNA are the factors deciding the extent of reassociation. Since the extent of reassociation is directly proportional to both Co and t, their arithmetic product (Cot) denotes the extent of reassociation. When $\log_{10}$Cot is plotted vs percent of ssDNA, it can give the idea of repetitive sequence present in the DNA.
- Out of the five DNA polymerases found in *E. coli*, polymerases I and III are the major enzymes which are involved in DNA replication. Rest are involved in repair of damaged DNA.
- DNA polymerase III is the principal enzyme involved in replication. The α subunit carries out the function of synthesizing phosphodiester linkage between adjacent nucleotides. It also has 3′ to 5′ exonucleolytic activity which is called proofreading activity.
- The number of bases added by the enzyme during replication is called the **processivity** of the enzyme. In DNA pol II, the two β-subunits are responsible for high processivity. α subunit is the catalytic subunit.

- DNA synthesis occurs with high amount of accuracy, and only one mismatch occurs per 100 million bases. That is called as **fidelity** of the enzyme.
- Three types of DNA sequences are found in origin of replication in *E. coli* (*OriC*); AT rich, DnaA box binding region and GATC methylation sites.
- DNA polymerase can extend polynucleotide only in 5′ to 3′ direction by adding new nucleotides at the 3′ end. In the leading strand, RNA primer is made by primase, and DNA polymerase III extends it towards the opening of the replication fork in the 5′ to 3′ direction. The other strand is called lagging strand, which is formed in 5′ to 3′. Several RNA primers are made by primase to facilitate synthesis of DNA fragments, also known as Okazaki fragments in the 5′ to 3′ direction.
- In eukaryotes there are multiple origins of replication. Depending on chromatin packaging, histone modifications also play significant roles in deciding origins of replication.
- Mammalian cells have over 12 different DNA polymerases. Four of these, alpha (α), epsilon (ε), delta (δ) and gamma (γ), have primary function of replicating the DNA.
- The 3′ end of DNA is replicated by telomerase enzyme. Telomerase has RNA and protein component. Telomerase RNA component contains sequence complementary to telomere repeat. Telomere lengthening is catalysed by the telomerase reverse transcriptase (TERT). This enzyme was discovered by Carol Greider and Elizabeth Blackburn in 1984. Cancer cells can undergo uncontrolled division which may not stop even if the telomere length is short. In them, the telomerase undergoes mutations which increase the activity of the enzyme.

References

Avery OT, Macleod CM, McCarty M (1944) Studies on the chemical nature of the substance inducing transformation of pneumococcal types induction of transformation by a desoxyribonucleic acid fraction isolated from pneumococcus type III. J Exp Med 179:137–158

Basu R, Lai LT, Meng Z, Wu J, Shao F, Zhang LF (2014) Using amino labeled nucleotide probes for simultaneous single molecule RNA-DNA FISH. PLoS One 9:e107425

Beadle GW, Tatum EL (1941) Genetic control of biochemical reactions in Neurospora. Proc Natl Acad Sci U S A 27:499–506

Brooker RJ (2018) Genetics: analysis and principle, 6th edn. McGraw-Hill Education, New York, pp 208–228

Graham GJ (2001) Cot analysis: single-copy versus repetitive DNA, Encyclopedia of life sciences. Wiley, pp 1–6

Khan SA et al (1997) Rolling-circle replication of bacterial plasmids. Microbiol Mol Biol Rev 61:442–455

Klug WS, Cummings MR, Spencer CA, Palladino MA (2012) Concepts of genetics, 10th edn. Pearson Education, California

Lamers MH, Georgescu RE, Lee SG, O'Donnell M, Kuriyan J (2006) Crystal structure of the catalytic alpha subunit of E. coli replicative DNA polymerase III. Cell 126:881–892

Lehman IR (1974) DNA ligase: structure, mechanism and function. Science 186:790–797

Maloy SR, Cronan JE, Freifelder D (1994) Microbial genetics Jones and Bartlett series in biology, 2nd edn. Jones & Bartlett Publishers, Inc.

Meselson M, Stahl FW (1958) The replication of DNA in Escherichia coli. Proc Natl Acad Sci U S A 44:671–682

Ruzicka M, Kulhanek P, Radova L, Cechova A, Spackova N, Fajkusova L, Reblova K (2017) DNA mutation motifs in the genes associated with inherited diseases. PLoS One 12:e0182377

Sanders MF, Bowman JL (2015) Genetic analysis: an intergrated approach, 2nd edn. Pearson Education, New Jersey, pp 227–266

Sharma-Kuinkel BK, Rude TH, Fowler VG Jr (2016) Pulse field gel electrophoresis. Methods Mol Biol 1373:117–130

Wetmur JG, Davidson N (1968) Kinetics of renaturation of DNA. J Mol Biol 31:349–370

Chromosomal Organization of DNA

8

Payal Gupta

8.1 Chromosome: Overview

If we look upon DNA as the blueprint of life, then chromosomes are the entities that hold the blueprint together. Waldeyer first coined the term "chromosome" in 1888 when he saw coloured bodies stained with an aniline dye under the microscope. Chromosomes are responsible for physically carrying the DNA and all associated proteins. Even the tiniest of beings, bacteria, can hold anywhere between 130 kbp and 14 Mbp of DNA, while the human genome holds over 3 billion bps which would measure ~2 m if placed in tandem. The Fig. 8.1 shows the sizes of the chromosomes of various organisms. It is bewildering how this massive amount of DNA is assorted and physically packaged in cells of microscopic sizes and yet made accessible for various life processes like replication, transcription, etc. Even though our understanding of the structure and organization of DNA packaging in chromosomes is constantly evolving, the last few decades have greatly enhanced our knowledge in this field.

There are many variations in the form of DNA packaging observed in different life forms. The bacterial genome comprises of a single DNA molecule which is present in the form of a single circular covalently closed chromosome and compacted with the help of packaging proteins. On the other hand, the haploid human genome has its DNA segregated into 23 units that form morphologically distinct linear chromosomes on compaction (Fig. 8.2). Then, there are specialized chromosomes like the polytene chromosomes that perform specialized functions.

However, chromosomes universally perform two fundamental functions: the precise transmission of genetic information and accurate control of gene expression. Specialized regions containing repetitive sequences help form structures such as telomeres, replication origin, etc. which help the chromosomes in the execution of its

P. Gupta (✉)
University of Calcutta, Kolkata, India

D. Kar, S. Sarkar (eds.), *Genetics Fundamentals Notes*,
https://doi.org/10.1007/978-981-16-7041-1_8

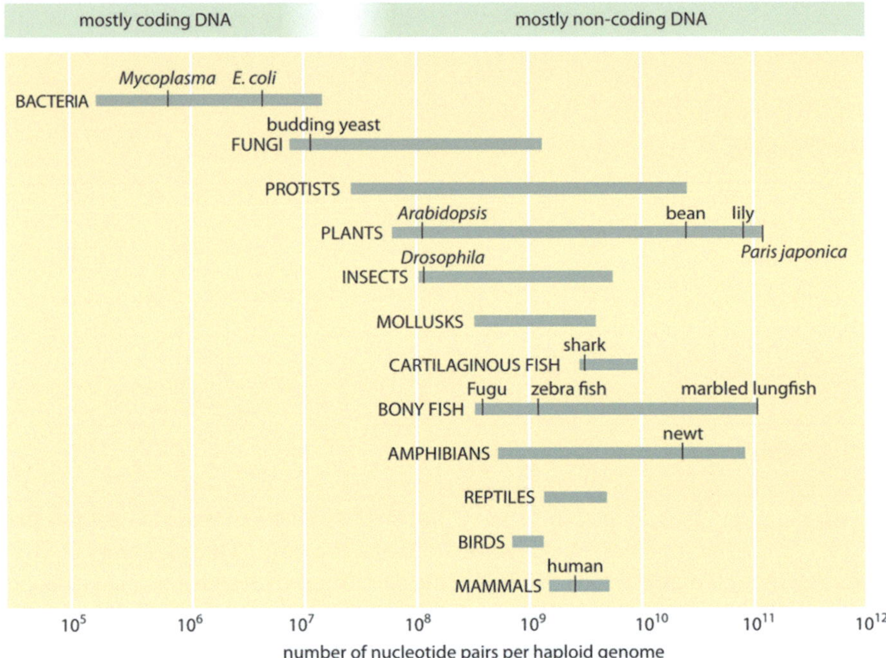

Fig. 8.1 The average size of the genome of organismal groups. The scheme depicts the average size of the genome of various groups of organisms and shows what proportion of the genome actually codes for peptides

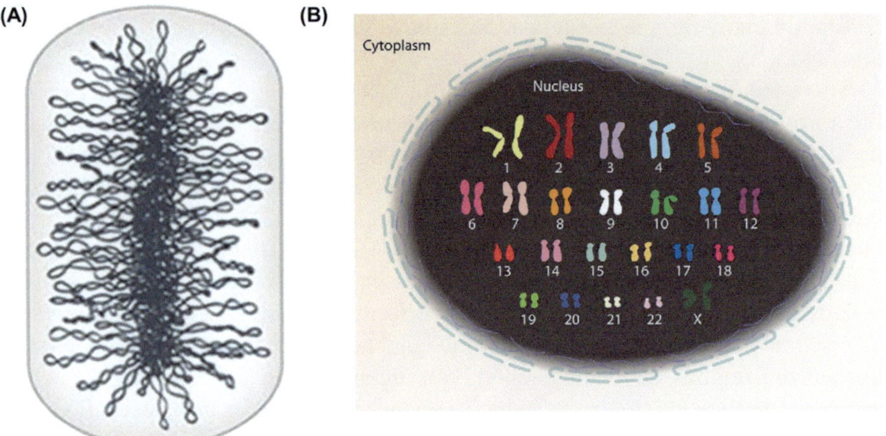

Fig. 8.2 Varying forms of chromosomes. (**a**) The diagram is a schematic representation of a bacterial nucleoid. It depicts the bottle brush model of the nucleoid with supercoiled loops that are interwound and radiate out of a dense core. (**b**) A schematic SKY image of a normal female karyotype

functions. The level of chromosome compaction also varies depending on which areas are needed for active transcription and which areas will remain inactive. Our current ideas of chromosomal structures revolve around understanding how chromatin is arranged in structural domain in the three-dimensional nuclear spaces and how this structural landscaping controls different facets of gene expression.

8.2 DNA Supercoiling

Now that we appreciate the gigantic size of the genetic material that must be packaged into a comparatively miniscule cellular space, it stands to reason that the DNA strings must be highly coiled and condensed. Yet, the coiling should still leave functionally important domains of the DNA accessible to proteins. This feat is achieved through the process of *supercoiling* which literally implies the coiling of a coil.

To understand the process, let us visualize the double helical DNA. An axis passes through the middle of the helix. When this axis is folded upon itself, it results in a supercoiled DNA (Fig. 8.3). This can be explained by using a small linear double helix with two to three turns. If both ends of the helix are twisted in direction

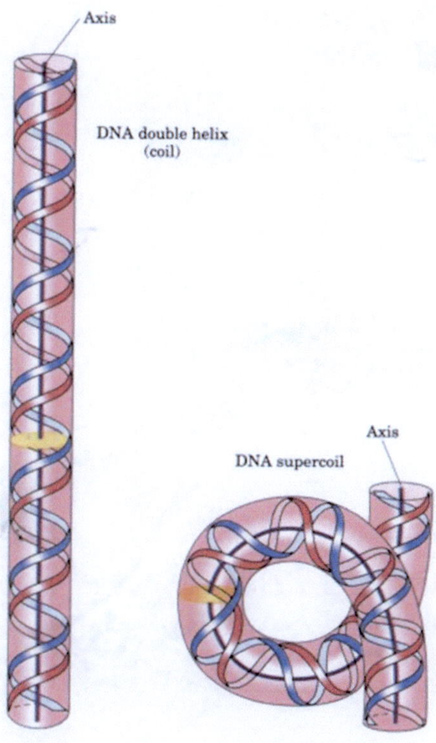

Fig. 8.3 The supercoiling of DNA. A linear double helix is shown on the left-hand side with an imaginary axis passing through it. The right-hand figure depicts a supercoiled DNA where the axis has folded on itself

of the helix winding, the number of turns in the helix will seem to increase. This supercoiled helix is over-twisted and under strain. If, however, the helix is twisted in the direction opposite to its coiling, it will appear to unwind. This state of the coil with lesser number of turns is a relaxed state. If the molecule is consistently over-twisted, it will relieve the molecular strain by twisting upon its own helical axis, thus creating a *positive supercoil.* Likewise an under-twisted moiety will result in a *negative supercoil.* Most basic processes like replication and transcription require the unwinding of the double helix. This is turn results in over-twisting of the domains lying ahead. Thus, supercoiling is an integral aspect of the tertiary structure of the DNA that is omnipotent in cellular DNA and tightly regulated by the cell.

To understand the physiological relevance of supercoiling, let us focus on closed circular B-form of DNA which has 10.5 bps per turn (Fig. 8.4). The underwinding of this DNA at any point will result in a strain. For example, if the DNA has 84 bps, in a

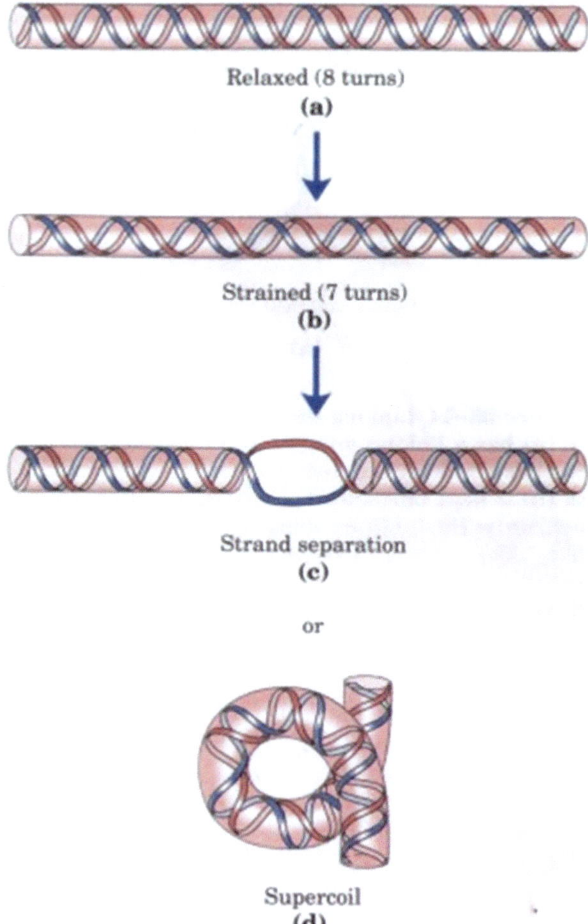

Fig. 8.4 DNA underwinding. (**a**) A relaxed 84 bps DNA segment with eight helical turns. (**b**) The underwinding of the DNA by removing one turn results in a strained DNA which is compensated by (**c**) separation of the strand over 10.5 bps or (**d**) by forming a supercoil

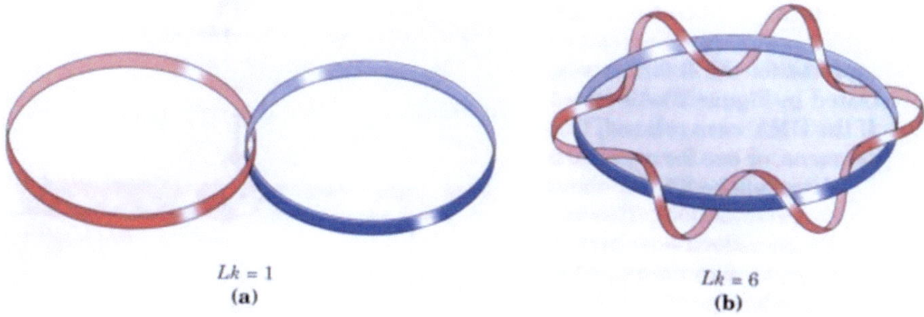

Fig. 8.5 Depiction of linking number (Lk). (**a**) The molecules shown physically interact at one junction and have an Lk of 1. (**b**) If one strand of DNA is kept unwound, then the number of times the second strand passes it defines the linking number for that molecule

relaxed state, it will consist of eight helical turns of 10.5 bps each. The DNA is now underwound, and one of these eight turns is removed, and then the 84 bps will be divided in seven helical turns containing 12 bps per turn instead of the 10.5 bps. This alteration will result in a thermodynamically strained structure. This strain can be compensated in two ways. Either the strands can simply separate over a stretch to resort to 10.5 bps per turn or the axis of the double helix can coil on itself to realign the base stacking to approximate of 10.5 bps per turn pattern.

DNA supercoiling could be better understood with the help of a branch of mathematics called topology. Of special interest is the concept of *linking number (Lk)* which denotes the number of helical turns in a closed circular DNA when no supercoiling is present (Fig. 8.5). This property does not vary even if the DNA is twisted or deformed as long as the two strands of DNA remain intact. By standards, if the DNA strands are twisted in a right-handed helix, the linking number is a positive integer; however, if the strands are twisted in a left-handed helix (as in Z-DNA), the linking number is a negative integer.

Two structural components make up the linking number, *twist (Tw)* and *writhe (Wr)*. Writhe denotes the coiling of the helix axis, while twist defines the local twisting or the spatial arrangement of the neighbouring base pairs. When supercoiling happens, the linking number of the DNA changes causing a strain which is compensated by variation in supercoiling (writhe) or by change in the twist patterns. This gives rise to the equation; $Lk = Tw + Wr$.

In all systems, the supercoiling is regulated by a class of enzymes called *topoisomerases* which play an important role in the processes of replication, DNA packaging, etc. The genre of topoisomerases is divided into two classes, *Type I* and *Type II*. Type I topoisomerases act by creating a nick on one strand, twisting it around the uncut strand and rejoining it, thus changing the Lk by 1. Type II topoisomerase acts by cutting both strands of the DNA, twisting them around one another and rejoining them, thus changing the Lk by 2 (Fig. 8.6).

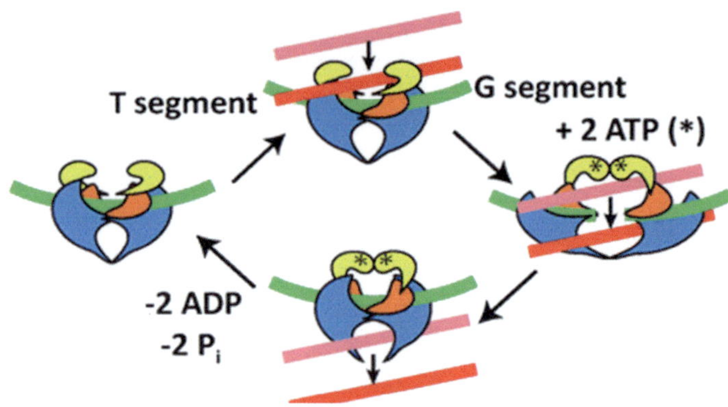

Fig. 8.6 The mechanism of strand passage through topoisomerase core. The type II topoisomerase (blue and yellow core) captures G-segment (gate) DNA (green). It then sequesters the transfer or T segment (indicated in shades of pink to denote movement of the DNA). Then they bind two ATP molecules thereby shutting the N-terminal gate (yellow) which is followed by the double-strand cleavage of the gate strand. The T segment passes across the cleaved DNA, and, together with the ATP hydrolysis products, it is then released

8.3 Organization of the Prokaryotic Chromosome

Having understood the basic dynamics of DNA supercoiling, we move onto the question of how the DNA is packaged in prokaryotes lacking a defined nucleus and how the stability of the coiled structures are maintained. Most of our knowledge of the organization of prokaryotic chromosomes comes from the studies conducted on *E. coli*. The single circular chromosome is packaged in the form of a nucleoid which is defined as the area of the cell where the chromosome DNA is located (Fig. 8.7).

The packaging of the bacterial genome into the small *nucleoid* space occurs via two processes; the first is the process of supercoiling discussed in Sect. 8.2, and the

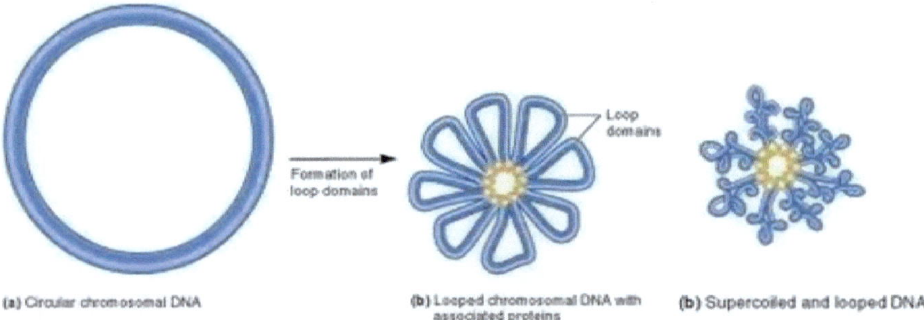

Fig. 8.7 Packaging of the bacterial chromosome. (**a**) A circular chromosome without any compaction. (**b**) The DNA-associated proteins form a central scaffold, and the bacterial DNA loops around them. (**c**) The loops are further supercoiled to form a condensed structure

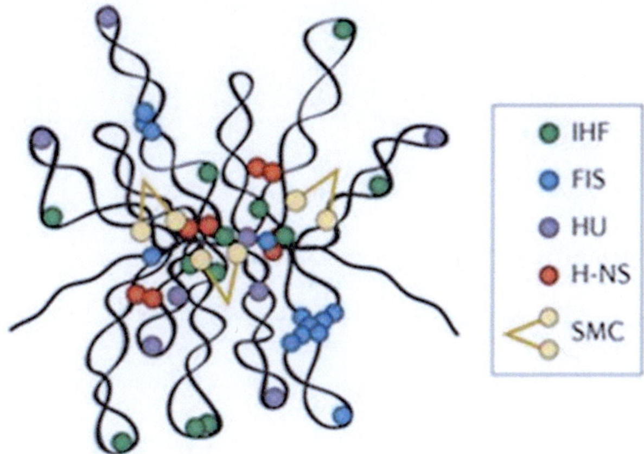

Fig. 8.8 Proteins associated with bacterial chromosomes. Several proteins like HU and H-NS are associated with the bacterial DNA to maintain its topological structure

other is the interaction of the DNA with packaging proteins. The main histone-like packaging proteins present in the *E. coli* are HU, IHF (integration host factor), Fis (factor for inversion stimulation) and H-NS. These positively charged proteins interact with the negatively charged DNA and, together with the topoisomerase and gyrase enzymes, maintain the supercoiling homeostasis of the bacterial genome. For example, the HU protein works with topoisomerase I and introduces acute bends in the chromosome which generates the tension required for the negative supercoiling. During normal growth most of the bacterial genome is negatively supercoiled. The negative supercoiling of DNA gives rise to plectonemic loops forming a network of supercoiled domains that are topologically insulated from each other. The current view is that these interwound loops are not rigid. They keep changing depending on the genetic transaction that occurs within and between them (Fig. 8.8).

8.4 Hierarchical Packaging of the Eukaryotic Chromosome

When DNA from eukaryotic cells is isolated using isotonic buffers like 0.15 M KCl, it is found to be associated with nearly equivalent proportions of protein in an extremely compacted complex called *chromatin*. The DNA in eukaryotes exists in the chromatin state throughout the interphase of the cell cycle. However, in order for the all the tangled DNA to accurately partition into two cells during mitosis, the DNA needs to condense into ordered structures called *chromosomes* during the prophase of the cell cycle. As discussed, the total length of total DNA of the diploid human cell is about 2 m. This is divided into 23 pairs of chromosomes each containing anywhere between 15 and 85 mm of DNA. The largest human chromosome, of ~85 mm long DNA, is packaged into a distinct mitotic chromosome 10 μm long and about 0.5 μm diameter. Even so, the chromosome has specialized

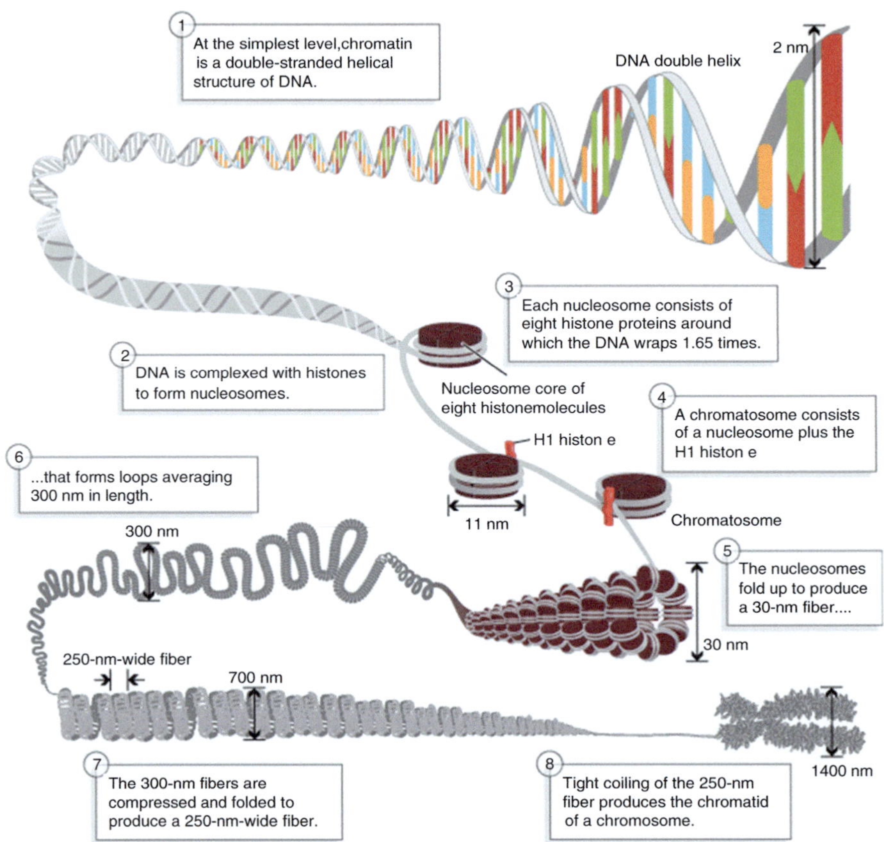

Fig. 8.9 The hierarchical packaging of eukaryotic chromosome. A simple dsDNA wounds around a nucleosome complex giving rise to a beads on string structure. This nucleosome forms a solenoid complex of 30 nm diameter. The 30 nm fibre further forms 300 nm long loops around the non-histone proteins undergoing several levels of compaction to form the chromosome

structures, and its dynamic topology allows access to proteins required for life functions. This is achieved by a hierarchical and orchestrated process of packaging with the help of different types of proteins (Fig. 8.9).

8.4.1 The Nucleosome Assembly

The most predominant proteins found attached to the eukaryotic DNA are the *histone* proteins. They represent a family of basic proteins, rich in positively charged amino acids such as lysine and arginine. This positive charge helps the protein interact with the negatively charged backbone of DNA. There are five principal types of histones: H1, H2A, H2B, H3 and H4. The histones form a highly structured

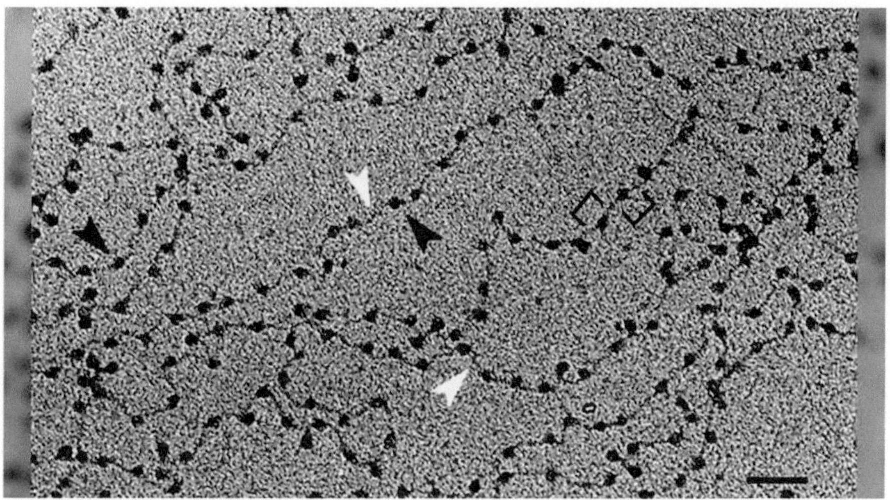

Fig. 8.10 An electron micrograph of nucleosome "beads-on-string" structure. The black brackets indicate nucleosome assembly, black arrowheads indicate the nucleosomal core, and white arrowheads indicate linker DNA. The scale bar indicates 50 nm. Image credit: Chris Woodcock

assembly, and the DNA loops around it to give rise to the beads on strings moieties known as *the nucleosome assembly.*

The nucleosome consists of a string of DNA wrapped around a protein core like a thread on a spool. The disc-shaped core is an octamer comprising of two copies of H2A, H2B, H3 and H4 each. Histone protein sub-assemblies come together to form the histone core with the H3 and H4 forming a tetrameric sub-assembly and the H2A-H2B dimeric sub-assembly joining it to complete the histone core. The DNA wraps around this core ~1.65 times using around 146 bps length. The H3 and H4 tetramers interact in the middle and the rear ends, while the rest of the DNA is bound to the H2A–H2B dimer via hydrogen bonds. Consecutive nucleosome cores are connected by short segments of linker DNA which harbours the linker histone, H1 (Figs. 8.10 and 8.11).

Each histone protein has an N-terminal tail that provides a guide for the DNA strand to wrap around the core, by creating grooves similar to those on a screw such that the DNA wraps in a specific pattern. Thus the beads form ellipsoidal bodies measuring 110 Å in diameter and 60 Å in height (Fig. 8.12).

8.4.2 The Solenoid Structure

The electron microscopy data show that the chromatin fibres have a diameter of ~30 nm. The 10 nm nucleosome fibre is wrapped into a highly structured *solenoid or a super-supercoil*. The solenoid model depicts the nucleosome as packaged into a spiral arrangement having six nucleosomes every turn. The nucleosomes fold on the

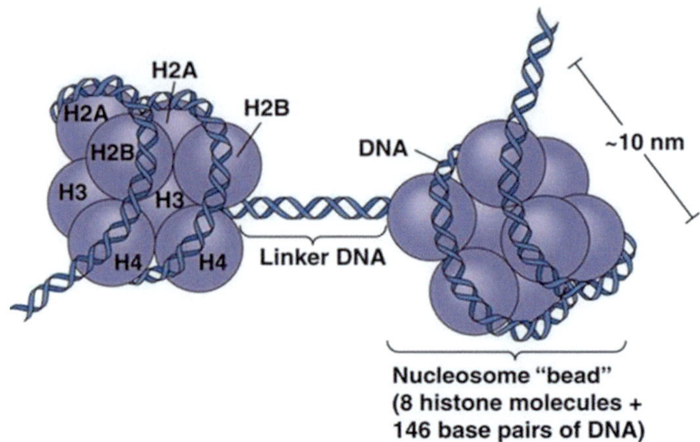

Fig. 8.11 The nucleosome assembly. The nucleosome consists of an octameric core comprised of the H2A, H2B, H3 and H4 units with about 146 bps of DNA wound around it

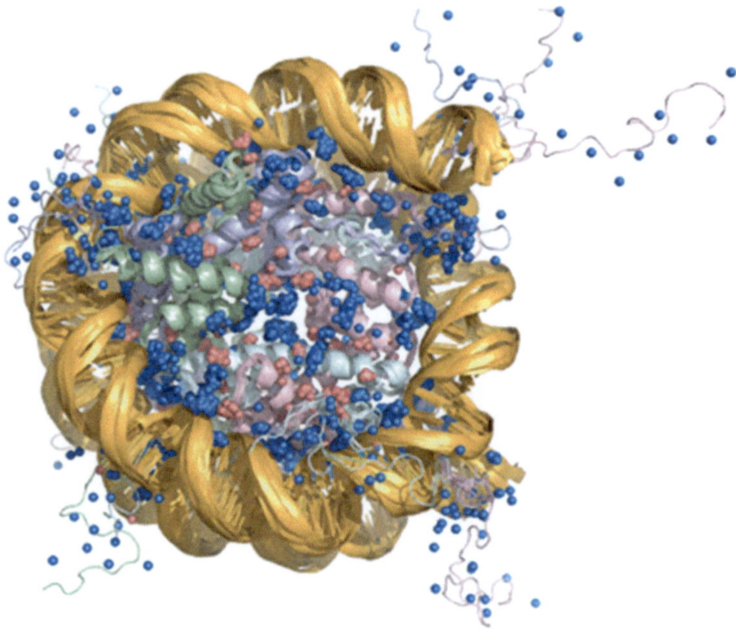

Fig. 8.12 Crystal structure of the nucleosome assemble. The image shows a schematic representation of the X-ray crystal structure of the nucleosome core. All reference frames are in alignment and computed by the PCA of the globular core of the histone octamer

inside with the help of the linker, H1, histone giving rise to a 30 nm chromatin (Fig. 8.13). A unit nucleosome and a unit of the H1 together are called the *chromatosomes*.

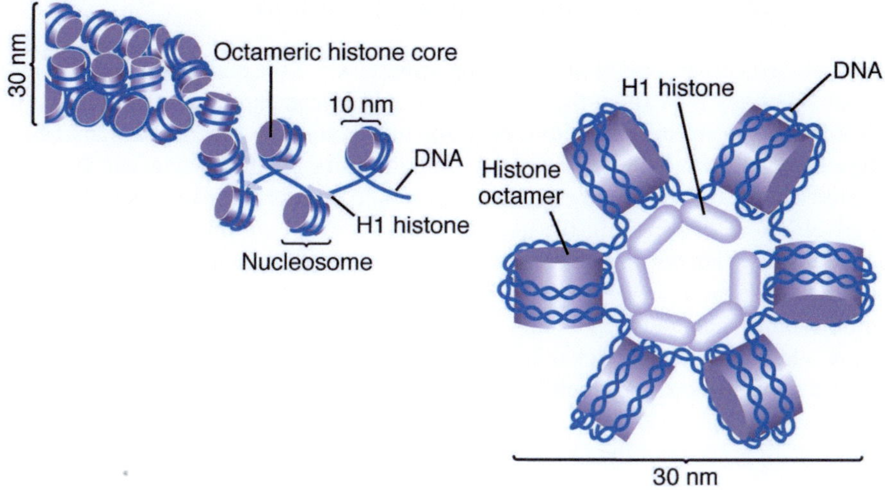

Fig. 8.13 The solenoid model. The nucleosomal core arrange in the form of a solenoid by bending over the H1 linker histone, thus giving rise to the 30 nm fibre

8.4.3 The Chromosomal Structure

The 30 nm solenoid chromatin again undergoes several orders of supercoiling to finally form the metaphase chromosome. The chromosome at the metaphase stage has the highest order of condensation. This is composed of loops of the 30 nm chromatin solenoid radiating from a central scaffold. The central scaffold contains non-histone proteins which anchor the chromatin at specific AT-rich regions called the *matrix-associated regions (MAR)* or *scaffold-associated region (SAR)*. Each loop emanating from the scaffold structure is about 100–300 nm (~60 kbps) long. It has been recently shown that this model of chromosome is neither a rigid model nor a random one. The chromosomes organize themselves in the three-dimensional nuclear space in the form of distinct territories with defined interactions with the nuclear subdomains. This gives rise to what is known as *topologically associated domains (TADs)*. TADs are essentially self-interacting regions of DNA which are far apart from each other in terms of their two-dimensional length but are brought together by the process of looping and three-dimensional organization of the chromosome. This helps to form two compartments: an open or transcriptionally active compartment and a closed or transcriptionally inactive compartment. So, TADs can be simply visualized as long stretches of DNA that come together in a common 3D space so that the genes that they house can be triggered or silenced in sync.

8.5 The Heterochromatin and Euchromatin

As discussed earlier, one of the most important tasks of the chromatin and chromosomal structures is to allow the accurate control of gene expression. Therefore, the chromatin fibre must make allowances for the proteins to reach and transcribe a segment of DNA required to be active while leaving other non-relevant segments inactive. In most regions the chromatin appears to be less compactly packaged and relatively more dispersed in the nucleus. These regions are transcriptionally more active and are known as the *euchromatin*. On the other hand, the more densely packaged regions of chromatin resembling the level of condensation seen in the chromosomes are transcriptionally inactive and known as the *heterochromatin*. These alternate states of packaging are achieved by various mechanisms including histone modification. Heterochromatin can be formed with the help of increased methylation of one histone, decreased acetylation of histones and hypermethylation of the cytidine bases of the DNA.

8.6 Chromosomal Banding

The study of the structure of chromosomes requires rigorous techniques to stain and visualize them under a microscope. This is done by enzymatic digestion or denaturation of the chromosome followed by staining with DNA binding dyes. Because of the variations in the extent of packaging, this procedure produces light and dark banding patterns on the chromosomes. Each chromosome has a unique banding pattern which helps in its identification (Fig. 8.14). Following are some of the banding techniques used routinely for the study of chromosomes:

G-banding—A controlled trypsin digestion of the chromosomes are performed followed by staining with Giemsa. The bands that take a dark stain are called G-bands, while the lightly stained bands are called G-negative. G-bands are mostly AT-rich segments of DNA.

Q-banding—This technique requires the staining of DNA with fluorescent dyes such as quinacrine, DAPI (4′,6-diamidino-2-phenylindole) or Hoechst 33258 and visualization under a fluorescence microscope. Since these dyes bind preferentially to the AT-rich regions of DNA, they produce a banding pattern similar to G-banding.

R-banding—This technique is performed by heat denaturing the DNA in high salt concentration before staining with Giemsa. This results in a banding pattern which is the reverse of G-banding. R-bands are also Q-negative. The same results can be achieved by using GC-specific binding dyes.

T-banding—This banding technique is a more severe form of R-banding used to identify the telomeric regions of DNA. This is achieved by either a more severe heat treatment of DNA prior to Giemsa staining or using a combination of standard and fluorescent dyes.

C-banding—This technique is used to stain constitutive heterochromatin of the centromere. The DNA is denatured with a barium hydroxide solution before being stained with Giemsa (Figs. 8.15 and 8.16).

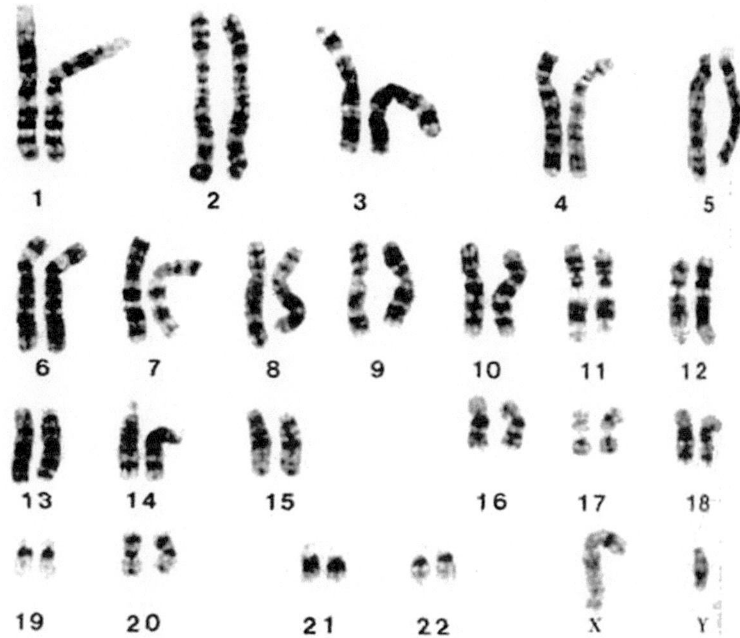

Fig. 8.14 G-banding pattern of the human chromosome as observed under a microscope. Giemsa is a protein stain that darkly stains the heterochromatic or transcriptionally inactive regions and lightly stains the euchromatic or transcriptionally active regions

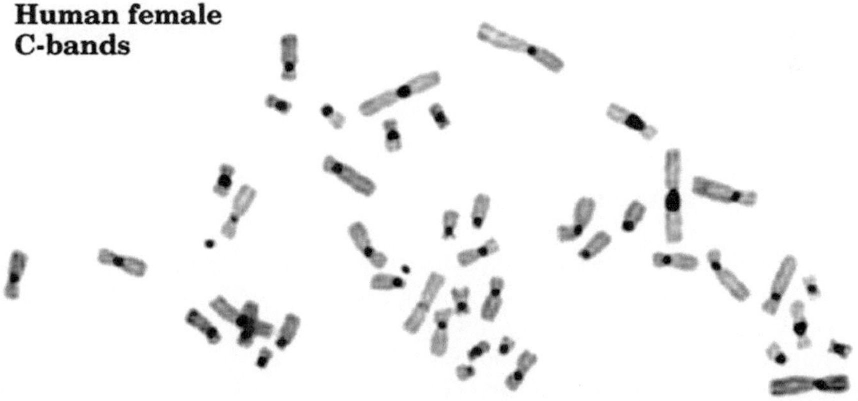

Fig. 8.15 C-banding patterns in human female chromosome. This technique specifically stains the centromeric region of the chromosome which is constitutively heterochromatinized

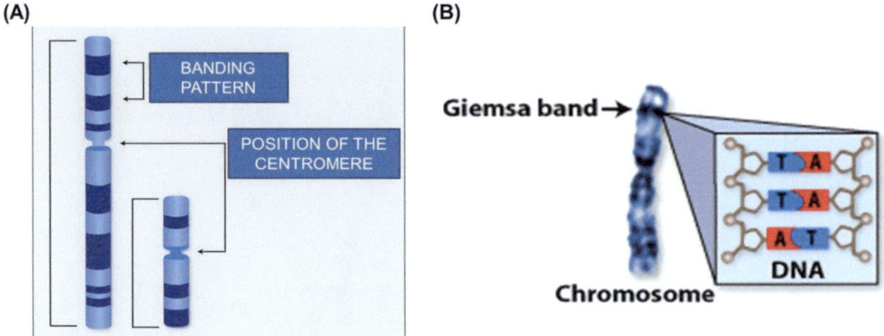

Fig. 8.16 Banding pattern of chromosomes. (**a**) Treatment and staining of chromosomes create alternating light and dark banding patterns which are unique for each chromosome. (**b**) The image depicts the G-banding pattern obtained after staining a chromosome with Giemsa which binds AT-rich regions

8.7 Morphology of the Eukaryotic Chromosome

At the metaphase stage of mitotic cell division, the chromosome consists of two molecules of DNA: one is parent DNA and the other is the DNA obtained from the replication of the parent DNA at the S-phase. These two molecules form two symmetrical structures called *sister chromatids* with each chromatid containing one DNA molecule. The chromatids are held together at the centromere which forms the point of attachment for the spindle fibres. When the two chromatids share a common centromere, they are called sister chromatids, but once they separate during the metaphase, each chromatid has its own centromere and is now known as a chromosome (Fig. 8.17).

8.7.1 The Centromere

The *centromere*, also known as the primary constriction, appears as a narrowed zone or a gap in the chromosome. The centromere harbours the *kinetochore complex* which has microtubules radiating towards the spindle pole of the cell. The kinetochore plays a crucial role in the movement of the chromosome towards the spindle pole during the cell cycle. The centromere is a heterochromatic region consisting of long stretches (~171 bps) of tandem repeats of short DNA sequences. Depending on the position of the centromere on the chromosome, the morphology is classified into the following groups:

- *Metacentric chromosome*: The centromere is present in the middle of the chromosome and divides it into two equal halves.

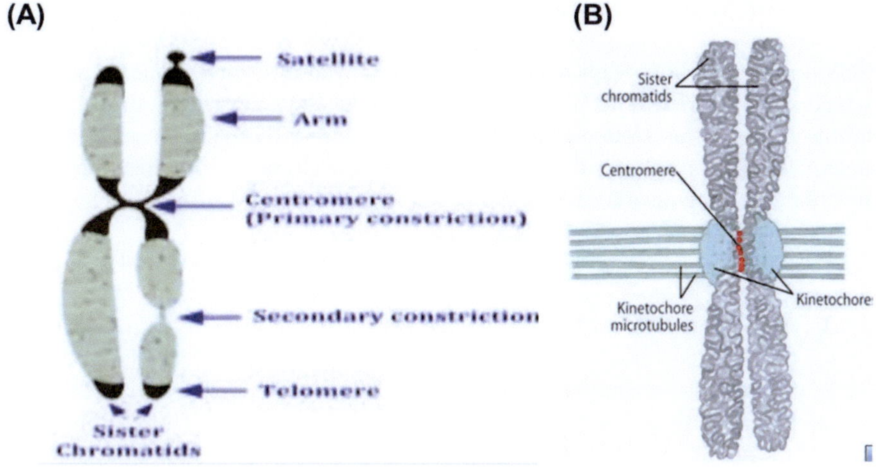

Fig. 8.17 The morphology of a metaphase chromosome. (**a**) The morphology of a eukaryotic metaphase chromosome shows the centromere, telomere, secondary constriction and satellite. (**b**) The centromere has the kinetochore assembly which connects to the microtubules and regulate separation of the sister chromatids during anaphase

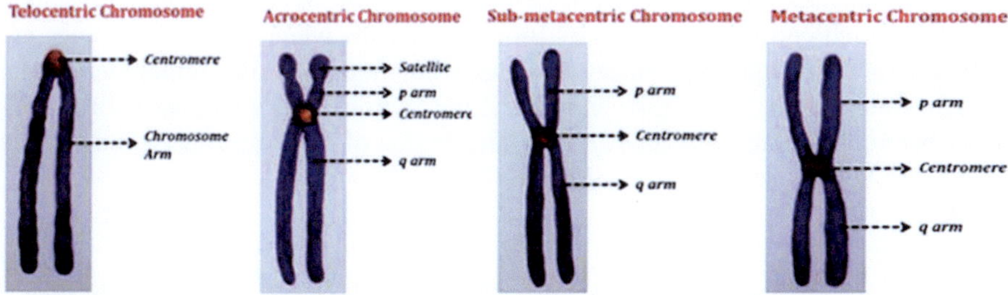

Fig. 8.18 Types of chromosomes based on the position of the centromere. The figure shows the variations of chromosomes based on the location of the centromere

- *Sub-metacentric chromosome*: A chromosome is said to be sub-metacentric when the centromere lies near the middle but not exactly in the centre of the chromosome. This results in two unequal arms. The shorter arm is called the “p-arm”, while the longer arm is known as the “q-arm”.
- *Acrocentric chromosome*: The centromere is located near the end of a chromosome resulting in unequal chromosome arms which has the shape of the letter “j”.
- *Telocentric chromosome*: The centromere is located at the end of the chromosome. The chromosome thus appears to have only one arm and has the shape of the letter “i” (Fig. 8.18).

8.7.2 The Secondary Constriction

Additional non-staining gaps may be seen in certain chromosomes. These are known as *secondary constriction (Sc)*. These are generally located towards the end of the chromosome arm and often contain genes encoding rRNAs or those that induce nucleoli formation. Since these serve as the epicentre for organization of the nucleolus, they are also known as the *nucleolus organising region (NOR)*.

8.7.3 Satellite DNA

The Sc constriction separates a segment of the chromosome from the rest of the arm. This rounded body is known as the *satellite* or the *trabant*. The satellite DNA consists of simple or moderately complex DNA sequence repeated multiple times over a long stretch of the DNA in tandem (end to end).

8.7.4 The Telomere

A distinctive feature of the eukaryotic chromosome is the *telomeric region* present at the end of each chromatid arm. They form sticky ends when the chromosomes are broken and differ markedly in structure and function from the remainder of the chromosome. As in case of the centromere, the telomeric region consists of constitutive heterochromatin. A span of 10–15 kbps region of highly conserved hexameric repeat $(TTAGG)_n$ makes up the telomere (Fig. 8.19).

Organism	Telomere sequence	RNA template sequence	RNA length
Tetrahymena	TTGGGG	CAACCCCAA	159
Euplotes	TTTTGGGG	CAAAACCCCAAAACC	190
Oxytricha	TTTTGGGG	CAAAACCCCAAAACC	190
Human	TTAGGG	CUAACCCUAAC	450
Mouse	TTAGGG	CCUAACCCU	450
S. cerevisiae	TG(1-3)	CACCACACCCACACAC	~1300
K. lactis	TTTGATTAGGTATG-TGGTGTACGGA	UCAAAUCCGUACACCA AUACCUAAUCAAA	~1300

Fig. 8.19 The telomeric sequence of various organisms. The sequence of telomerase RNA and the telomere repeat for different organisms have been depicted

8.8 Specialized Chromosomes

Certain eukaryotic organisms have tissues that harbour chromosomes with characteristic structures which differ greatly from the standard chromosome. These chromosomes, also known as the giant chromosomes, attain their largest size in the nuclei of the cells housing them. They are commonly found in the suspensors of the embryo of a few plants, in cells of the Malpighian tubules, in cells of the salivary glands of *Drosophila* and *Chironomus*, in oocytes of certain vertebrates, etc. These specialized chromosomes can be classified into two categories: *polytene chromosome* and the *lampbrush chromosome*.

8.8.1 The Polytene Chromosome

In these giant chromosomes, DNA replication is not followed by separation of the daughter chromatids giving rise to a giant chromosome. Essentially, the DNA replicates multiple times, but the nuclear division never happens. This phenomenon is known as *endomitosis*. The daughter chromatids never separate from each other and appear as a multistranded giant chromosome with the chromatids aligned side by side (Fig. 8.20).

They were first observed by E.G. Balbiani (1881) in the cells of the salivary glands of *Chironomus* larvae. Hence, these are also known as salivary gland chromosome. They are also found in the living cells of the gut, the cells of the Malpighian tubule and the salivary gland of *Drosophila* larvae. In *Drosophila*, they appear to have a length ~100 times greater than the length of the somatic metaphase

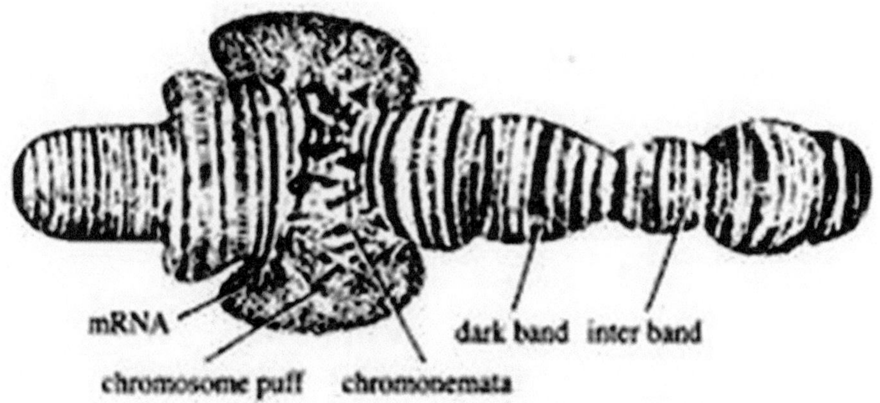

Fig. 8.20 A polytene chromosome. The figure depicts the polytene chromosome of an insect. The dark and light zones of bands and interbands are shown along with a puff or the Balbiani ring

chromosome. However, since these chromosomes pair during the interphase, their number appears to be half of the normal somatic cells.

The polytene chromosomes have alternating dark and light bands along its length when stained with Feulgen stain. The dark bands are heterochromatin regions, while the lighter bands, also known as the interbands, are the euchromatic regions.

8.8.2 Lampbrush Chromosome

Another type of giant diplotene chromosome found in the nuclei of oocyte of urodele amphibians is the lampbrush chromosome. They consist of a well-demarcated chromosomal axis, chromomeres and many looped extensions. Since the appearance of these chromosomes resembles the bristles of the brushes used to clean the chimneys of oil lamps, therefore they are known as the lampbrush chromosomes. A multitude of fine lateral loops gives them a "hairy" appearance. Visible in the meiotic prophase, the lampbrush chromosomes are found in the form of bivalents each having four chromatids. The homologous chromosomes are held together by chiasmata, and the axis of each homologue contains a row of chromomeres from which one to nine lateral loops emanate (Fig. 8.21).

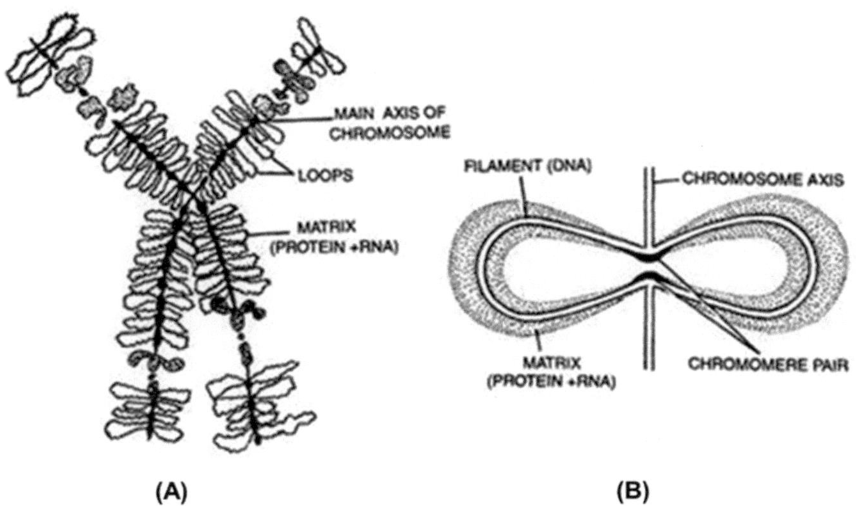

Fig. 8.21 The lampbrush chromosome. (**a**) Enlarged part of a lampbrush chromosome. (**b**) One loop of the chromosome enhanced

Box 8.1 Scientific Concept: Investigating DNA Supercoiling in Eukaryotic Genomes: Samuel Corless and Nick Gilbert

DNA supercoiling is an inherent and important property of the DNA. Supercoiling is often generated when polymerases or DNA-binding proteins unwind or bend the DNA double helix. Even though supercoiling has a pivotal role in genomic organization and function, its study is extremely challenging specially in the highly complex eukaryotic chromosome. Corless and Gilbert have described the following techniques for the study of the supercoiling dynamics.

1. Centrifugation or migration-based techniques
   The process of supercoiling results in a more compact DNA with a 3D structure remarkably different from that of a non-supercoiled DNA of the same length. Such compaction alters the property of migration of DNA during sucrose gradient centrifugation or agarose gel electrophoresis. When both supercoiled and non-supercoiled DNA molecules are subjected to sucrose gradient centrifugation, the supercoiled DNA molecule has a higher sedimentation rate compared to that of a non-supercoiled DNA molecule of the same molecular weight. Similarly, supercoiled, nicked and linear DNA molecules show differential rate of migration in agarose gel electrophoresis with the supercoiled DNA travelling the fastest. Further, using an intercalating agent like ethidium bromide, it is possible to differentiate between topoisomers with varying levels of supercoiling. The intercalating agents unwind the DNA in order to bind to them thereby inducing negative supercoils. Therefore, with the incorporation of the intercalating agent, the rate of migration for the DNA molecule will change as positive and negative supercoiling result in different rates of migration (Fig. 8.22).
2. The preferential intercalation of psoralen molecules in underwound DNA
   Psoralen is an intercalating molecule used to measure DNA supercoiling in normal eukaryotic chromosomes. A cell permeable psoralen derivative, 4,5′,8-trimethylpsoralen (TMP), intercalates in the DNA molecule and forms stable cross-links with pyrimidine nucleotides when exposed to UV light (365 nm). It binds preferentially to underwound segments in a probe-specific way. This is an invaluable technique as the TMP molecule can be conjugated with biotin and used for either a pull-down assay to enrich underwound DNA, or it can be used for immunofluorescence in vivo by treating the TMP bound DNA with a fluorescein-conjugated streptavidin moiety.

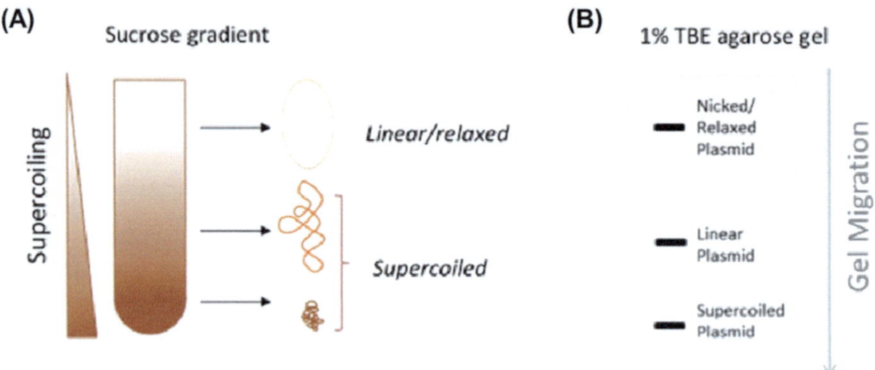

Fig. 8.22 Experimental evaluation of DNA supercoiling. (**a**) Differential partitioning of the supercoiled and linear DNA in sucrose gradient centrifugation. (**b**) Differential migration of nicked, linear and supercoiled DNA when subjected to agarose gel electrophoresis

8.9 Chapter Summary

- Given the size of the chromosome in proportion to the cells that harbour them, they need to be condensed manifold to be able to be physically and functionally accessible. For this purpose, DNA is often supercoiled. Supercoiled DNA appears as a coiled coil, and its topology is defined in mathematical of linking number and writher number. The topoisomerase enzymes function to introduce or relax these supercoils.
- The bacterial genome consists of a single nucleoid which is a single covalently closed circular DNA. The DNA is highly supercoiled forming coiled loops and is packaged with the help of non-histone basic proteins such as HU, HNS, etc.
- The packaging of the eukaryotic genome follows a distinct hierarchy. The first level of packaging is the nucleosome assembly which represents the "beads-on-string" structure. An octameric histone assembly forms the core with a segment of DNA bound around it. A linker histone binds the DNA which lies between two core histones.
- The solenoid structure or the "coiled coil" is the second level of packaging and forms the 30 nm fibre. The solenoid finally binds a scaffold matrix with the help of matrix-associated regions to form the final chromosomal structure.
- The highly condensed regions of the DNA are known as heterochromatin and are genetically inactive, while the relatively open regions known as euchromatin are genetically active.
- Each individual chromosome can be karyotyped and assigned a unique morphological signature based on its banding patterns. Giemsa is an important stain, and various treatments of DNA prior to staining can result in various types of banding patterns, viz. the G-band, C-band, R-band and C-band.

- The eukaryotic chromosome has key structural features. The centromere is a highly repetitive DNA segment that forms the attachment site for binding of spindle fibres. There is another constriction known as the secondary constriction evident on certain metaphase chromosomes.
- The telomere is a specialized region present at the end of a chromosome. An active telomerase enzyme can bind to extend this segment to prevent shortening of the chromosomal ends which do not replicate in somatic cells of higher eukaryotes.
- There are also specialized chromosomes such as the lampbrush and polytene chromosome which result from DNA that undergo multiple cycles of replication without segregating during mitosis. These have distinctive structures and are termed as the "giant chromosome".

Further Reading

Bates AD, Maxwell A (2005) DNA topology. Oxford University Press, New York

Bednar J, Horowitz RA, Grigoryev SA, Carruthers LM, Hansen JC, Koster AJ, Woodcock CL (1998) Nucleosomes, linker DNA, and linker histones form a unique structural motif that directs the higher-order folding and compaction of chromatin. Proc Natl Acad Sci 95:14173–14178

Bendich AJ, Drlica K (2000) Prokaryotic and eukaryotic chromosomes: what's the difference? Bioessays 22:481–486

Cairns J (1963) The chromosome of Escherichia coli. Cold Spring Harb Symp Quant Biol 28:43–46

Castle WE (1919) Is the arrangement of the genes in the chromosome linear? Proc Natl Acad Sci 5: 500–506

Cook PR (1995) A chromatic model for nuclear and chromosome structure. J Cell Sci 108:2927–2935

Corless S, Gilbert N (2017) Investigating DNA supercoiling in eukaryotic genomes. Brief Funct Genomics 16:379–389

Judd BH (1999) Genes and chromosomes: a puzzle in three dimensions. Genetics 150:1–9

Kornberg RD (1974) Chromatin structure: a repeating unit of histones and DNA. Science 184:868–871

Lewin B (2007) Genes IX. Jones and Bartlett, Sudbury

Noll M (1974) Subunit structure of chromatin. Nature 251:249–251

Pombo A, Dillon N (2015) Three-dimensional genome architecture: players and mechanisms. Nat Rev Mol Cell Biol 16:245–257

Snyder L, Champness W (2003) Molecular genetics of bacteria, 2nd edn. ASM Press, Washington, DC

Strachan T, Read AP (1999) Human molecular genetics. Wiley, New York

DNA Mutation, Repair, and Recombination 9

Atish Ray

A mutation is a process of alteration in genome sequence by mistakes during DNA replication or as the result of environmental stress including radiation, toxic chemicals, and smoke. Over a lifetime, the DNA of an organism can experience many alterations in the bases, (A, C, G, and T) sequence. The transient errors are corrected via a defined DNA repair mechanism. If the errors are not corrected in course of time, mutation takes place leading to a permanent impact on function, physiology, or phenotype mediated by awful translated proteins.

Commonly the term "mutation" has been associated with obvious detrimental outcomes. The harmful effect of mutation is undoubtedly a major issue but not always. In the natural selection process, positive mutations are inherited through the generations contributing to genetic variation within the species. This variation often facilitates the organism to survive and reproduce. On the other hand, "sickle cell anemia" is quite well discussed as an unfavorable outcome of mutations. This is the result of a mutation in the hemoglobin gene. Similarly, several forms of cancers, very common in the present day, are also known to be contributed by certain mutations. However, as mentioned above, unlike the harmful or disease-causing mutations, positive and neutral mutations are least discussed. For example, variation in blood types and eye color has neither good nor bad effects. Thus this kind of mutation can be generally regarded as neutral mutation. In such variation, it is often meaningless to distinguish between mutant and original alleles. Similarly, there can also be a positive effect of the mutation. Bacterial populations have certain mutations which can confer antibiotic resistance. Although this type of mutation is not at all positive for humans, however, logically they are beneficial (positive) mutations for bacteria.

There are some other examples of beneficial mutations. HIV resistance and malaria resistance are well documented. A 32 base pair deletion in a specific position of human CCR5 (CCR5-Δ32) confers HIV resistance to homozygotes and delays

A. Ray (✉)
Savil Technology and Business Incubator, Vadodara, India

D. Kar, S. Sarkar (eds.), *Genetics Fundamentals Notes*,
https://doi.org/10.1007/978-981-16-7041-1_9

AIDS outbreak in the European population. Surprisingly, there are certain instances where a single detrimental mutation has resulted in other beneficial effects. In sub-Saharan Africa, where malaria is common, significant numbers in the population carry the mutant allele of sickle cell trait. Such heterozygous individuals (carriers of sickle cell trait) are found to be resistant to malaria.

In this context, other terminology, "recombination," must be explained in contrast to "mutation." Genetic recombination (which is also known as genetic reshuffling) is the process of exchanging genetic material between different organisms leading to the production of offspring with combinations of parent's traits. In eukaryotes, a novel set of genetic information during genetic recombination can be naturally transmitted from the parents to the offspring in due course of time.

Both mutation and recombination can produce a novel trait. However recombinant traits can be assorted independently through the generation, but technically mutant traits are imprinted within the genome which can only be revered if it is corrected. In several cases, correction is possible through the precise mechanism known as the DNA repair mechanism, and this will be explored further in this chapter. In the following section, different facets of mutation, DNA recombination, and repair are elaborated.

9.1 Mutation: Overview of the Process

First of all, it's important to emphasize that mutation is one of the diverse alternatives to natural selection theory which was the subject of heated discussions both before and after the publication of *On the Origin of Species* by Charles Darwin in 1859. According to these conceptions, the mutation was taken as the preliminary considerable component which was believed to be acted as a source of novelty in producing new species. During the evolution of species, changes occur instantaneously through a sudden jump. Mutation plays two key roles in evolution: (1) it creates an evolutionary force that changes gene frequencies, and (2) it reserves the ultimate capacity of all genetic variation. With the progress of in-depth research on cell physiology, scientists concluded the molecular mechanism of mutagenesis. Not entering the history, lets us come directly to the overview of the mutation process. In this section, the entire process of mutation will be discussed thoroughly.

9.1.1 A Different Class of Mutation

The entire event of "mutation" can be classified in different categories depending on the different parameters including the size of the affected DNA, nature genetic alteration, impact on the translated peptide, and target cell types. It is important to remember that these parameters are not necessarily mutually exclusive. Terminologies related to the types of mutation are summarized in Table 9.1.

Table 9.1 Important terminologies and their definitions related to mutations

Term	Definition
Base substitution	Alteration of the base of a single DNA nucleotide
Transition	Base substitution of purine with purine or a pyrimidine with pyrimidine
Transversion	Base substitution where purine replaces a pyrimidine or a pyrimidine replaces a purine
Insertion	Addition of nucleotides
Deletion	Deletion of one or more nucleotides
Indel	Insertion or deletion
Frameshift mutation	Insertion or deletion that changes the reading frame of a gene
In-frame mutation	Deletion or insertion of a multiple of three nucleotides resulting in no shifting of reading frame
Reverse mutation	Changes of a mutant phenotype back to the wild-type phenotype
Missense mutation	Changes a sense codon into a different sense codon
Nonsense mutation	Changes a sense codon into a nonsense codon
Silent mutation	Changes a sense codon into a synonymous codon, leaving unchanged amino acid
Neutral mutation	Changes the amino acid sequence of a protein without altering its functional property
Loss-of-function mutation	A complete or partial loss of function
Gain-of-function mutation	The appearance of a new trait or function
Lethal mutation	The mutation that causes early infantile/premature death
Suppressor mutation	The mutation that suppresses the effect of the earlier mutations

9.1.2 Parameter 1: Size of the Affected DNA

Mutation can be broadly grouped into two types: (1) point mutation and (2) large-scale mutation. Point mutation refers to the change in a single (or few) bases in the DNA leading to alteration in the translated peptide. On the other hand, large-scale mutation can be demonstrated as the alteration/deletion/duplication of a comparatively larger part of the DNA. (Fig. 9.1a–c). However, in present days, the conception of large-scale mutation has been expanded and became more closely associated with chromosomal aberrations where abnormal ploidy (total number of complete sets of chromosomes in a cell) values are also found in certain cases (Fig. 9.1).

9.1.3 Parameter 2: Mode of Occurrence

The nature of mutation can be spontaneous or induced. Spontaneous mutations take place with a non-zero probability within a cell where several types of molecular mechanisms are involved such as depurination, deamination, mispairing, and tautomerism. These are discussed in the respective section in this chapter. On the other

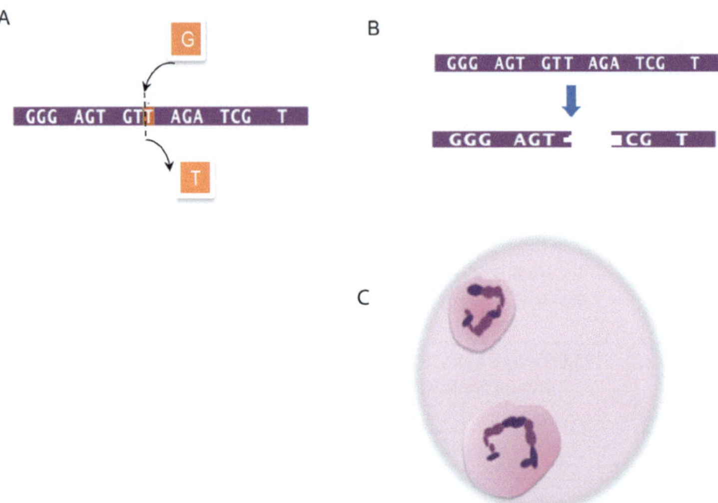

Fig. 9.1 Overview of mutation. Mutation can be accomplished by changes in (**a**) small portion of DNA (single or few) or by (**b**) deletion of the larger part. (**c**) Polyploidy, a type of chromosomal aberration well known in the plant *Rhoeo discolor* where chromosomes are associated to form a chain-like structure

hand, induced mutations are alterations in the DNA exposed to mutagenic chemical and/or other environmental stressors including ultraviolet light and ionizing radiation, base analogs, DNA intercalating agents, and DNA cross-linking agents.

9.1.4 Parameter 3: Target Cell Type

Mutation can occur both in somatic cells and in germ cells. Somatic mutations (also known as acquired mutations) occur in cells that are not part of a designated reproductive group and are not passed down to descendants. On the other hand, germ line mutations occur soon after fertilization and get transmitted through the offspring.

9.1.5 Parameter 4: Impact of Translated Peptide

Based on the impact on the translated peptide sequence, the mutation process can be classified as (1) frame shift mutation, (2) in-frame mutation, and (3) base substitution mutation. Base substitution can be either synonymous or non-synonymous. Furthermore, synonymous base substitution can also be silent, while non-synonymous base substitution can be divided into nonsense and missense type. Two important impacts of mutation—(a) frameshift and in-frame mutation and (b) synonymous and non-synonymous base substitution—are described especially in the following paragraphs.

9.1.6 Frameshift and In-Frame Mutation

Let us recall the triplet nature of the genetic code. The reading frame (from start to stop codon read during translation of peptide) can be shifted if the alteration (insertion or deletion) of bases takes place in such a position by which the combination of three bases is impaired. In contrast, if the alteration of bases takes place without impairing the three-base devisable combination, it creates no changes in the existing reading frame which is known as in-frame mutation. Let us consider a typical example. Proline, aspartic acid, tyrosine, and leucine are all encoded by CCU-GAC-UAC-CUA. Now if U is deleted from CCU, the entire peptide sequence will be altered. Instead of proline, aspartic acid, tyrosine, and leucine, the resulting peptide will be proline, threonine, threonine, and any other (any amino acid, stop codon, etc.). Usually, the frameshift mutation is the result of either insertion or deletion which is known together as "indel" mutation. It has been found that loss of proofreading activity is significantly associated with the frequency of UV-induced frameshift in bacteria. Investigation on novel frameshift mutation in search of the certain frameshift mutation is going on. Recently, a frameshift mutation in the *gon4l* gene is found to be associated with dwarfism in Fleckvieh cattle. Another study found a link between an autosomal recessive non-syndromic hearing loss and a new frameshift mutation (c.804delG) in the immunoglobulin-like domain containing receptor 1 (ildr1) (Fig. 9.2).

9.1.7 Synonymous and Non-synonymous Base Substitution

Sometimes a single base can be substituted spontaneously resulting in changes in the triplet codon without changing the reading frame due to the degeneracy of the genetic code. In some cases, this type of mutation results in an altered codon that encodes the same amino acid. This type of base substitution is known as synonymous substitution (Fig. 9.3). When synonymous substitution does not impact the phenotype, it is called a silent mutation. Similarly in non-synonymous base substitution, a codon is altered in such a way by a different codon that necessarily leads to modification of amino acid.

Two definite types of non-synonymous base substitution mutations are (a) nonsense and (b) missense mutation. We are familiar with the phrase nonsense concerning nucleotide biology. Nonsense codon (UAG, UAA, UGA) refers to the codon, the presence of which determines the termination of translation of a peptide. In nonsense mutation the single base substitution results in the addition of a stop/nonsense codon in the reading frame leading to premature termination of translation. This results in a truncated and non-functional protein. A nonsense mutation in the gene *Cyp11B1* is one of the mutations that result in affected steroid

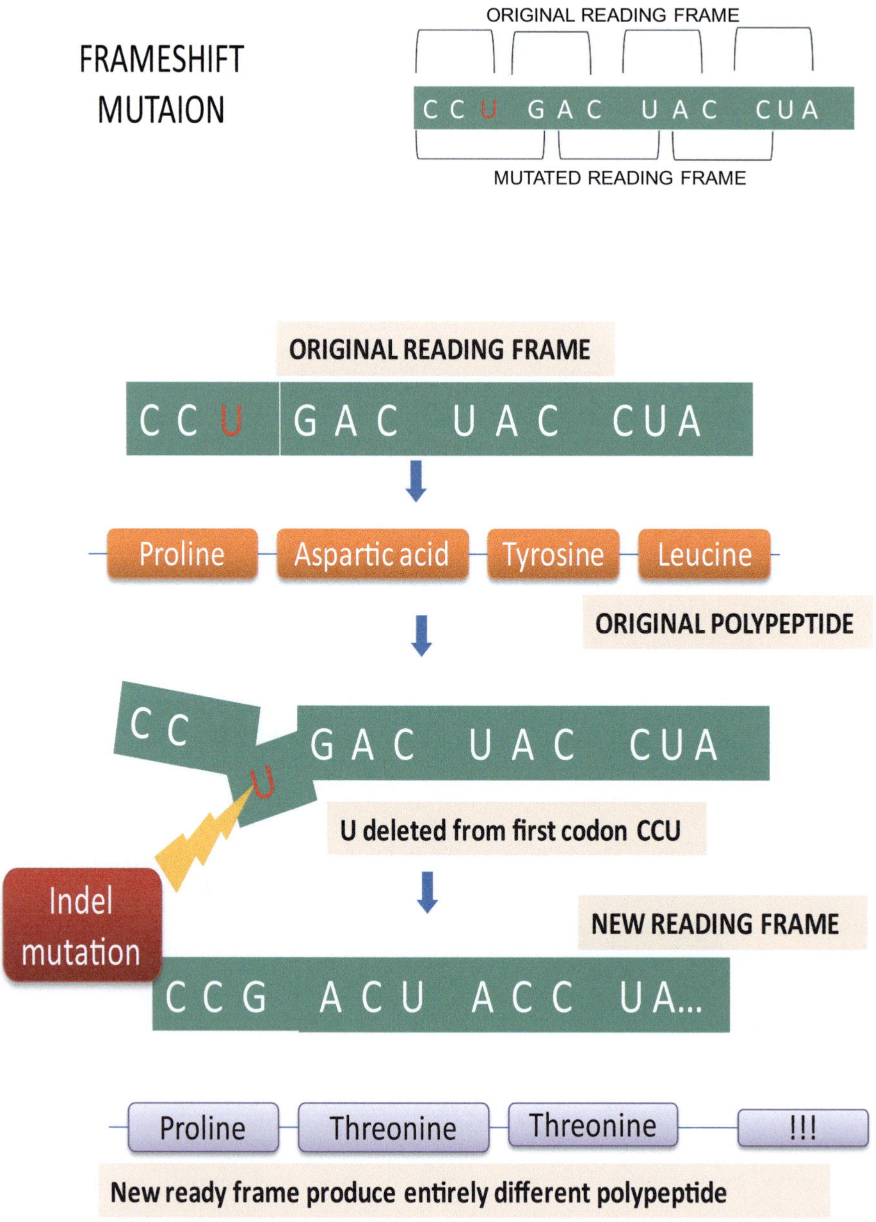

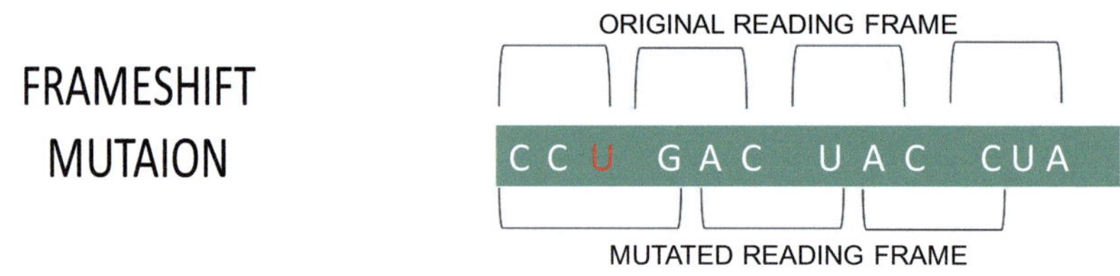

FRAMESHIFT
MUTAION
ORIGINAL READING FRAME
C C U G A C U A C C U A
MUTATED READING FRAME
ORIGINAL READING FRAME
C C U G A C U A C C U A
Proline
Aspartic acid
Tyrosine
Leucine
ORIGINAL POLYPEPTIDE
C C
U
G A C U A C C U A
U deleted from first codon CCU
Indel
mutation
NEW READING FRAME
C C G A C U A C C U A...
Proline
Threonine
Threonine
!!!
New ready frame produce entirely different polypeptide

Fig. 9.2 Mechanism of frameshift mutation. A single nucleotide insertion or deletion can lead to change in the entire reading frame of DNA known as frameshift mutation. Frameshift mutation results in different combinations of amino acid as compared to the original sequence

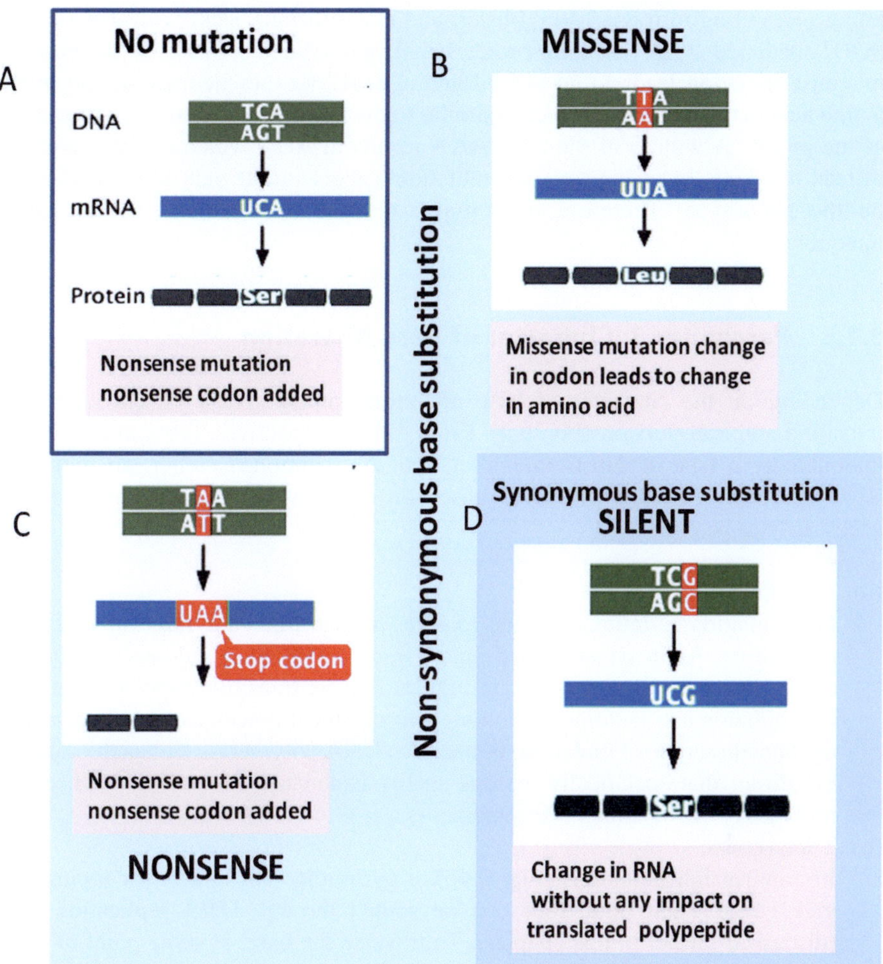

Fig. 9.3 Substitution mutation, a generalized conception. Substitution is commonly a point mutation. Substitution can be a synonymous or non-synonymous mutation. Figure (**a**) demonstrates non-mutated DNA. There are two types of non-synonymous mutation: (**b**) missense and (**c**) nonsense mutation. While (**d**) synonymous base substitution leads to changes in RNA without any impact on the translated peptide, nonsense mutation introduces a nonsense codon in the RNA leading to premature termination of the peptide

11beta-hydroxylase leading to a disease condition called congenital adrenal hyperplasia. This is an autosomal recessive disorder where deficiency of the sex steroid hormone impairs the primary and secondary characteristics of infants, children, or adults.

On the other hand, a missense mutation is a point base substitution mutation where the resulting altered amino acid leads to the non-functional protein. Such mutations are responsible for several diseases condition including epidermolysis bullosa (resulting in fragile skin, blister, and skin erosion), sickle cell disease, and SOD1-mediated ALS (an adult-onset, lethal, paralytic disorder). Sometimes the missense mutation leads to an altered amino acid, but the chemical nature of the amino acid in the mutant sequence is similar to the normal. For example, substitution of the middle A with G of the code AAA results in AGA which encodes arginine instead of lysine. In such a case, the mutation has eventually very little or no effect on the phenotype. Therefore these types of mutations are known as neutral mutations.

9.1.8 Parameter 3: Character of Base Alteration

Depending on the character of base alteration, mutation can be classified into transition and transversion (Fig. 9.4). From other perspectives, these mutations are undoubtedly a type of substitution mutation. As mentioned earlier, all mutations have been only logically classified based on different parameters which are not necessarily to be exclusive.

(a) Transitions

In a transition mutation, a purine replaces a purine, and a pyrimidine replaces a pyrimidine. Approximately two out of three single nucleotide polymorphisms (SNPs, we will discuss later) are expected to be transition types of mutation. Deamination and tautomerization are two chemical processes that contribute to the transition type of mutation. In bacteria, DNA polymerase III has the capacity of editing that specifically excises such mismatched bases. This reduces the probability of mutations (Fig. 9.4a).

(b) Transversion

In transversion mutations (Fig. 9.4b), a pyrimidine substitutes for a purine or vice versa. These mutations can be routed through DNA replication error (discussed further in this chapter). To traverse the base, at some point of DNA replication, purine-purine or a pyrimidine-pyrimidine mispair is critical. Although according to the dimensions of the DNA double helix this type of mispairs is energetically unfavorable, X-ray diffraction studies demonstrated that purine-purine pairs are possible. For instance, 8-oxo-2′-deoxyguanosine (8-oxodG), an oxidized deoxyguanosine derivative, can cause a spontaneous heritable G to T transversion mutations in germ line cells of mice. The example is one of the classical examples of the spontaneous molecular lesion discussed later (in the "route introduction of mutation section").

9.1.9 An Outline of Transition-Transversion Bias

Although not conserved, in general, the ratio of transitions to transversions is high in the naturally observed mutation. In many classical works, transition-transversion bias has been demonstrated as the ratio of differences. This has been expressed as a

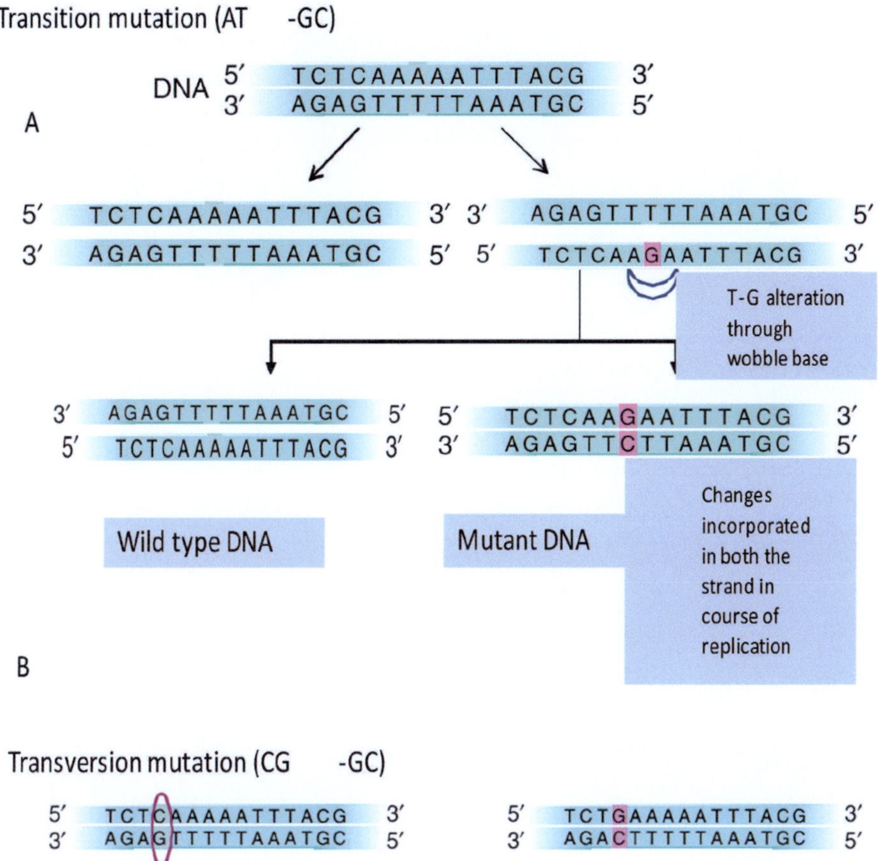

Fig. 9.4 Transition and transversion. AT to GC transition (**a**) is achieved through T-G alteration in the wobble base during DNA replication. Transversion (**b**) may take place during DNA replication via a similar mechanism as of transition causing a complete change of base in both the strands. In this figure, the fourth base pair (CG) is changed into GC

complex function of the degree of sequence divergence. In other words, transition bias was defined as bias in instantaneous rates. There is a possibility of two transversions while only one transition (Fig. 9.5). Each purine site may experience one type of transition (G-A) at rate α and two types of transversion (G-C, G-T) at rate β. Based on this consideration, probability of a transversion mutation should have a higher frequency than transition. However in reality that scenario is different. The conservation pattern of transition/transversion bias is not completely understood to date and is diverse among different species. For example, metazoan DNA sequences exhibit excess transition over transversion. A relatively high rate of mutation of methylated cytosine to thymine is thought to be one of the major causes of this bias. There are also several other reasons for this bias.

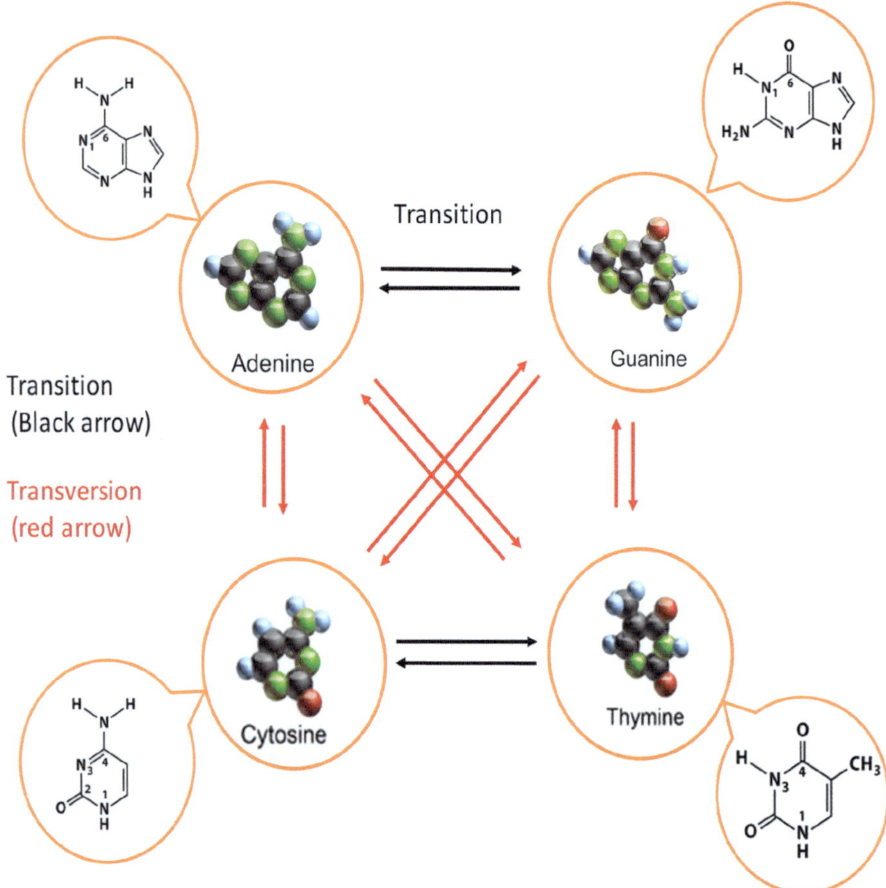

Fig. 9.5 Transversion versus translation. The probability of transversion is theoretically double than transition. However, in many cases, transition mutation is found in higher frequency than transition due to differential transition bias

Simply there are two major considerable facts behind this: (1) a single ring to single ring substitution (in transition) is energetically favorable as compared to double ring substitution for a single ring, and (2) transition mutation is more abundant in the population due to its possibility to reside as a silent mutation in the gene. Transition mutation is less likely to give rise to actual amino acid substitution due to its wobble base pair (refer to the silent mutation as described above). Therefore in the majority of the cases, transition mutation can reside as single nucleotide polymorphisms (SNPs) in a population without impacting the phenotypes. On the other hand, transversion mutations are more pronounced and have the potential to result in a catastrophic endpoint.

9.1.10 Similarity and Contrast: Synonymous Substitution Mutations, Silent Mutation, and Neutral Mutation

The terms "silent mutation," "neutral mutation," and "synonymous substitution mutations" are often used interchangeably in an oversimplified manner. Synonymous substitution mutation is technically defined as the substitution of base leading to different code but representative of the same amino acid as described above. However, synonymous mutations are not always silent, nor vice versa. Sometimes synonymous substitution mutation affects any of the processes including transcription, splicing, mRNA transport, and translation. From the evolutionary viewpoint, this implicates the fitness of the individual carrying the new gene to survive and reproduce. Several synonymous mutations in the fruit fly alcohol dehydrogenase gene, for example, were identified to cause suboptimal synonyms, resulting in a low enzyme yield. In this situation, the phenotypic effect is not silent. Similarly, there are other cases where synonymous substitution affects the protein function. (a) Intronic regions are often regulatory regions, and substitution mutation in this region may affect the protein function. (b) Timing of translation may be affected due to the substrate specificity of the tRNA to the infrequent codon which is reflected in codon bias. On the other hand, the mutations are neutral mutations where mutation replaces the amino acid of the same functionality. It is the type of non-synonymous substitution often missappropriately described as silent mutation.

9.1.11 Mutations Affect Gene Expression: "Loss of Function" Versus "Gain of Function"

Mutation can affect gene expression, and the consequence is either loss of function or gain of function. Loss-of-function mutations are achieved through the complete or partial absence of normal protein. Generally, in the loss-of-function mutation, the impairment occurs within the coding region of a gene so that the ultimate proteins are no longer able to work correctly. The mutation, on the other hand, can arise in regulatory areas that alter transcription, translation, or splicing events. In the gain-of-function mutation, alteration in the gene sequence produces the protein of a new molecular function. "Loss-of-function" mutations are often recessive. Therefore, homozygous individuals are only expressive mutant phenotype. There are several examples of loss-of-function mutation. The most widely discussed example of mutation is sickle cell anemia due to a mutation in β-globin gene which is a loss-of-function mutation resulting in the impaired oxygen-carrying capacity of hemoglobin. Similarly, a mutation in the insulin gene results in suboptimal production of insulin leading to diabetes type I. A more complex mutation is seen in PKU or phenylketonuria where metabolism of phenylalanine is reduced. Children seem normal at birth but begin to express with age leading to an underdeveloped brain.

In contrast, a "gain-of-function" mutation produces an entirely new trait. This mutation can cause a trait to arise at an inopportune moment and in an inopportune location. Gain-of-function mutations are frequently dominant or semi-dominant.

Gain-of-function mutations in scn11a, for example, have been reported in a small percentage of patients with painful peripheral small fiber neuropathy who have a congenital loss of pain sensibility. Certain mutations are conditional mutations in which the trait expresses only under certain conditions. On the other hand, certain mutations are lethal which can cause premature death or early infantile death in higher animals.

9.1.12 Mutations Are Instrumental in Evolution: "Gradual Change" Versus "Quick Jump"

Mutations are thought to be the "raw materials of evolution" and are essential to evolution. Every organism's genetic traits were initially the result of a mutation. The new genetic variant (allele) gets distributed via reproduction. Mutation improvises fundamental mechanisms of life including feeding, growing, or reproducing effectively. The mutant allele becomes increasingly abundant over time. As a result, the population diverges ecologically and physiologically from the original population that was unable to adapt. By removing a person bearing adaptive alleles at other genes, deleterious mutations also promote evolutionary change in small populations. In nature, both the evidence of gradual and quick evolutionary change are available as a result of mutation.

Most mutations causing evolutionary changes are single-point mutations affecting a single protein which remain less important, for example, genes that control the structure and effectiveness of the salivary glands. At a glance, mutations to salivary enzymes appeared to have less potential in impacting survival. However precise accumulation of slight mutations to saliva has an impact on evolving snake venom and snake evolution. Snake venom is a cocktail of proteins with varying effects. Other poisonous snake families have a distinct mix of genetically related species. Elapidae venom, which includes progenitors of sea snakes, coral snakes, and cobras, has evolved to be neurotoxic, but Viperidae venom, which includes rattlesnakes and bushmasters, acts on the circulatory system. Both families contain various species that, through mutations, inherited a minor edge in venom power from their ancestors, increasing the diversity of venoms and species over time.

Large-scale mutation usually affects quick evolutionary change within the population. For instance, there are certain organisms where chromosomal duplication (a type of large-scale mutation, broadly known as chromosomal aberration) takes place because their ancestor failed to undergo successful meiosis before sexual reproduction which resulted in the doubling of a chromosome. In North American grey tree frogs, this approach finally leads to "instant speciation."

As an animal, the process is also known in the plant kingdom which produces abnormally large seeds or fruits. These are having distinct traits with specific advantages. Most human edible cereals have enormous seeds as compared to other grasses. In most cases, this is due to genomic duplications in the ancestors, and the outcome of this error was successfully passed down to future generations. The phenomenon is also found in certain modern rice and wheat. In modern day's agricultural biotechnology practice, this process of Darwin's natural selection is often mimicked artificially to produce better offspring through plant interbreeding which is known as artificial selection. In no way, evolution is possible without random genetic change. However, the differential evolutionary fitness of mutant offspring and the probability of evolving by genetically modified individuals are the major issues in the successful production of the mutant population.

9.1.13 Mutation Rate

The mutation rate in genetics is defined as the number of new mutations in a single gene or organism over time. To put it in another way, it's the rate at which a gene switches from wild type to a certain mutation. Mutation rates are variable and diverse (Table 9.2). Therefore, mutation rates are determined for specific classes of mutations, and the spectrum of mutation rate is subdivided with subclasses. The mutation rates are usually expressed with different units including mutation base pair^{-1} per cell division^{-1}, mutation gene^{-1} generation^{-1}, and mutaiongenome^{-1} generation^{-1}. Several natural units of time for each of these rates are considered in practice. It is important to remember that only spontaneous mutations are considered to calculate the mutation rate of an organism. The genetic makeup of each organism, as well as environmental circumstances, has a significant impact on its mutation rate. The upper and lower bounds of mutation rates are still up for debate. However, it has been observed that certain health risks including cancer and other hereditary diseases in humans increase with an increase of the mutation rate. According to the recent estimate, the human mutation rate is approximately 0.5×10^{-9} per base pair per year (Fig. 9.6). In practice, mutation frequency and population history are used to compute the frequency (f) of the mutant organism in the overall organism. The mutation rate is usually denoted by μ often with subscripts to denote the type of

Table 9.2 Different mutation rates in different organisms

Organism	Mutation target/effect	Mutation rate	Unit
Bacteriophage T2	Lysis inhibition	1×10^8	Per replication
	Host range	3×10^9	
E. coli	Lactose fermentation	2×10^7	Per cell division
	Histidine requirement	2×10^8	
Neurospora crassa	Inositol requirement	8×10^8	Per asexual spore
	Adenine requirement	4×10^8	
Corn	Kernel color	2.2×10^6	Per gamete
Drosophila	Eye color	4×10^5	Per gamete
	Allozymes	5.14×10^6	
Mouse	Albino coat color	4.5×10^5	Per gamete
	Dilution coat color	3×10^5	
Human	Huntington disease	1×10^6	Per gamete
	Hemophilia A	3.2×10^5	

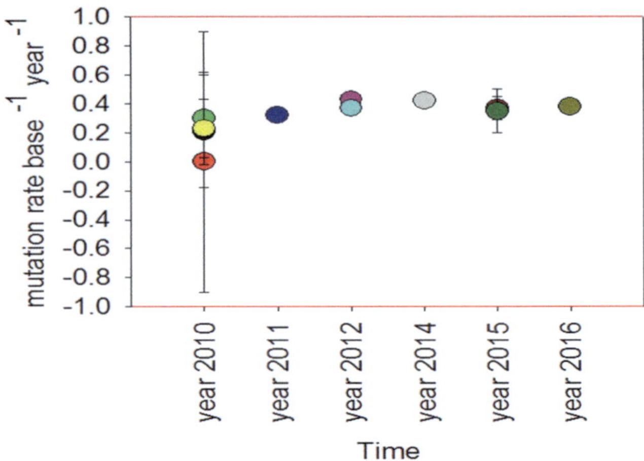

Fig. 9.6 Mutation rate. The mutation rate in humans is based on the recent data (Scally 2016). Mutation rate was calculated using family sequencing comparing genome samples across generations in one or more families. Within each, there are de novo mutations present in offspring and neither parent. The human mutation rate is estimated to be 0.5×10^{-9} per base pair per year, according to a recent report

mutation. However precise calculation of mutation rate is somehow challenging both theoretically and practically. There are certain mutations which have low and delayed phenotypic effect; hence, the frequency of mutant population remains intangible.

9.1.14 Mutation Hotspot

Based on the aspects discussed so far, we consider mutations in terms of the inactivation of the gene. For understanding mutation hotspots, some important feature of the mutation needs to be considered. (a) Most of the genes within a species show more or less similar rates of mutation relative to their size. As a result, vulnerability to mutation is basically proportional to the size of the genome. (b) Not necessarily all the base pairs in a single site are equally susceptible to mutation. (c) Majority of mutations reside at different sites, while few are found on the same site. (d) Due to (three) distinct point mutations, two independently isolated mutations at the same place may result in the same or different change in DNA. (e) While some sites may get multiple point mutations, others do not. This characteristic is demonstrated in the *lad* gene of *E. coli*.

Random hit kinetics can estimate the statistical probability of more than one mutation occurring at a given site, which is expressed as a Poisson distribution. From a random distribution expressed as 10×–100×, some sites are anticipated to gain more than the expected number of mutations. Hotspots are the term for these sites. Hotspots are known to cause spontaneous mutations; however, various mutagens may have variable hotspot selectivity. A modified base has evolved into mutational hotspots. Sites containing 5-methylcytosine, for example, are hotspots for spontaneous point mutations in *E. coli*, where the mutation occurs as a GC to AT transition. One of the most prominent reasons for the existence of such hotspots is the high frequency of spontaneous deamination of cytosine bases. While discussing the mechanism of spontaneous mutation, the process of deamination is covered in its own section of the chapter.

9.1.15 Somatic and Germinal Mutation

It is already discussed that mutations can occur in both somatic and germ cells. In this section, both somatic and the germinal mutation will be discussed with the typical examples.

Somatic mutation: As the name implies, a mutation is a somatic mutation when it takes place in any in the somatic cells. Somatic mutations cannot flow through the progeny and therefore do not impair the infant's genotype. The impact of somatic mutation fairly depends on the developmental stage when the mutation occurred. For instance, if the somatic mutation takes place in a single diploid cell at a very early stage of development, the mutant progenitor cell clonally multiplies producing mutant body parts. If mutation takes place in the later phase of life in the non-pluripotent cells, the impact mutations are closely restricted to a small portion (Fig. 9.7) known as the "mutation sector." Thus the formation of the "mutation sector" depends on the stage where the first mutation occurred. The phenotypic effect is often not observable if the mutation takes place in a single or a few cells. However, the mutation causing cancer is an exception where a mutation in few cells may lead to a significant phenotypic endpoint, because in these cases the mutation itself promotes cell division by impairing cell cycle regulation and reduced apoptosis.

One of the best examples of somatic mutation in the plant system is apple skin color. A rare form of somatic mutation is found in red delicious apple; half of the apple appears red, and half is green or golden. In these cases, a mutation occurred in the early developmental phase of the ovary wall cells which eventually differentiated into the fruit skin. The mutation changes the skin color from red to green or yellow. The exact mechanism and key mutations producing green or golden color are not completely characterized. However, according to recent reports, anthocyanin is found to be responsible for producing red color in the apple skin. Differential accumulation in different zone produces different colors, and MdMYB1-mediated MdGSTF6 expression acts as one of the important regulators in the anthocyanin production network.

Germinal mutation: Like somatic cell mutation, when the mutation occurs in germ cells, it is known as germinal mutation. In the germ line, if a specific tissue is programmed to generate sex cells, a germinal mutation occurs (Fig. 9.8). If mutant sex cells (either oocyte, spermatozoa, or both) fertilize, then the mutation will be passed down to the following generation. If the fertilization takes place between both

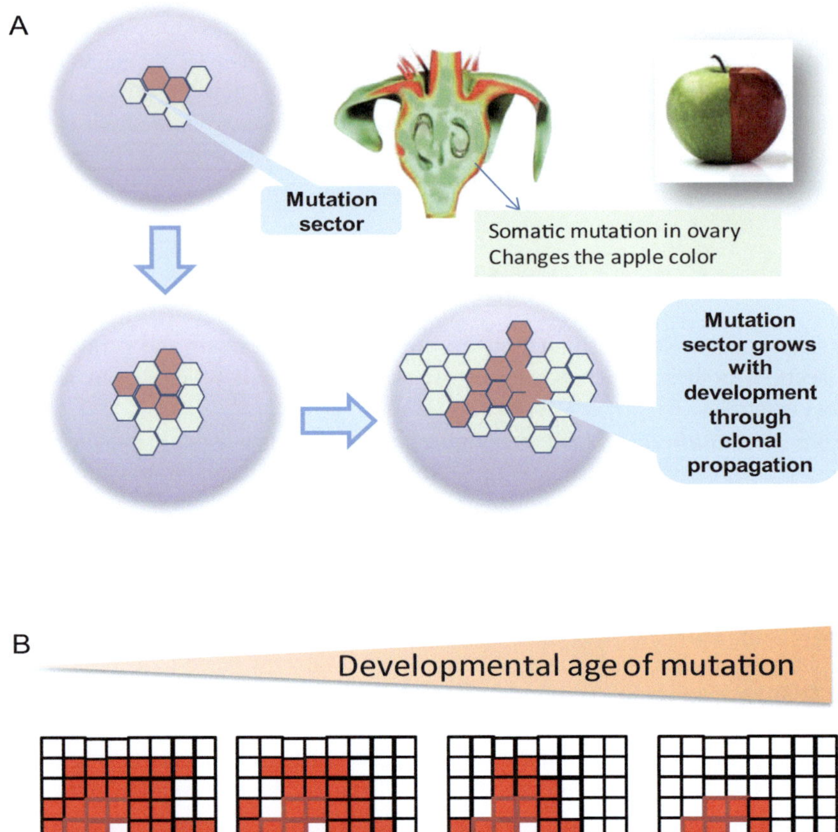

Fig. 9.7 Somatic mutation—a simplified representation: A specific example of somatic mutation in red delicious apple. Mutation in ovary cells produces a partially green or golden color. (**b**) Conceptual representation of the growth of mutation. The size of the mutation sector depends on the stage of development when the mutation occurred

the mutant counterpart and the progeny becomes homozygous mutant else it produces heterozygous mutant. An individual acquired mutation in the sex cells may otherwise remain perfectly normal until detected in the germ cell. For instance, X-linked mutation expressing hemophilia in European royal families is believed to

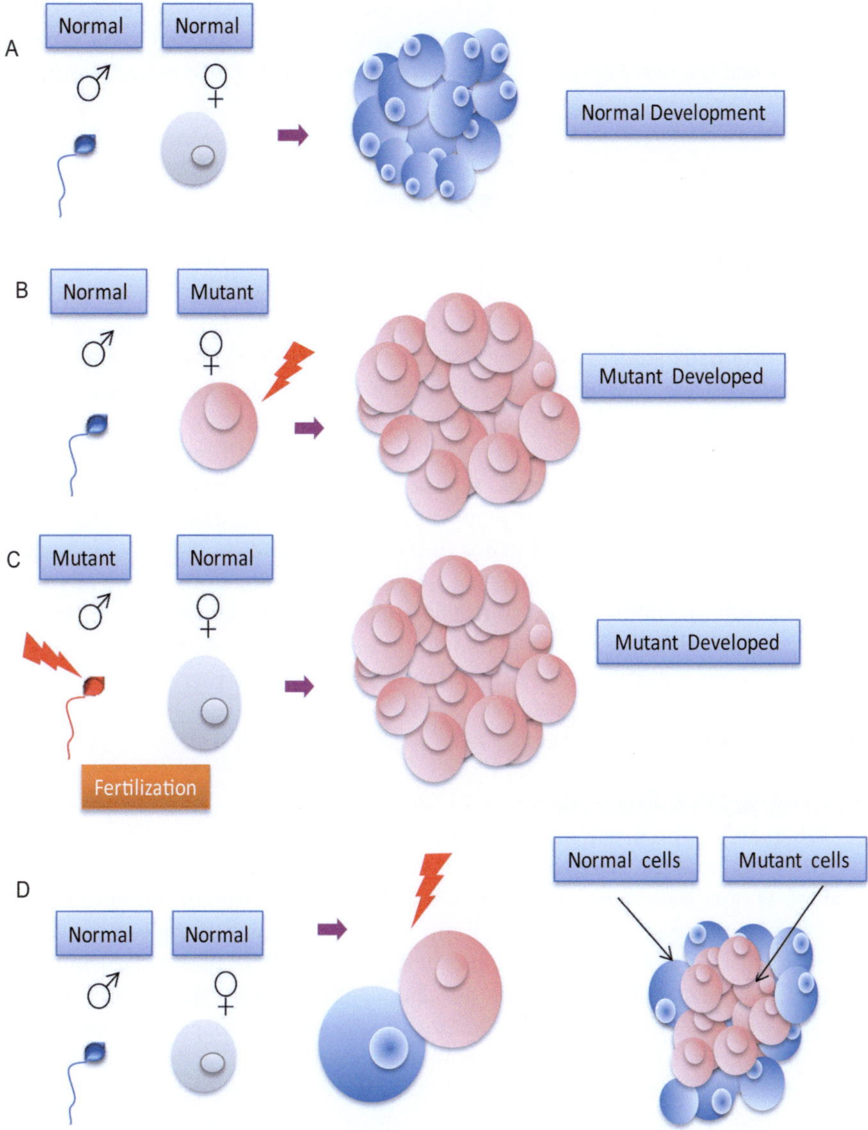

Fig. 9.8 Mutation in a different stage of development and their predicted result. Mutation in a different stage of life and different cells may result in different expected consequences. (**a**) No germinal or somatic mutation development of normal cells. (**b**) Mutation in oocytes may lead to the development of mutant fertilizing with the normal sperm or (**c**) vice versa. (**d**) Mutation in the very early stage somatic cells with high differentiation potency has a significant probability to affect the organism to a large extent

have aroused from the germinal mutation of Queen Victoria or one of her parents. However, as hemophilia (a bleeding disorder with impaired blood clotting capacity) is the X-linked recessive mutation, these are expressed only in the male descendants.

9.1.16 Spontaneous or Induced Mutation

9.1.16.1 Spontaneous Mutation

In general perception, the appearance of a new mutation is a rare event. In the majority of cases, geneticists were able to spot these mutations in nature. Mutations that were originally studied occurred spontaneously and are known as **spontaneous mutations**. The rate of spontaneous mutation is variable. A larger genome provides a large target; thus the probability of mutation frequency is high. A study of the five coat color loci in mice exhibited the range of mutation rate from 2×10^{-6} to 40×10^{-6} mutations per gamete per gene. A eukaryotic organism's spontaneous mutation rate is 2–12 × 10–6 mutations per gamete per gene, according to data from multiple studies. Similarly, with 100,000 genes in the human genome, 1–5 human gametes are projected to be mutated in some genes. Spontaneous mutations can be any sort of point mutation, including frame shift, substitution, and mismatch. There are three major routes by which spontaneous mutation has been introduced in a DNA such as (a) error in DNA replication, (b) deletion or duplication, and (c) molecular decay or spontaneous lesion. In this section, we will be discussing important routes of introducing spontaneous mutation.

(a) **Error in DNA Replication**
When an inappropriate nucleotide pair (e.g., A–C instead of A–T) develops in DNA synthesis, an error might occur, resulting in a base substitution. Let us now look at how this occurs. Every base in DNA exists in one of multiple isomeric forms known as tautomers, which differ in the positions of their atoms and the bonds that link them but remain in equilibrium. These forms are keto, imino, and enol. Among which keto forms are abundant (Fig. 9.9). At a certain point of DNA replication, one tautomer often shifts to another (inappropriate) resulting in mispairing of bases. This phenomenon is known as "tautomeric shift" first identified by Watson and Crick while they were formulating the model of DNA structure. The ability of the wrong tautomer of a standard base to mispair eventually potentiates the probability of mutation.
(b) **Deletions and Duplications**
Large deletions (deletions of more than a few base pairs) account for a significant portion of spontaneous mutations. The deletions occur in the majority of cases at repeated sequences. Several studies have shown that the longest repeated sequences are hotspots for deletions. Duplications of DNA segments have also been observed in several organisms, and they frequently occur at sequence repeats.
(c) **Molecular Decay or Spontaneous Lesion**

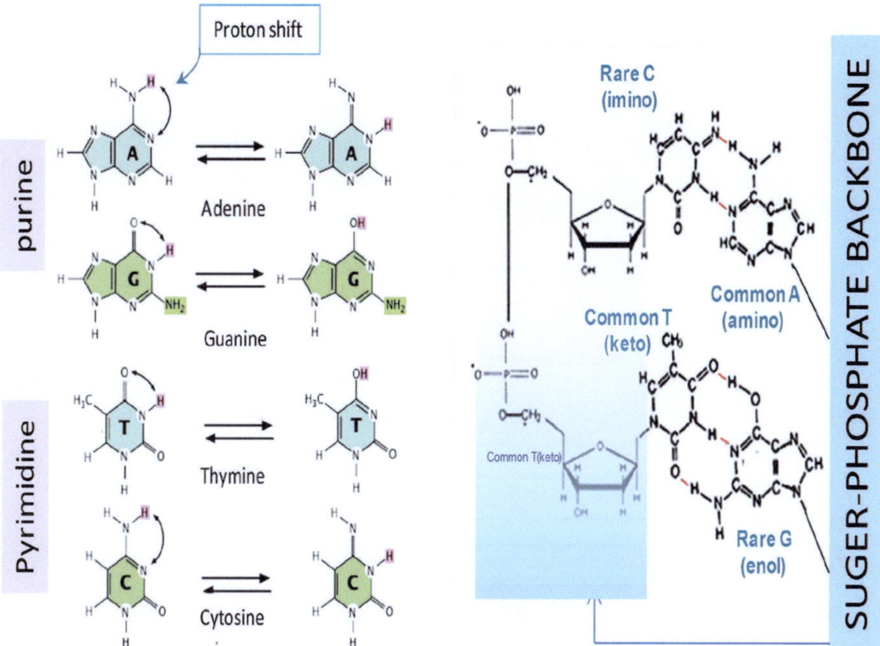

Fig. 9.9 Tautomerism in bases. Common and rare forms of bases participate in tautomeric shift leading to erroneous DNA replication and thereby resulting in point mutation in the nucleotide chain

Molecular decay or spontaneous lesions as the name implies constitute naturally occurring damage to the DNA. Major types of spontaneous lesions frequently result from depurination, deamination, and oxidative damage.

9.1.16.2 Depurination

Sometimes the glycosidic bond between the base and deoxyribose is interrupted, and subsequently, the purine residues (A and G) are lost from DNA. It has been shown and observed that a mammalian cell spontaneously loses ~10,000 purines from DNA in a single cell cycle period. Theoretically, this can lead to significant genetic damage. Base specificity is hindered at the apurinic sites. But that never happens because the efficient repair mechanism ((Fig. 9.10a) discussed later in this chapter) removes apurinic sites. However, under some circumstances, a base is frequently inserted across from an apurinic site, resulting in DNA mutation.

9.1.16.3 Deamination

Let us recall the chemical structure of cytosine and uracil. The sole structural difference is an amino group at C4. Uracil can be produced via the spontaneous deamination of cytosine. During replication, unrepaired uracil residues can pair with adenine, converting a G-C pair into an A-T pair (a GC AT) (Fig. 9.10b). This is one

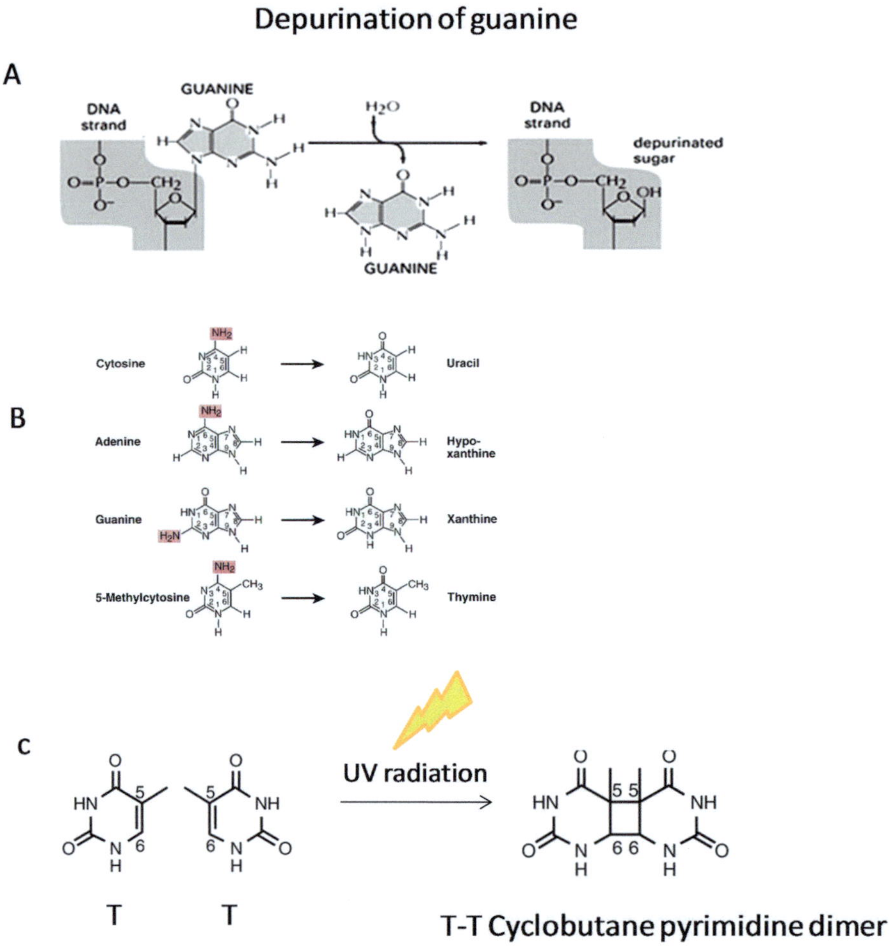

Fig. 9.10 The spontaneous lesion in bases. (**a**) Depurination in GTP produces depurination sugar leading to impaired specificity in base pairing. (**b**) Delaminated bases compromise the base-pairing specificity leading to mutation. (**c**) Formation of thymine dimer induced by ultraviolet radiation

of the classic examples of transition mutation. Among depurination and deamination, depurination is found more frequently.

(d) **Oxidatively Damaged Bases**

Several metabolic pathways produce reactive oxygen species (ROS), which include superoxide radicals (O2•), hydrogen peroxide (H2O2), and hydroxyl radicals (OH•). They have the potential to cause oxidative damage to DNA and DNA precursors (such as GTP). These can result in mutations, which have been linked to many human disorders. The 8-oxo-7-hydrodeoxyguanosine (8-oxodG, or GO) product, for example, frequently mispairs with A, leading to a high rate of G-T transversions. If left unrepaired, thymidine glycol prevents DNA replication, but it has not yet been linked to mutagenesis.

(e) **Pyrimidine Dimers**

Pyramiding dimers are a special case of spontaneous lesions formed from thymine, cytosine, or uracil bases in the nucleotide chain through a photochemical reaction. Ultraviolet (UV) light induces covalent linkages formation between two consecutive pyrimidine bases along a nucleotide chain (Fig. 9.10c). These pre-mutagenic lesions impair the structure and the base pairing. UV is the natural causative agent of these lesions; therefore, our skin cells exposed to sunlight can undergo up to 50–100 such reactions per second. However, these lesions are normally corrected immediately by a repair mechanism known as nucleotide excision repair (NER) (NER mechanism is later discussed in this chapter). In bacterial this correction takes place via a much simpler mechanism known as photolyase reactivation. Uncorrected lesions, beyond the threshold, can cause mutation due to misreading during transcription or replication. These lesions can also arrest the normal replication. This phenomenon is most commonly described in the bacterial system. However, pyrimidine dimers are thought to be the primary cause of melanomas in humans. The two most common UV products are CPDs (cyclobutane-pyrimidine dimers) and 6–4 photoproducts.

9.1.17 Mutation: A Reverse Phenomenon

9.1.17.1 Reverse Mutations and Suppressor Mutations

The impact of one mutation can be reversed by another mutation leading to restoration of the original phenotypic characteristic of an organism. These mutations are known as reverse mutations or back mutations, in contrast to the forward mutation where alteration of the original phenotypes takes place. For example, a nonsense mutation can be reversed by further changing the nonsense codon in the mRNA to a codon of amino acid. If these changes bring back the original amino acid sequence, then it is considered a true reversal. If the change of nonsense codon results in the addition of some amino acid other than the original, it is known as the partial reversion.

The phenotypic effect of an organism can also be restored by suppressor mutation. However, it is important to note that in this case restoration of the phenotypic effect is not true reversal; thus restoration is not robust. There must not be any confusion between the reverse mutations and the suppression mutation. A suppressor mutation balances out the effects of the first mutation, having left the organism with no evident phenotypic differences. Only the effects of the initial mutation are diminished by a suppressor mutation, but the first mutation is never reversed by the second (suppressor) mutation. Suppressor mutations may occur in the different sites of the same gene where the original mutations have taken place (intragenic suppressor), or they may occur in a different gene (intergenic suppressors). The mechanism of both types of the suppressor is different. Intragenic suppressors can change various nucleotides in the same codon as the original mutation, or they might change a nucleotide in a different codon entirely, whereas intergenic suppression is the consequence of the second mutation in another gene. For example, the effect of a nonsense mutation can be suppressed if the second mutation alters the anticodon in the respective tRNA which induced codon recognition capacity. For instance, if the anticodon 3′-AUG-5′ of a tyrosine tRNA gene is altered to 3′-AUC-5′, the suppressor tRNA still carries tyrosine but reads the nonsense codon 5′-UAG-3′. As a result, rather than terminating the chain, tyrosine is introduced into the polypeptide. Although the original mutation is not reversed in this case, it fairly suppresses the effect of the nonsense mutation.

9.1.18 Mutagenesis in Bacteria

Genomes of bacteria exist in circular DNA where the genes are arranged in an operon. The average size of a bacterial genome is around 4000 kb. Other than the circular DNA molecule, other elements including plasmids, transposons, integrons, or gene cassettes are also evident.

Bacterial genes with similar functions frequently share a promoter and are transcribed at the same time. An operon is a name for this type of system. The binding of transcription factors to the operator sequence of the DNA allows these operons to be regulated. Mutation in bacterial genome is introduced and inherited quickly through error-prone DNA replication due to having short generation turnover of the majority of the genes. Mutations in bacteria can produce alteration in structural or colony characteristics or loss in sensitivity to antibiotics. Bacterial spontaneous mutations occur at a rate of 1 in 105 to 108, resulting in random population variation. Few potential contributions in bacterial mutation are as follows:

(a) Auxotrophs: Auxotrophs have essential nutrient process dysfunctional.
(b) Resistant mutants: The population has developed a high level of tolerance to the stress of inhibitory chemicals or antibiotics.
(c) Regulatory mutants: Population having defective regulatory sequences such as promoter regions.
(d) Constitutive mutants: In constitutive mutants, the genes that normally turn on and off as in operons are expressed continuously.

Spontaneous mutations are most common in bacteria when DNA pol III synthesizes a new strand of DNA. In some cases, incorrect nucleotides are added or omitted during the procedure. In studies with *E. coli*, the lagging strand was found to be mutated 20 times more likely than the leading strand.

9.2 Mutation: Phenotypic Effects

9.2.1 Phenotypic Effects

Mutations have diverse phenotypic effects. These effects are inheritable when in the case of the germinal mutation. Only by comparing the mutant to the most prevalent phenotype in a natural population can the phenotypic effect of a mutation be realized. In contrast to the mutant, the original phenotype is known as the wild-type phenotype. As discussed above, a broader consequence phenotypic change in an organism may eventually lead to evolutionary change although in several cases of a slow propagating spontaneous point mutation in a population, it is difficult to identify the original phenotype. *Drosophila melanogaster*, for example, has red eyes by nature; hence flies with red eyes are considered wild type. Any other genetically determined eye color in fruit flies is regarded as a mutant phenotype, in contrast to the red eye color. Forwarding mutation refers to a mutation that affects the wild-type phenotype, whereas reverse mutation refers to a mutation that returns a mutant to its wild-type phenotype.

In nature, most mutations do not exhibit pronounced phenotypic effects and are believed to be silent mutations. There are two important theories behind these. (a) Only a small portion of the genome of a higher organism possesses coding functionality. For example, in humans with 3234.83 Mb of total haploid genome size, only ~2% possesses coding functionality. Therefore, a spontaneously occurring mutation has a higher probability to affect the non-coding region resulting now phenotypic effects. (b) All higher organisms have double sets of chromosomes that constitute the diploid genome. In case of several mutations, the phenotypic change is only expressed where genes in both the homologous chromosome are mutated; otherwise the mutations remain recessive. In this way, several diploid species increase the burden of genetic disease accumulating the large pool of mutation without expressing the same. Therefore, on the one hand, large portion of non-coding DNA may be advantageous in respect to the phenotypic effect of the mutation. However, the recessive character of a mutation is significantly disadvantageous increasing the mutation burden in the population without exhibiting phenotypic change.

9.2.2 Inheritance of Mutation

Most animals and several plants exhibit sexual dimorphism. There are two categories of chromosomes in this organism such as autosomes and sex chromosomes. Sex is determined by special sex chromosomes at the specific stage of development. The rules of inheritance which were primarily demonstrated with Mendel's example are the rules of autosomes. The sex chromosomes in diploid organisms are generally one pair only. We know that human has 46 chromosomes among which 44 autosomes (A) and 2 sex chromosomes (X and Y) in a diploid cell. XX combination determines the female, and XY combination determines the male progeny. A mutant trait can be

Table 9.3 Type of inheritance of mutational disorders

Types of mutant inheritance	Pedigree characteristics	Example
Autosomal dominant	1. Usually children of one affected parents are affected 2. Disease appears in every generation 3. Caused by one mutant allele 4. Vertical flow	Huntington disease, achondroplasia, myotonic dystrophy, Marfan syndrome
Autosomal recessive	1. Generally children are unaffected (heterozygotes); parents are affected 2. This is not typically found in every generation 3. Caused by two mutated alleles 4. Horizontal flow	Cystic fibrosis, Beta thalassemia, homocystinuria
X-linked dominant	1. Females are more frequently affected than the males 2. No male-to-male transmission	Hypophosphatemia, Rett syndrome
X-linked recessive	1. Males are more frequently affected than females 2. For a female to be affected, both her parents must be a carrier 3. Fathers cannot pass X-linked traits to their sons	Duchenne muscular dystrophy, hemophilia
Mitochondrial	1. Both males and female children can be affected 2. Can appear in every generation of a family	LHON: Leber hereditary optic neuropathy

inherited and either linked to autosomes or linked to a sex chromosome. Both types of inheritance can be either recessive or dominant. Different types of inheritance are summarized in Table 9.3. Recessive mutation produced haploinsufficiency that means the individual gets only 50% of the mutant product from one counterpart of the homologous chromosome. As a result of these, only homozygous genotype (having both the mutant counterpart) is only expressive of the trait. On the other hand, dominant mutation can be expressed where at least one counterpart of the homologous chromosome is mutated. Several diseases are inherited and either linked to autosomes or sex chromosomes, most X chromosomes. Some of the examples of different types of inheritance are as follows: autosomal dominant (Huntington disease, neurofibromatosis, and polycystic kidney disease); autosomal recessive (autosomal recessive cystic fibrosis, Tay-Sachs disease); X-linked dominant (X-linked hypophosphatemia, Rett syndrome); and X-linked recessive (color blindness, hemophilia). Like X-linked inheritance, there are certain characteristics not

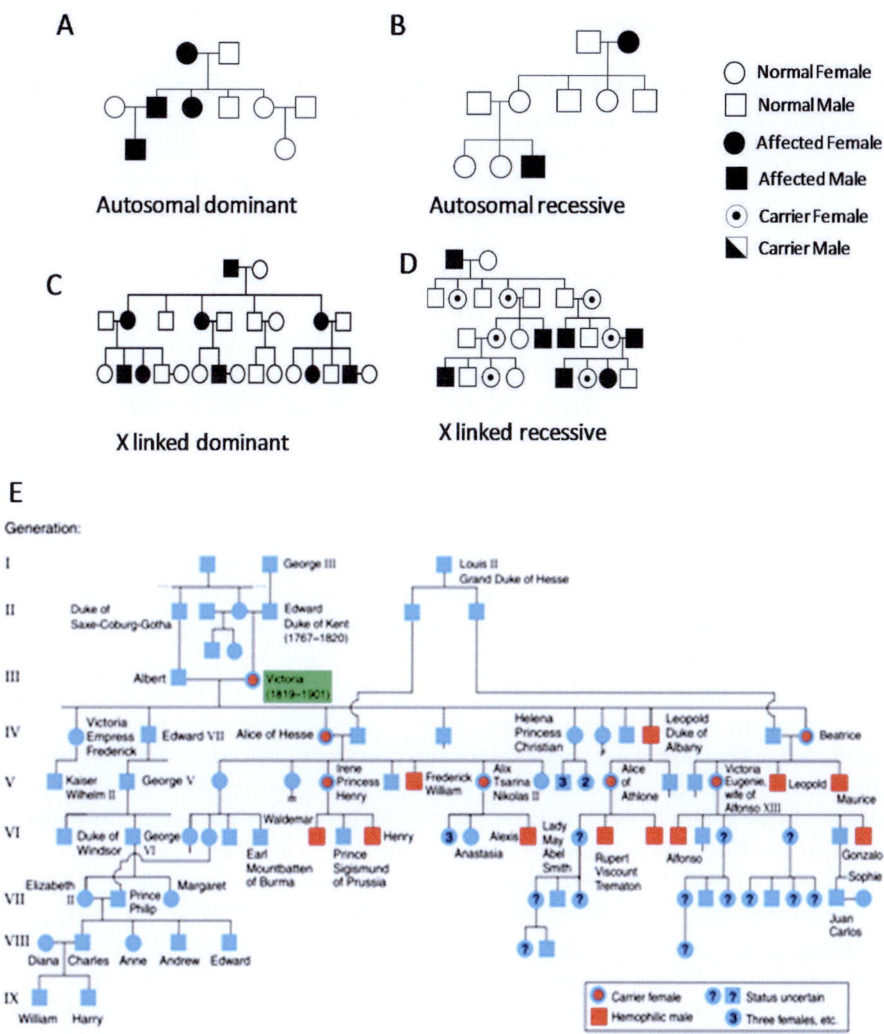

Fig. 9.11 Pedigree demonstrating a different type of inheritance. Basic pedigree pattern demonstrating inheritance of (**a**) autosomal dominant, (**b**) autosomal recessive, (**c**) X-linked dominant, and (**d**) X-linked recessive trait. (**e**) Pedigree of Queen Victoria and her descendants, showing inheritance of hemophilia trait

always linked to the Y chromosome. This is known as holandric inheritance. One of the classical examples of holandric inheritance in humans is a hairy ear. Figure 9.11 represents the typical pedigree demonstrating the flow of mutation through generations.

9.2.3 Morgan's Experiment

Thomas Hunt Morgan (1909) started working on the inheritance pattern of certain characteristics in fruit fly (*Drosophila melanogaste*r). During his experiment, Morgan found one fly in his laboratory colony with a white eye as compared to the red eyes of the normal population. This observation played a big role in the discovery and demonstration of inheritance mutation. Through a series of systematic genetic crosses, Morgan demonstrated the pattern of inheritance. As a result of the first cross between purebred red-eyed fly and a white-eyed male, he found almost all the individual in F1 had red eyes (only 4 was white among 1237 flies). From these results, he predicted that white eyes are a simple recessive trait. When F1 progenies were crossed with another, all of the females had red eyes, while half of the males had white eyes (Fig. 9.12). The result was a clear deviation of the simple recessive trait (expected value 25% of the total progeny). Thus it was hypothesized the locus of the mutation producing white eye in *Drosophila melanogaster* was X-linked recessive. Therefore males with the mutant X chromosome are haplosufficient expressing the trait. On the other hand, 50% of the females having only one mutant chromosome are compensated by the normal X counterpart, whereas the female having both the mutant X expresses the trait. Morgan's hypothesis was confirmed by a subsequent cross between a white-eyed female and a red-eyed male, which produced all red-eyed females and all white-eyed males, just as Morgan predicted. Morgan's hypothesis was statistically validated. Not all the individuals exhibited the expected result. A very small fraction (about 2–2.5%) deviated from the anticipated result. Later this deviation was explained by Bridges. In certain females, two X chromosomes fail to split during anaphase I of meiosis, resulting in some eggs receiving two copies of the X chromosome and others receiving none. This disorder is known as chromosomal nondisjunction. If these eggs are fertilized by sperm from a red-eyed male, three extra genotypes with different combinations of sex chromosomes are produced including X^wX^wY, X^+O, and $X^wX^wX^+$. Among these genotypes, $XwXwX^+$ is lethal. However, X^+O produce red-eyed male (because sex determination in *Drosophila* depends on X: A ratio and X^+O genotype only receive the male X chromosome counterpart), and X^wX^wY genotype produces a white-eyed female.

9.3 Mutation: Mechanism

The mechanism of mutation largely depends on the causative agents. The specific mechanism is studied in induced mutation, not in spontaneous mutation. In this part, different types of induced mutation and the associated mechanism are discussed.

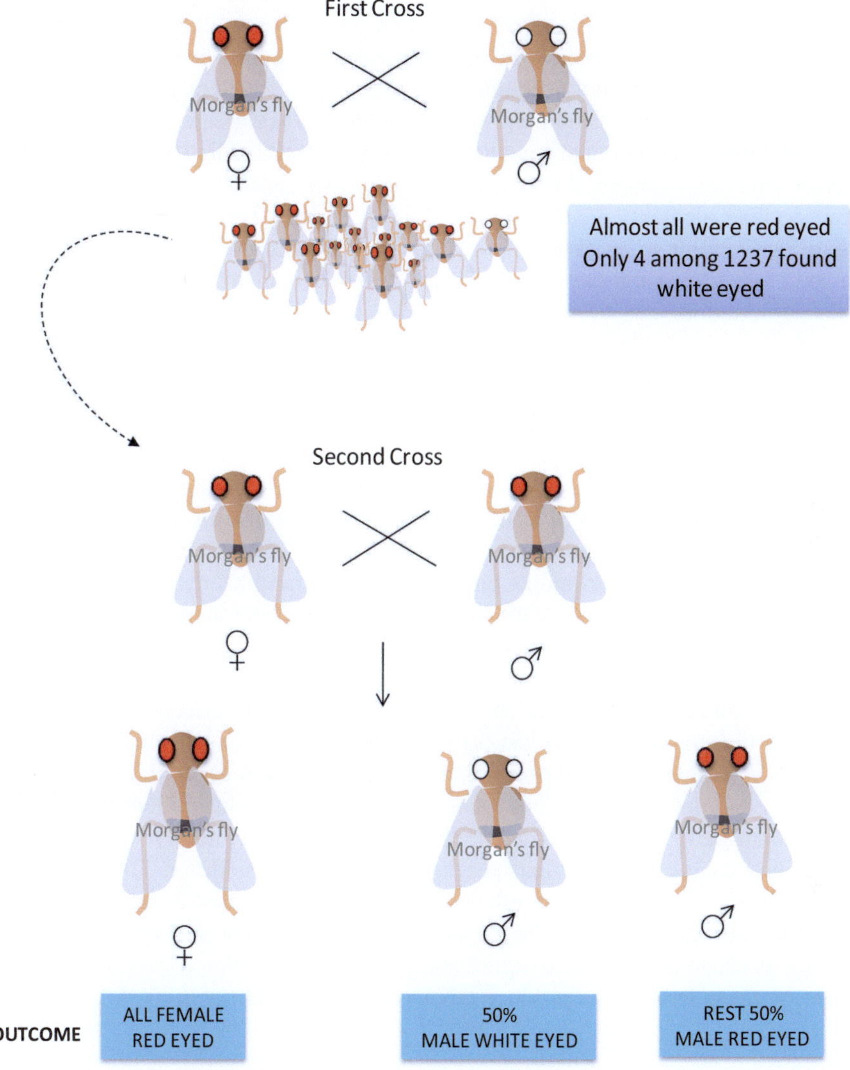

Fig. 9.12 Overview of Morgan's experiment on X-linked inheritance with *Drosophila melanogaster* of red and white eye color where red-eyed flies are wild type (purebred). In the first cross, almost all flies were red-eyed. In the second cross between red-eyed males and red-eyed females, 50% of males were red-eyed while the rest were white-eyed. On the other hand, 100% of females were found red-eyed

9.3.1 Induced Mutation

Mutations can be induced using different methods. The three most common approaches used to induce mutations are radiation, chemical, and transposon insertion. The early mutations in *Drosophila* were caused by radiation (X-rays), which Mueller used to cause fatal mutations. Other forms of radiation, including gamma radiation and fast neutron bombardment, have also been discovered to be proven mutagens. Similarly, chemical mutagens are frequently used to generate work point mutations. Alkylating agents and base analogs are the most routinely used mutagen in chemical mutagenesis. Other than radiation and chemicals, transposons or transposable elements are also used to generate mutants. Transposable elements are the mobile DNA fragments that can migrate from one site of the genome to another. As a result, new mutants are formed. In this section, we will be discussing the characteristics and mechanisms of different types of induced mutation.

9.3.2 Induced by Radiation

Ultraviolet light (UV) is the most common form of non-ionizing radiation. Cytosine and thymine, the two bases of DNA, are most susceptible to this radiation. Exposure to UV especially the longer wave UV light (UVA) can induce pyrimidine dimer between two adjacent pyrimidine bases in a DNA strand (as discussed earlier). UV exposure can also result in oxidative damage in DNA. Gamma radiation, on the other hand, is an ionizing radiation that can cause cancer.

As mentioned above, X-rays have enough potential to induce mutation depending on the magnitude of exposure. The differential mutagenic effects of X-ray are the matter of investigation in the last 50 years. In a recent genome-wide study, up to 6000 SNV (single nucleotide variation), CNV (copy number variation), and indel (insertion and deletion combined) mutation was found to be evident in the diploid mammalian germ line cells. These cells however have the ability to repair 30,000 similar endogenous DNA defects per day. As a result, an additional 6000 damage is less likely to have a long-term effect. Radiation-induced double-strand breaks and clustered damaged sites, on the other hand, are the most severe types of DNA damage because their repair is significantly delayed or hindered, and if not repaired properly, these lesions can produce DNA rearrangements.

9.3.3 Induced by Chemical

9.3.3.1 Incorporation of Base Analogs

Some chemical compounds are suitably similar to regular nitrogen bases, and such compounds are the base analogs. They are inserted occasionally into DNA to replace regular nucleotides. These bases lack the same stringent paring qualities as conventional bases, thus causing mutations during DNA replication. 5-bromouracil (5-BU), for example, is a thymine analog with bromine at the C5 position instead of the CH3 group. During transition mutation, 5-BU can frequently undergo a keto-enol tautomeric shift.

9.3.3.2 Specific Mispairing by Alkylating Agents

Certain alkylating agents, such as ethyl methanesulfonate (EMS) which used nitrosoguanidine (NG), can cause specific mispairing instead of getting incorporated into the DNA. These agents add alkyl groups to several positions on bases. The mutagenic effect is most pronounced when they add the oxygen at the six positions of guanine to produce an O6-alkylguanine. This addition leads to direct mispairing with thymine, resulting in GC $\rightarrow$ AT transitions during replication. Mutagenic specificity for EMS and NG exhibits a strong preference for GC $\rightarrow$ AT transitions. Alkylating agents can modify the bases in any dNTPs.

9.3.3.3 DNA Intercalating Agents

DNA intercalating agents interact non-covalently with the DNA molecule either through threading intercalation or through classical intercalation. Threading interaction is a groove-bound connection that is stabilized by hydrophobic, electrostatic, and hydrogen-bonding interactions. Classic interaction, on the other hand, is an intercalative association in which a planar, heteroaromatic moiety slides between the DNA base pairs.

Intercalation causes the DNA double helix to unwind. Depending on the intercalating agents, the degree of unwinding varies. For example, ethidium unwinds DNA by around 26° and proflavine by about 17°. These structural abnormalities eventually result in functional changes, such as suppression of transcription and replication, as well as mutations in DNA repair pathways (Fig. 9.13a).

9.3.3.4 DNA Cross-Linkers

DNA cross-linking agents have two independent reactive groups within the same molecule which can bind with a nucleotide residue of DNA. Both natural and synthetic chemicals can act as exogenous cross-linking agents, whereas endogenous cross-linking agents are the compounds that are the intermediary metabolites of biochemical pathways within a cell or organism. For example, cisplatin (cis-diaminedichloroplatinum (II)) and its derivatives act as exogenous cross-linking on the adjacent guanines at their N^7 positions. On the other hand, nitrous acid acts as the endogenous cross-linking agent via conversion of the amino group to the carbonyl group in the DNA (Fig. 9.13b).

9.3.3.5 Aflatoxin B_1

Aflatoxin B1 is a mycotoxin produced by *Aspergillus flavus* and *Aspergillus parasiticus*. Aflatoxin B1 is highly hepatotoxic and hepatocarcinogenic. It can generate apurinic sites resulting in the formation of an adduct at the N7 position of guanine (Fig. 9.13c).

9.3.4 Induced by Transposable Elements

Transposons are DNA elements that can move from one region to another region in the genome. When transposon shifts from a heterochromatin region to a

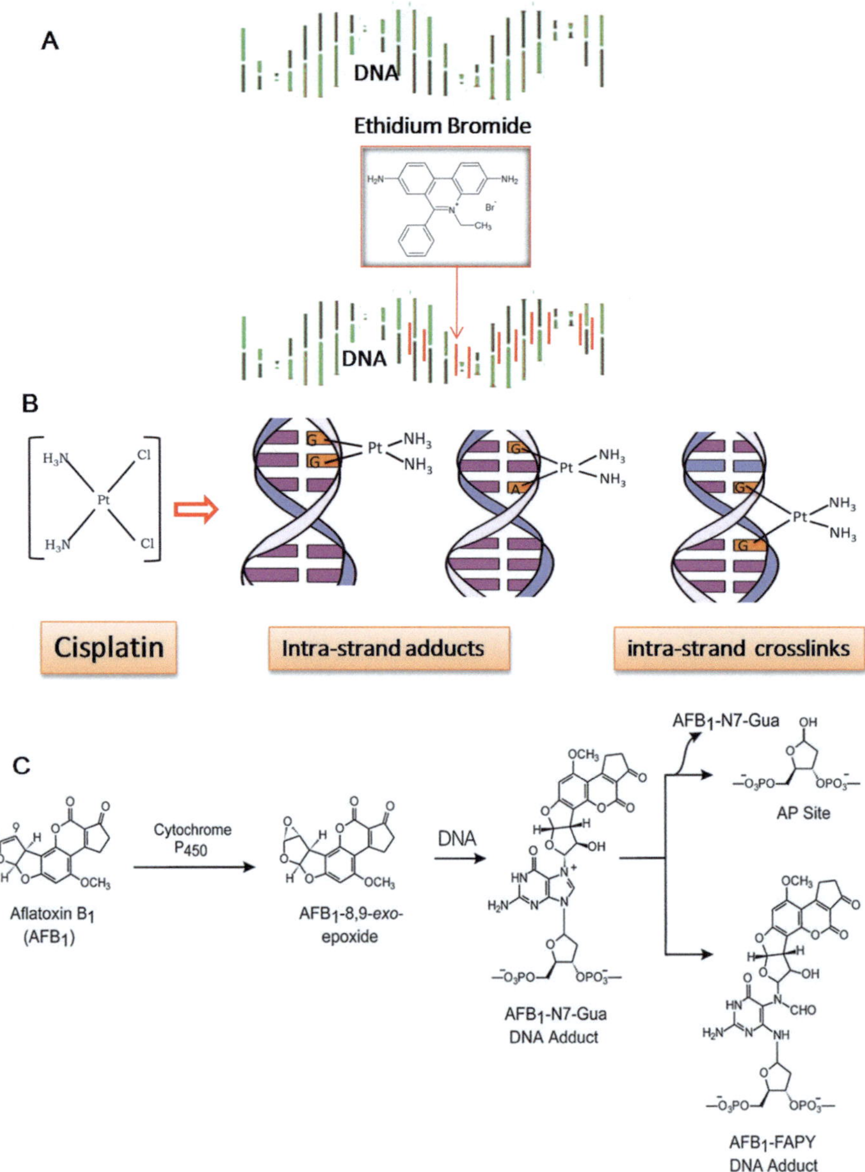

Fig. 9.13 Chemical-induced changes in DNA leading to mutation. (**a**) Ethidium bromide intercalation in DNA. (**b**) Cisplatin-mediated intra-strand adduct and intra-strand cross-links. (**c**) Aflatoxin-DNA interaction producing apurinic sites via formation of base aflatoxin adduct

transcriptionally active part of DNA, mutagenesis takes place. It is successfully used as a tool for mutagenesis and gene function. Transposons are most well demonstrated in *Drosophila melanogaster* (P elements), *Arabidopsis thaliana,* and

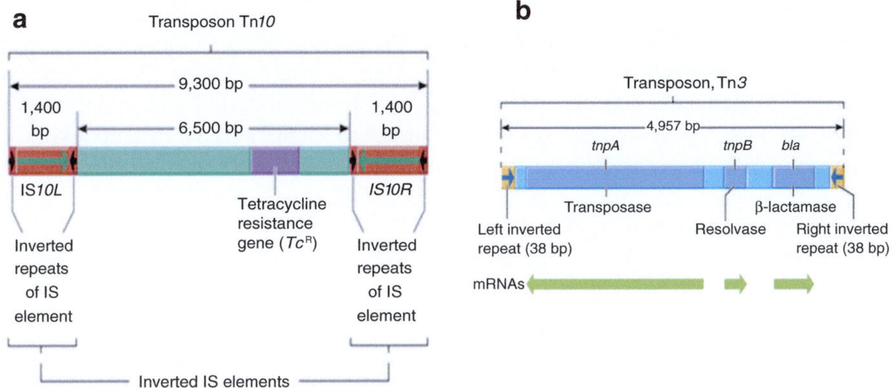

Fig. 9.14 The general features (structural) of transposons. (**a**) Composite and (**b**) non-composite transposons flanked by inverted IS sequence at both ends

Escherichia coli. Transposon mutagenesis was first discovered as "jumping genes" by *Barbara McClintock* in corn for which she was awarded Nobel Prize in 1983.

In bacteria, transposition mutagenesis is generally achieved by the plasmid from which transposons are extracted and placed into the host chromosome. Chemical mutagenesis is less efficient than transposon-mediated mutagenesis. It has a higher frequency of mutations. Furthermore, the organism's lethality is reduced. It can cause single hit mutations while also incorporating selectable markers in the construction. However, it is having lower transposition frequency inaccuracy in transposition. Initially, the transposon mutagenesis experiments are used to be achieved by bacteriophages and conjugative bacterial plasmids. Non-specificity was the key issue in this method. Shuttle mutagenesis is a relatively new approach that uses cloned genes from the host to introduce genetic elements. The transposon and transposition system is wide and complicated. Tn3, Tn5, Tn10, and many transposon systems have been developed. There are two types of bacterial transposons: composite transposons and non-composite transposons. Tn10 is an example of composite transposons (Fig. 9.14a), which have a core region with genes (e.g., genes that confer antibiotic resistance) flanked on both sides by IS elements known as IS modules. Tn3 (Fig. 9.14b) is an example of a non-composite transposon that contains antibiotic resistance genes but does not end with IS elements. In this context, the overview of the two most important transposition systems, the Tn5 transposon system, and the Sleeping Beauty transposon system are provided as follows.

9.3.4.1 Tn5 Transposon

Transposon mutagenesis is studied using the Tn5 transposon as a model system. Tn5 is a bacterial composite transposon with IS50R and IS50L insertion sequences flanking genes. IS50R codes for two proteins, Tnp and Inh. A transposase is represented by the letter Tnp. Transposase inhibitor (Inh) is a gene that is encoded by the Inh gene. Nineteen base pair components separate the IS50R and IS50L sequences. Mutation in these regions leads to impair the ability of transposase genes to bind to the sequences. Genes of interest are introduced into the transposon between IS50 sequences controlled by the host promoter. Incorporated genes usually include the target gene and the selectable marker to identify transformants. Almost all transposon systems share the most likely pathway for Tn5 transposition.

9.3.4.2 "Sleeping Beauty" Transposon

The SBTS ("Sleeping Beauty transposon system") is the first nonviral vector to successfully introduce a gene cassette into a vertebrate genome. Transposing the cassettes directly into the genome of the organism from the plasmid allows for long-term expression of transgenes. This transposition system has evolved over time as the transposon sequences have been improved and the utilization of transposase enzymes has increased, resulting in a transposition frequency of around 50%. Transposase enzymes can be expressed in both a cis and a trans manner in relation to the gene cassette. SBTS transposition has a similar underlying process to Tn5 transposition. Inverted repeat repeats, known as IR/DR sequences, flank the transposon. The gene sequences are the only eukaryotic components. Transposition allows for long-term transgene expression. This can be enhanced further by using transposase enzymes to improve transposition capacity. SB100X is a hyperactive mammalian transposase with a hundred times the efficiency of the first-generation transposase. Incorporating these enzymes into the cassette can lead to longer-lasting transgene expression (over 1 year). When pronuclear injection into mouse zygotes is used, transgenesis frequencies can reach 45%.

9.3.5 Trinucleotide Repeat: Mutation

9.3.5.1 CAG Repeats

Trinucleotide repeat expansion, also known as triplet repeat expansion (Fig. 9.15), is a form of DNA mutation that can result in a variety of disorders. In this type of mutation, the number of the trinucleotide repeat in gene or intron exceeds the normal stable threshold leading to microsatellite repeats. These mutations are dynamic and gradually increase the number of repeats in the gene. By generating spherical clusters known as RNA foci, trinucleotide repeats in introns can induce toxic effects in nucleases.

Slippage during DNA replication causes triplet expansion. During replication or DNA repair synthesis, instable tandem repetitions in the DNA sequence form "loop-out" structures in DNA. Endonuclease generates a nick on one side of the DNA strand, and DNA polymerase extends and seals a repeating triplet throughout this process. The number of repeats increases when the loop-out structure is produced from the sequence on the daughter strand, whereas the number of repeats reduces when the loop-out structure is formed on the parent strand. A fragile X syndrome is a classic example of an expanded trinucleotide repeat. Characteristically in the patient with this disease, the tip of the X chromosome is found attached only by a lean

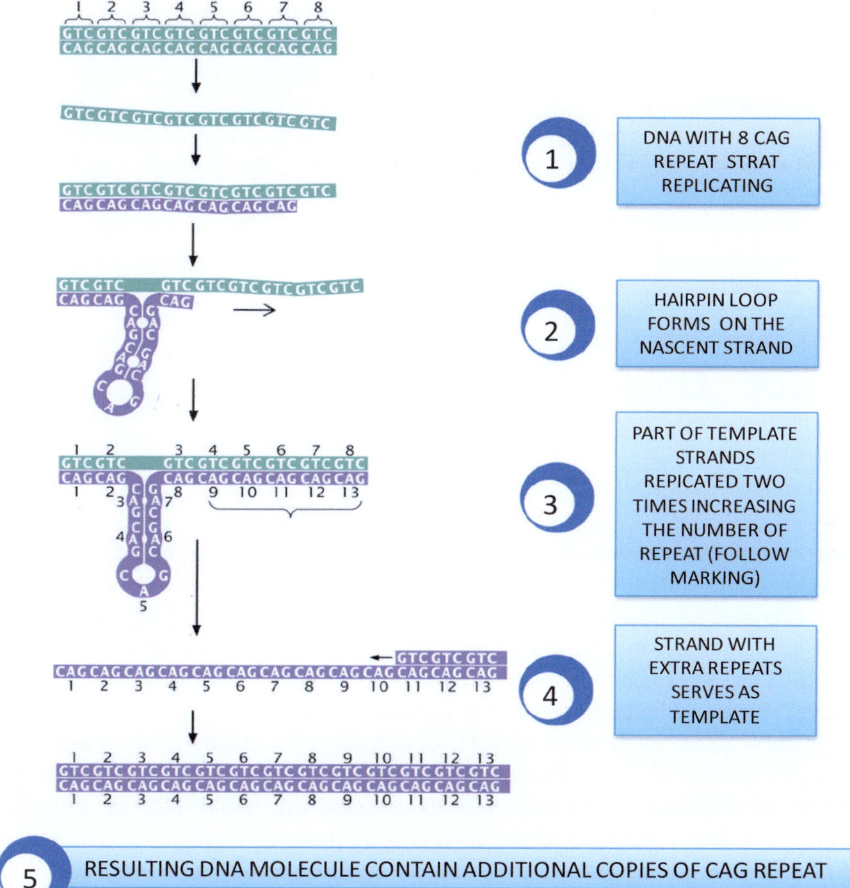

Fig. 9.15 Trinucleotide repeat expansion of CAG repeat via formation of hairpin loop. DNA with eight CAG repeat replicates and forms a hairpin loop in the nascent strand. As a result, the shortened complementary segment replicates again introducing the extra repeats. The strands with extra repeats serve as a template for further replication

thread. The disease is caused by several (around 60 or more as compared to the normal) CGG repeats in the fmr-*1* gene. Huntington's disease is another critical disease found to be caused by trinucleotide repeats.

9.3.5.2 Why 'Three' No Other Numbers in Nucleotide Repeats

An important question is why three, not the other numbers nucleotides are expanded? Dinucleotide repeats are a common feature of the genome as are larger repeats (VNTRs—variable number tandem repeats). One possible answer is any other numbers of nucleotide repeats may cause frame shift mutations causing obligation in developmental pathways and therefore which are not due to three

Table 9.4 Non-polyQ (non-polyglutamine) diseases

Type	Gene	Normal repeats	Pathogenic repeats	Codon
Fragile X syndrome	*fmr1*	6–53	230+	CGG (5′ UTR)
Fragile X-associated tremor/ataxia syndrome	*fmr1*	6–53	55–200	CGG (5′ UTR)
Fragile XE mental retardation	*aff2*	6–35	200+	CCG (5′ UTR)
Friedreich's ataxia	*fxn*	7–34	100+	GAA (Intron)
Myotonic dystrophy type 1	*dmpk*	5–34	50+	CTG (3′ UTR)
Spinocerebellar ataxia	*sca8*	16–37	110–250	CTG (RNA)
Spinocerebellar ataxia	*ppp2r2b*	7–28	66–78	nnn (5′ UTR)

nucleotide repeats may increase the probability of getting masked by early infantile or developmental lethality. On the other hand, mutation in three tandem nucleotides cannot produce catastrophic frameshift, and unless a stop codon is added, gene expression is not attenuated. Single nucleotide expansions to a coding region less likely to cause detrimental because a maximum of two amino acids are affected. However, the overwhelming number of repeats is deleterious.

Currently, an increased number of CAG repeats in the coding areas of unrelated proteins have been discovered to cause nine neurologic diseases. A polyglutamine tract is formed when enlarged CAG repeats are translated into a series of consistent glutamine residues ("polyQ"). This polyQ is more prone to aggregation. The activation of downstream regulatory processes can be disrupted by a prolong tract. The autosomal-dominant inheritance patterns define these disorders. Trinucleotide repeat diseases are characterized by a genetic anticipation phenomenon, in which the influence of a mutation grows with the individual's age but is not expressed at an earlier stage. The common symptoms of polyQ disease are progressive degeneration of nerve cells. Like polyQ repeats, there are certain non-polyQ diseases; however, they do not express specific symptoms. Some important polyQ and non-polyQ diseases are documented in Tables 9.4 and 9.5.

9.3.5.3 The Practical Implication of Mutagenesis: Site-Directed Mutagenesis

Mutagenesis is being rapidly used as an essential tool in several modern biological investigations. With the advancement of technology, mutagenesis can be achieved in specific sites via site-directed mutagenesis (SDM). To mutate plasmids, reverse PCR is employed. This approach amplifies the entire plasmid using two back-to-back primers, and the linear result is then ligated back to the circular form. Altering the primer sequences to include the desired mutation can affect the primer binding areas.

Table 9.5 PolyQ (polyglutamine) diseases

Type	Gene	Pathogenic polyQ repeats	Normal polyQ repeats
Dentatorubropallidoluysian atrophy	*atn1/ drpla*	49–88	6–35
Huntington's disease	*htt*	36–250	6–35
Spinal and bulbar muscular atrophy	*ar*	38–62	9–36
Spinocerebellar ataxia type 1	*atxn1*	49–88	6–35
Spinocerebellar ataxia type 2	*atxn2*	33–77	14–32
Spinocerebellar ataxia type 3 or Machado-Joseph disease	*atxn3*	55–86	12–40
Spinocerebellar ataxia type 6	*cacna1a*	21–30	4–18
Spinocerebellar ataxia type 7	*atxn7*	38–120	7–17
Spinocerebellar ataxia type 17	*tbp*	47–63	25–42

By adding flanking sequences to the primers, insertions can be made around the primer binding areas, and deletions can be made by simply leaving a space between the two primers. To change a target region, primer extension employs nested primers. Primers B and C contain the mismatched sequence to insert bases, as shown in the diagram. The mutant sequence is used in the first round of PCR, which uses primers A–B and C–D to create two products.

The smart thing happens in the second PCR round, and a new sequence is generated. Because primers B and C have complementary sequences, the first round's products will hybridize after being denatured after the first PCR cycle. The full-length product with the required mutation can then be amplified using primers A–D. Changes to this procedure could result in deletions or lengthy additions. The mechanism of SDM is depicted in Fig. 9.16a. Figure 9.17, on the other hand, depicts one of the potential applications of SDM.

9.4 Test for Mutagenicity: The Ames Test

- The Ames test is a widely employed method for testing mutagenic chemicals using bacteria (Fig. 9.18). A positive test predicts the carcinogenic potential of the chemical because cancer is one of the deadly outcomes of mutation. The test is a quick and easy way to screen a carcinogenic substance before performing animal tests.
- A histidine synthesis mutant (*his*$^-$) strain of *Salmonella typhimurium* is used in this test. These mutants are auxotrophic mutants requiring histidine for growth. The test asseses the capability of the bacteria to be converted in a "prototrophic" state which can grow in the histidine-free medium upon chemical exposure.
- The tester strains are developed to detect frameshift (e.g., strains TA-1537 and TA-1538) or point mutations in the genes that synthesize histidine (e.g., strain TA-1531). The tester strains have mutations in genes associated in lipopolysaccharide production, which makes the bacteria's cell wall more

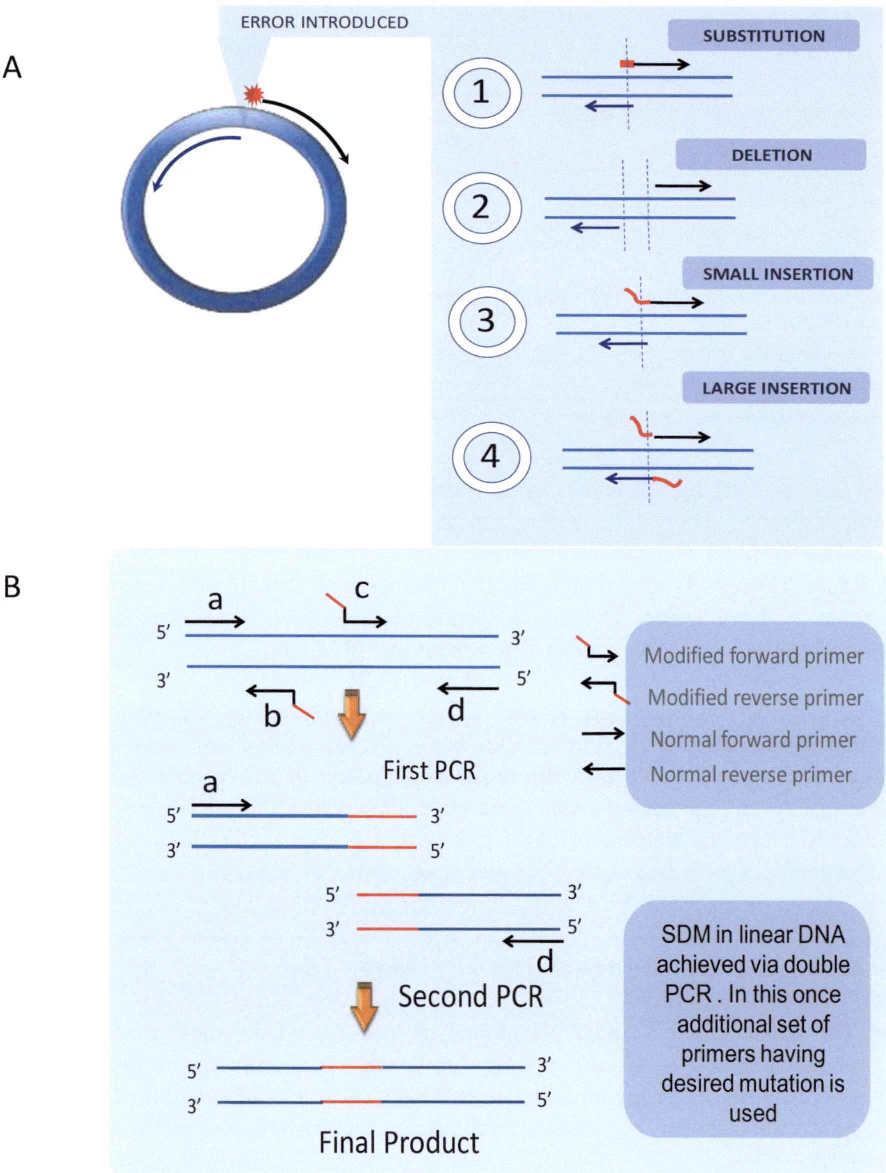

Fig. 9.16 Site-directed mutagenesis. Site-directed mutagenesis (SDM) can be achieved either in (**a**) plasmid and (**b**) liner DNA

permeable, in addition to his^{-} mutations. The tester strain also possesses a mutation that results in a flawed excision repair pathway, making the cells more sensitive.

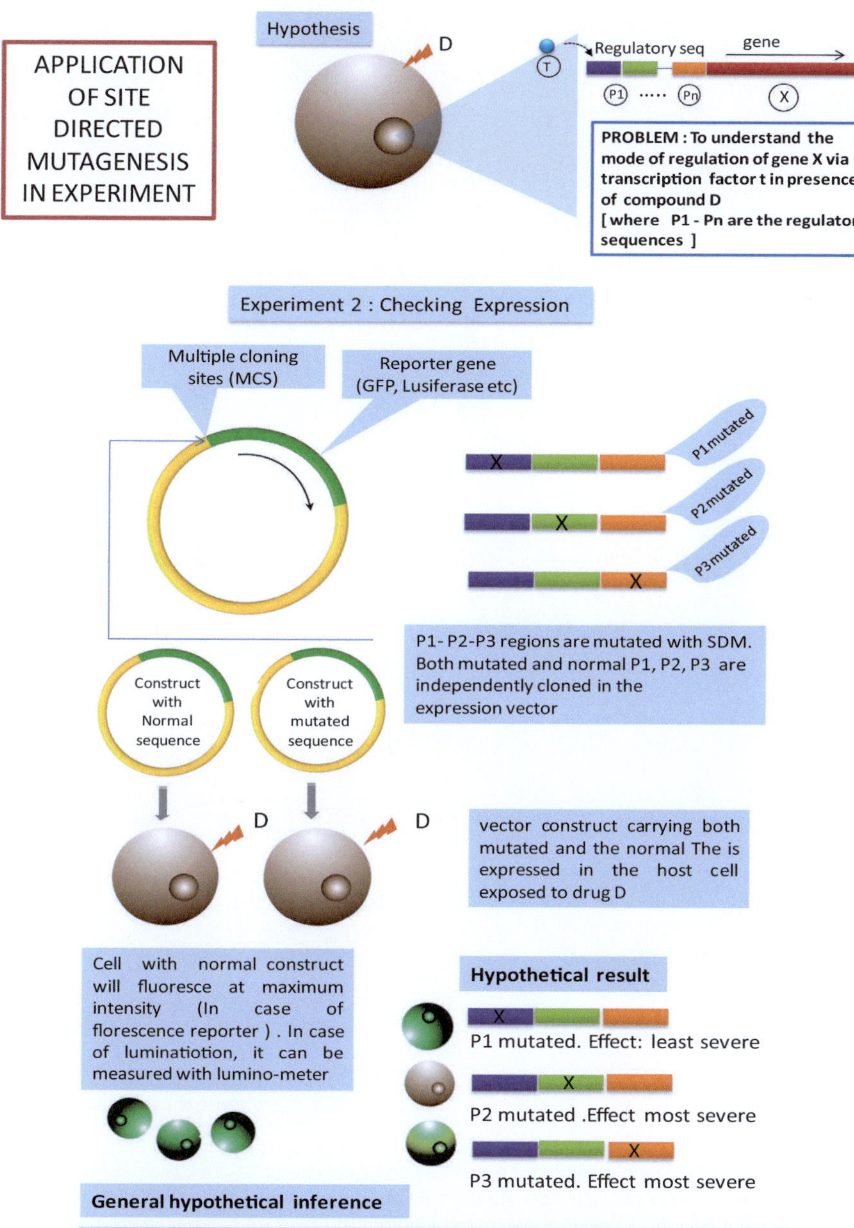

Fig. 9.17 A hypothetical experiment demonstrating the application of site-directed mutagenesis (SDM). Promoter activity can be assessed using SDM. The anticipated promoter sequence is mutated in a different position and cloned in an expression vector with a reporter gene. A positive correlation between SDM and affected expression of the reporter genes predict the involvement of the promoter a specific promoter sequence in driving its downstream genes

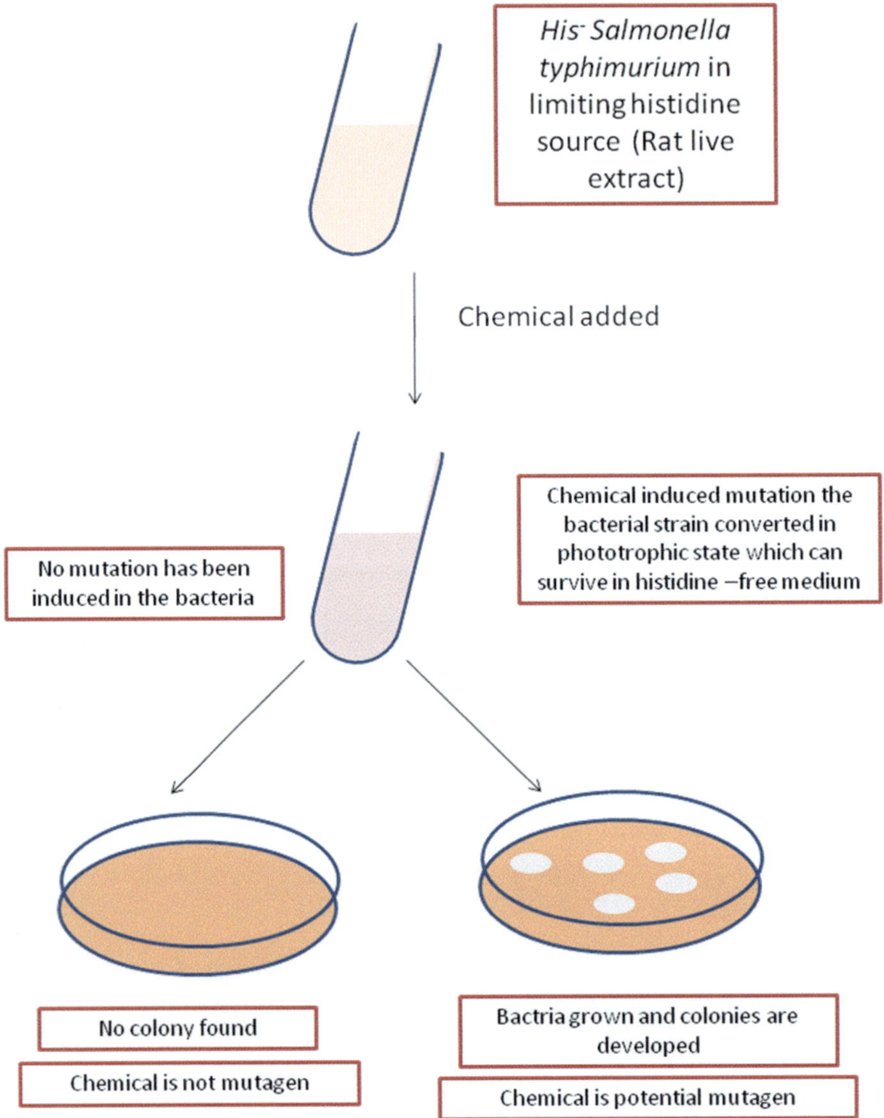

Fig. 9.18 Ames test. Chemical-induced mutation can be checked cost-effectively using the Ames test. Chemical-induced mutation in bacteria can convert it to a phototrophic state which can survive in histidine-free media. In colonies grown in the histidine-free media, it is assumed that the chemical added to the bacteria is a potential mutagen

- Initially, bacteria are grown on an agar plate with a limiting amount of histidine. On chemical exposure, when the medium histidine is depleted, bacteria will grow only if the exposed chemical has induced mutation and conferred survivability. Result is obtained by counting colonies after 48 h of incubation.

9.5 DNA Repair Mechanism

DNA repair is the process by which a cell detects the damage in its DNA and corrects it. The cell can tackle multiple assaults to DNA by a repair system. DNA damage in a human cell can occur both due to normal metabolic activities and in response to environmental stress. A probable estimate of about one million lesions day^{-1} cell^{-1} has can impair the cell physiology, affect cell growth and cell viability in humans. Among these, many of these lesions are mutation inducer. For example, malignancy can be induced by irreparable DNA damage (e.g., inter-strand cross-links or ICLs) due to failure of the DNA repair system. The rate of DNA repair is dependent majorly on three factors: (a) cell type, (b) environmental factor, and (c) age of the cell. A cell harboring a large amount of DNA damage is unable to repair completely resulting in either of the three major consequences: (1) senescence (an irreversible state of dormancy), (2) apoptosis (a type programmed cell death; a process through which cells commit suicide), and (3) tumorigenicity and/or malignancy (due to uncontrolled cell growth). The mechanism of DNA repair is a vital area of ongoing research in the present day. The 2015 Nobel Prize in Chemistry was awarded to Tomas Lindahl, Paul Modrich, and Aziz Sancar for their work on this area. In this section, we will be discussing the various mechanisms of DNA repair mechanism with diagrammatic representations.

There are mainly two types of DNA damage: (a) endogenous damage including replication errors and (b) exogenous damage. It is important to remember that the replication of damaged DNA before cell division can lead to the incorporation of wrong bases in the sister DNA. These alterations are irreversible by DNA repair mechanisms once they are passed down to daughter cells. Only back mutation can revert the changes. Many complex pathways are involved in repairing DNA. However, two general facts are important. (a) Most of the DNA repair mechanisms require both strands of DNA to replace whole nucleotides, and a template strand is essential to specify the base sequence in the DNA. (b) Redundancy of DNA repair events—several types of DNA damage can be corrected by several DNA repair pathways. This satisfies the significance of the DNA repair mechanism. DNA repair pathways are versatile and complex. Here, we will consider the general mechanisms of DNA repair. Commonly we can categorize DNA repair mechanisms into four types: (a) light-dependent repair, (b) base excision repair, (c) mismatch repair, and (d) nucleotide excision repair. The DNA repair mechanism is complex and depends on the cellular response due to the specific types of damage. These responses operate through complex molecular pathways. Let us first discuss the global response of DNA damage before discussing the specific types of DNA repair mechanisms.

9.5.1 An Overview of a Global Response to DNA Damage

Ionizing radiation, UV, or chemicals can cause multiple lesions in the cell at multiple sites. These include massive DNA lesions and double-strand breaks. Furthermore,

DNA-damaging agents can also damage other biomolecules including proteins, lipids, carbohydrates, and RNA. The common initiation signals for a "global response to DNA damage" are found among these double-strand breaks that halt replication forks. The cell's self-protective mechanism is an event-directed global response to any damage. This triggers multiple pathways determining the cell fate including lesion bypass, tolerance, or apoptosis cycle arrest and inhibition of cell division.

Chromatin relaxation and remodeling are the initial steps followed in response to DNA damage; these are achieved through several complex processes. The major three processes are mentioned briefly as follows:

1. Chromatin relaxation takes place rapidly at the damaged site of DNA. It can initially be accomplished by the JNK-SIRT6 axis where stress-activated protein kinase, c-Jun N-terminal kinase, phosphorylates SIRT6 on tenth serine residue in response to double-strand breaks. This facilitates recruitment of SIRT6 to damaged sites of DNA which in turn recruit PARP-1 (poly [ADP-ribose] polymerase 1) to DNA break sites immediately (time of half max accumulation at damage site is 1.6 s after the damage takes place). Eventually PARP-1 recruits DNA repair enzymes, MRE11, initiating DNA repair in around 13 s.
2. γH2AX (phosphorylated H2AX on serine 139 residue) is detected within 20 s of the formation of DNA double-strand break. (Half maximum accumulation of γH2AX occurs in 1 min.) Then RNF8 protein is associated with γH2AX mediating extensive chromatin decondensation.
3. 2 DDB1-DDB is the heterodimeric complex associated with CUL4A (the ubiquitin ligase protein) and PARP-1. This larger complex rapidly gets associated with UV-induced damage. The half maximum association time is 40 s. DDB1 and DDB2 both get attached with PARP-1 attracting the DNA remodeling protein ALC1. ALC1 relaxes the chromatin at the site of UV damage to DNA.

Cell cycle checkpoints are activated after rapid chromatin remodeling, allowing DNA repair to take place before the cell cycle proceeds. ATM and ATR kinases are the key contenders in this process. These are activated in 5–6 min after DNA damage. The cell cycle checkpoints—a protein detected in 10 min of DNA damage—are then phosphorylated as a result of this event.

[Further reading suggestion: DNA damage checkpoints, cell division cycle, cell division checkpoints, p53 pathways].

Two types of response to DNA damage are important (a) prokaryotic SOS response in bacteria and (b) eukaryotic transcriptional responses.

(a) **The prokaryotic SOS response**

Changes in gene expression in *E.coli* and other bacteria as a result of significant DNA damage are known as the SOS response. The bacterial SOS system is regulated by two proteins named LexA and RecA. LexA homodimer binds to SOS boxes, which are transcriptional repressors. LexA has been found to regulate roughly 48 genes in *E. coli*, including LexA and RecA. Although the SOS response mechanism is common in bacteria, it is absent in some taxa, such as spirochetes. Specific single-stranded DNA (ssDNA) regions resulting from stalled replication forks or double-strand breaks can activate the SOS response.

First, RecA binds to ssDNA mediated by an ATP-dependent reaction leading to the formation of RecA-ssDNA filaments which are then processed by DNA helicase producing separate DNA strands. After that, RecA-ssDNA filaments turn on the autoprotease activity of LexA leading to the cleavage and subsequent degradation of LexA dimer. Degradation of LexA repressor switches on the transcription of the genes involved in SOS response and allows for further signal induction. It is important to note that in *E coli*, SOS boxes are constituted of 20 nucleotide sequences close to the promoters having palindromic architecture. A high degree of sequence conservation has been found in these regions, whereas in other phyla, there is a considerable variation in structure and length of SOS box. The timing of the SOS response is fine regulated by the differential binding pattern of LexA to different promoters. At a very early stage, SOS response induces the lesion repair genes and induces the error-prone translesion polymerases including UmuCD'2 (DNA polymerase V) thereafter. Once the DNA is repaired, LexA cleavage activity is reduced restoring binding to the SOS box resulting in the normal homeostasis of gene expression.

(b) **Eukaryotic transcriptional responses to DNA damage**

Eukaryotic cells can activate critical defensive pathways in response to DNA damaging agents by activating various proteins involved in DNA repair, cell cycle checkpoint control, protein trafficking, and degradation. These transcriptional responses are exceedingly diverse, sophisticated, and well-coordinated across the genome. *Saccharomyces cerevisiae* demonstrates overlapping transcriptional profiles. Similarities to environmental assaults exhibit general stress-responsive pathways which are operated at the transcriptional level. The global pathway is lacking in a human probably because of heterogeneity of cell types in the higher organism.

Multiple genes involved in post-replication repair, homologous recombination, nucleotide excision repair, DNA damage checkpoint, transcriptional activation, regulating mRNA decay, and many more are expressed in response to DNA damage in general. A wide number of complicated signaling pathways triggered by stress and/or DNA damage share a common mechanism that leads to death or survival. Depending on the extent of stress in case the error is repaired adequately, the cell survives or else cell death takes place in the concourse of death-survival interplay.

9.5.2 Light-Dependent Repair

Light-dependent repair demonstrates the process through which the correction of erroneous DNA takes place in response to light (Fig. 9.19). Three types of damage can be eliminated by chemically reversing without using a DNA template. Therefore these types of DNA repair mechanisms are also known as "direct reversal." Bacteria and some eukaryotic cells can repair DNA through this mechanism. The most

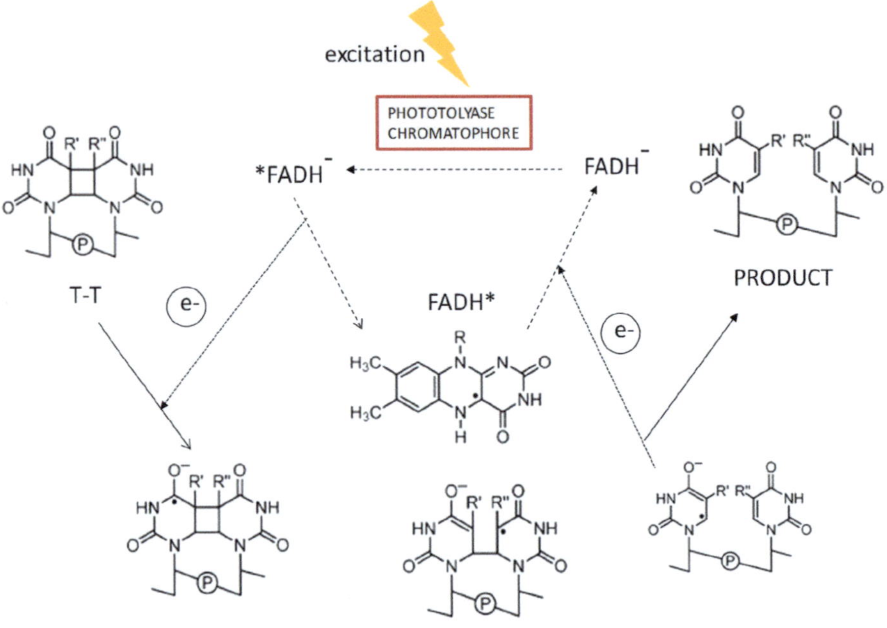

Fig. 9.19 Light-dependent DNA repair: Chemical reaction of photoreactivation of T-T dimer involving FAD as a cofactor

common form of this type of repair mechanism is known as the photoreactivation of pyrimidine dimers (y-y). The covalent bond of y-y is broken down in the photoreactivation process by the enzyme photolyase, whose activation is primarily dependent on energy absorbed from blue/UV light (300–500 nm wavelength) to enhance catalysis (Sancar 2003). The enzyme is a primitive type of enzyme which is lacking in humans. In humans, damage by UV radiation is usually repaired by nucleotide excision repair (will be discussed in the respective section). Another example of direct repair is a correction of O^6-methylguanine, an alkylation product of guanine that pairs with adenine, producing G-C ~ T-A transversions. The enzyme involved in this process is O^6-methylguanine-DNA methyltransferase. Because MGMT-mediated reactions are stoichiometric rather than catalytic, this method is expensive. The adaptive response in bacteria is a worldwide response to methylating compounds that confer resistance to alkylating agents when exposed to them consistently.

9.5.3 Base Excision Repair

In base excision repair (BER) (Fig. 9.20), a modified base is excised followed by the entire removal of the nucleotide. There are several enzymes involved in this cascade. The enzyme known as DNA glycosylases catalyzes the base removed from the DNA

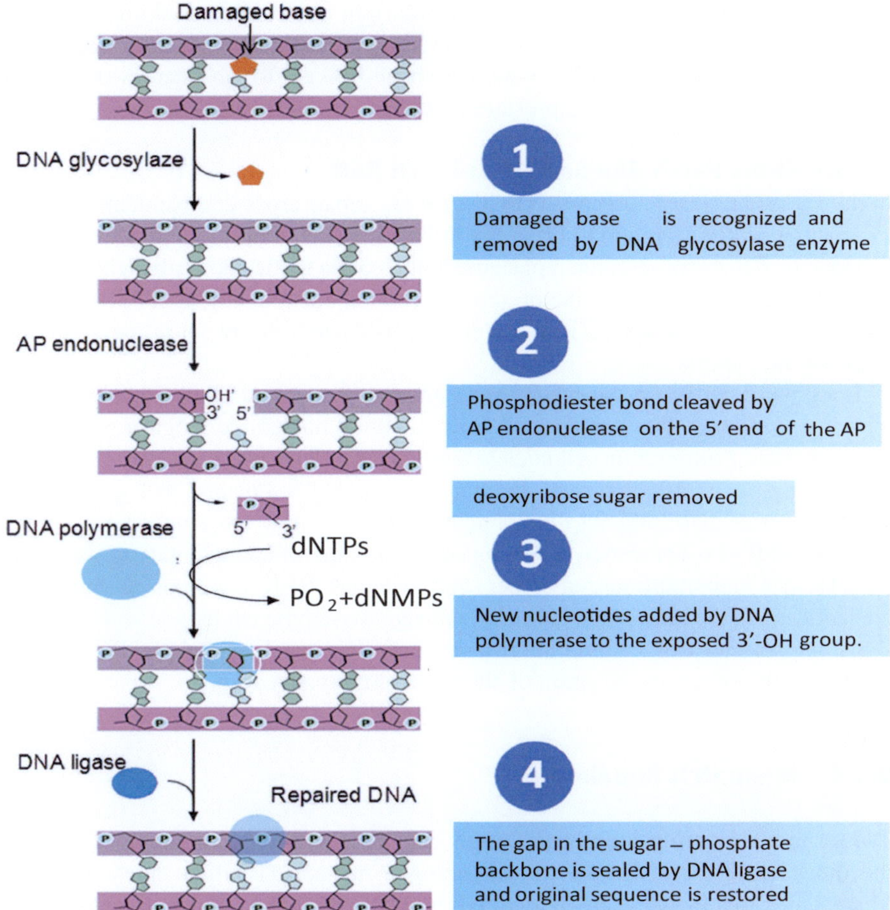

Fig. 9.20 Base excision repair (BER). BER is initiated by the enzyme DNA glycosylase. The step-by-step event is demonstrated in the figure. Once the damaged base is recognized by the DNA glycosylase enzyme, successive actions of AP endonuclease, DNA polymerase, and DNA ligase result in the repair of the gap

strand. The enzyme recognizes a specific type of base modification before removing those. For instance, uracil glycosylase recognizes and removes uracil produced by the deamination of cytosine. Other glycosylases can recognize hypoxanthine, 3-methyladenine, 7-methylguanine, and other modified bases. Endogenous oxidation and hydrolysis are used by BER enzymes to repair the damage. In the first step, the glycosylase enzyme, for example, 8-oxoguanine DNA glycosylase 1 (OGG1), cleaves the bonds between nucleotide base and ribose leaving intact ribose-phosphate chain resulting in apurinic or apyrimidinic (AP) site. In successive steps, these AP cites are repaired by AP endonuclease 1 (APE1). APE1 cleaves the phosphodiester chain 5′ to the AP site. Finally, through its associated AP lyase activity, DNA polymerase β (Polβ) inserts the right nucleotide based on the appropriate W-C pairing and eliminates the deoxyribose phosphate. A mutation (polymorphism) in the human OGG1 gene has been linked to an increased risk of malignancies such as lung and prostate cancer.

9.5.3.1 Short-Patch BER and Long-Patch BER

In the short-patch BER process, XRCC1 (X-ray repair cross-complementing group 1) acts as a scaffold protein to bring the Polβ and LIG3 (ligase III) enzymes (which connect the deoxyribose of the replacement nucleotide to the deoxyribosylphosphate backbone) together at the site of repair. The interaction among PARP-1 (poly [ADP-ribose] polymerase), XRCC1, and Polβ is a necessary event in the execution phase of the process.

Long-patch BER, on the other hand, substitutes a strand of nucleotides with a minimum length of two nucleotides. So far, 10–12 nucleotide length repairs have been reported. Proliferation cell nuclear antigen (PCNA) is used as a scaffold protein in the long-patch BER process. In this process, DNA Pol and Pol generate an oligonucleotide flap, and flap endonuclease 1 eliminates the existing nucleotide sequence (FEN1). Thereafter, DNA ligase I binds the oligonucleotide to the DNA (LIG1). It's important to remember that whereas BER can replace numerous nucleotides via the long-patch pathway, damage to a single nucleotide with minimal DNA impact can trigger both short-patch and long-patch pathways. Figure 9.21 represents the schematic diagram of short- and long-patch BER.

9.5.4 Mismatch Repair

The technique of replication is amazingly accurate. Each nascent copy of DNA, on the other hand, has a chance of containing a mistake (1 billion nucleotides 1). Mismatched nucleotides are integrated into a new DNA with a frequency of 104–105 without permanently affecting the freshly produced DNA during replication. Mismatched nucleotides are detected and corrected by DNA polymerases. Before executing the 5′ to 3′ polymerization, the DNA polymerase begins its proofreading activity and 3′–5′ exonuclease activity to delete the erroneous bases that have been introduced. Incorrectly paired bases distort the three-dimensional structure of DNA, and mismatch repair enzymes can identify these distortions. A mismatch repair system can detect and mend tiny unpaired loops in DNA generated by strand slippage during replication, in addition to improper base pairing (mentioned earlier in mutation section). Some trinucleotide repeats may make it easier for this repair system to miss them. Mismatch repair enzymes slice out the deformed region of the nascent strand and patch the gaps with new nucleotides using the original DNA strand as a template after the integrated error has been identified.

The presence of methyl groups on the unique sequences of the old strand helps mismatch repair proteins in *E. coli* discriminate between old and new strands. Dam methylase methylates adenine nucleotides in the sequence GATC after replication. Because the methylation process is delayed after replication, only the old strand is

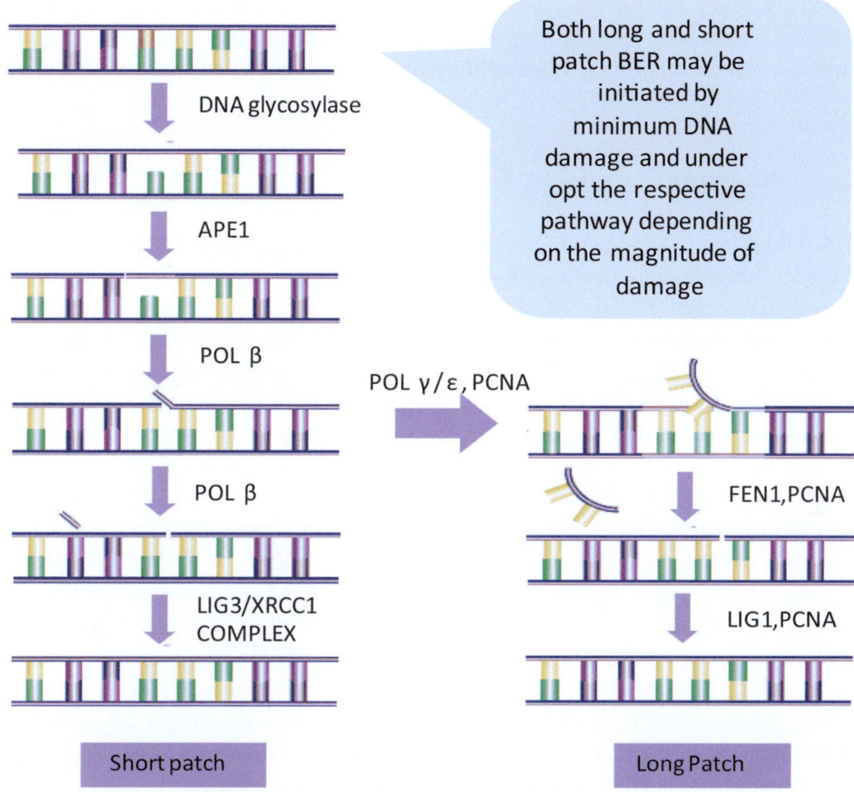

Fig. 9.21 Short-patch and long-patch base excision repair (BER). Short-patch and long-patch BER are both initiated by minimum DNA damage and opt the respective pathway depending on the magnitude of damage

found to be methylated. In E. *coli*, the proteins MutL, MutS, and MutH are the key elements of mismatch repair. MutS binds to the mismatched bases forming a complex with MutH and MutL. This complex takes the unmethylated GATC sequence to the proximity of mismatched bases. Unmethylated strands are nicked by MutH at the GATC site and degraded by the unmethylated strand. Finally, DNA polymerase and DNA ligase replace the missing nucleotides on the unmethylated strand.

In eukaryotic systems, nicks on the nascent lagging strand of DNA (before being sealed by DNA ligase) give a signal that directs mismatch proofreading of the proper strand. According to recent research, the nick allows for RFC-dependent orientation-specific loading of the replication sliding clamp, PCNA. At the nick, one face donut-shaped protein was located juxtaposed to the 3′-OH end. In the presence of a MutSbeta or MutSalpha, the loaded PCNA guides the MutL alpha endonuclease to function on the daughter strand. MutL and MutS have several eukaryotic homologs. Two major heterodimers—Msh2/Msh6 (MutSα) and Msh2/Msh3 (MutSβ)—are formed by MutS homologs, whereas five homologs of MutL have been found in eukaryotes including MLH1, MLH2, MLH3, PMS1, and PMS2. A simplified schematic representation of mismatch repair has been provided in Fig. 9.22.

9.5.5 Nucleotide Excision Repair

DNA damage in at least two bases which resulted in structural distortion is repaired by nucleotide excision repair (NER) (Fig. 9.23). NER is versatile and can repair many different types of DNA damage. It repairs single-strand breaks and serial damage from exogenous sources including bulky DNA adducts and UV radiation. Oxidative stress-induced DNA damage can also be repaired via the NER mechanism. NER system is one of the most important repair systems and ubiquitously found in all cells of all organisms from bacteria to humans and comprise of versatile enzymes. In bacteria, the system is comparatively simpler. For example, the *E coli* NER system is represented by four major proteins UVrA, UVB, UVrC, and UVrD.

XPA, XPC-hHR23B, replication protein A (RPA), transcription factor TFIIH, XPB, and XPD DNA helicases, ERCC1-XPF and XPG, Pol, Pol, PCNA, and replication factor C are among the more than 20 proteins involved in the NER pathway in mammalian cells. Overexpression of the excision repair cross-complementing 1 (ERCC1) gene has been linked to cisplatin resistance in non-small cell lung cancer cells and is associated with increased DNA repair ability.

NER mechanisms repair DNA damage in two ways: (a) global genomic NER (GGR) repairs damage across the genome, and (b) transcription-coupled repair (TCR) repairs genes while active RNA polymerase transcription.

9.5.6 Other Types of DNA Repair System

Apart from the four repair mechanisms discussed so far, there are more special two types of DNA repair mechanisms: (a) translesion synthesis (TLS) and (b) double-strand break repair system.

Translesion synthesis (TLS) is a DNA damage tolerance mechanism that prevents the DNA replication machinery from skipping replication of earlier DNA lesions like thymine dimers or apurinic sites. Instead of conventional DNA polymerases, it employs specialist translesion polymerases from the polymerase family, such as DNA polymerase IV or V. The active sites of this enzyme are well designed, making it easier to insert bases opposite damaged nucleotides. Different types of specialized DNA polymerases are engaged in bypassing or fixing various mistakes. For instance, Pol η arbitrates error-free bypass induced by UV irradiation, whereas Pol ι introduces mutations at these sites. Pol η, on the other hand, uses Watson-Crick base pairing to add the first adenine to the T-T dimer and Hoogsteen base pairing to add the second adenine syn conformation. Complex DNA lesions, like as G-T intra-strand cross-links, can be bypassed by human DNA pol η (G [8, 5-Me] T). A

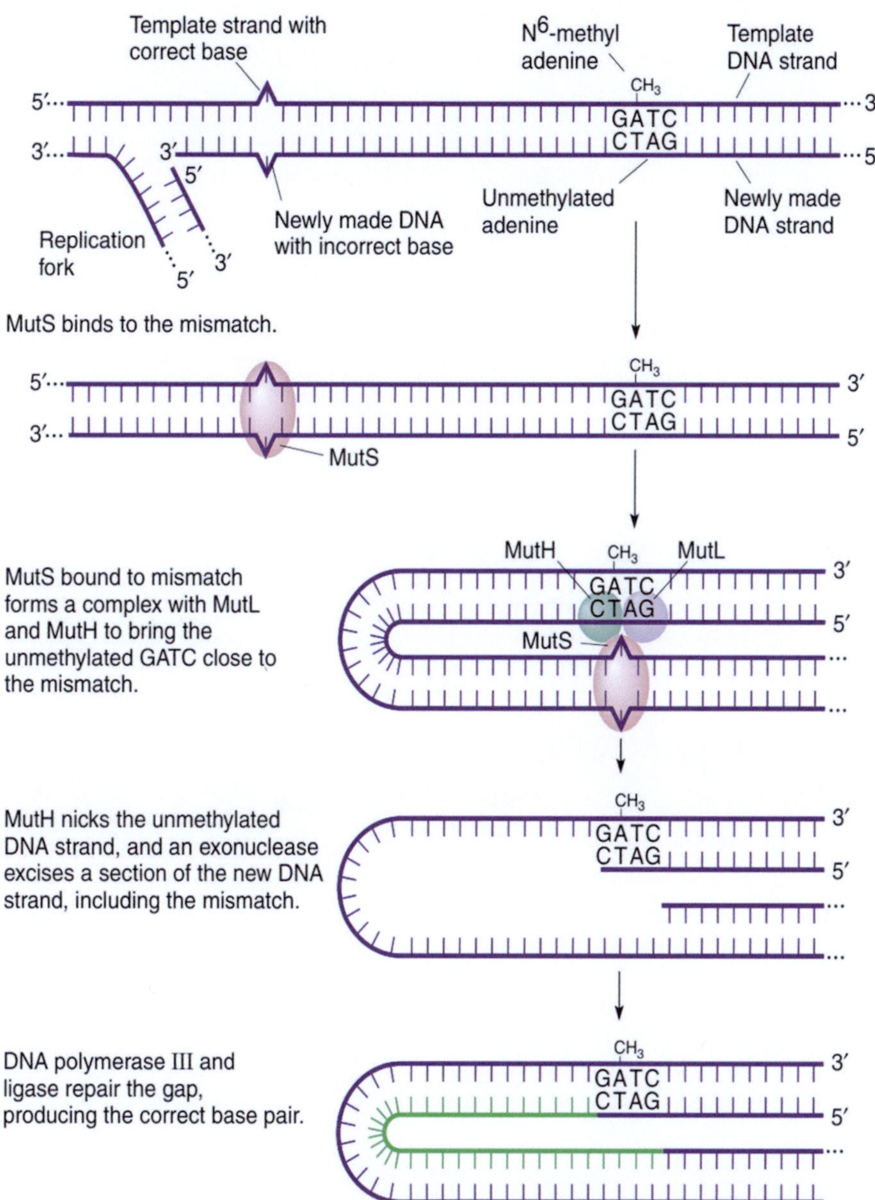

Fig. 9.22 Mismatch repair. Mismatch correction enzyme recognizes by reading the methylation state of a nearby GATC sequence. If the sequence is unmethylated, the mismatches are removed from the DNA strand, and new DNA is introduced. The figure shows the involvement of MutS, MutH, DNA pol III, and ligase in a step-by-step manner

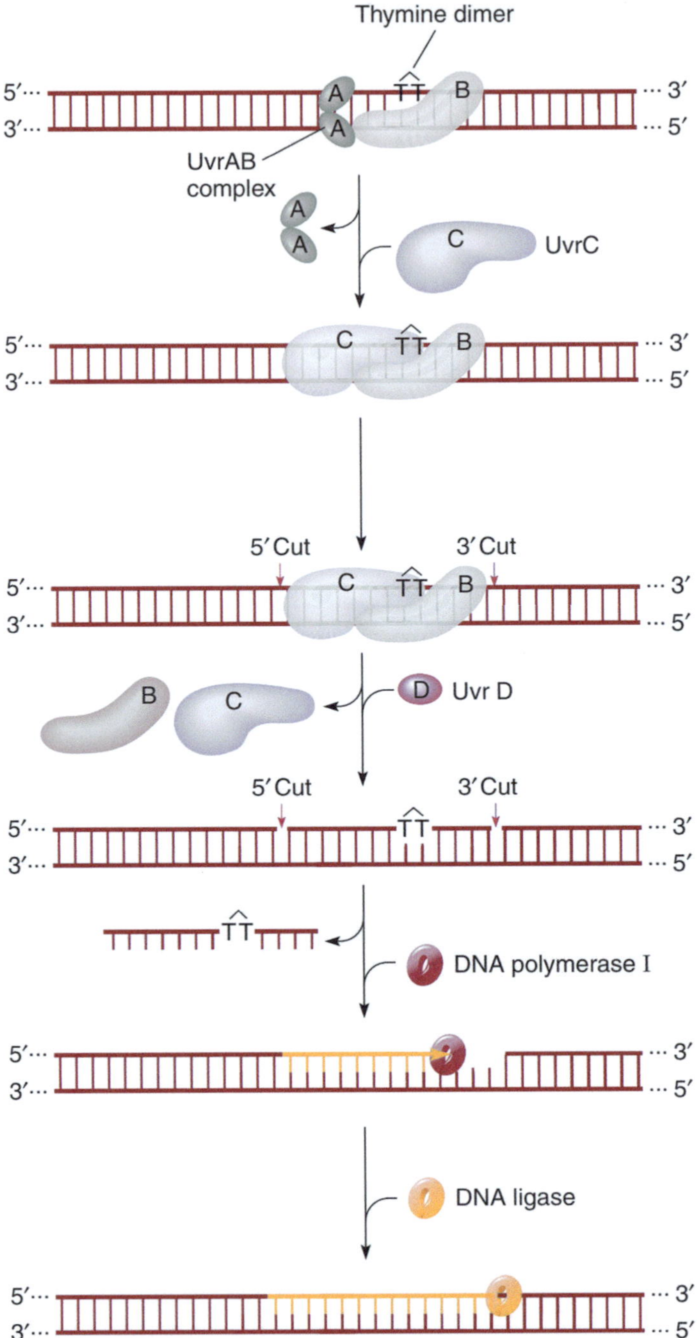

Fig. 9.23 Nucleotide excision repair mechanism. Thymine dimer can be repaired via NER with Uvr proteins. Briefly, the T-T dimer is recognized by the Uvr AB complex. Association of UvrC dissociates UvrA dimer from the site. Nick is formed at the 3′ and 5′ end of the Uvr complex. The nicked part is replicated with DNA pol I, and the end gap is filled up by DNA ligase

replicative polymerase performs extension after translesion synthesis. Finally, a specialized polymerase, Pol ζ, extends terminal mismatches.

On the other hand, sometimes DNA damage is more severe when breakage occurs at both the strand. This kind of DNA damage leads to genomic rearrangement often resulting in a gross chromosomal aberration. These types of DNA breaks are repaired by three mechanisms: (a) homologous end joining (NHEJ), (b) microhomology-mediated end joining (MMEJ), and (c) homologous recombination (HR). Important features of these repair mechanisms are discussed below.

(a) In NHEJ, DNA ligase IV, a specialized DNA ligase which forms a complex with the cofactor XRCC4, directly joins the two ends. The accuracy in repair is achieved by guidance of short homologous sequences known as microhomologies present on the single-stranded tails of the DNA ends to be joined.
(b) MMEJ begins with MRE11 nuclease excision of short range ends on either side of a double-strand break. MMEJ requires PARP-1 as an early component. In MMEJ, microhomology areas are paired, and then flap structure-specific endonuclease 1 (FEN1) is recruited to remove overhanging flaps. Following the recruitment of XRCC1-LIG3 to the location for ligating the DNA ends, undamaged DNA is produced.
(c) Homologous recombination requires an identical or substantially identical sequence to serve as a template for repairing a DNA break. The enzyme machinery that performs this repair is nearly identical to that which performs chromosomal crossover during meiosis.

9.5.7 Pathological Impact of Impaired DNA Repair Mechanism

According to experiments, the accumulation of mistakes can engulf a cell if the rate of DNA damage surpasses a threshold level of DNA repair potential, resulting in early apoptosis, various disorders including cancer, and increased susceptibility to carcinogens. As a result, several genetic illnesses linked to defective DNA repair pathways cause accelerated aging.

For instance, in NHEJ pathway and telomere maintenance mechanisms, deficient mice exhibited shorter life spans than wild-type mice. Similarly mice deficient in a key DNA unwinding protein, helicase has been shown to affect DNA repair mechanism and results in premature onset of aging.

Several individual genes have been identified which influence the life span of individuals. The effects of these genes are essentially environment-dependent, especially on the organism's diet. Caloric restriction causes nutrient-sensing mechanisms to extend life span and metabolic rate to decrease in a range of species. Although detail mechanism is still a matter of conjuncture, many DNA repair mechanisms are responded associated with caloric restriction. For example, several anti-aging agents have been shown to attenuate the constitutive level of mTOR signaling which is an evidence of reduction of metabolic activity. This results in a reduction of DNA damage by endogenous ROS. In an experiment, it has been shown

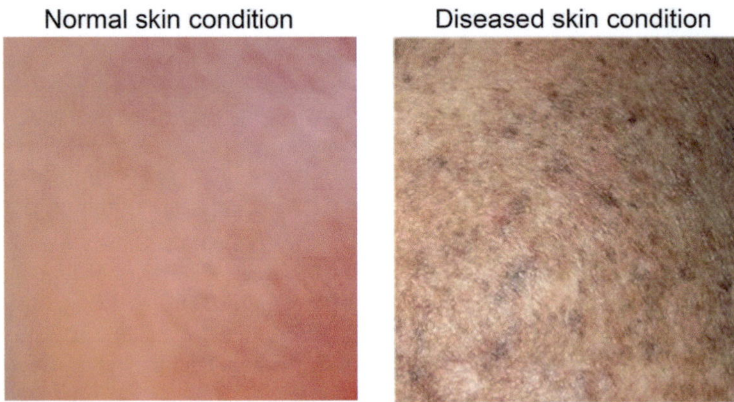

Fig. 9.24 Skin condition in xeroderma pigmentosum

that *C. elegans* extends its life span after the increase in gene dosage of SIR-2 that encodes the downstream DNA repair factor in NHEJ. The effect is promoted in caloric restriction conditions; there are several disorders associated with faulty DNA repair mechanism.

One of the best studied human DNA repair-related diseases is xeroderma pigmentosum. It is a rare autosomal disease responsible for anomalous skin pigmentation and acute sensitivity to sunlight (Fig. 9.24). Affected people are more likely to get skin cancer, with an incidence of 1000–2000 times that of unaffected people. Photolyase activity is absent in human cells (the enzyme which repairs pyrimidine dimers in bacteria). The NER system in humans corrects the majority of pyrimidine dimers. However, most persons with xeroderma pigmentosum have a deficiency in cellular NER, which leads to pyrimidine dimer buildup and, eventually, malignancy. Defects in numerous genes can cause xeroderma pigmentosum.

At least seven different xeroderma pigmentosum complementation groups have been experimentally identified. Two other genetic diseases associated with impaired NER system are trichothiodystrophy (brittle hair syndrome) and Cockayne syndrome. Persons having either of these diseases exhibit multiple developmental and neurological problems. Some genes commonly affect all three diseases. Two special cases associated with faulty DNA repair are HNPCC (hereditary nonpolyposis colon cancer) and Li-Fraumeni syndrome due to mutation in the p53 gene which exhibits a different form of cancer in different tissue.

9.6 Mechanism of DNA Recombination

Recombination is the process by which DNA molecules exchange its part with their counterpart. When this exchange occurs between homologous DNA molecules, it is known as homologous recombination. Homologous recombination takes place during crossing over where homologous regions of chromosomes are swapped

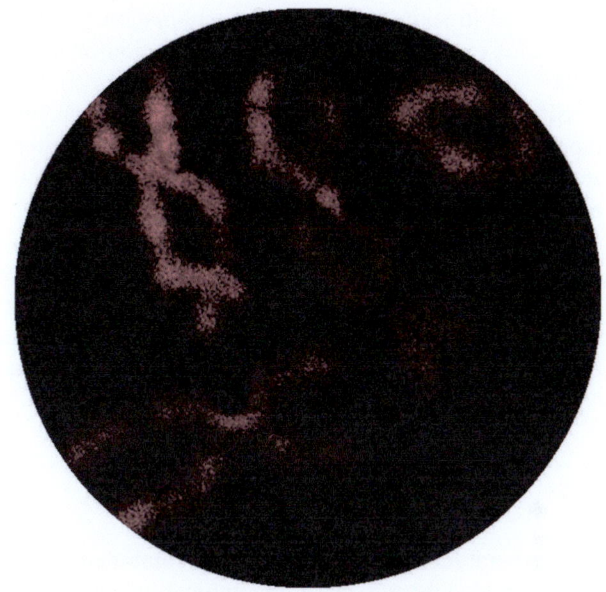

Fig. 9.25 Formation of chiasma. Recombination occurs during the crossing over of chromosomes via the formation of "X"-shaped chiasma

(Fig. 9.25). Through this process, genes are shuffled producing new combinations. Apart from mutation recombination is another vital genetic process implicating genetic variation. Rates of genetic recombination information are essential to creating genetic maps to deduce the linkage relations among genes. As mentioned earlier certain recombination process is essential for DNA repair.

Homologous recombination is a precise routine process during meiosis through multiple steps as follows: (1) one chromosome's DNA strand coincides with a homologous chromosome's nucleotide strand; (2) breaks appear in corresponding sections of DNA molecules; (3) parts of the molecules change their position accurately; and (4) all of the parts are securely attached. No genetic information is lost or acquired during this sequence. In this process, the exchange of DNA takes place via the formation of a heteroduplex. During meiosis, homologous recombination (crossing over) takes place in prophase forming an "X" structure known as chiasma that often remains microscopically visible till early anaphase. In this section, we will be discussing the different types and models of DNA recombination.

9.7 Types of DNA Recombination

Apart from general or homologous recombination, there are at least three more types of DNA recombination that have been identified in the living organism which are as follows:

Illegitimate or non-homologous recombination: It occurs in regions where there is no significant sequence similarity. However, when the DNA sequence at the breakpoints is examined in detail, small sequence similarity regions have been discovered in some situations. Recombination of two distantly placed comparable genes, for example, can result in the deletion of the intervening genes in somatic cells.

Site-specific recombination: Recombination between specific short sequences (approximately 12–24 bp) on otherwise distinct parental molecules is called site-specific recombination. Recombination at a given site necessitates specialized enzymes. The integration of certain bacteriophage DNA into a bacterial chromosome, for example, and the rearrangement of immunoglobulin genes in vertebrate animals are examples of site-specific recombination phenomena.

Replicative recombination: This type of recombination generates a new copy of a segment of DNA. Many transposable elements generate a new copy of the transposable element at a new location via replicative recombination.

9.8 Models of DNA Recombination

Two general models of homologous recombination have been proposed: (1) single-strand break initiated DNA recombination and (2) double-strand break initiated DNA recombination.

Double-stranded DNA molecules from two homologous chromosomes align precisely in the pathway began by a single-strand break in DNA molecule. A break in a single strand creates a free end that invades and connects with the other DNA molecule's end. Strand invasion and joining occur on both DNA molecules, resulting in the formation of two heteroduplex DNAs. A unique type of structure known as a Holliday junction is produced during this process, which is also known as the Holliday model of DNA recombination (discussed later in detail in this chapter).

In the recombination process initiated by double-strand breaks, the breaks occur in one of the two aligned DNA molecules. In this model, strand invasion is followed by the removal of certain nucleotides from the ends of the broken strands. Two heteroduplex DNA molecules are formed by successive displacement and replication, which are connected by two Holliday junctions and separated by additional cleavage. This model has been observed in yeast where double-strand breaks occur during meiosis prophase I.

(a) **Holliday Model**

The Holliday model of DNA recombination (Fig. 9.26) requires the (a) unwinding of DNA helices, (b) cleavage of nucleotide strands, (c) strand invasion, and (d) branch migration, followed by further (e) strand cleavage and (f) union to remove Holliday junctions. The important features are described as follows:

1. At the site of recombination, a heteroduplex is formed.
2. The recombinant joint has the ability to move with the duplex. Branch migration is the term for this type of movement. As one strand is displaced by the other, the branching point can migrate in either direction.

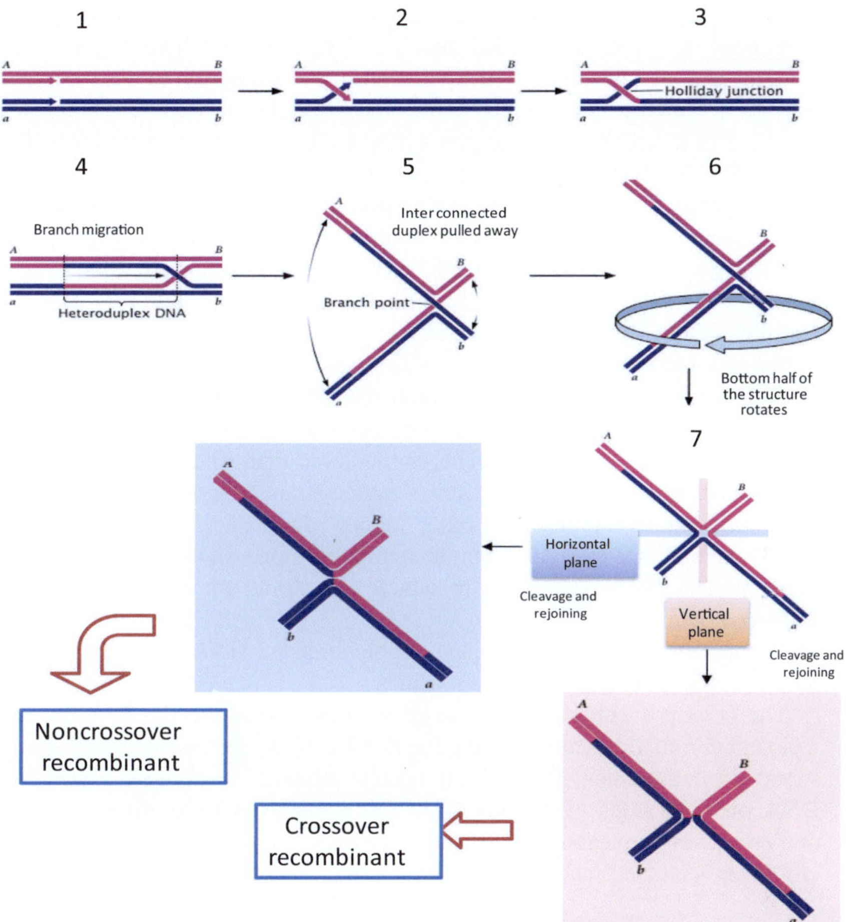

Fig. 9.26 Holliday model of DNA recombination. In the Holliday model of recombination, a nick is formed in one strand of pairing DNA, and subsequently, a heteroduplex is formed. The crossover junction is known as the Holliday junction which migrates to the desired location (branch point) of recombination along with the heteroduplex. The heteroduplex is separated by torsion, and recombination of DNA takes place at the branch point as demonstrated in the figure. Recombination of DNA can be crossed over or non-crossed over depending on the plane of rotation

3. Because the branch may have migrated since the molecule was isolated, branch migration confers a dynamic property on recombining structures that cannot be examined in vitro. The recombination enzymes catalyze branch migration.
4. The strand exchanged and joint must be resolved into two separate duplex molecules. A second set of nicks is required for the resolution.
5. The nick releases splice recombinant DNA molecules.

Experiments with short homologous sequences containing plasmid introduced to bacteria revealed that the length of the homologous region >75 bp and the distance >10 bp increase the efficiency of recombination. Important enzyme systems are associated with this procedure. Experiments with *E. coli* revealed that three genes play a vital role in recombination. These are *recB*, *recC*, and *recD*. These genes code for three distinct polypeptides that come together to generate the RecBCD protein. This protein cleaves the nucleotide strands as well as unwinds the double-stranded DNA. The RecA protein, which is produced by the recA gene, permits a single strand invasion to occur, causing one of the original strands to be displaced.

(b) **Double-Strand Break Initiated Model**

In this process, recombination (Fig. 9.27) is initiated by an endonuclease which cleaves one of the partner DNA duplexes. The cut in one DNA known as "recipient" DNA by the endonucleases becomes distended forming a gap by the action of the exonuclease. The exonuclease then dissolves one strand on each side of the break, resulting in a $3'$ single-stranded terminus.

A homologous region of the other "donor" duplex is invaded by one of the free $3'$ ends. This is called as a single-strand invasion. A D loop is formed when heteroduplex DNA is formed, in which one strand of the donor duplex is displaced.

Repair DNA synthesis generates double-stranded DNA by extending the D loop and using the free $3'$ end as a primer.

The D loop eventually grows large enough to match to the full gap on the recipient chromatid. After reaching the far side of the gap on the extruded single strand, the complementary single stranded is annealed, resulting in heteroduplex DNA on both sides of the joints, which is resolved by cutting. The initial non-crossover is released if both joints are resolved in the same way. A genetic crossover is created when the two joints are resolved in opposite directions.

Box 9.1 Scientific Concept: Testing the Mutagenicity Potential of Chemicals: Geert R Verheyen et al.

Testing of potentially hazardous chemicals is driven by the assessment of endpoints. Ames test using bacterial auxotrophic strain is a preliminary experiment that studies only the mutagenic potential of the chemical. However, it can't provide information regarding other harmful effects. The most important endpoints considered in testing the hazardous and/or genotoxic chemicals are cell death, apoptosis, mutagenicity, and DNA damage. The organization, OCED use to adopt the standard test guidelines set by the global scientific community. OCED has adopted the following tests as standard procedures for testing mutagenicity: (a) bacterial reverse mutation test, (b) mammalian cell gene mutation test, (c) transgenic rodent somatic and germ cell gene mutation

(continued)

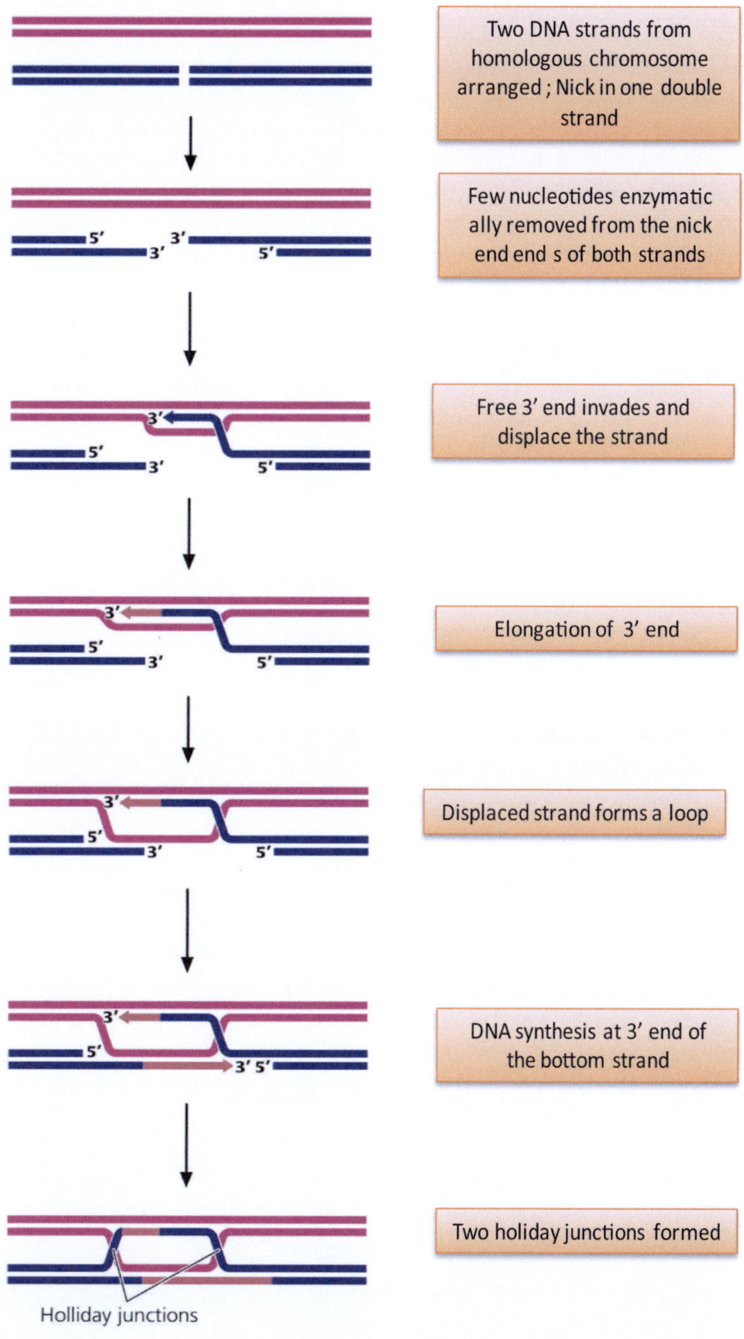

Fig. 9.27 Double-strand break model of DNA recombination. In the double-strand break model, nick is formed both in the strands of each pairing DNA molecule, and recombination takes place based on two Holliday junctions as demonstrated in the figure

Box 9.1 (continued)
assay, and (d) mammalian chromosome aberration test and mammalian micronucleus test.

However, the mode of toxic action is diverse. Therefore the genotoxic effect of a completely unknown chemical is a matter of in-depth toxicological research. Let us discuss the basic pattern of testing the genotoxicity of chemicals. Further reading is suggested in this regard.

Investigation system: (a) in silico, (b) in vitro, and (c) in vivo

1. In silico: Computer-generated program to compare the structural properties of the chemicals with the known chemical to interpret the probable effects in a predictive way.
2. In vitro: Cell types or cell lines are utilized to check the endpoints and mechanism of toxic action. Cell types are selected depending on the objective. For example, for investigating hepatotoxicity, hepatocytes culture or cell line is taken.

   Key Tools

   (a) Ames test: Mutagenic potential with bacterial strain is performed.
   (b) Checking DNA ladder: The DNA damage can be checked in agarose gel by the appearance of multiple bands or smears.
   (c) Micronucleus assay: The fragmented nucleus is assessed with the microscopic investigation.
   (d) Chromosomal aberration: Chromosomal break and/or damage is checked under a microscope.
   (e) PARP-1 cleavage: PARP-1 cleavage can be detected with immunoblots.
   (f) HPRT mutation: Mutation at hypoxanthine-guanine phosphoribosyltransferase (HPRT) and a transgene of xanthine-guanine phosphoribosyltransferase (XPRT) are studied.
3. In vivo: in silico prediction and subsequent in vitro is validated in the live animals. For different proposed mutations, different models are selected including, rats, mice, fish, etc. Zebrafish (*Danio rerio*) which resembles structural features of 70% of humans is a valuable model for genotoxicity study in the present days. Furthermore, apart from rats and mice, catfishes have been established models for testing immunotoxicity. Transgenic animals are used depending on the purpose
   - In present days, high-throughput gene expression profiling approach is considered a fruitful way to evaluate alteration in expressed gene and functional pathways due to chemical exposure.

9.9 Conclusion

From this chapter, we conclude changes in genetic material are essential for evolutionary change. Alteration in genetic information is accomplished either through mutation or via recombination. While recombination is the routine procedure for genetic alteration, mutation is incidental. Therefore mutation is more likely to impose the detrimental character in the genome although beneficial mutation is evident. Mutation can be achieved by several modes through which alteration in bases takes place. Several mutations are without significant phenotypic expression, or it remains subdued due to the countereffect of a second mutation. However many mutations are inherited either via generations, via autosomal linked fashion, or via sex chromosome linked to fashion. These produce the diseased phenotype. Small errors are corrected by the specialized repair mechanism. However, the errors accumulated beyond the critical tolerance potential eventually significant impact of mutation occurs. Therefore impairment in DNA repair mechanism machinery or mutation in the components of DNA repair machinery itself may act as the most severe consequence.

9.10 RuvC Enzyme and Holliday Junction

RuvC enzyme of *E. coli* is an important endonuclease that takes part in cleaving the four-way junctions in DNA structures during homologous recombination, and the enzyme system is also known as resolvase. It resolves the Holliday junction by collaborative action with the branch migration protein RuvAB. Two recombining DNA sections are eventually divided into discrete duplex molecules, which are subsequently joined together by ligation to form a continuous molecule. Certain bacteriophage three-strand or Y junctions, as well as four-strand Holliday junctions, have similar endonuclease to *E. coli*.

9.11 Summary

- Mutations can cause great suffering and are the sustainers of life. A mutation is the source of all genetic variation and the raw material of evolution.
- Without mutations and resulting variation, organisms are unable to adapt themselves to environmental change which may increase the risk of extinction of a species. In many cases mutation leads to detrimental effects resulting generation of a severe disorder.
- If a mutation occurs in gametes or somatic cells at an early point in development, the consequences are severe. Mutation can be inherited through the generation if takes place in the germ cells.
- Many disorder associated to the mutation either in autosome or in sex chromosome is inherited as a recessive or dominant fashion.

- There are several modes of mutation through which bases are altered and eventually alter the composition of the region of DNA. Mutation can be introduced in a DNA spontaneously or by induction of radiation and chemicals including alkylating agents and intercalating agents.
- Although mutation is the chief source of genetic variation, detrimental variation is noteworthy. Therefore our system possesses a specialized mechanism to correct the transient or temporary errors of DNA.
- The sustained impact of mutation occurs only when the DNA repair mechanism fails to correct the error.
- Apart from mutation, genetic recombination is the regular mode of reshuffling and alteration of genetic material.
- DNA of an organism is prone to different types of assaults (environmental and chemical). However, due to certain characteristics of genetic codes including wobble base pairing and transition-transversion bias, an organism acquires a higher probability to defend the mutational change. Silent mutation and reverse mutation potentially subdued the effect of the mutation.
- In present days, the phenomenon "mutation" is being widely used as an important genetic tool to investigate a gene function known as site-directed mutagenesis (SDM).
- Mutagenesis in bacteria has been tactfully utilized to screen the mutagenic potential of chemicals cost-effectively through the Ames test.

References

Sancar A (2003) Structure and function of DNA photolyase and cryptochrome blue-light photoreceptors. Chem Rev 103:2203–2237

Scally A (2016) The mutation rate in human evolution and demographic inference. Curr Opin Genet Dev 41:36–43

RNA Transcription 10

Manasa G. Sharma

10.1 RNA Transcription: Overview

The process of producing proteins from nucleotides is termed as gene expression. Genetic information present in the DNA, is first rewritten, generating RNA, the process of which, is termed as transcription. During transcription, starting with a DNA template, RNA is essentially synthesized (Fig. 10.1). The three major types of RNA generated during transcription are tRNA, mRNA, and rRNA. Other non-coding RNAs (ncRNAs) are also generated. The term "coding RNA" refers to RNAs that are translated to proteins, whereas "non-coding RNA" refers to RNAs that do not code for proteins. Messenger RNAs (mRNAs) serve as templates for translation, and ncRNAs are involved in many regulatory functions in the cell. Recent emerging trends have suggested that many proteins play dual roles, and these are referred to as bifunctional (bifRNAs). Every single gene is transcribed into several copies of mRNA, and each mRNA molecule is capable of generating identical copies of a single protein. The amount of a protein synthesized from a gene is regulated during transcription and translation.

The first step in gene expression where the enzyme RNA polymerase converts a DNA segment into RNA is called transcription. DNA and RNA both make use of nucleotide base pairing as a complementary language.

The direction of flow of genetic material was established as the central dogma by Francis Crick in 1957. This states that the flow of genetic material is in the direction of DNA to RNA to protein and is fundamental to all organisms like complex multicellular organisms as well as unicellular prokaryotes.

This phenomenon came into the picture after the discovery of reverse transcription in retroviruses, which does not follow central dogma. Exceptions also include prions, where information flow is from protein to protein (Fig. 10.2).

M. G. Sharma (✉)
Ramaiah University of Applied Sciences, Bangalore, India

D. Kar, S. Sarkar (eds.), *Genetics Fundamentals Notes*,
https://doi.org/10.1007/978-981-16-7041-1_10

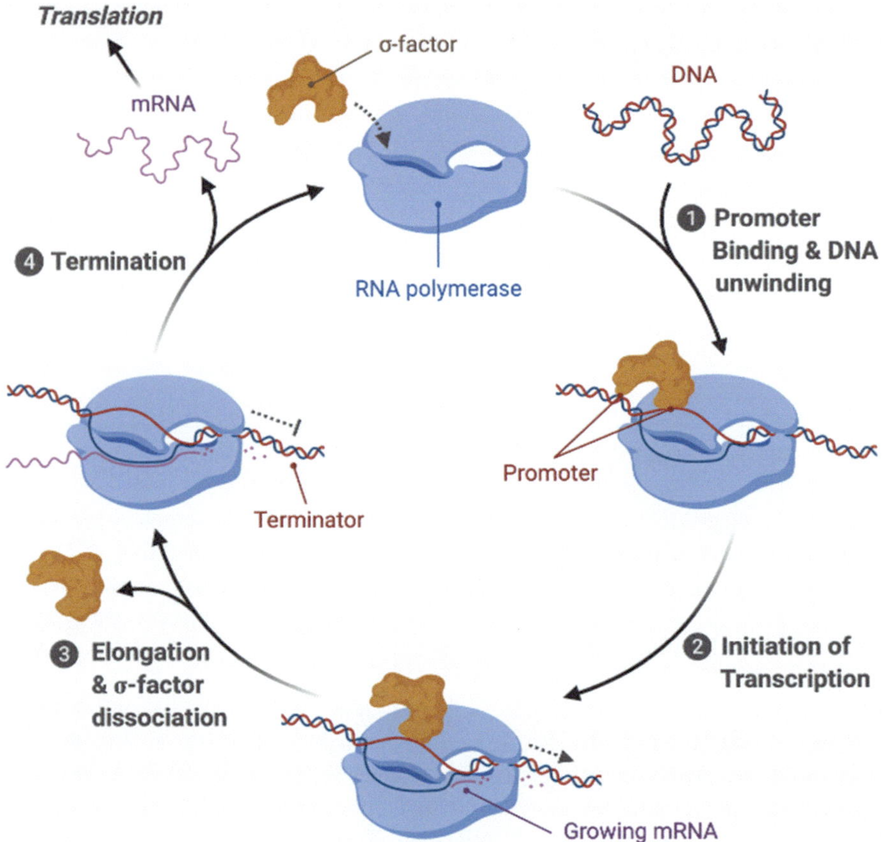

Fig. 10.1 RNA transcription overview. All living organisms carry genes, in the form of nucleotide sequences that contain the information required to synthesize proteins. To carry out essential functions such as survival, growth, and proliferation, these proteins are essential

10.2 Types of RNA: Overview

The product of transcription is RNA. Four kinds of nucleotides make up the RNA, and phosphodiester bonds link them linearly. The difference between DNA and RNA lies in their structures, mainly the sugar moiety and the type of nucleotides that make up the respective nucleic acids. DNA primarily contains deoxyribose sugar, whereas RNA contains ribose sugar—with an extra –OH group. These structural differences have contributed to the nomenclature of the biomolecules. Also, thiamine (T) is one base pair that is part of DNA, whereas it is replaced by uracil in RNA. The other three bases—adenine (A), cytosine (C), and guanine (G)—are in both of these nucleic acids.

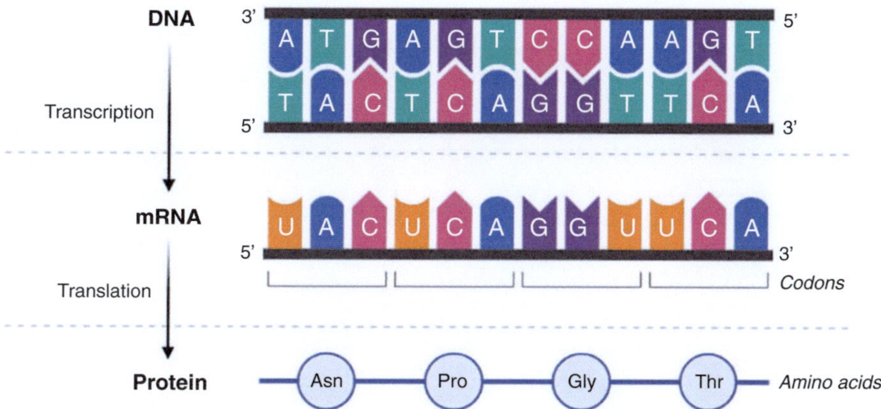

Fig. 10.2 Central dogma. The central dogma is a concept in molecular biology that explains how the residue-by-residue transfer of genetic information happens in living organisms. It states that such information transfer follows a particular order and that it does not move from protein to nucleic acid or protein to protein

The substitution of thiamine with uracil in RNA leaves us with very few chemical differences between the two molecules, and the base complementary properties remain the same as DNA except the U base pairing with A. However, the structure differs a lot. This can be attributed to the fact that RNA molecules are single-stranded and they have different confirmations. Upon completion of the human genome project, a major discovery was made where studies suggested that only 2% of the entire human genome coded for proteins. But, the amount of the genome that was being transcribed was exceptionally high (62%) indicating that there were RNA molecules in huge amounts that did not code for proteins. These non-coding RNAs were found to be involved in cell cycle regulation, maintaining structural integrity, and other cellular and molecular functions.

Many different types of RNAs are responsible for various functions. The important ones include (Fig. 10.3):

- mRNA—They are protein-coding messenger RNAs.
- rRNA—These are ribosomal RNAs that are the main structural components of ribosomes. They are essential for the catalysis of translation.
- tRNA—These clover leaf-shaped molecules are adapter molecules that facilitate the transfer of amino acids to the site of translation.

Others include:

- snRNA—The processing of pre-mRNA is mediated by small nuclear RNAs.
- snoRNA—Processing and modification of pre-rRNAs are facilitated by small nucleolar RNAs.
- miRNA—MicroRNAs are responsible for post-transcriptional gene regulation.

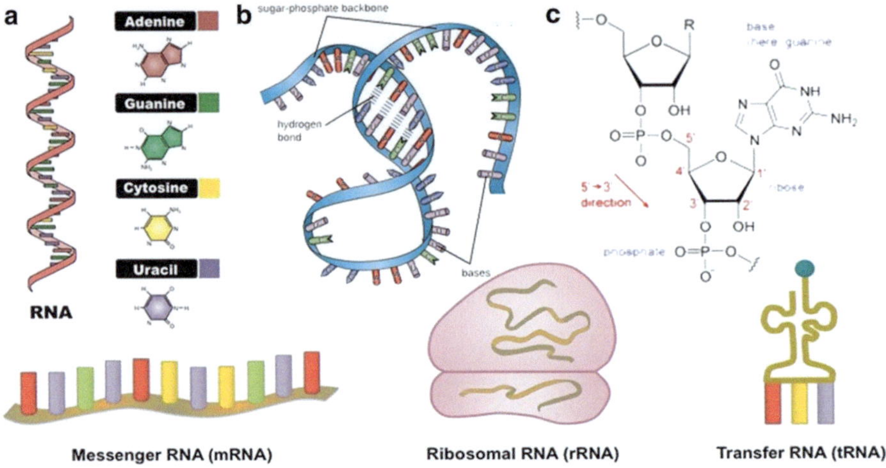

Fig. 10.3 The three major types of RNA: (**a**) mRNA or messenger RNA, (**b**) rRNA or ribosomal RNA, and (**c**) tRNA or transfer RNA

- piRNA—These are known as Piwi-interacting RNAs, and they are known to regulate epigenetic and post-transcriptional silencing of transposons that protect germlines from certain transposable elements.
- lncRNA—Chromatin remodeling is influenced by these long non-coding RNAs. They also regulate gene expression at the transcriptional and post-transcriptional level.
- circRNAs—These are circular RNAs that regulate gene expression and alternative splicing.

10.3 Process of RNA Transcription

10.3.1 Prokaryotes

The mechanism of transcription has remained the same in all living species. Prokaryotic transcription has been extensively studied in bacteria such as *E. coli*. The transcription of RNA involves three steps: initiation, chain elongation, and termination.

10.3.2 Promoter Recognition by RNA Polymerase

RNA polymerase locates the target DNA by recognizing the promoter region. Promoter sequences are usually located in the region preceding the transcription start site (TSS) on DNA. The TSS is usually referred to as +1, and the nucleotides towards the 3′ end of the template strand from this point are said to be downstream of

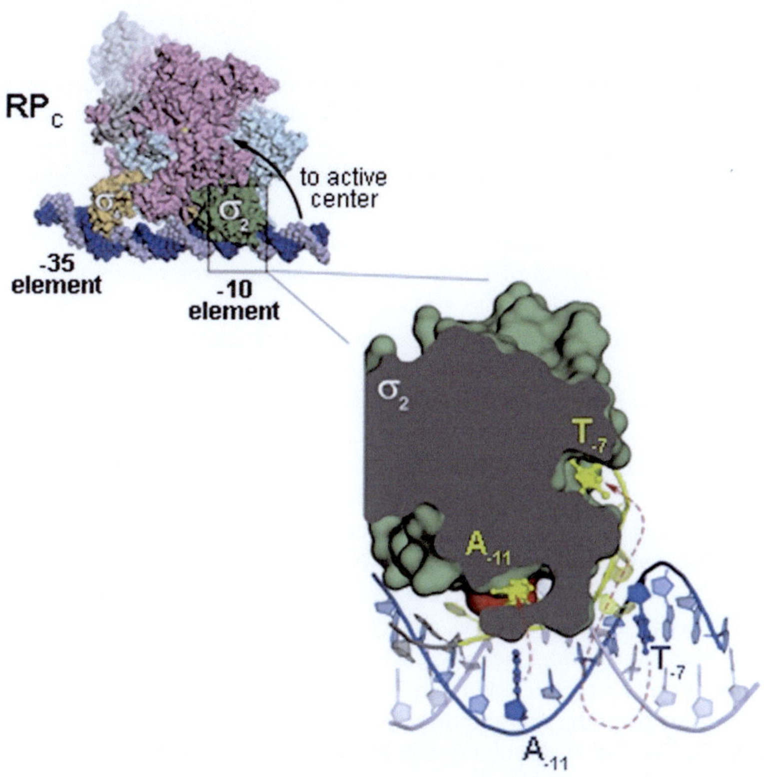

Fig. 10.4 Feklistov et al. study the −10 promoter element recognition "$T_{-12}A_{-11}T_{-10}A_{-9}A_{-8}T_{-7}$ consensus sequence," i.e., the major process involved in the opening of the bacterial promoter by the RNA Pol "σ subunit"

the transcript start site. The nucleotides preceding +1 are referred to as −1, and the bases towards the 5′ direction are upstream. Promoters are characterized by the presence of two hexameric AT-rich signature sequences at −10 and −35 bases and are referred to as −10 and −35 elements, respectively (Fig. 10.4) The −10 element is commonly known as the Pribnow box. The −10 and −35 elements are separated by a nonspecific stretch of 17–19 bases. This region is known as the spacer region. Several studies analyzing promoter regions of a large number of bacteria suggest that there are conserved sequences (TATAAT and TTGACA) at −10 and −35 elements, respectively. The activity of a promoter is highly dependent on its ability to bind strongly to the RNA polymerase, thereby increasing its efficiency of inducing conformational changes in the DNA-polymerase close complex leading to opening the DNA duplex and quick disassociation of the promoter. Heterogeneity and variance in the promoter sequences lead to differential levels of gene expression. In eukaryotes, transcription initiation by Pol II is triggered when the core promoters are recognized through the interactions with DNA, transcriptional (co)activators, and modified histones.

Initiation: The core enzyme is aided by the σ factor to locate the transcription binding site. This activity is mediated by specific nucleotide sequences on the DNA known as the promoter regions. Promoter recognition is crucial for the initiation of transcription. Initially, the holoenzyme is attached to the DNA weakly and rapidly slides across it. Once the σ factor locates the promoter region, the holoenzyme adheres tightly to the DNA and specifically interacts with the outer edges of the bases of the DNA.

The tightly bound holoenzyme then initiates transcription and unwinds around 12–14 bases of the DNA duplex forming a transcription-competent open promoter complex. The σ factor binds to the unpaired bases of any one strand of the transcription complex, and the core enzyme starts assembling the complementary ribonucleotides.

This process is tedious and generates a lot of stress. This is because the RNA polymerase stays attached to the promoter region pulling the upstream DNA region into its active site expanding the transcription bubble. This is known as scrunching (Fig. 10.5). This stress coupled along with the steric clashes between the RNA undergoing elongation and the σ element set up abortive transcription in which the newly synthesized short stretches of the RNA are discarded from the transcription machinery, while the RNA polymerase remains in the same place and starts again. After the abortive initiation rounds, and synthesis of around 17 bases, the initiation transitions into elongation. Due to scrunching pressure, the RNA polymerase breaks free off the promoter region leaving behind the σ factor.

Elongation: After the transcription cycle is set up, the elongation process has to be stabilized. The Pol II machinery equips additional factors to stop the premature dissociation of Pol II. These factors are known as elongation factors, and they associate with Pol II just after initiation. They help the polymerase move along the DNA template on the chromatin network. Pol II goes through something known as "promoter-proximal pausing" after the transcription of 30–50 nucleotides downstream to the TSS (transcription start site). This is depicted in Fig. 10.6. Pol II is released by cyclin-dependent kinase 9 (CDK9) a subunit of positive transcription elongation factor b (P-TEFb). This is facilitated by phosphorylation of several components that make up the transcribing elongation complex (TEC). The P-TEFb works on the core promoters by the COF bromodomain-containing protein 4 (BRD4). A negative elongation factor (NELF) complex is also responsible for promoting Pol II pausing with the help of the DRB sensitivity-inducing factor (DSIF). DSIF directly binds to Pol II, but NELF has a preferential binding activity, and it binds to the assembled DSIF/RNAP II complex.

The NELF complex functionally competes with TFIIF. TFIIF subsequently binds to Pol II and induces conformational changes in the polymerase which are ideal for elongation. Phosphorylation of DSIF, Pol II, and NELF causes dissociation of NELF from Pol II which creates optimum conditions for it to move on to the elongation phase. Certain histone chaperones also help by rapidly disassembling and assembling nucleosomes ahead of the moving Pol II. In eukaryotes, transcription elongation goes hand in hand with RNA processing, and the newly formed RNA molecules

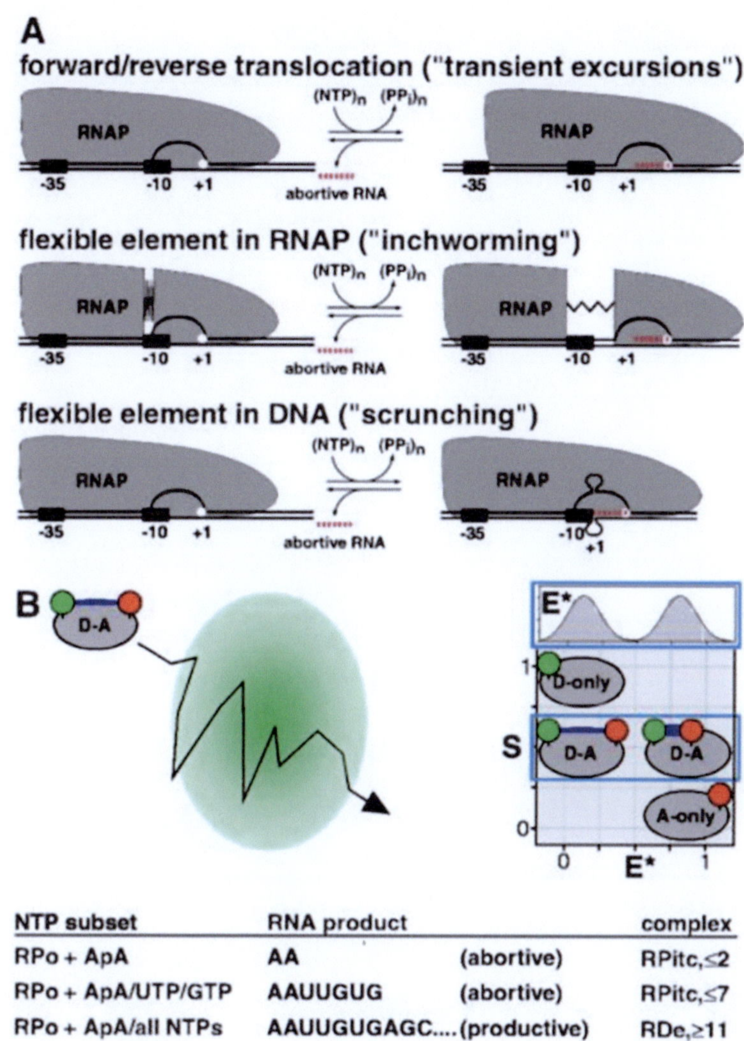

NTP subset	RNA product		complex
RPo + ApA	AA	(abortive)	RPitc,≤2
RPo + ApA/UTP/GTP	AAUUGUG	(abortive)	RPitc,≤7
RPo + ApA/all NTPs	AAUUGUGAGC....	(productive)	RDe,≥11

Fig. 10.5 Studies establishing that transcription initially follows a "scrunching" mechanism. RNA polymerase stays on the promoter region of the DNA and directs the downstream DNA past its active center and into itself (Kapanidis et al. 2006)

are processed in the nucleus, before being exported to the cytoplasm as mature RNAs.

Termination: The terminator signal triggers cascades that cause the core enzyme to dissociate from the template, which releases the newly synthesized RNA transcript and re-associates with the σ factor so that it can start a new round of transcription. The major two kinds of transcription termination mechanisms studied

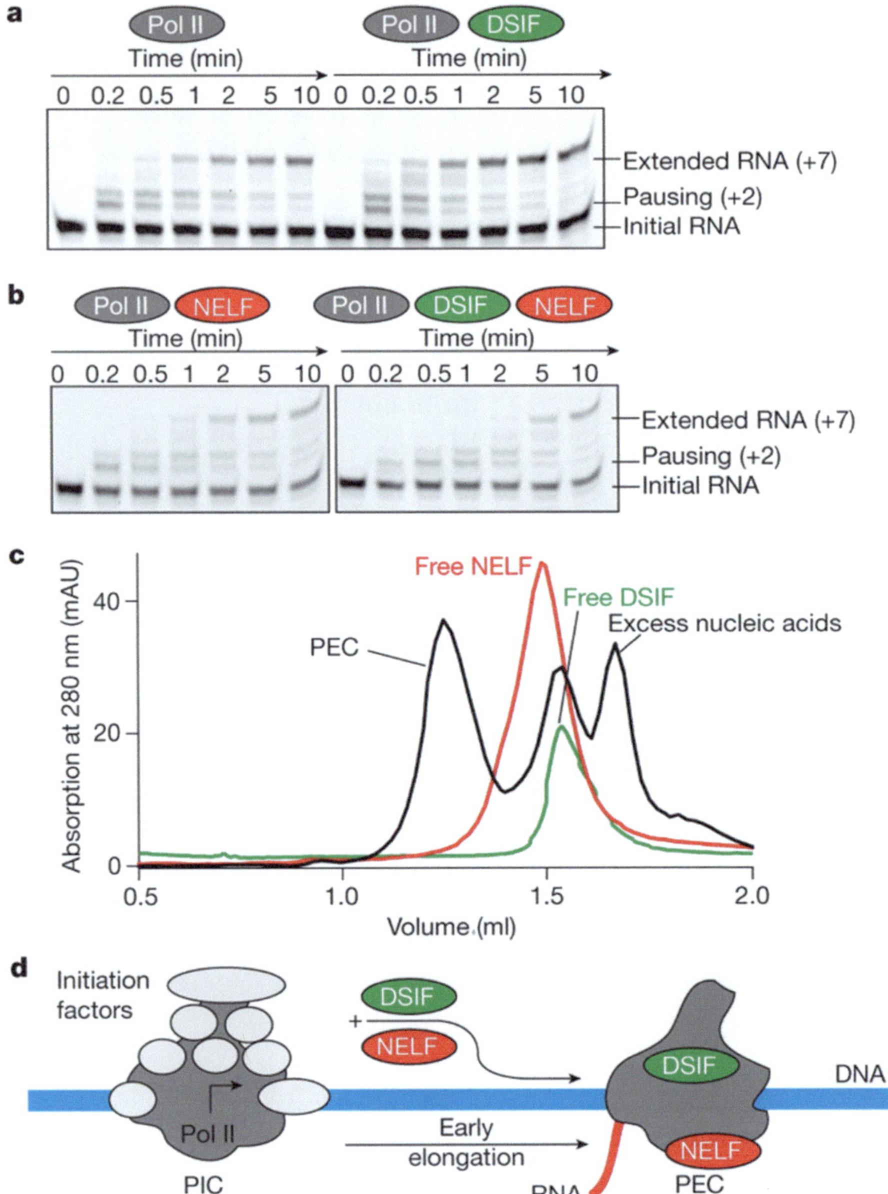

Fig. 10.6 Pol II-DSIF-NELF: Paused transcription complex (Vos et al. 2018). The negative elongation factor (NELF) as well as the DRB sensitivity-inducing factor (DSIF) protein complexes help stabilize the paused Pol II. NELF binds to the polymerase funnel and forms a bridge between the polymerase units that are mobile, thereafter contacting the trigger loop. This restrains the mobility of Pol II which is necessary for pause release

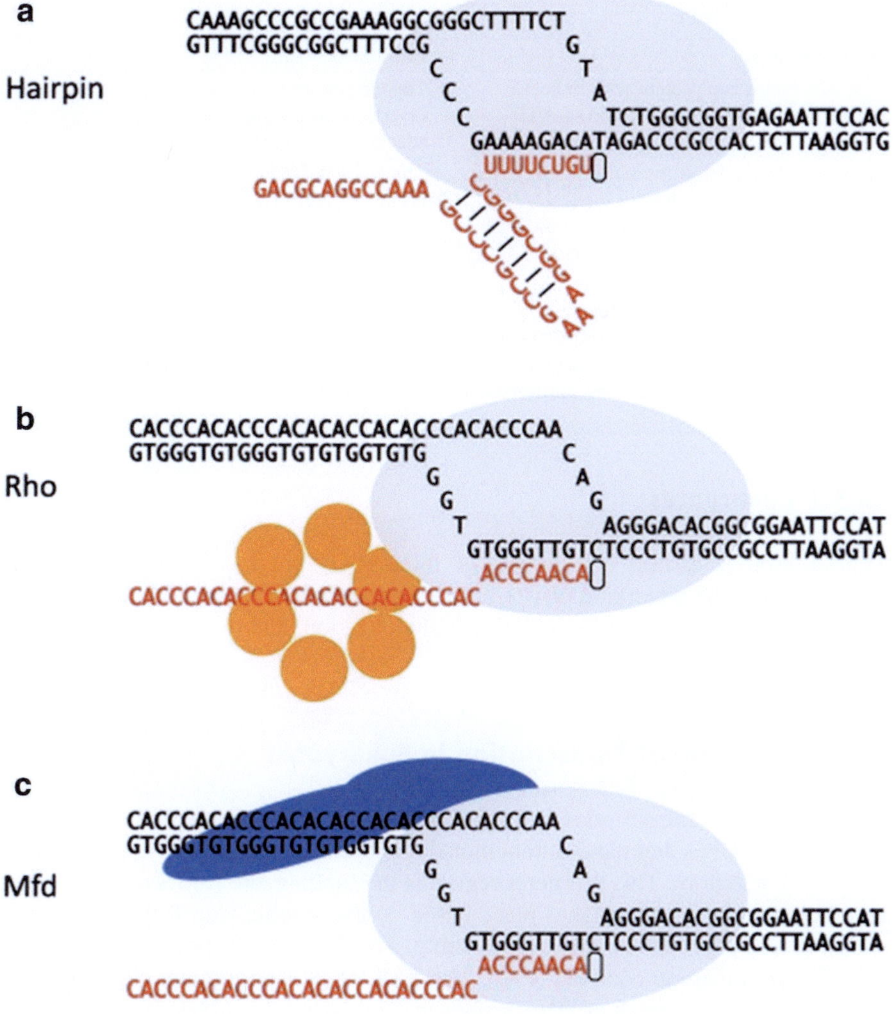

Fig. 10.7 Mechanisms of bacterial transcription termination. (**a**) Transcription is induced by RNA hairpin formation; (**b**, **c**) DNA translocase Mfd and RNA translocase Rho move towards the RNAP, engaging it and forcing dissociation of the elongating complex

in bacteria are factor-dependent termination (Rho-dependent) and intrinsic (Rho-independent) termination (Fig. 10.7, Table 10.1).

Table 10.1 Key differences between the Rho-independent and Rho-dependent transcription termination mechanisms

Rho-dependent	Rho-independent
The Rho factor is a protein with helicase activity and works by utilization of ATP energy	The terminator region is characterized by an inverted repeat sequence followed by an adenine-rich region (AAAA)
Rho binds to the RNA template and moves in the 5′-3′ direction along with the RNA polymerase. It breaks the hydrogen bonds between the DNA and RNA transcript	The RNA polymerase moves on the transcript producing mRNA. Due to hydrogen bonding, the complementary regions in the inverted repeat sequence region form a hairpin loop structure
The DNA/RNA hybrid is pulled apart when the Rho factor reaches the transcription bubble, and this releases the transcript from the transcription bubble, terminating transcription	Hairpin structure stops the RNA polymerase activity. Weak interactions between adenine-uracil bonds in the U-rich region also destabilize the DNA template and the RNA transcript separating them from each other

10.3.3 Eukaryotes

Transcription is carried out by one of three RNA polymerases. The RNAP is recruited based on what kind of RNA is being transcribed. Eukaryotic transcription also involves three stages: (a) initiation, (b) elongation, and (c) termination.

10.3.4 Initiation of Transcription in Eukaryotes

Prokaryotic RNA transcription is facilitated by RNA polymerases that is capable of binding to a DNA template autonomously, but eukaryotic RNAPs make use of transcription factors. This promoter region is the binding site for these factors, and then polymerase selection takes place based on the requirement. RNAPs and transcription factors form a completed assembly. This binds to the promoter in order to form the pre-initiation complex (PIC). The TATA box is a promoter element which is basically a short stretch of DNA sequence. This found 25–30 base pairs upstream from the transcription start site is the best characterized and most studied core promoter element in eukaryotes. However, the TATA box is only present in about one in ten mammalian genes. Various other core promoter regions are also known to exist.

TBP binds to its binding site "the TATA box." The binding of TFIID containing the subunit TBP to the TATA box results in the assembly of five more transcription factors. Around the TATA box (Fig. 10.8), these complexes and the RNA polymerase combine forming the pre-initiation complex. Transcription factor II H (TFIIH) drives the unwinding of the DNA double helix in order to provide a DNA template that is single stranded to the moving RNA polymerase. The pre-initiation complex alone is not entirely responsible for transcription initiation. Other than activators, coactivators, repressors, and corepressor proteins are also responsible for regulating

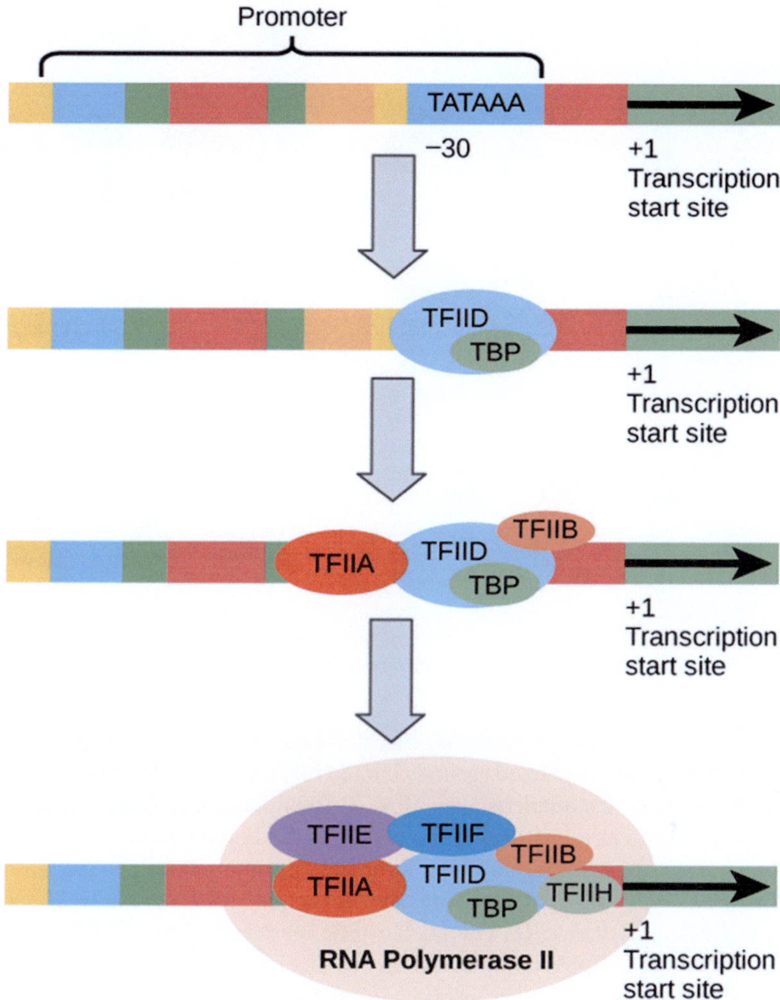

Fig. 10.8 The TATA box region is known to be the major core promoter element in eukaryotic transcription. TATA-binding protein (TBP), a transcription factor, binds to this region. Transcription factor II D (TFIID) contains the subunit TBP

transcription. Activators increase the rate of transcription, and repressors decrease the transcription rate.

10.3.5 Transcription Through Nucleosomes

The release of the polymerase is triggered by the formation of pre-initiation complex, and elongation begins. This phase is characterized as soon as the RNA is synthesized

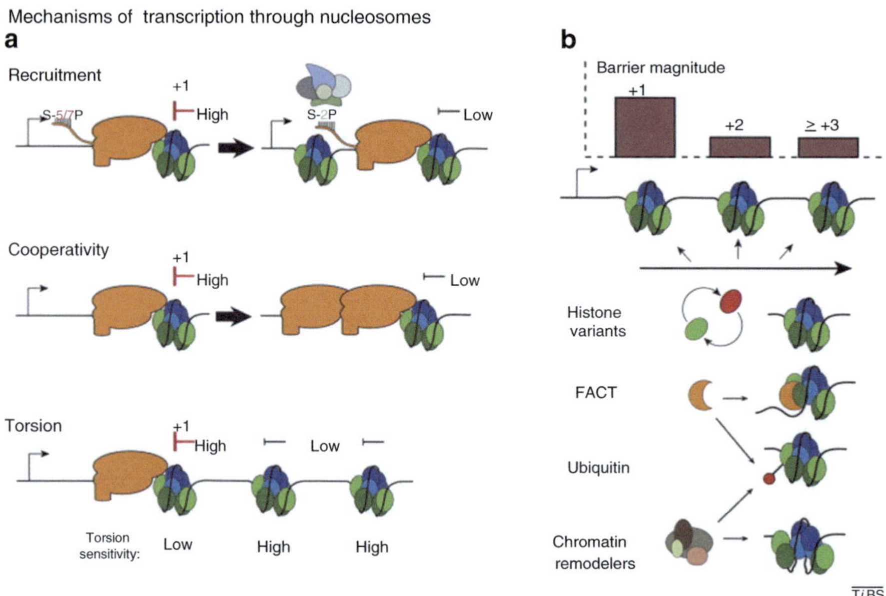

Fig. 10.9 Teves et al. discuss the mechanism of transcription through nucleosomes and suggest that the stability and the dynamics of transcribed nucleosomes are affected by Pol II transit

by the polymerase in the direction of 5′ to 3′. In eukaryotes, DNA is present in the form of chromatin and is packed around charged histone molecules. DNA along with histones makes up the nucleosomes. FACT (facilitates chromatin transcription) protein helps in the removal of histones which is mediated by. The disassembly of nucleosomes is also carried out by FACT ahead of a moving Pol II, by the removal of eight histones (Fig. 10.9). This unwinds the chromatin and gives the DNA template access to RNA polymerase II. The nucleosome reassembles the FACT behind the moving and mobile RNA polymerase II. Pol II elongates the newly synthesized RNA until transcription termination signals are encountered.

10.3.6 Elongation

Twelve protein subunits make up the dynamic RNA polymerase II. This enzyme is a sliding clamp, and single-stranded DNA-binding protein possesses helicase activity. New RNA strands synthesis does not require extra proteins due to this unique multifunctionality of Pol II when compared to replication by the DNA polymerase. But, RNA Pol II requires several accessory proteins for transcription initiation until it is positioned at the +1 initiation nucleotide. After the elongation process has begun, Pol II leaves behind the initiation proteins by a process called “escaping.” This is shown in Fig. 10.10.

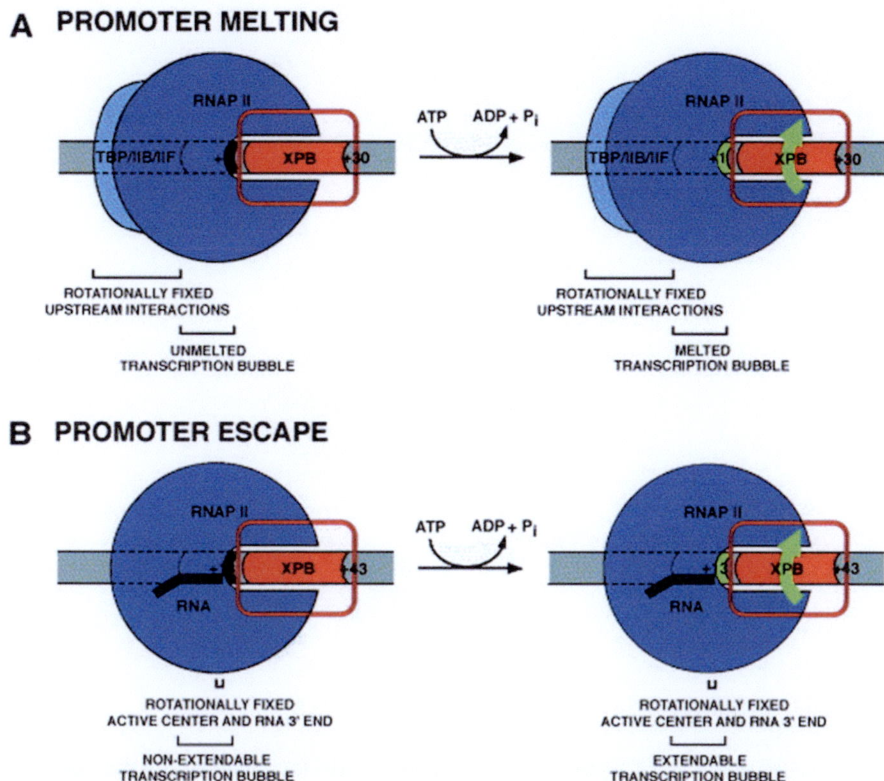

Fig. 10.10 Promoter escape is triggered by the generation of the nascent mRNA strands. This stage can be recognized by abortive transcripts formation and also by the functionally and physically unstable transcription complex

The template DNA guides the RNA polymerase to move in the 3′ to 5′ direction. New nucleotides are added by the RNAPs to the 3′ end of the RNA strand and synthesize new RNA strands in the 5′ to 3′ direction. Ahead of the moving RNA Pol, DNA double helix is unwound and simultaneously rewound behind it. When synthesis is happening, 25 unwound DNA base pairs are known to be attached along with new RNA strands which are about 8 nucleotides long.

10.3.7 Termination

Transcription termination differs based on which RNA polymerase is recruited. A specific (11 bp long in humans; 18 bp in mice) sequence is transcribed by RNA polymerase I that transcribes ribosomal rRNA which contains the sequence. This sequence is recognized by a termination protein called TTF-1 (transcription termination factor for RNA polymerase I) and begins the termination of transcription.

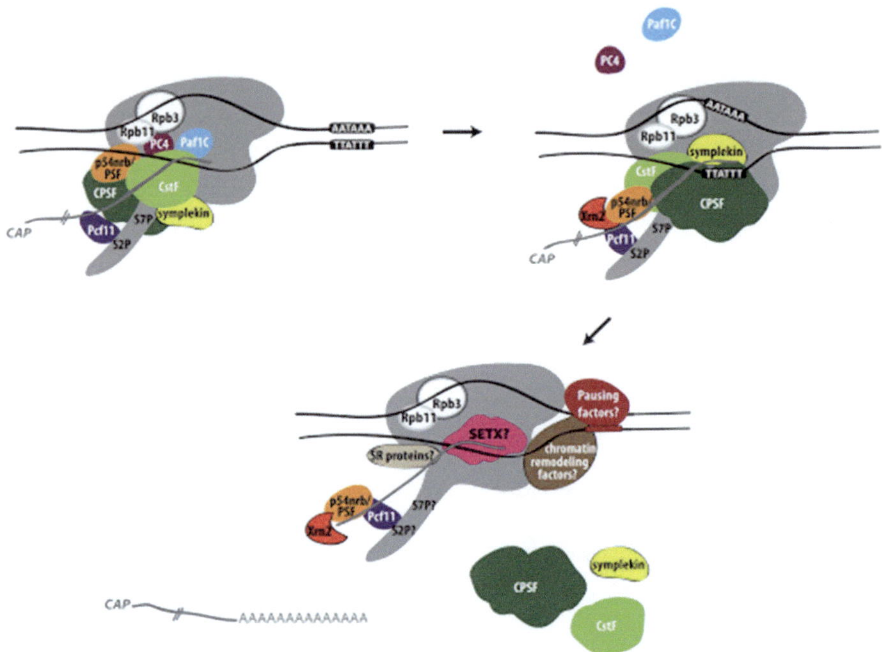

Fig. 10.11 Mammalian RNAPII termination at protein-coding genes

From the template DNA strand, RNA polymerase I dissociates causing the release of the new RNA that has just been synthesized. Before the transcription is complete, RNA Pol II cleaves the transcript and is cleaved at an internal site, releasing the upstream portion of the preliminary transcript. This acts as the initial RNA (or pre-mRNA) before further processing can take place. Upon encountering the cleavage site, the end of the gene is reached. A 5′-exonuclease (Xrn2 in humans) digests the remaining transcript as it is being transcribed by Pol II. After the overhanging RNA is digested by Pol II, the 5′-exonuclease catches up to the polymerase II and helps it dissociate from the DNA template strand and concluding transcription.

Where pre-mRNA synthesis is involved, the end of the gene is determined again by the cleavage site. This site is located between an upstream AAUAAA sequence and a downstream GU-rich sequence separated by about 40–60 nucleotides. After transcription of both of these sequences, the CPSF protein binds to the AAUAAA sequence, and the CstF protein binds to the GU-rich sequence (Fig. 10.11).

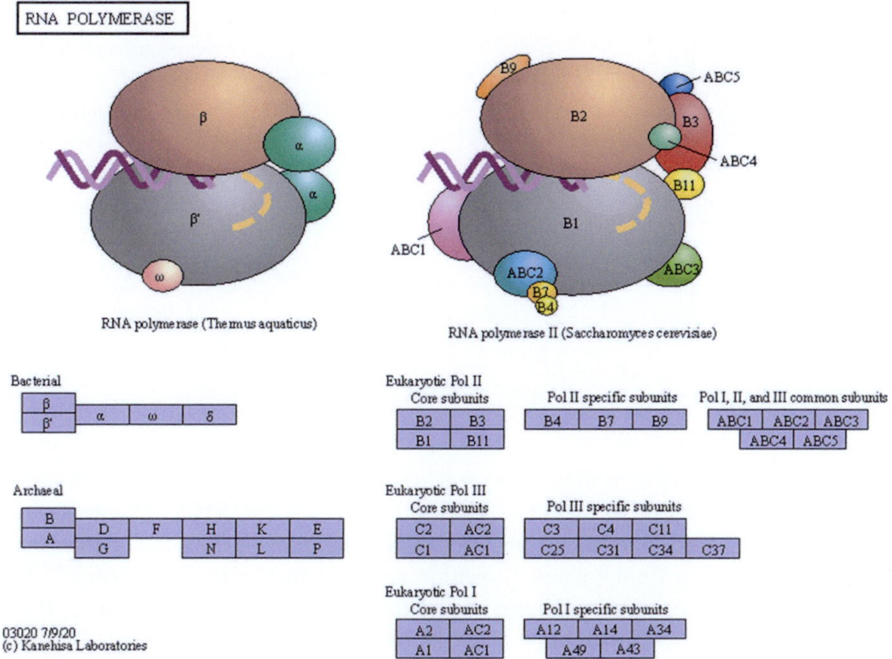

Fig. 10.12 Illustration of eukaryotic and prokaryotic bacterial RNA polymerase holoenzyme

10.4 RNA Polymerase: Mechanism

Bacterial RNA polymerase is made up of a core complex consisting of multiple subunits and an initiation factor called the sigma (σ) factor (Fig. 10.12). The core complex has nonspecific polymerase enzyme activity and can bind to DNA and nicks in a nonspecific manner and is known as the core enzyme (E). The σ factor is essential for the sequence-specific transcription activity. The σ factor along with the core enzymes make up the RNA polymerase holoenzyme (Eσ). Most bacterial RNA polymerases are functionally similar, but there are significant structural differences. The *E. coli* core complex is made of five subunits $-\alpha_2$, β, β′, and ω that possess different functions. Bacteria express several forms of σ factor that recognize and bind different promoter sequences in response to various signals and environmental triggers. In *E. coli*, the main σ factor is σ^{70} or σ^{D}. It expresses housekeeping genes.

The active site and the DNA-binding cleft of the bacterial RNA polymerase structurally look like a crab claw that possesses two pincers. The β and β′ subunits occupy more than 80% mass of the core enzyme and form the pincers generating a cleft for the entry of the template DNA into the active site of the enzyme. The pincer formed by the β′ subunit is known as the clamp which changes its orientation by swinging between the open and closed conformations. The β and β′ subunits form two double-psi beta-barrel (DPBB) domains. The DPBB domain interacts with the

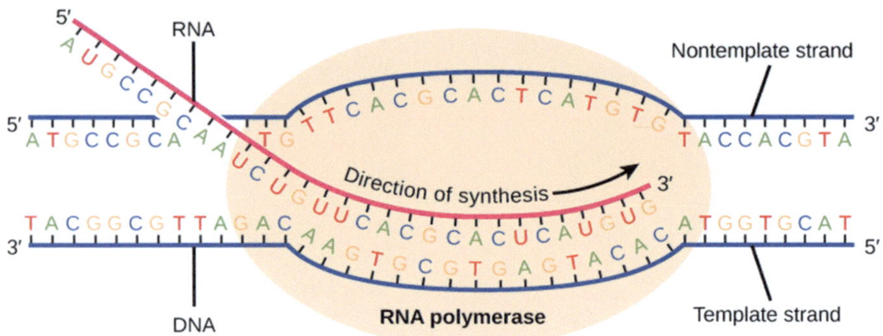

Fig. 10.13 Direction of synthesis of RNA transcript in a transcription bubble

incoming nucleotides with their basic residues on the surface. The DPBB domains are used by most of the cellular RNA polymerases for RNA synthesis. The ribonucleosides enter into the active site of RNA polymerases through a secondary channel, which is a funnel-shaped opening separate from the main channel. The secondary channel contains a binding cavity for accessory factors that control RNA polymerase activity. Two additional motifs from the β′ subunit play very important roles in the RNA synthesis reaction. One motif is the trigger loop involved in catalysis, and the other is the bridge helix used for the translocation of DNA and RNA during the nucleotide addition cycle.

RNA polymerase carries out transcription which is also enzymatically catalyzed. Transcription happens in three distinct phases: initiation, elongation, and termination, which constitute the transcription cycle. In the initiation phase, RNA polymerase recognizes and binds to the DNA at the promoter region which lies upstream of the DNA template. RNA polymerase unwinds the DNA double helix and creates a transcription bubble and exposes 12–14 bases on each strand. One of the helices of DNA acts as the template strand for transcription for the complementary ribonucleotide bases to align. The template strand is commonly referred to as the non-coding strand. The other strand of the DNA double helix will have the same base sequence as the RNA (except uracil instead of thiamine) and is known as the coding strand. RNA polymerase covalently links the base pairs on the template strand. After this step, nine to ten bases of the newly synthesized RNA and the template DNA remain attached forming a temporary DNA-RNA duplex structure. After the synthesis of ten bases, RNA polymerase proceeds to enter the elongation phase. It also involved in the unwinding of the double helical DNA in front and rewinding it from behind. The RNA polymerase moves in the 3′ to 5′ direction of the template, but the direction of chain elongation is in the 5′ to 3′ direction (Fig. 10.13).

At the molecular level, it is understood that the RNA polymerase works by creating phosphodiester bonds between the incoming ribonucleotide triphosphates and the growing chain of RNA. This is a thermodynamically feasible and irreversible reaction. The RNA polymerase adds around 10–100 bases every second. It does not dissociate from the DNA until the transcript is completely formed, and this characteristic is known as processivity. The movement of the RNA polymerase across the template strand happens in such a way that the enzyme is capable of detecting mismatches and other errors. The fidelity of transcription is extensively taken care of because a misincorporated ribonucleotide base leads to disastrous consequences. In case an error is detected, the enzyme moves back onto the template and excises the misincorporated base at the 3′ end exhibiting proofreading activity and replaces it with the correct base. The binding of the RNA polymerase is lenient and allows it to move on the DNA template at different rates. Transcription is terminated at the site where the RNA polymerase recognizes a terminator sequence. The transcript is then released from the transcription bubble. In eukaryotes, RNA polymerases are also involved in the modification of transcripts, a process known as the post-transcriptional modification in which primary transcripts, the firsthand product of the transcription; undergo certain modifications to become functional.

10.4.1 The Three Eukaryotic RNA Polymerases (RNAPs)

Compared to prokaryotes, mRNA synthesis in eukaryotes is significantly complicated. Eukaryotic transcription involves three polymerases that are comprised of more than ten subunits.

RNA polymerase I: It is a characteristic nuclear substructure in which ribosomal RNA (rRNA) is transcribed, processed, and subsequently assembled into ribosomes. It is found in the nucleolus region. The rRNA molecules are structural RNAs because they offer structural maintenance and support, but they are not translated into proteins but are essential to carry out translation. Majority of the rRNAs, except the 5S rRNA, are synthesized by RNA Pol I.

RNA polymerase II: Main polymerase involved in the synthesis of protein-coding nuclear pre-mRNAs. Eukaryotic pre-mRNAs are subjected to extensive post-transcriptional after. RNA polymerase II transcribes a majority of eukaryotic genes, which includes protein-encoding genes as well as genes that encode for various regulatory RNAs, like microRNAs (miRNAs) and long non-coding RNAs (lncRNAs).

RNA polymerase III: Transcribes structural RNAs including the 5S pre-rRNA, transfer pre-RNAs (pre-tRNAs), and small nuclear pre-RNAs. Small nuclear RNAs are involved in "splicing" pre-mRNAs and regulating transcription factors.

10.5 Prokaryotic vs Eukaryotic RNA Transcription

The process of transcription is essentially the same in both prokaryotes and eukaryotes. But more steps are involved in eukaryotic transcription (Fig. 10.14). Bacteria and species belonging to archaea require only one type of RNA polymerase, whereas eukaryotes require at least three main enzymes—RNA polymerases I, II, and III (Pol I, II, III), along with polymerases IV and V which are present in plants

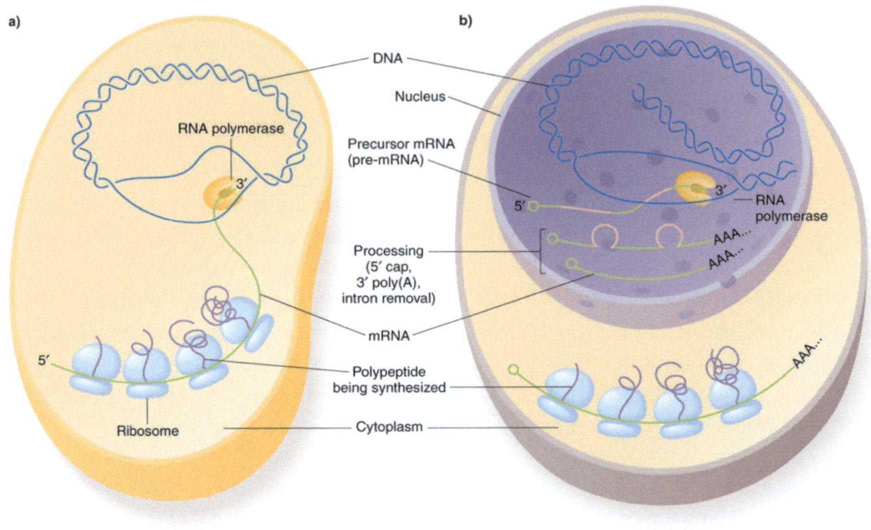

Fig. 10.14 Transcription in prokaryotes vs transcription in eukaryotes. (**a**) Bacterium. (**b**) Eukaryote

that transcribe different subsets of RNA. In all three RNA polymerases, the core enzyme is structurally conserved and comprises ten subunits. Additional subunits are located on the periphery. Out of the three, Pol II is known to transcribe the maximum number of genes.

DNA is replicated, and RNA is translated in the same shared space in prokaryotes because of the absence of a nuclear membrane. In eukaryotes, the nucleus is the site of DNA replication and transcription, whereas protein synthesis occurs in the cytoplasm. RNA is exported across the nuclear membrane before it can undergo translation. Transcription and translation are separated by physical barriers. The primary transcript in eukaryotes, which is also known as "heterogeneous nuclear RNA (hnRNA)," is subjected to post-transcriptional processing in order to make a messenger RNA (mRNA) molecule that can pass through the nuclear membrane.

10.6 Regulation of RNA Transcription

In regulation, transcription factors are key players. Transcription factors are DNA-binding proteins that work by repressing or activating gene transcription. Preferential activity is displayed by some transcription factors. These bind to each other, cis-acting DNA sequences as well as to both DNA and other transcription factors. In order to promote repression or activation, specific promoters act as binding sites for these transcription factors. Certain transcription factors act exclusively as activators or repressors, and some others function as either activators or repressors. The genome of the bacteria *E. coli* is comprised of around 300 genes that code for proteins that function as transcription factors that up- or downregulate transcription. The functional properties of most of these proteins are still unknown. For the most part, they are known to regulate a large number of genes. Half of all regulated genes are controlled transcription factors such as CRP, FNR, IHF, Fis, ArcA, NarL, and Lrp. A single promoter is known to be controlled by 60 transcription factors. Data inferred from sequence analysis suggest that bacterial transcription factors can be classified into various families and based on these studies. Among these, 12 groups of families have been extensively analyzed and characterized including the LacI, AraC, LysR, CRP, and OmpR families. Bacterial promoter activity also depends on multiple environmental factors and seldom on one signal. Multiple signals are necessary for promoter response. Various transcription factors mediate these events. Many promoters are controlled by two or more transcription factors, with each factor responding to a particular environmental signal.

10.6.1 Repression of Transcription Initiation

A repressor protein negatively regulates gene expression, by binding to the DNA in order to inhibit the initiation of transcription. An effector molecule essentially binds to the repressor which decides whether or not the repressor is capable of binding to the DNA. Transcriptional control can be best explained by the lac operon expression in *E. coli*. In this case, the initiation of transcription is brought about from the lac operon promoter by a repressor protein known as LacI. A site known as operator on the DNA acts as the binding site for LacI. This region shares a stretch of sequences with the promoter. Because of the region of similarity, competitive binding of RNAP and LacI repressor takes place, which makes it a requirement that the repressor should be released from the operator for RNAP to bind to the promoter (Fig. 10.15).

The repressor is subsequently compounded by the binding galactoside, which causes the binding of the RNAP to the promoter region and the initiation of transcription due to the destabilization of the repressor-operator complex (Fig. 10.16). LacI has two additional binding sites in the lac, an upstream site and a downstream site, located in the first gene of the operon. These sites have a lower affinity for the repressor protein, and, likely, they do not directly inhibit transcription initiation.

10.6.2 Small RNAs

In bacteria several studies have indicated that a subset of small RNAs has been found to regulate transcription in bacteria. An important example is the 6S RNA that inhibits transcription at 70-dependent promoters by binding to the active site of 70-RNAP and competing for DNA binding. It has been proposed that, in the conserved secondary structure of 6S RNA, a single-stranded central bulge within a highly double-stranded molecule that is essential for 6S RNA function is present

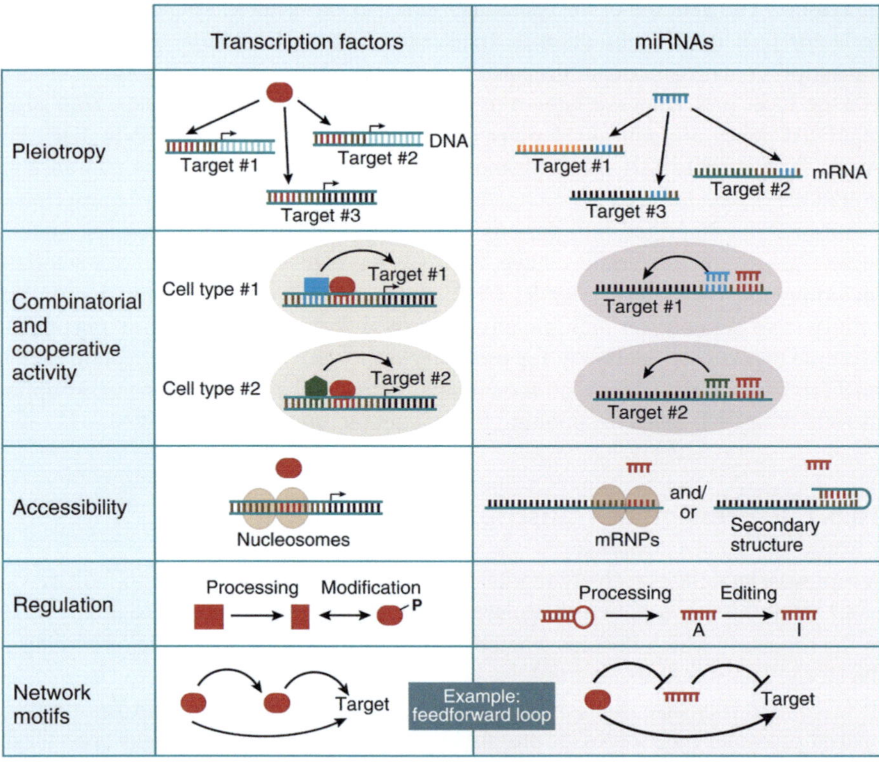

Fig. 10.15 Gene regulation by transcription factors and microRNAs (Hobert 2008)

from which it can be hypothesized that 6S RNA mimics the open conformation of promoter DNA. 6S RNA blocks access to the promoter DNA, and in some cases, it is also used as a template for RNA synthesis.

10.6.3 Regulation of Transcription Initiation via Changes in DNA Topology

Negatively supercoiled DNA acts as the template for transcription. DNA melting is necessary for the open transcription complex assembly. The degree of supercoiling influences and affects the efficiency of some of the promoters (Fig. 10.17). They are also stimulated by negative supercoiling. The effect of superhelicity on transcription initiation has been demonstrated in several in vitro studies and in in vivo models by gyrase inhibitors, which introduce negative supercoils. Some promoters are also sensitive to the degree of supercoiling, and some are not; the reason for this lies in the fact that the sequence of some promoters is easier to melt.

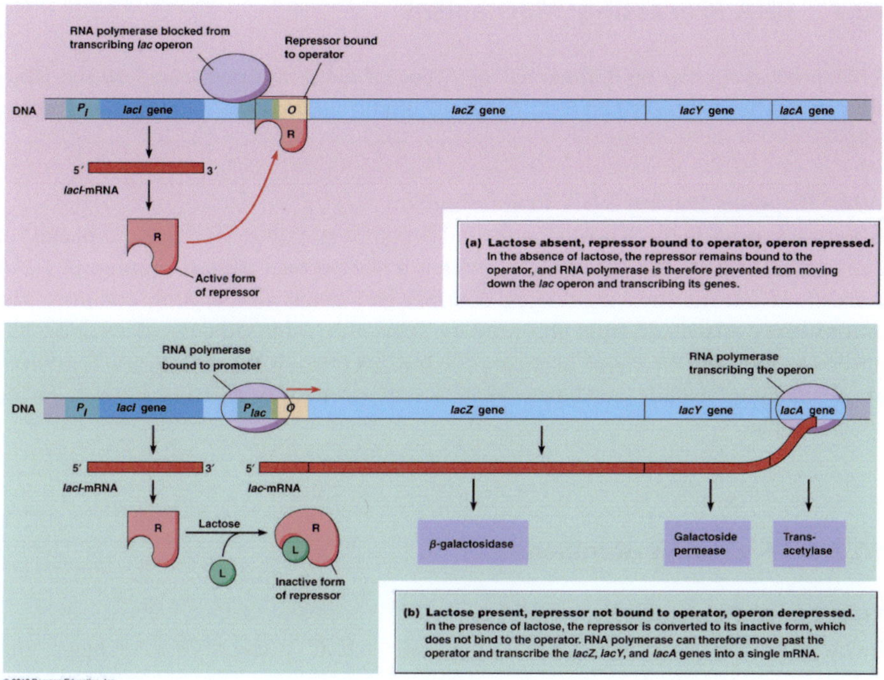

Fig. 10.16 The Lac operon concept and the regulation of gene expression in bacteria

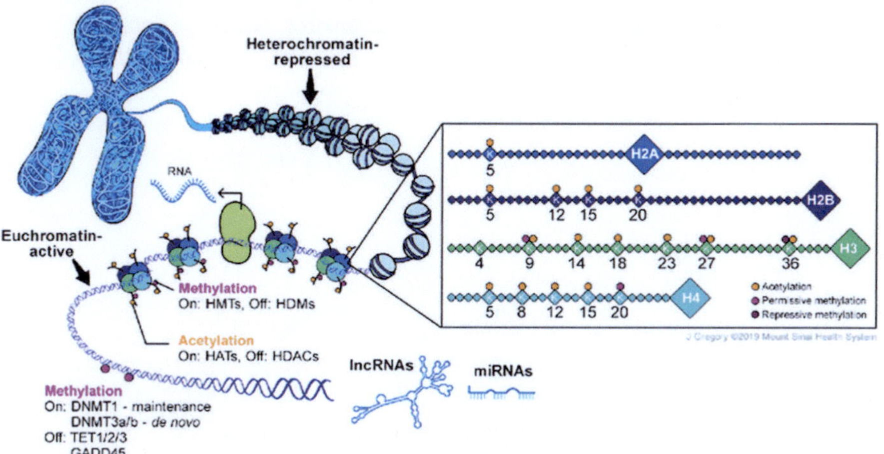

Fig. 10.17 Epigenetic regulation of gene regulation. Gene expression is regulated by DNA methylation, histone post-translational modifications (PTMs), and the actions of non-coding RNAs, among other mechanisms. To fit within the nucleus, DNA is wrapped around histone proteins creating a higher-order chromatin structure, which can facilitate or prevent access to gene regulatory machinery through steric mechanisms (Torres-Berrío et al. 2019)

10.7 RNA Processing: Mechanism

RNA processing can be defined as "any type of alteration performed on the RNA after it has been transcribed from DNA to obtain its complete functionality in the cell."

What Happens During RNA Processing?
After transcription, the RNA is processed before it is exported to the cytoplasm for translation (Fig. 10.18). The hnRNA which is the product of transcription of DNA consists of introns (non-expressing codons) and exons (expressing codons). The introns are excised off from the primary transcript. This process of excising the introns and joining the exons in eukaryotic mRNA (also tRNA and rRNA) is referred to as splicing, usually mediated by a set of protein complexes known as the spliceosomes.

10.7.1 Processing of mRNA

10.7.1.1 5′ Capping

Eukaryotic mRNA is not stable at the ends and is susceptible to damage thus requiring modification to protect it from ribonucleases. The pre-mRNA hence undergoes capping at the 5′ end immediately after transcription and is then released by Pol II. GTP condensation with triphosphates at the 5′ end is an event that triggers the capping reaction followed by guanine methylation at N-7. This methylation produces the modified guanine or 7-methylguanosine which is attached to the triphosphates of the first transcribed base. Capping of the nascent mRNA protects

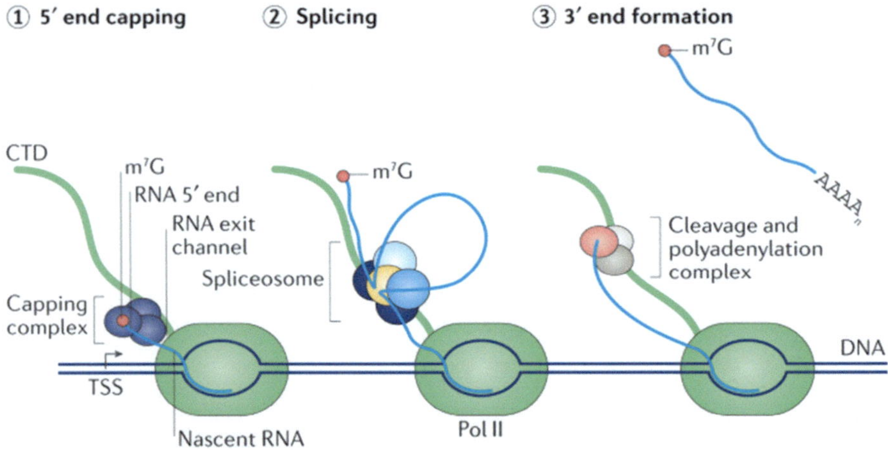

Fig. 10.18 Overview of mRNA processing (Desterro et al. 2020)

it from enzymatic degradation by RNAse and also helps in the identification of mRNA by the eukaryotic factors and initiates translation by ribosomes.

It is already known that a protein called m7G cap-binding protein eIF4E binds to the 5′ cap of mRNA and recruits the 40S ribosome subunit to the 5′ end of the mRNA. The ϒ phosphate is removed from the first nucleotide by the enzyme 5′-triphosphatase. The enzyme guanylyl transferase facilitates the attachment of guanosine nucleotide to the first nucleotide of the pre-mRNA. The beta phosphate of the RNA transcript displaces the pyrophosphate group at the 5′ position of the GTP molecule. The cap formation involves a 5′-5′ linkage between the two substrates.

To the terminal end of the RNA, and in the opposite direction, a G residue is added. 5′ diphosphate RNA acts as a substrate for the specific enzyme guanylyltransferase. However, GMP transfer is not catalyzed to give rise to a 5′ monophosphate RNA. Only 5′ end of the pre-mRNAs contain caps, whereas processed 5′ ends followed by endolytic mRNA cleavage do not contain caps.

In the terminal guanine, seven positions are methylated. This event is the first methylation event. The methyl groups are obtained from *S*-adenosylmethionine in the presence of the enzyme methyltransferase. A cap known as cap 0 carries a single methyl group. The enzyme usually adds methyl groups to 2′ hydroxyl groups on the ribose sugar of the next two nucleotides in the mRNA adjacent to the cap. The addition of methyl group to 2′OH of ribose of the first nucleotide is known as cap-1. The addition of methyl group to 2′OH of the ribose of the second nucleotide and third nucleotide is called cap-2, cap-3, and so on. The decapping enzyme removes the cap and the cap-binding complex, and the mRNA is subjected to degradation after translation (Fig. 10.19).

10.7.1.2 3′ Polyadenylation

Polyadenylation is a post-transcriptional mechanism in which the addition of poly (A) tail to the messenger RNA at the 3′ end takes place. The poly(A) tail is around 100–250 residues long. The mechanism takes place by endonucleolytic RNA cleavage coupled with the synthesis of polyadenosine monophosphate on the newly formed 3′ end also known as the polyadenylation site. A poly(A) tail is added to the 3′ UTR of newly synthesized pre-mRNAs by the enzyme poly(A) polymerase, which is in turn followed by the recognition of the poly(A) signal and endonucleolytic cleavage of the pre-mRNA at the poly(A) site. Polyadenylation increases the efficacy of mRNA by protecting the 3′ downstream sequences against several nucleases and also plays important roles in mRNA export to the cytosol, its localization, stability, as well as translation. A set of proteins cleave the 3′ segment of the newly synthesized pre-mRNA and then the poly(A) tail. Another important function of the poly(A) tail is to recruit RNases that cleave the RNA. Almost all eukaryotic mRNAs except animal replication-dependent histone mRNAs are polyadenylated. Other important functions of the poly(A) tail include the export of mature mRNA from the nucleus to the cytoplasm, increasing the stability of mRNA and offering protection from cleavage, and signal recognition for the binding of translational factors (Fig. 10.20).

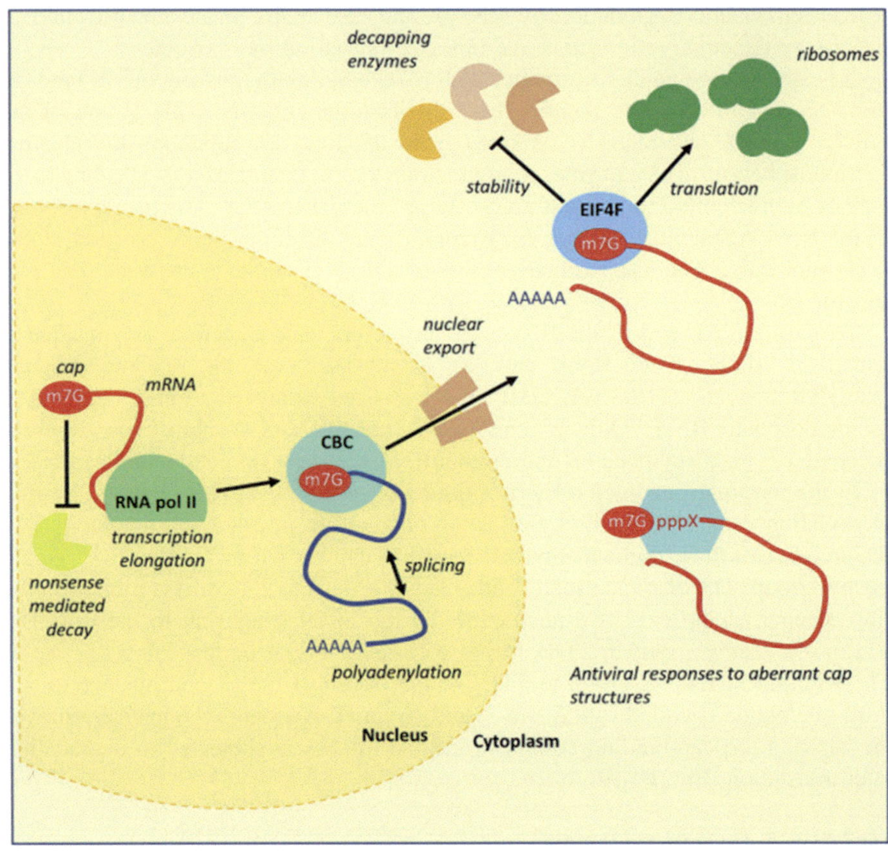

Fig. 10.19 The mRNA cap is a methylated modification of the 5′ terminus of mRNA. RNA processing and translation factors are recruited to the mRNA cap. The mRNA cap protects transcripts from degradation and defines mRNA as "self." The formation of the mRNA cap is regulated by cellular signaling pathways. mRNA cap regulation results in changes in gene expression and cell function (Galloway and Cowling 2019)

A multiprotein complex present in the nucleus of eukaryotes primarily targets precursor mRNA. This multi-protein complex excises the 3′ end and adds polyadenyl groups to the cleaved end. The enzyme CPSF (cleavage/polyadenylation specificity factor) specifically binds to the recognition site with the following sequence—5′AAUAAA3′ also known as the polyadenylation signal (PAS) is recognized by the RNA cleavage complex.

When RNA polymerase II recognizes the termination sequence 5′ TTTATT 3′ on the DNA template, transcription is terminated. The polyadenylation machinery is also linked to the spliceosomes.

Alternative polyadenylation (APA) is yet another mechanism of RNA processing which produces distinct 3′ ends on mRNAs. APA is also a gene regulation mechanism in eukaryotes. It is tissue-specific and is extensively studied to understand

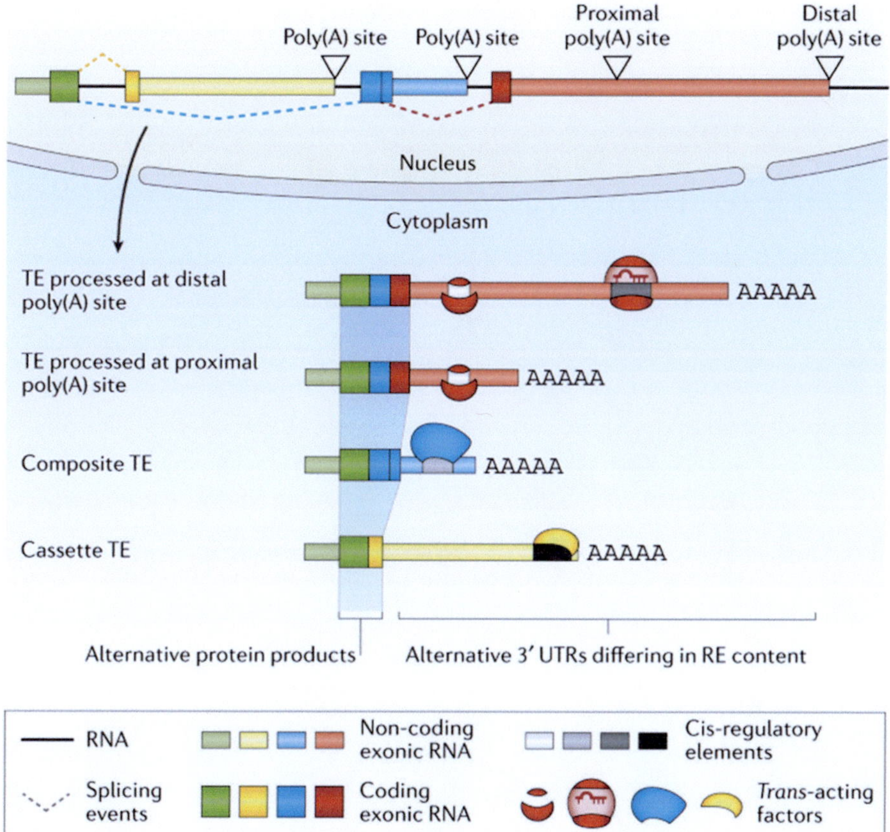

Fig. 10.20 The process of alternative polyadenylation (Gruber and Zavolan 2019)

proliferation and differentiation in cells. Alternative polyadenylation is sometimes used to reduce the length of the coding region which can lead to the expression of different proteins.

10.7.1.3 RNA Splicing

RNA splicing is an important post-transcriptional process where the non-coding intron sequences are removed from the transcript and the exons are subjected to processing and rejoining. The splicing complex is similar to restriction enzymes that recognize specific sites within the RNA and cleave and ligate the RNA at the cleaved sites. Splicing of pre-mRNA takes place in the nucleus before export.

There has been a significant amount of progress in computational analyses and sequencing methods, and they have led to the discovery of novel splicing isoforms and non-canonical splicing mechanisms. One such example is co-transcriptional splicing. This allows for the epigenetic and epitranscriptomic fine-tuning of gene expression. Studies have revealed that intrinsically disordered domains of RNA Pol

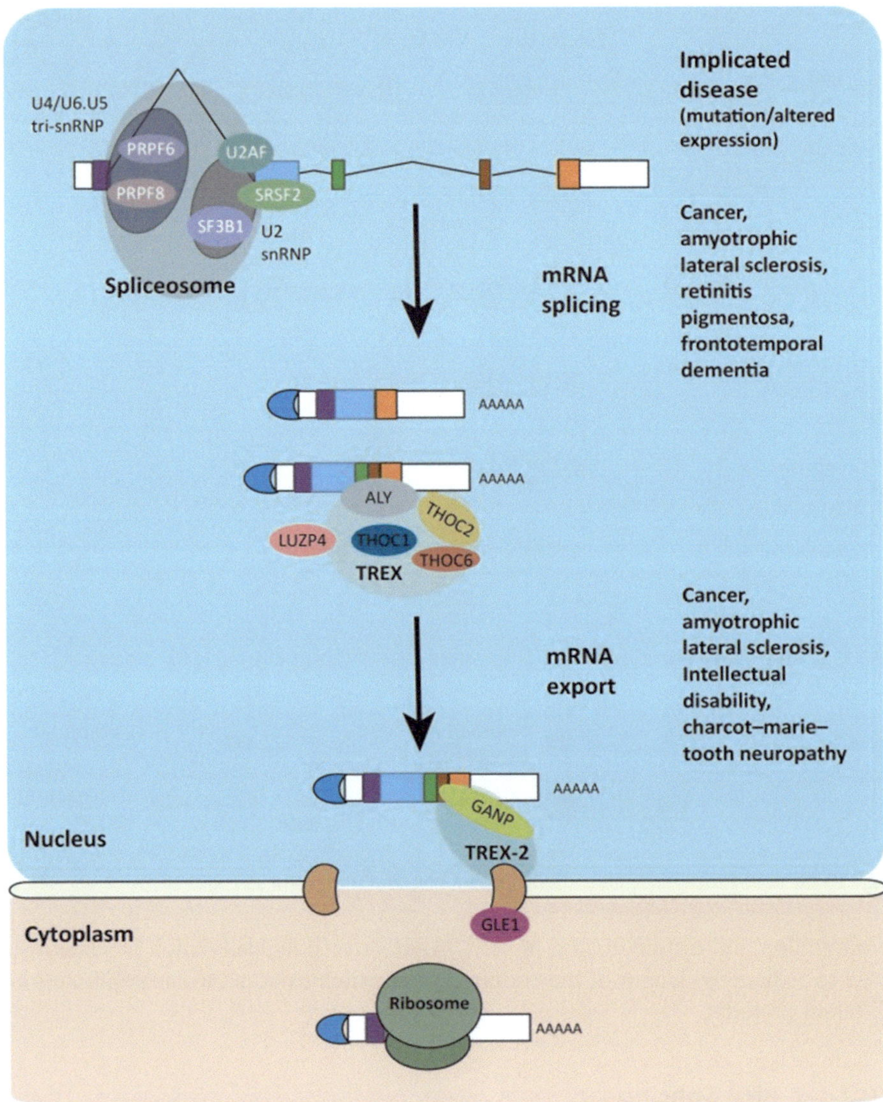

Fig. 10.21 Emerging evidence highlights that the RNA splicing and export machinery can display regulatory potential. Core spliceosome components can display regulatory potential if their levels become limiting for the function of complexes. These findings have important implications for the contribution of selective mRNA processing and export to the development of human cancers and neurodegenerative disorders (Carey and Wickramasinghe 2018)

II form local condensates, and several splicing factors are known to optimize splicing reactions (Fig. 10.21).

Exon junction complexes facilitate recursive splicing and also inhibit cryptic splice sites. Circular RNA splicing efficiency is enhanced by the low-efficiency splicing of the flanking introns. Pre-mRNA splicing is crucial in eukaryotic gene expression. Identification of exact splice sites and the accurate removal of introns are also essential for the generation of mRNA and its isoforms. Splicing regulation is mostly well understood. Emerging studies have also revealed that certain non-canonical splicing mechanisms exist. These are important in the regulation of gene expression.

10.7.1.4 Alternative Splicing

Alternative splicing is a process that allows a messenger RNA (mRNA) to express different forms of proteins (Fig. 10.22). Alternate splicing occurs by the reorganization of the intron and exon sequences in various combinations. This alters the coding sequence of the mRNA. Alternative splicing of the precursor mRNA increases the complexity of gene expression and plays an important role in cell differentiation and growth. Alternative processing is tightly regulated, and this regulation is supervised by the regulatory elements linking both transcription and splicing.

10.7.1.5 Sequestration as RNP

After nuclear pre-mRNA undergoes complete processing, it is recognized due to the absence of association with splicing factors, and the RNA associated with the functional spliceosome complex is retained in the nucleus. As splicing begins from the cap site and moves towards the polyadenylation site, the hnRNP A1 protein factor binds to the single-stranded RNA molecules that are already exposed. The

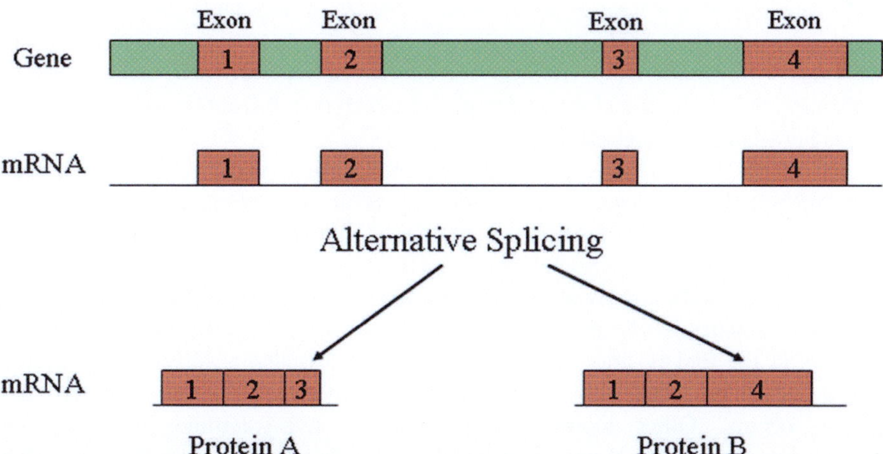

Fig. 10.22 A gene that contains numerous exons and introns can be spliced together in various ways. For example, in a gene containing eight exons, the mRNA transcribed from that gene can contain exons 1–7

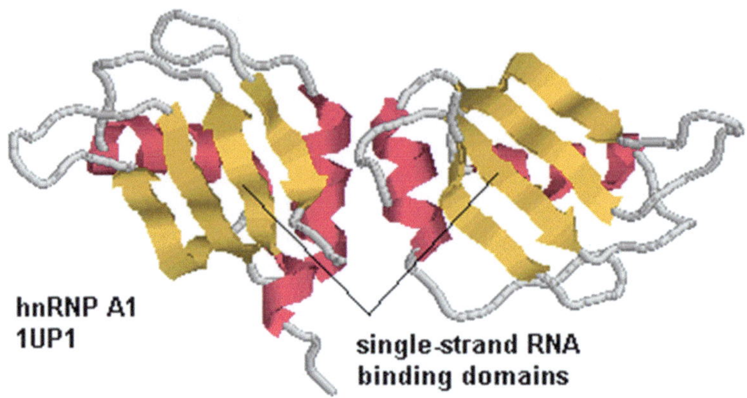

Fig. 10.23 Crystal structure of hnRNP A1

final processed and fully mature pre-RNA molecules are devoid of any bound splicing factors. Some of the sequences on the proteins are markers for nuclear export signals (NES) and nuclear localization signals (NLS). hnRNP A1 protein also acts as a carrier molecule for mature pre-mRNA (Fig. 10.23).

10.7.2 Processing of tRNA

Transfer RNA or the tRNA is the primary molecule that facilitates the process of translation. It consists of a single RNA strand made up of 75–95 nucleotides. tRNA is the smallest of the three types of RNA. The 20 amino acids that make up the primary peptide chain all have a specific tRNA that binds to it and transfers it to the growing polypeptide chain during translation. tRNAs are also called adapter molecules. tRNAs have a cloverleaf structure which is stabilized by strong hydrogen bonds between the nucleotides.

All the tRNA molecules have a 3′ end with a conserved 5′-CCA-3′ sequence. Some tRNAs have unusual and modified bases in their primary structure. These unusual bases are mostly a result of post-transcriptional enzymatic modifications of the normal bases in the polynucleotide chain. Two common modifications include pseudouridine (ψU), a derivative of uridine, in which uridine is modified such that the uracil attaches to the ribose to the carbon at the fifth position instead of the nitrogen in the first position, and dihydrouridine (D), also a derivative of uridine where enzymatic reduction of the double bonds between the fifth and the sixth carbon occurs. Other modified bases include hypoxanthine, thymine, and methylguanine. Studies have suggested that cells that do not have these modified bases have shown retarded growth leading to the conclusion that the modified bases have a role in enhanced and better tRNA function.

Parts of tRNA	Function
Acceptor stem	The incoming amino acids are attached to the acceptor stem by the paring of the 5′ and 3′ ends of the tRNA molecule. 5′-CCA-3′ end is present which is a single-stranded region protruding from the double-stranded stem
ψU loop (T loop)	It is named as T loop due to the presence of the modified base pseudouracil in this region or loop. The sequence 5′-TψUCG-3′ contains the modified base
D loop	It is named as D loop because of the presence of the modified base dihydrouridine
Anticodon loop	It is named as anticodon loop because it contains the anticodon which recognizes the codon in the mRNA and the corresponding specific amino acid is attached to the tRNA. The anticodon loop always has uracil in its 5′ end
Variable loop	It is present between the anticodon loop and the T loop, also called the pseudouracil loop, and it is so called because of its variation in the base stretch between 3 and 21 bases

10.7.2.1 Secondary and Three-Dimensional Structure of tRNA

There are regions of complementarity within an RNA molecule that enables RAN to form small stretches of double helical patterns which are subsequently stabilized by base pairing. tRNA molecules have a unique conserved pattern showing both single- and double-stranded regions which are also commonly referred to as the cloverleaf model. The cloverleaf model includes significant structures such as the acceptor stem, the ψU loop, the D loop, and the anticodon loop and a fourth variable loop (Fig. 10.24).

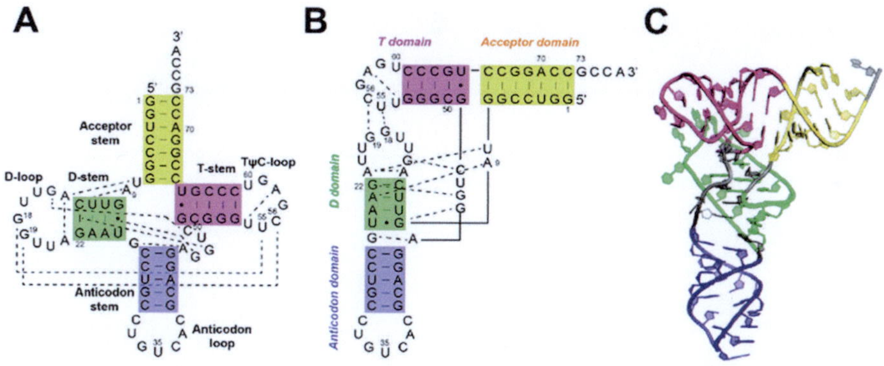

Fig. 10.24 (**a**, **b**) Secondary and tertiary structures of tRNA. (**c**) Crystal structure of tRNA (Liu et al. 2015)

X-ray crystallography studies revealed the tertiary structure which takes the shape of the letter L. This structure enabled us to better understand that the orientation of the acceptor stem and anticodon loop and that they are at opposite ends of the adaptor molecule. The acceptor stem and the pseudouracil loop form an extended continuous helix. The anticodon stem associates with the D loop stem to form an extended second helix. The two helices are perpendicular to each other bringing the D loop and the ψU loop together. Interactions such as base stacking, hydrogen bond formation between the bases, and the interaction between bases and the sugar-phosphate backbone stabilize the three-dimensional L-shaped structure and the final confirmation of the tRNA molecule.

10.7.3 Processing of rRNA

rRNAs account for around 80% of the total RNA present in cells, and they are the main components of ribosomes. Ribosomes are made up of two subunits, a large subunit (the 50S) and a small subunit (30S). Each subunit is made up of specific rRNA molecules. The rRNAs along with proteins and enzymes combine to form ribosomes, which are sites of protein synthesis. The small and large rRNAs contain around 1500 and 3000 nucleotides in prokaryotes such as bacteria and 1800 and 5000 nucleotides in eukaryotes such as humans. The 16S rRNA is the only rRNA in the small subunit of the ribosome and is also called the small subunit rRNA or ss-rRNA. The 5S and 23S are both components of the large subunit of the ribosome. Ribosomes are denoted by the sedimentation unit "S." In eukaryotes and archaea, four rRNAs are present: 18S in the small subunit and 5S, 5.8S, and 28S in the large subunit. Mitochondria contain 12S and 16S rRNAs. The processing of rRNA is depicted in Fig. 10.25.

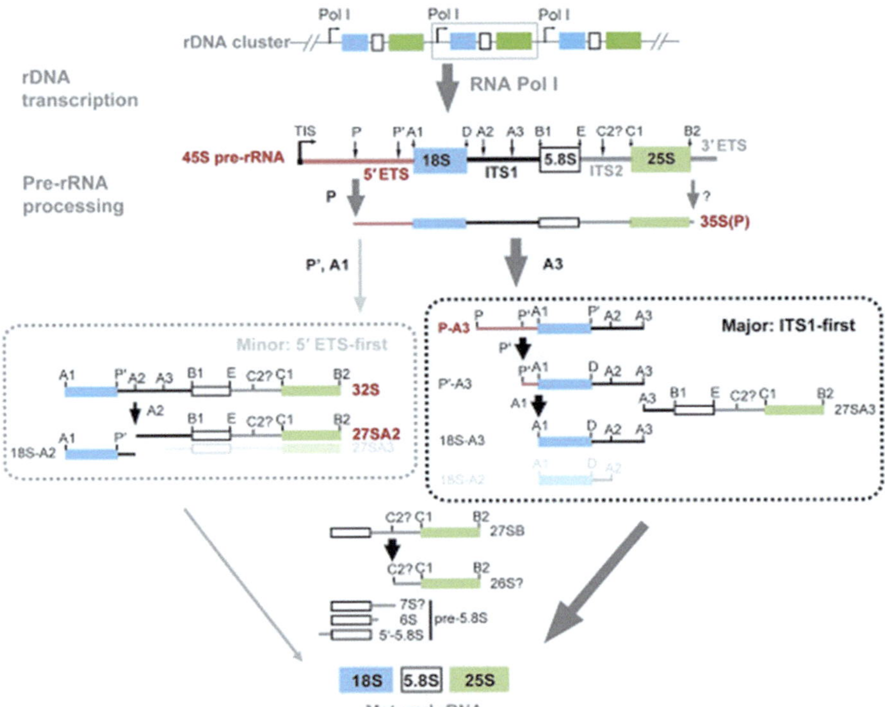

Fig. 10.25 The processing of rRNA

10.8 RNA Editing: Mechanism

RNA editing involves series of molecular processes where the RNA sequence is altered to allow the mature RNA to show variance from the RNA that is encoded by the genomic DNA. Editing includes processes like deletion, insertion, and substitution of the nucleotides. The variation observed in the messenger RNA (mRNA), ribosomal RNA (rRNA), transfer RNA (tRNA), and microRNAs (miRNA) can be attributed to RNA editing. The process of RNA editing occurs in the time interval between the transcription of DNA into mRNA and the translation of this mRNA to protein.

With the discovery of RNA editing, more light is being shed on novel post-transcriptional modifications. RNA editing is facilitated by adenosine and cytidine deaminases acting on DNA and RNA (Fig. 10.26). Adenosine to inosine (A-to-I) editors are members of the ADAR and ADAT protein families. They are important molecules that are crucial in the regulation of alternative splicing and transcription. Other kinds of editors such as cytidine to uridine (C-to-U) editors are members of the

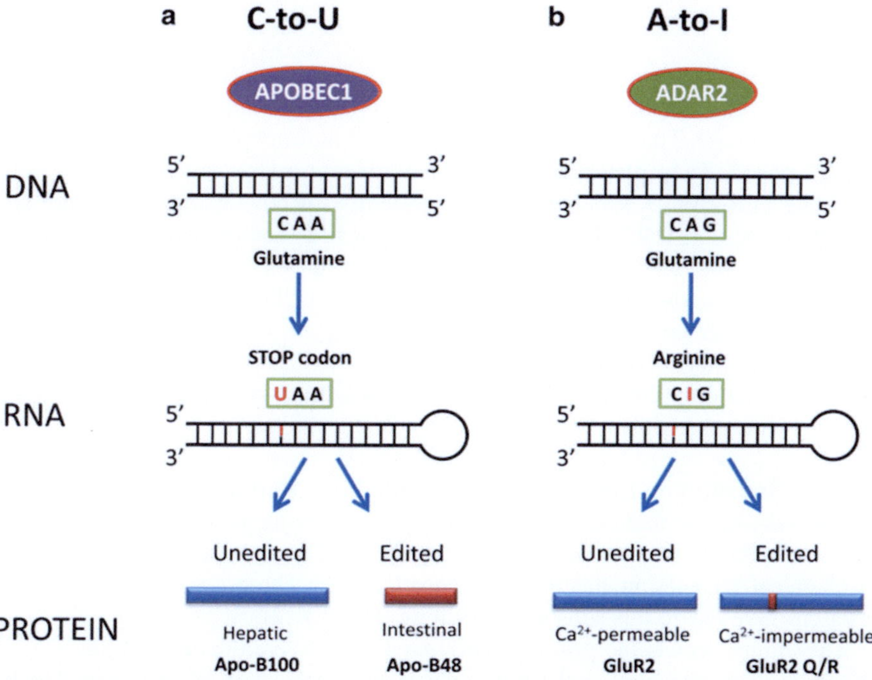

Fig. 10.26 RNA editors such as cytidine and adenosine deaminases are functionally important in regulating cellular processes. (**a**) Apolipoprotein B is produced in the gut which is mediated by APOBEC1 editing. Glutamate is transformed to a stop codon by C-to-U editing at residue 2153 of hepatic Apo-B100, and a truncated protein Apo-B48 is produced in intestinal cells. (**b**) The glutamate receptor 2 (GluR2) mRNA at position 607 is edited by ADAR2 in neurons, resulting in change of adenosine to inosine (Christofi and Zaravinos 2019)

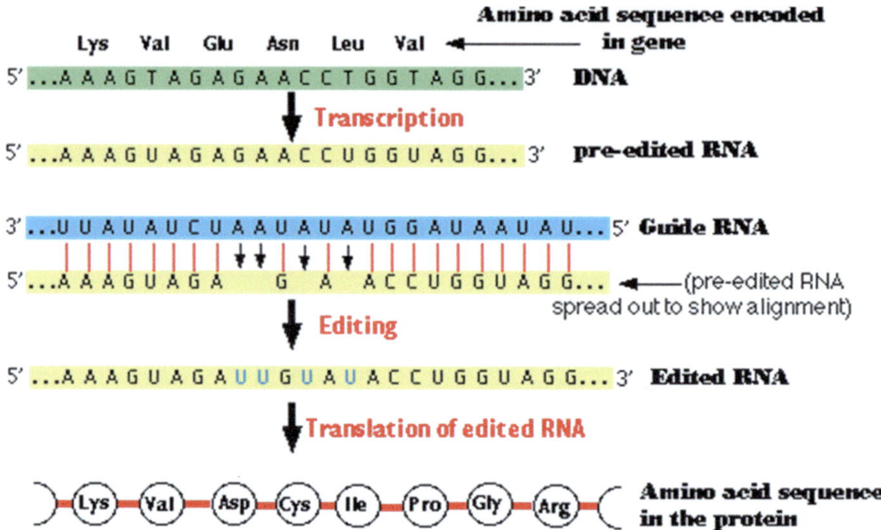

Fig. 10.27 RNA editing: overview

AID/APOBEC family and are key players that mediate innate and adaptive immunity and are also responsible for antibody diversification, antibody generation, and antiviral response. These editors are enzymes, and they are present in the nucleus or the cytoplasm. They play a role in the modification of several RNA molecules, including miRNAs, tRNAs, and most importantly mRNAs. Some editors are also capable of editing DNA. Latest technologies such as next-generation sequencing (NGS) have provided us with a large amount of data regarding these post-transcriptional modifications. RNA editing is often implicated in disorders such as cancer and other neurological diseases concerning the brain and the CNS. RNA editing is directly affected by cancer heterogeneity, carcinogenesis, response to treatment, and drug efficacy. Research on RNA editing will lead to the discovery of novel biomarkers identification and diagnostic techniques.

What Is Substitution Editing or Site-Specific Base Modification Editing?
A base substitution that causes a significant change in the coding properties or the structure of RNAs is known as substitution editing. These alterations often arise due to chemical changes in the individual nucleotides. Three common deamination reactions give rise to this substitution editing (Fig. 10.27).

This figure depicts the process of RNA editing, where the original guide RNA undergoes post-transcriptional editing which upon translation gives rise to proteins with different functional properties.

10.8.1 A-to-I Editing

Conversion of A to I: Adenosine deaminases convert an A to inosine (I), which is translated in the form of G by the ribosomes. The abundantly seen type of RNA editing system is A-to-I editing by double-stranded RNA-specific adenosine deaminase (ADAR) enzymes. Data gathered from transcriptomic studies has revealed several "recoding" sites supposedly at which A-to-I editing results in substitutions of bases in protein-coding sequences (Fig. 10.28). The recoding sites are also

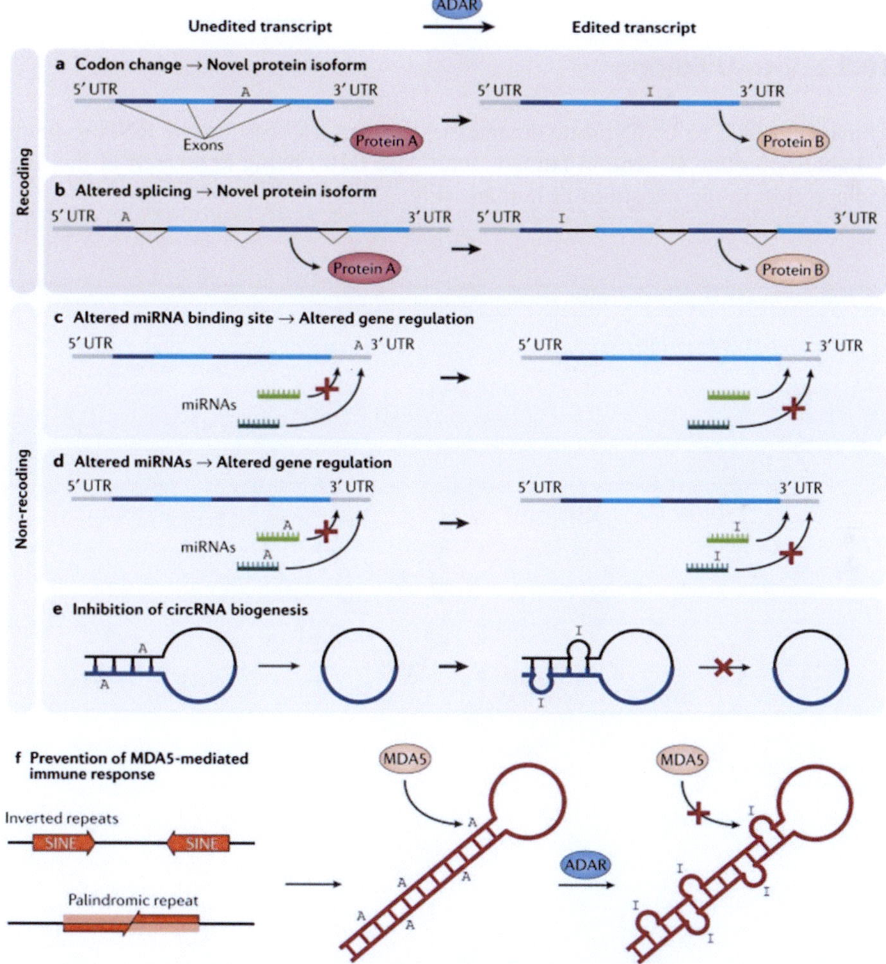

Fig. 10.28 A-to-I editing by double-stranded RNA-specific adenosine deaminase (ADAR) enzymes (Eisenberg and Levanon 2018). The deamination of adenosine and formation of inosine, leading to unstable double-stranded RNA base pairing. This alteration leads to reduced production of siRNA, by the dsRNA, and this disrupts the RNAi pathways

conserved within lineages and are subjected to positive selection, and they have functional and evolutionary importance. Mapping studies of the editosome complex in various species of the animal kingdom has suggested that most A-to-I editing sites are present within mobile genetic elements in the non-coding parts of the genome and evidence points to the fact that editing of these non-coding sites might have a critical role to play in protection against innate immunity activation by the self-transcripts. Recoding, as well as non-coding events, has been implicated in genome evolution and their deregulation, which could lead to diseased conditions. ADARs are being extensively studied and being adapted for RNA engineering.

10.8.2 C-to-U Editing

Conversion of C to U: Cytidine deaminases convert a C base in the RNA to uracil (U) apolipoprotein B gene in humans with Apo B100 being expressed in the liver and apo B48 in the intestines of humans (Fig. 10.29).

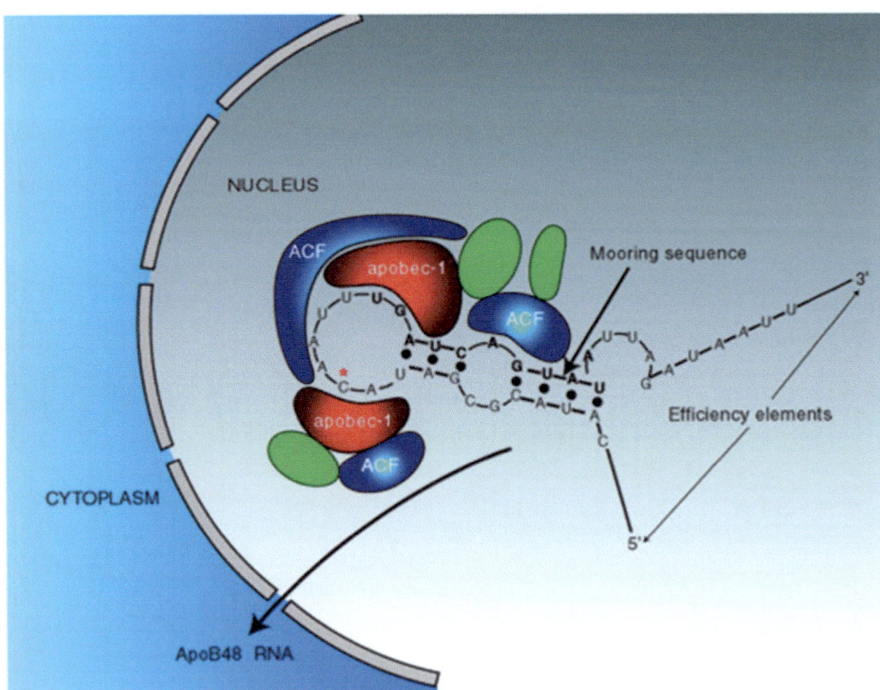

Fig. 10.29 C-to-U RNA editing of apolipoprotein B. In this figure a ~35-nucleotide region of apoB RNA flanking the edited base is shown also highlighting apobec-1 and ACF binding to RNA both 5′ and 3′ of the edited base and depicts the presence of additional proteins that may modulate assembly of the holoenzyme

RNA editing by cytidine deamination, the extent to which editing takes place, its regulation, and enzymatic and molecular basis have not been properly established. Hundreds of gene transcripts are known to undergo site-specific C-to-U RNA editing in macrophages and monocytes during M1 polarization and in response to hypoxia and interferons, respectively. This allows for the altering of the amino acid sequences of proteins, especially those that are involved in the viral disease pathogenesis. In single-stranded DNA, cytidines are deaminated by APOBEC3A and also inhibit retrotransposons and viruses. Amino acid residues of APOBEC3A involved in anti-retrotransposition and DNA deamination were also found to affect its RNA deamination activity. In plants, C-to-U editing is seen in the mitochondrial RNA of flowering plants.

10.8.3 Editing in Mitochondria

10.8.3.1 Types of RNA Editing in Mitochondria

Protein-coding transcripts (mRNA) are always edited, but introns, rRNA, and tRNA species are also edited in a few cases. In a mitochondrial system, editing restructures many different mRNA transcripts and generates a translatable reading frame. This structured re-tailoring is usually seen in insertion/deletion editing. Editing also gives rise to translation initiation or termination codons. RNA editing affects internal codons; however, substitution editing specifically occurs at first or second positions of codons. Editing often changes the protein properties and causes the protein to significantly deviate from the predicted gene sequence.

In mitochondria, RNA editing systems are known to be mechanistically distinct, and several RNA editing systems are not related mechanistically (Fig. 10.30). The molecular events that occur during editing are phosphodiester bond cleavage and re-ligation. In some cases, editing happens by direct base conversion methods such as deamination. These unique characteristics are discussed along with the concept of the evolution of RNA editing processes. Mitochondrial editing systems are also said to be phylogenetically isolated.

The phylogenetic distribution of most mitochondrial RNA editing systems is restricted, and evidence suggests that these intrinsic mechanisms in the editing systems have evolved from discrete eukaryotic lineages during evolution. Considering U insertion/deletion editing, it has been observed only in kinetoplastid protozoan (Kinetoplastida), in the early-diverging and free-living bodonids (Bodonida), and the late-diverging and parasitic trypanosomatids (Trypanosomatina). This type of editing has not been observed within diplonemids or euglenids. Mitochondrial mRNA editing that occurs in dinoflagellate mitochondria has not been observed in the apicomplexans or ciliates. A different type of editing has been observed in plant organelles (C-to-U and U-to-C) which are not found in any of the green algal mitochondrial systems.

In this way, the mitochondrial editing systems are narrowly restricted and deflect suspicions away from the hypothesis that the editing systems were present in a common eukaryotic ancestor.

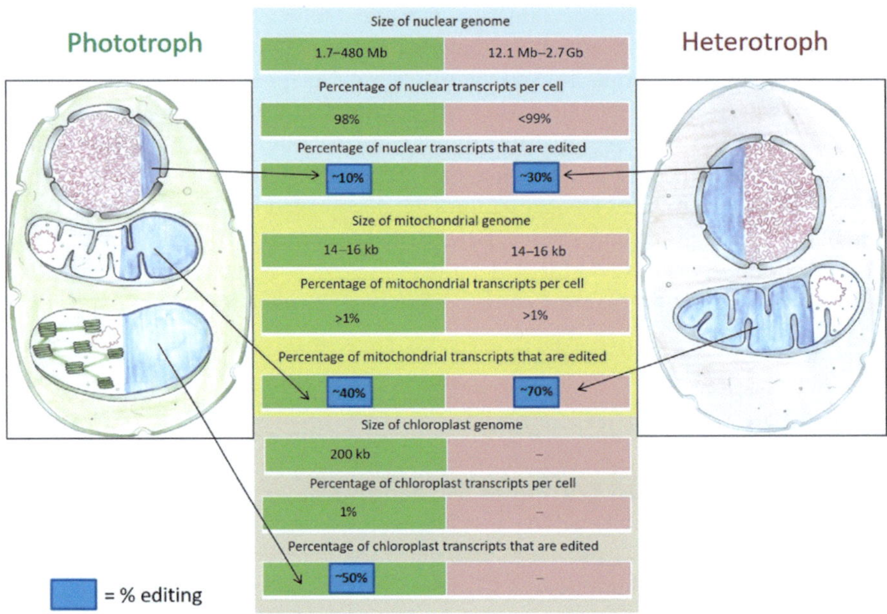

Fig. 10.30 Varying levels of RNA editing is depicted in this schematic illustration. Heterotrophs are known to display higher levels of RNA editing

10.8.4 Editing in Plastid

Mitochondria and plastids in terrestrial plants show evidence for post-transcriptional editing from C to U as well as U to C in several transcript sequences. In plastids such as chloroplasts, around 6–20 C bases were found to be deaminated to U bases. RNA analysis studies in mitochondria reveal that in some plants, around 500–1000 U to C conversations is common. Two important factors—(a) PLS-type pentatricopeptide repeat (PPR) protein and (b) multiple organellar RNA editing factors (MORFs, also known as RNA-editing factor interacting protein (RIP))—are trans-acting factors that are involved in this process (Fig. 10.31). MORF9 binding induces significant compressed conformational changes of $(PLS)_3$PPR, revealing the molecular mechanisms by which MORF9-bound $(PLS)_3$PPR has increased RNA-binding activity. Similarly, increased RNA-binding activity is observed for the natural PLS-type PPR protein, LPA66, in the presence of MORF9.

Most plant species do not support U-to-C conversions, and these edits are rare events, specifically two to three edits in the entire mitochondrial RNA sequence. Sites for RNA editing are usually found in the coding regions of mRNAs and rather less in the introns and non-translated regions. It has been observed that in certain situations, RNA editing in tRNA molecules corrects errors and restores normal base pairings. Here, RNA editing becomes crucial because only proper editing can ensure further maturation, accurate folding, and processing of the precursors of tRNA.

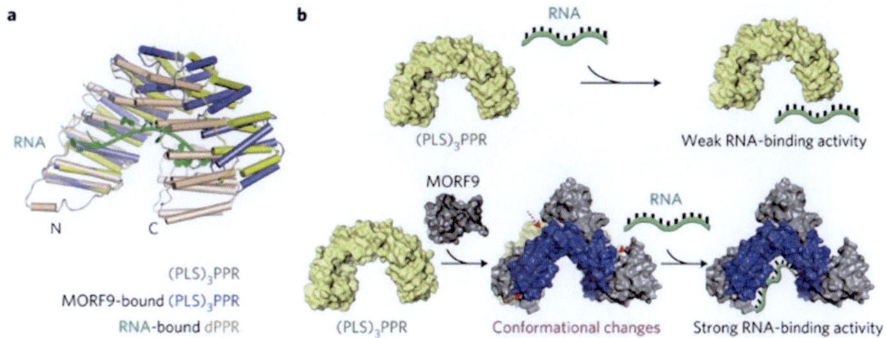

Fig. 10.31 Yan, J., Zhang, Q., Guan, Z. et al. characterize the interactions between a designer PLS-type PPR protein (PLS)$_3$PPR and MORF9 and strongly suggest that RNA-binding activity of (PLS)$_3$PPR is drastically increased on MORF9 binding. Crystal structures of (PLS)$_3$PPR, MORF9, and the (PLS)$_3$PPR-MORF9 complex are shown in the figure

Figure 10.32 depicts the molecular interactions that influence both RNA editing and chloroplast signaling, essentially suggesting that RNA editing is crucial for normal functioning. In plant organelles, rRNAs are subjected to minimal RNA editing. The mechanisms underlying the recognition of these editing sites, the enzymatic action, and the molecular pathways involved are yet to be determined.

In plants, other nucleotide insertions or edits have not been observed. One hypothesis suggests that one reason for RNA editing is that it exists to trigger the activity of a particular RNA-specific C deaminase. These deaminases, however, do not catalyze reverse U-to-C reactions.

Research points to the idea that RNA sequences are involved in guiding the "editosome" editing complexes to specific sites. Cis- or trans-acting RNA molecules facilitate this guiding function but are not native to the sequence regions of the edit sites, but there are no common sequence motifs that have been identified around the different C-to-U conversion sites. In positions preceding the edited Cs, a low amount of G residues has been found. Downstream nucleotides in both mitochondria and plastids are not involved in the editing site specifications, whereas the upstream sequences play a significant role. However, in both the organelles, the upstream region differs in various editing sites, while some sites require only about 5–20 nucleotides and others require around 200 nucleotides. Sequence duplications are also seen in mitochondria. Here as if enough number of upstream sequences are present, RNA editing is accurately maintained. By experimenting in vivo with transgenic plastids, it was proved through upstream and downstream sequence insertions.

Identification of potential RNA editing intermediates suggests that RNA editing in plant organelles is a post-transcriptional process. Partially edited transcripts contain some C bases that have been deaminated to Us. The Cs encoded by the genome exist, in other potential editing sites. Partially edited RNA molecules do not follow a particular order of editing. This means that the hypothetical "editosome"

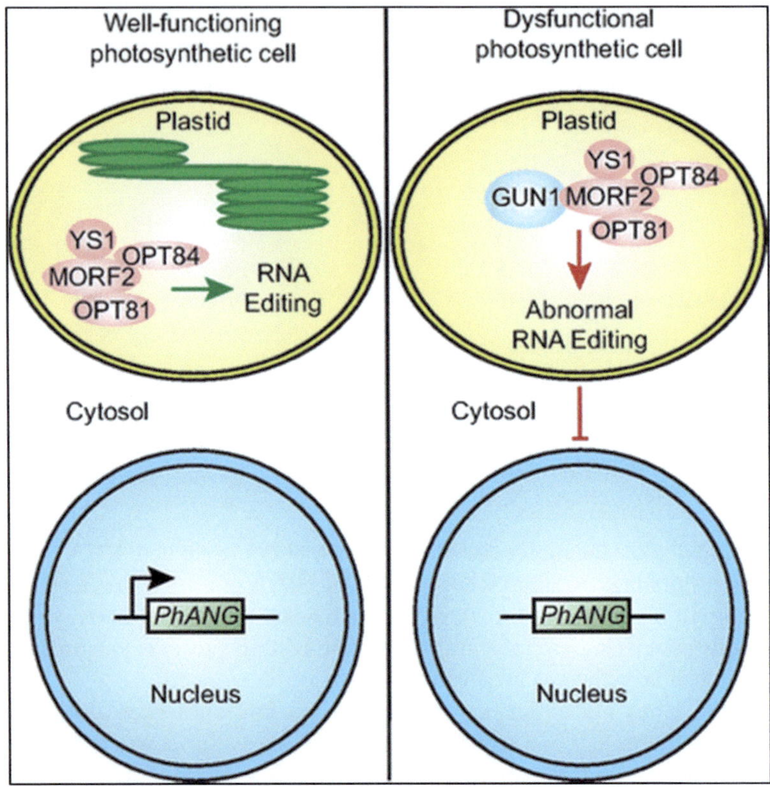

Fig. 10.32 Proposed interactions between RNA editing and chloroplast signaling. In well-functioning photosynthetic cells (left), chlorophyll accumulates in the thylakoid membranes (green), and chloroplasts perform photosynthesis. The GUN1 protein does not accumulate, and thus the chloroplast-to-nucleus signaling that depends on GUN1 is not active. MORF2 contributes to RNA editing (green arrow) and interacts with OPT81, OPT84, and YS1. Light signaling, tissue-specific signals, and the circadian rhythm (not shown) drive high-level expression of PhANGs (black arrow), which promotes chloroplast function (Vo et al. 2019)

complex does not linearly scan the RNA molecule and the selection of editing sites or regions is arbitrary. These partially edited mRNAs are found in minimal amounts in the plant mitochondria, and they are translated into a family of different proteins. But this assumption only holds for one type of protein sequence, and it is said to be present in the protein complexes of the respiratory chain. These sequences generally are the polypeptide sequences that are best conserved with their respective homologs in other organisms and are selected by their physiological and biochemical functionality. So, it is likely that polypeptides synthesized from unedited RNA molecules would not function properly and such proteins, for example, would hinder the efficiency of respiration in mitochondria.

10.8.5 Coediting in Virus

In RNA viruses, during transcription of mRNA, the transcription machinery incorporates additional nucleotides that are not specific to the viral genome. For example, in certain paramyxoviruses such as measles, Sendai, parainfluenza, and mumps, viruses around one to ten G residues are inserted at specific editing sites (Fig. 10.33). In Ebola viruses, it is observed that additional A residues are incorporated during transcription.

So, it is observed that co-transcriptional editing of RNA in RNA viruses happens by insertion of non-templated nucleotides by a mechanism known as "stuttering of the RNA-dependent RNA polymerase." Specific sequence motifs are present in the viral RNA that can induce the RNA polymerase complex to stutter and repeat the last transcribed nucleotide of the template RNA before resuming transcription.

The replication of Ebola viruses, paramyxoviruses, and other RNA viruses is facilitated by the transcription of RNA-dependent RNA polymerase that is virus

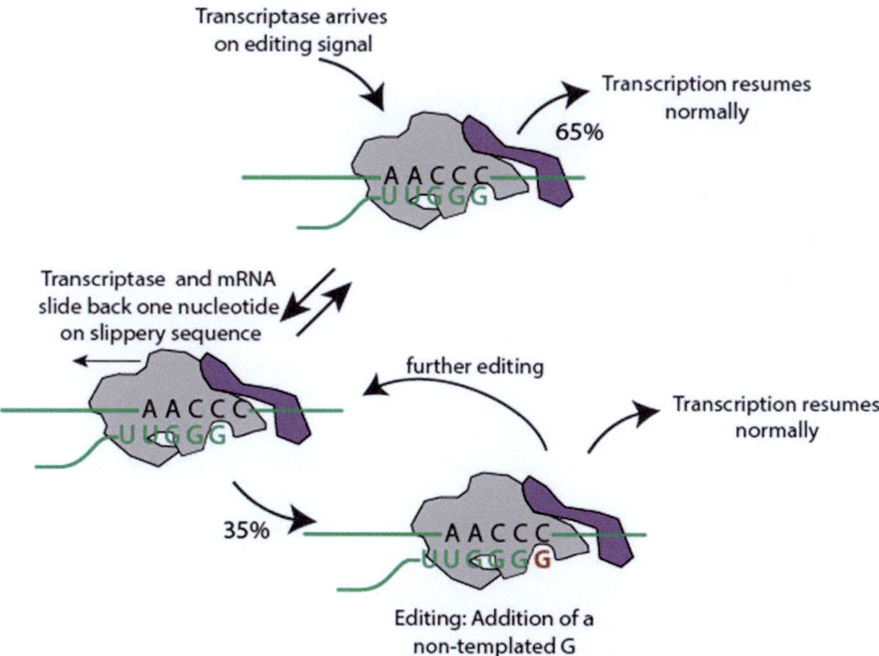

Fig. 10.33 Certain viruses like the Sendai viruses encode genes to express multiple proteins. They do this with the help of overlapping open reading frames (ORFs) by RNA editing. In viruses like these, the RNA polymerase is capable of reading the same template base more than once, creating insertions that subsequently lead to different mRNAs and generating different types of proteins

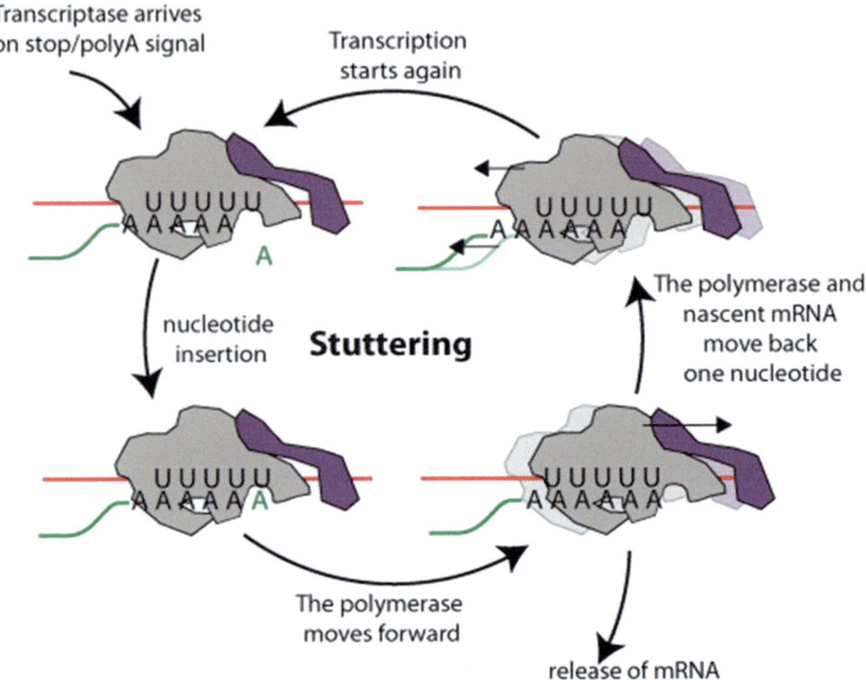

Fig. 10.34 Negative-strand RNA viruses belonging to paramyxoviridae and ebolaviridae are known to polyadenylate mRNA during transcription through a polymerase stuttering mechanism. The viral polymerase acquires a stuttering behavior upon encountering the stop signal present at the end of each gene comprising a stretch of U bases. After each adenine insertion, the RNA polymerase moves back one nucleotide with the mRNA, copying U hundreds of times at the end of viral mRNA thereby producing a poly(A) tail and releasing the polyAdenylated mRNA to stop transcription or scan to restart on the next gene

encoded resulting in antisense RNA. Stuttering and pausing mechanisms are mostly seen in these types of polymerases at certain nucleotide base combinations, mostly mono- or oligonucleotide tracts (Fig. 10.34). At the nascent mRNA, 3′ ends up to several hundred As are added by the same polymerase, although they are not templated. These As stabilize the mRNA by using a mechanism similar to polyadenylation. mRNA polymerase (complex) pauses at these positions, while the RNA replicase (complex) synthesizes the replication intermediate RNA from the virion RNA. The same RNA polymerase is influenced in a differential manner by additional cofactors, and replication takes place in the virion, while transcription usually occurs in the cytoplasm.

RNA-dependent RNA polymerase encoded by viruses also pauses and incorporates non-templated nucleotides by the same "stuttering" mechanism around genomic stop codon of the first open reading frame in the unedited mRNA. The reading frame is shifted by insertion of one to two Gs or A bases upstream of the translational stop codon. Upon translation this results in the generation of different proteins with different carboxy-terminal sequences. Compared to the genomically predicted first open reading frame, the extra amino acid sequences in the mumps viruses are known to be double the size.

Transcription termination in *E.coli* and sequence-specific RNA polymerase pausing and "stuttering" are comparable. Unstable transcription complexes are also induced similarly by sequences at the template sites. Pyrimidine-rich sequences are present along with long U stretches preceding Cs can cause the viral RNA polymerase to slow down, slip back one nucleotide on the template, and incorporate another G nucleotide opposite the same C again. Due to stuttering progression, a sequence of three Gs is altered as initially predicted by the genomic RNA to four Gs in the edited mRNA. RNA viruses facilitate dissociation and realignment of this polymerase-product complex to the previously transcribed nucleotide on the RNA template which can be explained by the specific rate constants of dissociation and RNA-binding protein.

Box 10.1 Scientific Concept: High-Throughput Detection of RNA Processing in Bacteria: Erin E. Gill et al.

A γ-proteobacterium *Pseudomonas aeruginosa* is a causative agent of opportunistic infections in hospitalized immunocompromised patients and chronic lung infections in patients suffering from cystic fibrosis. *P. aeruginosa* has been extensively studied because it has devastatingly contributed to human morbidity and mortality. *P. aeruginosa* is of considerable medical importance due to its metabolic diversity, motility, quorum sensing, ability to produce biofilm, adaptive responses to evade antibiotic stress, and extreme virulence. The main *P. aeruginosa* strain that is implicated in opportunistic infections is "PAO1." However, accurate information regarding the transcription start site (TSS) is unavailable to date. The molecular pathways underlying post-transcriptional modifications of RNA transcripts remain elusive in *Pseudomonas* as well as other organisms. E. Gill et al. suggest that understanding the biochemical makeup of *P. aeruginosa*, and obtaining the detailed map of TSS and subsequent RNA processing of transcripts and influences virulence, antimicrobial resistance, and other essential cellular functions, is crucial to figuring out the mystery behind the regulation of pathogenesis and drug resistance and the identification of novel drug targets.

Maintaining an inventory of RNA processing sites Transcription Start Sites is necessary to understand cellular processes. RNA sequence-based analysis helps map the set of post-transcriptional modifications occurring in the transcriptomes of organisms. This is a crucial and challenging objective. The completion of transcription results in the occurrence of a series of tightly regulated secondary modifications ultimately leading to the maturation of the RNA transcript. These processes are fundamental to the functionality of many RNAs and also strongly influence the overall behavior of the RNA molecule.

(continued)

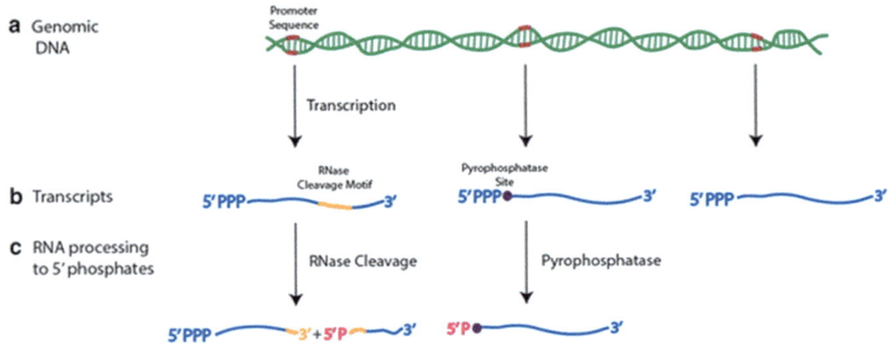

Fig. 10.35 RNA transcription and processing. (**Erin E. Gill et al.**) (**a**) Initiation of RNA transcription from a promoter sequence (indicated in red) within the genome and subsequent polymerization of ribonucleoside triphosphate resulting in a 5′ triphosphate at the 5′ end of the nascent mRNA transcript and a 3′ hydroxyl at its 3′ terminus. (**b**) mRNAs undergoing cleaving by endonucleases to giving rise to two fragments of RNA or can undergoing degradation by exonucleases from 5′ or 3′ termini. (**c**) RNA processing events that result in either a 5′ triphosphate (dRNA-Seq) or 5′ monophosphate (pRNA-Seq) and that simultaneously contain a terminal 3′ hydroxyl

Box 10.1 (continued)

A terminal 5′ triphosphate is present in the primary transcript contains. The conserved pyrophosphatase RppH (YgdP in *P. aeruginosa*) selectively removes the 5′ triphosphate and leaves a 5′ monophosphate in bacteria. This 5′ monophosphate causes destabilization of the mRNAs by making them susceptible to degradation (Fig. 10.35). The multi-subunit degradosome is an important complex involved in this process. At its core, it contains the 5′ phosphate-sensitive exonuclease/endonuclease RNase E. 5′ monophosphate is known to significantly increase RNase E's endonuclease activity. Endonucleases that can cleave RNA and leave behind a 5′ phosphate can also result in the production of stable RNAs and activate RNA degradation via degradosome-mediated pathways required for cellular function.

10.9 Summary

- The process of producing proteins from nucleotides is termed as gene expression. Genetic information present in the DNA, is first rewritten, generating RNA, the process of which, is termed as transcription. During transcription, starting with a DNA template, RNA is essentially synthesized. Every single gene is transcribed into several copies of mRNA, and each mRNA molecule is capable of generating identical copies of a single protein.

- The first step in gene expression where the enzyme RNA polymerase converts a DNA segment into RNA is called transcription. DNA and RNA both make use of nucleotide base pairing as a complementary language.
- Many different types of RNAs are responsible for various functions. The important ones include mRNA, tRNA, and rRNA.
- Prokaryotic transcription has been extensively studied in bacteria such as *E. coli*. The transcription of RNA involves three steps: initiation, chain elongation, and termination. RNA polymerase locates the target DNA by recognizing the promoter region.
- The core enzyme is aided by the σ factor to locate the transcription binding site. This activity is mediated by specific nucleotide sequences on the DNA known as the promoter regions. Promoter recognition is crucial for the initiation of transcription.
- After the transcription cycle is set up, the elongation process has to be stabilized. The Pol II machinery equips additional factors to stop the premature dissociation of Pol II. These factors are known as elongation factors, and they associate with Pol II just after initiation.
- The terminator signal triggers cascades that cause the core enzyme to dissociate from the template, which releases the newly synthesized RNA transcript and re-associates with the σ factor so that it can start a new round of transcription.
- Bacterial RNA polymerase is made up of a core complex consisting of multiple subunits and an initiation factor called the sigma (σ) factor. The core complex has nonspecific polymerase enzyme activity and can bind to DNA and nicks in a nonspecific manner and is known as the core enzyme (E).
- At the molecular level, it is understood that the RNA polymerase works by creating phosphodiester bonds between the incoming ribonucleotide triphosphates and the growing chain of RNA. This is a thermodynamically feasible and irreversible reaction. The RNA polymerase adds around 10–100 bases every second.
- The process of transcription is essentially the same in both prokaryotes and eukaryotes. But more steps are involved in eukaryotic transcription. Bacteria and species belonging to archaea require only one type of RNA polymerase, whereas eukaryotes require at least three main enzymes—RNA polymerases I, II, and III (Pol I, II, III), along with polymerases IV and V which are present in plants that transcribe different subsets of RNA.
- Transcription factors are DNA-binding proteins that work by repressing or activating gene transcription. Preferential activity is displayed by some transcription factors. These bind to each other, cis-acting DNA sequences as well as to both DNA and other transcription factors. In order to promote repression or activation, specific promoters act as binding sites for these transcription factors.
- The degree of supercoiling influences and affects the efficiency of some of the promoters. Some promoters are also sensitive to the degree of supercoiling, and some are not; the reason for this lies in the fact that the sequence of some promoters is easier to melt.

- After transcription, the RNA is processed before it is exported to the cytoplasm for translation. Eukaryotic mRNA is not stable at the ends and is susceptible to damage thus requiring modification to protect it from ribonucleases. The pre-mRNA hence undergoes capping at the 5′ end immediately after transcription and is then released by Pol II.
- Polyadenylation is a post-transcriptional mechanism in which the addition of poly (A) tail to the messenger RNA at the 3′ end takes place. The poly(A) tail is around 100–250 residues long.
- RNA splicing is an important post-transcriptional process where the non-coding intron sequences are removed from the transcript and the exons are subjected to processing and rejoining. Alternative splicing is a process that allows a messenger RNA (mRNA) to express different forms of proteins.
- Transfer RNA or the tRNA is the primary molecule that facilitates the process of translation. It consists of a single RNA strand made up of 75–95 nucleotides. tRNA is the smallest of the three types of RNA.
- RNA editing involves series of molecular processes where the RNA sequence is altered to allow the mature RNA to show variance from the RNA that is encoded by the genomic DNA. Editing includes processes like deletion, insertion, and substitution of the nucleotides. Conversion of A to I: Adenosine deaminases convert an A to inosine (I), which is translated in the form of G by the ribosomes. Conversion of C to U: Cytidine deaminases convert a C base in the RNA to uracil (U).
- In a mitochondrial system, editing restructures many different mRNA transcripts and generates a translatable reading frame. This structured re-tailoring is usually seen in insertion/deletion editing. Mitochondria and plastids in terrestrial plants show evidence for post-transcriptional editing from C to U as well as U to C in several transcript sequences.
- In RNA viruses, during transcription of mRNA, the transcription machinery incorporates additional nucleotides that are not specific to the viral genome.

References

Carey KT, Wickramasinghe VO (2018) Regulatory potential of the RNA processing machinery: implications for human disease. Trends Genet 34(4):279–290. https://doi.org/10.1016/j.tig.2017.12.012. Elsevier Ltd

Christofi T, Zaravinos A (2019) RNA editing in the forefront of epitranscriptomics and human health. J Transl Med 17(1):319. https://doi.org/10.1186/s12967-019-2071-4. BioMed Central Ltd

Desterro J, Bak-Gordon P, Carmo-Fonseca M (2020) Targeting mRNA processing as an anticancer strategy. Nat Rev Drug Discov 19(2):112–129. https://doi.org/10.1038/s41573-019-0042-3. Epub 2019 Sep 25

Eisenberg E, Levanon EY (2018) A-to-I RNA editing—immune protector and transcriptome diversifier. Nat Rev Genet 19(8):473–490. https://doi.org/10.1038/s41576-018-0006-1. Nature Publishing Group

Galloway A, Cowling VH (2019) mRNA cap regulation in mammalian cell function and fate. Biochim Biophys Acta Gene Regul Mech 1862(3):270–279. https://doi.org/10.1016/j.bbagrm.2018.09.011. Elsevier B.V.

Gruber AJ, Zavolan M (2019) Alternative cleavage and polyadenylation in health and disease. Nat Rev Genet 20(10):599–614. https://doi.org/10.1038/s41576-019-0145-z. Nature Publishing Group

Hobert O (2008) Gene regulation by transcription factors and MicroRNAs. Science 319(5871): 1785–1786. https://doi.org/10.1126/science.1151651. American Association for the Advancement of Science

Kapanidis AN, Margeat E, Ho SO, Kortkhonjia E, Weiss S, Ebright RH (2006) Initial transcription by RNA polymerase proceeds through a DNA-scrunching mechanism. Science 314 (5802):1144–1147. https://doi.org/10.1126/science.1131399

Liu J, Osbourn A, Ma P (2015) MYB transcription factors as regulators of phenylpropanoid metabolism in plants. Mol Plant 8:689–708. https://doi.org/10.1016/j.molp.2015.03.012

Torres-Berrío A et al (2019) Unraveling the epigenetic landscape of depression: focus on early life stress. Dialogues Clin Neurosci 21(4):341–357. https://doi.org/10.31887/DCNS.2019.21.4/enestler. Les Laboratoires Seriver

Vo TV, Dhakshnamoorthy J, Larkin M, Zofall M, Thillainadesan G, Balachandran V, Holla S, Wheeler D, Grewal SIS (2019) CPF recruitment to non-canonical transcription termination sites triggers heterochromatin assembly and gene silencing. Cell Rep 28(1):267–281. e5. https://doi.org/10.1016/j.celrep.2019.05.107

Vos SM, Farnung L, Boehning M, Wigge C, Linden A, Urlaub H, Cramer P (2018) Structure of activated transcription complex Pol II-DSIF-PAF-SPT6. Nature 560(7720):607–612. https://doi.org/10.1038/s41586-018-0440-4. Epub 2018 Aug 22

Protein Translation 11

Tanushree Banerjee

As we have studied in the previous chapter, DNA is the carrier of genetic information. There are four bases, and the permutation and combination of these bases store this enormous information. The information stored in DNA is transferred to RNA, and this process is called as transcription. The messenger RNA after being transcribed moves out of the nucleus carrying the genetic information. Hence, it is called messenger RNA or mRNA. The information present in mRNA is then read and is translated into proteins. The process of converting the genetic information stored in mRNA into proteins is called translation. In this chapter we will learn about the genetic code and how it is being translated to proteins. We will also learn about the various regulatory mechanisms involved in translation.

11.1 Genetic Code

The information present in the RNA is in the form of ribonucleotide bases organized in a pattern. In this pattern the bases serve as "letters," and the combination of three bases serves as "word." Each three-letter words made of three bases code for an amino acid. This three base code is called as **triplet code** or **genetic code** or **codon**. There are certain features of the genetic code which are universal.

1. Code is unambiguous—Each codon specifies only a single amino acid. More than one amino acid cannot be coded by the same codon.
2. Code is degenerate—One amino acid can be specified by more than one codon. Out of 20 amino acids, 18 amino acids are coded by more than one codon.

T. Banerjee (✉)
Molecular Neuroscience Research Laboratory, Dr. D. Y. Patil Biotechnology and Bioinformatics Institute, Dr. D. Y. Patil Vidyapeeth, Pune, India
e-mail: tanushree.banerjee@dpu.edu.in

D. Kar, S. Sarkar (eds.), *Genetics Fundamentals Notes*,
https://doi.org/10.1007/978-981-16-7041-1_11

3. Code is contiguous—There are no breaks, no comma, and no punctuation in it. It is read one after the other.
4. Code is non-overlapping—A ribonucleotide can be a part of only one codon.
5. Start and stop signal: AUG and GUG act as signal to start the translation process, and UAA (ochre), UAG (amber), and UGA(opal) act as stop signals.
6. The code is linear and universal with minor exceptions.

Initially, it was assumed that DNA directly codes for proteins using ribosomes. Later, Jacob and Monod postulated the existence of a less stable RNA intermediate, mRNA, that is translated into proteins.

In the early 1960s, Sydney Brenner proposed that the code must be a triplet since three bases are the minimal requirement for forming at least 20 combinations coding for 20 amino acids. If the code would have been a two-letter word, 16 unique combinations were possible (4^2). A triplet code gives 64 combinations (4^3), whereas 4 letter words gives 256 combinations (4^4), which is far more than the required number of combinations for generating 20 amino acids.

Existence of triplet code was first experimentally shown by Francis Crick, Leslie Barnett, Brenner, and R. J. Watts-Tobin. They took bacteriophage T4 which is known to cause lysis of *E. coli* strains *B* and *K12*. Mutant strain of phage T4, *rII*, was unable to infect *K12* but could still infect strain *B*. Crick and colleagues used acridine dye proflavine which is a DNA-intercalating agent and causes one or more insertion and deletion mutation during replication. An insertion of a single nucleotide causes the entire reading frame to shift changing all subsequent downstream codons. These mutations are called **frameshift mutations**. Crick and colleague hypothesized that if rII is again treated with proflavine, random frameshift mutations would occur. That might induce new mutations or could even revert the earlier mutation by causing some insertion near the original mutation. The revertant could then be selected by its ability to cause the *K12* infection. They observed many mutants where there was one insertion and one deletion which could cause reversion to the wild-type behavior. However, when two deletions and two insertions occurred, wild-type behavior was not restored.

11.1.1 Non-overlapping Nature of the Code

Sydney Brenner imagined a 5-nucleotide sequence, GTACA; parts of the central codon TAC were shared by the outer codons GTA and ACA. When any central amino acid is considered, only 16 combinations of tripeptide sequences were possible. Brenner had actually found many more tripeptide combinations. This observation supported non-overlapping code.

Another argument against overlapping code was given by Francis Crick in 1957. He predicted the existence of an adaptor molecule that will have hydrogen-bonding ability with the nucleotide sequence and form covalent bond with the amino acids. If genetic code would have been overlapping, various adaptors would have tried to occupy overlapping sites making the translation process complex and error prone. It was later found that Crick's prediction about the adaptor molecule was correct. Transfer RNA (tRNA) plays the role of adaptor molecules.

11.1.2 Degenerate Nature of Code

Crick also proposed that the code is contiguous and lacks any punctuation or breaks. He hypothesized that out of the 64 codons, only 20 codes for amino acids, and the rest 44 codes for nothing and hence were referred to as blank or non-sense codons. However, his experiments later with phage T4 *rII* point mutants proved that 44 remaining codons were not blank. He observed that cases where wild type sequence was restored, had mutations like (+) with (−); (++) with (−−); and (+++) with (−−−). However, in between the deletion and insertions, there were numerous codons out of frame. It was very likely that out of those wrong codons, some of them will be among those 44 non-sense codons. In that case the translation would terminate prematurely, and restoration of wild type would not be possible. As the mutant combinations were able to infect *E. coli K12*, Crick and colleague concluded that remaining 44 codons must not be blank.

11.1.3 Deciphering Genetic Code

In 1968 Nirenberg, Har Gobind Khorana, and Robert Holley won the Nobel Prize in Physiology or Medicine for their seminal work on the genetic code. Nirenberg characterized specific coding sequence; Har Gobind Khorana developed the process for the synthesis of nucleic acids; and Robert Holley discovered the chemical structure of transfer RNA. Their contribution is recognized for their "interpretation of the genetic code and its function in protein synthesis." The experimental model was a cell-free protein-synthesizing system in a test tube with the enzyme polynucleotide phosphorylase, which allowed the production of synthetic mRNAs.

11.1.4 Synthesizing Polypeptides in a Cell-Free System

The in vitro system is made by adding all the essential factors like ribosomes, tRNAs, amino acids, and mRNA template for translation. Few of the amino acids were radioactively labeled. In 1961, mRNA was yet to be isolated. Hence, enzyme, the polynucleotide phosphorylase, was used to synthesize artificial RNA template catalyzing the reaction shown in the template.

This enzyme is known to degrade RNA in vivo. However, in vitro in the presence of high concentration of ribonucleoside diphosphates, the backward reaction rate increases, leading to RNA synthesis. This enzyme does not require any DNA template and inserts ribonucleotides depending on their concentration (Fig. 11.1) (Klug, W.S., et al. Concepts of Genetics, 10th ed. Pearson Education, California, 2012).

Initially, Nirenberg started synthesizing RNA with only one type of ribonucleotide generating either poly U, poly A, poly C, or poly G. In all their experiments, they made all amino acids available but radiolabeled only one amino acid per experiment. When they labeled C^{14}-phenylalanine, they could track its incorporation when the RNA sequence was poly U. It proved that UUU codon codes for phenylalanine. Using similar experiments they found AAA codes for lysine and CCC codes for proline. Poly G could not form a functional template as it gets folded back on itself.

Next, they used RNA heteropolymers for protein synthesis. In their next experiments, they used combinations of two different nucleotides. The relative proportion of each ribonucleoside was known to them. Hence, they could predict the frequency of each of the possible triplet codons. They could ascertain the percentage of each amino acids in the resulting polypeptide. Upon analyzing the sequence of the polypeptide, they could predict the composition of the triplet codons (Tables 11.1 and 11.2).

Let's assume only A and C are used for the synthesis of RNA in the ratio 1A:5C. There is 1/6 possibility for A and 5/6 possibility for C to occupy a position. Based on this assumption, the frequency for AAA will be $(1/6)^3$ or about 0.4%. For AAC, ACA, and CAA, the frequencies will be same as the number of A is occurring twice

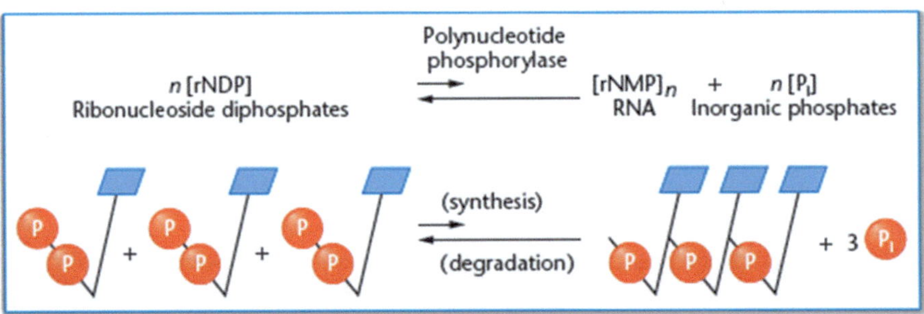

Fig. 11.1 The reaction catalyzed by polynucleotide phosphorylase. The equilibrium favors degradation of RNA, but the reaction can be "forced" towards forming RNA (Adapted from: Klug, W.S., et al. Concepts of Genetics, 10th ed. Pearson Education, California, 2012)

Table 11.1 Polynucleotide Phosphorylase Catalyzed Reaction

Possible compositions	Possible triplets	Probability of occurrence of any triplet	Final %
3A	AAA	$(1/6)^3 = 1/216 = 0.4\%$	0.4
1C:2A	AAC, ACA, CAA	$(5/6)(1/6)^2 = 5/216 = 2.3\%$	$3 \times 2.3 = 6.9$
2C:1A	ACC CAC CCA	$(5/6)^2(1/6) = 25/216 = 11.6\%$	$3 \times 11.6 = 34.8$
3C	CCC	$(5/6)^3 = 125/216 = 57.99\%$	57.9
Total			100

Table 11.2 Mixed Copolymer Experiment

Percentage of amino acids in protein	Probable base composition assignments
Lysine <1%	AAA
Glutamine 2%	1C:2A
Asparagine 2%	1C:2A
Threonine 12%	2C:1A
Histidine 14%	2C:1A, 1C:2A
Proline 69%	CCC, 2C:1A

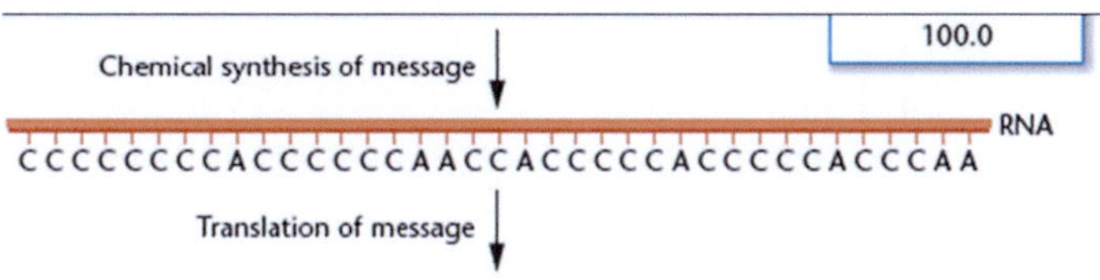

Fig. 11.2 Mixed copolymer experiment. Results and interpretation of a mixed copolymer experiment in which a ratio of 1A:5C is used (Adapted from: Klug, W.S. et al. Concepts of Genetics, 10th Edition, Pearson Education, California, 2012)

and C is occurring once. The frequency will be $(1/6)^2(5/6)$ or about 2.3% for each triplet. Similarly, each 1A and 2C will have $(1/6)(5/6)^2$ or 11.6%. CCC will have $(5/6)^3$ or 57.9%. Because proline appeared 69% of the time, it could be proposed that proline is encoded by CCC and 2C:1A (11.6%). Histidine appeared 14%, so it is probably encoded by 2C:1A (11.6%) and 1C:2A (2.3%). Threonine at 12% is likely to be coded by only one 2C:1A. Asparagine and glutamine each appears to be coded by one of the 1C:2A and lysine by AAA. Together these form 100% of the amino acids (Fig. 11.2). So, their theoretical calculation of amino acid composition and codon assignment was found logical. However, the actual codon sequence for heterogeneous bases could not be predicted. Subsequently, they used all four ribonucleotides in various proportions to determine the codon usage. However, the specific sequences remained unknown.

Har Gobind Khorana synthesized polynucleotides without a template to decipher the genetic code in the early 1960s. He repeatedly synthesized di, tri, and tetra nucleotide sequences and then ligated those short stretches enzymatically. This method was later known as **repeating copolymer synthesis**. By this approach the actual sequence of the codons could be deciphered. Let's understand Khorana's experiment: (1) Trinucleotide made of only U and C and then joined together (UUCUUCUUC) could be read as UUC, UCU, and CUU depending on the initiation point. In cell-free translation system, it gave rise to polypeptide of three amino acids, phenylalanine, serine, or leucine. So we know that three triplets codes for these three amino acids. (2) Then synthesizing dinucleotide sequence of U and C and ligating them together will create UCUCUC, which can be read in only two ways, UCU and CUC, generating only leucine and serine. Therefore, from the two experiments, we

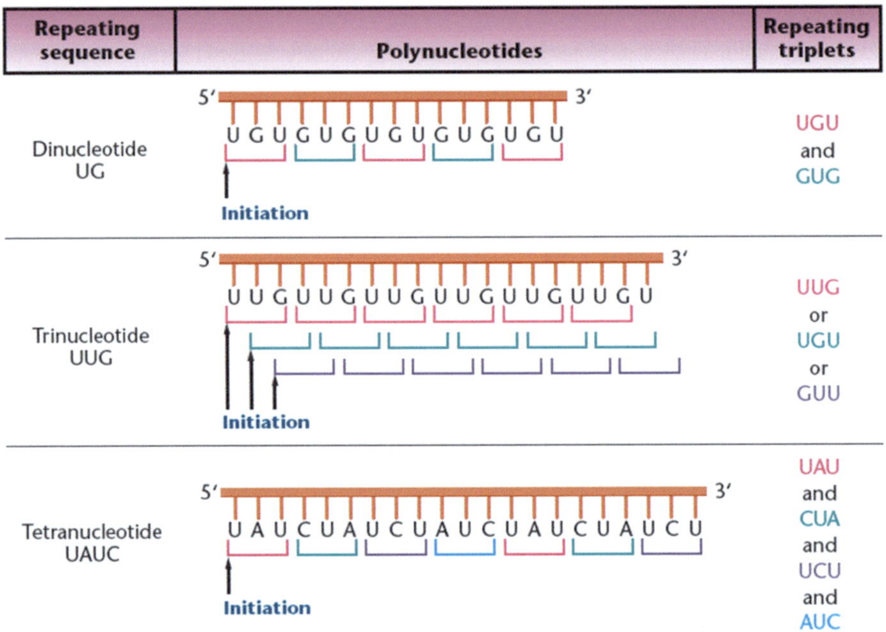

Fig. 11.3 Repeating copolymer synthesis. Synthesis of repeating copolymers from di-, tri, and tetranucleotides (Adapted from: Klug, W.S. et al. Concepts of Genetics, 10th ed. Pearson Education, California, 2012)

can conclude that UCU codon, which is common in both, must be coding for either leucine or serine but not phenylalanine. (3) For further information tetranucleotide sequence was created, UUAC, which produces triplets UUA, UAC, ACU, and CUU. Three amino acids were observed to be incorporated: leucine, threonine, and tyrosine. CUU is the common codon in the three experiments, and leucine is the common amino acids produced. Hence, it could be concluded that CUU codes for leucine (Fig. 11.3) (Klug, W.S. et al. Concepts of Genetics, 10th ed. Pearson Education, California, 2012). Therefore, Khorana's repeating copolymer approach could decipher the actual sequence of the codons. His approach could also prove the degeneracy of the code. As we can see in the third experiment, there are four unique codons, but only three amino acids were coded. So, two of the codons must have coded for the same amino acid.

Another method for deciphering genetic code was developed by Leder, known as triplet binding assay. In this process the amino acid was radioactively labeled. The labeled amino acid was then incubated with tRNA to create charged tRNA. By that time, codon compositions were already known although not the exact codon sequences. Therefore, it was possible to select a few amino acids that should be tested for each triplet. The radioactively charged tRNA, the RNA triplet, and ribosomes were used to carry out the assay. Those were incubated on a nitrocellulose filter (Fig. 11.4). The nitrocellulose filter retained the ribosomes due to their large

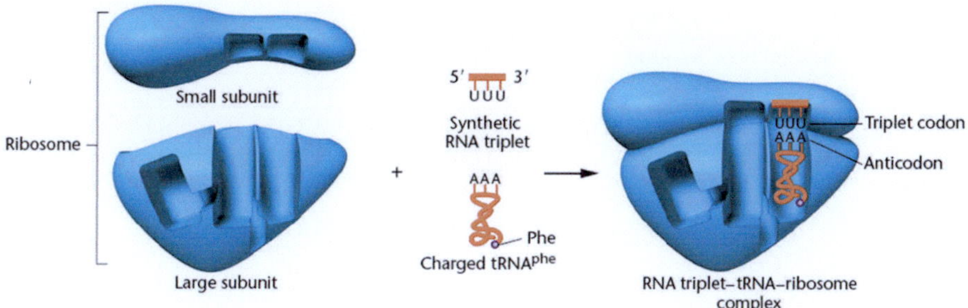

Fig. 11.4 Triplet binding Assay. Using UUU triplet in the ribosome to act as a codon, it attracts AAA anti-codon of the charged tRNA^{phe}. If A is radiolabeled, then the incorporation of A can be confirmed by the retention of radioactivity in the filter membrane (Adapted from: Klug, W.S. et al. Concepts of Genetics, 10th ed. Pearson Education, California, 2012)

size. If radioactivity remained in the filter, it indicated that correct amino acid has been incorporated, and hence the corresponding codon sequence could be predicted.

11.2 Codon: tRNA Interaction

Information stored in mRNA in the language of 4 nucleotide bases is translated into a polypeptide made of 20 amino acids (Table 11.3). This process of protein synthesis is therefore called as **translation**. In every translation process, an adaptor is required which can understand both the languages. In protein synthesis, this adaptor is tRNA. It can form chemical bonds with both mRNA and amino acids.

Each amino acid is carried by a particular tRNA which has a specific anti-codon sequence. The tRNA charged with amino acid and having anti-codon complementary to the codon present in the RNA base pairs with it. The base pairing between codon and anti-codon is also anti-parallel like that of DNA. So, if codon is 5′ AUG 3′, then anti-codon will be 3′ UAC 5′. However, it was observed that tRNAs bind to several alternative codons apart from the codon to which it is complementary. These alternative codons have a different base at the third position. The first two bases from 5′ end of the codon should form perfect base pairing as per AU: GC rule. At the third position of the codon, certain amount of flexibility was allowed (Fig. 11.5) (Griffiths, A. J. F, Introduction to Genetic Analysis. New York, NY :W.H. Freeman & Company, 2015). Therefore, it was hypothesized that base pairing between codon and anti-codon at the first two bases (from 5′ end in codon and 3′ end in anti-codon) is stringent which ensures accuracy of translation but less stringent at the third position due to wobbling. Wobble is a situation where there could be more than one spatial alignment of the base leading to lose base pairing. This was known as **wobble hypothesis** (Table 11.4). Wobble hypothesis supported the degeneracy of the genetic code where codons having same bases at the first two positions and different bases at the third position could code for the same amino acid.

Table 11.3 Genetic code tabulation

	Second base position										
		U		C		A		G			
First base position	U	UUU	[1]P	UCU	S	UAU	Y	UGU	C	U	Third base position
		UUC		UCC		UAC		UGC		C	
		UUA	L	UCA		UAA	Stop	UGA	Stop	A	
		UUG		UCG		UAG		UGG	W	G	
	C	CUU	L	CCU	P	CAU	H	CGU	R	U	
		CUC		CCC		CAC		CGC		C	
		CUA		CCA		CAA	Q	CGA		A	
		CUG		CCG		CAG		CGG		G	
	A	AUU	I	ACU	T	AGU	N	AGU	S	U	
		AUC		ACC		AGC		AGC		C	
		AUA		ACA		AGA	K	AGA	R	A	
		AUG	M	ACG		AGG		AGG		G	
	G	GUU	V	GCU	A	GAU	D	GGU	G	U	
		GUC		GCC		GAC		GGC		C	
		GUA		GCA		GAA	E	GGA		A	
		GUG		GCG		GAG		GGG		G	

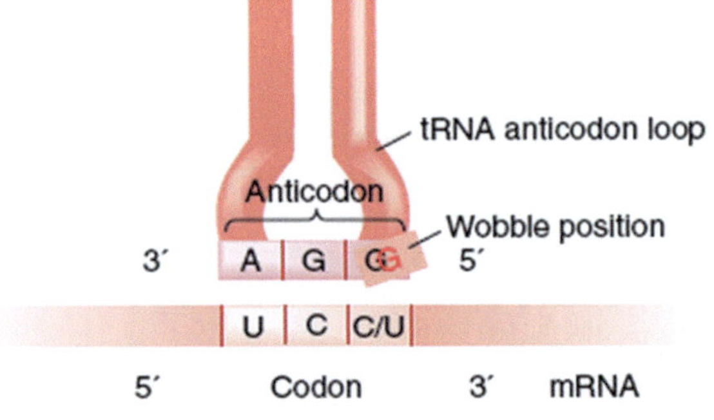

Fig. 11.5 tRNA and codon interaction. In the third position (5′ end) of the anti-codon, G can take two positions due to wobbling. Thus, it can pair with U and C. Therefore, single tRNA species can recognize two codons—UCU and UCC (Adapted from: Griffiths, A. J. F, Introduction to Genetic Analysis. New York, NY :W.H. Freeman & Company, 2015)

Table 11.4 Codon-anti-codon base pairing allowed under wobble rule

5′ end of anti-codon	3′ end of codon
G	C or U
C	G only
A	U only
U	A or G
I	U, C, or A

Therefore, 20 different tRNAs are present for carrying 20 different amino acids. Some organisms have extra tRNAs which have different anti-codon regions but gets charged by the same amino acids. These are called as isoaccepting tRNAs.

11.3 Structure of Ribosome

The site of protein synthesis is the ribosome in both prokaryotes and eukaryotes. These are large ribonucleoprotein complexes present in the cytosol. The complete ribosome is made of two subunits, the larger subunit and the smaller subunit. The ribosomal subunits are assembled together only during protein synthesis. At other times the two subunits exist separately in cytosol. Each subunit of ribosome is composed of several rRNA types and many proteins. Ribosomal subunits were isolated and centrifuged in ultracentrifuge. Depending upon their sedimentation coefficients in Svedberg (S) units which is an indication of molecular size, their names were derived. In prokaryotes, the small subunit is 30S, and large subunit is 50S, and they associate to form 70S particle. In eukaryotes, the small subunit is 40S and large subunit is 60S. Small and large subunits together form 80S particle. Almost two-third of the ribosomes in both prokaryotes and eukaryotes are RNA, and only one-third are proteins. Although there is difference in the size of ribosome in prokaryotes and eukaryotes, the overall composition pattern and process of translation is very similar. It shows that translation is an evolutionarily conserved process.

In prokaryotes, the large subunit is made of 5S rRNA (120 nucleotides) and 23 SrRNA (2900) nucleotides. The small subunit is made of 16S rRNA (1540 nucleotides). The large subunit has 34 proteins, and small subunit has 21 proteins. In eukaryotes, the large subunit is made of 5S (120 nucleotides), 5.8S (160 nucleotides), and 28S rRNA (4700 nucleotides). The small subunit is made of 18SrRNA (1900 nucleotides). The large subunit has 49 proteins, and small subunit has 33 proteins (Fig. 11.6).

Ribosome provides binding site for mRNA and tRNA. The mRNA binding site lies in the small subunit. Ribosome has three tRNA binding sites (A site for acceptor tRNA, P site occupied by the nascent peptide, and E site occupied by the uncharged tRNA which has transferred its amino acid to the peptide) in the ribosomes. It spans both the ribosomal subunits. The anti-codon end is present at the small subunit, and the amino acylated end is at the large subunit. The small subunit is responsible for reading the genetic information, while the large subunit is responsible for peptide bond synthesis, elongation, and protein release. Translocation is brought about by interplay between both subunits.

The crystal structure of the 50S ribosomal subunit, with the peptidyl transferase center (PTC) located with the help of model substrates, has been solved by Nissen et al. in 2000. The PTC consists only of RNA which forms the cavity of PTC. Another long cavity starts near the PTC and passes through the large ribosomal subunit and emerges on its back side. This is the ribosomal tunnel which provides path for the exit of nascent peptides. The inner side of the tunnel is lined with RNA and nonglobular parts of ribosomal proteins. Proteins like L4, L22, and L39e are also present in the inner wall of the tunnel. These walls largely consist of hydrophilic non-charged groups. Hence, it helps the nascent peptide to pass through without having any strong hydrophobic interactions with the globular proteins lining the tunnel. Presence of hydrated ions, water molecules, and sugar phosphate backbone gives negative potential to the tunnel.

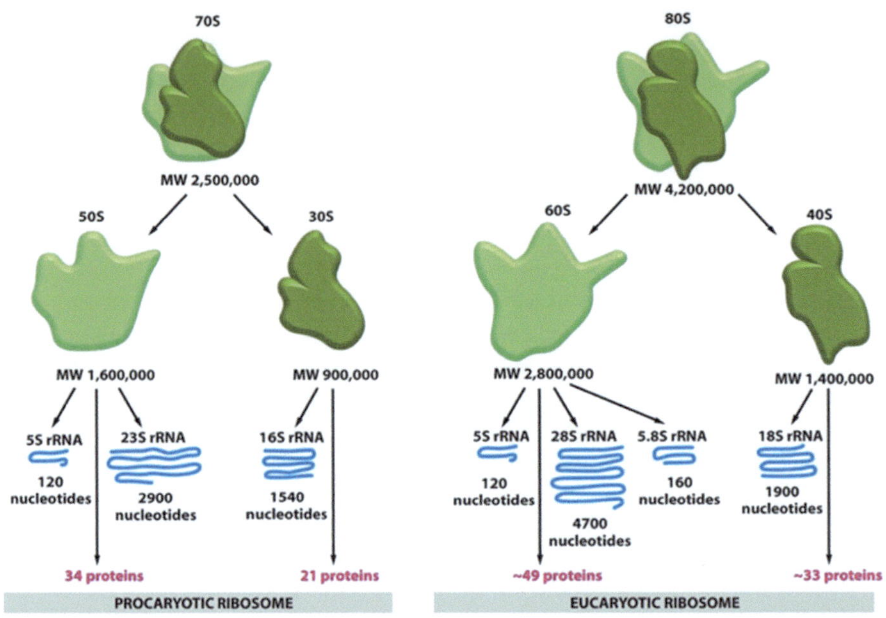

Fig. 11.6 Subunit composition of prokaryotic and eukaryotic ribosome. The large ribosomal subunits have been shown in light green, and small ribosomal subunits have been shown in dark green

Crystallographic studies have revealed that the A, P, and E sites are at least 20 Å, and perhaps as much as 50 Å, wide, thus defining the atomic distance that the tRNA molecules must shift during each translocation event. This is considered a fairly large distance relative to the size of the tRNAs themselves. The complete translational complex of ribosomes with associated mRNA and tRNA was crystallized, and structure was solved at the atomic level by Ramakrishnan and Noller using ribosomes from bacterium *Thermus thermophilus (Schmeing TM, et al. The crystal structure of the ribosome bound to EF-Tu and aminoacyl-tRNA. Science. 326, 688-694, 2009;26. Desai N, Brown A, Amunts A, Ramakrishnan V. The structure of the yeast mitochondrial ribosome. Science. 2017 Feb 3;355(6324):528-531. 27. Desai N, Yang H, Chandrasekaran V, Kazi R, Minczuk M, Ramakrishnan V. Elongational stalling activates mitoribosome-associated quality control. Science. 2020 Nov 27;370(6520):1105-1110 ivates mitoribosome-associated quality control. Science. 2020 Nov 27;370(6520):1105-1110*). In 2009 Nobel Prize in Chemistry was awarded to Venkatraman Ramakrishnan, Thomas Steitz, and Ada Yonath "for studies on the structure and function of the ribosome." The three groups deciphered the structure of ribosomes up to a resolution of 3 Å.

Electron microscopic structure of polyribosome was observed in early 1960s (Fig. 11.7) (Slayter, Henry S. et al. The visualization of polyribosomal structure, 7, 652-657, 1963). Recently, using a unique high-resolution approach—the technique of **time-resolved single-particle cryo-electron microscopy** (cryo-EM)—the 70S *E. coli* ribosome was captured and examined while in the process of translation

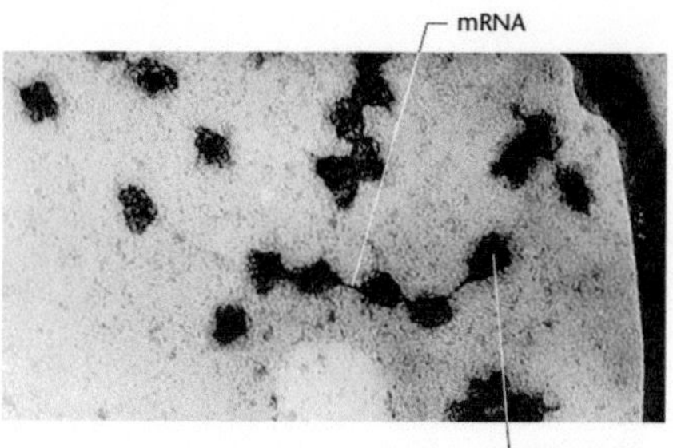

Fig. 11.7 Electron microscopic structure of polyribosomes. (Adapted from: Slayter, Henry S. et al. The visualization of polyribosomal structure. , 7, 652-657, 1963) (**a**) Translational complex of rabbit reticulocytes synthesizing hemoglobin mRNA

at a resolution of 5.5 Å by Niel Fisher's group. The study demonstrated the trajectories followed by tRNA during the process of translocation. They also showed the conformational changes occurring in the ribosomes during the process of translation using time-resolved single-particle cryo-electron microscopy.

Two additional regions of ribosomes which are important for protein synthesis are the decoding center in the 30S subunit and the peptidyl transferase center in the 50S subunit. The decoding center ensures that only the correct tRNA having anticodon which perfectly base pairs with the codon gets accepted. This site is referred to as A site. Peptidyl transferase center is formed by a pocket-like symmetrical 23SrRNA dimer that is composed of two RNA units which have L-shaped structure. The symmetrical association of the two units is important for providing the proper positioning of the two aminoacyl-tRNAs leading to appropriate stereochemistry. The size of each symmetrical dimer is approximately the same as that of tRNA. 23S rRNA interacts with the CCA terminus of peptidyl-tRNA in both the P site and A site.

11.4 Structure of tRNA

The existence of an adaptor molecule was predicted by Francis Crick in 1957. tRNA was later discovered. It is a small molecule with a stable compact structure. It is composed of 75 to 90 nucleotides and is well conserved throughout evolution. In 1965, Robert Holley and colleagues isolated and sequenced tRNA from yeast. Several unique nucleotides were present in tRNA like inosinic acid (contains purine hypoxanthine), ribothymidilic acid, pseudouridine, etc. Initially it was difficult to comprehend the presence of these unique bases. Later it was hypothesized that presence of these bases increases the probability of hydrogen bonding and helps in formation of compact conformation of tRNA.

Holley's analysis of tRNA sequence led him to propose the two-dimensional cloverleaf model of tRNA (Fig. 11.8) (Griffiths, A. J. F, Introduction to Genetic Analysis. New York, NY :W.H. Freeman & Company, 2015). tRNA has a characteristics secondary structure created by hydrogen bonding between base pairs. His model showed that tRNA sequence is double stranded at certain regions leaving single-stranded interspersed regions in between them forming a loop

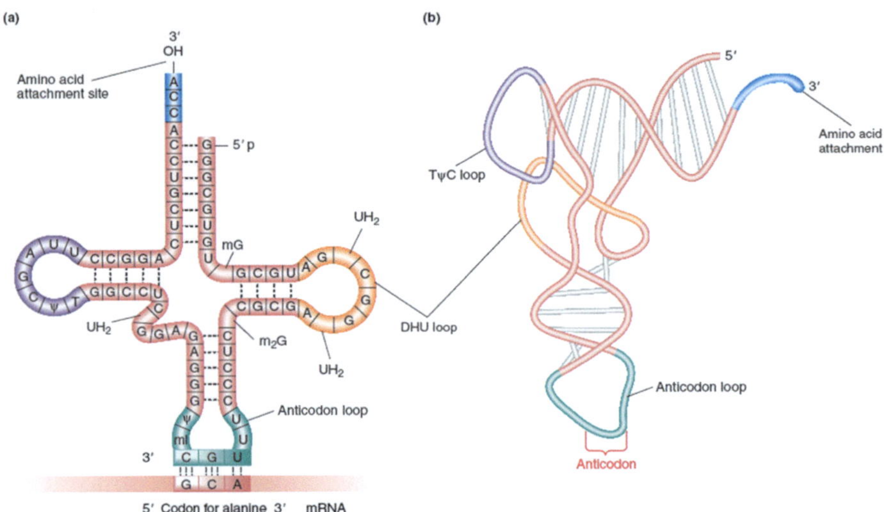

Fig. 11.8 Structure of tRNA. (**a**) The schematic structure of yeast alanine tRNA. The tRNA-codon base pairing has been shown. (**b**) Diagram of 3D structure of yeast phenylalanine tRNA. The abbreviations ψ, $_{m}$G, m$_2$G, $_{m}$I, and UH$_2$ refer to pseudouridine, methylguanosine, dimethylguanosine, methylionosine, and dihydrouridine, respectively (Adapted from: Griffiths, A. J. F, Introduction to Genetic Analysis. New York, NY :W.H. Freeman & Company, 2015)

structure. Hence, it was called as stem-loop structure. Genetic code was already deciphered, and hence Holley searched for complementary sequence of bases for each codon. He observed that the complementary bases are present in one of the loop regions. Hence, it was called anti-codon loop because codons in mRNA are read in the 5′ to 3′ direction; anti-codons are oriented and written in the 3′ to 5′ direction.

Studies on other tRNA species revealed many common features. First, the 3′ end of tRNA molecule has CCA. At this end the amino acid is covalently joined to the terminal adenosine residue. Second, it has four double helical stems and three single-stranded loops forming the L-shaped cloverleaf L-shaped flattened two-dimensional structure. Third is the presence of an anti-codon loop which base pairs with mRNA codon. Fourth is the presence of a loop rich in dihydrouridine, known as dihydrouridine loop (DHU). Apart from these, there is a loop whose length varies, as is known as variable loop. Due to the variation of length of this loop, the length of tRNA sequence varies. Later X-ray crystal structure of tRNA was solved.

11.5 Charging of tRNA

The freely available tRNA present in the cytosol needs to attach itself to the appropriate amino acid so that it can carry the amino acid to the small ribosome subunit and mRNA complex. The class of enzymes known as amino acyl tRNA synthetases attaches the amino acids to the tRNA. This process is called charging of

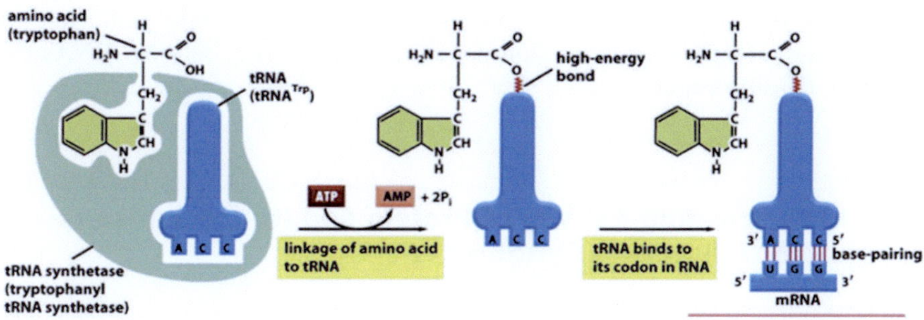

Fig. 11.9 Charging of tRNA. The tRNA synthetase enzyme has binding site for tRNA and the amino acid. ATP is hydrolyzed into AMP and 2Pi. This energy is used to synthesize high-energy ester bond which links the amino acid with the tRNA (Adapted from: Alberts, B. et al. Molecular Biology of the Cell, 4th edition New York: Garland Science, 2002)

tRNA. There are 20 different aminoacyl-tRNA synthetases recognizing one amino acid each and joining those to the compatible tRNA. The adenine residue at the 3′ end has free 2′ and 3′ hydroxyl groups. The free amino acid is first activated by ATP, forming adenylated amino acid by reacting with AMP and releasing two inorganic phosphates. The adenylated amino acid is then linked with tRNA by forming high-energy ester bond between the free 2′ or 3′ hydroxyl of 3′ adenine of tRNA and carboxyl group of amino acid (Alberts, B. et al. Molecular Biology of the Cell, 4th edition New York: Garland Science, 2002). Adenylation of amino acid and the linking of adenylated amino acid to tRNA occur as coupled reaction, and the energy release from phosphoanhydride bond hydrolysis present in ATP is used for the synthesis of high-energy ester bond. The energy of this bond is subsequently used for the synthesis of peptide bonds (Fig. 11.9). The aminoacylation reaction proceeds towards activation of the amino acid due to the hydrolysis of released pyrophosphate. In case the tRNA gets charged with the wrong amino acid, it is removed by the aminoacyl tRNA synthetase enzyme. This enzyme has an amino acid-binding pocket and an editing pocket. The amino acid-binding pocket allows only the correct amino acid by creating a spatial constraint. Any amino acid larger or smaller than the correct amino acid is not able to bind at the binding pocket. However, in cases wrong amino acid of approximately the same size to that of the correct amino acid binds, then the enzyme sends it to editing pocket. Here, it is hydrolyzed from tRNA and released from the enzyme. Hence, these enzymes possess proofreading activity increasing translational accuracy.

Base pairing between codon and anti-codon leads to incorporation of amino acids in the nascent peptide chain. This base pairing is specific in the first and second nucleotide positions of the codon (5′–3′ direction). However, at the third base of the codon, the base pairing may or may not follow Watson-Crick rule of base pairing. If the first base of the anti-codon from 5′ to 3′ direction is C, the base pairing is specific. However, if it is U or G, it may not follow Watson and Crick base pair rules. This occurs because of a wobble in tRNA positioning. Hence, it is called as **Wobble**

hypothesis. It allows a single tRNA to bring more than one amino acid against a single codon. Hence, it allows codon degeneracy. Therefore, only 32 tRNAs can recognize 61 different codons.

11.6 Proofreading Activity of Protein and Its Comparison to RNA and DNA Synthesis

During DNA and RNA synthesis, the fidelity is achieved majorly by two strategies: correct substrate selection and removal of mismatch called proofreading. In DNA polymerase (DNAP), the active site selects only dNTPs over rNTPs by the help of steric gate. Side chains of two amino acids of the DNAP active site exclude the 2′OH group and hence exclude rNTPs. For RNA polymerase (RNAP), rNTPs are selected over dNTPs as 2′OH of rNTP interacts with RNAP active site which is important for entering the active site. Hence, dNTPs are excluded.

In case there is a mismatch, it causes disruption in the active site of the enzyme. The DNAP slows down, and the wrong incorporation in DNA is removed by 3′ to 5′ exonucleolytic activity of DNAP. In DNAP I 5′ to 3′ exonucleolytic activity is also present which can remove up to ten nucleotides at a time. In RNA synthesis, if misincorporation occurs, then mismatched nucleotide at position +1 of the RNA is displaced away from the template. Due to this displacement, RNAP is paused. RNAP then moves back by one position and takes the mismatched RNA nucleotide to the proofreading site. The misincorporated nucleotide is cleaved off, and RNA synthesis resumes.

During protein synthesis, also the fidelity is maintained based on correct selection of charged tRNA and removal of incorrect tRNA. Both these activities are brought about by amino-acyl tRNA synthetases (aaTS). These aaTS have two catalytic sites, one for selection of tRNA and another for editing. Correct amino acids have the highest affinity for the active site pocket of the synthetase, and based on the affinity of bonding, the correct amino acid is preferred. As there are specific aaTS for every amino acid, the active site excludes the amino acids larger than the correct one. In case of misincorporation, aaTS forces the incorrect amino acid to the editing pocket where it is removed by hydrolysis from AMP and released from the enzyme. This is called hydrolytic editing (Beuning, P.J. et al. Hydrolytic editing by class II aminoacyl-tRNA synthetase. Proc Natl Acad Sci USA. 97, 8916-20, 2000).

11.7 Process of Protein Translation

The process of translation occurs in complex dynamic and continuous process. However, to understand the step-by-step mechanisms, we will divide the process into three steps: **initiation, elongation, and termination**.

11.7.1 Initiation

We will discuss the process of translation in prokaryotes. Although the basic steps of initiation, elongation, and termination remain the same in both prokaryotes and eukaryotes, the factors involved in them are different. Eukaryotic translation process will be discussed in Sect. 11.8.

When the translation is not occurring, the ribosomal subunits exist separately in the cytosol. The small subunit binds to a protein molecule known as **initiation factor 3 (IF3)**. The small ribosomal subunit along with IF3 binds to the mRNA near its 5′end and moves along the mRNA from 5′ to 3′ direction searching for the start codon AUG. The search halts at a specific sequence, AGGAGG, known as **Shine-Dalgarno** sequence (Fig. 11.10) (Arakawa, K et al. Computational Genome Analysis Using The G-language System. Genes, Genomes and Genomics. 21–13, 2008). It was discovered by John Shine and Lynn Dalgarno in 1974. The exact sequence varies slightly from species to species. This sequence is located about three to nine base pairs upstream of the initiation codon in the 5′ UTR of the mRNA. This sequence base pairs with a pyrimidine-rich sequence, UCCUCC, present at 3′ end of the 16S rRNA of the small ribosomal subunit. This interaction between the small ribosomal subunit rRNA and mRNA helps the small ribosomal subunit to dock on the mRNA and form the **preinitiation complex**. Subsequently, the initiator tRNA binds the start codon. The amino acid carried by the initiator tRNA in prokaryotes is *N*-formylmethionine (fMet). *N*-formyl methionine is synthesized by the enzyme methionine tRNA formyltransferase. An initiation factor IF2 with a GTP molecule facilitates binding of charged initiator tRNA. Then the 50S large subunit of ribosome is attached to the preinitiation complex with the help of IF1. The assembly of ribosome is accomplished by the hydrolysis of a GTP molecule to GDP. The initiation factors IF1, IF2, and IF3 leave once the ribosome assembly is complete. This completes the formation of **70S initiation complex** (Fig. 11.10). When the ribosome assembly is completed, the A (acceptor), P (peptidyl), and E (exit) sites, PTC and ribosome tunnel are formed in the ribosome. The initiator tRNA in the initiation complex occupies the designated P site of the ribosome (Fig. 11.11)

Fig. 11.10 Representation of Shine-Dalgarno sequence. Sequence logo for Shine-Dalgarno sequence in *Escherichia coli*. Bases present at each position are represented by stacked alphabets. Height of each nucleotide corresponds to its contributing frequency (Adapted from: Arakawa, K et al. Computational Genome Analysis Using The G-language System. Genes, Genomes and Genomics. 21–13, 2008)

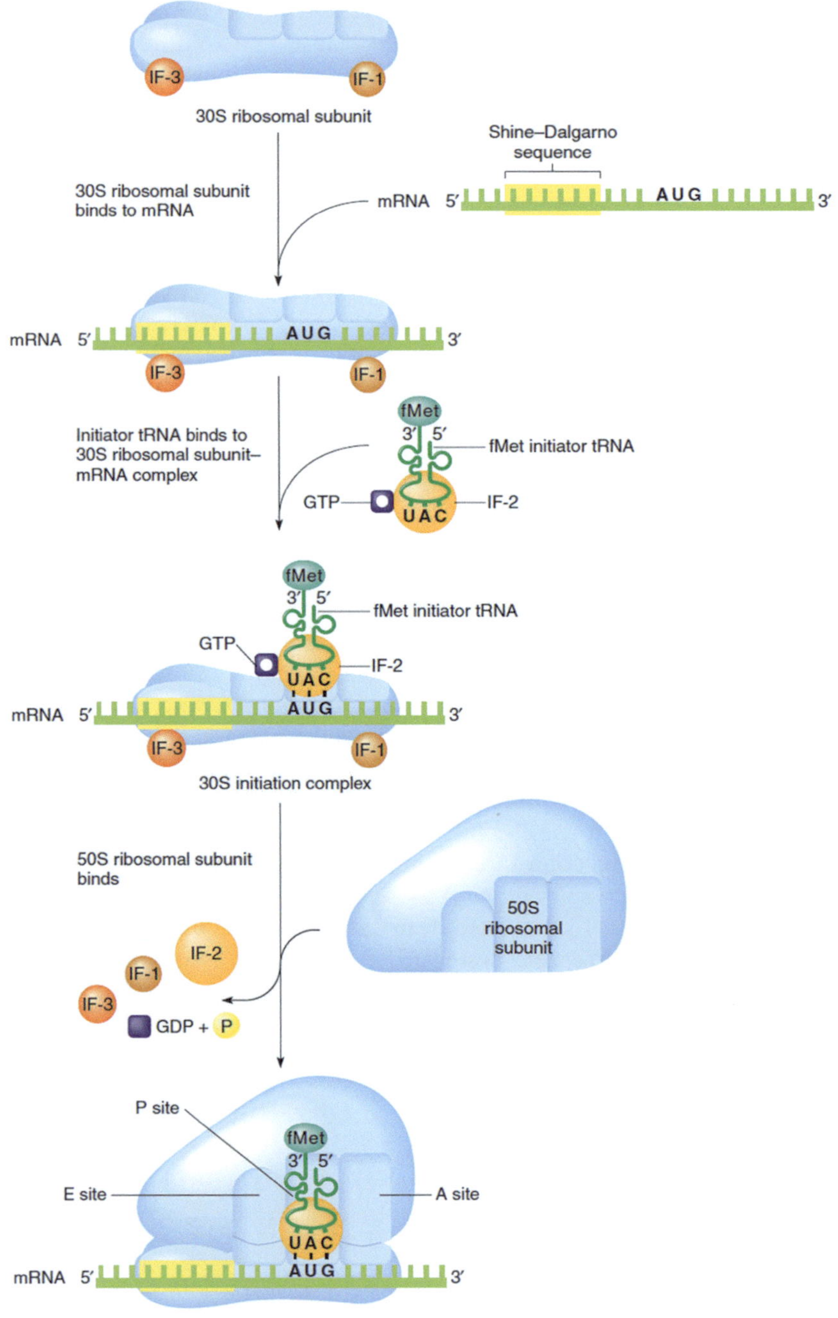

Fig. 11.11 Initiation of protein translation in prokaryotes. Ribosomes have been shown in light blue. Shine-Dalgarno sequence has been highlighted in yellow. tRNAs have been shown in green. The initiation process has been depicted in three steps: (1) formation of preinitiation complex, (2) formation of 30S initiation complex, and (3) ribosome assembly (Adapted from: Russell, P.J.: iGenetics: A molecular Approach, 3rd ed. Benjamin Cummings, New York, 2010)

(Russell, P.J.: iGenetics: A molecular Approach, 3rd ed. Benjamin Cummings, New York, 2010).

11.7.2 Elongation

Elongation phase consists of three steps:

1. Bringing the charged tRNAs to join the A site
2. Formation of peptide bond between the incoming amino acid and the most recently added amino acid to the nascent peptide
3. Translocation of ribosome in the 3′ direction

These steps require the help of several elongation factors (EFs) and energy. A charged tRNA is bound by EF-Tu and GTP. EF-Tu is a monomeric G protein whose active form (bound to GTP) binds aminoacyl-tRNA. The EF-Tu-GTP-aminoacyl-tRNA trimeric complex binds to the ribosome A site. Codon-anti-codon recognition leads to change in ribosome conformation, hydrolysis of GTP and tRNA with the anti-codon complementary to the codon at the A site enters the A site, and GTP linked to EF-Tu is hydrolyzed to form EF-Tu-GDP. EF-Tu-GDP complex then gets released from the tRNA (Lewin, B. Genes VIII. 15th Edition. Upper Saddle River, NJ: Pearson Prentice Hall, 2004). Released EF-Tu forms a complex with another protein Ts in the cytosol. Ts releases Tu for replenishing the levels of EF-TU-GTP (Fig. 11.12).

Then the enzyme peptidyl transferase catalyzes the peptide bond synthesis between the amino acid of the P site and that of the A site. This leads to addition of an amino acid to the nascent peptide chain, and the peptide chain now gets transferred from the P site to the A site. The tRNA at the P site is no longer carrying the peptide chain. It now occupies the E site from where the uncharged tRNA leaves the ribosome. The 23SrRNA of the large ribosomal subunit, which forms the **peptidyl transferase center** (PTC), catalyzes the peptide bond synthesis. Peptide bond formation occurs by nucleophilic attack of the α-amino group of the aminoacyl-tRNA on the carbonyl carbon of peptidyl-tRNA (Nissen, P., Hansen, J., Ban, N., Moore, P.B., Steitz, T.A.: The structural basis of ribosome activity in peptide bond synthesis. Science. 289, 920-930, 2000). The N3 of A2451 of 23SrRNA abstracts a proton from the α-amino group, facilitating the nucleophilic attack of the nitrogen on the carbonyl carbon of the peptidyl-tRNA. When the ester bond of the peptidyl-tRNA is cleaved, a proton is delivered from the adjacent 2′-OH group, which in turn receives a proton from the α-amino group of the aminoacyl-tRNA. Due to this catalytic ability, 23SrRNA belongs to the category of ribozymes. In the next step, the elongation factor EF-TG uses the energy released from GTP hydrolysis to translocate the ribosome by moving it in the 3′ direction on mRNA. This translocation is exactly one codon in length. It places the ribosome over a new codon which is not occupied by tRNA. This site becomes the new acceptor site. The tRNA which is attached to the nascent peptide is now occupying the new P site

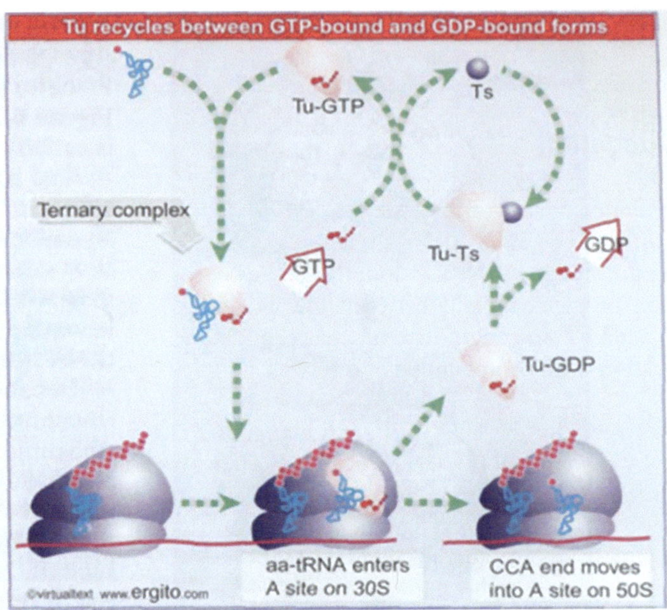

Fig. 11.12 Elongation of peptide chain during synthesis in prokaryotes. Ribosomes have been shown in purple, tRNAs have been indicated in light blue, and nascent peptide chain has been shown in pink. Tu and Ts proteins present in the cytosol get dissociated, and Tu reacts with elongation factors (EF) to form EF-Tu. EF-Tu combines with GTP to form EF-Tu-GTP complex (Adapted from: Lewin, B. Genes VIII. 15th Edition. Upper Saddle River, NJ: Pearson Prentice Hall, 2004)

which was earlier A site. The earlier P site now becomes the E site (Fig. 11.13) (Kathrin, L. et al. The Role of 23S Ribosomal RNA Residue A2451 in Peptide Bond Synthesis Revealed by Atomic Mutagenesis, Chemistry & Biology, 15, 485-492, 2008).

The process of elongation and translocation keeps repeating over and over again. An additional amino acid is added to the nascent peptide chain after each cycle. Once the nascent polypeptide reaches a certain length (about 30 amino acids), it exits the ribosome through the ribosome tunnel.

11.7.3 Termination

Termination is the final phase of translation which occurs when the acceptor site codons are any of these three codons: UAG, UAA, or UGA. These codons do not specify any amino acids and hence do not invite any tRNA. These are known as stop codons, or nonsense codons. The termination codon calls upon the GTP-dependent release factors instead of elongation factors.

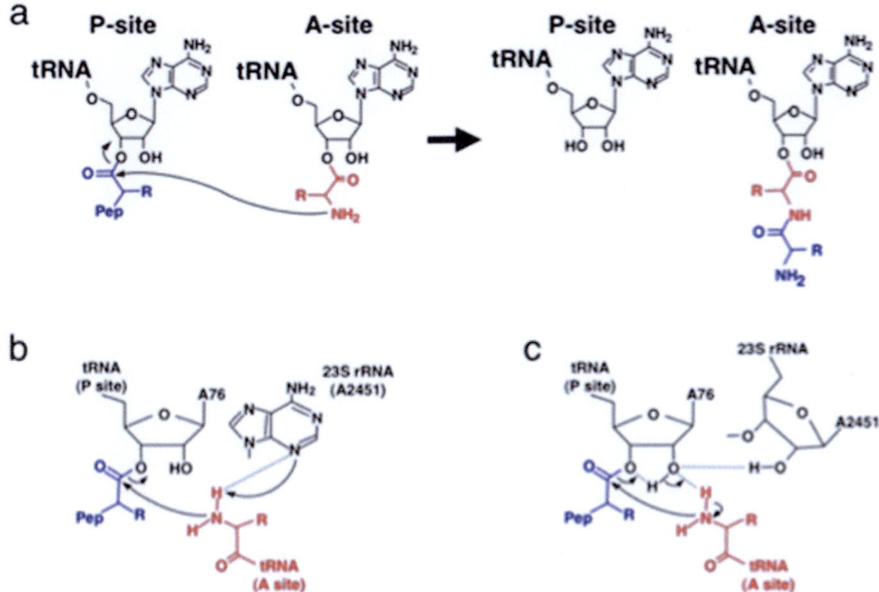

Fig. 11.13 Catalytic role of 23SrRNA in synthesis of peptide bond. A2451 exists as tautomer. It can have a negative unprotonated N3 and a neutral protonated N3. (**a**) N3 of A2451 of 23SrRNA abstracts a proton from the NH_2 group of amino acid of acceptor tRNA. Deprotonated amine group of amino acid present with acceptor tRNA attacks the carbonyl carbon of the peptidyl-tRNA. (**b**) A protonated N3 stabilizes the tetrahedral carbon intermediate by hydrogen bonding to the oxyanion. (**c**) The proton from N3 is then transferred to the peptidyl tRNA as the newly formed peptide deacylates (Kathrin, L. et al. The Role of 23S Ribosomal RNA Residue A2451 in Peptide Bond Synthesis Revealed by Atomic Mutagenesis, Chemistry & Biology, 15, 485-492, 2008)

In bacteria, two release factors, RF1 and RF2, recognize stop codons. RF1 recognizes UAG and UAA, and RF2 recognizes UAA and UGA. RF1/2 resembles C-terminal domain of EF-G and tRNA. Hence, it can occupy the A site just like tRNA. Binding of RRFs causes the PTC to hydrolyze the nascent peptide chains from tRNA and to leave the ribosome. Subsequently, ribosome release factors (RRFs) bind to the A site which mimics the anti-codon and acceptor end of tRNA (Fig. 11.14). EF-G then causes translocation of the ribosome leading to disassembly of larger and smaller subunits (Russell, P.J.: iGenetics: A molecular Approach, 3rd ed. Benjamin Cummings, New York, 2010).

11.7.4 Polyribosome

A cell needs to constantly make proteins from hundreds to thousands of mRNA transcripts. The process of translation must be very fast, accurate, and efficient. A single bacterial cell might contain 20,000 ribosomes, collectively contributing to one-fourth of the cell mass. So, a single mRNA is translated by multiple ribosomes

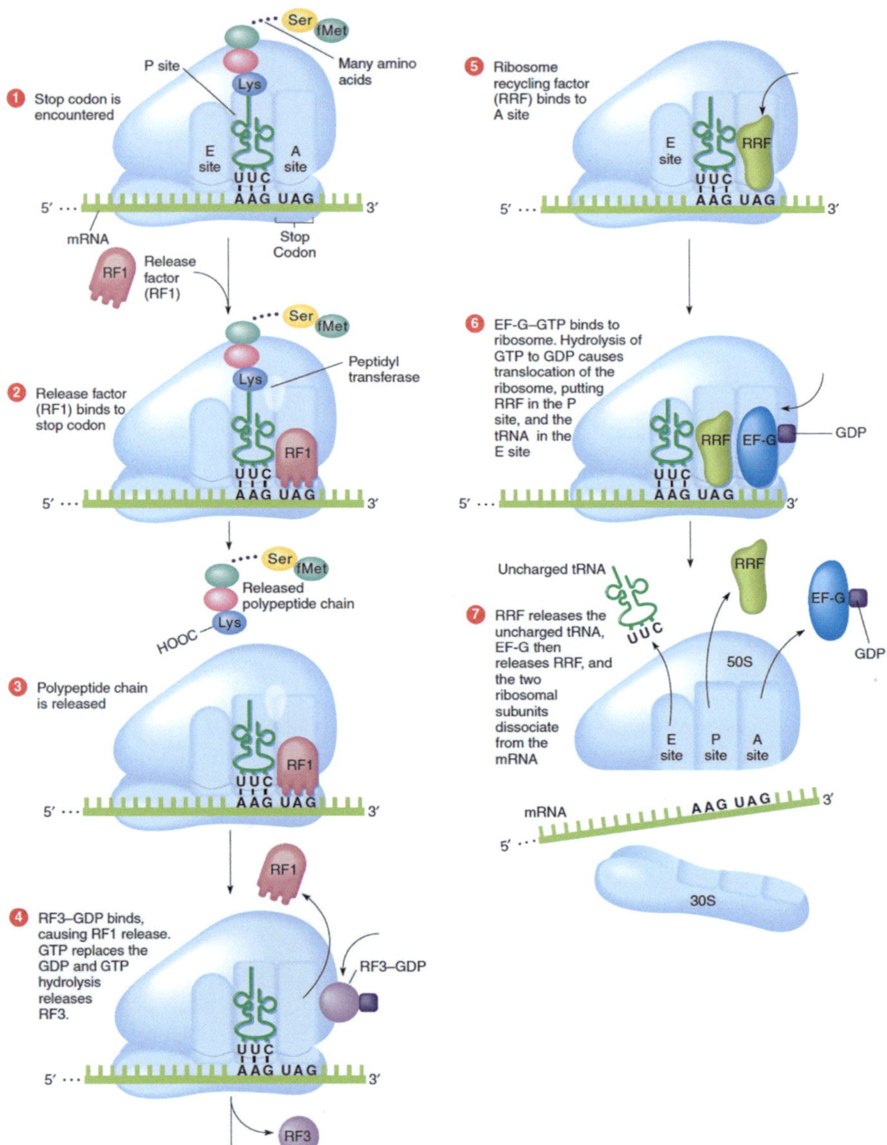

Fig. 11.14 Termination of peptide synthesis in prokaryotes. (1) Stop codon is at A site. (2) Release factor binds. (3) Polypeptide chain is released. (4) RF3-GDP binds causing RF1 release. (5) Ribosome recycling factor binds to A site. (6) EF-G GTP binds to A site, causing ribosome translocation and disassembly. (7) Ribosome disassembles, and RRF releases uncharged tRNA (Adapted from: Russell, P.J.: iGenetics: A molecular Approach, 3rd ed. Benjamin Cummings, New York, 2010)

making multiple copies of the same proteins. Electron micrographs of actively translating mRNA have revealed beaded necklace kind of a structure of mRNA being translated by multiple ribosomes. That translation complex is called **polyribosomes** or **polysomes** (Fig. 11.15). Each ribosome independently translates the complete polypeptide. Hence, polyribosome increases the efficiency of translation.

In bacteria, since there is no separate nucleus, both transcription and translation occur in the cytosol. Hence, translation starts in the 5′ end of mRNA which is still being transcribed at the 3′ end. Therefore, in prokaryotes, translation and transcription are coupled process, whereas in eukaryotes, transcription occurs inside nucleus, and only the completely synthesized RNA is exported out of nucleus and gets translated in the cytosol.

11.7.5 Translation of Polycistronic mRNA

Each polypeptide producing gene in eukaryotes produces monocistronic mRNA, meaning mRNA that directs the synthesis of a single kind of polypeptide. In eukaryotes, a single start codon is identified in mRNA to initiate the synthesis of

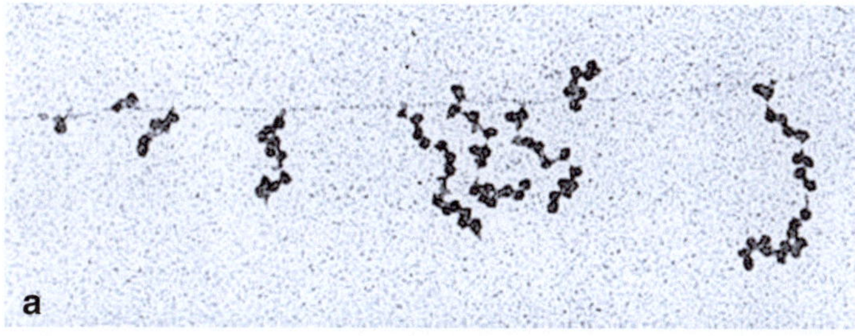

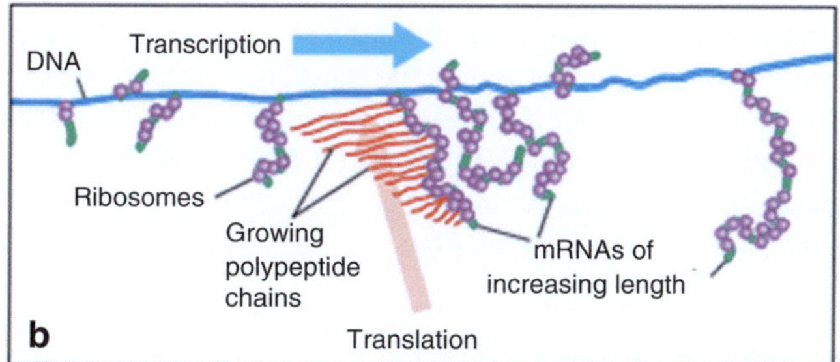

Fig. 11.15 Electron micrograph of a polyribosome. (**a**) Multiple ribosomes translating a single mRNA. The ribosomes closest to the stop codon have the longest polypeptide [Adapted from O. L. Miller et al., 1970, *Science*; 169:392–5]. (**b**) Schematic rendition of the polyribosome electron micrograph (Adapted from: Sanders, M.F., Bowman, J.L.: Genetic Analysis: An Integrated Approach, 2nd ed. Pearson Education, New Jersey, 2015)

one kind of polypeptide chain. In contrast, bacterial and archaeal genes often share a single promoter, and the resulting mRNA transcript can lead to the synthesis of multiple polypeptide chains. In bacteria, genes participating in a single metabolic pathway are part of a single operon, and one operon produces single polycistronic mRNA. Each cistron contains sequence information for translation initiation. In between the start and stop codons of the adjacent cistrons, there are intercistronic spacer sequence that is not translated.

11.8 Prokaryotic vs Eukaryotic Protein Translation

Although the overall steps of translation are very similar in prokaryotes and eukaryotes, there are differences in the factors involved; the sequences that act as signals for translation initiation are different. In this section we will discuss eukaryotic translation and how it is different from prokaryotic translation process.

11.8.1 Eukaryotic Translation Initiation

For translation initiation in prokaryotes, Shine-Dalgarno sequence acts as the docking site for initiator tRNA, and the preinitiation complex forms there. However, in certain prokaryotes like proteobacteria and mitochondrial genome, Shine-Dalgarno sequence is absent. In them the small subunit of ribosomes scans the mRNA from the 5′ end, and depending upon RNA, unfoldedness and accessibility of the initiation codon protein translation can start. If recognition site is different from consensus recognition sequence, then it will sometimes ignore the first AUG to produce two or more proteins differing in N terminus. This process is called as **leaky scanning**. Hence, a single mRNA can lead to synthesis of different proteins having variable length at their 5′ end. Apart from that, there are structural and sequence motifs of mRNA, which also contribute to **Shine-Dalgarno independent initiation**.

The size of eukaryotic ribosomes is larger than that of the prokaryotes, and the mRNA has expansion sequences apart from the core sequence required for translation. There are other major differences in the translation process. Eukaryotic mRNA contains a purine-rich sequence (A and G), three bases upstream from that of the AUG initiator codon which is followed by a G (A/GNN**AUG**G) where N can be any nucleotide base. This was discovered by Marilyn Kozak and hence is called Kozak sequence (Fig. 11.16). It is analogous to the Shine-Dalgarno sequence of prokaryotes.

11.8.1.1 5′ mRNA Capping

5′ end of mRNA has a 7-methylguanosine cap (7-mG). Capping occurs while transcription is occurring. The enzymes Cet1, Ceg1, and Abd1, an RNA 5′triphosphatase, a GTP-mRNA guanylyltransferase, and an RNA guanine-7-methyltransferase, bring about the process of capping. One of the terminal phosphate groups is removed by RNA triphosphatase. A bisphosphate group (i.e., 5′(ppN)[pN]

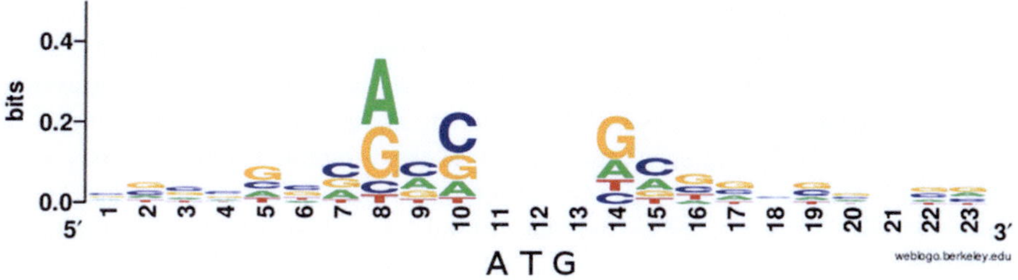

Fig. 11.16 Representation of Kozak sequence. A sequence logo showing the most conserved bases around the initiation codon from 10,000 human mRNAs. Height of each nucleotide corresponds to frequency of its occurrence. More conserved residues have more height

n) remains at the terminus. GTP is then added to mRNA by guanylyltransferase. In this process the GTP loses the pyrophosphate. This results in the 5′–5′ triphosphate linkage, producing 5′(Gp)(ppN)[pN]n. The methylation of the 7-nitrogen of guanine is then carried out by mRNA (guanine-N7-)-methyltransferase, leading to demethylation of *S*-adenosyl-L-methionine.

11.8.1.2 Circularization of mRNA During Translation

During eukaryotic translation, the 5′ and 3′ ends of the mRNA come together with the help of eukaryotic initiation factors. Circularization may serve a regulatory role to ensure translation of only intact (capped and polyadenylated) mRNAs. eIF4F complex is composed of eIF4E, eIF4A, and eIF4G proteins. This eIF4F binds to the 5′-cap of an mRNA and prepares an mRNA for recruitment of a 40S subunit. The small subunit of ribosome binds at the 5′ UTR. The eukaryotic initiation factor (eIF2) along with GTP brings the methionine charged initiator tRNA to the 5′cap. Small subunit along with eIF2-GTP, $_{\text{met}}$tRNA, eIF1, eIF1A, eIF5B with bound GTP, eIF3 large subunit, and 40S ribosomal subunit makes 43S preinitiation complex. eIF4B promotes the ATP-dependent RNA helicase activity of eIF4A, and eIF4F is needed to unwind secondary structure present in a 5′-UTR that would otherwise impede scanning of the 40S subunit during initiation. 43S preinitiation complex starts scanning along 5′ to 3′ direction in search of start codon. The scanning process is dependent on ATP-powered helicases but independent of mRNA secondary structure or sequence. 43S complex is recruited to the Kozak sequence by eIF4F. The poly(A) binding protein (PABP), which binds the poly(A) tail, interacts with eIF4G and eIF4B which are present at the 5′cap. This leads to circularization of mRNA (Gallie, D.R.: The role of polyA binding protein in the assembly of cap binding complex during translation initiation in plants. Translation (Austin). 2, e959378, 2014). When 43S complex binds the mRNA at the start site, it forms the 48S preinitiation complex. IF2 and eIF5B are structurally similar GTPases that catalyze joining of the large subunit to the preinitiation complex.

There are certain eukaryotic mRNAs which are not capped like plus-strand RNA virus mRNAs (poliovirus, human rhinovirus, and hepatitis A and hepatitis C viruses); growth factors VEGF, FGF2, and PDGF; transcription factors c-myc;

apoptosis-associated gene products Apaf-1; the potassium channel Kv1.4; etc. In addition, many plant messengers are uncapped. Their 5′ UTRs are mostly 200 bases long or have more than one AUG codon. They are translated by **cap-independent internal initiation**. The ribosome binds to an internal RNA sequence near the initiator AUG. Such sequences are known as the **internal ribosome entry sites (IRES)**. IRES structure has conserved structural motifs but less conserved sequence motifs. IRES have been categorized into different types based on their requirement. Type I requires eIF4G and associated factors, Type II requires eIF2 and associated factors but not eIF4G, Type III doesn't require either eIF2 or GTP, and Type IV seems to interact directly with rRNA. For example, hepatitis C virus can bind to mRNA at an IRES without any initiation factors. Another major difference is that eukaryotes have methionine as initiator amino acid and not formylated methionine. The eukaryotic mRNA has stretches of multiple adenine residues at its 3′ end, known as **poly(A) tail** which is unique to eukaryotes.

Archaeal translation shares features of both eukaryotic and prokaryotic translation with certain differences from both the systems. Archaeal ribosomal subunit composition is similar to that of bacteria than of eukaryotes. The process of translation initiation is similar to that of eukaryotes. It uses methionine as first amino acid and not *N*-formylated methionine. Unlike bacteria, most of archaeal mRNA is leaderless, lacking 5′UTR and lacking Shine-Dalgarno sequence. The process of mRNA start site recognition in leaderless sequences is not yet known. Both bacteria and eukaryotes have 5′ UTR unlike most archaea. Therefore, it is hypothesized based on observations, and in vitro experiments that last common ancestor (LUCA) of bacteria, archaea, and eukaryotes had leaderless sequence lacking 5′UTR. 5′ UTR is a more recent development. Though in a few archaeal species Shine-Dalgarno sequence have been found. Archaeal initiation factors (aIFs) are homologous in structure and function to eukaryotic initiation factors (eIFs).

11.8.1.3 Eukaryotic Translation Elongation

The biochemistry of polypeptide elongation is highly conserved in all three domains of life. They all use two elongation factors: one for bringing the charged tRNA to the acceptor site and the other for translocation of ribosome to the 3′ direction. The names of archaeal and eukaryotic homologs of EF-Tu and EF-G present in bacteria are given in the table below (Table 11.5). Based on sequence comparison, archaeal elongation factors are more similar to that of eukaryotes; hence eukaryotes are evolutionary closer to archaea, as proposed by Carl Woese. Elongation rate in bacteria is 20 amino acids per second, and in eukaryotes, it is 15 amino acids per second.

Table 11.5 Translation elongation factor homologs

Function	Bacterial homolog	Archaeal homolog	Eukaryotic homolog
Adjusts tRNA in A site	EFT	aEF1	eEF1
Promotes translocation	EFG	aEF2	eEF2

11.8.1.4 Eukaryotic Translation Termination

The elongation cycle continues until one of the three stop codons UAA, UAG, or UGA is encountered at the A site. Instead of charged tRNA, the release factors (RF) bind and catalyze the release of polypeptide bound to tRNA at the P site. The eukaryotic release factor, eRF3, is a GTP-binding protein. The eRF3-GTP acts in concert with eRF1 to promote cleavage of the peptidyl-tRNA, thus releasing the completed protein. The uncharged tRNA now occupies the P site. After the release of polypeptide, **ribosome release factors** bind and cause translocation of ribosome. As there is no peptide, the tRNA occupying the earlier P site now exits the ribosome from the E site, and ribosome disassembles.

11.9 Regulation of Protein Translation

Translation process is majorly regulation at the initiation step by various mechanisms which we are going to discuss in this section. The rate of assembly of multifactor preinitiation complex and its interaction with eIF4F complex interacting with the mRNA cap form the first step in the regulation of translation initiation (Sonenberg, N., Hinnebusch, A.G.: Regulation of translation initiation in eukaryotes: mechanisms and biological targets. Cell. 136, 731-745, 2009). The rate of selection of the start codon by the scanning mechanism is the next step which can regulate the process of eukaryotic translation. The scanning process is in turn dependent upon the activity of helicases. One of the principal mechanisms of translation control is by the phosphorylation of eIF2α on Ser51. There are four different eIF2α kinases in mammals activated by different stresses, PKR (double-stranded RNA in virus infection), PERK (unfolded proteins in the ER), HRI (heme deprivation), and GCN2 (amino acid starvation), that phosphorylate the same residue in eIF2α. eIF2α-GTP brings initiator $_{\mathrm{met}}$tRNA. Once tRNA base pairs with mRNA, eIF2α-GTP is hydrolyzed to eIF2α-GDP which then leaves the PIC. The guanine exchange factor eIF2B then removes GDP and replenishes eIF2-α-GTP for the next round of initiation. However, during cellular stress, the eIF2 kinases phosphorylate eIF2α which remain tightly bound to eIF2B. Hence, eIF2B is not able to exchange GDP to GTP. eIF2α-GTP is not replenished, delivery of $_{\mathrm{imet}}$tRNA is stopped, and hence translation initiation stops. Repression of translation initiation is an efficient mechanism to conserve energy and nutrients during stress.

Recently it has been shown that during stress conditions when the cells need to produce specific proteins related to stress management and reduce the production of other proteins that are involved in cell growth and proliferation, the abundance of tRNA pool varies. The tRNA pool gets skewed towards those which can recognize the rare codons. Thus, tRNA pool adjustment is an important mechanism for regulating the translation of stress-related proteins.

Another important mechanism for translation regulation is by controlling the 5′7mGcap. Binding of eIF4F can be hindered by eIF4E homolog, 4E-HP. The interaction between eIF4G and eIF4E in the eIF4F complex is hindered by eIF4E-binding proteins (4E-BPs). Hence, eIF4E-BPs can inhibit cap-dependent translation. miRNAs (approximately 22 nt long oligonucleotides) are another significant factor for controlling protein synthesis. miRNA can destabilize mRNA or inhibit translation by being a part of protein complex known as **RNA-induced silencing complex (RISC)**. miRNA-loaded RISC base pairs with complementary sites located at the 3′ UTR of many mRNA, regulating their translation.

11.9.1 Translation Pause

During translation, it is observed that at certain points of translation, complex ribosomes are close together as if they halted. This process is called as ribosome stalling which leads to a pause in the process of translation. This process is known to occur in both prokaryotes and eukaryotes. The stalling process is required for ensuring the fidelity of translation process. Hence, it is a part of ribosome-associated quality control process. Ribosome stalling promotes ribosome disassembly and releases the mRNA for degradation.

Box 11.1 Scientific Concept: Protein Synthesis by Single Ribosomes: Francesco Vanzi et al.

We have learned that the energy required for peptide bond comes from the high-energy ester bond which links the amino acid to the 3′ adenine residue of tRNA. Energy required for ribosome translocation, while maintaining the fidelity of amino acid selection and the correct reading frame, comes from the GTPase activities of two G-protein elongation factors, EF-Tu and EF-G. To understand the production of force and displacement of ribosome, single-molecule technique was needed.

In 1999, Sytnik et al. showed that ribosomes adsorbed on the mica surface are able to bind amino-acyl-tRNA at the peptidyl transfer site (P site) of the ribosome. Francesco et al. extended their work by measuring polyphenylalanine polypeptide synthesis on poly(U)-programmed mica-bound ribosomes. Ribosomes were adsorbed on mica surface at different concentrations of 70S ribosome. Surface images of ribosomes adsorbed on mica were obtained by tapping mode atomic force microscopy. Synthesis of poly(Phe) was observed to be dependent on elongation factor and temperature but not on the concentration of charged tRNA.

Changes in the range of motion of 3′ end of mRNA was done by microscopic techniques called tethered particle motion (TPM). Ribosomes (1 μM), N-Ac[14C]-Phe-tRNAPhe (1 μM), and biotinylated long-chain poly (U) (2.6 μg/μL) were preincubated at 30 °C for 15 min before being applied to a mica flow chamber (10–15 μL) for 5 min. The chamber was then washed,

(continued)

Box 11.1 (continued)
and blocking solution (15 μL) was added. Thereafter, neutravidin-labeled fluorescent beads were added and incubated for 30–60 min. The 3′ biotinylated ends of mRNA were then fluorescently labeled by neutravidin fluorospheres (Fig. 11.17). After the removal of excess beads by washing, three different groups of beads were visualized by epi-fluorescence with an inverted microscope: one freely diffusing under Brownian motion; second, immobilized beads; and third group with diffusive motion with restricted volume (i.e., the root means squared horizontal displacement from the average position of the bead, Drms >100 nm). The restricted Brownian motion was around a central nodal point, mostly in a radial pattern. They observed a time-dependent reduction in the Drms (Vanzi, F. et al. Protein synthesis by single ribosomes. RNA. 9, 1174-1179, 2003).

TPM data pooled from six beads displaying a reduction of Drms with time and applying the physical properties of poly(U) led to calculation of peptide synthesis rate by a single ribosome.

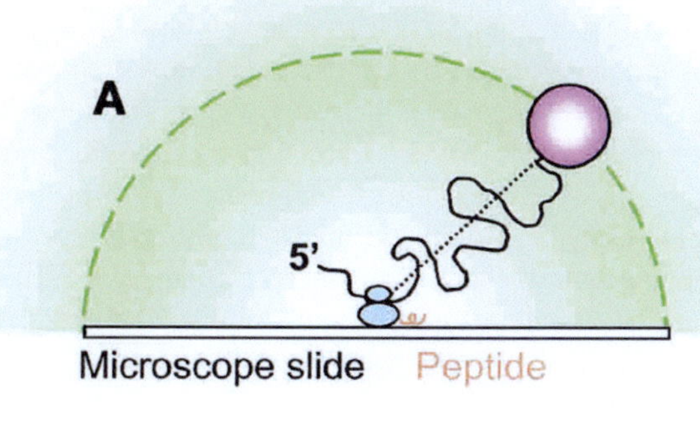

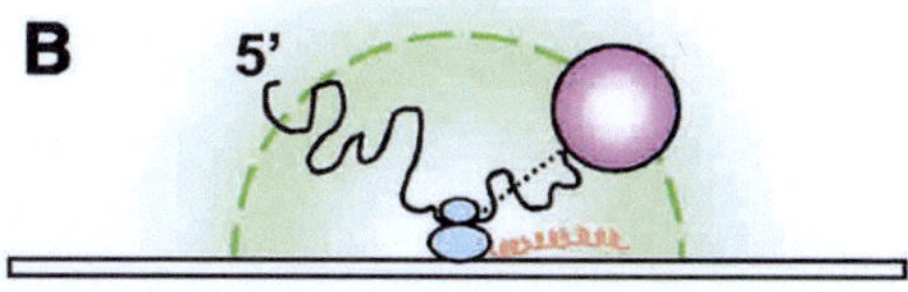

Fig. 11.17 Principle of the tethered particle method. (**a**) A microsphere (shown in pink) is attached to the 3′ end of immobilized ribosome bound mRNA molecule (curvy black line). End-to-end length of the tether is shown as a black dotted line. Nascent peptide is shown in orange. The microsphere can diffuse only in the radial area marked in green. (**b**) As peptide synthesis proceeds, the ribosome pulls the 3′ end of the mRNA towards itself, reducing the range of restricted diffusion

Box 11.2 Scientific Concept: Errors in Protein Synthesis Increase the Level of Saturated Fatty Acids and Affect the Overall Lipid Profiles of Yeast: Ana Rita D et al.

Protein synthesis is a highly accurate mechanism like DNA replication and transcription. Under normal conditions, the error rate of amino acid misincorporation is only 10^{-3} to 10^{-4}. Cell has its protein quality control mechanism like molecular chaperones, proteasomes, autophagy, etc. which ensures the normal functioning of the cell. However, there exist certain mutations which lead to increase in error rate causing significant deleterious effect on the cell.

In yeast, codon misreading decreases growth rate, alters cell morphology, increases reactive oxygen species, and causes deregulation of gene expression, proteotoxic stress, etc. Lipids play significant role in cell physiology by controlling membrane fluidity and participating in signaling cascades and energy metabolism. They are also important for protein functions and protein-protein interactions. Therefore, it is important to understand that errors in protein synthesis can have significant effect on lipid profiles and metabolism. Ana Rita D et al. constructed yeast strains to express a recombinant Ser-tRNA that decodes Ala and Gly codons (Araújo, A.R.D. et al. Errors in protein synthesis increase the level of saturated fatty acids and affect the overall lipid profiles of yeast. Plos One. 13, e0202402, 2018). The profiles of phospholipid (PL), triglyceride (TG), and fatty acids (FA) were then evaluated by various techniques. Fatty acid profiles were assessed by analyzing fatty acid methyl ester by gas chromatography. The analysis showed reduction in unsaturated fatty acids and increase in saturated fatty acids. PL classes were separated by thin layer chromatography (TLC). Phosphatidylcholine was the most abundant PL in the growth phase. The profile of triglycerides was analyzed by LC-MS (liquid chromatography-mass spectrometry). The most abundant TG species had either three C16 or two C16 and one C18 fatty acyl chains. The mechanism by which amino acid misincorporation leads to changes in lipid profile remains to be understood.

Box 11.3 Scientific Concept: Protein Synthesis by Single Ribosomes: Francesco Vanzi et al.

We have learned that the energy required for peptide bond comes from the high-energy ester bond which links the amino acid to the 3′ adenine residue of tRNA. Energy required for ribosome translocation, while maintaining the fidelity of amino acid selection and the correct reading frame, comes from the GTPase activities of two G-protein elongation factors, EF-Tu and EF-G.

(continued)

Box 11.3 (continued)

To understand the production of force and displacement of ribosome, single-molecule technique was needed.

In 1999, Sytnik et al. showed that ribosomes adsorbed on the mica surface are able to bind amino-acyl-tRNA at the peptidyl transfer site (P site) of the ribosome. Francesco et al. extended their work by measuring polyphenylalanine polypeptide synthesis on poly(U)-programmed mica-bound ribosomes. Ribosomes were adsorbed on mica surface at different concentrations of 70S ribosome. Surface images of ribosomes adsorbed on mica were obtained by tapping mode atomic force microscopy. Synthesis of poly(Phe) was observed to be dependent on elongation factor and temperature but not on the concentration of charged tRNA.

Changes in the range of motion of 3′ end of mRNA was done by microscopic techniques called tethered particle motion (TPM). Ribosomes (1 μM), N-Ac[14C]-Phe-tRNAPhe (1 μM), and biotinylated long-chain poly (U) (2.6 μg/μL) were preincubated at 30 °C for 15 min before being applied to a mica flow chamber (10–15 μL) for 5 min. The chamber was then washed, and blocking solution (15 μL) was added. Thereafter, neutravidin-labeled fluorescent beads were added and incubated for 30–60 min. The 3′ biotinylated ends of mRNA were then fluorescently labeled by neutravidin fluorospheres. After the removal of excess beads by washing, three different groups of beads were visualized by epi-fluorescence with an inverted microscope: one freely diffusing under Brownian motion; second, immobilized beads; and third group with diffusive motion with restricted volume (i.e., the root means squared horizontal displacement from the average position of the bead, Drms >100 nm). The restricted Brownian motion was around a central nodal point, mostly in a radial pattern. They observed a time-dependent reduction in the Drms.

TPM data pooled from six beads displaying a reduction of Drms with time and applying the physical properties of poly(U) led to calculation of peptide synthesis rate by a single ribosome.

11.10 Summary

- Genetic code was deciphered by Nirenberg, Khorana, and Holley. Har Gobind Khorana synthesized polynucleotides without a template to decipher the genetic code.
- Genetic code is triplet code formed by three nucleotides specifying an amino acid. Genetic code is universal, degenerate that is one amino acid can be coded by more than one codon, unambiguous that is one codon can code for only one amino acid, and contiguous as there are no gaps and no overlapping.

- Triplet codon gives 64 combinations (4^3) to be recognized by 20tRNA, coding for 20 amino acids. The reading frame is marked by an initiation codon that is either AUG or GUG. There are three stop codons: UAA, UAG, and UGA.
- Protein synthesis comprises of four steps: (1) binding of amino acids by appropriate tRNAs, (2) initiation, (3) elongation, and (4) termination.
- Amino acyl tRNA synthetase enzymes bring about charging of tRNA with appropriate amino acids. Charged tRNA interacts with the codon of the mRNA with the help of its anti-codon loop. This binding allows flexibility at 3′ end of codon which is called wobble hypothesis. Wobble rule allows one tRNA to recognize more than one codon.
- Protein synthesis occurs in the ribosomes; ribosomes have two subunits: small and large. Prokaryotes have 70S ribosome, and eukaryotes have 80S ribosome.
- Amino acid is linked to the 3′ end adenine by high-energy ester bond. The high-energy ester bond is used for peptide bond synthesis.
- Process of translation is divided into three steps: initiation, elongation, and termination. In bacterial translation initiation, the small subunit of ribosome gets positioned over initiation codon. The first tRNA brings in *N*-formylmethionine in bacterial cells, and subsequently the large subunit of ribosome binds with the help of initiation factors and GTP.
- In prokaryotes, Shine-Dalgarno sequence is recognized by the small ribosome subunit during scanning, and in eukaryotes, mRNA has Kozak sequence which is recognized by small ribosomal subunit. Some eukaryotic mRNAs lack consensus initiation sequence but have internal ribosome entry site.
- 7-methylguanosine cap (7-mG) is present in eukaryotic mRNA. Circularization of mRNA occurs during translation in eukaryotes.
- Elongation step is dependent on two GTP-bound translation elongation factors. During elongation, charged tRNA enters A site, and peptide bond is formed between amino acid at A site and P site. The ribosome moves to the next codon, and earlier P site becomes the E site; A site becomes the new P site.
- Termination occurs when either of the three stop codons occupies the A site. The stop codons are recognized by release factors, and ribosome is disassembled by ribosome release factors.
- Translation is tightly regulated in the initiation step by phosphorylation of eIF2α by eIF2α kinase.

References

Alberts B et al (2002) Molecular biology of the cell, 4th edn. Garland Science, New York

Arakawa K, Suzuki H, Tomita M (2008) Computational genome analysis using the G-language system, Genes, genomes and genomics. Global Science Books, pp 21–13

Araújo ARD, Melo T, Maciel EA, Pereira C, Morais CM, Santinha DR, Tavares JF, Oliveira H, Jurado AS, Costa V, Domingues P, Domingues MRM, Santos MAS, Witt SN (2018) Errors in protein synthesis increase the level of saturated fatty acids and affect the overall lipid profiles of yeast. PloS one 13(8):e0202402. https://doi.org/10.1371/journal.pone.0202402

Beuning PJ, Musier-Forsyth K (2000) Hydrolytic editing by class II aminoacyl-tRNA synthetase. Proc Natl Acad Sci U S A 97:8916–8920

Desai N, Brown A, Amunts A, Ramakrishnan V (2017) The structure of the yeast mitochondrial ribosome. Science 355(6324):528–531. https://doi.org/10.1126/science.aal2415

Desai N, Yang H, Chandrasekaran V, Kazi R, Minczuk M, Ramakrishnan V (2020) Elongational stalling activates mitoribosome-associated quality control. Science 370(6520):1105–1110. https://doi.org/10.1126/science.abc7782

Gallie DR (2014) The role of polyA binding protein in the assembly of cap binding complex during translation initiation in plants. Translation (Austin) 2:e959378

Griffiths AJF (2015) Introduction to genetic analysis. W.H. Freeman & Company, New York

Klug WS, Cummings MR, Spencer CA, Palladino MA (2012) Concepts of genetics, 10th edn. Pearson Education, California, pp 344–373

Lang K, Erlacher M, Wilson DN, Micura R, Polacek N (2008) The role of 23S ribosomal RNA residue A2451 in peptide bond synthesis revealed by atomic mutagenesis. Chem Biol 15 (5):485–492. https://doi.org/10.1016/j.chembiol.2008.03.014

Lewin B (2004) Genes VIII Chicago, 15th edn. Pearson Prentice Hall, Upper Saddle River

Miller OL Jr, Hamkalo BA, Thomas CA Jr (1970) Visualization of bacterial genes in action. Science 169:392–395

Nissen P, Hansen J, Ban N, Moore PB, Steitz TA (2000) The structural basis of ribosome activity in peptide bond synthesis. Science 289:920–930

Russell PJ (2010) iGenetics: a molecular approach, 3rd edn. Benjamin Cummings, New York, pp 102–128

Sanders MF, Bowman JL (2015) Genetic analysis: an integrated approach, 2nd edn. Pearson Education, New Jersey, pp 305–337

Schmeing TM, Voorhees RM, Kelley AC, Gao Y-G, Murphy FV, Weir JR, Ramakrishnan V (2009) The crystal structure of the ribosome bound to EF-Tu and aminoacyl-tRNA. Science 326 (5953):688–694. https://doi.org/10.1126/science.1179700

Slayter HS, Warner JR, Rich A, Hall CE (1963) The visualization of polyribosomal structure. J Mol Biol 7(6):652–657, IN5. https://doi.org/10.1016/S0022-2836(63)80112-6

Sonenberg N, Hinnebusch AG (2009) Regulation of translation initiation in eukaryotes: mechanisms and biological targets. Cell 136:731–745

Vanzi F, Vadimirov S, Knudsen CR, Goldman YE, Cooperman BS (2003) Protein synthesis by single ribosomes. RNA 9:1174–1179

Regulation of Gene Expression in Prokaryotes

12

Tanushree Banerjee

We have learned how genetic information is encoded in DNA and how it is organized into genes. We have also learned that the information encoded in genes gets translated into proteins. We must understand that not all proteins are required at equal quantity at all times. Some are required more, and some are required less in amount. Depending upon environmental conditions, stress factors, growth phase, and the requirement of a cell keep changing. Therefore, the protein synthesis rates also keep changing. The process of producing gene products is called as gene expression, and it is a tightly regulated process so that it can be modified depending upon the requirement of the cell. In this chapter, we will learn about the mechanism of gene regulation specifically in prokaryotes.

Bacteria serves as an important model organism for understanding gene regulation. It has short life cycle, multiple generations can be studied easily over a short span of time, and hence it is easy to identify mutants. Therefore, we now discuss the changes in gene expression in response to environmental changes.

12.1 Concept of Gene Regulation

Escherichia coli is an excellent model organism for studying gene regulation. They can switch on and switch off expression of certain genes depending upon environment or phase of life cycle like gene replication, cell division, etc.

It was observed that bacteria synthesize lactose-metabolizing enzymes only when lactose was present in the medium. These enzymes were therefore called as adaptive or facultative enzymes. Later, that terminology was replaced with inducible enzyme

T. Banerjee (✉)
Molecular Neuroscience Research Laboratory, Dr. D. Y. Patil Biotechnology and Bioinformatics Institute, Dr. D. Y. Patil Vidyapeeth, Pune, India
e-mail: tanushree.banerjee@dpu.edu.in

D. Kar, S. Sarkar (eds.), *Genetics Fundamentals Notes*,
https://doi.org/10.1007/978-981-16-7041-1_12

as the production gets induced only when the inducer-like lactose is present. The pathway is then said to be inducible. The other enzymes which are always present are called constitutive enzymes.

A contrasting system also exists, where the presence of gene product inhibits the gene expression. Tryptophan is an amino acid which gets synthesized by the cell. If tryptophan is present in sufficient amount, the cell does not need to synthesize it anymore. Therefore, it inhibits the anabolic pathway which leads to tryptophan synthesis. Therefore, the end product of the pathway tryptophan acts as a repressor. Therefore, the pathway is then said to be repressible.

Inducible and repressible pathways can be controlled by positive and negative regulation. In positive regulation mode, the gene expression continues when a regulator molecule directly stimulates RNA production. In this case, the regulator molecule is called as activator. In negative regulation mode, the RNA production continues as default unless it is being shut off by a regulator molecule. That regulator molecule is called as repressor. Sometimes, another small molecule binds to the repressor and enables it to bind to DNA causing repression of gene expression (Fig. 12.1). This small molecule is called as corepressor. Other mechanism for gene repression is an inhibitor binding to an activator so that positive regulation by activator does not occur. Both corepressor and inhibitor lead to gene repression. Therefore, it can be said that negative regulation is important where default pathway of expression is "on" and positive regulation occurs where default pathway is "off." Both inducible and repressible systems could be controlled by the combination of positive and negative control. Both lactose metabolism pathway and tryptophan synthesis pathway are under negative control mechanism.

12.2 Concept of Operon

It was at the turn of the twentieth century that Francois Jacob and Jaques Monod gave the concept of adaptive enzymes. Regulation of the lac operon was first described by François Jacob and Jacques Monod. Their research demonstrated how enzyme quantities can be controlled directly at the level of transcription. They hypothesized that certain metabolic enzymes are expressed only when cell is exposed to their respective substrate. When a bacterium does not have the substrate, it does not need to synthesize the enzyme which metabolizes it. They further investigated this hypothesis by studying lactose metabolism in *E. coli*. They observed that presence of lactose increased the enzymatic activity of the lactose-metabolizing enzymes by 1000–10,000 times. They also observed that lactose-metabolizing enzyme activity increased as there was enhancement in the expression of the enzyme. Removal of lactose leads to immediate lowering in the gene expression of the entire lactose metabolism pathway. Hence, it became evident that all the genes involved in lactose metabolism are being controlled for transcription together. The group of two or more genes which get transcribed from a single promoter are called as **operon**. Therefore, operon could be called as a DNA which gets transcribed by a single promoter synthesizing polycistronic mRNA (two or more genes). An

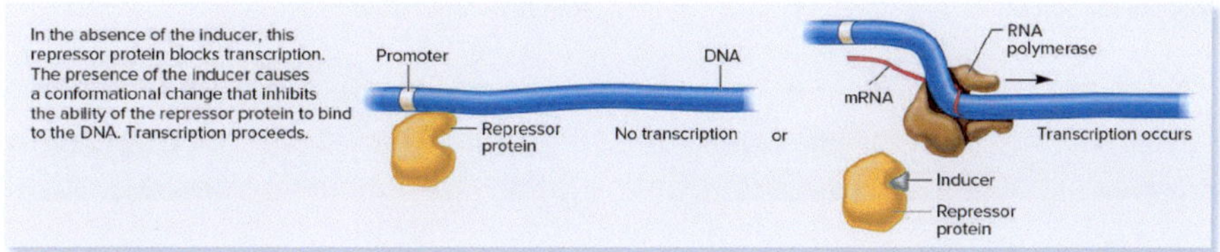

(a) Repressor protein, inducer molecule, inducible gene

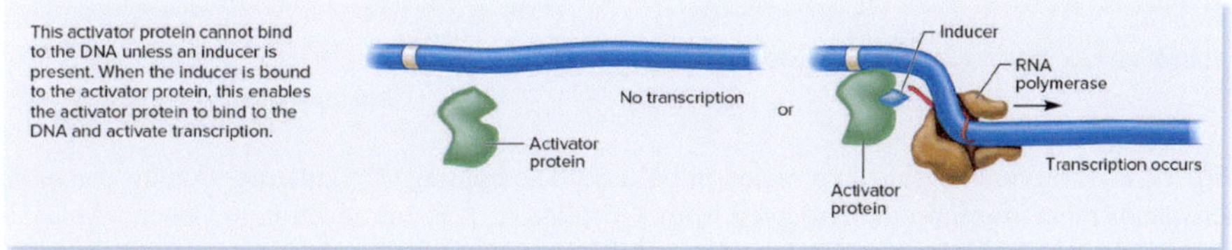

(b) Activator protein, inducer molecule, inducible gene

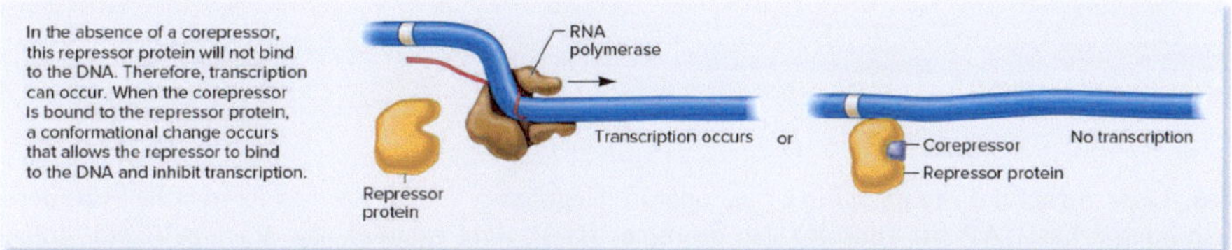

(c) Repressor protein, corepressor molecule, repressible gene

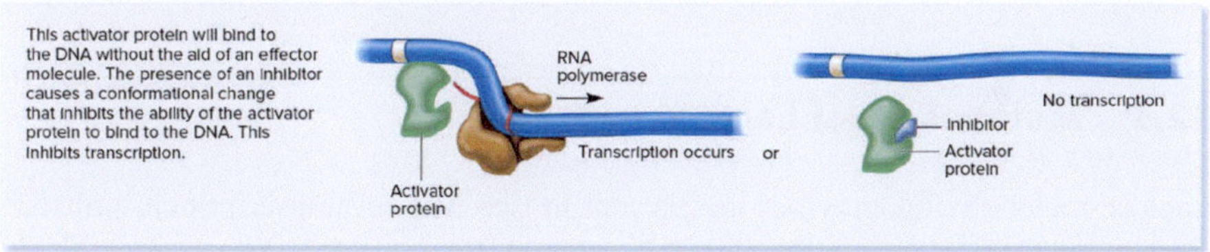

(d) Activator protein, inhibitor molecule, repressible gene

Fig. 12.1 Binding sites on a genetic regulatory protein. In these examples, a regulatory protein has two binding sites: one for DNA and the other for a small effector molecule. The conformational changes in the regulatory protein are brought about by the binding of small effector molecule, leading to changes in its DNA-binding site. **a)** Regulation of operon by repressor in presence and absence of inducer **b)** Regulation of operon by activator in presence and absence of inducer **c)** Regulation of operon by repressor in presence and absence of co-repressor **d)** Regulation of operon by activator in presence and absence of inhibitor (Adapted from: Brooker, R.J.: Genetics: Analysis and Principle, 6th ed. pp 336–360. McGraw-Hill Education, New York, 2018)

operon is flanked by a promoter for initiating mRNA synthesis and a terminator indicating end of mRNA synthesis (Fig. 12.2).

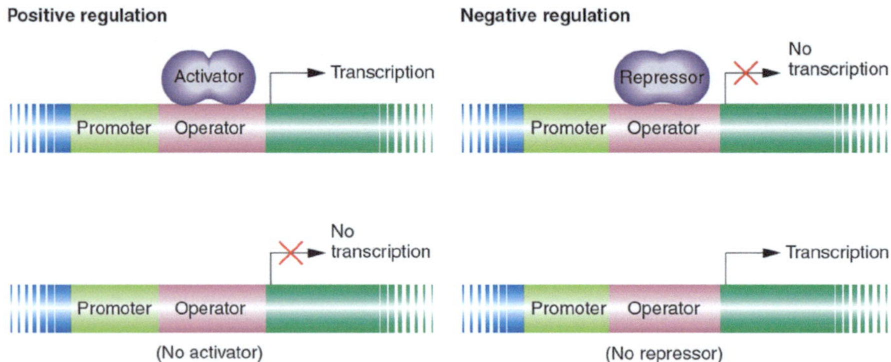

Fig. 12.2 Overview of gene expression in bacteria. The binding of regulatory proteins can either activate or block transcription (Adapted from: Griffiths, A. J. F. Introduction to Genetic Analysis. New York, NY :W.H. Freeman & Company, 2015)

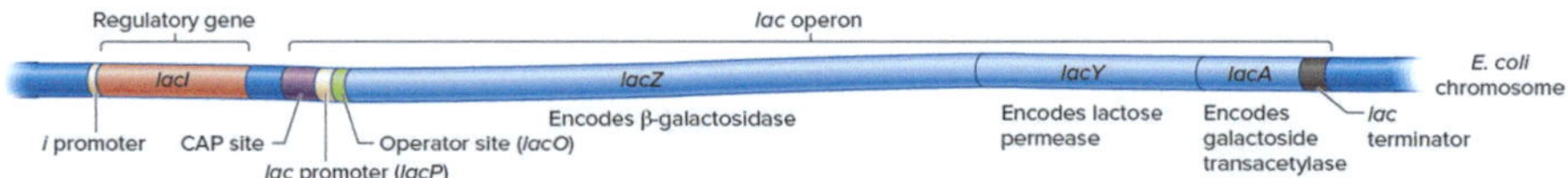

Fig. 12.3 Structural organization of lac operon. Regulatory *lacI* genes has its own lacI promoter. Lac operon has CAP site (purple), lac promoter (lacP, light orange), lac operator (lacO, green), structural genes (*lacZYA*, blue), and lac terminator (gray) (Adapted from: Brooker, R.J.: Genetics: Analysis and Principle, 6th ed. pp 336–360. McGraw-Hill Education, New York, 2018)

12.3 Metabolism of Lactose in *E. coli*

Lactose-metabolizing enzymes are present in two distinct transcriptional units: *lac* operon and *lacI*. *lacI* is not the part of lac operon. First, we will discuss about the lac operon. It has a catabolite activator protein (CAP) site, promoter (lacP), operator (lacO), and three protein-encoding genes *lacZ, lacY,* and l*acA*. *lacZ* encodes for β-galactosidase (Fig. 12.3). This enzyme cleaves lactose into glucose and galactose. *lacY* encodes for lactose permease. This protein helps in active uptake of lactose. *lacA* encodes galactoside transacetylase which modifies lactose and other lactose analogues by attaching hydrophobic acetyl groups. Attachment of acetyl group helps lactose to diffuse out of the cell, preventing toxicity caused by excessive lactose. CAP site is regulatory DNA sequence present in lac operon. *C*atabolite *a*ctivator *p*rotein (CAP) binds to this site, hence called CAP site. Operator site is another DNA regulatory element where the lac repressor binds (Beckwith 1967).

lacI has its own promoter which is constitutively expressed at a low level. It encodes for lac repressor which binds to the operator of the *lac* operon. It is a homotetrameric protein which is required in small amounts to repress lac operon (Fig. 12.4). Negative supercoiling of DNA enhances lacI binding to the operator;

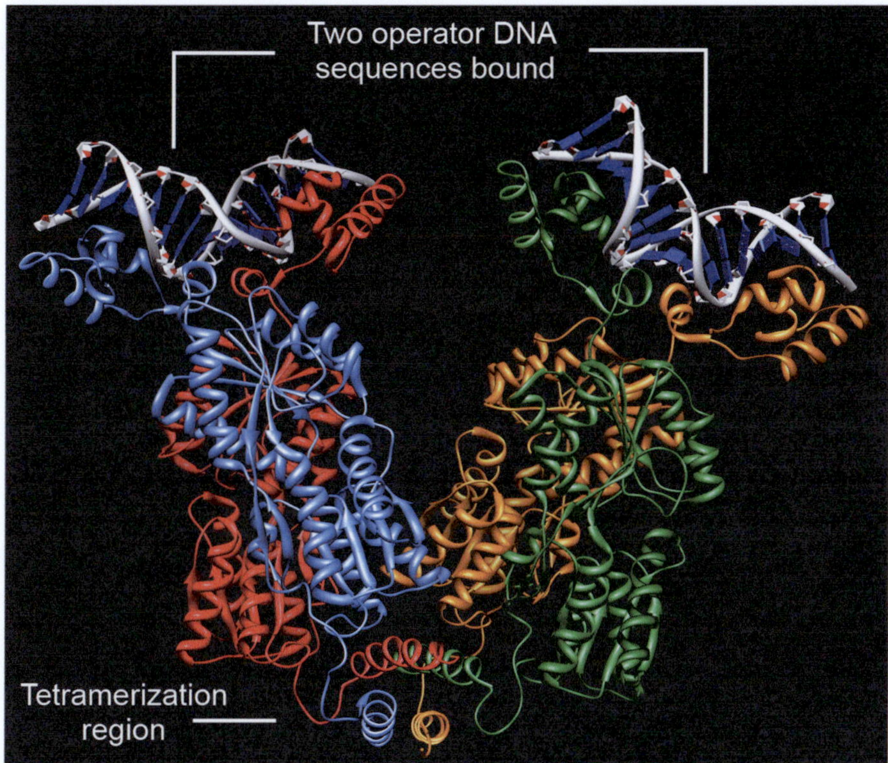

Fig. 12.4 LacI binds to two operator sequences in tetrameric form leading to looping of DNA. There are two dimeric *lacI* functional subunits (red+blue and green+orange). Each subunit binds to two DNA operator sequences (labeled). Tetrameric *lacI* thus induces DNA looping (Rutkauskas D. et al. Proc Natl Acad Sci U S A. 106, 16627–16632, 2009)

hence, it is known to act as a topological barrier which is superhelically modulated to control the expression of *lacZYA* under various growth conditions.

12.3.1 Lactose Uptake in *E. coli*

Lactose is taken up by lactose permease. It is encoded by *lacY* gene. It is a transmembrane protein which actively takes up lactose from the environment. It cotransports lactose with H^+. Since bacteria maintains a high proton gradient across its membrane, it actively accumulates lactose in the cytosol against its concentration gradient by utilizing the free energy released from downhill translocation of H^+. *LacY* also catalyzes the converse reaction, utilizing free energy released from downhill translocation of sugar to drive uphill translocation of H^+. It has 12 helices that forms a hydrophilic cavity creating a single binding site that is alternatively

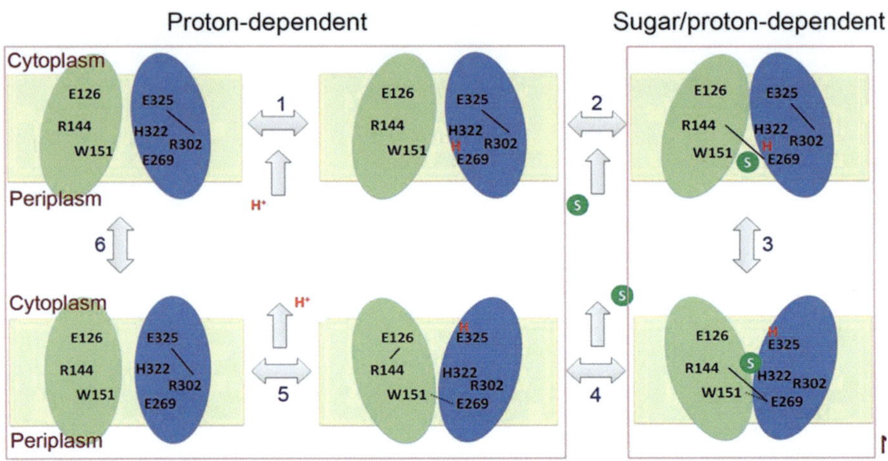

Fig. 12.5 A possible lactose/H$^+$ symport mechanism. The key residues involved in changing the symporter conformation are indicated. H-bonds are shown in black dotted lines. The protons are shown in red, and the substrates are shown in green (Kaback, H.R. C R Biol. 328, 557–67, 2005)

Table 12.1 lac operon genes and regulatory sequences

Gene/Sequence	Product/Sequence Type	Function	Important Mutants
Protein-Producing Genes			
lacI	Repressor protein	Contains two binding sites, one for the operator and one for allolactose, the inducer	I$^-$: Unable to bind to operator I^S: Unable to bind the inducer (allolactose)
lacZ	β-galactosidase	Cleaves lactose into two monosaccharides (glucose and galactose)	Z$^-$: No functional β-galactosidase
lacy	Permease	Facilitates lactose transport across the cell membrane	Y$^-$: No functional permease
lacA	Transacetylase	Protects against harmful by-products of lactose metabolism	A$^-$: No transacetylase
Regulatory Sequences			
lacO	Operator	Binds repressor protein to block transcription of operon genes	O^C: Fails to bind repressor protein
lacP	Promoter	Binds RNA polymerase	P$^-$: Fails to bind RNA polymerase or does so weakly

accessible to either side of the membrane. Binding of substrate leads to major conformational shift in the enzyme known as inward and outward conformations. Influx consists of six steps: starting from the outward-facing conformation (A), protonation of LacY (B), binding of lactose (C), a conformational change that results in the inward conformation (D), release of substrate (E), release of the H+ (F), and return to the outward conformation (Fig. 12.5).

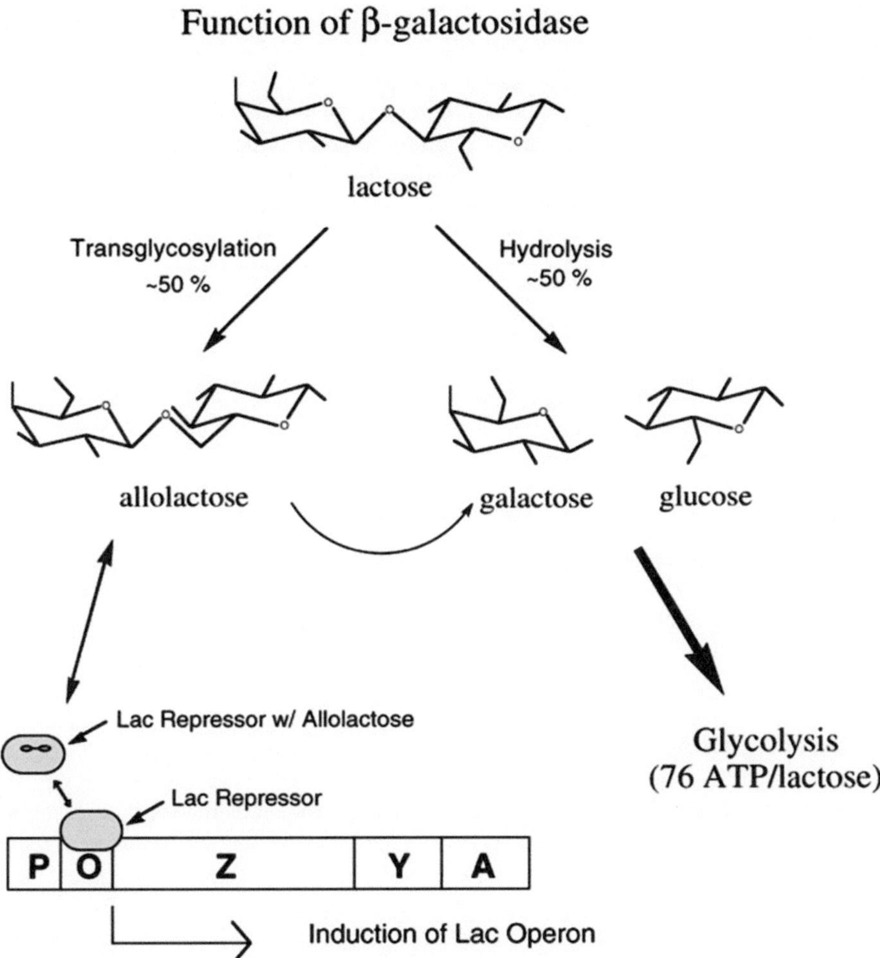

Fig. 12.6 Schematic summarizing of the roles of β-galactosidase in the cell. The enzyme β-galactosidase can cause hydrolysis of lactose to galactose plus glucose and transgalactosylate lactose to form allolactose, and it can hydrolyze allolactose. The presence of lactose results in the synthesis of allolactose which binds to the lac repressor and reduces its affinity for the lac operon. This in turn allows the synthesis of β-galactosidase, the product of the *lacZ* gene (Adapted from: Juers, D.H. et al. Protein Sci. 21, 1792–807, 2012)

12.3.2 Structural Gene

There are three structural genes present in the lac operon, *lacZ, lacY,* and *lacA*. *lacZ* codes for the enzyme β-galactosidase (Table 12.1). β-galactosidase is a tetramer of four identical polypeptide chains, each of 1023 amino acids. This enzyme has three catalytic activity: It can cleave the disaccharide lactose to form glucose and galactose, which can then enter glycolysis; the enzyme can catalyze the transgalactosylation of lactose to allolactose, and the allolactose can be cleaved to the monosaccharides. It is allolactose that binds to lacZ repressor and creates the positive feedback loop that regulates the amount of β-galactosidase in the cell. It hydrolyzes X-gal (5-bromo-4-chloro-3-indoyl-β-D-galactopyranoside), a soluble colorless compound. X-Gal consists of galactose linked to a substituted indole (Fig. 12.6). Its hydrolysis releases the substituted indole. Indole then spontaneously dimerizes to give an insoluble, intensely blue product.

lacY gene codes for an integral cytoplasmic membrane protein, lactose permease. It belongs to a conserved transporter family known as Major Facilitator Superfamily. LacY protein is made of 417 amino acid residues. It has 12 helices. It acts as symporter of H^+ and lactose. Helices are connected with hydrophilic loops with both N and C termini facing the cytosol.

lacA gene codes for galactoside transacetylase. It acetylates lactose, and acetylated lactose can diffuse out of the cell membrane reducing lactose-induced toxicity. Unlike β-galactosidase and lactose permease, the function of galactoside transacetylase remains debatable. The action of acetylation of lactose was extended to other galactoside molecules like isopropyl-1-thio-βD-galactoside (IPTG), with the identification of 6-O acetyl-IPTG as the chemically altered compound. However, acetylated-IPTG did not revert to the original IPTG and did not act as an inducer of the lac operon or as a substrate of lactose permease.

12.3.3 Regulatory Mutation

In the 1950s, Jacob, Monod, and Pardee identified rare mutant strains of bacteria which had abnormal lactose metabolism. One type of mutant-designated *lacI*$^-$ resulted in the constitutive expression of lac operon. Even when the lactose was absent, lac operon in these mutants continued to express. The exact mode of action of lac repressor was not known at that time. It was thought to produce an activator for the operon which kept it transcriptionally active throughout. To understand the nature of this mutation, they applied a genetic approach in 1961. It involved bacterial mating by conjugation. The circular segments of DNA are called F factors. Sometimes, the F factors carry the genes that were originally present in the chromosome. Then they are called as F′ factors. They identified F′ factors which carried the *lacI* gene and lac operon. These F′ factors can get transferred from one cell to another by conjugation. The strain of bacteria containing F′ factors are called as **merozygote or partial diploid**.

Mutations in Regulator Gene: Lac repressor encoded by the *lacI* gene has two binding sites: one for binding the DNA at the operator site and other for binding allolactose. There are lacI$^-$ mutants which either fail to produce lac repressor or the repressor is unable to bind to the operator site. Hence, the lac operon remains constitutively active both in presence and absence of lactose. There are other regulatory mutations known for the *lacI* gene. *lacI*^{s} mutation produces repressor that cannot bind allolactose. Hence, it always remains bound to the operator

Table 12.2 Synthesis of beta-galactosidase and permease by haploids and partial diploids with regulatory gene mutation

Genotype	β-Galactosidaise		Permease		Description
	Lactose	No Lactose	Lactose	No Lactose	
1. $I^- P^+ O^+ Z^+ Y^+$	+	+	+	+	Constitutive transcription due to lacI⁻ mutation.
2. $I^+ P^+ O^C Z^+ Y^+$	+	+	+	+	Constitutive transcription due to lacO^C mutation.
3. $I^S P^+ O^+ Z^+ Y^+$	–	–	–	–	Transcription is not inducible, due to lacI^S mutation.
4. $I^+ P^- O^+ Z^+ Y^+$	–	–	–	–	No effective transcription, due to lacP⁻ mutation.

Table 12.3 Synthesis of beta-galactosidase and permease by haploids and partial diploids with structural gene mutation

Genotype	β-Galactosidaise[a]		Permease[a]		Description
	Lactose	No Lactose	Lactose	No Lactose	
1. $I^+ P^+ O^+ Z^+ Y^+$	+	–	+	–	Wild-type (lac$^+$)
2. $I^+ P^+ O^+ Z^- Y^+$	–	–	+	–	No functional β-galactosidase (lac$^-$)
3. $I^+ P^+ O^+ Z^+ Y^-$	+	–	–	–	No functional permease (lac$^-$)
4. $I^+ P^+ O^+ Z^+ Y^- / I^+ P^+ O^+ Z^- Y^+$	+	–	+	–	Wild-type response by complementation (lac$^+$)

[a] Symbols + and – indicate production and no production, respectively, of functional enzymes.

(Table 12.2). Even if lactose is present, the operon cannot be induced creating a **super-repressed state.**

Mutation in Structural Genes: Mutant strains which had lost the ability to synthesize β-galactosidase or permease were identified and were mapped to *lacZ* and *lacY* structural genes, respectively. These mutations led to changes in the amino acid sequence and structure of these proteins which caused loss of function in most cases. From the study of partial diploids, they could infer that *lacZ* and lacy mutations are independent of each other. Partial diploids with *lacZ* + *lacY*- on plasmid and *lacZ-lacY+* on bacterial chromosome produced normal enzymes (Table 12.3). Hence, they could conclude that single functional gene either in chromosome or in plasmid is capable of producing normal lac operon phenotype (Juers et al. 2012).

Mutation at DNA-Binding Site: Jacob and Monod could identify mutants having defective promoter regions. Those were called *lacP*- mutations. RNA polymerase is not able to bind to the defective promoter. These mutant strains don't produce any protein of the lac operon either in the absence or presence of lactose. lacP- mutations are *cis* acting and hence inhibit the synthesis of genes present in the same DNA molecule. Therefore, *lacI+lacP-lacZ+/lacI+lacP+lacZ-* will not produce functional β-galactosidase.

Mutations in the Operator Site: They identified certain mutations where β-galactosidase was produced even in the absence of lactose. The mutation was mapped to the operator region. It was called constitutive operator or lacOc. Such mutation doesn't allow the binding of repressor. Hence, RNA polymerase can constitutively bind to the promoter, and transcription continues even in absence of lactose. A partial diploid *lacI + lacOclacZ+/lacI + lacO + lacZ+* shows constitutive production of β-galactosidase. It showed that constitutive mutation *lacOc* is dominant over *lacO+*. *lacI + lacO + lacZ+/lacI + lacOclacZ-* did not show constitutive β-galactosidase production. Therefore, the operator mutations affect genes which are present on the same molecules.

12.3.4 Positive and Negative Control

In positive control mode, the presence of an activator directly stimulates the operon transcription. For lac operon, presence of lactose acts as an activator, and hence, it is a positive regulator of the operon. Lactose gets converted to allolactose by β-galactosidase. Each of the four subunits of the lac repressor has single bonding site for allolactose. Allolactose then binds to the repressor, induces a major conformational change in the repressor, and reduces the affinity of the repressor for the operator. Hence, when allolactose is present, the repressor can no longer bind to the operator site, and the operon is actively transcribed. The action of a small effector molecule such as allolactose is called as **allosteric regulation**. Repressor molecule acts as the allosteric protein, and the binding sites for allosteric molecule is called **allosteric site**.

In negative control, the operon stays in actively transcribing mode unless bound by the repressor. For lac operon, repressor is transcribed by *lacI* gene in a separate transcriptional unit. This repressor binds to the operator site and keeps the operon in transcriptionally inactive state (*off* state). When the bacteria are not having any lactose in its system, it does not need to synthesize the lactose metabolizing enzymes. Hence, the operon is kept switched off by the repressor. Binding of repressor to the lac operator to keep it in default *off* state is considered to be the **negative control** of the operon.

12.3.5 Genetic Evidence for Operon Model

Generation of merozygotes described in the previous section could lead to identification of *lacI* gene function. There are two main points in their experiment: First, two *lacI* genes may be different alleles. For example, the chromosomal DNA might be carrying the mutant *lacI*$^-$ allele, and the F factor might carry the normal *lacI* allele. Second, the genes on F factor and chromosomal DNA are not physically adjacent to each other. Their group used the *lacI*$^-$ mutant strain which constitutively expresses lac operon and compared it with various merozygotes. The merozygote had a *lacI*$^-$ mutant gene on the chromosome and normal *lacI* allele on the F′ factor. Each strain was grown and divided into two tubes. One tube did not contain lactose, and the other tube contained lactose. Cells were sonicated which led to release of the enzymes present inside the cell. If lac operon was expressed, then among the released enzymes β-galactosidase would have been present. β-galactosidase was known to breakdown the substrate β-*o*-nitrophenylgalactoside (β-ONPG). β-ONPG is a colorless molecule, but upon being cleaved by β-galactosidase, it gives yellow color. Therefore, the amount of yellow color generated was the indication of presence of β-galactosidase and expression of lac operon.

In the *lacI*$^-$ mutant strain, yellow color due to β-ONPG breakdown was obtained whether or not lactose was present. It was expected as the mutant lacks the lac repressor, the operon is constitutively expressed, and β-galactosidase gets produced even in the absence of lactose. It means that *lacI*$^-$ mutant loses the inducible nature of lac operon. However, for the merozygote, β-galactosidase was observed to be expressed only in the presence of lactose as evident from the yellow color produced by the breakdown of β-ONPG. Therefore, it could be concluded that one normal copy of *lacI* gene is present in the F′ factor although the chromosome contained the mutant *lacI*. *lacI* therefore was predicted to encode repressor protein which could diffuse throughout the cell and bind to lac operon. It was also inferred that *lacI* gene need not be physically adjacent to the lac operon or present within the operon. Repressor, once present inside the cell, can bind to the operator either on chromosome lac operon or F′ factor lac operon. Therefore, the repressor is known as **trans-acting factor,** and this effect brought about by the repressor is called **trans effect** (Fig. 12.7).

In contrast, when a normal lac operon and *lacI* gene containing F′ factor is introduced into a cell with defective operator site on the chromosome, the lac operon on the chromosome continues to be expressed without lactose. This occurs because repressor cannot bind to the defective operator. Having a normal operator site in the F′ factor did not lead to inducible nature of chromosomal lac operon. Hence, operator did not show any trans effect. It, therefore, is known to have **cis-effect** or in other words, **operator is a cis-acting element**.

12.4 Mechanism of Lac Operon

As we learned in the previous sections, lac operon has the organization as CAP site, promoter site, operator site, *lacZ* gene, *lacY* gene, *lacA* gene, and the terminator site. *LacI* is not part of the operon and has a separate promoter site. We know that lac operon needs to get expressed only when lactose is present. In the absence of lactose, its analog allolactose is also absent. In the absence of allolactose, Lac repressor remains bound to the operator, preventing operon activation. However, in presence of lactose, allolactose becomes available. It binds to the repressor, changing its conformation so that it can no longer bind to the operator (Beckwith 1967). This leads to induction of lac operon expression (Fig. 12.8).

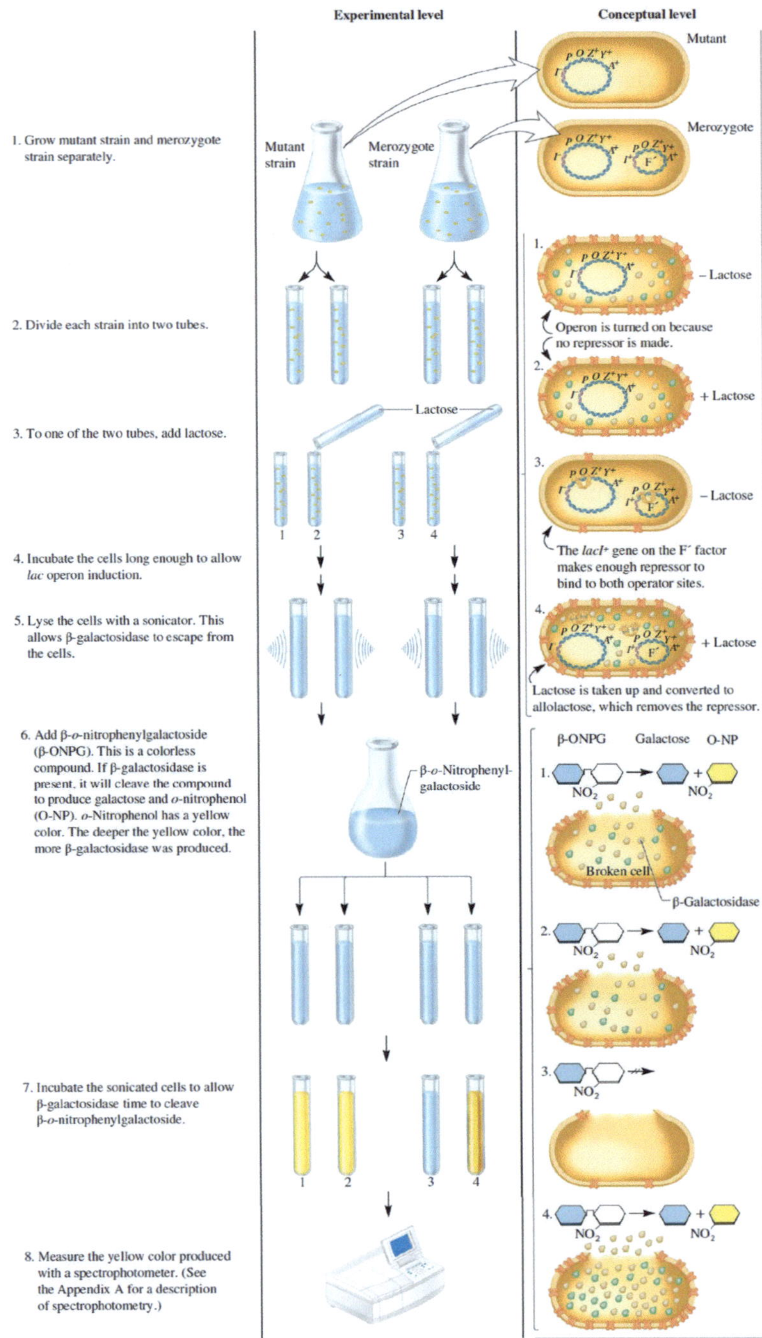

Fig. 12.7 Evidence that the *lacI* gene encodes a diffusible repressor protein. Starting material: The genotype of the mutant strain was *lacI*$^-$ *lacZ*$^+$ *lacY*$^+$ *lacA*$^+$. The merozygote strain had an F′ factor that was *lacI*$^+$ *lacZ*$^+$ *lacY*$^+$ *lacA*$^+$. The F′ factor had been introduced into the mutant strain via conjugation (Adapted from: Brooker, R.J.: Genetics: Analysis and Principle, 6th ed. pp 336–360. McGraw-Hill Education, New York, 2018).

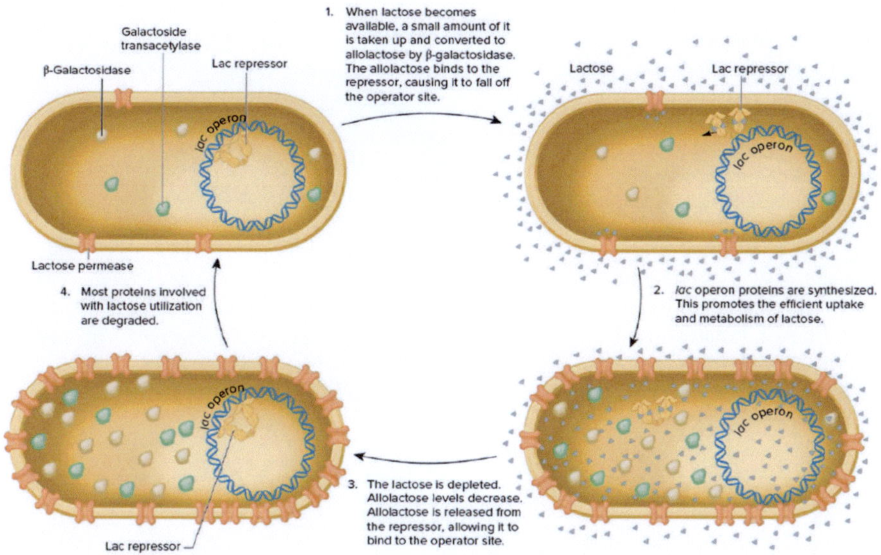

Fig. 12.8 The cycle of lac operon induction and repression. Lac operon codes for genes that metabolize lactose. When lactose is present, genes of the lac operon is induced, and proteins involved in lactose uptake and metabolism are synthesized. When the lactose is absent, the lac operon is repressed, blocking the transcription of lactose-metabolizing genes (Adapted from: Brooker, R.J.: Genetics: Analysis and Principle, 6th ed. pp 336-360. McGraw-Hill Education, New York, 2018)

In reality, the repressor is not able to inhibit the transcription of the lac operon completely. Basal levels of all three enzymes encoded by the lac operon are present although the level is too low and is not enough to enable the bacterium to utilize minimal lactose present in the environment. However, when lactose concentration rises in the environment, it needs lactose permease to go inside. At that point, the basal level of lactose permease present before the actual induction of lac operon helps the cell to take up lactose. Availability of lactose inside the cell then induces lac operon expression. Therefore, it is important that the repressor does not completely inhibit transcription of lac operon.

When lactose gets depleted in the environment, the concentration of allolactose inside the cell also keeps reducing due to the action of metabolic enzymes. After a certain point, allolactose concentration decreases below the minimal level required for repressor binding. Therefore, despite the high affinity of allolactose for the repressor, it is not able to bind to it.

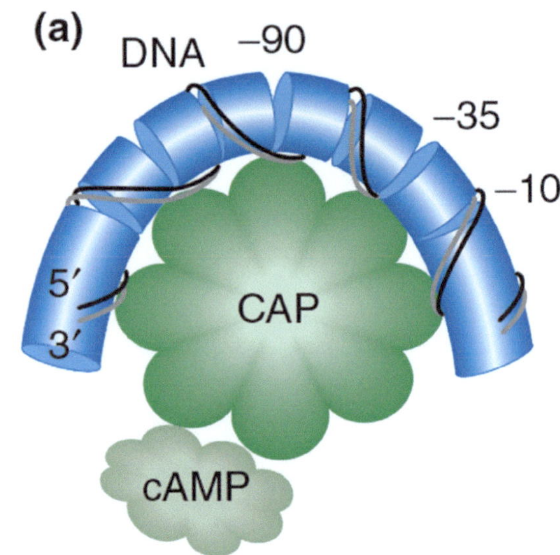

Fig. 12.9 CAP-cAMP complex. CAP-cAMP complex binding bends the DNA at an approximate angle of 90°. The bending enables the RNA polymerase to bind and activate transcription of lac operon (Chakerian, A.E., Matthews, K.S. Mol. Microbiol. 6, 963–968, 1992, Fulcrand, G. et al. Sci Rep. 14, 19243, 2016)(Adapted from: Griffiths, A. J. F, Introduction to Genetic Analysis. New York, NY :W.H. Freeman & Company, 2015)

Table 12.4 Transcription condition for the lac operon

Glucose	Lactose	cAMP	Allolactose	*lac* Operon Transcription	Explanation
Present	Absent	Absent	Absent	None	Glucose is present to provide energy. There is no allolactose to bind repressor. There is no CAP–cAMP complex to bind CAP site.
Present	Present	Absent	Present	Basal	Glucose is present to provide energy; absence of cAMP prevents positive transcription regulation, but allolactose is present and acts as an inducer to allow a small amount of transcription.
Absent	Absent	Present	Absent	None	CAP–cAMP forms, but no allolactose is present to block repressor binding at operator.
Absent	Present	Present	Present	High	Inducer and CAP–cAMP are available to induce and positively regulate transcription.

12.5 Regulation of Lac Operon

Apart from regulation of lac operon by lactose and repressor protein, it is also regulated by another way that is known as **catabolite repression**. It is brought about by the presence of glucose, which is a catabolite. We all know that glucose is the more readily available form of sugar which is easily catabolized to produce more energy in the form of adenosine triphosphate (ATP) than lactose. Considering the efficiency of energy production, the bacteria need not utilize lactose in the presence of glucose. Therefore, presence of glucose represses the transcriptional activation of

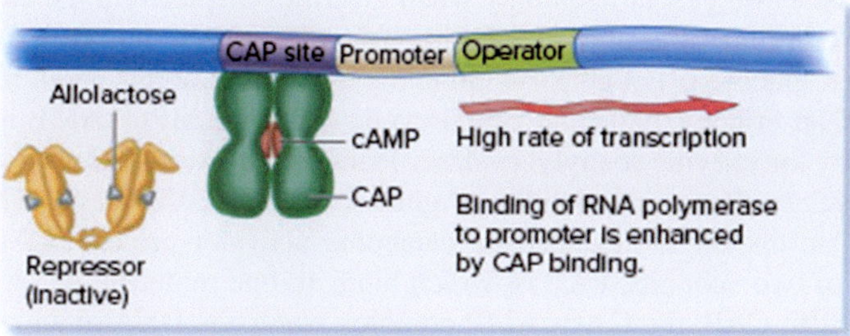

(a) Lactose, no glucose (high cAMP)

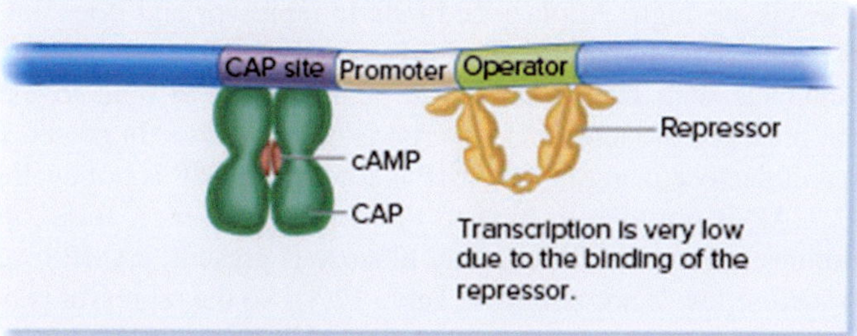

(b) No lactose or glucose (high cAMP)

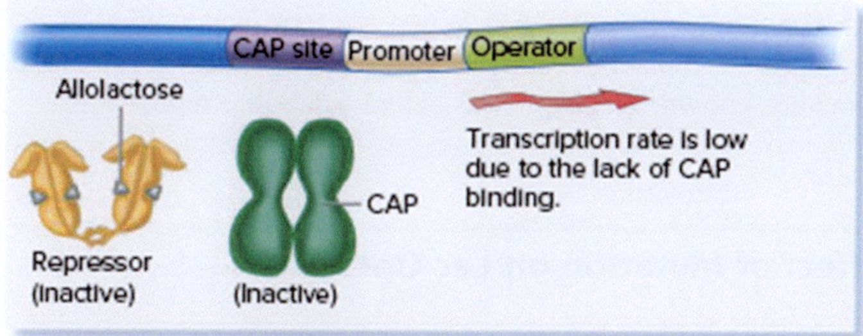

(c) Lactose and glucose (low cAMP)

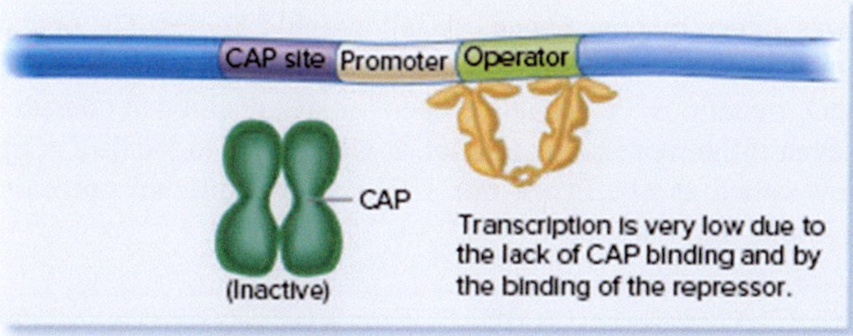

(d) Glucose, no lactose (low cAMP)

Fig. 12.10 Regulation of lac operon by catabolite repression. (**a**) High rate of transcription when lactose is present and glucose is absent. (**b**) Low rate of transcription in absence of lactose and glucose as CAP is active but repressor is bound to operator. (**c**) Both lactose and glucose are present.

lac operon. Lactose gets utilized only when glucose is depleted in the environment. The use of one sugar after the other by the bacteria is called **diauxic growth**.

However, glucose is not able to bind to the operon but brings about its action by the help of an effector molecule called cyclic-AMP (cAMP). cAMP is produced from ATP by the enzyme adenylyl cyclase. Transport of glucose into the cell inhibits adenylyl cyclase. Hence, availability of glucose reduces cAMP level. The effect of cAMP on lac operon is mediated by catabolite activator protein (CAP). CAP is composed of two subunits, each of which binds to one molecule of cAMP. When cAMP binds to CAP, the CAP-cAMP complex bends the DNA at an approximate angle of 90°. This bending enables the RNA polymerase to bind and activate transcription of lac operon (Fig. 12.9). When only lactose is present, allolactose and cAMP levels are high. Allolactose binds to repressor and does not allow it to bind to the operator. cAMP binds to the CAP. cAMP-bound CAP interacts with CAP site. CAP interacts with RNA polymerase and helps it to bind to the promoter. Therefore, in presence of lactose, lac operon is highly active. In presence of lactose and glucose, allolactose is high but cAMP is low. As cAMP is not available to bind to the CAP, CAP is not able to bind to the CAP site. Hence, transcription of lac operon continues at low rate. When only glucose is present, cAMP levels are very low and allolactose levels are also low (Table 12.4). So the repressor is bound to the operator site, and CAP site is not occupied by the CAP protein. Hence, transcription of lac operon continues at very low basal rate (Fig. 12.10).

The term catabolite repression may be confusing as it involves action of an inducer cAMP and an activator CAP. When the term was coined, it was observed that glucose represses transcription of lac operon. The mode of interaction of cAMP and CAP was not known. Hence, it was called catabolite repression.

12.6 Effect of Mutation on Lac Operon

In the previous section, we have learned about the lac operon mutation, *lacI*$^-$. It was the study on *lacI*$^-$ mutant that proved lac repressor is a diffusible *trans*-acting protein whereas operator is a *cis*-acting element. Later, molecular understanding of lac operon was gained by genetic and crystallographic studies. The operator site was identified due to mutations that prevented repressor binding. These mutations, called lacO$^-$ or lacO$_c$ mutations located at the operator site, resulted in constitutive operon expression even if the repressor is normal. It was shown to localize at operator site, which is now called as O_1. In the late 1970s, two additional operator sites were identified, O_2 and O_3.

Fig. 12.10 (continued) Transcription is low due to the lack of CAP binding. (**d**) Glucose is present and lactose is absent. Transcription rate is very low due to lack of CAP binding and binding of repressor (Adapted from: Brooker, R.J.: Genetics: Analysis and Principle, 6th ed. pp 336–360. McGraw-Hill Education, New York, 2018)

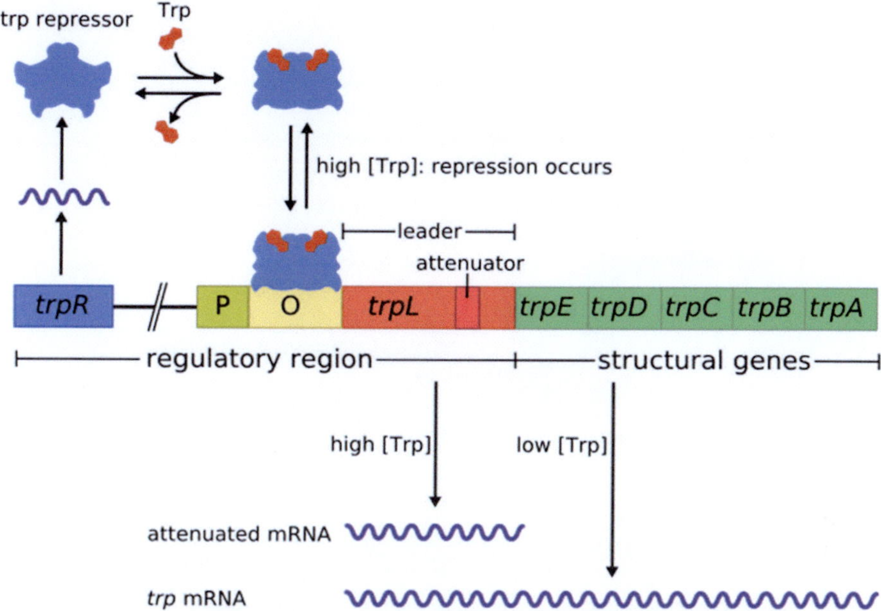

Fig. 12.11 Overview of trp operon. In absence of tryptophan, operon transcription occurs. In presence of high tryptophan, operon is repressed (Adapted from: https://en.wikipedia.org/wiki/Trp_operon)

12.7 Trp Operon in *E. coli*

We will now learn about another important operon called the trp (pronounced as "trip") operon. The genes present in this operon are responsible for encoding enzymes involved in the synthesis of tryptophan. This operon contains five enzyme-encoding genes called as *trpE, trpD, trpC, trpB,* and *trpA,* which are involved in tryptophan biosynthesis. Apart from these, there are two genes, *trpR* and *trpL*, that play a role in regulation of trp operon. *TrpL* is part of trp operon, and *trpR* acts as a separate transcriptional unit having its own promoter. *trpR* encodes for **trp repressor** protein. When tryptophan levels in the cell are very low, trp repressor cannot bind to the operator site and its transcription continues. When the tryptophan levels in the cell becomes high, it acts as a corepressor and binds to the repressor (Fig. 12.11). Binding of corepressor to the repressor brings a conformational change in the repressor, and hence it can bind to the operator. Binding of repressor hinders the binding of RNA polymerase to the promoter, and hence trp operon is shut off (Fig. 12.12). The anabolic end product of the trp operon, tryptophan, itself acts as the repressor for the operon (Crawford and Stauffer et al. 1980).

(a) Low tryptophan levels, transcription of the entire *trp* operon occurs

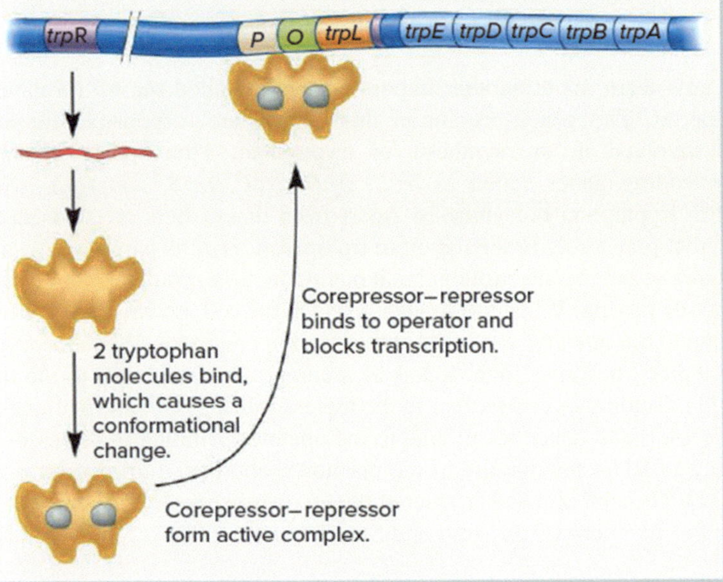

(b) High tryptophan levels, repression occurs

Fig. 12.12 Regulation of *trp* operon by end product tryptophan. (**a**) When tryptophan levels are low, it cannot bind to trp repressor protein, repressor is not able to bind to the operator site, and transcription continues. (**b**) When tryptophan levels are high, it binds to the repressor. The repressor

In a gram-positive bacteria *Bacillus subtilis,* the trp operon is present within a supraoperon. There are six genes of trp operon which are flanked by three genes of supraoperon on each side.

12.7.1 Trp Operon Structural Genes

The full-length polycistronic trp mRNA is about 6800 nucleotides in length. It codes for five trp polypeptides. The operon codes for two bifunctional polypeptides, that is, trpD and trpC. TrpD consists of glutamine amidotransferase domain (designated as trpG domain as the analogous monofuctional trpG protein is present in other organisms). The other domain is an anthranilate phosphoribosyltransferase, designated as trpD domain. Similarly, trpC polypeptide also has two domains, and the amino terminal has trpC domain. This domain is responsible for the indole-3-glycerol phosphate synthetase reaction. The distal half is responsible for isomerization of phosphoribosylanthranilate. TrpE and TrpD polypeptides form a tetrameric functional complex (Yanofsky 1971). These complexes catalyze the reaction of chorismate + glutamine + anthranilate and anthranilate + PRPP + phosphoribosylanthranilate. TrpB and Trp A polypeptides form a complex that catalyzes the reaction of indole-3-glycerol phosphate + L-serine leading to synthesis of L-tryptophan. The structural genes are organized in the same order as the functions of their corresponding polypeptide domains in tryptophan synthesis. The trpC region preceding the trpF region and trpB preceding trpA are the only exceptions.

12.7.2 Concept of Attenuation

We learned that trp repressor binds to the operator site only in the presence of corepressor tryptophan. Hence, in the presence of tryptophan, the operon for tryptophan synthesis is shut off. In 1970s, Yanofsky and colleagues observed mutant strains of bacteria which lacked trp repressor but could still inhibit trp operon in the presence of high tryptophan concentration. They also found mutations where trpL gene which codes for the leader sequence was missing and high expression of trp operon was observed. Studies on these mutant strains led to the understanding of another mode of trp operon regulation called **attenuation**.

During attenuation, transcription begins but is terminated before the complete mRNA is made. A short stretch of mRNA is transcribed, and it terminates shortly past the *trpL* gene. The transcription is terminated before the transcription of

Fig. 12.12 (continued) then binds to the operator to inhibit transcription (Adapted from: Brooker, R.J.: Genetics: Analysis and Principle, 6th ed. pp 336–360. McGraw-Hill Education, New York, 2018)

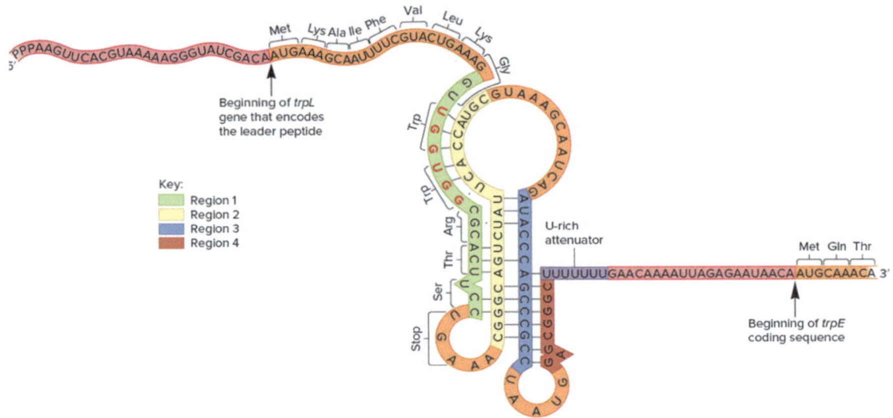

Fig. 12.13 Sequence of the *trpL* mRNA produced during attenuation. trpL mRNA has self-complementary regions which base-pairs with each other. Regions 1 and 2, 2 and 3, and 3 and 4 have possible base pairing. One region can base-pair with the other only once (Adapted from: Brooker, R.J.: Genetics: Analysis and Principle, 6th ed. pp 336–360. McGraw-Hill Education, New York, 2018). So if 2 has hydrogen bonded with 1, it cannot base-pair with 3. Similarly, if region 3 has hydrogen bonded with 4, it cannot base-pair with 2. The last U in the purple attenuator sequence is the last nucleotide that would be transcribed if attenuation is occurring

structural genes; hence, attenuation stops production of tryptophan. The attenuation occurs due to the presence of **attenuator sequence.** This sequence is present immediately downstream from the operator site. The first gene in trp operon is *trpL* gene. The mRNA of *trpL* gene codes for 14 amino acids that form the leader peptide. The leader peptide contains two tryptophans. These two tryptophan acts as sensor for tryptophan levels in the cell. If tryptophan levels are high, then only the tRNA gets charged with tryptophan, and leader peptide is synthesized. In case tryptophan is less, the tRNA doesn't get charged with tryptophan, and the leader peptide synthesis slows down.

Another important feature of leader mRNA is that it has self-complementary regions which base-pairs with each other to form stem loops. Region 2 is complementary to region 1 and also to region 3. Region 3 is complementary to region 2 as well as to region 4. Therefore, three stem loops are possible: 1–2, 2–3, and 3–4. One region can take part in only one stem loop formation. So if region 2 pairs with 1, it cannot base-pair with region 3. Alternatively, if region 2 base-pairs with 3, then region 3 cannot base-pair with region 4 (Fig. 12.13). The 3–4 stem loop together with U-rich attenuator forms an intrinsic terminator and leads to **ρ-independent termination**.

The formation of 3–4 stem loop with U-rich attenuator needs to form for attenuation. The amount of tryptophan regulates the formation of 3–4 stem loop. In bacteria, generally, translation and transcription occur together. As the transcription of *trpL* gene continues, region 1 rapidly hydrogen bond with region 2, and region 3 is not able to base-pair with region 2 and can therefore hydrogen bond with region

Fig. 12.14 Mechanism of attenuation of trp operon by stem loop formation of trpL mRNA. Attenuation occurs in part (**a**) and (**c**) due to formation of a 3–4 stem loop. (**a**) No Translation, 1-2 and 3-4 stem loops form **b**) Low Tryptophan Levels, 2-3 stem-loop forms **c**) High Tryptophan levels, 3-4 stem-loop form) (Adapted from: Brooker, R.J.: Genetics: Analysis and Principle, 6th ed. pp 336–360. McGraw-Hill Education, New York, 2018)

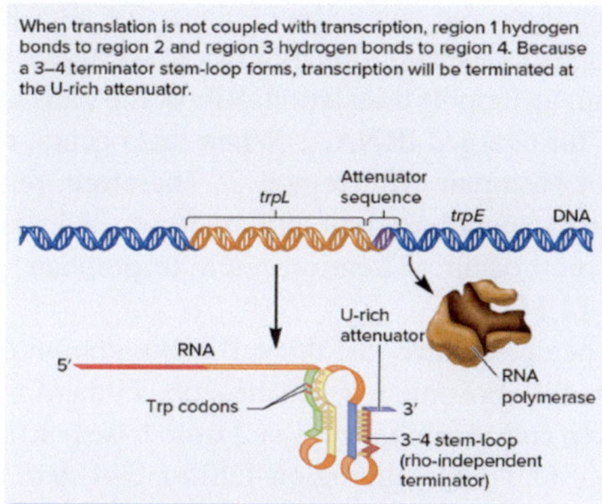

(a) No translation, 1–2 and 3–4 stem-loops form

Coupled transcription and translation occur under conditions in which the tryptophan concentration is very low. The ribosome pauses at the Trp codons in the *trpL* gene because insufficient amounts of charged $tRNA^{Trp}$ are present. This pause blocks region 1 of the mRNA, so region 2 can hydrogen bond only with region 3. When this happens, the 3–4 stem-loop structure cannot form. Transcriptional termination does not occur, and RNA polymerase transcribes the rest of the operon.

Attenuator sequence
trpL
mRNA for *trpE*
Leader peptide
DNA
Trp codon
Ribosome stalls
2 3 4
1
trpE
5′

(b) Low tryptophan levels, 2–3 stem-loop forms

Coupled transcription and translation occur under conditions in which a sufficient amount of tryptophan is present in the cell. Translation of the *trpL* gene progresses to its stop codon, where the ribosome pauses. This blocks region 2 from hydrogen bonding with any region and thereby enables region 3 to hydrogen bond with region 4. This terminates transcription at the U-rich attenuator.

Attenuator sequence
trpL
trpE
DNA
Stop codon
U-rich attenuator
5′
1 2
3′
3 4
3–4 stem-loop

(c) High tryptophan levels, 3–4 stem-loop forms

4. Therefore, the attenuation occurs just after the transcription of *trpL* gene. When tryptophan concentration is low, trp-charged tRNA is not formed and is enough amount to support translation. Ribosome halts at Trp codons of leader *trpL* gene as it waits for charged $tRNA_{trp}$. When this occurs, region 1 is covered by ribosome and cannot base-pair with region 2. Therefore, region 2 base-pairs with region 3. As region 3 cannot base-pair with region 4, 3–4 stem loop is not formed and attenuation does not occur. Therefore, the tryptophan operon is successfully transcribed (Fig. 12.14).

When cells have just the sufficient amount of tryptophan, the mRNA translation occurs till ribosome reaches the stop codon of *trpL* gene. The pausing of ribosome at the stop codon inhibits region 2 from base pairing with region 3, allowing region 3 to form a stem loop with region 4. Since 3–4 stem loop gets formed, the transcription of the rest of the tryptophan operon is not transcribed (Fig. 12.14).

In *Bacillus subtilis*, Trp operon is regulated by attenuation where two alternative RNA secondary structures are formed: anti-terminator and terminator. A tryptophan-activated RNA-binding protein, TRAP, can exist in either of these alternative structures. When tryptophan availability is less, TRAP is inactive, and anti-terminator RNA structure persists leading to transcription of tryptophan structural genes.

Attenuation regulates transcription of many other amino acid-synthesizing operon. In all these operons, the leader peptide contains the amino acid which is synthesized by the enzymes coded by the operon. For example, histidine operon has seven histidine codons in the leader sequence.

12.7.3 Mechanism of Trp Operon

We learned till now that operons can be regulated by either positive and negative regulation and sometimes both. Catabolic enzymes are mostly inducible, and anabolic enzymes are repressible. Trp operon is repressible as it is an anabolic operon and the end product, and tryptophan acts as a corepressor which binds to the repressor molecule. Binding of repressor-corepressor complex to the operator shuts off the operon transcription. Tryptophan availability inside the cell controls the formation of charged $tRNA_{trp}$ which in turn regulates attenuation process. Hence, genetic regulation provides bacteria with an energy-efficient mechanism for preventing the overproduction of any anabolic product more than the cell's requirement.

12.8 Regulation of Gene Expression in Lambda Phage

Bacteriophages are the class of viruses which uses bacteria as their host. We know that viruses can invade host cells and utilize host cell machinery to produce viral gene products. Viruses can then exist within the host cell (lysogeny) by integrating its DNA into host chromosome or can form progeny viruses, lyse the host cell, and

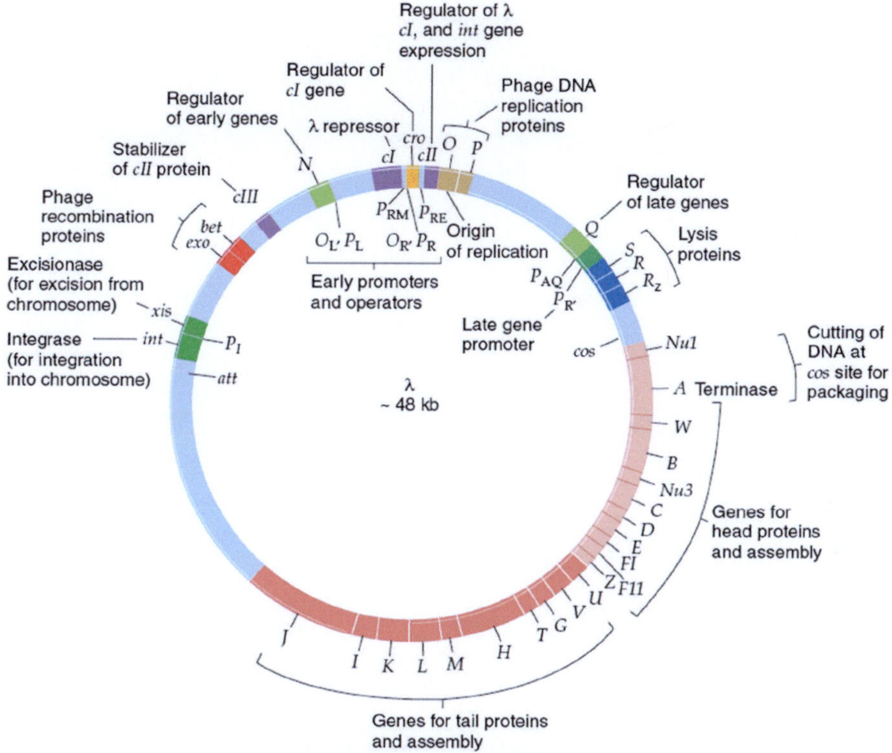

Fig. 12.15 A map of phage, showing the major genes. P_L promoter for leftward transcription of the left early operon, P_R promoter for rightward transcription of the right early operon, P_{RE} promoter for repressor establishment, and P_{RM} promoter for repressor maintenance (Adapted from: Russell, P.J.: iGenetics: A molecular Approach, 3rd ed. pp 36–59. Benjamin Cummings, New York, 2010)

invade new host cells (lysis). In the lysogenic phase, the expression of the viral genome is repressed, and the virus is said to exist as prophage.

Bacteriophage λ has emerged as important genetic model as it shares gene regulatory mechanism with both prokaryotes and eukaryotes. Once it adsorbs onto the bacterium and inserts its DNA into host cytoplasm, the phage DNA circularizes. At that point, the decision is made whether to follow **lysogenic or lytic cycle**. The decision leads a highly regulated genetic switch. There are two promoters, P_L and P_R. P_L leads to transcription toward the left-hand side, and P_R leads to transcription in the right-hand side (Fig. 12.15). In rightward transcription, *cro* (control of repressor and other) is the first gene to be transcribed. Cro protein plays an important role in initiating the lytic phase. In the leftward transcription, *N* is the first gene to be transcribed. The resulting N protein is a transcription anti-terminator. It helps transcription to proceed past the termination sites. As a result of N protein synthesis, other genes are transcribed; those are cII, O, P, and Q. cII protein can turn on

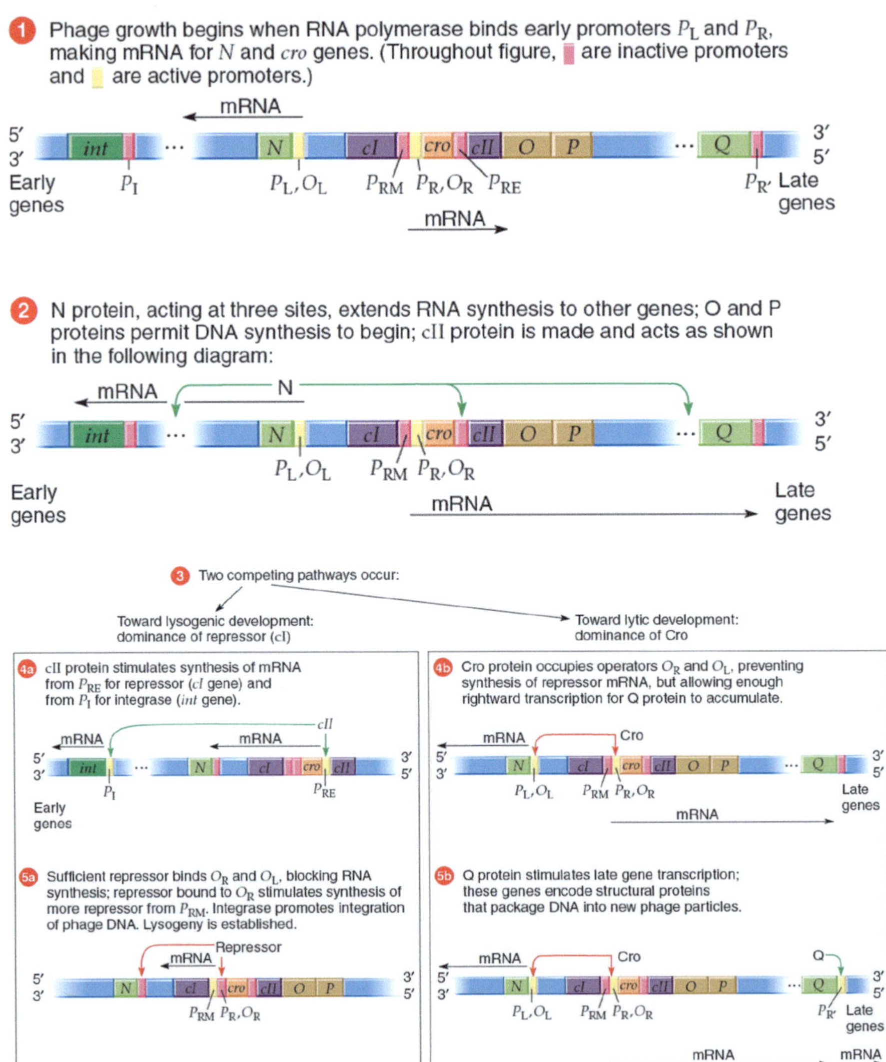

Fig. 12.16 Regulation of phage lambda genes for deciding lysogeny and lysis. Expression of genes after infection of *E. coli* and the transcriptional events that occur when either the lysogenic or lytic pathway is followed. In the figure, stimulation of transcription is indicated by green arrows, and repression of transcription by red arrows (Adapted from: Russell, P.J.: iGenetics: A molecular Approach, 3rd ed. pp 36–59. Benjamin Cummings, New York, 2010)

production of cI protein (repressor) and int protein (product of integrase gene, required for inserting phage DNA into host chromosome). O and P genes code for DNA replication proteins. Protein Q turns on late genes involved in lytic phase.

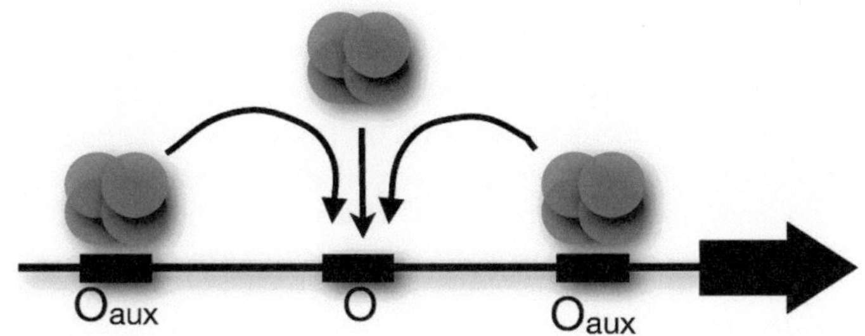

Fig. 12.17 Hypothetical contributions to local repressor concentration at a bacterial operator (O) include contributions because of free repressor and contributions because of DNA-bound repressor at auxiliary operators (O_{aux}) through DNA looping (Becker, N.A. et al. Nucleic Acids Res. 41, 156–166, 2013)

12.8.1 Lysogenic Pathway

The establishment of lysogenic pathway requires cII protein (translated from *cII* gene of early right operon) and cIII (translated from *cIII* gene of early left operon). cII is stabilized by interacting with cIII. The stabilized cII protein then activates the transcription of *cI*. The product of *cI* gene is called the **λ repressor**. It binds to operator regions O_L and O_R. The binding of λ repressor inhibits further transcription of early operons under the control of P_L and P_R. N and cro proteins translated from the operons under P_L and P_R can no longer take place. N and cro are highly unstable proteins, and hence their cellular levels come down quickly. There is another promoter known as **promoter for repressor maintenance, P_{RM}**. Repressor gene is transcribed from that promoter to maintain the repressor levels. Thus, if enough repressor is available, then it binds to O_L and O_R, shutting down the transcription from early promoters. It also leads to production of int protein by the action of cII, establishing lysogeny.

12.8.2 Lytic Pathway

The lytic pathway is induced by DNA damage caused by ultraviolet light. Bacteria has RecA proteins which functions in DNA recombination and repair. When DNA is damaged, RecA stimulates the repressor polypeptides to cleave themselves into two and gets inactivated. In the absence of repressor, *cro* gene is transcribed. Cro protein reduces RNA synthesis from P_L and P_R. This leads to reduction of cII protein synthesis. Reduction in cII blocks λ repressor synthesis from P_{RM}. Synthesis from P_R is also reduced, but if enough Q protein is available, the synthesis of late genes continues for establishing the lytic phase (Fig. 12.16). Therefore, to summarize, if repressor dominates, lysogeny ensues, and if cro dominates, lysis occurs.

Box 12.1: Scientific Concept: Mechanism of Promoter Repression by Lac Repressor-DNA Loops – Nicole A. Becker et al.

We have learned that lac repressor is both positively and negatively regulated. It stays repressed when the operator is bound by the repressor. The repressor binding becomes weak in presence of allolactose, and as a result, RNA polymerase can access the promoter transcribing the lac operon genes. Apart from the promoter proximal operator, there are two more auxiliary operators present just upstream and far downstream of the promoter proximal operator. The authors of this paper studied the role of these auxiliary operators in regulation of lac operon. The authors of this paper hypothesized that local repressor concentration at regulatory promoter proximal operator is necessary for lac operon repression. The auxiliary operators help in increasing the local repressor concentration at the regulatory operator in distance-dependent manner (Fig. 12.17).

The authors created many artificial promoter constructs in bacterial cells with artificial auxiliary operators at varied distances from the regulatory operator. They observed that repressor binding increases in a distance-dependent manner by strong auxiliary operators. They also observed that DNA looping occurring in the repressor traps the promoter and occludes RNA polymerase binding. They also observed that downstream auxiliary operator increased local repressor concentration but did not trap the promoter in the repressor loops. Hence, they concluded that promoter repression is equivalent to DNA loops of comparable size whether the promoter lies within the loop or adjacent to the loop.

Their results also showed that DNA looping can repress T7 transcription initiation but not T7 transcription elongation. It could be possible that RNA polymerase cannot initiate from the tightly bent DNA due to lack of sufficient energy required for untwisting the DNA.

12.9 Summary

- Gene expression is tightly regulated in prokaryotes depending upon the requirement of the cell. For anabolic pathways, genes are expressed only when the final product of the pathway is either absent or very low than the required amount. For proteins involved in catabolism, the corresponding genes are expressed only when the required catabolite is present in the system.
- The genes participating in a metabolic pathway are organized together which are called as operons. They are mostly organized as promoter, operator, structural genes, and terminator. Structural genes are transcribed as single mRNA.
- For inducible operons, transcription is normally shut off, unless the inducer is present. In repressible operons, transcription is normally on unless shut off by the

presence of repressor. When repressor binds and inhibits transcription, it is called negative control. When inducer/activator binds and initiates transcription, it is called positive control.

- Repressor may or may not be coded by the operon. Therefore, it might act as *trans* element. Repressor protein often act in *trans,* but operators are mostly *cis*-acting elements.
- The lac operon is an inducible operon. It gets induced by the presence of lactose. Lactose gets converted to allolactose, which then binds to repressor and inactivates it. Once the repressor becomes inactive, it leaves the operator and RNA polymerase can then bind to the promoter to transcribe its structural genes.
- The lac operon is also controlled by positive regulation through catabolite repression. cAMP is the catabolite activator protein for lac operon. Its levels are inversely related to glucose. Therefore, when glucose is low, cAMP is high and it activates the transcription of lac operon.
- *lac* operon has three operator sites: O_1 is the regulatory operator, and O_2 and O_3 are the auxiliary operator. Auxiliary operators increase the local concentration of repressor and hence facilitate binding of repressor to the regulatory operator.
- The *trp* operon of E. coli is a repressible operon. It normally transcribes the structural genes unless tryptophan is present in high amount in the cell. Tryptophan acts as corepressor. Hence, in presence of corepressor, the repressor binds to the operator and shuts off transcription.
- The trp operon is also regulated by attenuation. The first gene of *trp* operon is the leader sequence. It has four self-complementary regions. Stem loop formed between regions 3 and 4 leading to attenuation of transcription. Tryptophan residues present in the leader sequence also regulates trp operon by attenuation process.
- Jacob, Monod, and Pardee's experiment on partial diploid or merozygotes showed that lac repressor is a diffusible trans-acting element, whereas operator only acts in cis form.
- Viruses can integrate its DNA into host chromosome and undergo lysogenic life cyle or can form progeny viruses, lyse the host cell, and invade new host cells, which is called lytic life cycle. Bacteriophage λ can choose between lysogenic and lytic life cycle by the help of genetic switches.
- λ-phage has two principal promoters, P_L and P_R, for transcription of early operons. P_L leads to transcription toward the left-hand side, and P_R leads to transcription in the right-hand side. In rightward transcription, *cro* (control of repressor and other) is the first gene to be transcribed. In the leftward transcription, *N* is the first gene to be transcribed. The resulting N protein is a transcription anti-terminator.
- The establishment of lysogenic pathway requires cII protein and cIII. cII is stabilized by interacting with cIII. The stabilized cII protein then activates the transcription of *cI.*
- *cI* codes for λ repressor which binds to operator regions O_L and O_R. The binding of λ repressor inhibits further transcription of early operons under the control of

P_L and P_R. N and cro proteins translated from the operons under P_L and P_R can no longer take place.

- There are two more promoters, known as P_{RE} and P_{RM}, known for their function of repressor establishment and repressor maintenance, respectively. If repressor dominates, lysogeny ensues, and if cro dominates, lysis occurs.

References

Becker NA, Peters JP, Lionberger TA, Maher LJ III (2013) Mechanism of promoter repression by Lac repressor–DNA loops. Nucleic Acids Res 41:156–166

Beckwith JR (1967) Regulation of the lac operon. Recent studies on the regulation of lactose metabolism in Escherichia coli support the operon model. Science 156:597–604

Brooker RJ (2018) Genetics: analysis and principle, 6th edn. McGraw-Hill Education, New York, pp 336–360

Chakerian AE, Matthews KS (1992) Effect of lac repressor oligomerization on regulatory outcome. Mol Microbiol 6:963–968

Crawford IP, Stauffer GV (1980) Regulation of tryptophan biosynthesis. Annu Rev Biochem 49: 163–195

Fulcrand G, Dages S, Zhi X, Chapagain P, Gerstman BS, Dunlap D, Leng F (2016) DNA supercoiling, a critical signal regulating the basal expression of the lac operon in Escherichia coli. Sci Rep 14:19243

Griffiths AJF, Wessler SR, Carroll SB, Doebley J (2015) Introduction to genetic analysis. Freeman & Company, New York, NY :W.H

Juers DH, Matthews BW, Huber RE (2012) LacZ β-galactosidase: structure and function of an enzyme of historical and molecular biological importance. Protein Sci 21:1792–1807

Kaback HR (2005) Structure and mechanism of lactose permease. C R Biol 328:557–567

Russell PJ (2010) iGenetics: a molecular approach, 3rd edn. Benjamin Cummings, New York, pp 36–59

Rutkauskas D et al (2009) Proc Natl Acad Sci U S A 106(39):16627–16632

Yanofsky C (1971) Tryptophan biosynthesis in Escherichia coli. Genetic determination of the proteins involved. JAMA 218:1026–1035

Regulation of Gene Expression in Eukaryotes

13

Aathmaja Anandhi Rangarajan

Similar to prokaryotes, eukaryotes have also evolved with complex gene regulation mechanisms to control the expression of variety of genes in different cell types. Thousands of proteins orchestrate in a timely manner to ensure the activation and repression of genes in response to internal and external environment. This spatial and temporal precision of gene expression is critical in all biological processes from cell differentiation to cell death, while deregulation of gene expression often leads to disease. The regulation of gene expression happens in several stages including transcription, posttranscription, translation, and posttranslation. This complex regulation involves both cis-acting and trans-acting elements. Cis-acting elements are present in the coding and noncoding regions of the DNA itself involving remodeling or modifications of the DNA. Trans-acting elements are proteins such as transcription factors and other DNA-binding proteins that enhance or suppress gene expression.

Most of the gene regulation in eukaryotes happens at the level of transcription initiation. Transcription is mainly performed by RNA polymerase II along with several cis- and trans-acting factors. The eukaryotic transcriptions dictated by two distinct sets are cis-acting DNA elements: (i) proximal regulatory elements which contain the core promoter and (ii) distal regulatory element which includes enhancers, silencers, insulators, and locus control regions (Fig. 13.1). Trans-acting factors are usually regulatory proteins like transcription factors that binds to cis-acting elements to activate or repress transcription. In humans, the transcription factors are present far less in numbers (~1850) when compared to the number of genes (~20,000 to 25,000) (Venter et al., 2001). To compensate for this skewed proportion, the cell has evolved with multiple cis-acting elements that can work along in various combinations with trans-acting factors to control gene expression.

A. A. Rangarajan (✉)
University of Michigan, Ann Arbor, MI, USA
e-mail: anandhir@msu.edu

D. Kar, S. Sarkar (eds.), *Genetics Fundamentals Notes*,
https://doi.org/10.1007/978-981-16-7041-1_13

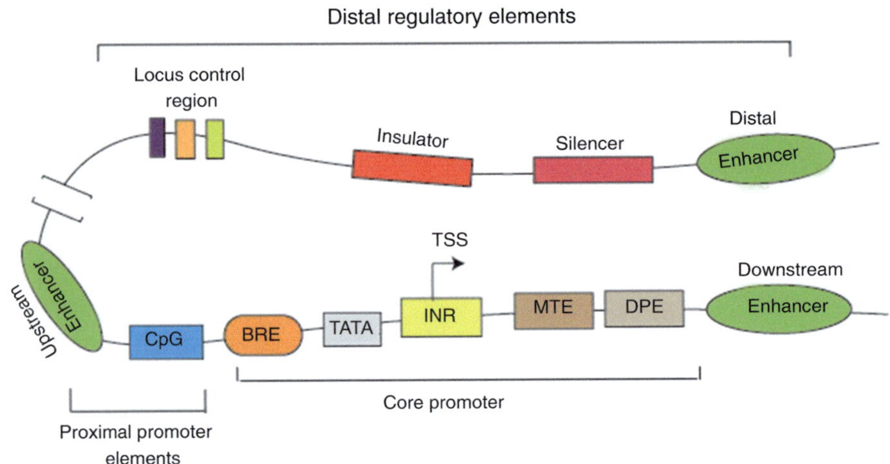

Fig. 13.1 Figure depicting several *cis*-acting elements involved in the regulation of gene expression during transcription. The core promoter which spans about 1 kb is surrounded by promoter elements such as CpG islands and upstream and downstream enhancers. The core promoter contains initiator element (INR) which also possesses transcription start site (TSS) surrounded by TFIIB recognition element (BRE), TATA box (TATA), motif ten element (MTE), and downstream promoter element (DPE). Distal regulatory elements such as insulator, silencer, and locus control region are located proximal or distal to the promoter region

During transcription initiation, as a first step, RNA polymerase II binds to the core promoter region together with some transcription factors. The first identified core promoter region was TATA box which possesses consensus sequence of TATAa(t)AAg(a) located 25 to 30 bp upstream of the transcription start site. Later, it was found that TATA box only consists of 32% of human promoters. Apart from TATA box, core promoter elements can contain other elements such as initiator element (Inr), downstream promoter element (DPE), downstream core element (DCE), TFIIB recognition element (BRE), and motif ten element (MTE) (Maston et al., 2006). These core promoter elements are present at specific distance from the transcription initiation site, and they possess distinct consensus sequences, which are described in Fig. 13.2. These elements in combination or alone recruits specific transcription factors to initiate or repress transcription.

Core promoter elements are surrounded by proximal promoter elements that spans from −50 to −500 region of the transcription start site and alters the rate and level of transcription. These regions typically contain multiple sites for binding activators. An example for the proximal promoter elements is the CpG islands which are 500 bp–2 Kb in length and are highly GC rich. 60% of human genes promoter region contains CpG islands. Most of the CpG dinucleotides present in the genome are methylated at the cytosine base whose fifth carbon is methylated, whereas the one in the CpG island is unmethylated. These are involved in regulation of many housekeeping genes and other regulated genes. When methylation proteins such as MeCP2 bind the cytosine of CpG island and also recruit histone-modifying

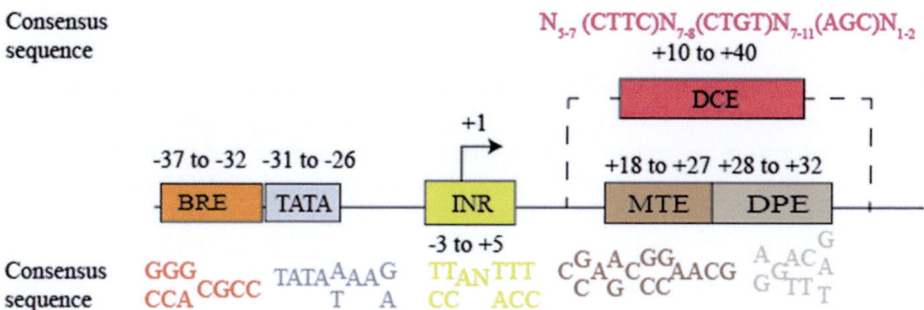

Fig. 13.2 RNA polymerase II core promoter elements. The core promoter elements are represented in colored boxes. TFIIB recognition element (BRE), TATA box, initiator element (INR), motif ten element (MTE), downstream promoter element (DPE), and downstream core element (DCE) are represented with respect to the transcription start site (+1). Consensus sequence of these elements are shown with the corresponding color of each element

complexes, it results in the repressive chromatin structure leading to silencing of transcription.

Apart from core and proximal promoter elements, there are several distal elements such as enhancers, insulators, silencers, and locus control regions that repress or activate the genes. Enhancers typically contain multiple binding sites for the binding of specific transcription factors that work together to enhance transcription. Enhancers are independent of orientation and distance and independent and can be present several Kb away from the transcription start site. It is proposed that by "looping out," the intervening DNA sequence enhancer region comes in contact with the core promoter region. In contrast to enhancers, silencers repress the transcriptional activity. The function of enhancers is independent of distance and orientation from the promoter region. Silencers provide binding sites for repressors which sometimes act in the presence of corepressor. These repressors and corepressors may block the activator from binding to the enhancer sites or block the binding of RNA polymerase II to start the transcription. In some cases, repressors can also recruit chromatin remodeling complexes which perturbs the binding of transcription complexes. Insulators or boundary elements limit the activity of transcriptional regulatory elements. They are of typically 0.5–3 Kb in length and possess roles such as blocking promoter-enhancer activity or block the repressive chromatin region (Recillas-Targa et al., 2002). They function in position-dependent and orientation-independent manner. Locus control regions (LCR) consists of groups of several regulatory elements which collectively work to control the gene expression. LCRs consist of several cis-acting elements such as insulators, enhancers, silencers, chromosome scaffold attachment regions, and nuclear matrix. They are general bound by transcription factors, activators, and repressors. Each of the component affects the gene expression differentially, and the collective action of these results in the overall effect on the gene expression. These LCRs confer spatial and differential gene expression depending on the stimuli (Li et al., 2002).

13.1 Control of Transcription by Protein

Apart from the cis-acting elements, several trans-acting proteins affect the transcription. The trans-acting proteins involved in the initiation of transcription can be categorized into three, namely, general transcription factors and promoter specific activators and coactivators. General transcription factors include RNA polymerase II itself and transcription factors such as TFIIA, TFIIB, TFIID, TFIIE, TFIIF, and TFIIH. The general transcription factors assemble in the core promoter in the orchestrated order to enable the formation of pre-initiation complex (PIC) which enables the RNA polymerase II to initiate the transcription at the transcription start site. In the first step, TFIID comprised of multi-subunit TATA box binding proteins (TBPs) and tightly bound TBP-associated factors (TAF). TBP makes direct contact with RNA polymerase along with the transcription factors such as TFIIA and TFIIB. TBP also provides binding sites for certain activators which stimulate transcription at the stage of pre-initiation complex. The binding of TFIIA further enhances the binding of TBP to the TATA box. TFIIB interacts with multiple proteins TFIID, TFIIF, and RNA polymerase II along with BRE elements downstream of TATA box. Binding of TFIIB is the rate-limiting step in the recruitment of RNA polymerase II which is stimulated by activators. In the next step, once the pre-initiation complex is formed, TFIIE binds to RNA polymerase II and TFIIH. TFIIF then binds to this complex and interacts with RNA polymerase II and TFIIB. It helps in nonspecific binding of DNA sequences and helps stabilizing pre-initiation complex. In the last step, a multi-subunit complex TFIIH binds and initiates the transcription. This complex contains ATPase, protein kinase, and helicase activity which unwinds the DNA around the initiation site (Orphanides et al., 1996).

This entire assembly of pre-initiation complex and the activation in the core promoter region allow merely basal level of transcription from the transcription start site. Transcriptional activity from the promoter is enhanced severalfold by other special class of proteins named activators.

13.1.1 By Activators

Transcriptional activity of many genes is greatly stimulated by a set of promoter-specific factors known as activators. Activator proteins contains two domains, namely, DNA-binding domain (DBD) and activator domain (AD) which is crucial for the activation of transcription. The important role of activators is to bring the RNA polymerase II along with the general transcription factors to the core promoter. It does that by binding to the general transcription factors and changes its confirmation which activates the pre-initiation complex formation. Activators can also help in the subsequent transcription process such as initiation and elongation after the formation of pre-initiation complex by helping in the clearance of pre-initiation complex and releasing paused RNA polymerase II (Green, 2005). Examples of activator proteins include zinc finger proteins, homeobox, fork head, helix-loop-helix (HLH), ETS, basic leucine zipper (bZIP), and PIT-Oct-Unc (POU)

DNA-binding domain (Pabo, 1992). DNA-binding sites of activators are generally small of about 6–12 bp. Most of the activators form hetero- or homodimers; hence, their binding regions usually comprise of two identical half sites. Activators undergo numerous posttranslational modifications including phosphorylation, acetylation, and glycosylation. In many cases, these posttranslational modifications have positive role in the transcriptional activation. For example, acetylation of p53 enhances its binding to DNA and thereby positively affect the transcription. Transcriptional activators which interact with regulatory proteins to enhance transcription are referred to as coactivators. Coactivators generally cannot bind DNA and needs the help of other DNA-binding proteins. Coactivators act similar to activators in stimulating pre-initiation complex or by remodeling repressive chromatin complexes. Since most of the histone-modifying enzymes and chromatin remodeling enzymes do not have sequence specificity, these enzymes are brought to the target gene sites by the coactivators. The most important effect of the activators is the synergistic activation of transcription. Synergy refers to the higher activation level by the combination of multiple regulatory factors due to the action of several factors acting together. In some specialized scenario, multiple activators bind with each other to form a stable nucleoprotein complex on the DNA called enhanceosome. These multiple activators cooperatively activate the target promoter. One of the best studied examples of enhanceosome is the IFN-β promoter in which all the activator comes in contact with the cofactor p300. The recruitment of the cofactor p300 is only possible when all of the activation domains are present. Enhanceosomes are generally found in tightly regulated genes such as those that are involved in wound healing or pathogen defense (Ma, 2011).

13.1.2 By Repressors

Similar to activation, repression is also an important gene regulatory mechanism which is mediated by specific set of proteins called as repressors. The mechanism of repression is important to avoid nonessential transcription from several genes which are detrimental and costly to the cell if not regulated. There are two types of repression observed in eukaryotes, general or global repression or gene-specific repression. Transcriptional repression occurs when repressors sequester regions or proteins involved in pre-initiation complex such as RNA polymerase II which results in overall decrease in the transcription. Gene-specific repression takes place when specific gene or gene sets is targeted by repressor or corepressor. Gene-specific repression is mediated by repressor or corepressor which binds to activator or coactivator and inhibits their activity or decreases their level. In some cases, it could also bind to promoter and does not allow the formation of pre-initiation complex. In short-range repression, repressor acts locally by decreasing the activity of activators or members of the pre-initiation complex that is present proximally. On the other hand, in the long-range repression, distally located repressor proteins affect the activity of the promoters by remodeling chromatin or by looping out the intervening DNA sequence to access the activators or coactivators. These short- and long-range repressions are not mutually exclusive, and both can occur at the same time. For example, hairy protein can perform both short-range and long-range interactions at the same time.

Repressor proteins are categorized into three categories: Class I, Class II, and Other types. Class I repressor proteins all possess DNA-binding activity. They could be sequence specific or non-sequence specific. Class II repressor proteins cannot bind to DNA themselves but act with other proteins to enable repression. They are also termed as corepressors. Some proteins of Class II can also have dual role which act as both activators and repressors at different contexts. Class III proteins does not bind DNA directly or indirectly but usually act on activators, coactivators, and pre-initiation complex to reduce the levels of these proteins (Gaston and Jayaraman 2003).

In some cases, relatively small amount of variation in the gene expression might significantly change the cell fate. Some corepressors are involved in the fine of such genes. In yeast, it was found that such corepressors act as chromatin-modifying enzymes and reduce the transcriptional noise while not shutting off the transcriptional machinery.

13.1.3 By Steroid Hormones

Regulation of gene expression can also be regulated by steroid hormones which act on intracellular transcription factors which positively or negatively regulate specific sets of gene. Steroid hormones are synthesized from cholesterol and are lipophilic in nature. They are produced in several regions of the body such as adrenal cortex (glucocorticoids, mineralocorticoids, adrenal androgens), testes (testicular androgen, estrogen), and ovary (estrogen and progestins). Steroid hormones reach the target cell via blood, cross the membrane by diffusion because of the innate lipophilic property, and bind to steroid hormone receptors (SHR). Usually, steroid hormone receptors are bound by chaperons which prevent them from folding and aggregation. Binding of steroid hormones to the steroid hormone receptor relives it from the chaperone, changes its confirmation to active form, enters the nucleus where it binds to the target genes (Fig. 13.3). These active forms of steroid hormone receptors can regulate gene transcription in several ways: (i) by binding to the hormonal-responsive elements (HRE) and initiating chromatin remodeling which activates or represses the target gene machinery, (ii) by binding to other transcription factors that can modulate the activity of other genes, and (iii) by cross-talking with signal transduction pathway to activate nuclear transcription factors which in turn regulate the target genes. SHRs are characterized by DNA-binding domain which helps in binding to HRE and ligand-binding domain (LBD) that binds to ligand.

Androgens (testosterone and dihydrotestosterone) are male sex steroid hormones that are responsible for the proper functioning of male reproductive system. Testosterone and dihydrotestosterone bind to androgen receptors which mediate transcription. Androgen receptors are present in male reproductive tract, in female reproductive tract, and in various other diverse tissues (Davey and Grossman, 2016). Estrogens influence many processes including reproduction, cardiovascular

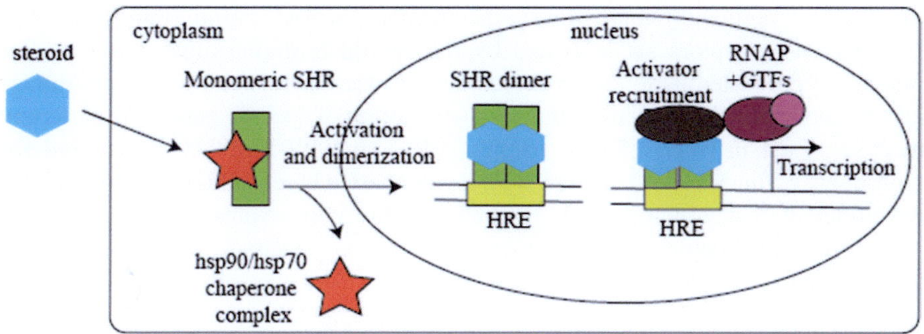

Fig. 13.3 General mode of activation by steroid hormones. Steroid hormones (blue) enters the cytoplasm and binds to monomeric steroid hormone receptor (SHR) (green) which leads to the release of hsp70/hsp90 chaperone complex (red). SHR dimerizes and translocates into the nucleus followed by binding to hormone-responsive element (HRE). This recruits activator proteins (brown) which enable chromosome remodeling to enable the RNA polymerase binding (magenta) with general transcription factors (pink) to activate the transcription

health, bone integrity, behaviors, and cognition. Estrogen binds to estrogen receptor in the nucleus. This complex further binds to sequences containing estrogen response element along with activator protein SP1 in the promoter region. This binding results in the recruitment of co-regulatory proteins that results in increased or decreased levels of regulatory proteins which results in response to particular stimuli. Estrogen receptors exist in two different forms: ERα and ERβ. ERα is expressed in the liver, mammary gland, pituitary, hypothalamus, uterus, and vagina. ERβ is highly expressed in prostrate, ovary, colonic epithelium, and lung (Deroo and Korach, 2006). The activation of glucocorticoid hormones is mediated by cortisol that is secreted from the adrenal gland which in turn binds to glucocorticoid receptors. After the glucocorticoid receptor is activated by the hormone, it undergoes homodimerization, translocates into nucleus in which it binds to specific DNA-responsive elements, and activate transcription. Activated glucocorticoid receptor can also bind to transcription factors such as NF-κB or AP-1 and repress them from acting as transcription factors. Thus, it plays a dual role in activating the anti-inflammatory protein in the nucleus and repressing pro-inflammatory protein such as NF-κB or AP-1 in the cytosol. Glucocorticoid receptors are expressed in most of the cell types in the body which takes place in regulation of several biological processes such as development, metabolism, and immune response (Vandevyver et al., 2014). Mineralocorticoid receptors possess similar affinity for both mineralocorticoids such as deoxycorticosterone and glucocorticoids such as cortisol. Upon activation by the binding of ligand, it translocates inside the nucleus, homodimerizes, and further binds to hormone-responsive elements present in the promoter, thereby activating transcription. Mineralocorticoid receptors are expressed in several types of tissues such as heart, central nervous system, brown adipose tissue, kidney, sweat glands, colon, and immune cells. It helps in maintaining the normal salt concentration in the body by activating the proteins involved in regulating ionic transports such as Na^+/K^+ pump (Funder, 2005). Progesterone receptors are activated by the steroid hormone progesterone. After binding of progesterone to the progesterone receptor, the receptor undergoes dimerization and enters the nucleus to induce transcription of specific genes. Progesterone receptors exist in two isoforms, namely, PR-A and PR-B. PR-A is required for reproductive function and uterine development of whereas PR-B is required for development of mammary gland. Most PR-positive cells express both the isoforms including uterus, mammary gland, brain, ovary, testes, pancreas, and tissues in lower urinary tract (Daniel et al., 2011).

13.2 Regulation of Transcription by Chromatin

Nearly 99% of the DNA in the eukaryotes is packed into highly complex, condensed structure known as chromatin. Chromatin dynamics plays a crucial role in the transcriptional regulation and chromosome function. The basic structural subunit of chromatin consists of nucleosome. Each nucleosomes consists of octamer of histones made up of two copies of four core histones H2A, H2B, H3 and H4 that is wrapped around the DNA as shown in the Fig. 13.4. The core octamer with the linker DNA and the histone H1 comprises the basic repeating subunit of chromatin complexes. Histone octamers help in maintaining the nucleosomal stability by making numerous contacts with the nucleosomal DNA. The N-terminal region of the histones protrudes outside of this core complex since it undergoes several covalent modifications which regulate the chromatin structure and function. Chains of nucleosomes are placed in zigzag confirmation which forms complex, highly organized, condensed structure.

Chromatin dynamics is intertwined with the gene regulation. This condensed state of chromatin itself poses structural barrier for the binding of transcription factors and other proteins to bind and initiate transcription. In another level of regulation, the histones are removed or modified by the chromatin-modifying complexes which render the chromatin in actively transcribing state of inactive state. These chromatin-modifying complexes fall under two categories based on

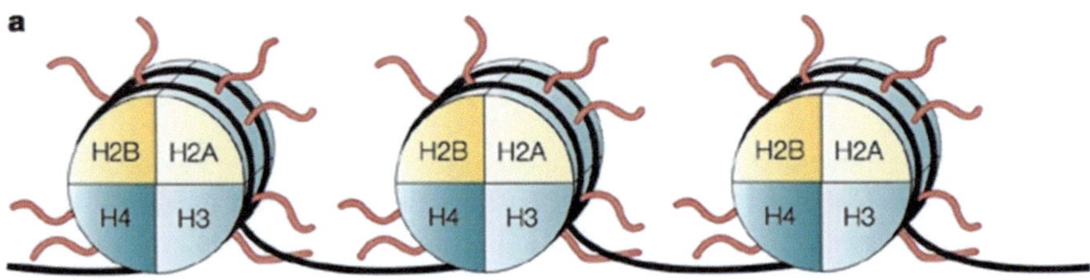

Fig. 13.4 Organization of histone core proteins. Core proteins of histones are depicted in the figure as H2A, H2B, H3, and H4. Each of the histone is present in two copies forming an octamer which wraps the DNA (black)

their type of regulation: (i) ATP-dependent chromatin-modifying complexes which use energy from ATP hydrolysis to disrupt or alter the histones, and (ii) histone-modifying enzymes which are responsible of posttranslational modification of histone molecules thereby regulating the gene expression (Li et al 2007).

13.2.1 Chromatin Remodeling

Chromatin dynamics is shaped by several ATP-dependent chromatin remodeling enzymes. These remodeling enzymes take part in several processes of chromosomal dynamics. They enable the formation of properly spaced nucleosomes, move or eject histones facilitating transcription factor binding, and participate in exchanging the canonical histones with histone variants in specific chromosomal regions.

All of the ATP-dependent chromatin remodeling enzymes are categorized into four types depending on the type of ATPase subunit they possess: (i) imitation switch (ISWI), (ii) chromodomain helicase DNA binding (CHD), (iii) switch/sucrose non-fermentable (SWI/SNF), and (iv) INO80. ISWI and CHD complexes help in initial DNA histone complexes to mature into nucleosomes. The important role of ISWI is to create arrays of nucleosomes by placing nucleosomes at specific distance from each other, thereby limiting chromosome accessibility and gene expression. SWI/SNF family of proteins is involved in the ejection or sliding of histones from the DNA, thereby promoting the binding of transcription factors and other specific proteins. INO80 subfamily is involved in replacement of canonical histone with variant histones. For example, they are involved in replacing canonical histones H2A and H3 with other histone variants mediated by Snf2-related CBP activator protein (SRCAP) and p400. This inclusion of variant histone may affect the transcription factor recruitment, exclusion of proteins, and activity (Fig. 13.5).

Nucleosomal density has been found to be lower in the regions of functional transcriptional factor binding sites. Once activators bind to the promoters, they recruit other coactivators and general transcription factors to assemble pre-initiation complex (PIC). Histone acetylation has also been important for gene activation at the promoter region, and histone acetylases are also recruited by activator proteins. Activators and coactivators also interact with chromatin remodelers SWI/SNF complex to free the DNA from histones for the recruitment of pre-initiation complex. Histones which are bound at the promoter region that are active are lost after pre-initiation complex assembly which reassemble in the promoter region once the genes are turned off. In large number of genes where partial pre-initiation complex is present, the nucleosomes were found to be intact, whereas when the RNA polymerase II binds and subsequently transcription is activated, nucleosomes are removed from the DNA indicating the role of RNA polymerase II itself as chromatin remodeler.

During transcription elongation, RNA polymerase is released from the general transcription factors and travels into the coding region of the genome where it encounters nucleosomal barrier. These barriers are overcome by transcription machinery along with the help of chromatin remodeling complexes. During

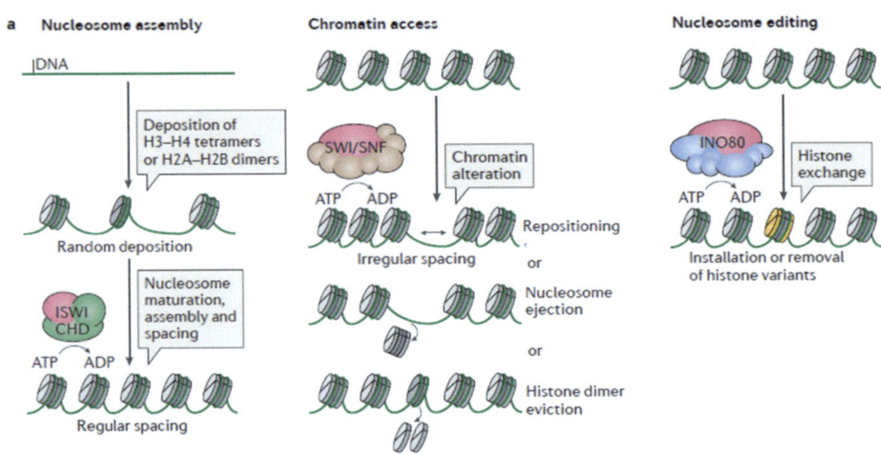

Fig. 13.5 Different functions of chromatin remodeling complexes. The ATPase-translocase subunit of all remodelers is depicted in pink; additional subunits are depicted in green (imitation switch (ISWI) and chromodomain helicase DNA binding (CHD)), brown (switch/sucrose non-fermentable (SWI/SNF)), and blue (INO80). Nucleosome assembly: Particular ISWI and CHD subfamily of chromatin remodeling complex participates in the deposition of histones, nucleosome maturation, and spacing. Chromatin access: SWI/SNF subfamily remodelers alter chromatin by repositioning nucleosomes, ejecting octamers, or evicting histone dimers. Nucleosome editing: Remodelers of the INO80 subfamily (INO80C) change nucleosome composition by exchanging canonical and variant histones, for example, and installing H2A.Z variants (yellow) (Figure taken with permission from Clapier et al. 2017)

elongation, RNA polymerase II C-terminal domain undergoes two important phosphorylation events in which Ser5 and Ser2 are phosphorylated. These phosphorylation are important for recruitment of elongation complex along with chromatin remodeling enzymes. PAF/RTF complex with the help of Spt4/5 binds phosphorylated Ser5 and recruits chromatin remodeling complexes. PAF helps in recruiting H3K4 methyltransferase Set1 complex (COMPASS) for elongating RNA polymerase II. PAF complex associated with phosphorylated Ser5 is also important for recruiting other chromatin regulators Chd1 and histone chaperone-like factors Spt6 and FACT. These chromatin regulators help in clearance and sequestering histones downstream of the elongating RNA polymerase II complex paving way for the elongation.

13.2.2 Histone Modification

Histones are not only involved in the formation of nucleosomes, but histone modifications play a crucial role in the regulation of gene expression. These modifications are done by histone-modifying enzymes which bind to N-terminal region of histone molecule that are exposed from the nucleosome. There are seven major posttranslational modifications that are described to take place on tails of histone: (1) acetylation, (2) methylation, (3) phosphorylation, (4) ADP ribosylation, (5) glycosylation, (6) sumoylation, and (7) ubiquitylation. Each of these modification results in distinct effect on gene regulation (Bannister and Kouzarides, 2011).

13.2.2.1 Histone Acetylation

Histone acetylation is one of the most well studied and first discovered posttranslational modification in histone. All of the histone proteins (H2A, H2B, H3, and H4) can be acetylated at several lysine residues. Acetylation cancels the positive charge of histone molecules, thereby reducing the binding of histones to the DNA, which helps in exposing the DNA for the binding of transcription factors and other DNA-binding proteins. Thus, histone acetylation helps in activation of transcription. The acetylation of several lysine residues results in combinatorial enhanced effect on the structure of the nucleosome when compared to the single residue acetylation. These are performed by histone acetyltransferase (HAT). Humans possess five groups of HATs: HAT1, CBP/p300, GNAT, MYST, and other HATs. In contrast to HATs, histone deacetyl transferase (HDAT) deacetylates the histone core and restores the positive charge and binding of histone, resulting in transcriptional repression. There are four classes of HDATs (class I, class II, class III, class IV). Different classes of HDATs are associated with multiple complexes and have functional redundancy (Luo and Dean, 1999).

13.2.2.2 Histone Methylation

Unlike histone acetylation, histone methylation leaves the charge of the histone molecule unchanged and therefore does not affect the binding of histones to the DNA directly. Histone methylation affects the binding of other chromatin remodeling enzymes to the DNA which in turn has a positive or negative effect on transcription. Histones are usually methylated at lysine or arginine residues. At lysine residues, mono-, di-, or trimethylation can occur, and at arginine residues, mono- and dimethylation can occur. Methylation has diverse effect on the status of the transcription. H3K4 and H3K79 mono-methylation leads to transcription activation, whereas H3K9 and H3K27 dimethylations are associated with transcription repression. Like acetylation, methylation of histones is regulated by histone methyltransferase (HMTs) and histone demethylases (HDMs). HMTs can be categorized into two categories: (i) lysine methyltransferases (KMTs) and protein arginine methyltransferases (PAMTs). HDMs are further categorized into two groups: Jumonji domain containing HDMs and lysine-specific demethylase 1 or 2. The former removes the methyl group using Jumonji domain, while the latter removes methyl group with other complexes. Methylation regulates chromatin structure by interacting with other chromatin remodelers. It was shown that trimethylation of H3K4 recruits chromatin remodeling complexes such as ING protein and CHD protein which in turn activates the transcription (Zentner and Henikoff, 2013).

13.2.2.3 Histone Phosphorylation

Histone phosphorylation leads to reduction of positive charge of histones eventually leading to destabilization of histone-DNA binding, thereby enabling DNA access to the protein complexes. Histone phosphorylation has been found to impact nucleosomal structure, open the chromatin, and enable the binding of transcription factors and DNA repair proteins. After the double-stranded break, the histones H2A and H2A.X variant are phosphorylated which recruits the DNA repair machinery and increases the accessibility of double-stranded break to the repair proteins. In some cases, the phosphorylated histone protein can decrease the binding of proximal methylated lysine with its binding partner. For example, phosphorylated H3S10 blocks the binding of H3K9me3 and K3K27 with HP1 and S28, respectively (Banarjee and Chakravarti 2011).

13.2.2.4 Histone ADP Ribosylation

Histones can be mono- or poly-ADP ribosylated on glutamate and arginine residues. ADP ribosylation adds negative charge to the histones, thereby loosening the contact of histones with DNA leading to loosened chromatin state. Mono- and poly-ADP ribosylation is mediated by ADP ribosyl transferases (ART) and ADP-ribose-protein hydrolases (ARHs) or poly-ADP-ribose glycohydrolases removes the ADP ribose subunit by hydrolysis. All four core histone proteins (H2A, H2B, H3, and H4) along with the linker H1 can be ADP ribosylated. Poly-ADP ribosylation of H3, H2B, H4, and H1 is induced by DNA damage, indicating their possible involvement in DNA damage response mechanism (Messner and Hottiger, 2011).

13.2.2.5 Histone Sumoylation and Ubiquitylation

Unlike other posttranslational modifications in histones which result in smaller modifications in the amino acid side chains, sumoylation or ubiquitylation leads to large covalent modification by the addition of large peptides. In histone sumoylation or ubiquitylation modifications, small ubiquitin-like molecules (SUMO) (100 or fewer amino acids) or ubiquitin (76 amino acids) are added to the lysine residue of the histones by the consecutive action of three enzymes: E1-activating enzyme, E2-conjugating enzyme, and E3-ligating enzymes. Ubiquitylation can be reversed by the action of isopeptidases. Histone ubiquitylation is usually observed on H2A and H2B tail lysine residues. H2AK119 mono-ubiquitylation was shown to repress transcription by its interaction with polycomb complexes. DNA double-stranded breaks induce H2BK123 mono-ubiquitylation which plays an important role in recruiting DNA repair proteins. All the core histone proteins undergo sumoylation which antagonizes acetylation or ubiquitylation that can take place at the same lysine residue. The molecular mechanism by which sumoylation affects the chromatin has yet to be elucidated.

13.2.2.6 Other Histone Modifications

Other noncanonical histone tail modifications are histone deimination, addition of histone O-GlcNAcylation, tail clipping, and protein isomerization. In histone deimination, PADI4, which is a peptidyl arginine deiminase, converts arginine to citrulline that cancels the positive charge of arginine. In histone O-GlcNAcylation, some histones like H2A, H2B, and H4 undergoes modifications by β-linked N-acetylglucosamine (O-GlcNAc) at serine or threonine residues. O-GlcNAc catalyzes this reaction, whereas β-N-acetylglucosaminidase removes the monosaccharide from the histone. In tail clipping, certain residues at the N-terminal tail region of the histone are clipped. For example, the mouse enzyme Cathepsin L was shown to remove the H3 histone N-terminal region during embryonic state cell differentiation. Histone proline isomerization is mediated by proline isomerases which catalyze the interconversion of cis and trans conformations in the peptide bonds of proline. One of the known examples of proline isomerase is scFpr4 from yeast which isomerizes H3P38. This modification is shown to be linked to H3K36 methylation which could probably affect the recognition of methyltransferase.

13.3 Gene Silencing by DNA Methylation

DNA methylation is a major epigenetic modification of eukaryotic genome, which has profound effect on eukaryotic genome. DNA methylation usually results in silencing of transcription. DNA methylation also plays a crucial role in important processes involved in cell differentiation and developmental processes such as genomic imprinting, spermatogenesis, and X-chromosome inactivation. DNA methylation also plays a role in silencing retrotransposons by preventing deleterious recombination during embryonic development. DNA methylation is prominent in lower eukaryotes to higher mammals, and the enzymes involved in the process are highly conserved during evolution. DNA methylation is mediated by DNA methyltransferases (Dnmt) which catalyze the transfer of methyl group from S-adenyl methionine to the fifth carbon of cytosine residue to form 5-methyl cytosine. There are two classes of DNMTs described according to their role: (i) Dnmt1 and (ii) Dnmt3a and Dnmt3b. Dnmt1 takes part in DNA replication, and methylation pattern from the parent DNA is copied to the daughter DNA strand (Fig. 13.6a). Dnmt3a and Dnmt3b are involved in the formation of new methylation pattern in the genome; thus, they are termed as de novo Dnmts. (Fig. 13.6b).

13.3.1 Mechanism of Gene Silencing by DNA Methylation

Nearly 4% of the human genome is methylated. The majority of DNA methylation happens in the cytosine residue that precedes a guanidine or CpG island. CpG refers to cytosine and guanine residue separated by a phosphate group. CpG islands are found primarily near the promoter regions than at the other regions of the genome. However, most of the CpG sites in the early embryonic stage and the somatic cells involved in the cell differentiation are unmethylated to allow the expression of those genes. The methyl group of methyl cytosine protrudes into the major groove of the DNA, thereby directly repressing transcription by affecting the binding of several DNA-binding proteins and transcription factors. Methyl-CpG binding proteins

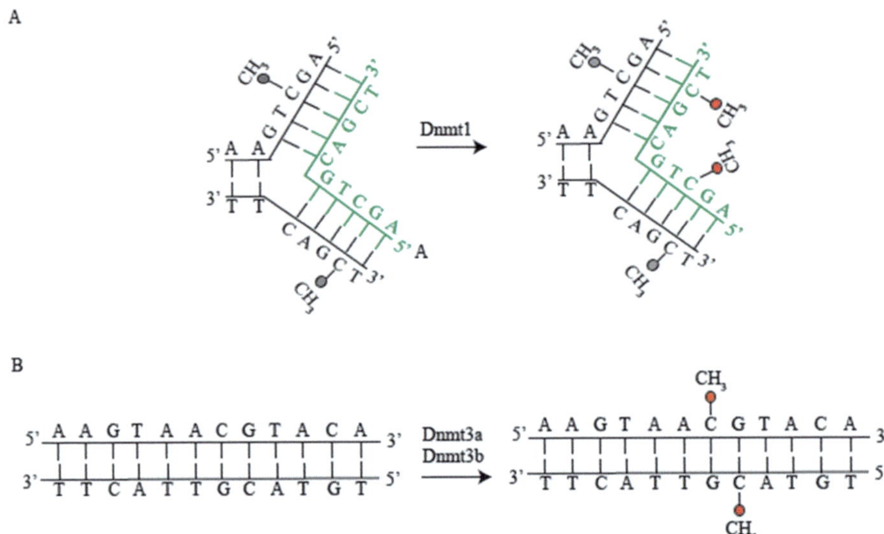

Fig. 13.6 Types of DNA methylation by DNA methyltransferases (Dnmts). (**a**) Dnmt1 maintains the methylation pattern in the daughter strand during replication. Dnmt1 duplicates the methylation pattern (gray) of the parent strand and makes the copy of methylation (red) onto the CpG sites in the daughter strand (green). (**b**) Dnmt3a and Dnmt3b are the de novo methyltransferases that transfers methyl group to the genomic DNA on the CpG sites

(MBDs) bind to methyl groups in the CpG-binding domains and recruit repressor complexes resulting in transcriptional repression. This family of proteins includes MeCP2, MBD1, MBD2, MBD3, and MBD4. Three of the MBD proteins, MBD1, MBD2, and MeCP2, involved inhibition of transcription in methylation-dependent manner. All the four MBD proteins associate with different corepressor complexes. MeCp2 binds to mSin3a corepressor complex to achieve repression. MBD2 is a part of MeCp1 repressor complex, and it is involved in DNA binding. MeCp1 is a multiprotein complex that contains Mi-2-NuRD (nucleosome remodeling histone deacetylase) repressor, comprising of histone deacetylases and chromatin remodeling protein Mi-2. MBD3 also interacts with Mi2-NuRD repressor complex (Newell-Price et al., 2000 and Moore et al. 2013)

13.3.2 DNA Methylation and Its Interaction with Other Chromatin Modifiers

However, the binding of this complex needs deacetylation of histones. DNA methylation and the associated factors also act together with the histone modification system. Histone acetylation usually results in activation of transcription, and DNA methylation pattern is negatively correlated with histone acetylation. Dnmt1 and Dnmt3b binds to histone deacetylases and deacetylating histones which results in condensed chromatin and transcriptional silencing. Dnmt1 and Dnmt3a bind to histone methyltransferase SUV39H1 which inhibits transcription by methylating H3K9. H3K9me-2-specific methyltransferases G9a/GLP binds to Dnmt1 enzyme and enables de novo methylation of retrotransposons in embryonic stem cells. MBD proteins also interact with several histone deacetylases (HDACs) to achieve corepression. UHRF1 (ubiquitin-like PHD and RING finger domain containing 1) protein binds to methylated H3K9 and maintains the activity of Dnmt1 enzyme. H3K4 methylation inhibits the CGI methylation, while most of the CpG sites are subjected to methylation by Dnmts. H3K4 methyltransferase Setd1 is recruited to CGI sites though its interaction with Cfp1 that binds to unmethylated CpG sites. This recruitment enhances the H3K4 methylation at these sites in CpG sites that are unmethylated. Apart from histone-modifying enzymes, DNA methylation also plays a role with ATP-dependent chromatin-remodeling enzymes. MBD3 protein interacts with Mi2-NuRD repressor complex. This complex also contains Mi2β which is a part of SWI/SNF family of chromatin remodelers that destabilize histone DNA interaction.

13.3.3 Demethylation of DNA

Although DNA methylation patterns can be transferred from mother cell to daughter cells, the methylation pattern changes during different processes of development. It changes due to physiological cues during development or as a pathological response to cancer and aging processes. DNA demethylation can take place via active and passive mechanisms. Active demethylation involves Ten-eleven translocation enzymes (tet) that add a hydroxyl group to methyl group of 5-methyl cytosine which converts it to 5-methylhydroxycytosine. Tet enzyme can also mediate oxidation of 5-methylcytosine to form 5-formylcytosine or 5-carboxylcytosine. In passive mechanism, Dnmt1 enzyme is either absent or inhibited, thereby inhibiting the methylation of cytosine during cell division. Active DNA demethylation takes place in germ cell during development and also in somatic cells in gene-specific manner. During embryonic development at stage E7.5, posterior epiblast cells are set to be formed as primordial germ cells (PGCs). During migration to the genital ridge, it has a similar epigenetic pattern to that of epiblast cells. Incidentally, by the time it reaches the genital ridge at stage E11.5, most of the epigenetic markers are changed including the deletion of DNA methylation pattern which induces transcription of several genes. In somatic cells, demethylation happens in locus-specific manner at brain-derived neurotrophic factor (*Bdnf*) when the neurons are depolarized. When T cells are active and when there is no DNA replication, the enhancer region of interleukin-2 gets demethylated.

13.3.4 DNA Methylation and Diseases

DNA methylation is crucial for regulation of gene expression in several physiological and developmental processes. Alteration in the DNA methylation pattern results in several diseases including cancer since methylation is an epigenetic modification and 5-methylcytosine, in which cytosine can be deaminated to thymine, is passed to the daughter cells. Thus, methylated CpG island acts as a hotspot for mutations to occur. MBD4 contains thymine DNA glycosylase that recognizes T-G mismatch and corrects 5mc deamination. Methylation of the CG1 sites at the FMRI gene 5′ end causes fragile X syndrome. Fragile X syndrome patients have 200 copies of CGG sites at their 5′ end, whereas normal individual possesses 50 sites. Because of the presence of 200 CGG sites, these regions are silenced by the de novo methyltransferases resulting in silencing of FMRI gene. Mutation of Dnmt1 is associated with T-cell lymphomas, whereas Dnmt3 mutation results in chromosomal aberrations including chromosomal fusions and aneuploidy. Mutations in *MECP2* results in Rett syndrome which is a neurological disorder. Rett syndrome is associated with females who are heterozygous for mutations in the X-linked *MECP2* gene. Rett syndrome is characterized by impaired motor skills, microcephaly, abnormal breathing, and other associated neurological complications. Global DNA hypermethylation is also involved in the development and progression of cancers. When DNA methylation is reduced with DNA methyltransferase inhibitors, it resulted in suppression of some forms of cancer. On the other hand, hypomethylation could also result in some forms of cancer such as cervical cancer, ovarian cancer, colon cancer, and hepatocellular carcinomas (Li and Zhang, 2014).

13.4 Alternative Splicing of mRNA

In eukaryotes, single gene can code for multiple proteins, and the eukaryotic mRNA contains both introns (noncoding region) and exons (coding region). However, before translation into proteins, the noncoding introns should be removed from the mRNA, whereas the coding exon must be joined. This process of trimming introns and combining several combinations of exon regions of mRNA to generate different proteins is called splicing. This process increases the functional diversity of a gene by the formation of several proteins and adds to another level of gene regulation. When splicing process is tampered, it results in disruption of cellular functions leading to disease. RNA splicing was first discovered in adenovirus and later was discovered in other eukaryotes. The mechanism and the proteins involved in splicing reaction are highly conserved among different genera of eukaryotes.

13.4.1 Molecular Mechanism of Splicing

The process of splicing is mediated by dynamic macromolecular ribonucleoprotein (RNP) complex known as the spliceosome. There are two types of spliceosome

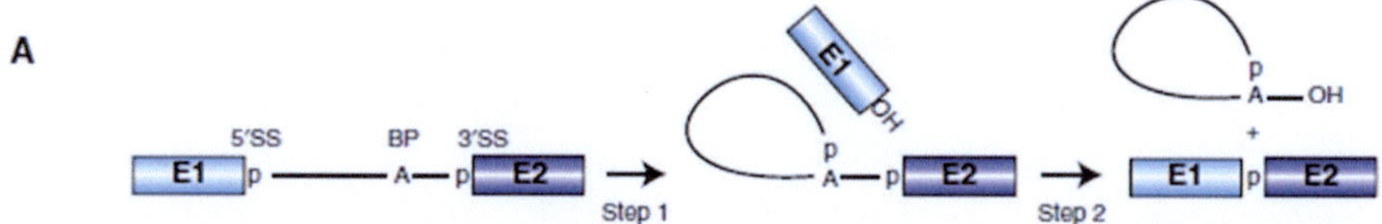

Fig. 13.7 Schematic representation of esterification reaction of pre-mRNA splicing. E1 and E2 represent exons. The solid line represents introns. The branch site adenosine is represented by letter A. The phosphate group is represented as (p) in 5′ and 3′ splice sites. (Figure adapted with permission from Will and Lührmann 2011)

complex prevalent in eukaryotes: U2-dependent spliceosome that catalyzes the eviction of U2 type of introns and the less common U-12-type spliceosome. There are several cis-acting and trans-acting factors which act together to ensure precise splicing reaction. Intron is recognized by 3′ splice site, branch site, and 5′ splice site. The 5′ splice site is characterized by GU at the 5′ end of intron which is followed by the branch site. Branch site is located usually 18–40 nucleotides upstream of 3′ splice site which is usually characterized by A nucleotide. The branch site is succeeded by several pyrimidine residues called polypyrimidine tract. This is flowed by 3′ splice site at the end which is characterized by AG nucleotide. Additionally, there exist other cis-acting regulatory elements such as exonic and intronic splicing enhancers (ESE and ISE) and exonic and intronic splicing silencers (ESS and ISS). They facilitate the binding of regulatory proteins which enhances or represses the action of the spliceosome complex at the nearby splice site. The most prominent splicing repressors consist of heterogeneous ribonucleoproteins (hnRNPs) family and polypyrimidine tract-binding protein (PTB) family. The introns are generally cleaved by two esterification reaction. In the first reaction, the 2′OH located at the branch site adenosine performs nucleophilic attack on the 5′ splice site and cuts them. The 5′ end region of the phosphate is ligated to the branch adenosine which forms a lariat structure. In the second step, the 3′OH of the 3′ splice site is cleaved which leads to ligation of 5′ and 3′ exons forming the mature mRNA (Fig. 13.7).

Spliceosome is assembled in the ordered fashion by snRNPS (small nuclear ribonucleoproteins) and other associated proteins to precisely recognize the splice sites and perform the splicing reaction. In the first reaction, U1 snRNP recognizes and binds the 5′ splice site forming complex E. In the next step, SF1 and U2AF proteins interact with branch site and polypyrimidine tract. This is succeeded by the binding of U2snRNP binding to branch site forming A complex. The preassembled U4/U6.U5 tri-snRNP is recruited which generates precatalytic B complex. After this, protein and RNA rearrangements lead to dissociation of U1 and U4 snRNPs which leads to activated B complex. Prp2 which is an RNA helicase catalytically activates B* complex which performs the first step of splicing reaction. Finally, the process leads to the formation of C complex which performs the next catalytic step. Finally, the spliceosome complex disassembles and further performs next rounds of splicing reaction (Fig. 13.8).

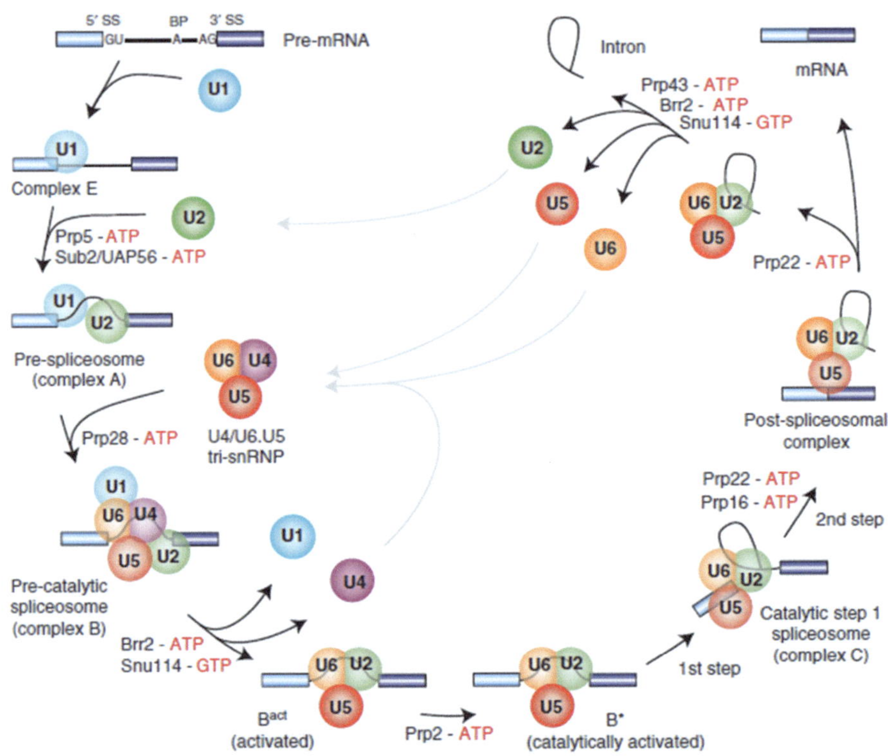

Fig. 13.8 Steps involved and assembly and disassembly of spliceosome complex. Various snRNPS are indicated in circles. Exons are indicated by colored boxes, and introns are represented by solid lines. The other associated proteins which facilitate conformational changes in various complexes, such as Prp5, Sub2/UAP56, Prp28, Brr2, Prp2, Prp16, Prp22, Prp43, and Snu114, are indicated. (Figure adapted with permission from Will and Lührmann 2011)

13.4.2 Modes of Alternative Splicing

In many cases, the mRNA can be alternatively spliced by including or excluding specific introns or exons leading to wide variety of proteins from single mRNA. This mechanism is known as alternative splicing. More than 95% of human mRNA undergoes alternative splicing in tissue-, development-, or signal transduction-specific manner. This accounts for the formation of more than 90,000 proteins from approximately 25,000 coding region.

There are five different modes of alternative splicing that were observed. (a) Exon skipping: In exon skipping, coding exon is spliced out from the primary transcript. Exon skipping is the usual type of alternative splicing that occurs in mammalian cells. (b) Mutually exclusive exons: In this type, one of the two exons is retained while the other is skipped. (c) Alternative 5′ splice site: In this type, noncanonical 5′ splice site is used which modifies the 3′ boundary of upstream exon. (d) Alternative

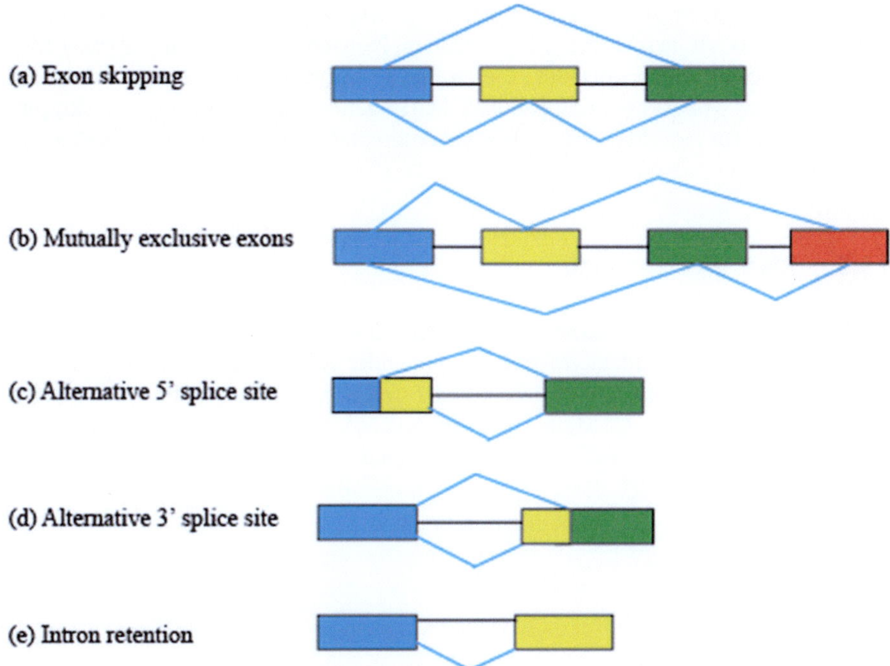

Fig. 13.9 Types of alternative splicing: In this figure, introns are indicated by solid black line between the exons. Exons are represented in different colored boxes, and the splicing events are represented in blue

3′ splice site: In this type, noncanonical 3′ splice site is used which modifies the 5′ boundary of downstream exon. (e) Intron retention: In intron retention, some introns are retained without being spliced in the messenger RNA which gets translated to proteins along with other exons (Wang et al., 2015) (Fig. 13.9).

13.4.3 Alternative Splicing and Disease

When splicing machinery is disrupted, it leads to mis-splicing of several transcripts leading to mutant gene transcripts and aberrant proteins. Thus, changes in the splicing mechanism lead to several diseases including cancer. Many mutations associated with human in disease conditions are found in splicing components. Spinal muscular atrophy, a neurodegenerative disease, is caused due to the mutation of SMN1 gene-coding SMN protein which regulates snRNP assembly. Hutchinson-Gilford progeria is a genetic disorder whose symptoms include premature aging. This is caused by the mutation in the intronic splice site which leads to the formation of cryptic splice site. Thalassemia is caused by abnormality in the hemoglobin gene which leads to abnormal levels of red blood cells. This is caused due to the mutation at the 5′ splice site of hemoglobin gene. Apart from these, alterations in the splicing causes several forms of cancer. Changes in splicing was observed in cancer cells in CD44, which is a transmembrane glycoprotein. Nonsense mutation in the exon 18 of the *BRCA1* gene which interrupts the exonic splicing enhancer causes breast and ovarian cancer. Mutation of 659 codon of MlH1 gene which results in exon skipping is responsible for hereditary nonpolyposis colorectal cancer (Tazi et al., 2009).

13.5 RNA Interference

RNA interference is a phenomenon in which 20–30 nt noncoding RNA inhibits translation by binding to target mRNA molecule. RNA interference was initially discovered in *C. elegans* for which Nobel Prize was awarded in the year 2006 to Andrew Fire and Craig C. Mello. Both microRNA (miRNA) and short interfering RNA (siRNA) play a key role in RNA interference. Although RNA interference was initially discovered in *C. elegans*, the phenomenon was also later discovered in plants, fungi, *Drosophila,* and other mammals.

13.5.1 MicroRNA

MicroRNAs (miRNAs) are genetically encoded noncoding RNA that involves in regulation of genes. Approximately 5% of human genome produce more than 1000 miRNAs which regulate at least 30% of the genes. MicroRNAs regulate several processes including cell growth, differentiation, heterochromatin regulation, and cell proliferation. The dysfunction of this process is implicated in neurological disorders, cardiovascular diseases, and cancer. A miRNA is processed from primary miRNA (pri-mRNA) in the nucleus. It is typically 1000 nt long consisting of single or clustered double-stranded hairpins. This pri-mRNA is further processed by the microprocessor complex that comprises of Drosha and dsRNA-binding protein DGCR8. After processing 65–70 nt, stem loop structure containing miRNA is produced.

13.5.2 Small Interfering RNA

Small interfering RNA (siRNA) is generally exogenously produced: synthetic- or viral-induced interfering RNA. siRNAs have well-defined structure containing 20–24 nt long dsRNA phosphorylated at 5′ ends and hydroxylated at 3′ end with two overhanging nucleotides. In the next step, Dicer enzyme catalyzes the formation of siRNAs from long dsRNAs. siRNAs are of immense important for studying gene function by knocking out specific genes. siRNA can be easily delivered into the cell by transfection.

13.5.3 Molecular Mechanism of RNA Interference

RNA interference on mRNA targets by exogenous (siRNA) and endogenous (miRNA) origin is regulated by RNA-induced silencing complex (RISC). The first step of RNA interference involves processing of RNAs which can be loaded onto the RISC complex. Endogenous miRNA is processed from pre-miRNA which leads to formation of characteristic stem loop structure in the nucleus, and further is transported to cytoplasm. Exogenous siRNA (of viral origin or synthetically made) is transported to cytoplasm. At cytoplasm, it is cleaved into short RNA fragments by Dicer enzyme. Dicers are large endonucleases containing a helicase domain, dimerized pair of RNase III domain that cleaves the RNA and PAZ domains (from the names of the proteins PIWI, Argonaute, and Zwille proteins), which specifically binds to 3′ overhangs of RNA. After processing with Dicer, dsRNA is formed which is a duplex of 20 to 24 nt long with 2 nt overhang at 3′ and a phosphate group at the 5′ end.

In the subsequent step, Dicer is aided by dsRNA-binding protein (dsRBP) to load the dsRNA complex to Argonaute protein of RISC. Dicer, Argonaute, and TRBP constitute minimum RISC loading which is responsible for generating double-stranded RNA and loading onto the Argonaute protein. After presenting the dsRNA to the Argonaute, the 3′ and 5′ end of the guide strand is bound by the PAZ proteins and MID domain of the protein forming the RISC. While the complex is loaded onto RISC, only one strand of RNA is attached to which is necessary silencing while the other strand is removed. The former is known as guide strand, and the latter is known as passenger stand, respectively. In the case of miRNA, the loaded strand from the duplex is referred to as miRNA, and the opposite strand is known as miRNA*. In the next step, RISC performs searching for the target mRNAs and binds to single-stranded mRNA which is complementary to the guide strand. The seed sequence constituting 2–6 bp of the guide strand initializes the binding to the target. Once it is bound to the target, RISC cuts the mRNA between 10 and 11 nts in a non-ATP-dependent manner. The target mRNA is further degraded preventing translation of it into proteins. In case of miRNAs, rather cleaving miRNAs bind to the 3′ untranslated regions of the target mRNA with imperfect complementarity which blocks the binding of ribosome preventing translation (Wilson and Doudna, 2013) (Fig. 13.10).

13.5.4 Applications of RNA Interference

RNA interference is a very useful biological tool to study the function of genes. siRNAs are designed for gene of interest and transfected into the cells, thereby decreasing the level of gene expression. The siRNAs may not completely abolish the expression of gene but can decrease it to a significant level. The length and type of siRNA used vary according to the species. In mammals, short RNAs are used since long RNAs elicit immune response. Functional genomics applications of RNA interference have been extensively used in studying the gene functions of

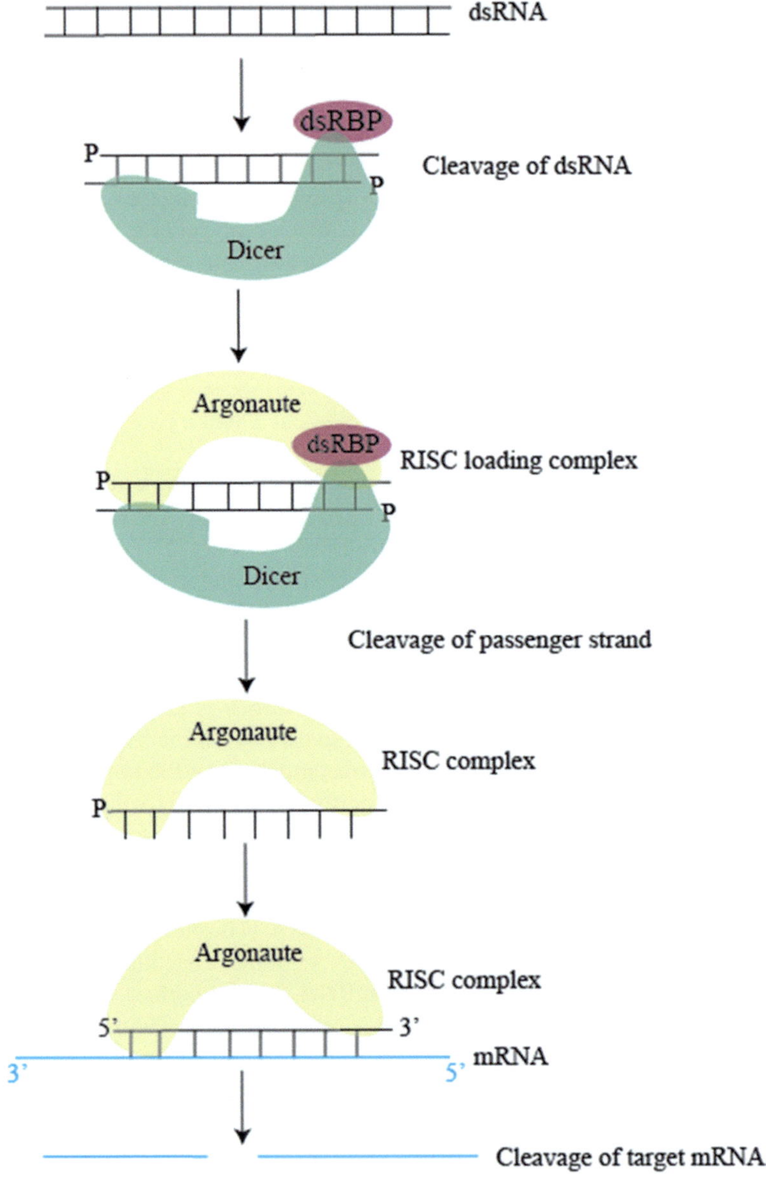

Fig. 13.10 Molecular mechanism of RNA interference. Double-stranded RNA (dsRNA) is cleaved by Dicer (green) and dsRBP-like TRBP (magenta). The resulting RNA duplex is bound by Argonaute protein. In the subsequent step, the passenger strand is removed, and the guide strand attached to Argonaute protein which forms the RISC. The RISC binds to complementary mRNA (blue) and degrades them by the action of Argonaute protein

C. elegans and *Drosophila melanogaster* since characteristic phenotype generated by RNA interference is heritable, and simple delivery of siRNA makes it attractive tool for studying gene functions in *C. elegans.* RNA interference has also been extensively used for mapping genome and annotation. RNA interference has also been used in functional genomic studies in plants such as wheat, *Arabidopsis,* and maize.

RNA interference technology also has many therapeutic applications. Many preclinical studies have shown efficient suppression of target genes, making it an attractive tool in clinical therapy. RNA interference treatment was initially proposed as antiviral strategy targeting several viral RNAs such as hPV, HIV, hepatitis B, hepatitis A, influenza virus, SARS coronavirus, and Adenovirus. Many clinical trials have shown successful deregulation of oncogenes (cancer-causing genes), making it as an attractive option for treatment of various cancer. Since RNA interference can target-specific genes, it does not affect the function of other genes, thereby increasing the target specificity of cancer treatments which is a major roadblock with chemotherapeutic approach. RNA interference has also shown potential in treating neurological diseases such as Alzheimer's, Parkinson's, and neuroglutamine diseases. RNA interference has also played a significant role in biotechnology in the development of wide variety of crops. It has resulted in the invention of unique type of crops such as tobacco free from nicotine, coffee plants free from caffeine, and nutrient-enhanced plants. By targeting polyphenol oxidase gene by RNA interference in apples, browning of apples after cutting was abolished, since polyphenol oxidase-silenced apples cannot convert chlorogenic acid into quinone product which causes browning of apples (Agrawal et al., 2003).

13.6 Posttranscriptional Regulation of Gene Expression

Posttranscriptional control plays a crucial role in the control of gene expression. It usually represents control of RNA level between the stages of protein transcription and translation. The stability, the distribution, and the level of mRNA determine the quality and level of various proteins. Thus, the stability and maturation of mRNA is highly regulated, and it varies during development or different environmental conditions.

13.6.1 Control of mRNA Degradation

The mRNA molecules are usually synthesized by RNA polymerase II and are relatively less stable molecules. The typical structure of mRNA contains cap structure at the 5′ region and polyadenylation site at the 3′ region which protects them from degradation. However, many mRNAs do not result in the translation into proteins, and after the translation of proteins, the mRNA has to be degraded; otherwise, it would result in overproduction of proteins. The mRNA degradation pathway in eukaryotes begins with the process of deadenylation of 3′ end. The

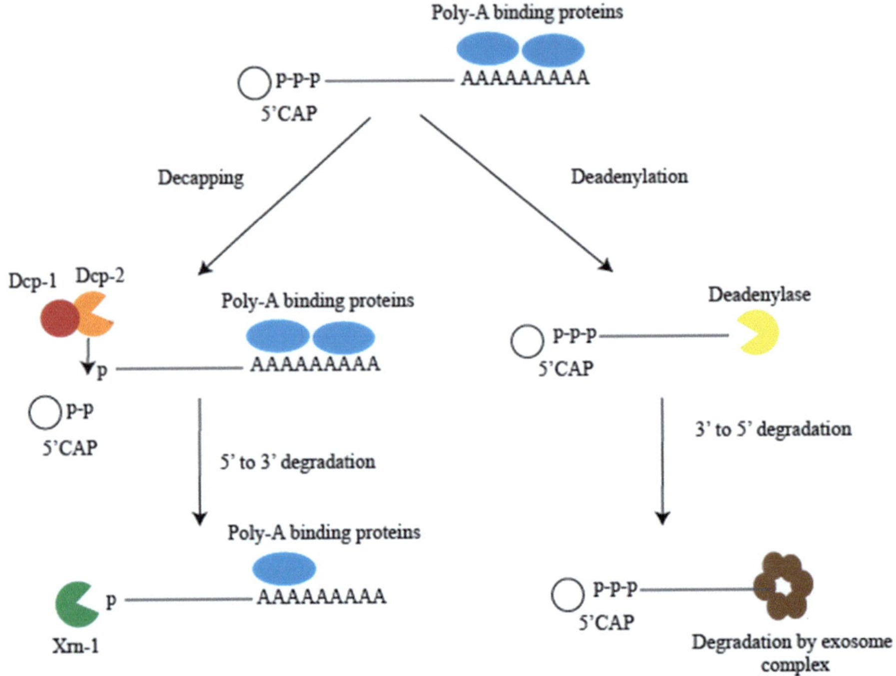

Fig. 13.11 Depiction of different routes of mRNA degradation. Poly-A-binding protein (blue) is bound to 3′ polyadenylation region, and 5′ region is bound by cap. Enzyme Dcp-1 (red) and Dcp-2 (orange) complex cleaves the 5′ cap region leaving the monophosphate at 5′ end. This is attacked by Xrn-1 endonuclease which cleaves the mRNA in 5′ to 3′ direction. Deadenylases (yellow) cleave the polyadenine residues at the 3′ end. This is then targeted by multiprotein exosomal degradation complex (yellow) which cleaves the mRNA in 3′ to 5′ direction

process deadenylation is the most critical step in the mRNA degradation. Later, the 5′ cap will be removed by the process of decapping, and the RNA will be degraded in 5′ to 3′ direction by Xrn-1 endonuclease and 3′ to 5′ direction by exosomal complex (Fig. 13.11).

13.6.1.1 3′ Deadenylation

The poly-A tail length at the 3′ region of the mRNA determines the stability and mRNA transport and translation. The poly-A tail region is usually bound by poly-A-binding proteins (PABP) which facilitate translation and shortening of the poly-A tails enabling 3′ to 5′ exonuclease activity. Poly-A tail reduction sends signal for mRNA degradation to several pathways such as nonsense-mediated mRNA decay (degradation because of premature stop codon), ARE-mediated decay (mRNAs possessing AU-rich elements), and miRNA-mediated decay. In eukaryotes, PAN2-PAN3 and CCR4-NOT are two important protein complexes involved in deadenylation.

13.6.1.2 Decapping

Another way of destabilizing and getting access to the mRNA is by decapping the 5′ end of the mRNA. The process of decapping is performed by Dcp-2 enzyme belonging to Nudix hydrolase family which is conserved among the eukaryotes. Dcp-2 cleaves the mRNA using the Nudix motif, which is triggered by Dcp-1 enzyme, resulting in m^7GDP and 5′ monophosphate RNA. Dcp-2 enzyme prefers enzyme that are at least 25 nt in length.

13.6.1.3 5′ to 3′ mRNA Degradation by Xrn-1 Endonuclease

Decapped mRNA with the exposed 5′ phosphate end is amenable to degradation by Xrn-1 endonuclease. Xrn-1 processively cleaves the mRNA from 5′ to 3′ direction. Decapping and Xrn-1 endonuclease activity are coupled, in which Xrn-1 interacts with one of the decapping complex protein Dcp-1 or its homologues.

13.6.1.4 3′ to 5′ Degradation by Exosomes

In some cases, mRNAs can also be degraded in 3′ to 5′ direction. Typically, this type of 3′ to 5′ mRNA degradation is usually mediated by exosome machinery. The 3′ end of the RNA is targeted by multi-subunit enzyme complex exosome for degradation and processing. The components of exosome are highly conserved in eukaryotes. Six subunits, Rrp41, Rrp42, Rrp43, Rrp45, Rrp46, and Mtr3, forms the hexameric core ring. Rpr4, Rpr40, and Csl4 bind to RNA though their KH motif-binding domain, and Rpr44 contains the PIN domain which attaches the exosome ring and performs 3′ to 5′ endonucleolytic cleavage. After the degradation by exosome complex, the remaining m^7GpppN is hydrolyzed by m^7G-specific pyrophosphatase DcpS. DcpS is called scavenger decapping enzyme which has specificity for small RNA fragments hydrolyzing the remaining cap region.

13.6.1.5 Other Types of Aberrant mRNA Degradation

When mRNA was aberrantly synthesized, it results in the formation of aberrant proteins; therefore, such mRNAs have to be identified and degraded. One way of removing these aberrant transcripts is nonsense-mediated mRNA decay (NMD) in which mRNAs containing premature termination codon is identified and degraded. NMD also degrades RNA with long 3′ untranslated region (UTR), mRNAs possessing an upstream ORF at the 5′UTR, and mRNAs containing introns at the 3′UTR. The effect of NMD is dependent on many cis-acting elements in the mRNA and the exon junction complex (EJC) components. The NMD factors are highly conserved among the eukaryotes. The key protein of NMD pathway is UPF1 which is an ATP-dependent RNA helicase. UPF1 interacts with UPF2, an acidic protein which in turn interacts with UPF3, an RNA-binding protein. UPF1 with UPF2 and UPF3 enhances degradation of mRNA by endonucleolytic cleavage, 5′ to 3′ or 3′ to 5 mRNA degradation.

In some cases, error in the mRNA causes stalling of the ribosome blocking further translation. It may lead to No-go decay (NGD), nonstop decay (NSD), or nonfunctional 18srRNA decay (NRD). Dom34-Hbs1 protein takes part in all of these pathways. When elongation is blocked by secondary mRNA structures or while translating several positively charged amino acids, it results in No-go decay (NGD). Consequently, the aberrantly formed protein is degraded by ubiquitin-based protein degradation. When transcripts lack in-frame stop codon, it results in continues translation, and the nonstop decay (NSD) is triggered by cryptic polyadenylation signal in which ribosome is stalled. Defects in the ribosomal 18srRNA caused by mutations or by improper biogenesis can lead to translational stalls which lead to nonfunctional 18srRNA decay (NRD) (Newbury, 2006 and Siwaszek, et al., 2014).

13.6.2 Control of Protein Degradation

Protein degradation is one of the important cellular machinery which determines the half-life of proteins. The half-life of protein varies between different kinds of proteins ranging from minutes to hours. The proteins that are turned over rapidly act as regulatory protein in other processes such as transcription factors. This rapid turning-over process enables to increase or decrease the levels of the protein in accordance with the external stimuli. Mistranslated and aggregate proteins are also recognized and degraded rapidly by the protein control machinery, the accumulation of which otherwise is detrimental to the cellular processes (Glick et al, 2010). Protein degradation in eukaryotes takes place by two well-studied systems, namely, lysosome-mediated autophagy and ubiquitin proteasome pathway.

13.6.2.1 Autophagy

Autophagy plays an important role in removal of misfolded and aggregate proteins; degrades damaged organelles such as peroxisomes, mitochondria, and endoplasmic reticulum; and targets intracellular pathogens. The process of autophagy involves lysosomal-mediated protein degradation by several autophagy-related proteins (ATG). Autophagy is classified into three types, namely, macro-autophagy (autophagy), micro-autophagy, and chaperone-mediated autophagy. In all of these types, proteolytic degradation is mediated by cytosolic machineries of the lysosome. (i) In macro-autophagy (autophagy), the degradation of cytosolic membrane takes place by sequestering into membrane-bound vesicle called autophagosome which is then fused to the lysosome. (ii) In micro-autophagy, the lysosomal components are directly taken into the lysosome by lysosomal invaginations. In chaperone-mediated autophagy, chaperone proteins such as Hsc-70 help in translocating the proteins across the lysosomal membrane. These chaperone proteins are recognized by the lysosomal-associated protein 2A (LAMP-2A) which binds to target protein and degrades them. The macro-autophagy (also referred to as autophagy) is a complex process orchestrated by several proteins on consequential manner. It involves (i) formation of phagophore, (ii) autophagosome formation, and (iii) lysosome fusion and degradation (Fig. 13.12).

(i) Formation of phagophore: The formation of phagophore involves activation by conserved nutrient sensors such as mTOR and AMP-activated kinase (AMPK) in which mTOR acts as an inhibitor and AMPK acts as an activator. It activates the two assembly complexes, ALGI/PLKI and P13K complex, which help in the formation of phagophore.

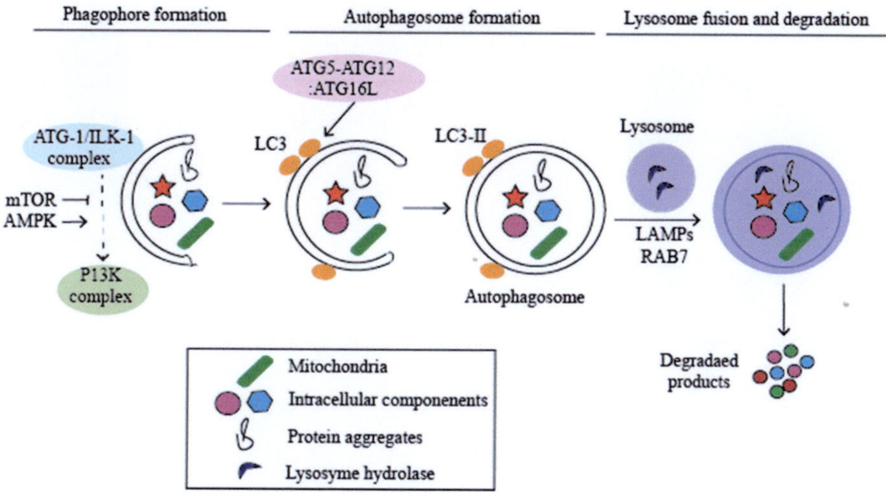

Fig. 13.12 Molecular mechanism of autophagy. Autophagy is controlled by signals from mTOR and AMPK pathway which depends on the activity of ATG-1/ILK-1 and P13K complexes. In the subsequent steps, ATG5-ATG12:ATG16L conjugates LC3 to lipid phosphatidylethanolamine to form LC3-II which enables the anchoring to autophagosomal membrane. It fuses to the lysosome via LAMPs and RAB7, and hydrolases of lysosome degrades the molecules and releases the products for further recycling

(ii) Autophagosome formation: Autophagosome formation consists of two conjugation steps. In the first step, ATG7 and ATG10 mediate the conjugation of ATG5, and ATG12 proteins interact and bind to ATG16L. In the next step, ATG7 and ATG3 along with ATG5-ATG12:ATG16L complex conjugate LC3 to lipid phosphatidylethanolamine (PE) to form LC3-II that allows the binding of the complex to the autophagosomal membrane.

(iii) Lysosome fusion and degradation: The mature autophagosome then fuses with the lysosome mediated by lysosomal proteins such as LAMPs and RAB7. Once it is fused, the targeted protein or organelle is digested by several lysosomal hydrolases into final products consisting of amino acids, lipids, and nucleotides which translocate to the cytoplasm and get incorporated into the cellular processes (Fig. 13.12).

13.6.2.2 Ubiquitin-Proteasome Pathway

Selective protein hydrolysis in eukaryotic cells takes place via ubiquitin-proteasome pathway in which cytosolic and nuclear proteins are marked with ubiquitin which is targeted for rapid proteolysis. Ubiquitin is an 8.6 kDA regulatory protein that is highly conserved among the eukaryotes. Ubiquitin attaches to the lysine residues of the protein via an isopeptide bond. Additional ubiquitin residues are added to the ubiquitin residue forming polyubiquitin chain. Such polyubiquitylated chains are recognized by the 26S proteasome complex that degrades the proteins and releases the ubiquitins which are used in the next cycle.

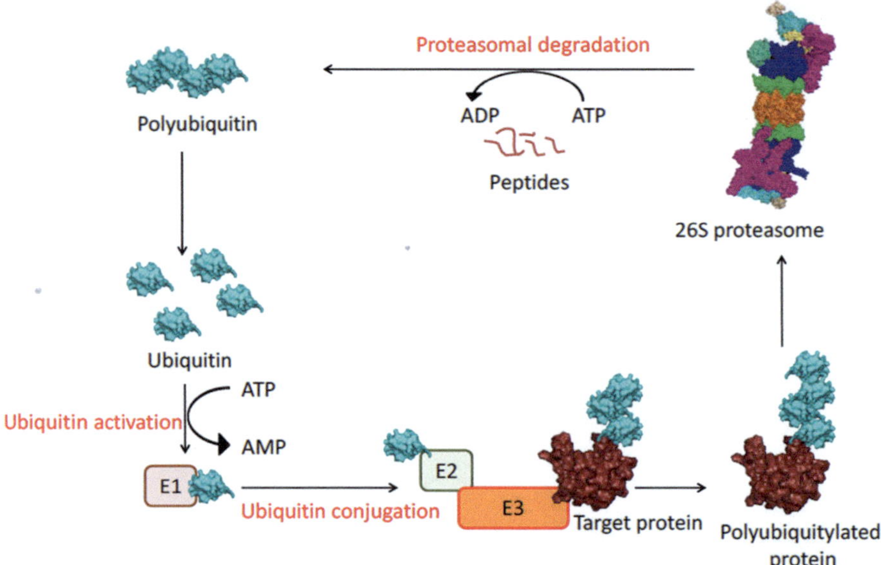

Fig. 13.13 The ubiquitin-proteasome pathway: The ubiquitin residues are activated by ATP-dependent manner by E1 ubiquitin-activating enzyme. It is then conjugated to E2 via E2 ubiquitin-conjugating enzyme. This complex interacts with E3 which interacts with E2 and target protein and transfers the ubiquitin to the target protein. After polyubiquitylation by addition of ubiquitin chains, the substrate is targeted to 26S proteasome where the proteins are deubiquitylated and the target protein is degraded and the released ubiquitin is recycled for the next round

The process of ubiquitination consists of three steps: (i) activation, (ii) conjugation, and (iii) ligation.

(i) Activation: This process involves two steps and is catalyzed by E1 ubiquitin-activating enzyme. E1 enzyme binds to both ATP and ubiquitin and catalyzes the reaction in which acyl adenyl group is transferred to C-terminus of ubiquitin. In the next step, ubiquitin is transferred to cysteine residue with the release of AMP.
(ii) Conjugation: This step is catalyzed by the E2 ubiquitin-conjugating enzymes. In this process, the E2 enzyme catalyzes trans(thio)esterification reaction in which ubiquitin is transferred from E1 to E2.
(iii) Ligation: This step is catalyzed by E3 ubiquitin ligases. In this step, E3 ligase interacts with the E2 and target protein. It then mediates isopeptide bond formation between lysine residue of the target protein and C-terminal glycine residue of ubiquitin. There are several hundreds to ubiquitin ligases present. Polyubiquitylation takes place by the addition of additional ubiquitin residues via one of the seven lysines that is present in the ubiquitin.

The 26 s proteasome is about 2000 kDa consisting of one 20S protein subunit and two regulatory cap subunits. The hollow core in the central region enables forming cavity in which protein degradation takes place. Each end of the proteasome contains 19S regulatory cap proteins which possess multiple ubiquitin-binding sites. These cap proteins recognize the ubiquitin-tagged proteins that enable transferring them to the catalytic core where the protein hydrolysis takes place (Fig. 13.13).

13.6.2.3 Mechanism of Protein Degradation by Ubiquitin Proteosome

Polyubiquitylated substrate is recognized by the intrinsic ubiquitin receptors proteins of 19S cap proteins. The ubiquitin receptor proteins contain N-terminal ubiquitin-like domain (UBL) and few or more ubiquitin-associated domain (UBA). A 19S proteasome cap recognizes the UBL domains, and the UBA domains bind ubiquitin through three helix bundles. The ubiquitin chain is removed from the substrate by deubiquitylase enzyme, Rpn11, and the ubiquitin chains are recycled. The substrate is further extended by the Rpt1–6 ATPases and transferred to the 20S proteasomal complex for degradation. The 20S core particle mediates proteolysis by the β-subunits via threonine-dependent nucleophilic attack. Degradation of protein takes place in the central chamber by the action of two β rings and degrades the substrate to short polypeptides of 7–9 residues long. In some cases, it forms biologically active and functional molecule such as NF-κB which plays a crucial role in immune response.

13.6.2.4 Ubiquitination and Diseases

Ubiquitination process is an important posttranslational modification which is involved in processes such as cell cycle progression, proliferation, and development. When ubiquitin pathway is disrupted, it results in several diseases. High levels of ubiquitin in the brains cause decreased level of amyloid precursor protein which plays a key role in triggering Alzheimer's disease. Mutation in ubiquitin B resulting in truncated peptides have also been observed in Alzheimer's disease and tauopathies. Mutations in ubiquitin E3 ligase cause several diseases such as Angelman syndrome, Fanconi anemia, and 3-M syndrome. Mutation of E3 ligase is also implicated in several types of cancer such as renal cell carcinoma, breast cancer, colorectal cancer, and glioblastoma (Dikic, 2017 and Saeki, 2017).

Box 13.1: Scientific Concept: Quantifying Site-Specific Chromatin Mechanics and DNA Damage Response

DNA is wound around the histones forming highly condensed structure called chromatin. The chromatin is found to be highly dynamic with several condensation states rather than more rigidly defined euchromatin and heterochromatin region. Whitefield et al. (2018) studied the dynamics of transcriptionally active and inactive region using single particle tracking and also analyzed the chromosome dynamics during DNA damage. They used 96 tetracycline

(continued)

Box 13.1 (continued)

response elements (TRE) named U2OS TRE that allows controlling gene expression be the binding of transcription activator (TA) or transcription repressor (TetR) tagged with mCherry. GFP-tagged fibrillarin was transfected to visualize bulk chromatin motion. The tracking revealed that transcriptionally active regions were indistinguishable from chromosomal bulk dynamics motion, whereas transcriptionally repressed regions showed decreased mobility compared to transcriptionally active region and chromosomal bulk dynamics (Fig. 13.14a and b).

During DNA damage such as double-stranded breaks, the nucleosome must reorganize itself for DNA, and histones have to be displaced in order to splice back the DNA by repair proteins. Whitefield et al. (2018) studied the effect of DNA damage by using KillerRed-labeled cells to induce DNA damage with transcriptional activator (TA) or repressor (TetR). In transcriptionally repressed regions, induction of DNA damage increased the chromosomal mobility similar to that of the bulk chromosome dynamics (Fig. 13.14d). In contrast, upon DNA damage, transcriptionally active regions undergo dynamic changes in a time-dependent manner which results in chromatin relaxation and reduction in propagation at these regions (Fig. 13.14c). The overall effect of it decreases the large-scale chromatin motions, thereby reducing improper repair and translocations in the chromosome. These conclusions give us the concept that the human genome is present in variety of chromatin states rather than previously determined two states of euchromatin and heterochromatin.

Box 13.2: Scientific Concept: Transcriptional Gene Silencing in Humans

Transcriptional gene silencing by small noncoding RNAs was extensively studied in yeast (*Saccharomyces pombe*), plants (*Arabidopsis thaliana*), and worms (*C. elegans*) and was later found to also control gene transcription and epigenetic states in human cells. Transcriptional gene silencing (TGS) can cause long-term epigenetic effect on the genes that can be transferred to the daughter cells in contrast to RNA interference. Endogenous transcriptional gene silencing was shown to be facilitated by microRNAs (miRNAs). In 2005, FANTOME and ENCODE consortia revealed that human genome encodes large number of long noncoding RNAs (lncRNAs) which are more than 200 nt in length, and many of them are antisense to the protein coding counterparts. Later, the biological role of lncRNA was revealed in the X-chromosome inactivation and regulation of p15 and p21 tumor-suppressor genes and several other genes such as BDNF, MYCN, DHRS4, KCNQ1, NBAT, and HIV-1.

(continued)

Box 13.2 (continued)

Small RNA mediates epigenetic transcriptional gene silencing (TGS) by making use of template mRNA at the specific promoter and further recruits epigenetic silencing complex which consists of DNMT3a, Ago1, EZH2, and HDAC1. Long noncoding RNAs also work with similar epigenetic silencing complexes which are targeted to the specific loci for epigenetic silencing. In human cells, both small and long noncoding RNAs mediate its action via DNMT3a which in turn regulates the epigenetic state of the target genes (Fig. 13.15). This is of particular interest because of its potential to be long-lasting and heritable which could play a significant role in therapeutics. These new findings provide us the hope of using small and long noncoding RNAs in therapeutic applications. It is a promising therapeutic approach because of strand-specific targeting of gene, stability of small RNAs, and its prolonged effect on the gene expression that can be used for a specific gene of interest and possess high efficiency because endogenous pathway of RNA-directed transcriptional gene silencing overlaps with the small and long noncoding RNA silencing mechanism.

13.7 Chapter Summary

- Regulation of eukaryotic gene takes place at the level of transcription, posttranscription, translation, and posttranslation levels. The transcription levels are regulated by cis- and trans-acting regulatory elements. Cis-acting elements include enhancers, silencers, and locus control regions, and trans-acting factors consist of activators, repressors, and specific transcription factors.
- Chromatin complexes consisting of nucleosomes also play an important role in controlling transcription. Nucleosomes consist of octamer of histones around which the DNA is wrapped. Chromatin remodeling complexes and histone-modifying enzymes positively or negatively influence the nucleosomal structure to activate or repress transcription.
- DNA methylation is a major epigenetic modification resulting in silencing of transcription. The majority of DNA methylation happens in the CpG islands that are found near the promoter elements.
- Splicing is an important posttranscriptional modification in which introns are spliced out to produce mature mRNA with exons for translation into proteins. Splicing is mediated by macromolecular complex called proteasome. In alternative splicing, combinations of exons are used to get different proteins.
- RNA interference is a phenomenon in which noncoding consisting of 20–30 nt inhibits translation by binding to target mRNA. This includes endogenous microRNA (miRNAs), long noncoding RNA (lncRNA), and exogenous or viral origin small interfering RNA (siRNA).

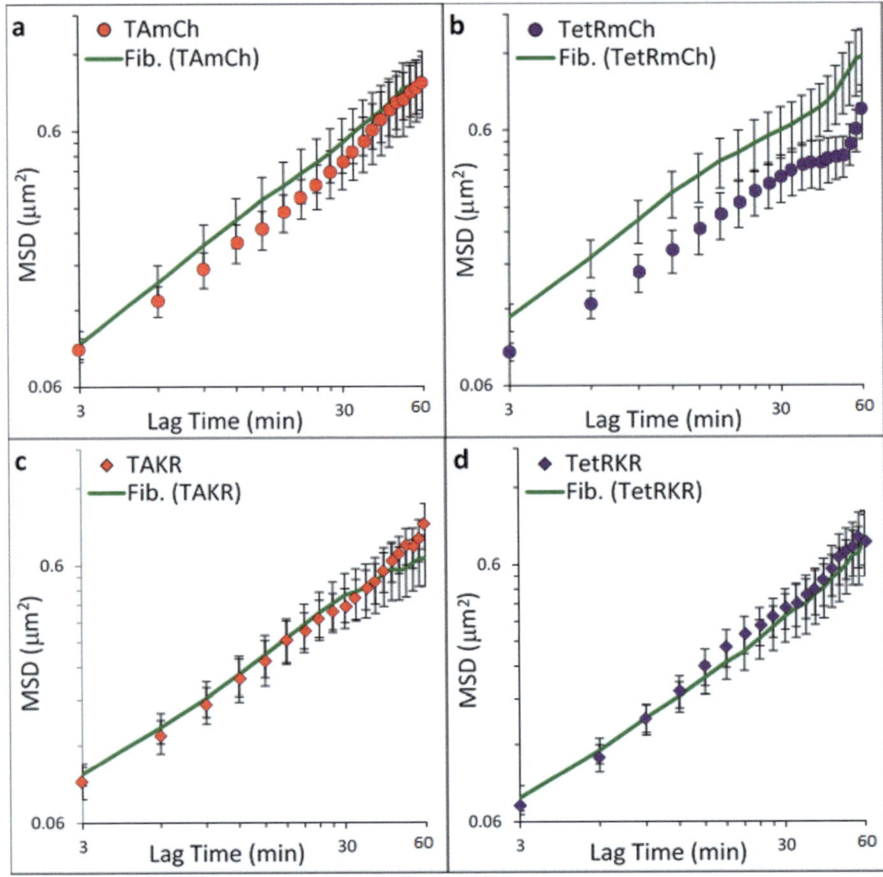

Fig. 13.14 Mean square displacement (MSD) comparing mobility of the bulk chromatin motion to the four tracers. Bulk chromatin motion measured as cotransfected GFP fibrillarin. (**a**) TA-mCherry (TAmCh) is denoted by red circles, and cotransfected GFP fibrillarin (Fib. (TAmCh)) is denoted by green line. (**b**) TetR-mCherry (TetRmCh) is denoted by purple circles, and cotransfected GFP fibrillarin (Fib. (TetRmCh)) is denoted by green line. (**c**) TA-KillerRed (TAKR) is denoted by red diamonds, and cotransfected GFP fibrillarin (Fib. (TAKR)) is denoted by green line. (**d**) TetR-KillerRed (TetRKR) is denoted by purple diamonds, and cotransfected GFP fibrillarin (Fib. (TetRKR)) is denoted by green. Error bars are SEM. (Figure taken from Whitefield et al. 2018)

- The stability of mRNA is determined by the action of several mRNA degrading enzymes that play an important role in posttranscriptional regulation. The mRNAs can be decapped at the 5′ end and deanylated at the 3′ end. It can then be attacked by the Xrn-1 endonuclease or degraded by the exosome.
- Posttranslational control involves the regulation of protein stability and degradation. Along with the bulk proteins, mistranslated and aggregated proteins are also rapidly degraded. This is performed by the process of autophagy in which lysosomal hydrolases perform protein degradation. In ubiquitin-proteasome

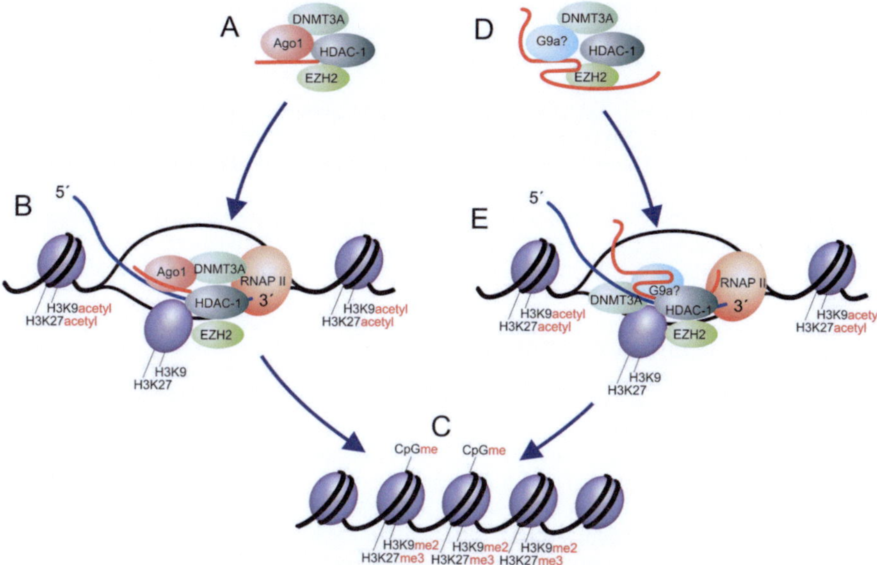

Fig. 13.15 Antisense RNA-directed transcriptional gene silencing (TGS). Small antisense noncoding RNAs can be (**A**) introduced into the nucleus and (**B**) interact with and recruit epigenetic silencing complexes consisting of DNMT3a, Ago1, EZH2, and HDAC1 to homology containing targeted loci by interactions with low-copy promoter-associated transcripts resulting in (**C**) epigenetic silencing consisting of histone and DNA methylation and ultimately chromatin compaction of the targeted locus. (**D**) Long antisense noncoding RNAs have also been observed to interact with similar epigenetic silencing complexes and (**E**) localize with these complexes at targeted loci resulting in (C) epigenetic silencing of the lncRNA-targeted locus. (Figure taken from Weinberg and Morris 2016)

pathway, proteins are polyubiquitylated that are recognized by the proteasome complex which degrades the proteins.

References

Agrawal N et al (2003) RNA interference: biology, mechanism, and applications. Microbiol Mole Biol Rev 67:657–685. https://doi.org/10.1038/561

Banerjee T, Chakravarti D (2011) A peek into the complex realm of histone phosphorylation. Mol Cell Biol 31(24):4858–4873. https://doi.org/10.1128/MCB.05631-11

Bannister AJ, Kouzarides T (2011) Regulation of chromatin by histone modifications. Cell Res 21(3):381–395. https://doi.org/10.1038/cr.2011.22

Clapier CR et al (2017) 'Mechanisms of action and regulation of ATP-dependent chromatin-remodelling complexes. Nat Rev Mol Cell Biol 18(7):407–422. https://doi.org/10.1038/nrm.2017.26

Daniel AR, Hagan CR, Lange CA (2011) Progesterone receptor action: defining a role in breast cancer. Expert Rev Endocrinol Metab 6(3):359–369. https://doi.org/10.1586/eem.11.25

Davey AR, Grossman M (2016) Androgen receptor structure, function and biology: from bench to bedside. Clin Biochem Rev 37(1):3–15. https://doi.org/10.1038/sc.1992.29
Deroo BJ, Korach KS (2006) Review series Estrogen receptors and human disease. J Clin Invest 116(3):561–570. https://doi.org/10.1172/JCI27987
Dikic I (2017) Proteasomal and autophagic degradation systems. Annu Rev Biochem 86(1): 193–224. https://doi.org/10.1146/annurev-biochem-061516-044908
Funder JW (2005) Mineralocorticoid receptors: distribution and activation'. Heart Fail Rev 10:15–22
Gaston K, Jayaraman PS (2003) Transcriptional repression in eukaryotes: repressors and repression mechanisms. Cell Mol Life Sci 60(4):721–741. https://doi.org/10.1007/s00018-003-2260-3
Glick D, Barth S, Macleod KF (2010) Autophagy: cellular and molecular mechanisms. J Pathol 221:3–12
Green MR (2005) Eukaryotic transcription activation: right on target. Mol Cell 18(4):399–402. https://doi.org/10.1016/j.molcel.2005.04.017
Li E, Zhang Y (2014) Dna methylation in Mammals. Cold Spring Harb Perspect Biol 6(5):a019133
Li Q et al (2002) Review article Locus control regions. Blood 100(9):3077–3086. https://doi.org/10.1182/blood-2002-04-1104
Li B, Carey M, Workman JL (2007) The role of chromatin during transcription. Cell 128(4): 707–719. https://doi.org/10.1016/j.cell.2007.01.015
Luo RX, Dean DC (1999) Chromatin remodeling and transcriptional regulation. J Natl Cancer Inst 91(15):1288–1294. https://doi.org/10.1093/jnci/91.15.1288
Ma J (2011) Transcriptional activators and activation mechanisms. Protein Cell 2(11):879–888. https://doi.org/10.1007/s13238-011-1101-7
Maston GA, Evans SK, Green MR (2006) Transcriptional regulatory elements in the human genome. Annu Rev Genomics Hum Genet 7(1):29–59. https://doi.org/10.1146/annurev.genom.7.080505.115623
Messner S, Hottiger MO (2011) Histone ADP-ribosylation in DNA repair, replication and transcription. Trends Cell Biol 21(9):534–542. https://doi.org/10.1016/j.tcb.2011.06.001
Moore LD, Le T, Fan G (2013) DNA methylation and its basic function. Neuropsychopharmacology 38(1):23–38. https://doi.org/10.1038/npp.2012.112
Newbury SF (2006) Control of mRNA stability in eukaryotes. Biochem Soc Trans 34(1):30–34. https://doi.org/10.1042/bst0340030
Newell-Price J, Clark AJL, King P (2000) DNA methylation and silencing of gene expression. Trends Endocrinol Metab 11(4):142–148. https://doi.org/10.1016/S1043-2760(00)00248-4
Orphanides G, Lagrange T, Reinberg D (1996) The general transcription machinery of RNA polymerase II. Genes Dev 10(7):2657–2683
Pabo C (1992) Transcription factors: structural families and principles of DNA recognition. Annu Rev Biochem 61(1):1053–1095. https://doi.org/10.1146/annurev.biochem.61.1.1053
Recillas-Targa F et al (2002) Position-effect protection and enhancer blocking by the chicken-globin insulator are separable activities. Proc Natl Acad Sci 99(10):6883–6888. https://doi.org/10.1073/pnas.102179399
Saeki Y (2017) JB special review - Recent topics in ubiquitin-proteasome system and autophagy: Ubiquitin recognition by the proteasome. J Biochem 161(2):113–124. https://doi.org/10.1093/jb/mvw091
Siwaszek A, Ukleja M, Dziembowski A (2014) Proteins involved in the degradation of cytoplasmic mRNA in the major eukaryotic model systems. RNA Biol 11(9):1122–1139. https://doi.org/10.4161/rna.34406
Tazi J, Bakkour N, Stamm S (2009) 'Alternative splicing and disease. ', Biochim Biophys Acta 1792(1):14–26. https://doi.org/10.1016/j.bbadis.2008.09.017
Vandevyver S, Dejager L, Libert C (2014) Comprehensive overview of the structure and regulation of the glucocorticoid receptor. Endocr Rev 35(4):671–693. https://doi.org/10.1210/er.2014-1010
Venter JC et al (2001) The sequence of the human genome. Science 291(5507):2001

Wang Y et al (2015) Mechanism of alternative splicing and its regulation. Biomed Rep 3(2): 152–158. https://doi.org/10.3892/br.2014.407

Weinberg MS, Morris KV (2016) Transcriptional gene silencing in humans. Nucleic Acids Res 44(14):6505–6517. https://doi.org/10.1093/nar/gkw139

Whitefield DB et al (2018) Quantifying site-specific chromatin mechanics and DNA damage response. Sci Rep 8(1):1–9. https://doi.org/10.1038/s41598-018-36343-x

Will CL, Lührmann R (2011) Spliceosome structure and function. Cold Spring Harb Perspect Biol 3:a003707. https://doi.org/10.3858/emm.2008.40.6.686

Wilson RC, Doudna JA (2013) Molecular mechanisms of RNA interference. Annu Rev Biophys 42(1):217–239. https://doi.org/10.1146/annurev-biophys-083012-130404

Zentner GE, Henikoff S (2013) Regulation of nucleosome dynamics by histone modifications. Nat Struct Mol Biol 20(3):259–266. https://doi.org/10.1038/nsmb.2470

Part III

Molecular Genetics II: Analysis of Genomes

Techniques of Molecular Genetics 14

Nidhi Sharma and Shrish Tiwari

14.1 Recombinant DNA Technology

In advanced molecular biology and genetics, recombinant DNA technology appears as a boon that facilitates and opens up new challenges in the age of biotechnology. This technology has overtaken the conventional genetics and given rise to the opportunities to edit a genetic makeup of any organism. This has become possible to identify a particulate target, cut, and insert into another organism. The very first man-made recombinant product in the biotechnology industry was human insulin, and it was only possible to develop by taking advantage of this technology. How this recombinant technology story came in existence? It has begun with the discovery of restriction enzyme. Werner Arber first time reported that bacteria has the machinery which cuts the foreign viral DNA into fragments, and thus, he isolated this enzyme known as "restriction enzyme." With his discovery, geneticists came to know how to "cut" and "paste" a DNA molecule and were encouraged to find novel restriction enzymes. In a simple definition, recombinant technology is sum up into three main tools as follows: (1) enzymes exploit for cutting, polymerization, and ligation, (2) vectors as a vehicle for DNA transfer, and (3) host organism as a production factory.

With the discovery of recombinant technology, it was further advanced by the collaboration of Stanley Cohen and Herbert Boyer in 1972, and they successfully established their first ever company, Gentech, in 1976 based on recombinant DNA technology.

N. Sharma
La Sapienza University of Rome, Rome, Italy

S. Tiwari (✉)
Aligarh Muslim University, Aligarh, Uttar Pradesh, India

D. Kar, S. Sarkar (eds.), *Genetics Fundamentals Notes*,
https://doi.org/10.1007/978-981-16-7041-1_14

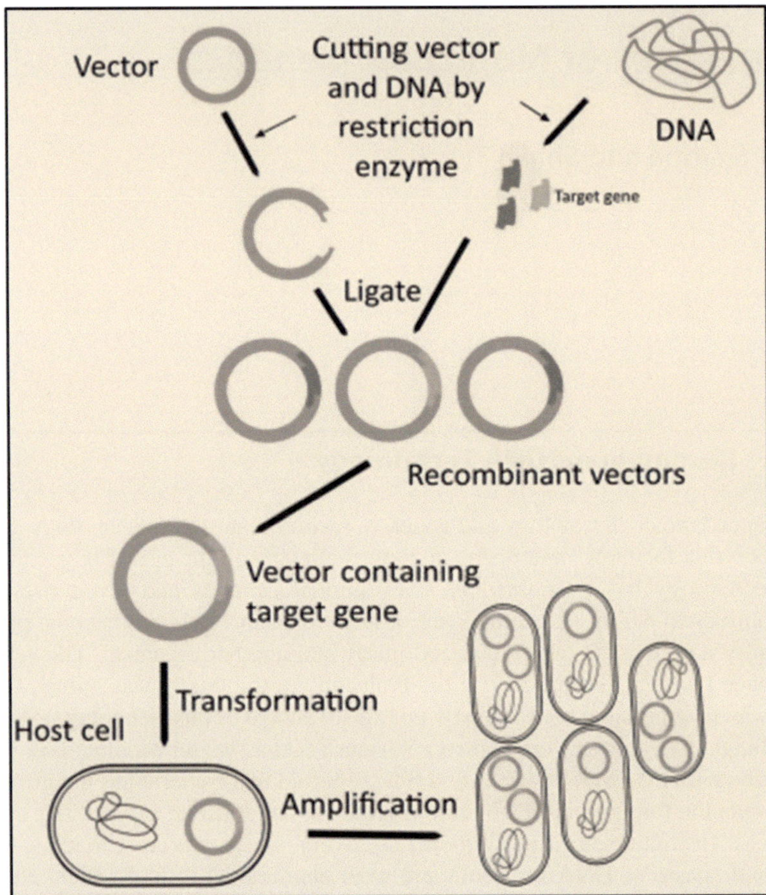

Fig. 14.1 Principle of recombinant DNA technology. Recombinant DNA technology includes few stepwise processes where a DNA fragment is associated with vector-mediated (i) ligation, (ii) recombination, (iii) replication, (iv) transformation, and (v) amplification to produce a recombinant clone of desired DNA fragment (gene) in host cell

A recombinant DNA technology has five major steps: (1) cutting the DNA of interest by the site-specific restriction enzyme, (2) amplification of DNA copies by polymerase chain reaction (PCR), (3) insertion of amplified DNA into the appropriate vectors, (4) incorporating the vectors in the desired host organism, and (5) cultivating and harvesting the desired recombinant product (Fig. 14.1).

Currently, applications of recombinant DNA technology are not only limited in recombinant protein production but also has been applied in gene therapy, clinical diagnosis, and animal and plant transgenesis.

14.1.1 Vector-Vehicle of Carrying DNA

Vector is a DNA molecule other than genetic DNA that functions as a carrier to transfer/insert a foreign gene, replicate, and express into another cell. Vectors are one of the most essential and powerful tool for gen cloning. Vector also helps in the detection whether insertion and expression of desired gene is done successfully in the host organism. This is because vector encodes some biomarker-associated gene, those only expressed in definitive growth condition. Cloning vector can be taken from bacteria, virus, and cells (some higher organism like yeast) to be incorporated to a foreign DNA particle for cloning purposes. As cloning vector adopts the atmospheres in an organism, this vector too shows features that target a convenient insertion and removal of gene or DNA. Cloning in vectors further can be followed by subcloning in another vector that would show more specificity such as expression vector, etc. Two most common used vector in recombinant DNA technology are *E. coli* plasmid vector and bacteriophage λ vectors.

Mainly six major types of vectors are being used in recombinant technology so far which are as follows:

Plasmid: Plasmid is an extrachromosomal bacterial DNA that contains the ability of autonomous replication. Plasmid's machinery allows it to invade the bacterial cell and undergo self-replication inside the host cell. Plasmid does not kill host cell on invasion, rather a plasmid containing antibiotic resistance gene helps the host to survive in the presence of that antibiotic. Plasmid vector is structurally divided into restriction enzyme site, origin of replication, and gene insertion site along with reporter gene. Reporter gene distinguishes recombinant plasmid to nonrecombinant plasmid (Fig. 14.2). Reporter gene can be either an antibiotic resistance gene or a gene which produces colorimetric substance or a luciferase activity. All these features will show up in cloning culture if the plasmid have received our gene of interest. All plasmid does not have similar copy number, but usually, they have high copy number, and pUC19 is among one of the plasmids that has highest copy number of 500–700 per cell.

Phage: Foreign DNA can be delivered as an insertion into linear DNA derived from the bacteriophage such as lambda phage. This insertion doesn't diminish the life cycle of phage. Genetically engineered lambda (λ) phage has the capacity to carry an insert about 9–25 kb in size. At the structure point of view, this phage shows two restriction sites at both ends of linear DNA – "stuffer fragment." Stuffer fragment is replaced by DNA insertion of interest through digestion with restriction of endonuclease (Fig. 14.3). Larger fragments of DNA can easily be incorporated in phage vectors than the plasmids. This is why phage is more often being used in recombinant technology.

Cosmids: Cosmid vector is a pure hybrid construct that contains both the feature of plasmid and phage to increase the capacity to carry a foreign DNA inside the phage head. Cosmid is generally created by incorporating an antibiotic resistance site from the plasmid and cos site from the phage. Antibiotic resistance site provides sites for digestion, while cos helps in packaging of phage. Cosmids are deprived of lambda genes responsible for progeny phage particle production after infection;

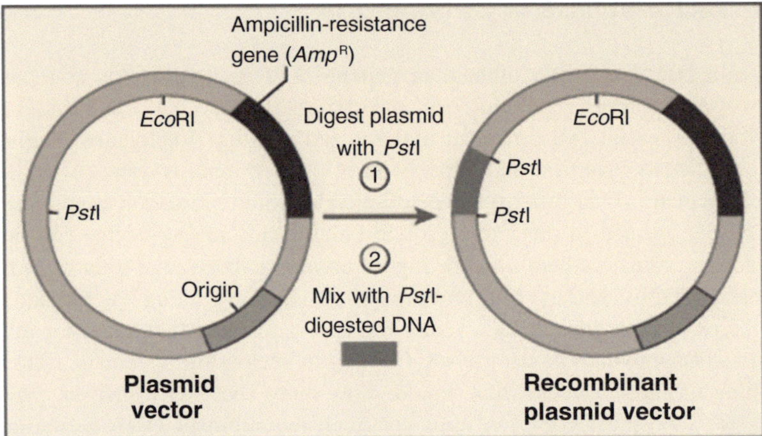

Fig. 14.2 Structural characteristics of a typical plasmid pUC19. A basic structure of plasmid and its recombinant forms after insertion. The ampicillin resistance site in plasmid will allow the host bacterium to grow on ampicillin-containing medium. Origin site is a point at which replication will start as immediate as host bacteria will start to replicate in the culture

thus, cosmid has more capacity for acquiring more foreign DNA than a single bacteriophage. As a result, a cosmid is able to receive 40–60 kb foreign DNA fragments. Cloning through cosmid vector begins first with the insertion of foreign DNA into cosmid and packaging of cosmid into phage head, as shown in Fig. 14.4. Once packaging is done, cosmid-containing phages are allowed to infect *E. coli* cells, cosmid injects its DNA fragment into the cultured *E. coli* where cosmid replicates by using plasmid replication system, and positive-infected clones will be identified by the presence of antibiotic resistance marker from the plasmid vector. Cosmid vector has the advantage over others because it is the most desirable vector in construction of the genomic library of higher organism that has the large genomic DNA. Cosmid can accept up to 40 kb DNA, while phage is only able to receive DNA of 20 kb, a literally half number of cosmid. However, cosmid also has a disadvantage that some cosmid does not maintain stability on the propagation of *E. coli* in the culture because *E. coli* has high copy number of plasmid replication system.

Bacterial Artificial Chromosomes: BAC is a slightly advanced vector from the previous ones with new additional features. BAC has constructed with circular bacterial F factor plasmid integrated with foreign DNA fragment of 180–200 kb. BAC has predominantly been used for constructing *genomic library* of large DNA such as plant genome. In the past decades, BAC is being more acknowledged for *engineered transgenic mice*. These Tg mice had developed through the delivery of a DNA fragment of interest via direct injection to the fertilized single-cell mouse embryo. This gene delivery will result in a stable integration of transgene in mouse embryo; however, this integration is a random process. A transgene construct includes a gene fragment that upholds a *eukaryotic promoter*; associated regulator elements, i.e., enhancer, suppressor, and locus control region, etc.; an *open reading*

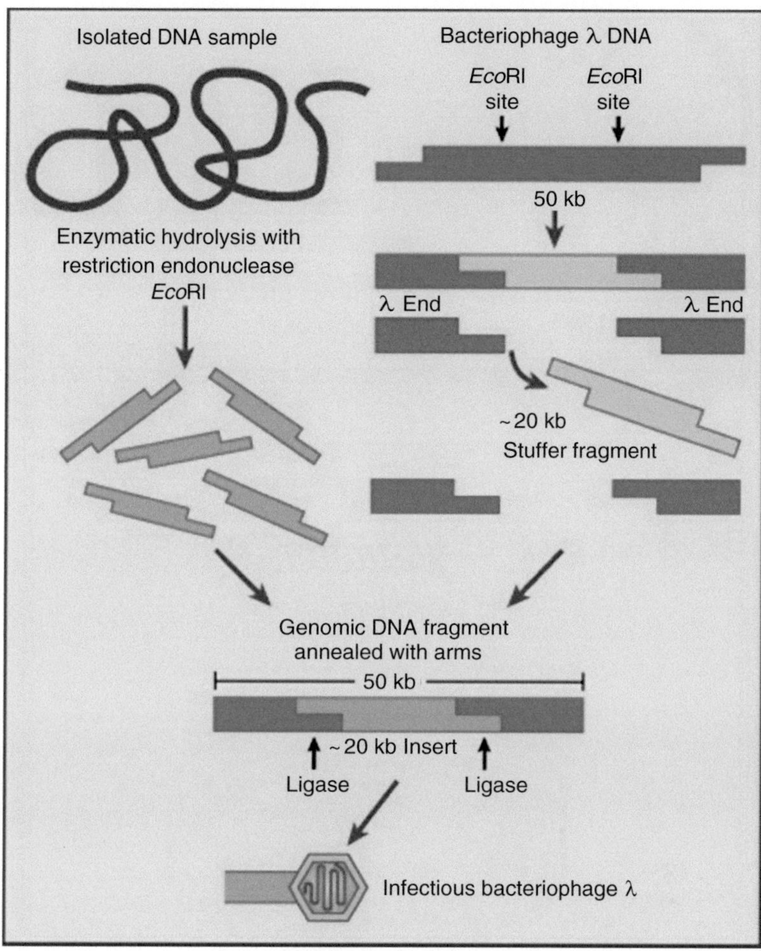

Fig. 14.3 Structure of a typical bacteriophage such as lambda (λ). The image depicts the structure and the recombination process of a λ bacteriophage (a typical bacteriophage mostly used for cloning). A phage DNA with large linear DNA fragment (stuffer fragment) flanked with two restriction will be removed by endonuclease. Always, target DNA is a replacement of a stuffer fragment. Digestion with endonucleases produces arms with sticky ends possessing complementary sequences that will help the target DNA to attach. Target DNA and bacteriophage arms put in a mixture that also contains a ligase enzyme to ligate this arm with target DNA will also promote the assembly of phage protein that begins the head formation

frame (ORF); and a eukaryotic polyadenylation (polyA) sequence. ORF contains reporter genes that encode the essential marker gene such as lacZ, green fluorescent protein, Cre recombinase, etc., to recognize the recombinants during the cloning.

BAC has been extensively studied in understanding virus genetics and its role in human pathogenesis. For example, molecular function of most human viral genes is not known. Recently, many of them including human cytomegalovirus

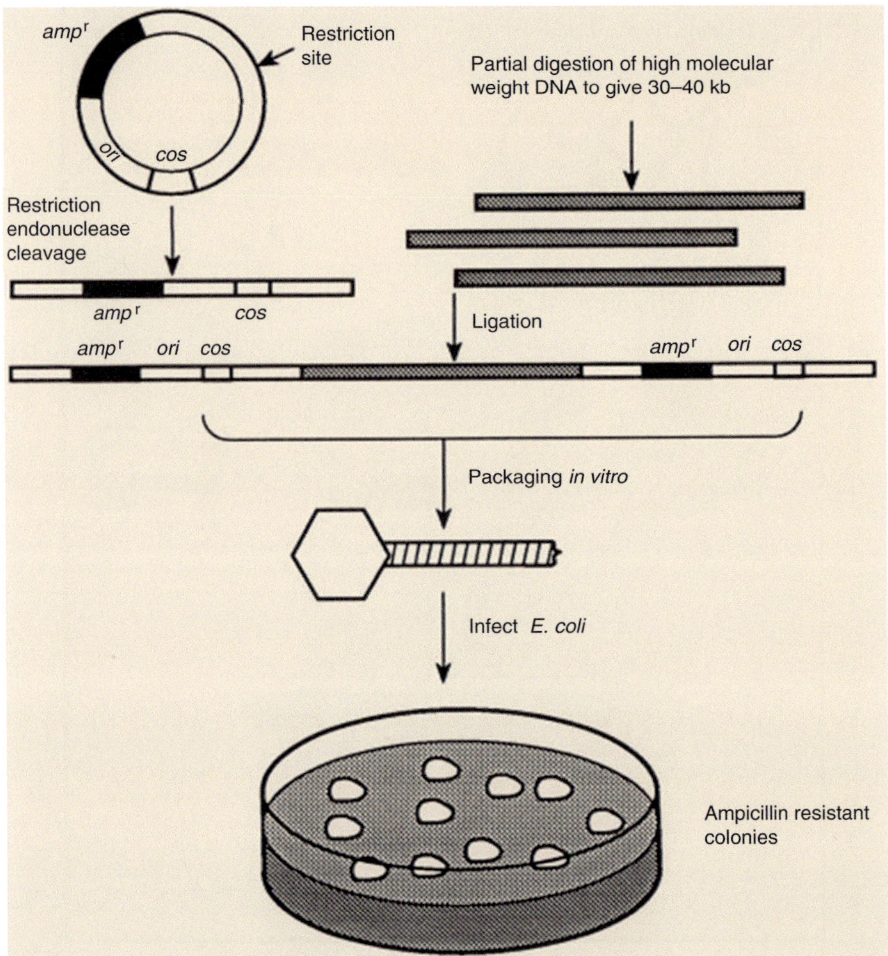

Fig. 14.4 Cloning method using a cosmid vector. This vector contains a cos site, a restriction site for inserting exogenous DNA, and a gene for ampicillin resistance. Exogenous DNA is cut with an appropriate restriction enzyme, as is the vector. The vector and exogenous DNA are ligated together, producing a recombinant molecule of 37–52 kb that can be packaged in λ by in vitro packaging. The packaged vector infects E. coli, injecting its DNA into the host, where it circularizes and multiplies. Escherichia coli cells that receive the cosmid are distinguished from cells that are not infected by their ability to survive on media containing ampicillin

(herpesvirus), mouse CMV, pig CMV, pseudorabies virus, mouse gamma herpesvirus, herpes simplex virus, and Epstein-Barr virus have been cloned and studied by using BAC construct. In each case, BAC has replicated and cloned in mammalian cell.

However, over the advantages, BAC has some limitations as well such as it cannot produce recombinants at large scale due to its low copy number.

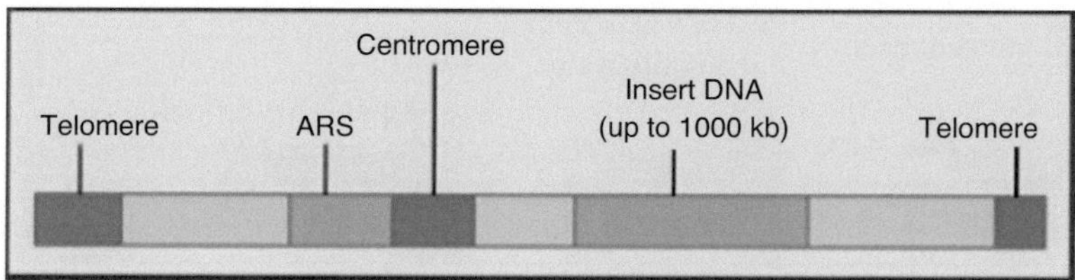

Fig. 14.5 YAC vector construct. A typical YAC consists of telomeres at both end as a part of chromosome structure, autonomous replication sequence (ARS), and insertion part where foreign DNA can be integrated

Yeast Artificial Chromosomes (YAC): This artificial chromosome engineered in yeast cell has the capacity to clone genomic DNA of 1000 kb. YAC is an artificial vector containing (i) telomeres at both end, (ii) origin of replication denoted as autonomous replicating sequence (ARS), and (iii) centromere (center of the chromosome) and a selectable marker along with DNA insert of 1000 kb (Fig. 14.5).

Above, we explained the BAC cloning system, and if we compare YAC and BAC, it has considered that BAC is more advantageous over YAC because of clonal instability, high chimerism, low quality of DNA, etc.

The first-time transgenesis was successfully achieved with YAC that cloned approximately 500–600 lengthy genomic DNA and was cultivated in yeast as a linear chromosome; however, this system has some limitations like it produces high amount of chimeric (clone contains a genomic DNA from different region and produces chimeric), instable clone, and it requires excess efforts to isolate relatively large YAC, intact genomic DNA, and relatively high probability of YAC transgene fragmentation.

Human Artificial Chromosome: HAC is an artificial chromosome that supports the cloning of any gene into human cell. This vector is useful for expression study of chromosomal genes and also is capable to clone large DNA fragment.

14.1.2 Cloning Strategies

Cloning is an engineered cellular machinery that has been used to create multiple copies of certain genes which later undergo a study of their expression, functions, etc. The first step in cloning is fragmentation of DNA and insertion into the vector to be expressed and copied. In cloning, vector works as a carrier or vehicle to carry DNA fragment as an insertion. Constructed plasmid with desired DNA insert is a modified form which is reintroduced to the host cell for replication and multiplication. Plasmids divide and copy their DNA as bacteria dived and grow in the culture. Bacterial DNA (host DNA) is being divided and copied along with the inserted DNA in the plasmid. This inserted DNA (in cloning, it's a human genome) in the plasmid is usually referred to as "foreign DNA" to distinguish it from host DNA. A notable characteristic of plasmid is ease with handling and integration of foreign DNA into it. These vectors are characterized with specific DNA sequences to be recognized by

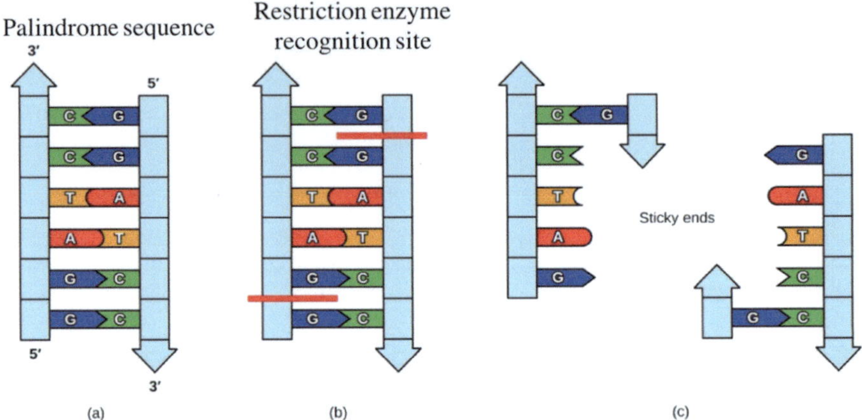

Fig. 14.6 Palindrome structure of restriction enzyme. (**a**) A 6-nucleotide long sequence specific to restriction enzyme recognition site is always palindrome in nature; sequence of nucleotide on 5′ to 3′ is the same as 5′ to 3′ on the complementary sequence, and (**b**) restriction enzyme cut the DNA and (**c**) results into "sticky ends"

digestive enzyme like restriction endonuclease to cut the vector into short fragments. This site-specific enzymes are usually produced by bacteria in the defense mechanism against foreign DNA intervention.

Restriction enzyme cut both DNA strands in a way that each end has 2–4 nucleotide overhang after each cut, and such type of cut is known as "staggered cut." Restriction enzyme-specific sequence in dsDNA is a 4–8 nucleotide sequence called "palindrome" sequence. Palindrome means a word can be read the same in both forward and backward direction; likewise, nucleotide sequence in palindrome reads in same order in both directions.

After stagger cut by restriction enzyme, overhangs on each strand are capable to make complementary strand by hydrogen bonding in both forward and reverse direction, and thus such ends are called **"sticky end"** (Fig. 14.6).

Sticky ends on single strand initiate making hydrogen bond to generate complementary strand. This process is annealing process and accompanied by the action of additional enzyme such as ligase enzyme. DNA ligase helps in replication when these two sticky ends come together, and ligase enzyme joins the DNA fragments permanently (Fig. 14.6).

Cloned plasmid now contains two types of genomic information, one of itself and other of foreign DNA, and therefore, as a result of cloning, this plasmid is known as recombinant DNA molecule, and proteins produced through this cloning procedure are known as recombinant protein (Fig. 14.7) However, in cloning, not all vectors used for cloning will necessary express the respective protein, and thus, only expression vectors are capable for protein expression and are genetically engineered accordingly that in this organism this particular protein has to be expressed on some stimulation. Expression of protein on the stimulation can be controlled by a scientist if it is needed; therefore, this cloning strategy is safe.

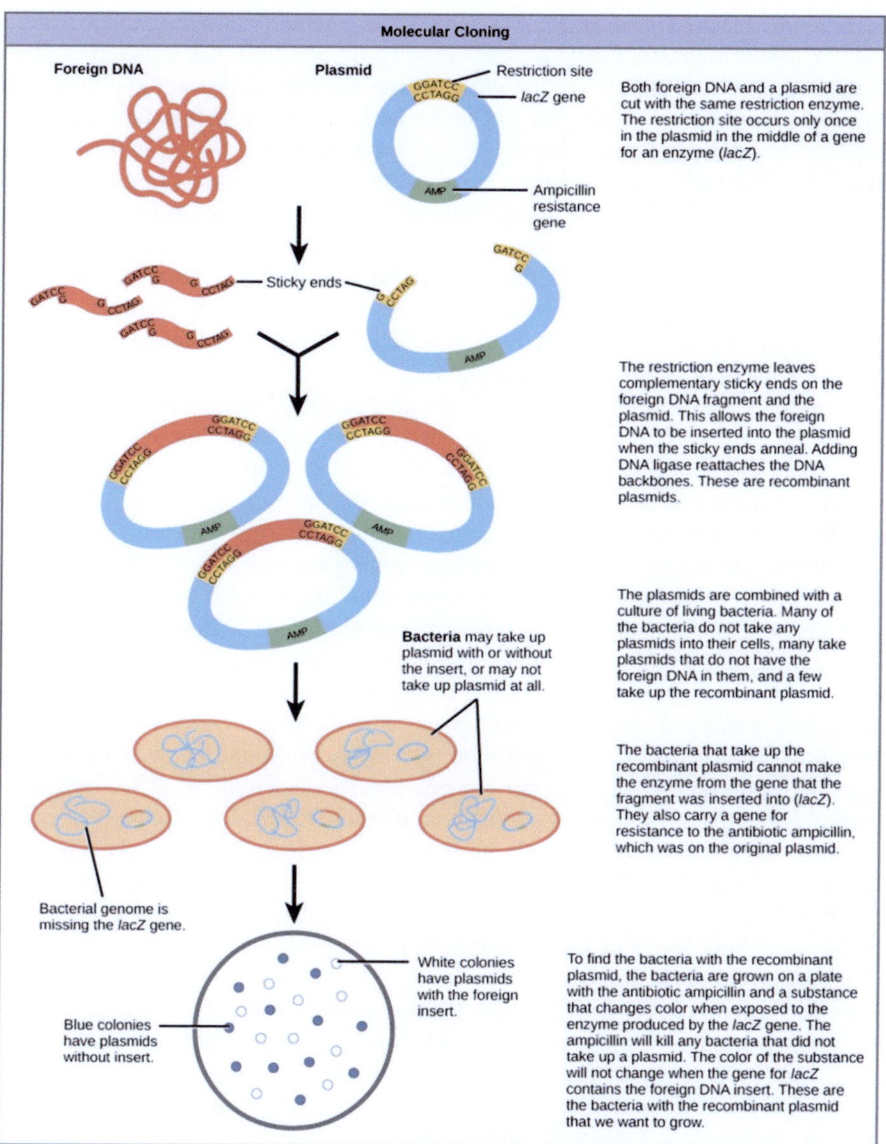

Fig. 14.7 A schematic diagram of cloning strategies. A molecular cloning includes (i) DNA cleavage by restriction enzyme, (ii) insertion of foreign DNA, (iii) ligation of DNA, (iv) transformation in bacterial cell, and (v) selection of cloned DNA of interest

In the history of recombinant DNA technology, many plasmids have been engineered to clone 1000–30,000 nucleotide pairs till date; however, larger DNA weren't possible to clone in conventional plasmid vector, and so researchers introduced yeast artificial chromosome (YAC) in which large DNA can be inserted.

Earlier, only plasmid is being used for gene delivery which was able to clone smaller size of gene fragment; however, the capacity of plasmids has evolved, and its derivative like YAC and BAC are able to clone 100,000 to one million nucleotide pairs of foreign gene. F plasmid has been used in BAC which is far better than conventional plasmid vector, whereas yeast chromosome has been converted into a vector for mammalian gene cloning. In fact, BACs have low copy number, so it can maintain the largely cloned sequence with stability in the E. coli. Also, it can avoid the scramble issue; cloned gene sequence can recombine with other sequence carried by other copies of plasmid. Therefore, having various adapting features like stability, clone large fragment, and easy to handle, BACs have become the preferable choice for constructing DNA libraries of complex organism, i.e., human and mouse genome, as mentioned in previous section.

In recent years, researchers have developed a modified cloning system known as "***seamless cloning.***" This strategy has become in existence to overcome those limitations like (i) low efficiency and time-consuming restriction and ligation steps and (ii) unwanted nucleotide insertion in desired sequence, which might result in abruptly translated product of desired gene. Consequently, seamless cloning is an enzyme-free, sequence-free, and vector-free cloning method. This method includes a compatible set of tailed and *non-tailed primers* and linear vector with cohesive ends. This innovative technique evades prolonged cloning steps such as cutting through restriction enzyme and insertion into the vector, and rather, it allows an insert of DNA fragment into linear vector. For example, in Gibson assembly, which was created by Daniel G. Gibson in collaboration with the J. Craig Venter Institute, plasmids and primers are designed with two identical sequences of about 40 base pairs on each end. An exonuclease enzyme digests one strand of DNA back from each 5′ end, creating a single-stranded region that can anneal to its complementary sequence on the vector or plasmid. DNA polymerase is used to close the gaps, and DNA ligase links the joined segments together to create continuous sequence. The entire process is carried out in a single *isothermal reaction*. Golden Gate Assembly by New England Biolabs allows the insertion of multiple gene inserts into a vector using a type IIS restriction enzyme (which recognizes sites outside of its recognition sequence and cut them) and T4 DNA ligase enzyme. When the cleavage sites are designed correctly, the plasmid is assembled without the original restriction site.

14.1.3 Polymeric Chain Reaction (PCR)

Kary Mullis in 1980 has invented a groundbreaking method to amplify the DNA. He named this process a polymerase chain reaction (PCR). PCR technique is about the ability of DNA polymerase enzyme to synthesize new strand on complementary site of DNA template in vitro. Polymerization is carried out by heat-sensitive DNA polymerase: *Taq polymerase*. *Taq polymerase* enzyme had been isolated from the thermal bacterium "*Thermus aquaticus*" and is the only enzyme used for DNA replication. Since this bacterium grow in extreme thermal condition, it has evolved the protein which is temperature resistant.

PCR is a cyclic process that has a series of 20–40 thermal cycle of heating and cooling that allows the enzymatic reaction to be carried out to amplify the target DNA. The principle of PCR leads to the fact that target sequence doubles in each thermal cycle, and this doubling is an exponential process, represented by 2^N (here, N is the number of cycles; if the cycle has 30 repeats, then it will generate $2^{30} = 1073741824$ copies of DNA from a single template sequence).

The components of PCR are usually a mixture of small fraction of template DNA (in few micrograms), forward and reverse primers that bind to the flank of the target sequence, nucleotides (dNTPs or deoxynucleotide triphosphates), and small amount of heat-resistant Taq polymerase. After performing the reaction, this mixture results in a large amount of DNA as a product.

PCR reaction steps are as follow:

Initiation: This process allows the reaction to be first heated to 94–96 ° C (or sometimes 98 degree if polymerase is highly thermostable) for 10–20 minutes. Heating the reaction mixture activates the thermostable polymerase and denature the other contaminants (if any). Initiation step on higher temperature will lyse the cell and denature the other cellular components like unnecessary proteins and DNase (enzyme which destroys the DNA). To avoid nonspecific amplification and primer dimer, some reagent or antibody such as Hot Start DNA polymerase can be used. Initiation step on high temperature is necessary to heat activate the polymerase enzyme.

Denaturation: Reaction mixture heated on 94–98 ° C for 20–30 seconds separates the double-stranded DNA into single-strand DNA as hydrogen bonds break on high temperature. This process is sometimes known as melting process.

Annealing: Annealing process requires target-specific primer that is oligonucleotide stretch, binds to template sequence, and guides the DNA polymerase to replicate the DNA. After denature process, reaction has to be cooled down till 50–60 °C to allow the primer annealing. This process lasts for 20–40 seconds. It is important to notice that optimal temperature for primer annealing depends on primer melting temperature T^m: a temperature at which half of the duplex DNA is dissociated into single strand. It has been observed that if temperature is too high, primer does not anneal, and if temperature is too low, nonspecific priming and nonspecific DNA amplification do occur. Therefore, an ideal annealing temperature can be considered as 3–5 °C below the T^m of the primers. This difference is high enough for specific primer annealing and low enough for nonspecific priming. Primer concentration in the mixture should always be higher than the DNA template, so once reaction will start, the primer-template hybridization will replace the reannealing of the templates. Once annealing between primer and template is done, polymerase starts incorporating the oligonucleotide on the template strand. As soon as the primers anneal to the template, DNA polymerase can start incorporating dNTPs onto the template strand.

Elongation/Extension: Once primer anneals to the template, polymerase enzyme incorporates dNTPs in 5′ to 3′ direction, and elongation of synthesis strand will continue. The synthesis strand is complementary to the template. Optimum temperature for extension step varies for different DNA polymerase used in the reaction.

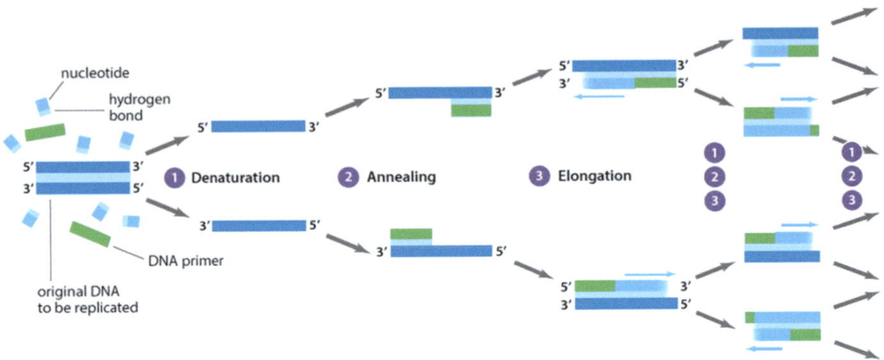

Fig. 14.8 A detailed mechanism of PCR. A PCR mechanism includes the following steps: (1) Denaturation covers the breaking of hydrogen bonds of dsDNA on the temperature of 94–98 °C for 20–30 seconds. (2) Annealing takes place between 50 and 65 °C for 20–40 seconds that allows primers to anneal to the template strands. (3) Elongation allows dNTPs to be added continuously to the template strand, and exponentially amplified DNA will be the end product

The ideal temperature for most of DNA polymerase is 72–78 °C. How much time a reaction will take to finish the extension of the synthesis strand completely depends on how longer the template strand is and on what speed DNA polymerase enzyme adds dNTPs to the template strand. A common polymerization speed lies between 1 and 1.5 kb/min under optimal condition.

Final elongation: reaction mixture is kept on 72–78 °C (which is an optimum temperature for all types of DNA polymerase) for 5–15 min. This holding time will ensure that any remaining single strand would get enough time to elongate in the end of the PCR cycles (Fig. 14.8).

Final Hold: The final product can be stored in 4–15 °C for an indefinite time for short-term storage.

PCR – Primer Design

Potential primer design will ensure the fidelity of PCR product and avoid the nonspecificity of primers. A few points that are described below should be taken into account:

- Primer length should always be between 18 and 30 nucleotides. Lengthy primers promote specific annealing and high melting temperature, while shorter primers can bind easily but with less affinity.
- Optimal melting temperate (T_m) for primers are 50 °C–65 °C. T_m for forward and reverse primers should have difference of 5 °C so that both primers can anneal simultaneously.
- During PCR, annealing temperature (Ta) should be set to ±5 °C of melting temperature T_m. This difference lies between higher enough for specific binding and lower enough for efficient annealing.

- The GC content of primer should be 40%–60% for efficient binding. The GC content of target sequence affects the melting temperature as higher GC content has higher melting temperature.
- If primer is ending with Gs and Cs on 3′ terminal, it will promote target-specific binding. However, Gs and Cs should not be more than three bases in the last five bases.
- Secondary structure in target sequence will result in poor primer annealing and low yields. Thus, it should be avoided if possible.
- Nucleotide repeats of four or more should be avoided (e.g., ATATATAT).
- Complementary sequence in either primers should be avoided; otherwise, that may result in self-dimer or primer dimer. Moreover, excessive concentration of primers than the template in a mixture will result in primer dimer and self-dimer, and such dimers can compete template-primer annealing.

  There are some primer designing tools available online, i.e., PrimerBLAST, Primer3, and OligoCalc which can simplify the primer design process.

Technical Issues

PCR requires low amount of DNA sample, i.e., 5 μL to 100 μL, and such low volume can face the issues of evaporation and insufficient pipetting. The second most challenging issue is the amount of mixture solution. Large volume requires long holding time for thermal equilibrium. Let's say bigger amount of solution takes longer time for an external temperature to be transmitted to the center of the solution. Much solution will take longer time for thermal equilibrium at each cycle, and therefore, longer holding time is required for each cycle. Thus, volume of solution is linearly correlated to timing of entire thermal cycle. A standardized volume for PCR mixture is 20 μL to 50 μL.

The sample mixture is pipetted into reaction PCR tubes which are thermostable. Volume of reaction mixture is 20 μL to 50 μL and usually pipetted in thin wall 0.2 mL PCR reaction tube. PCR reaction tubes can be purchased as individual tubes with or without caps or 8–12 tubes connected together called "strip tubes." High-throughput labs usually used 96- or 384-well plate for routine PCR. Other than theses commercial tubes, PCR also has been performed in *microscopic plate* by pipetting the sample spot on it, covered by coverslip and closed with mineral oil.

14.2 Construction of Library

14.2.1 Construction of Genomic Library

Molecular cloning is one of the promising tools these days which allow researchers to study protein function and structure. Cloning also provides a platform to reproduce recombinant proteins. In this technique, genes for a particular protein of interest will be cloned with the advantage of vector, PCR restriction enzymes, etc. Construction of genomic library is being associated with the range of multiple application in the field of molecular cloning. The genomic library provides us the useful information about the source of DNA fragment which has been cloned and stored for various use. The bright side of molecular cloning is that scientist can create and store DNA fragments obtained from the different sources in a suitable host organism. This cloned DNA is kept restored in a suitable microorganism which is protected by indigenous cellular machinery to protect and replicate exogenous DNA. This type of libraries can be a source for various genetic material such as cDNA, alleles, mutants, mRNA, mitochondrial DNA, etc.

Genomic library construction takes a few steps as follows: Cells are collected or grow to crush, and genetic DNA has isolated them from other proteins. In a cellular extract when DNA is isolated from other cellular components, DNA/RNA will have appeared in the aqueous phase since DNA is dissolved in water while other elements will appear in the solvent phase (like phenol) and be taken out by a pipette. There are several methods for precipitating, of which one is adding a solvent like alcohol to the diluted DNA and precipitate it. However, DNA/RNA remains together, and additional RNase will degrade the RNA but not the DNA. Extracted DNA will be followed by fragmentation through restriction enzyme on particular sites and inserted to the vectors like plasmid, phage, and cosmid. These vectors containing the DNA fragment will now be transferred to the bacterial host to replicate and multiply to produce more copies. After a couple of cycle of bacterial growth, the bacterial cell will produce phage particles or plasmid copies, contacting overlapped genomic fragments. As a result, few clones contain the entire gene of interest, some contains partial, and few contains no gene of interest. Therefore, the marker gene will help to recognize the bacteria that contains a full genome of interest. The genomic library must include the positive clones that have the entire genome integrated into the vector, a library for human genome constructed by using cosmid, and each cosmid is carrying the random gene of 30,000 to 40,000 bp long fragment. Because of this feature (clone large fragments), cosmid provides 99% chances for every gene to be presented in the human genome library. Establishment of the genomic library will provide the sources for subsequent experiments or for the primary purpose to which genomic library has been developed. To this end, the genomic library should be stored carefully and safely for future purposes. For example, a random genomic library produced with phage particle will contain a suspension in the test tube and should be kept safe. At large scale, most of the genomic libraries are stored at − 80 °C. Other examples including bacterial cells containing plasmids are stored and protected from the adverse effect of freezing by adding glycerol as a cryoprotectant during freezing while phage particles are protected by dimethyl sulfoxide (DMSO) which has the cryoprotective properties (Fig. 14.9).

However, the main restriction of genomic library is that such kind of library is accessible to create comparatively small genome like prokaryotic organism. A genomic library of any eukaryotes seems difficult to construct and maintain since it contains very large genome and small fraction of noncoding genome; therefore, cDNA library omits this issue when it comes to eukaryotic genome.

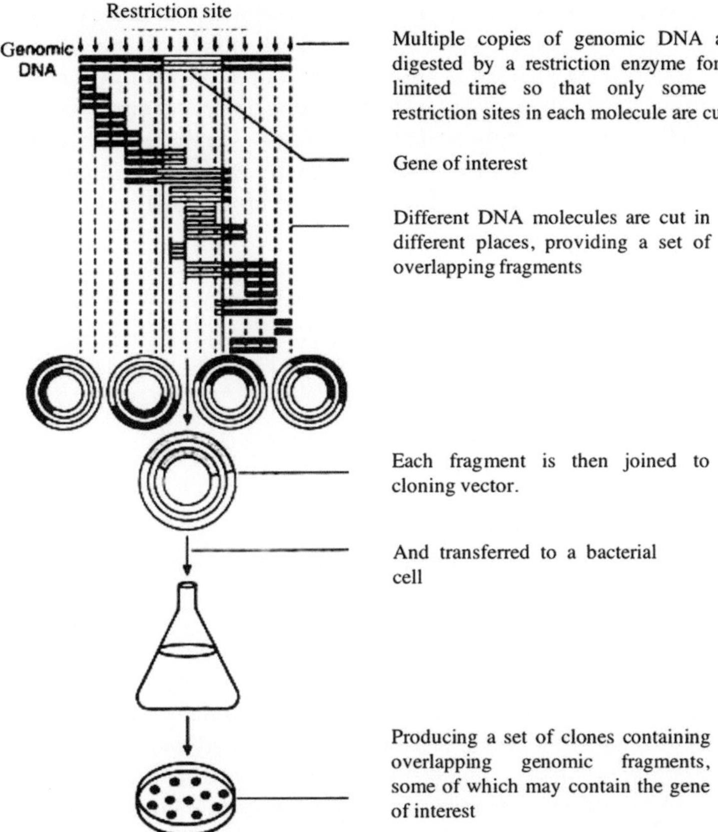

Fig. 14.9 Schematic diagram of steps involved in construction of genomic library. DNA of interest were first digested with restriction enzymes which give segmented DNA containing set of overlapping fragments. This fragment and cut DNA molecules were joined to cloning vector. Cloning vector were further cloned in the bacterial medium and screened for positive cloned DNA of interest

14.2.2 Construction of cDNA library

After constructing a genomic library where the entire genome is settled down in the form of a library which contains all the genomic information of an organism, we will discuss the importance of cDNA library in this section. A cDNA library contains many nucleotide sequences that are a complementary sequence to mRNA of the same species. In simple words, cDNA library is constructed by using mRNA rather than the whole genome, and mRNA further develops cDNA (complementary DNA) which is kept stored as a form of a library for future purpose.

Most of eukaryote genome consists of repetitive sequences that are not transcribed into mRNA, such as noncoding sequences, and cDNA library is deprived of such noncoding sequences. Construction of cDNA library is applicable only for higher eukaryotic organism because prokaryote and lower eukaryotic organism genome does not contain introns, and thus cDNA library is necessary for these organisms. Hence, cDNA libraries have been produced only for higher eukaryotes, and cDNA library is an alternate to create a genomic library for eukaryote organisms.

cDNA library has two additional advantages: (i) It is enriched with genes actively transcribed into functional proteins, and (ii) cloning of eukaryote genes in bacteria would not cause problem since it does not contain introns, and either bacterial genome does not contain introns.

However, cDNA library also has some disadvantages over genomic library. The major one is that it only stored sequences that are present in mature mRNA. Sequences that are altered after transcription are not involved in the cDNA library. Some sequences such as promoters and enhancers do not transcribe into RNA so that cDNA of such sequences are not present in the library. Another important thing to be noticed here is that cDNA library represents those mRNA sequence that is expressed in the tissue, and RNA isolation is possible at the same time. Moreover, it has been observed that the frequency of all genes in genome library is abundant and in equal distribution, while the frequency of DNA in cDNA library corresponds to the abundance of mRNA in a given tissue.

Method

cDNA library construction initiates with first step to isolate mRNA from the rest of cellular RNA, i.e., tRNA, rRNA, rnRNA, etc. There are many methods available for RNA isolation, but the most common is TRIZOL method till the time. As poly adenine (poly A) tail at the 3′ end is a prominent feature of most eukaryotic mRNA, this long stretch provides a convenient hook for separating mRNA from the rest. Oligo dT column, an oligonucleotide stretch with thymine (oligo dT chains), is a very common technique which has been used for mRNA separation, based on the concept that when mRNA pass through the column, adenine will pair with thymine and be retained to the column while the rest of RNA will elute out from the column (Fig. 14.10). Later in the steps, intact mRNA can be washed away through eluting buffer which will break the hydrogen bonding between adenine and thymine, and mRNA will be detached from the column.

Extracted mRNA will now be transcribed into cDNA upon the action of an important enzyme – reverse transcriptase. Reverse transcriptase enzyme triggers single-strand DNA synthesis from the template RNA (a reverse transcription). This DNA synthesis process needs additional DNA nucleotide in the mixture to bind on 3′ OH group of primer (similar to normal transcription). Reverse transcriptase is a retrovirus (i.e., HIV) enzyme, where the genetic material is RNA and is transcribed into ssDNA. During this process, a short fragment of oligo dT will be added to the mixture that works as a promoter. Primer will bind to poly (a) tail on 3′ end of the mRNA and will provide free OH group for initiation of DNA synthesis.

Resulting RNA-DNA hybrid molecule is then interrupted by partial digestion by RNase to separate the ssDNA from the RNA. Partial digestion will leave some gaps on the hybrid strand on which the DNA polymerase can bind and initiate the synthesis of complementary DNA strand. Undigested small RNA fragments will

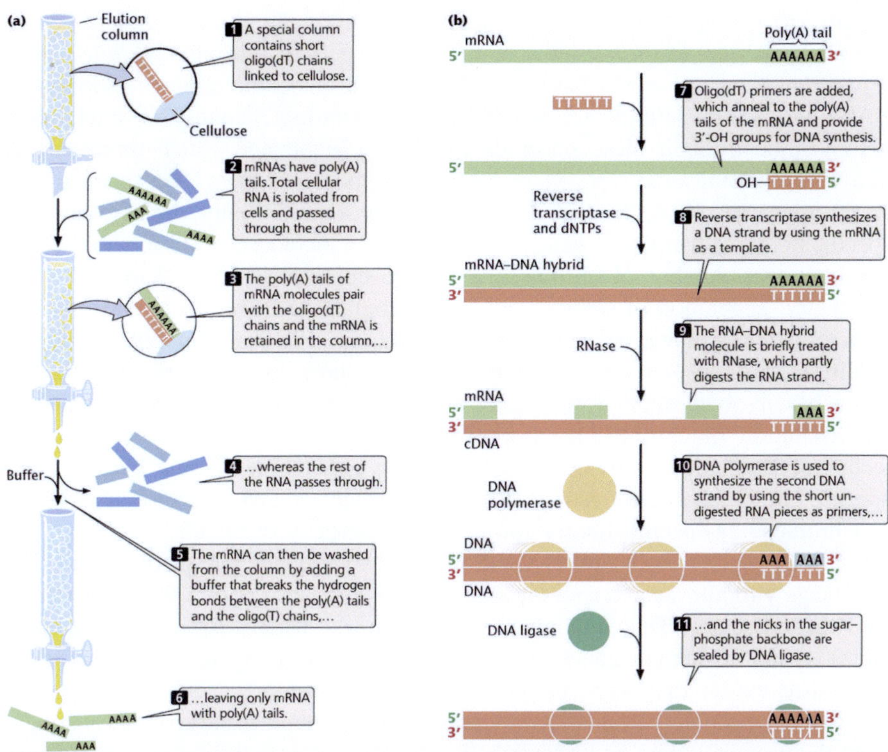

Fig. 14.10 Schematic diagram of cDNA library construction. cDNA construction involves (**a**) mRNA isolation through elution column of oligo dT and (**b**) cDNA construction from isolated mRNA using reverse transcriptase enzyme. Reverse transcriptase synthesizes DNA strand on the mRNA templates which were isolated by dT column. This mRNA-DNA hybrid lysed with RNase to release the mRNA and to intact the DNA for DNA polymerase as a template that further synthesizes the cDNA. DNA ligase used for ligation of cDNA. (Benjamin. A. Pierce., Genetics: A conceptual approach)

be used as a primer, and DNA from the RNA-DNA hybrid will be used as a template. While DNA is synthesizing, all the RNA fragments will be displaced eventually by DNA polymerase, and nicks will be scaled by DNA damage machinery.

14.2.3 Construction of Chromosome-Specific Library

Large and complex genome in eukaryotes creates more difficulties in map saturation. It is clear that disintegration of full genome into smaller parts will be easier to study separately and could accelerate its analysis easily rather than the entire complicated genome in once. Thus, chromosome-specific library would be ideal to construct library from the subset of the genome instead of lengthy complexed genome. Chromosome-specific library can be applied in (i) cytogenic genome mapping studies, (ii) region-specific marker isolation, and (iii) study of integration of genetic and physical map. Chromosome-specific region has been separated by *flow-sorting separation* (flow cytometry-based separation of chromosome regions) followed by BAC, cosmid, bacteriophage, and YAC-based cloning. Cloning of fractionated chromosome through these plasmids can be an advantage which represents an individual chromosome type. Many human chromosomes have been mapped such as chromosomes 19, 6, 21, and 22 and Y chromosome. Initially, the chromosome-specific library construction was labor intensive and includes tedious methods which require larger number of chromosomes. Pure chromosome was sorted by flow cytometry, generation of somatic cell hybrid containing targeted chromosomes, or a combination of both procedures. Thus, to eliminate such obstacles and improve the quality of chromosome purity, many researchers have developed the new method which has involved single flow-sorted chromosome which is also favorable and even the resolution of chromosome population is poor. Single sorted chromosomes technique has unique prevalence for the rapid generation of pure chromosome-specific libraries for many *genetic disorder* or cancer-related chromosomes.

Chromosome-specific library (*physical mapping*) is eventually useful to map transcribed sequence, i.e., mRNA and *heterogenous nuclear RNA* (hnRNA). This advantage will allow us to spot the protein coding gene or transcribed sequence on the chromosome. This mapping finally shows where those all estimated 50,000–100,000 human genes are located on the chromosome. However, it has mentioned earlier that some unique sequence for promoters, enhancers, protein-recognizing proteins, etc., does not present in genomic library; therefore, apart from the protein coding and transcribed sequences instead, these unique sequences also can be easily positioned on the human chromosome. Ultimately, chromosome mapping or chromosome library will be advantageous for positioning sequences on the chromosome and could significantly heighten the scope of human genome functional map among others.

14.3 Screening of Gene Library

Once genomic library is prepared, it can be stored, can be used for purification of proteins, and can be reanalyzed to check further whether a fragment of desired sequence is present in our genomic library or not. Multiple strategies are developed for screening purposes. Among all, probing is one of the most common method, and the rest are involved in hybridization, colony hybridization, PCR, immunological assay and protein functional analysis, etc. Here, we will understand the screening strategies for any genomic or DNA library as follow:

14.3.1 Hybridization

Hybridization is a type of probing where the target sequence in DNA or cDNA library can be determined with labeled DNA probe (Fig. 14.11). Probe must be a

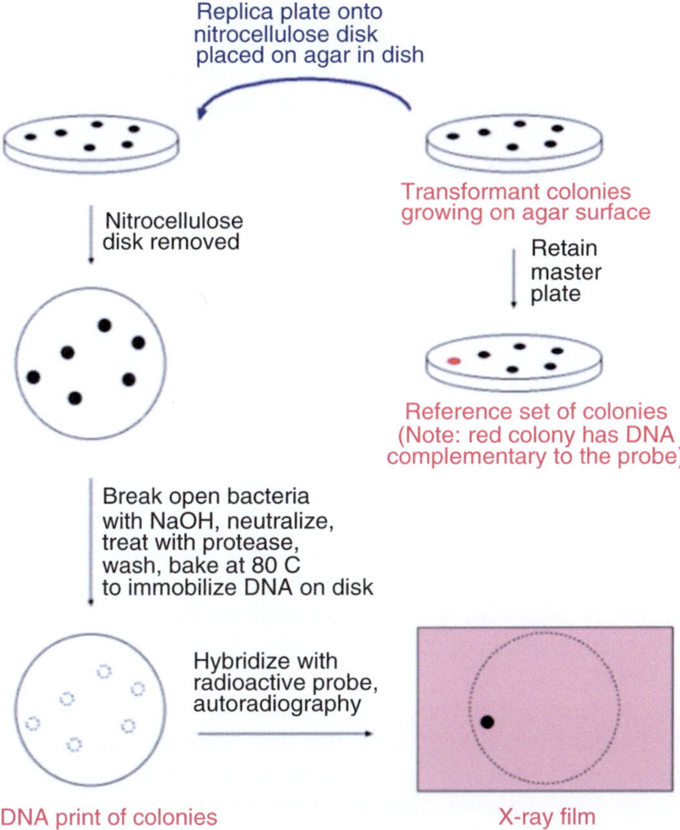

Fig. 14.11 Hybridization method to detect DNA probe from the library. Schematic diagram of important steps involved in hybridization method to detect DNA probe hybridization on the nylon/nitrocellulose membrane. The target complementary DNA for probe is identified on the master plate of samples detected as red dot which further supposed to be isolated from the master plate. The probe is radiolabeled so that autoradiography is being used on for detection

stretch single-stranded oligonucleotide or a double-stranded cDNA or a PCR product of the same gene. The DNA probe can be either radioactively labeled or tagged with fluorescent protein. Labeled probe is a key step for hybridization that will produce signal or colored precipitation on the hybridization with target sequence. Radioactively labeled probe gives radioactive signal by replacing phosphate backbone with radioactive phosphate 35 (p^{35}). Whereas nonradioactive probe system of which biotin has widely used up to now, for example, biotin is tagged to the probe and, when it is hybridized with the target sequence followed by addition of substrate antibody or solution, will result in colored precipitates that are produced after efficient hybridization. Probe in some cases may be an RNA sequence.

The first step is labeling a nylon membrane with our DNA candidates from the library and adding certain buffer or NaOH to denature the double-stranded DNA.

Labeled membrane is exposed to the probes. Upon the hybridization between probe and its complementary fragment, we get the signal of either fluorescence or coloration. In case of radioactive probe signal, X-ray or autoradiography can be used to measure the radioactivity coming from the hybridization which clearly indicates that hybridization occurs successfully. This membrane further can be compared with master plate to know where signal is indicating and where our desired fragment is located on the master plate (Fig. 14.11).

14.3.2 Colony Hybridization

A successful cloning of desired gene in the E. coli bacteria further requires screening to pick up cloned colony from the culture plate. Target DNA sequence is present in the transformed colony that will be further detected by hybridization method with radioactive DNA probes (sometimes labeled RNA probes can also be used). Colony hybridization technique is also referred to as replica plating by some authors. The technique is depicted in Fig. 14.12 and is briefly described. The transformed cells are grown as colonies on a master plate. Samples of each colony are transferred to a solid matrix such as nitrocellulose or nylon membrane. The transfer is carefully carried out to retain the pattern of the colonies on the master plate. Thus, the nitrocellulose paper contains a photocopy pattern of the master plate colonies. The colony cells are lysed and deproteinized.

The DNA is denatured and irreversibly bound to matrix. Now, a radiolabeled DNA probe is added which hybridizes with the complementary target DNA. The non-hybridized probe molecules are washed away. The colony with hybridized probe can be identified on autoradiograph. The cells of this colony (from the master plate) can be isolated and cultured.

Many a times, multiple colonies are detected on hybridization by a DNA probe. This is due to overlapping sequences. To identify which colony has the complete sequence of the target gene, corresponding colonies can isolate from the master plate and digested with the restriction enzyme, and data observed from the restriction endonuclease analysis will be helpful to detect the exact transformed or cloned colonies.

Moreover, sometimes overgrown colonies interfere with the signal, and therefore, some published journal had been spotted that *overgrown colonies* can lead to interference with background and are difficult to distinguish the labeled/hybridized colonies from the negative ones. To prevent the variation in size of colonies, it is advised that inoculation loop should be touched lightly to the agar plate instead rubbing it forcefully.

The other thing to be noticed is accuracy in the placement of *nitrocellulose membrane* on the agar plate. Nitrocellulose membrane usually adhere to the agar plate because of a light layer of water that kept the membrane enough moist. However, excess water makes movement of bacterial colonies along with displacement of membrane, and thus, cultured colony that appears in fuzzy edges can be seen in X-ray film.

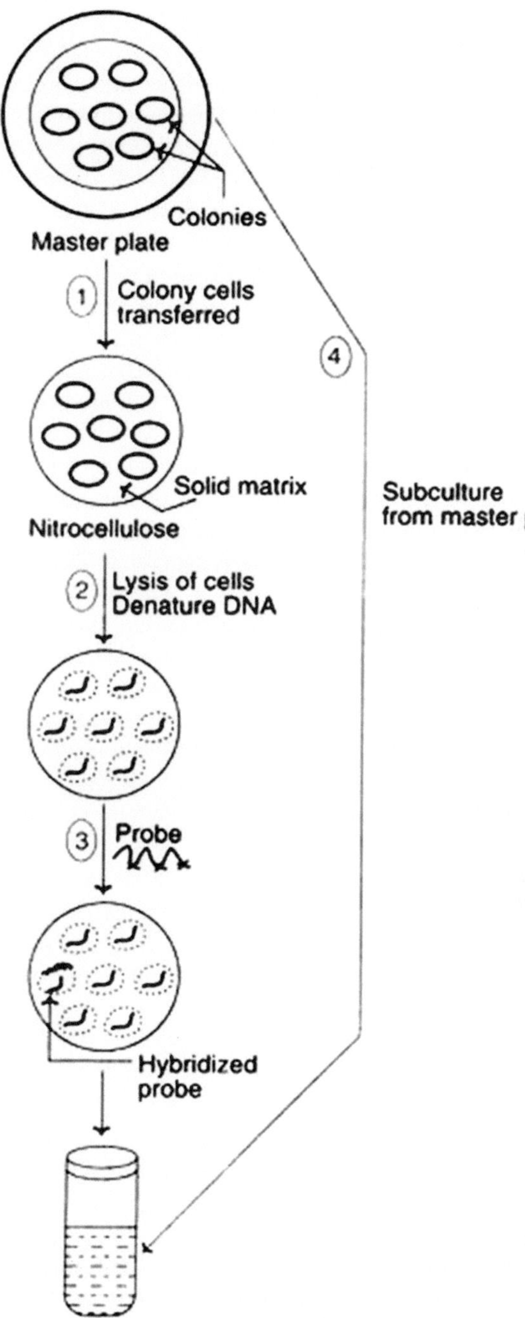

Fig. 14.12 A flowchart diagram of colony hybridization. Transformants containing DNA targets were grown on nitrocellulose membrane in the colonies. Targeted DNA were identified by radiolabeled DNA and probed with labeled DNA, mentioned in step 3. In step 4, identified hybrid probe matches with master plate and then subcultured to isolate them from others for future purpose

14.3.3 PCR

Polymerase chain reaction (PCR) is as good as the hybridization technique for screening DNA libraries. PCR offers an opportunity to identify complex and rare sequences in a clone that has been produced by a complex bacteriophage, plasmid, or cosmids. PCR facilitates the abundance of a unique and complexed sequence by amplification, and thus it gives particular and sensitive technique in short time to isolate a specific gene of interest which might not be possible by simple hybridization. But adequate information on the flanking sequences of target DNA must be available to prepare primers for this method. This method has been first developed by Takumi and Lodish in 1994 for the screening of cDNA libraries.

PCR protocol first follows the culture of bacterial colonies in multiwell plates same as they usually grow in L broth. Cultured bacterial colonies in the multiwell plate contain phage particles or plasmid with a cloned target sequence. The phage releases its DNA into the medium; that's why the extraction of DNA is not needed here. This phage culture is now ready to be used as PCR template for screening. The rest of the following procedures are the same as usual, and we use them for a conventional PCR. PCR primer is specific to the target sequence and will run for a typical PCR reaction cycle protocol. After PCR run, the product should run on the agarose gel, stained with ethidium bromide, and take the image. Later to this, the gel is dried on 70 °C to denature the DNA and further probed with complementary oligonucleotide sequence for hybridization process. In this way, each well is screened by PCR, and the positive wells are identified.

14.3.4 Immunological Assay

Immunological techniques can be used for the detection of a protein or a polypeptide, synthesized by a gene (through transcription followed by translation). The procedures adopted for immunological assay and hybridization technique (described already) are quite comparable. Screening procedure by immunological assay is depicted in Fig. 14.13.

The protocol includes grown colonies on the master plate will be transferred on the solid matrix such as nitrocellulose membrane. These colonies are then lysed, and extracted protein will bind to the surface of the matrix. Fixed protein will be treated with a primary antibody that is specific to the protein and will interact with the protein (protein is the antigen here). The target DNA encodes a protein. The molecular target of primary antibody is an epitope – a short amino acid sequence that folds into a three-dimensional structure as a secondary protein structure on the surface of antigen (protein). Incubation of antibody-antigen follows by removing the unbound antibody by washings, and then a secondary antibody specific to primary antibody is added to the membrane.

Again, the unbound antibodies are removed by washings. The second antibody carries an enzyme label (e.g., horseradish peroxidase or alkaline phosphatase) bound to it. The detection process is so devised that as a transparent substrate, it is acted

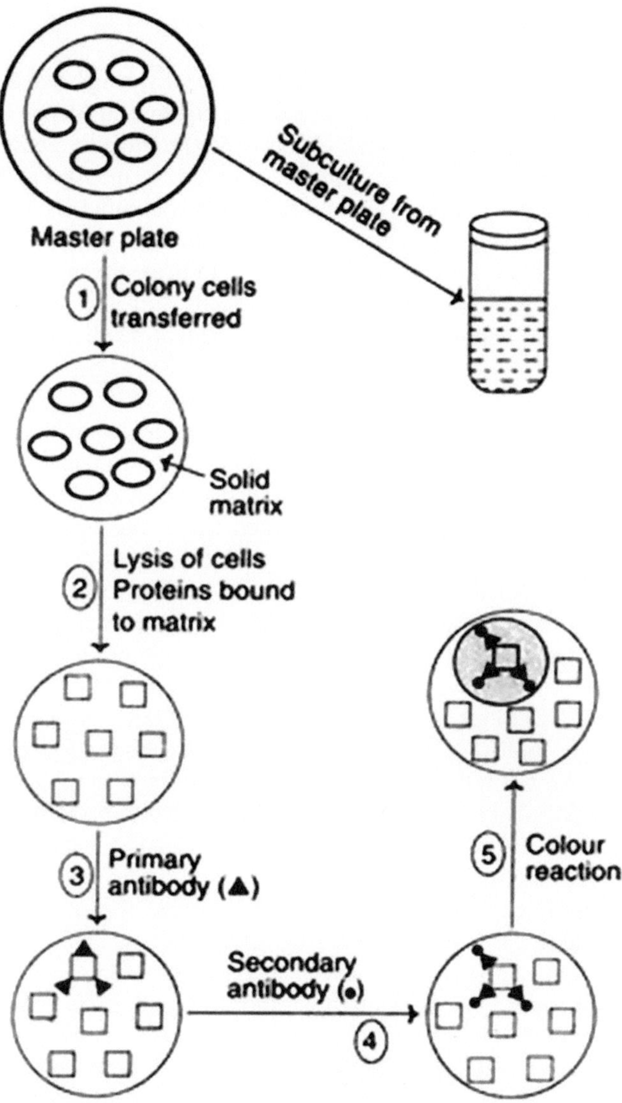

Fig. 14.13 Immunological assay for screening a gene library. Transformants from the gene library were grown on the master plate and transferred to the solid matrix. Colonies were lysed on the solid matrix using lysis enzyme. Exposed proteins are probed with primary antibody and secondary antibodies. Signals on the solid matrix indicates the targeted protein or protein of interest which further picks up from the master replica and subculture for further uses

upon by this enzyme, and a colored product is formed. The colonies which give positive result (i.e., colored spots) are identified. The cells of a specific colony can be subcultured from the master plate.

14.3.5 Protein Function

Sometimes, gene library contains a few genes that directly synthesize protein straight into the host cell and released in the form of the enzyme. It means that the target sequence product is an active enzyme which is not produce by endogenous genes (bacterial genes). In this case, screening is possible by measuring the enzyme's activity in the host cell. Enzyme activity is mainly measured by identification of released product after enzyme-substrate reaction when the substrate is added onto the membrane. The enzyme-substrate product will indicate the presence of the target sequence/functional protein/enzyme in the host cells. For instance, α-amylase and β-glucosidase are two natural enzymes that can be identified by this technique.

Chapter Summary

- Recombinant DNA technology (RDT) is a set of molecular techniques that enable the possibilities to alter the genetic makeup of an organism by modifying or altering the composition of genetic material like DNA or RNA. RDT provides tools for locating, cutting, joining, analyzing, and changing DNA sequence and for inserting the sequence into the cell.
- Restriction endonuclease enzyme plays a vital role that makes double-stranded cuts in DNA at the specific sequence site. DNA fragments as a result of cuts by a restriction enzyme can be separated on the gel through electrophoresis technique and visualized by labeling the fragments with radioactive or chemical tag.
- Plasmids and bacteriophage are very essential and basic vectors (extrachromosomal DNA) in molecular techniques. Many other modified vectors have been developed that contain the features of many in one and can replicate larger DNA such as cosmid, phage, and plasmid-generated cosmid, bacterial artificial chromosome (BAC), and yeast artificial chromosome (YAC). These modified vectors have larger capacity to hold big DNA.
- Cloning is a strategy to generate identical copies of small DNA fragments (molecular cloning) or entire organisms (reproductive cloning). In molecular cloning, the desired DNA fragment is inserted into a bacterial plasmid using restriction enzymes and transferred to the host cell to further multiply and express.
- Polymerase chain reaction (PCR) is a method used for amplification of DNA enzymatically without cloning. A solution for containing DNA is heated, so it breaks down into single-stranded, primer (a complementary sequence of the DNA template) bind to 5′ and 3′ end of the single-stranded DNA. High temperature activates Taq polymerase to synthesize new strand from the primer. Each time cycle is repeated, the amount of DNA doubles.
- Genes can be isolated and stored by creating a DNA library utilizing bacterial colonies and viral plaque that has incorporated DNA fragment within. A genomic library contains the entire genome of an organism, while a cDNA library contains DNA fragment complementary to all different mRNA expressed in a cell.

- Screening of gene library requires a sensitive and powerful technique which can be hybridization, colony hybridization, PCR, immunological assay, and protein function.
- The immunological assay is one of the modified techniques of hybridization where primary and secondary antibodies are hybridized with interest of DNA and produce the signals to detect whether cloning is done or not. Protein function refers to another method where enzymatic reaction signals as a function of protein if the target DNA is specific to synthesize a protein which can bind with substrate enzyme and produce signals.

14.4 Screening of the Genomic Library

Screening of the genomic library is a process that involves the identification of clones carrying the gene of interest. Screening depends on the characteristic property of a clone in a library. The genomic DNA library consists of a thousand clones in the form of plaques or colonies on the plate.

14.4.1 Screening by Hybridization

- Screening with the help of the hybridization technique uses a gene probe (a short nucleotide sequence specific for that particular gene) to identify the gene of interest.
- The process begins with (Fig. 14.14) the production of a replica filter for each plate that uses “colony lift” as a procedure. This procedure is quite similar to southern blotting as discussed later in the book. A nitrocellulose or nylon membrane is placed on the top of a Petri plate that contains the cells which carry the desired gene of interest. This plate is kept for a short time (approximately 1 minute). When the membrane is kept on the top of the cell culture plate, a part of each bacterial colony from the plate will bind to the membrane. The membrane is removed and is soaked in a solution containing sodium hydroxide (NaOH) that will help to release the bound DNA with the membrane and additionally will denature it.
- In the neutralization step that uses buffer solution, the single-stranded DNA molecules will be immobilized on the nitrocellulose membrane with either heat or UV irradiation. The membrane is then hybridized with a suitable probe under specific conditions.
- This is followed by autoradiography procedure which shows black spots representative of the desired clones.
- The final step in the procedure is to align the X-ray film (autoradiograph result) image with that of the plate and subculture the desired colony.

A diagram showing the process of screening by hybridization. The process starts with the replication of a colony of cells from the library onto a filter. This filter is

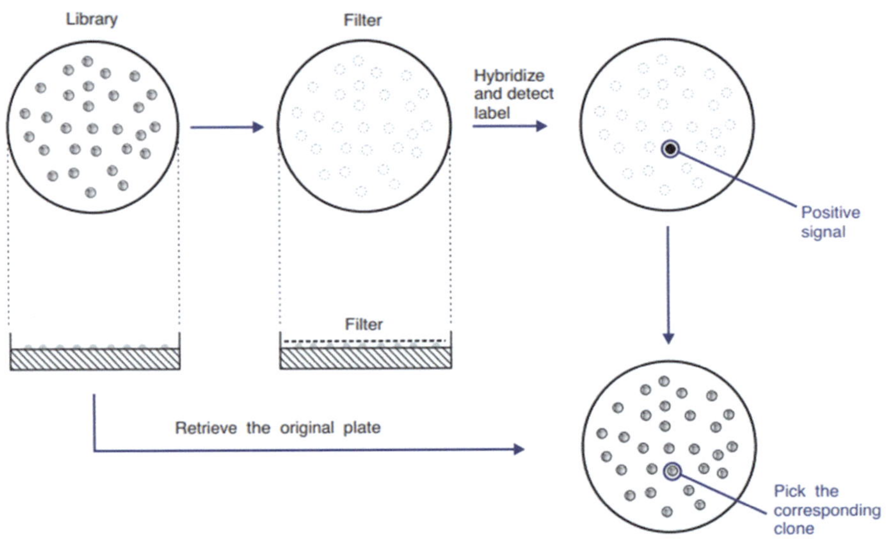

Fig. 14.14 A Schematic diagram representing the process of screening of clones by hybridization

nothing but a membrane made up of nitrocellulose or nylon fiber. A suitable probe is then hybridized to the gene of interest for a short duration of time. This hybridization makes a portion of the DNA colony to transfer from the library to the filter plate. The filter plate is thus then washed with NaOH to release the binding of ssDNA. This is then put under X-ray film to observe the cloned gene of interest. A black dot here shows that the gene is being cloned. This gene is marked and compared with the original Petri plate to screen the gene of interest.

14.4.2 Expression

Identification of clone within a cDNA library is an important task to achieve. This is often done by immunological screening which uses antibody against purified virus or a Coat protein. Antibody screening can be carried out on a cDNA-cloned gene on a wide range of vectors that include plasmid and phage-based vectors. A wide range of vectors λgt11 and λZAP are now available commercially that can be used for the generation of expression library. These vectors include antibiotic screening. Antibody screening can also be done on a simple plasmid vector based on the blue-white screening principle. cDNA inserts are expressed as fusion proteins with beta-galactosidase after being induced by IPTG. The following methods of expression are described in brief in this regard.

14.4.2.1 Library Plating

In this method, cDNA libraries are transformed into a suitable bacterial host, and colonies are grown on plates at an optimum density to visualize single-cell colonies at 37 °C overnight.

A grid is drawn and numbered on a nitrocellulose circle that is a size of a Petri plate which is then soaked in IPTG solution of 10 mM strength. This is followed by air-drying and carefully placing it on a solid media plate. A similar grid is drawn on the bottom of another solid media plate. Individual colonies are picked up from the original cDNA library. They are then streaked at a numbered position of a nitrocellulose grid; this streaking process is repeated on an agar plate using a sterile loop. After enough colonies have been streaked out on to the plates are inverted and incubated overnight at 37 °C.

14.4.2.2 Western Blot Analysis

An antibody-positive colony is grown overnight in a liquid broth at 37 °C in the presence of 5 mM IPTG. The cells are harvested by microcentrifugation, resuspended in SDS-PAGE-loading buffer, and boiled for 2 minutes. A microcentrifuge is spined briefly to remove cell debris. The supernatant is also removed into a fresh tube.

The sample is boiled for 2 minutes to separate it on 12.5% polyacrylamide gels. This is followed by Western blotting onto a suitable membrane.

14.4.3 Hybrid Arrest and Release

14.4.3.1 HRT and HART (Hybrid Release Translation and Hybrid Arrest Translation)

These are a set of two related techniques that find their usage in the identification of translated protein products that are encoded by a cloned gene of interest in a cell-free system. The frequently used cell-free systems are usually prepared from germinating weed seeds or from rabbit reticulocyte cells since these cell-free systems are highly active in protein synthesis. Cell extracts contain all the prerequisites needed for protein synthesis including ribosomes, tRNA, and all other types of essential machinery.

The process involves the addition of mRNA to a cell-free translation system with a mixture of 20 amino acids that are common to all proteins. One of the most commonly labeled amino acids is S^{35} methionine.

The mRNA molecules undergo a process of translation to produce a mixture of radioactive proteins that can be separated by using gel electrophoresis and are then visualized by autoradiography.

Each band on the autoradiogram represents a single protein coded by one of the mRNA molecules present in the sample.

This technique works best as clone obtained from a cDNA library.

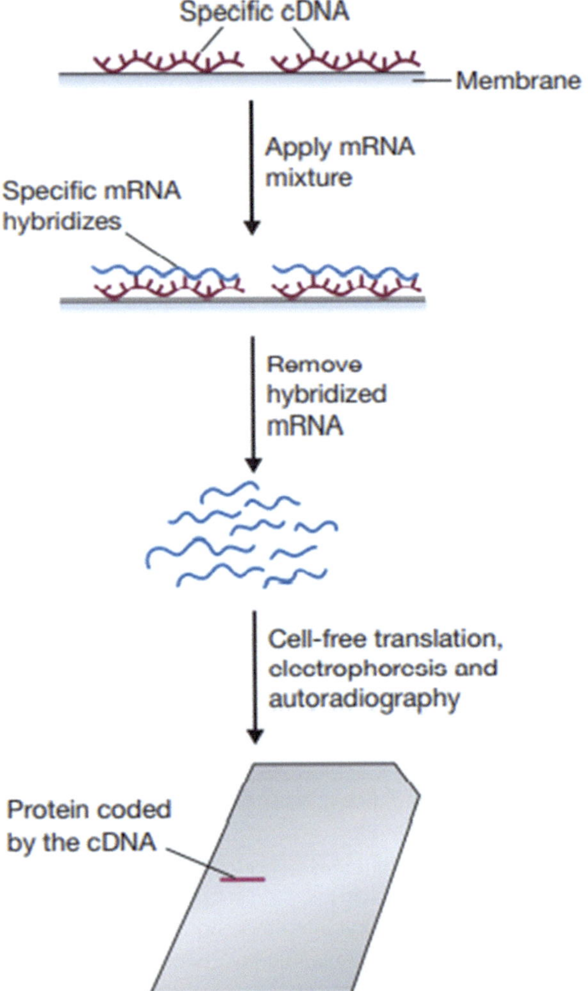

Fig. 14.15 A diagram representing the process of hybrid release transition. The processes include binding of specific cDNA complementary to a gene of interest on a nitrocellulose membrane. This is followed by the addition of an mRNA mixture that binds to cDNA. The hybrid mRNA is then recovered and is translated into protein in the cell-free system. The method thus provides a pure sample of protein products

HRT (Hybrid Release Translation)

This technique involves the following steps:

- cDNA is denatured (Fig. 14.15) and is then fixed on a nitrocellulose membrane followed by an incubation with mRNA sample.
- This mRNA complementary to the cDNA hybridizes and remains attached to the membrane.
- After removing the unbound molecules, the hybridized mRNA is recovered and is translated in the cell-free system for protein synthesis.
- This method will provide a pure sample of a protein coded by the cDNA.

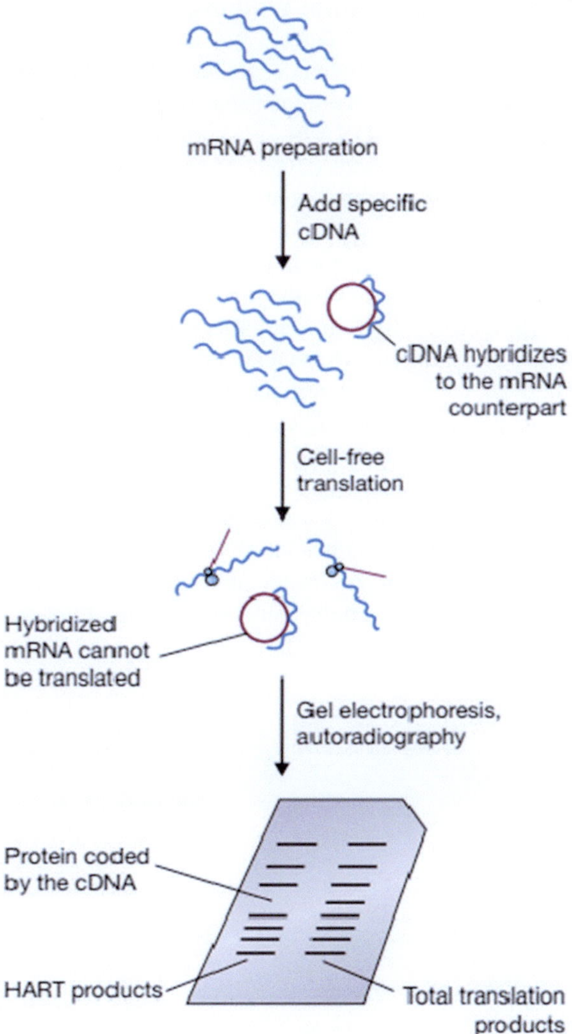

Fig. 14.16 A diagram showing the process of hybrid arrest translation. The steps involved include mRNA preparation which is complementary to specific cDNA. The hybridized mRNA cannot be translated to the protein product. Gel electrophoresis is done that shows protein bands. The gel then shows results that could represent protein coded by cDNA and HART products

HART (Hybrid Arrest Translation)

- Hybrid arrest translation (Fig. 14.16) is a little different from the hybrid release translation in the sense that in this technique, there is no need to denature the cDNA, and the cDNA is directly added to the mRNA sample.
- This results in hybridization between the cDNA and the mRNA that are complementary to each other. The unbound mRNA is not discarded; rather, the whole sample is translated in a cell-free system.
- As the hybridized mRNA is unable to translate, all the proteins except the ones coded by the cloned gene are synthesized. The absent protein from the radiograph is nothing but the protein product of the cloned gene of interest.

14.4.4 Chromosomal Walking

It is developed by Welcome, Bender, Pierre Spierer, and David S. Hogness in the early 1980s. It is a technique that is used for identifying DNA regions next to a known segment of DNA by sequential hybridization of the clones (Fig. 14.17). It is one of the first methods developed for assembly of clone contigs (set of overlapping DNAs that together represent a consensus region of DNA).

Here in this example, the library contains 96 clones each containing different inserts. The walks start with one of the clones that are used as a hybridization probe against all other clones in the library. The clone in this example is the A1 probe, and it hybridizes to itself and the other probes E7 and F6. The insert from the last two clones must therefore have an overlapping sequence with the insert from clone A1. The walk is continued, and the probing is repeated, but this time, the insert is from F6, Thus, the hybridizing clones are A1, F6, and B12, showing that the insert from the B12 insert has overlapped with that of the F6 fragment.

A schematic diagram representing chromosomal walking by PCR (Fig. 14.18). The two oligonucleotides annealed within the end region of insert number 1. They are used in PCR clones with all other clones in the library. Only clone 15 gives a PCR product, and this result indicated that insert clone 1 and clone 15 show an overlap. The walk could then be continued by sequencing the fragment from the

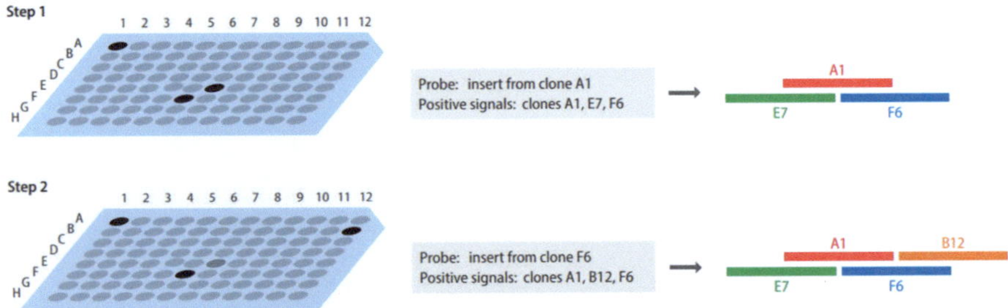

Fig. 14.17 A schematic diagram representing the process of chromosomal walking

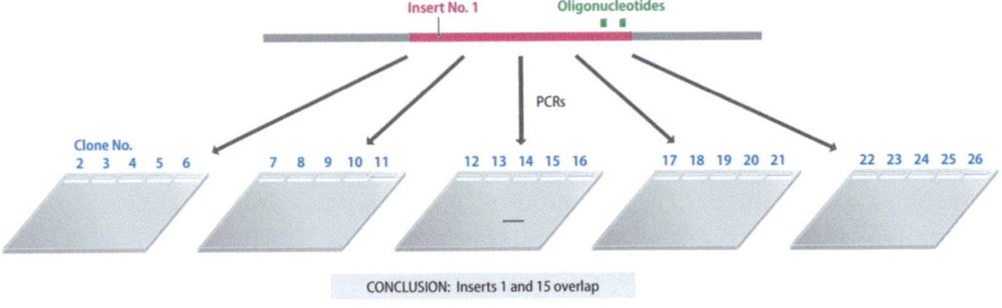

Fig. 14.18 Chromosomal walking by PCR method

other clone 15, using them to design a second pair of oligonucleotides and using them in a new set of PCR with other clones.

Limitations of Chromosomal Walking

- The drawback with chromosomal walking is that it can only sequence and map small lengths of chromosomes.
- Chromosomal walking is a slow process, and it becomes very challenging to assemble contigs of more than 15–20 clones by this method.
- To add the difficulty of walking through a repeated sequence that is scattered through genes.
- Chromosomal walking can be stopped by unclonable sections of DNA.

Applications of Chromosomal Walking

- This disease can be used for the analysis of genetically transmitted disease. It can also be used to look for mutations.
- It is used in the discovery of SNPs.
- One of the best applications of chromosomal walking was the identification of cystic fibrosis gene.

Chapter Summary

- Genomic DNA library is a representative of the total genomic DNA of an organism.
- Screening of genomic DNA library by hybridization is one of the most commonly used technique to identify the gene of interest. This technique works on the same principle that uses Southern blot. The autoradiograph is a representative of the cloned gene of interest.
- Expression of libraries helps us to identify a clone within a gene library that is of particular interest. Currently, a wide range of vectors is commercially available for library screening. Immunological screening with the help of antibodies can also provide great assistance to identify the gene of interest. This can be achieved with the help of plasmid vectors that use blue-white screening strategies.
- Library screening and Western blot analysis are currently used techniques for identifying a gene of interest from gene libraries.
- Hybrid arrest and release is a set of technique that is used to identify translated protein products in cell-free systems. HRT (hybrid release translation) and HART (hybrid arrest translation) use slightly two different approaches to identify a specific protein of interest. These techniques work best with the cDNA library.
- Chromosomal walking is a technique that is used to map regions of DNA next to an unknown sequence. It is the first method developed for assembly of clone contigs.

14.5 Site-Directed Mutagenesis by PCR

Site-directed mutagenesis is a technique that purposefully introduces a mutation at the specific site of a DNA to monitor the effect of changed protein. A basic understanding of the central dogma of biology says that DNA holds the genetic information to synthesize any protein. Changes in the sequence of DNA can alter the structure and function of the targeted protein. Site-directed mutagenesis allows us to create a mutation in the segment of DNA. This technique was first reported to be used by Charles Weissmann in 1974. With the advancement in science, this technique is developed, and it currently offers different methods of mutagenesis, namely, cassette mutagenesis, primer extension method, and PCR method of site-directed mutagenesis.

Cassette Mutagenesis: This is one of the early methods of site-directed mutagenesis in which an artificially created DNA fragment containing the desired mutation replaces the corresponding sequence in the wild-type gene (Fig. 14.19). Wells et al. originally introduced this method to generate improved variants of enzyme subtilisin. The efficiency of mutagenesis in this method was close to 100%. A major drawback of this method was the requirement of unique restriction sites that flank the region of interest.

Site-Directed Mutagenesis by Primer Extension

The single primer method of site-directed mutagenesis developed by Gillam et al. and Zoller and Smith is the simplest method available for introducing mutation in the DNA segment (Fig. 14.20). The first step of the process involves synthesizing a DNA oligonucleotide that serves as a starting material for the polymerization reaction. The oligonucleotide synthesized is typically 20 nucleotides long that carries a base mismatch with a complementary sequence. The target DNA must be single stranded, and cloning the gene in an M13-based vector makes the process easy. A modification to this method was introduced by Dalbadie-McFarland et al. in1982 where they clone DNA in a plasmid, obtained the duplex form, and converted it to a partially single-stranded molecule that is suitable for the process.

Limitations of Primer Extension Method

- The double-stranded duplex molecules generated by this method have a high chance of getting contaminated by any single-stranded nonmutant template DNA that was not copied during this process.
- The presence of this single-stranded non-template DNA reduces the proportion of mutant progeny. This can be corrected by sucrose gradient centrifugation or by agarose gel electrophoresis but is a time-consuming process. A mixed population of mutant and nonmutant progeny can be seen because the in vivo DNA synthesis can lead to segregation of the heteroduplex molecule.
- Separating mutant progeny by cell's machinery is a complex process and is dependent on several factors which cannot be controlled.
- The requirement of a single-stranded template is an added disadvantage of this method. This is because of two factors. First is the fact that single-stranded

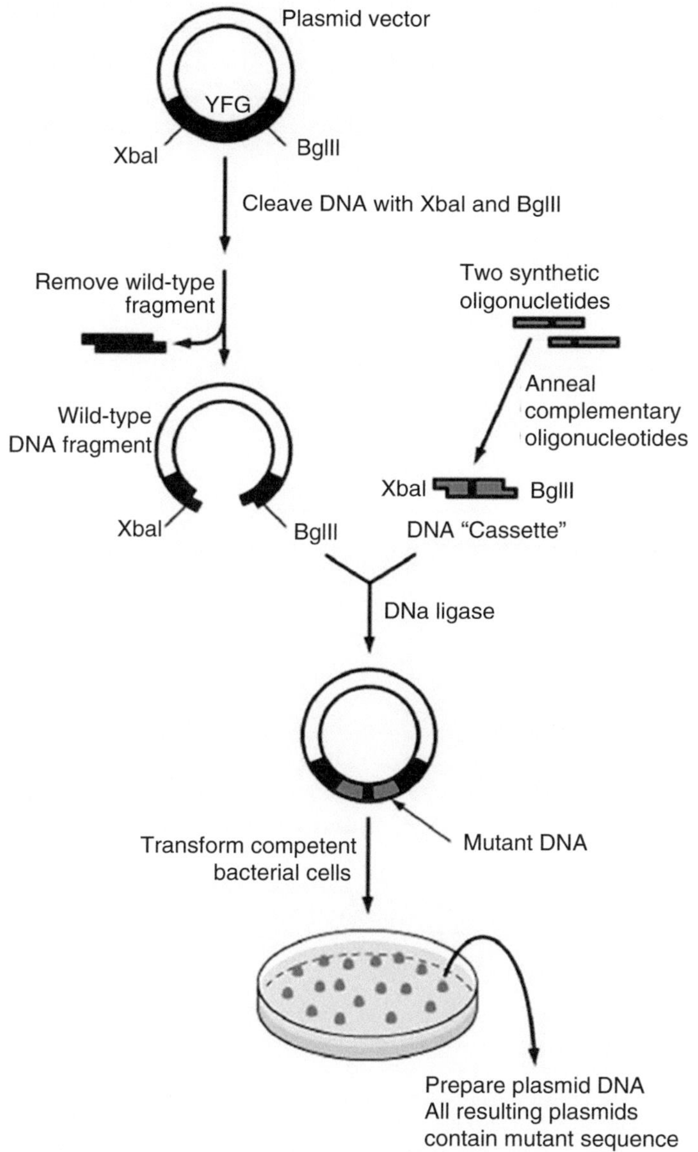

Fig. 14.19 The schematic diagram explaining the procedure of cassette mutagenesis. The plasmid containing a copy of the desired gene (YFP; black segment) is digested with two well-known restriction enzymes, for example, Xbal and BglII. Both of them have only one unique restriction site in the entire plasmid. The reaction mixture is separation on agarose gel by electrophoresis. The larger fragment is purified from the gel. A pair of single-stranded oligonucleotide is synthesized by automated DNA synthesis. These two strands are complementary to each other and differ from the original sequence at only the single position containing the desired changes followed by hybridization of the two strands because of their complementary nature. The vector and the two strands are then ligated using the enzyme DNA ligase. The transformed cells contain mutation at the desired location

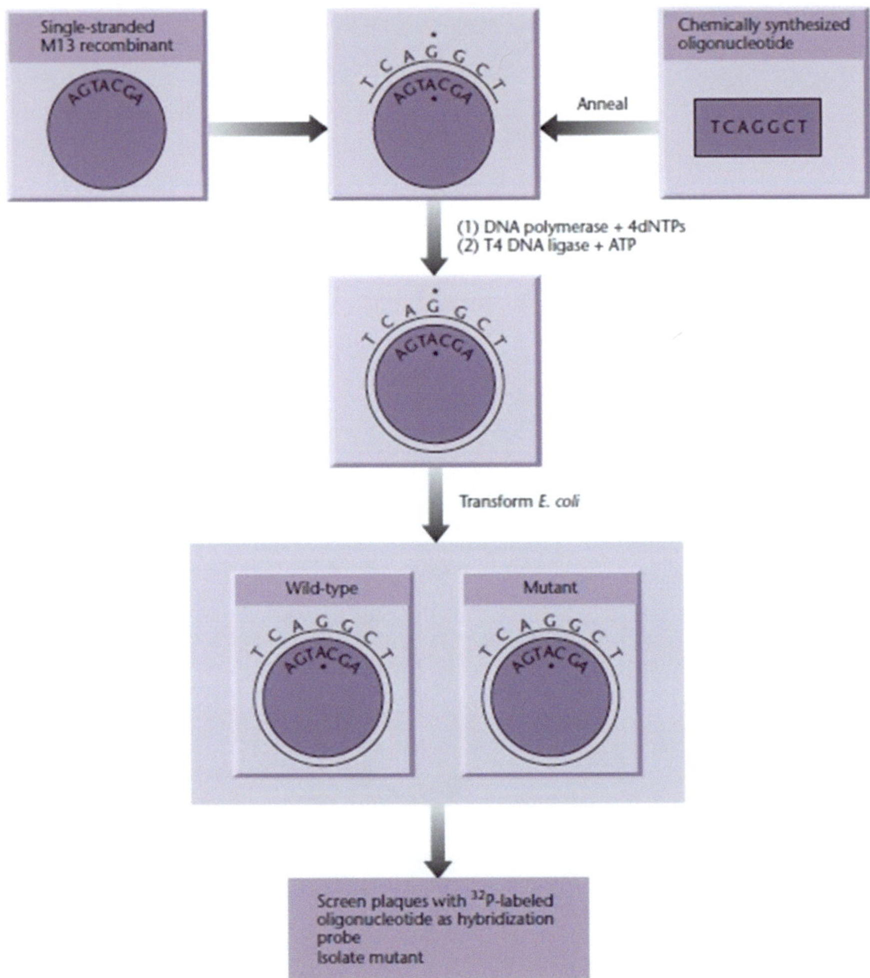

Fig. 14.20 The figure represents site-directed mutagenesis by primer extension method. Single-stranded DNA is first prepared. It is then annealed to a synthetically made oligonucleotide. The sequence is complementary to the wild-type sequence of the template except for the places that have mismatched nucleotides that contain a mutant DNA sequence. The remaining strand is synthesized by DNA polymerase using multigenic oligonucleotide as a primer followed by end join ligation with the help of ligase enzyme. The resulting product contains one wild-type strand and another mutant strand. This DNA is then introduced to E. coli cells. The DNA is sequenced and analyzed for mutation

molecules are hard to prepare as compared to double-stranded molecules, and secondly, gene insertion is less stable in single-stranded DNA and can be easily degraded due to several factors.

14.5.1 PCR Method of Site-Directed Mutagenesis

PCR is considered one of the most revolutionary discoveries in science. A lot has already been talked about PCR in the preceding sections. The main focus of this topic will be to give readers an insight into site-directed mutagenesis by PCR-based techniques. The key requirement of all PCR-based methods is to use high-fidelity polymerase which is achieved by *Taq* polymerase with the error rate of 10^{-5} order error/bp/duplication. This enzyme can be thus utilized to introduce nonselective additional mutations in the gene of interest. Other enzymes such as *Pfu* with 10^{-6} error/bp/duplication and Phusion with an error rate of magnitude of 10^{-7} errors/bp/duplication are preferred choices for site-directed mutagenesis over *Taq* polymerase.

PCR finds wide range of applications in the field of molecular biology, and site-directed mutagenesis is not an exception to this. The central objective of PCR-based site-directed mutagenesis technique is to separate/remove the template-strand DNA from amplified strand to increase the efficiency of mutant clone after transformation. If not removed, this can lead to a generation of false-positive results; therefore, it is necessary to remove the template strand after PCR amplification.

The early works that used PCR-based site-directed mutagenesis are dated back to the year 1986 when Scharf et al. showed the potential use of PCR in this technique.

PCR Conditions for PCR-Based Site-Directed Mutagenesis
PCR conditions for site-directed mutagenesis is quite unconventional, and the number of cycles is reduced to 12–18 cycles only. The standardization of PCR conditions is required for obtaining optimal amplification. This includes the use of DMSO or glycerol as they aid in melting the strands whose T_m is close to 80 °C.

Dpn1 Digestion
After the product amplification confirmation by agarose gel electrophoresis, the template stranded is given Dpn1 treatment. This enzyme specifically recognizes 5′-GATC-3′ as the sequence where adenine is methylated on both the strands, thus specifically digesting parental plasmid, and will not digest unmethylated PCR-amplified product.

14.5.2 Overlap Extension Method

Higuchi et al. have depicted a variety of the fundamental strategy which empowers a transformation in a PCR-created DNA segment to be presented anyplace along its length. His technique used two primary PCR reactions that produce two overlapping DNA fragments both bearing the same mutation in the overlapping region (Fig. 14.21). The overlapping in the sequence allows the fragments to hybridize. This is followed by the extension of one of the two possible hybrids carried out by DNA polymerase, and the other hybrid is degraded in the reaction mixture. One can introduce addition/substitution and deletion by this method. The drawback of this

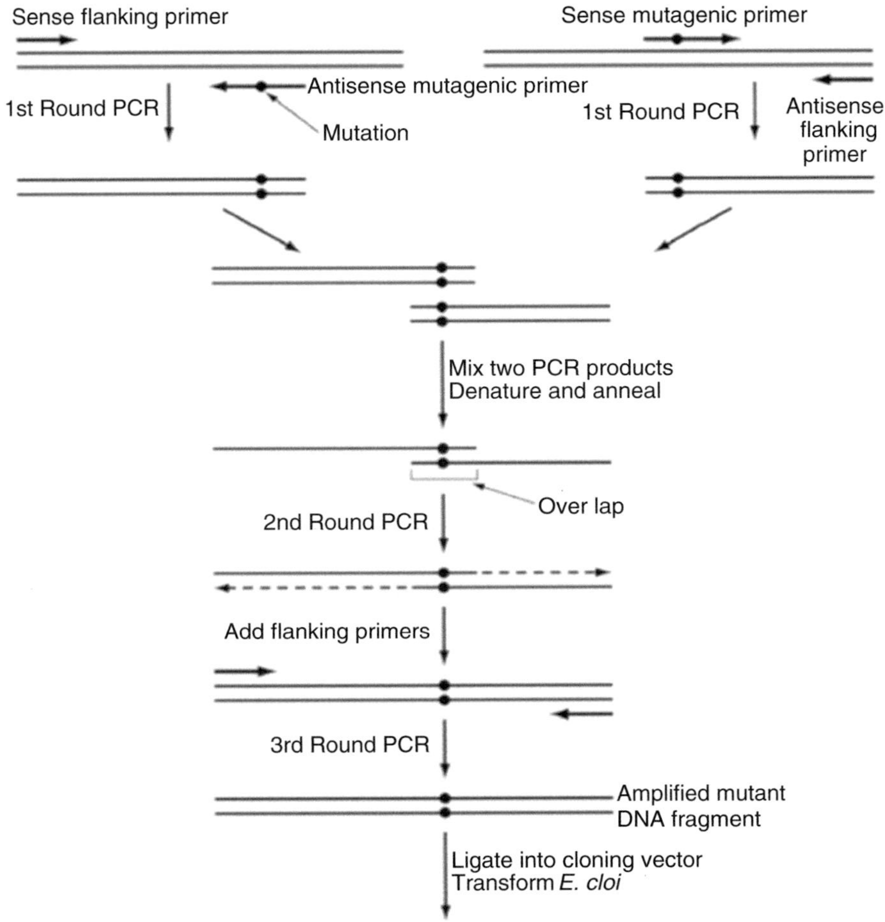

Fig. 14.21 The above schematic diagram shows site-directed mutagenesis by overlap extension PCR. The initial two rounds of PCR created two covering segments of the original template, both containing the mutation at the overlapping region. The two PCR products are annealed and then subjected to the second round of PCR to create the entire segment with a mutation. The flanking primers contain the restriction site for joining the segment back to the original vector

method was that it required four primers and three PCRs (two PCR cycles to amplify the overlapping segments and the final PCR cycle to fuse these two segments).

14.5.3 Megaprimer Method

A relatively simpler method was developed by Sarkar and Sommer in 1990 which uses three primers and two rounds of PCR. This modification used the product of the first PCR as a megaprimer (Fig. 14.22) for the second PCR. This technique uses a

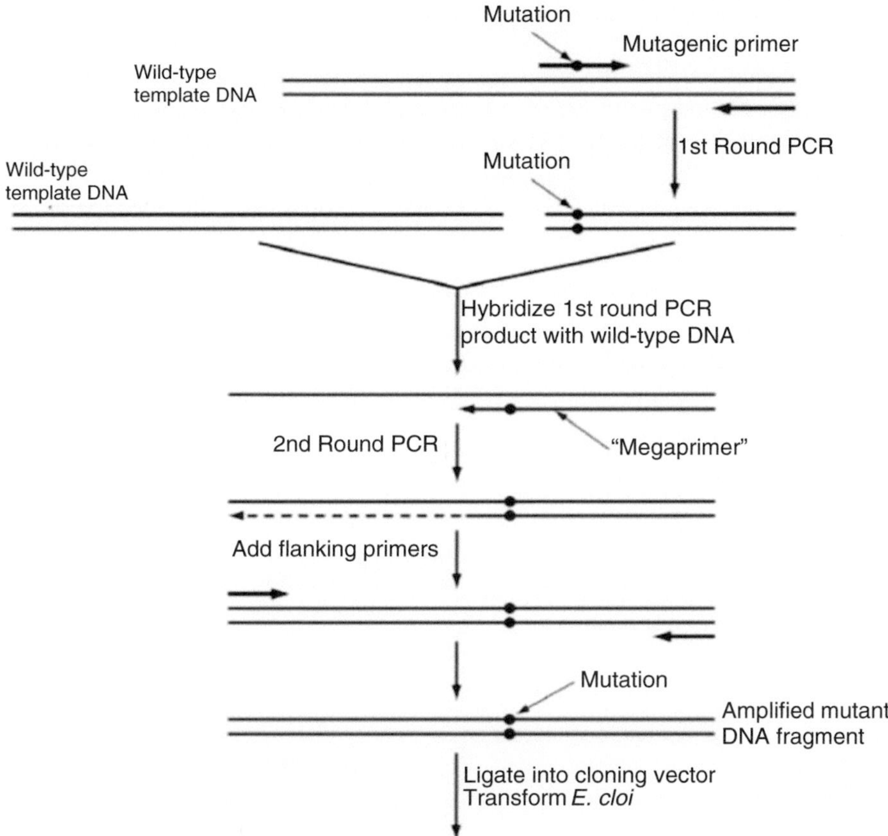

Fig. 14.22 The schematic diagram here represents site-directed mutagenesis by the megaprimer PCR method. This method involves the use of two PCR cycles, and the first cycle makes the fragment of the template DNA containing the desired mutation. The megaprimer thus formed is hybridized to wild-type template DNA, and the second round of PCR is used to generate an entire molecule with a mutation. Restriction sites present in the flanking primer is used to clone the fragment back to the vector

single mutagenic primer to create changes in the target template. The first round of amplification involves the wild-type template using a sense or an antisense mutagenic primer and an appropriate flanking primer. The amplified product is utilized in the second round of the PCR cycle with wild-type template and the other flanking primer to create a fragment that is of the same length as the original target DNA containing the mutation. The key to this technique uses the fact that an amplified product of the first PCR cycle is used as the primer for the second cycle. The overlapping of the template and the mutagenic strand is more extensive in the megaprimer method.

14.5.4 Inverse PCR-Based Technique

This technique requires the use of two primers to create the desired mutation (Fig. 14.23). The unique feature of this method is that the entire vector gets amplified while making the mutation. The two primers the one that contains the desired mutation extend the circular DNA template in opposite direction. The amplification results in a linear double-stranded DNA molecule containing the mutation at one end. After amplification, the ends are ligated back, and the circular DNA molecule is transformed in E. coli.

PCR-based site-directed mutagenesis offers an advantage of simplicity and speed over other forms of site-directed mutagenesis; however, the possibility of introducing unwanted mutations due to the error-prone nature of some thermostable

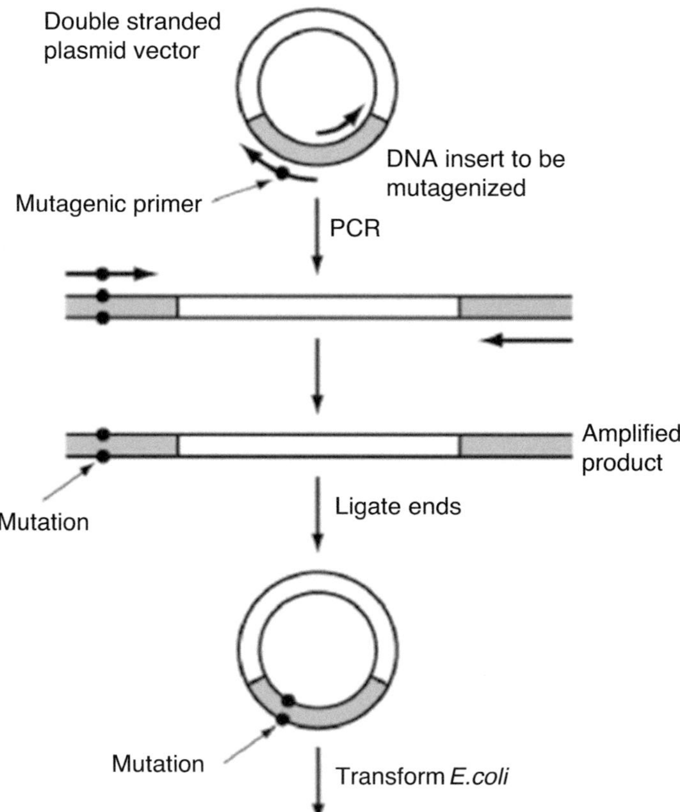

Fig. 14.23 The figure shows a process of site-directed mutagenesis using an inverse primer. As depicted, this method uses two primers, one of them being mutagenic in nature and the other being the normal simple primer. These primers extend the target DNA in opposite direction. After the completion of the PCR cycle, the circular vector becomes linearized which contains the desired mutation. The resulting linear DNA is religated and made circular. The DNA is then transformed into E. coli cells

DNA polymers might offer some limitation to this method. With advancements in science, a large number of thermostable polymerases with high fidelity are now available that can improve this limitation, thus enhancing the potential use of PCR-based site-directed mutagenesis.

Chapter Summary

- Site-directed mutagenesis is a technique that allows us to introduce mutations in the DNA sequence for a better understanding of the protein structure-function relationship.
- The early method involved in this technique could produce close to 100% efficiency but have its own limitations.
- The use of unique restriction sites and lack of control in in vivo DNA synthesis and the use of special single-stranded DNA molecules limited the use of non-PCR-based site-directed mutagenesis.
- PCR-based site-directed mutagenesis used PCR reactions to introduce mutations in the target DNA fragment.
- This includes techniques, namely, overlap extension method and the megaprimer method inverse PCR method.
- The overlap extension method uses four primers, and three PCR cycles limited the use of this method for site-directed mutagenesis.
- The megaprimer method was a bit advance than the overlap extension method and used two PCR cycles and three primers.
- The inverse primer method used a set of two primers to create the desired mutation.
- All these new advancements in PCR-based site-directed mutagenesis techniques made it possible to introduce mutations in a specific part of DNA and allow us to study how these mutations affect the structure and function of the resulted protein.
- PCR-based site-directed mutagenesis allows us to engineer proteins and enhance their properties.

14.6 Visualization of Biomolecules

14.6.1 Visualizing DNA by Southern Blot

- Southern blot is a technique that is used to detect DNA molecule.
- E. M. Southern in 1975 is credited for the discovery of this useful technique.
- This technique aims at finding out a specific fragment of DNA in a sample.
- The basic principle of this technique is hybridization. This technique is used to separate DNA fragments by size.

14.6.1.1 Procedure of Southern Blotting

Following electrophoresis (Fig. 14.24), the test DNA fragments are denatured in strong alkali. As the electrophoretic gels are fragile and the DNA can diffuse within

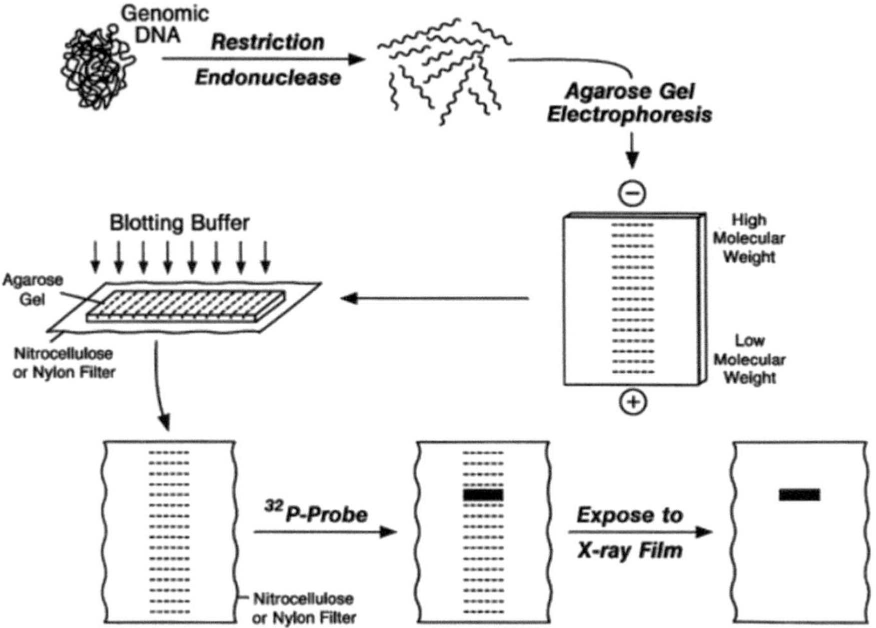

Fig. 14.24 A schematic diagram showing the basic structure of the Southern blotting technique. The process involves the isolation of genomic DNA from the cell of bacterial, plant, or animal origin. Once the genomic DNA is isolated, restriction endonuclease digests the sample of genomic DNA. This is followed by agarose gel electrophoresis which separates DNA bands according to size. These fragments are then transferred onto a nitrocellulose membrane or a nylon membrane. The nitrocellulose membrane which now has DNA fragments is then probed with radioactive phosphorus. After probing the DNA fragment with radioactive material, the nitrocellulose membrane is exposed to X-rays. This is then viewed on an autoradiogram, and thus, the DNA of interest is identified by the process of Southern blotting

the gel, it is usual to transfer denatured DNA fragments by blotting onto a durable nitrocellulose membrane to which the single-stranded DNA binds readily.

After transfer, the DNA fragments need to be fixed to the membrane so that they cannot detach. In case of nitrocellulose paper, nucleic acid immobilization occurs noncovalently after baking for 2 hrs at 80°C. In case of nylon membrane, either it is baking for 1 hour at 70°C or UV irradiation at 254 nm. Nucleic acid binds covalently with nylon membrane after UV irradiation for 5 minutes. The individual DNA fragments become immobilized on the membrane at positions which are a faithful record of the size separation achieved by gel electrophoresis. Following the fixation step, the membrane is placed in a solution of labeled (radioactive or nonradioactive) RNA, single-stranded DNA or oligodeoxy nucleotide which is complementary in sequence to the blot transferred DNA band or bands to be detected. Since this labeled nucleic acid is used to detect and locate the complementary sequence, it is called the probe. The probe is allowed to hybridize to its complementary single-stranded target DNA sequences on the membrane. Conditions are chosen which maximize the rate of hybridization compatible with a low background of nonspecific binding of the membrane. After the hybridization, reaction has been carried out, and the membrane is then washed extensively to remove nonspecifically bound probe. If the probe is radioisotope labeled, then the membrane is exposed to photographic film. If the probe is nonisotopically labeled with biotin or digoxigenin, the membrane may be with the chemiluminescent substrate to detect the labeled probe and then exposed to photographic film. The probe will form a band on the film at a position corresponding to the complementary sequence on the membrane.

14.6.1.2 Applications of Southern Blotting

- Identifying specific DNA in a DNA sample.
- It is used to construct RFLP maps. These maps are valuable in detecting and studying various kinds of mutations including deletions, additions, substitutions, and gene rearrangement.
- Diseases such as sickle cell anemia, cystic fibrosis, Duchenne muscular dystrophy, and others can be detected by Southern blot studies.
- Southern blotting finds its application in forensics particularly for criminal identification and DNA fingerprinting.
- This technique is of high value in identification of genetically engineered organisms.
- Prognosis of cancer and prenatal diagnosis of genetic diseases.
- Determination of the molecular weight of a restriction fragment and to measure relative amounts in different samples.
- It is used for the confirmation of DNA cloning results.

14.6.2 Visualizing RNA by Nothern Blot

This technique is used to study the level of gene expression. This study can be tissue specific or condition specific. It is often seen that genes are transcribed in a tissue-specific manner. A gene may have limited expression in normal circumstances, but it may be highly expressed in diseased condition. The comparison of this healthy and diseased could be made by Northern blot analysis. The level of gene expression can be easily detected if one can find the amount of RNA transcribed from the gene of interest.

The process of Nothern blotting (Fig. 14.25) involves the following steps:

- This technique measures the amount and size of RNA transcribed from genes and estimates their abundance.
- Firstly, an RNA extract is electrophoresed in an agarose gel, using a denaturing buffer such as formaldehyde that ensures that the RNA transcript does not form any secondary structures.
- The agarose gel is blotted onto a reactive DBM (diazobenzyloxymethyl) paper and hybridized with a labeled probe so as to detect RNA of specific interest.

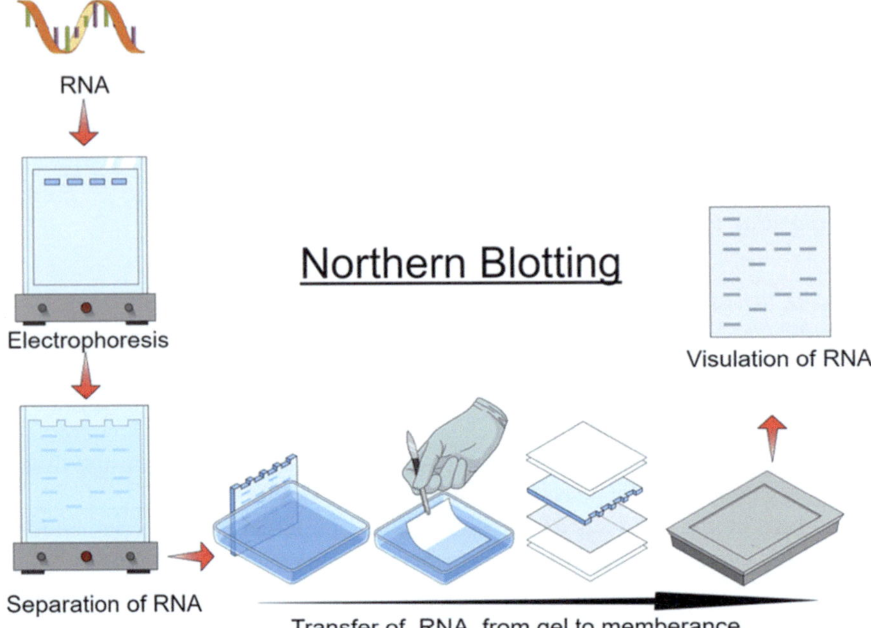

Fig. 14.25 A flow diagram representing the process of Nothern blotting. Nothern blotting is quite similar to the above discussed Southern blotting. The only difference lies in the fact that in Nothern blotting, the detected molecule is RNA and not DNA. A key point to mention during the process of Northern blotting is the use of formaldehyde that removes any kind of secondary structures formed. The secondary structures in RNA are quite common and may be formed because of inter- or intramolecular hydrogen bonding. The process starts with collection of RNA sample which is followed by electrophoresis that leads to separation of RNA fragments according to size. The RNA fragments separated on agarose gel are transferred to a membrane, and then specific RNA is visualized on an autoradiogram

- RNA bands can also be blotted onto nitrocellulose paper under appropriate conditions and suitable nylon membranes.

14.6.2.1 Applications of Nothern Blot

- This technique is use to study the process of RNA degradation and alternative splicing.
- Nothern blotting finds its use in the study of differential gene expression that are tissue specific and condition specific that might offer an early detection of disease.
- Diagnosis of several diseases (Crohn's diseases) including viral infection.
- This technique is widely used to study the overexpression of oncogenes and downregulation of tumor suppressor genes in cancerous cells.
- Detect specific mRNA molecular weights and contents in a sample.

14.6.3 Visualizing RNA by RT-PCR

RT-PCR stands for Reverse Transcriptase Polymerase Chain Reaction. Another way of detecting a specific mRNA is through PCR that enables to amplify the specific message. This process requires copying of mRNA to cDNA with the aid of reverse transcriptase. This is a highly specific method, and it becomes possible to detect specific RNA species in a single cell (Fig. 14.26). This process also helps in detecting low levels of mRNA and also makes it easy to analyze gene expression that is difficult to obtain in huge quantity, for example, gene expression in cells from tumors in order to pinpoint those genes that are expressed in such conditions. Tissue-specific expression or expression of cell under stress conditions can be monitored by this technique. RT-PCR also requires less quantity of mRNA and hence requires fewer cells to achieve same goal as compared to conventional Northern blot.

RT-PCR is a boon during the current times of SARS-COV2 (coronavirus pandemic). It has gained a lot of repute in recent times and has helped researchers and medical doctors to know the exact viral load in the patient's body. This helps in proper diagnosis and relevant treatment of the diseased person.

14.6.4 Visualizing Proteins by Western Blotting

Western blotting (also called **immunoblotting**) is a technique used for analysis of individual proteins in a protein mixture (e.g., a cell lysate). In Western blotting (immunoblotting), the protein mixture is applied to a gel electrophoresis in a carrier matrix (SDS-PAGE, native PAGE, isoelectric focusing, 2D gel electrophoresis, etc.) to sort the proteins by size, charge, or other differences in individual protein bands (Fig. 14.27). The separated protein bands are then transferred to a carrier membrane (e.g., nitrocellulose, nylon, or PVDF) (**blotting**). The proteins adhere to the membrane in the same pattern as they have been separated due to interactions of charges. The proteins on this immunoblot are then accessible for antibody binding for detection.

Antibodies are used to detect target proteins on the Western blot (immunoblot). The antibodies are conjugated with fluorescent or radioactive labels or enzymes that give a subsequent reaction with an applied reagent, leading to a coloring or emission of light, enabling detection.

The term Western Blotting is also based on a play of words. The Southern blot, which is a method to detect specific DNA sequences, is named after Ed Southern, who first described this procedure. The Western blot (immunoblot), as well as the Northern blot (for RNA detection), play on the meaning of this name.

Applications of Western Blotting

- The confirmatory HIV test.
- Used as the definitive test for Bovine spongiform encephalopathy (mad cow disease).
- Some forms of Lyme disease testing employ Western blotting.

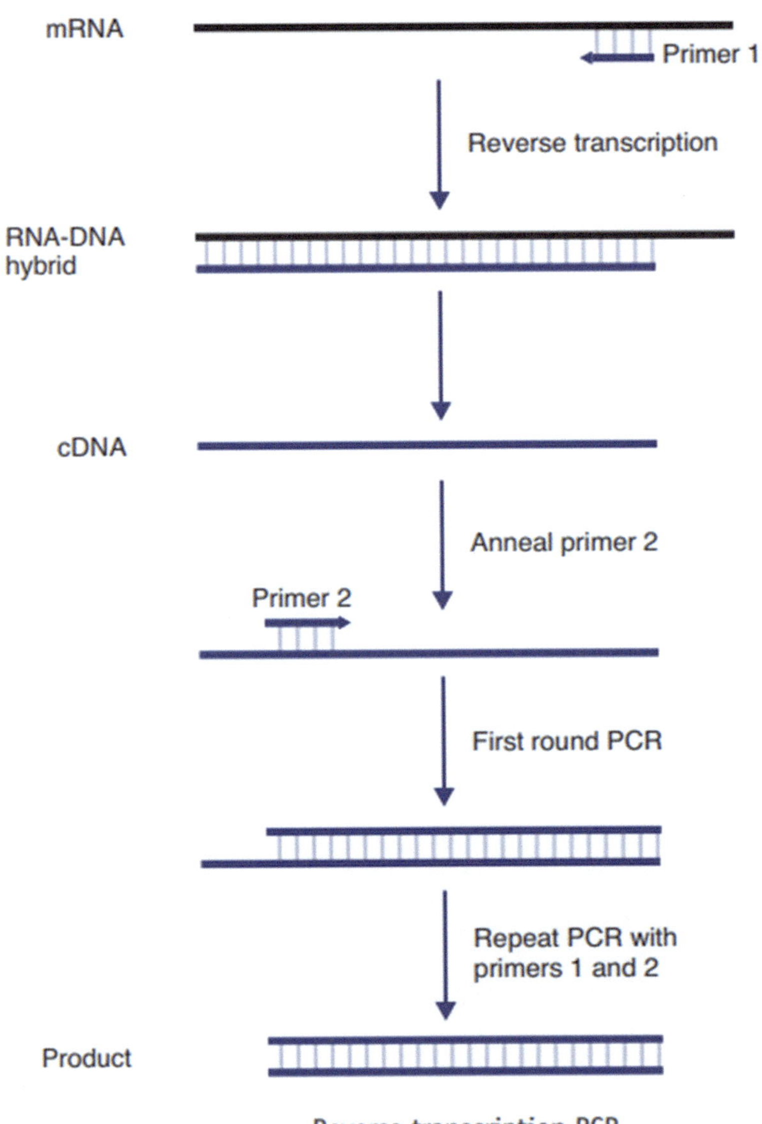

Fig. 14.26 A schematic diagram representing the process of reverse transcription. The primer base-pairs with the mRNA, and it extends along the length of mRNA molecule with the help of enzyme reverse transcriptase. The RNA-DNA hybrid thus formed is cleaved with the help of RNAse H, an enzyme that specifically digests the RNA segment. The DNA thus left is actually called cDNA (c, complementary). Another primer anneals to this cDNA and adds nucleotides that are complementary to this cDNA. This completes the first round of PCR cycle. The cycle is repeated multiple times by using primers 1 and 2. This generates the PCR product. This technique has recently become an advantage in dealing with coronavirus pandemic

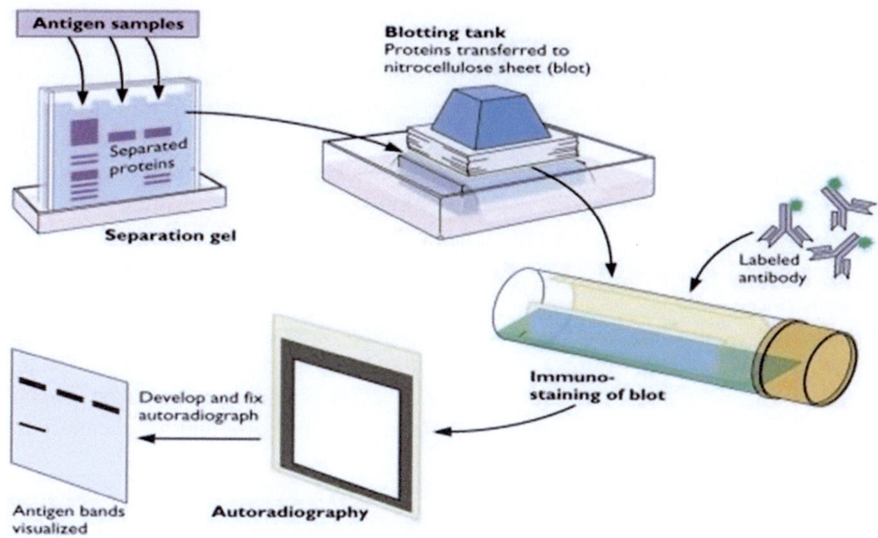

Fig. 14.27 A schematic showing the process of Western (immunoblotting) blotting

Table 14.1 Difference between southern blot, northern blot, and western blot

Characteristics	Southern blot	Northern blot	Western blot
Molecule to be detected	DNA	RNA	Protein
Extraction	Alcohol precipitation	Cellulose chromatography	Differential configuration
Separation (gel used)	Agarose gel electrophoresis	Agarose gel electrophoresis	SDS-PAGE
Denaturation	Alkali (NaOH)	Not required	Not required
Blotting method	Capillary blotting	Capillary blotting	Electroblotting
Membrane used	NC or nylon	Nylon or DBM	NC or PVDF
Blocking	Pretreatment	Not required	BSA/milk powder
Probe used	Radiolabeled ssDNA	Radiolabeled ssDNA	One or two antibody
Hybridization	DNA-DNA	DNA-RNA	Ag-Ab complex
Detection	Autoradiography	Autoradiography	Colorimetric
Application	DNA fingerprinting	Disease diagnosis	HIV and hepatitis B

- Western blot can also be used as a confirmatory test for hepatitis B infection.
- In veterinary medicine, Western blot is sometimes used to confirm FIV+ status in cats.
- This technique is also employed in the gene expression studies.
- It is used in the definitive test for BSE.**Chapter Summary**
- Working with DNA, RNA, and proteins is very important part in the field of molecular biology.

- ***Blot:*** A membrane on which biological molecules such as protein and nucleic acids are adsorbed or immobilized.
- The process of transferring these molecules (proteins and nucleic acids) from a gel to the membrane followed by their detection on the membrane is known as blotting.
- Generally, the process of transferring molecules from agarose gel to nitrocellulose membrane is often done by capillary blotting which uses capillary action to transfer DNA, RNA, or protein onto a nitrocellulose membrane; however, in some cases, electroblotting is also observed that uses current electricity for the same purpose.
- Dot blot: A technique often used to detect proteins. DNA/RNA can also be detected. This is performed to screen the binding capabilities of antibodies. Simplification of Western blot with the exception that proteins to be detected are not first separated by electrophoresis. Fast method and is cheaper than Western blotting techniques.
- A method of Eastern blotting is also present. This is nothing but a modification of Western blotting that could detect posttranslational modifications.

14.7 Analyses of Gene

14.7.1 Physical Mapping by Restriction Endonuclease

Physical mapping is a technique used in molecular biology that is used to determine the physical structure of the genome (Fig. 14.28). This includes the exact position of restriction sites, the positions of specific clones, and the complete sequence of gene. Restriction endonucleases are the enzymes that recognize specific sequence of

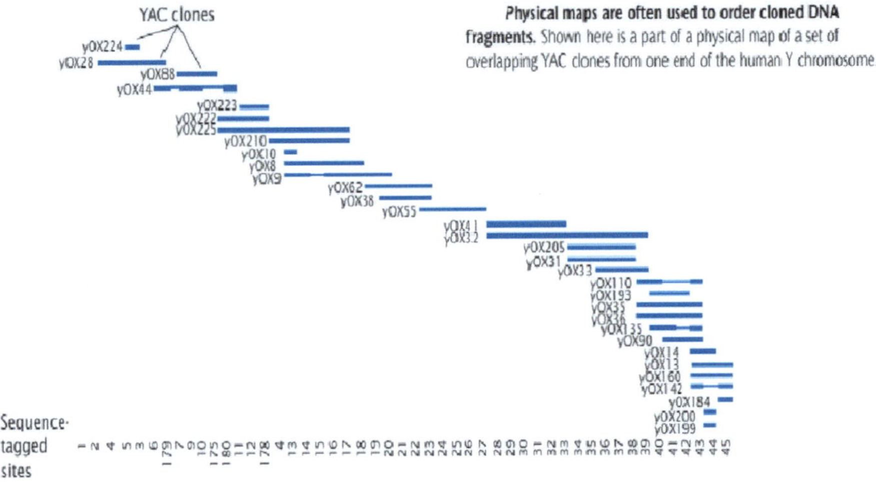

Fig. 14.28 A diagram showing practical application of physical mapping

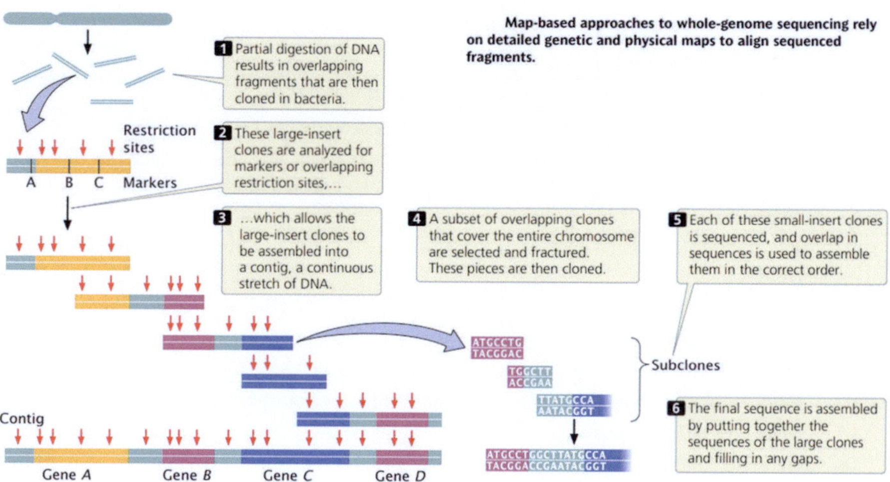

Fig. 14.29 A schematic diagram representing the procedure for construction of a physical map

nucleotide and cut at a specific position. Restriction endonuclease can either produce blunt ends or they can produce sticky ends. Physical maps are used to arrange fragments of a cloned DNA.

Mapping of a Sequence with the Help of Restriction Endonuclease

This is one of the most common used techniques to create a physical map. In order to know the exact location of restriction sites, a sample of DNA is cleaved with a restriction enzyme, and another sample is cut with a different restriction enzyme (Fig. 14.29). A third sample is cut with both the restriction enzymes that are previously used to cleave sample of DNA. After restriction endonuclease digestion, the DNA segments produced are separated by gel electrophoresis, and their relative sizes are compared. Overlapping in the size of the fragments produced can be used to position restriction sites on the original DNA molecule. These maps are often created for genomic analysis. The basic difference between a **genetic map** and a **physical map** is that genetic maps are based on recombination rates and are measured in percent recombination, whereas physical maps are based on physical distances and measured in base pairs.

14.8 Sequencing of Genes

Biomolecules are basic structural and functional unit of a living cell. Nucleotides (DNA and RNA) form the genetic basis of inheritance. We now know that proteins are in fact the translated message of the RNA which in turn is transcribed from DNA. A gene by definition is a set of three consecutive nucleotides that codes for a functional polypeptide chain or an RNA molecule. The functional properties of a gene can only be decoded once we know the gene sequence. DNA sequence is

Table 14.2 A brief history of DNA sequencing

Year	Milestone achievement
1963	The alanine tRNA of yeast was sequenced by Robert Holley
1972	MS2 bacteriophage was sequenced using RNA sequencing by Walter Fiers
1973	With wandering spot analysis, Walter Gilbert and Allan Maxam sequenced lac operator
1975	The plus-minus sequencing of FX174 bacteriophage was developed by Sanger and Coulson
1977	The chemical cleavage method of DNA sequencing was discovered by Maxam and Gilbert
1977	The chain termination method of sequencing was discovered by Frederick Sanger
1980	The Nobel Prize for DNA sequencing was awarded to Sanger and Gilbert
1986	The semiautomated DNA sequencing machine was invented by Leroy E. Hood
1997	Fully automatic sequencing machine with capillary electrophoresis
2005	Next-generation sequencers like Genome Sequencer GS20, 454 Life Sciences, Roche
2006	Genome Analyzer, Solexa/Illumina
2007	SOLID Applied Biosystem

perhaps the most crucial technique that is currently available to molecular biologist that allows to determine the precise order of nucleotide in a piece of DNA. DNA sequencing nowadays has become a vital activity of most labs. These methods are about 50 years old. Gene sequencing is defined as a method of determining the order and arrangement of nucleotides in DNA fragments. Any segment of RNA or DNA can be used to derive the sequence of the gene. The DNA sequence provides valuable information about the presence of regulatory regions, coding regions, homologous sequence, and sequence variation in two forms of genes or alleles. In the mid-1970s, rapid and efficient DNA sequencing was made possible, but earlier, these techniques were restricted to individual genes, but with advancement in technologies, it has been made possible to sequence the entire genome since 1990s. A number of different methods are devised to sequence DNA and are broadly classified into the chain termination method and the next-generation sequencing method The chain termination sequencing method will be discussed in detail, and for next-generation sequencing, please refer to high-throughput sequencing.

Sanger Dideoxy Sequencing Method of Gene Sequencing

This method of DNA sequencing was developed by Frederick Sanger and his colleagues in the mid of 1970s. This technique found wide range of application in sequencing the DNA segment. This is also known as chain termination method or dideoxy method of DNA sequencing. The basic principle behind this technique is that single-stranded DNA molecules that differ in size by a single nucleotide can be separated from one another by polyacrylamide gel electrophoresis.

Chain termination sequencing (Fig. 14.30) is performed by DNA polymerase. It makes a copy of DNA molecule that is being sequenced. This method begins with a short nucleotide being annealed to the template DNA oligonucleotide acts as a primer for the synthesis of new DNA strand that is complementary to the template.

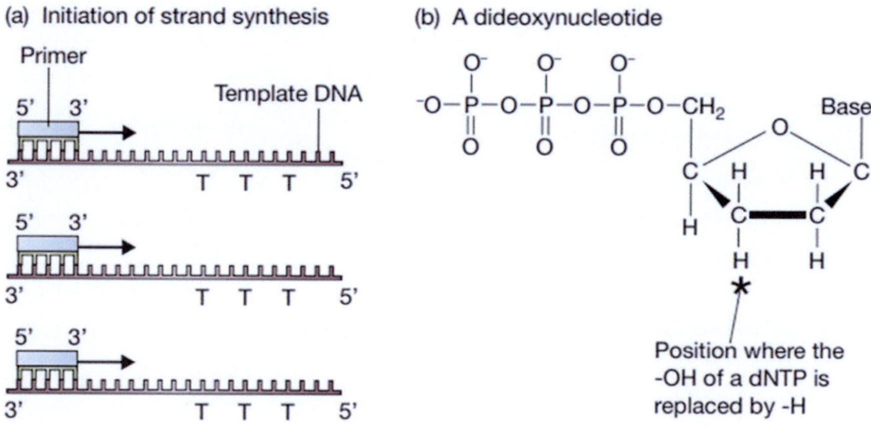

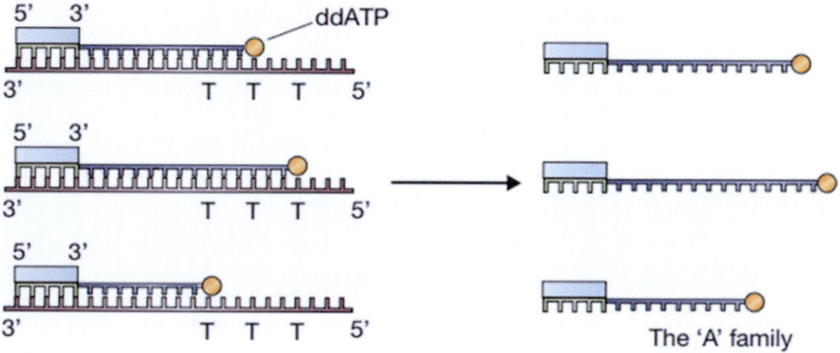

Fig. 14.30 A schematic diagram representing the process of chain termination

The basic science behind the chain termination reaction is that DNA polymerase cannot discriminate between the deoxy and dideoxynucleotide. Once incorporated, a dideoxynucleotide blocks further elongation because it lacks the 3′ OH group needed to form a linkage with the next nucleotide. Since the ratio of deoxyribonucleotide to dideoxynucleotide is relatively higher, strand synthesis does not always terminate close to primer, and a DNA may be extended to several hundred nucleotides before the incorporation of dideoxynucleotide. As the consequence to this formation of new molecule takes place, these new molecules are of different length, each of them ending in a dideoxynucleotide.

A schematic diagram representing the process of chain termination. Figure 14.30a represents a primer that is annealed to a template DNA that extends from 5′ to 3′ direction. Figure 14.30b represents a deoxynucleotide which lacks a 3′ OH crucial for elongation of DNA. This molecule halts the process of DNA synthesis. The DNA polymerase is unable to distinguish between deoxynucleotide and

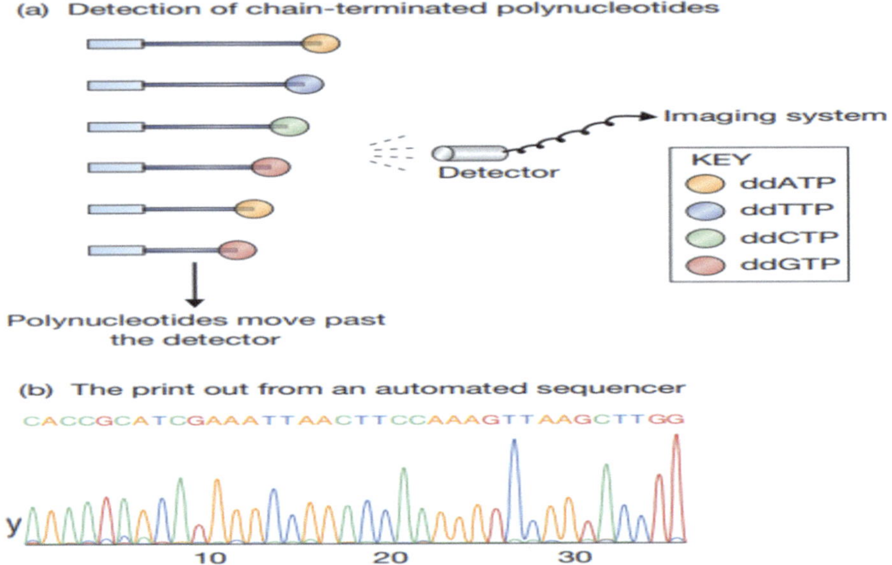

Fig. 14.31 A diagram representing the detailed sequencing of a DNA fragment with the help of a detector. Usually, this detection is achieved by a fluorescently labeled molecule. Here is this diagram: (**a**) A is labeled as orange color, T is blue, C is green, and G is red. So where there is a green signal, one can say the nucleotide is adenine. (**b**) The diagram (Fig. 14.32) represents a sequence of nucleotide in the form of a printout. This sequence could also be stored for future reference

dideoxynucleotide. The dideoxynucleoside is purposely added to terminate chain elongation. Figure 14.30c represents the incorporation of dideoxynucleotide that results in the formation of DNA fragments of different chain lengths.

To correctly sequence the DNA fragments, it is important to identify the dideoxynucleotide at the end of each chain-terminated molecule. This step is done by electrophoresis in polyacrylamide gel. This is done to separate the molecules according to their length.

After separation, molecules are run through a fluorescence detector that helps in discriminating labels attached to deoxynucleotide (Fig. 14.31). The detector therefore determines that each molecule ends in A, C, G, or T. The sequence can be printed by the operator or could be saved for future reference.

A unique feature of chain termination reaction is that not all DNA polymerases can be used in this procedure. This is due to the fact that a DNA polymerase have diverse enzyme activities. It can degrade as well as synthesize DNA. DNA degradation can occur in 5′–3′ direction and can occur in 3′–5′ direction. If it occurs in 3′–5 direction, this could adversely affect chain-terminating sequence. This is due to the fact that 3′–5′ activity prevents chain termination from occurring as this activity removes the dideoxynucleotide immediately after it has been added at the 3′ end of strand of this being synthesized.

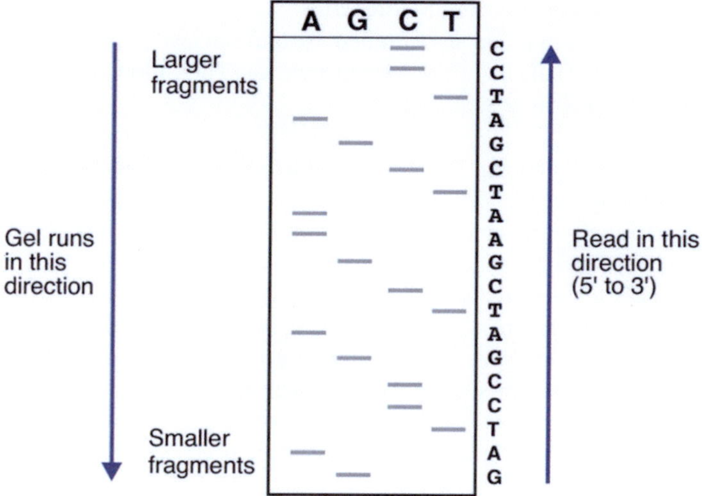

Fig. 14.32 Determining the DNA sequence through Sanger method. This figure represents the method to determine the exact sequence of DNA through Sanger sequencing method. It is clearly visible that gel runs from top to bottom where top represents larger- or bigger-sized fragments, and the bottom represents the smaller size fragments. However, the DNA sequence is determined from bottom to top and read in 5′–3′ direction. One key point to mention here is that the sequence that is determined here is of the non-template strand. It is complementary to the template strand

In the actual method of chain-terminating sequence, the Klenow polymerase was used as a sequencing enzyme. This enzyme was a modified version of DNA polymerase I in the sense that its 5′–3′ exonuclease activity was removed, and a mutation was done so as to hamper its 3′–5′ exonuclease activity.

However, Kelow fragment is not suitable for DNA sequencing due to its low processivity, i.e., it can only synthesize a short DNA strand before dissociation from the template due to natural causes. This limits the length of sequence to about 250 bp in a single experiment.

These drawbacks in Klenow fragments lead to the use of Taq DNA polymerase with high processivity and no exonuclease activity, making it suitable for chain termination sequence.

Chain termination sequence method that uses Taq polymerase is called thermal cycle sequence as it is quite similar to PCR with the slight modification that uses just single primer, and the reaction mixture includes four dideoxynucleotides. The use of single primer makes the amplification linear and not exponential as in the case of real PCR.

Limitations of Chain Termination Sequencing

Sanger dideoxy chain termination has a limited use and can lead to problems that can negatively affect result analysis of the sequenced data. The following are the most common drawbacks of Sanger sequencing:

1. Sanger method of sequencing is mainly used to sequence short pieces of DNA about 300–1000 bp.
2. The quality of sanger sequencing is often compromised for the first 20–40 bases because it is where primer binds.
3. Sequence quality tends to reduce after 700–900 bases.
4. Some unwanted DNA segments might also be sequenced. This problem may arise due to the fact that if a DNA fragment being sequenced has been cloned, one might also find some vector sequence in the final result.
5. Sanger sequence is relatively expensive in terms of the fact that it takes about 500 $ per 1000 bases. The Human Genome Project used Sanger sequencing to sequence human genome.

14.8.1 High-Throughput Sequencing (HST)

High-throughput sequencing is a set of techniques that uses modern machines to sequence DNA. This has greatly influenced our understanding of biology, human diversity, and disease. Advancement in DNA sequencing technologies have opened up new opportunities and introduced us to the era of personal genomes and genomic medicine. These technologies are quickly reliable and often automated.

Human Genome Project presented its first draft in 2001 which was followed by genomic sequencing of several other organisms. Sanger sequencing allowed all these projects to accomplish but was not desirable due to limited throughput and large cost investments. It is estimated that the first Human Genome Project costed around 0.5 to 1 billion dollars. Due to this high-cost investment, the National Human Genome Research Institute (NGHRI) created a DNA technology which cut down the original cost to 70 million dollars, aiming to achieve a 1000 dollars genome sequencing in a decade. HST thus emerged to fulfil this ambitious dream.

To begin, some improvements were made in conventional Sanger sequencing techniques which lead to a cost reduction by 100 folds; however, to reach a 1000-dollar genome threshold, an additional jump of five orders was much needed. The cost of genome sequencing (without interpretation) is less than 2000 dollars, all thanks to HST technologies. The following paragraph will be discussing some of the most popular HST technologies just to give an overview about recent trends in genome sequencing.

14.8.1.1 Illumina

This biotech giant released Genome Analyzer II in 2006, and advancements in Illumina's technologies over the past years have set a pace for huge profits in terms of output and reductions in cost. As an outcome, Illumina technologies rule the HST business. This technology uses a fluorescently labeled molecule 3′-O-azidomethyl-dNTPs to halt the polymerization reaction, thus enabling the removal of bases that didn't incorporate and allow fluorescent imaging to determine the added nucleotide. An added advantage common to all Illumina models is that overall error rates are below 1% and the most common type of error encountered is substitution type.

Illumina currently provides a variety of sequencing machines optimized for various uses. The most common sequencers are MiSeq, NextSeq500, and the HiSeq series. The MiSeq and HiSeq are the more accepted platforms. The MiSeq is a fast personal benchtop-sized sequencer and could sequence small genomes in just 4 hrs. HiSeq, on the other hand, is designed for high-throughput applications which can generate about 1 Tb outputs in 6 days (pretty fast!). Illumina also launched NextSeq and HiSeq X Ten in 2014. NextSeq 500 was also designed as a benchtop sequencer for individual labs. NextSeq is capable of producing 120 Gb of data in less than 30 hrs. NextSeq also uses a unique two-channel sequencing strategy in which cytosine is labeled red, thymine is labeled green, and adenine is labeled yellow (labeled with a mixture of red and green), and guanine remains unlabeled. The four-channel sequencing strategy is used in HiSeq and MiSeq platform. Two-channel sequencing strategy is favored because it reduces data processing times and increases the throughput.

14.8.1.2 Life Technologies/ThermoFisher/Ion Torrent

Ion Torrent's semiconductor sequencing technology was commercialized by Life Technologies in 2010 in the form of benchtop Ion PGM sequencer. This sequencer senses pH changes induced by hydrogen ions during DNA extension. These changes are monitored by a sensor placed at the bottom of a microwell and changed into a voltage signal. The voltage signal is directly proportional to the number of bases that are incorporated during sequencing. In addition to this, Ion Torrent avoids optical scanning to distinguish nucleotides during cycles of sequencing. This speeds up the sequencing process and is cost-effective.

Ion Torrent launched another machine in 2012 named the Ion Proton that has an advantage to increase the output over the PGM by a magnitude of the order of 1 Gb vs. 10 Gb. Proton is currently limited to 200 bp reads as opposed to 400 bp read by PGM. PGM found its use in targeted resequencing and small genome analysis, whereas Proton has the ability to sequence exomes and perform whole transcriptome analysis.

The speed of these sequencers can range from 2 to 8 hrs that depends on factors like the machine and chip used for sequencing. This property makes these sequencers quite useful in terms of clinical applications. The most common types of errors that could occur during this type of sequencing include additions and deletions. A homopolymer tail of more than 6 bp could increase the error rates.

14.8.1.3 Pacific Biosciences

They are credited for pioneering single-molecule real-time sequencing (SMRT). They used strand displacing polymers that allowed the original DNA molecule to be sequenced multiple times, which leads to an increase in accuracy. This method involves direct sequencing of native and potentially modified DNA. DNA synthesis takes place in zeptoliter-sized chamber termed as zero-mode waveguides (ZMWs) in which a single polymerase is fixed at the bottom of the chamber. The science behind these chambers adds value to this technique by reducing background noise such as phosphate-labeled versions of all four nucleotide as a result of this polymerization can occur continuously, and DNA sequence can be read in real time from fluorescent signals recorded in the video.

RS II is the only sequencing machine made commercially available by Pacific Biosciences. This was launched in 2010. This machine can have enhanced performance when the chemistry involved in the systems gets altered. SMRT cell can produce up to 1 Gb reads of data in just 4 hrs. The error rates in these machines can be as large as 11%, which is a disadvantage to single-molecule sequencing platforms. This machine finds useful application in projects involving de novo assembly of small bacterial and viral genomes. Reconstruction of structural variation (SV) in genomes and isoform usage in transcriptome is an important area where SMRT sequencing is advantageous over short-read technologies.

Lower throughput and higher per-base sequencing cost can limit the scope of most genome-wide studies. SMRT sequencing has a distinguishing feature of polymerization monitored in real time. This technology can also map 6-methyladenine and 5-methylcytosine genome-wide in bacteria.

14.8.1.4 Oxford Nanopore Technologies

Nanopore-based sequencing is a recent single-molecule strategy that has made remarkable progress in recent years. It majorly relies on the movement of DNA nucleotides through a small channel. The sequencing is done by measuring characteristic changes in the current. The first available device for nanopore sequencing is MinION, which was a USB-powered portable sequencer. The error rates reported in this sequencing were 4.9%, 7.8%, and 5.1% for insertion, deletion, and substitution, respectively. MinION reads have been successfully used to figure out the positional and structural aspects of bacterial resistance island. With lower throughputs and higher error rates, this sequencing tool is not a preferred choice in current times.

Use of HTS Application

1. Genome sequencing and variation: The first de novo sequencing was done for *Acinetobacter baumannii.* HTS has also been able to characterize SV in the human genome. HTS are frequently used to give an insight about human diversity and disease.
2. Mapping regulatory information of genome: HTS can map regulatory elements of DNA at high resolution. This mapping of sequences back to the genome may reveal location of chromatin modification.
3. Mapping the 3-D organization of the genome: Global organization and compartmentalization of chromosomes are advanced by HST technologies. Hi-C was one of the first technique to allow impartial genome-wide probing of chromatin organization and revealed open and closed genome states.

4. Characterizing the transcriptome: HTS technologies could characterize various classes of RNA. This also includes characterization of RNA structure, RNA-protein interactions, and genomic localization. HST has led to the development of microRNA target discovery. HTS application have also made it possible to determine transcript structure both in vitro and in vivo.
5. Microbiome sequencing: HST technologies have enabled classifying metagenomic samples. This can provide a knowledge of the microbial diversity from wide variety of sources. In fact, this has also given rise to the concept of personal microbiome.
6. Genome sequencing of rare disease: The ability to sequence large genome, exomes, and transcriptomes has added to our current knowledge about human health and disease. This is particularly relevant for rare mendelian disorders and cancer. To quote an example, exome sequencing of a child with severe inflammatory bowel disease uncovered a mutation in an important regulator of inflammation, X-linked inhibitor of apoptosis (XIAP). Based on the severity of the child's symptoms as well as the molecular diagnosis, a bone marrow transplant was given to the patient, which subsequently reduced his symptoms.
7. Cancer genome sequencing: The Cancer Genome Atlas and International Cancer Genome Consortium performed sequencing on thousands of cancer and normal cell pairs. These studies thus provide a better understanding of cancers at molecular level. **Limitations of HTS:** Although HTS has a lot to offer, its limitation includes low hit rate due to incompatible libraries and potential false-positive results.

14.8.2 Whole Genome Sequencing

The genome of each individual organism contains its entire genetic information. Whole genome sequencing is a powerful technology that helps researchers to obtain the entire genetic information of the genome and reveals the complexity and diversity of the genome. Whole genome sequencing can detect variants including single nucleotide variants (single nucleotide polymorphism, SNPs), and insertions, deletions, and copy number changes are large-scale structure variance. Whole genome sequencing can be divided into de novo sequencing and resequencing based on whether there is a reference genome. A reference genome can make genome assembly easy and rapid.

Methods of Whole Genome Sequencing: In the early 1980s, Sanger successfully completed the genome sequencing of the lambda phage by using shotgun method, and the method was successfully applied to large viral DNA, organelle DNA, and bacterial genome. Shotgun sequencing is the classic strategy for genome sequencing.

Shotgun strategy provides a technical guarantee for large-scale sequencing. This technology first randomly interrupts a complete target sequence into small fragment sequence, separating and then splicing them into a consistent sequence by using the overlapping relationship of this small fragments. For large genomes, it mainly includes two methods: the hierarchical shotgun sequencing with the clone-by-clone method and the other is the whole genome shotgun sequencing.

Clone-by-Clone Method: This method was once adapted by the Human Genome Project consortium, and this method can generate high-density maps, making the genome assembly easier. It generally includes four steps:

- Preparation of BAC clone library.
- Preparation of clone fragment.
- BAC clone sequencing.
- Sequencing assembly.

However, this method is time-consuming and costly, so it is seldom used at present.

Whole Genome Sequencing

This generally involves four steps: random fragmentation of genomic DNA, size selection using electrophoresis, library construction, and sequencing sequence genome assembly. Two different sizes of DNA fragments including longer insert of 2–2.5 kb and short insert of 0.5 to 1.2 kb are selected from the agarose gel. The longer inserts are cloned in phage or cosmid vectors. The short inserts are cloned in plasmid vectors.

Advantages of Whole Genome Sequencing

- Do not require genome map.
- Less time-consuming.
- Cost-effective.

Disadvantages Include

- Genome assembly of eukaryotic genome is difficult to abundant repetitive sequences.
- Genome sequence using this method is not a reliable source and might produce false results.

Advanced Sequencing Technologies Promote Whole Genome Sequencing

In this method, genomic DNA is first randomly fragmented using sonication or nebulization. They are then ligated to the platform specific to sets of double-stranded adapters to generate shotgun library. Subsequently, these library fragments are sequenced by next-generation sequencing instruments. All NGS instruments utilize microfluidic device to contain the amplified fragment of the shotgun library followed by an imaging step that collects data from fragments being actively sequenced. The data analyses workflow for whole genome sequencing includes raw read quality control, data preprocessing, alignment, variant calling, genome assembly, genome annotation, and other analyses. NGS instruments require the PCR amplification step before sequencing, while minION and PacBio RSII rely on single molecule for sequencing. This sequencing technology omits the amplification step and can perform direct sequencing. Although NGS has enabled population scale analyses of small variants, it is difficult to identify larger structural variations. De novo assembly using next-generation sequencing is often at lower quality compared to early methods. The single-molecule sequence technology can get over these difficulties which span nearly the entire chromosome and not sensitive to GC content. These have been used to produce highly accurate *de novo* and reference assemblies for microorganisms, plants, animals, and humans, enabling new insights to revolution and sequence diversity.

Box 14.1: Scientific Concept: Efficient Method for Site-Directed Mutagenesis in Large Plasmids Without Subcloning – Louay K. Hallak et al.

Conventional site-directed mutagenesis (SDM) techniques allow easy manipulation of DNA sequence in small plasmid; however, introduction of mutation in larger plasmids is often considered a daunting task. This is so because introduction of mutation in larger plasmids involves the necessity of subcloning. Subcloning is a time-consuming and labor-intensive technique. In addition to this, subcloning method is not reliable for large plasmids. A relatively new DNA mutagenesis technique has been developed in this regard and named URMAC (Unrestricted Mutagenesis and Cloning). The basis of this technique is to use quick biochemical reactions and replace it with conventional subcloning technique. UMRAC allows the use of a wide range of plasmid size and allows simultaneous introduction of multiple mutations. UMRAC uses two PCR reactions each followed by ligation step that allows it to circularize the product with an optional third enrichment PCR step which is followed by traditional cloning step that requires two restriction sites.

URMAC Method

The basic principle involved in this method is to convert a linear PCR product generated from a plasmid template to a circular DNA that can be opened at second site. This involves the amplifications with primers that contain required mutations circularized by religation again and amplified with original primers to replicate the DNA that contains the required mutations. URMAC uses a set of two primers, namely, the starter primers and the opener mutagenic primers. These primers are phosphorylated at the 5′ end so that they can participate in the ligation step.

Starter primers are used twice, firstly in the initiation step that amplifies the modification target sequence and secondly in the enrichment step.

The open primers are used to introduce the mutation of interest.

The URMAC process requires six sequential steps (Fig. 14.33) that include a PCR reaction followed by ligation followed by another PCR reaction. This is followed by another ligation which is often succeeded by another PCR enrichment step. The final step in this procedure involves digestion with restriction enzymes (Fig. 14.34).

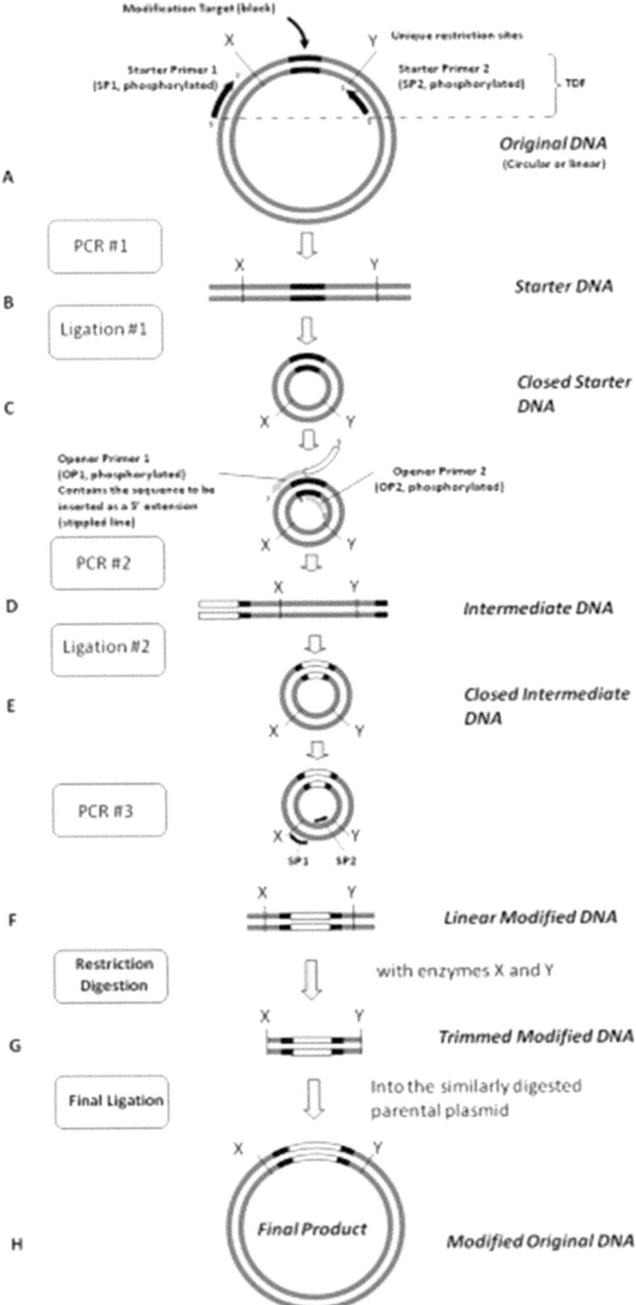

Fig. 14.33 A schematic diagram representing the process of URMAC method. This diagram represents an imaginary insertion within the modification target (black lines) in the original DNA plasmid. The first PCR generates the starter DNA copy of the modification target that is positioned between specific restriction sites X and Y in this figure. This is achieved by the use of thermostable DNA polymerase using starter primers S1 and S2 represented by black arrows. T4 DNA ligase then

URMAC offers the following advantages over other conventional SDM techniques:

1. PCR reactions have a higher success rate in small fragments rather than full-length plasmids.
2. URMAC tends to avoid the errors of introducing polymerase errors as reported in the QuickChange Method and inverse PCR method because the fragment being amplified is much smaller.
3. URMAC do not require sequence verification as any region of plasmid that is not the direct target of DNA mutagenesis remained unaffected by DNA polymerase.
4. URMAC is very fast as compared to traditional SDM subcloning techniques with an average of a single day to complete URMAC and an additional 3 days to clone the final product into the original plasmid. It's quite fast when compared to 3–4 weeks required for subcloning.
5. This technique can reduce the challenge of high GC-containing plasmids by avoiding PCR amplification of those parts of plasmids. URMAC is cost-effective when compared to subcloning as it requires less labor and materials.
6. It also offers versatility for handling any combinations of deletions, additions, and substitutions.

Box 14.2: Scientific Concept: The pPSU Plasmids for Generating DNA Molecular Weight Markers

Nucleic acid visualization by gel electrophoresis is one of the most common techniques in molecular biology. The visualization aims to know the exact size of the DNA bands, and molecular weight markers or ladders are commonly used for this purpose. With the advancement in the basic understanding of plasmid creation, a group of researchers constructed a pair of cost-effective plasmids. The pPSU1 and pPSU2 pair of molecular weight marker plasmids can produce both 100 bp and 1 Kb DNA ladders when digested with two common restrictions, enzymes Pst1 and EcoRV, respectively. The 100 bp ladder fragments have been optimized in such a way that they can be migrated appropriately on both agarose and native polyacrylamide. The pPSU molecular weight marker plasmids were constructed in such a way to provide low

(continued)

Fig. 14.33 (continued) circularizes the starter DNA in the next step that forms closed starter DNA. The closed starter DNA acts as a template for the next PCR cycle that uses opener primers OP1 and OP2 yielding a mutated intermediate DNA. OP1 incorporates an insertion mutation that have the sequence of interest attached to its 5′ terminal end. The intermediate DNA is circularized by the aid of T4 DNA ligase. The SP1 and SP2 primers are used in the enrichment step amplifying the linear modified DNA. The linear modified DNA and the original plasmid are digested with restriction endonuclease that cleaves at unique sites X and Y. The appropriate fragments are joined to produce the modified original DNA

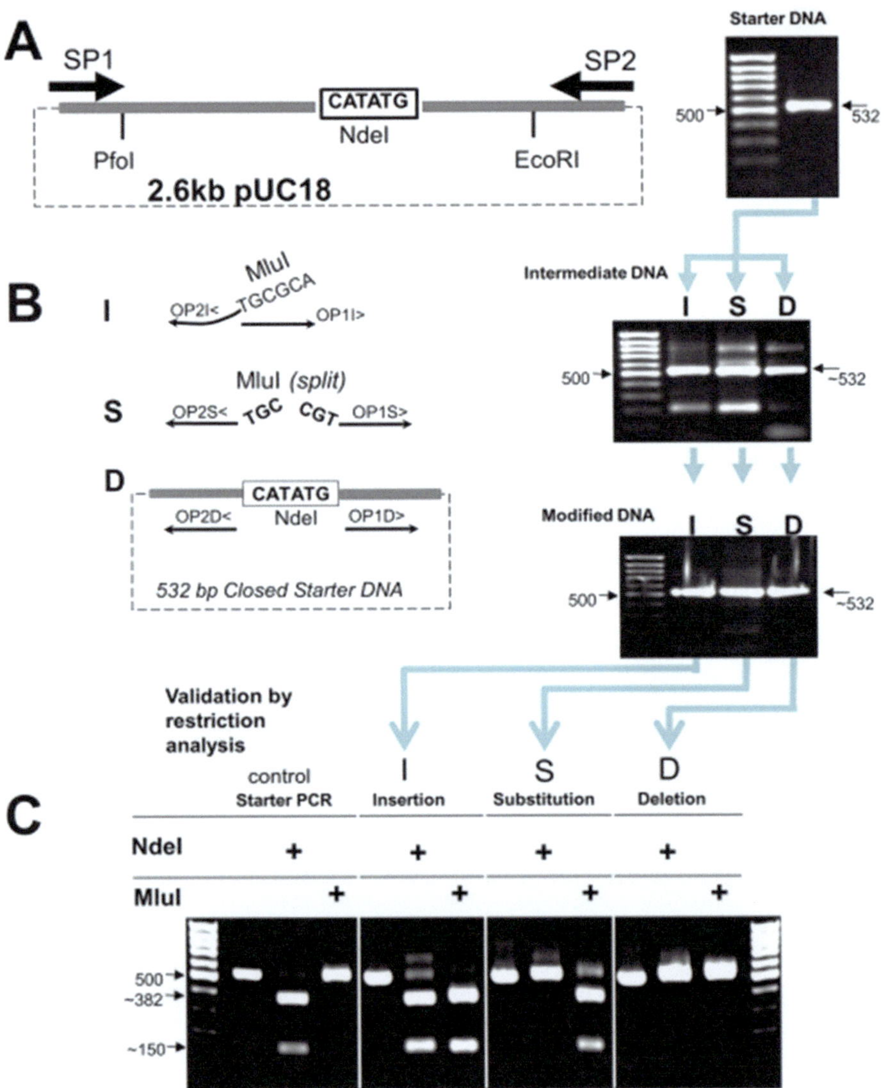

Fig. 14.34 A schematic diagram representing the validation of URMAC method by insertion, substitution, and deletion of some restriction site in pUC 18 plasmid. An illustration (**a**) showing modification target relative to the positions of restriction sites that occur between the starter primers S1 and S2. As a result of first PCR reaction, the starter DNA migrated as expected, 532 bp on a 1% agarose gel. A DNA ladder of 100 bp size is shown for comparison. (**b**) is showing a diagram that introduces different kinds of mutations using a closed starter DNA from a PCR product. **a** is used as a template and is used to create opener/mutagenic primers. The top picture at the right corner shows intermediate DNA which contains mutations. The figure at the bottom shows modified DNA after enrichment with SP1 and SP2 primers. (**c**) Validation of URMAC mutagenesis for the three different types of mutations by restriction analysis

Box 14.2 (continued)

molecular weight markers when digested with Pst1 restriction enzyme. Once digested with Pst1, both plasmids produce a ladder of 50, 100, 200, 300, 400, 500, 600, 700, 800, 900, 1000, 1500, 2000, and 4100 bp reference fragments.

An additional site of 500 bp Pst1 fragment is deliberately added to provide a visual landmark for the fragment on the electrophoresis gel. EcoRV digestion of the same pair of plasmids produced a 1 Kb ladder which gives 500, 750, 1000, 1500, 2000, 3000, 4000, and 5000 bp reference fragment by making the 10 kb pPSU plasmid linear. The pPSU1 and pPSU2 plasmids can be digested individually with appropriate restriction enzyme, or they can be combined and digested in a single reaction. These two plasmids can be linearized with Nco1 and BglII to yield bands that can be up to 10 kb in size. Ppsu1 alone can produce 0.5, 1.0, 1.5, 2.0, 3.0, 5.0, 7.0, and 10 kb fragments by combining EcoRV, EcoRI, and Nco1 digest. The plasmid hence digested can empower us with large number of bands to be visualized simultaneously. In addition to this, these plasmids also confer ampicillin resistance and can be grown on standard E. coli strains and media which adds to its benefits of being a cost-effective and multipurpose use plasmid.

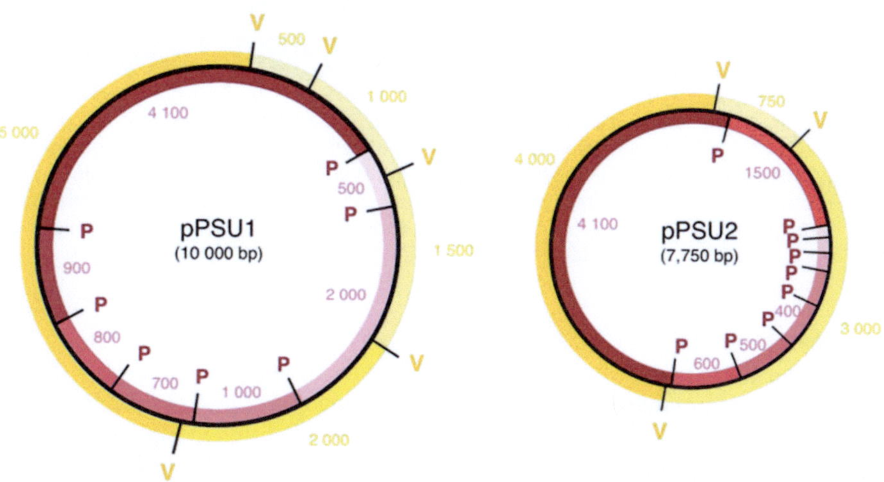

Fig. 14.35 A figure representing the plasmid maps of pPSU1 and pPSU2. A nesting approach is used to create a 64 kb of 100 bp and 1 kb ladder fragments generated by the use of Pst1 enzyme of each plasmid shown in red and 1 Kb fragments produced by EcoRV digestions. pPSU1 contains the 500, 700, 800, 900, 1000, 2000, and 4100 bp PstI fragments, and 500, 1000, 1500, 2000, and 5000 bp EcoRV fragments. pPSU2 contains 50–600, 1500, and 4100 bp PstI fragments, and 750, 3000, and 4000 bp EcoRV fragments

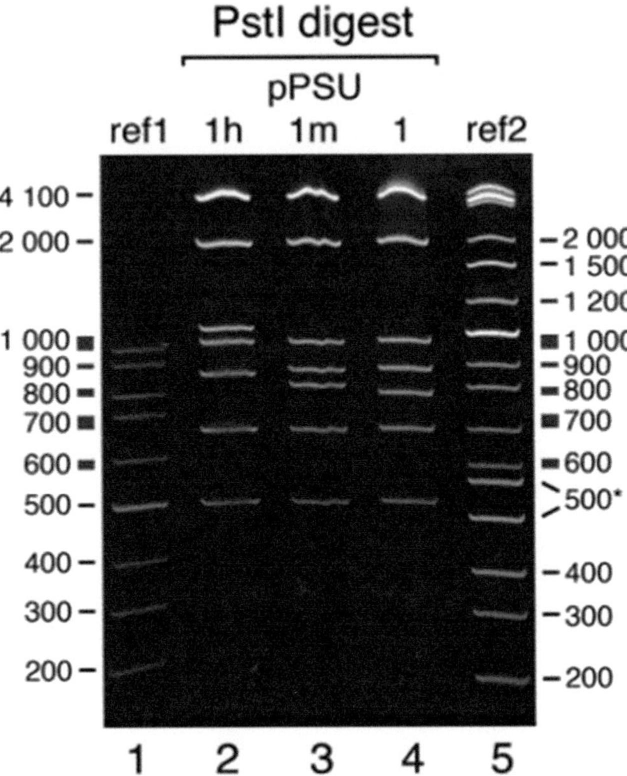

Fig. 14.36 The figure here represents an actual ladder on a polyacrylamide gel (10%) after electrophoresis. Lane 1: reference Thermo Scientific GeneRuler 100 bp ladder (ref1). Lane 2: PstI digestion of the intermediate pPSU1 plasmid contains 800, 900, and 1000 bp lambda DNA fragments which migrate anomalously slowly. Lane 3: pPSU1m contains replacements for the 800 and 900 bp fragments from lambda DNA and the 1000 bp fragment from the human Bmi1 RING domain gene. Lane 4: PstI digestion of the final pPSU1 plasmid containing a replacement 800 bp fragment from the human G9a histone methyltransferase gene. Lane 5: reference New England Biolabs 2-Log DNA ladder (ref2)

Figures 14.36, 14.35, 14.36, 14.37, and 14.38

The above three figures from 14.36 to 14.38 are the representatives of different ladders that are produced by the restriction endonuclease digestion under different conditions. The results of the ladder generated with different set of restriction enzyme are compared with the standard (Thermo Scientific GeneRuler) in this case. This helps easy identification of the DNA segment and thus helps in determining the actual size of the DNA segment.

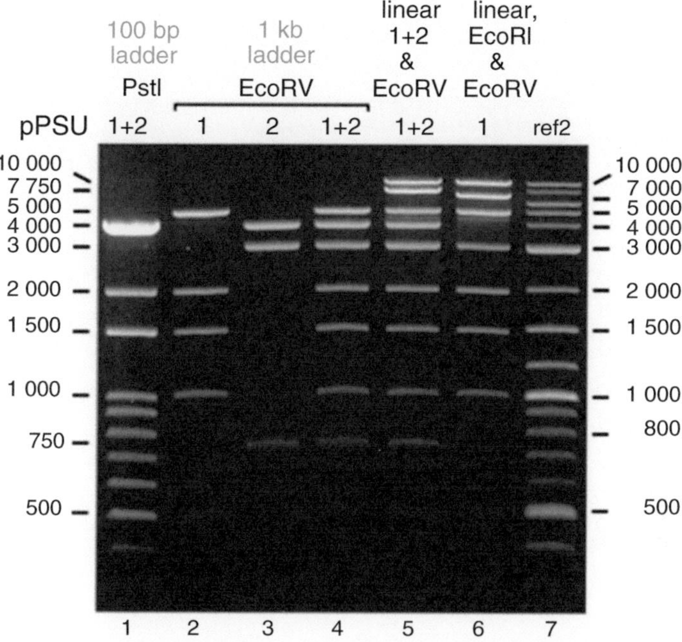

Fig. 14.37 This is an actual representation of the ladder in Lane 1: when the pPSU plasmids (pPSU1 and pPSU2) is cleaved by Pst1 restriction enzyme. Lane 2: pPSU 1 digested with EcoRV. Lane 3: pPSU 2 digested with EcoRV. Lane 4: shows when both the plasmids are joined and digested simultaneously. Lane 5: linear pPSU plasmids (pPSU1 and pPSU2) digested with EcoRV. Lane 6: linear set of plasmids digested with EcoRV and EcoRI. Lane 7: represents the reference ladder that is used to know the exact size of DNA

14.9 Summary

- Sequencing is a process that determines the correct order of nucleotides in a piece of DNA.
- The sequencing technology was introduced by Fedrick Sanger in mid-1970s (1977) with the aim to sequence genomes of small organisms.
- His technique used chain termination strategy also known as dideoxy method which uses dideoxynucleotide molecule which stops the elongation reaction. This generates a variety of different sized fragments that are detected radiolabelly to sequence a DNA fragment.
- However, the chain termination method of sequencing had its own limitations. It was a very expensive technology to use. The sequencing process was often very time-consuming, and sometimes, these sequencing produced false-positive results.
- All these limitations lead to the advancements of next-generation sequencing technologies.

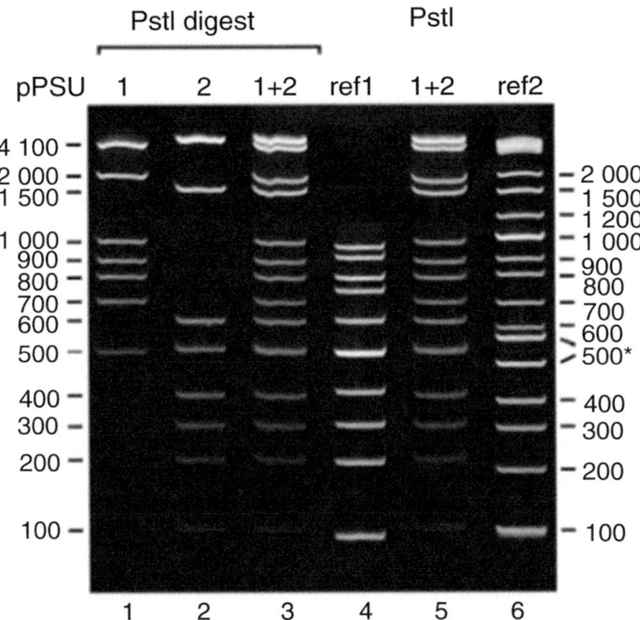

Fig. 14.38 The ladder pattern is shown in this diagram. Lane 1: pPSU1 digested with Pst1. Lane 2: pPSU 2 digested with Pst1. Lane 3: pPSU plasmids (pPSU 1 and pPSU 2) digested with Pst1. Lane 4–6: Thermo Scientific GeneRuler lane

- The next-generation sequencing technologies used a set of advanced machines that produced better sequencing results at a faster rate and were cost-effective.
- Some recent advancements in next-generation sequencing technologies resulted in formation of high-throughput sequencing.
- The use of high-throughput sequencing made human health and disease simpler to understand, thus creating a branch of personalized medicine and pharmacogenomics.
- Whole genome sequencing provides a thorough insight of the genome-wide studies that could lead to a new era of medical and health research.
- DNA sequencing developed around the 50s is now an advanced area of study and research and can potentially reveal the secrets of biology.

Genomics 15

Sai Krishna AVS, Sonali Patle, Parampreet Kaur, Shama Omkumar, and Aarti Sharma

15.1 Overview of Genomic Analysis

The field of genomics is interdisciplinary, and it is defined as the study of genes present in an organism along with the various techniques involved. It studies the functions, structures, mapping, and various techniques that are used to edit genomes. Genome, on the other hand, is defined as the information repository of an organism. It is the entire set of genetic information that is important for the proper functioning of the living being over its period of life. Genomes include the genes that are coding in nature as well as the noncoding DNA, mitochondrial DNA, as well as the DNA of the chloroplast.

The word genome was first formulated in 1920 to explain "the haploid chromosome set, which, together with the pertinent protoplasm, specifies the material foundations of the species" (Fig. 15.1). The discovery of Mendelian genetics was done in the year 1900, and the identification of chromosomes as the carrier of genetic information was done in the year 1902. The fact that DNA carries genetic information did not come into light till the late 1920s.

When referred to the Google ngram analysis, it states the case-dependent episodes of the words "gene," "genome," and "chromosome" in the aggregation of books written in English from 1920 to 2008.

Genome analysis necessitates the prediction of genes, and a vast range of both animals and plants are sequenced, including microorganisms (Table 15.1), but the

S. KrishnaAVS (✉) · P. Kaur · S. Omkumar
Ramaiah University of Applied Sciences, Bangalore, India

S. Patle
Ajay Kumar Garg Engineering College, Ghaziabad, India

A. Sharma
NIMHANS, Bangalore, India

D. Kar, S. Sarkar (eds.), *Genetics Fundamentals Notes*,
https://doi.org/10.1007/978-981-16-7041-1_15

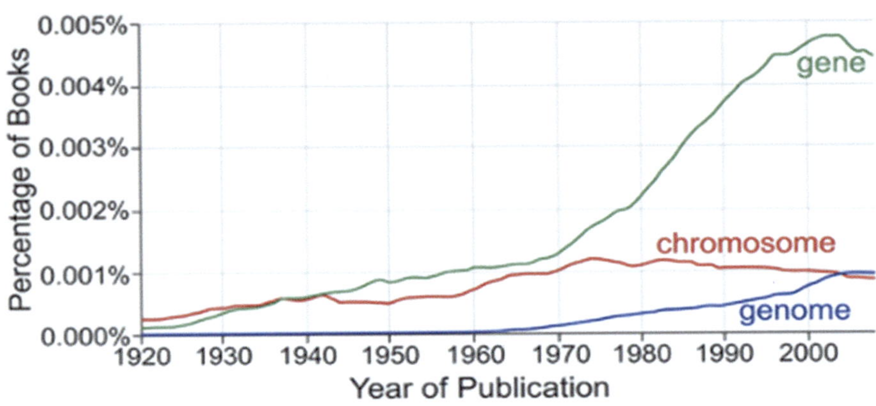

Fig. 15.1 The interchange of the word "genome" compared to similar expressions.

Table 15.1 Estimated sizes of certain genomes and the number of genes in them (SCFBio, IIT Delhi)

Species	Genome size (mb)	Number of genes
Mycoplasma genitalium	0.58	500
Streptococcus pneumoniae	2.2	2500
Escherichia coli	4.6	4400
Saccharomyces cerevisiae	12	5800
Caenorhabditis elegans	97	19,000
Arabidopsis thaliana	125	25,500
Drosophila melanogaster	180	13,700
Oryza sativa	466	45–55,000
Mus musculus	2500	29,000
Homo sapiens	3300	27,000

pace at which genome annotation is done is not at par with the pace of genome sequencing. There is a high demand for making computational tools to predict genes. The main question that fascinates scientists regarding the genome is to identify what part of the genome codes for proteins and what part of it is junk and the classification of junk DNA.

To successfully sequence a genome, firstly, an organism is selected and the sequencing of DNA is done, followed by the assembly of the sequence (also known as sequence compilation) to represent the actual chromosome, and finally, annotation and analysis of the sequence are done.

DNA Sequencing

The first step in exploring the genome is to determine the DNA sequence. The term refers to the biochemical methods utilized to determine the accurate representation of adenine, guanine, cytosine, and thymine, respectively.

There are three main types of methodologies that are used for sequencing DNA:

1. Chain termination method.
2. Chemical degradation method.
3. Pyrosequencing method.

Chain Termination Method

This method was developed by Frederick Sanger and his coworkers in the year 1977. It is also called the dideoxy method, where the complementary DNA is produced by the DNA polymerase along with dNTPs and ddNTPs. The termination of the reaction occurs at random sites due to the addition of ddNTP with the oligonucleotide chain.

Chemical Degradation Method

The establishment of the procedure was done by two scientists named Allan Maxam and Walter Gilbert in the year 1977. The sequence of a dsDNA molecule is determined following chemical treatment. The chemicals make cuts at specific nucleotide sequences.

Pyrosequencing

In the year 1996, Mostafa Ronaghi and Pal Nyren invented the procedure of pyrosequencing. The procedure necessitates the addition of a deoxynucleotide to the end of the growing strand, and this is detected due to the emitted light.

DNA Sequencing Methods

High-Throughput Sequencing Methods

A wide range of methods called next-generation or second-generation sequencing methods were invented in the later part of 1990s and implemented in the 2000s. All of the methods were named second generation or next generation to create a clear differentiation from the previous techniques used for sequencing including Sanger sequencing. This includes next-generation short-read and the second-generation long-read sequencing methods.

Long-Read Sequencing

1. Single-molecule real-time sequencing—This is a synthesis-dependent approach where DNA is produced in zero-mode waveguides by using an unmodified polymerase and freely floating fluorescently labeled nucleotides.
2. Nanopore sequencing—This approach works by monitoring changes to an electric current when the nucleic acid is passed through the nanopore. The resulting signal is then decoded to provide the specific sequence.

Short-Read Sequencing

1. Polony sequencing—In 2005, this technique was first used to sequence the genome of *Escherichia coli*. This involves an automated microscope, in vitro coupled tagged library, and ligation-based sequencing chemistry.

2. 454 pyrosequencing—It was invented by "454 Life Sciences," after which "Roche Diagnostics" took its possession. This technique necessitates utilization of emulsion PCR and pyrosequencing.
3. Illumina (Solexa) sequencing—This method was developed by Shankar Balasubramanian and David Klenerma, and it involves the use of polymerases that are engineered and reversible dye terminators technology.
4. SOLiD sequencing—This procedure was invented by Applied Biosystems. It was an approach that used sequencing by ligation.
5. Ion Torrent sequencing—The technique is developed using the basic chemical reactions in sequencing, although it has a unique detection system which utilizes semiconductors.
6. DNA nanoball sequencing—This technique amplifies the little pieces of genetic material to a DNA nanoball utilizing rolling circle replication mechanism, and it is followed by sequencing by ligation to determine the nucleotide sequence.

15.1.1 Sequence Compilation

The second step in genome analysis is the assembly or compilation of the fragment of DNA to recreate the original sequence. The process of DNA sequencing cannot read the whole-genome fragment at once. Instead, it reads shorter fragments around 50–30,000 bp long, and the size of the fragment depends on which technology of sequencing is being used.

To explain what sequence compilation means, take the example of a paper being shredded and rejoined again to form the original one by just looking at the shredded pieces.

History

The very first assemblers or compilers were developed in the late 1980s and 1990s. They were developed to join the large number of fragments generated by using automated sequencers.

Scientist faced tons of problems to sequence the first eukaryotic genome—*Drosophila melanogaster* in 2000 and a year later techniques such as Celera assembler and Arachne. These assemblers can handle genomes from 130 million to 3 billion base pairs in size.

The basic definition of sequence compilation is to align and merge fragments, and there are two approaches to assemble the genome.

i. Mapping and assembly—If a previously sequence genome is available as a reference, then the newly sequenced genome is first plotted by referencing genome by alignment, followed by assembling in a proper order.
   Example: Bowtie—It is a read aligner which maps and assembles the genome.
ii. De novo assembly—When there is no reference genome available, the de novo approach of assembling is used as explained in Fig. 15.2. In this technique, paired reads are used instead of single reads as they help in generating scaffolds.

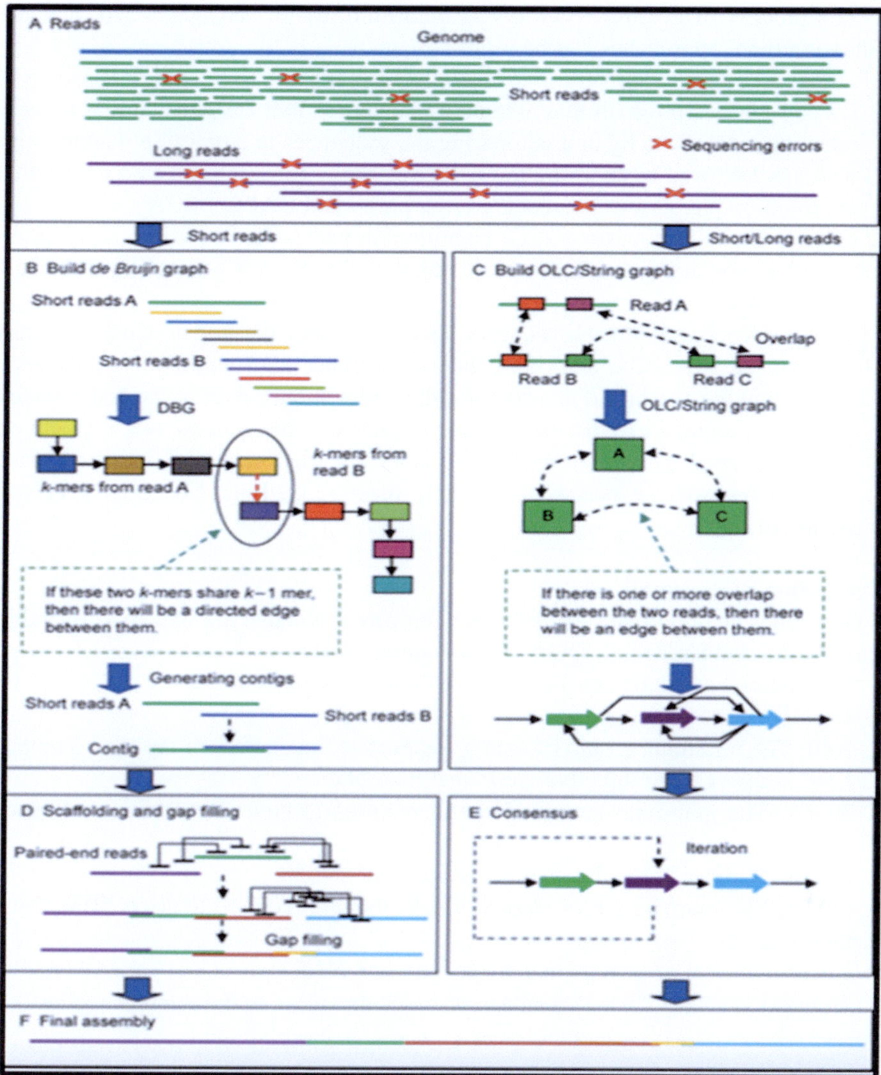

Fig. 15.2 The illustration of the pipeline of de novo assembly (Liao et al. 2019). "The subgraph (**a**) shows all reads. The subgraph (**b**) shows the principle of building de Bruijn graphs. The subgraph (**c**) shows the principle of building OLC/String graph. The subgraph (**d**) shows the principle of scaffolding and gap filling. The subgraph (**e**) shows the consensus operation. The subgraph (**f**) shows the final genome sequence"

Sequence assembly is a systematic approach where the "assembly of the sequence reads into contigs, assembly of the contigs into scaffolds (supercontigs), and assembly of the scaffolds into chromosomes."

The process of genome assembly is algorithm driven and automated, and the following three approaches can be used:

1. Greedy—It is a program that assembles sequences that are most similar to each other, and it does so by first comparing the sequences in a pairwise manner with overlaps followed by merging the best overlaps. The gaps generated due to this are filled by paired-end sequencing (e.g., phrap and CAP).
2. Overlap-layout-consensus (OLC) Hamiltonian path—This approach is based on pairwise comparisons, and after the comparison, a graph is generated using reads and overlaps.
   The graph represents each sequence as a node, and the nodes which overlaps generate an edge. The algorithm then determines the Hamiltonian transversal pathway in the graph, and this contains the nodes and the overlaying nodes which are then combined together to form the sequence of the genome (e.g., Arachne).
3. De Bruijn graph and Eulerian path—It is mostly used in assembling short reads but also has some experience with long reads (e.g., Euler-SR, Oases, Velvet, ALLPATH)

Evaluation of Assembly

Evaluation is an important step as this will inform us whether the sequence assembly has met the standards (e.g., QUAST, evaluation tool).

The following criteria can be used to evaluate assemblies:

1. **N50:** The minimum length of contig required to cover 50% of the total length.
2. **L50:** Number of contigs that are longer than N50.
3. **NG50:** The minimum length of contig required to cover 50% of the reference genome.
4. **LG50:** Number of contigs that are longer than NG50.
5. **NA50:** Minimum length of aligned blocks that are required to cover 50% of the total length.
6. **LA50:** Number of contigs that are longer than NA5.
7. **Genome fraction (%):** Percentage of bases that align to the reference genome.

15.1.2 Sequence Annotation

Annotation is defined as the procedure which marks the genes in order to identify their functions, locations, and the coding regions, and the steps are shown in Fig. 15.3.

The process of sequence annotation comprises of three steps:

1. Recognition of the part inside the genome which is noncoding for a protein.
2. Recognition of genome elements.
3. Addition of elements and the biological information.

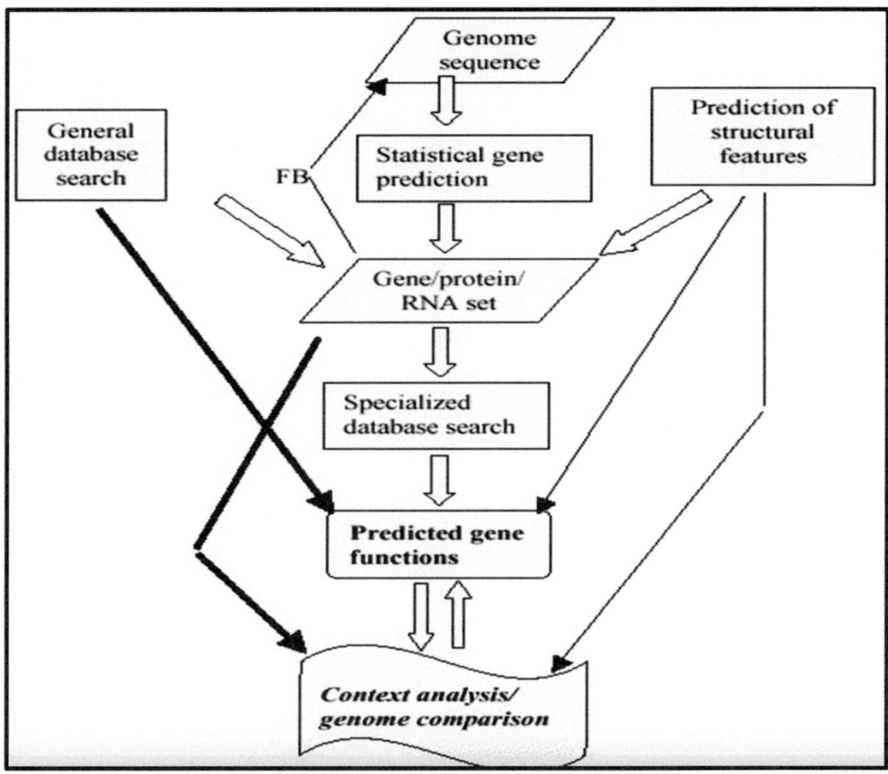

Fig. 15.3 A genome annotation flowchart. The flowchart represents the steps in genome annotation, locating the positions of all the genes, feedback from gene identification, general database search, specialized database search, statistical gene prediction, and prediction of structural features

These steps are performed by automatic annotation tools via computer analysis. One such very simple method which is most widely used is BLAST. A BLAST search for homology, and that data helps in annotating the genomes and genes. Any discrepancies can be handled by manual annotators. A few other annotations can be done via a database that utilizes the context information of the genome along with similarity scores and experimental information through their various subsystems approach.

Annotation can be broadly classified into two types:

1. Structural.
2. Functional.

The structural annotation part deals with recognition of elements in the genome such as:

- The localization of ORFs.
- Gene structure.
- Coding locations.
- Regulatory motifs regions.

The second part involving functions deals with addition of genomic elements and the biological information such as:

- Information regarding biochemical function.
- Information regarding biological function.
- The required interactions and regulation.
- Expression.

A wide range of ongoing projects are going on related to annotating the genome, for example, ENCODE, Entrez gene, Ensembl, GENCODE, and GeneRIF.

*Project **GeneQuiz** was the first automatic system for genome analysis. It performed similarity searches followed by automatic evaluation of results and generation of functional annotation.*

15.2 Functional Genomics

This branch aims in determining the functions and interactions of genes. It uses the huge amount of data that is generated by a variety of transcriptome and genomic projects.

The main focus of functional genomics is transcription, translation, and gene expression along with their protein-protein interactions.

The function genomic analysis measures the various changes in DNA, i.e., the genome as well as the epigenome, RNA, and different interaction between protein, DNA, and RNA which in turn influence the sample's phenotype.

The branches of functional genomics are:

1. Genotyping
2. Transcription profiling.
3. Epigenetic profiling.
4. Nucleic acid-protein interactions.
5. Meta-analysis.

Techniques

A variety of techniques are used at DNA, RNA, and protein level, and these techniques are as shown in Fig. 15.4.

These techniques at different levels are defined as follows:

- At the DNA level, genomics and epigenomics.
- At the RNA level, transcriptomics.

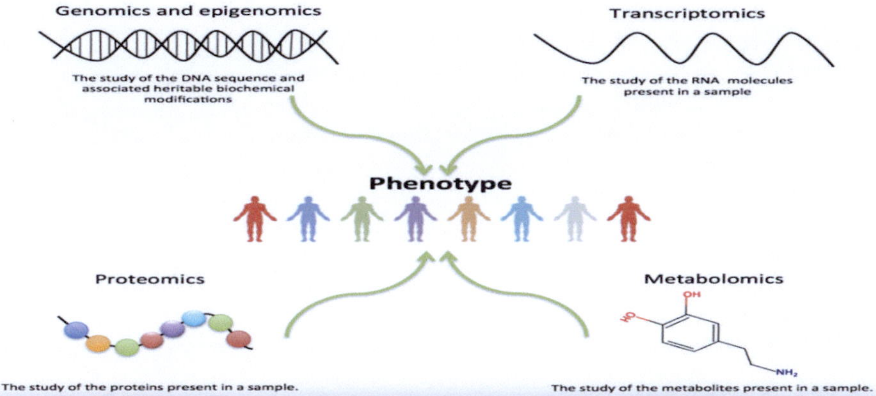

Fig. 15.4 Functional genomics is the study of how the genome, transcripts (genes), proteins, and metabolites work together to produce a particular phenotype (EMBL-EBI)

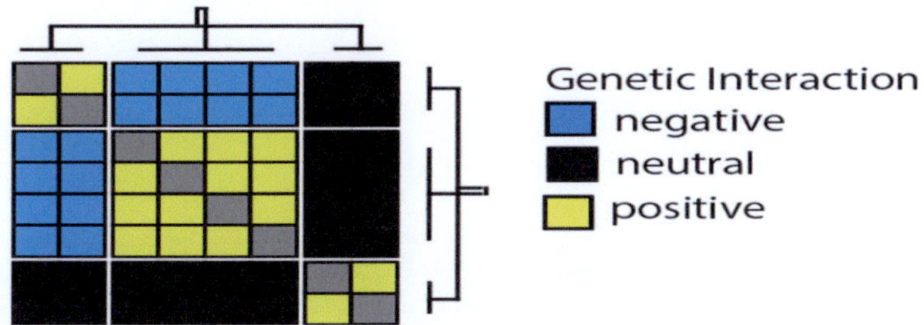

Fig. 15.5 Genetic interaction mapping. GI mapping entails perturbing genes in pairs (e.g., knockout, knockdown, or overexpression) to see how one gene influences the phenotype of the other. Cell viability is commonly used as a phenotypic readout, with GIs that increase cellular fitness being labeled "positive" and GIs that decrease cellular fitness being labeled "negative" (Krogan Lab, UCSF)

- At the protein level, proteomics.
- At the metabolite level, metabolomics.

Genetic interaction mapping—A systematic approach for deleting genes or inhibiting their expression is one way to identify genes with related functions even if there is no interaction between them physically.

The technique involves pairwise perturbation (e.g., knockout, knockdown, or overexpression) of genes in order to determine how the genes modulate each other's phenotypes as explained in Fig. 15.5.

DNA/protein interactions—Techniques such as ChIP sequencing and CUT&RUN sequencing are developed in order to identify the interaction between DNA and protein, and this is due to the fact that proteins help in regulating expression of genes (Fig. 15.6).

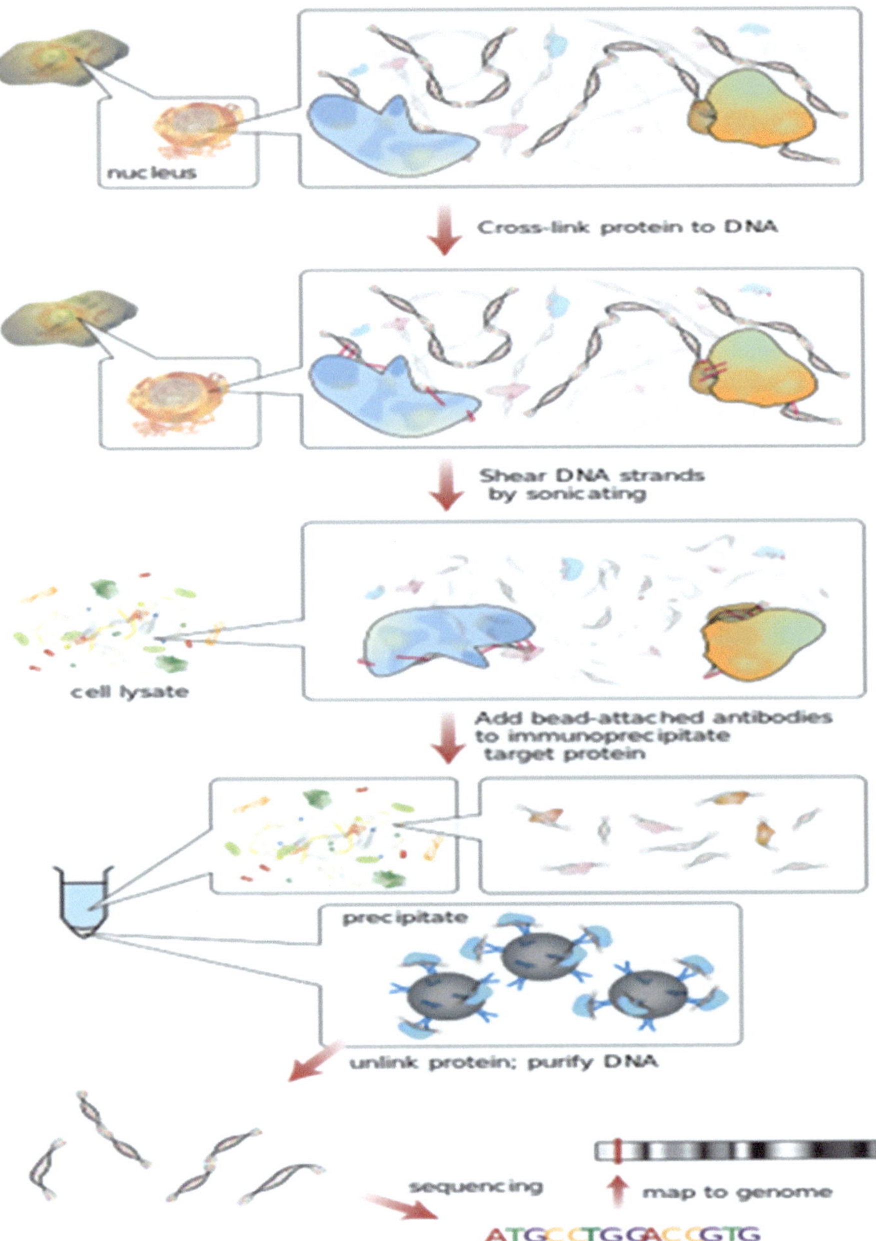

Fig. 15.6 ChIP sequencing workflow**.** ChIP-Seq can be used to map global binding sites for a protein by identifying the binding sites of DNA-associated proteins. Cross-linking of DNA-protein complexes is usually the first step in ChIP-Seq. The fragmented samples are then treated with an exonuclease to remove any unbound oligonucleotides. The DNA-protein complex is immunoprecipitated using protein-specific antibodies. The DNA is extracted and sequenced, yielding high-resolution protein-binding site sequences

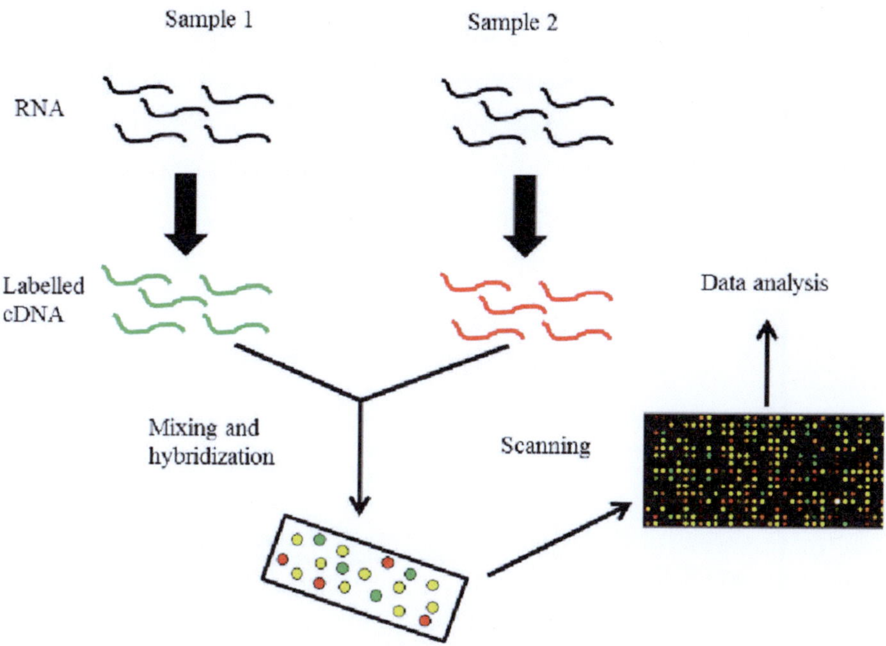

Fig. 15.7 An overview of DNA microarray technology. RNA is isolated from the control and the target samples, and labeled cDNA is then hybridized (Muhammad Afzal, Researchgate)

DNA accessibility assays—ATAC-seq, DNase-Seq, and FAIRE-Seq are few such assays that help in identifying the accessible regions on the DNA such as open chromatin region, etc.

Microarray—It helps in detection of a thousand gene expressions at a single time. They are microscopic slides with thousands of spots indicating a defined position, and each spot contains a known amount of DNA or gene (Fig. 15.7).

SAGE—It stands for Serial Analysis of Gene Expression. This approach is established on RNA sequencing as explained in Fig. 15.8. The transcriptomic technique generates a small overview of mRNA present in the sample in the form of small tags, and these tags correspond to fragments of those transcriptomes.

From the desired samples, RNA is extracted and purified. Reverse transcriptase uses biotinylated poly-T tails as primers to transform mRNA to double-stranded cDNA. Anchoring enzymes recognize and cut at their respective restriction sites, which are found naturally every few hundred base pairs, to fragment biotinylated cDNA. The fragmented cDNA's biotinylated ends are bound to streptavidin beads and separated into two parts. At the cleaved end of each cDNA fragment, appropriate adaptors with an endonuclease site for tagging enzymes are ligated. Tagging enzymes are added to release fragmented cDNA (up to 26 bp) from the beads by cleaving a small distance from their endonuclease sites found at each adaptor. The blunt ends of each of the two fragmented cDNA segments are ligated together to form a ditag, which is then amplified by PCR. Ditags are cleaved at their adaptors,

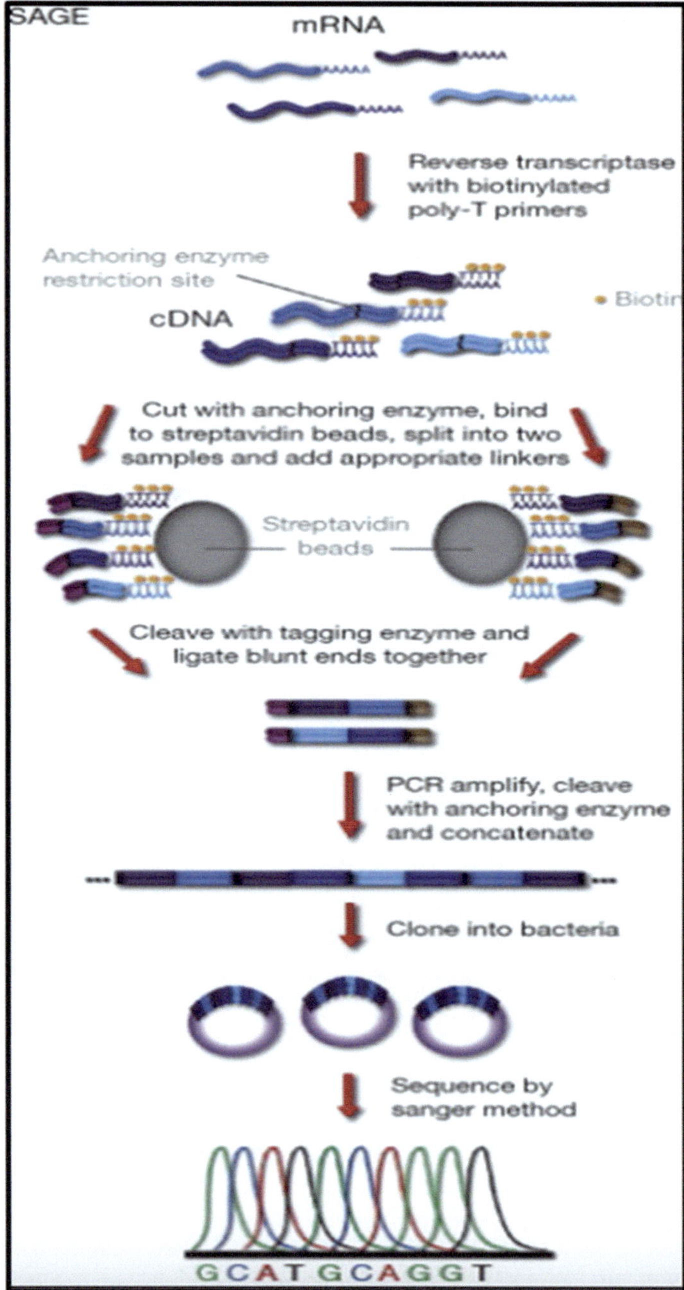

Fig. 15.8 SAGE. Gene expression analyses in sequence (SAGE)

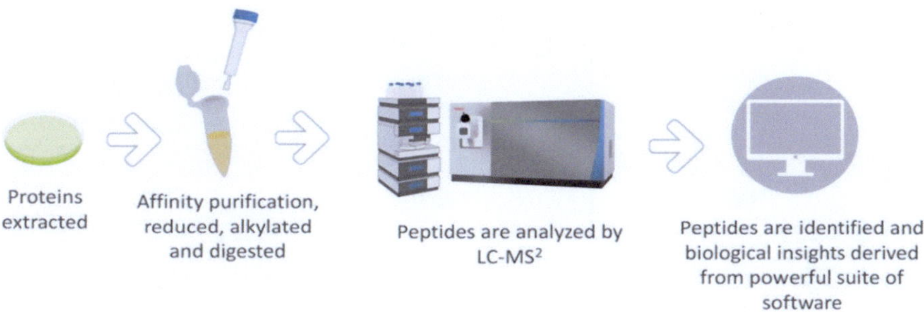

Fig. 15.9 Affinity purification and mass spectrometry (Thermo Fisher Scientific)

concatenated, and cloned into plasmids following PCR. Plasmids are sequenced after cloning, mapped back to the gene, and then analyzed for gene expression.

RNA sequencing—This is the most efficient way to study gene expression and transcription; it is widely used as compared to microarray and RNA sequencing. In this technique, the quantity and sequence of RNA can be analyzed using next-generation sequencing.

Yeast two hybrid systems—In order to determine protein-protein interaction, this tests one "test" protein against a variety of potential proteins. This approach is mostly based on the GAL-4 transcription factor. If two proteins are being tested, one of their genes will bind with the DNA binding domain of GAL-4 (GAL4-DBD), while the other will bind with the activation domain of GAL-4 (GAL4-AD).

Affinity purification and mass spectrometry—This technique is used to determine the protein that interacts with one another, present in complexes. The affinity enrichment approach uses specific binding property between molecules to successfully isolate the protein of interest as explained in Fig. 15.9.

Using the proximity biotinylation approach, affinity purification mass spectrometry can be used to look at particular protein-protein interactions within protein complexes or to look at protein complexes more broadly at the interactome level. Interactions can be studied under different conditions by combining affinity purification with quantitative MS, resulting in a much more dynamic view of protein-protein interactions. This workflow also allows for PTM analysis, allowing researchers to investigate the role of posttranslational modifications (PTMs) in facilitating protein-protein interactions.

RNAi—This method is being used to silence or knock down genes. It uses siRNA or shRNA.

CRISPR Screens—As explained in Fig. 15.10, it deletes genes in a multiplexed manner and quantifies the amount of guide RNA in order to determine essential genes. The enzymes "*Cas9 and dCas9*" are required in deletion and inhibition of gene and its expression.

The Cas9 nuclease modifies a specific sequence by repairing the susceptible DNA through the pathway of NHEJ or homologous recombination accompanied by a customized template of DNA. Then the dCas9 binds the DNA and hinders the RNA

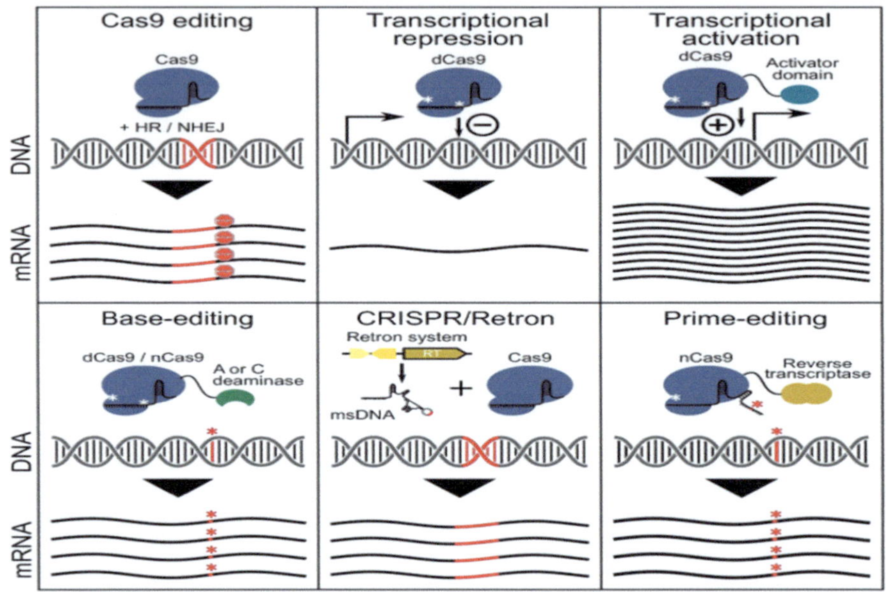

Fig. 15.10 CRISPR-based tools are usually compliant to high-throughput screening

polymerase, thereby obstructing the gene expression. The fusion of dCas9 and a transcription activator domain starts the transcription. The mutation of nucleotides happens by new base-editing technologies without the insertion of DNA nicks by using the integration of Cas9 nickase (nCas9) and in some cases dCas9 into adenine or cytosine deaminase. The utilization of retron systems to give rise to multi-copy single strands of DNA (msDNA) like a template for editing is another way. Lastly, a contrived guide RNA and a merged nCas9 and reverse transcriptase introduce changes encrypted in the guide sequence.

Bioinformatics Approaches for Functional Genomics

As the data produced is very large in quantity, the bioinformatics method is very useful in the analysis of genomic data.

Data clustering, unsupervised machine learning, as well as artificial intelligence techniques are used. A variety of computational analysis methods are developed in the last decade to understand deep mutational scanning experiments, and phydms is one such software, and the mechanism is explained in Fig. 15.11.

Phydms runs the results of the experiment with the phylogenetic tree in order to provide accurate information.

Deep mutational scanning synthesizes every amino acid change in a protein, their activity is then assayed, and the comparison to the wild-type variant determines the mutation.

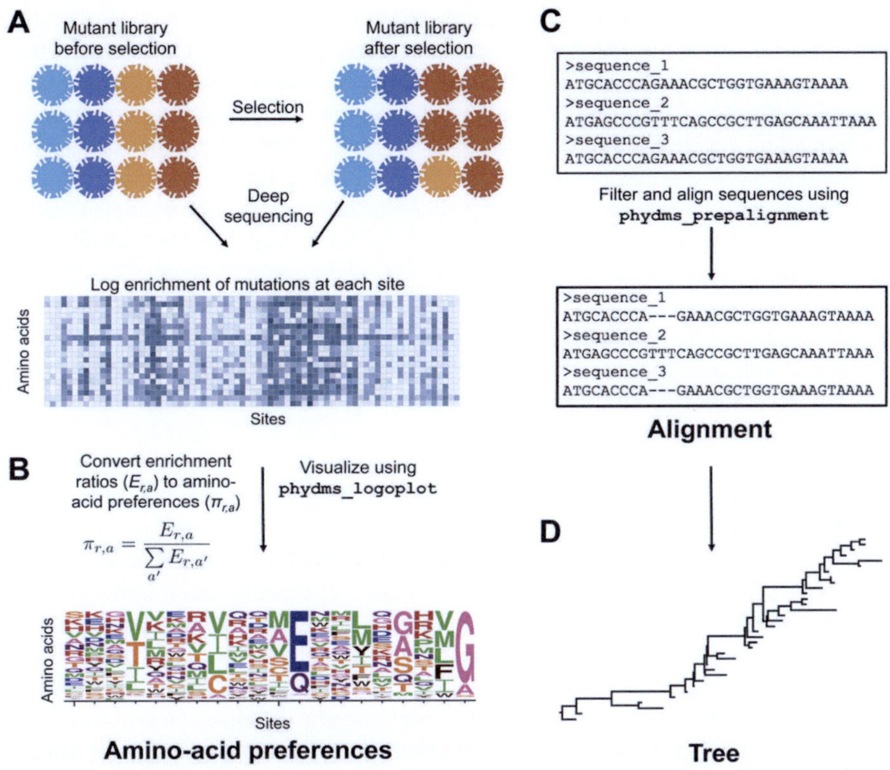

Fig. 15.11 Phydms workflow (A-D). Phydms is a program that allows you to compare the results of deep mutational scanning experiments to natural selection on genes. Phydms allows rigorous comparison of how well different experiments on the same gene capture real natural selection when given a phylogenetic tree topology inferred with another program

The technique is also useful in determining the structure of protein and protein-protein interactions. Phydms use experimentally informed codon models, and it is written in Python by Bloom laboratory.

15.3 Features of Prokaryotic Genome

15.3.1 Introduction

The prokaryotic genome is fundamentally very different from eukaryotic genome. There is a lack of physical separation in prokaryotic eel between DNA and cytoplasm. The size of prokaryote genome is much smaller; however, it is observed that an overlay in size extends between one of the largest prokaryotic genomes and one of the smallest eukaryotic genomes. *E. coli* K12 is known to have a genomic size of 4639 kb, i.e., 2/fifth the extent of the genome of yeast with only 4405 genes. A small

DNA molecule, i.e., circular in shape is known to carry the whole prokaryotic genome in addition to which there may be genes that occur in plasmids. These genes are highly efficient and are known to express genes responsible for resistance toward antibiotics. They are also able to make use of toluene, i.e., a complex compound as a carbon source. In prokaryotes, there is diversity in the organization of genome with few that possess unipartite genomic components such as *E. coli.* Other organisms have a genome which is multipartite in nature. An example of this is the organism *Borrelia burgdorferi B31* that is found to contain a chromosome, i.e., straight and linear of the length 911 kb while consisting of 853 genes. This genome may also be accompanied by 17–18 circular and linear molecules that altogether generates around 533 kb and 430 genes.

John Cairns, in 1963, has proved that the genomic constituents of *E. coli* consist of a chromosome, i.e., single and circular. Later, in 1979, the pioneering plasmid in Streptomyces, which is linear, was identified, and in 1989, scientific evidences pointed out that the chromosome of *Borrelia burgdorferi* is straight and linear. This formed an evidence that bacterial DNA molecules need not be circular. Furthermore, the "megaplasmid" was identified in *Sinorhizobium meliloti* in the early 1980s, disputing the hypothesis wherein almost the complete genome of bacteria is present on the chromosome. Eventually, a scientific development regarding this emerged that the organism *Rhodobacter sphaeroides* was composed of a "second chromosome." This development led to many more fundamental cell function theories revolving around the fact that a critical cell function can be expressed by several replicons present in the bacterial genome. Nearly 9–10% of the genome in bacteria do not possess chromosome, i.e., single and circular, such as *E. coli.* Rather, it has been found to contain multiple critical and large replicons that may be straight or circular. This type of genome composition regarding a chromosome with one or more than one big extra replicon constitutes to be a multipartite genome (also known as divided genome).

15.3.2 Concept and Classification of Replicons

The genome of prokaryotes is composed of a single replicon which consists of its own replicator and initiator. Several terms have been used to establish the classification of DNA molecules which might be present in the genome that is multipartite in nature.

Replicon and Secondary Replicon: Irrespective of its exact nature, any DNA molecule is in general referred to as replicon, and a single replicon is again branched into further subtypes on the basis of exact features. A secondary replicon is one that is not the primary chromosome belonging to the cell. Replicon can be subcategorized as such: chromosomes, second chromosomes, chromid, megaplasmid, and plasmid (Fig. 15.12).

Chromosome: This is the primary replicon and is also the largest. It consists of the critical genes. It is found that the chromosome forms the core for nearly all genetic material in prokaryotes. However, there are a few anomalies such as the

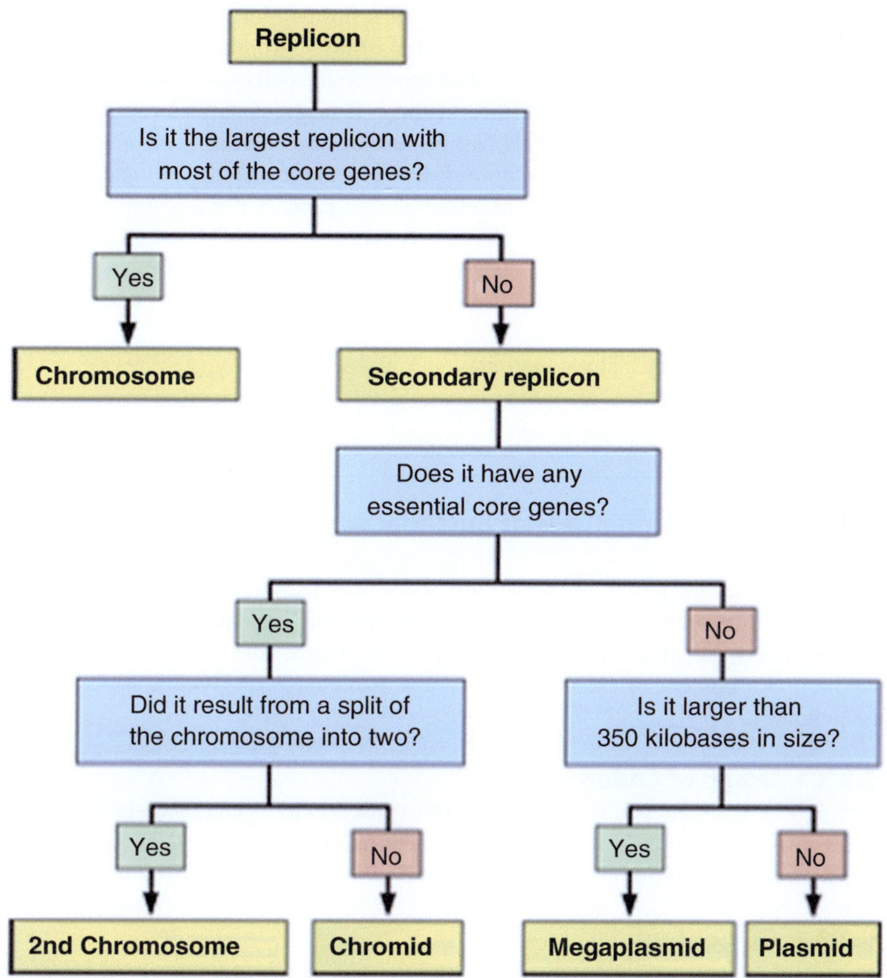

Fig. 15.12 Flowchart representing the classification of bacterial replicons. (Adapted from "The Divided Bacterial Genome: Structure, Function, and Evolution" authored by George C. diCenzo, Turlough M. Finan Department of Biology, McMaster University, Hamilton, Ontario, Canada)

chromosome present in *Burkholderia xenovorans* LB400 and *Sinorhizobium meliloti* 1021 are responsible for around 50% of their respective genomic constituents.

Plasmid and Megaplasmid: They are defined as type of replicons that lack the core genes. The majority of secondary replicons in bacterial genomes are redundant for cell viability as they do not carry core genes and are nonessential. There is significant amount of difference from chromosome in terms of GC percentage and composition of dinucleotides. Megaplasmid is nothing but a plasmid of larger size.

Chromid: It is an arrangement of plasmid with chromosome. In terms of replicon, it forms a common link for chromosome and plasmid. Chromids contain regulatory controls that help in integrating their replications into the cell cycle. They also carry a core gene that is important for viability of the cell wherein its loss could result in possible death of the cell. Chromids can be further classified as primary chromids and secondary chromids. Primary chromids are critical for viability of the cell, while chromids that are secondary in nature might be unnecessary but are required by the organism for survival in its natural habitat. A replicon is considered as secondary replicon when it is important for its native habitat. The tRNAarg and engA genes on the pSymB replicon of *S. meliloti* represent few of the critical genes present on the chromid.

Second Chromosome: This constitutes a term used in describing a rare occurrence wherein a replicon which is secondary is composed when the ancestral chromosomal is divided into further two replicons.

15.3.3 Genomic Signatures of Bacterial Replicons

Multiple prokaryotic genomes differ in their features such as usage of codons (ratio at which similar codons occur in the genome), relative abundance of dinucleotides (the frequency at which a certain base-pair of nucleotides can occur within a genome), and content of GC nucleotides (the percentage of guanine and cytosine in a genome).

15.3.3.1 Usage of Codon

The usage of codon depends on the level of expression in a gene wherein genes that are expressed in high levels constitute a codon usage which is similar to the comparative tRNA presence. This is not the case with low expression of genes.

15.3.3.2 GC Content

GC content differs quite relatively in prokaryotes and can lie in the range of 15–75%. The GC content of an organism can also be influenced by factors such as environmental adaptation and recombination. Furthermore, GC content can differ quite notably within a genome and can thereby be used to recognize and identify genes recently obtained by horizontal gene transfer. Adequate studies and research have depicted that the process of replication starts by a bound, fixed origin of replication (ori) which further sets forth in a bidirectional manner replicating the two arms known as replichores till the replication complex extends unto the replication terminus (ter) site, i.e., positioned directly opposite to the ori (Fig. 15.13).

Replication, one of the most basic and critical phenomena of bacterial cell cycle, employs selection and mutational pressure that spans across the genome. This aids in deciding the polarity of the genome by addition of nucleotides that are biased in an asymmetrical manner within the lagging and leading strand. This constitutes a certain skew in the composition that is identified with ease while marking a graph of standardized abundance value containing guanine content (G) with respect to

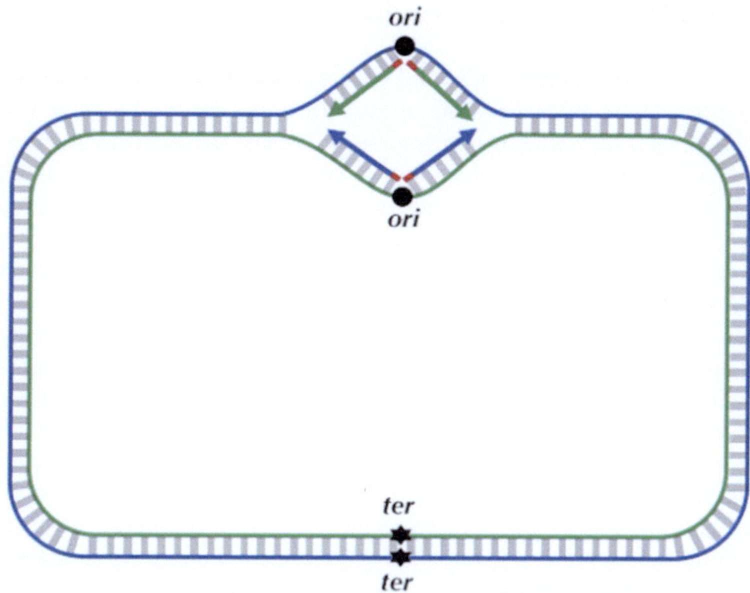

Fig. 15.13 DNA polymerases at work replicating a chromosome as the replication fork extends from *ori* to *ter*. The blue strand is directed clockwise, and the green strand is directed counterclockwise. (Phillip Compeau, Programming for lovers, Chapter 1: finding replication origins in bacterial genomes)

cytosine content (C). This is called as a **GC skew graph,** and it aids in segregating a genome into further two subregions, i.e.:

(a) One having an abundance of guanine above cytosine that corresponds to the base composition of the leading strand.
(b) One having an abundance of cytosine above guanine that corresponds to the base composition of the lagging strand.

The shift points that are marked in the graphs of the GC skew are found to correspond to the loci region of *ter* and *ori.* Several studies have indicated that the GC skew can be identified only in bacterial organisms that have a circular chromosome and not in organisms with a straight or linear chromosome.

15.3.3.3 Relative Quantity of Dinucleotide

Studies regarding the topic of quantity of dinucleotide of the organizational genome reveal that they are distinguished for every single genome of bacterial origin and are representative of phylogeny of bacteria. It has also been illustrated by several studies that relative quantity of dinucleotide has the ability to distinguish plasmids and chromids from chromosomes.

15.3.4 Physical Structure of the Prokaryotic Genome

- Size: The size of majority of prokaryotic genome is less than 5 Mb which is present in a DNA molecule which is adapted to the nucleoid with an exception being *B. megaterium* which has a genome of 30 Mb. The most commonly studied prokaryote is the *E. coli* which has a size of just 1.0 × 2.0 μm with the chromosome having a circumference of 1.6 mm to sequence the entire genome. Packaging of the genome is done with help of DNA binding proteins. Studies of *E. coli* chromosome revealed that it is negatively supercoiled. Supercoiling happens in positive manner or negative manner, i.e., if surplus turns are made to enter in the double helix of the DNA, it is known as positive supercoiling. If the turns are excised or removed, it is known as negative supercoiling (Fig. 15.14). Because the circular chromosome of *E. coli* has no free ends, the strain or torsional stress cannot be released by rotation. Instead, the molecule responds to the strain by writhing itself in a more compact structure. The enzymes involved in supercoiling are DNA gyrase and DNA topoisomerase I. There is restricted freedom to rotate in *E. coli* DNA molecule if a break is introduced at any point in the chromosome. However, there is loss of supercoiling in a limited region of DNA when a break is made.
- The genome of prokaryotes has to fit into a comparatively small space. For example, the circular chromosome of E. coli is known to have a circumference of 1.6 mm while the cell size of E. coli is around 1.0 × 2.0 um. The organization of DNA in the nucleoid is essentially studied from E. coli. Supercoiling has been identified and studied to be an ideal technique of packaging the circular DNA molecule into a relatively smaller space. Supercoiling of the circular DNA molecule structure winds over and around itself to lead to the formation of a more closed and compact structure, and this can be aided by two important enzymes. Those enzymes are DNA gyrase and DNA topoisomerase I.
- Further studies regarding nucleoids and cells that are carefully isolated and living have revealed the DNA of *E. coli* to be restricted in their ability to rotate once a break is introduced. This has been found to be caused due to the attachment of proteins to the DNA molecule, thereby restraining its capacity in relaxing. Upon a single rotation at the DSB site, the supercoiling is lost from a section of the DNA. The present profile of the DNA molecule of *E. coli* has been found to be attached to a protein core from which over 39–50 loops that are supercoiled in nature are known to emerge radially outward to the cell. A separate DNA loop is found to comprise around 90–100 kb of DNA, i.e., supercoiled. This constitutes the quantity of DNA that is separated as a consequence of the double-stranded single
- The enzymes DNA gyrase and DNA topoisomerase I are critical in sustaining the supercoiling of the DNA molecule. A set of around four proteins are also found to accompany the enzymes with a special ability to aid in the packaging of the DNA. Hu, a packaging protein, is morphologically very distinguished from histones of eukaryotes. However, they perform their function in an identical manner by making a tetramer leading to unwinding of over 60 bp of DNA (Fig. 15.15).

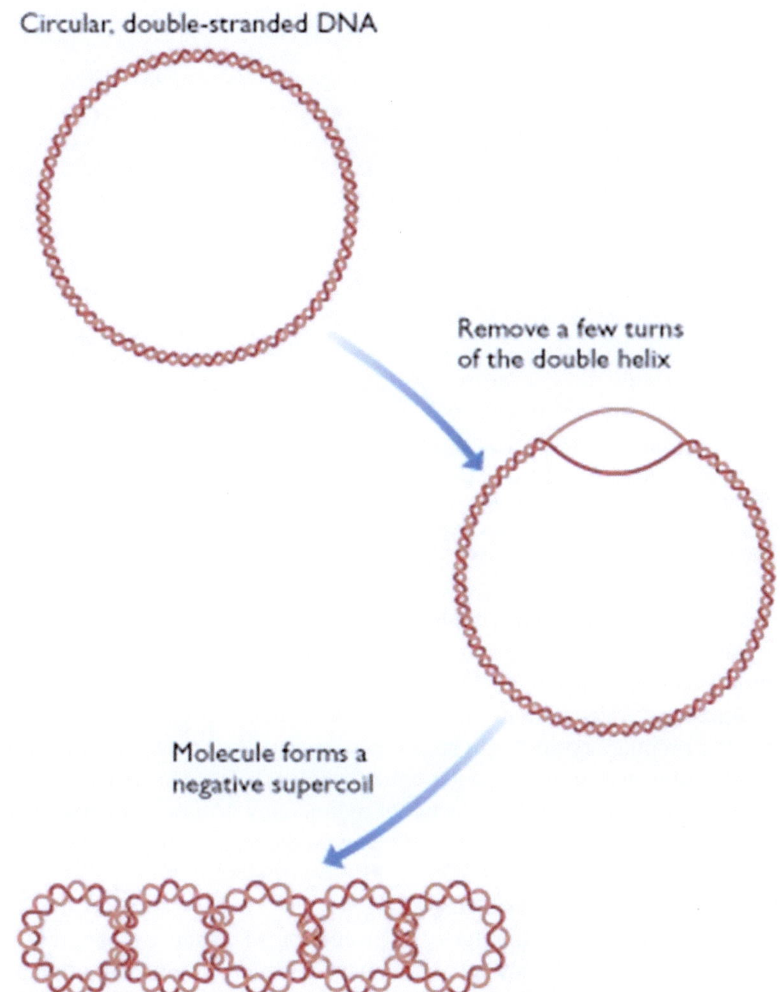

Fig. 15.14 This figure depicts the unwinding of a double-stranded circular molecule of DNA leading to the formation of supercoiling that is negative. (Chapter 2, Genome Anatomies, Genomes, TA Brown 2002)

15.3.5 Repeats in Prokaryotic Genome

A repeat in a prokaryotic genome is mathematically represented by a subsequence of a given genome which bears strong resemblance to another subsequence in the same genome. In order to correctly define a genomic repeat, one has to examine biological and mathematical factors that influence the occurrence of the repeat. Repeat that occur in an unexpected and random assembly out of the genetic context cumulates interest as such repeats have clearly not occurred as a consequence of certain phenomena and generally may discover a critical significance of biological origin.

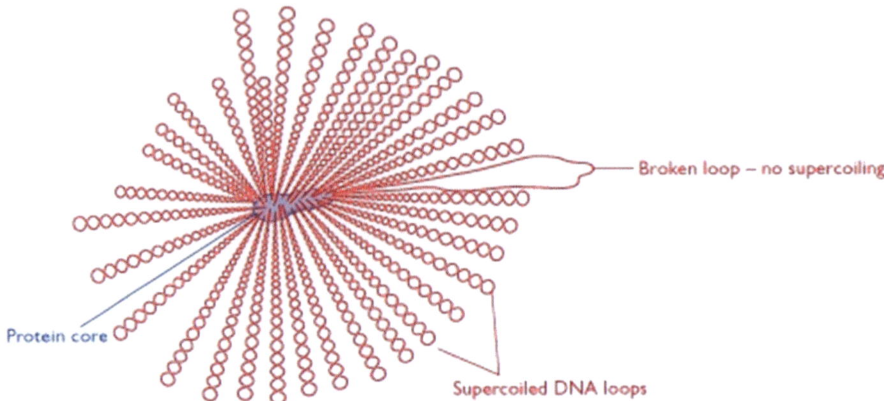

Fig. 15.15 A diagram depicting the formation of structure of a nucleoid of *E. coli*. Around 39–50 loops of DNA that are supercoiled emerge radially outward with the protein core in the center. A loop is depicted in its unwound circular form representing a break that has formed in that section of DNA, thereby causing the loss of supercoiling. (Chapter 2, Genome Anatomies, Genomes, TA Brown 2002)

It has been found that the repeats are of biological significance as they are formed as a direct consequence of significant overlapping and sequence recombination. An association between genetic elements can lead to the overlapping of the genes.

Co-expression of genes may also occur when identical copies of a repeat are present in their regulatory genes. In such cases, repeats are known to efficiently give rise to associations within genes in genomic complexes. Repeats can also constitute operons or entire genes, wherein functional redundancy occurs as a subsequence of intersection by overlapping of functional genes in the two copies.

15.3.5.1 Recombination and Repeats

Homologous recombination involves the promotion of strand exchange by RecA between homeologous and homologous sequences which are primary critical steps. RecA is considered to be the critical component of the process. However, it is absent in certain small genomes of obligatory endomutualists, like *Buchnera,* which lack repeat elements and thereby cannot repair themselves by the common repair mechanisms.

15.3.5.2 Origin of Repeats

Evidences have pointed out that repeats can occur in several different possible methods wherein generic repeats are known to make an appearance by the common recombination processes; however, more special families of families are known to originate through specialized recombination processes like site-specific integration for prophages, transposition for insertion sequences (ISs), or poorly understood mechanisms for elements like retrons or clustered regularly interspaced short palindromic repeats (CRISPR). After their genesis, repeats can associate in general recombination processes independently of their origin (Fig. 15.16).

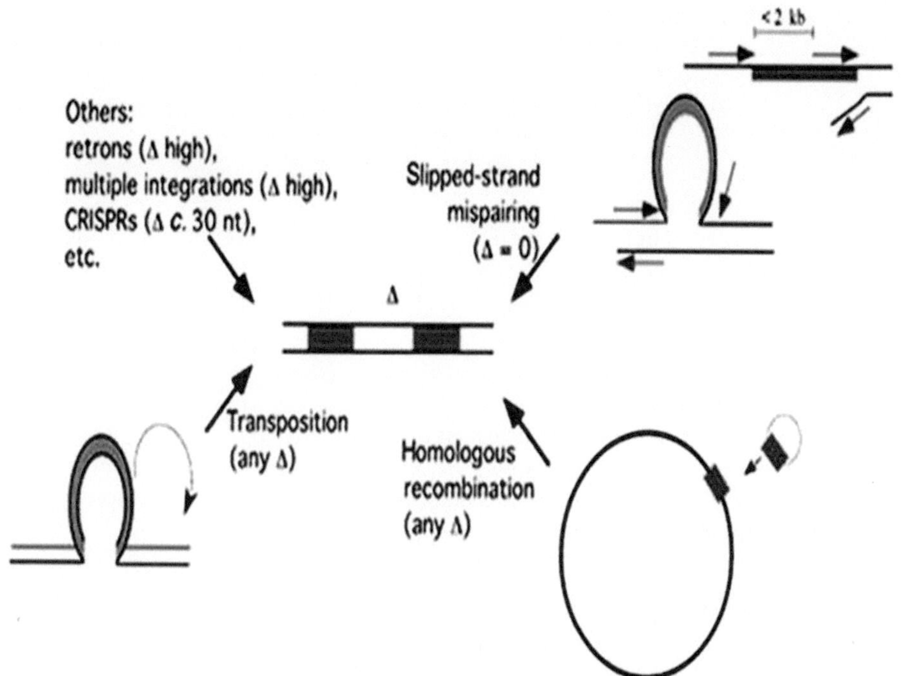

Fig. 15.16 Different methods by which a repeat is created. Δ stands for the size of the spacer between the two repeats expected under the different mechanisms. (*FEMS Microbiology Reviews*, Volume 33, Issue 3, May 2009, Pages 539–571, Genesis, effects and fates of repeats in prokaryotic genomes- Todd J. Treangen, Anne-Laure Abraham, Marie Touchon, Eduardo P.C. Rocha)

15.3.5.3 Repeats of Generic Nature

Horizontal gene transfer and homologous recombination between shorter degenerate repeats are known to form generic repeats. Repeats are created as a consequence of horizontal gene transfer whenever the incoming DNA is comprised of elements of elements that are identical to preexisting information.

Integration of information, i.e., laterally transformed, can occur through site-specific recombination. Homologous recombination that occurs in between degenerate and short repeats can lead to conversion of the gene and thereby efficiently enlarge the area of likeness in between the elements, thus generating larger repeats. This conversion of the gene can allow the integration of similarity between sequences intact.

Illegitimate recombination can occur between small degenerate closely spaced repeats that may cause "Dislocated Mutagenesis" as a consequence of which the assembly of a constricted repeat is formed by a degenerate one. Owing to the properties of slipped strand mispairing, the repeats have to be in close proximity in the genome with a length of only few nucleotides.

15.3.5.4 Selfish Elements

The production of SE in a genome is known to be a major source of repeats by which TEs are undoubtedly the best observed and ubiquitous in presence. Based upon their structure and transposition mechanism, they can be classified into two major groups:

(a) Class I elements that constitute retrotransposons that are known to transpose by an RNA intermediate.
(b) Class II elements that constitute DNA transposons that are known to transpose by a DNA intermediate.
(c) Furthermore, retrotransposons have been subclassified into three main categories.

15.3.5.4.1 Group II introns

The only transposable elements to be distributed to all three domains of life belong to Group II introns. They have the ability to self-splice introns, can grow by reverse transcription, and are efficient in maintaining both retro mobility reactions and splicing of introns by itself.

15.3.5.4.2 Retrons

Genetic elements that can generate several DNA, i.e., single stranded and linked covalently to RNA via transcriptase enzyme, are known as retrons.

15.3.5.4.3 Diversity-generating retroelements

A recently discovered family of genetic elements, diversity-generating retroelements can distribute specific benefits in certain conditions by injecting large quantities of diversities in the sequence of the target gene.

15.3.5.5 Interspersed Repeats

Interspersed repeats are isolation of highly recognized families of DNA repeat elements from a vast classification of eukaryotic and prokaryotic organisms. However, the exact function and origin are unknown for these elements. They are of ubiquitous nature in archaeal and bacterial plasmids and chromosomes. These repeats are known to perform important functions in genetic evolution, organization of the genome, and genome plasticity. Interspersed repeat sequences in prokaryotes are highly intergenic (Table 15.2).

15.4 Eukaryotic Genome

15.4.1 Introduction

The typical genome that has been studied quite often with regard to the eukaryotic genome is the human genome. The human system forms a good model for studies regarding the eukaryotic genome. The eukaryotic nuclear genome can be divided

Table 15.2 Families of interspersed repeats (Adapted from: Todd J. Treangen, Anne-Laure Abraham, Marie Touchon, Eduardo P.C. Rocha, Genesis, effects and fates of repeats in prokaryotic genomes *FEMS Microbiology Reviews*, Volume 33, Issue 3, May 2009, Pages 539–571)

Repeats	Acronyms	Features
Repetitive extragenic palindromic or palindromic units	REP or PU	21–65 bp imperfect palindrome, extragenic sequence, potential stem-loop structure probably transcribed
Bacterial interspersed mosaic elements	BIME	40–500 bp mosaic combination of REP separated by other sequence motifs
Clustered regularly interspaced short palindromic repeats	CRISPR	Noncontiguous direct repeats (DR, 24–47 bp) separated by stretches of similarly sized unique spacers (26–72 bp) Potential stem-loop structure, i.e., probably transcribed
Miniature inverted repeat transposable elements	MITE	100–400 bp nonautonomous element (mobilizable in *trans* by full-length transposase) Probably derived from IS by internal deletion
Intergenic repeat unit or Enterobacterial repetitive intergenic consensus	IRU or ERIC	69–127 bp large palindromic sequence
Insertion sequence	IS	0.7–3.5 kbp autonomous element
Transposons	Tn	Autonomous element that codes for transposase and a number of gene products (e.g., antibiotic resistance, virulence factor)

into two or more linear DNA molecules, each of which is stored in a chromosome which is distinguished.

All eukaryotes are known to also contain shorter mitochondrial genomes that are generally circular in nature. Plants are observed to have an additional genomic structure, i.e., present in their chloroplast known as the chloroplast DNA. This is unique to the eukaryotic genome of plants alone and cannot be identified in any human or animal eukaryotic model.

Despite the similarity in the fundamental morphology of all eukaryotic genomes, they differ in their genomic size and can thus be distinguished from one another. The smallest eukaryotic genome has been found to be shorter than 10 Mb in length, and the largest eukaryotic genome is larger than 100,000 Mb. By observing the table below, the range of eukaryotic genome size is known to correspond to a specific level to the organism's complexity.

Fungi, a simple eukaryotic system, can be observed to have the smallest genome with respect to its highly generic and simplified nature, while higher eukaryotes like vertebrates and flowering plants have larger genomic sizes. One of the well-known exceptions to such an understanding is the nuclear genome of *S. cerevisiae* that at 12 Mb is of 0.004× the size of a human genome. It would be estimated to contain 0.004 × 35,000 genes, i.e., 140 genes. However, *S. cerevisiae* is comprised of 5800 genes.

Table 15.3 Sizes of eukaryotic genome (Chapter 2, Genome Anatomies, Genomes, TA Brown, 2002)

Species	Genome size (in Mb)
Fungi	
Saccharomyces cerevisiae	12.1
Aspergillus nidulans	25.4
Protozoans	
Tetrahymena pyriformis	190
Invertebrates	
Caenorhabditis elegans	97
Drosophila melanogaster	180
Bombyx mori	490
Locusta migratoria	500
Vertebrates	
Homo sapiens	3200
Mus musculus (mouse)	3300
Plants	
Arabidopsis thaliana (vetch)	125
Oryza sativa (rice)	430
Zea mays (maize)	2500
Triticum aestivum (wheat)	16,000

The complete absence of corresponding link between the organism's genome complexity and its genomic size led to the theory of the C-value paradox. *S. cerevisiae* is a common illustration that depicts this point (Table 15.3).

15.4.2 Packaging of DNA into Chromosome

Scientific evidences have suggested and pointed out that the chromosomes are generally smaller and shorter than the DNA molecules that they are comprised of. Therefore, a specific organized packaging system is required to package a DNA molecule into its chromosome. The most critical findings regarding the packaging of DNA were formed in the early 1970s by a combination of analysis by biochemical methods and electron microscopy techniques. The presence of DNA-binding proteins known as histones is very crucial in packaging the DNA molecules in a chromosome. In the 1970s, several groups of scientists carried forth nuclease protection on chromatin (DNA-histone complexes) which were carefully removed from the nuclei via methods fabricated to preserve the chromatin structure as much as possible. In a nuclease protection assay, the complex is subjected to treatment with an enzyme that cleaves the DNA at certain sites of the DNA that is not bound to the protein. The sizes of the obtained DNA fragments depict the location of the protein complexes on the original DNA molecule. It has been found that the bulk of DNA fragments varies in lengths of nearly 200 bp and multiples of 200 bp, i.e., suggestive of a common spacing of histone proteins along the DNA (Fig. 15.17).

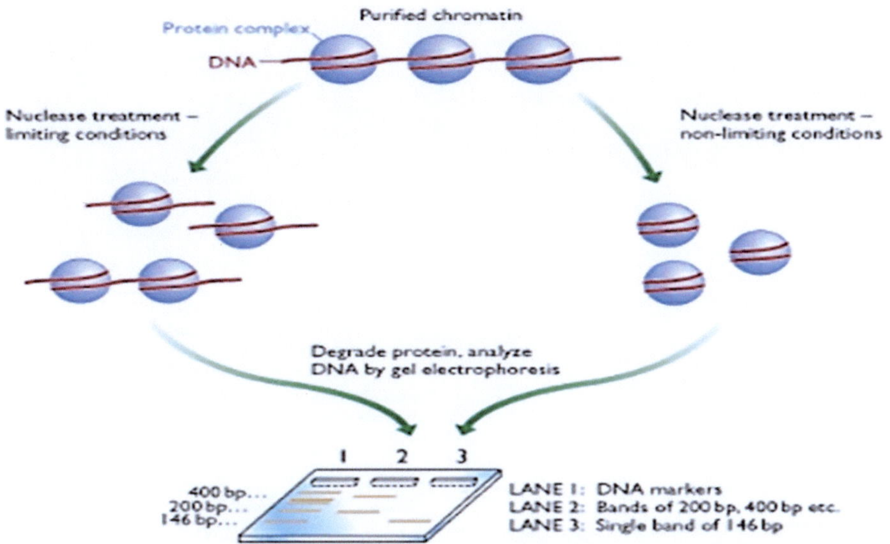

Fig. 15.17 Nuclease protection analysis of chromatin from human nuclei. (Chapter 2, Genome Anatomies, Genomes, TA Brown 2002)

Electron micrographs depict the biochemical results, and a subsequent inference was made with regard to the purified chromatin. The spacing of histone binding along the length of the DNA at regular intervals of distance was drawn as conclusion. The histone protein was found to bind to the DNA molecule in a manner similar to the presence of "beads on a string." More biochemical assays have revealed that a single bead, known as nucleosome, consists of eight histone protein molecules, i.e., two each of histones H2A, H2B, H3, and H4.

However, it was observed that the "beads on a string" appearance is known to depict an unpacked form of chromatin which forms less frequently in nuclei that are living. Techniques to break the cell have been conducted in a very gentle manner and have revealed a much compact version of the complex known as the **30-nm fiber** (approximately 30 nm in width). Many developments have been proposed regarding the 30-nm fiber among which the solenoid structure is one of the most popular.

The formation of nucleosomes as a form of DNA packaging can shorten the fiber length around sevenfold. This means that a stretch of DNA, i.e., a meter in length, will lead to formation of a "string of beads" chromatin fiber of just 14 centimeters length (approx. 6 inches) long (Fig. 15.18). The chromatin is coiled further to lead to generation of a much thicker and shorter fiber, i.e., termed as the 30-nm fiber (the size is nearly 30 nanometers in diameter).

Histone H1 has been observed to perform a critical function of stabilizing the chromatin of complex structures, and 30-nm fibers are generated with ease due to the presence of H1 protein. Modification of histones by enzymes is performed to acetylate, methylate, or phosphorylate the protein molecule to enable transcription and replication (two of the major fundamental mechanisms) to occur without any

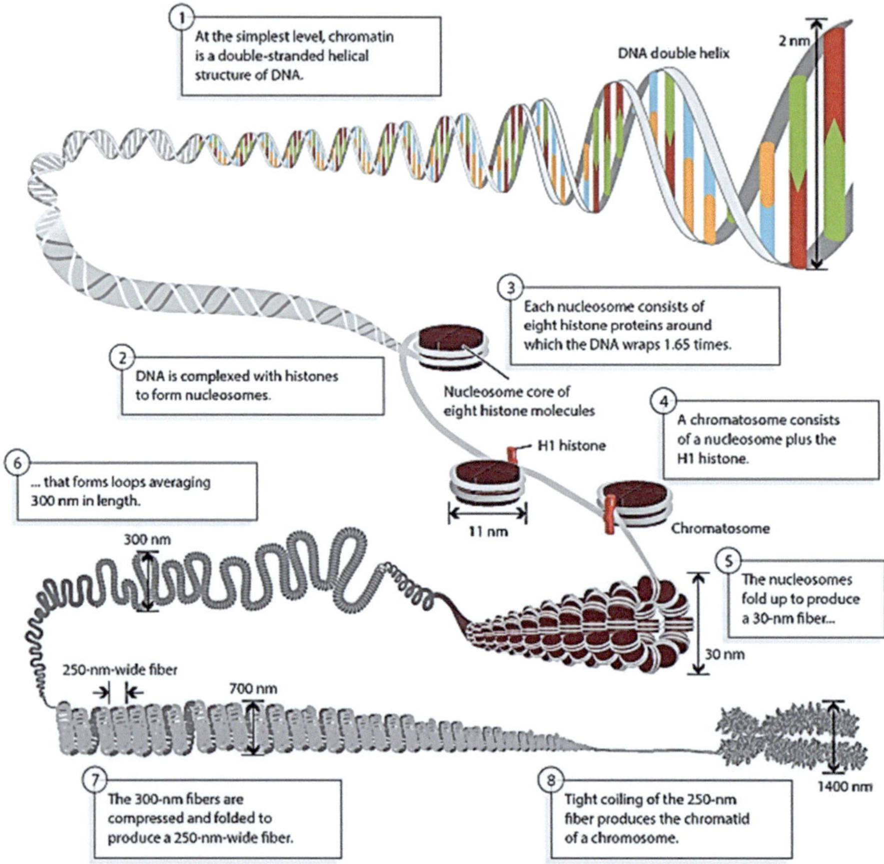

Fig. 15.18 Chromosomes are made up of DNA tightly wound around histones. (Adapted from Pierce, Benjamin. *Genetics: A Conceptual Approach, 2nd ed.)*

hindrance. Chromatin remodeling complexes can replace histones to reveal the DNA sequences and allow the binding of polymerase enzymes to the DNA sequence.

15.5 Human Genome Project

The Human Genome Project, a global research study, aims to interpret the human genome sequence, define all of the genes found within it, and develop the research insights to explore all genetic data that has been generated. This groundbreaking project is focused on the matter of fact that isolating and analyzing the genetic component found within the DNA will allow scientists with new powerful keys to address the progression of the disease and develop modern prevention and treatment strategies. Except for physical injuries, almost all medical conditions are linked with

Table 15.4 Disease-related genes identified using positional cloning (Collins and Fink 1995)

Year	Disease
1986	Chronic granulomatous disease Duchenne's muscular dystrophy
1989	Cystic fibrosis
1991	Fragile X syndrome Familial polyposis coli
1992	Myotonic dystrophy Norris's syndrome
1993	X-linked A gammaglobulinemia Neurofibromatosis type 2 Huntington's disease Wilson's disease
1994	McLeod's syndrome Polycystic kidney disease Diastrophic dysplasia
1995	Spinal muscular atrophy Ataxia telangiectasia Alzheimer's disease Hereditary multiple exostoses

a change in the structure as well as the function of DNA due to mutations. These disorders comprise heritable "Mendelian" diseases associated with single mutations, complex and common disorders caused by genetic alterations in multiple genes, and disorders caused by DNA mutations acquired during childhood, such as several cancers. Although many of these activities and research have been carried out by scientists for decades, the Human Genome Project stands out because of its efforts. The human genome is made up of three billion nucleotides, enough for it to cover 1000-thousand-page phone books if every other nucleotide is defined by a single letter. To process large quantities of data rapidly, affordably, and precisely, researchers must develop new methods, given the size of the human genome. These methods will be used to classify DNA for disease studies in families, to construct genomic maps, and to identify the gene sequences and other large significant fragments of DNA.

The Human Genome Project's primary objective is to create three important research facilities that will enable researchers to recognize genes that are relevant in both typical biology and rare and common diseases. Advanced techniques like positional cloning enable researchers to scan the genome for disease-linked genes without first determining its function of the corresponding protein. Since 1986, when investigators used positional cloning to discover the gene for the chronic granulomatous disease, this approach has resulted in the sequestration of nearly 40 genes that are disease related and will enable the identification of several additional genes forthcoming (Table 15.4).

All the three methods that are being created by this Genome Project help to narrow down the gene that is being searched. The genetic map, for example, is made up of thousands of distinctive features—small, distinct pieces of DNA—that are pretty much equally distributed along the length of chromosomes. Researchers may use this technique to pinpoint the position of a gene on a chromosome. Investigators use another method, the physical map, to locate a particular gene once this area has been established. Physical maps are considered as imbricating DNA sequences that can cover throughout the chromosome. These collections have been cloned and preserved in anticipation of potential testing. Instead of digging through the chromosomes all over again, researchers will indeed be allowed to retrieve the exact segment of DNA required once the physical map is complete. The development of a detailed sequence map of the DNA, which will include the perfect sequence of all the nucleotides that form up the human genome, will be the final tool.

15.5.1 Origin of the Project

The Human Genome Project (HGP) has radically altered the field of biology and is accelerating the medical revolution. Renato Dulbecco proposed the HGP in a review issued in 1984, arguing that the sequence of the human genome would aid in the knowledge of cancer. In May 1985, Robert Sinsheimer convened a conference solely based on the HGP, with 12 experts debating the project's merits. The session finished that the plan was technically feasible but would be incredibly difficult to implement. However, there was some debate on whether it was a good idea, with six people in attendance in favor and six against. The skeptics contended that big science could be bad as it can divert the assets away from "true" small science, that the genome is all junk and its sequencing is worthless, that they are not ready to take over such a massive task and should await once the technology is up to the task, and also that sequencing and mapping the genome is a straightforward and necessary task. Approximately 80% of biologists, as well as the National Institutes of Health, were opposed to the HGP during its early years of advocacy (mid to late 1980s). The US Department of Energy (DOE) is driven for the project at first, claiming that understanding the sequence of the genome would facilitate us to better comprehend the impact of radiation upon the genome caused by atomic explosions as well as other forms of transmitted energy. This DOE activism was crucial in igniting the controversy and, eventually, the HGP's approval. Surprisingly, the US Congress was more supportive than most biologists. The attraction of foreign competition in medicine and biology, the prospective for economic gains and technological spin-offs, and the capability for more innovative ways to treat the disorders were all acknowledged by many in Congress. In 1988, a National Academy of Science committee approved the HGP, and the flow of thought shifted: The initiative was launched in 1990, and the completed series was written ahead of time and on a budget in 2004.

As genomics technology advanced, this three-billion-dollar, 15-year initiative progressed significantly. The HGP's initial goal was to create a human genetic map, followed by a physical map of the human genome, and eventually a sequence map. The HGP was a driving force behind the advancement of high-efficiency DNA planning, sequencing, and mapping technologies throughout its life. There was hope when the HGP was established in the 1990s that the then-current technologies used for sequencing could get a replacement. The whole process, now known as "first-generation sequencing," used agarose gel electrophoresis to build up the ladders as well as fluorescent or radioactive-based tagging techniques to execute base calling. It seemed a too time-consuming and low degree of efficiency for effective sequencing. In the end, a 96-capillary version of first-generation sequencing technology was used to decode the original human genome reference sequence. Alternative methods, such as multiplexing and hybridization sequencing, were tried but failed to scale up. Meantime, the HGP saw gradual enhancements in the momentum, cost, production, as well as precision of first-generation fluorescent-based preprogrammed strategies for sequencing. The objective of the acquisition of generating a full-blown physical map was disposed of in the subsequent stages of the project in favor of establishing the complete sequence earlier than expected because researchers were clamoring for sequence results. Craig Venter's ambitious idea to build a corporation (Celera) to use the technique of the whole-genome shotgun approach to decode the entire sequence rather than the fragmentary approach that uses the clone-by-clone method, used by the International Consortium, utilizing the vectors of bacterial artificial chromosome (BAC) intensified this push. Government funding organizations backed Venter's proposal, which called for the development of a clone-based prototype sequence to every chromosome, with the final version coming later. These concurrent efforts sped up the process of creating a genome sequence that would be invaluable to biologists.

The Human Genome Project generated a carefully selected and precise sequence as reference for every other different chromosome, with only a few discrepancies and major heterochromatic regions excluded. The sequence that has been used as a reference has proved important for both the advancement as well as eventual extensive use of second-generation sequencing technology, started in the mid-2000s, in the supplement to provide a basis for the succeeding research in the variation of the human genome. Second-generation cyclic array platforms of sequencing generate many hundreds of millions of short reads (initially 30–70 bases, even several hundreds of bases) in such a single cycle, which are usually mapped against the genome of reference maintaining extreme redundancy throughout the coverage. The HGP paved the way for several sequencing techniques (ChIP-Seq, RNA-Seq, and bisulfite sequencing) that have greatly sophisticated research surveys of gene transcription and control and genomics.

15.5.2 Mapping of Human Genome

The process of determining human genome starts with mapping or characterizing the chromosomes that lead to the development of a physical map followed by its sequencing of the order of DNA bases on chromosomes that will ultimately generate a genetic map. The order of genes or other markers on each chromosome, as well as the spacing between them, is described by a genome map. Human genome maps have been generated on a variety of dimensions and resolution levels. Genetic linkage maps, which represent the comparative positions of a chromosome with

DNA markers (functional genes as well as other recognizable DNA sequences) based on their pattern of inheritance, have the lowest resolution. The biochemical properties of DNA molecules are defined by physical maps.

15.5.2.1 Mapping Strategies

15.5.2.1.1 Genetic Linkage Maps

The relative positions of DNA markers that are specific throughout the entire chromosome are depicted on a genetic linkage map. Any inherited molecular or physical trait that varies between different persons and can be accessed in the research lab is a probable genetic marker. You can use fragments of DNA that have been expressed (genes) or any segment of DNA as markers with no defined coding mechanism but a predictable pattern of inheritance. Since DNA sequence variations are abundant and to be successful in mapping, the DNA markers should be polymorphic, i.e., different patterns should prevail between different individuals and that they can be detected within various members of the family in an investigation. Polymorphisms are variants in the DNA sequence that appear every 300 to 500 base pairs on average. Exon sequence variations can cause visible changes including variations in blood type, the color of eyes, and susceptibility toward a disease. The majority of differences exist among the noncoding DNA sequences and have very little to no impact on the nature or functionality of an entity, but they can be detected at the nucleotide level and can be utilized as markers. Two examples for these kinds of markers comprise (1) restriction fragment length polymorphisms (RFLPs), which represent variations within different DNA sequence sites that are able to be cleaved by restriction enzymes, and (2) variable number of tandem repeat sequences, which are very short repetitive fragments that differ in the number of repeated units and, thus, in length (an easily measurable characteristic feature). The genetic linkage map of humans is derived from the frequency at which the inheritance of two markers is together.

The markers within the same chromosome that are close to each other are more likely to be passed down from parent to infant. DNA strands split and rejoin in various positions on the same chromosome or the other copy of the same chromosome during normal sperm and egg cell development (i.e., the homologous chromosome). Recombination during meiosis is a mechanism that allows two separate markers that have been previously on the very same chromosome to be separated (Fig. 15.19). Centimorgans (cM), called for geneticist Thomas Hunt Morgan, are used to calculate the distance between markers within the genetic map. If recombination separates two different markers 1% of the time, they are said to be 1 cM apart. A genetic distance of 1 cM is approximately equivalent to a physical distance covering up to 1 Mb. Most human genetic map regions have a resolution of around 10 megabytes. A genetic map with high resolution (2 to 5 cM) is one of the genome project's short-term goals; modern consensus maps of certain chromosomes have ranged from 7 to 10 cM between genetic markers. The use of rDNA technology, such as chromosomal fragmentation induced by radiation and fusion of cells to build

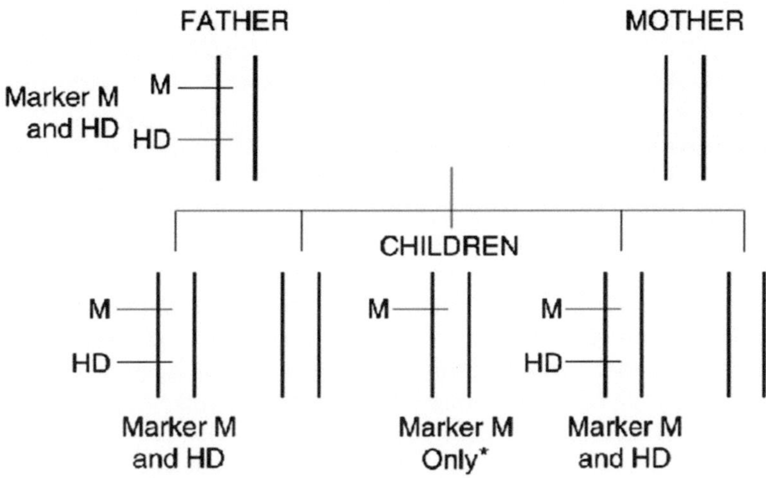

Fig. 15.19 Constructing a genetic linkage map. The vertical lines in this diagram represent pairs of chromosomes 4 for each member of a family. A short-established DNA sequence used as a genetic marker (M) and Huntington's disease (HD) are two characteristics that can be identified in any child who inherits them from the father. The implication that only one infant inherited a single trait (M) from that chromosome suggests that the father's hereditary material recombined during the sperm development process. This event's frequency aids in determining the distance between two DNA sequences on a genetic map

a group of cells that is particular and variable elements of a chromosome, has improved genetic mapping precision.

15.5.2.2 Restriction Enzymes: Microscopic Scalpels

Restriction enzymes are bacteria-derived enzymes that identify very small sequences of DNA and cleave them at particular locations. (One of these enzymes' common biological functions is to defend bacteria by targeting genetic material of the virus and other foreign DNA.) Few enzymes cleave the DNA occasionally, resulting in the formation of few huge fragments (several thousand to a million bp). The majority of the enzymes cleave DNA at a higher rate, resulting in a huge number of small fragments (few hundreds to more than a thousand base pairs).

On average, restriction enzymes yield:

- Pieces of 256 bases (4-base recognition sites).
- Pieces of 4000 bases (6-base recognition sites).
- Pieces of 64,000 bases (8-base recognition sites).

15.5.2.3 Physical Maps

The degree of resolution varies between different types of physical maps. The chromosomal (also known as cytogenetic) map is the lowest-resolution physical map, and it is based on the distinct banding patterns observed by light microscopy of stained chromosomes. On a chromosomal map, a cDNA map depicts the locations of expressed DNA regions (exons). The cosmid contig map shows the order of overlapping DNA fragments across the genome in greater detail. The order and distance between enzyme cleavage sites are described by a macrorestriction map.

15.5.2.4 Low-Resolution Physical Mapping

Chromosomal Map—Genes or other recognizable DNA fragments are attributed to their corresponding chromosomes in a chromosomal map, with distances measured in base pairs. An approach which is known as in situ hybridization includes labeling the markers with a specific identifiable label that may be used to physically associate these markers with specific bands (identified by cytogenetic staining). Soon after the probe that is labeled binds with one of the DNA strands that is complementary in an integral chromosome, its position can be determined. Even the finest chromosomal maps could only identify a DNA fragment to a region of about 10 Mb, the size of a standard chromosome band, until recently. FISH techniques have improved, allowing DNA sequences as near as 2 to 5 Mb to still be aligned. In situ hybridization techniques have been tweaked while using chromosomes during the interphase phase of cell division, because they're less compact, resulting in a resolution of about 100 kb. Additional advancements in banding may enable chromosomal bands to be linked to unique fragments of DNA that have been amplified, which could help assess recognizable physical characteristics linked to chromosomal abnormalities.

Complimentary DNA Map—The locations of exons concerning unique regions of chromosomes are shown on a cDNA map (those transcribed into mRNA are known as expressed DNA regions). When cDNA is being generated in the lab while utilizing the mRNA strand as a guiding template, base-pairing rules are followed (a nucleotide G on the mRNA will pair only with a C on the newly synthesized DNA strand). The cDNA would then be traced back to regions within the genome. Since cDNAs reflect genomic regions that have been expressed, they are believed to classify the sections of their genome with the most medical and biological relevance. The chromosomal position of genes whose roles are unknown can be calculated using a cDNA map. When the estimated position of a disease gene has been identified using genetic linkage techniques, the map may also indicate a range of candidate genes to test for disease gene hunters.

15.5.2.5 High-Resolution Physical Mapping

Top-down (producing a macrorestriction map) and bottom-up (producing a microrestriction map) are the only two approaches that are currently being used for high-resolution physical mapping (generating a contig map). Maps illustrate organized DNA fragment sets formed by restriction enzymes cleaving genomic DNA with any of the strategies. Cloning or polymerase chain reaction (PCR) methods are used to amplify the fragments.

Macrorestriction Maps: Top-Down Mapping—A single chromosome is split into large pieces (with rare-cutter restriction enzymes) and then sorted and divided up; the small fragments will then be mapped further. The corresponding macrorestriction maps show the sequence in which rare-cutter enzymes cleave and the distance between them (Fig. 15.20). This method produces maps with far more consistency

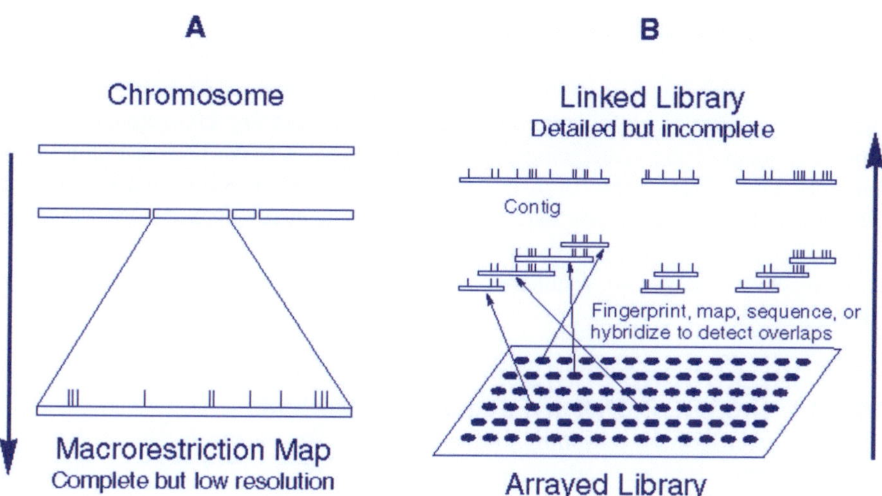

Fig. 15.20 Physical mapping strategies (P. R. Billings et al.). (**a**) Top-down physical map is generating fewer gaps within the map, but the position of particular genes may not be allowed by the resolution of the map. (**b**) Bottom-up strategies produce highly detailed maps with smaller areas, but they also have a lot of gaps. Both methods are being used in tandem

with smaller gaps separating the fragments compared to contig maps; however, the resolution of the map being lower might not be suitable for identifying a unique set of genes. Additionally, this method seldom produces long stretches of mapped locations. Currently, this method can locate DNA fragments in regions ranging from 100,000 to 1 million bytes.

Contig Maps: Bottom-Up Mapping—Breaking the chromosomes into smaller parts, in which every fragment is ordered and cloned, is the bottom-up method. Adjacent blocks of DNA are created by the arranged fragments (contigs). The resultant clone library presently ranges in size from 10,000 bp to 1 Mb. The fact that these stable clones are available to other researchers is a benefit of this approach. The FISH technique may verify contig construction by localizing cosmids with a particular segment within chromosomal bands.

Large DNA fragments can now be cloned synthetically using vectors that can hold the fragments of human DNA as big as 1 MB, thanks to technological advances. These vectors are kept as artificial chromosomes in yeast cells (YACs). (See DNA amplification for more information.) Before the creation of YACs, the largest cloning vectors (cosmids) only had inserts of 20 to 40 kb. The YAC method significantly decreases the number of clones that must be ordered; several YACs span whole human genes. Subcloning, a method in which actual inserts are cloned in the form of fragments into relatively small vectors, will create a much more elaborated map consisting of a huge YAC insert. High-capacity bacterial vectors (those that can handle large inserts) are indeed being established since some YAC regions are unstable.

15.5.3 Sequence of Human Genome

The use of paired-end sequences (also known as mate pairs) originating from subclone repositories with distinctive insert sizes and cloning characteristics was a crucial element of the sequencing approach used in this mega base-size and larger genome. The effectiveness of using end sequences from long segments (18 to 20 kbp) of DNA cloned into bacteriophage lambda in the assembly of microbial genomes led to the suggestion of using end sequences from 150-kbp bacterial artificial chromosomes to simultaneously map and sequence the human genome (BACs).

15.5.3.1 Sources of DNA and Sequencing Methods

This part examines the reasoning and moral principles overseeing giver choice to guarantee ethnic and sex variety alongside the procedures for DNA extraction and library development. The plasmid library development is the principal basic advance in shotgun sequencing. On the off chance that the DNA libraries are not uniform in size, are nonchimeric, and don't arbitrarily address the genome, at that point, the ensuing advances can't precisely remake the genome grouping.

1. Library construction and sequencing—The organization of excellent plasmid libraries in a variety of supplement estimates is vital to the entire genome shotgun sequencing process, as it allows for the acquisition of sets of grouping peruses (mates): one read from each plasmid embeds two closures. Great libraries have an equal representation of all genome parts, few clones without embeds, and no contamination from sources like the mitochondrial genome or Escherichia coli genomic DNA. Every giver's DNA was used to build plasmid libraries in at least one of three sizes: 2 kbp, 10 kbp, and 50 kbp. The major focus is on creating a simple method that could be executed in a consistent and repeatable manner and monitored effectively when developing the DNA-sequencing process (Fig. 15.21).
2. Trace processing—An automated trace-processing pipeline was built to process each sequence file. After consistency and vector trimming, the average trimmed sequence length was 543 bp, and sequencing accuracy was exponentially distributed, with a mean of 99.5% and fewer than 1 in 1000 reads being less than 98% accurate.
3. Quality assessment and control—Each sequence read must be put in the genome in a specific way, and even a small error rate will reduce the assembly's effectiveness. Furthermore, for the algorithms discussed below, preserving the validity of mate-pair information is important. As sequencing reactions progressed through the process, procedural controls were developed, including strict rules built into the laboratory information management system, to ensure the validity of sequence mate pairs.

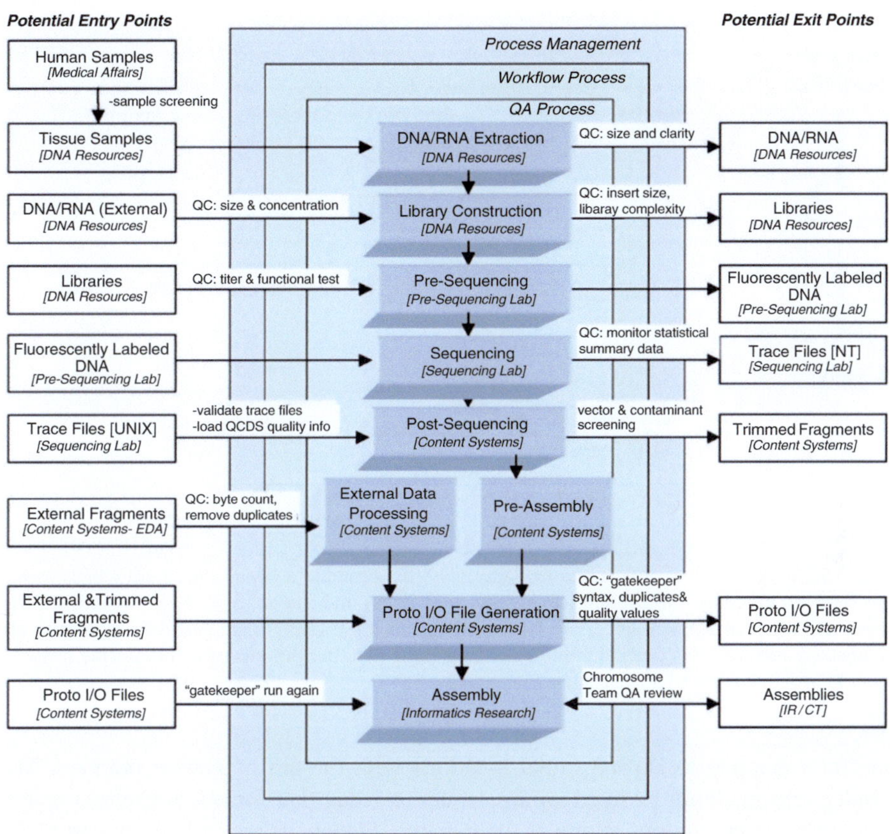

Fig. 15.21 Flow diagram for sequencing pipeline (J. Craig Venter et al.). With an emphasis on consistency inside and across divisions, samples are collected, selected, and processed by standard operating procedures. Each method has its collection of inputs and outputs, as well as the ability to share samples and data with both internal and external organizations when adhering to quality standards

15.5.3.2 Genome Assembly Strategy and Characterization

The two main approaches that are being used to assemble the genome are discussed in this section. To create an independent, nonbiased view of the genome, one approach involves computationally combining all sequence reads with shredded data from GenBank. Based on mapping details, the second method involves clustering all of the fragments to a region or chromosome. After that, the clustered data was shredded and placed through a computational assembly process. Both methods yielded a virtually identical reconstruction of the assembled DNA sequence, complete with proper order and orientation. Shotgun sequence assembly is a classic inverse problem: Given a random sample of reads from a target sequence, reconstruct the order and location of those reads in the target sequence. Celera assemblies are made up of a series of contigs that are grouped and oriented into scaffolds, which

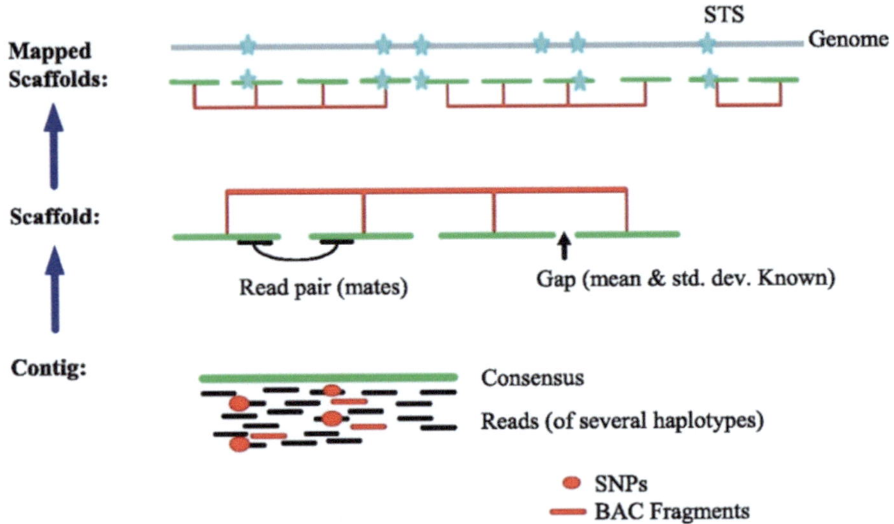

Fig. 15.22 Anatomy of whole-genome assembly. To generate a contig and a consensus series, overlapping shredded BAC contig fragments (red lines) and internally derived reads from five separate individuals (black lines) are combined (green line). Using mate-pair details, contigs are connected into scaffolds (red). Scaffolds are then mapped to the genome (gray line) using physical map knowledge from STS (blue star)

are then mapped to chromosomal locations with the aid of known markers. The contigs are made up of overlapping sequence reads that form a consensus reconstruction of a contiguous region of the genome. Mate pairs are an integral part of the assembly process. They're used to create scaffolds in which the size of gaps between contigs is known to a reasonable degree of accuracy. This is achieved by noting that a pair of reads, one of which is in one contig and the other in another, means a distance and orientation between the two contigs (Fig. 15.22).

Assembly strategies—It was decided to take two separate approaches to assembly. The first was a whole-genome assembly process that used Celera data and Publicly Funded Human Genome Project (PFP) data as synthetic shotgun data, and the second was a compartmentalized assembly process that partitioned Celera and PFP data into sets localized to large chromosomal segments and then conducted ab initio shotgun assembly on each set (Fig. 15.23).

1. Whole-genome assembly (WGA)—The WGA assembler is made up of five stages: Screener, Overlapper, Unitigger, Scaffolder, and Repeat Resolver. All microsatellite repeats with less than a 6-bp element are found and labeled, and all recognized interspersed repeat elements, such as Alu, Line, and ribosomal DNA, are screened out. Marked regions are checked for overlaps, while screened regions are not but maybe included in an overlap with unscreened matching segments.

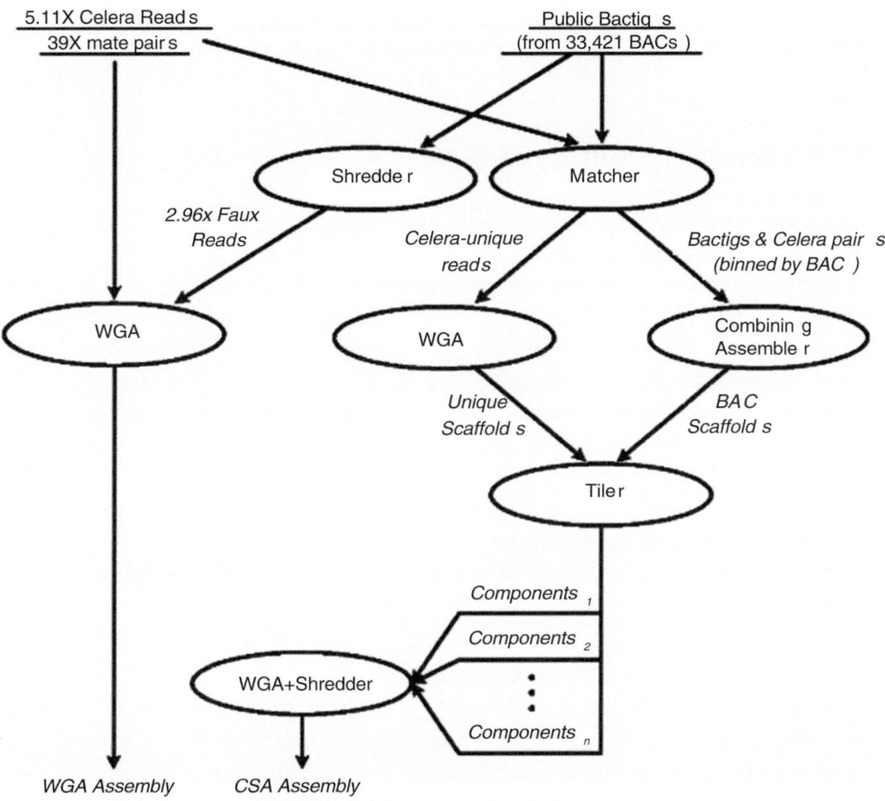

Fig. 15.23 Architecture of Celera's two-pronged assembly strategy. Each oval represents a computation process that performs the function indicated by its mark, with labels on arcs between ovals representing the types of objects generated and/or consumed by the process. The discussion in the text that describes the words and phrases used is summarized in this diagram

Every read is compared to every other read in the Overlapper to find full end-to-end overlaps of at least 40 bp and no more than 6% differences in the match. Early in the process, the assembler must avoid selecting repeat-induced overlaps. Unitigger helps in achieving this aim. Unitigs are the contigs produced from these subassemblies (for uniquely assembled contigs). These unitigs are uncontested interval subgraphs of the graph of all overlaps in formal terms.

The Unitigger generated a series of correctly assembled subcontigs that covered an estimated 73.6% of the human genome. The Scaffolder then proceeded to connect these into scaffolds using mate-pair details. When two or more mate pairs mean that a pair of U-unitigs are at a certain distance and orientation concerning each other, the likelihood of this being incorrect is approximately 1 in 1010, assuming that mate pairs are false less than 2% of the time. A consensus sequence of each contig is produced at the end of the assembly process, as well as at several intermediate points. The theory of maximum parsimony guides an algorithm, which uses quality-value-weighted measures to evaluate each foundation.

15.5.4 Features of Human Genome

The human haploid genome comprises approximately 30,000 genes and is and around three billion bp in length. Since each base pair can be encoded with two bits, this amounts to approximately 750 M of data. A single somatic cell comprises two times as many base pairs or around six billion. Males have comparatively lesser base pairs than a female since the Y chromosome has around 57 million while X had around 156 million. Since human genomes differ by only about 1% in sequence, the differences of any given genome of a human from a standard reference perhaps reduced approximately to 4 MB without losing any data. The genome's entropy rate varies greatly among noncoding and coding sequences. It is similar to the limit of 2 bits per base pair for coding sequences (approximately 45 Mbp) but less for noncoding pieces. The entropy rate for each chromosome varies from 1.5 to 1.9 bits per base pair. Y chromosome is an exception, which has 0.9 bits of entropy rate per base pair.

15.5.4.1 Noncoding DNA vs. Coding DNA

Coding and noncoding DNA sequences are widely used to separate the human genome's material. Coding DNA refers to DNA that has the potential to synthesize mRNA by transcription and ultimately leading to the production of proteins; these sequences account for just 2% of the entire genome. The sequences that don't encode for any proteins are considered as noncoding DNA (roughly 98%). RNA molecules with essential roles can be found in certain noncoding DNA (rRNA and tRNA). The ENCODE (Encyclopedia of DNA Elements) initiative, which attempts to explore the entire genome of the human by utilizing a variety of analytical methods whose findings are representative of biological activity, is one example of contemporary genome research focusing on the role and evolutionary origins of noncoding DNA. The analyzed genome framework has evolved into a much more focused analytical category than the conventional concept of DNA-coding genes since noncoding DNA exceeds the number of coding DNA.

15.5.4.2 Coding Sequences (Protein-Coding Genes)

Protein-coding sequences are the most researched and well-understood part of the human genome. Though many biological processes (such as rearrangement of DNA and alternate splicing of pre-mRNA) can result in the development of far more unique proteins than that of the genes encoding for the proteins, these sequences eventually contribute to generating all the proteins. The exome contains the genome's entire functional protein-coding ability and is made up of nucleotide sequences that are encoded by coding sequences and can then be translated to functional proteins. The exome was considered as the first big achievement of the

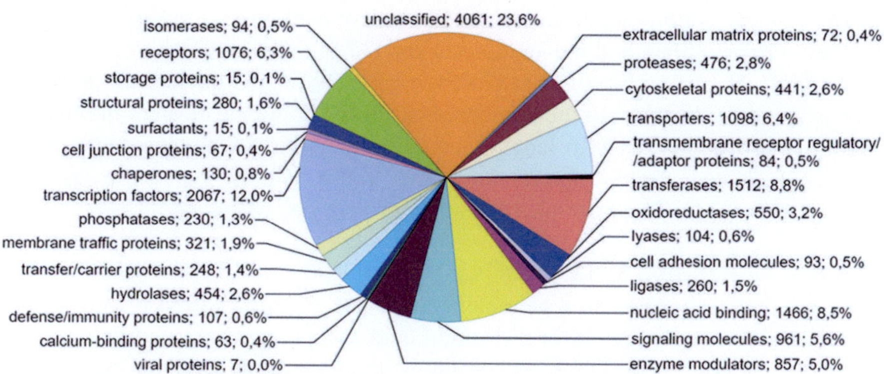

Fig. 15.24 Human genes classified according to the function of transcribed proteins, expressed as many encoding genes and as a percentage of all genes (PANTHER Pie Chart)

Genome Project for its biological importance as well as the reality that it makes up only around 2% of the entire genome (Fig. 15.24).

1. **Number of protein-coding genes**—In databases like UniProt, around 20,000 human proteins have been annotated. Historically, estimates for the number of protein-coding genes have ranged from 200,000 in the late 1960s to 2000,000 in the early 1970s, and many scientists noted in the early 1970s that only about 40,000 for the total number of functional loci are present which includes functional noncoding as well as protein-coding genes. This is because mutational load arising through deleterious mutations sets a maximum limit. The number of genes encoding the protein is comparable to that of several fewer complex species such as roundworms and fruit flies. This distinction may be due to humans' widespread utilization of alternative splicing of pre-mRNA, which helps them to create a wide range of functional proteins by selectively adding exons.
2. **Chromosome's protein-coding capacity—**Genes encoding the proteins are unequally scattered throughout the chromosomes, varying from several hundred to over 2000, with chromosomes 1, 11, and 19 having the highest gene abundance. Gene-abundant and gene-poor regions can be located on each chromosome, which can be connected to bands on the chromosome and GC material. The importance of these established gene density arrangements is still unknown.
3. **The magnitude of protein-coding gene**—Within the human genome, the scale of protein-coding genes varies greatly. For instance, the HIST1HIA gene, which codes for the histone H1a, is small and straightforward. A 781-nucleotide mRNA, lacking introns, encodes a protein with 215 amino acids arising from 648 bp ORF. According to the 2001 human reference genome, dystrophin (DMD) was considered as the biggest protein-coding gene, spanning 2.2 million nucleotides, while a fox-1 homolog 1, which is an RNA binding protein, RBFOX1, which spans 2.47 million nucleotides, was discovered in a much more recently published statistical analysis of modified human genome results.

The longest coding sequence (114,414 nucleotides), consisting of the highest number of exons (363), and the lengthiest exon (17,106 nucleotides) are all found in Titin (TTN). The mean size of a protein-coding gene is 66,577 nucleotides, the mean exon size is 309, the mean of the number of exons is 11, and the mean of the encoded protein is 425 amino acids in length, according to estimates in accordance with a carefully selected range of protein-coding genes across the entire genome. Table 15.5 provides few examples of human protein-coding genes.

15.5.4.3 Noncoding DNA (ncDNA)

The nucleotide sequences in a genome that are not included within protein-coding exons and thus never appear in the amino acid composition of gene products are referred to as noncoding DNA. According to this definition, ncDNA makes up as much as 98% of entire human genomes. Genes that encode for noncoding RNA (e.g., transfer RNA and ribosomal RNA), untranslated regions of mRNA, pseudogenes, introns, repetitive DNA sequences, regulatory DNA sequences, and sequences linked to transposons are all examples of noncoding DNA. Noncoding DNA refers to a variety of sequences that are found within genomes.

The human genome contains less than 1.5% protein-coding sequences (specifically, coding exons). Furthermore, introns make up about 26% of the human genome. The human genome includes noncoding DNA regions in addition to genes comprising both introns and exons and regions that regulate gene expression (8–20%). The precise amount of noncoding DNA that contributes to cell physiology has been a point of contention. According to a recent report by the ENCODE project, 80% of the human genome is either transcribed, binds to regulatory proteins, or is involved in some other biochemical activity.

Most of these sequences control chromosome structure by restricting heterochromatin formation regions and controlling chromosome characteristics like centromeres and telomeres. Few other noncoding regions function as origins of replication. Finally, some regions of the genome are transcribed into functional noncoding RNA, which controls protein-coding gene expression, translation of mRNA and its stability, and chromatin remodeling [including DNA binding protein modifications, methylation of DNA, DNA recombination]. It's also probable that many of the transcribed noncoding regions have no function and are the outcome of the nonspecific activity of RNA polymerase.

15.5.4.3.1 Pseudogenes

Pseudogenes are nonfunctional transcripts of protein-coding genes that have accumulated due to inactivating mutations. They are also produced by gene duplication. The total number of pseudogenes with the human genome is on the order of 13,000, and the number of functional protein-coding genes in certain chromosomes is approximately the same. During molecular evolution, gene replication is a popular process for generating new genetic material. The gene family of the olfactory receptor is one of the most well-studied models of a pseudogene. In humans, as many as 60% of the genes in this given family are dysfunctional pseudogenes. In contrast, just 20% of the genes within the olfactory receptor gene family of the

Table 15.5 Common examples of human protein-coding genes (Ensembl genome browser, July 2012)

Protein	Chromosome	Gene	Length	Exons	Exon length	Intron length	Alt splicing
Breast cancer type 2 susceptibility protein	13	BRCA2	83,736	27	11,386	72,350	Yes
Cystic fibrosis transmembrane conductance regulator	7	CFTR	202,881	27	4440	198,441	Yes
Cytochrome b	MT	MTCYB	1140	1	1140	0	No
Dystrophin	X	DMD	2,220,381	79	10,500	2,209,881	Yes
Glyceraldehyde-3-phosphate dehydrogenase	12	GAPDH	4444	9	1425	3019	Yes
Hemoglobin beta subunit	11	HBB	1605	3	626	979	No
Histone H1A	6	HIST1H1A	781	1	781	0	No
Titin	2	TTN	281,434	364	104,301	177,133	Yes

mouse are pseudogenes. According to research, the most closely related primates all have proportionally fewer pseudogenes, implying that this is a species-specific trait. This evolutionary observation explains why humans have a less acute sense of smell than other mammals.

15.5.4.3.2 Untranslated Regions and Introns of mRNA

The preliminary transcripts of protein-coding genes normally include substantial sequences which don't code for any functional product in the form of 5′- and 3′-untranslated regions, introns in addition to the few ncRNA transcripts encoded by distinct genes. Introns are 10 to 100 times longer than the exons of most protein-coding genes in the human genome.

15.5.4.3.3 Regulatory DNA Sequences

Many different regulatory sequences exist in the human genome, all of which are essential for regulating gene expression. According to conservative figures, these sequences make up 8% of the genome, but extrapolations from the ENCODE project say that gene regulatory sequence makes up 20–40% of the genome. Noncoding DNA is a form of genetic "switch" that does not code for functional proteins but does control the temporal and spatial expression of genes (enhancers). Since the late 1960s, regulatory sequences have been identified. Recombinant DNA technology was used to identify the first regulatory sequences within the genome. The identity of these sequences might later be deduced through evolutionary conservation, thanks to the advent of genomic sequencing. For example, the evolutionary split between primates and mice occurred 70–90 million years ago. As a result, computerized evaluations of gene sequences will reveal noncoding sequences that are being conserved, which will indicate their role in functions such as gene regulation.

15.5.4.3.4 Repetitive DNA Sequences

Approximately half of the human genome is composed of repetitive DNA sequences. Low complexity repeat sequences with many adjacent copies, tandem repeats, make up around 8% of the human genome (e.g., "CAGCAGCAG..."). Tandem sequences can vary in length from a few nucleotides to tens of thousands of nucleotides. These sequences are used for forensic DNA and genealogical DNA analysis because they are extremely inconstant among nearly connected persons. Microsatellite sequences (e.g., the dinucleotide repeat (AC)n) are recurring sequences with less than 10 nucleotides. Trinucleotide repeats are especially important among microsatellites because they can occur in the exonic regions of protein-coded genes, causing genetic disorders. Huntington's disease, for example, is caused by the expansion of a trinucleotide repeat (CAG)n within the gene on human chromosome 4. A hexanucleotide repetition of the sequence (TTAGGG)n, which ends the telomeres, is also a microsatellite. Tandem repeats of comparatively longer sequences (10 to 60 nucleotides in length) are termed as Minisatellites.

15.5.4.3.5 Transposons (Mobile Genetic Elements) and Their Remnants

Transposable genetic materials, or DNA sequences that can duplicate and introduce clones of themselves in the other sites within the same host genome, are prevalent in the human genome. Alu, the most common transposon line of descent, has around 50,000 operating clones and can inject itself into both intergenic and intragenic locations. LINE-1 is another lineage with about a hundred active clones within a genome (the number differs among various people). They make up more than half of total human DNA, along with the nonfunctioning remnants of obsolete transposons. Transposons, also known as "jumping genes," have been playing a significant role in shaping the human genome. A few of these sequences are retroviruses that have been generated endogenously, which are copies of viral DNA that have been incorporated permanently within the genome and are crossed down the generations. LTR retrotransposons (8.3% of the total genome), SINEs (13.1% of the total genome) including Alu elements, LINEs (20.4% of the total genome), SVAs, and Class II DNA transposons are the various types of mobile elements present in the human genome (2.9% of the total genome) (Fig. 15.25).

A diagram depicting the general structure of various types of mobile components. The proportion of the human genome that is predicted to be present is shown on the right. ITR stands for the inverted terminal repeat, LTR stands for the long terminal repeat, and UTR stands for the untranslated region. Gag, which codes for viral structural proteins; pol, which codes for viral enzymes; and env, which codes for the viral envelope protein are all HERV genes. The sense and antisense promoter regions are indicated by black arrows. DNA is represented by black and gray double helices. AAAA stands for polyA tail.

15.5.5 Human HapMap Genome

The International HapMap Project's mission is to identify the general characteristics of variations of a DNA sequence within the human genome and enable this

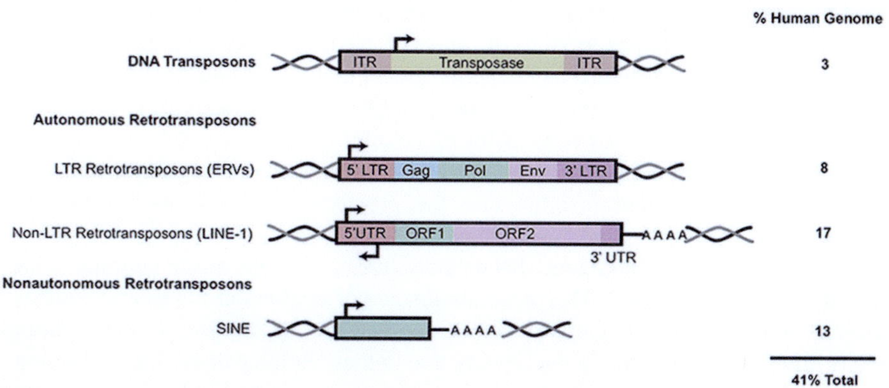

Fig. 15.25 Transposable elements in the human genome

knowledge publicly accessible. By identifying the genetic traits of one million or even more expression levels, their patterns, as well as the direction of the association among them in genetic data from communities with ancestors from areas of Asia, Africa, and Europe, global collaboration is constructing a map of such variations across the genomes. The HapMap will aid in the identification of genetic variations that influence common diseases, as well as the development of screening methods and the selection of targeted therapies. The project aims to classify variation in the sequence, their prevalence, and associations within them in DNA specimens from inhabitants with lineage from Asia, Africa, and Europe to establish common schematics of DNA sequence variation in the human genome. As a result, the project will develop tools that will make it simple to apply the indirect association method to any operational candidate gene, any area proposed by familial linkage assessment, and, eventually, the entire genome for disease risk factor scans.

Variation in the human DNA sequence—Approximately 0.1% of nucleotide positions vary between each replica of the human genome (on average, one per thousand bases). An SNP, the most common type of variant, is a variation in the base present at a specific location in the DNA sequence between chromosomes (Fig. 15.26). Few chromosomes in a community, for example, can have a C allele at a particular location, while other people have a T allele. It is assumed that approximately ten million locations (on average, one alternative per 300 bases) differ in the human population in such a way that both the alleles are ascertained at an incidence of 1%. These ten million standard SNPs account for 90% of the divergence in the population. The remaining 10% is due to a wide variety of variations, each of which is extremely rare in the population. A genomic DNA sample is tested ("genotyped") to determine the existence of specific SNP alleles in an organism.

DNA samples and populations—The project will analyze 270 DNA samples, including 90 samples from a US Utah population, gathered by the "Centre d'Etude du Polymorphisme Humain" (CEPH) and new samples gathered from 90 Yoruba individuals, 30 trios of Nigerian, 45 distinct Japanese in Yokohama, and 90 Yoruba people. A survey of 45 irrelevant persons must be enough to identify 99% of haplotypes in a population with a prevalence of 5% or higher. Random individual samples, trios, or larger pedigrees may be used in Linkage Disequilibrium (LD) studies; every structure has benefits (easier sampling) and drawbacks (larger pedigrees) (with a rising number of related individuals, productivity is dwindling). Irrelevant persons and trios have a lot of credibility when it comes to predicting localized LD patterns, according to established smart computing algorithms. The trios may have valuable insights into the project's genetic analysis platforms' precision.

To accurately explain genetic discrepancy throughout the whole genome, a high density of SNPs is needed. The mean abundance of markers in the shared database, dbSNP, when the task began was about one per kb (2.8 million SNPs), although many domains had a much lower density due to their variable distribution. There are three methods for determining the consistency of a genotype. To begin, each center received the same arbitrarily chosen set of 1500 SNPs for analysis and genotyping

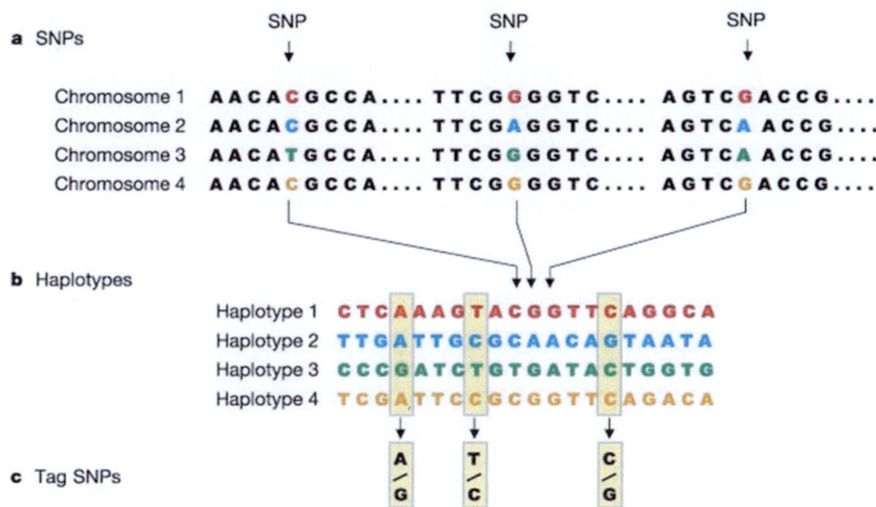

Fig. 15.26 SNPs, haplotypes, and tag SNPs (the International HapMap Project). (**a**) SNPs are single nucleotide polymorphisms. A small piece of DNA from four separate people's variety of the exact chromosomal area is seen. The majority of the DNA sequence in these chromosomes is similar, but three bases are seen where there is the difference. Each SNP has two potential alleles; the alleles C and T are contained in the first SNP. (**b**) A haplotype is composed of a specific set of alleles found at neighboring SNPs. The identified genetic makeups for 20 SNPs spanning 6000 bases of DNA are depicted here. Just the varying bases, like the 3 SNPs in section a, are displayed. Demographic data revealed that many of the chromosomes in this zone have haplotypes 1–4. (**c**) It is adequate to genotype only the three tag SNPs out from the 20 SNPs to distinguish such four haplotypes. For example, if these three tag SNPs on a specific chromosome have the sequence A–T–C, this configuration represents the pattern defined for haplotype1. Multiple chromosomes in the populations bear the standard haplotypes

with the 90 CEPH DNA samples used for the study. Genotyping centers provided data that was beyond 99.2% complete and 99.5% accurate on average (comparative with the concordance of at least two other forums). Second, samples for convergent validity checks are used in each genotyping test, with replicates of five different samples with a blank in every 96-well plate. Also, data from trios can be used to verify that SNP alleles are inherited in a consistent Mendelian manner. The information from the independent samples serves as a review to ensure that perhaps the SNPs seem to be in Hardy-Weinberg equilibrium across all communities (a test of genetic mating patterns). Although there are a tiny proportion of SNPs that failed such assessments due to biological factors, they seem to be more prone to failure even when a genotyping system creates recurrent errors, like under-calling heterozygotes. Third, a sample of SNP genotypes from every center will indeed be selected at random and re-genotyped by several other centers. These comprehensive third-party quality assessments will ensure that the data that has been generated during the project is complete and reliable.

While using the HapMap to analyze vast genomic areas, several comparative studies can occur when evaluating tens to hundreds of thousands of SNPs as well as haplotypes for disease correlations. It would be difficult to distinguish true from false-positive outcomes as a result of this. To validate the findings and classify the functionally significant SNPs, functional analyses, analytical techniques, and validation tests of variants would be needed. The HapMap has a lot of potential as a modern method for discovery—to help us better understand the genetic factors that influence health and disease. Fundamental science researchers, population geneticists, physicians, epidemiologists, sociologists, theologians, and the public, in general, will work together to reap the maximum benefits.

15.5.6 Human Genome Vs. Chimpanzee Genome

Chimpanzees are humans' nearest living relatives. The split between human and chimp ancestors took place about 6.5–7.5 million years ago. The genetic characteristics that differentiate us from chimps and distinguish us as humans are still a source of fascination. Human and chimp genomes underwent several changes after their ancestral lineages diverged, including substitutions of a single nucleotide, duplications and deletions of DNA fragments of various sizes, addition of mobile genetic elements, and chromosomal rearrangements.

The analogy of the genomes of chimpanzee and human demonstrates remarkable resemblance, substantial discrepancies, and new insights for biomedical research:

1. It demonstrates unequivocally that humans and chimps share a similar and recent evolutionary origin, as predicted by Charles Darwin in 1871.
2. It reveals essential properties of the human genome for human medicine, such as the types of genes that have developed the fastest over thousands of years and unique regions of chromosomes that have experienced significant positive selection throughout the history of moderns. Ever since some of these represent reactions to current pathogens or evolutionary changes important to human well-being, this reflects light on human biology and, in particular, human diseases.
3. It shows that humans and chimps have been able to handle more genetic mutations than many other animals, such as rats. This supports a significant evolutionary assumption, and it could explain why primates have more creativity than rats, as well as a higher rate of genetic disorders.

The major outcomes of the consortium league are as follows:

1. The genomes of chimps and humans are remarkably identical, and they encode for proteins whose functions are highly similar. The DNA sequences of the two genomes are almost identical and can be directly compared. Also, after accounting for DNA insertions and deletions, humans and chimps share 96% sequence similarity. In chimps and humans, 29% of genes encode the exact amino

sequences at the protein level. Since chimps and humans separated about six million years ago from a common ancestor, the modern human protein has only undergone one unique shift.

2. When compared to other animals, some groups of genes change extremely rapidly in both chimps as well as humans. Genes involved in sound perception, nerve signal transmission, sperm formation, and cellular ion transport are among these groups. The expeditious course of evolution of the above genes might have affected primates' unique characteristics.
3. Throughout evolution, humans and chimps probably acquired more potentially dangerous mutations throughout their genomes than rats, mice, and other rodent species. Though such modifications can result in disorders that reduce the overall fitness of a species, they might have enabled primates highly adaptable to drastic changes in the environment and allowed them to develop specific evolutionary acclimatization.
4. The mutual sections of the two genomes differ by around 35 million DNA base pairs. There are an additional five million sites that vary due to an insertion or deletion of one of the lineages, as well as a much smaller number of chromosomal rearrangements. The majority of these discrepancies are thought to be nonfunctional DNA. However, up to three million variations are present in important genes encoding structural proteins that include other functional areas within the genome. The biological foundation for the peculiar features of the human race, comprising human-specific disorders such as Alzheimer's, few cancers, and HIV, can be found somewhere within these relatively few variations.
5. Despite the lack of statistical evidence, a few groups of genes seem to be developing more quickly in humans than in chimps. Genes that encode transcription factors, elements that control the function of few other genes and perform important functions during embryonic development, are the single largest outlier.
6. Perhaps more drastic shifts have occurred in a limited number of other genes. The chimp genome lacks or has partial deletions of more than 50 genes found in the human genome. The exact number of gene deletions in the human genome is unclear at this time. Three primary inflammation genes appear to be missing from genome of the chimp, which may explain a few of the reported variations among chimps and human beings in terms of immune responses. Humans, but from the other hand, tend to already have impaired the role of the caspase-12 gene, which expresses an enzyme which may well influence Alzheimer's disease progression.
7. Over the past 250,000 years, six regions of the human genome have provided strong signs of specific sweeps. (Particular sweeps appear when a mutation comes up within a population and is so beneficial that it propagates across the population to a few hundreds of generations and eventually becomes "normal.") One region includes as many as 50 genes, whereas others contain no known genes and are referred to as a "gene desert" by scientists. Surprisingly, components in this gene desert can regulate the transcription of a nearby protocadherin gene that has been linked to nervous system patterning.

Another study group looked at non-repetitive non-reference sequences (NRNR) in 15,219 Icelanders' genomes. An aggregate of 326,596 bp of NRNR DNA was discovered, with only 244 inserts longer than 200 bp accounting for 84% of the total. Notably, as compared to the chimpanzee genome, it was discovered that more than 95% of the NRNRs longer than 200 bp were also existing in the genome assembly of the chimpanzee, implying that they were ancestral. As a result, misinterpretation of polymorphic sequences may affect the total degree of human-chimp interspecies divergence due to a lack of knowledge on genome populational diversity. It does not, however, negate any of the assumptions and evidence presented in this study. Nonetheless, these results eventually contribute to the notion that human and chimp pan-genomes must be produced and compared.

15.6 Features of the Chloroplast Genome

Tobacco and Marchantia polymorpha, a liverwort, were the foremost plants in which the chloroplast genome (cpDNA) was sequenced successfully. Ever since, the cpDNA of 3721 species of plants, comprising green algae and that both terrestrial and freshwater plants, has been discovered. The organelle genome database at the National Center for Biotechnology Information (NCBI) has them. The advent of high-throughput sequencing technologies has allowed for such substantial advancement in the domain of chloroplast genetics in recent years. The completely sequenced genome of *Synechocystis* sp. PCC6803, a cyanobacterial species, represents a real advance in cpDNA science. It is a well-known primordial photosynthetic organism that is commonly used in research on photosynthesis, carbon and nitrogen assimilation, and the evolution of plastids. *Synechocystis* was quickly dubbed a model microorganism due to its unusual characteristics, which include a fully sequenced genome and the ability to feed in both autotrophic and heterotrophic modes. Finally, phylogenetic comparative studies among *Arabidopsis thaliana* and different microorganisms reported that more than 4500 genes encoding plant proteins are derived from Cyanobacteria. It is significant to note that the proteome of the chloroplast is made up of about 3000 proteins, the major part of which is transcribed by the genomic DNA in the nucleus.

However, the nuclear genome does not contain any of the chloroplast proteins. The chloroplast genome encodes some of them. The number of genes coded by cpDNA varies from 0 to 315 in different plant species (NCBI). Significantly, proteins encoded by a few of these genes have an important role in chloroplast operations, especially photosynthesis. The majority of cpDNA-encoded proteins are generated by their internal expression mechanism. Furthermore, both chloroplast-derived transcription factors and non-chloroplast-derived transcription factors control the translation of proteins that are being coded by cpDNA at different levels of the process. Furthermore, when a plant is exposed to multiple forms of abiotic pressures, the chloroplast proteome undergoes major changes. The majority of the time, it's about proteins involved in photosynthesis.

15.6.1 Structure and Organization of the cpDNA

At the onset of the plastom organization studies, chloroplast DNA was expected to represent just a circular molecule inside living plant tissue. Many angiosperm species have multibranched linear structures of cpDNA, according to recent microscopy studies. However, it is still unknown in which form the cpDNA will exist as its most common. It varies by plant species and is affected by several factors such as cell growth stage, type of tissue, and experimental model. The DNA of chloroplasts is circular and has a length of 120,000–170,000 bp. They usually have a contour length of 30–60 micrometers and a mass of 80–130 million daltons. The genomes of most chloroplasts are fused into a single wide ring, except for dinophyte algae, whose genome is split up into nearly 40 tiny plasmids, each 2000–10,000 bp in length. There are 1–3 genes in each minicircle, but empty plasmids without any coding DNA have also been discovered.

A combination of the two inverted repeats divides a long single copy section (LSC) from a short single copy section (SSC) in many chloroplast DNAs. The length of the inverted repeats varies greatly, from 4000 to 25,000 bp. Inverted repeats of plants, which are each 25,000 bp long, are at the upper end of this scale. Inverted repeat regions typically have three rRNA genes and two transfer RNA genes, and they can be modified or contracted to have as few as 4 or as many as 150 genes (Fig. 15.27). Although any two inverted repeats are usually unidentical, they can often be quite similar, suggesting that they evolved in concert. The inverted repeat regions of land plants are strongly conserved and accumulate few mutations. Similar inverted repeats have been found in the cyanobacterial genome and two other lineages of the chloroplast (Rhodophyceae and Glaucophyta), while few DNAs of the chloroplast, such as some red algae and peas, have lost them. Few members, such as Porphyra, made the direct repeats by reversing one of their inverted repeats. The inverted repeats probably help to stabilize the remainder of the genome because chloroplast DNAs that have lacked a few of the inverted repeat segments appear to get rearranged more.

Nucleoids—In young leaves, every chloroplast comprises about 100 copies of the DNA, which decreases in aged leaves to about 20 copies. They're generally stuffed in the form of nucleoids that can have multiple similar rings of chloroplast DNA. Each chloroplast contains a large number of nucleoids. Although chloroplast DNA isn't identified with authentic histones, red algae have been found with a histone-like chloroplast protein (HC) encoded by the chloroplast DNA that firmly packs every chloroplast DNA rings in the form of a nucleoid. The chloroplast DNA nucleoids in primitive red algae are concentrated in the middle of the chloroplast, while the nucleoids in green plants and green algae are scattered across the stroma.

Transfer of genes between chloroplast and cell nucleus—Most of the genes from the ancestral chloroplasts have been passed into the cell nucleus from chloroplast DNA during evolution. Endosymbiotic gene transfer is a mechanism that also took place in the second semiautonomous organelle, mitochondria. Around 18% of the genes in the nuclear genome of *A. thaliana* are thought to be of cyanobacterial origin. A gene encoding the chloroplast translation initiation factor 1 has been "acquired"

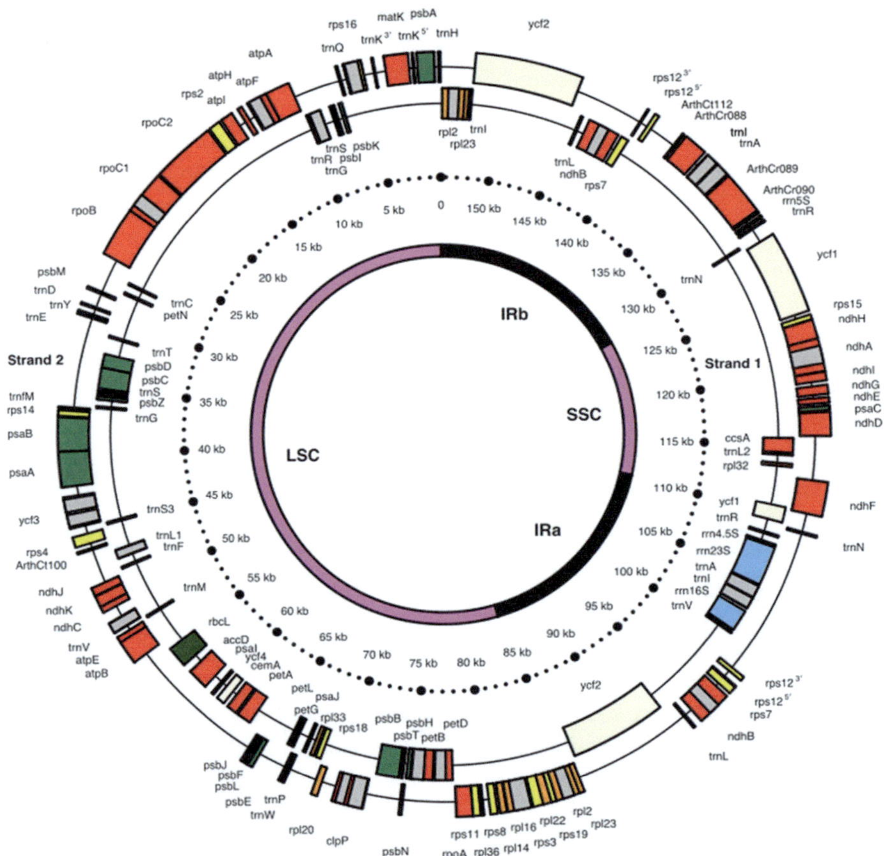

Fig. 15.27 A graphic representation showing the distances of two inverted repeat regions (IRs). A long unique sequence (LSC) and a short unique sequence (SSC) are marked on the *Arabidopsis thaliana* cpDNA map (Prof. Emmanuel Douzery). Small ribosomal subunit proteins are indicated by yellow; large subunit ribosomal proteins, orange; hypothetical chloroplast open reading frames, lemon; protein-coding genes involved in photosynthetic reactions, green, or other functions, red; ribosomal RNAs, blue; and transfer RNAs, red and black. Introns are indicated by a gray tint

by the cellular genome from the chloroplast DNA is one example. The RAR genes, a significant number of genes involved in DNA damage repair as well as recombination, were also transferred in large numbers from chloroplast DNA to the cellular genome. Because there is a high degree of similarity seen between multiple genes that are associated with recombination as well as repair of damaged DNA found in *Arabidopsis thaliana* and a human whose mutations are believed to cause abnormalities like many cancers including non-lipid colon cancer, or Cockayne syndrome, breast cancer research in this area is particularly important to medicine.

15.7 Features of Mitochondrial Genome

Mitochondria are cellular organelles with an extrachromosomal genome that is derived from and distinct from the genome of the nucleus. Sylvan and Margit Nass first described and isolated mitochondrial DNA (mtDNA) in 1963, when they identified that few mitochondrial fibers appeared to be DNA-linked based on their fixation, stabilization, and staining actions. Eighteen years later, in 1981, the first mtDNA complete sequence was published and developed as the mtDNA Cambridge Reference Sequence (CRS). The mitochondrial DNA molecule is a circular dsDNA molecule consisting of about 16,569 bp and a weight of 10^7 daltons that is 5 millimeters in diameter and free of histones. Due to differences in G + T base composition, mtDNA strands have different densities. The strong (H) strand contains genes for 2 rRNAs (12S and 16S), 12 polypeptides, and 14 tRNAs, while the light (L) strand contains genes for only 8 tRNAs and 1 polypeptide. Both 13 protein products are components of the oxidative phosphorylation system's enzyme complexes. Intron less genes and small, or even absent, intergenic sequences, except in one regulatory region, are other distinguishing features of mtDNA. The D-loop of the mitochondria region is found in many mitochondrial genome mainly in the noncoding region (NCR), created by the stable inclusion of a third 680 base DNA strand, also known as 7S DNA. The D-loop or noncoding area, a 1121 bp section situated between 16,024 and 576 positions in accordance with the CRS numeration, is where replication begins (Fig. 15.28). There are two transcription promoters in the D-loop region, one for each strand. The convention that was originally proposed was moderately changed gradually, deciding the substitution of CRS for rCRS (revised Cambridge Reference Sequence), which is used to position the nucleotide numbers in the mitochondrial genome. More specifically, each base pair's numerical designation starts at an unspecified location on the H strand and continues for roughly about 16,569 bp around the molecule.

A set of comparable haplotypes characterized by the mixture of single nucleotide polymorphisms (SNPs) within mitochondrial DNA derived from a common origin is known as a haplogroup, formed as a result of mutations that have been accumulated simultaneously through maternal lineage. Every somatic cell might just have up to 1000 mitochondria, with each comprising up to 10 copies of mitochondrial DNA. Mitochondrial DNA typing seems to be more likely to yield a DNA typing result instead of a polymorphic region typing in the nuclear genome when the quantity of the retrieved DNA is too little or dissolute. In contrast to genomic DNA, mitochondrial DNA is solely transmitted from the mother, which explains why, apart from mutation, the mitochondrial DNA sequences of siblings, as well as maternal relatives, are similar. In forensic investigations, such as in the analysis of a deceased individual's remains, identified maternal relatives could contribute some samples as a reference for a comparative analysis with the mitochondrial DNA type, and this functionality can be extremely beneficial. Due to the lack of recombination, reference samples may be taken from maternal families from many generations away from the origin of proof. The fact that most people's mtDNA is haploid and monoclonal streamlines the procedure of deciphering the sequenced DNA data.

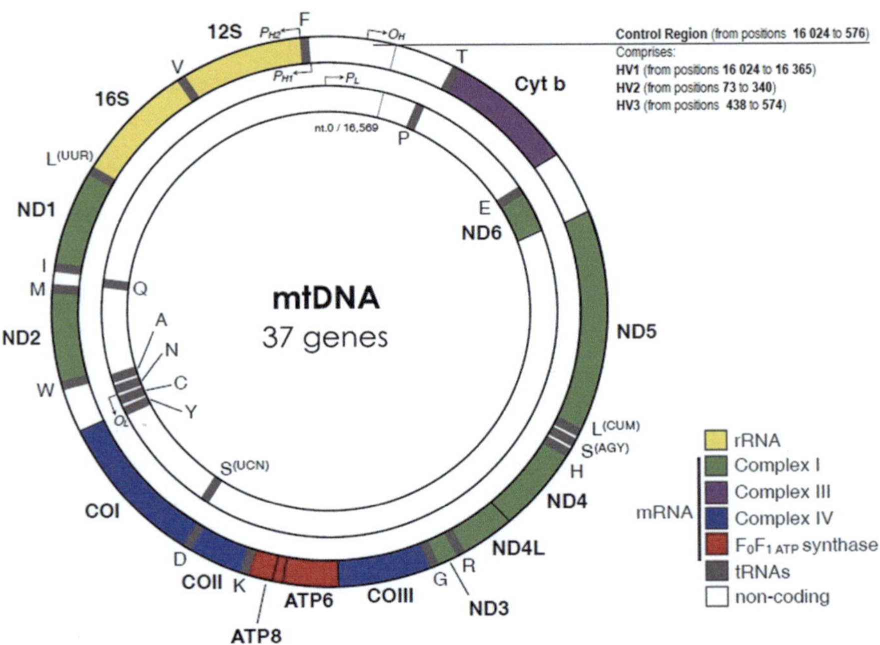

Fig. 15.28 Representation of the human mitochondrial genome including the genes and their control regions labeled. The mitochondrial genome, consisting of 37 genes that are crucial to respiratory chain assembly and function

Even so, heteroplasmy may be discovered on rare occasions. If an individual has more than one detectable mtDNA type, he or she is said to be heteroplasmic. Heteroplasmy is classified into two categories: length polymorphisms and point substitutions. For forensic human identification, only the latter is important. The majority of forensic laboratories around the world do not record polymorphism lengths, and the standards for the identification of humans with mitochondrial DNA don't require them.

Mitochondrial DNA in Forensic Human Identification—The evidence sample (s) and the mitochondrial DNA sequences of a referral sample are allowed for comparison in forensic analysis. If the two sequences are distinct, it is possible to rule out the possibility that they came from the same source. Although any further nucleotide differences seen between the two sequences are not specified in any scientific report or advisory publication, criminal routine departments regard them as an exemption situation. The specimens cannot be rejected out if their sequences of mitochondrial DNA are similar because they must be from the very same maternal lineage or have a similar origin. Whenever heteroplasmy is found in both cases around the same nucleotide positions, they shouldn't be ruled out. If one of the samples is homoplasmic in contrast to the other which is heteroplasmic, yet both share exactly one mitochondrial DNA species, the samples should not be ruled out because they may have come from the same source. Many scholars have reported that samples with one-base differences in mtDNA be investigated again, especially in terms of mutation rate.

15.8 Comparative Genomics

The analysis of the variations and resemblance within the framework of the genome and the arrangement of various species is known as comparative genomics. How do the distinctions between humans and other species manifest themselves in our genomes? For example, how similar are the different kinds and the number of proteins in different species like bacteria, yeast, worms, fruit flies, and humans? Comparative genomics is essentially the application of bioinformatics techniques to whole-genome sequence analysis to define biological concepts, i.e., biology in silico. Two factors influence comparative genetics. The first is an attempt to increase a far more in-depth comprehension of the macroevolutionary process (the root of all main classes of organisms) and the other one on a local scale (factors that cause the uniqueness of related species). The necessity to translate DNA sequences is the second engine data into known-function proteins. The justification DNA sequences encoding essential information can be found here. It's more likely that cellular roles will be conserved than sequences encoding nonessential information between species noncoding sequences or functions.

The development of paralogs and orthologs is a crucial step in the evolution of genes. It's important to differentiate between orthologs and paralogs when comparing genome organization in different species. Orthologs are homologous genes in distinct species that code for the proteins with the exact functionality. They have evolved through direct vertical descent. Paralogs are homologous genes that encode proteins with similar but not identical functions within an organism. These concepts mean that orthologs have simply arisen by the progressive mutation accumulation, while paralogs originate from gene duplication followed up by the accumulation of mutations (Fig. 15.29).

Many bacterial genomes are smaller than that of the minimal eukaryotic genome. The genome of the obligate intracellular parasite *Encephalitozoon cuniculi* is the smallest eukaryotic genome ever sequenced having a genome size of 2.9 Mb, while in *E. intestinalis,* its closest relative may even have a smaller genome of only 2.3 Mb. The length of intergenic spacers is reduced in these species, and most putative proteins are shorter than their orthologs in other eukaryotes, resulting in genome compression. *Neurospora crassa*, a multicellular fungus, has about 10,000 ORFs, which is about 25% less than the fruit fly *Drosophila melanogaster*. Most of the abovementioned genes have no homologs in either *Saccharomyces pombe* or *Saccharomyces cerevisiae*.

Genes and regulatory sequences can be identified using comparative genomics. It can be difficult to accurately classify genes in a full genome sequence, and identifying regulatory elements can be even more difficult. Aligning orthologous genomic sequences from various organisms and searching for sequence conservation regions is an effective tool for identifying functional components such as genes and

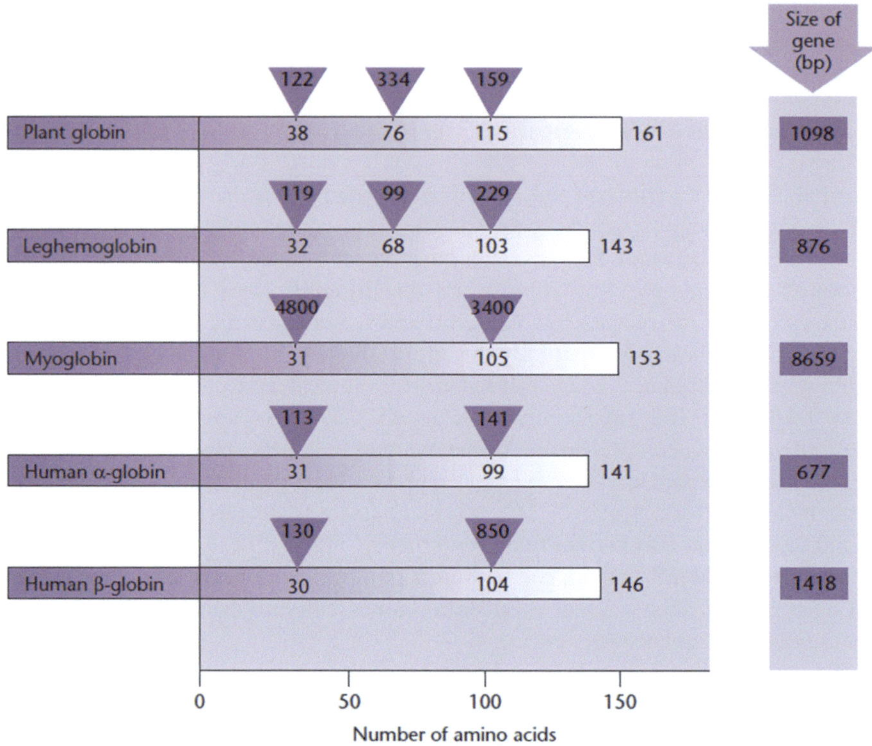

Fig. 15.29 Example of paralogs are the protein superfamilies. The location of introns in different globin superfamily members. Inside the inverted triangles, the proportion of the introns in base pairs is mentioned. It's worth noting that the sizes of the polypeptides and the locations of the various introns are fairly consistent

the regions regulating them. The rationale behind this strategy is that causes of mutations within the functional DNA are deleterious and therefore counterselected, followed by a slower rate of evolution in those functional elements. The amount of divergence captured and the phylogenetic scope of the aligned sequences are the two most significant factors influencing the outcome of a comparative study. The strength and resolution of the tests are affected by the amount of divergence. The scope, which is defined as the narrowest phylogenetic category that includes all interpreted sequences, has an impact on the potential application of insights and the generalizability of the results. A dipteran scope, for example, could be utilized to look for elements that are involved in their common ancestor along with those present before metazoa, arthropods, and hexapods diverged (Fig. 15.30).

The driving force of the evolution of species can be studied at the genome level. *Saccharomyces paradoxus*, *Saccharomyces mikatae*, and *Saccharomyces bayanus* are thought to have split from *Saccharomyces cerevisiae* about 5–20 million years ago. A comparative study has been conducted upon sequencing all four genomes.

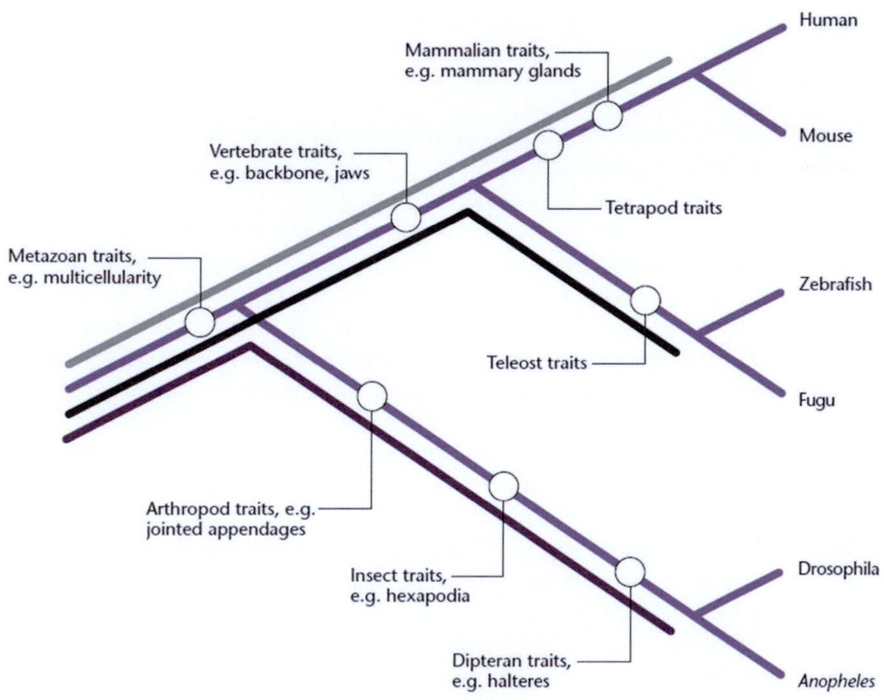

Fig. 15.30 The importance of scope and the impact of shared ancestry on comparative sequence analysis. The light purple tree depicts the relationships between six genomes that are currently being studied (not to scale). Every line that has been colored represents the scope of the phylogeny that shall be applied to each pair of species at the terminal nodes: Gray represents placental mammals, black represents teleosts, and dark purple represents dipterans. Colored line overlaps suggest common heritage and catch characteristics that have been shared by the mentioned scopes and, by extension, functional components.

Table 15.6 Genomic rearrangement in three yeast species in comparison with *S. cerevisiae*. S.B.

Species	Reciprocal translocations	Inversions	Segmental duplications
S. paradoxus	0	4	3
S. mikatae	4	13	0
S. bayanus	5	3	0

They discovered a high degree of "genomic churning" near the telomeres, with gene families showing major changes in the number, order, and orientation. Outside of the telomeric regions, only a few rearrangements were observed and are listed in Table 15.6. All the 20 inversions which generally belong to the same iso-acceptor form were transcribed in the opposite direction that flanked by tRNA genes. The importance of tRNA genes in genomic inversion has previously gone unnoticed.

Seven of the nine translocations happened among Ty components, and two resulted among closely related ribosomal gene pairs.

Box 15.1: A HapMap Harvest of Insights into the Genetics of Common Disease—Teri A. Manolio et al.

In order to study the genetic variation in humans and the patterns associated with it, as well as create a database of the same, the International HapMap Project was designed. These patterns in genetic variation would be established with the intent of studying common diseases and their genetic associations.

Forty common diseases were characterized, and 100 loci were identified for these diseases. These preliminary results included genome-wide association studies providing novel strategies for the development of several fields such as pathophysiology. Undiscovered etiological pathways were identified for certain commonly occurring diseases suggesting that novel drug targets could be recognized based on inheritance patterns and risks. The human genome is often subjected to certain evolutionary pressures that have also been analyzed through HapMap-based studies. These analyses have suggested that the disease-causing pathogens and their adaptation to new environments is influenced by multiple loci.

The goal of the International HapMap Project was essentially to provide a foundation for genetic studies of human health and disease and to curate a public, genome-wide database of patterns of common human sequence variation. The first draft of the human genome sequence was published in the year 2001, and the final draft was published in 2003. Immediately after this, the HapMap Project came into picture as there was necessity to study and characterize human genomic variation, including the millions of naturally occurring single nucleotide polymorphisms (SNPs), and quantitate the frequencies of these associations.

Genome-wide association (GWA) studies are advantageous because genetic variations that occur in common diseases can be studied quickly while also being cost-effective. The HapMap Project brought about a revolution in the field of GWA studies. There is a growing body of evidence that suggests the need to investigate the role of evolutionary pressures on the human genome and how it affects the inheritance of various diseases. It has also led to important advances in estimation of genotypes specifically relating to untyped SNPs in a nonredundant and unique fashion, while also aiding in studies about population substructure (Fig. 15.31). Another major benefit of HapMap is that the DNA and cell line samples are openly available along with the consult and the process through which they were collected. This has continuously contributed to genomic research and GWA studies.

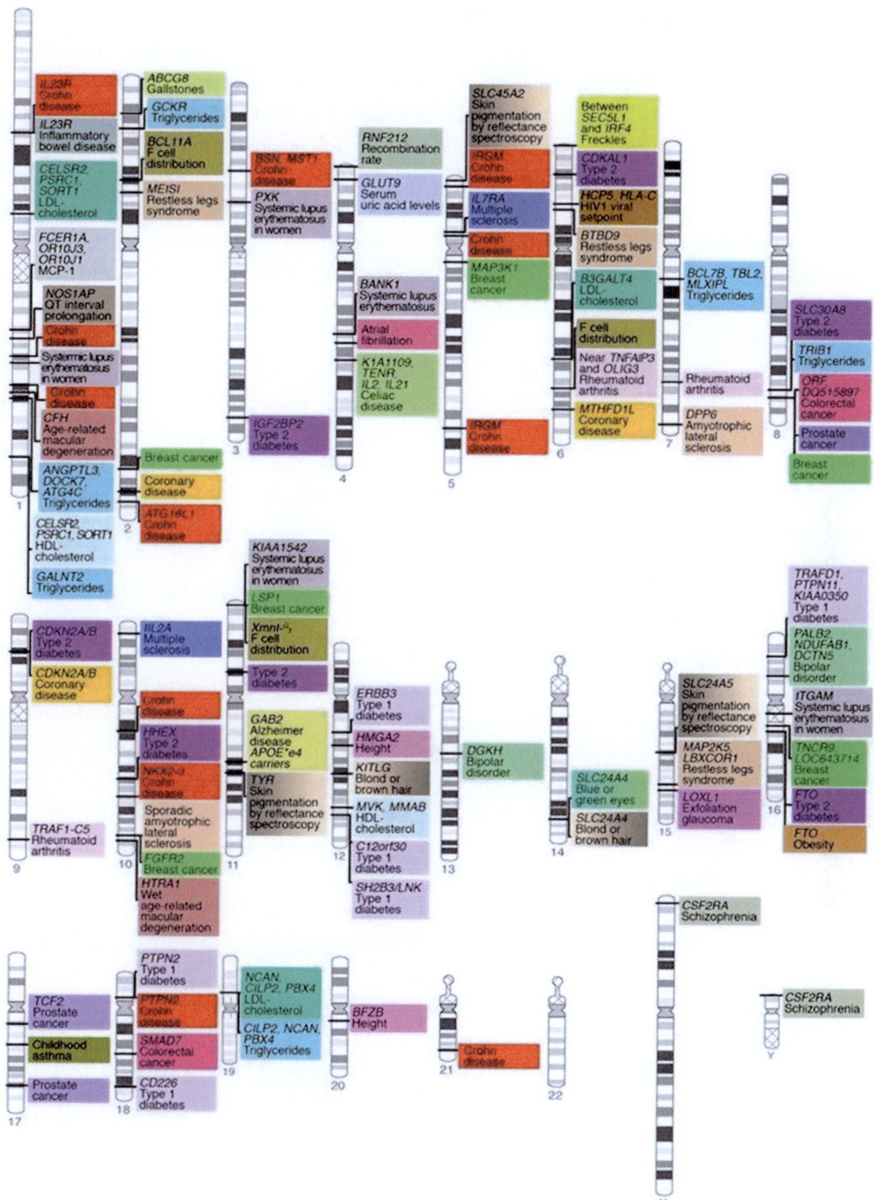

Fig. 15.31 In GWA studies, the various SNP-trait associations that were detected are depicted in the figure. Involved neighboring genes, according to chromosomal location and associations that showed a significance value of $P < 9.9 \times 10^{-7}$

15.8.1 Building a HapMap for the Human Genome

In order to consider the ethical issues, carry out data analysis and genotyping, formulate a scientific plan, select suitable diseases and SNPs that need to be typed, and make the data publicly available, a consortium was created among researchers hailing from various countries including the UK, the USA, China, Japan, Canada, and Nigeria. This came to be known as the International HapMap Project. From four populations with varying genetic ancestry, a human haplotype map was produced as a product of this consortium by genotyping of 270 samples. These samples were collected from people from whom specific consent was taken for the purpose of the project and research on the same.

In the year 2005, after the completion of the phase 1 of the project, a description of the one million SNPs that were sequenced was published. Subsequently, the HapMap Project proceeded to phase II during which three million SNPs were sequenced and the data published in 2007. Due to absence of polymorphism in about 1.3 million SNPs out the 4.4 million originally selected SNPs, the former could not be genotyped. Some did not pass quality control assessment. Centromeric regions, telomeric regions, gaps in sequences, duplications, and insertions were found to be quite challenging to study. These regions came to be known as "not HapMap-able."

This project ultimately led to the discovery of association patterns among SNPs in the human genome, and the variation of these patterns across genomes was determined. In the four populations that were studied, the variation patterns showed a level of similarity to a certain extent. Some populations such as the Yoruba population (sampled from Nigeria) had relatively short haplotype blocks and less overall LD. The regions that showed a higher LD value were similar in all the four populations. The haplotypes displayed various degrees of diversity with the blocks which also showed variation across the four populations.

Evidence from the data gathered strongly suggested that the selection of the tag SNPs using HapMap played an important role and that they were transferrable across other populations.

Exceptions to this include rarer SNPs. Due to small sample sizes, there were various levels of differences in the determination of LD and allele frequency which also proved to be a major limiting factor for the transferability of the HapMap-derived tag SNPs. To avoid these errors and increase the accuracy, HapMap is being developed with seven additional populations.

15.9 Summary

- The process of identifying functional elements along a genome's sequence and thus giving it meaning is known as genome annotation. It's required because DNA sequencing generates sequences with unknown functions.
- The study of how genes and intergenic regions of the genome contribute to various biological processes is known as functional genomics. A researcher in

this field studies genes or regions on a "genome-wide" scale (i.e., all or multiple genes/regions at the same time) in the hopes of narrowing down a list of candidate genes or regions to investigate further.

- The entire genetic composition of an organism constitutes to form its genome which is classified into prokaryotic and eukaryotic genome based on the type of organism and features of the genome. A prokaryotic genome is considerably smaller in size than a eukaryotic genome and has a less defined nucleus. A prokaryotic genome can be classified into several categories and are physically diverse. The genome of prokaryotes is composed of a single replicon which consists of its own replicator and initiator. Several terms have been used to establish the classification of DNA molecules which might be present in the genome that is multipartite in nature. Repeats in a prokaryotic genome are known to influence important functions and are of high biological significance.
- The eukaryotic nuclear genome can be divided into two or more linear DNA molecules, each of which is stored in a chromosome which is distinguished. All eukaryotes are known to also contain shorter mitochondrial genomes that are generally circular in nature. Plants are observed to have an additional genomic structure, i.e., present in their chloroplast known as the chloroplast DNA. This is unique to the eukaryotic genome of plants alone and cannot be identified in any human or animal eukaryotic model.
- The smallest eukaryotic genome has been found to be shorter than 10 Mb in length, and the largest eukaryotic genome is larger than 100,000 Mb. The complete absence of corresponding link between the organism's genome complexity and its genomic size led to the theory of the **C-value paradox.** *S. cerevisiae* is a common illustration that depicts this point.
- The presence of DNA-binding proteins known as histones is very crucial in packaging the DNA molecules in a chromosome. The histone protein was found to bind to the DNA molecule in a manner similar to the presence of "beads on a string."
- The Human Genome Project aims to interpret the human genome sequence, define all of the genes found within it, and develop the research insights to explore all genetic data that has been generated. As genomics technology advanced, this three-billion-dollar, 15-year initiative progressed significantly with the initial goal to create a human genetic map, followed by a physical map of the human genome, and eventually a sequence map.
- Human genome maps have been generated on a variety of dimensions and resolution levels among which genetic linkage and physical maps are utilized to order the genes on each chromosome.
- The effectiveness of using end sequences from long segments (18 to 20 kbp) of DNA cloned into bacteriophage lambda in the assembly of microbial genomes led to the suggestion of using end sequences from 150-kbp bacterial artificial chromosomes to simultaneously map and sequence the human genome (BACs).
- The human haploid genome comprises approximately 30,000 genes and is and around three billion bp in length. Protein-coding sequences are the most researched and well-understood part of the human genome. Genes that encode

for noncoding RNA (e.g., transfer RNA and ribosomal RNA), untranslated regions of mRNA, pseudogenes, introns, repetitive DNA sequences, regulatory DNA sequences, and sequences linked to transposons are all examples of noncoding DNA.

- The International HapMap Project's mission is to identify the general characteristics of variations of a DNA sequence within the human genome and enable this knowledge publicly accessible. The HapMap will aid in the identification of genetic variations that influence common diseases, as well as the development of screening methods and the selection of targeted therapies.
- Human and chimp genomes underwent several changes after their ancestral lineages diverged, including substitutions of a single nucleotide, duplications and deletions of DNA fragments of various sizes, addition of mobile genetic elements, and chromosomal rearrangements. As compared to the chimpanzee genome, it was discovered that more than 95% of the NRNRs longer than 200 bp were also existing in the genome assembly of the chimpanzee, implying that they were ancestral.
- Mitochondria are cellular organelles with an extrachromosomal genome that is derived from and distinct from the genome of the nucleus. The mitochondrial DNA molecule is a circular dsDNA molecule consisting of about 16,569 bp and a weight of 10^7 daltons that is five millimeters in diameter and free of histones. Mitochondrial DNA studies can be employed in forensic human identification.
- The analysis of the variations and resemblance within the framework of the genome and the arrangement of various species is known as comparative genomics. The development of paralogs and orthologs is a crucial step in the evolution of genes. It's important to differentiate between orthologs and paralogs when comparing genome organization in different species.

References

Collins FS, Fink L (1995) The human genome project. Alcohol Health Res World 19(3):190–195

Liao X, Li M, Zou Y, Wu F-X, Pan Y, Jianxin W (2019) Current challenges and solutions of de novo assembly. Quantit Biol 7:90–109. https://doi.org/10.1007/s40484-019-0166-9

Application of Molecular Genetics

16

Dhruti Patwardhan and Nidhi Sharma

16.1 Biotechnology to Study Human Gene

Biotechnology plays an important role in the study of human genes. Techniques from molecular biology can be used to identify genes associated with diseases and the mutations involved in them. These techniques can be used for prenatal testing as well. A sample can be taken from the amniotic fluid or fetal tissue to obtain DNA which can be screened for mutations using recombinant DNA technology. This is especially useful in cases where the affected protein cannot be detected in the early stages because it is either produced in lower quantities or not expressed in fetal stages. This allows us to predict if the newborn will be affected by certain diseases that it may be susceptible to on the basis of family history.

16.1.1 Huntington's Disease

Huntington's disease (HD) is a neurodegenerative disorder which involves loss of motor control, jerky movements, and decline in cognition accompanied by change in personality and psychiatric symptoms. It is a rare disease, and symptoms of the disease appear later in life between 35 and 40 years of age. It is inherited in an autosomal-dominant manner. George Huntington was the first to describe this disorder in detail in 1872. Since the symptoms appear later, many affected individuals already have children and have passed on the genes to their offspring.

D. Patwardhan
Indian Institute of Science, Bangalore, India

N. Sharma (✉)
La Sapienza University of Rome, Rome, Italy

D. Kar, S. Sarkar (eds.), *Genetics Fundamentals Notes*,
https://doi.org/10.1007/978-981-16-7041-1_16

There is a 50% chance of the offspring getting affected with the disorder as it shows autosomal-dominant inheritance.

Huntington (*HTT*) was the first disease-associated gene to be mapped to the human chromosome in 1983. It started with a group of scientists who worked on finding a DNA probe that showed a specific restriction fragment length polymorphism (RFLP) pattern for HD. They tested 12 probes on Southern blot of chromosomal DNA digested with HindIII. One of the probes showed a specific RFLP pattern for DNA from two families which had a history of HD. A large amount of effort over 20 years was devoted to identifying these families and obtaining their pedigree and medical histories. To identify where in the human DNA this HD-specific probe was binding, researchers made use of a series of mouse cell lines called human-mouse somatic cell hybrids. These cell lines were engineered to contain a specific subset of human chromosomes. On hybridizing the probe to a number of these cell lines, it was found that the probe recognized a region on the fourth chromosome. They therefore concluded that the gene responsible for HD was present on the human chromosome 4 and in the region to which the probe was binding.

Over the next 10 years, efforts were put in to identify this gene and the nature of its mutations. It was found that the gene had a trinucleotide repeat of CAG at the beginning of the gene. In normal individuals, number of this repeat varied from 6 to 21. In affected individuals, this number was found to be greater than 40, even up to 100. The trinucleotide expansion was identified as the cause of the disease. Today, molecular genetic approaches can be used to detect the number of CAG repeats in the *HTT* gene. We can therefore predict if an individual will suffer from the disease later in life and the chances of passing on the disease to their offspring.

16.1.2 Cystic Fibrosis

Cystic fibrosis is caused by a mutation in the cystic fibrosis transmembrane conductance regulator (*CFTR*) gene. Cystic fibrosis affects cells that produce mucous, sweat, and digestive juices. It causes the fluids to become thick and blocks ducts and passageways adversely affecting the lung and digestive system. It may cause difficulty in breathing and persistent lung infections. It is inherited in an autosomal-recessive pattern.

One of the most common mutations in CFTR is a 3 bp DNA deletion known as 508. To detect mutations in *CFTR* for the purpose of diagnosis, allele-specific oligonucleotides (ASO) are used. These are probes constructed as an exact match against the mutated allele or the normal allele. This probe will only bind to the allele which is an exact match and will not bind even if there a single nucleotide mismatch. To do the test, DNA is isolated from the white blood cells of the individual to be tested. This is spotted on a nylon membrane and allowed to hybridize with both the ASOs against normal and mutated allele under specific conditions which do not allow a mismatch. In unaffected homozygotes, only the ASO against the normal allele will hybridize. In case of carriers, both ASO against normal as well as mutated

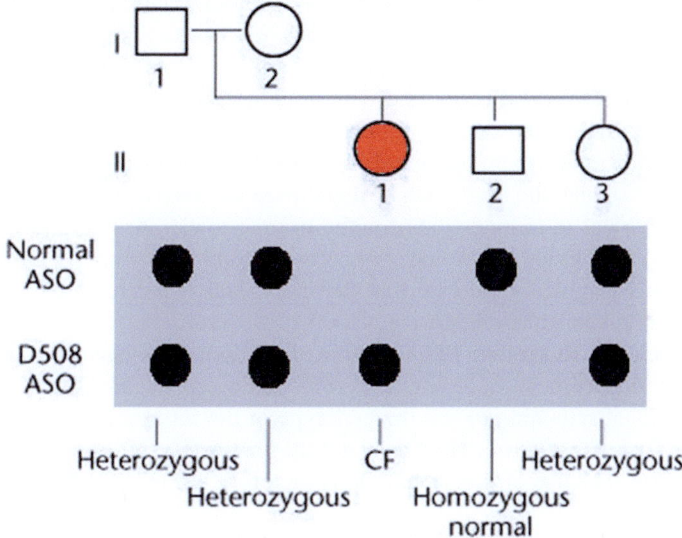

Fig. 16.1 Genetic diagnosis of cystic fibrosis. Allele-specific oligonucleotides are prepared for both normal allele (normal ASO) and allele having Δ508 deletion (D508 ASO). DNA is extracted and blotted for parents and three children. Both ASOs hybridize for parents both of whom are heterozygous. The affected child (child 1) shows hybridization only with D508 ASO. Child 2 is homozygous for the normal allele and shows hybridization with only normal ASO. Child 3 is a carrier and shows hybridization with both ASOs

allele will hybridize. In case of affected individuals, only the ASO against mutant allele will hybridize. This is illustrated in Fig. 16.1. Although a powerful technique, it suffers from an obvious limitation. The *CFTR* gene may have mutations other than 508, which will not be detected by this technique. Thus, a negative result on this does not necessarily mean that the individual has no mutations in this gene. However, as more mutations are identified and more genomic data is available, we will be able to identify most if not all mutations that are present in the population and provide a screening that has a better coverage.

16.1.3 Sickle Cell Anemia

Sickle cell anemia is a disorder in which the shape of erythrocytes is affected due to a substitution mutation in the β globin gene. The abnormal erythrocytes become elongated and curved resembling a sickle due to polymerization at low oxygen tension. The normal erythrocytes are disc shaped. Aggregation of red blood cells leads to oxygen deprivation to many tissues which might severely damage them.

The mutation in β globin gene eliminates restriction site for the enzymes *Mst*II and *Cvn*I. These differences lead to a different restriction pattern seen on the Southern blot for mutated allele and normal allele. This distinguishing feature can be used to diagnose individuals having the mutated allele. The MstII restriction enzyme has three sites in the normal β globin gene cleaving the gene into two fragments. In the mutated allele, the site in the middle is lost cleaving the gene into a single fragment. DNA can be extracted from an individual exposed to *MstII* digestion, and the fragments can be separated by gel electrophoresis. This can be transferred to a nylon membrane and visualized by Southern hybridization using probes that recognize the fragments of β globin gene. Two small fragments indicate that the individual is homozygous for the normal allele. A single large fragment denotes that the individual is homozygous for the abnormal allele. A large fragment and two small fragments indicate that the individual is heterozygous, having one normal and one abnormal allele.

The difference in pattern of restriction digestion fragments produced can be utilized here to diagnose individuals having sickle cell anemia. It can be used to perform prenatal screening to test the genotype of the fetus and determine if he/she will suffer from the disease. However, not all mutations will eliminate or create a restriction site. The use of this technique is therefore limited.

16.2 Biotechnology to Study Plants

Plant breeders often select plants having favorable characteristics and use their seeds for further breeding. This is nothing but manual selection of traits to ensure a more robust crop with desirable traits. Biotechnology can be used for the same. Genes that confer the desirable trait from different organisms can be isolated and introduced in the genome of the plant of interest. We can increase nutritional value of crops, obtain insect- and herbicide-resistant crops, and also increase their yield.

16.2.1 Transgenic Crops

Plants whose genome have been modified by genetic engineering techniques to introduce a desirable trait are called transgenic plants. Vitamin A deficiency is a major health concern especially in Africa and Southeast Asia. Vitamin A deficiency may lead to blindness and also weakens the immune system. In these countries, vitamin A-rich foods like milk, eggs, and fish are generally expensive and not within the reach of the poorer sectors. Rice is relatively cheaper and is a staple diet in these parts. Though rice is a good source of carbohydrates, it lacks micronutrients like vitamins and minerals. Golden rice was created in 1999 with an intention to fortify rice crop with vitamin A. This is called biofortification. The technology was provided free of cost by its investors for use in the public sector rice varieties without any limitations on use of recombinant crops and seeds.

Genetic engineering in rice plant allows it to biosynthesize β carotene which is a precursor of vitamin A. β carotene is produced in the endosperm, giving the rice its characteristic golden-yellow color. For this, two genes were introduced in the rice plant, Phytoene synthase (*psy*) from daffodil and phytoene desaturase (*crtl*) from soil bacterium *Erwinia uredovora*. These genes are placed under endosperm-specific promoter. The introduction of these gene leads to the synthesis of lycopene which is naturally converted to β carotene by the plant's endogenous enzymes. In 2005, golden rice 2 was announced which produced significantly higher amounts of β carotene than the original golden rice. Golden rice received its first approval in 2018 from Australia, New Zealand, Canada, and the USA. Golden rice is a cost-effective way to provide essential nutrients to a large population. It is also safe for consumption.

16.2.2 Herbicide Resistance

A large portion of the crop yield is destroyed due to weed infestation. These unwanted weeds can be removed by herbicides. The herbicides, however, may also affect the crop plants. They may get washed into the water and deposited in the soil which also adversely affects the environment. Herbicide-resistant plants will protect plants from herbicides while allowing surrounding weeds to be destroyed.

Glyphosate is an herbicide which has the ability to kill plants by inhibiting an enzyme called EPSP synthase which is present in the chloroplast. This enzyme is important for the synthesis of essential amino acids, and without this enzyme, plants are unable to survive. This herbicide does not affect humans and is effective at low concentrations. It is also rapidly destroyed by soil microorganisms. EPSP synthase is also present in bacteria and essential for their survival. There is a strain of *E. coli* that is however resistant to glyphosate. The *EPSP synthase* gene of this resistant strain can be used to confer resistance against glyphosate in crop plants. To do this, the *EPSP synthase* gene from glyphosate-resistant *E. coli* was cloned into a vector under a plant viral promoter sequence and upstream of the plant's transcription termination sequence. This recombinant vector was introduced into the bacterium *Agrobacterium tumefaciens*.

Discs were cut out from plant leaves and infected with *Agrobacterium tumefaciens* carrying the vector. Due to infection with bacteria, the plant tissue developed calluses which consists of unorganized plant parenchymal cells. These callus cells were tested for their ability to resist glyphosate. The calluses which were able to survive were further cultured and grown into transgenic plants. These plants were exposed to high concentration of glyphosate, and only those plants which were able to produce the resistant EPSP synthase gene in high quantities were able to survive and the others died.

Glyphosate-resistant corn and soybean have been created in this manner and are available in the USA and other countries since its introduction in 1996.

Unfortunately, persistent use of herbicides due to resistant crops have now led to the evolution of herbicide-resistant weeds. Weeds are developing resistance mechanisms for a large number of herbicides. We therefore need more studies pertaining to understanding evolution of resistance and sustainable solutions addressing these issues.

16.2.3 Insect Resistance

Just like weeds, insects also impact crop production to a large extent. With growing human population, it is essential to optimize our food production to meet the demand. Creating insect-resistant crops have led to an increase in overall yield of crops like corn, potato, and cotton. It has also cut down on the use of insecticides which may have harmful effects on humans.

Genetically modified crops containing δ-endotoxin also known as Cry proteins from *Bacillus thuringiensis* (Bt) were introduced in the mid-1990s. Most of the recently produced insect-resistant strains contain multiple Cry proteins which are toxic to Lepidoptera and Coleoptera species. The Bt toxin when ingested by the insect is solubilized in the midgut where it gets proteolytically cleaved at the N terminal to an active form. The active molecules bind to a receptor in the epithelial cells of the midgut. This induces formation of pores in the membrane resulting in osmotic lysis and cell death which eventually kills the insect. In the transgenic plants, Bt toxin is expressed directly in its active form. Transgenic insect-resistant crops have had a major beneficial impact on agriculture by improving crop yield and reduction of pests.

16.2.4 Production of Biofuels

With rapidly depleting sources of fossil fuels and its negative impact on the environment, it is essential to look for alternate sources of energy. Biofuels are one such source which can fully or partially replace the use of fossil fuels. The first generation of biofuel utilized sugarcane or sugar beet for production of ethanol through fermentation. Different alcohols, for example, butanol, can be produced by applying different fermentation process. The first generation of biofuels utilized crops which can also be used as food or feed, thereby increasing their demand and creating shortage of food crops. With the advancements in biotechnology, second generation of biofuels were developed which reduced reliability on food crops. Lignocellulosic biofeedstock was used for this which was a less expensive biomass. The cellulose and hemicellulose present in this biomass is broken down into its constituent simple sugars through enzyme-catalyzed hydrolysis. These are then fermented with the use of microorganisms to produce alcohol. These fuels are also known as cellulosic ethanol or cellulosic biobutanol. Newer biofuels now referred to as the third generation of biofuels are produced by using algae biomass with microbial enzymes. Microalgae offer many advantages because they are able to rapidly double their biomass. They also are rich in oil, and some of them have oil content about 80% of their dry weight. They can also be grown in waste or non-potable water. This solution also has its own problems. It requires a high density of algae culture over large surfaces which is not very optimum. Apart from algae, other prokaryotes or eukaryotes which show high accumulation of oils are also being considered for production of biofuels.

There are multiple stages at which use of biotechnology can increase efficiency of production of biofuels. Microbial enzyme activity can be enhanced and improved via biotechnology to achieve better microbial digestion and fermentation of biomass. The use of genetically modified organisms for pretreatment or conversion to ethanol can boost productivity. The cell wall and composition of lignocellulose in plants being used for production of biofuels can be modified using biotechnology to increase the yield of ethanol. Using herbicide- and insecticide-resistant plants can improve the biomass produced. Creating plants that can survive and grow in harsh soil or weather conditions will allow plants meant for biofuel production to be grown on nonarable land. This will allow arable land to be used only for crop production.

16.3 Biotechnology to Study Pharma Product

Biotechnology can make use of the fact that all organisms contain the same nucleotides and follow the same code for its conversion to proteins. This allows a human gene to be inserted into a plant or a bacteria to produce the protein of interest. The bacteria or plants can then be cultured on a large scale, and protein of interest can be isolated and purified. Thus, the bacteria are acting as molecular factories producing the protein of interest on a large scale.

16.3.1 Recombinant Insulin

Insulin became one of the first human protein to be produced using recombinant DNA technology and licensed for therapeutic use in 1982. Insulin is required in the body for glucose metabolism, and lack of insulin leads to diabetes. Insulin can be given to diabetics externally which allows them to maintain their blood glucose levels. Insulin is produced by cells in the pancreas in the form of a precursor peptide known as preproinsulin. This gets cleaved, and some amino acids are removed from its center and at its end. This leads to formation of two polypeptide chains called A and B which are held together by disulfide bonds.

Before the use of rDNA technology, insulin was produced by extracting it from porcine or bovine pancreas. There were two issues associated with this method. One was that although animal insulin was chemically similar to human insulin, it is not identical. This difference led to immune reaction from patients causing inactivation of insulin and inflammation in many patients. Secondly, production of insulin from animals was very expensive and difficult to obtain in large quantities. Both these issues were prevented by using bacteria to produce human insulin. Since posttranslational modifications differ between bacteria and humans, the insulin gene was not inserted as is in *E. coli.* Instead, the polypeptides were synthesized separately using two different plasmids. Polypeptide chain A has 21 amino acids, and polypeptide chain B has 30 amino acids. The genes for these two chains were constructed using oligonucleotide synthesis (Fig 16.2). The genes were inserted in the vector adjacent to a lacZ gene to produce fusion proteins. The fusion proteins consisted of polypeptide A or B fused to β galactosidase (product of *lacZ* gene). The vector also contained a gene for antibiotic resistance which was useful in selection of bacteria containing the vectors. These recombinant bacteria were cultured in large-scale fermenters. From the bacterial extracts, fusion proteins were isolated and treated with cyanogen bromide to remove the β galactosidase (Fig 16.2). The insulin chains were then purified and mixed. The chains were able to spontaneously unite to form the active molecule. This insulin was capable of being purified, packaged, and used in therapy.

16.3.2 Recombinant Growth Hormone

Human growth hormone is produced in the body by the pituitary gland and released into the blood. It performs a host of biological functions like metabolism of proteins, carbohydrates, and lipids as well as cell proliferation and immune regulation. Growth hormone is essential to ensure proper growth and stature, and its deficiency may lead to dwarfism. Dwarfism can be treated by administering growth hormone. It is also used in the treatment of burns, bone fractures and disintegration, and gastric burns. Till mid-1980s, the only source of human growth hormone was human cadaver tissue. The supply for this hormone was therefore limited. There were also reports that associated pituitary-derived growth hormone with Creutzfeldt-Jakob disease. Recombinant DNA technology provided a means of safely producing abundant amounts of human growth hormone (hGH).

hGH is produced by the pituitary as a prehormone. It contains a hydrophobic leader peptide of 20 amino acids. During secretion, this leader peptide is removed to produce the mature hormone of 191 amino acids in length. To facilitate the direct expression of mature hormone, cDNA coding for the leader peptide was removed. The cDNA for growth hormone was partially chemically synthesized and partially derived from the actual mRNA of human pituitary. This was significant because unlike insulin, which is only 51 amino acids long, chemically synthesizing entire mRNA for such a large protein like hGH would have been difficult. This cDNA was cloned into a plasmid which was introduced into a strain of *E. coli.* Since nonglycosylated form of the hormone was active, prokaryotic system was preferred for its production. The recombinant bacteria is grown in large quantities, and growth hormone produced was isolated and purified.

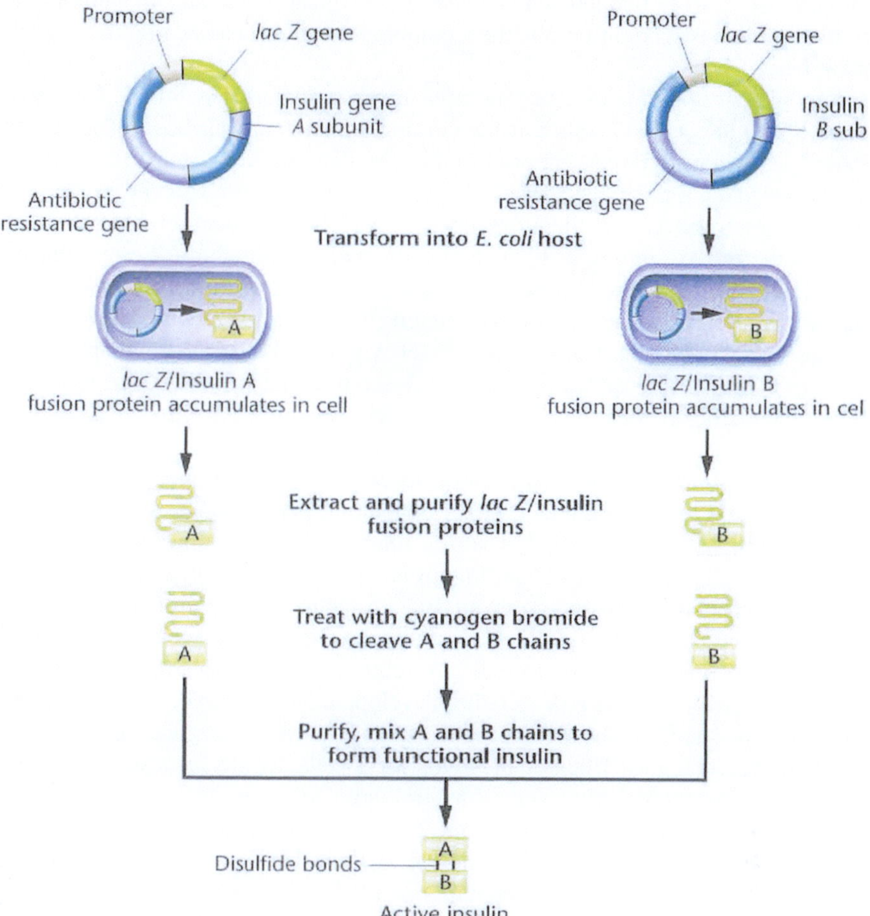

Fig. 16.2 Production of recombinant insulin. The genes coding for polypeptide chains A and B of insulin are inserted in a bacterial plasmid fused to the *lacZ* gene. The plasmid is transformed into E. coli and cultured in large-scale fermenters. The fused lacZ/insulin A or lacZ/insulin B fusion protein accumulates in the cell from where it is extracted and purified. It is further treated with cyanogen bromide to separate insulin from β galactosidase. The A and B chains are purified and mixed to form the active insulin protein

16.3.3 Recombinant Vaccine

Traditionally, vaccines have been produced by either killing or attenuating the pathogen. This inactivated pathogen is no longer able to cause the disease but stimulates the immune system to produce antibodies against it, providing protection against future infections from the pathogen. Using recombinant DNA technology, only the peptide that acts as the antigen can be produced which can elicit an immune response. These are called subunit vaccines. They provide protection against possible risk of live or attenuated vaccines like reversal of attenuation and virulence in susceptible hosts. Additionally, the recombinant proteins can be produced in large quantities.

Recombinant protein vaccine currently in use is against hepatitis B. Hepatitis B virus (HBV) infects liver cells causing chronic infection and cirrhosis. The hepatitis B surface antigens (HBsAg) are produced in yeast expression system. Yeast cells are capable of making posttranslational modifications in proteins. Protein products that require glycosylation can therefore be produced in the eukaryotic yeast. It also secretes the HBsAg into the supernatant of the culture allowing for easier purification. The HBsAG when administered assemble into viruslike particles which are highly immunogenic and capable of eliciting an immune response. Recombinant vaccine against human papillomavirus (HPV) has also been developed which contains the L1 major capsid protein. Many subunit vaccines, however, have weak immunogenicity on their own and need to be administered with an adjuvant to promote long-lasting and strong protective immune response.

Genetic engineering can also be used to create live recombinant vaccines. The idea is to use a live recombinant vector containing heterologous antigen encoding genes. The live vector can elicit a strong immunological reaction against its own antigens as well as toward the heterologous antigens being expressed. An example is the work being done on recombinant BCG vaccine. The vector *M. bovis* BCG provides many advantages. It is safe and can elicit T-cell-mediated immunity. Recombinant BCG (rBCG) expressing foreign antigens for various diseases like malaria, tuberculosis, and HIV is being developed. For example, rBCG expressing HIV antigens has been shown to produce specific antibodies against HIV, produce interferon γ, and induce T helper and cytotoxic T cells. Efforts are also being used to utilize viral vectors for expression of heterologous antigens.

Direct injection of DNA plasmid into the muscle to induce immune response is also another approach that is being studied as a vaccine system. In a DNA vaccine system, the antigen can directly be expressed by host cells in a manner similar to viral infection. They have been shown to elicit both humoral and cell-mediated immunity. DNA vaccines avoid problems associated with producing recombinant proteins like inaccurate folding of protein and purification costs. DNA vaccines, however, have their own set of problems like low efficiency of transfection of cells in vivo, production of anti-DNA antibodies, and possible integration into host genome. Although successful in animal models, DNA vaccines have shown limited immunogenicity in primates. Ongoing efforts in increasing its effectiveness include strategies like augmenting gene expression, co-expression of cytokines and other molecules that boost immune response, and formulations to protect DNA from degradation.

Even after production of purified vaccine, there are challenges involved in administering the vaccines especially in developing countries. Absence of facilities for manufacturing, transportation, and storage pose challenges for vaccination in remote places. To circumvent these issues, creation of edible vaccines was proposed. Edible vaccines are transgenic plants or animals that express the antigen of a pathogen and, when consumed, can elicit an immune response in the body. These vaccines would provide the advantage of being inexpensive, not require special storage conditions, as well as not require trained medical personnel for administration. Transgenic tobacco plants having leaves expressing antigenic subunit of hepatitis B virus have been produced. This is just a model system, and for actual use, the gene for HBV would be transferred into a food plant. In another example, rabies antigen was expressed in spinach and fed to volunteers. Eight of the fourteen volunteers showed high expression of rabies-specific antibodies. Edible vaccines are undergoing further studies and clinical trials.

16.3.4 Recombinant Protein

The proteins mentioned above including recombinant insulin, recombinant growth hormone, as well as the vaccines are examples of recombinant proteins. The first recombinant protein was somatostatin produced in 1976. From there, a host of other proteins have been cloned and expressed in heterologous systems from bacteria to plants which are widely used in therapy today. Recombinant proteins provide the advantage of producing large number of products at lower cost. It is also safer as it avoids transmission of infection from animal- or cadaver-derived products.

The X-linked disorder hemophilia occurs due to a lack of clotting factors. Hemophilia A is caused by a defect in factor VIII, and hemophilia B is caused by a defect in factor IX. For a long time, plasma-derived clotting factor concentrates were used for the treatment of hemophilia. However, due to lack of proper screening methods, they were discovered to be also transmitting blood-borne viruses including HIV and hepatitis. In 1984, the gene for factor VIII was successfully cloned, paving the way for its production using recombinant technology. By 1992, factor VIII was commercially produced and licensed for therapeutic use. Factor IX was commercially available for people with hemophilia B in 1997. These recombinant proteins provided a safe and effective method for treatment of hemophilia. Table 16.1 lists the recombinant proteins used in clinical treatments.

A range of recombinant proteins including hormones, antibodies, enzymes, and vaccines have been produced which have been used for therapy in certain diseases.

16.4 Biotechnology to Study Animals

Applying novelty of molecular genetics in the field of biotechnology, the study of animals has become one of the important branches of study that includes a vast area of related topics such as animal health, animal in research, transgenic animal, gene pharming, etc. Animal biotechnology is an important branch of biotechnology that includes a wide range of topics such as use of animal in research, clones, transgenic animal, gene pharming, and animal health. In the past history of human beings, animals have been bred to the enormous purposes over 1000 years, such as breeding of (i) working dogs to flock pasture animals, (ii) cows produces more tender and tasty meat, (iii) horses that are easier to tame, and so on. All these things have been

Table 16.1 List of recombinant proteins and their therapeutic applications

Proteins	Therapeutic application
DNase I	Cystic fibrosis
Coagulants factors	Hemophilia A and B
Erythropoietin	Anemia in chronic renal disease
Glucocerebrosidase	Gaucher disease
Growth hormone	Pituitary dwarfism
Insulin	Diabetes
Alpha interferon	Some leukemias, Kaposi's sarcoma, and hepatitis B and C
Gamma-1b interferon	Chronic granulomatous disease
Interleukin-2, interleukin-3, and interleukin-4	Immunotherapy of cancer
Tissue-type plasminogen activator	Acute myocardial infarction and massive pulmonary embolism
Antibodies for cellular immunotherapy	Neoplastic processes
Vaccines	Influenza and hepatitis A and B
Monoclonal antibodies anti-antibodies	Lupus and rheumatoid arthritis

possible by the use of biotechnology. Notably, animal biotechnology is the core result of genetic engineering which includes recombinant technology for transferring exogenous or foreign DNA into germ line. Genetic engineering is a modification of an organism's characteristic by adjusting its genetic material.

Use of Animals in Research

Animal models are an essential part of primary research. The need of medical assessment of various drugs or products produced for disease treatment is supported by the use of animal models. It has been noticed that working on computer models and in vitro cell studies would not reproduce the result as an organism does. Thus, working with computer models and cells will be a supplement to animal research. Regardless of initial failure, recent development in animal biotechnology has changed the fate of agriculture, medicine, and animal breeding, and a prompt effort to conserve the endangered animals to save the flora and fauna is remarkable.

It's very well known that for any new product to be approved for human use, the manufacturer must first demonstrate that it's safe, and therefore, trials on cell culture, in live animals, and on human subjects are facilitated by virtue of this field. The most required animal used for testing is a pure-bred rat, mice, primates, etc., prior to trial on human subjects. Zebra fish, a hardy aquarium fish, is extremely used for valuable research, while dogs are enormously used for study of cancer, heart disease, and lung disorders. Importantly, HIV and AIDS like lethal studies are always conducted on monkey and chimpanzees.

16.4.1 Transgenic Animals

Transgenesis (transfer of significant genes) or transgenic animal is one of the most significant and exciting research tools in biotechnology. Through this advanced technology, inserting new genes in livestock for economical important characteristics such as fertility, resistance, or tolerance to the environmental stress has become a major revolution in animal breeding. Another significant application that transgenic animal brought to us would be producing clotting factors in the milk of domestic livestock.

To elaborate the literal definition of transgenic animal is a deliberate modification of genome of one animal by introducing DNA through recombinant DNA technology and then must be transmitted to the germ line that homogeneously spread to every cell and pass over through the next generations. It means that gene can be altered artificially so that some characters of animals are changed and can pass to all offspring.

In modern history, fundamental to these techniques is an ability to culture early embryos in vitro which allowed a variety of manipulations in the genome to be performed. One of the first genetic manipulations of the embryo is the production of chimeric mice achieved by mixing an early stage of development (eight cells) to develop a single form of embryo carrying chimeric characteristics in adult. Likewise, knocked-out mouse is also another great achievement in the history of biotechnology. First ever knocked was created by Mario R. Capecchi, Martin Evans, and Oliver Smithies in the year 1989 who won the Nobel Prize in 2007 for introducing knockout mice technique. Knockout generally refers to a deletion (knockout) of a gene to inactivate its function.

Reasons that Promote the Production of Transgenic Animal

1. Some transgenic animals are being produced, keeping the goal in mind for specific economic traits which can produce milk containing particular human proteins like protein C (a potential coagulant) and fibrinogen (plasma glycoprotein).
2. The primary goal was to produce advanced cattle breed which is subsequently followed by the second goal of production of disease model. This disease model shows disease symptoms as the result of transgenesis. This model has been manufactured to study several disease studies. For example, OncoMouse® and 3× Tg AD mice are produced to study various human cancer and Alzheimer's disease. These animals successfully show the symptoms of the disease alike humans.
3. One ultimate purpose is to study the interaction between environment and genome. This can be possible with transgenesis that provides a unique opportunity to design unique models for such study.

Methodology for Producing Transgenic Animal

To date, there are several methods in used, but the most efficient methods are as follows:

(a) Recombinant retrovirus.
(b) Embryonic stem cell.
(c) Pronuclear DNA microinjection.

Recombinant Retrovirus: Recombinant retrovirus technique is mostly useful for eukaryotic genome. Retrovirus is a family of virus in which genetic material is RNA and transcribed into DNA by specific enzyme "Reverse Transcriptase (RT)." Of this significant characteristic, RT becomes an essential tool for cloning or transgenesis of eukaryote genome. Eukaryote genome cannot clone in directly bacterial cell because it contains additional introns, which make the genome heavy and lengthy which cannot transfer to the bacterial cells. Thus, mRNA from the desired gene will transfer to this virus which later will be transcribe into single-strand complementary DNA (cDNA) by RT. DNA polymerase will transcribe the second strand, complementary to the first strand to complete the DNA amplification. A direct gene transfer to retrovirus will be followed by injecting virus to the host cell and cultured to maintain the cell line, and cultured cell will develop to a mature embryo. Retrovirus method is mainly used to transfer genetic material into host cell to produce the chimeric characteristics of transgenic animal containing diverse genetic constitution from the donor.

Embryonic Stem Cell

This is one of most common techniques and often being used to produce successful transgenic animal. This method is first initiated with isolation of totipotent cell (a cell that has a capacity to develop into any specialized cell) from the embryo. The gene of interest is being inserted in vitro into these cells (totipotent) and then incorporated into the host embryo resulting in a chimeric animal (Fig. 16.3).

DNA of interest is isolated and injected to the embryonic stem cells by gene delivery method, and cells were cultured and maintained carefully. This embryonic stem cell containing DNA is injected to the blastocysts and implanted to the uterus of foster mother. The first generation will be heterozygous, and mating with wild type will produce the homozygous transgenic strains.

Pronuclear DNA Microinjection

First time, this technique was described by Gordon et al., and mice were the first transgenic animals produced through this method. Male and female pronuclei (nucleus of sperm or egg cell) are microscopically spotted after immediate hours of sperm entry into the cell.

The gene of interest or cluster of recombinant genes in a construct will be subjected to microinject into either of these pronuclei (Fig. 16.4). This manipulated cell is first needed to be cultured in vitro (in a lab, not in an alive organism) to develop to a specific embryonic phase, which is further transferred to the recipient female. The recipient female is always a pseudopregnant that promotes the hormone stimulation in the body to make her uterus receptive to the embryo that has to be transplanted in it. However, transfer of gene by DNA microinjection would not be

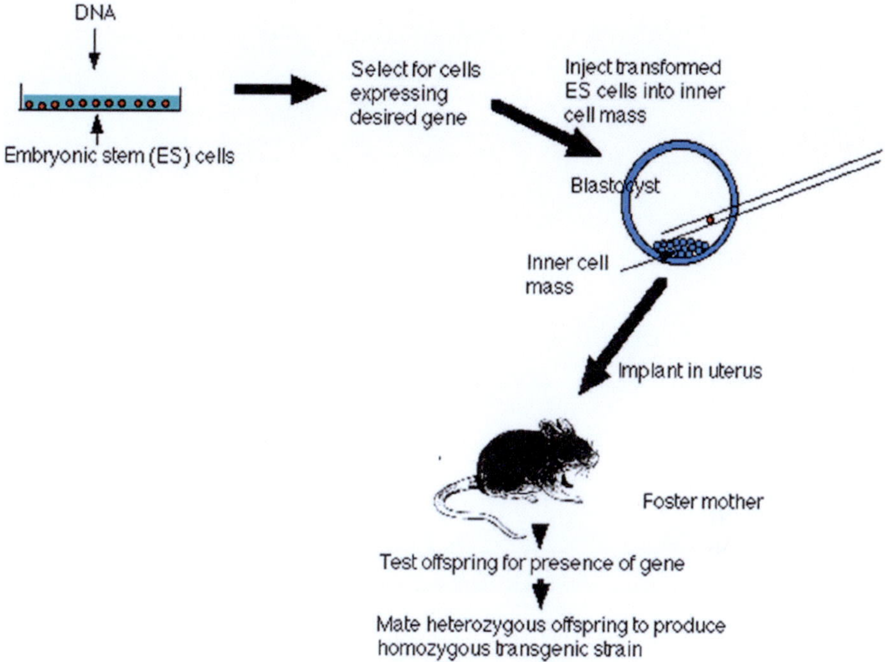

Fig. 16.3 Embryonic stem cell-based transgenic method

homogeneous since it is a random process, so it's not necessary that all pups will be born with expressed desired gene.

Application of Transgenic Animals in Human Welfare

Disease Model: For ages, mice model has been depicted as potential source for studying human disease since mice shares physiological, anatomical, and genomic characteristics with human. To study most effective disease such as cancer, AIDS, and Alzheimer, transgenic animals have been produced to exhibit similar symptoms in order to understand the basic fundamentals of these disease. In short, transgenic animals enable scientists to understand the role of genes in specific diseases.

Growth. Transgenic pigs and sheep are being produced with increased growth and heavy body composition. This attempt had been achieved by transfer of gene that regulates growth hormone. Delivery of proteohormone (peptide hormone) made it possible.

Quality of Animal Products. Improved quality and composition of animal products can be achieved by respective gene transfer. A model for this hypothesis is being proposed by Mercier in the year 1987. He proposed a model that is lacking lactose content in the milk on purpose. Practically, sheep and cattle had been produced carrying lactose gene but combined with udder specific promoter. This construct results lactose degradation into the end product. Thus, milk lacking lactose will be useful for a large population that are suffering with lactose intolerance.

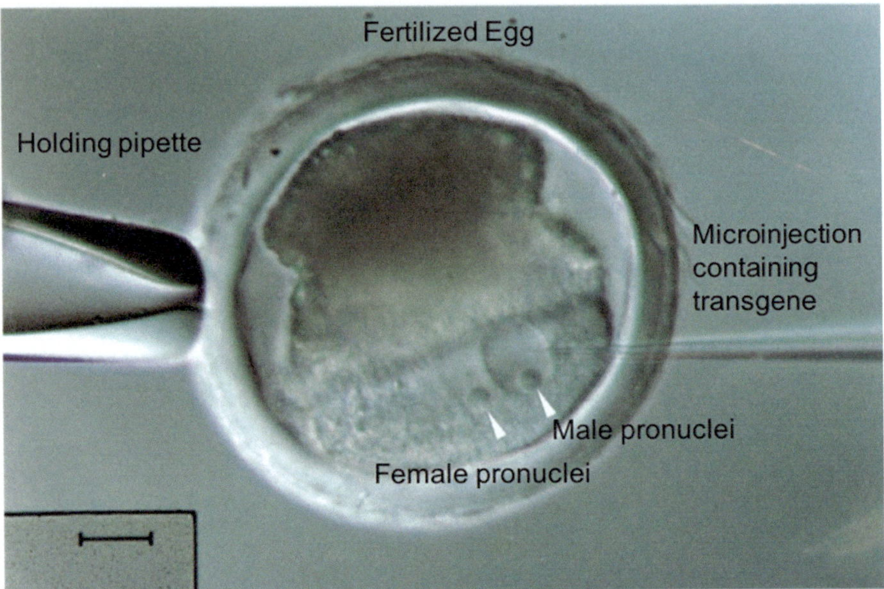

Fig. 16.4 DNA microinjection in pig embryo to inject transgene (image taken on scale bar 20 μm). A DNA construct or recombinant construct of transgene was prepared and injected to the pronuclei (nucleus either egg or sperm cell) with needle or microinjector. This is further followed by cultivating into matured embryo and implantation to the foster mother

Researchers also have produced transgenic animals that promote cysteine synthesis (an essential amino acid) in the animals that enhance the woo growth particularly.

Gene Pharming. Pharming seems a misspelled word for farming, but it isn't at all. This word comes from two different word "farming" and "pharmaceutical." Thus, pharming denoting here is the production or farming of significant genes or proteins by means of secretion in the transgenic animal's blood, milk, saliva, eggs, etc. Altering the gene makeup (modifying its own DNA or splicing) of an animal through transgenesis or transfer of a particular gene for production of valuable proteins for human purpose leads to the idea of gene pharming. In this direction, tissue-specific promoter inducing the protein production in domestic animal is a reliable source for human needs. Therefore, remarkable efforts have been made by scientists, in particular direction to use animals as bioconversion system.

In the year 1987, Gordon et al. and Simons, McClenaghan, and Clark successfully demonstrated that human T-PA (tissue-specific plasminogen activator, to treat the blood clotting) and sheep beta-lactoglobulin were, respectively, expressed in the milk of transgenic mice.

Industrial Applications

Two scientists from Nexia Biotechnologies, Canada (2001), had spliced a spider gene into the cells of lactating goats. Eventually, they observed that goats started to produce silk in a form of tiny strands from their body along with their milk. The amount of silk was quite enough to commercialize it. This strand was subjected to be extracted and weaved into thread that would be useful for manufacturing objects like military uniform, tennis racket strings, etc.

In 1997, the first cow "Rosie" was produced as a transgenic cow that secreted protein-enriched milk at 2.4 grams per liter. This cow was more nutritional than normal bovine milk. The milk was containing human gene "alpha-lactalbumin."

Ethical Issue

Beside its application in human welfare, transgenic animals had faced many bioethical issues raised by environmentalist and activist and cannot be ignored by scientist, biotechnology industry, policy maker, and public domain. Those ethical issues and doubts we tried to sum up here are as follows:

1. Why transgenesis does not have any universal or standardized protocol?
2. Should only promising research permit to demand such protocol?
3. Why transgenic animals have to be taken into consideration of human welfare only? What about lab animal welfare or other forms of life?
4. Should transgenic method first needed to be examined on in vitro (cultured cells in the laboratory) before using live animals to reduce animal surfing?
5. Do transgenic animals provide an evolutionary benefit in the right direction or instead will result in a drastic consequence for nature and humans?
6. Should patent policies be legitimate to restrict the free exchange of scientific research?

16.4.2 Improved Reproductive Rate

Animal reproduction through animal biotechnologies have endowed many improvements in agriculturally important traits and livestock in which production of development of genetically improved animals for farming is the primary concern for researchers. The very primitive and first biotechnology tool applied to improve the production was artificial insemination as it is well known that reproductive success is of topmost importance for economic efficiency of cattle production in animal pharming. High productive efficiency is required for efficient milk production and meat production and therefore has an influence on herd profitability. Among the most recent of these emerging technologies, reproductive cloning and production of transgenic animals are the best choice to improve the reproductive rate in animals. Nevertheless, the other novel techniques include synchronization of estrus, in vitro fertilization, multiple ovulation (female released more than one egg in a month by intrauterine insemination), embryo transfer, and cloning, and they are all important potential tool for improvement of livestock reproductivity rate. Researchers have found that it is important to avoid the risk and challenges that impede productivity, reproduction rate, and health with adverse environment conditions. This concern has been prioritized in animal reproduction system that results in encouragement of the practice of techniques mentioned below.

Estrus Synchronization

This is a technique related to regulate the estrus synchronization of female. Estrus synchronization is basically a manipulation of heating time to reduce for a short period (36 to 96 hr). Such synchronization can be achieved by using of one or more hormones. This technique is one of the competent methods that increase the possibility of animals to breed at the beginning of breeding season.

Artificial Insemination (AI)

AI has been practiced worldwide on large scale from more than half of the century. This is one of the most common and efficient method among all. The purpose of this technique is to determine an efficient bull with maximum fertility rate. It has been proven by several studies that seminal plasma contains fertility-associated antigen which defines the differences in fertility rate between males whose seminal composition is the same though. To this reason, even every bull comprised the same amount and composition of semen but may affect the sperm capacitating, fertilization, or related events. Because of technology boom, in recent years, sperm collection and AI have been improved by the advent of sperm sexing or selection which here refers to a possibility to select the sperm carrying X chromosome or Y chromosome with 85–95% of accuracy. This method is common in some domestic animals such as buffalos. At practical ground, this technique is facilitated by using cutting-edge flow cytometric principle, where fluorescent-labeled X chromosome-bearing spermatozoa can separate from the fluorescent-labeled Y chromosome-bearing spermatozoa. In addition to this, the production of more male progeny and the reduction of sex-linked disease are the most beneficial outcomes of this approach. However, this technique has some limitations including reduced number of sexed sperms and variety of damage to the sperm cells, viz., sperm membrane destabilization. Changes in the competence, thereby, life span of spermatozoa would reduce after fertilization in female genital tract.

Embryo Transfer

This technique is one of the tools which provides a faster rate to livestock and an opportunity in which both male and female has contributed equally. This method involves superovulation which is an important step to increase the oocyte number from the superior donor. The first mammalian embryo transfer was reported by Walter Heape in 1890, while the first birth of calf through this method was reported by Betteridge. The first live calves developed from bubaline embryos (using embryo transfer method) were born in 1983 in the USA and later in India.

The essential stages for this method are as follows:

- Donor cow of good pedigree animals and treated with hormones (FSH and LH) to stimulate ovulation and release eggs in large number—multiple ovulation (MO).
- Insemination is performed using semen of a chosen bull.

- Embryos are flushed nonsurgically after 6–7 days of insemination.
- Collected embryos are implanted to the recipient cows whose estrus cycle is at the correct receptive stage due to hormone manipulation.
- Embryo may be frozen and stored for a long time.

In India, embryo transfer and protocols of ET technology are being standardized for cattle, buffalo, sheep, goat, camel, and other species of animals. Embryo transfer is an effective method, and genetic improvement is higher in embryo transfer than AI alone.

In Vitro Fertilization (IVF)
In vitro fertilization is an artificial technique to fertilize the oocytes outside and transfer the developed embryo to the recipient mother. The oocytes can be collected either from slaughterhouse or live animals. This oocyte undergoes the maturation and fertilization in vitro and later will develop into mature and viable embryo. "Pratham" was the first IVF buffalo calf produced in India in 1990. Through this technique, it is now possible to study development process, gene expression, epigenetic modification, and cytogenetic disorder in various species and also provide a model for study of embryogenesis in human. In this method, unfertilized eggs are first fertilized in the laboratory and cultured for a few days until they have successfully developed into early embryos. This embryo is now ready to be implanted in the recipient uterus through a long syringe. It is important to notice that the recipient mother should be in receptive stage of estrus cycle before the implantation. The first live offspring achieved from IVF technology was a rabbit.

Cloning
The literal meaning of clone is an identical copy of any organism or tissue or cell which contains identical copies of genetic information. In cloning, it is possible to reproduce an entire organism from any cell taken from parent organism, and resulted clone is an identical copy of parental organism in every means. The genetic composition of clone is identical to its donor (or parental organism). The main objective of cloning is to increase the number of identical copies of superior livestock to produce high-quality end product although cloning does not change the genetic makeup of the animal.

In nature, cloning is quite usual and frequent. For instance, naturally occurring asexual reproduction in some lower eukaryotes and prokaryotes is similar to those twins reproduced from one fertilized egg.

The major breakthrough in cloning world had appeared in 1996 when Ian Wilmut and his colleagues have successfully produced a clone of sheep named "Dolly" (Fig. 16.5). Dolly was produced from fertilization between cultured adult somatic cell as a donor and enucleated oocyte recipient cell (lacking chromosomal DNA).

The cloning process includes the following steps: (i) Chromosomal DNA will be removed from a mature oocyte. (ii) This egg nucleus will be replaced by the somatic cell nucleus from the donor which is supposed to be cloned. The donor cell now will be fused with enucleated oocyte, and reprogramming of somatic cell genome will be

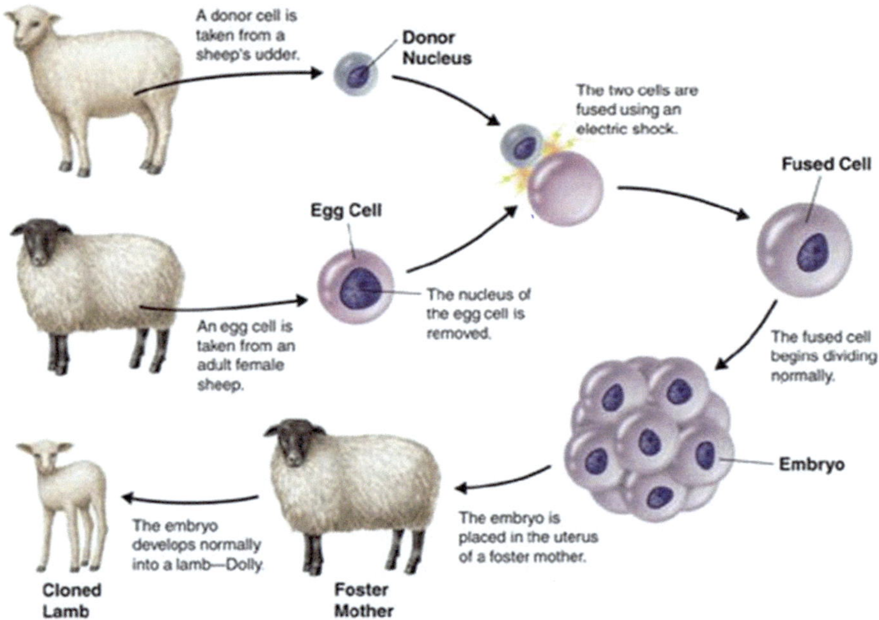

Fig. 16.5 Cloning method to produce "Dolly" sheep. Enucleated oocyte is fused with somatic cell by electric shock. Fused cell developed into fertilized egg and matured embryo, followed by implantation into foster mother. This embryo has a genetic information identical to the donor cell and appeared into Dolly sheep which is a clone

activated as of embryonic genome. Activation of reprogramming can be induced by chemical or electrical pulse. Reconstructed cloned embryo is further cultured, and this viable embryo is transferred to synchronized recipient which carries the live cloned offspring until parturition.

In addition to the genetic cloning of valuable animals, it is also possible to generate the genocopies of exceptional animal from those who are incapable of reproduction. For example, in the case of steers or animals that are dead, cells can be stored.

16.4.3 Improved Health

At the global level, increasing figures of population simultaneously are adding to the requirements of overproduction of high-quality protein, meat, milk, and eggs as essential needs for life, but sustaining a good health and protection from adverse environmental factors which can affect the livestock's longevity are quite questionable and the primary concerns to be taken. Apparently, biotechnology plays a significant role in the diagnosis of livestock diseases and genetically transmitted conditions which reduce animal's health and productivity drastically. Nowadays, advanced biological technique produces cheaper and more efficient drugs because natural drugs from natural source materials are excessively expensive. In this scenario, drug production utilizing applications of genetic engineering in either microbial or tissue culture system has become a wise decision eventually in favor of human and animal's health. Largely produced human insulin, human growth hormone, and plasminogen activator (used in treating heart disease) are the biggest success in animal biotechnology.

16.4.3.1 Vaccines

Immune system of animals is induced by arrays of vaccine to produce antibodies targeting disease or infection. Emerging recombinant DNA technology has introduced the possibilities to develop a recombinant antibody and vaccines commercially available at low cost. Empirical knowledge in vaccine development and relationship with immune system makes scientist and industries to produce massive range of vaccine that can perform better to boost the body's immune system than the conventional vaccines. Data from the trial history suggested that these engineered vaccines are way safer than the traditional vaccines which may develop the "revert effects" (inactive non-virulent can revert into virulent and cause disease). Therefore, such genetically engineered vaccines have been developed to eliminate this threat to animal health.

Biotechnology industry is booming day by day to produce entirely new engineered vaccines and their new ways of uses. Vaccines have been developed for many purposes, i.e., modulator for growth hormone to increase the growth rate, additives as a feed conversion, stimulator in milk production, enhancer to improve animal carcass and meat quality, and modulator in reproduction system to enhance or to suppress the reproduction rate.

Recombinant or engineered vaccines are useful for those diseases for which vaccine has been not developed. These vaccines do not contain the dangerous infectious agent unlike the traditional vaccines, and this property makes these vaccines efficiently safer. Production of vaccines are considered as less expensive with mass production, and maintenance cost is negligible since it can be stored even at room temperature.

16.4.3.2 Diagnosis

Examination of poor health in cattle, pets, and other domestic animals is an additional responsibility for farmers and biotechnologist. Improvement in diagnosis methods or in tools makes the situation under control for many poultry firm running around. Nearly a decade, scientists from Japan and Taiwan became in spotlight when they invented the DNA to detect the hereditary weakness in poultry pigs during transportation or in slaughterhouse. This test has identified the gene expression associated with "porcine stress syndrome" in pigs. They observed that pig with this gene in active state produce pale and poor quality of meat. Now, it has become an easy job for poultry people to use DNA testing to identify the pig with this active gene and can eliminate during the breeding program to reduce the risk in the offspring.

16.4.4 Feed Additives

The main objective to use feed additives is to enhance the quality of feed for the animal to improve animal's performance and health. Feed additives can be available in many forms, relatively concentrated form, such as vitamins produced by animal or vegetable origin, amino acids, enzymes, minerals, antibiotics and probiotics, and single-cell proteins. For example, yeast products high in protein have been used as a feed additive for many animals: cattle, pigs, and poultry. Rich in nutrition and highly edible, these products also help in creating a healthy balance of bacteria in the digestive tract and prevent bacterial diarrhea. A beneficial bacterial product "phytase," commercially named as "TRANSPHOS," had been used as an inexpensive feed additive. The wide use is to substitute the costly mineral phosphate used as an additive in the feed of monogastric animals. Similar to this, bacteriocin is another feed additive that had been produced and used to fight against livestock pathogen like Listeria monocytogenes, Staphylococcus aureus, etc. Lysine is the most essential supplement for animal growth, and in routine life, animals hardly get this supplement in enough amount. This L-lysine monohydrate is safe, stable, and edible, being produced in many countries from bacteria through fermentation and added to the feed material to increase the quality of nutrition in feed.

Feed additives are categorized as follows:

16.4.4.1 Antibiotics

Antibiotics contain antimicrobial and antifungal properties, usually of plant or fungal origin produced for pharmaceutical purposes, and can be synthesized in laboratories. Antibiotics are meant to be used for the treatment of infections, but there are a few antibiotics available in the market that can improve the growth of animals and increase the feed conversion efficiency. The most common antibiotic used as feed additive is "ionophore." The function of ionophore includes metabolic role in improvising the production efficiency. These ionophores have general metabolic role within the animal to improve the production efficiency.

16.4.4.2 Enzymes

Applying advance biotechnology, many enzymes are produced at large scale and relatively inexpensive (McDonald et al. 2010). These enzymes are widely being used as a feed additive in a nonruminant and ruminant diet. The primary goal was to improve the nutrition value when poor quality and inexpensive ingredients are incorporated during feeding the animal. Many enzymes are available commercially including phytase (phosphorus digestion), hemicellulose (plant cell wall digestion), and cellulase/xylanases as a feed additive. Also, digestibility of amino acid can be improved with phytase supplementation.

16.4.4.3 Probiotics

While antibiotics are designed to be involved as a feed additive to treat any bacterial infection, on the other hand, probiotics are being used to improve the strength of certain strains of bacteria in the gut. Probiotics basically are a microbial population, which enhance the activity of the digestive system. Apart from all, these probiotics (microbial population) also have been observed to produce vitamin B complex and many digestive enzymes, for protection against toxins, to increase intestinal mucosa immunity, etc.

16.4.4.4 Beta-agonists

Beta-agonist is a natural or synthetic organic compound that shares a common chemical structure with phenethanolamines. Therapeutically, this compound is involved in massive use to maintain smooth muscle mass. Beta-agonist is a type of metabolic modifier which means such compounds modify the metabolism in specific and directed way. These compounds show overall effect on productive efficiency (weight gain or milk production), improving carcass composition (lean vs. fat ratio), increasing milk yield in lactating animals, and decreasing animal waste per production unit. Two main compounds that are popular and commercially available are somatotropins and beta-adrenergic. Such compounds are widely used as feed additive to improve the nutrient amounts in feeds and the productivity of livestock.

16.5 Genetically Modified Organisms (GMOs)

Genetically modified organisms (GMOs) are completely transformed animals whose genetic material has been commodified in a way that it could not occur naturally. It is now possible by cutting-edge application of genetic engineering, and molecular biology together has been involved in modification of genetic makeup of any organism.

In the history of GMs, Europe is the leading place that was producing cloned and GM animals throughout the 90s. The most famous example of genetically modified sheep named “Dolly” had been developed through the cell nucleus transfer from the differentiated cell. This first genetically modified animal was produced at Roslin Institute in Scotland. Afterward, a genetically modified bull named “Herman” was produced by a Dutch biotechnology company “Gene Pharming Europe.” The purpose of producing genetic modification was aimed to generate a subsequent trait of female offspring that would be able to produce milk proteins like lactoferrin in their milk and would be an excellent source for food, nutraceutical, and pharmaceutical purposes. Other examples including some experimental animals like genetically modified mouse, pig, fish, and chicken have been developed within European institutions. These animals have been produced with specific advantages and benefits to food production and other areas of application.

Genetic modification of organisms can be understood by the following categories:

- Green genetic engineering (or agrogenetic engineering): Established to develop genetically modified plants in agricultural firms and food sector.
- Red/yellow genetic engineering: Aim to develop medical diagnostic tool, gene therapy, and essential drugs like insulin and vaccines.

- Gray/white genetic engineering: Involved in production of enzyme or chemicals by utilizing genetically modified microorganism.
- Genetically modified animals: Produced genetically modified organism to the purpose of improvement in the quality of food, milk, meat, and dairy product.

The list of few commercially available genetically modified organisms (GMOs) is as follows:

16.5.1 Glowing Fish

Glowing fish, commercially known as GloFish, is a genetically modified fish and initially had been developed for pollution detection instead for consumption. Naturally bioluminescent fishes are found in the sea and inhabitant of darkest part of sea. Scientists from Singapore adapted this idea and developed a fluorescence transgene that is sensitive to sea pollutant. They had transferred the naturally derived fluorescence gene-incorporated pollutant-sensitive biosensor that will detect the environmental pollutant. Other than its scientific role, GloFish also had captured the public interest for decoration purpose, and so it has become commercial for public to keep in home aquarium. Since 2003, it has become commercial under the license of Yorktown Technologies.

Currently, there are 12 lines of GloFish in the market, including tetras, zebra fish, and barbs, in such colors as electric green, moonrise pink, and cosmic blue.

16.5.2 GM Salmon

Besides insects, fishes had been genetically modified to provide good source of dietary consumption to the consumers. The first genetically modified fish in the market is the AquAdvantage salmon. After three decades of its production, in August 2017, it has become available in Canada. This GM salmon is produced by AquaBounty Technologies which is twice in size and grown in same period of non-GE salmon. GE salmon is produced by recombinant gene containing growth hormone gene from Chinook which is activated by adjacent gene from ocean pout (a fish). AquAdvantage had been approved by US FDA and declared safe to eat. This fish also contains same nutrition as other non-GE Atlantic salmon. There had been no biological side effects observed according to the FDA reports.

16.5.3 GM Mosquito

Mosquitoes are worldwide known to be vector for detrimental diseases such as malaria, dengue, chikungunya, and Zika. In earlier past, to combat these epidemic diseases, scientists were on the mission to reduce the female population.

Scientists at Imperial College London had successfully reduced female population by means of genetic engineering. An enzyme known as I-PpoI is an endonuclease that cut the ribosomal gene sequence (rDNA) in the mosquito. This enzyme specifically destroys the rDNA cluster located on X chromosome. Thus, scientist at Imperial College London had developed the mosquito strain in which sperm cell possesses this enzyme to cut the rDNA cluster located on X chromosome and produced only male offspring. Applying this technology, they were successfully able to produce 95% male offspring which will further inherit this enzyme to the generations.

Another example of GE mosquito is Friendly™ Aedes that had been manufactured at an Oxford-based biotechnology company, Oxitech. This mosquito had been produced by inserting the gene that kills the insect at larval stage. So when male Aedes, which does not consume human blood, mate with wild-type Aedes female, then progeny will be unviable and killed at larval stage before the adulthood.

These mosquitoes had been tested in fields located in different places. First time, Friendly™ Aedes were launched in Brazil (Piracicaba) in April 2015, and a 91% reduction in dengue fever were observed in Dorado district of Brazil after immediate launching in 2016. In total, 12 cases were observed in 2015/2016 that was a huge reduction from 133 in previous years.

16.5.4 Eco-friendly Pig

GM pig had been produced at University of Guelph in Ontario, Canada, in 1999. They develop GM pig that contains phytase enzyme in their saliva. Phytase enzyme digests the plant phytate into phosphorus which is useful for algal growth when the phosphorus is released from the manure of GM pig into the water stream. The origin breed was Yorkshire pigs that genitivally modified for phytase enzyme.

16.5.5 Bird Flu-Resistant Chicken

To fight against most devastating disease for chicken such as bird flu, scientists in the UK had developed transgenic flu-resistant chicken. Thereafter, in another attempt, scientists from University of Cambridge had developed GM chicken with short hairpin RNA. This structure somehow blocks the spread of the influenza virus (mechanism is unknown). Thus, this technology had improved the poultry chicken and environment as well as human health which is prevented by flu infection.

16.5.6 Human Gene Therapy

Gene therapy is referred here as an incorporation of gene to replace a defective gene that might have been involved in many genetic disorders. This is a technique to correct the genetic error which occurs in some genetic disorder like cystic fibrosis, cancer, ADA, immunodeficiency, etc. First-time gene therapy in human history was performed successfully by William French Anderson, Michael Biase, and Ken Culver in 1990. These guys showed that a severe immunodeficiency, adenosine deaminase (ADA) deficiency, also known as "boy in a bubble disease," can be treated with gene therapy. To spot a bit light over here, ADA is a recessive disease carrying two copies of recessive allele of ADA gene. Normally, two copies of ADA gene promote the production of adenosine deaminase in cells throughout, but error in even one gene will inhibit the conversion of deoxyadenosine (a waste product) into inosine and thus will lead to heavy buildup of deoxyadenosine in the body. This accumulated buildups later undergo phosphorylation, convert into toxic triphosphate responsible for killing T cells, and eventually result in failure of immune system and early death.

Gene therapy can be done in two possible ways:

1. Somatic gene therapy is taking into account that gene transfers into the body cell by means of somatic cell rather than to germ cell (egg or sperm cell). The aim to somatic gene therapy is not to let the gene pass to the offspring in the future but just stay in patient's body till its effective state. Study and trials on somatic therapy has prevailed its success as clinically effective. Gene therapy for ADA that has been discussed above was the first somatic gene therapy in 1990 and 1991 with two patients of ages 4 and 11 years old. Both kids are growing well with the continuity of the treatment. Later in 1992, a 29-year-old woman experiencing familial *hypercholesterolemia*, a genetic condition (defect in the chromosome 19) that is associated with increased cholesterol in the blood due to defective LDL receptor on the liver, was treated with somatic gene therapy. This woman was treated with homozygous FH ex vivo delivery to the liver. This treatment was carried out for 18 months, and liver biopsies demonstrated no discernible abnormalities.

   Consequently, five more patients had been successfully treated with gene therapy since then. After the success of above stories, scientists are focusing on the clinical trial for many other diseases especially chronic genetic disorder and cystic fibrosis.
2. In addition to somatic gene therapy, gene transfer would have been done with germ line cells (eggs and sperm) as well. Gene delivery to the germ cells would modify the genetic makeup of germ line and would definitely pass on the future generations. Germ line gene therapy would be capable to vanish the risk for inherited genetic disorder from the family forever. This type of assurance could be achieved by another method like diagnosis during the IVF if there is any known risk before the implantation. Germ line therapy is a distant prospect and have negative opinion; such therapy is illegal in most of the Europe. However, germ line gene therapy and somatic gene therapy raise different issues. Only somatic gene therapy brings the effective prospect of treatment and have provided a promising cure for few genetic disorders although treatment is complex and success rate is uncertain.

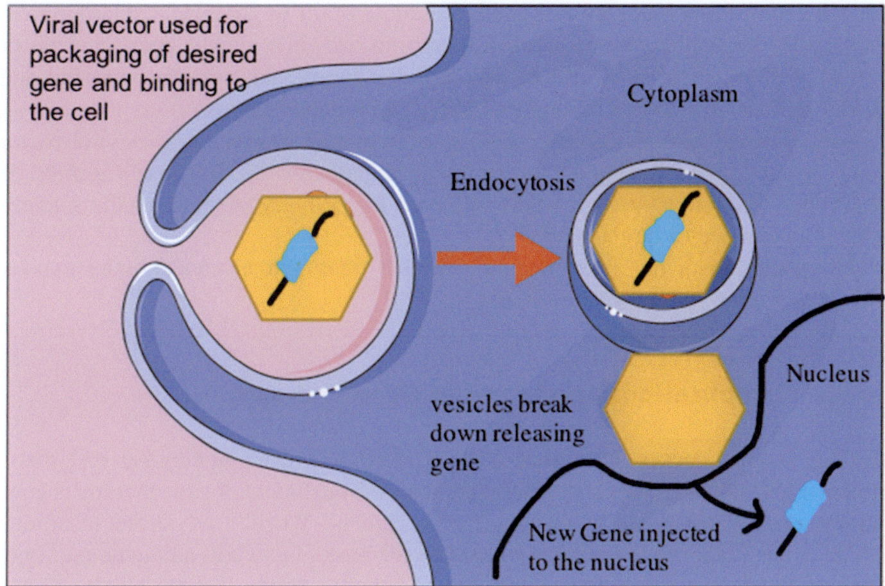

Fig. 16.6 Gene delivery system through a viral vector used in cancer therapy

Despite that techniques are not advanced in gene therapy, researchers are still attempting to develop the methods for gene transfer into the cells in the culture, animals and humans. Within the effect, viral genome was first ever reported as an efficient method for gene delivery into the mammalian cells in the culture. In the beginning of 1980 with the development of retroviral vectors, gene delivery into cultured mammalian cells became widely accepted (Fig. 16.6).

Genome editing has been established as a powerful and efficient tool as a part of gene therapy. Compared to the earlier, current techniques are much more efficient and advanced for modification of DNA. These days, researchers are able to investigate the gene editing in plants, insects, zebra fish, mice, and human cell line in vitro. In theory, gene editing is capable to introduce point mutation to investigate transcription regulation and epigenetic modification. Hence, this technique has a promising contribution in medicine. In recent years, genome engineering is advanced by introducing a powerful and efficient tool: cluster regulatory interspaced short palindromic repeat (CRISPR) nuclease Cas 9. *CRISPR* chops the DNA sequences identified by guide RNA. CRISPR technique is undergoing a widespread use in the research and has already been used for genome engineering of more than dozen species.

Interestingly, a recent study from Chinese research group have demonstrated human embryo genome editing by using *CRISPR/Cas9* system. Unfortunately, this experiment arises significant questions from the scientific and technical point of view upon the risk of these technology over the future. The Chinese research team claimed that the embryo was "*mosaic*" in nature, meaning that only few cells had desired gene editing but there was enough number of off-target effects or mutation in nontargeted genes that can be harmful if embryo had been viable. Thus, in terms of human welfare, their work further points out a significant concern on social and ethical policies of genome editing especially in human embryos.

Gene delivery system through viral vector in human gene therapy. A viral vector had been used for packaging of gene of interest being absorbed by the cell membrane or through the endocytosis method followed by delivery of gene to the nucleus which is the place of target gene.

This mechanism has been widely used for treatment of cancer, lung cancer, immune deficiency, cystic fibrosis, etc.

16.5.7 Complications and Issues in Gene Therapy

Besides its great success in various disease treatments, gene therapy has rose many ethical issues in the society. One of the great accomplishment of gene therapy is gene editing is now no more an obstacle for scientists. World's advanced technique CRISPR has made this opportunity possible for biologist to edit any gene required. Thus, it had become a grave concern that the day is not far for parents to achieve when they will be desiring a customized baby, and they can decide a list of new features such as redhead, blue eyes, and extrovert to be added in their child's genome. While everyone is in the race to make their kids smarter, so why wouldn't you?

It's ethically and economically a dilemma, questioning upon fate and fairness, about vanity and values. Considering all the issues of grave concern, a vigilance committee including not only scientist but also lawyers, doctors, religious, and ethicist have decided to permit somatic cell gene therapy to cure genetic disease but should not extend to the germ line gene transfer which will cause the gene editing for preferred child appearances, which would further pass on to offspring. This action may lead to breach the policies of availing gene therapy and will reason for harmful outcome for human society.

There is also sensitive concern about safety highlighted in 1999 after an incident when a patient was participating as volunteer for gene therapy trial through viral vector, but shockingly, this patient had a fatal immune reaction after injecting viral vector for treatment of his metabolic disorder.

16.6 Molecular Markers

A molecular marker is referred to as a tool to detect variation at DNA or gene level among the individuals. Variation occurs due to base pair changes, rearrangements/ translocation/inversion of sequence, insertion or deletion of sequence, and variation in the number of tandem repeats. Molecular markers can be categorized into protein markers such as allozymes, DNA marker such as mitochondrial DNA (mtDNA), and nuclear DNA marker such as microsatellites, tandem repeats, RAPD, RFLP, AFLP, etc. Molecular markers have been widely used to study genetic variation because of its dynamic properties such as ubiquitous, stably inherited, contain multiple alleles for each marker, devoid of pleiotropic effects, detectable in all tissues, and long shelf life of DNA sample.

16.6.1 Types of Nuclear Molecular Marker

16.6.1.1 Restriction Fragment Length Polymorphism (RFLP)

RFLP represents a molecular marker that utilizes the information about restriction enzyme recognition sites in different individual. However, this technique is not so often being used nowadays, but it was the only technique being used for DNA analysis in forensic science widely. The principle of this technique is that any small mutation like base substitution, insertion, deletion, duplication, and inversion within the whole genome can remove or incorporate new restriction site. Thus, any two individuals will show differences by a great as many as RFLPs.

RFLP analysis distinguishes that where the disease gene lies on the chromosome. In RFLP assay, a cloned DNA fragment is used as a probe and must bind to one site alone which reveals a different-sized restriction fragments as an RFLP locus on Southern blot.

In model organism, a set of strain or individual that provides standard RFLP loci has been utilized for RFLP mapping of that particular species. Human RFLP mapping has been established on a set of individuals in 61 families worldwide and an average of eight children per family. Mapping of human RFLP loci was taken into consideration that disease allele may be linked to the restriction site, and this linkage information of each individual may provide a sustainable approach to assess the probability of a person getting disease in next generation. RFLP mapping is a useful tool to predict any genetic disease linkage in pedigree analysis. Despite the fact that RFLP is one of the primitive and efficient method, it is still a very time-consuming and tedious method compared to the new advanced DNA analysis techniques. Moreover, it requires comparatively large sample size to determine RFLP mapping. Remember that RFLP is always considered as a "single locus marker" since we investigate the single restriction enzyme recognition site on a DNA fragment which reflects similar to the single locus on an allele.

16.6.1.2 Random Amplification of Polymorphic DNA (RAPD)

RAPD is a PCR-based molecular technique to develop DNA markers. This technique has been established by Welsh and Maclelland in 1991. RAPD was designed with the concept of the DNA sequence-based polymorphism at a very large number of loci among the individual. RAPD is a pre-DNA sequencing free technique that makes it more conceivable. This technique has been beneficial for certain purposes in animal breeding such as to identify and classify the accessions of the breed, for identification of the breed, and to study genetic diversity.

RAPD principle works on the virtue of a set of primers, a short oligonucleotide sequence that binds to many different loci. This binding results in the amplification of *random sequence* in the complexed DNA sample. Amplification is done in the PCR and amplified PCR product influenced by the size of both primer and template DNA. Let's consider that two different genomic DNA from two different individuals create a different pattern of the amplified PCR product. A specific fragment found in individual 1 not present in individual 2 will represent a *DNA polymorphism*, and this difference can be used as a genetic marker. An advantage of RAPDs is that it is technically simplified irrespective of demanding any prior DNA sequence information. It has been observed that this method is more advanced and constructive compared to *fingerprinting* and RFLP though this technique has a significant disadvantage of the fact that polymorphism can only detect the presence or absence of a particular amplified DNA sequence in the sample that appears a band of a specific molecular weight. This information is regardless of homozygosity or heterozygosity of that particular DNA fragment. Also, reproducibility of the data is a technical demerit in RAPD.

16.6.1.3 Amplified Fragment Length Polymorphism (AFLP)

AFLP is another PCR-based molecular marker and very specific to the species and subspecies. The common use is determining the close relationship between species or subpopulation in mostly plants, humans, animals, fungi, and bacteria. This marker prevails the application in investigation of genetic variation, population structure, and differentiation. AFLP is capable to generate hundreds of replicable markers from small size DNA sample. In this way, AFLP provides high-resolution genotyping quality. Pertaining its advantageous features, AFLP is known as a substantial genetic marker, used in many broad applications, for example, molecular systematic studies, phylogenic studies, DNA fingerprinting, quantitative trait loci (QTL) mapping, and population genetics. AFLP analysis is more convenient because it generates large numbers of marker fragment for any organism without having prior knowledge of genomic sequence. Additionally, AFLP requires only small fraction of starting template, and it shows much higher reproducibility compared to RAPD. On the basis of factors like time consumption, cost-effectiveness, resolution, and reproducibility, AFLP is equal to the other markers, i.e., RAPD, RFLP, and microsatellite.

Template fragments are generated by (1) digestion with combination of two restriction enzymes and (2) ligation of restriction enzyme-specific adaptors (red and blue) to each end (3). After adaptor ligation, a preamplification step is completed by using combination of primers that matches any of these adapter sequences, and each primer carries selective nucleotide that is represented here with N. (4) Selective PCR amplification step completed by adding additional nucleotides to the primers (specific to two restriction sites) (5). Gel electrophoresis image shows amplified AFLP product.

However, the major drawback of this technique is that AFLPs cannot distinguish dominant homozygous individuals from dominant heterozygous individuals as it is a dominant biallelic marker (i.e., SNP, two possible nucleotide variation for a single position). AFLP is widely used in population genetics and genome typing and subsequently more suitable for detecting genetic polymorphism and tracing animal genetic resources.

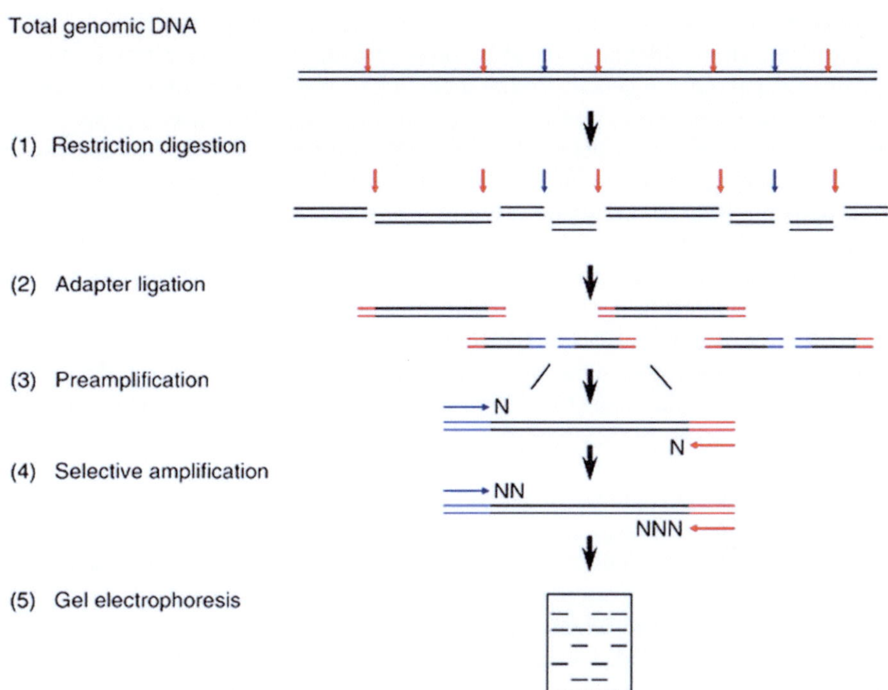

Fig. 16.7 Schematic representation of AFLP method

For a high-impact AFLP, researcher should choose an appropriate primer combination that will generate the sufficient polymorphic marker to study. Literature suggested that at least minimum combination of three primers from the dozens of combinations (that have been used for examining polymorphism in a few individuals, and the results enable them (primers) to identify optimal pairs) will be used for a sufficient gain. AFLP success from the practical point of view requires four major factors including (i) standardized reaction condition, (ii) optimized reagents, (iii) a robust and reliable electrophoresis platform, and (iv) accurate analysis software (Fig. 16.7).

16.6.1.4 Sequence-Tagged Microsatellite

Microsatellites or *simple sequence repeated* (SSR) loci are also known as variable number of *tandem repeats* (VNTRs) and simple sequence length polymorphisms (SSLPs). These repeats are very common in eukaryotes genome and, to some extent, in prokaryotes genome. *Microsatellites* vary from one to six nucleotides in length. Depending upon the numbers of nucleotide in the repeats, they are categorized as mono (A), di (AA), tri (AAA), tetra (AAAA), penta (AAAAA), and hexa (AAAAAA). Di-, tri-, and tetranucleotide repeats are so often used for genetic studies. These repeats are present in repeated manner at least 5–20 times with the minimum length of 12 nucleotide.

Microsatellite provides ease to perform high score, high polymorphism, and a strong automation for thousands of markers at a time and makes this technique suitable for a number of genetic studies including (i) DNA fingerprinting, (ii) genetic mapping, and (iii) paternity analysis and genetic diversity. Microsatellite sequence at specific loci can be easily detected by PCR. Primer for PCR should be specific to conserve flanking region of repeats. A sequence tag is used as a primer to amplify the corresponding SSR and resulted to amplified sequences known as sequence-tagged microsatellite (STMs). These sequence tags must have been taken from the group of enormous amplified restriction fragments or from the genomic DNA. Primer for STM amplification is designed by combination of universal primers that have been used for microsatellite repeats and anchored at $5'$ end of the targeted microsatellite repeat. Thus, primer designing for STM is cost-effective and cut down the cost of developing SSR marker (constructing a microsatellite marker) direct to the synthesis of a single primer specific to a conserved flanking region (conserved DNA sequence) of SSR.

When we have to apply STMPs in large complex DNA, complications have been reported in isolation of restriction fragments containing target SSRs from the pool of amplicon which has been used to construct a sequence tag profile. This problem has been solved by enriching the pool of SSR-rich restriction fragments available from the same DNA template which were used to construct the sequence tag profile, reported in Hayden and M. J. et al. In spite of many advantages, STMSs also possess some limitations including extensive use of high cost of cloning, sequencing, and primer synthesis. Still standard protocol uses radioisotope, and primer selection procedure suffers from a number of problems like redundancy of clones and the occurrence of artificial chimeras.

16.7 DNA Fingerprinting

This technique is also known as DNA profiling and DNA typing. DNA fingerprinting is usually collections of fragmented DNAs from individuals to be compared for particular purpose and generate a DNA-specific profile in a term of fingerprinting. DNA fingerprinting is nonetheless but a distinctive pattern of DNA fragments according to the length isolated by gel electrophoresis. In a forensic field, DNA sample is first isolated and purified from the suspects and victim in order to suspect a crime scene. These samples are further digested by restriction enzyme, amplified by PCR, and profiled using electrophoriesis. DNA fingerprinting is first invented by Alec Jeffreys in 1985 in England. He used restriction enzyme to cut the DNA into fragments because PCR had not been developed that time. Initially, fragments were used with radioactivity-labeled DNA, but now this technique is improvised with the discovery of PCR and fluorescent dye. Routine fingerprinting testing is accommodated with repeated sequence or short tandem repeats which allow to distinguish DNA fragments more effectively. Nowadays, DNA fingerprinting is always performed by PCR assay.

To understand the analysis result of fingerprinting, we can consider the DNA fragments on autoradiograph image below. This radiograph contains five essential lanes describing (i) markers, a known size of DNA fragments; (ii) control lane (TS), a sample from positive source and expected to bind with DNA probes favorably; (iii) experimental lane which denoted the blood collection from defendant, defendant's shirt, and jeans; and (iv) blood sample from victim's shirt and victim's blood.

The given example shows that blood on defendant's shirt and jeans are similar to the blood of victim. Defendant was lying here because his own blood doesn't match with victim's blood and blood on his clothes (Fig. 16.8).

DNA analysis has become the routine job in many forensic labs for solving the crime scene and identifying the suspects. In the beginning of DNA technology, scientists have realized that DNA database from the crime scene would provide a criminal justice in no time, and with more efficiency also, some local DNA database have been created on purpose. Before the DNA fingerprinting, protein assay was being used for several years for solving the cases, but this technique was not reliable because of less availability of technical support and protein degradation was the major issue. DNA collection was much easier, and DNA is more resistant to physical damage compared to the protein. DNA analysis can be performed from any tissue, i.e., blood, hair, saliva, semen, skin, and bones, while protein markers are restricted to the cells where these proteins are expressed. Besides, this technique has also been used to analyze the relationships and source of another organism in the population. For example, DNA fingerprinting has revealed that anthrax has been transmitted to the population having same source of origin.

16.8 Fluorescent in Situ Hybridization (FISH)

If a part of genome is possible to clone, then it can be used to make a labeled probe for hybridization to chromosome in situ. The logic of this approach is identical to Western or Southern blot just in case this probe does not bind with any DNA or protein instead binds to largely intact chromosome since the probe is cloned for chromosome specific. This method involves a few steps to be performed such as isolation of chromosome by tearing cells chemically or mechanically and spread on the microscopic glass. The chromosome on slide is supposed to be denatured so that the double-stranded long DNA can convert into single stranded. Thereafter, denatured labeled probe is added to this mixture. In result, the probe will be hybridized to homologous sequence in situ with the chromosomal DNA on the slide and apparently location of hybridization on the chromosome will be detected by bright fluoroscent spot on the chromosome DNA under the fluoroscent microscope (Fig. 16.9). With the advance technology, FISH can also be used to localize and detect various RNA target (mRNA and miRNA) within the cells and tissue.

Now, the probe sequence will be used to map the position of hybridization on the chromosome by observing the banding pattern related to centromere or any other cytological feature. Unfortunately, this technique does not allow to observe recombinational mapping due to low resolving power, as in example of two genes that

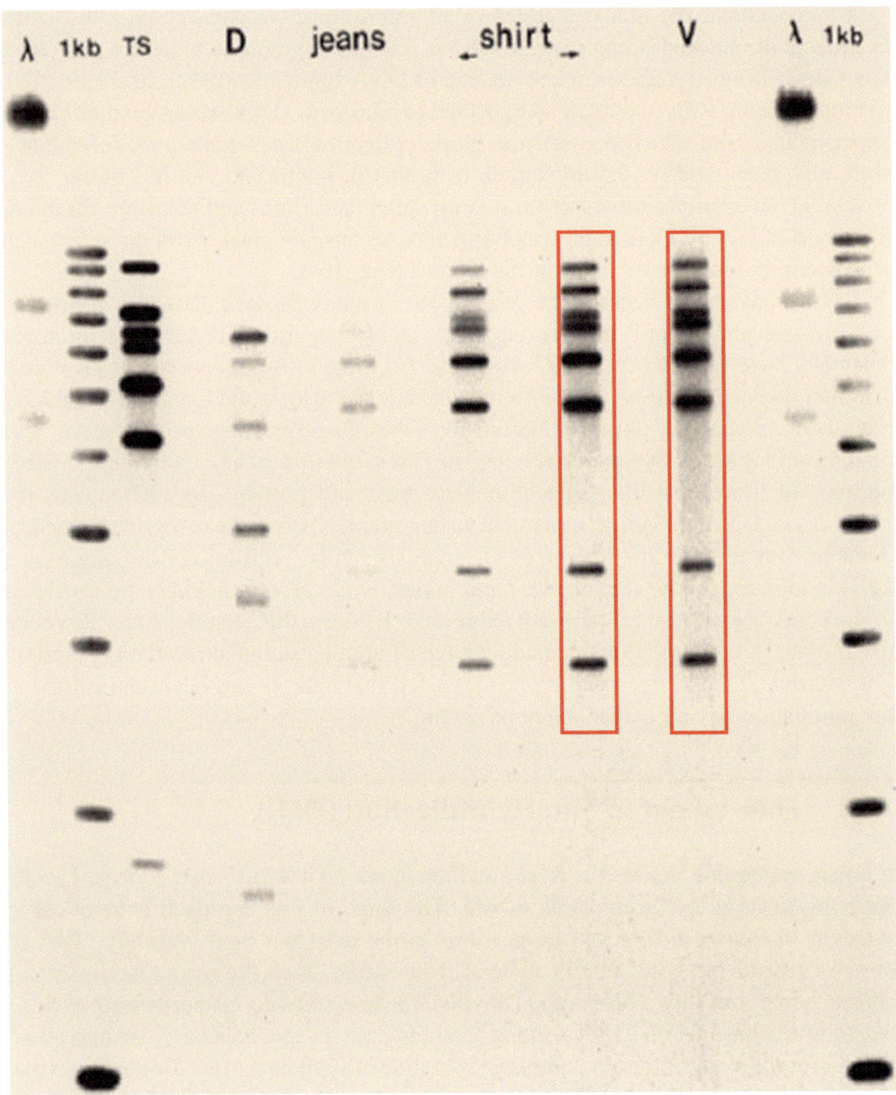

Fig. 16.8 DNA fingerprinting/profiling from a crime scene. The DNA samples collected from the victim (V) were found on defendant's (D) clothing (jeans/shirt). First lane shows DNA ladder (λ), and second lane shows positive control (TS). This profile shows a successful DNA fingerprint assay from a crime scene resulting in prediction of criminal

positioned 5 cM apart to each other on the human chromosome, but in situ hybridization mapping technique would not be able to distinguish this recombinational-like feature.

In advance in situ hybridization technique, probe can be a cDNA, cRNA, or synthetic oligonucleotide. While choosing the probe, researchers should consider the

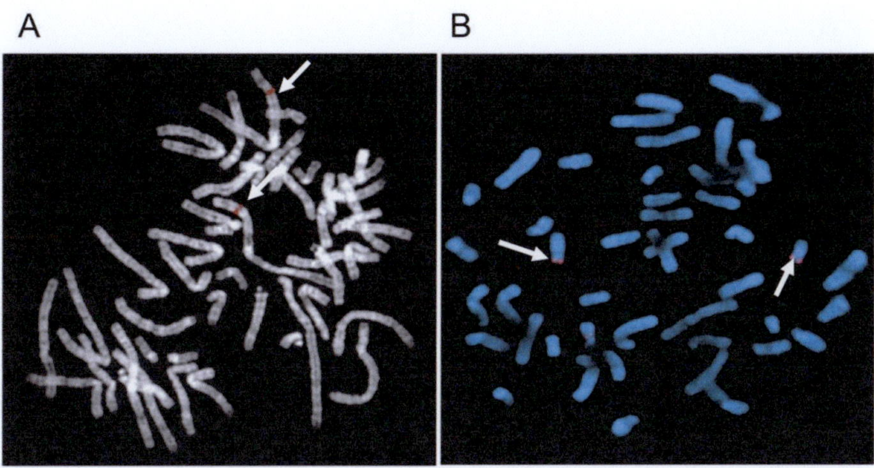

Fig. 16.9 FISH image of chromosome shows locus-specific fluorescent signals. (**a**) Cytogenetic bands (gray) with a hybridized probe with chromosome spotted in red. (**b**) A clone selected from patient suffering with multiple congenital malformations and mental retardation. FISH analysis is used for locating the break point of a translocation on chromosome 11 or 19, and FISH was able to show the red signals split between chromosomes 11 and 19 where the translocation took place

binding sensitivity, specificity, production facility probe penetration strength, and stability of hybrid. These probes are basically conjugated with fluorescein, biotin, or digoxigenin.

FISH probes can be categorized (Fig. 16.9) as (i) locus-specific probe or (ii) chromosome paint probes. Locus-specific probes are particular to detect a distinct gene or chromosomal area, and such probes help detect any deletion or amplification of DNA sequence (if any occur), while *chromosome paint* (Fig. 16.10) probes are completely specific for detecting structural abnormalities within the chromosome structure or rearrangement of chromosome. The optimal size of probe should be of length 50–300 bases.

Box 16.1: Production of Recombinant Human Proinsulin in the Milk of Transgenic Mice (Qian X et al.)

Diabetes is the third most worldwide health issue among the people and characterized with high blood sugar which can lead to a number of serious complications such as heart disease, stroke, kidney failure, blindness, nerve damage, Alzheimer's, etc. The main causes of diabetes are lacking insulin production (Type I) and failure of its use after production (Type II). After success of medical trial, insulin treatment has become one of the promising treatments for diabetes. At molecular level, human insulin gene in vivo is first transcribed and translated in single-chain precursor known as "proinsulin"

(continued)

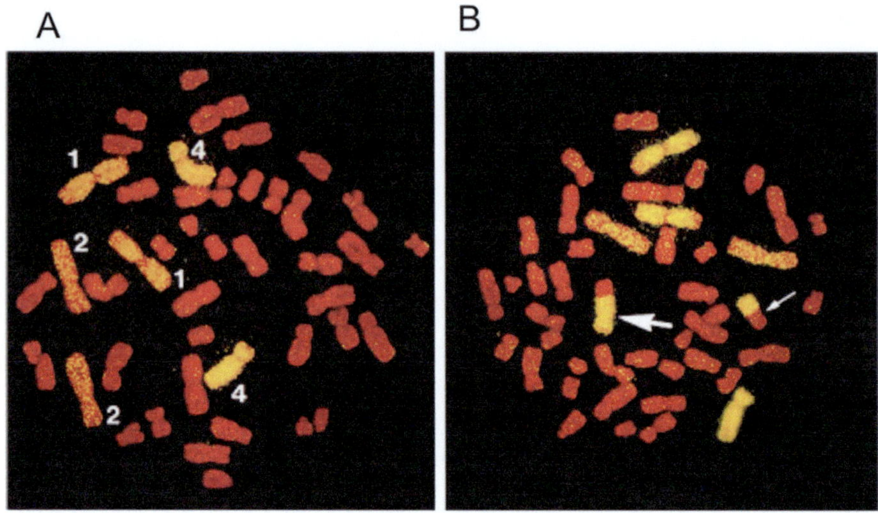

Fig. 16.10 FISH image of human metaphase chromosome painting. (a) Chromosomes 1, 2, and 4 were labeled yellow and rest painted red. (b) Image shows reciprocal translocation between chromosoms appeared in bicolor chromosom indicated with white arrow. FISH technique used to detect translocation between chromosomes is stained yellow, while chromosome is stained red

Box 16.1 (continued)

which is 110-amino acid (aa)-long peptide in the ß cells of the islet of Langerhans in the pancreas. This signal peptide, first 23–24 aa at the N terminus, is removed during the protein folding in the endoplasmic reticulum which results in proinsulin (86 aa, 9.5 kDa). Molecular structure of protein has three domains: an amino terminal B chain of 30 aa, 3.4 kDa, a carboxy terminal A chain (21 aa, 2.4 kDa), and a connecting C chain (34 aa, 3.0 kDa). This structure is generally considered as a premature protein structure, but a cut by neuroendocrine cell-specific prohormone convertase (PC1 and PC2) in order to remove C chain and remaining A and B chain together is bound with a sulfide bond known as mature insulin (5.8 kDa). With an alarming need of insulin production for diabetes treatment, recombinant DNA technology has become a boom for the medical field. Human insulin now biosynthetically can be possible to manufacture at large scale for clinical use worldwide. Primitively, chain A and chain B successfully were produced in two separate bacterial strains and followed by purification and sulfide bond formation by air oxidation. Furthermore, insulin and its analogues were produced in a yeast as inactive proinsulin that further goes for enzymatic cleavage of C chain by using enzymes such as trypsin and carboxypeptidase

(continued)

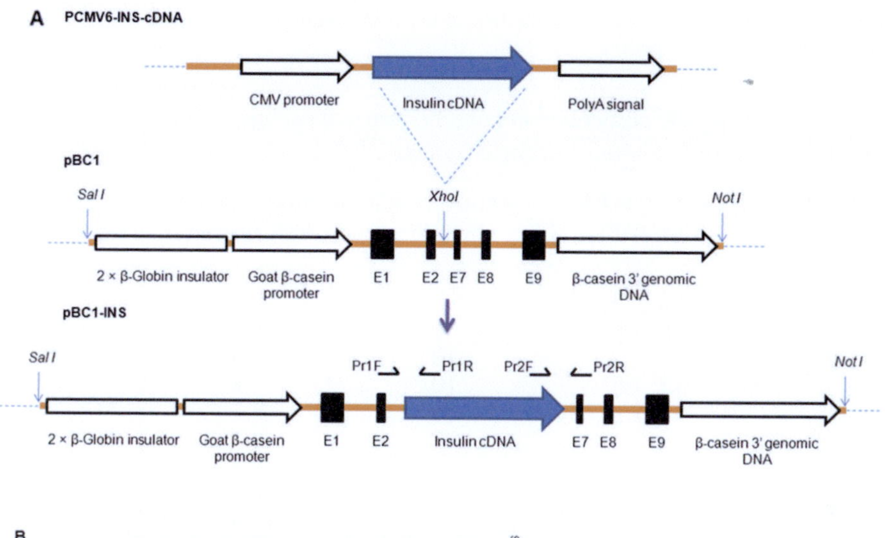

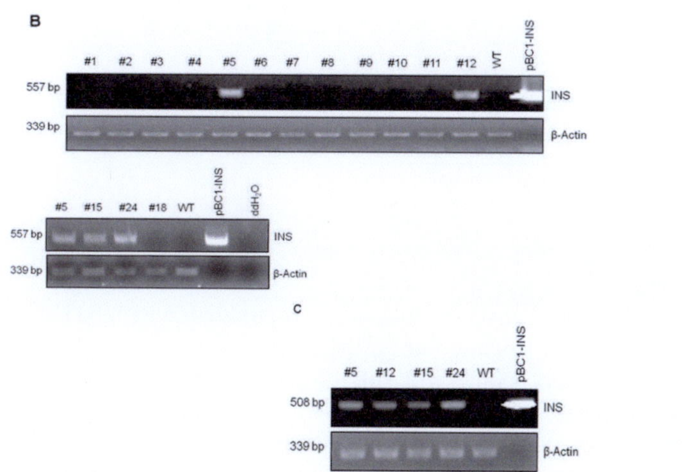

Fig. 16.11 Transgene construction and the identification of transgenic mice. (**a**) Schematic representation of the transgene construction. The full length of insulin cDNA in the pCMV6-XL5-INS-cDNA was amplified by PCR and inserted into the pBC1 vector at the Xho I site, generating the pBC1-INS construct. Before microinjection, the pBC1-INS construct was excised with Sal I and Not I. From left to right, the linearized pBC1-INS comprises the 2 × β globin insulator; the goat β-casein promoter and untranslated exons E1 and E2; human insulin cDNA; untranslated goat b-casein exons E7, E8, and E9; and 39 genomic DNAs. Pr1F, Pr1R, Pr2F, and Pr2R primers were used in PCR for the identification of the transgenic mice. (**b** and **c**) Identification of the transgenic mice by PCR using the Pr1 primer pair (**b**) and Pr2 primer pair (**c**). Non-transgenic wild-type (WT) mouse DNA was used as a negative control, and the DNA used for microinjection served as a positive control. b-actin was amplified to show the same amount of DNA used in each PCR reaction

Table 16.2 List of proinsulin concentration in the collected milk of transgenic mice at the midlactation period

Transgene copy numbers and human proinsulin expression levels in multiple lines of transgenic mice			
Line FO (gender[a])	Offspring F1 (gender[a])	Transgene copy number	Expression level in milk[b] (μg/μl)
5 (F)		13	1.4
	5_1 (F)	7	1.2
	5_2 (F)	6	2.0
12 (F)		20	7.8
	12_1 (F)	10	7.2
	12_2 (F)	6	8.1
15 (F)		7	4.1
	15_1 (F)	5	2.1
	15_2 (F)	3	1.6
24 (M)		9	NA
	24_1 (F)	5	5.7

[a]F = female, M = male.
[b]Milk samples were collected at mid-lactation, and the concentration of human proinsulin was quantified by ELISA.

Box 16.1 (continued)

B to obtain full potential mature insulin. However, this production method has some limitations on the net production rate, and according to diabetic federation, in 2030, one out of ten people will suffer with insulin deficiency or diabetes which gives rise to the possibility of increased demand of insulin production. Current insulin production methods are insufficient to cover the targeted demands to cure the diabetes. Thus, the biopharmaceutical protein production in the mammary glands of transgenic animals is one of the exclusive and extensive trial to increase the proficiency of insulin production for medical uses. This approach has achieved the heights of current expectations in the medical field as it promises the production of high-quality therapeutic medicine for human at very efficient cost. In the current scenario, production of human proinsulin in the milk of transgenic mice provides a possibility and feasibility for scaling up the potential proinsulin production by using transgenic dairy animals.

To generate the transgenic mice expressing the human proinsulin milk, full length of human insulin cDNA was amplified in the pCMV6 vector andinserted into the mammary gland-specific expression vector pBC1

(continued)

Box 16.1 (continued)
(Fig. 16.11a). pBC1 vector contains goat ß-casein promoter and flanked by 5′ to 3′ untranslated sequence of goat ß-casein gene (Fig. 16.11a) This transgene was first linearized in the vector and injected into the fertilized mouse eggs followed by transfer into the recipients. Recipient female mice were crossed with wild type male, which gives F1 transgenic mice pups. However, qPCR for identification of copy number of transgenes in the founder and offspring reveals a slight difference in the numbers of transgenes in the offspring from its founder (Table 16.2) which can be due to the loss of transgene transfer efficiency. However, loss of transgenes is a common phenomenon that explains that transgenes are usually exogenous fragmented DNA supposed to be introduced to some random sites into the host genome and will not result in 100% transfection capability of vectors (Fig. 16.11)

Secretion of proinsulin is targeted only in the milk product of transgenic mice, but to confirm the negligible secretion of insulin in the blood or plasma of these animals, further ELISA test was performed. ELISA test confirmed that there is no difference between the insulin levels in the blood of transgenic and non-transgenic mice. Also, no glucose amount was observed in the plasma of non-transgenic mice which was similar to the transgenic mice. To further obtain the bioactivity of proinsulin or conversion into mature insulin, this proinsulin which was collected from the milk was further digested with enzymes. Digested product was targeted to the CHO cells which express the enormous insulin receptor on its surface. Treated CHO cells were harvested and analyzed for tyrosine phosphorylation of insulin receptor. Levels of total phosphorylation were compared with positive control, undigested transgenic milk, and digested and undigested non-transgenic milk. Results from these samples do not show any detectable activity compared to the digested transgenic milk which confirms the bioactivity of proinsulin produced from the milk of transgenic mice. This observation further clarifies that mammary gland does not produce the mature insulin since the digestion of proinsulin was done exogenously, and therefore, no insulin bioactivity was observed in undigested proinsulin. This mammary gland may not express PC1 and PC2, enzymes required to release the C chain from the proinsulin, but mammary gland can recognize the insulin signal peptide and able to secret the proinsulin protein in the milk. This proinsulin production is empirically favorable in the human health since it does not affect the blood glucose and blood insulin levels that reduce the systemic side effects on diabetic patients.

16.9 Summary

- Molecular genetics is being used for a long time, which has been engaged in enormous applications in the field of animal biotechnology, transgenic animal, production of genetically modified organisms, human gene therapy, development of molecular markers, and forensic science.
- The most advantageous contribution of molecular genetics is in development of transgenic animal. Transgenic animal is produced by injecting DNA into fertile egg that contains foreign DNA (of desired requirement or trait) that is integrated into a chromosome. Knockout mice is a transgenic mouse in which genes are disabled of particular role.
- In molecular genetics, it is now possible to improve the reproduction system and health of cattle and domestic animals by enrichment of diet, improved diagnosis tool, providing feed additives, adding nutrient supplements, etc.
- In human gene therapy, detrimental diseases are now possible to treat by altering the disease-associated gene in human cells.
- Variation in DNA sequence of individuals or studying polymorphism in a population can be assessed by analyzing molecular markers such as RFLP, RAPD, AFLP, and microsatellite.
- DNA fingerprinting and in situ hybridization are mostly used in forensic science. Both techniques are being used in understanding the crime scene and analyzing the samples through these methods to identify criminals.

Further Reading

Amos B, Schlotterer C, Tautz D (1993) Social structure of pilot whales revealed by analytical DNA profiling. Science 260:670–672

Baguisi A, Behboodi E, Melican DT, Pollock JS, Destrempes MM, Cammuso C et al (1999) Production of goats by somatic cell nuclear transfer. Nat Biotechnol 17(5):456

Brinster RL (1974) The effect of cells transferred into the mouse blastocyst on subsequent development. J Exp Med 140(4):1049–1056

Brumlop S, Finckh MR (2011) Applications and potentials of marker assisted selection (MAS) in plant breeding. BfN-Skripten (Bundesamt für Naturschutz) 298

Campbell KH, McWhir J, Ritchie WA, Wilmut I (1996) Sheep cloned by nuclear transfer from a cultured cell line. Nature 380(6569):64

Cao D, Oard JH (1997) Pedigree and RAPD-based DNA analysis of commercial US rice cultivars. Crop Sci 37(5):1630–1635

Chakravarthi PV, Sri Balaji N (2010) Use of assisted reproductive technologies for livestock development. Vet World 3(5)

Charters YM, Robertson A, Wilkinson MJ, Ramsay G (1996) PCR analysis of oilseed rape cultivars (Brassica napus L. ssp. oleifera) using 5′-anchored simple sequence repeat (SSR) primers. Theor Appl Genet 92(3–4):442–447

Cheung VG, Nowak N, Jang W, Kirsch IR, Zhao S, Chen XN et al (2001) Integration of cytogenetic landmarks into the draft sequence of the human genome. Nature 409(6822):953

Chial H (2008) Rare genetic disorders: learning about genetic disease through gene mapping, SNPs, and microarray data. Nature. Education 1(1):192

Choudhary KK, Kavya KM, Jerome A, Sharma RK (2016) Advances in reproductive biotechnologies. Vet World 9(4):388

Chuang CK, Chen CH, Huang CL, Su YH, Peng SH, Lin TY et al (2017) Generation of GGTA1 mutant pigs by direct pronuclear microinjection of CRISPR/Cas9 plasmid vectors. Anim Biotechnol 28(3):174–181

Clark AJ, Simons P, Wilmut I, Lathe R (1987) Pharmaceuticals from transgenic livestock. Trends Biotechnol 5(1):20–24

Dubock A (2019) Golden rice: to combat vitamin a deficiency for public health. In: Leila Queiroz Zepka, Veridiana Vera de Rosso and Eduardo Jacob-Lopes. IntechOpen, London

Dzau VJ, Ralph J (2015) Cicerone. Responsible use of human gene-editing technologies. 411–412

Fuller R (1989) Probiotics in man and animals. J Appl Bacteriol 66(5):365–378

Gatehouse AM, Ferry N, Edwards MG, Bell HA (2011) Insect-resistant biotech crops and their impacts on beneficial arthropods. Philos Trans R Soc Lond Ser B Biol Sci 366(1569): 1438–1452

Gordon K, Lee E, Vitale JA, Smith AE, Westphal H, Hennighausen L (1987) Production of human tissue plasminogen activator in transgenic mouse milk. Bio/Technology 5(11):1183

Grossman M, Raper SE, Kozarsky K, Stein EA, Engelhardt JF, Muller D et al (1994) Successful ex vivo gene therapy directed to liver in a patient with familial hypercholesterolaemia. Nat Gen 6(4):335

Gupta PK, Varshney RK (2000) The development and use of microsatellite markers for genetic analysis and plant breeding with emphasis on bread wheat. Euphytica 113(3):163–185

Hammer RE, Pursel VG, Rexroad CE, Wall RJ, Bolt DJ, Ebert KM et al (1985) Production of transgenic rabbits, sheep and pigs by microinjection. Nature 315(6021):680

Hanahan D, Wagner EF, Palmiter RD (2007) The origins of oncomice: a history of the first transgenic mice genetically engineered to develop cancer. Genes Dev 21(18):2258–2270

Hayden MJ, Good G, Sharp PJ (2002) Sequence tagged microsatellite profiling (STMP): improved isolation of DNA sequence flanking target SSRs. Nucleic Acids Res 30(23):e129–e129

Hernandez Gifford JA, Gifford CA (2013) Role of reproductive biotechnologies in enhancing food security and sustainability. Anim Front 3(3):14–19

Howlett DR (2011) APP transgenic mice and their application to drug discovery. Histol Histopathol 26(10):1611

Jensen E (2014) Technical review: in situ hybridization. Anat Rec 297(8):1349–1353

Lathe R, Clark AJ, Archibald AL, Bishop JO, Simons P, Wilmut I (1986) Novel products from livestock. Oxford University Press, Oxford

Liras A (2008) Recombinant proteins in therapeutics: haemophilia treatment as an example. Int Arch Med 1(1):4

Maksimenko OG, Deykin AV, Georgiev PG (2013) Use of transgenic animals in biotechnology: prospects and problems. Acta Nat 5.1(16)

Manmohan S, Niraj K (2010) Transgenic animals: production and application. Int J Pharmac Sci Res 2:12–22

McCarthy C (n.d.) What are GloFish? https://www.petmd.com/fish/what-are-glofish

McNicol AM, Farquharson MA (1997) In situ hybridization and its diagnostic applications in pathology. J Pathol 182(3):250–261

Mercier JC (1986) Genetic engineering applied to milk producing animals: some expectations. Exploiting new technologies in animal breeding: genetic developments/edited by C. Smith, JWB King, and JC McKay

Nandani K, Thakur SK (2014) Randomly amplified polymorphic DNA-a brief review. Am J Anim Vet Sci 9(1):6–13

Nascimento IP, Leite LC (2012) Recombinant vaccines and the development of new vaccine strategies. Braz J Med Biol Res 45(12):1102–1111

Ormandy EH, Dale J, Griffin G (2011) Genetic engineering of animals: ethical issues, including welfare concerns. Can Vet J 52(5):544

Paun O, Schönswetter P (2012) Amplified fragment length polymorphism: an invaluable fingerprinting technique for genomic, transcriptomic, and epigenetic studies. In: Plant DNA fingerprinting and barcoding. Humana Press, New York, pp 75–87

Pierce BA (2008) Genetics: a conceptual approach, vol 1. Freeman, New York, NY

Plaschke J, Ganal MW, Röder MS (1995) Detection of genetic diversity in closely related bread wheat using microsatellite markers. Theor Appl Genet 91(6–7):1001–1007

Qian QIAN, Cheng H, Sun Z, Zhu L (1996) The study on determining true and false hybrid rice II you 63 using RAPD molecular markers. Zhongguo Shuidao Kexue 10(4):241–242

Rezaei M, Zarkesh-Esfahani SH (2012) Optimization of production of recombinant human growth hormone in Escherichia coli. J Res Med Sci 17(7):681–685

Röder MS, Korzun V, Wendehake K, Plaschke J, Tixier MH, Leroy P, Ganal MW (1998) A microsatellite map of wheat. Genetics 149(4):2007–2023

Ross MJ, Olson KC, Geier MD, O'Connor JV, Jones AJS (1986) Recombinant DNA synthesis of human growth hormone. In: Raiti S, Tolman RA (eds) Human growth hormone. Springer, Boston, MA

Selden RC, Springman K, Hondele J, Meyer J, Winnacker EL, Kräußlich H et al (1985) Production of transgenic mice, rabbits and pigs by microinjection into pronuclei. Reprod Domest Anim 20(4):251–252

Singh B, Chauhan MS, Singla SK, Gautam SK, Verma V, Manik RS et al (2009) Reproductive biotechniques in buffaloes (Bubalus bubalis): status, prospects and challenges. Reprod Fertil Dev 21(4):499–510

Smith C, Gibson JP (1987) On the use of transgenics in livestock improvement. Anim Breed Abstr 55:1–10

Tseten T, Murthy K (2014) Advances and biotechnological applications in biofuel production: a review. Open J Renew Sustain Energ:29–34

Virk PS, Ford-Lloyd BV, Jackson MT, Newbury HJ (1995) Use of RAPD for the study of diversity within plant germplasm collections. Heredity 74(2):170

Vize PD, Michalska A, Ashman R, Seamark RF, Wells JR (1987) Improving growth in transgenic farm animals. In EMBO Workshop Germline Manipulation of Animals

Ward KA, Franklin IR, Murray JD, Nancarrow CD, Raphael KA, Rigby NW, et al. (1986a) The direct transfer of DNA by embryo microinjection

Ward KA, Franklin IR, Murray JD, Nancarrow CD, Raphael KA, Rigby NW, Byrne CR, Wilson BW, Hunt CL (1986b) The direct transfer of DNA by embryo microinjection. Proceeding of 3rd World Congress on Genetics Applied to Livestock Production Lincoln, Nebraska

Weber JL, May PE (1989) Abundant class of human DNA polymorphisms which can be typed using the polymerase chain reaction. Am J Hum Genet 44(3):388

Winter P, Pfaff T, Udupa SM, Hüttel B, Sharma PC, Sahi S et al (1999) Characterization and mapping of sequence-tagged microsatellite sites in the chickpea (Cicer arietinum L.) genome. Mol Gen Genet MGG 262(1):90–101

Wolff JA, Lederberg J (1994) An early history of gene transfer and therapy. Hum Gene Ther 5(4): 469–480

Genetic Analysis of Development

17

Tapodhara Datta Majumdar and Atrayee Dey

17.1 Model Organism for Genetic Study

17.1.1 Yeast (*Saccharomyces cerevisiae*) (Fig. 17.1)

Yeast is a saprophytic and unicellular fungi. It has a wide application in food production especially brewer's yeast or baker's yeast. Yeasts are eukaryotes and are characteristically different than prokaryotic bacteria. They have a dimension much greater than bacteria which are 5×10 μm and 0.5×5 μm, respectively. They have many defined features that make them unique and specific, like they have an innate resistance toward numerous antibiotics, sulfamides, and also antibacterial agents. It's a property which cannot be transferred to other organisms. Yeasts are widely found in soil, water, plants, animals, and insects. Plant tissues (i.e., leaves, flowers, and fruits) are very common yeast habitats, but a few species are also found in animals. Some yeast varieties can also be isolated from extreme environment conditions like high salinity, low water potential, low temperature, and also low oxygen availability (Table 17.1).

17.1.1.1 Saccharomyces cerevisiae

Saccharomyces cerevisiae (*S. cerevisiae*), widely known as brewer's yeast, can survive either as single-celled or as pseudo-mycelia. *S. cerevisiae* differs from

---

Both authors have contributed equally.

T. D. Majumdar (✉)
Indian Statistical Institute, Kolkata, India

A. Dey
Department of Zoology, Banwarilal Bhalotia College, Asansol, West Bengal, India

D. Kar, S. Sarkar (eds.), *Genetics Fundamentals Notes*,
https://doi.org/10.1007/978-981-16-7041-1_17

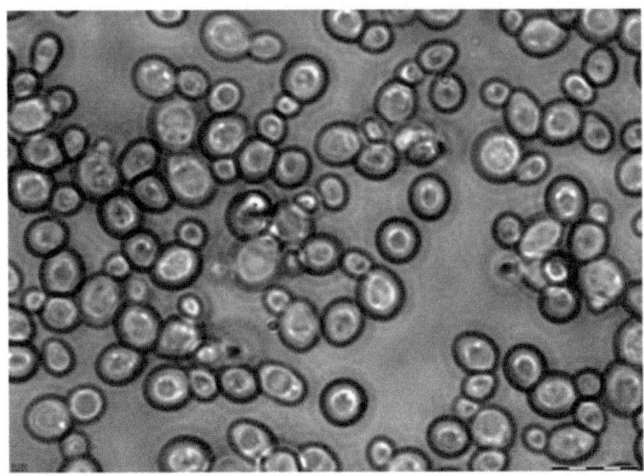

Fig. 17.1 The yeast culture. High-resolution microscopic image showing yeast culture under high magnification depicting both budding and vegetative yeast cells

Table 17.1 Yeast habitats

Habitat	Yeast
1. Plants	*Debaryomyces hansenii*, *Hanseniaspora uvarum*, *Wickerhamomyces anomalus*, *Metschnikowia pulcherrima*, *Magnaporthe oryzae*, *Roesleria subterranea*
2. Animal and insects	*Sugiyamaella mastotermitis*, *Saitozyma flava*, *Apiotrichum mycotoxinovorans*, *Candida albicans*, *Cryptococcus neoformans*
3. Soil	*Lipomyces*, *Schwanniomyces*, *Apiotrichum dulcitum*, *A. porosum*, *Cyberlindnera saturnus*, *Lipomyces starkeyi*
4. Water	*Rhodotorula*, *Debaryomyces hansenii*, *Candida lipolytica*, *Candida tropicalis*
5. Extreme environment	*Glaciozyma*, *Leucosporidium* (extreme cold), *Methanopyrus kandleri* (extreme hot), *Picrophilus torridus* (acidophile), *Halobacterium*, and *Natronobacteria* (high saline conditions)

other yeasts mainly in their ability to ferment sugars. Due to this it finds wide application in food and dairy industry (Table 17.2).

17.1.1.2 *Saccharomyces cerevisiae*: Structure and Components (Fig. 17.2)

A budding cell of *S. cerevisiae* shows the cellular components under electron microscope. Being a eukaryote *S. cerevisiae* has membrane embedded distinct organelles like mitochondria, nucleus, Golgi, and endoplasmic reticulum. The diagram is a part of vegetative reproduction where a budding cell arises from the mother cell and separates as daughter cell under normal growth conditions.

1. Cell wall: Yeast falls under the category of eukaryotes that possesses an external cell wall. It is the outermost cell organelle and provides external protection and maintains the osmotic balance of the cell. It has an outer and an inner layer. It protects the cellular components from foreign particles and other cell wall-

Table 17.2 Taxonomy of *S. cerevisiae*

Domain	Classification
1. Kingdom	Fungi
2. Division	*Ascomycota*
3. Subdivision	*Saccharomycotina*
4. Class	*Saccharomycetes*
5. Order	*Saccharomycetales*
6. Family	*Saccharomycetaceae*
7. Subfamily	*Saccharomyetoideae*
8. Genus	*Saccharomyces*
9. Species	*cerevisiae*

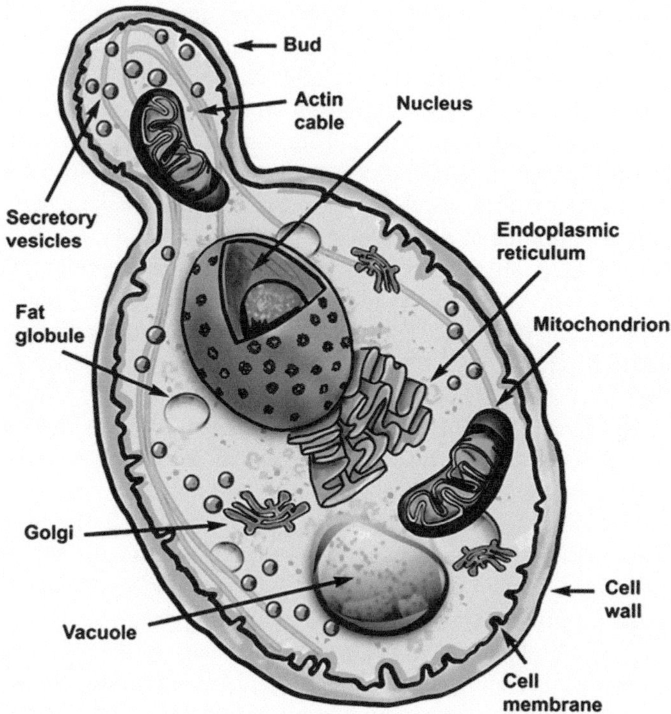

Fig. 17.2 *S. cerevisiae* cell structure and components

degrading enzymes of different hosts like plants. The outer layer is build up of chains of densely populated glycosylated mannoproteins and functions in cell-cell communication process. The inner layer of the cell wall provides firmness to the wall and is 70–100 nm to 200 nm thick depending on the growth conditions. The main building components of the cell wall are L1,3-glucan and chitin and contribute to about 50–60% of the wall dry weight. The glucan-chitin network provides elasticity to the wall. Cell wall proteins are covalently linked to this network either indirectly through a L1,6-glucan moiety or directly. Some proteins are bonded to cell wall via disulfide bonds. The phosphodiester bridges of the carbohydrate chains of the cell surface proteins make the wall hydrophilic and help in water retention during drought conditions. An estimated 1200 genes of *S. cerevisiae* build up the cell wall. Morphological and compositional changes in the wall occur depending on the phase of the cell cycle, nutrient availability, and environmental conditions such as pH, temperature, and the availability of oxygen.

2. Cell membrane: The next organelle after cell wall is the cell or plasma membrane, which is 7.5 nm wide. It is composed of polar lipids and proteins which, by their interactions, construct the famous lipid-protein bilayer of eukaryotes. The proteins are located asymmetrically and are either intrinsic (present all throughout the membrane) or extrinsic (partially embedded in the membrane). The proteins are involved in transport of solutes, signal transduction, anchoring of the cytoskeleton, and synthesis of outer membrane components. The inner side consists mainly of phospholipids like phosphatidylethanolamine, phosphatidylinositol, and phosphatidylserine. The major lipid classes are glycerophospholipids, sphingolipids, and sterols. Glycerophospholipids consist of two fatty acid acyl chains ester-linked to glycerol-3-phosphate; various substituents such as choline, ethanolamine, serine, myoinositol, and glycerol are linked to the phosphoryl group. Cardiolipin is also widely present. Sphingolipids have a ceramide backbone which is composed of a long-chain base phytosphingosine that is N-acylated with a hydroxy C26 fatty acid. *S. cerevisiae* contains three sphingolipids: inositol phosphate ceramide, mannosyl-inositolphosphate-ceramide, and mannosyl-diinositolphosphate-ceramide. The membrane also constitutes sterols (hydrophobic molecules with a polar hydroxyl group) mainly ergosterol and some zymosterol.
3. Nucleus: The genetic assembly of a eukaryote is present in its membrane-bound distinctive nucleus. Due to this it is considered as one of the most essential organelles of the cell. The DNA-containing chromosomes are present in the nucleus. It endorses formation, expression, and functioning of chromosomes. It also is the site of transcription of DNA to mRNA and aids in the formation of rRNA in the nucleolus. After formation, mRNA and rRNA export to cytoplasm and proceed for protein synthesis. Large pores are present in the nuclear membrane that regulates the traffic of macromolecules into and out of the nucleus. The DNA is held together in the chromosome via histone proteins, which are present in the nucleus and help in chromatin folding. In budding yeast, the main structural elements are the nuclear envelope, the nuclear pore complex, and the nucleolus. The nuclear envelope consists of chromatin anchorage sites, spindle pole body, heterochromatin, and the ribosomal DNA (rDNA). The nuclear pore complex takes part in repairing damaged DNA. As a eukaryote the yeast nucleus is functionally compartmentalized. The formation and functioning of these compartments and subcompartments are not due to the intranuclear membranes but depend instead on sequence elements, protein-protein interactions, specific anchorage sites at the nuclear envelope or at pores, and long-range contacts between specific chromosomal loci, such as telomeres. Finally, long-range interaction of loci in trans, such as the clustering of telomeres or of transfer RNA (tRNA) genes, influences the nuclear order.

4. Vacuole: The vacuole of *S. cerevisiae* plays an important role in the physiology of this organism. Vacuoles are the largest compartment in yeast cells, occupying about 25% of the total cell volumes. Vacuoles contain mainly low molecular weight solutes and ions, but quite small amount of proteins. Among the many roles of this organelle are pH homeostasis and osmoregulation, protein degradation, and storage of amino acids, small ions, and polyphosphates. It also plays precious role in sporulation, protein turnover, and ion transport. Besides, it functions in maintaining homeostasis of cytosol in many respects.
5. Golgi apparatus and mitochondria: The Golgi apparatus is responsible for the proper functioning of secretory proteins in eukaryotic cells. It helps in modifying and ultimate sorting of the newly synthesized secretory proteins, which are called cargo proteins. The cargo proteins are synthesized in endoplasmic reticulum (ER) and transported to the Golgi apparatus into cisternae (membrane sacs). After the cargo proteins enter the Golgi apparatus, they are modified and processed by Golgi enzymes. These are then transferred to the trans-Golgi network and transported to their destinations via transport proteins. Mitochondria are the other essential components of eukaryotic cells and are endosymbiotic in origin. They have their own genetic machinery. Mitochondria consist of two membranes, the inner and outer membrane. The intermembrane space lies between the two membranes, and together they guard the mitochondrial matrix. They consist of tubular networks. Mitochondria are known as the powerhouses of eukaryotic cells and are one of the most important organelles in respiratory active cells as they take an active part in synthesis of cellular ATP and also play an important role in tricarboxylic acid (TCA) cycle and oxidative phosphorylation. Its function also includes amino acid and lipid metabolism as well as synthesis of iron-sulfur clusters and heme.
6. Endoplasmic reticulum (ER): The ER originates from the nuclear envelope and forms a membranous tubular network that exudes and branches all along through the cytoplasm. It mainly takes part in protein exportation. An amino-terminal signal peptide targets the secreted proteins to the ER. The signal peptide gets removed upon entry into the ER lumen, and the protein assumes its final, folded conformation for the first time. Those proteins that do not fold properly or fail to form correct three-dimensional structures usually get associated with an ER-binding protein known as BiP. BiP may also serve to catalyze the folding or assembly of normal protein structures. Besides this, synthesis and assembly of lipid bilayers are also initiated in ER. The ER also helps in assembly of dolichol-linked oligosaccharides and their transfer.

17.1.1.3 Life Cycle and Reproduction (Fig. 17.3)

Life cycle of *S. cerevisiae* can be divided into two phases: (1) mitotic phase in which the mother cell undergoes budding and goes through G1, S, G2, and M phases to form an identical daughter cell and (2) meiotic phase, which forms four haploid spores held together in ascus. The spores come out of ascus under favorable conditions and undergo mitosis to form new diploid.

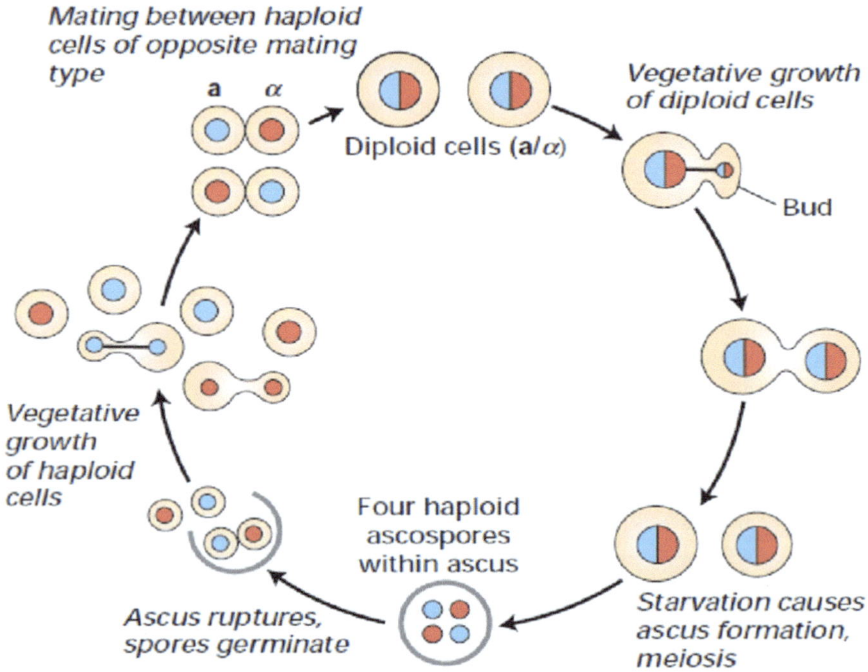

Fig. 17.3 Life cycle and reproduction of *S. cerevisiae*

The genome size of *S. cerevisiae* is approximately 1.2 x 10^6 base pairs comprising of 16 chromosomes, with about 5700 protein-coding genes. The yeast cell divides vegetatively under normal conditions and gives birth to one daughter cell. Being eukaryotes it has four phases of life cycle, i.e., G1 (gap 1), S (stationary), G2 (gap 2), and M (mitotic) phases. At the end a genetically identical daughter cell is formed. But before the daughter is released, each sister chromatid is duplicated and segregated by mitosis. The G1 phase consists of round yeast which still remains unbudded. It then develops a small bud from one side and reaches the S phase. The chromosomes which remain diffused and indistinguishable in the nucleus during the G1 gets duplicated, yielding pairs of sister chromatids in the S phase. In the G2 phase, the bud grows, and the nucleus is found adjacent to the bud. The pair of sister chromatids remain attached to each other and are still diffused in the nucleus. In the M phase, chromosome goes through a twofold condensation and separates, as a result of which the sister chromatids are segregated between the mother and bud. These chromatids are pulled apart by spindle fibers. Finally, cytokinesis divides the cytoplasm and leads to the formation of two genetically identical nuclei, one mother and one daughter cell. Both of them again return in G1 phase, thus completing the mitotic life cycle.

When two haploid cells mate and fuse to form a diploid cell, it leads to formation of homologous chromosomes. The diploid cell contains two of each chromosome. Diploid grows either by budding (vegetative mitosis) or undergoes meiosis. There are two mating types of the yeast, termed as a and α. Haploids of a can only mate with that of α and vice versa. The haploids produce pheromones to induce mating, which attracts the haploids of the opposite mating type. This stimulates fertilization. During fertilization the cells move toward each other by growing projections, a process called shmooing. The cells then fuse to form a diploid cell, which is α/a and cannot mate. The newly formed zygote can divide either mitotically or meiotically. Mitotically they generate diploid cells, whereas meiotically they generate four haploid cells of which two will be α and two will be a.

Yeast usually divides vegetatively. Under conditions when diploid budding yeasts are starved of nutrition, meiosis takes place. Meiosis leads to production of four haploid spores. Spores are more resistant to the environment than vegetatively dividing cells. The four spores of a single meiosis are held together in an ascus, or tetrad, surrounded by a thick wall. When favorable conditions return, the haploids get released from the ascus. They then divide by mitosis to again form new diploid. Two closed nuclear divisions characterize meiosis in yeast. The first is meiosis 1, where homologous chromosome associate lengthwise and then segregate into daughter nuclei. In meiosis 1, chromosome number gets halved in the daughter nuclei. The maternal and paternal genes get mixed together through independent assortment of the individual chromosomes. So it is also referred as reductional division. Here crossing over takes place between the paired homologues by homologous recombination. The second meiosis or meiosis 2 is similar to mitosis and gives rise to four haploids. The meiosis of yeast makes it one of the most widely studied model organisms for eukaryotes. Researchers can easily control the process and mediate crosses between strains.

17.1.1.4 *S. cerevisiae* as an Effective Model Organism

S. cerevisiae was the first eukaryotic organism whose whole genome was sequenced. Various mammalian homologues were found in it. About 31% of important mammalian protein sequences have homology to potential protein-encoding genes of yeast. Two close homologues of mammalian ras proto-oncogene, *RAS1* and *RAS2*, were studied to be present in it. Among the best matches are the human genes that cause hereditary nonpolyposis colon cancer (MSH2 and MLH1 in yeast), neurofibromatosis type 1 (IRA2 in yeast), ataxia telangiectasia (TEL1 in yeast), and Werner's syndrome (SGS1 in yeast). It has been successfully used as a model in regulation of gene expression, signal transduction, cell cycle, metabolism, apoptosis, and neurodegenerative disorders. *S. cerevisiae*, being a facultative anaerobe, can satisfy its energy requirements with adenosine triphosphate (ATP) generated by fermentation. So it doesn't need much amount of other mitochondrial proteins for its survival. Due to this various knockout mutations can be done in yeast mitochondria for studying mitochondrial functions. This makes yeast an ideal organism for identifying the molecular processes required for biogenesis of respiratory-competent mitochondria. The budding yeast of *S. cerevisiae* has been nowadays extensively used in aging research. Two major approaches, "budding life span" and "stationary phase," have been used to determine "senescence" and "life spa" in yeast. Role of genes like Sir2, histone deacetylases, gene silencing, rDNA circles, and stress genes has also been extensively studied in the aging process in yeast. Even, integrated analysis of existing regulatory and metabolic networks, by studying about 55 transcription factors regulating 750 metabolic genes, and novel regulatory mechanisms was also put forward in a recent study.

17.1.1.5 Uses and Threats

The widest use of *S. cerevisiae* is probably in the bakery industry in baker's and brewer's yeast, used in bread and other bakery products. The enzymes produced during the fermentation process are used in formation of different alcohols, organic acids, and esters. These impact flavor and vary from strain to strain. In addition yeast may be utilized as food. They contain a good concentration of protein as well as most of the B complex vitamins. At present days *S. cerevisiae* is also used in producing biodiesel and biofuel.

S. cerevisiae is usually non-harming to humans. But recently a certain fungemia has been detected which is caused by the yeast. Besides, a very rare and an acute case of pyelonephritis was also diagnosed.

17.1.2 Fruit Fly (*Drosophila melanogaster*) (Fig. 17.4)

Drosophila melanogaster (*D. melanogaster*), the common fruit fly, has its ancestral home in Africa. It was first put forward by Thomas Hunt Morgan as a model organism, over 100 years ago, when he discovered the white gene in the fly. Since then it has been used as a model organism in both medical and scientific research for over a century. One of the main advantages of the fly is that it shares 60% homology to humans and has 75% genes homologous to human diseases. Due to this it finds ample application in studies including gene biology, cell biology, developmental

Fig. 17.4 Fruit fly (*Drosophila melanogaster*)

Table 17.3 Taxonomy of *D. melanogaster*

Domain	Classification
1. Kingdom	*Animalia*
2. Phylum	*Arthropoda*
3. Class	*Diptera*
4. Family	*Drosophilidae*
5. Genus	*Drosophila*
6. Subgenus	*Sophophora*
7. Species	*melanogaster*

biology, population genetics, and most importantly neuroscience. *Drosophila* has a simple nervous system, consisting of around 100,000 neurons that can carry out complicated neuronal tasks, similar to that in humans, such as learning and memory. Due to this it has find immense popularity in the study of a wide range of human neurodegenerative diseases including Huntington's, Alzheimer's, and Parkinson's disease. *Drosophila* has minimal generation time, low maintenance costs, and well-developed genetic structure. They produce large numbers of externally laid embryos, and they can be genetically modified in numerous ways. Due to the above properties, the fruit fly can be used to study complicated cycles relevant in biomedical research, including cancer. Another emerging use of the fly is in the field of regenerative biology (Table 17.3).

17.1.2.1 *Drosophila melanogaster*: Structure and Components (Fig. 17.5)

D. melanogaster's body is divided primarily into three segments, namely, the head, thorax, and abdomen.

Some characteristic features of *Drosophila* play an important role in its taxonomy classification and species identification. Likely, the shape of its head and face, the number of relative size of oral setae, and compound eye are some most relevant distinguished features. The head is complex in structure and has a different morphology in different species. Next to the head comes the thorax. Prothorax, mesothorax, and metathorax are the three main segments of thorax. Humeral callus of prothorax bears a pair of setae. This pair of setae is very useful in identifying the species. Mesothorax is the largest segment of fruit fly and provides the wings the required strength for flight. The number and disposition of acrostichals or bristles on the mesonotum (the dorsal portion of the mesothoracic integument of insects) are important characteristics for species identification. Each thorax segment consists of a pair of legs, which are further segmented into coxa, trochanter, femur, tibia, and tarsus. The tarsus region is further divided into five small tarsal segments. Chaetotaxy (arrangement of bristles on the leg of the flies), body coloration, length of the leg segments, and sex combs are widely used in species identification and classification. The abdomen of *Drosophila* is covered by chitin. The dorsal region of the abdomen segment is known as tergite, while the ventral region is called sternite. These two regions are widely useful in taxonomy studies and also help in differentiating between male and female. The pigmentation patterns of tergites and

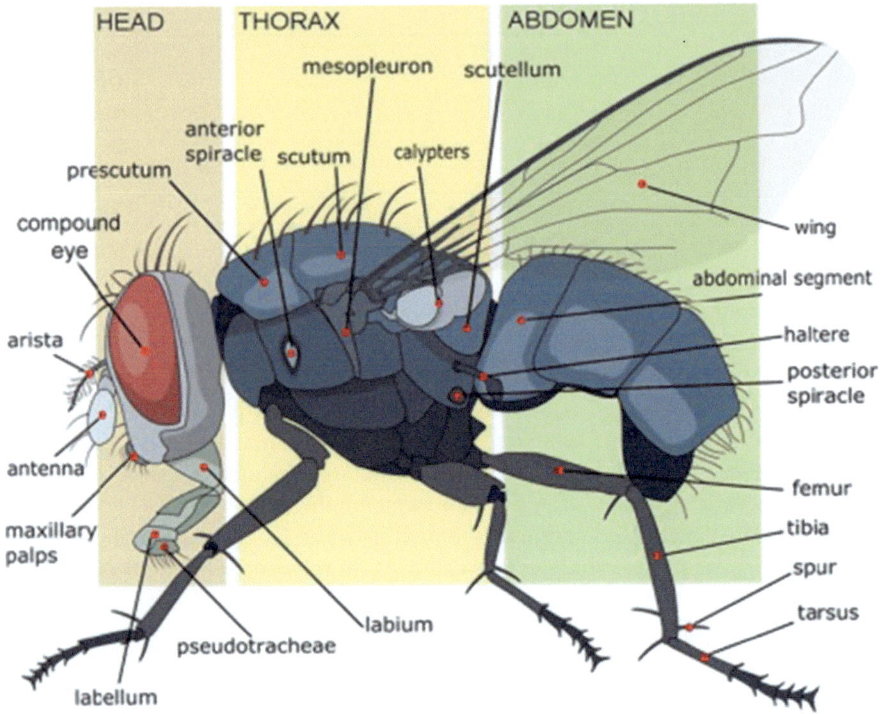

Fig. 17.5 Anatomy of *D. melanogaster*

the six quadrilaterally shaped sternites in female compared to only four sternites in male are widely used for identifying the species and gender.

The anatomy of the fly is very conserved and interesting. Having homology to human genome, it has analogy to various developmental process and signaling pathways of humans. Over the years it has helped researchers tremendously in getting a picture of the development and signaling processes in humans by serving as a very efficient model organism. Some selected functional anatomy are the brain and nervous system, muscle, kidney, gastrointestinal tract, and cardiovascular system.

Ectodermal brain of *Drosophila* and its development provide an immense understanding of neurobiology. Neuroblast is first formed in the embryo which then grows into the ventral nerve cord. The brain turns into a much mature and complex structure from larvae to pupa. It develops defined lobes having distinct functions. Fly muscle tissue is much conserved in evolution and provides the main support during flight. The kidney consists of three parts, i.e., the garland cells, the nephrocyte, and the Malpighian tubule. *Drosophila* nephrocytes share evolutionarily conserved domains of the vertebrate glomerular slit diaphragm. Nephrocytes and garland cells of fruit fly are a classic example of primitive reticuloendothelial and renal system. Fluid balance is maintained by the Malpighian tubule. The gastrointestinal tract of *Drosophila* is a very complex structure and over the years has been used to study host-pathogen interactions. It's heart is primitive and is morphologically different from the multichambered heart of vertebrates, though the functioning and genetic paradigm are very much similar. Detailed studies of the fly's heart have provided insights into studies regarding cardiac aging and also diabetes-related cardiac diseases.

17.1.2.2 Life Cycle and Reproduction (Fig. 17.6)

The life cycle of *Drosophila* is divided into egg, larval, pupal, and adult stages. The first larva comes out in 24 h and completes the instar stages by the next 72 h, reaching the pupal stage. Within the next 3–4 days, adult fly comes out and can live up to 70–90 days under favorable conditions.

D. melanogaster has a very short life cycle. A female fly lays huge number of eggs and will continue to do so for weeks. Larva hatches out from eggs in a span of 24 h at 25 °C and passes through three "instar" stages of development by the next 72 h. Pupa is formed at the end of the third instar stage. At this time various organs of the body provide the required energy for growth and metamorphosis into adult fly,

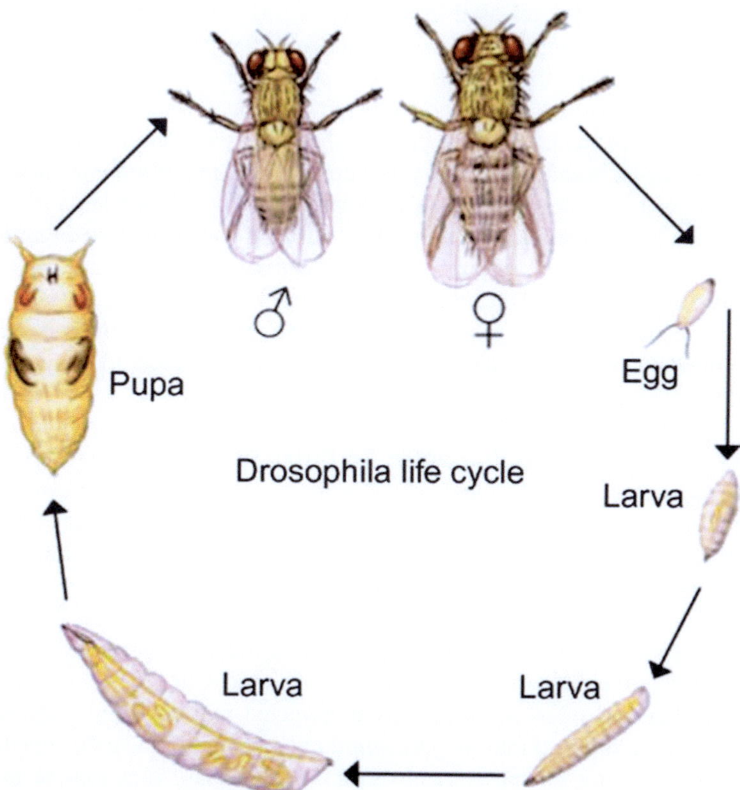

Fig. 17.6 Life cycle of *D. melanogaster*

for instance, the imaginal discs that ultimately develop into adult structures such as wing, eye, and limbs and consist mostly of undifferentiated epithelial cells. After 3–4 days eclosion occurs and the adult fly emerges. Female flies do not lay eggs in the first days of life. After the completion of a life cycle of 10 days, adult flies can live up to 70–90 days.

During the growth cycle, a number of genes play their part. The collaborative expression of these genes produces various proteins which lead to the entire larval setup of the fly. These proteins form the anterior-posterior axis and the dorsal-ventral axis called the bicoid and the dorsal, respectively. The synchronization of these proteins activates the transcription of specific cascades of genes. These genes include gap genes, pair-wise genes, segment polarity genes, and hox genes that ultimately divide the embryo into structural segments and functional regions. Once the embryo is completely developed, the first instar larva hatches out and starts eating. The larva eats a lot this time, and this food not only helps for its growth but also as future storage food (fats and sugars) during the later stages of metamorphosis, mainly during pupal stage. During the growth process, molting takes place when the larvae shed the exoskeleton. This process involves the chain-wise functioning of a number of hormones like ecdysone, juvenile, and most importantly the prothoracicotropic hormone (PTTH). PTTH, released from neurosecretory cells in the brain, stimulates the release of the molting hormone ecdysone from the prothoracic glands into the hemolymph. Ecdysone then forms a new cuticle or exoskeleton. Ecdysone is complemented by another hormone, eclosion hormone, which helps in initiating the molting process by allowing the larva to shed the exoskeleton and enter the next instar stage. Larvae wander for a place to pupate at the end of the third instar stage. At this time the larvae are appropriately referred to as "wandering larvae." Once the flies reach the final stage of larva, they attain a critical size, where the organs should be properly developed before pupation starts. During the pupal stage, most of the larval structures are lost. During this transition time, autophagy, a "self-eating" signaling mechanism, takes place. This converts the stored nutrients (fats and sugars), uptaken during the larval stage, into necessary food to provide energy required for the animal survival.

17.1.2.3 *Drosophila melanogaster*: An Efficient Model Organism

Drosophila melanogaster has served for years as a very important and efficient model organism for various genetic studies by researchers. The genome of the fly has been manipulated to study the function of the genes by generating transgenes by various methods. Some of them are mentioned below.

1. P-element transposons—Transposons are either linear or circular DNA that can be inserted into chromosomal DNA at specific sites with the help of transposase enzyme. Plasmid DNA are primarily constructed with the modified P-element inserts and are injected into the *Drosophila* embryo germline cells together with plasmids encoding for the P transposase. P transposase functions to insert the P-element transposon into a random point in the genome of *Drosophila*. The fly then produces a progeny that will carry the transgene. To determine the successful

integration of the transgene in the genome, sometimes a marker DNA is linked to it.
2. Homologous recombination—P-element transposons have many disadvantages, including lack of target site specificity. Therefore homologous recombination was put forward. This method uses the cell's own genetic makeup to insert foreign DNA or even for knockout genes by the process of homologous recombination.
3. Bacterial artificial chromosome (BACs)—BACs and recombination engineering are gaining attraction in introduction of transgenes in *Drosophila*. The technique uses recombination to insert genes or other DNA fragments into BACs. The process is becoming more popular because it is faster and easier than using restriction enzymes and ligases to insert DNA.
4. CRISPR/Cas 9—Small sets of repetitive nucleotide sequences with interspaced non-repetitive sequences are present in the bacteria. These are called Clustered Regularly Interspersed Short Palindromic Repeats (CRISPR) and function in bacterial immunity against viruses. When a phage attacks the bacteria, these CRISPR transcribe into shorter CRISPR RNA (crRNA) and noncoding trans-activating crRNA (tracrRNA). Cas enzymes help the bacteria to keep viral DNA copies in the form of CRISPR arrays, after the virus attacks the first time. These DNA copies get transcribed and processed into crRNA and tracrRNA when the virus attacks again. They together target the virus to be cleaved by other cas enzymes like Cas9. This technique of cleavage of viral DNA by cas 9 in association with crRNA and tracrRNA is used to insert artificial constructs into the fly genome. The cleavage caused by cas 9 is then repaired by techniques such as homologous recombination.

Besides transgenesis, mutagenesis is also widely used to study altered genetic functions in *Drosophila*. There are many ways to generate mutants, two most important being P-element mutagenesis (where P-element transposons are used to disrupt gene functions by inserting or removing them from certain positions in the genome) and chemical mutagens like ethyl methanesulfonate (produces a form of guanine, O6-methylguanine, that incorrectly base pairs with thymine during DNA replication).

Also *Drosophila* is used nowadays for genetic screening, which is aimed in identification of noble genes that are important to understand new and important biological cycles and mechanisms.

Outside the importance in studying the gene functioning, *Drosophila* also finds application in studying other fields like behavioral scoring, signaling pathways, and also metabolic cycles.

1. Behavioral scoring—Complex behavior like visual learning, courtship, and olfactory learning are well-established behavior in the fly and can be well assayed by studying their behavioral pattern.
2. Signaling pathways—*Drosophila* has helped in studying various signaling pathways, which play important role in humans. The most important being the Notch pathway, which is involved in many diseases like alagille syndrome, aortic

valve disease, cerebrovascular dementia, and leukemia. Other signaling pathways which are well studied in *Drosophila* includes JAK/STAT, TGF-β, Wingless, Hedgehog, Toll, Hippo, and MAPK.
3. Metabolism—*Drosophila* has also gained recent importance in putting light into the mechanisms underlying complex conditions like diabetes, obesity, hyperlipidemia, and inborn errors of metabolism.

Due to the abovementioned uses of *Drosophila*, a variety of applications in disease research have been found. Below are examples of some such diseases where the fly has given significant contribution.

1. Cancer—Over the years the fly has implied a lot in cancer biology. Hippo signaling, which controls organ size, proliferation, and apoptosis, was first characterized in *Drosophila*. Other human cancerous genes had also find homologue in *Drosophila*.
2. Insulin signaling peptides—*Drosophila* insulin-like peptides (Dilps) were identified in the flies and were found to act as sensors that regulate energy balance and growth.
3. Neurodegeneration—This is one field where *Drosophila* has an immense contribution. It has been used in the study of a wide range of human neurodegenerative diseases including Huntington's, Alzheimer's, and Parkinson's disease.

17.1.2.4 Threats

Drosophila melanogaster till date is not known to cause any harm to humans. However in two of very recent articles in *Nature* journal, the researchers have found a change in aggressive and courtship behavior of the fly with change in environment and climate. The other group of researchers investigated a primitive cannibalism in the fly behavior, where the younger larvae attacked and consumed the later-stage larvae. Though these are not threats to the humans, they can affect the laboratory preservation of the flies.

17.1.3 Nematode Worm (*Caenorhabditis elegans*) (Fig. 17.7)

An electron microscope image showing the thread-like structure of free-living nematode, *Caenorhabditis elegans*, under high magnification.

17.1.3.1 Introduction

Discovered by Sydney Brenner in the mid-1960s, *Caenorhabditis elegans* (*C. elegans*) is a free-living nematode found mainly in soil and compost heaps and has since become one of the most established model organisms. It has advantages of a very short life cycle which lasts only for 3 days, is small size (1.5 mm in adults), and can be easily handled under laboratory conditions. Due to these properties, this soil nematode offered great potential for genetic analysis. The population mainly consists of self-fertilizing hermaphrodites (XX) with a rare and very small

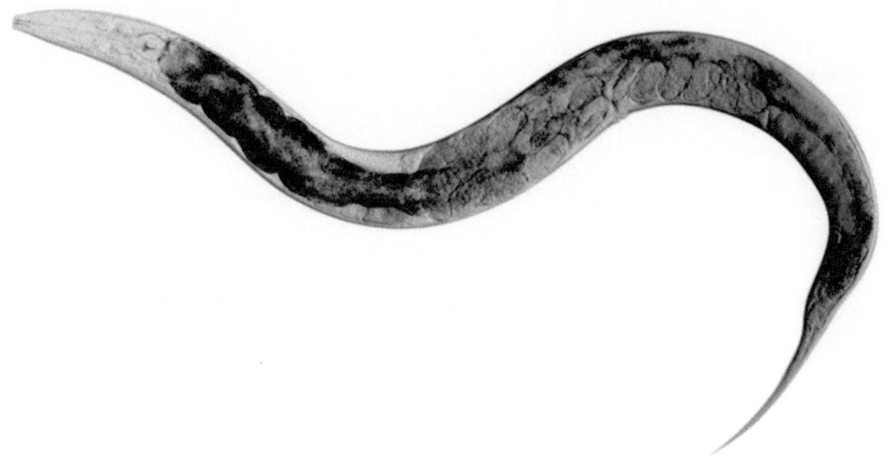

Fig. 17.7 *Caenorhabditis elegans*

Table 17.4 Taxonomy of *C. elegans* (ITIS Report)

Domain	Classification
1. Kingdom	*Animalia*
2. Phylum	*Nematoda*
3. Class	*Chromadorea*
4. Order	*Rhabditida*
5. Family	*Rhabditidae*
6. Genus	*Caenorhabditis*
7. Species	*elegans*

percentage of males (X0), who have a distinct morphology. The other distinguished features of *C. elegans* are the nematode's small but compact genome which is only 20 times that of *E. coli* and a simple cellular system which consist slightly greater than 1000 cells, of which 302 cells make the nervous system (Table 17.4).

17.1.3.2 *Caenorhabditis elegans*: Structure and Components (Fig. 17.8)

The body of *C. elegans* is divided into anterior, posterior, dorsal, and ventral regions. They are primarily hermaphrodites having both male and female reproductive organs in the same body.

1. Feeding—They are predators and possess grasping mouth parts and spines to catch and eat their prey.
2. Respiration, circulation, and excretion—They exchange gases and excrete metabolic wastes through their body wall. The nutrients and waste are carried out throughout their body by diffusion.

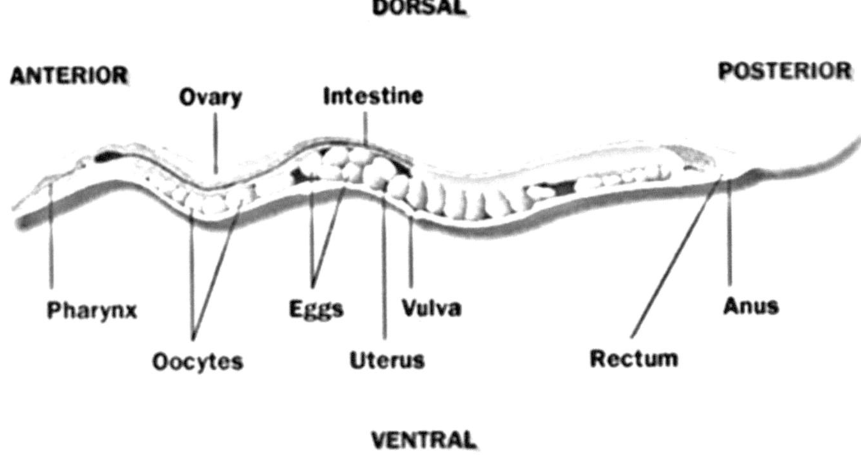

Fig. 17.8 Structure and components of *C. elegans*

3. Nervous system—They have a very simple nervous system. Nerves start from ganglia in the head and run through the whole body. They have several sense organs which can sense hosts and preys.
4. Movement—The muscles extend throughout the entire body. The pseudocoelom has a fluid that works together with the muscles as a hydrostatic skeleton.
5. Reproduction—*C. elegans* has a well-developed and simple reproductive system. Being maximum hermaphrodites they undergo self-reproduction; in very rare cases male progeny is found. A special mention is the presence of the organ vulva, which is required during mating, as males inject sperm through it and for deposition of embryos after internal fertilization. A further detail about the reproduction in *C. elegans* is given in the next section.

17.1.3.3 Life Cycle and Reproduction (Fig. 17.9)

The life cycle of *C. elegans* is divided into egg, larval, pupal, and adult stages. There are three instar larval stages. Improper food and environment leads to these larval stages entering the diapauses (e.g., the dauer stage shown in the diagram above, which instar larva L2 can enter under inappropriate growth conditions). After the last instar L4, *C. elegans* enters the pupal stage and finally the adult stage.

C. elegans adults lay embryos that pass through stages like gastrulation, comma stage, and two- to threefold changes in embryo formation before hatching into the first larval stage. The first-stage larvae, L1, comes out of the eggshell and begins feeding. When adequate amount of food is there with low population density of maturing larvae, the larvae go through a definite genetic clock and produce four larval stages, namely, L1, L2, L3, and L4, before molting into adults. The cycle takes about 3.5 days at 20 °C under rich nutritional conditions. But harsh and unfavorable environmental conditions with little food and crowded population can trigger L1s to

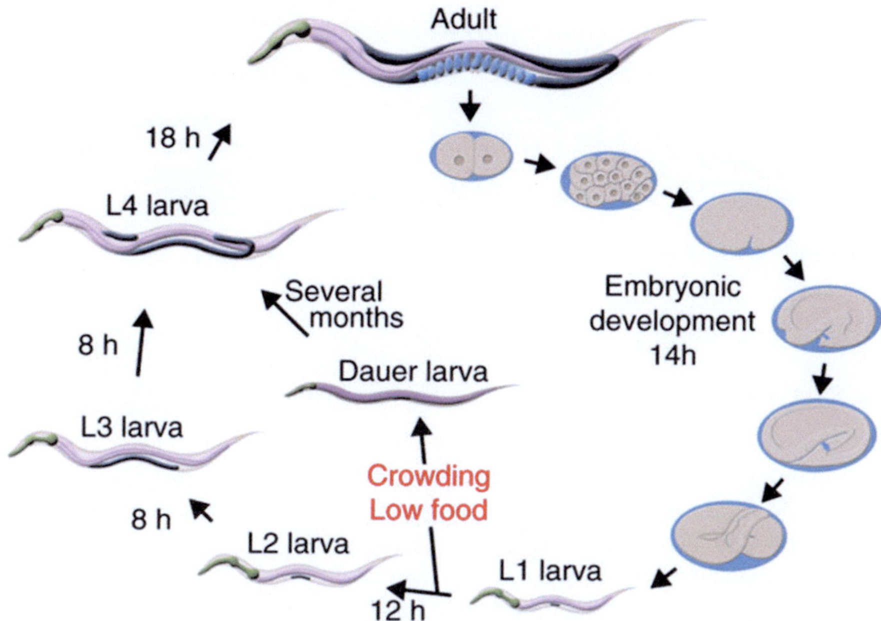

Fig. 17.9 Life cycle of *C. elegans*

enter diapause state, blocking further development. When the scarcity of food reduces, they regain development and reach the L2 stage. If in the L2 stage the larvae again face inappropriate growth conditions, they enter into a stage called dauer, where development of the reproductive system is arrested. If the dauer find food, then the third larvae, L3, come and proceed to adulthood. The larvae can enter the third diapause when L4 adults are again left without food. At this diapause the reproduction is halted by the degradation of most of the germline, except proliferating zone cells which remain cell cycle arrested. Upon refeeding, worms in adult reproductive diapause resume germ cell proliferation, meiotic development, and oogenesis and can become fertile.

C. elegans primarily disseminate as hermaphrodites, with two X chromosomes and a diploid set of autosomes (2X, 2A). Up to L2 the nematode lives as hermaphrodites. In the L3 stage, some of the hermaphrodites produce the male germ cells. Nondisjunction occurs in about 0.1–0.2% cases which lead to male progeny (X0) production. On the other hand, 50% of cross-progeny becomes male if mating occurs. These male germ cells then get differentiated into sperm between L3 and L4 stage. The sperms are produced both by the hermaphrodite individuals and from mating. They stay in the spermatheca and wait for the ovulation of the very first oocyte. Female germ cells are produced in the L4 stage which finally differentiate into oocytes. Once the ovulation is over, the sperm meets the oocyte and fertilization occurs marking the initiation of embryogenesis. When the hermaphrodite is unmated, it can produce about 300 embryos, but in the mated condition the

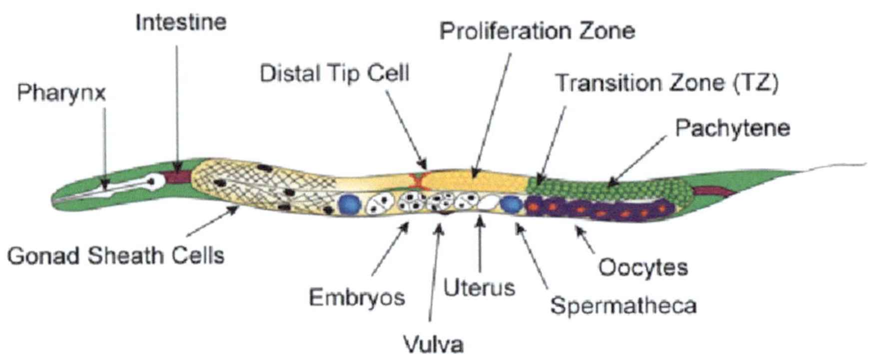

Fig. 17.10 Adult *C. elegans* highlighting the reproductive system

hermaphrodite can produce up to 1000 embryos. This shows that the main factor on which self-fertility depends is not oocyte production but rather the amount of self-sperm formed by the hermaphrodite (Fig. 17.10).

The reproductive system of *C. elegans* carries the major organs like the uterus, vulva, spermatheca, and distal tip cell along with embryos and oocytes.

Gonadogenesis or formation of gonads in *C. elegans* depends on the presence of four gonad precursors Z1, Z2, Z3, and Z4. It begins at L1 stage of larval development. Z2 and Z3 are the germline progenitors, and Z1 and Z4 are the somatic precursors. The germline progenitors are present in between the somatic precursors. The germline precursors give rise to germline gonad cells, whereas the somatic precursors give rise to distal tip cell (DTC) and the other somatic gonad cells. From the L4 stage, the hermaphrodite takes the shape of the adult, with the gonad arm on the dorsal side of each U-shaped tube capped by the DTC. Spermatogenesis completes within a short span of 24 h in both the hermaphrodite and male. Four spermatids are formed from spermatogenesis which produces motile and amoeboid sperms. The spermatids are generated by two sequential meiotic divisions of meiotic germ cells. The prophase arrest stage is absent in the cell cycle of spermatogenesis. Oocyte, on the other hand, is large and nutrient-rich. In contrary to prophase of sperm formation, meiotic prophase of oogenesis takes a considerable elongated time span of about 54–60 h. The oocyte is then released for fertilization. There is an extended pachytene in oogenesis, where germ cells synthesize RNAs and proteins that are donated to the oocyte. The oocyte most adjacent to the spermatheca (proximal oocyte) undergoes ovulation into the spermatheca and takes part in fertilization. Before this it undergoes meiotic maturation which includes breakdown of nuclear envelope, progression to metaphase of meiosis I, and rearrangement of the oocyte cortex and cytoplasm. In the absence of sperm, oocytes get arrested in diakinesis and ongoing oogenesis is inhibited. Sperms are not available when adult hermaphrodites are depleted off their self-sperms or when the female undergoes any mutation. Thus, oocyte production and utilization only occur in the presence of sperm.

17.1.3.4 *Caenorhabditis elegans*: An Efficient Model Organism

The nematode *C. elegans* has emerged as an important animal model in various fields including neurobiology, developmental biology, and genetics. The characteristics of the nematode that have made it a highly acceptable model organism include its well-characterized genome which can be genetically manipulated, vividly described developmental program, ease of maintenance, short life cycle, and small body size. It has been widely accepted as a model organism, with a number of toxicological, host-pathogen interaction and metabolic pathways studies being carried out on it over the years. Some of them are described below. Since reverse genetic and transgenic experiments are much easier and less expensive to conduct in *C. elegans* as compared to many other model systems, it is a useful model for studying effect of chemical exposures on metabolic pathways. *C. elegans* can act as both in vitro and in vivo mammalian models in toxicology. *C. elegans* also provides a good platform for the study of host-pathogen interactions. It has been discovered that the nematode has an inducible defense response and that the defense mechanism shares similarity to that of other organisms. This makes the worm a potential model for the study of innate immunity. The interaction between prokaryotes and eukaryotes is pervasive and has important medical and environmental significance. But despite this importance, the existing data on the mechanisms and pathogenic consequences of bacterial-fungal interactions within a living host is very scarce. *C. elegans* was utilized as a host to study the interactions between two ecologically related and clinically troublesome pathogens. One is the bacteria, *Acinetobacter baumannii*, and the other is the fungi, *Candida albicans*. It was observed from the results that *Acinetobacter baumannii* inhibits filamentation, a key virulence determinant of *Candida albicans*. The nematode has also proved efficient in studying substrate-inhibitor interaction mechanisms. It was found in another study that the interaction of a lead compound (named lead compound-76) had an inhibitory effect on MPK-1, needed by the worm in vulva development and egg laying and the established homologue of the human ERK-2 (extracellular signal-regulated kinase-2). Compound 76 showed selectivity in inhibition of phosphorylation of LIN-1, a MPK-1 substrate essential for vulva precursor cell formation. Thus *C. elegans* served as a model organism to study specificity of compounds targeting the ERK protein. It also helped in studying mammalian bacterial pathogenesis, as it was also shown to be killed by *Pseudomonas aeruginosa*. Besides these *C. elegans* has also proved to be a useful organism for investigating molecular and cellular aspects of numerous human diseases.

17.1.3.5 Threats

There are no known threats caused by *C .elegans*, till date, to humans. Except it can sometimes be host to pathogenic bacteria and fungi, which should be handled and taken care of.

Fig. 17.11 *Xenopus* frog. Female *Xenopus laevis* (left) and female *Xenopus tropicalis* (right)

17.1.4 Western Clawed Frog (*Xenopus tropicalis*) (Fig. 17.11)

Xenopus tropicalis (*X. tropicalis*), also known as *Silurana tropicalis* or the Western clawed frog, is a small, aquatic frog that is found all along the west coast of equatorial Africa. It is a close neighbor of the widely used model organism *Xenopus laevis* (*X. laevis*). It complements *X. laevis* in taxonomy, anatomy, and life cycle. *X. tropicalis* bears a diploid genome and a high conservation of gene synteny with the human genome. It was first used as a model organism for genomic research in the early 1990s. After this it has been used as an acceptable and interesting model for biomedical studies, developmental and cell biology, biochemistry, functional genomics, and immunology especially those metabolic cycles and systems that can influence vertebrate development from embryonic stages through adulthood. The embryos of the frog develop externally and the tadpoles are transparent. These features facilitate experimental manipulation and post-factum analysis of animals. Another important feature which makes *X. tropicalis* more acceptable model organism than *X. laevis* is its short life span of 6 months as compared to 18 months in *X. laevis*. Raising and culturing them are also relatively easy as they can be easily grown in water tanks or in recirculating aquatic systems (Table 17.5).

17.1.4.1 *Xenopus tropicalis*: Structure and Components (Fig. 17.12)

The anatomy of frog has similarity with that of humans but is much simpler.

The body of the frog is divided into three portions, i.e., head, short neck, and trunk. The neck is short and rigid, permitting only limited head movement. The frog body consists internally of a single cavity called the coelum. The trunk is thick and

Table 17.5 Taxonomy of *X. tropicalis*

Domain	Classification
1. Kingdom	*Animalia*
2. Phylum	*Chordata*
3. Class	*Amphibia*
4. Order	*Anura*
5. Family	*Pipidae*
6. Genus	*Xenopus*
7. Species	*tropicalis*

a)

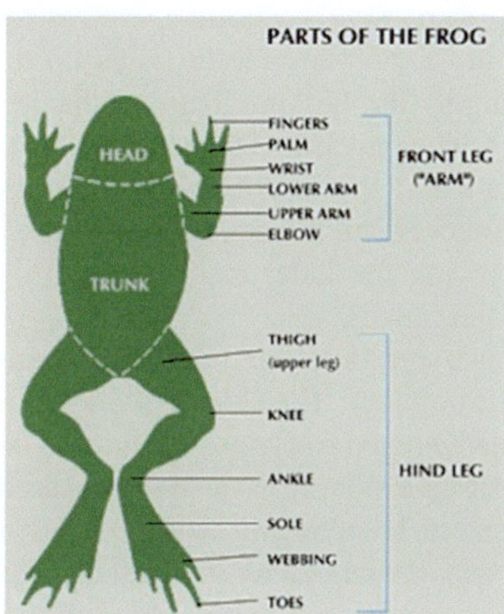

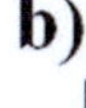

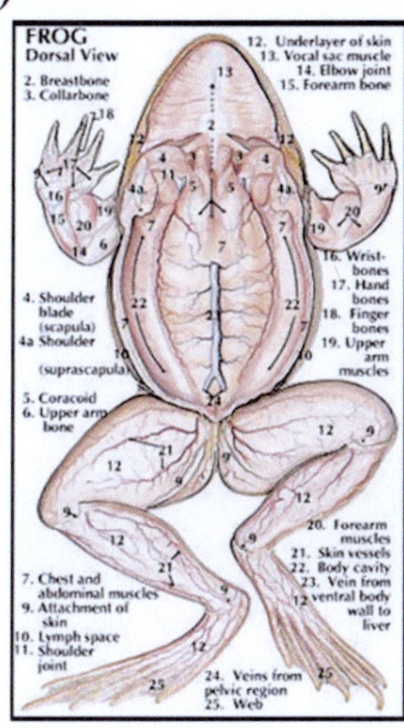

Fig. 17.12 Anatomy of *Xenopus tropicalis.* The structure and components of *Xenopus tropicalis.* (**a**) Overall structural components of the frog. (**b**) Dorsal view of the frog anatomy

forms walls of the coelom. All the frog's internal organs are held in the coelum, and it is a continuous hollow space with no such partition as diaphragm in man.

The frog skull is flat with an expanded area that encloses the brain. The skeleton is bony and provides support and protection to the frog's body. The skeleton consists of nine vertebrae in the vertebral column and no ribs. The forearm and hind legs of frog perform important functions in the frog's locomotion. They have similarity to that of humans in terms of structure and nature of bones. Radio-ulna is the only forearm bone in frogs, whereas humans have two forearm bones, the radius and the ulna. But both have a single upper arm bone, the humerus. The hind legs of the frog are highly specialized for leaping and consist of a single leg bone, the tibiofibula, in contrast to man who has two lower leg bones, tibia and the fibula. Femur is the single thigh bone found in both man and frog. Frog's leg also consists of two elongated anklebones, or tarsals. These are the astragalus and the calcaneus. The astragalus and calcaneus correspond to the human talus and heel bone, respectively. The frog does not have a tail, except ruminents of a primitive tail, called urostyle. The skeleton movement is supported by muscles, mainly striated muscle, whereas internal organs contain smooth muscle tissue.

The frog's heart is three chambered and protected by the pericardium. It is made up of two upper chambers, the right and left atrium and a one single ventricle as the lower chamber. In contrast men contain two lower compartments, the right ventricle and the left ventricle. The pure (oxygen-laden) and impure (oxygen-poor) blood are always present together in the frog ventricle, but the blood however never mixes. The right atrium passages toward the bottom of the ventricle and allows the impure blood that enters into it from the body, to pass to the bottom. On the other hand, the pure blood from the left atrium also enters the same single ventricle. But as the oxygen-poor blood is present toward the bottom of the ventricle, it holds up the oxygen-laden blood and protects it from flowing to the bottom. This mechanism prevents both the blood from coming into contact. The pure blood then moves out of the heart along with impure blood, when the latter leaves the ventricles and enters the vessels leading to the lungs. The lung vessels, however, are filled with oxygen-poor blood, blocking the oxygen-laden blood and forcing the latter to detour into the arteries. This carries the oxygen-laden blood to the tissues.

Frog principally respires through its skin. It has a soft, thin, and moist skin having an extensive network of blood vessels running throughout. The skin is divided into two layers, an outer epidermis and an inner dermis. Oxygen freely passes through the membranous skin and enters into the blood. Other than skin, frogs also have paired, simple, saclike lungs to breathe. The mechanism of breathing in frog is different from that in man. It has no ribs or diaphragm like men, and neither its chest muscles are involved in breathing. The frog simply opens its mouth to breathe and let the air flow into the windpipe. It can also breathe when its mouth is closed. It does this by lowering the floor of the mouth keeping the nostrils open. This causes the air to enter the enlarged mouth. Then, with nostrils closed, the air in the mouth is forced into the lungs by contraction of the floor of the mouth.

The process of digestion in frog starts in the mouth. The teeth in frog have no function in digestion and are present only in the upper jaw. The tongue is highly specialized and helps in catching prey. Whenever the frog encounters a prey, it can flick open the tongue from its folded position in the throat. The prey gets attached to the tongue due to its sticky texture. The food then moves to the stomach through the esophagus. From here the food moves into the small intestine, where most of the digestion occurs. Large digestive glands, the liver and the pancreas, are attached to the digestive system by ducts. Ureters carry the liquid wastes from the kidneys to the urinary bladder, and the solid wastes from the large intestine pass into the cloaca. Both liquid and solid wastes leave the body by way of the cloaca and the cloacal vent.

The nervous system in frog is well developed and is divided into the brain, spinal cord, and nerves. The brain is made of medulla, cerebellum, and cerebrum. The automatic functions like digestion and respiration are controlled by the medulla, whereas body posture and muscular co-ordination are regulated by cerebellum. The cerebrum is very small as compared to human. The brain consists of ten cranial nerves and ten pairs of spinal nerves. Olfactory lobes, present in the forepart of brain, monitor the sense of smell in frogs. The eye has fixed lens that cannot change its focus. The eyelids are poorly developed and cannot move, so to close its eye, the frog draws the organ into its socket. A third eyelid called the nictitating membrane is present but is almost rudimentary. Only in some instances that it may be drawn over the pulled-in eyeball. External ear is absent and both eardrums (tympanic membranes) are exposed. There is only one bone in the frog's middle ear, and semicircular canals help to maintain body balance.

17.1.4.2 Life Cycle and Reproduction (Fig. 17.13)

There are different stages of development in *Xenopus* frog. The fertilized egg from ovum and sperm goes through zygotal, blastulation, and gastrulation stages and finally reaches the adulthood. *Xenopus tropicalis* completes the life cycle in 6 months, whereas *Xenopus laevis* takes 12 months.

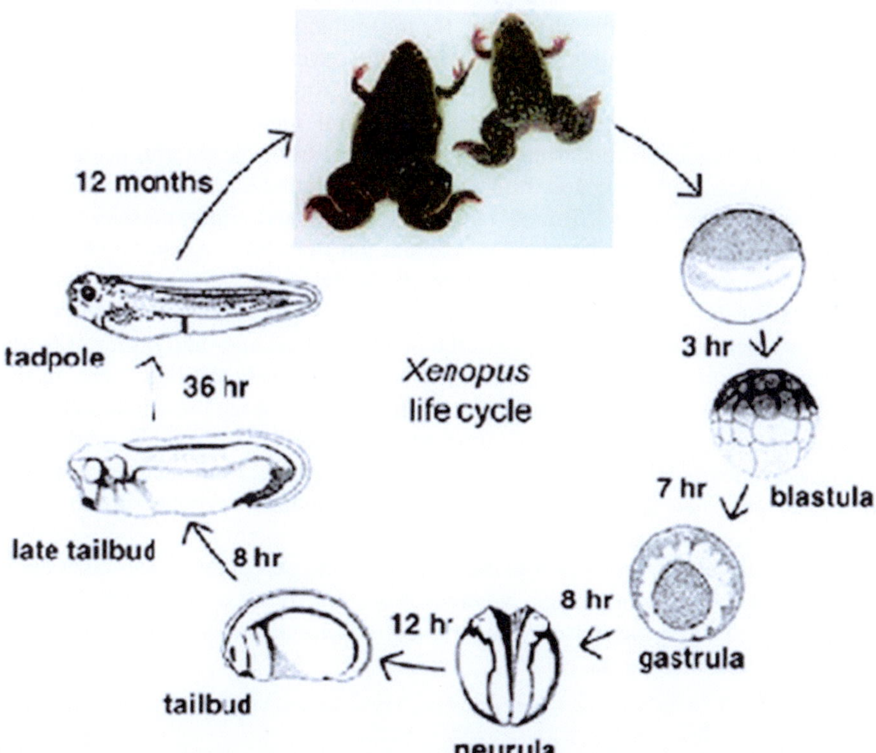

Fig. 17.13 Life cycle of *Xenopus*

Like any other amphibians, *X. tropicalis* starts its life as an embryo. The embryos can be used for various biomedical studies as they are able to tolerate considerable manipulation like single cell, germ layer dissections, and tissue transplantations. The eggs and the embryos are widely used in targeted gene knockout, knockdown, and overexpression studies and also serve as a source for high-throughput biochemical studies. Besides, the cell-free extracts made from *Xenopus* oocytes are used as a coherent in vitro system for cell and molecular biological studies. Oocytes are also widely used for studies of ion transport, channel physiology, and environmental toxicology. Due to all the above properties, the eggs and the embryos can serve as an outstanding tool in biomedical research.

The egg is composed of an animal and a vegetal region, which are covered by a vitelline membrane. After fertilization, the cortex determines the future dorsal region at a position opposite to the site of sperm entry. The blastulation and the gastrulation phases start soon after fertilization within a few hours. The blastula follows a radial symmetry. The three germ layers mesoderm, endoderm, and ectoderm are next formed. Mesoderm and endoderm are formed from the marginal zone and are then internalized at the blastopore. Ectoderm spreads to cover the embryo, a process called the epiboly. Archenteron (future gut cavity) is formed from the dorsal endoderm after it separates from mesoderm. The lateral mesoderm then spreads ventrally to cover inside of archenteron. By the end of gastrulation, the archenteron formation is completed, and mesoderm completely covers the gut internally. Epiboly gets accomplished by this time, with ectoderm covering the embryo. Yolk cells are internalized and serve as food source. The dorsal mesoderm during this time develops into notochord and somites. The somites, from mesoderm, at this stage form dermatome, vertebrae, and trunk muscles. The dermatome turns into the future dermis and the limbs are formed by the vertebrae and trunk muscles. Lateral mesoderm becomes heart, kidney, gonads and gut muscles and ventral mesoderm form blood forming tissues. The endoderm gives rise to the lining of gut, liver and lungs. Part of the ectoderm forms the neural plate, which further forms the neural tube. The anterior neural tube becomes the brain, whereas mid- and posterior neural tube becomes spinal cord. Neurulation is followed by early tail bud stage. In this stage the brain is divided, ears and eyes are formed, and three bronchial arches are formed. The tail is also formed as an extension of notochord, somites, and neural tube. Lastly, neural tube, from the edges of the neural folds, forms neural crest cells. This crest cell then develops the sensory and autonomic nervous system, skull, pigment cells, and cartilage.

X. tropicalis and *X. laevis* follow the same life cycle except that *laevis* takes about a year to reach the adult stage, whereas *tropicalis* reaches adulthood in about 5–6 months. This is the reason why *tropicalis* is more preferred nowadays as a model organism than *laevis*.

17.1.5 *Xenopus tropicalis*: An Efficient Model Organism

As mentioned earlier *X. tropicalis* is a close relative of the widely used model organism *X. laevis*, with the advantage of a short life span as compared to the latter. So, *X. tropicalis* is nowadays gaining rapid importance as a model organism. It has an established genome sequence which can be manipulated for various biomedical, cell biology, genetics, developmental biology, and also drug discovery and delivery studies. Some of its applications as an emerging model organism are enlisted below. Systematic chemical genetics screens (a term used to describe the difference in chemical manifestation between mutant strains of various organisms) had been successfully carried out in *X. tropicalis* and *X. laevis*. *X. tropicalis* is a fast-breeding and diploid relative of *X. laevis* but much smaller in size. It is now successfully adopted for research in developmental genetics and functional genomics. Its embryogenesis cycle is almost similar to that of *X. laevis* but has relatively simpler genome and shorter generation time. This makes it more convenient for use in genomic and transgenic practices. Its embryos closely resemble those of *X. laevis*, except for their smaller size. Therefore, assays and molecular probes developed in *X. laevis* can be readily adapted for use in *X. tropicalis*. It can also be a key model to host-pathogen interaction studies as it was found in a recent study that *Chlamydia pneumoniae* (a human pathogenic bacteria) is pathogenic to *X. tropicalis*. It is also used for transcription, translation, and targeted mutagenesis studies by applying various recent and advanced technologies like transcription activator-like effector nucleases (TALENs), zinc-finger nucleases (ZFNs), and CRISPR/Cas system. The usefulness of *X. tropicalis* as a model species to investigate endocrine disruption and developmental reproductive toxicity was assessed. *X. tropicalis* is reported to have similarity in reproductive system with that of fish, birds, and mammals after developmental exposure to estrogens. This makes *X. tropicalis* a promising model for research on endocrine disruption and developmental reproductive toxicity. *X. tropicalis* has also entered the world of cancer, as it was very first studied as a test organism for blood cancer. Vertebrate biology, particularly the genetic, biochemical, and environmental factors that influence the development of a vertebrate from embryonic stages through adulthood, can be well studied in *X. tropicalis*. It is also finding uses as an important test species for estimating the impact on the health of rare and endangered species of amphibians, which are exposed to environmental toxins and disease due to waterborne pollutants and infectious agents such as the chytrid fungus. It is also finding wide applications in functional genomics. Successful understanding of the activity of genes in this easy model system will be of a significant help in understanding of similar genes in other vertebrate systems.

17.1.5.1 Threats

There are no known threats caused by the Western clawed frog to humans, till date, except the hygiene issues. It should be handled with care and aseptically.

Fig. 17.14 Mouse (*Mus musculus*)

17.1.6 Mouse (*Mus musculus*) (Fig. 17.14)

The house mouse, *Mus musculus*, was established in the early 1900s as one of the first genetic model organisms. The major advantage of it is being a mammal, so it shares genetic and physiological similarities with humans, therefore efficiently serves as models for human phenotypes and disease. Other than this it has a short generation time, produces comparatively large litters, is easily handled and grown, and shows visible phenotypic variation. There are many biological processes in the development process of rodent and primate lineages, which are conserved during the years of evolution, and mice had served invaluable in studying these processes. Besides this over the years mice had also helped in investigating the developmental mechanisms by which the conserved mammalian genome develops and differentiates to give rise to a variety of different species. Today, *Mus musculus* is widely known as an excellent mammalian model for studying a wide variety of traits and diseases, including those involved in metabolism, development, neurological disorders, immunity, male sterility, adaptive evolution, comparative genomics, and non-Mendelian inheritance. Mouse (2.5 Gb) has almost 99% similarity in the coding region genes with humans (2.9 Gb). Despite this they can differ strikingly in experimental results from humans. The experimental results and the knowledge developed during a century of research on the mouse present an opportunity of interpretation of human genes and their functions with experimental studies of corresponding mouse genes (Table 17.6).

Table 17.6 Taxonomy of *Mus musculus*

Domain	Classification
1. Kingdom	*Animalia*
2. Phylum	*Chordata*
3. Class	*Mammalia*
4. Order	*Rodentia*
5. Family	*Muridae*
6. Genus	*Mus*
7. Species	*musculus*

17.1.6.1 *Mus musculus*: Structure and Components (Fig. 17.15)

Mice being a mammal has similar anatomy to humans. The diagram depicts the internal organs of the mice starting from the esophagus to the anus.

Mus musculus is a mammal and has an anatomy like any other mammal. The anatomical characteristics of *Mus musculus* are as follows.

Mammals are endothermic and require necessary insulation to maintain their body temperature. This is provided by the body hair, which is also a signature of being a mammal. These animals are warm blooded and have mammary or milk-producing glands. Oil glands or sebaceous glands are located almost all over the body. They produce sebum (a lipid mixture) that is secreted onto the hair and skin and provides water resistance and lubrication. Sweat glands or sudoriferous glands are absent in mice.

Mice have the dental formula as follows: incisors 1/1, canine 0/0, premolars 0/1, and molars 3/3. Mice have a total of 18 teeth, including 4 incisors, 2 premolars, and 6 molars. Incisors are very important for mice. Mice have a natural yellow pigment in their teeth. Mice are monophyodont, i.e., they have only one set of teeth throughout life.

The skeletal structure can be classified into axial and appendicular. The axial skeleton consists of the bones which make up the skull, vertebral column, sternum, and ribs, whereas the appendicular skeleton consists of bones which make up the appendages and girdles. The dorsal portion of the skull is made up of a series of bones, namely, unpaired occipital and interparietal, the paired parietals, frontals, nasals, premaxillae, maxillae, zygomatics, and squamosals. The occipital, interparietal, and parietals form the cranium of the skull. Its posterior wall is made up of the occipital, interparietal forms a part of the top dome, and the parietals form the rest of the top and a portion of the sides of the cranial cavity. Frontals and nasals form the anterior portions of the brain and along with premaxillae, maxillae, zygomatics, and squamosals form the face and upper jaw. The vertebral column or the spinal cord is formed as an extension of the occipital bone. Occipital continues anteriorly as the spinal cord and forms a support of the thoracic and abdominal cavities, further extending up to the tail. The spinal cord consists of the following vertebrae: 7 cervical, 12–14 thoracic, 5 or 6 lumbar, 4 sacral, and 27 to 30 caudal. The most anterior of the seven cervical are the atlas and axis. Each thoracic vertebra articulates with a pair of ribs and some sacral articulates with the innominate bones of the pelvic girdle. The appendicular skeleton consists of the paired pectoral and

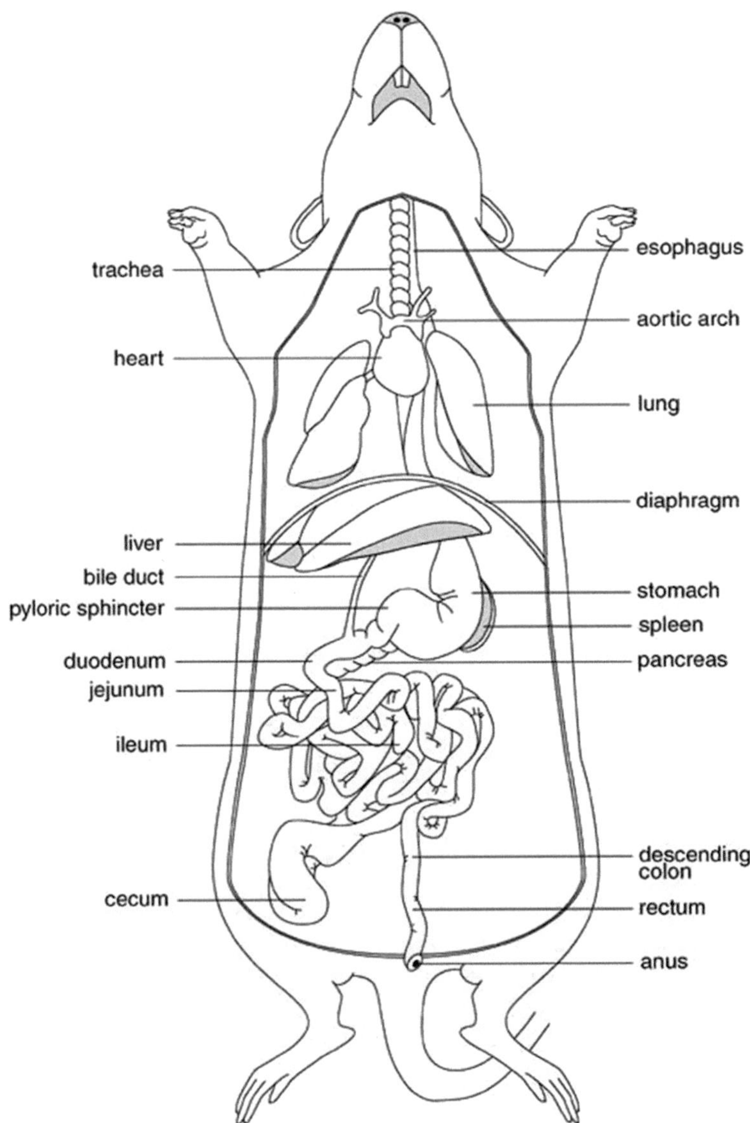

Fig. 17.15 Anatomy of *Mus musculus*

pelvic girdles and the bones of the limbs. Each pectoral girdle consists of dorsal scapula and ventral clavicle. The adult pelvic girdle is made up of right and left innominate bones. These bones are further attached dorsally to one or more sacral vertebrae and ventrally at the pubic symphysis. The bones of the forelimb are humerus, radius and ulna, eight carpals, five metacarpals, five first phalanges, four second phalanges, and five third phalanges with their tips. The bones of the hind limb are femur, tibia and fibula, seven tarsals, five metatarsals, and the same number of phalanges as in the forelimb.

The heart of the mice is a closed circulatory system, i.e., blood remains within the vessels. It is enclosed by the pericardial cavity, which is a division of the thoracic cavity. Being a mammal a mice's heart is four chambered. The chambers are muscular-walled and are divided into the left and right atria and the left and right ventricle. The right side (right atria and right ventricle) receives the impure blood from the veins and pumps it to the lungs for oxygenation. The left side (left atria and left ventricle) receives oxygenated blood from the lungs and pumps it to the body via the arteries. There are two principal vessels that carry blood from the heart. They are the pulmonary artery from the right ventricle and the aorta from the left ventricle. Branches of these arteries supply all parts of the body. There are three principal veins entering the atria of the heart: the pulmonary veins from the lungs; the superior vena cava from the head, neck, chest, and forelimbs; and the inferior vena cava from regions of the body posterior to the diaphragm.

The lungs are large and fill the entire thoracic cavity along with the heart. Muscular diaphragm controls the volume and movement of air to and fro the lungs.

The lymphatic system consists of vessels which transport the lymph to the blood, from lymphatic tissues. The lymph gets distributed, on its way to the blood channels, partially to the nodes that lie in the course of the lymphatic vessels and also to the peripheral nodules present at the beginning of lymph channels in the digestive tube. Mice do not possess palatine and pharyngeal tonsils.

The spleen is present in the left anterior quadrant of the abdominal cavity. It is the largest secondary immune organ in the body and is slightly curved, elongated, and oval. It initiates immune reactions against blood-borne antigens and purifies the blood of foreign and toxic matters. It also removes old or damaged red blood cells from circulation. It is composed chiefly of lymphatic tissue but has no lymph vessel connections. The thymus was regarded as a vestigial organ with little or no function in the adult animal. But now it has been categorized as an endocrine gland made up of reticular tissue. In rodents the thymus has mainly two functions. It serves as the site for lymphopoiesis in the embryo and newborn. The formed lymphocyte migrates from the thymus to the spleen, lymph nodes, and other lymphoid organs. In addition, the thymus produces an immunotrophic hormone, one which helps in enhancing the immunological potential of cells.

Mice contain three pairs of salivary glands which are located in the subcutaneous tissue of the face and neck: the parotid, submandibular, and sublingual. Each of them is connected with the oral cavity through a single excretory duct. They are not mixed glands and each has only one type of secretory cell. Parotid and submandibular secrete serous and sublingual secretes mucous. The digestive tube extends from the pharynx to the anus and includes the esophagus, stomach, small intestine, and large intestine. The small intestine consists of duodenum, jejunum, and ileum. The large intestine consists of caecum, colon, and rectum. The liver is a large gland occupying the anterior third of the abdominal cavity. It touches the arch of the diaphragm anteriorly and partially meets the stomach and duodenum posteriorly. The liver functions include producing bile which digests fat, storing glycogen and transforming wastes into less harmful substances. The mouse have a gall bladder which is absent in rats. The pancreas is a diffuse gland with a pinkish color. It is suspended in the mesenteries between the stomach, duodenum, and ascending and transverse colons. It has both endocrine and exocrine functions. As an exocrine organ, it secretes enzymes or digestive juices into the small intestine. There, it continues breaking down food that has left the stomach. As an endocrine organ, it produces the hormone insulin and secretes it into the bloodstream, where it regulates the body's glucose or sugar level. The urinary system includes the kidneys, ureters, urinary bladder, and urethra. The principal function of the urinary system is the maintenance of water and electrolyte homeostasis. The female genital system is composed of ovaries, oviducts, uterus, and vagina. The male genital system consists of testis, excretory ducts, accessory glands, urethra, and penis.

The reproductive system is well developed. Both ovaries are equally functional, and the ova are fertilized in the oviducts. Embryo develops within the uterus and lies within a fluid-filled amniotic sac, which provides necessary nourishment to the embryo and also protects the embryo from any friction. It has a well-built and connected placenta which connects the embryo with the mother's system and provides nourishment and immunity to the embryo. Male testes are contained within the scrotum outside the body cavity.

The mouse has a complex brain like other mammals, smaller in size then rat. Mouse brain consists primarily of water, protein, and lipids. It consists of enzymes like alkaline phosphatase and glutamic acid decarboxylase, but their expressions vary with development. Alkaline phosphatase is present in large amounts during fetus stage, whereas glutamic acid decarboxylase is present all throughout but its amount increases with maturation up to 30 days. Thus maturation to adulthood involves shifts in enzyme activity which presumably is related to morphological and functional changes.

17.1.6.2 Life Cycle and Reproduction (Fig. 17.16)

Mice reach sexual maturity in about 5 weeks and can mate within 2 months of birth. Mice have a short gestation period of 21 days, and a female can give birth to about seven litters in a year.

The mouse life cycle will vary depending on their habitant but typically lasts about 2 years indoors. The first stage of a mouse's life cycle lasts about a week. It then reaches its adolescence which lasts a few weeks and is marked by big personality changes. Over the next several weeks, it eats plenty of food to provide fuel to his growing body. The male becomes sexually mature once he is 5 weeks to 10 weeks old, and females are 8–12 weeks old before they're sexually mature adults. Once they reach maturity, they undergo mating. Once mating completes, females go into heat for 4–5 days. Once pregnant, mouse gives birth after 3 weeks, and five to eight litters (baby mice) are produced at a time. The females can undergo mating and reproduce up to ten times per year. At the start of the mouse life cycle, newborns are hairless and blind but develop thin fur and gain sight and mobility after 2 weeks. Mice reach sexual maturity in about 5 weeks and are ready to mate about 2 months after birth. Mice can breed throughout the year if they get suitable conditions, and

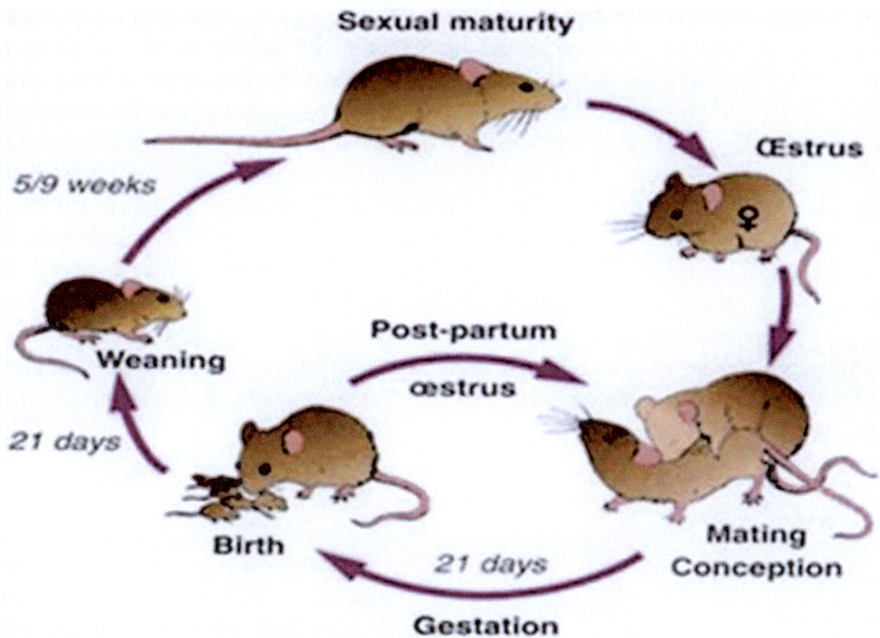

Fig. 17.16 Life cycle of *Mus musculus*

with a short gestation period of only 21 days, they can give up to seven litters a year. With the birth of each individual mouse and reaching sexual maturity, the total population of a particular habitat of mice under ideal conditions can reach up to 15,000 in a year. The maximum life span of a mouse is about 3 years. Females are capable of becoming pregnant immediately after giving birth. They can even give birth and raise two healthy litters of normal size and weight simultaneously without significantly changing their own food intake. However when there is scarcity of food, the females can extend their pregnancy by over 2 weeks and give birth to young ones of normal number and weight.

Males can ejaculate multiple times in a row and can mate with multiple females, especially when there are several estrous females present. Males also have the capacity to copulate at much shorter intervals than females. Dominant males have higher efficiency of mating, and the females are more likely to use the sperm of dominant males for fertilization. In group mating, females often switch partners and show a clear choosing preference between an unknown male and ones with whom they have already mated (a phenomenon called the Coolidge effect).

17.1.6.3 *Mus musculus*: As a Model Organism

Mice surpasses yeasts, worms, and flies in being excellent model system to studies regarding immune, endocrine, nervous, cardiovascular, skeletal, and other complex physiological systems that mammals share. Being a mammal, mice can easily develop diseases like cancer, atherosclerosis, hypertension, diabetes, osteoporosis, and glaucoma and therefore can act as an efficient model for their study and effective cure. Other diseases which affect humans but not mice like cystic fibrosis and Alzheimer's can be induced by manipulating the mouse genome and environment. It has relatively low cost of maintenance and has a higher multiplication rate with reproducing once every 9 weeks. These make it advantageous in biomedical research. Also it is nowadays finding more reliability as a robust tool in the field of aging. Since laboratory mice have a life expectancy of only a few years, genetic manipulation can be done on life span and aging parameters during the relatively short period. Moreover, experiments on mice with an extended life span and those showing premature aging, together with genetic mapping strategies, have provided novel views toward fundamentals of aging. Again with recent advances in mapping neural circuitry governing behavior, mouse has also evolved to be a model for studying anesthetic action.

17.1.6.4 Threats

Mouse and other rodents can cause and spread many diseases, mostly viral. Some of them are as follows:

1. Hantavirus—A deadly viral disease spread from rodents' feces and urine to humans.
2. Salmonellosis—A food poisoning spread by rodents' feces.
3. Lymphocytic choriomeningitis—Rodent-borne viral disease which can cause serious neurological problems.
4. Rat-bite fever—A fatal infectious disease spread by infected rodents or consumption of food contaminated by them.
5. Bubonic plaque—Also known as "black death," though almost eradicated these days. In the Middle Ages, it was one of the most deadliest rodent-borne diseases creating huge epidemics.

17.1.7 Zebrafish (*Danio rerio*) (Fig. 17.17)

Zebrafish (*Danio rerio*) is a teleost and a member of the minnow family that lives in freshwater and is no bigger than 2 inches in length. In the 1970s, George Streisinger first used it as a model organism, because of its simplicity as compared to mouse and ease to manipulate genetically. They have various distinctive features which led to its widespread use as a model system. These include their ease of care, fecundity, rapid development, small size, and ease of manipulation. The brain, digestive tract, musculature, vasculature, and innate immune system of zebrafish have physiological and genetic similarity to humans, having genes which have functional similarity with 70% of human disease genes. They have proved to be a huge help to scientists and had changed their way of understanding treatments for cancer, spinal cord injuries, and potentially the regeneration of limbs in humans. It is a powerful vertebrate model

Fig. 17.17 Zebrafish (*Danio rerio*)

Table 17.7 Taxonomy of *Danio rerio*

Domain	Classification
1. Kingdom	*Animalia*
2. Phylum	*Chordata*
3. Class	*Teleostei*
4. Order	*Cypriniformes*
5. Family	*Cyprinidae*
6. Genus	*Danio*
7. Species	rerio

system for studying developmental biology and serves as a model for various diseases and screening for novel therapeutics (Table 17.7).

17.1.7.1 *Danio rerio*: Structure and Components (Fig. 17.18)

The most important organs of fish are the gills and swim bladder which help in its respiration and maintain buoyancy, respectively. The fins help in swimming. The other important organs are the kidney, heart, spleen, pancreas, intestine, urogenital pore, and the reproductive system.

Many of the organs of zebrafish are analogous with mammalian counterparts, including the brain, heart, liver, spleen, pancreas, gallbladder, intestines, kidney, testis, and ovaries.

The very first thing important for a fish's survival is its ability to swim and maintain its buoyancy and oxygen supply during its movement. Gills and the gas bladder help to maintain the fish's buoyancy. The gills play an important role in the oxygenation of the blood. Water flows via the gills and brings in oxygen for the blood. Blood, from the body, flows via the filament arteries and releases carbon dioxide to the water. Besides this fish's acid-base balance and osmoregulation are also regulated partly by gills. They also help in excretion of waste products. The gas bladder, or swim bladder, plays the key role in maintaining buoyancy. In the embryo

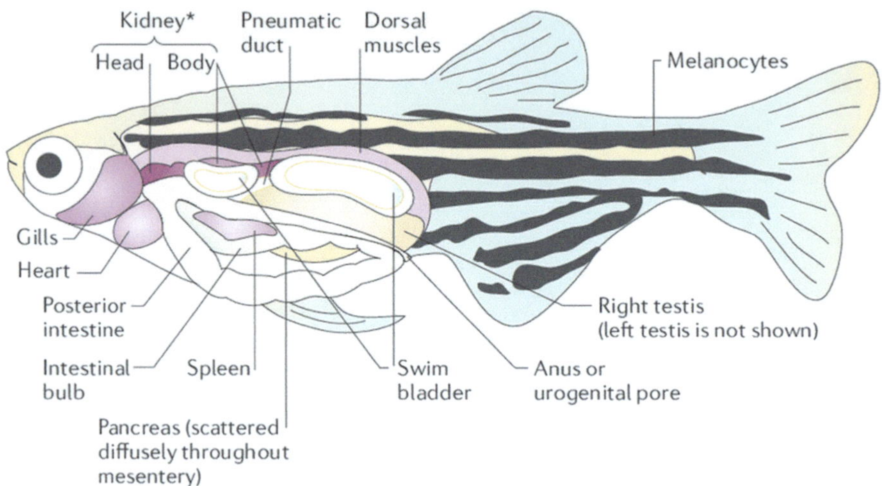

Fig. 17.18 Anatomy of *Danio rerio*

stage, it is derived from the upper digestive tract but practically has no digestive function. In the zebrafish, the swim bladder and the esophagus are connected by a pneumatic duct which allows the fish to fill up the swim bladder by gulping in air.

Fish's heart is one of the simplest in the developmental hierarchy of animals. It consists of only two chambers, i.e., one atrium and one ventricle. In zebrafish, the heart is situated anterior of the main body cavity and ventral to the esophagus. Deoxygenated blood from the body is carried by sinus venosus, and oxygenated blood from the heart is distributed by the ventral aorta to the gills via the afferent branchial arteries. The blood present in the sinus venosus subsequently passes through the sinoatrial valve into the atrium. The atrium then contracts to force the blood into the ventricle, via the atrioventricular valve. The ventricle first dilates to let the blood in and then contracts to pump the blood into the bulbus arteriosus via the ventricular-bulbar valve. From here the blood is distributed to the gills.

Zebrafish body is not distinctively divided into stomach, small intestine, and large intestine. Instead one single long tube is present as the intestine which folds twice in the abdominal cavity. The difference can be traced in the morphological anatomy of the mucosa columnar epithelial cells and the number of goblet cells which depicts that both the cells have functional differentiation between them. The intestinal epithelium consists mainly of columnar-shaped absorptive enterocytes and secondly of the goblet cells.

The zebrafish kidney lies on the ventral side of the vertebral column, distinctly divided into head and trunk regions. It is further divided into nephrons with a glomerulus, proximal tubules, distal tubules, and collecting ducts, as found in mammals. The kidney along with the spleen filters out the foreign particles and defective blood cells from the body. The spleen is made up of parenchyma cells, which in turn consists mainly of erythrocytes and thrombocytes (red pulp). Bacterial cells and other foreign bodies are trapped in splenic ellipsoids.

Though the fish possess distinct exocrine and endocrine functioning pancreatic tissues, a discrete pancreas is absent in it. The exocrine pancreatic tissue is scattered along the intestinal tract, while the endocrine pancreatic tissue encompasses α-cells, β-cells, and δ-cells. The α-, β-, and δ-cells have distinctive functions of producing glucagon-like peptide, insulin, and somatostatin, respectively.

Zebrafish testes are lateral, paired organs that comprise a series of tubules or blind sacs. Testis contains the Sertoli cells which supports the sperm formation. Cytoplasmic projections of Sertoli cells lead to the formation of cysts. Spermatogenesis occurs in these cysts which completely surrounds a single spermatogonium. Mature spermatozoa are carried to the genital orifice by two caudally merged ducti. Ovaries are paired, elongated structures. A short oviduct conducts the eggs to the outside.

Fish liver consists of three lobes that lie along the intestinal tract. The liver maintains the metabolic homeostasis of the body. This includes the processing of carbohydrates, proteins, lipids, and vitamins. It also detoxifies and synthesizes serum proteins such as albumin, fibrinogen, complement factors, and acute phase proteins. The gall bladder is composed of the bile ducts. The gall bladder carries the greenish bile that reaches the intestine via the common bile duct.

Lastly, the basic components of the zebrafish brain have similarity with that of higher animals. It is divided into five regions: the telencephalon, the diencephalon, the mesencephalon, the metencephalon, and the myelencephalon. Senses of smell, reproductive behavior, feeding behavior, color vision, and some other aspects of memory are controlled by the telencephalon. The olfactory bulb connects the olfactory (smell) organ to the telencephalon. The diencephalon is further divided into three components: the epithalamus, the thalamus, and the hypothalamus. The mesencephalon mainly deals with the normal sight.

17.1.7.2 Life Cycle and Reproduction (Fig. 17.19)

Zebrafish grows from egg, cleavage, gastrula, to adult. The gastrulation is seen to be completed within the initial 16 h. The full-grown adult emerges about 4–5 days after fertilization.

A fertile zebrafish lays approximately 200–300 embryos every 5–7 days. The embryo develops very fast and is transparent, so the stages of development can be viewed, which makes the study easy. The very first stage of the embryo is blastulation which lasts only for 3 h; this is followed by gastrulation which ends up within the next 5–6 h. Segmentation of the body and formation of most primary organs like the ears, eyes, segmenting muscles, and brain are observed within the first 20–24 h. Within the next 72 h, the embryo hatches out from the egg and, after an initial acclimatization period, goes in search of its food by the next 48 h. In a period of just 4 days, the embryo starts developing and showing features of adulthood. The developmental stages can be explained in brief as follows.

Zygote: It is the fertilized egg which is formed after completion of the first zygotic cell cycle.

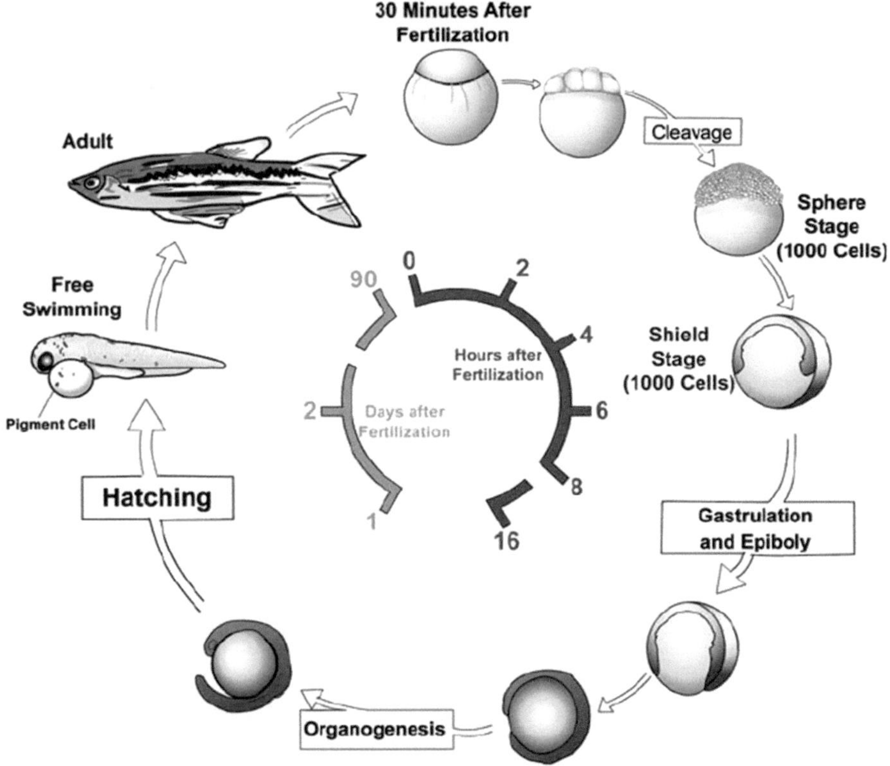

Fig. 17.19 Life cycle of *Danio rerio*

Cleavage: Zygote cleaves to two to six cells very rapidly in a synchronous manner.

Blastula: It is characterized by rapid and metasynchronous cell cycles, which is followed by a lengthened and asynchronous division, the midblastula stage. This asynchronous division is followed by the cells of one side of the blastula spreading and surrounding the remaining cells and the yolk, in a process called epiboly. This is the first coordinated cell movement in zebrafish embryos and begins before gastrulation.

Gastrula: Involution, convergence, and extension from the epiblast, hypoblast, and embryonic axis are the various morphogenetic changes which come during gastrulation. Beginning of gastrulation also marks the end of epiboly.

Bud: Epiboly totally ends and the yolk plug gets completely covered.

Segmentation: At the end of the epiboly, the first and second somites appear which marks the beginning of the body segmentation. Primary organogenesis and earliest movements take place at this time. Pharyngeal arch primordia, early neuromeres, and the tail develop.

Pharyngula: During this phase the body axis straightens from its early curvature around the yolk sac and marks the phylotypic stage of the embryo. Metabolic processes like circulation and pigmentation start. Fins begin to develop.

Hatching: Morphogenesis of primary organ systems completes. Cartilage development in the head and pectoral fin occurs. With this the egg hatches asynchronously.

Larval: Then finally comes the larval stage when swim bladder inflates. The fish starts food-seeking with active avoidance behaviors.

After the larval stage, finally the fish reaches its sexually matured adult form. Once reaching the adult stage, they allocate to suitable places for reproduction. The shape of the belly is a distinguishable character between the genders as the belly of the males is sleeker, while those of females are fuller and rounder. Zebrafish show signs of aging when reaching 2 years of life, although they can live for 3–4 years. The average life span of zebrafish depends on strain and rearing.

17.1.7.3 *Danio rerio*: As a Model Organism

Zebrafish has become much popular as a model organism in recent decades mainly because of its certain advantages over other model organisms like ease of care, prolific breeding, rapid external development, requirement of less space to culture, and low maintenance cost. These features make this fish an attractive model organism for developmental, toxicological, transgenic studies, human disease, drug discovery, cancer, etc. Some of the recent developments of use of the fish as a model organism are enlisted here. Much of the preliminary work on zebra fish was on developmental biology, in which it has done a distinctive contribution over the years. When Streisinger and Kimmel first used it as a model system in early twentieth century, it was for the study of developmental biology like embryonic development, muscle development, nervous system, etc. Since then it has been used for various similar studies, like that on development of enteric nervous system, angiogenesis, and regeneration, mainly because it has 70% homologue in disease genes with that of humans. It has also been used in investigating aquatic endocrine-disrupting compounds that can cause alterations in development, physiological homeostasis, and health of vertebrates. Complex behavior like that of learning and memory, aggression, anxiety, and sleep had been tried to be studied on the fish. The studies strongly suggest that conserved regulatory processes function in both zebrafish and mammals and result in common underlying behavior between them. Apoptosis is known to play an important role in mechanisms like embryogenesis, tissue homeostasis, and immune system regulation. The zebrafish serves a powerful model organism for studying apoptotic cell death in vertebrates, both during normal development and under cellular stress. Zebrafish are well suited in exploring ocular development, function, and disease as they have a well-developed eyesight and their retina and lens show morphological similarity to other vertebrates including humans. The most advanced use of zebrafish is in cancer research and drug testing and targeting. It has also been extensively used in examining other human disorders like hematopoietic disorders, kidney disorders, and cardiovascular disorders. Being an aquaculture fish, *Danio rerio* can help in studying growth behavior of other aquaculture fish. Research on its nutrition and growth, stress, and disease resistance can be expected to improve husbandry and formulated feeds of aquaculture species.

17.1.7.4 Threats

It has no reported threats till date except maintaining proper hygiene while handling the fish.

17.2 Mechanism of Gene Control in Development

17.2.1 Maternal Effect

Multicellular organisms are formed by the synchronization of the male and female gametes, i.e., ovum and sperm. The conjunction of the gametes gives rise to a single fertilized cell. This cell then generates more cells that will interact with one another via highly conserved signaling pathways to build an embryo with defined axes and organ systems. The development of the embryo continues and ultimately gives rise to a mature adult capable of producing the cells necessary to form the next generation. On the other hand, hermaphrodites can produce both sperm and ovum. This egg or embryo, during its development, gets a major contribution of DNA, nutrition, mRNA, and proteins from the ovum, i.e., the mother gamete. This legacy which it obtains from its mother is called the maternal effect. A huge number of genes together contribute to this effect. These maternal gene products are responsible for regulating meiosis, transitions between meiotic and mitotic cell cycles, and oocyte development. They also contribute to the important development cycles of the embryo including fertilization and help in activation of the embryo's own gene products after the preliminary utilization of mRNAs and proteins provided by the mother, during zygotic genome activation. Besides the genetic contribution, a mother can also pass on their behavioral traits to their offspring; for instance, nursing, grooming, predator defense, and "decisions" on when and where to lay eggs can all affect the offspring and its survival. Maternal effects can influence selection of a population of offspring into a particular environment. It plays an important role in determining various ecological and evolutionary processes like population dynamics, phenotypic plasticity, niche construction, life history evolution, and the evolutionary response to selection. The traits expressed due to maternal effect affect the fitness of the progeny, which in turn can play a role in selection of the offspring in the environment and therefore govern the evolutionary dynamics of the population. Additionally, maternal effects can also affect the change in responses of the offspring toward adaptation with changing environments.

The maternal effect is not the same in all organisms. In some animals, only the first few initial cleavage cycles are dependent on maternal RNAs and proteins, like that in mice and nematodes (*C. elegans*), before their zygotic genes get activated. Other organisms, such as *Drosophila*, *Xenopus*, and zebrafish, rely on maternal RNAs and proteins for much of the developmental period before activation of the embryonic genome. Several genes and their corresponding effect play an important part in synchronizing the maternal effect in the above model organisms. Some of them are discussed below.

The soil nematode *C. elegans* is one of the most highly studied organisms for maternal gene effect. The fertilization and the development of the nematode require the simultaneous action of a number of maternal genes. Some of them are mes-2, mes-3, mes-4, mes-6, skn-1, pie-7, and mex-7. Among them mes-2, mes-3, mes-4, and mes-6 are classified as sterile genes. It has been found that the mutant phenotypes of these sterile genes are similar and thus were considered to have a common origin. They are involved in a common process of encoding nuclear proteins that are essential for germline development. MES-2 and MES-6 have homology to members of the Polycomb group of chromatin regulators, found in insects and vertebrates. They are homologous to enhancer of zeste and extra sex combs. MES-3 is a novel protein, and MES-4 has function in growth control and is a SET-domain protein. In nematodes maternal determinants are found to play a role in the autonomous or cell-intrinsic developments of the early embryonic blastomeres. Pharyngeal cells are produced by only the posterior blastomere P1 in *C. elegans* after the first cleavage. From a study it was found that P1 can produce pharyngeal cells with the help of the maternal gene skn-1. *C. elegans* embryogenesis bears a stage called MS, which is an 8-cell blastomere. Cleavage of embryo at this stage produces pharyngeal cells, body wall muscles, and also cell deaths. Studies have shown that two maternal effect genes of *C. elegans*, pie-7 and mex-7, are involved in the division of MS. MS can show similarities to the wild-type blastomere stage if these two genes undergo mutations. In pie-7 mutants one additional posterior blastomere adopts an MS-like fate, and in mex-7 mutants four additional anterior blastomeres adopt an MS-like fate. A key molecular component of the duplication process of centrosome/centriole during the cell cycle is a protein kinase called ZYG-1. The paternal product of it regulates duplication and bipolar spindle assembly during the first cell cycle, whereas its maternal product regulates these processes thereafter.

In *Drosophila* embryo, the pattern of segmentation is controlled by both zygotic and maternal genes. About 25 zygotically active genes and 20 maternally active genes jointly synchronize for the process. The expression of fushi tarazu (ftz), an active zygote segmentation gene, at the cellular blastoderm stage in progeny which has mutations in six maternal effect genes, namely, exuperantia (a member of the anterior class of maternal effect segmentation genes); staufen and vasa (members of the posterior class); and torso, trunk, and fs(l)N (members of the terminal class), was observed. The gene ftz showed disruption in its normal functioning in the mutant varieties, depicting the importance of maternal effect on growth of *Drosophila*. Staufen gene was found responsible for localization of both anterior and posterior maternal determinants. In another study, it was found that three genes, dorsal (dl), twist (mi), and snail (ma), undergo maternal-zygotic interactions to produce the dorsoventral pattern of the *Drosophila* embryo. It was also seen that introduction of a new maternal genetic element, named the selfish-genetic element, in wild-type *Drosophila*, made them resistant to insect-borne pathogens.

In mouse embryo development, involvement of around 14 MEGs (maternal effect genes) have been studied, namely, maternal antigen that embryos require (Mater), heat shock factor 1 (Hsf1), oocyte-specific deoxyribonucleic acid (DNA) methyltransferase-1 (Dnmt1o), DNA-cytosine-5-methyltransferase 3-like (Dnmt3l), formin-2 (Fmn2), postmeiotic segregation increased 2 (Pms2), T-cell leukemia/lymphoma 1a (Tcl1a), nucleophosmin/nucleoplasmin 2 (Npm2), germ and embryonic stem cell enriched protein (Stella), zygote arrest 1 (Zar1), and ubiquitin-conjugating enzyme.

1. The gene which transcripts Mater protein is present in the cytoplasm of growing oocytes but not in other tissues. It is expressed till late blastocyst stages. A single copy of gene encodes the Mater protein.
2. Stress stimuli like heat, etc. activates Hsf1, which then activates stress-inducible genes.
3. Both the nuclei of early stage oocytes and the cytoplasm of mature oocytes contain the Dnmt1o gene. It helps in maintaining genomic methylation patterns in mammalian somatic cells.
4. Dnmt3l is associated with catalyzation of de novo methylation of CpG islands.
5. Fmn2 is expressed in the central nervous system, spinal cord, and brain from development to maturation stages.
6. Pms2A is required for methyl-directed post-replication mismatch repair. It is a DNA mismatch repair gene homologue and was identified as a MEG using null mutation mice.
7. Tcl1a helps in the transduction of anti-apoptotic and proliferative signals in T-cells and increases AKT kinase activity by interacting with AKT.
8. Npm2 transcripts are specified to growing oocytes. Its protein is expressed in the nucleus of oocytes before germinal vesicle breakdown (GVBD) and in the cytoplasm after GVBD.
9. Stella is a novel gene expressed in primordial germ cells (PGCs), cytoplasm of oocytes and embryos, and early developmental stage pre-implantation embryos especially in pronuclei of zygotes.
10. The transcripts of Zar1 are detected and expressed in high amount from early oocytes to the initial zygotic embryo stages.
11. Ube2a is a ubiquitin-conjugating enzyme and associated with DNA repair. It is a homologue to RAD-6, which is a radiation-repair gene.
12. Zfp36l2 binds to mRNAs having class II AU-rich elements and aids in its degradation.
13. Uchl1 is a deubiquitinating enzyme which is expressed in organs like the ovary, testis, placenta, and neuron.
14. Filia binds to Mater and is crucial for pre-implantation embryo development.

Dorsal axis in amphibians consists of notochord, somites, neural tissue, and head structures. In a study it was found that β-catenin encoded by maternal mRNA is needed for the development of dorsal axis of *Xenopus*. Xwnt-5A transcripts, a maternal gene, are expressed throughout development of *Xenopus* and are well active in the anterior and posterior regions of embryos at late stages of development. Overexpression of Xwnt-5A in *Xenopus* embryos leads to complex malformations. *Xotx2*, a *Xenopus* homeobox maternal gene, is expressed at low levels throughout the development of the frog from unfertilized egg to late blastula, when its expression increases and has a function in the formation of the frog's brain. Foxi1e is a zygotic transcription factor that is essential for the expression of early ectodermal genes. Foxi1e mRNA is maternally encoded and highly enriched in hemisphere cells of the blastula of the frog.

Over the years, advanced researches in zebrafish have made this organism impeccable in detailed understanding of the role of maternal factors in early vertebrate development. A large number of maternal effect genes have been discovered in this teleost fish. Some of them are janus (contributes in cell adhesion), yobo (has a role in axis convergence and extension), ichabod (functions in dorsal organizer induction), alk8 (plays a role in development of ventral cell fates), pbx4 (helps in hindbrain segmentation and rhombomere identity), etc. A gene named mission impossible (mis) was identified in zebrafish. The gene is a maternal gene and plays an active part during gastrulation, contributing in activities like cell movement and the activation of some endodermal target genes.

17.2.2 Zygotic Gene Activity

17.2.2.1 Body Segmentation (Fig. 17.20)

The genes show a coordinative and chain-wise activation of gap genes by maternal genes, followed by activation of pair-rule genes by gap genes and then finally activation of segment polarity genes by pair-rule genes.

Segmentation in vertebrates and arthropods is a process in which body axis is formed by serial repetition and differentiation in patterns of similar anatomical modules. The process involves simultaneous activation of a series of transcription factors. The activation of these transcription factors is a chain-wise process in which many genes are involved. These genes initiate the pathway of these transcription factors that, in turn, induce the expression of other transcription factors, thus creating cascades of gene expression. In this way a multistep signal pathway is initiated, which ultimately amplifies the initial signal and results in the segmentation of the body. In vertebrates, segmentation results in the development of the spine, head, and limbs, whereas in insects, it results in formation of the abdomen, head, and thorax.

The segmentation is a result of the synchronous activity of a series of genes. Most of the maternal genes taking part in segmentation are present in the eggs during the time of fertilization. Their genetic regulation starts before the onset of the cellular blastoderm stage. Firstly the proteins transcripted by the maternal coordinate genes, namely, *bicoid*, *nanos*, *caudal*, etc. gets diffused from the anterior and posterior poles of the embryo. The maternal transcripts also have definitive activation sites during embryogenesis like *caudal* (*cad*) and *hunchback* (*hb*) which are ubiquitously distributed in the fruit fly embryo, whereas transcripts *bicoid* (*bcd*) and *nanos* (*nan*) are found only in the anterior and posterior parts of the embryo, respectively. Once

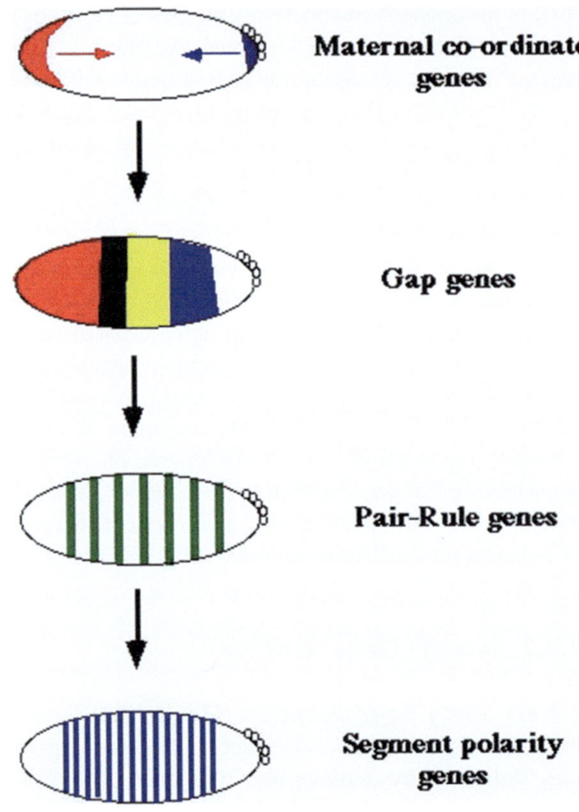

Fig. 17.20 Gene control in body segmentation of *Drosophila*

translated, the signals from this factors control the expressions of the zygotic genes in the developing fruit fly. The first zygotic genes to be activated by the maternal genes are the zygotic gap genes (the very first zygotic genes responsible for segmentation), i.e., *hunchback*, *Krüppel*, *knirps*, etc. They are expressed all along the anterior to posterior axis of the fly embryo. The gap gene domains circumscribe the progenitors of several adjoining segments and sub-divide the embryo into anterior, middle, and posterior regions. The gap genes regulate each other and the next set of genes in the hierarchy, the pair-rule genes, namely, *even-skipped*, *hairy*, *fushi tarazu*, etc. Once activated the pair-rule genes generate periodic gene expression events called pair-rule stripes. The strips decide the position of the boundaries between the segments to be formed and are expressed in seven stripes of cells corresponding to every other segment. Their formation and regulation are controlled and governed by the pair-rule genes. Pair-rule gene expression then induces the final set of transcription factors, i.e., the segment polarity genes, namely, *wingless*, *hedgehog*, *engrailed*, etc. This set of genes is expressed in 14 segmentally repeated stripes. The segment polarity genes, unlike the other classes of segmentation genes, require regulatory proteins other than transcription factors (i.e., secreted signaling molecules, receptors, kinases, etc.). These proteins mediate interactions between cells. At the end the body is divided into identical repeated pattern of segments expressed by the segment polarity genes. The final group of genes are the homeotic genes which control the character (i.e., head, thorax, abdomen) of each segment.

In situ RNA and protein expression patterns of these genes gave the idea of where the genes are first expressed:

- Maternal coordinate genes: The maternal genes which make maximum contribution to the process of segmentation are the bicoid, localized at the anterior end of the embryo and nanos, located at the posterior end of the embryo. Both of them are primarily expressed during oogenesis.
- Gap genes: Gap genes are first expressed at the syncytial blastoderm stage. They are the Krüppel, hunchback, and knirps. Krüppel mRNA is found in parasegments 4–6, whereas hunchback and knirps mRNA are found in the anterior and posterior half of the embryo, respectively.
- Pair-rule genes: They are first expressed during syncytial blastoderm. They decide the boundary of future segmentation around each embryo by forming the seven stripes.
- Segment polarity genes: Primarily expressed during cellular blastoderm, segment polarity genes play the most important part during the final segmentation process. They form the 14 stripes of transcription around each embryo which decides the ultimate segmentation of the body axis.

The maternal factors which are found to play an important role in segmentation in fruit flies are studied not to be as important for other insects and neither in vertebrates. However, the hox genes, which are said to control the ultimate fate of each segment in the body, are found to be universal. They define what part of the body each segment will become later in the cascades of nearly all insects and vertebrates. Importance of hox genes in fruit flies was identified by a number of mutation studies. Mutations in specific hox genes caused legs to grow out of a fly's head, wings to grow where they shouldn't be, and a number of other physical deformities.

17.2.2.2 Organ Formation (Fig. 17.21)

The cells during the process of embryogenesis get genetically programmed to form the organs which make up the body. The organs are specialized tissues which synergistically work together to maintain various functioning of the body and also contribute to environmental adaptation (Fig. 17.21).

Flowchart depicting the cells of the embryo undergoing morphogenesis, specification, proliferation, and differentiation to undergo organogenesis.

A number of genes work together in an organized manner to ultimately form the mature organs. Some of the genes studied in various model organisms which contribute in organogenesis are listed below:

1. A homeotic gene, sex combs reduced (Scr), was found to be involved in the formation of salivary glands in *Drosophila melanogaster*. It was found that the

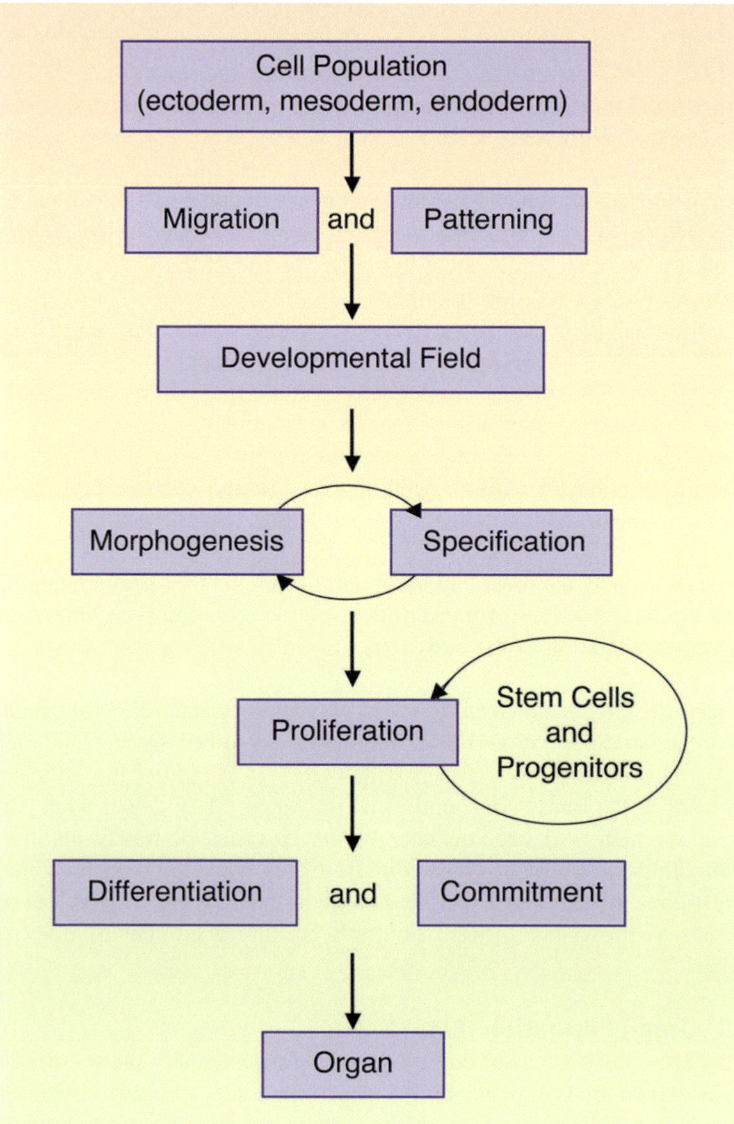

Fig. 17.21 Organogenesis

anterior-posterior extent of the salivary gland primordium, a placode of columnar epithelial cells derived from parasegment 2, is established by the positive action of the gene.

2. The development of *Drosophila* hindgut is a good model system for studying establishment of epithelial cell primordium, its internalization by gastrulation, patterning in the anteroposterior and dorsoventral axes, interaction of epithelium

with mesoderm and mesenchymal layers, cell shape change, and cell rearrangement. It was found that caudal/Cdx, brachyenteron, fork head/HNF-3, and wingless/Wnt genes are actively involved in posterior patterning, cell rearrangement, and gut maintenance in the fly.
3. A zygotically active gene called pha-4 is studied to have function in proper development of the pharynx of *C. elegans*. The PHA-4 protein is present in nuclei of almost all pharyngeal cells. Mutation in the gene can block the proper formation of the pharynx.
4. A protein complex named Chromatin Assembly Factor 1 (CAF-1) aids in chromatin assembly during DNA replication. It was found that a mutation in CAF-1b in zebrafish caused hindrance in cell cycle and led to defects in its progression. It also interrupted the differentiation of several organs, including the retina, optic tectum, pectoral fins, and head skeleton.
5. A detailed analysis of the global gene expression of mouse organogenesis from egg to the adult was studied. Comparative analysis identified both conserved and divergent gene regulations in mouse and human organogenesis.
6. Hox genes (regulates body segmentation) require coordination of various downstream target genes for its function. These genes are called the "realizator" genes. Abdominal-b, a hox gene in *Drosophila*, involved in the organogenesis of the external respiratory organ of the larva, was found to activate four intermediate signaling molecules and transcription factors.

17.2.3 Analysis of Development in Vertebrates

17.2.3.1 Vertebrate Homologues

Homologues in vertebrates are usually referred to as genes or structures in different organisms in different taxa, which share a common ancestry. Some of the genes in model organisms which share vertebrate homology are discussed below:

1. A certain group of proteins are responsible for guiding the migrating cells and axons to their targets in the nervous system of the developing embryo under extracellular environment. One such protein family found in vertebrates is the netrins. They are involved in guiding axons and cells to their targets by functioning as diffusible attractants and repellants. In *C. elegans* one such netrin is UNC-5. Loss of proper action of unc-5 gene causes migration defects. Two vertebrate homologues of UNC-5 were found in a study. It was seen that these two homologues along with UNC-5 define a new family of immunoglobulin superfamily. Their mRNA shows prominent expression in various classes of differentiating neurons.
2. Polycomb group (PcG) and trithorax group (trxG) proteins are chromatin-mediated regulators of a number of developmentally important genes including the homeotic genes. Most of these genes are conserved from flies and humans. Trithorax-like (Trl), a trxG member gene, is present in *Drosophila*. It encodes the essential multifunctional DNA binding protein called GAGA factor (GAF). A

homologue of this Trl-GAF was known missing in vertebrates, until recently a vertebrate homologue to GAF, c-krox/Th-POK, was reported on the basis of sequence similarity and comparative structural analysis.
3. Apterous gene defines dorsal cell fate and directs the outgrowth of the wing during *Drosophila* wing development. The homologue of the wings in insects is the forelimbs of vertebrates. It was seen in a study that the function of the apterous gene is carried out by two separate proteins in vertebrates. Lmx-1 specifies dorsal identity and Lhx-2 regulates limb outgrowth. From the study it was found that Lhx-2 is a more closer homologue to apterous than Lmx-1.
4. The Polycomb group (PcG) of proteins maintain expression pattern of genes during the preliminary stages of development. They have an important role in epigenesis and maintains the expression of a huge number of genes, most important being the homeotic genes. Whether different vertebrates have homology in PcG still remains much unclear. Based on the fact that all PC homologues consist of an N-terminal chromodomain and a C-terminal Polycomb repressor box, 100 PC homologues of different organisms were identified by their protein and genome sequence database.
5. The nervous system and limb evolution of the vertebrates and invertebrates have been well studied and are established to have a vast amount of gene and protein homology.
6. Over the years the vertebrate jaw evolution has been extensively studied. The evolution of the vertebrate jaw can be viewed as a change of developmental program for the functioning and specification of crest cells. Various homeobox genes like Otx2 and other Hox genes participate in the formation and homology of the jaw.

17.2.3.2 Mouse: Insertion Mutation

Mouse has been used as a model system for various mutation studies. Various gene mutations have been carried out in mice for the sake of studying complex disease formation like that of cancer and tumorigenesis. Some of the gene mutations are listed below:

1. A transgenic mice model of Alport syndrome (a syndrome of progressive loss in kidney function) was developed. Some exons of *Col4a4* and *Col4a3* and the intergenic promoter region were deleted in the mouse line (OVE250) by transgenic insertion. This was confirmed by genetic and molecular analysis which identified deletion of genomic DNA at the transgenic insertion site. This insertion led to severe progressive glomerulonephritis in the mice. It also lacked both α3 and α4 chains of collagen IV.
2. Mouse mammary tumor virus (MMTV) was used to insert mutation in both *int*-1 and *int*-2 loci of premalignant and malignant neoplasms from the GR mouse strain. The mutations were able to determine the earliest stages of neoplastic development as well as progression of premalignant lesions to tumors.
3. Mutations were inserted in genes like cellular proto-oncogenes or tumor suppressor genes with the help of murine leukemia retroviruses (MuLVs). This led the

mice to undergo leukemia and lymphoma as seen from studies. In a study >200 viral insertion sites was sequenced, from which >35 genes was identified that were altered by viral insertion in 4 AKXD mouse strains. It was found that the mutations were strain specific as each AKXD strain displays a unique mutation profile. Even some of these mutations identified genes that had no previous reports of being involved in causing cancer.

4. Insertion of a novel transposon ETnII-β, between the genes for Dusp9 and Pnck, causes mutation which leads to multiple malformations in polypodia mice. Mutation due to ETn insertion includes dysregulation of nearby interval gene expression at early stages of development.

17.2.3.3 Mouse: Gene Knockout Mutation

Many of the studies regarding mechanism behind human biological process and diseases have been carried out in mice. Mice have many advantages for the above purpose like they are small, have relatively short life spans, and are prolific. Most importantly it shares 99% gene analogy to human. Target-specific gene mutation in mice has been used recently for understanding the function of various genes responsible for development, metabolism, and diseases. Gene targeting is most widely used for designing gene knockout mice. In this a drug resistance marker replaces an essential coding region in a genetic locus, to understand the function of the gene in embryogenesis and in normal physiological homeostasis. Gene knockout mutations in specific targeted genes can mold and change it into a more useful form and discern its biological role. Homologous recombination is the most commonly used technique for knockout. It takes the advantage of the organism's own DNA repair machinery to replace the targeted gene with the engineered homologous genomic sequence, in a specific locus. Only a small percentage of cells actually take part in the process as homologous recombination takes place at a very lower rate, where only 6 to 14 kb of homology is typical for targeting constructs.

Recombination vector and Cre/loxP technology are the two most important techniques used for knockout. Recombinant vectors are widely used and are successful means of knockout mutations. A drug selection marker is required for its construction and positive selection of the susceptible variety. The most common markers are neomycin (mostly used), puromycin, and hygromycin. The marker is placed in the genetic frame of an open reading frame. Two homologous recombination events function to insert the targeting construct containing the drug resistance gene into a homologous genetic locus. The most widely used technique to inject the vectors into large number of stem cells is electroporation, as it provided better targeting ratio (1:2400) than microinjection (1:15 targeted recombinants to random integrants).

The Cre/loxP technology works almost similar to that of the recombinant vectors. They are also designed with a drug selection marker, flanked by two homology arms in the construct. But unlike recombination vectors, Cre/loxP requires both positive and negative selection to isolate cells with the properly recombined mutation, and the entire exon is not replaced by the drug selection marker. The positive selection is done by Cre recombinase, which is used to replace the normal coding sequence in a targeted allele with the mutated version. One homology arm will carry the planned point mutation, micro-deletion, or insertion to be introduced into the targeted gene.

To design a vector for gene targeting, the following steps are suggested:

1. Firstly, the allele of the target gene should be well studied. Its genomic structure, exon/intron sequence information, size, chromosomal location to be targeted, and the location of most major restriction enzyme sites for subcloning should be researched and identified prior to designing the vector.
2. Homology arm formation requires a mouse genomic fragment which contains a large and relatable portion of the required gene. A 129/Sv genomic clone is most commonly used for this purpose, as most stem cells were derived from this mouse strain.
3. Both positive and negative selection markers needed for constructing the targeting vector, especially in the Cre/loxP method, should be relatable and properly chosen. The most common positive and negative selection markers are neomycin phosphotransferase (neo^r) gene and HSV thymidine kinase (HSV-tk), respectively. Besides, puromycin and hygromycin are also common positive markers. The neo^r gene is often inserted in the opposite orientation to transcription in the target allele, while designing the vector.

Several reports had proved that knockouts in mice can lead to malfunctioning and also prenatal deaths of the embryo. The death can be due to a number of disturbances including failure of proper vascular circulation and transition from yolk sac-based to liver-based hematopoiesis. Several other reasons like improper implantation and formation of a yolk sac vascular circulation, defective chorioallantoic placenta formation, etc. leads to cutoff of connections with maternal system. These can also lead to fetal destruction. Gene knockouts can also lead to malfunctioning and underdevelopment of a number of embryonic organ and body systems, including the central nervous system, gut, lungs, urogenital system, and musculoskeletal system.

17.2.3.4 Zebrafish: Gene Knockdown Mutation

Similar to mouse gene knockout, zebrafish has also been extensively used for gene knockout studies. Some of these extensive experiments are listed below:

1. Zebrafish is a powerful model for studying forward genetics. In a study the fish was used to study reverse genetics that produced null phenotypes in G0 zebrafish. CRISPR/Cas9 was used to target four ribonucleoproteins in a single yolk. In this process early embryonic and stable phenotype was produced. This helped in rapid screening of genes which are helpful in development, physiology, and disease models in zebrafish.
2. RNA-guided endonucleases (RGENs), in the form of Cas9 protein-guide RNA complexes (as they were derived from the prokaryotic type II CRISPR-Cas system), were used for the establishment of gene knockout mice and zebrafish. To achieve this RGENs were injected into the embryo of both the species when it

was in the one-cell stage. In the process germline transmittable mutations in up to 93% of newborn mice were achieved with minimal toxicity.

3. Proviral insertions were widely used to mutagenize genes in the zebrafish genome, using high-throughput sequencing and mapping technologies. Using in vitro fertilization, about 0.5% of the predicted and established mutations was understood and characterized in this study. From the results a high mutation efficiency was predicted, and also phenotypes relevant to both developmental processes and human genetic diseases were put forward.
4. Zinc finger-based knockout of zebrafish was also put forward by a group of researchers, which notably enhanced the chances of the fish to be established as an efficient experimental organism for understanding human diseases.

17.2.3.5 Mammalian Stem Cell (Fig. 17.22)

Stem cells are cells capable of growing into a specialized cell. They can develop to various specific adult organ cells like muscle cells, intestinal cells, liver cells, blood cells, nerve cells, cardiac cells, etc.

A stem cell is a cell that has the inherent remarkable potential of developing into a specialized new cell. This specialized new cell can be cells of other tissues or self-renewal to produce more of the same stem cells. A stem cell is capable of converting into more specialized cell like muscle cell, a red blood cell, or a brain cell. They also can internally repair damaged cells of many tissues and organs like the gut and bone marrow, dividing in order to replace damaged cells or multiply cells in mammals. In

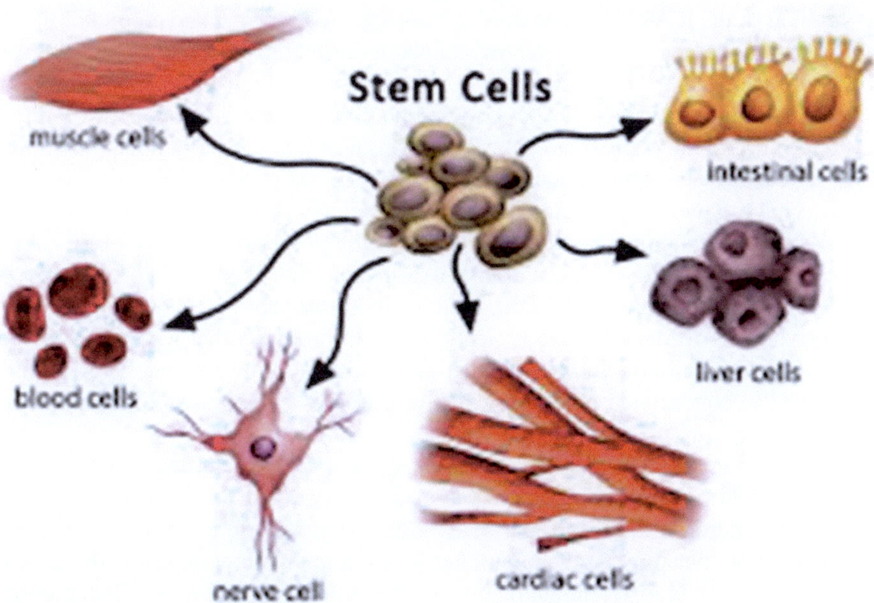

Fig. 17.22 Stem cells

certain other cases, stem cells divide only under special conditions like in the pancreas and heart. Stem cells are distinguished from other cell types by two important characteristics. First, they are primarily unspecialized and can renew themselves when required, even after elongated dormant stages. Second, under certain physiologic or experimental conditions, they can be made tissue- or organ-specific with special functions.

Two kinds of mammalian stem cells have been proposed: embryonic stem cells and non-embryonic "somatic" or "adult" stem cells. In the late 1990s, human stem cells were derived and successfully grown under laboratory conditions. They were named human embryonic stem cells (hESCs). The embryos were produced through in vitro fertilization procedures.

1. Embryonic stem cells (ESCs)—ESCs are produced from the inner cell mass (ICM) of the embryo. After fertilization the mammalian embryo undergoes a series of cell divisions, without any growth in total volume. These cells called blastomeres get progressively smaller. At a point of time, they stop dividing and rearrange to form a hollow sphere of cells called the blastocyst, encircling a fluid-filled cavity called the blastocoel. These cells then form the outermost epithelial layer called the trophectoderm and the ICM. The cells of the trophectoderm or trophoblast form the fetal part of the placenta, while the cells of the ICM give rise to the embryo proper. Around days 5–6 after fertilization in mouse and days 8–9 in humans, the cells of the inner mass can be isolated and put in culture. In culture only the ICM is plated and the trophectoderm is removed. ICM is plated on to a feeder layer of mouse or human embryonic fibroblasts, which is essential for the survival of the ICM. The ICM then develops into a close-packed colony of ESCs, which can then be cultured and maintained to ultimately produce stable cell line. ESCs differ from the original ICM cells, notably in their pattern of epigenetic modifications. It was found by experimentation that the generated ESCs retain their pluripotency and can generate all tissues in chimeric mice, including germline tissues, when injected into the blastocyst. When the adult chimeric mice have gametes generated from cultured ESCs, breeding of such chimeras will produce an animal composed entirely of the progeny of cultured ESCs. This event is called the germline transmission. But the ESCs cannot themselves generate the entire embryo. They require the presence of feeder cells for their survival. The feeder cells are typically mouse fibroblasts that have been treated with mitotic inhibitors to prevent their proliferation. Human feeder cells can also be used in conditioned medium, which presumably contains appropriate growth factors.
2. Somatic stem cells—Many mammalian somatic cells including those of bone marrow, skin, gut lining, blood vessels, endocrine glands, mammary gland, prostate, lung, retina, and parts of the nervous system contain stem cell populations that might self-renew and generate somatic cells normally as well as proliferate and differentiate in response to wounding or disease. This multiplication of the somatic stem cells is tissue specific, and therefore these stem cells are addressed as stem cell niche. They are defined as a specialized subset of tissue

cells and extracellular matrix that produces one or more somatic stem cells and control their self-renewal and differentiation throughout the life of an organism. Some niches like those of CNS are said to retain multi-potency, i.e., the ability to differentiate into major cell types appropriate to their originality.

Mainly two kinds of stem cells make up the bone marrow. They are the hematopoietic and the non-hematopoietic stem cells. The hematopoietic stem cells produce the blood cells in the body. The non-hematopoietic stem cells (also called mesenchymal stem cells, or skeletal stem cells) are a class of stromal cells in the bone marrow and generates bone, cartilage, fat, and cells that support the formation of blood and fibrous connective tissue.

Other than bone marrow, there are also many other organ-specific stem cells. For instance, the neural stem cells have three major categories: nerve cells (neurons), astrocytes, and oligodendrocytes (non-neuronal cells). The epithelial layer in the digestive tract also bears stem cells like absorptive cells, goblet cells, Paneth cells, and enteroendocrine cells. The skin stem cells are present in the basal layer of the epidermis and at the base of hair follicles. The follicular stem cells give rise to hair follicle and to the epidermis, and the epidermis in turn has stem cells which give rise to keratinocytes, which migrate to the surface of the skin and form a protective layer.

Sometimes a phenomenon known as transdifferentiation can take place. Here some adult stem cells differentiate into organs different from their inherent cell lineage, for example, brain stem cells that differentiate into blood cells or blood-forming cells that differentiate into cardiac muscle cells and so forth. This process can be used for the differentiation of one cell type into another under a well-controlled programmed condition of genetic modification. Nowadays research on reprogramming adult somatic cells to become like embryonic stem cells (induced pluripotent stem cells, iPSCs) through the introduction of embryonic genes are also being carried out.

17.2.3.5.1 Application of Stem Cells

1. Blood and immune system disorders—Blood and immune system disorders are treated by stem cell therapy by transplantation of bone marrow containing human somatic cells (HSCs) or purified cell fractions from bone marrow.
2. Metabolic diseases—Lysosomal storage disorders are also recently being treated with HSC transplantation, including the mucopolysaccharidoses such as Hurler's syndrome.
3. Autoimmune diseases/multiple sclerosis (MS)—Multiple sclerosis, a neurologic autoimmune disorder, can be cured by stem cell therapy. The disease affects CNS and causes multiple impairments of motor, sensory, and cognitive functions. Besides this, ESCs can also be a good option for treatment of demyelinating diseases.
4. Brain tumors—Neural stem cells have a unique capacity of migrating through tissues which can be both neural and non-neural. This property can be used extensively to treat brain tumors especially that cannot be effectively removed by surgery or chemotherapy.

17.2.3.5.2 Negative Concerns Regarding Stem Cell Therapy

1. Use of animal cells or products—According to current FDA regulations, the use of feeder cells from other mammals to culture ESCs may limit the use of the stem cells.
2. Tumor-forming potential of ESCs—In the undifferentiated state, ESCs have a tendency of forming teratoma, which can lead to a substantial risk during its implantation.

17.2.4 Analysis of Developmental Result

17.2.4.1 Sex Determination in Drosophila

The sex of individuals is the result of balance between the combined expression of the X and Y chromosomes. But in *Drosophila*, sex is primarily determined by balance of the X-A ratio, i.e., balance between female-determining factors encoded on the X chromosome and male-determining factors encoded on the autosomes. The process of sex determination involves the coordination of many genes. As autosomes do not carry the sex-determining genes, the copies of the X chromosome are primarily responsible for the purpose. However the expression of these genes is not equal in both the sexes due to the presence of two X chromosomes in female and only one in male. This problem is sorted by dosage compensation. It is a process by which the gene products generated by the X-linked genes in males and females are equalized. This is accomplished in the flies by adjusting the levels of X-linked gene products by doubling expression from the X chromosome in males. In flies sex determination begins immediately at fertilization with no specific time period. The transcription for the male and female gender specificity is triggered on sensing the definite X-A ratio and does not require the involvement of any hormones. The number of X chromosomes with respect to the autosomes plays an important part in defining the gender type. When one X chromosome is present in a diploid cell (1X-2A), the fly is male, whereas two X chromosomes in a diploid cell (2X-2A) leads to female progeny. The XO variety is sterile male. *Drosophila* can generate gynandromorphs, i.e., animals in which certain regions of the body are male and other regions are female. Gynandromorphs are formed when an X chromosome is lost from one embryonic nucleus and the cells which are formed from such cell are XO male instead of XX female. Sex hormones are absent in insects and such traits are controlled by body itself.

The sex determination in *Drosophila* is governed by the integrated action of a number of genes like *Sex-lethal* (*Sxl*), *transformer* (*tra*), *transformer-2* (*tra2*), *intersex* (*ix*), and *doublesex* (*dsx*) genes. XX progeny can convert into males when these genes undergo mutations, whereas XY males remain unaffected. Under homozygous condition the *intersex* (*ix*) gene leads the XX flies to develop an intersex phenotype having portions of male and female tissue in the same organ, whereas absence of the *doublesex* (*dsx*) gene leads both XX and XY flies to turn into intersexes.

1. *Sex-lethal* (*Sxl*)—Present in both male and female, *sxl* is a switch gene for female progeny. It is active in XX female from early stages of development. Its activation requires high X-A ratio, a series of transcriptional and posttranscriptional factors, and a number of regulatory proteins. It becomes active in XX females from the initial 2 h of fertilization, and once activated it remains in this state as its protein product is able to bind to and activate its own promoter. Immediately after activation the gene transcribes *Sxl* mRNA (an embryonic mRNA) that is found for only about 2 h more. The resulting SXL female-specific RNA-binding protein modulates the expression of a set of downstream genes, ultimately leading to sexually dimorphic structures and behaviors. In contrary in XY cells, *Sxl* remains inactive during the early stages of development. A certain class of proteins called the numerator (encoded by the X chromosome) is said to be responsible behind this female-specific activation of *Sxl*. The other group of proteins called the denominator block the binding or activity of the numerator proteins. These are autosomally encoded proteins such as Deadpan and Extramacrochaetae.
2. The transformer genes—*Transformer* (*tra*) is one of the somatic genes that take an active part in sex determination. It is also a switch gene and can be both female and male specific. It regulates sexual dimorphism based on RNA splicing in many insects. The female transcript encodes a functional TRA protein, and the male transcript encodes a nonfunctional truncated TRA protein. The gene activates sex-specific splicing of doublesex (dsx) pre-mRNA, along with TRANSFORMER-2 by binding to dsx repeat elements. Dsx is the final gene in the genetic cascade of sex determination and promotes female sexual development. The activation of the *tra* gene is also responsible for the activation of *Sxl* gene, and when *Sxl* is switched "off," a nonfunctional TRA protein is formed. This results a switch to male specificity of dsx, which then generates the male DSX-M protein.
3. *Double sex gene* (*dsx*)—The final and most important gene in the sex determination gene cascade is the *doublesex* (*dsx*) gene. It is the final gene to be activated in the series and needs the consecutive activation of both *srl* and *tra* genes for its functionalization. It is active in both males and females, but its primary transcript is processed in a sex-specific manner. When the X-A ratio equals to 1, the *sxl* gene leads to activation of *tra* gene in a female-specific manner. A female-specific splicing factor gets initiated which causes the splicing of the *tra* gene transcript. This *tra* gene product interacts with the Tra2 splicing factor to cause the *doublesex* pre-mRNA to be spliced in a female-specific manner. When transcribed in this manner, the *dsx* transcript promotes female development and inhibits male development. If the *doublesex* transcript is not acted on in this way, it will be processed in a manner that will be male specific.

17.2.4.2 Dosage Compensation in Drosophila

The presence of Y chromosome in *Drosophila* is necessary just for the specification of male gender and lacks any gene related to phenotypic development or sexual maturity. Instead the sex-related genes and proteins are associated with the X chromosome and its number. As males consist of only one X chromosome, it is necessary to equalize the products of male X chromosome with that of female. This is done by a process called dosage compensation. In this animals equalize the amount of gene products released by the X-linked genes in both male and female. This phenomenon was first observed in the fly by H.J. Muller in the early 1930s who also coined the name "dosage compensation."

The X chromosomes in *Drosophila* are identical in shape and genetic content and are active in all somatic cells. They carry many housekeeping genes and other genes which take active part in developmental pathways. Males have one X and a Y chromosome. The Y chromosome differs from the X in morphology and genetic information.

The X-autosome (A) ratio mainly controls both sex and dosage compensation and not the X-Y ratio. Y chromosome is only required for male fertility. For proper maintenance of dosage compensation, it is very necessary to have a control on the number and function of X chromosomes, failure to which can be lethal. The first gene which gets activated during initiation of dosage compensation is a critical binary switch gene called sex lethal (*Sxl*). It is present on the X chromosome and is regulated by transcription factors encoded by the chromosome. XX chromosomes are able to initiate *Sxl* expression from promoter, Pe, whereas embryos with XY fail to express *Sxl* from Pe. In flies, dosage compensation is mediated by the dosage compensation complex (DCC) also known as male-specific lethal complex (MSL), as loss in proper function of the complex can lead to lethality of the male phenotype. The *Drosophila* MSL is a ribonucleoprotein complex and is composed of at least five proteins, namely, MSL-1 (male-specific lethal 1, scaffolding protein), MSL-2 (male-specific lethal 2, RING finger protein), MSL-3 (male-specific lethal 3, chromodomain protein), MOF (males absent on the first, histone acetyltransferase), and MLE (maleless, RNA helicase). The SXL protein, present only in females, works to suppress the activity of MSL complex by repressing the translation of msl2 mRNA by binding in both the 5′ and 3′ untranslated regions (UTRs) of the mRNA. If SXL is absent in females, dosage compensation is aberrantly turned on, and these females die. Conversely, if SXL is expressed in males, dosage compensation is turned off and males die.

The five components of the DCC complex have their specific functions and combinedly lead to the activity of the complex. The MLE protein binds single-stranded RNA or DNA and is an ATP-dependent RNA-DNA helicase. MSL2 takes part in ubiquitination of itself and the other MSLs and targets them for proteolysis when required. It also binds DNA through its CXC domain—a stretch of 37 amino acids rich in cysteine. MSL3 contains a chromodomain that targets the MSLs to active X-chromosome genes in association with nucleosomes which contain histone H3 methylated at lysine 36 (H3K36me). MSL3 like MSL2 can also bind to DNA and methylated histone H4 at lysine 20. All the MSL proteins are male-specific and absent in females. It was found from various studies that the presence of MSL1 or MSL2 was compulsory for all the other proteins in the complex to bind to X chromosome. These sites are therefore considered nucleation sites for MSL targeting and spreading. DNA sequence motifs and histone acetylation are largely responsible for the targeting of MSL to the X chromosome and dosage compensation. Gene activation for dosage compensation involves the MSL-associated MOF acetyl transferase activity on H4K16 (histone H4 lysine 16), which represents a hallmark of the male X chromosome.

Two non-coding RNAs (ncRNAs), called RNA on X (roX), are responsible for targeting the MSL complex to the male X chromosome in *Drosophila*. They lack a significant open reading frame, are dissimilar in size, and are not colocalized with the MSL complex along the length of the X. roX RNA function was understood by mutation of X chromosome carrying both roX1 and roX2. It was found from the studies that MSL complex becomes mislocalized when both the RNAs underwent mutation, and the males showed an unknown phenotype when any one of them faced mutation. Partial purification of the complex suggests the presence of a tight core consisting of MSL1, MSL2, MSL3, and MOF proteins, with roX RNA and the MLE helicase lost except under very low salt concentrations. It was seen from studies that MSL complex can bind with acetylate histone H4 on lysine within nucleosomes in vitro, even if it lacks roX RNAs. From this it was put forward that the MSL complex possessed every components essential for dosage compensation and only need the RNAs to stimulate assembly and spreading. Besides this overexpression of MSL proteins can partially overcome the lack of roX RNAs.

MSL1 and MSL2 play the primary role in the MSL complex, as their interaction marks the initiation of the complex. Two subunits of MSL2 interact with an MSL1 dimer for the initiation. The interaction occurs near the RING finger of MSL2 (a C_3HC_4 zinc-binding domain) and an amino-terminal coiled coil domain in MSL. MSL1 further associates with MSL3 and MOF. It leads to chromatin binding and scaffold formation for interaction with MSL3 and MOF via adjacent conserved carboxy-terminal domains. MSL2 also functions in ubiquitination of itself and other members of the MSL complex components, including MSL1, MSL3, and MOF, but not MLE in vitro. The MSL3 has an active chromodomain-bearing histone acetyl transferases. It uses this domain in locating target genes for the MSL complex by interacting with active chromatin marks such as H3K36me3. The next member of the MSL complex is the MOF, which is the most important part in the MSL complex as it is the principal component which mainly controls the gene regulation by the MSL. It is a part of the MYST subfamily of HATs and is characterized by the presence of a chromodomain and enzymes that specifically acetylate lysine 16 in vivo. The vital role of the rest of the complex is to support MOF and to localize it to its targets on the X chromosome. MOF recruitment is particularly important as MOF also participates in the nonspecific lethal (NSL) or MBD-R2 complex in both sexes. The NSL complex is necessary for mortality in both the sexes. It is found at 5′ ends of most active genes. The last member of the complex, MLE, shows RNA/DNA helicase, adenosine triphosphatase (ATPase), and single-stranded RNA/single-stranded DNA binding activities in vitro. The function of MLE denotes that RNA has a potential role in MSL function. MLE most probably interacts with RNA or alters its structure, particularly the roX RNAs. Other than male-specific factors, many general factors are also responsible for dosage compensation. They are involved in chromatin organization and transcription in both sexes. For example, JIL-1, a tandem kinase, is found along all chromosomes in both males and females but is more highly concentrated on the male X chromosome. JIL-1 functions in modifying nucleosomes, specifically on the male X chromosome.

17.2.4.3 Sex Determination in Mammals

In placental mammals, the presence or absence of a Y chromosome determines the sex of the individual as the X chromosome is common in both male and female (XX in females and XY in males). Even in cases of aneuploidy, the sex determination depends upon presence of the Y chromosome.

Painter, in 1923, showed that the presence of X and Y chromosomes in the karyotype is characteristic of the male sex of humans. Sex of an individual is determined by the presence or absence of the Y chromosome. Sex chromosome aneuploids are males if they retain the Y chromosome and females if they lack it, irrespective of the number of the X chromosomes in the chromosome set. Different systems of chromosomal sex determination were put forward which studied the various combinations of sex chromosomes which ultimately determined the sex: females with XX and males with XY in mammals and females with ZW and males with ZZ in birds and snakes. The most ancient part of the X chromosome was found in marsupial and eutherian species. In various studies it was found that genes in the ancient conserved region of humans and also in the recently discovered regions have similarities with chromosomal arrangement of chickens and other birds. The sex chromosomal arrangement of humans has also resemblance with that of turtles, fish, marsupials, and eutherians.

In mammals gonads develop within the urogenital system and are formed from the intermediate mesoderm. The urogenital system is divided into three regions: pronephros, mesonephros, and metanephros, which develop anterior to posterior along the nephric or Wolffian duct. Gonads arise from the ventrolateral surface of each mesonephros, which serve as primitive kidney during embryogenesis. The pronephros is vestigial. The genital ridges are composed of somatic cells derived from the mesonephros and primordial germ cells.

The process is carried out by a series of genetic regulation (Fig. 17.23).

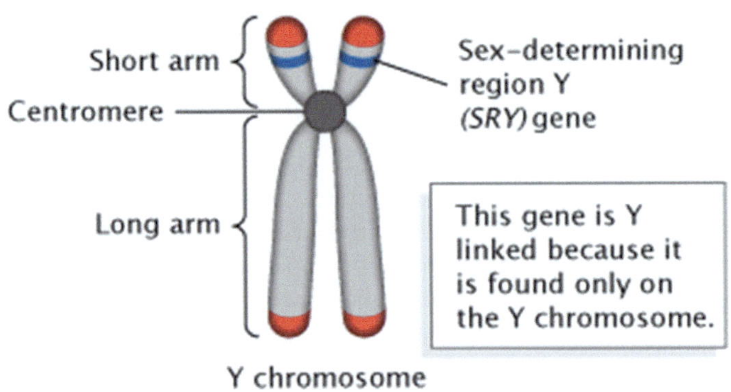

Fig. 17.23 Location of the SRY gene on Y chromosome

The SRY gene (blue) dictates the sex determination in mammals. It is located on the Y chromosome:

1. Sex-determining region Y (SRY) gene—The SRY is the most important gene in sex determination in mammals. It is located on the Y chromosome and functions in development of testis, the male-specific organ which defines the preliminary stage in sex determination and gender differentiation. Initially the embryo has the capacity of differentiating into both testis and ovary. It possesses two ducts, namely, Wolffian and Müllerian ducts, which develops into the male and female reproductive tracts, respectively. At about seventh week of embryo development, the SRY gene activates and encodes a unique transcription factor that activates a testis-forming pathway. Once testis is formed, it produces testosterone and anti-Müllerian (AMH) hormone. Testosterone and one of its derivatives, dihydrotestosterone, induce formation of other organs in the male reproductive system, while AMH causes the degeneration of the Müllerian duct. SRY protein is however absent in females, and a different set of proteins activate the ovary-forming pathway. Developed ovary then produces estrogen, which triggers development of the uterus, oviducts, and cervix from the Müllerian duct. Müllerian duct plays important role in females and is not degenerated as in males.
2. Steroidogenic factor 1 (SF1)—SF1 functions along with SRY and is a member of the orphan receptors, nuclear receptors which have no known activating ligand. This transcription factor has two highly conserved regions which are specific to mammals, a DNA-binding domain composed of two zinc fingers and a carboxyl terminal domain in the zinc fingers. It is studied that SF1 is involved in steroid biosynthesis and was found to be associated with endocrine function such as gonads, adrenals, pituitary, and hypothalamus.
3. Lim 1 and Pax 2—These genes help in the differentiation of the intermediate mesoderm. They take part in early gonad development and the final formation of the urogenital system. They also help in kidney development during the early stages of the organ.
4. Wilms' tumor 1 (WT1)-associated gene—This also takes an active part in early gonad and kidney development. Mutations in this gene are found to be associated with three different but related syndromes in humans. The first are childhood tumors in the kidney which can develop when the gene undergoes heterozygous deletions. Second is Denys-Drash syndrome in which urogenital malformations are caused when WT1 undergoes heterozygous missense mutations in the zinc finger DNA-binding domain. Third is Frasier syndrome, specific to females where they display urogenital malformations.
5. M33 gene—It was discovered in mouse. It also takes part in the early development of the gonads but in a different way than the above genes.

Three different cell lineages altogether make up the gonads and germ cells. They are the supporting cell lineage, steroidogenic cell lineage, and connective cell lineage. The supporting cell lineage gives rise to Sertoli cells in the testis and follicle cells in the ovary. The Sertoli and follicle cells provide protection and the required growth environment to the germ cells. The steroidogenic cell lineage gives rise to the Leydig and the theca cells in male and female, respectively. These cells are responsible for producing the sexual hormones during the maturity and development of the gonads. They contribute to the development of the secondary sexual characteristics of the embryo. The connective cell lineage leads to the formation of the gonads and germ cells as a whole. During the early stages of development of testis, Sertoli and germ cells are produced in the testicular cords, whereas Leydig cells are excluded to the interstitium. Basal lamina in testis is collectively formed by the cord and Sertoli cells. The action of SRY gene (remains active for a day and a half during early gonad development) triggers differentiation of the Sertoli cell lineage in the testis. After activation the Sertoli cells direct the differentiation of the rest of the cell types in the testis. Ovary is less structured than testis, and its development and organization take place later than that of the testis. The connective tissue lineage, in the case of ovary, gives rise to stromal cells with no myoid cell equivalent.

17.2.4.4 Dosage Compensation in Mammals

As most of the sexual development related genetic loci is present in the X chromosome, it is necessary that the output of the genes is equal in both male and female. Disproportionation in the level of expression of sex chromosomes can lead to different copy numbers of X-linked genes in males and females. Due to this there can be dissimilarity in the level of gene products in terms of RNA and proteins, which would, in turn, require differences in metabolic control and other cellular processes. In order to maintain the equality in genetic output of both the sex chromosomes X and Y, dosage compensation is carried out in the genes. It balances the level of X-linked gene products between the sexes. Dosage compensation is carried out in three ways in mammals: by twofold upregulation or twofold downregulation in the expression of X-linked genes in males and females, respectively, and by the complete inactivation of one of the two X chromosomes in females.

Many genes are involved in the sex determination process, and consecutive activation and inactivation of these genes control the sexual differentiation in mammals. The product of these genes then initiates the activation of another cascade of genes and the consecutive regulatory cycles that mediate the progression of the various pathways which ultimately leads to sex identification. In humans, it is the protein product of the SRY gene on the Y chromosome that directs the embryo toward the male pathway. In the absence of dosage compensation, chromosome degeneration takes place which has two major problems: first, a high amount of chromosomal difference generates between the male and female progeny and consequently among the members of the same species, and second, the heterogametic sex is monosomic for a large chromosome and thus monoallelic for a large number of genes.

Males of both mammals and drosophila have one copy of X and females have two copies. The Y chromosome only functions in sex determination in both the organisms. It has a relatively poor number of genes and is largely heterochromatic. But the X is much larger and has much more number of genes. This difference in the number of X chromosomes (2:1) can cause a twofold difference in many gene products between the sexes. XY males are monoallelic for the great majority of X-linked genes. This problem faced by males was addressed by Susumu Ohno about 40 years ago, who hypothesized that this problem could be solved if the X in males produces twofold products and matched the female. This led to the studies in the fruit fly *Drosophila melanogaster* using high-throughput gene expressions, which confirmed the occurrence of dosage compensation through upregulation of X-linked genes in XY males.

The evolutionary development of both the sex chromosomes X and Y are co-related. They differ significantly in their genetic content where the human X chromosome contains about 1100 genes, whereas the Y chromosome contains only 100 genes. Over the evolutionary period, the Y chromosome went through many changes in function and genetic composition to maintain the difference between the male and female progeny, including controlling the activity of SRY gene to develop the testis and mark the difference between the genders. The X chromosome, on the other hand, acquired the maximum percentage of sexual related genes. Genes that enhance male sexual reproduction and spermatogonia also accumulate on the X chromosome. But genes of later spermatogenesis are not expressed on X.

The other important mechanism of dosage compensation is X inactivation. During the early period of mammalian development, the differentiation in both the X chromosomes is governed and controlled by the Xic gene. Only the chromosomes which have the Xic gene can induce X-chromosome inactivation (XCI), as they contain a region of about 1 Mb occupying the Xic gene.

There are several genes responsible for dosage compensation. Some of them are sdc-1, sdc-2, sdc-3, dpy-21, dpy-26, dpy-27, dpy-28, and dpy-30. Mutations in them reduce the viability of XX but not XO animals. The mutations cause stoppage of dosage compensation, and the genes in XX individuals undergo twofold upregulation in their expression, though it does not affect the XO individuals in any way.

DPY-26, DPY-27, and DPY-28 proteins showed similarity to 13S condensin, a complex essential for control and proper functioning of mitotic and meiotic chromosomes from yeast to man and hence is believed to have function in the above mechanisms.

SDC-2 functions in chromosome specificity and hermaphrodite specificity during dosage compensation. It is the switch gene that activates dosage compensation in hermaphrodites.

The hermaphrodite formation is controlled by the activation of xol-1. Under active conditions xol-1 leads to male development in XO embryos, whereas it leads to hermaphrodite formation in XX embryos when in inactive state. Also when inactive in XX it leads to activation of dosage compensation. The ratio of the X chromosomes to sets of autosomes (X-A signal) regulates the activation of the xol-1. The twofold differentiation in expression in X-chromosome dose between males and hermaphrodites is dominated by the on/off state of xol-1.

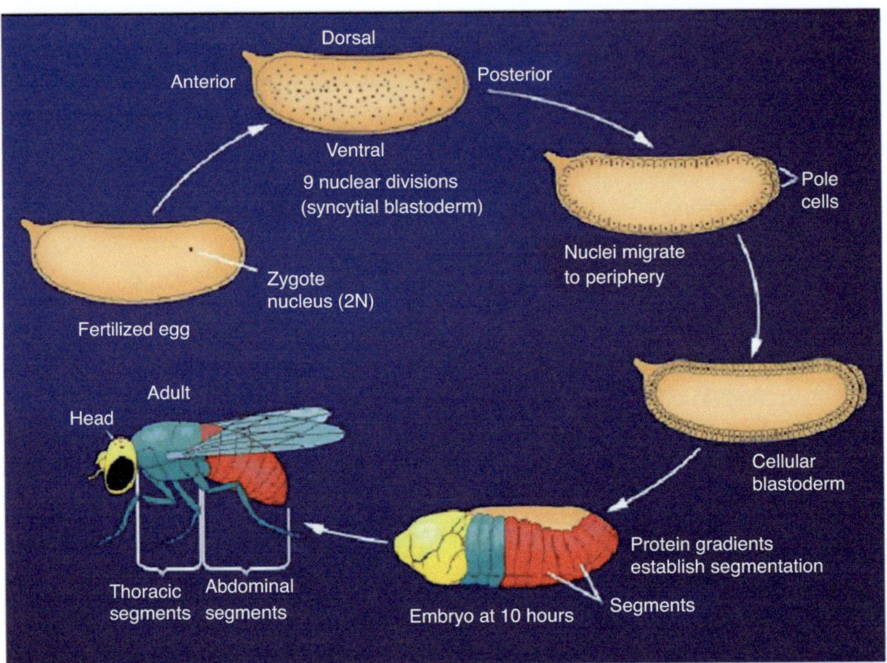

Fig. 17.24 Developmental stages in the life cycle of *Drosophila*

17.2.4.5 Developmental Stages of Drosophila (Fig. 17.24)

The *Drosophila* life cycle starts from fertilized egg to the syncytial blastoderm, cellular blastoderm, and finally the embryo. The adult fly hatches out from the embryo and undergoes various developmental stages to finally result the segmented body.

The *Drosophila* life cycle consists of a number of stages: embryogenesis, three larval stages, a pupal stage, and the adult stage.

The nuclear division (mitosis) starts soon after fertilization. Cytokinesis (division of cytoplasm) does not take place in the early stages of embryogenesis; instead a multinucleate cluster of cells is formed called syncytial blastoderm. By the tenth nuclear division, the nuclei formed during syncytial blastoderm migrates toward the periphery of embryo and gets partitioned into separate cells (cellular blastoderm) at the 13th division. The major body axes and segment boundaries start developing and are characterized by formation of morphogen gradients, which play a key role in pattern formation. Subsequent development results in an embryo with morphologically distinct segments.

Early embryogenesis—The *Drosophila* egg is the shape of a sausage. The sperm enters through a micropyle which is present at the anterior end. After fertilization, mitotic division occurs within the first 90 min. No cleavage is formed which results in a common cytoplasm and many nuclei. This is called a syncytium. After about nine to ten divisions, nuclei move to the periphery to form the syncytial blastoderm (2 h).

Embryogenesis—By the 13th mitosis division, individual cells begin to be formed when the membranes start surrounding the nuclei. This stage is called the cellular blastoderm. After this about 15 cells move to the posterior as pole cells and ultimately become the germline. After blastoderm formation, gastrulation starts at about 3 h.

Gastrulation—Cells do not divide during gastrulation, instead they separate from one another and move to internal locations under the ectoderm. Formation of mesodermal tube and nerve cord takes place at this stage. Mesodermal tube forms from ventral tissue and ultimately becomes muscle and connective tissues, whereas nerve cord occupies the ventral region. Neuroblasts lie between mesoderm and outer ectoderm. Both posterior and anterior midgut fuse and ectoderm becomes epidermis.

Segmentation—The ventral blastoderm or the germ band develops into the trunk region. It further pushes the posterior end of the body over the dorsal side and marks the beginning of segmentation. The very initial segmentation grooves begin to appear from the posterior of one parasegment and the anterior of the next. There are 14 parasegments: 3 mouth, 3 thorax, and 8 abdominal.

Larvae—After 24 h of fertilization, the larva comes out. It gets divided into anterior (acron) and posterior (telson) ends. It is segmentally divided into the head, three thoracic segments, and eight abdominal segments. The ventral side of the larvae possesses denticle belts, alternating patches of denticle hairs, and cuticle on each segment, used for locomotion.

Metamorphosis—Once the egg moves into larval stage, the cycle of the fly goes through three instar stages, separated by molts. Pupae come out from the third instar larvae and undergo metamorphosis. The metamorphosis includes all the developmental processes which take place after the pupal stage to the adult body structure, including the final segmentation. Small sheets of epidermis called the imaginal discs develop into the adult tissues. The discs grow throughout life cycle. In addition to imaginal discs, tissues are also formed from histoblasts, especially the abdominal segments. The adult fly ultimately develops six legs, two wings, two halteres, and two eye antennas, plus genital, head discs, and about ten histoblasts.

Maternal effect genes play an important part during zygotic development. They take an active part in oogenesis as well as fertilization. Two such genes are bicoid and nanos whose protein product plays important function during embryo development and are present in the egg at fertilization. Besides these genes also have a significant role in the process of segmentation in the fly. They activate the transcription of the zygotic genes, namely, gap genes, pair-rule genes, and segment polarity genes, which control the segment patterning in the fly. The gap genes roughly subdivide the embryo along the anterior/posterior axis, the pair-rule genes divide the embryo into pairs of segments, and the segment polarity genes set the anterior/posterior axis of each segment. The segmentation takes place in a synchronized manner by the consecutive activation of these three genes. Firstly, the maternal genes encode transcription factors that regulate the expression of the gap genes, and the gap genes then encode the transcription factors of the pair-rule genes, which in turn encode the transcription factors of the segment polarity genes. After the division of the body into segments, another set of genes called the homeotic genes gets activated. These genes then control the formation of anatomical structures like legs, wings, and antennae on the segments. The homeotic genes include a 180 nucleotide sequence called the homeobox, which is translated into a 60 amino acid domain, called the homeodomain.

Besides the maternal and zygotic genes, other growth factors also take part in the development of *Drosophila.* For instance, Torpedo (an epidermal growth factor (EGF) receptor homolog) is expressed in the dorsal follicle cells and takes part in the dorsoventral patterning of the embryo. Besides this another transforming growth factor (TGF) alpha homologue called gurken is also required for follicle cells to adopt dorsally.

Anterior-posterior axis specification in *Drosophila* is controlled during oogenesis by localization of bicoid and oskar mRNA, more precisely, localization of bicoid mRNA at the anterior pole and oskar mRNA at the posterior pole. This transfer of the mRNAs depends on microtubules. Initially, oocyte is present at one end of the nurse cells (cells that provide food, helps and provides stability to other cells). The adjacent follicle cells then move to the posterior side. Microtubules then aid toward the posterior side and help the necessary nutrients to flow from nurse cells to posterior. The microtubules are then directed in an opposite direction by a signal from the follicle cells, and the oocyte nucleus moves anteriorly and to one corner.

17.2.4.6 Microarray Study of Drosophila Development

The developmental pathway of *Drosophila* needs the conjoint activity of thousands of genes, starting from embryogenesis to adult and metamorphosis. In order to know the mechanism behind the growth cycles of the fly, it is necessary to understand, identify, and analyze the function and expression of these genes. In recent times, advanced approaches like microarray are used to understand the expression of various genes. A microarray basically consists of an array of microscopic slides, each containing a particular kind of gene which acts as probes to detect the expression of the respective genes. This technique is used now for studying the functioning of the genes of *Drosophila.* Many genes have been discovered over the years. Among them some take part in developmental process and are active during metamorphosis, whereas some others are not directly associated with metamorphosis. Additionally, many genes of unknown function were identified that may be involved in the control and execution of metamorphosis.

DNA microarray and mRNA in situ hybridization are two widely used techniques for epigenetic and other genomic studies. Time-related assessment of differences in gene expressions from mRNA extracted at various stages of embryonic development can be efficiently done by microarray studies. Integrative analysis of in situ expression patterns and microarray gene expression data is a very efficient way to find out the roles that genes play during development and to identify sets of genes that contribute somehow to developmental processes.

Advantages of microarray (expression screens)—These experiments are frequently performed in order to screen for molecules regulating a specific biological

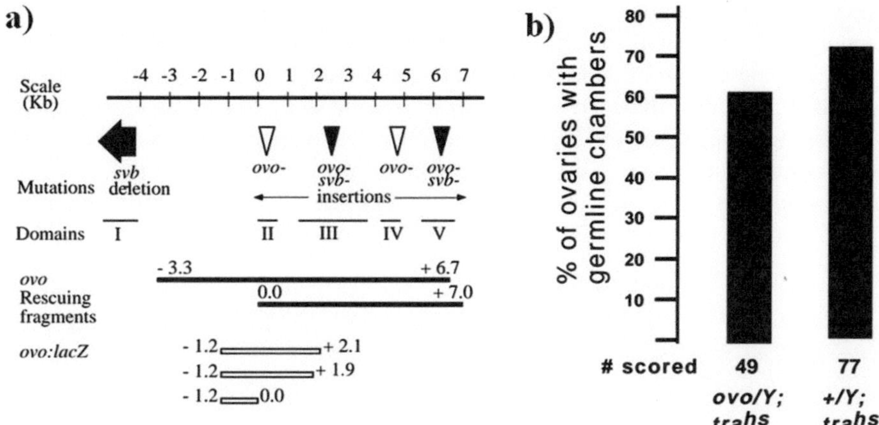

Fig. 17.25 The insertion and function of *ovo*$^+$ gene in germline sex determination. (**a**) Molecular map depicting the *ovo* locus with the consecutive insertion and deletion sites of the *ovo* and *svb* genes. The bold arrows represent the deletions, whereas the arrowheads represent the insertions (filled in (insertion of *ovo*$^-$ and *svb*$^-$) and unfilled (insertion of *ovo*$^-$)) sites. These are the sites where multiple insertions and deletions were carried out in order to understand the function of both the genes. Below the domains the wild-type fragments are shown in which the mutation studies have been carried out. At the bottom the three reporter gene fragments of *ovo* gene, used for building the gene constructs for mutation, are indicated. The transcription proceeds from left to right. (**b**) Map showing that the activity of *ovo* gene is not required in XY female germ cells. Females carrying both *ovo*$^+$ and *ovo*$^-$ genes were crossed with males to obtain the XY female germline. The siblings were then scored for the activity of the *ovo* gene, and no significant difference was found

process. There are many genes whose regulations were studied by microarray. Some of them are as follows:

1. Target genes induced by b-catenin, a key component of the WNT signaling pathway—A human renal carcinoma cell line was used for the purpose. The cells did not express catenins and were stably transfected with a b-catenin construct. From microarray studies it was found that LEF/TCF binding sites were necessary for transcription of b-catenin by Nr-CAM promoter and their proper functioning.
2. Similar screens have been performed in order to identify the targets of other WNT signaling pathways, GDNF signaling, Mitf-dependent SCF/Kit signaling, and the targets of the transcription factors TBX2, SIX5, PAX3, WHN, HOXA11, and HOXA13. While studying HOXA11 transcription, one gene, Integrin a8, was upregulated 20-fold after induction of Hoxa11 expression.

Technical issues: The microarray till date has many technical limitations. In order to perform genome-wide expression data in detail, it is necessary to have the genome information. The complete genomic data of various important model organisms like frog and chicken is missing. This hinders the proper utilization of the technique.

Another disadvantage and perhaps one of the most important ones is the limiting amount of RNA available from standard embryonic dissections. Lastly in the case of mammals, the main limitation is their complexity both at transcriptional and cellular level that limits the efficient use of microarray in the study of mammalian genomics.

Imaging gene expression in four dimensions: Microarray in the near future can be used to study the gene expression patterns and their activity in four dimensions. It can lead to excellent studies of cell differentiation, communication, death, migration, and division in time and three- and four-dimensional space.

Box 17.1: Scientific Concept: Function of Drosophila ovo + in Germline Sex Determination Depends on X-Chromosome Number (Brian Oliver et al.) (Fig. 17.25)

The differentiation of the gonads and the development of the germline are mainly characterized by the process of spermatogenesis or oogenesis. It is known that in drosophila after fertilization and formation of embryo, the sex of the individual is not determined by the X-Y ratio but by the X-A ratio. One of the principal steps in this process is counting the number of X-chromosomes. Some studies have shown in somatic cells that this function is carried out and controlled by the *Sxl*$^+$ gene, but nothing much is known about how the process is carried out in germ cells. It is often thought that both the germ and somatic cells follow the same pathway, which is studied not to be true. Though both the cell types have the functioning of the *Sxl*$^+$ gene in common, the germline sex determination also needs the association of some other X-chromosome genes. These are the *ovo*$^+$, *otu*$^+$, and *snf*$^+$ genes. These genes act in association with the *Sxl*$^+$ gene and helps in its transcription and activation in a female-specific manner. It is studied that *Snf*$^+$ is an RNA-binding protein and may therefore act directly on *Sxl* + pre-mRNA. But the function of *ovo* + and *otu* + was not known for a long time. In this work, Oliver et al. first estimated a function of *ovo*$^+$ in the germline sex determination and that it is highly active in the females. They made *ovo* reporter genes and found that they show high activity in the germline of females and low activity in the germline of males. They saw that when XY flies are transformed into somatic females, they do not show high levels of reporter activity, but when XX flies were transformed into somatic males, they showed high level of reporter activity. This shows that high-level *ovo*$^+$ expression depends on the number of X chromosomes, not on the X-A ratio. It was found that the *ovo*$^+$ functioning is only required for XX flies which showed that the gene played an important role in sensing the number of X chromosomes. Mutations in *ovo* have no effect on XY males, X0 males, or XY females but have pronounced effects on germ cell viability in XX females, XX females with sex transformed germlines, and XX males indicating that *ovo*$^+$ gene products are required for events occurring only in flies with two X chromosomes.

17.3 Summary

- "Genetics is about how information is stored and transmitted between generations," John Maynard Smith.
- A gene is a functional unit of heredity. These genes control the growth and development of an organism throughout its life cycle. The study of function and mechanism of genes in the evolution of a living being is called developmental genetics or genetic analysis of development.
- When life begins from a single cell, it involves three processes: cell division (production of cells), cell differentiation (diverging of cells to different cell types), and morphogenesis (forming structure of organism). This whole phenomenon of conversion of a single cell to an organism is due to the orchestrated synchronization of genes which forms the basis of development.
- During the onset of life from an egg to an embryo, two groups of genes come to play. They are the maternal genes (in development of egg to embryo) and zygotic genes (in further embryonic development). These two sets of genes are among the earliest, governing the onset of life from egg to embryo to infant. Maternal genes are genes that are present in the ovum during fertilization and have an immense contribution in the initial development of the zygote in terms of DNA, nutrition, mRNA, and proteins. Zygotic genes play an important role in segmentation and differentiation of the body in organisms and also in various organ formations. Many model organisms are used as important tools to study various developmental processes and the role of genes in it.
- A model organism is a species that has been widely studied, because it is easy to maintain and breed under laboratory setting and has many experimental advantages. It is used to enlighten many biochemical, genetic, and signaling pathways and serves as disease models, like that of cancer and neurological disorders, etc. *Saccharomyces cerevisiae*, *Drosophila melanogaster*, *Caenorhabditis elegans*, *Xenopus tropicalis*, *Mus musculus*, and *Danio rerio* are the most widely used model organisms. These organisms also serve as vertebrate homologues for analyzing the various developmental stages in vertebrates.
- Studying the phases of development is very important for the diagnosis of diseases and studying various signaling and biochemical pathways. These homologues have a much simpler life cycle and genetic framework than higher vertebrates, which helps researchers in giving an idea of the activity and importance of any gene. For this many gene knockout and gene mutation studies have also been carried out in organisms like mouse and zebrafish. *Drosophila melanogaster* or the fruit fly is one of the most studied model organisms. It has many gene homologues in humans. So, developmental stages of the fruit fly and the genes associated with it give an idea of both insects and vertebrate developments. Many robust and advanced techniques are also available to study the interaction of these genes like microarray.
- Besides, other important areas of research in developmental genetics are studying the process of sex determination of any individual. The interdependency and the

cumulative functioning of the sex chromosomes are major factors of research in developmental biology and are well studied from the last century.

- Another field which is gaining immense importance in present-day research is mammalian stem cells. Stem cells denote certain primary cells which have the pluripotent ability to give rise to a cell lineage and ultimately a tissue. They are finding wide applications in various immune disorders, hematopoiesis, metabolic disorders, tumors, etc.

References

Abolaji AO, Kamdem JP, Farombi EO, Rocha JBT (2013) Arch Bas App Med 1:33

Allocca M, Zola S, Bellosta P (2018) Drosophila melanogaster: Model for Recent Advances in Genetics and Therapeutics, p 113

Altmann K, Durr M, Westermann B (2007) Methods Mol Biol 372:81

Angeles-Albores D, Leighton DH, Tsou T, Khaw TH, Antoshechkin I, Sternberg PW (2017) G3 Genes Genomes Genet 7(9):2969

Arenas A, Fernández A, Gómez S (2009) Handbook on biological networks, vol. 10. InTech, Rijeka, Croatia. p. 243

Augustine S (2012) Doctoral dissertation. Aix-Marseille

Austin CP, Battey JF, Bradley A, Bucan M, Capecchi M, Collins FS, Dove WF, Duyk G, Dymecki S, Eppig JT, Grieder FB (2004) Nat Genet 36(9):921

Bakloushinskaya IY (2009) Biol Bull 36(2):167

Bieler J, Pozzorini C, Naef F (2011) Biophys J 101(2):287

Blaxter M (2011) PLoS Biol 9(4):1001050

Blum M, Ott T (2018) Cells Tissues Organs 205(5–6):303

Bock J, Fukuyo Y, Kang S, Phipps ML, Alexandrov LB, Rasmussen K, Bishop AR, Rosen ED, Martinez JS, Chen HT, Rodriguez G (2010) PLoS One 5(12):e15806

Brakebusch C, Pihlajaniemi T (2011) Mouse as a model organism: from animals to cells. Springer, Dordrecht

Briggs JP (2002) Am J Physiol Regul Integr Comp Physiol 282(1):R3

Brockdorff N, Turner BM (2015) Cold Spring Harb Perspect Biol 7(3):a019406

Buzzini P, Turchetti B, Yurkov A (2018) Yeast 35(8):487

Carroll SB, Winslow GM, Schupbach T, Scott MP (1986) Nature 323(6085):278

Chege PM, McColl G (2014) F ront Aging Neurosci 6:89

Chen F, MacKerell AD, Luo Y, Shapiro P (2008) J Cell Communic Signal 2(3–4):81

Cho JH, Bandyopadhyay J (2012) Salmonella-A Diversified Superbug. IntechOpen, Rijeka

Copp AJ (1995) Trends Genet 11(3):87

D'Costa A, Shepherd IT (2009) Zebrafish 6(2):169

Dahm R, Geisler R (2006) Mar Biotechnol 8(4):329

Dooley K, Zon LI (2000) Curr Opin Genet Dev 10(3):252

Eimon PM, Ashkenazi A (2010) Apoptosis 15(3):331

Ekker SC (2008) Zebrafish 5(2):121

Fischer S, Prijkhozhij S, Rau MJ, Neumann CJ (2007) Cell Cycle 6(23):2962

Gayatri PN (2012) Rev Lit

Gelbart ME, Kuroda MI (2009) Development 136(9):1399

Georgiev P, Chlamydas S, Akhtar A (2011) Fly 5(2):147

Gershon H, Gershon D (2000) Mech Ageing Dev 120:1

Glass AS, Dahm R (2004) Ophthalmic Res 36(1):4

Grainger RM (2012) Xenopus protocols. Humana Press, Totowa, NJ, p 3

Gravato-Nobre MJ, Hodgkin J (2005) Cell Microbiol 7(6):741

Hall B, Limaye A, Kulkarni AB (2009) Curr Protoc Cell Biol 44(1):19
Hansen GM, Skapura D, Justice MJ (2000) Genome Res 10(2):237
Heard E, Disteche CM (2006) Genes Dev 20(14):1848
Herndon LA, Wolkow CA, Driscoll M, Hall DH (2017) Effects of ageing on the basic biology and anatomy of C. elegans. Springer, Cham. p. 9
Herrgard Markus J., Lee Baek-Seok, Portnoy V., Palsson B.: 16, 627 (2006)
Honigberg SM (2011) Eukaryot Cell 10(4):466
Hosen MJ, Vanakker OM, Willaert A, Huysseune A, Coucke P, DePaepe A (2013) Front Genet 4: 74
Jennings BH (2011) Mater Today 14(5):190
Justice MJ, Dhillon P (2016) Dis Model Mechan 9:101
Kalb JM, Lau KK, Goszczynski B, Fukushige T, Moons D, Okkema PG, McGhee JD (1998) Development 125(12):2171
Kaletta T, Hengartner MO (2006) Nat Rev Drug Discov 5(5):387
Khan FR, Alhewairini S (2018) Current trends in cancer management. IntechOpen, London
Kim H, Ishidate T, Ghanta KS, Seth M, Conte D, Shirayama M, Mello CC (2014) Genetics 197(4): 1069
Klis FM, Mol P, Hellingwerf K, Brul S (2002) FEMS Microbiol Rev 26(3):239
Krishnan KG, Milionis A, Tetteh F, Loth E (2017) Aerosp Sci Technol 69:181
Kuratani S (2003) Paleontol Res 7(1):89
Kuroda MI, Hilfiker A, Lucchesi JC (2016) Genetics 204(2):435
Langheinrich U, Hennen E, Stott G, Vacun G (2002) Curr Biol 12(23):2023
Lee KY, Jang GH, Byun CH, Jeun M, Searson PC, Lee KH (2017) Biosci Rep 37:3
Lehoczky JA, Thomas PE, Patrie KM, Owens KM, Villarreal LM, Galbraith K, Washburn J, Johnson CN, Gavino B, Borowsky AD, Millen KJ (2013) PLoS Genet 9(12):e1003967
Lengyel JA, Iwaki DD (2002) Dev Biol 243(1):1
Leonardo ED, Hinck L, Masu M, Keino-Masu K, Ackerman SL, Tessier-Lavigne M (1997) Nature 386(6627):833
Leung MC, Williams PL, Benedetto A, Au C, Helmcke KJ, Aschner M, Meyer JN (2008) Toxicol Sci 106(1):5
Levine M (2008) Genome Biol 9(2):207
Lisa R. Girard, Tristan J. Fiedler, Todd W. Harris, Felicia Carvalho, Igor Antoshechkin, Michael Han, Paul W. Sternberg, Lincoln D. Stein, and Martin Chalfie (2006) WormBook: the online review of Caenorhabditiselegans biology, Nucleic Acids Research, 35 (2007)
Lovegrove B, Simoes S, Rivas ML, Sotillos S, Johnson K, Knust E, Jacinto A, Hombría JCG (2006) Curr Biol 16(22):2206
Lu W, Phillips CL, Killen PD, Hlaing T, Harrison WR, Elder FFB, Miner JH, Overbeek PA, Meisler MH (1999) Genomics 61(2):113
Lucchesi JC, Kuroda MI (2015) Cold Spring Harb Perspect Biol 7(5):a019398
MacWilliam IC (1970) J Inst Brew 76(6):524
Marsh EK, May RC (2012) Appl Environ Microbiol 78(7):2075
Matharu NK, Hussain T, Sankaranarayanan R, Mishra RK (2010) J Mol Biol 400(3):434
Menke AL, Spitsbergen JM, Wolterbeek AP, Woutersen RA (2011) Toxicol Pathol 39(5):759
Meyers JR (2018) Curr Protoc Essent Lab Techn 16(1):e19
Mirzoyan Z, Sollazzo M, Allocca M, Valenza AM, Grifoni D, Bellosta P (2019) Front Genet 10:51
Morris DW, Barry PA, Bradshaw HD, Cardiff RD (1990) J Virol 64(4):1794
Murgatroyd C, Spengler D (2010) Genome Biol 11(2):105
Nguyen KD, Disteche CM (2006) Nat Genet 38:47
Norton W, Bally-Cuif L (2010) BMC Neurosci 11(1):90
O'Reilly LP, Luke CJ, Perlmutter DH, Silverman GA, Pak SC (2014) Adv Drug Deliv Rev 69:247
Panzer S, Weigel D, Beckendorf SK (1992) Development 114(1):49
Parng C, Seng WL, Semino C, McGrath P (2002) Assay Drug Dev Technol 1(1):41

Pazdernik N, Schedl T (2013) Introduction to germ cell development in Caenorhabditis elegans. Adv Exp Med Biol 757:1–16
Peleg AY, Tampakakis E, Fuchs BB, Eliopoulos GM, Moellering RC, Mylonakis E (2008) Proc Natl Acad Sci 105(38):14585
Perlman RL (2016) Evol Med Public Health 2016(1):170
Phifer-Rixey M, Nachman MW (2015) elife 4:e05959
Pomiankowski A, Nothiger R, Wilkins A (2004) Genetics 166(4):1761
Pretscher J, Fischkal T, Branscheidt S, Jager L, Kahl S, Schlander M, Thines E, Claus H (2018) Fermentation 4(2):31
Rodriguez-Esteban C, Schwabe JW, Pena JD, Rincon-Limas DE, Magallón J, Botas J, Belmonte JC (1998) Development 125(20):3925
Ruvinsky I, Gibson-Brown JJ (2000) Development 127(24):5233
Montes de Oca R, Salem AZM, Kholif AE, Monroy H, Pérez LS, Zamora JL, Gutiérez A (2016) Yeast: description and structure. Yeast Addit Anim Prod
Salz H, Erickson JW (2010) Fly 4(1):60
Samsonova AA, Niranjan M, Russell S, Brazma A (2007) PLoS Comput Biol 3(7):e144
Sandeman D (1999) Naturwissenschaften 86(8):378
Schlegel A (2012) Cell Mol Life Sci 69(23):3953
Segner H (2009) Toxicol Pharmacol 149(2):187
Senthilkumar R, Mishra RK (2009) BMC Genomics 10(1):549
Sheikh K, Forster J, Nielsen LK (2005) Biotechnol Prog 21(1):112
Showell C, Conlon FL (2009) Cold Spring Harb Protoc 2009(9):131
Sibirny AA (2016) FEMS Yeast Res 16(4):1
Smith L, Greenfield A (2003) Hum Mol Genet 12(1):R1
Stem Cell Basics, NIH Stem Cell Information, NIH 2015
Stephenson R, Metcalfe NH (2013) J R Coll Physicians Edinb 43(1):70
Stewart GG (2017) Brewing and distilling yeasts. The yeast handbook. Springer, Champions. p. 55
Sung YH, Kim JM, Kim HT, Lee J, Jeon J, Jin Y, Choi JH, Ban YH, Ha SJ, Kim CH, Lee HW (2014) Genome Res 24(1):125
Swain A, Lovell-Badge R (1999) Genes Dev 13(7):755
Taddei A, Gasser SM (2012) Genetics 192(1):107
Tan MW, Mahajan-Miklos S, Ausubel FM (1999) Proc Natl Acad Sci 96(2):715
Tee SY (2011) Doctoral dissertation, UTAR
Terskikh AV, Bryant PJ, Schwartz PH (2006) Pediatr Res 59(S4):13R
Thisse C, Zon LI (2002) Science 295(5554):457
Tyler MS (2000) Developmental biology, A guide for experimental study, 2nd. ed. p. 85
Vanhooren V, Libert C (2012) Ageing Res Rev 12(1):8
Varshney GK, Lu J, Gildea DE, Huang H, Pei W, Yang Z, Huang SC, Schoenfeld D, Pho NH, Casero D, Hirase T (2013) Genome Res 23(4):727
Vijayalakshmi M (n.d.) Drosophila melanogaster-Life Cycle
Vleminckx K, Dimitrakopoulou D, Tulkens D, Van Vlierberghe P (2019) Front Physiol 10:48
Wangler MF, Bellen HJ (2017) Basic science methods for clinical researchers, Elsevier, London. pp 211–234
Wasilczuk AZ, Maier KL, Kelz MB (2018) Methods Enzymol 602:211
White KP, Rifkin SA, Hurban P, Hogness DS (1999) Science 286(5447):2179
White R, Rose K, Zon L (2013) Nat Rev Cancer 13(9):624
Wieschaus E, Nusslein-Volhard C, Kluding H (1984) Dev Biol 104(1):172
Wu RS, Lam II, Clay H, Duong DN, Deo RC, Coughlin SR (2018) Dev Cell 46(1):112
Xue L, Yi H, Huang Z, Shi YB, Li WX (2011) Int J Biol Sci 7(7):1068
Xue L, Cai JY, Ma J, Huang Z, Guo MX, Fu LZ, Shi YB, Li WX (2013) BMC Genomics 14(1):568
Yurkov AM (2018) Yeast 35(5):369

Molecular Genetics of Cancer

18

Bhawna Chuphal

18.1 Cancer: Overview (Fig. 18.1)

Coined by Hippocrates in the fifth century BC, cancer has proven to be a leading cause of death in Western countries mostly. The tale of cancer goes back to 1761 when Giovanni Morgagni did autopsies for the first time to relating after death pathogenic findings to illness of patients and laid the foundation for study of cancer scientifically known as oncology, the study of cancer. Over a million people are diagnosed each year in the USA (Table 18.1) with at least half of them died and the treatment being often costs a fortune as well. Looking at the statistics provided by a study published in the journal of *American Cancer Society*, it is quite evident that the USA reported approximately 1.7 lakhs new cancer cases and more than 6 lakhs death by cancer toward the end of 2019. When it comes to cancer medical expenses, it has been estimated by the Agency for Healthcare Research and Quality (AHRQ) that the USA alone has spent up to 80.2 billion dollars in the year 2015. Patients suffering from cancer feel as if their bodies have been invaded by an extraterrestrial force; however, the malignancies arise from the self. Cancer is considered a group of disorders in which the normal regulation of cell cycle is lost. In fact, a series of genetic mutations are well established to be the leading cause of cancer arising from a single cell (Cavenee & White 1995). A healthy cell comprises a portion of an ordered array of other cells around it and undergoes cell division only when the stimulatory and inhibitory signals from external environment balance out and favor the event. This cell is replaced by new ones if worn out or damaged. However, with replication or growth comes an inevitable hazard of genetic mutations impairing the regulatory circuits inside a cell, occasionally leading to unscheduled cell division (Fig. 18.1). The growth of cancer cells is a consistent process, producing new cells, thereby crowding out normal cells and creating ruckus at the onset site of cancer.

B. Chuphal (✉)
University of Delhi, Delhi, India

D. Kar, S. Sarkar (eds.), *Genetics Fundamentals Notes*,
https://doi.org/10.1007/978-981-16-7041-1_18

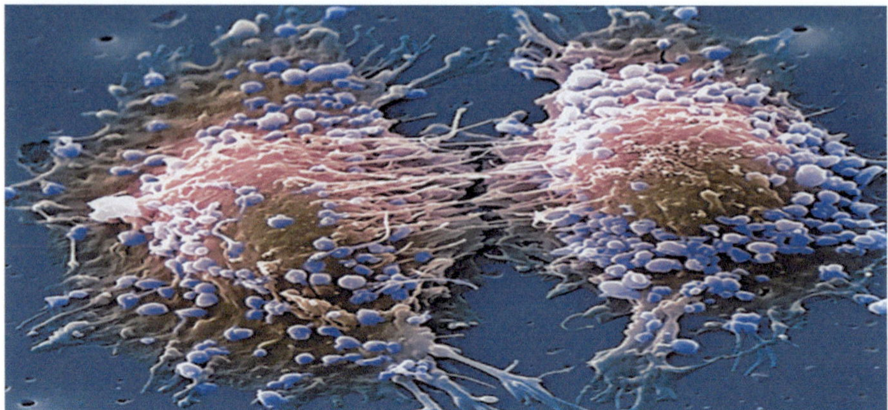

Fig. 18.1 SEM of two dividing prostate cancer cells. Two cancer cells shown here are undergoing cytokinesis

Cells having cancer are well known to possess the ability of spreading via a process known as metastasis (meh-TAS-tuh-sis) and reaching to various other body parts. For instance, lung cancer cells can spread to the bones and divide there, but it is still called lung cancer unless it started in the bones. No matter where a cancer originates, the cell loses its native shape and boundary, ceases responding to inhibitory signals, and goes haywire in case of division. The resulting mass of cells, in turn, can crowd out and rob healthy tissue of nutrients. The worst scenario is when it can invade the barriers separating organs and metastasize to distant sites.

18.2 Cancer Types

There exist several different cancer types, and it may start in several tissues including the lungs, the skin, the eye, or even in the blood (Table 18.2). Although alike in some criteria, cancers differ in manner of growth and invasion. With loss in normal functioning, cancer cells lose their regular shape forming a distinct mass known as tumor. Tumor is generally of two types: benign, where tumor is localized, and malignant in which the cells invade other tissues via metastasis.

18.2.1 Gonadal Tumors

There are various classes of tumors under this category that may arise from totipotent cells giving rise to a variety of tissue-type tumors, often within "germ cell" tumors and mainly consist of endodermal sinus tumor, choriocarcinoma, seminoma/dysgerminoma, teratocarcinoma, and embryonal carcinoma. Although common in the gonads (ovaries/testes), these germ cell tumors might sometimes occur at sites

Table 18.1 Top 10 cancer types along with reported new cases and deaths in males and females in the USA, 2016 (Siegel et al. 2016)

In males	**Estimated new cases**		**Estimated deaths**	
Cancer sites	**Percentage of cases**	**Number of cases**	**Percentage of cases**	**Number of cases**
Brain and other	–	–	3%	9440
nervous system	8%	70,820	8%	26,020
Colon and rectum	–	–	4%	12,720
Esophagus	5%	39,650	–	–
Kidney and renal	4%	34,090	4%	14,130
pelvis	3%	28,410	6%	18,280
Leukemia	14%	117,920	27%	85,920
Liver and intrahepatic	6%	46,870	–	–
bile duct	5%	40,170	4%	11,520
Lung and bronchus	4%	34,780	–	–
Melanoma of the skin	21%	180,890	8%	26,120
Non-Hodgkin	–	–	7%	21,450
lymphoma	7%	58,950	4%	11,820
Oral cavity and				
pharynx				
Prostrate				
Pancreas				
Urinary bladder				
Total	100%	841,390	100%	314,290
In females	**Estimated new cases**		**Estimated deaths**	
Cancer sites	**Number of cases**	**Percentage of cases**	**Number of cases**	**Percentage of cases**
Brain and other	–	–	6610	2%
nervous system	246,680	29%	40,450	14%
Breast	63,670	8%	23,170	8%
Colon and rectum	23,050	3%	–	–
Kidney and renal	26,050	3%	10,270	4%
pelvis	–	–	8890	3%
Leukemia	106,470	13%	72,160	26%
Liver and intrahepatic	29,510	4%	–	–
bile duct	32,410	3%	8630	3%
Lung and bronchus	–	–	14,240	5%
Melanoma of skin	25,400	3%	20,330	7%
Non-Hodgkin	49,350	6%	–	–
lymphoma	60,050	7%	10,470	4%
Ovary				
Pancreas				
Thyroid				
Uterine corpus				
Total	662,640	100%	281,400	100%

which are extragonadal. There exists another type of gonadal tumors which may arise from stroma of connective tissue, for instance, tumors of granulosa-theca cell, hilar cell, and lipid cell in females and Sertoli-Leydig cell tumors in males, depending upon the nature and function of stromal cells.

Table 18.2 The different kinds of tumors associated with tissue type

Tissue	Malignant tumors	Benign tumors
Adult fibrous tissue	Fibrosarcoma	Fibroma
Bile duct	Cholangiocarcinoma	Bile duct adenoma
Blood vessels	Angiosarcoma, hemangiosarcoma	Hemangiopericytoma, hemangioma
Bone	Osteosarcoma	Osteoma
Breast	Cystosarcoma phyllodes	Fibroadenoma
Fat	Liposarcoma	Lipoma
Glandular epithelium	Adenocarcinoma	Adenoma
Hematopoietic cells	Aleukemic leukemia, Leukemia (various types)	Myeloproliferative disorders, preleukemias
Kidney	Hypernephroma; renal cell carcinoma	Renal tubular adenoma
Liver	Hepatocellular carcinoma	Hepatic adenoma
Lymph vessels	Lymphangiosarcoma	Lymphangioma
Lymphoid tissue	Hodgkin lymphoma and Non-Hodgkin lymphoma, multiple myeloma, plasmacytoma	Plasmacytosis
Placenta	Choriocarcinoma	Hydatidiform mole
Smooth muscle	Leiomyosarcoma	Leiomyoma
Stratified squamous epithelium	Malignant skin adnexal tumors, epidermoid carcinoma and squamous cell carcinoma	Skin adnexal tumors, seborrheic keratosis, and papilloma
Nerve cells	Medulloblastoma, neuroblastoma	Ganglioneuroma
Nerve sheath	Malignant schwannoma Neurofibrosarcoma Malignant meningioma	Neurilemmoma, neurofibroma, schwannoma
APUD system		
• Adrenal medulla • Pancreas • Pituitary • Parathyroid • Thyroid (C cells) • Stomach and intestines • Carotid body and chemo-receptor system	Islet cell carcinoma Malignant carcinoid Malignant carcinoid Parathyroid carcinoma Medullary carcinoma Malignant Pheochromocytoma	Islet cell adenoma, gastrinoma Insulinoma Chemodectoma Paraganglioma Basophilic adenoma Chromophobe adenoma Eosinophilic adenoma Parathyroid adenoma C cell hyperplasia Pheochromocytoma

18.2.2 Classification of Cancer

Cancers are broadly classified based on tissue type in which they originate and primary site where they develop (Table 18.3). Tissue type also known as histological type as established by the International Classification of Diseases for Oncology includes hundreds of different cancer types which have been primarily grouped

Table 18.3 Some common types of cancers

	Types
1.	Anal cancer, adrenal cancer, acute myeloid leukemia (AML), acute lymphocytic leukemia (ALL)
2.	Basal and squamous cell skin cancer, bile duct cancer, bladder cancer Bone cancer, brain and spinal cord tumors, breast cancer
3.	Colorectal cancer, Castleman disease, chronic myeloid leukemia (CML), cervical cancer, chronic myelomonocytic leukemia (CMML), chronic lymphocytic leukemia (CLL)
4.	Ewing family of tumors, eye cancer (ocular melanoma), esophagus cancer, endometrial cancer
5.	Gestational trophoblastic disease, gastrointestinal neuroendocrine (carcinoid) tumors, gallbladder cancer, gastrointestinal stromal tumor (GIST)
6.	Hodgkin lymphoma
7.	Kidney cancer, hypopharyngeal cancer, Kaposi sarcoma, laryngeal cancer
8.	Lung cancer, lymphoma, liver cancer, lung carcinoid tumor, leukemia
9.	Myelodysplastic syndromes, Merkel cell skin cancer, multiple myeloma, melanoma skin cancer, malignant mesothelioma
10.	Non-small cell lung cancer, neuroblastoma, non-Hodgkin lymphoma, nasopharyngeal cancer, cavity and paranasal sinuses cancer
11.	Ovarian cancer, osteosarcoma, oropharyngeal cancer, oral cavity
12.	Prostate cancer, pituitary tumors, penile cancer, pancreatic neuroendocrine tumor (NET), pancreatic cancer
13.	Rhabdomyosarcoma, retinoblastoma
14.	Stomach cancer, soft tissue sarcoma, small intestine cancer, small cell lung cancer, skin cancer, salivary gland cancer
15.	Thyroid cancer, thymus cancer, testicular cancer
16.	Uterine sarcoma
17.	Vulvar cancer, vaginal cancer
18.	Wilms' tumor, Waldenstrom macroglobulinemia

into six major categories, namely, carcinoma, leukemia, lymphoma, myeloma, sarcoma, and mixed types. Here four of them are dealt with in detail.

18.2.3 Carcinoma

Carcinoma often refers to cancer related to body lining (internal or external) or epithelial malignant neoplasm and accounts for majority of the cancer cases (80–90%) since the involved tissue type is found throughout the body ranging from the skin to internal passages such as alimentary canal tract or gastrointestinal tract and exterior or interior lining of organs. Carcinomas are again classified into two major subtypes: adenocarcinoma and squamous cell carcinoma (SSC). Adenocarcinomas develop inside a gland or organ and are often seen as thickened mucosa at first. It generally occurs in mucus membranes and easily spreads throughout the soft tissue at their place of origin. SCCs mainly originate from squamous epithelium, chiefly found in lining of body cavities, lung, and skin. It has been long

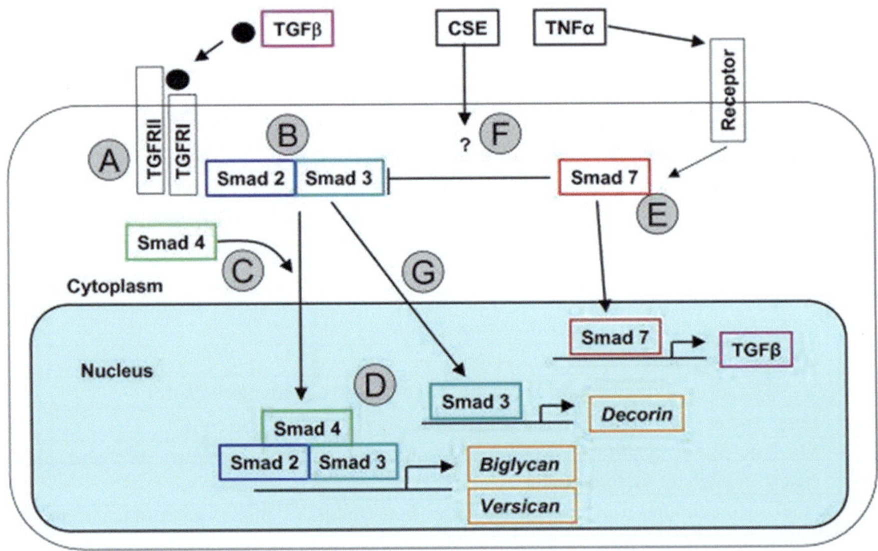

Fig. 18.2 A schematic view of TGFβ signaling pathway. Ligand TGF-β binds to its type II and type I receptor (**a**) leading to complex formation and type I receptor phosphorylation which subsequently phosphorylates Smad2 or 3 (**b**), binding with Smad4 and translocating to the nucleus shown in **c**. The complex associates with enhancers in target genes (**d**) inside the nucleus. TNF upregulates Smad7 inhibiting the signaling pathway (**e**) and CSE might also interfere with the Smad pathway (**f**)

established that SCC caused by mutation increases transforming growth factor-beta (TGFβ) secretion levels in the surrounding microenvironment. Briefly, after binding with ligand, TGFβ receptor complex takes part in phosphorylation of Smad2 and Smad3 resulting in complex formation with Smad4 translocating into nucleus, binding to protein binding elements specific to Smad, and regulating various downstream transcriptional pathways. The activated complex in turn also activates a number of non-canonical signaling pathways which are independent of Smad proteins (Fig. 18.2).

18.2.4 Immunotherapy

TGFβ promotes immune suppression in a number of ways and hence provides a target for cancer immunotherapy. In addition, it also inhibits growth of tumor cells at early stages of cancer, thereby rendering them unresponsive to its underlying signaling. In the recent past, various TGFβ receptor inhibitors have undergone development, for instance, galunisertib (LY2157299), with focus on their impact on patients with heart diseases. The current range of TGFβ inhibitors has been proven effective in combination with other agents as well as a single agent. For instance, in a preclinical trial, TGFβ has been demonstrated to drive away response related to immune suppression in mice with prior EGFR inhibitor treatment. Perhaps in combination with immune checkpoint blockade, inhibition therapy targeting TGFβ will have the most potential as the signaling is closely related to immune checkpoint signaling. Moreover, in murine SCC, it has been shown that addition of TGFβ depleting antibody abrogates the elevated TGFβ signaling and Treg expansion induced by anti-PD-1 treatment. Reports including bifunctional antibodies such as TGFβ/CTLA-4 and TGFβ/PD-1 which have shown antitumor response in certain breast cancer lines and melanoma in both clinical and preclinical trials show that bifunctional antibodies which work by combining antibodies for immune checkpoints blockade with ligand binding domain are also well effective. Another such example is M7824, antibody against PD-L1 and bifunctional antibody of TGFβRII ligand trap, which has shown positive results in colon and breast cancer in the murine model system by activation of T cell and NK cell inside the tumors.

18.2.5 Sarcoma

The words *sarco* and *oma* have Greek origin meaning fleshy and tumor, respectively. Although relatively rare and often malignant, sarcomas are derived from embryonic mesodermal layers of mesenchymal tissues (Table 18.4). They constitute less than 10% of all cancer types and have been shown to have high morbidity as well as mortality rate among children and young adults rather than older adults. Associated with sarcoma, there are certain environmental exposures and genetic predisposition syndromes even though the majority of these cancer types are considered sporadic with unknown etiology.

Most of the sarcomas are reported to have alterations in either retinoblastoma (RB) pathway or p53 pathway; hence hereditary retinoblastoma patients are often at higher risk of developing sarcomas. Since these two pathways have members which

Table 18.4 Examples of sarcomas

S. no.	Name	Origin
1.	Osteosarcoma or osteogenic sarcoma	Bone
2.	Chondrosarcoma	Cartilage
3.	Leiomyosarcoma	Smooth muscle
4.	Rhabdomyosarcoma	Skeletal muscle
5.	Mesothelial sarcoma or mesothelioma	Membranous lining of body cavities
6.	Liposarcoma	Adipose tissue
7.	Hemangioendothelioma or angiosarcoma	Blood vessels
8.	Fibrosarcoma	Fibrous tissue
9.	Glioma or astrocytoma	Neurogenic connective tissue found in the brain
10.	Myxosarcoma	Primitive connective tissue
11.	Mesenchymous or mixed mesodermal tumor	Mixed connective tissue

are viewed as either brakes or accelerators of cell cycle, usually in tumors, loss of proteins acting as brakes takes place, whereas those acting as accelerators get amplified. For instance, when protein such as INK4A is lost, it results in amplified phosphorylation of RB protein through cyclin D1/CDK4 inhibitory loss, thereby preventing cells from undergoing cell division by disarming RB protein. On the other hand, in cases of sarcomas, increased expression of CDK4 or cyclin D1 has often been demonstrated and leads to same results as during loss of INK4A protein. In addition, with sarcomas, RB protein loss has been associated, rendering its inability to block the process of cell division. Another example is ARF protein loss which leads to HDM2 inhibitory loss in turn enhancing its ability to block normal functioning of p53 protein. Further, amplified HDM2 in case of sarcomas enhances its ability to inhibit normal functioning of p53 protein. It has been well documented that p53 loss of function in sarcomas impairs the ability of affected cells with already damaged DNA to undergo apoptosis (Helman & Meltzer 2003).

18.2.6 Therapeutic Implications

At the present, treatment strategy is based largely on histological grade of tumor, ease of resecting, along with absence or presence of metastases. This grading is entirely dependent on features often determined medically such as degree of nuclear atypia, mitoses, necrosis, and others. The main purpose behind histological grading is to be able to predict most likely biological behavior of a tumor under consideration, which specifically includes spread due to metastasis or recurrence of the tumor locally after its surgical removal. Low-grade tumors have high possibility of getting cured with complete surgical resection though they do occasionally metastasize, whereas high-grade ones are likely to recur and spread distantly following local therapy. However, it has been suggested that further expression of genes along with proteomic analysis in case of such tumors may in turn improve our ability to predict future behavior of various sarcomas. Interestingly, there have been documented successes in recent development of therapies targeting specific genetic alteration involved in these tumors. Gastrointestinal stromal tumors (GISTs) represent such examples involving improved diagnosis as well as treatment strategy depending on mechanisms driving the process of tumorigenesis. Before this improved diagnosis and treatment, these stromal tumors were thought to be tumors of smooth-muscle cells, i.e., leiomyosarcomas of the GI tract (gastrointestinal tract) and hence never expected to give positive response to chemotherapeutic agents. Subsequently, after careful histological grading, these tumors were considered to be derived instead from Cajal's interstitial cells and thereafter majorly characterized by activation in tyrosine kinase of c-KIT receptor through point mutations. Luckily, imatinib, a drug well known to inhibit the activity of c-KIT tyrosine kinase, led to pronounced effects when tested in GIST patients, leading to major regression in tumor activity in previously thought unresponsive patients. However, in the recent past, GISTs having lack of KIT mutation were found to be associated with an activating mutation in PDGFα receptor (PDGFRA) in many cases. The interesting part discovered was that mutations in either of these receptors led to a similar downstream signaling pathway which included activation of MAPK (mitogen-activated protein kinase) as well as AKT pathway. By far, mutated tumor-specific target blockage such as c-KIT in case of GISTs has been most successful at molecularly targeting therapeutic interventions, but further therapeutic advances are required as many more kinases are involved in sarcomas which seem to be activated and not yet mutated. Although potential kinase inhibitors being developed in cancer therapeutics have been recently listed, it still remains to be largely determined as to whether any of the drugs/ inhibitors will have any pronounced effect in sarcomas having non-mutated activation of kinase signaling pathways.

18.2.7 Leukemia

Increased numbers of leucocytes in our blood and/or bone marrow (the production site for blood cells) lead to development of several malignant disorders commonly called as leukemia. Leukemia often renders patients prone to infection as it associated with the overproduction of inefficiently functioning immature white blood cells. However, there are cases as seen in chronic lymphocytic leukemia (CLL) where either leukemia cells present in dominance have been reported to be matured, or like in the acute leukemias where precursor cells of various lineage are found, or as in chronic myeloid leukemia (CML) where both precursor and mature cells have been reported. While leukemias may be found in all age groups, each type is reported to have specific age distributions, with acute myeloid leukemia (AML) progressively common in people with older ages, whereas acute lymphoblastic leukemia (ALL) is most commonly seen in early childhood. CML on the other hand is reported to be very rare among young children, and being most common in the West, CLL is considered almost exclusive to people above 40 years. Few examples of different leukemia are granulocytic, lymphatic, lymphoblastic, and polycythemia vera.

Studies in AML have characterized several genes with recurring mutations with both prognostic and biologic implications, especially within normal functioning karyotype and/or cytogenetic subset having intermediate risk involved of the host cells. The genes involved with leukemia via mutations include *CEBPA*, *FLT3*, *MLL*, *NPM*, and *NRAS*. Two major approaches based on cytarabine and anthracycline are currently being followed for AML treatment, but the associated outcomes still remain poor and unsatisfactory, especially for high-risk patients or patients of older age. Hence one promising treatment strategy would be study and development of novel agents having diverse mechanisms to target AML by targeted therapeutics including oligonucleotide constructs and certain kinase inhibitors along with histone deacetylase inhibitors, leading to arrest of cell growth and thus apoptosis via conformational changes and histone acetylation. Furthermore, new class of therapies are more concerned with targeting repair of DNA, replication of DNA, along with cell cycle and underlying signaling. While a few of this class of therapies are under early study and development phases, others have given away promising results at preclinical and clinical trials. For instance, immunohistochemical investigation of p53 protein expression in patients with leukemia has shown positive expression of said protein in 7 out of 24 AML cases, in 3 out of 15 CLL cases, while in 1 of 11 CML cases and in 4 out of 8 ALL cases. From the study, it was concluded that abnormalities in p53 protein expression may have a vital role in leukemogenesis as well as development of chronic leukemia's more malignant clones because in all, leukemia cases with p53 protein expression was 26%.

18.2.8 Lymphoma

Lymphomas are defined as clonal neoplasms arising from subsets of innate and adaptive immune cells such as natural killer (NK) cell, T cell, and B cell, respectively, at their different maturational stages. With almost 4% of all new reported malignancies, lymphomas are the fifth most common cancer type with highest mortality rate in Western countries with B-cell deprivation representing more than 80% of mortality cases. In addition, major pathogenetic mechanisms have been reported for B-cell lymphomas opposed to derived from other two cell types, and till date, the genome level characterization has been done for mostly a huge subclass of B- and T-cell lymphomas (TCLs). These studies unraveled into genetic mechanism's novel insights leading to immune escape along with various cellular mechanisms such as activation of B-cell receptor, epigenetic and spliceosome alterations, GTPase families, non-coding sequences, TCR signaling, certain regulators and their dependencies, unregulated proteolysis, and altered metabolism of tumor cell. Immunotherapies related to cancer have further highlighted that immune escape depends on genetic mechanisms (Table 18.5). For lymphomas, excellent platforms are provided by regulatory checkpoints leading to circumvention of antitumor response for exploitation of mechanisms involved in initiation, progression, and therapy resistance (Elenitoba-Johnson & Lim 2018).

Table 18.5 Immune evasion and contributing genetic mechanisms in lymphoma

S. no.	Mechanism	Genes involved	Lymphoma
1.	3′ Untranslated-region (UTR) disruption	*PD-L1*	Adult T-cell leukemia/lymphoma, diffuse large B-cell lymphoma
2.	Transcriptional regulation	*STAT3* regulation of *PD-L1*	Anaplastic lymphoma kinase-positive anaplastic large cell lymphoma
		MYC regulation of *PD-L1*, *CD47*	Burkitt lymphoma
		MEK/ERK regulation of *PD-L1*	Anaplastic lymphoma kinase-positive anaplastic large cell lymphoma
3.	Mutations	*B2M, CD58*	Hodgkin lymphoma, diffuse large B-cell lymphoma
4.	Structural rearrangements	*X/PD-L1*, *X/PDCD1LG1*	Primary mediastinal large B-cell lymphoma, Hodgkin lymphoma

18.2.9 Therapeutic Targets

Therapy development for lymphoma has employed the E3 ubiquitin ligase's substrates along with insights into the regulatory mechanism. For example, cereblon encoded by *CRBN* gene, one of the E3 ubiquitin ligase (CRL4) substrate receptors, is directly targeted by thalidomide and IMiDs which is a collective of all related immunomodulatory drugs. After binding, these drugs alter the specificity of cereblon to CRBN-CRL4 complex, leading to lymphoid transcription factor's degradation, thereby suggesting that drug collective, IMiDs, may prove effective in cases of multiple myeloma and primary effusion lymphomas which are types of B-cell tumors. Comparable applications of these drugs have further been proposed for both T-cell and B-cell Hodgkin lymphoma as well as non-Hodgkin lymphomas. In addition, studies have been done in a B-cell neoplasm, Waldenstrom macroglobulinemia, which is dependent for sustenance of BCR-driven cell growth on aberrant proteasome functioning, activation of MYD88, and BCR signaling. These studies used a combinatorial targeting approach in clinical trials wherein BTK/MYD88 node was targeted using ibrutinib and 20S proteasome was inhibited using bortezomib. Further extending the same approach, various inhibitors with small molecular weight were used for targeting ubiquitin carboxyl-terminal hydrolase isozyme L5 (UCHL5) and ubiquitin-specific protease 14 (USP14). Both being deubiquitinating enzymes, the inhibitors resulted in tumor-specific apoptosis of otherwise resistant lymphoma cells. Further, exposure to an inhibitor of both UCHL5 and USP14, known as VLX1570 residing in the cap of 19S proteasome, resulted in downregulation of proteins associated with BCR signaling such as BTK, CXCR4, MYD88, NFAT, and NF-κB suggesting its potential to inhibit components of the proteasome for clinical translation.

In spite of the fact that histological type classifies the cancer in a more efficient way, it is the cancer name based on primary sites that the general public is more familiar to. The common sites at which development of cancer takes place include the following.

18.2.10 Lungs

Lung cancer (Fig. 18.3), considered to be the world's most common cancer type, can be of various types, namely, lung carcinoid tumor, small cell lung cancer (SCLC), and non-small cell lung cancer (NSCL) with NSCL being common with prevalence rate of over 85%. SCLC averages about 10–15% of lung cancers and tends to quickly spread, whereas less than 5% of cancers are lung carcinoid cancer which rarely spread due to slow growth. Till recently, translocation of anaplastic lymphoma kinase and mutation in epidermal growth factor receptor are among numerous molecular events having been identified offering hopes to patients with metastatic lung cancer.

There have been many oncogenes related to lung cancer and include RAS, MYC, and HER-2/NEU dominant oncogenes which act by overtaking normal cell growth

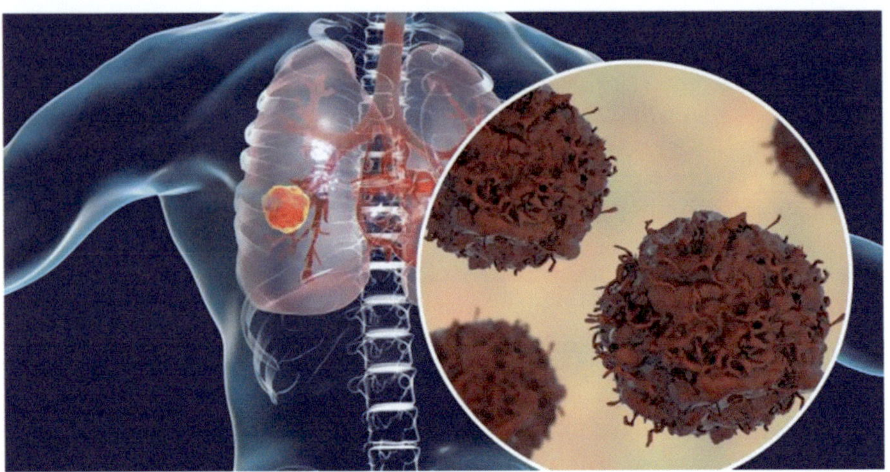

Fig. 18.3 Pictograph showing lung cancer. The enlarged view depicts the metastatic form of lung cancer

and functions and RB, p15, p16, and p53 among various tumor-suppressor genes (TSGs) that act in controlling further cellular growth. For cancer development and progression, molecular alterations in either proto-oncogene or TSG are involved. Lung cancers have been shown to exhibit multiple genetic lesions involving mutations leading to dominant cellular proto-oncogene activation or activation of those involved in inactivation of TSGs (Singh & Kathiresan 2014). These alterations prime to cellular capabilities often acquired and can be grouped further based on function in sets of six: self-sufficiency in growth signals, antiproliferative signal insensitivity, apoptosis evasion by anti-apoptotic molecule upregulation or pro-apoptotic molecule downregulation, boundless replicative potential as a result of telomerase activation, sustained angiogenesis, and metastasis.

18.2.11 Therapeutics

Cancer treatment generally includes procedures such as surgery, chemical therapy, and radiation therapy or therapy including drug targeting. Chemotherapy often uses a combination of drugs which can be given orally or intravenously to kill the cancer-affected cells and is generally a pursued treatment after the surgery for lung cancer. Application of chemotherapy along with cisplatin and etoposide is reported to have pronounced positive results in case localization of the tumor is within the irradiation field, while for a longer time, the standard treatment is mentioned to be chemotherapy along with etoposide and cisplatin. For advanced lung cancer treatment, current trend includes efficacy of several treatments often combining antineoplastic and platinum-based antineoplastic agents which have been recently developed. In addition, the marine flora has displayed a plethora of pharmacological properties including antioxidant, immunostimulatory, and antitumor activities and is often considered a rich source of medicinal polyphenols and sulfated polysaccharides. The mechanism of action of these marine florae includes prevention of oxidative damage to DNA by the phytochemicals released, induction of apoptosis, and activation of macrophages, thereby possibly controlling the process of carcinogenesis. Another example of marine flora exhibiting anticancer properties are the mangrove resources which are enriched in anticancer lead compounds. Associated species and around 45 mangroves have been reported to possess anti-cancerous properties, yet very scarce studies have been done on cancer types such as lung cancer.

18.2.12 Female Breasts

Breast cancers (Fig. 18.4) mostly begin either in the ducts carrying milk, also known as ductal cancer, or in the glands producing the breast milk, called lobular cancer.

Breast cancer may be detected even before it causes lump formation or develops the symptoms. The important part is that many of the lumps developing in the breasts are benign. Non-cancerous breast tumors are often abnormal growths but do not metastasize and hence are not life threatening.

Accounting for ~10% of all breast cancers, hereditary breast cancer is caused by deletion of a tumor suppressor gene rather than mutation leading to an oncogene gain. A plethora of mutations in a variety of genes are known for susceptibility to breast cancer, the most significant being BRCA2 and BRCA1 (Deng & Scott 2000; Osborne et al. 2004) genes typically responsible for about 80–90% of high-risk type,

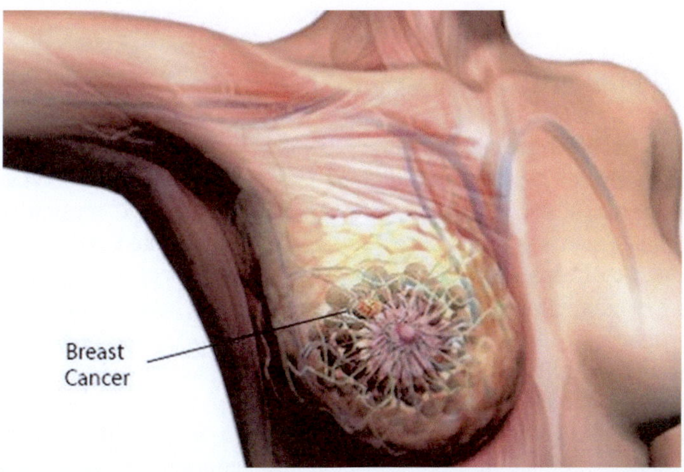

Fig. 18.4 Breast cancer. Breast cancer either begins in the ducts or the lobules. The figure shows breast anatomy and development of breast cancer. (Image ©WebMD, https://www.emedicinehealth.com/breast_cancer/article_em.htm)

genetically determined tumors. These genes have been reported to be sporadically seldom mutated, highlighting their role in pathogenetic promotion of carcinogenesis. Many of these genes are reported in breast cancer in case of males. For instance, men carrying germline mutations in either BRCA2 or BRCA1 gene are usually at a higher risk of carcinoma development than the rest of the general population not carrying the same mutations. Additional genes supposedly involved in the breast cancer genesis include TSGs, namely, p53 and PTEN. Studies have shown that mutation in the p53 coding region is the most common factor contributing to the inactivation of p53 signaling pathway in human cancers as it has a vital role as a transcription factor in regulating DNA repair and cell cycle. The level of protein is reported to be increased inside cells during cellular stress periods, and hence the cell cycle arrest gives time for DNA damage repair or for cellular death. In case of mutation, one or all of these mechanisms fail leading to an increase in abnormal protein production, thereby allowing cells with DNA damage to proceed with replication. Li-Fraumeni syndrome often associated with various tumors such as sarcomas, lymphomas, melanomas, carcinomas, leukemia, etc. is also caused by p53 germline mutations. Another tumor suppressor gene involved in cell division, proliferation, and death, located at chromosome 10q23, is phosphatase and tensin homologue (PTEN), and PTEN with germline mutations is associated with syndromes such as Cowden syndrome (includes aberrant protein expression and increased breast cancer frequency), rare Mendelian syndrome, certain high-risk cancer types such as thyroid and endometrial cancers, and benign manifestations. PTEN acts in tumor suppression which regulates chromosomal disruption, and its integrity leads to extensive breakage at centromere breakage as well as chromosomal translocations. Other pathways often associated with cancer including breast cancer are PI3K/AkT and Ras/Raf/MEK/ERK signal transduction pathways generally involved in regulation of apoptosis and cell cycle process in various cellular types.

18.2.13 Therapeutic Strategies

18.2.13.1 Chemotherapy

Cyclophosphamide. It is a derivative of nitrogen mustard, an alkylating agent, initially synthesized and used for improving the nitrogenated mustard selectivity. Over time it has been a cytotoxic agent, clinically implemented and proven effective on a wide range of tumors, including breast cancer. Most of the tumoral cells have enzymatic systems, phosphamidasis, and phosphatases, responsible for the activation of cyclophosphamide. The conversion of cyclophosphamide to 4-hydroxycyclophosphamide begins in the liver, a tautomerization process that yields aldophosphamide. One of the byproducts of this cleavage process is N, N-bis-2-(2 chloroethyl)phosphorodiamidate, which is a bifunctional alkylating agent, an active product of cyclophosphamide reported to act as an alkylating agent on DNA with N7 position of particularly susceptible guanine. The ability of this drug to disrupt mitosis and cell differentiation in rapidly proliferating cells takes the main focus. It has been used as an adjuvant therapy and in numerous combination therapies along with fluorouracil and methotrexate (MTX) for treating patients with high risk for relapse.

Methotrexate (MTX). Belonging to antimetabolites and folic acid analog class of drugs, MTX is known to prevent cell division as it gets embedded in prerequisite material for nuclear neosynthesis, or because it combines with the life necessary enzymes in irreversible manner, hence preventing normal cellular division. Treatment with this drug results in blocking the synthesis of thymidine 5′-monophosphate (TMP) by prevention of N5,N10-methylenetetrahydrofolate synthesis required for DNA synthesis.

5-Fluorouracil (5-FU). This drug also belongs to antimetabolite class and is known to successfully prevent biosynthesis of nucleotide pyrimidines. 5-FU is considered inactive in normal as well as tumor cells and acquires its cytotoxic activity only after the cell's bioregulation is disrupted. The mechanism of action of 5-FU is similar to that of MTX and prevents DNA synthesis leading to arrest of cell cycle.

Anthracyclines. Anthracyclines, belonging to cytotoxic antibiotics, are anticancer agents whose antineoplastic action is mainly due to interaction with genetic material eventually triggering cell death. Although very effective, these antibiotics are also known to be toxic as they act as intercalating agents and often fail to discriminate between malignant and healthy cells. The mechanism of actions leading to tumor cell death are (a) p53-independent and/or p53-dependent DNA damage, (b) DNA topoisomerase II inhibition, (c) apoptosis induction mediated through cytochrome c, (d) proteasome interactions, and (e) free radicals' generation which results in oxidative damage. Doxorubicin (Adriamycin) and epirubicin are the major anthracyclines used in the treatment of breast cancer.

Taxanes. Taxanes consist a group of drugs which include docetaxel and paclitaxel, known by their trade names Taxotere and Taxol, respectively. These drugs are known to treat various types of malignancies, including lymphomas and leukemias, and many solid tumor types, such as breast, brain, lung, neck, prostate, and ovarian cancer, and are progressively being used at early stages of the disease. They function by disrupting microtubules' structure which play vital role in several important cellular functions. During the growth of normal cell, microtubules are responsible for division of cell, and once a cell stops dividing, these microtubules disintegrate. Taxanes, however, stop the disintegration of these microtubules, and as a result, cancer cells cannot divide and grow because these are clogged with the intact microtubules. Epothilones are one such example which possess anti-tubulin activity hence inducing tubulin polymerization and microtubule stability causing the cell cycle arrest at the G2/M transition (Ligresti et al. 2008).

18.2.13.2 Hormone Therapy

Aromatase inhibitors. The hormone estrogen is well known to promote the growth in many breast cancers. After menopause, the main source of estrogen acquirement is the conversion of androgens from adrenal cortex into estrogen via an enzyme called aromatase cytochrome P450 (CYP19) mainly found in the adipose-filled tissues in the body. Aromatase inhibitors work via prevention of cell proliferation in breast cancer by blocking CYP19 and thereby estrogen synthesis. Two major classes of these inhibitors exist which differ in chemical composition and mechanism of function, namely, Type 2 and Type 1 inhibitors. Type 1 inhibitors are steroids like exemestane which bind to the aromatase enzyme in irreversible manner, whereas in nature, Type 2 inhibitors are non-steroidal such as letrozole and anastrozole and bind to enzyme in a reversible manner. These inhibitors, especially anastrozole, provide an alternative to tamoxifen (TAM) due to their ability to significantly increase thymidine 5′-triphosphate (TTP) and lower rate of incidences such as vaginal bleeding and thromboembolic events.

Tamoxifen. Hormonal intervention has been implemented for breast cancer treatment as early as the 1800s, when Beatson treated it by performing an ovariectomy. In years that follow, the underlying mechanisms to this treatment were demonstrated to be related to hormonal responsiveness of these cancers and their dependence on estrogen receptor (ER) signaling. The antiestrogen with non-steroidal, TAM, is now being used for breast cancer treatment in postmenopausal women given its attribute to both the cell growth arrest and cell death induction. In spite of the belief that the primary mechanism of action of TAM is related to the ER signaling inhibition, studies in the recent past have indicated the presence of non-ER-mediated mechanisms which seem to include signaling proteins, such as c-myc, transforming growth factor-ß (TGF-ß), calmodulin, and protein kinase C (PKC). Research has also showed vital roles played by caspases and MAPK, including p38 and c-Jun N-terminal kinase, in its apoptotic signaling. TAM has significantly been able to decrease the mortality rate in case of breast cancer having ER-positive cells in both postmenopausal and premenopausal women.

18.2.13.3 Therapy Based on Antibody

The term "targeted therapy" came into existence due to a newly discovered generation of cancer drugs which were specifically designed to interfere with targeted molecules and were alleged to play a crucial role in either tumor progression or growth. Antibodies are considered as another type of targeted therapy by many scientists. They are often used when conjugated with cytotoxic cellular poisons, drugs, or radioisotopes, killing the malignant cells with target antigen. The role and inherited nature of genetic polymorphisms in prolonging the effect of drugs and their disposition is being rapidly elucidated by pharmacogenomic studies. These studies thus greatly enhance the process of drug discovery, thereby providing a stronger scientific basis for optimizing drug therapy on the genetic constitution basis respective of the patient.

Trastuzumab. Being the pioneer approach concerning neoplastic cell growth, trastuzumab (Herceptin®, Genentech Inc., San Francisco, CA) is a recombinant human monoclonal antibody which targets the human epidermal growth factor receptor 2 (HER2 or ErbB-2) protein extracellular ligand binding domain, generally expressed in all epithelial cell types and belongs to the human epidermal growth factor receptor (HER) family. The receptor gene seems to be overexpressed in 30% of breast cancers, compromising an unfavorable prognosis and the tumor's aggressive behavior. The most significant predictor of relapse time and overall survival in breast cancer patients is HER-2/neu gene amplification which promises to distinguish between patients who will not show any response and those who may have registered a significant response to the antibody. Nonetheless, this high expression of HER2 receptor protein is suitable as antibody therapy target using trastuzumab. Even though most patients usually tolerate trastuzumab, however, its adverse effects include chills and/or acute fever and are responsible for the cardiac dysfunction (CD) among patients having anthracycline prior exposure as well as death and anaphylaxis in patients receiving anthracycline and cyclophosphamide treatment combined with trastuzumab, and therefore treatment of patients with trastuzumab previously are at a higher risk of cardiomyopathy development, suggestive of the cardiomyopathy associated with anthracyclines. Till today, the molecular mechanisms associated with trastuzumab-associated cardiomyopathy are vague, and the risk can be substantially lowered by using trastuzumab with concurrent paclitaxel and trastuzumab or as a single agent.

18.2.13.4 Colon and Rectum

Colorectal cancer as the name suggests starts as a growth on the inner lining of either the colon or the rectum. These growths are termed polyps. Only some polyps can turn cancerous over time depending upon the type. There are mainly two types of polyp:

1. Adenomatous polyps (adenomas): Adenomas are often called a precancerous condition as there are chances that these polyps may cause cancer.
2. Hyperplastic polyps and inflammatory polyps: Being very common, this type is not precancerous.

If development of polyp takes place, it can further penetrate and grow into the colon or rectum wall eventually. Colorectal cancer usually develops in the innermost layer (the mucosa) and further grows outward via the other layers.

Once inside the colorectal wall, cancer cells can metastasize into blood or lymph vessels and may grow into nearby lymph nodes or spread further to distant parts of the body. The stage of the cancer depends on the extent of how much it has metastasized outside and growth into the wall of the colon or rectum. Owning majority of colorectal cancers (96%), adenocarcinomas tend to produce the mucus for lubrication of inside of colon and rectal wall.

18.3 Cancer and Apoptosis

Lying under the idiopathy and complexity of each cancer type is a restricted set of "mission-critical" events which propel the tumor and the associated progeny cells into unrestrained division and invasion. One of the key events is known to be unregulated cell proliferation, which, along with compensatory suppression of apoptosis required for support, gives a minimal "stage" inevitable for the support of neoplastic advancement. Inside our body, many surplus and unwanted cells are usually disposed of via a mechanism that is already programmed into them. Programmed cell death, which is also known as apoptosis (Greek roots meaning falling off), is vital among animals as its absence might impair the proper functioning of organs. Apoptosis is also equally important for prevention of tumor formation because if a cancer cell is killed, it cannot multiply to form a dangerous mass of cells. Hence, apoptosis keeps a proper check against the mutated cells which would otherwise divide uncontrollably leading to the formation of HULK! A detailed analysis of apoptosis mechanism is important in a way that it is essential in many of the disease pathologies. In some cases, too much apoptosis is associated with some major problems, such as the case of degenerative diseases, whereas in other cases, the culprit is too little apoptosis, cancer being the perfect model. Kerr et al. described apoptosis in the late 1900s for its morphological characteristics, including chromatin condensation, membrane blebbing, cell shrinkage, and nuclear fragmentation. Our understanding of development biology has had insightful implications through the discovery of the fact that apoptosis is a gene-directed program, which implies that factors distressing cell proliferation and survival and differentiation of cells can regulate cell numbers.

Moreover, any mutations in the genes controlling apoptosis lead to the disruption in the process. This process is the most investigated one in biological research. During apoptosis, a family of proteases called caspases play an inevitable role. These proteases cleave peptide bonds, thereby inactivating the target proteins. The target consists of a plethora of proteins including components of cytoskeleton, and lamins, which are the building blocks of the inner lining of the nucleus. The proteolytic cleavage results in chromatin fragmentation, formation of cytoplasmic blebs at cell surfaces, and shrinkage of cells themselves. Cells with these integrated defects are usually engulfed by host phagocytes, one of the key players of the innate immunity, and are processed for destruction. If the machinery of programmed cell death is impaired or inactivated, a defective cell would have the potential to survive and divide uncontrollably and hence become cancerous.

18.3.1 Morphological Changes in Apoptosis

Across all the cell types and species, these changes remain quite the same and alarm the cytoplasm as well as nucleus. For nucleus, the hallmarks of death by apoptosis are nuclear fragmentation and chromatin condensation often with cell rounding up; reduced cellular volume, i.e., pyknosis; and pseudopod retraction. During the condensation of genetic material, i.e., chromatin, periphery of the nuclear membranes starts to form ring-like or crescent structure. Further, the chromatin condenses till it breaks up inside the cell with intact plasma membrane throughout the process. The morphological features at the later stages of apoptosis include membrane blebbing, a loss of membrane integrity, and ultrastructural cytoplasmic organelle modifications. Under normal circumstances, the phagocytic cells engulf cells undergoing apoptosis (Elmore 2007).

18.3.2 Biochemical Changes in Apoptosis

The main biological changes observed during apoptosis: (a) caspase activation, (b) DNA and protein breakdown, and (c) phagocytic cells recognize the cell membrane changes. The expression of phosphatidylserine (PS) during early apoptosis is **flipped out** to the outer cell membrane from the inner membrane, which eventually allows macrophages for early recognition of the dead cell, followed by phagocytosis without pro-inflammatory cellular component release. For the detection of apoptosis, a recombinant phosphatidylserine-binding protein, Annexin V, has been developed, which has higher affinity for the phosphatidylserine amino acid residues, thus used as recognition signal for apoptosis. Another protein, Calreticulin, has the ability to bind with the LDL receptor-related proteins, thereby indicating the engulfing cells. In the microvascular endothelial cells, expression of an adhesive glycoprotein, thrombospondin-1 and CD36, has been observed. Then DNA breaks into 50–300 kilo bp pieces and eventual inter-nucleosomal cleavage by endonucleases into oligonucleosomes into 180–200 bp fragments. An essential step of apoptosis includes caspase (c-caspase, caspase-aspartic acid, a group of cysteine proteases) activation which has signature cleaving site after aspartic acid residues. Once caspases are activated, they break down the cytoskeleton and nuclear scaffold, cleave vital cellular proteins, and activate DNAse, hence degrading nuclear DNA.

18.3.3 Mechanism of Apoptosis

Two initiation pathways, (a) mitochondrial (intrinsic) and (b) death receptor (extrinsic) pathways (Fig. 18.5), exist eventually leading to execution phase of apoptosis known as common pathway. A third pathway is the perforin/granzyme pathway; however a fourth lesser known intrinsic endoplamic reticulum pathway is also present for apoptosis. The extrinsic death receptor pathway starts when an appropriate ligand, Fas (fatty acid synthetase) ligand (FasL) and TNF-α (tumor necrosis factor alpha), is bound to the receptor, Fas (CD95) and type 1 TNF receptor (TNFR1), respectively. The TNFR has an extracellular domain rich in cysteine amino acids and a cytoplasmic domain consisting of 80 amino acid residues (known as death domain). These receptors are well known for having cytosolic death domain that plays a critical role in recruitment of proteins such as Fas-associated death domain (FADD), pro-caspase-8, and TNF receptor-associated death domain (TRADD). The death effector domain (DED) at the amino-terminal of pro-caspase-1 is seen to be involved in the interaction with DED of FADD.

The complex so formed by ligand-receptor-adaptor known as death-inducing signaling complex (DISC) leads to autocatalysis of pro-caspase-8 and hence activation of caspase-8. In addition to the FasL/FasR and TNF-α/TNFR ligand/receptor combination, other lesser known ligand/receptor combinations have also been discovered such as Apo2L/DR4, Apo2L/DR5, and Apo3L/DR3 (Apo3 ligand/death receptor 3).

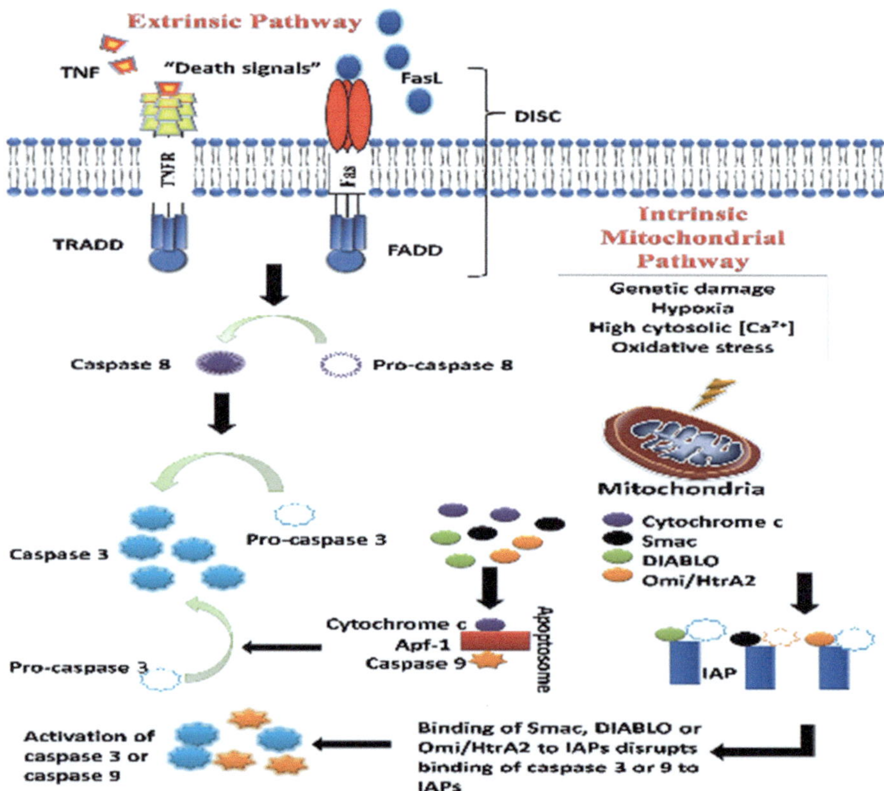

Fig. 18.5 Schematic representation of the apoptotic intrinsic and extrinsic pathways

Several internal stimuli such as high cytosolic Ca^{2+} concentration, permanent genetic damage, and severe stress by oxidative species are known to trigger the intrinsic mitochondrial pathway. There are basically two types of stimuli, negative and positive, that control the intrinsic mitochondrial pathway. Absence of cytokines, hormones, and growth factors leads to apoptosis suppression, which acts as negative stimuli, thereby leading to apoptosis. The positive stimuli include hypoxia, radiation, viral infection, hyperthermia, and reactive oxygen species which directly trigger apoptosis events. Irrespective of the stimuli, increase in mitochondrial permeability and pro-apoptotic molecule release in the cytoplasm are the ultimate result of this pathway, which is under Bcl-2 family protein regulation. There are two main groups of the B-cell lymphoma protein 2 (Bcl-2) proteins: the first one involves pro-apoptotic proteins (Bad, Bcl-Xs, Bak, Bax, Bid, Bim, and Hrk), and the second one consists of anti-apoptotic proteins (Bcl-2, Bcl-W, Bcl-XL, Bfl-1, and Mcl-1). Interaction of caspase recruitment domain (CARD) with Apaf-1 (apoptotic protease activating factor) is essential for the formation of apoptosome from pro-caspase-9, Apaf-1, and cytochrome and leads to the caspase-3 activation. On the other hand, DIABLO (direct IAP binding protein with low PI), Omi/HtrA2 (high-temperature requirement), or Smac (second mitochondrial activator of caspases) disrupts the interaction of caspase-3 or caspase-9 with the IAPs by binding to inhibitor of apoptosis proteins (IAPs) and results into caspase activation. When the cell has committed for apoptosis, AIF (apoptosis induce factor), CAD (caspase-activated DNase), and endonuclease G released from the mitochondria at a later stage promote the apoptotic events. All these pro-apoptotic signal molecules, CAD, endonuclease G, and AIF translocate to the nucleus and result in the DNA fragmentation and peripheral nuclear chromatin condensation.

The perforin/granzyme pathway involved the release of granzyme enzyme (Granzyme A and Granzyme B) and perforin from the cells, which are specifically linked to the cytotoxic T-lymphocytes (CTL) and natural killer (NK) cells. The release of perforin from CTL and NK cells results in formation of pores in the membrane of targeted cells, through which influx of cytoplasmic granules occurs. Granzyme A and Granzyme B are serine proteases and have the ability to cleave at the aspartate residues. By interaction with the mannose-6-phosphate receptors (IGFR2), Granzyme B enters the target cells and also through the pores formed by the perforin enzyme. By cleaving the ICAD (inhibitor of caspase-activated DNase), Granzyme B has potential to activate the pro-caspase-10. Granzyme B cleaves the anti-apoptotic signal and leads to mitochondrial cytochrome c release, subsequently inducing apoptosis. In addition, Granzyme B possesses the capacity of direct activation of caspase-3 and leads to the execution pathway of apoptosis. Since Granzyme B activates caspase-dependent pathway, Granzyme A leads to apoptosis in caspase-independent pathway. Through the activation of DNase NM23-H1, a product of tumor suppressor gene, Granzyme A, leads to DNA nicking. This DNase has the ability to prevent cancer in the CTL, thereby essential in the immune surveillance. The HM23-H1 gene is under the control of a nucleosome assembly protein, SET. The SET protein complex comprises of SET, Ape1, pp32, and HMG2 proteins which protects the structure of DNA and chromatin. Granzyme A abolishes the functions of SET complex by cleaving the SET proteins and induces apoptosis in the target cell. However, a nuclear pathway has also been proposed which depends on the Pml oncogenic domains (PODs) or nuclear bodies. It has been reported that mice with defective pml gene develop resistance against apoptosis through unknown mechanisms. The PODS includes Daax, Zip kinase, and Par4 proteins, which has a role in apoptosis induction, and defects in these have been reported in cancers.

Caspase-3 acts as a converging point for the intrinsic and extrinsic pathways and is responsible for cleaving the inhibitor of the caspase-activated deoxyribonuclease (ICAD), thereby activating CAD, known to have indispensable role in nuclear apoptosis. In addition, downstream caspases (caspase-6 and caspase-7) cleave cytoskeletal proteins, protein kinases, inhibitory subunits of endonuclease family, and DNA repair proteins. These processes result in the abovementioned morphological changes involved in apoptosis. Caspase-3 cleaves the gelsolin protein, which is involved in the polymerization of actin protein and serves as the binding site for phosphatidylinositol phosphate, resulting in the disruption of the cytoskeleton and intracellular signaling. Due to evasion of cell death, cancer is often seen as an outcome of a series of genetic changes which involve transformation of normal cell into malignant one.

Generally reduced caspase function or impaired death receptor signaling or disruption of pro- and anti-apoptotic protein balance due to an overexpression or an underexpression or both leads to cell death evasion. The caspases are placed in two groups: (1) those related to caspase-1 (e.g., caspase-1, caspase-4, caspase-5, caspase-13, and caspase-14) and are involved in inflammation and cytokine processing and (2) those involved in apoptosis (e.g., caspase-2, caspase-3, caspase-6, caspase-7, caspase-8, caspase-9, and caspase-10). The other group has been divided into (a) effector caspases (caspase-3, caspase-6, and caspase-7) essential in cleavage of cellular components and (b) initiator caspases (e.g., caspase-2, caspase-8, caspase-9, and caspase-10) involved in initiation of apoptotic pathway. Thus, decrease in apoptosis and carcinogenesis is the result of low concentration or impairment in caspases. However, few reports suggest that caspase disruption is only responsible for cell death delay. On the other hand, protease loss leads to pathological increases in cell numbers. The cellular FADD-like interleukin-1 beta-converting enzyme inhibitory protein, i.e., c-FLIP, acts as an inhibitor of extrinsic death receptor pathway by binding to caspase-8 and FADD. Another inhibitor of the death receptor mediated apoptosis is Toso which regulates T-cell apoptosis by inhibiting the activation of caspase-8. Endometrial, colon and prostate cancer, non-small cell lung carcinoma (NSCLC), hepatocellular carcinoma, and melanoma reported higher expression of c-FLIP protein, whereas in some cancers (ovarian, prostate, endometrial, and colon cancer) increased expression of c-FLIP reported poor prognosis and increased progression of cancer. To date, only few reports are available which show the caspase mutation involvement in cancers. Noonan's syndrome is a result of disruption in Apaf1, autoimmune lymphoproliferative syndrome type II is caused by mutations in caspase-10, and mutations in caspase-5 including frameshift mutations are reported to occur in endometrial tumors, gastrointestinal tumors, and hereditary non-polyposis colorectal cancers. Abnormalities in death receptor and ligand signaling pathway lead to dodging of apoptosis extrinsic pathway as they play a crucial role in the same. Regardless of the type or mechanism of defects, these abnormalities may include receptor function impairment or downregulation of receptor expression and reduction of the levels of death signals eventually contributing to reduction of apoptosis. CD95 reduced expression is associated with neuroblastoma cells or treatment-resistant leukemia. Reesink-Peters et al. (2004) reported that the factors responsible for carcinogenesis in cervical cancer include loss of Fas and dysregulation of DR5, DR4, FasL, and TRAIL (tumor necrosis factor-related apoptosis-inducing ligand) in the cervical cancer sequence of cervical intraepithelial neoplasia (CIN). TRAIL (also known as Apo2L) is essential for the tumor suppression through the NK-mediated apoptosis (Delbridge et al. 2012), and FasL is critical for killing the cancerous cells through the CTL. However, with progression of tumor cells, these cells develop resistance against the FasL and TRAIL either through expression of nonfunctional or downregulation of Fas receptor, thereby escaping the immune attack and preventing resistance against apoptosis. The c-FLIP and RIP are crucial in the activation of

NF-κB and ERK pathway in the TRAIL-resistant NSCLC. The constitutive expression of NF-κB and upregulation of c-FLIP by the NF-κB lead to various human cancers. The apoptosis-inducing proteins (IAPs) are essential for caspase activation. IAPs consist of at least one copy of BIR (baculovirus IAP repeat) and one to three copies of zinc-bonding fold which are essential for the anti-apoptotic activity of IAPs. The IAP family includes cIAP1, cIAP2, and XIAP (X-linked mammalian inhibitor of apoptosis protein), member that binds via BIR domain and results in suppression of the caspase-3, caspase-7, and caspase-9 activity, thereby evasion of cell from the apoptosis. Survivin and livin (ML-IAP) are other members of IAP family involved in the inhibition of caspase-9 only. In several cancers, overexpression of IAPs has been documented, and IAPs are also associated with the resistance of cancer against chemotherapy and maintains the survival and growth of cancer cells. These IAP molecules suggest a potential target to fight against cancer. The development of synthetic peptides of IAP antagonists which mimics the natural IAP antagonists (Smac and DIABLO) present in the mitochondria may help to curb the cancer development. The upregulation of anti-apoptotic proteins such as Bcl-2 and downregulation of pro-apoptotic proteins such as Bax are observed in several cancers that leads to inhibition of intrinsic apoptotic pathway. The tumor suppressor protein, p53, regulates Bcl-2 and Bax expression, and p53 mutation accounts for 50% of human cancers. Upon sensing the DNA damage, ataxia telangiectasia-mutated gene (ATM) activates the p53 pathway of apoptosis, and ATM gene mutation is reported in the several cancers. In addition, several signaling pathways lead to tumor development. For the development of cancer, phosphatidylinositol 3-kinase/AKT pathway activation without the requirement of ligand (cell survival signals) is responsible. Several researches are ongoing to develop the specific molecular target therapy against the cancer. For the treatment of cancer, the pro-apoptotic and anti-apoptotic proteins, p53 protein, caspases, and several signaling components present potential molecular targets. These molecular target therapies are less toxic with less side effects compared to the chemotherapy.

18.3.4 Apoptosis During Cancer

The question now appears as to what triggers apoptosis during tumor development. Extracellular triggers include loss of cell-matrix interactions, radiation, hypoxia, and depletion of growth/survival factor. Disruption in proliferative signals produced by oncogenic mutations, malfunction of telomeres, and DNA damage are among the various factors misbalancing the internal system and triggering apoptosis. Although very unlikely, in some cases, apoptotic "trigger" leads to the alleviation of an anti-apoptotic signal. For instance, IGF-1 promotes cell survival by activating the PI-3 pathway; however, "death by default" can be triggered by survival factors such as IGF-1. In contrast, p53 activation under stress and stimuli forms pro-apoptotic factors promotes apoptosis through the involvement of pro-apoptotic molecules such as Bax.

Apoptotic trigger identification may provide valuable insights into the tumor evolution. Excessive exposure of skin to UV radiation leads to apoptosis induction,

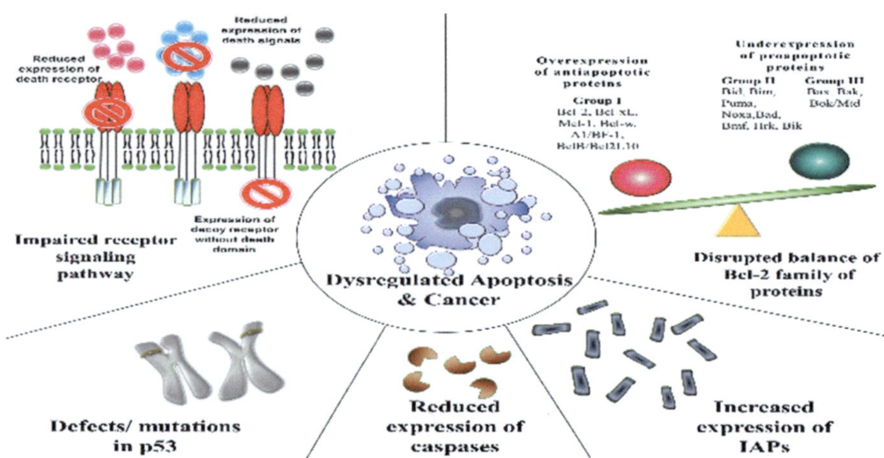

Fig. 18.6 Evasion of apoptosis and carcinogenesis: the contributing mechanism

and p53 function loss leads to damaged cell survival, thereby initiating tumor development (Fig. 18.6). The developing tumors encounter hypoxia as they outgrow the blood supply which then activates p53 eventually promoting apoptosis. Cells which have underlying apoptotic defects can survive the hypoxic stress, which leads to clonal expansion. For telomere malfunction-induced apoptosis, not only hypoxia but p53 is also required. Thus, cells with p53 mutation survive and are unstable genomically where loss of p53 and telomerase stimulates the development of tumor. Every target of cancer treatment arose from every abnormality or defect along the apoptotic pathways like a double-edged sword. Treatment strategies and drugs represent the potential approach in eliminating cancer cells and have the ability for restoring normal apoptotic signaling pathways because cancer cells in a way depend on these defects for thriving (Hassan et al. 2014; Lowe & Lin 2000; Wong 2011). Potential classes of anticancer drugs opened by recent advancement and important discoveries are summarized in Table 18.6.

18.4 Telomere Shortening, Telomerase, and Cancer

Present toward the end of chromosomes and considered as protective structures, telomeres contain repetitive nucleotide sequences (TTAGGG in case of vertebrates) along with linked proteins, commonly termed as shelterin consisting of six proteins: TRF1, TRF2 (telomere double-stranded DNA-binding proteins), RAP1, TIN2, TPP1, and single-stranded DNA-binding protein (POT1) (Table 18.7).

By preventing nucleolytic degradation, irregular recombination, and chromosomal end-to-end fusion, telomeres bring about genomic stability and chromosome integrity. At a rate of 50–200 bp per cell replication, the telomeric DNA shortens and has an average length of 10–15 kb in humans (Okamoto & Seimiya 2019).

Table 18.6 List of potential new anticancer drugs

S. no.	Strategies of treatment	Remarks
1.	Target the proteins of Bcl-2 family	
	(a) Bcl-2 family proteins targeting agents i. Oblimersen sodium ii. Bcl-2 family protein inhibitors iii. BH3 mimetics (b) Silencing the Bcl family anti-apoptotic proteins/genes	In combined action with the conventional anticancer drugs, chemosensitizing effects have been shown in chronic myeloid leukemia patients and improved their survival Sodium butyrate, depsipeptide, flavopiridol, and fenretinide affect gene and protein expression. Molecules like HA14–1, gossypol, ABT-737, GX15–070, and ABT-263 act on the proteins ABT-737 inhibits anti-apoptotic proteins and exhibits cytotoxicity in SCLC cell line, lymphoma, and primary patient-derived cells ATF4, NOXA, and ATF3 bind and inhibit Mcl-1 Bcl-2 specific siRNA inhibits the target gene expression with antiproliferative effect and promotes pro-apoptotic effect in pancreatic carcinoma cells under both in vivo and in vitro conditions Bmi-1 silencing in MCF breast cancer cells downregulates Bcl-2 and pAkt expression and increases apoptosis in vivo and in vitro along with increased sensitivity of these cells to doxorubicin
2.	Targeting the p53	
	(a) Gene therapy based on p53 (b) Drug therapy based on p53 i. Small molecules ii. Other agents: Nutilins, MI-219, and Tenovins (c). Immunotherapy based on p53	Retroviral vectors containing normal p53 gene were injected into tumor cells derived from NSCLC patients, and this therapy is feasible Wild-type p53 introduction sensitize the colorectal and prostate and neck and head cancers and glioma to the ionizing radiation ONYX-015, a genetically engineered oncolytic adenovirus, selectively replicates in and lyses tumor cells, which are p53 deficient Phikan083 binds and restores mutant p53 CP-31398 destabilizes the DNA-p53 core domain complex and intercalates with DNA, thereby restoring unstable p53 mutant Nutilins inhibit MSM2-p53 interaction and stabilize p53, resulting in induction of cancer cell senescence selectively MI-219 disrupts the MDM2-p53 interaction, thereby inducing selective apoptosis in tumor cells and inhibits tumor growth and proliferation Tenovins under in vivo condition decrease tumor growth A recombinant replication-defective adenoviral vector containing vaccine with normal p53 gene given to advanced-stage cancer patients and reported stable disease

(continued)

Table 18.6 (continued)

S. no.	Strategies of treatment	Remarks
3.	Targeting IAPs	
	(a) Targeting XIAP i. Antisense approach ii. siRNA approach (b) Targeting Survivin i. Antisense approach ii. siRNA approach (c) Other IAP antagonists i. Small molecule antagonists	In vivo improvement in tumor control by the radiotherapy Antisense oligonucleotide use with chemotherapy in vivo and in vitro exhibits enhanced activity of chemotherapeutic in lung cancer cells Using this method increased the sensitivity of human cancer cells toward radiation The sensitized hepatoma cells lead to death receptor and chemotherapeutic agent-induced cell death by siRNAs Spontaneous apoptosis is induced in LOX and YUSAC-2 malignant melanoma cells when transfected with anti-sense Survivin It sensitizes squamous cells of the neck and head to chemotherapy and induces apoptosis Inhibits medullary thyroid carcinoma cell proliferation and growth Diminishes pancreatic cancer cell's radioresistance through downregulation of Survivin Induces apoptosis in human lung adenocarcinoma cells (SH77 and SPCA1) and inhibits proliferation Enhances apoptosis, inhibits cell proliferation, and suppresses expression in ovarian cancer cells (SKOV3/DDP) Enhances the sensitivity of NSCLC cells Gene therapy and CDK and Hsp90 inhibitors are attempted in targeting survivin Cyclopeptidic Smac mimetics 2 and 3 bind to XIAP and restore the caspase-9/caspase-3/caspase-7 activity, which is inhibited by XIAP SM-164 enhances activity of TRAIL by targeting XIAP and IAP1
4.	Targeting the caspases	
	(a) Drug therapy based on caspases (b) Gene therapy based on caspases	Apoptin induces apoptosis in malignant cells selectively Small molecule caspase activators lower the caspase threshold activation and increase the cancer cells sensitivity toward drugs In addition with etoposide, human caspase-3 gene therapy induces apoptosis and reduces volume of tumor in AH130 liver tumor model Constitutively active caspase-3 gene transfer into HuH7 human hepatoma cells results in selective apoptosis A recombinant adenovirus with immunocaspase-3 promotes in vivo and in vitro anticancer effect in hepatocellular carcinoma

(continued)

Table 18.6 (continued)

S. no.	Strategies of treatment	Remarks
5.	Targeting signal transduction pathways	
	(a) Targeting NF-κB (b) Targeting PI-3/Akt pathways (c) Targeting Ras-GAP	In response to TNF and chemotherapeutic agents, NF-κB signaling is inhibited, resulting in inhibition of cell growth I-κB or proteasome inhibitor inhibits NF-κB activity and induces tumor cell death in association with radiation and chemotherapy Small molecule inhibition of enzymes involved in PI-3/Akt pathways leads to tumor cell apoptosis Farnesyltransferase is responsible for active Ras. Farnesyltransferase inhibitor, as an antitumor agent, leads to massive apoptosis and mammary carcinoma regression in *ras* transgenic mice

Table 18.7 List of shelterin proteins and their function

Function	Shelterin proteins involved
End protection from ATM- and ATR-dependent DNA damage responses End protection from DNA repair pathways: non-homologous end joining and homologous recombination	TRF2, POT1
Higher-order telomere loop structure called t-loop	TRF2
DNA bending activity, telomere replication at the S phase of the cell cycle	TRF1
Telomere localization	TPP1
Negative regulation of telomerase	TIN2

NB: The t-loop is formed with the help of TRF1 DNA bending ability which prevents the telomerase and machinery involved in response to DNA damage to be recognized by the telomere ends resulting in a single-stranded G-overhang (G-tail, 3′-overhang) invasion at the ends to double-stranded telomeric DNA

Inside normal somatic cells of human, the semi-conservative mode of replication leads to the gradual shortening of telomeric ends, and cells with awfully short telomeres undergo apoptosis. However, there are cells with self-renewal powers, which surprisingly maintain telomeres to overpower the programmed cell death caused by short telomeres and become immortal such as embryonic stem cells and almost all cancer cells. Generally, a critically short length of telomere is the trigger for a cell entering subsequent replicative senescence. The ultimate fate of the triggered cell is cell death; alternatively, on the other hand, if death doesn't take place, cells continue to divide, often resulting in genomic instability and chromosomal abnormality.

The literature available up to now suggests that telomere maintenance occurs through two important pathways: activation of telomerase-independent mechanism employing homologous recombination DNA repair pathway also known as

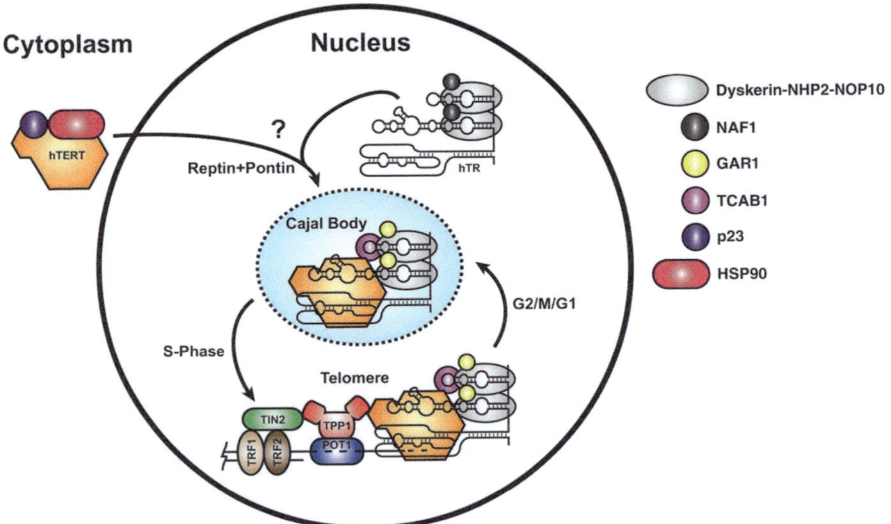

Fig. 18.7 Telomerase assembly, maturation, and recruitment to telomere. The synthesis of hTERT takes place in the cytoplasm. The hTR and hTERT assembly into functionally active telomerase ("?" subcellular location still unknown) is assisted by reptin and pontin (AAA+ ATPases). The recruitment of telomerase to telomeres takes place via interaction of TPP1 with TEN domain of hTERT in the S phase of cell cycle

alternative lengthening of telomeres (ALT) and telomerase transcriptional activation, telomerase reverse transcriptase (TERT). Out of all cancer cells, 85–95% express telomerase; the rest undergo ALT pathway activation. Telomerase in general consists of two main components, TERT and telomerase RNA (TERC/TR), along with some proteins such as pontin/reptin, NOP10, DKC1, TCAB1, and NHP2.

The recruitment of telomerase to telomeric single-stranded DNA takes place via interaction with TPP1, a telomerase-localizing protein. Using TERC as a template, telomeric sequences are synthesized by TERT. The mechanism taking place is shown in Fig. 18.7. Studies have reported that TERC is constitutively expressed in human somatic cells under normal condition, whereas TERT expression is silenced epigenetically. Interestingly, most cancer cells re-express the limiting factor, TERT, and hence acquire telomerase activity.

Many groups of scientists have studied the mechanism regulating TERT transcription over the past years. In 1999, in order to isolate the TERT gene 5′ promoter region, three independent groups were successful. The core promoter region essential for the transcription of the gene was demonstrated to fall upstream of the transcription start site by 260 bp. Further, the E-box having the signature sequence -CACGTG- is reported to bind with C-MYC, SP1, and other transcription factors at −165 bp and + 44 bp along with five GC boxes with conserved sequence -GGGCGG-, respectively, resulting in the induction of TERT mRNA expression. The promoter region also possesses binding sites for other transcription factors such as AP-1, E2F, and ERE (estrogen response element) which are essential for TERT mRNA expression. Another important factor which creates loops in the chromatin and functions as insulator across the genome is CTCF (CCCTC binding factor). It is also involved in positive and negative regulation of gene expression by either promoting the promoter-enhancer association or blocking it in a manner dependent on the position, respectively. In addition, phosphatidylinositol-3 kinase (PI3K)/AKT kinase pathway enhances TERT activity via phosphorylation of TERT at the post-translational level. Hence, the regulation of expression of TERT takes place at multiple levels through several factors. TRF1 promotes telomere replication at the S phase and in contrast negatively regulates telomerase through recruitment of TIN2. Telomerase is actively also regulated by TPP1-POT1 wherein TPP1 interacts with telomerase and promotes telomerase processivity, whereas POT1 limits the access of G-overhangs to telomerase by binding to single-stranded DNA. In addition, for TPP1-TERT interaction, phosphorylation of TPP1 is required, which is dependent on cell cycle.

18.4.1 TERT Expression: Epigenetic Regulation

Studies have reported that TERT expression levels correlate with DNA methylation levels at promoter region of the gene, which is in contrast to the role of methylation in gene silencing. The levels of methylation in TERT are reformed between explicit positions in the promoter region.

For instance, the highly methylated regions lie between −600 and − 200 bp, and − 200 to +150 bp region corresponds to relatively low methylation rate. The −200 to −100 bp area has the presence of GC box and E-box, collectively called core promoter region known to activate TERT mRNA expression by binding to SP1 and C-MYC as mentioned before. Hence, the methylation levels are reduced as compared to +1 to +100 bp though both the areas fall under low methylation rate. Also,

the positive correlation between TERT expression and methylation level in the promoter region has been demonstrated by whole genome sequencing which also shows that the expression negatively correlates to methylation level of gene body. To further support this, Sterna and group have shown that −600 bp methylation level inversely correlates to TERT transcription. In addition, when compared to monoallelic mutant cancer cells, wild-type allele had higher level of DNA methylation.

18.4.2 Regulation of Gene Expression by Telomeres

The length of telomere affects the expression of genes located near the region, and this phenomenon is known as telomere position effect (TPE). This phenomenon was first discovered by Gottschling and group in *Saccharomyces cerevisiae* and reported that the expression of RNA polymerase II was suppressed when inserted next to telomere locus. The reason behind TPE was suggested to be silent chromatin conformation in the telomeric region. An example includes ISG15 which shows higher expression in the aged cells. ISG15 is located at 1 Mb from the telomeric end and on the chromosome number 1p36.33. Interestingly, Robin and group demonstrated that not only adjacent genes but those far from telomeric region (upto 10 Mb) are also affected by TPE by a process known as long-range loop formation. This phenomenon involving atypical TPE is known as TPE-OLD (telomere position effect over long distances). TRF2 is a component of shelterin and falls under TPE-OLD. This protein tethers interstitial telomere repeats to telomere and generates the so-called long-range loop of chromatin. These repeats are present 100 kb downstream of TERT gene and suppress its expression by forming chromatin loop in a TRF2-dependent manner. This suppression is aided by LDS1 (lysine-specific demethylase 1) and REST (RE-1 silencing transcription factor) which bind to promoter region bound to TRF2 and add silencing histone marks. The regulation of gene expression in telomere length-dependent manner might involve TERRA, telomeric non-coding RNA associated with telomere length regulation, telomere end protection, and maintenance of chromatin structure. TERT overexpression leads to increase in number of telomeric tracts and TERRA signal enhancement, the reason being telomere elongation. Therefore, in telomere-elongated cancer cells, decreased expression of ISGs is associated with TERRA signal enhancement. This is further validated by the fact that ISG expression in cells having short telomeres is repressed by oligonucleotides which mimic TERRA. All these observations point toward the possibility that downregulation of ISG expression is dependent on upregulated TERRA signals in cancer cells having longer telomeres (Zhu et al. 2016).

18.4.3 TERT Promoter Mutations in Cancer

In sporadic melanoma, point mutation in the TERT promoter at −146 base pairs (C > T) and − 124 (C > T) from TSS (transcription start site) was discovered by Horn's group and Huang's group. Furthermore, in familial melanoma, Horn and group reported point mutation at −57 bp (T > G) from TERT transcription start site. These mutations lead to upregulation of TERT mRNA expression by providing E-twenty-six (ETS) transcription factor (GGAA, reverse complement) novel consensus binding motifs in the promoter region. These mutations being common type in noncoding somatic cells are present in different types of cancers (Table 18.8).

Except in myxoid liposarcoma, there seems to be a low percentage of mutation in TERT promoter as activation of ALT and not telomerase occurs in about 60% of sarcoma. In light of these observations, TERT promoter mutations are likely to be mutually exclusive with death domain-associated protein (DAXX), the α-thalassemia/mental retardation syndrome X-linked (ATRX), and other ALT pathway-associated chromatin modeling proteins. In case of mutations in TERT promoter in the case of ovarian carcinoma, the telomerase activity is regulated by ARID1A and PIK3CA.

Table 18.8 TERT promoter mutations involved in cancer types

Type of cancer	% Occurrence of TERT promoter mutation
Melanoma	67
Glioma	51.1
Primary glioblastoma	83.3
Myxoid liposarcoma	79
Bladder cancer	59
Urothelial carcinoma	50.8
Hepatocellular carcinoma	44
Medulloblastomas	21
Ovarian clear cell carcinoma	15.9
Squamous cell carcinoma	14.4
Thyroid cancer	10
Osteosarcoma	4.3

18.4.4 Telomere Shortening in Cancer

In contrast to the acquired ability such as elongation of telomere involved in cancer having activated telomerase, the length of telomere appears to be shorter in cancers such as prostate cancer. The fact was validated through whole genome/exome sequencing data from TCGA (The Cancer Genome Atlas) wherein Barthel and group examined 31 cancer cohort types for length of telomere. They demonstrated that while 30% of cohorts were suspected to be regulated by ALT, a majority (70%) exhibit shorter telomeres compared with normal samples.

Nevus and melanoma section staining using fluorescence in situ hybridization provided further validation. Chiba and group suspected and examined that telomere maintenance through cell division in human embryonic stem (ES) cells involved the TERT point mutants. Upon inactivating checkpoint proteins along with TERT promoter mutation, immortalization phenotype was achieved. Interestingly, in approximately 70 cell cycles, telomere length was shortened despite TERT mutation, followed by emergence of TERT expression and telomerase activity. As compared to non-cancerous tissues, cancerous ones show high expression of shelterin genes such as POT1, TRF (1–2), and TIN2. Interestingly, telomere length is reported to be inversely correlated with elevated expression of these genes/TERT/telomerase activity (Fig. 18.8). In normal telomerase silent cells, hTERT or its ectopic introduction activates telomerase activity and bypasses senescence leading to cell immortalization, demonstrating that for both crisis and senescence, the telomeres are mechanistically important (Shay & Wright 2011).

18.4.5 Telomerase Therapeutics

In general, immunotherapy, gene therapy, and small molecule inhibitors are among three classes of agents that have been developed for targeting telomerase molecular

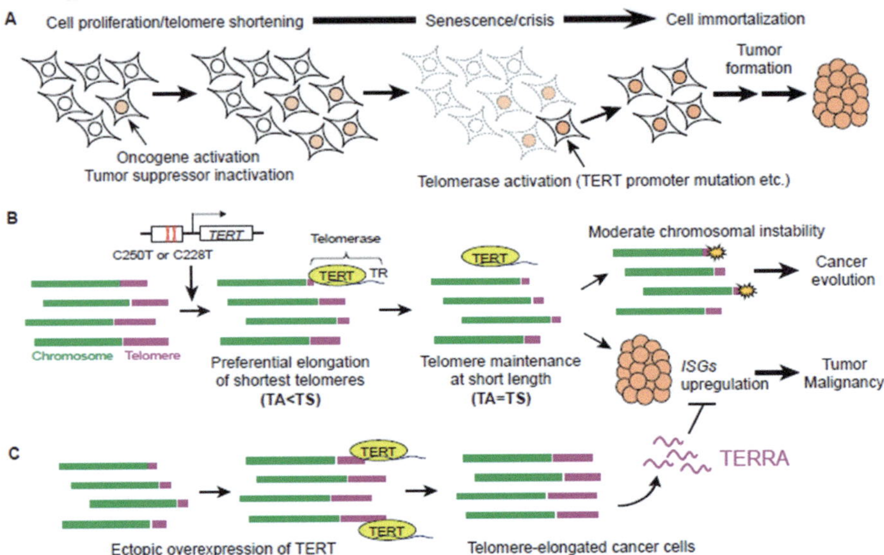

Fig. 18.8 Implication of telomere shortening in cancer development. Telomere length inversely correlates with TERT or even telomerase activity (**b**). In normal telomerase silent cells, hTERT or its ectopic introduction (**c**) activates telomerase activity and bypasses senescence leading to cell immortalization (**a**)

biology. Among the more advanced approaches, immunotherapy and small molecule inhibitor therapy are prevalent. Gene therapy targets a suicide vector by using either hTR (template RNA component) or hTERT promoter specific for oncolytic virus. Immunotherapy is currently being employed for advanced cases of pancreatic cancer and is undergoing phase III clinical trials, whereas small molecule therapy is being pursued for non-small cell lung cancer and breast cancer currently in phase II trials. In the case of lung cancer (non-small cell), use of small molecule therapy under controlled manner involving telomerase inhibitor, namely, imetelstat, is prevalent for prolonging the remissions after chemotherapy. Under this treatment method, the patients of lung cancer are randomized such that few of them receive bevacizumab, an angiogenic inhibitor, some receive imetelstat, while others receive imetelstat along with the angiogenic inhibitor. In case of multiple myeloma, small molecule therapy involving imetelstat is being verified as a cancer biomarker for depletion of stem cell.

18.5 Carcinogens

Carcinogens are the substances that have the capacity to induce cancer. Carcinogens, genotoxic agents, have the potential to directly bind the DNA and result in DNA damage, impairment of DNA repair machinery, and alteration in the proto-oncogene and tumor suppressor gene and hence tumor development (Moschel 2001; Sugimura 2000). In addition, carcinogen alters expression of genes through epigenetic effects.

In the 1950s an idea was put forward that a carcinogen causes cancer by causing some mutations, but a proof was not available initially. The first proof was put forward by Bruce Ames. The name of the test which he introduced is Ames test which assesses the mutagenic capability of a chemical. A special strain of bacteria which could not synthesize histidine was taken, which is an essential amino acid, and since these bacteria were made to grow in a medium lacking histidine, the cells couldn't survive. They added a chemical whose mutagenic capability was to be tested to the media; that caused mutations in the bacterial cells: some of them were back mutations, and in some bacterial cells, the capability of synthesizing histidine was restored, and those bacterial cells started dividing.

Carcinogen has been classified under six groups by the International Agency of Research on Cancer:

1. Biological agents.
2. Arsenic, fibers, metals, and dust.
3. Pharmaceuticals.
4. Chemical agents and related occupation.
5. Radiation.
6. Personal habits and indoor combustions.

Biological carcinogens include several viruses (HBV, HTLV, KSHV, EBV, and HPV) (Butel 2000), bacteria (*Chlamydia trachomatis*, *Helicobacter pylori*), and animals (*Schistosoma haematobium*, *Opisthorchis viverrini*, *Clonorchis sinensis*). The ultraviolet radiations cause cataract and skin cancer. Several pharmaceuticals including anticancer drugs, analgesics, and estrogen are documented as carcinogens. Chemical carcinogen directly acting on DNA includes alkyl and aryl epoxides, nitrosoureas, sulfonate, sulfate, and nitrosamides, while indirectly acting chemical carcinogens include hydrocarbons with aromatic rings and amines, alkyl nitrosamines, or aflatoxin B1. The effluents released from the industries (vinyl chloride, benzene, chromium compounds, and aromatic amines) and through combustion of fossil fuels (polycyclic aromatic hydrocarbons) are causing various cancers. The chemicals used as pesticides, fungicides, and pesticides are potential carcinogens. Chemical compounds dysregulate the cytochrome P-450 cellular enzyme in the liver and induce cancer development. Because of cigarette smoking habits, nicotine chemicals cause lung, esophagus, mouth, pharynx, bladder, pancreas, and larynx cancer. Several assays are discovered to check carcinogenic potential of the particular substance which includes Ames test for the early detection of carcinogen, single-cell gel assay for DNA strand break, FISH, DNA adducts assay, and Syrian hamster embryo cell transformation assay. Ames test can detect the mutagen agent but does not confirm its carcinogenic capacity. To detect the carcinogen, differential genetic expression assay, cDNA hybridization method, and checking protein synthesis pattern are used.

18.6 Oncogenes

The oncogenes are initially reported as tumor-inducing genes encoded by retrovirus in rodents and birds. Retroviruses picked up these genes from the host genes (proto-oncogenes) as the dominated mutated forms of gene (Lee & Muller 2010). The signals usually regulating cell division are molecules which either stimulate the cell division process or inhibit it. Oncogenes cause cancer development when present, and they can be either introduced via viruses into the host cells that cause cancer or some genes in the host genome itself can be mutated to form oncogenes. These genes can code for proteins which result in stimulation of excessive proliferation of cells and can lead to enhanced cell survival. The activation of oncogene is the result of epigenetic modifications, mutation, gene rearrangement or gene amplification, and subsequently tumor development (Botezatu et al. 2016). The early stages of tumor are associated with the gene mutation and translocation, while gene amplification is associated with the late stage of tumor. The proteins encoded by oncogenes discovered till date are classified among four major classes: growth factors, GFs (*v-sis*), GF receptors (*v-erbB*, *v-fms*, *v-kit*), GF response transducers (*v-src*, r-raf, v-ras), and transcription factors; TFs which mediates gene expression of growth factors induced genes (*v-jun*, *v-fos*). The identification and sequencing of the simian sarcoma virus resulted in the discovery *v-sis* oncogene. The cellular homologue of PDGFB protein encoded by *v-sis* gene is PDGF (platelet-derived growth factor), which initially thought to be responsible for the proliferation of arterial smooth muscle cells. PDGF binds to the PDGFR (PDGF receptor) and Src family of tyrosine kinases (SFKs) and activates PI3K-Akt pathway, therefore, involved in regulation of proliferation and cellular growth (Cantley et al. 1991; Shih & Holland 2006). The overexpression of PDGF and PDGFR is documented in the cancer of the central nervous system, sarcoma, glial cell tumor, germ-cell tumor, and gastrointestinal sarcoma. The GF receptor tyrosine kinase, epidermal growth factor receptor (EGFR), overexpression is reported in the tumor growth progression and involved in NSCLC, esophageal, ovary, cervix, breast, endometrium, and bladder cancer (Nicholson et al. 2001). EGFR family includes HER4, HER3, and HER2, of which overexpression of HER2 is associated with numerous cancers. The *v-kit* oncogene has a cellular homologue, *c-kit*, that encodes a KIT protein, belonging to the family of receptor tyrosine kinase. Mutation of the *c-kit* gene has been reported in the exon numbers 11, 13, and 17, and valine (V) to an alanine (A) amino acid substitution at 599 position (V599A) in the KIT is the most common mutation found. These mutations are observed in the acral, mucosal, conjunctival and cutaneous melanoma, and gastrointestinal stromal cancer. Another oncoprotein belonging to tyrosine kinase family, c-ErbB, has been discovered in humans, along with identification of v-ErbB protein in avian erythroblastosis virus, which is encoded by oncogene, *v-erbB*. The *c-erbB* gene amplification is found in gliomas and many other carcinomas, while point mutation in the *c-erbB* gene results in lung carcinomas. With the discovery of McDonough feline sarcoma virus, *v-fms* oncogene has been identified, which is analogous to the colony-stimulating growth factor receptor and associated with various cancers. In Rous sarcoma virus, *v-src* is derived from *c-src* of the host genome, thereby responsible for the transformation of fibroblasts. Oncogene *ras* of Kristen and Harvey sarcoma virus encodes for oncoprotein Kristen-ras (Ki-ras) and Harvey-ras (Ha-ras) which results into the transformation of the fibroblasts, hematopoietic cells, and epithelial cells into cancerous cells. The cellular homologue of *v-ras* is *c-ras*, and mutation in the *c-ras* leads to thyroid, colon, lung, and pancreatic carcinoma and acute myeloid and lymphocytic leukemia. The cellular homologue in humans of *v-jun* oncogene from the avian sarcoma virus (ASV17) is *c-jun* gene, which encodes for the c-Jun protein with 39 kDa molecular weight, consisting of leucine zipper DNA-binding domain at the C-terminus with basic in nature and transcription activator domain present at the N-terminal region. Under the stress, c-Jun promotes apoptosis through JNK signaling. Mutation in the c-Jun leads to the transformation of fibroblasts and tumor development in mice. Detailed analyses of viral oncogenes, proto-oncogenes, and tumor suppressor genes are discussed below.

18.7 Viral Oncogenes

In 1909, Dr. Francis Peyton Rous provided the first evidence of an infectious etiologic agent that causes cancer, and the RNA virus inducing chicken sarcoma was named Rous sarcoma virus in honor of Dr. Rous. He injected the cells, extracted from the hen breast tumor, into other hens who developed sarcoma later. For this discovery, he was awarded with Nobel Prize in 1966. Now approximately 20–25% human cancers in the world have a viral etiology (Table 18.9). The first

Table 18.9 Human DNA and RNA virus

Taxonomic grouping	Examples	Oncogenes	Tumor types
1. DNA viruses			
Adenoviridae *Hepadnaviridae* *Herpesviridae* *Papovaviridae* *Papillomaviridae*	Adenovirus types 12, 18, 31 HBV EBV KSHV Merkel cell polyomavirus BK virus, JC virus HPV 16, 18, 31, 45	E1A, E2B HBx LMP-1, BARF-1 vGPCR T antigens E6, E7	Various solid tumors only in rodents Hepatocellular carcinoma Burkitt lymphoma, B-cell lymphoma, NPC Kaposi sarcoma, primary effusion lymphoma Merkel cell carcinoma Solid tumors in rodents and primates Cervical and anal cancer, oral cancer
2. RNA viruses			
Flaviviridae *Hepacivirus* *Retroviridae* HTLV	Hepatitis C virus Human T-cell leukemia virus type I	– Tax	Hepatocellular carcinoma Adult T-cell leukemia/ lymphoma

demonstration of human retrovirus was human T-cell leukemia virus type 1 (HTLV-1), which causes adult T-cell leukemia. Another group of scientists were identifying the involvement of human DNA viruses in the transformation and development of cancer. This led to the discovery of herpes simplex virus type 2 (HSV2) and association of human papillomaviruses (HPVs) in cervical cancer development.

Some viruses possess an oncogene which is derived from cellular proto-oncogene of host cell, known as viral oncogene. These viral oncogenes are integrated into cells, while a particular virus is infecting cells, resulting into cancer development (Bishop 1985).

18.7.1 Human DNA Tumor Viruses

18.7.1.1 Human Papillomavirus (HPVs)

The integration of viral genome into the genome of cancer cell represents the most common characteristics of infection by these HPVs. Being a DNA virus, the genome size is 8 kb and is known to cause tumors, both benign and malignant in animals including humans. The two main viral oncoproteins of HPVs, E7 and E6, are involved in cervical cancer induction, which destabilize p53 and pRb, the two most important tumor suppressors of the cell. Sexual contact is the primary mode of transmission of these viruses. In view of this, Gardasil, the first cancer vaccine, was approved by the US Food and Drug Administration in 2006. Females in age group 9–26 years use this vaccine for cervical cancer, genital warts, and precancerous genital lesion prevention, which are caused by HPVs (HPV18, HPV16, HPV11, and HPV6).

HPVs consist of high-risk (oncogenic) HPVs and low-risk (non-oncogenic) HPVs. High-risk HPVs are considered to be oncogenic, which include the HPV16 and HPV18 and cause various cancers including cervical cancer. The HPV oncoproteins, E6 and E7, from high-risk HPV have the ability to transform the normal keratinocytes and epithelial cells into carcinogenic cells. Only the high-risk E6 and E7 and not the low-risk E6 and E7 have the capacity to interact with the p53 and pRb, tumor suppressor proteins. In order to comprehend the molecular mechanism behind E6 and E7 oncogene interaction with the tumor suppressor genes, first E6 and E7 structure needs to be known (Fig. 18.9). The E6 oncoprotein contains approximately 150 amino acids, having molecular weight of ~18 kDa, and is a nuclear protein with basic nature. Because of high-risk HPV16, 16E6 oncoprotein encoded by the HPV16 has been analyzed in detail. Structurally, 16E6 contains three nuclear localization signals (NLS), a PDZ domain-binding motif at the C-terminal along with four zinc-binding motifs (Cys-X-X-Cys), which results into formation of two Cys/Cys fingers leading to direct zinc binding. The active form, E6AP (E6-associated protein), interacts with the p53 and leads to the ubiquitination mediated p53 proteasomal degradation. 16E6 has higher binding affinity for p53 and efficiently causes the degradation, compared to the 18E6 (encoded by the HPV18), whereas 11E6 oncoprotein, encoded by the low-risk HPV11, has the least binding affinity for p53 and weakly influences the p53 degradation. The

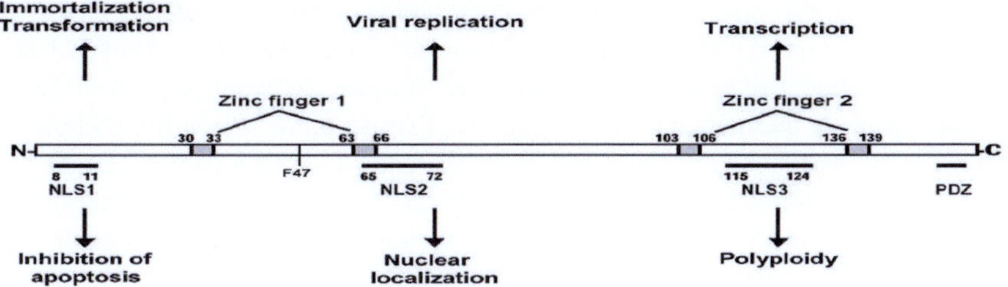

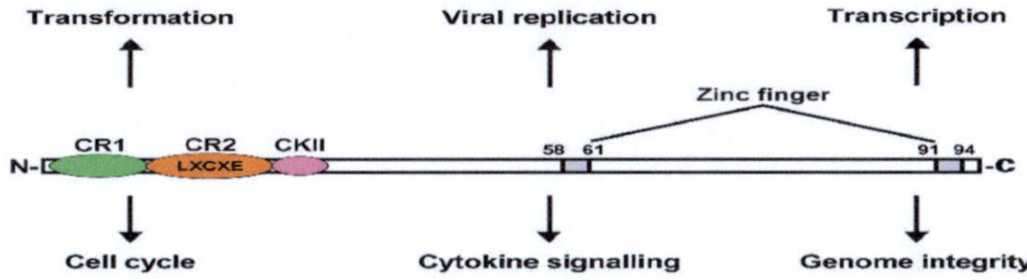

Fig. 18.9 Schematic representation of HPV16 oncoproteins E6 (**a**) and E7 (**b**)

amino acid present at 47 position (F47) in the E6 oncoprotein is essential for interaction with p53. Any mutation in the E6 F47 amino acid leads to the abrupt ubiquitination process and p53 degradation. The E6 from high-risk HPVs has shown to inactivate the CYLD (cyclindromatosis tumor suppressor protein) tumor suppressor, under hypoxia condition, by interacting with CYLD deubiquitinase and causing unstoppable NF-κB activation. 16E6 is known to regulate the gene transcription by interacting with various transcription factors and coactivators, such as p300/CBP (EP300/cAMP response element-binding protein-binding protein), IRF-3 (interferon regulatory factor 3), and c-Myc. 16E6 oncoprotein interacts with the c-Myc and leads to immortal cells by inducing the transcription of h-TERT. With the help of nuclear localization signal present near C-terminal, 16E6 is known to regulate the E6E7 bicistronic RNAs splicing by interacting with the RNA. In addition to above functions, 16E6 is reported to interact with the protein phosphatase H1, TNFR1 (tumor necrosis factor receptor 1), and PDZ proteins, thereby regulating the signal transduction (Zheng 2010).

Similar to the E6, high-risk E7 has a higher affinity for interaction with the pRb compared to the low-risk E7. E7 oncoprotein has approximately 100 amino acids and is a nuclear protein like E6. The detailed analysis of 16E7 of HPV16 leads to the findings that 16E7 contains nuclear localization signal and casein kinase II phosphorylation sites in the N-terminal domain and two zinc-binding motifs near the C-terminal domain. The N-terminus of 16E7 has a sequence similarity with the complete CR1 (conserved region 1) and CR2 (conserved region 2) of the adenovirus E1A oncoprotein and also related sequences found in T antigen of SV40. In the CR2 region, LXCXE motif is present, which upon interaction with the LXCXE-binding cleft of the p107, p130, and pRb results into the degradation of these proteins through interaction with the ubiquitin ligase complex, cullin-2. E7 not only degrades the pRb but is also essential for the viral life cycle and HPV-infected cells for various cellular functions. E7 has been reported to upregulate the expression of p21 and p16 (cyclin-dependent kinase inhibitor), thereby resulting in dysregulation of cell cycle. E7 oncoprotein interacts with the regulator of centromere, γ-tubulin, and inhibits its recruitment at the centromere and thereby leads to chromosomal abnormalities, resulting in mitotic defects and aneuploidy. In addition to above functions, E7 is reported to interact with various other proteins also, such as steroid receptor coactivator 1, p300, p600, and PCAF (P300/CBP-associated factor). These interactions of E7 and E6 oncoprotein with the tumor suppressor proteins and cell cycle regulator result in induction of the malignancies.

18.7.1.2 Epstein-Barr Virus (EBV)

The main target of this virus is B lymphocytes; the virus transforms B lymphocytes into lymphoblasts, instead of replicating within them, resulting in immortality of cells. The EBV contains oncogene, *LMP1* (latent membrane protein-1 or BNLF1), which is mainly expressed in lymphoma associated with the EBV and is crucial for B-cell transformation as well as for signal transduction disruption inside a cell. EBV is also involved in nasopharyngeal carcinoma and some types of Hodgkin disease. In nasopharyngeal carcinoma (NPC), early gene *BARF1* (BamHI-A reading frame-1) is essential for carcinogenesis, which is reported to express as latent gene in some NPC samples.

Depending on the latency forms, different expressions of EBNA (EBV nuclear antigen) and LMP1 are observed in B cells. EBNA-2 and EBNA-5 are crucial for the expression of LMP1 gene. Expression of only EBNA1 and LMPs is observed in NPC, T and NK lymphoma, and Hodgkin lymphoma with type II latency. With type I latency in Burkitt lymphoma, expression of only EBNA1 is observed. It has been reported that cytokines (IL-10 and IL-21) are involved in the expression of the LMPs.

The molecular weight of EBV integral membrane oncoprotein LMP1 is 62 kDa. LMP1 contains 386 amino acid residues with 24 amino acids present at the cytoplasmic N-terminal, 186 amino acids are responsible for six transmembrane domains, and 200 amino acids present at the cytoplasmic C-terminal. The full-length LMP1 with 386 amino acids is required for B cell transformation into cancerous cells. LMP1 interacts with the TRAF (TNFR-associated factor) and JAK3 (Janus kinase 3) through the C-terminal and induces the signaling molecules activation in B cells, resulting in the proliferation of B cells. The levels of LMP1 oncoprotein in the clonal population of B cell vary largely, and this leads to different functions of oncoprotein in clonal population such as promotion of cell proliferation by NF-κB in cells with intermediate expression of LMP1. However, with high expression of LMP1, protein synthesis is inhibited by activation of PERK (protein kinase R-like endoplasmic reticulum kinase) kinase which prevent phosphorylation of eIF2α, leading to higher expression of activating transcription factor 4 (ATF4) and, thereby, transactivation of its own promoter. In addition, LMP1 leads to the upregulation of anti-apoptotic genes, prevention of p53-mediated apoptosis through the inhibition of pro-apoptotic proteins, and activation of A20 gene.

The *BARF1* gene which is involved in NPC is located at the BamHI-A region and encodes a 31 kDa protein. The 54 amino acid regions at the N-terminus of BARF1 protein are responsible for the expression of anti-apoptotic proteins, resulting in the transformation of rodent's fibroblasts and B cells. It has been reported that, in B cells of Burkitt lymphoma with EBV-negative, BARF1 has potential to induce tumor growth in these cells. BARF1 expressing B cells reported to have higher expression of c-myc, CD23, and CD2. The secreted form of BARF1 acts as receptor for the CSF (colony-stimulating factor) and is also involved in the inhibition of IF-α (interferon-α) from the mononuclear cells. The secreted form of both BARF1 and LMP1 is detected in NPC patient's serum and acts as mitogenic factor.

18.7.1.3 Kaposi Sarcoma-Associated Herpesvirus (KSHV)

KSHV contains 165–170 kb genome size with double-stranded DNA. Several viral oncoproteins encoded by the viral oncogenes are involved in the tumorigenesis through the inhibition of apoptosis, immune modification, cell cycle regulation, and regulation of signal transduction pathway. KSHV is responsible for KS (Kaposi sarcoma), PEL (primary effusion lymphoma), and MCD (multicentric Castleman disease). Virus encodes for various oncoproteins including latency-associated nuclear antigen (LANA); v-FLIP (viral-FLIP) also known as K13; Kaposin C, B, and A; K1 also as named VIP (variable ITAM-containing protein); viral interleukin-6 (vIL6); viral interferon regulatory factor-1 (vIRF-1); and viral G protein-coupled receptor (vGPCR). The LANA is responsible for the inhibition of p53 and pRb proteins and interaction with the bromodomain of RING3/Brd2 protein complex, thereby progresses the cell from G_1 to S transition. LANA protein is essential for the upregulation of hTERT, cooperates with H-ras oncogene, and leads to the expression of Myc gene through β-catenin by preventing its degradation from the glycogen synthase kinase-3β (GSK-3β), resulting in cancer development. The K13 oncoprotein interacts with the FADD and caspase-8, thereby inhibiting the death receptor-mediated apoptosis. The K13 also activates the NF-κB, responsible for the viral latency and Kaposi sarcoma. Through maintaining the levels of IL-6 and GM-CSF, Kaposin protein is reported to be involved in PEL, and high levels of Kaposin protein are found in PEL cancer cell lines. K1 is a 46 kDa glycoprotein which contains an immunoreceptor tyrosine-based activation motif (ITAM). K1 leads to the constitutive expression of NF-κB and PI-3 K/Akt pathway, thereafter resulting in immortalization of the cells. Abundant levels of K1 have been reported in the KS, PEL, and MCD. The vGPCR encoded by KSHV is involved in human endothelial cell immortalization and angio-proliferative tumors. Through JAK/STAT activation and MAPK pathway, vIL-6 is involved in the pathogenesis of the KS and MCD. The vIRF-1 inhibits the host system antiviral response by blocking the IRF3/CBP/p300 complex formation and suppressing the interferon response. vIRF-1 promotes cancer development through inhibiting the TNF-α- and p53-mediated apoptosis.

18.7.1.4 Human Polyomaviruses

Simian virus 40 (SV40) from the rhesus monkeys is the most important member of this family. Merkel cell polyomavirus (MCV) is a newly discovered human polyomavirus which causes approximately 80% of Merkel cell carcinoma (MCC). The other polyomaviruses, Polyomavirus-6 (HPyV6) and HPyV7, along with the MCV inhabit the human skin.

During the early viral infection, the mRNA of human polyomavirus undergoes alternative splicing mechanism, resulting in the translation of two viral nonstructural oncoproteins: large T (LT) antigen and small t (st) antigen. LT antigen interacts with p53 and pRb, tumor suppressor proteins. In addition to the role of st antigen in interaction with the protein phosphatase 2 (PP2A), it also promotes the transforming capacity of LT. To understand the molecular mechanism behind the interactions of LT antigen with the p53 and pRb, LT antigen structure has been analyzed in detail. LT antigen protein contains two conserved regions (CR1 and CR2) in the J domain of the N-terminal, DNA origin-binding domain (OBD) at the middle, and ATPase/DNA helicase region in the C-terminal domain (Fig. 18.10). The J domain of LT N-terminal contains Hsc70 chaperone binding sites and is essential for the proper protein folding. The LXCXE motif is also present near the N-terminal of LT antigen, which interacts with the LXCXE-binding motif of the pRb protein, thereby regulating the pRb activity. The ATPase/DNA helicase region at C-terminus of LT antigen is crucial for its interaction with the p53 binding region and regulating the p53 activity. Further, malignant cell growth requires activation of insulin-like growth factor I (IGF-I), which is done by the LT-p53 complex. For viral DNA replication, DNA origin-binding domain of LT antigen is required. Similar to LT antigen, the J domain is also present in the st antigen because of common derivation from the exon 1 region of mRNA, while other parts are different from st antigen. With the presence of a unique PP2A-binding site, st antigen interacts with the PP2A, which is required for various signaling pathways and leads to inhibition of its phosphatase activity.

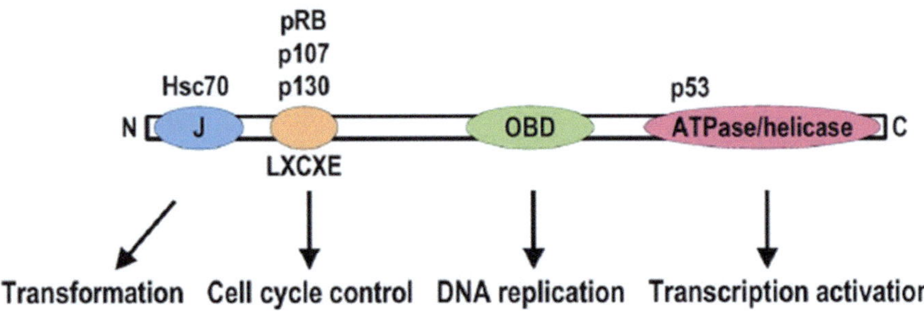

Fig. 18.10 Schematic representation of the SV40 large T antigen

18.7.1.5 Human Adenoviruses

The human adenoviruses are the DNA viruses mainly responsible for the infections of the respiratory system. The two main oncogenes of the adenoviruses are E1A and E1B, which are responsible for tumor development and help to understand the cellular processes of the tumor developing cells.

Till now, no reports are present in which adenovirus caused the tumor in humans, but many adenovirus serotypes (adenovirus types 2, 5, 12, 18, and 31) are known to cause tumors in rodents and hamsters. The main oncoproteins encoded by the adenovirus include E1A and E1B. The E1A protein (289R) is a 289 amino acid residue nuclear protein, which comprises of four conserved regions: CR1 located in the N-terminal domain, CR2 and CR3 located on the middle, and CR4 present at the C-terminal domain (Fig. 18.11). The E1A protein is responsible for the transactivation of various early genes including E1B genes. Because of the alternative splicing in the E1A mRNA, 243 amino acid residue protein (243R) has been encoded, which lacks the CR3 region. During the early viral infection, levels of both 289R and 243R E1A proteins are high, which are involved in the stimulation of DNA synthesis in cells and transformation, inactivation of the pRb, and induction of apoptosis through the interaction with various proteins. The CR2 region of E1A is known to interact with higher affinity to the pRb, p107, and p130 proteins, and along with CR1 region (which has low binding affinity for pRb and its related proteins) E1A inhibits the binding of pRb with the E2F target genes, thereby continuous expression of E2F transcriptional factors, which is responsible for expression of various cell cycle genes. The residues of the CR1 and N-terminus domain of E1A are required for transformation and interact with p300/CBP protein and lead to reduction in the acetylation of the H3 histone protein at lysine 18, thereby regulating the chromatin structure. For the acetylation of H3K18, the CBP and p300 are required. By interacting with the promoter of various genes temporarily and in time-dependent manner, E1A regulates the genes essential for cell cycle and growth, development, differentiation, and antiviral activities.

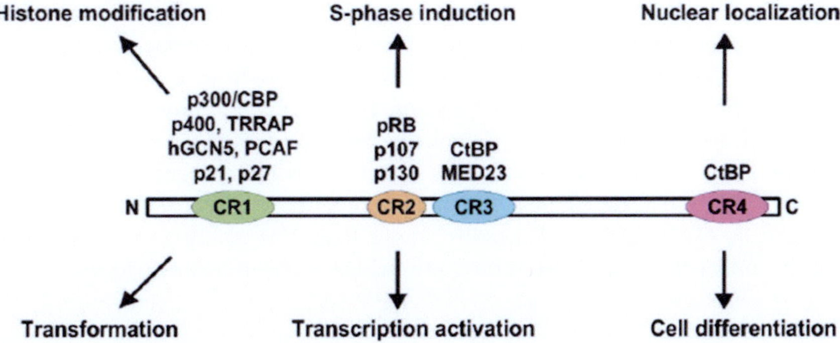

Fig. 18.11 Schematic representation of the adenovirus E1A-289R protein

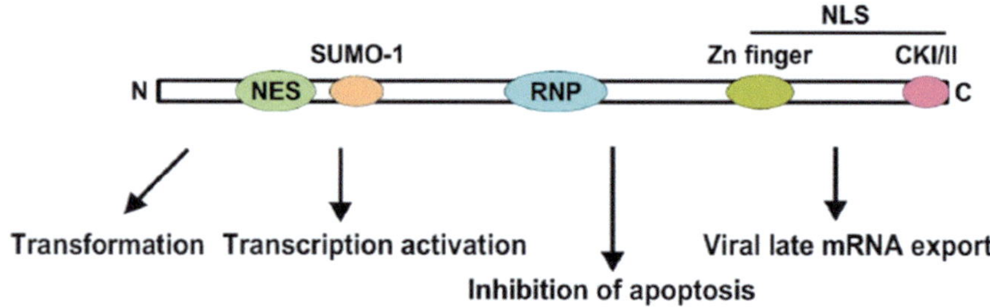

Fig. 18.12 Schematic representation of the adenovirus E1B-496R protein

The pre-mRNA of E1B under alternative splicing and encodes two main proteins of 176 amino acid residues (176R/19 K) and 496 amino acid residues (496R/55 K). There is no sequence homology between the 176R and 496R, but both have the capacity to upregulate the cell growth and transformation. The E1B-496R protein contains nuclear export signal (NES) near the N-terminal domain, ribonucleoprotein motif (RNP) in the middle and casein kinase I/II phosphorylating sites, nuclear localization signal (NLS), and zinc-binding motif near the C-terminal domain (Fig. 18.12). During viral infection, prevention of Bcl-2 family proteins oligomerization results in inhibition of caspase-mediated apoptosis of host cells, and transformation of rodent cells is promoted by the E1B-19 K and E1B-55 K. Both E1B-19 K and E1B-55 K along with the E1A protein is required for cancer induction in the lungs of transgenic mice, which is expressing both E1A and E1B transgenes. With the presence of nuclear export signal, transport of viral late mRNA is carried out by the E1B-55 K protein. E1B-55 K protein is also essential for the interaction of p53 and causing its degradation, thereby hampering p53 functions.

18.7.1.6 Hepatitis B Virus (HBV)

HBV contains DNA genome and is of only 3 kb size. This virus can cause cell transformation by the presence of a gene called gene X, which codes for transcriptional regulator HBx protein, through which it can cause hepatocellular carcinoma. However, HBx itself cannot cause tumors as evident from its inability to induce cancer in X gene transgenic mouse strains. HBV is mainly found in Southeast Asia and sub-Saharan Africa. HBx gene has the ability to disrupt the signal transduction, thereby affecting the cell growth; thus it can be a viral oncogene. The first human cancer vaccine to be developed, with high immunogenic effect and lasting protective efficacy, is HBV vaccine. Currently, more than 170 countries include the HBV vaccine under their vaccination program, which is cheap and safe to use.

18.7.2 Human RNA Tumor Viruses

18.7.2.1 HTLV-1

HTLV-1 causes T-cell lymphoma, mainly affecting the $CD4^+$ lymphocytes, and this can be transmitted through blood transfusion, breastfeeding, or sexual contact. Only 1% of people carrying this virus develop leukemia after decades. This virus is endemic in the Caribbean Basin, South Africa, and Japan. This retrovirus does not integrate into host genome and does not contain the viral homologue of the cellular proto-oncogenes. Tax oncoprotein of this virus is involved in the progression of cell cycle in the T cells and also sets up the continuous proliferating system.

Tax is also known as p40tax/Tax1 nuclear phosphoprotein, which comprises 353 amino acids with 40 kDa molecular weight. Tax1 acts as transcriptional activator of the virus promoter region and transformation of the T-cells, but not involved in maintaining T-cell transformation. Transgenic mice with Tax1 expression are prone to develop various type of tumors, making the concept of Tax1 as oncogenic clear. Recent reports indicate that Tax1 is essential for the transformation, but for maintaining the T-cell leukemia, another viral protein HBZ (HTLV1 basic leucine zipper factor) or expression of microRNA is required. In T-cell leukemia samples, detectable amounts of Tax1 are not observed.

The structure of tax1 protein consists of zinc finger motif and nuclear localization signal in the N-terminal, leucine zipper-like motif and nuclear export signal at the middle, and PDZ binding domain, activation domain, Golgi localization motif, and secretion motif at the C-terminal domain. The N-terminal domain of Tax1 acts as transcriptional factor for various cellular genes through interaction with the CREB and is also crucial for the transport of various RNAs and proteins within the cell. The leucine zipper motif is essential for the interaction with PP2A and NF-κB, thereby regulating the expression of various genes. In addition, Tax1 is essential for the regulation of various cell cycle genes, chromatin remodeling proteins, as well as activation of viral and cellular transcription. Another protein, Tax2/p37tax, encoded by the HPLV-2, shows similar properties to the Tax1 such as lymphocyte transformation.

18.7.2.2 Xenotropic Murine Leukemia Virus-Related Virus (XMRV)

XMRV is first identified from the patients with prostate cancer. This virus is also involved with chronic fatigue syndrome. The receptor, XPR1 (xenotropic and polytropic retrovirus receptor 1), is essential for the infection of XMRV. However, recent studies are controversial related to the XMRV roles in chronic fatigue syndrome and in prostate cancer.

18.7.2.3 Hepatitis C Virus (HCV)

The *Flavivirus* family contains HCV, having RNA genome which carries a long open reading frame involved in encoding a 3010 amino acid precursor polypeptide, which results in ten different proteins, seven nonstructural proteins, and three structural proteins. However, HCV's role in oncogenesis is not properly understood; liver cancer associated with HCV requires persistent infection. No oncogenes are discovered in the viral genome, but for the HCV replication, liver-specific miR-122 is involved. Till now, there is no vaccine available for the HCV, but various researches are ongoing to develop a safe vaccine with long-lasting effects against HCV infection.

18.7.2.4 Rous Sarcoma Virus (RSV)

Rous sarcoma virus is an avian sarcoma virus, involved in the transformation of fibroblasts. Using the RNA fingerprinting technique of RSV genome, identification of *gag*, *pol*, *env* (genes required for the virus reproduction), and *src* gene (required for the oncogenic activity of virus) can be carried out. The *src* gene comprises 20% of the virus genome and is present in the C-terminal of the virus RNA. The first oncogene to be discovered is *src* oncogene and encodes v-src protein with 60 kDa molecular weight which belongs to the tyrosine kinase protein family. The *src* gene is essential for the viral oncogenic activity but not responsible for the virus replication and reproduction. The sequencing of *src* gene leads to the finding of its source of origin, as it is a cellular gene, not a viral gene. The *src* DNA sequence of the host is incorporated into the virus DNA during viral infection, since loss-of-function or gain-of-function mutation in certain genes converts them to oncogenes. The viral *src* gene has deletion at carboxy-terminal, which includes essential phosphorylation sites and also several point mutations in comparison to the cellular *SRC* gene. The SRC protein kinase activity is under the regulation of a polyproline-binding SH3 domain and a phosphotyrosine-binding SH2 domain, two modular domains of SRC protein, and disruption in the intramolecular interactions in these SH2 and SH3 domains is essential for the activation of catalytic domain of SRC protein. Because of these regulations, SRC protein has lower kinase activity and no oncogenic activity in comparison to the viral src protein, since, because of point mutations in the SH3 domain and deletion in the C-terminal, these regulations are absent in viral src protein, resulting in the constitutive expression of viral src protein with higher kinase activity and oncogenic activity.

18.7.2.5 Avian Myelocytomatosis Virus MC29

Avian myelocytomatosis virus MC29 is the cause for the development of Burkitt lymphoma, B-cell lymphoma, neuroblastoma, and medulloblastoma because of the presence of *myc* gene in its genome. Basically, *myc* gene is responsible for several human cancers. With the development of RNA fingerprinting technique and DNA sequencing techniques, it has been reported that *myc* is not a viral gene but incorporated into viral RNA as a contiguous stretch. The *myc* is the second oncogene to be discovered. The avian retroviruses CMII, OK10, and MH2 also contain *myc* gene. *MYC* gene is also reported in the human genome. Presence of three *MYC* genes in the human genome has been discovered: *c-MYC* (or *MYC*), *MYCN*, and *MYCL1*. The homologous gene of viral *myc* gene is cellular *MYC* gene. With the discovery of MYC protein location in the cell nucleus, it is represented as a transcription regulator. But MYC protein alone has lower gene-binding affinity and is not able to regulate the gene expression, but in association with the MAX protein, MYC-MAX heterodimer has higher gene-binding affinity and has potential to regulate gene expression. With regard to *c-MYC* gene chromosomal translocation at the immunoglobulin gene region and under the immunoglobulin enhancer, overexpression of *c*-MYC protein was observed in Burkitt lymphoma (Boxer & Dang 2001). The *MYCL1* and *MYCN* genes are responsible for several human cancers. The amplification and expression of *MYCN* gene was observed in the neuroblastoma. The *c-MYC* gene is not only essential for the cancer induction but also involved in providing resistance against inhibitors of PI3K and inhibition of *RAS*-induced tumors in mice.

18.7.2.6 Harvey Sarcoma Virus and Kirsten Sarcoma Virus

During the murine leukemia virus infection in rats, recombination between the host genome and viral genome leads to the origin of two new retroviruses: Harvey sarcoma virus and Kirsten sarcoma virus. Both the retroviruses contain *ras* gene in their genome and encode 21 kDa protein, Ras. The Kristen-*ras* (*Ki-ras*) from the Kristen sarcoma virus and Harvey-*ras* (*Ha-ras*) from the Harvey sarcoma virus have similar properties. These oncogenes are responsible for the transformation of the fibroblasts, hematopoietic cells, and epithelial cells into cancerous cells. The *ras* gene is also involved in pancreatic cancer and its homologue is also found in the human genome. The Ras protein is a GTPase, bound to the GTP in its active form, and located at the cytoplasmic side of plasma membrane and interacts with lipid membrane through isoprenylation post-translational modification of the protein. The GTPase-activating protein (GAP) stimulates the GTPase activity of Ras. Mutations in the 12 and 61 amino acid residues of RAS protein affect its binding with the GAP protein, and GTP-bound Ras protein level increases with the decrease in GTP hydrolysis rate, thereby stimulating the oncogenic activity of Ras. The Ras protein interacts with various signaling components such as with the PI3K catalytic subunit and affects the MAPK pathway through binding with the RAF protein.

18.7.2.7 Abelson Murine Leukemia Virus

Abelson murine leukemia virus contains an oncogene, *v-abl*, which is responsible for the pre-B-cell transformation into the cancerous cells by the two-step method. Initially, during viral infection in the bone marrow, progressive proliferation of the pre-B cells is observed which marked the first step. The second step includes a period of apoptosis and abrupt growth marked as crisis phase. Some pre-B cells escape this crisis period and give rise to fully malignant cells. The v-abl oncoprotein negatively regulates the p53 tumor suppressor protein and p19Arf protein expression. The p19Arf protein is essential for the activation of p53 through interaction with the Mdm2 protein. The v-abl oncoprotein interacts with the c-myc promoter region, resulting in cellular Myc protein expression and is also involved in the cellular Ras expression, which are essential for virus stimulated transformation and growth. The v-abl activates the PI-3 K and JAK pathway which promote the growth and survival of cancerous cells. The Akt pathway is also under the regulation of v-abl oncoprotein, which interacts with the Mdm2 protein, thereby suppressing the p53 activity. Through regulating various signaling proteins and tumor suppressor protein p53, v-abl leads to oncogenesis of pre-B cells.

18.7.2.8 Avian Erythroblastosis Virus

The avian erythroblastosis virus causes leukemia of erythrocytes, known as erythroblastosis because of the presence of oncogene *erbB*. In the virus-infected cells, presence of 74 kDa ErbB protein has been detected which is encoded from the fusion of cellular ErbB mRNA and viral erbB mRNA. The ErbB viral protein is phosphorylated and glycosylated and belongs to the tyrosine kinase family which functions as integral plasma membrane protein. The sequence analysis shows the close similarity of ErbB protein with the cellular EGFR. The viral ErbB protein has large deletion in the extracellular domain compared to EGFR, modification at N-terminal domain, and mutation at the C-terminal that comprises the kinase domain. Because of the mutation at the kinase domain of ErbB protein, it works in ligand-independent manner with constitutive expression. These mutations lead to the disruption in the signaling component of the host system and result in oncogenesis. Another viral oncogene from the avian erythroblastosis virus has been sequenced, named *erbA* (Table 18.10), and encodes an ErbA protein, which belongs to the hormone receptor family and shows close similarity with the cellular THRA.

18.7.2.9 Mode of Action of Viral Oncogenes

Presently there are only three strategies known through which viral oncogene products might act.

1. Phosphorylation: The primary activity occurs on the plasma membrane surface. The transforming protein which acts as potential substrate strikes the surface, ultimately spanning the whole plasma membrane or residing at the inner surface.
2. DNA synthesis: Cancer cells exhibit unrestrained cell division by involving some transforming proteins eliciting the property to initiate DNA synthesis.
3. Transcription regulation: Many transforming proteins may interact with enhancers or promoters to influence transcription of cellular genes.

18.8 Proto-oncogenes

The normal cellular oncogenes (*c-onc*) are the host genes, whose product form important constituents of various signaling pathways which control the proliferation, division, and growth of cells. Viral oncogenes are homologues of these cellular oncogenes, such as *c-src* which is the cellular homologue of *v-src*. Proto-oncogenes are the representatives of normal genes of cells, which show similarity with the nucleotide or protein sequences that are tumorigenic or have the potential of transforming. A plethora of circumstantial evidence now points to the fact that alteration in either copy number, expression, or structure of one of these genes is responsible for the several malignancies of humans. The proteins encoded by these genes include transcription factors, signal transducers, growth factors, and growth factor receptors. Epigenetic and genetic changes are responsible for the conversion of these proto-oncogenes to oncogenes. Scientists have isolated many *c-onc* from different organisms including humans, by probing them using *v-onc*. The

Table 18.10 List of retroviral oncogenes (Vogt 2012)

S. no.	General class	Oncogene	Virus	Host species	Predicted function
1.	Non-receptor protein tyrosine kinase	*abl* *fes* *fps* *src* *fgr* *ros* *yes*	Abelson murine leukemia ST feline sarcoma Fujinami sarcoma Rous sarcoma Gardner-Rasheed feline sarcoma URII avian sarcoma Y73 sarcoma	Mouse Cat Chicken Chicken Cat Chicken Chicken	Tyrosine-specific protein kinase Tyrosine-specific protein kinase Tyrosine-specific protein kinase Tyrosine-specific protein kinase Tyrosine-specific protein kinase Tyrosine-specific protein kinase Tyrosine-specific protein kinase
2.	Receptor protein tyrosine kinase	*erbB* *fms*	Avian erythroblastosis McDonough feline sarcoma	Chicken Cat	Truncated version of epidermal growth factor receptor Analog of colony-stimulating growth factor receptor
3.	Serine/ threonine protein kinase	*mil (mht)* *mos* *raf*	MH2 Moloney sarcoma *3611* Murine sarcoma	Chicken Mouse Mouse	Serine/threonine protein kinase Transcription factor Serine/threonine protein kinase
4.	Growth factor G protein	*H-ras* *K-ras* *sis*	Harvey murine sarcoma Kirsten murine sarcoma Simian sarcoma	Rat Rat Monkey	GTP-binding protein GTP-binding protein Analog of platelet-derived growth factor
5.	Transcription factor	*erbA* *fos* *jun* *myb* *myc* *rel*	Avian erythroblastosis FJB osteosarcoma Avian sarcoma Avian myeloblastosis MC29 myelocytomatosis Reticuloendotheliosis	Chicken Mouse Chicken Chicken Chicken Turkey	Analog of thyroid hormone receptor Transcriptional activator protein Transcriptional activator protein Transcription factor Transcription factor Transcription factor

conservation of these genes across different species suggests that the proteins encoded by them are involved in vital cellular events. HMGA1 proto-oncogene reactivation has been reported in several cancers. The protein encoded by these genes is responsible for carcinogenesis under both in vitro and in vivo conditions (Choudhuri & Chanderbhan 2007). After their expression was discovered, Rat1a fibroblasts and lymphoid cells were proved to be transformed by their overexpression. Under in vitro transgenic animal models, HMGA1 overexpression results in pituitary tumor, benign mesenchymal tumors, and lymphomas. Importantly, these proteins may also disrupt the tumor suppressive pathways such as pRb and p53 pathway.

18.9 Mutant-oncogenes

The proteins encoded by proto-oncogenes regulate key cellular activities, and, therefore, a single mutation in any of these genes can disturb the normal functioning inside a cell pushing it onto the highway of becoming cancerous. The first ever proof linking mutant c-onc to cancer was given in case of human bladder cancer by Robert Weinberg and colleagues using transfection test (Fig. 18.13).

The team identified a DNA fragment that carried an allele of *c-H-ras* oncogene from the bladder cancer which had the reproducible capacity for transformation of cells into the cancer cells. Mutant c-ras oncogene involvement has been reported in a variety of tumors in humans (Fig. 18.14) (Bos 1989).

Isolate DNA from tumor cells and add marker to each strand of DNA

↓

Transfer DNA into the normal cells

↓

Oncogene integrates into the normal cells and transform its descendants into cancerous cells

↓

Cancerous cells form colony in the culture

↓

Isolate the specific DNA from the cancerous cells, with the marker added

↓

Repeat the procedure to verify the transforming ability

Fig. 18.13 Flowchart showing steps involved in transfection test for identifying nucleotide sequences capable of making cells cancerous. DNA from tumor cell is transferred into normal cells leading to its integration. The DNA is then isolated based on its marker

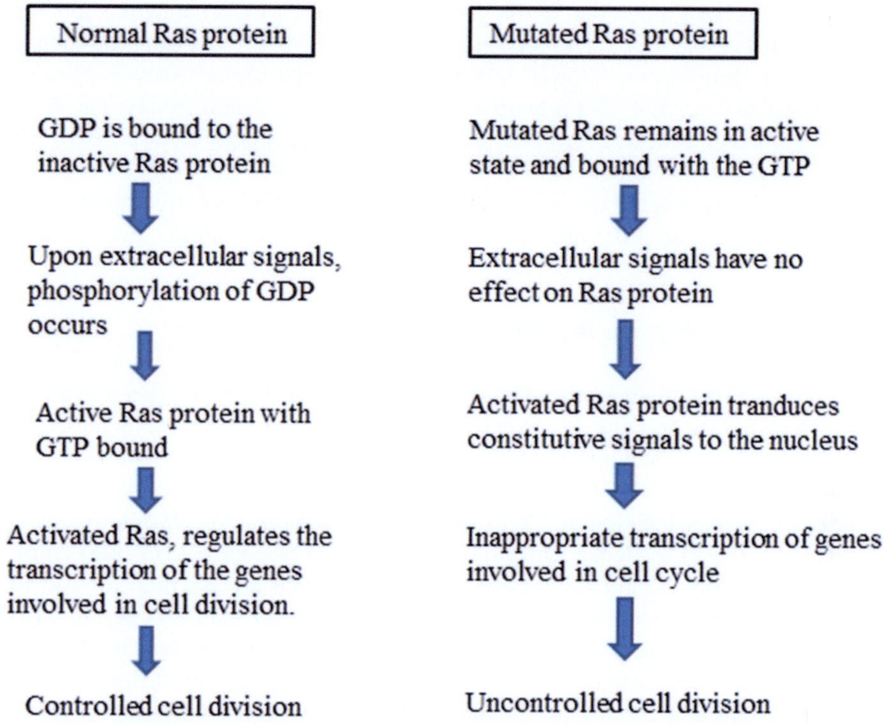

Normal Ras protein	Mutated Ras protein
GDP is bound to the inactive Ras protein	Mutated Ras remains in active state and bound with the GTP
↓	↓
Upon extracellular signals, phosphorylation of GDP occurs	Extracellular signals have no effect on Ras protein
↓	↓
Active Ras protein with GTP bound	Activated Ras protein tranduces constitutive signals to the nucleus
↓	↓
Activated Ras, regulates the transcription of the genes involved in cell division.	Inappropriate transcription of genes involved in cell cycle
↓	↓
Controlled cell division	Uncontrolled cell division

Fig. 18.14 Ras protein signaling and cancer. Normal Ras protein is regulated in presence of extracellular signal and the mutated Ras protein leads to uncontrolled cell division

The amino acid change in one of the three positions, 12, 59, and 61, impairs the mutant Ras protein to come out of its activated state, thereby stimulating cells to divide continuously.

18.9.1 Phosphatidylinositol 3-Kinase (*PI3K*)

The lipid kinase PI3K constitute a heterodimer of adaptor/regulatory (p85) and catalytic subunit (p110), resulting from the transcription from different genes and undergoing alternative splicing during the translation. The interactive binding of growth factors with the receptor tyrosine kinase is responsible for PI3K activation, subsequently production of phosphatidylinositol 3,4,5 trisphosphate (PIP3) and diacylglycerol (DAG) from phosphatidylinositol 4,5 bisphosphate (PIP2), leading to the Akt pathway activation, thereby regulating the cell survival, cellular growth and motility, apoptosis, and cell adhesion and transformation. With the discovery of avian hemangiosarcoma virus (belongs to avian retrovirus family), ASV16, the kinase activity and oncogenic activity of the PI3K have been reported. Many people were found to have mutations in the gene encoding the p110α catalytic subunit of PI3K (Karakas et al. 2006). The gene encoding the PIK3CA (PI3K catalytic subunit) is located on chromosome 3 (3q26.3) with 34 kb gene size and consists of 20 exons, which translate into 124 kDa protein having 1068 amino acid residues. The B-cell defects, colorectal cancer, liver necrosis, and embryonic lethality are associated with the mouse model with knockout of both PI3K subunits. The PTEN (phosphatase and tensin homologue deleted on chromosome ten) acts as PI3K negative regulator, by dephosphorylating the PIP3 and thus disrupting the PI3K signaling.

18.9.2 Protein Kinase B (*PKB*)

The PKB is also known as the Akt serine/threonine kinase and hyperactivated in several human cancers. The Akt hyperactivation is responsible for the increased cell growth, proliferation and energy metabolism, and development of resistance against apoptosis. Activated PI3K generates PIP3, to which Akt PH domain binds, thereby stimulating the translocation of Akt to the plasma membrane. Akt phosphorylation is mediated by the PI3K-dependent kinase-1 (PDK1) at threonine (Thr 308) and by PDK2 at the serine amino acid residue (Ser 473), which are essential for Akt full activation. There is still a contradiction between FOXO and mTOR, as which downstream effector molecule of Akt is responsible for the cancer development. Akt inhibits the stimulation of the forkhead family of transcription factors (FOXO), which is responsible for cell proliferation inhibition, whereas activation of mTOR stimulates cell proliferation (Hay 2005). The TSC2/TSC1 (tuberous sclerosis complex 2/tuberous sclerosis complex 1) acts as a negative regulator of Akt by inhibiting the mTOR activity. The development of benign tumor has been reported with the germline mutation in the TSC2 and TSC1 encoding genes. Several Akt-mTOR target molecules have been under development which can be used in cancer therapy.

18.9.2.1 B-RAF

The oncogene *B-RAF* encodes a serine/threonine protein kinase B-RAF. Three paralogs of RAF serine/threonine protein kinase are C-RAF, B-RAF, and A-RAF. Binding of cytokines, hormones, and growth factors to the RAS, a membrane-bound G protein, leads to activation of RAF kinase, subsequently MEK (mitogen-activated protein kinase) activation, which activate ERK (extracellular signal-regulated protein kinase), thereby regulating the apoptosis, cell proliferation and differentiation via cytoskeleton rearrangement, metabolism regulation and gene expression. RAF protein structure constitutes the N-terminal domain with two conserved regions, CR1 and CR2, and C-terminal containing the third conserved region, CR3 and kinase domain. It has been reported that 7% of cancers contain B-RAF mutation. The incidence of mutation in the B-RAF as shown in Fig. 18.15 is highest among the malignant melanoma (27–70%), serous ovarian cancer (~30%), papillary thyroid cancer (36–53%), and colorectal cancer (5–22%), while a variety of cancers has low frequency of B-RAF mutation. The most frequent mutation found in *B-RAF* gene is the transversion of thymidine to adenosine at 1796 position of nucleotide, which

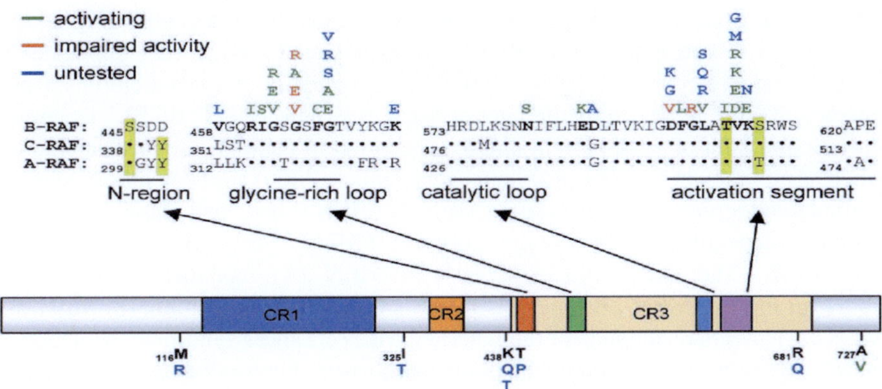

Fig. 18.15 Schematic representation of *B-RAF* oncogene mutations associated with the cancer

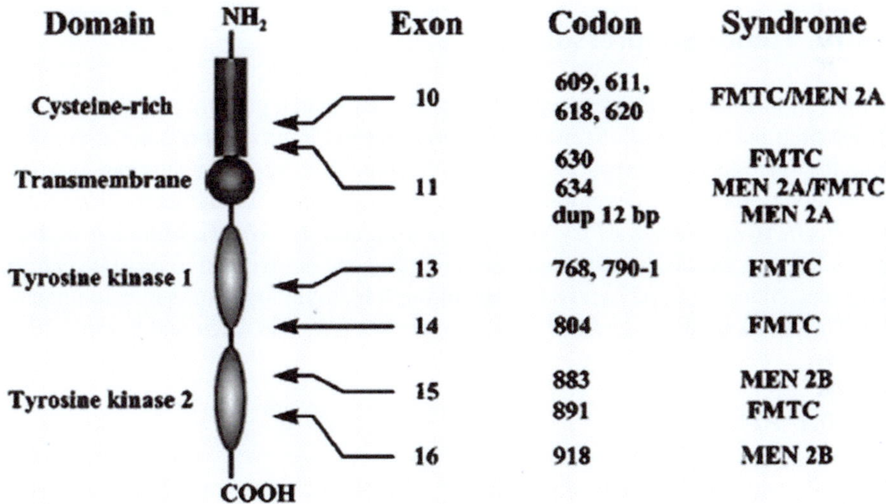

Fig. 18.16 Diagram showing RET gene with codons identified in MEN 2 families

leads to conversion of valine at 599 amino acid position to the glutamate in the B-RAF protein. This mutation, ^{V599E}B-RAF, is linked to 90% mutations of melanoma and thyroid cancer, while in non-small cell lung cancer, the rate of mutation is very low. However, several other sites of mutation are also reported with potential to cause cancer (Garnett & Marais 2004).

18.9.2.2 RET

The *RET* oncogene encodes a RET protein (Fig. 18.16), which is a membrane-bound tyrosine kinase having intracellular tyrosine kinase, transmembrane, and extracellular domain. On binding with the artemin, neurturin, and GDNF (glial cell line-derived neurotrophic factor), RET kinase is activated and has been implicated in multiple endocrine neoplasia type 2 (MEN 2) and papillary thyroid carcinoma. Upon interaction with various activating genes (such as ELKS, H4, HTIF, ELE1), constitutive expression of RET protein kinase is reported in papillary thyroid carcinoma, whereas mutation in the *RET* proto-oncogene (Jhiang 2000) has been observed in the multiple endocrine neoplasia type 2 (MEN 2). MEN 2 has been classified in three subtypes depending on the organ affected: FMTC, MEN 2B, and MEN 2A (Fig. 18.16). The parathyroid hyperplasia (in MEN 2A), medullary thyroid carcinoma (in FMTC, MEN 2B, and MEN 2A), and pheochromocytoma (in MEN 2B, MEN 2A) are characteristics of the inherited MEN 2 cancer syndrome. Mutation in the *RET* gene region, which encodes for the tyrosine kinase domain and cysteine-rich domain of RET protein are reported, thereby responsible for the constitutive activity of tyrosine kinase and results in transformation of the cell.

18.10 Tumor Suppressor Gene

Tumor suppressor genes are responsible for regulating cell growth, proliferation, and differentiation. The gain-of-function mutation results in proto-oncogene conversion into the oncogene, whereas loss-of-function mutation in tumor suppressor genes leads to various cancer developments. The tumor suppressor gene includes *p53*, *Rb*, *PTEN*, *BRAC1*, *BRAC2*, *APC*, *NF1*, *p27* (*Kip1*), and *p16* (*Ink4*). Mutation in these genes leads to cancer development. Loss of *p16* gene is involved in prostate cancer. The gene silencing of *p27* and *p16* gene through methylation results in carcinogenesis. The detailed analysis of few tumor suppressor genes is discussed below.

18.10.1 PTEN

PTEN (phosphatase and tensin homologue deleted on chromosome 10) or *MMAC-1* gene is located on chromosome 10 (10q23), which encodes a PTEN protein with 403 amino acid residues and responsible for the downregulation of PI-3 K-Akt pathway (Fig. 18.17) (Carnero et al. 2008). Under the mitogenic factor influence, Ras protein kinase is activated, resulting in PI-3 K activation. The PI-3 K is responsible for the conversion of PIP2 into PIP3, thereafter activating the Akt signaling, resulting in cell growth and proliferation. *PTEN*, a tumor suppressor gene, regulates this pathway, through dephosphorylation of PIP3 and inhibiting the activation of the Akt signaling, thus preventing the excessive proliferation of cells. Inactivation of *PTEN* either through mutation or homozygous deletion at both the alleles results in loss of PTEN protein function (Wang et al. 1998). The familial neoplastic syndrome, Cowden disease, is associated with the germline mutation in the *PTEN* tumor suppressor gene. The large deletion at 10q region of *PTEN* gene is observed in a variety of cancers such as endometrium melanoma and glioblastoma, breast cancer, prostate cancer, renal cancer, and NSCLC (Chen et al. 2018; Leslie & Downes 2004).

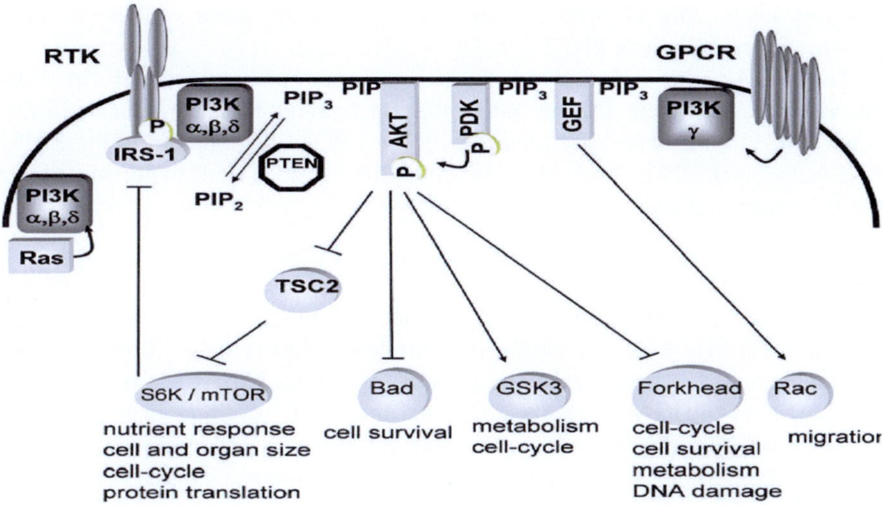

Fig. 18.17 Schematic representation of the interaction of PTEN and PI3K-Akt pathway

18.10.2 NF1

NF1 (neurofibromatosis type 1) gene is located on the chromosome number 17 (17q11.2) and encodes a tumor suppressor protein, Neurofibromin (Cichowski & Jacks 2001). Loss of both alleles of *NF1* gene results in familial tumor syndrome; Neurofibromatosis type 1 is well known as von Recklinghausen, because of its discovery by Friedrich von Recklinghausen in 1882, with a frequency of 1 in 3500 individuals. Defects in the *NF1* gene are associated with tumors in the nervous system, glioblastoma, and myeloid leukemia. In addition to cancer syndrome, neurofibromatosis type 1 disease is associated with abnormal pigmentation, bone deformations, and mental retardation. Neurofibromatosis acts as a GTPase-activating protein and results in inhibition of Ras signaling activation. Overexpression of the Ras pathway is observed in neurofibromatosis type 1.

18.10.3 *BRCA1* and *BRCA2*

BRCA1 (*br*east *ca*ncer 1) gene is located on the chromosome number 17 (17q21.13) and encodes a 1863 amino acid residue long protein. The structure analysis of BRCA1 protein revealed the presence of the RING domain at the N-terminus and BRCT domain on the C-terminus. The interaction of RING domain with the BARD1 (BRCA1-associated RING domain) and also by binding to the Rad51 mediate the DNA repair through homologous recombination. The interaction of p53 protein with the BRCT domain of BRCA1 and BRCA1 regulates the p53 transcription. Mutation in the N-terminal and C-terminal of BRCA1 protein leads to cancer of the breast, ovary, prostate, and colon. With the increase in age, chances of breast cancer through mutation in the *BRCA1* gene increase. BRCA1 also regulates various cell cycle proteins (Baer & Ludwig 2002).

The *BRCA2* gene is located on the chromosome number 13 (13q12-q13) and encodes BRCA2 protein. Despite having the structural dissimilarity from the *BRCA1* gene, *BRCA2* gene share may feature with the *BRCA1*. By binding to the Rad51, BRCA2 helps in DNA repair. Mutation in the *BRCA2* gene leads to prostate cancer, gastric cancer, and melanoma.

18.10.3.1 APC

APC (adenomatous polyposis coli) gene is located on chromosome number 5 and encodes a 312 kDa protein consisting of multiple domains (Fig. 18.18), which serves as a binding site for various proteins, including β-catenin, CtBP (C-terminal binding protein), Asefs, IQGAP, axin, microtubule, and EB1, resulting in regulation of spindle formation, chromosome segregation, cell adhesion and migration, and cytoskeleton organization. Germline mutation in *APC* gene causes familial adenomatous polyposis (FAP), which is characterized by the presence of various polyps in the intestine. Mutation in *APC* gene leads to colon cancer and lung cancer. Two *APC* genes are present in humans: *APC* and *APC2*, which encodes for APC (containing 2843 amino acid) and APC2 (containing 2303 amino acid). The APC protein consists of oligomerization domain, an armadillo domain, mutational cluster region with 15–20 amino acid residue repeats essential for β-catenin binding and SAMP repeats involved in axin binding, and CtBP binding region with basic region, EB1 binding region, and DLG binding domain. As regards binding of various proteins to the APC protein, functions of APC are mediated. APC plays the suppressive effects

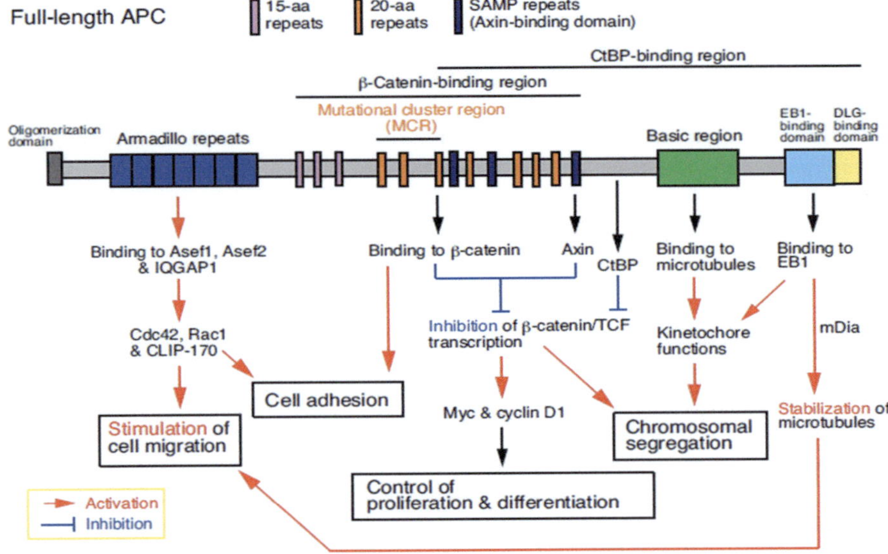

Fig. 18.18 Schematic representation of APC protein and its functions

on the Wnt signaling, which is responsible for the cell proliferation and differentiation (Aoki & Taketo 2007; Senda et al. 2005).

18.11 Two-Hit Hypothesis

Analysis of cancer following a dominant pattern of inheritance helped researchers to identify many tumor suppressor genes. Like all genes, mutations occur in tumor suppressor genes, although almost all loss-of-function mutations occurring are rather recessive in nature. Cancer is mostly bound to happen if only a second mutation knocks out the normal functioning when it hits the somatic cells. Hence, cancer development requires two loss-of-function mutations, "two hits," one for each of the two copies of the tumor suppressor gene (Fig. 18.19) (Chial 2008). In 1971, Alfred Knudson explained this idea for the occurrence of retinoblastoma.

18.12 RB1 Tumor Suppressor Gene

The two-hit hypothesis was idealized way before the human genome project was completed. Among the firstly isolated tumor suppressor genes was the RB1 gene (Dyson 2016) responsible for causing retinoblastoma in 1985 by a pair of scientists, Raymond White and Webster Cavenne. They showed that chromosome 13 was missing huge segments in retinoblastoma cells and eventually the gene was isolated.

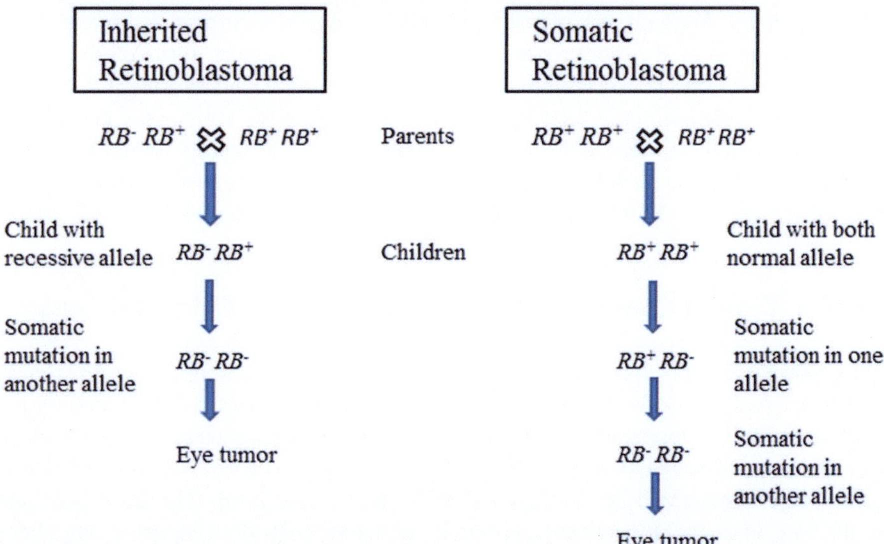

Fig. 18.19 Knudson's two-hit hypothesis in case of retinoblastoma. Inherited and sporadic retinoblastoma are the two types of conditions which lead to eye tumor formation

The gene causing retinoblastoma is located on chromosome number 13 (13q14.2). Over the decades, scientists trying to characterize and identify the function by deleting the Rb1 gene using mouse models have shown that RB1 can block cell division and also regulate cell death. It has the ability to interact with various proteins for carrying out many other functions. This gene encodes a protein product pRb, having 948 amino acids, belonging to the RB gene family along with the retinoblastoma-like 2 gene (RBL2) and retinoblastoma-like 1 gene (RBL1). RBL1 and RBL2 are present on chromosome 20 (20q11.2) and 16 (16q12.2) encoding p107 and p130, respectively, which are involved in slowing the various neoplastic growths. Till now, it is not clear as to why the retina is so susceptible for RB1 gene mutation. There are ~300 cases arising every year in the USA. The mortality rate being very low in developed countries is reported to be very high (approximately 70%) in developing countries.

Two forms of retinoblastoma are reported: hereditary form and non-hereditary form. In the hereditary form of retinoblastoma, retinoblastoma is present in the family history, through which one mutated allele is passed on to the child, while the other allele is mutated during gamete formation or in developing retina. In the non-hereditary form of retinoblastoma, mutation in one allele occurs either during gamete formation or in developing retina, and the other allele is mutated at any time during the life because of exposure to any radiation or genetic alterations. In other case of non-hereditary retinoblastoma, the allele is mutated during the lifetime resulting from epigenetic effects or genetic mutations. Hereditary form of retinoblastoma includes ~40% of the cases, while non-hereditary form accounts for ~60% of retinoblastoma cases. Retinoblastoma can be unilateral or bilateral. Bilateral retinoblastoma is generally hereditary and unilateral retinoblastoma is usually non-hereditary. Bilateral retinoblastoma leads to multifocal tumors and malignancy to various parts of the body; benign tumor with less severe form of disease is observed in the unilateral retinoblastoma. Early detection and treatment of retinoblastoma are essential, otherwise the tumor can spread to the optic nerve and further to the brain and can metastasize to the bones also. "Trilateral retinoblastoma," characterized by the development of primary midline intracranial neoplasm, has been observed in 10% children with hereditary retinoblastoma (Dimaras et al. 2015).

18.12.1 Clinical Characteristics and Diagnosis of Retinoblastoma

Leukocoria is the most common presentation of the retinoblastoma. It is the first sign which is usually observed by parents/guardians. Leukocoria is characterized by the presence of white appearance in one or both pupils. It results when a large tumor or smaller tumor associated with the retinal detachment is present. Retinoblastoma is generally characterized by strabismus (misalignment of eye). The least common signs may also include orbital cellulitis, hyphema (blood in anterior chamber), glaucoma (increased intraocular pressure), heterochromia (different color of pupils), visible extraocular growth, and decreased vision. Patients with advanced stages of retinoblastoma also show symptoms such as orbital swelling and proptosis because of extraocular invasion.

Ophthalmoscope is the simplest test to detect leukocoria. Under the anesthesia effect, portable slit lamp examination is done to check hyphema and iris neovascularization; indirect ophthalmoscopy and fundoscopy are used to detect various lesions. Further imaging methods of diagnosis include ultrasonography, computed tomography (CT), and magnetic resonance imaging (MRI). MRI is preferred over CT scan to reduce the exposure effects of ionizing radiations on the patient. MRI is done for the brain and eye orbitals to detect any malignancy. Trilateral retinoblastoma has been only detected through the MRI. Early diagnosis of retinoblastoma increases the survival chances of patients.

18.12.2 Structure and Function of RB1 Gene

Human RB1 gene consists of 178,143 bp with 27 exons and 26 introns. The RB1 gene encodes 928 amino acid protein, pRb. pRb protein contains three domains: N-terminal domain, A/B pocket domain, and C-terminal domain (Fig. 18.20). RB gene family proteins, pRb, p130, and p107, belong to the "pocket" proteins family, because of the presence of large pocket domains. These pocket domains contain binding sites for many of their interacting proteins. In addition to the presence of binding sites, A/B pocket domain of pRb also contains a LXCXE-binding cleft, which serves as a separate binding site, and through this cleft pRb interacts with

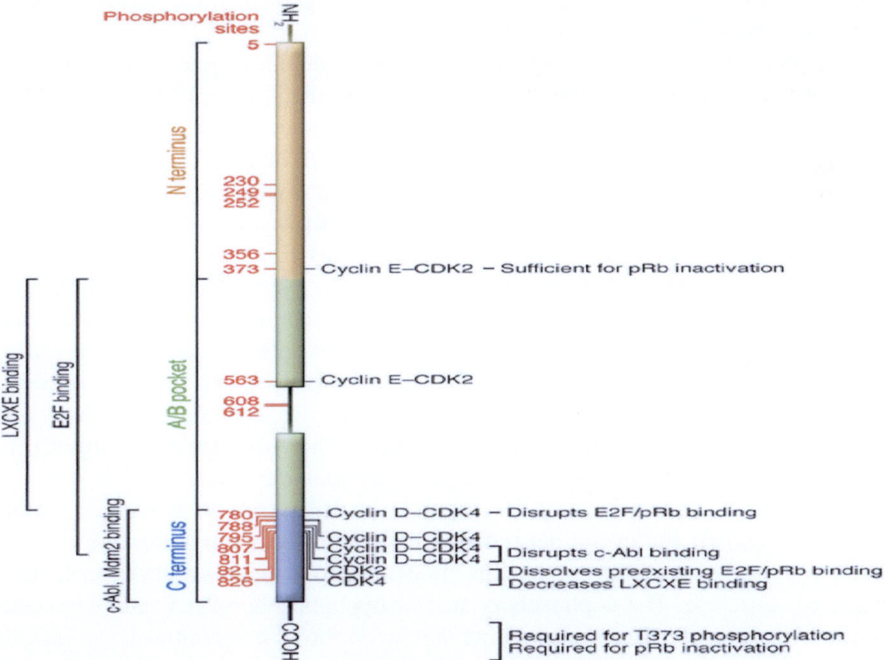

Fig. 18.20 Schematic representation of pRb protein domains

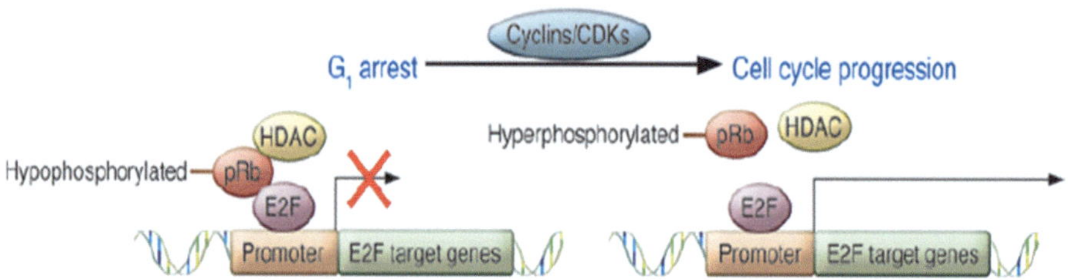

Fig. 18.21 Regulation of E2F target genes by the pRb

proteins that share an LXCXE-binding motif. Being highly evolutionary conserved, LXCXE-binding cleft also interacts with LXCXE sequence of various viral oncoproteins (human papilloma virus E6 and E7, adenovirus E1A protein, and simian virus 40 large T antigen). These viral oncoproteins bind to the LXCXE-binding motif of pRb protein and lead to its inactivation. Discovery of these viral oncoproteins plays important roles in the understanding of retinoblastoma (DeCaprio 2009). pRb contains 16 CDK (cyclin-dependent kinase) phosphorylating sites, through which activation and inhibition of pRb activity are controlled. Since the dephosphorylated/hypophosphorylated state of pRb is the active state, which is controlled by the CDKI (cyclin-dependent kinase inhibitors) of the Cip/Kip families and Ink4, phosphorylated/hyperphosphorylated state of pRb is the inactive state of pRb. Ink4 family contains p19, p18, p16, and p15 proteins, and Cip/Kip family consists of the p57, p21, and p27 proteins.

The product of RB1 gene, pRb protein, is responsible for the terminal differentiation of the myocytes by promoting the functional interaction of MyoD (myoblast determination protein 1) with the coactivator MEF2 (myocyte enhancer factor 2). By binding with the histone deacetylase 1 (HDAC1), pRb inhibits its association with the MyoD. HDAC1 is involved in gene silencing by deacetylation of histone proteins of target gene. As binding of HDAC1 with MyoD is prevented, MyoD indirectly promotes the expression of MyoD transcriptional target genes. The LXCXE motif of A/B pocket domain of pRb protein is responsible for the interaction with HDAC1. pRb plays an indispensable role in the fate determination, differentiation, survival, and migration of developing neurons. pRb protein is involved in regulation of arresting the cell at G_0 phase of cell cycle, G_1 cell cycle arrest, senescence, apoptosis, and transcriptional control. pRb regulates these functions through E2F transcription factor and Skp protein. The A/B pocket domain of pRb has the binding site for E2F target genes and Skp protein.

E2 promoter binding factor (E2F) regulates the transcription of various cyclins (cyclins A and E) and cyclin-dependent kinases (CDKs), by binding to their promoter region. These CDKs and cyclins are essential for the transition from G_1 to S phase of cell cycle. The unphosphorylated or hypophosphorylated state represents the active state of pRb, which either alone or through recruitment of histone deacetylase 1(HDAC1), ATPases, and DNA methyltransferases binds to E2F gene promoter region, thereby blocking the E2F target gene transcription (Fig. 18.21).

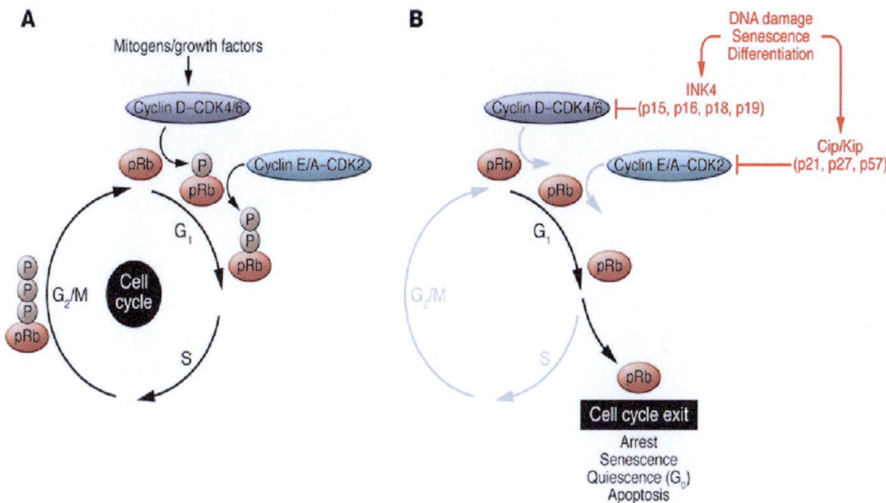

Fig. 18.22 Regulation of pRb. (**a**) Mitogens/growth factors lead to phsophorylation of pRB thus progressing the cell in S-phase. (**b**) Growth inhibitory factors result in cell cycle arrest at G1 phase

The phosphorylated or hyperphosphorylated state of pRb is its inactive state (Nevins 2001).

Skp2 (S phase kinase associated protein), human F-box protein, regulates the transition from G_1 to S phase of cell cycle and cellular senescence through interaction with cell cycle inhibitory protein, p27. p27 inhibits the cyclin E-CDK2 complex, which are responsible for phosphorylation of pRb. In growth-promoting signals (EGF, PDGF, TGF-β), Skp2 binds with the p27 and leads to its degradation, thereby activating the cyclin E-CDK2, which phosphorylates pRb, resulting in S phase transition of cell cycle. When growth-promoting signals are absent, pRb binds with the Skp2, which further interacts with the APC/C^{cdh1} and results in Skp2 degradation. p27 will accumulate in the cell, which inhibits pRb phosphorylation by inhibiting the cyclin E-CDK2 complex. The active form pRb (unphosphorylated form) leads to G_1 cell cycle arrest.

Since activity of pRb is controlled by the CDKs and CDKIs; these CDKs and CDKIs respond to various mitogen factors, DNA damage, senescence, and differentiation signals, thereby regulating the pRb activity. Growth stimulatory signals lead to the phosphorylation of pRb through CDKs, thereby inactivating the pRb, resulting in S phase of cell cycle. When growth inhibitory signals such as DNA damage, senescence, and antiproliferative signals are present, CDKIs inhibit the CDKs, thereby dephosphorylating the pRb and cell arrest at G_1 phase, or undergo senescence, quiescence, and apoptosis (Fig. 18.22).

Additional functions of RB1 include its involvement in heterochromatin formation, thus ensuring the chromosomal integrity by stabilizing the histone methylation through interaction with two LXCXE containing histone methyltransferase, Suv4-20 h1 and Suv4-20 h2. Any disruption in these interactions results in delayed mitosis process and centromere fusion, abrupt chromosomal segregation, and genomic instability. pRb inhibits the intrinsic kinase activity of TAF1 (TATA binding protein-associated factor 1). pRb promotes the centromeric localization of the CAP-D3/condensing II (Condensin-2 complex subunit D3) protein complex, thereby regulating the chromatin condensation, cohesion, and stability. In heritable retinoblastoma samples, it has been found that there is aneuploidy of 6p and 1q chromosome. Deletion in the 16q chromosome has been also reported in the retinoblastoma samples. Genes present in these regions may also involve retinoblastoma development. The Mdm2 (Mouse double minute 2) related protein, MdmX, is responsible for inhibition of p53 functions. The MDMX gene is located in the 1q32 region. In 65% and 10% of retinoblastoma cases, MDMX and MDM2 have been upregulated, since these proteins inhibit the p53-mediated apoptosis, resulting in tumor development. Deletion or inactivation of pRb leads to abrupt expression of E2F target genes, such as MAD2 (mitotic arrest deficient 2), which is responsible for encoding mitotic spindle checkpoint protein. Thus, deregulated expression of MAD2 leads to tumorigenesis (Di Fiore et al. 2013).

18.12.3 Treatment Therapy

On the basis of International Retinoblastoma Staging System, retinoblastoma has been classified among five groups: E (enucleation, advanced), D (diffuse, large) C (confined, medium), B (medium), and A (small) (Table 18.11). Till now various therapies have been used to treat retinoblastoma, like focal therapy (external beam radiation, brachytherapy, cryotherapy, thermotherapy), chemotherapy (vincristine + low-dose carboplatin, vincristine + high dose of carboplatin + etoposide + G-CSF, subtenon carboplatin, and prophylactic 3-agent chemotherapy), and enucleation. These treatments result in severe side effects. Various research have been undertaken to develop a targeted approach for treatment of retinoblastoma with less side effects (Lin & O'Brien 2009; Sachdeva & O'Brien 2012; Teixo et al. 2015).

18.12.3.1 External Beam Radiation

It is the most effective method to treat the retinoblastoma. It is generally advisable to the patients with bilateral retinoblastoma. When chemotherapy fails and there is recurrence of disease, then EBR is used to treat the tumor. EBR is effective in cases of small tumors in the macula, where chemotherapy does not work. The range of dose from 42Gy to 46 Gy is prescribed, with the 58–88% protection of the eye. Patients with heritable retinoblastoma receiving EBR treatment lead to several secondary malignancies, such as osteosarcoma and cancer of the nasal cavities and the brain. Children below 1 year of age have high risk of developing undesired effects like cataract, facial and temporal bone hypoplasia, vitreous hemorrhage, and orbital hypoplasia.

18.12.3.2 Brachytherapy

Brachytherapy involves the implantation of a radioactive agent in the sclera, near the tumor base. In this therapeutic approach, several radioactive agents have been used, such as ruthenium-109 (^{109}Ru), iridium-192 (^{192}Ir), iodine-125 (^{125}I), gold-198

Table 18.11 Different stages of retinoblastoma

Group	Characteristics	Treatment	Prognosis	Toxicities
E (advanced, enucleation)	≥1 mm tumor in anterior segment and ciliary body, no vision, neovascular glaucoma, vitreous hemorrhage, phthisical eye, orbital cellulitis, extraocular disease	1. Enucleation 2. Chemotherapy with prophylactic-3 agent	Morbidity high from the treatment and no visual potential	Neurotoxicity, hyponatremia, nephrotoxicity, ototoxicity, secondary leukemias
D (large, diffuse)	>3 mm from tumor margin with diffuse vitreous or subretinal seeding, and subretinal fluid	1. Chemotherapy with three agents (vincristine, carboplatin, etoposide) for six cycles 2. With or without subtenon or carboplatin 3. External beam radiation (EBR)	Variable visual prognosis; high morbidity from treatment	Neurotoxicity, hyponatremia, nephrotoxicity, ototoxicity, secondary leukemias Decreased ocular motility, optic atrophy with ischemic necrosis, pseudo-preseptal cellulitis Midfacial hypoplasia, soft tissue and osteosarcoma, brain tumors
C (medium, confined)	≤3 mm from tumor margins with vitreous or subretinal seeding	1. Chemotherapy with three agents (vincristine, carboplatin, etoposide) for six cycles 2. Treatment with or subtenon or carboplatin 3. Focal therapy	Visual prognosis variable	Neurotoxicity, hyponatremia, nephrotoxicity, ototoxicity, secondary leukemias Decreased ocular motility, pseudo preseptal cellulitis, optic atrophy with ischemic necrosis Vitreous seeding, radiation retinopathy, retinal tears
B (medium)	>3 mm height, subretinal fluid clear ≤3 mm from tumor margin	1. Vincristine + low-dose carboplatin up to six cycles 2. Focal therapy from two to six cycles	Good visual prognosis	Neurotoxicity, hyponatremia, nephrotoxicity, ototoxicity, Vitreous seeding, radiation retinotherapy, retinal tears

(continued)

Table 18.11 (continued)

Group	Characteristics	Treatment	Prognosis	Toxicities
A (small)	$\leq$3 mm height, $\geq$2 mm disk diameter from fovea, $\geq$1 mm disk diameter from optic nerve	Focal therapy (argon/YAG laser, brachytherapy, hyperthermia, or cryotherapy)	Good visual and overall prognosis; usually eradicated	Tissue damage, vitreous seeding, retinal tears, chorioretinal atrophy, radiation retinotherapy

(^{198}Au), cobalt-60 (^{60}Co), yttrium-90/strontium-90 (^{90}Y/^{90}Sr), palladium-103 (^{103}Pd), and rhodium-106/ruthenium-106 (^{106}Rh/^{106}Ru). In this therapy for 2–4 days, the tumor is exposed to 40–45 Gy ionizing radiation. Brachytherapy is used in patients who stop responding to chemotherapy and EBR. Brachytherapy is only advisable to the small tumor; however it is not recommended to large tumor size or which is associated with the macula. This therapy has lesser side effects compared to the EBR and includes cataract formation, radiation retinopathy, and optic neuropathy.

18.12.3.3 Thermotherapy

Thermotherapy process consists of applying heat directly to the tumor through the use of infrared radiation. In this therapeutic approach, temperature ranges from 45 °C to 60 °C that does not result into blood coagulation in retinal vessels. It is advisable for the small dimensions of retinoblastoma.

18.12.3.4 Laser Photocoagulation

Laser photocoagulation is associated with the delivery of diode laser or argon laser or a xenon arc to the tumor. This technique is recommended to the small posterior tumor, in which blood supply to the tumor is prevented by coagulation of blood. Normally at monthly intervals, two or three sessions are required for the effective treatment. Side effects of this treatment include the preretinal fibrosis, retinal vascular occlusion, and retinal detachment.

18.12.3.5 Cryotherapy

The principle of cryotherapy is the complete destruction of the vascular endothelium that supplies the tumor, through the freezing process. This therapy is recommended for the small peripheral tumors. On a monthly interval, one to two sessions with three times per session of cryotherapy are done to treat the tumors. The side effects of this therapy include retinal detachment and vitreous hemorrhage.

18.12.3.6 Chemothermotherapy

In this therapeutic approach, chemotherapy and thermotherapy are applied at the interval of a few hours to combat the large-sized tumors. This treatment is most effective in small-sized tumors present near the optic nerve and fovea. The major disadvantage of this treatment is the atrophy of focal iris, retinal detachment, and optic disk and corneal edema.

18.12.3.7 Chemotherapy

Chemotherapy consists of the administration of drugs through different routes such as intravenous, intra-arterial, periocular, or intravitreal to reduce the tumor size so that disease can be eradicated completely using other therapies. For intraocular retinoblastoma or unilateral retinoblastoma, intravenous chemotherapy is recommended. Carboplatin, vincristine, and etoposide are the main chemotherapeutic agents, which are used in combination for six cycles based on the patient's body weight. The major side effects of these drugs are high risk of bacterial infections and development of new tumors in various body parts. In view of side effects, less toxic chemotherapeutic agents, such as topotecan and 2-deoxy-D-glucose (2-DG), are used. Topotecan is an inhibitor of DNA topoisomerase-1 and 2-DG is a glycolytic inhibitor. In intra-arterial chemotherapy, infusion of melphalan drug into the ophthalmic artery is done. It is safe and very effective for the unilateral retinoblastoma. This technique is recommended for medium- and large-sized tumors. This approach leads to local ocular toxicity. Injections of carboplatin are given in periocular chemotherapy with systemic chemotherapy to increase the dose in the vitreous for controlling retinoblastoma. The side effects of this approach include optic atrophy, strabismus, and orbital and eyelid edema. Melphalan drug is used in intravitreal chemotherapy against the retinoblastoma. It is one of the most effective methods, but numerous injections in the eye for a period of 1 year lead to vitreous seeding.

18.12.3.8 Enucleation

Enucleation is applied in the advanced cases of retinoblastoma. In this approach the eye is replaced with orbital implants of plastic, silicone, or hydroxyapatite. Enucleation leads to complete vision loss. It is only effective when timely done, otherwise chances of malignancies increase.

18.12.3.9 Targeted Approach with Small Molecules

Since all the above therapeutic approaches lead to severe side effects, there is a need for the development of targeted approaches with less side effects. With the improvement in the understanding of the different interacting molecular pathways, new approaches have been developed. It is a well-known fact that p53 is involved in apoptosis, so by stimulating the p53-mediated apoptosis pathway, tumor growth can be inhibited. A small molecule, Nutilin-3, inhibits the Mdm2 and MdmX interaction with the p53, resulting in apoptosis in retinoblastoma through p53-mediated pathway. When nutilin-3 is used in combination with the topotecan, a p53 inducer, it leads to synergistic effect on the reduction of retinoblastoma cells. The side effect of nutilin-3 and topotecan is negligible. It is reported that the cells with upregulation of E2F1 activity are responsive to the HDAC inhibitors. Thereby, another targeted approach in which HDAC inhibitors (HDACi) are used leads to apoptosis in retinoblastoma cells. These molecules are still under clinical trials. To combat the retinoblastoma, targeted approaches with less side effects need to be developed.

18.13 p53 Tumor Suppressor Gene

The 20 kb gene encodes for a nuclear phosphoprotein, p53, with 53 kDa molecular weight and containing 11 exons and 10 introns, located on chromosome 17pl3. The other members of the family to which p53 gene belongs include p63 and p73. Although being related structurally and functionally, p53 has evolved as a tumor suppressor gene in higher vertebrates, whereas p73 and p63 are involved in developmental biology. p53 was first discovered in 1979, as bound to viral oncoprotein in SV40 transfected cells. p53 has DNA-binding protein property and is usually found in low quantities inside a normal cell's nucleus, whereas 5–100X quantities can be localized in both transformed and tumor cells.

Human p53 protein consists of 393 amino acids and comprises of various domains (Fig. 18.23): an amino-terminal domain (1–42 amino acids) and proline-rich region (61–94 amino acids) at N-terminal, a middle domain (102–292 residues), and an oligomerization domain (324–355 residues), strongly basic regulatory domain (363–393) at C-terminal (301–393 residues) along with nuclear export signal sequence and nuclear localization signal (Fig. 18.23). The amino-terminal region regulates the interaction with transcription factors (such as acetyltransferase and MDM2) and transactivation activity. The stability of p53 is controlled by the proline-rich region. Any mutation or deletion in this region increased the susceptibility of p53 to degradation by MDM2. The central region of p53 is evolutionarily highly conserved. The negative regulatory role is played by basic C-terminus, which is involved in cell death induction. Various structural studies have revealed that the majority of p53 mutations are missense mutations in the central DNA-binding domain found in cancers, and mostly 126–130 residues are focused to study p53 mutation (Bai & Zhu 2006; Choisy-Rossi et al. 1999; Harris 1996; Joerger & Fersht 2010; Kamaraj & Bogaerts 2015; Sigal & Rotter 2000).

18.13.1 The Physiological Functions of p53

Following genotoxic stress, p53 functions as a tumor suppressor gene, thereby preventing unregulated cell proliferation and maintaining the genome

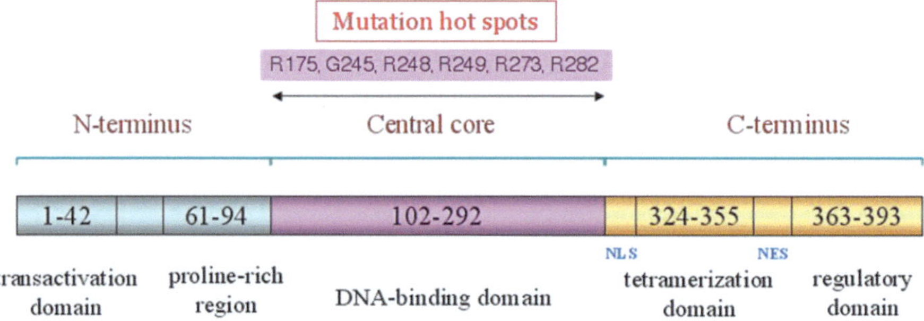

Fig. 18.23 Diagrammatic representation of the p53 protein structure

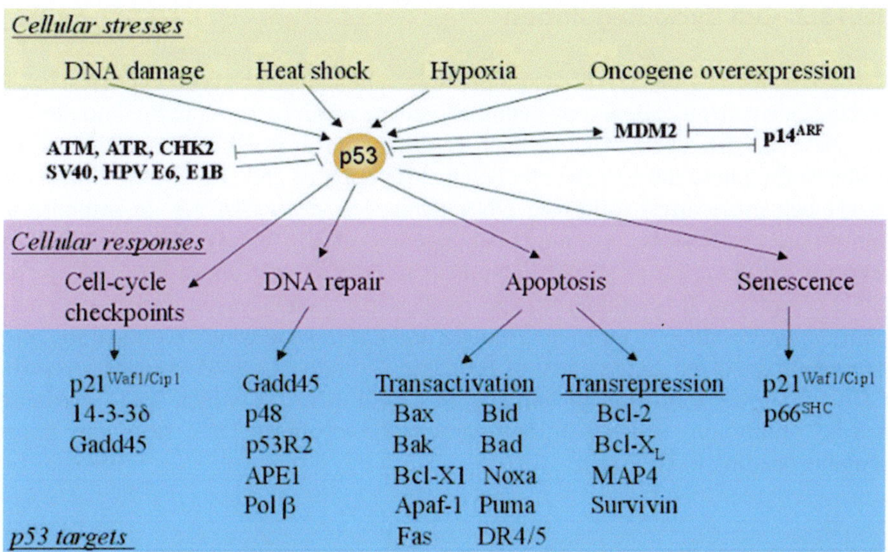

Fig. 18.24 Schematic representation of p53 at the middle of a complex signaling network under stress

integrity (Finlay et al. 1989). Following various stimuli both extracellular and intracellular in nature, such as DNA, hypoxia, heat shock, and overexpression of oncogene, activation of p53 occurs and hence triggers various biological responses. The activation of this protein involves overall fold-change in protein level as well as via extensive post-translational modification, eventually activating the several p53-targeted genes. For instance, in case of double-strand DNA damage, activation of protein kinase ATM (ataxia-telangiectasia mutated) results in Chk2 kinase activation. In turn, both the proteins lead to phosphorylation of p53, thereby resulting in p53-dependent apoptosis or cell cycle arrest. Damage to DNA leads to blocking of replication process and activates the ATR (ATM and Rad3-related) and Chk1, subsequently resulting in p53 phosphorylation and activation. Activation of genes by wild-type p53 results in DNA repair, senescence, cell cycle arrest, and apoptosis, by regulating the downstream signaling molecules involved in these processes (Fig. 18.24), such as $p21^{waf1/Cip1}$, Bcl-2 family, and Gadd45 (growth arrest and DNA damage inducible protein 45). In addition, p53 elevated level leads to inhibition of various gene expressions such as cyclin B1, bcl-X, bcl-2, MAP4, and survivin. Interestingly, it was seen that in ovarian cancer cells having p53-expressing adenovirus infection, 80% of putative p53-responsive genes are repressed.

18.13.2 Cell Cycle Regulation

Cell cycle arrest in the G1, G2, and S phases can be induced by p53. The arrest at G1 and G2 subsequently allows cell genomic damage repair before entering into the S & M phase of cell cycle. Once repair is done, the arrested cells enter into the proliferating phase through the biochemical function of p53. The primary mediator of G1 cell cycle arrest following DNA damage is p21waf1/Cip1. In response to stimuli such as stress, upregulation of endogenous Cip1/p21waf1 mRNA and protein levels takes place via p53. ZRXL motif of p21waf1/Cip1 in turn binds to cyclin-CDK complexes. It is reported that p21waf1/Cip1 overexpression blocks Rb phosphorylation, resulting in arrest at G1 phase and E2F release which is critical for gene expression involved in the entry of S phase. Similarly, Gadd45 prevents cyclinB/CDK1 complex formation by binding to CDK1 (CDC2), which leads to kinase activity inhibition, and 14–3-3δ separates the cyclinB/CDC2 from the target proteins, leading to G2 arrest.

18.13.3 Induction of Apoptosis

p53-induced apoptotic genes include p53AIP1 (p53-regulated apoptosis-inducing protein 1), Scotin, Bax (Bcl-2-associated X protein), DRAL, PIDD (p53-induced protein with death domain), PIG3 (p53-inducible gene 3), Fas/CD95 (cell death signaling receptor), Noxa (from the Latin word for "harm" or "damage"), Puma (p53-upregulated modulator of apoptosis), Apaf-1 (apoptotic protease-activating factor-1), PERP (p53 apoptosis effector related to PMP-22), DR5/KILLER (death receptor 5), etc. (Fig. 18.25). Bax plays a critical role in apoptosis mediated through Puma and also reportedly participates in response to death indirectly via Puma. However, p53-mediated apoptosis seems to require Bax in a cell-dependent manner. Loss of Bax results in 50% of accelerated brain tumor growth which are related to the loss of function in p53. In addition to this, Bax is involved in the apoptosis in intestinal epithelial cells and thymocytes. In addition, several Bcl-2 family members and mitochondrial proteins are also involved in apoptosis induced by p53. During apoptosis, p53 is seen to regulate the genes encoding Apaf- 1, apoptosome component, and PIG3, involved in depolarization of mitochondria and can also promote apoptosis through death receptor activation located at the plasma membrane, including DR5, DR4 and Fas/CD95, and often IAP (inhibitor of apoptosis proteins) inhibition. Apoptosis is also induced by the ER-dependent mechanism by transactivating Scotin expression, a protein present in the ER and nuclear membrane. p53 also promotes apoptosis in transcription-independent manner, in certain cell types, and under certain conditions. For example, in mitochondria, p53 is reported to directly bind Bcl-XL/Bcl-2 and thus displace Bax or BH3 domain-only pro-apoptotic proteins, thereby facilitating Bax-dependent apoptotic changes in mitochondria. p53 induces oligomerization by binding to Bak in mitochondria, which facilitates plasma membrane permeabilization and release of cytochrome

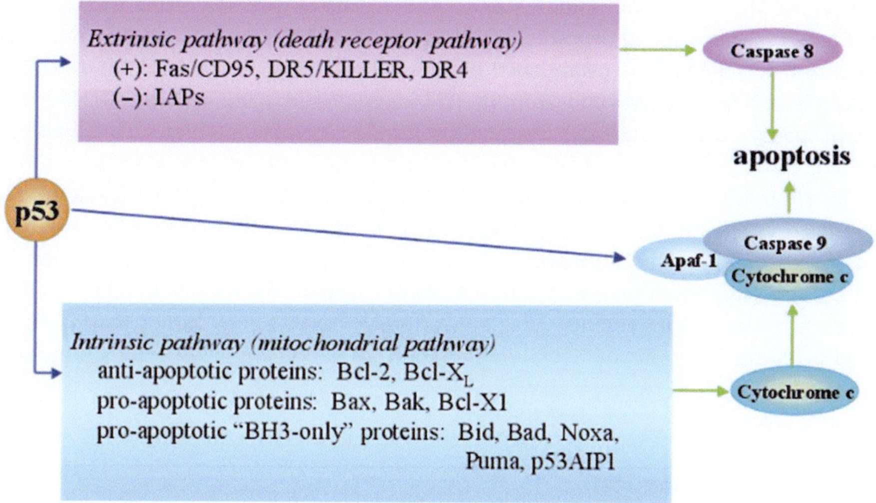

Fig. 18.25 Schematic representation of p53-associated genes and pathways involved in apoptosis

c. This is a tissue-specific response. This is a very fast response (30 min), which takes over the transcriptional response (taking more than 2 h).

Studies have shown that tumorigenesis is accelerated in the mouse brain when there is loss of p53-dependent apoptosis. Transcriptional factors such as ASPP (apoptotic-stimulating protein of p53) family, JMY (junction-mediating and regulatory protein), c-Myc, p73, and p63 affect the balance between the cell cycle arrest and apoptosis. The balance between p21Waf1/Cip1 and Puma identified in human colorectal cancer cells regulates the apoptosis and cell cycle arrest in response to p53.

18.13.4 p53 Level and Activity Regulation

p53 exists mainly in an inactive form and is maintained at a very low concentration under normal conditions. Crucially, p53 low basal level has to be even tightly controlled during cell cycle progression. A complex cellular protein network is required for regulation of p53 level including PARP-1, MDM2, JNK, HPV16 E6, SV40 T-antigen, E1B/E4, Pirh2, and WT-1. p53 stability increases when it binds to E1B/E4, WT1, or SV40 T antigen, whereas its degradation accelerates when it associated with MDM2 or E6. MDM2 blocks p53 transcription or stimulates its export to the nucleus, or its degradation, thereby, inhibits p53 activity. MDM2 protein interacts with transactivation domain of p53 at N-terminal and blocks its interaction with the transcription components. Because of its intrinsic E3 ubiquitin-ligase activity, it mediates p53 ubiquitylated-dependent degradation. It recruits the histone deacetylase 1 (HDAC1) at the p53 C-terminal, thereby marking it for ubiquitination. However, under stress, p53 regulates MDM2 gene transcription through p53-dependent promoter present in MDM2 gene. In this manner, an autoregulated network is established between the activity of p53 and MDM2 gene expression. Under genotoxic stimuli, p53 degradation mediated by MDM2 was abolished by various mechanisms, leading to p53 response maintenance (Shaikh & Niranjan 2015).

In addition to the protein complexes, stability and activity of p53 are also highly regulated via post-translational modifications, such as phosphorylation, acetylation, ADP-ribosylation, ubiquitylation, sumoylation, etc. Acetylation and phosphorylation are two important post-translational modifications, which leads to p53 stabilization and promotes its activity as a transcription factor for its target genes. These modifications also prevent p53 degradation by abolishing its interaction with the MDM2.

18.13.5 p53 Mutation

A few common missense mutations in the DNA-binding domain (102–292 amino acids) of p53 subsequently results in both the loss of tumor suppressor function and accumulation of a mutant protein, having its own oncogenic life. The p53 loss of function in cancers is brought about by several mechanisms, including lesions, gene mutations, or mutations in the downstream mediators of p53's function. Approximately 18,000 p53 mutations are identified in the majority of human cancers. Majority of these tumors are either because of loss or mutation in p53, resulting in its activation. For example, 70% of lung cancer, 45% of stomach cancer, and 60% head and neck, colon, bladder, and ovary cancer are because of p53 mutation. In addition, mutations in the downstream signaling molecules of p53 function result in 50% of human cancers. In Li-Fraumeni syndrome, inherited germline mutation is present in one allele of p53, and people with this syndrome are more susceptible for cancer development during lifetime. Brain, breast, adrenal gland, hematological system, and connective tissue tumors develop because of loss of p53 alleles. It has been reported that mice deficient in p53 allele are associated with tumorigenesis. In addition, p53 inactivation is also associated with the tumor's development, because of its inactivation by defects in upstream signals or oncoproteins.

18.14 Cell Cycle and Cancer

There exist a confined number of vital events which propel tumor cells and its daughter cells into uncontrolled proliferation and invasion throughout the body. One such vital event is cell proliferation gone haywire which along with suppressed apoptosis gives a platform to tumor cells for further neoplastic progression. Normally, a cell cycle consists of three important events, namely, growth, DNA synthesis, and division, where the duration of each event is tightly controlled by chemical signals provided to or by the cells. In addition, the transition between each phase

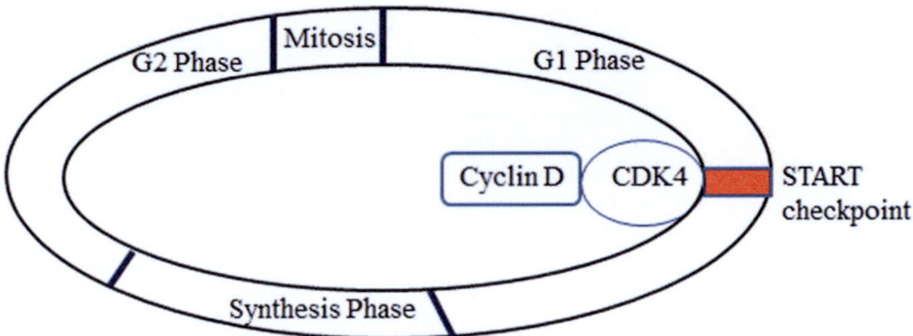

Fig. 18.26 Schematic representation of START checkpoint in the cell cycle

demands selective chemical signals and timely responses which if gone wrong such as in cases where signals are not correctly sensed or when the cell is not prepared for the response can give rise to cancerous tissue. The major phases involved in the cell cycle are G1 phase, S phase, G2 phase, and M phase which are regulated at various checkpoints and halt the progression of cell until it has all the necessary machinery such as DNA synthesis or repair of faulty DNA. A cell can only progress further into the division process when all the checkpoints are satisfied. Cyclins and cyclin-dependent kinases (CDKs) are critical during cell cycle. CDKs are catalytically active cell cycle components, which can transfer the phosphate group, thus regulating the activity of other proteins (Malumbres & barbacid 2009). However, their activity majorly depends on their interaction with cyclins forming cyclin/CDK complexes leading to CDK activation. Normal cell cycle requires cyclic pattern in the formation and degradation of cyclin/CDK complexes.

G1 phase involves the most important checkpoint, START (Fig. 18.26), to decide the appropriate time to proceed to S phase. The cell is said to be committed to cell division only if it passes this checkpoint after which DNA replication is initiated. There exist inhibitory proteins which while sensing issues in G1 phase such as DNA damage can halt the cyclin/CDK complex, thereby preventing the cell to enter S phase.

In tumor-inflicted cells, these checkpoints are deregulated often due to defects in genetic machinery such as mutation in genes encoding cyclin or CDKs (Table 18.12) or by modified proteins during cell cycle disruption (Table 18.13).

Normal cells have been programmed in such a way that they pause at the START to ensure all the machinery is working before proceeding for DNA replication. On the other hand, cells in which this checkpoint is faulty move to S phase without repairing the DNA damage. Over a period of time, these accumulated mutations cause further cell cycle deregulation, thus leading to the formation of aggressive cancerous cells.

Table 18.12 List of cyclins and CDKs along with their function involved during cell cycle

Protein	Chromosome	Function	Relevance in human cancer
Cyclin A	4(q25-q31)	Regulation of S phase and G_2-M transition by forming complex with CDK2 and 1	Hepatocellular carcinoma and breast carcinoma showed overexpression
Cyclin B1	5(q13-qter)	Regulation of G_2-M transition by forming complex with CDK1	Breast carcinoma showed overexpression
Cyclin D1	11q13	Early G_1 phase regulation by forming complex with CDK4/6	Overexpression in various tumors including lymphoma, breast cancer, and parathyroid adenoma is reported
Cyclin D2	12q13	Early G_1 phase regulation (in some cells) by forming complex with CDK4/6	Overexpression in few colorectal cancers
Cyclin E	19q12	Regulation of late G_1 and G_1-S transition by forming complex with CDK2	Overexpression is reported in various cancers including colon, breast, prostate carcinomas, and leukemia
CDK1	10	Regulation of G_2-M transition by forming complex with cyclin B1	Overexpression in breast cancer
CDK4	12q13	Early G_1 phase regulation by forming complex with cyclin D	Mutation in case of melanomas and overexpression in brain tumors

18.14.1 Cell Cycle and Signal Transduction

Cells at quiescent stage are stimulated by appropriate signals, so that they enter into the cell cycle. These signals enter the cell by molecules like growth factors and hormones that have the ability to bind to receptors present on the cell surface relaying the signals into the cytoplasm via a process called signal transduction. The final result is certain gene expression activation which can propel a cell out of a quiescent phase into cell cycle. By contrast to normal cells, cancer cells are reported to have defects in signal transduction pathways ranging from abnormal growth signals or molecules present at downstream signaling. In addition, cancerous cells have also been shown to stop responding to external growth inhibitory signals.

18.14.2 Cell Cycle Checkpoints

To deal with exogenous and endogenous sources of DNA damage, cells have cell cycle checkpoint and conserved DNA repair mechanisms (Funk 2001; Kastan & Bartek 2004; Visconti et al. 2016). In order to regulate events in cycle progression such as case of genome damage, complex signal transduction pathways play a role which together constitute "checkpoints." These checkpoints allow cell DNA damage repair by arresting cell cycle. In order to avoid risking generation of altered progeny, some of the cell types prefer to undergo the programmed cell death (apoptosis). When damage to the genome is detected, cell cycle checkpoints delay the cell

Table 18.13 Types of proteins involved in cancer due to cell cycle disruption

Protein	Chromosome	Function	Role in human cancer	Mouse knockout models
p21CIP1	6p21	Block G_1 and S phase by binding to multiple cyclin/CDK complexes and proliferating cell nuclear antigen (PCNA); induced by p53	Rare mutations in the breast, bladder, and prostrate carcinomas	Defect in G1-S checkpoint, no spontaneous tumors, no tumor suppressor
p27KIP1	12p13	Induce G_1 arrest by binding to multiple cyclin/CDK complexes and inhibit them	Variable loss in expression of protein in several malignancies and heterozygosity loss	Pituitary hyperplasia/ adenoma, organomegaly, gigantism; haplo-insufficient tumor suppressor
p57KIP2	11p15.5	Induce G_1 arrest by binding to multiple cyclin/CDK complexes and inhibit them	Mutation found in Beckwith-Wiedemann syndrome patients, few inactivation identified	Adrenal hyperplasia, developmental defects, neonatal lethality; no spontaneous tumors
p61NK4a	9p21	Induce G_1 arrest by binding to CDK4/6 and inhibit its function	Often inactivated in bladder and lung carcinomas, pancreatic adenocarcinomas, melanoma	Carcinogen-induced increase in melanomas, low chances of spontaneous mutations; cooperative effects with haplo-insufficient p14ARF status
p14ARF	9p21	G_1 and G_2 arrest by blocking MDM2 inhibition of p53	Mutation in gliomas, melanoma cell lines; targeted in acute T-cell leukemia	High incidence of induced and spontaneous mutations; p16INK4a$^{-/-}$ p19ARF$^{-/-}$ show very similar phenotype

progression, by affecting the critical cell cycle regulator activity. The checkpoints are essential for the genetic stability maintenance, and therefore any mutations in their components result in the aberrant cell cycle progression under perturbing stimuli. How a cell responds to any DNA damage type is one essential constituent of the cancer biology field. From in vivo and in vitro studies using animal models, and from the fact that mutations in genes implicated in DNA damage responses, it can be well concluded that damage to cellular DNA leads to cancer. Also,

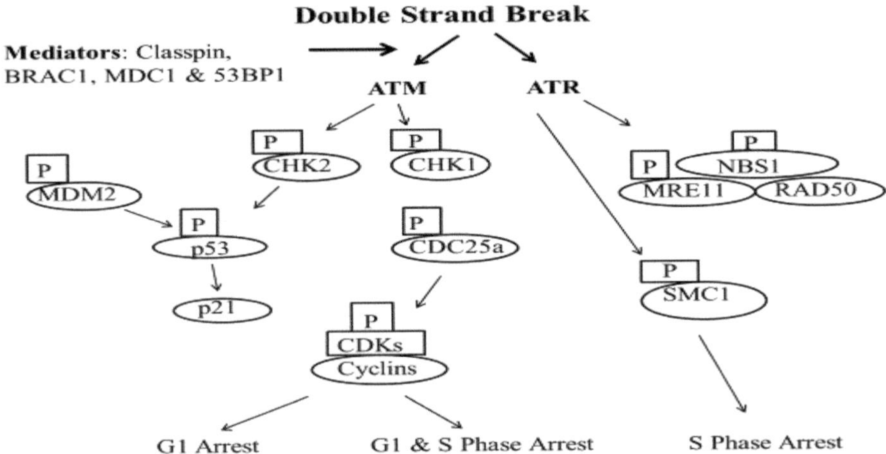

Fig. 18.27 Cell cycle checkpoint pathways. ATM(ATR)/CHK2(CHK1)–p53/MDM2-p21 pathway is shown to stop the G1-S phase transition of cell. *P* phosphorylation

surprisingly DNA damage is often used for cancer treatment. Many of the therapeutic approaches used to cure malignancies by targeting the DNA includes chemo agents and radiation therapy. In addition, DNA damage itself leads to various side effects, for example, hair loss, gastrointestinal toxicities, and bone marrow suppression. So, weirdly, DNA damage is crucial for the disease cause as well as treatment of disease and also for the toxicities for the same disease. A variety of different repair mechanisms exists for the plethora of DNA lesions. The cell, besides undergoing toward DNA damage repair, undergoes apoptosis or blocks the progression of cell proliferation. Although we still have a limited understanding about the coordination of cell cycle arrest or programmed cell death with DNA repair, this coordination is of utmost importance for cell outcome optimization. In addition to the DNA damage, cells are also required to cope with other stresses such as deficiency in nutrients or oxygen levels. The term "cell cycle checkpoint" referred to the process in which cell cycle progression is halted until the cell ensures that earlier steps, such as mitosis or DNA replication, are completed.

Cell cycle progression is a highly regulated process in normal cells to ensure that each step should be completed before proceeding to the next. There are three well-known cell cycle checkpoints, the G1/S, the G2/M, and M checkpoints (Fig. 18.27), in the whole cell cycle where equilibrium between external and internal signals is checked before a cell enters the next stage.

18.14.2.1 The G1 and G1/S Checkpoint Responses

G1/S checkpoint monitors the cell size and DNA damage and halts the cell cycle progression until these faults are corrected. After the cell size and DNA integrity pass the normality test, the checkpoint is traversed, and cell proceeds for S phase. In mammalian cells, the most essential response toward the DNA damage while traversing through the G1 phase is ATM(ATR)/CHK2(CHK1)–p53/MDM2-p21 pathway, resulting in induction of G1 arrest.

Although there appears to be a relatively constant ATM and CHK2 expression during cell cycle, it is the concentration of CHK1 and ATR that fluctuates with low levels till mid-G1 and starts their important activities once G1/S transition approaches. ATR/ATM acts by phosphorylating the p53 at serine 15 which falls under the amino-terminal transactivation domain. Threonine 18 and serine 20 within the same domain of p53 are reported to be targets of CHK1/CHK2. In addition, MDM2, ubiquitin ligase which normally binds to p53, is targeted after the DNA damage by both ATM/ATR and CHK2/CHK1. These modifications aid stabilization and accumulation of p53 protein and increase in its activity. The main target of p53 is the transcription of p21CIP1/WAF1 inhibitor of CDKs, which inhibits the G1/S-promoting cyclin E/Cdk2 kinase causing G1 arrest of the cell. This results not only in the inhibition of DNA synthesis initiation but also leads to safeguard of RB/E2F pathway in its active, growth-suppression mode and eventually causes G1 blockade sustain. In humans, the most commonly deregulated tumor suppressor pathways, governed by p53 and pRB, are targeted by G1 checkpoint. Increased CHK1 and ATR expression along with induction of Cyclins E and A, and CDC25A phosphatase, cyclin E(A)/CDK2 kinase activator occurred during late G1. Under genotoxic stress, enhancement in the physiological mechanism via CHK1 and CHK2 increased activity and leads to CDC25A downregulation and eventually cyclin E(A)/CDK2 complex inhibition. Interestingly, despite the p53 and CDC25A constitutive phosphorylation by checkpoint kinases, CDC25A-degradation cascade is faster as compared to the p53 pathway. Thus, the CHK1/CHK2-CDC25A checkpoint activated quickly, independent of p53, and results in delay of G1/S transition, unless p53-dependent mechanism sustains and leads to prolonged G1 arrest.

18.14.2.2 S Phase Checkpoint

Activated by genotoxic insults, the intra-S phase causes transient and reversible inhibition of those DNA replication origins that have not been initiated yet. Two independent, ATM/ATR controlled pathways are responsible for the entry into S phase. One of the mechanisms is operated through the CDC25A-degradation cascade. CDC45 protein required for DNA polymerase recruitment into pre-replication complexes is blocked by CDK2 activity inhibition downstream of CDC25A pathway; hence CDK2 activity inhibition leads to prevention of the new origin initiation. The other part of checkpoint involves the ATM-mediated NBS1 phosphorylation, in particular at serine 343 and serines 966 and 957 of the SMC1 cohesin protein. The concept of these two branches of intra-S phase checkpoint reported for ionizing radiation and to ultraviolet light responses. Apart from origin of replication inhibition, another essential function of S phase checkpoints (replication checkpoint) is maintenance of the stalled replication fork integrity. Such fork stability maintenance, by undiscovered mechanisms, helps the prevention of primary lesions converting into DNA breaks and facilitates recovery of DNA replication subsequently. The primary function of G1/S CDKs is activation of E2F. For the S phase induction and to overcome a G1 arrest, cyclin E-associated kinase and E2F overexpression are required.

18.14.2.3 The G2 Checkpoint

When a cell experiences DNA damage during G2 phase or when it progresses into G2 with unrepaired damage during S or G2 phases, G2 checkpoint (G2/M checkpoint) prevents the cell from mitosis initiation. In addition, the G2 arrest reflects the S/M checkpoint contribution, after sensing some persisting DNA lesions from the previous S phase. The important target of the G2 checkpoint is the cyclin B/CDK1 kinase and its mitosis-promoting activity specifically whose activation during stress is inhibited by CHK1/CHK2, ATM/ATR, and/or sequestration, degradation of subcellular components mediated by p38-kinase, and/or CDC25 phosphatase inhibition which under normal conditions activate CDK1 at the boundary of G2/M. BP1 and BRC1 are involved in the G2 checkpoint response regulation as being analogous to S phase checkpoint mediator role. Cell cycle inhibitors such as CDK inhibitor p21, 14–3–3 sigma proteins, and GADD45a (growth arrest and DNA damage inducible 45 alpha) upregulation are controlled by BRCA1 and p53 which maintain the G2 checkpoint. The fact that tumors with p53 mutation tend to accumulate in G2 due to DNA damage and indicate sustenance of G2 arrest is brought about by p53-independent mechanisms. Moreover, this phenomenon has put forward new scope to interfere with this checkpoint to sensitize cancer cells as a potential treatment, with G1/S checkpoint deficient, to DNA damage due to irradiation or drug usage.

Mutation of any kind in the genes involved in checkpoints of the cell cycle contributes to cancer development. For example, mutation in genes controlling the first two checkpoints of the cell cycle might result in continuous growth and division without DNA repair. Likewise, if genes for cyclins are expressed at a wrong time or at incorrect levels, the cell may never be able to exit the cell cycle and enter the quiescent phase, the result being the cancerous cell.

18.14.3 Cyclin D1 and Cyclin E Proto-oncogenes

Three homologues of Cyclin D are found in mammalian cells: Cyclin D1 (Bates & Peters 1995; Diehl 2002), cyclin D2, and Cyclin D3, which are responsible for the formation of active protein kinase by binding with either CDK6 or CDK4. The structure analysis of Cyclin D revealed the presence of LXCXE-binding cleft near N-terminus that binds with the LXCXE motif of pRb and cyclin box in the middle that interacts with the CDKs and PEST sequence at the C-terminal. The induction and assembly of Cyclin D1 with CDK4 are regulated by the growth factors via Ras-mediated pathways. Under the influence of mitogenic factors, Ras kinase is activated, which further activate downstream molecule Raf, subsequently activation of mitogen-activated protein kinases (MEK1 and MEK2), resulting in continuous extracellular signal-regulated protein kinase (ERK) activation, which regulates the transcription of Cyclin D1 and its association with CDKs. In another signaling, Ras leads to the activation of PI3K (phosphatidylinositol 3-kinase), thereafter Akt activation, which results in glycogen synthase kinase-3β (GSK-3β) inhibition. The GSK-3β leads to phosphorylation of the specific threonine (Thr-286) residue present

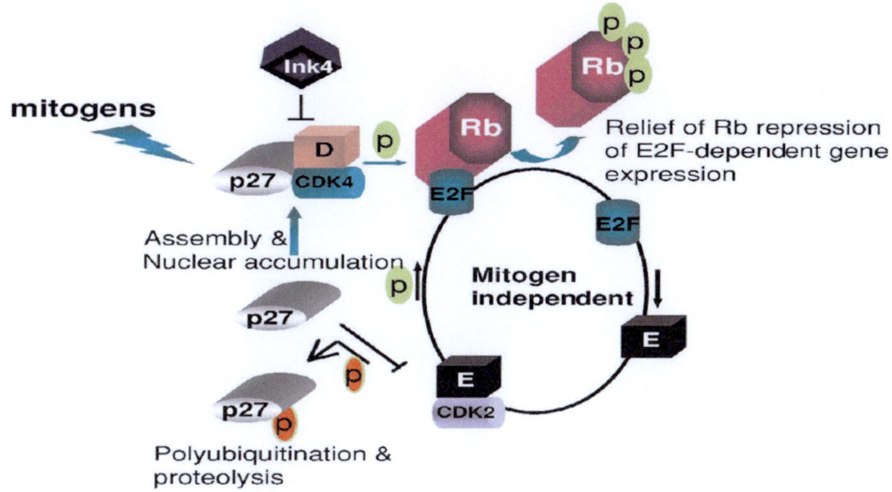

Fig. 18.28 Schematic representation of restriction point control

near the C-terminal of Cyclin D, which is essential for Cyclin D polyubiquitination-mediated proteolysis and nuclear export. The accumulation of Cyclin D in the nucleus induces a cell cycle G1 phase. To cross the G1 phase checkpoint, Cyclin D leads to the pRb protein inhibition through its phosphorylation, thereafter release of E2F transcription factor that regulates Cyclin E (Hwang & Clurman 2005) and other proteins involved in the G1-S phase transition (Fig. 18.28). In addition, Cyclin D1/CDK4 complex bind with the Cip/Kip family (CDK inhibitory protein) proteins and stabilize Cyclin D1 in the nucleus, inhibits the cyclin D1 nuclear export, and induces activation of the Cyclin E/CDK2.

PRAD1 is considered a putative proto-oncogene usually activated via gene arrangement or gene amplification in various cancer types. It was isolated as gene rearranged in parathyroid tumors by Motokura et al. in 1991. This gene (aka *CCND1*) encodes for cyclin D1 which acts at the G1/S cell cycle transition leading to pRB phosphorylation by sequestering to CDK4. The gene encoding the cyclin D1 is located on the chromosome number 11 (11q13). The inversion between the 11q13 region and the gene encoding region (11p15) for the parathyroid hormone leads to the parathyroid adenoma. Because of the inversion between these genes, an alternative name for the cyclin D1 encoding gene (*CCND1*) is named *PRAD1*. In addition to being overexpressed in various neoplasms, Cyclin D1 has also been identified in acral melanomas (hand and feet occurrence). The translocation between the 11q13 region and gene region encoding for the heavy chain of immunoglobulin present on chromosome 14 (14q32) accounts for the 70% B-cell lymphoma, because of the upregulation of Cyclin D1 gene. This translocation has been reported in 15–20% multiple myeloma. However, amplification of the Cyclin D1 encoding gene is documented in 30–46% of NSCLC, 30–50% of neck and head squamous cell carcinoma, 25% of pancreatic tumor, 49–54% of pituitary adenoma, 13% of breast

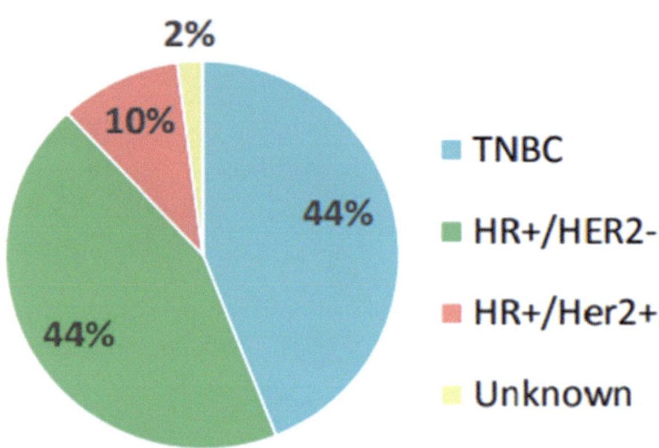

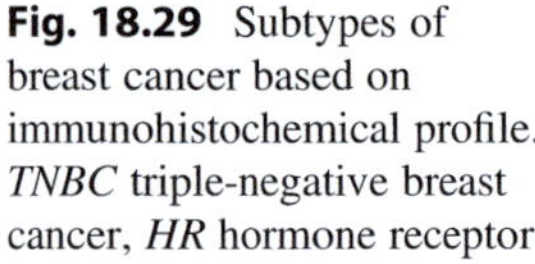
Fig. 18.29 Subtypes of breast cancer based on immunohistochemical profile. *TNBC* triple-negative breast cancer, *HR* hormone receptor

cancer, and 15% of bladder cancer. Mutation in the p16 also results in dysregulation of Cyclin D and induction of tumor development. The Cip/Kip family is associated with the tumor suppression; however, loss of one allele in *Kip1* gene results in tumor progression via overexpression of Cyclin D. The expression of *Wnt-1* transgene in association with the heterozygosity at the *Cip1* locus is demonstrated with upregulation of Cyclin D and increased tumor progression. The truncated form of Cyclin D resulted from the A/G single nucleotide polymorphism (A870G) in the Cyclin D encoding gene is reported in the development of lung colon and cancer. Further, insertion of retroviral gene near the cyclin D gene region is also associated with numerous cancers. Treatment with the synthetic inhibitor of Cyclin D and its associated kinase, knockdown of Cyclin D, and ectopic expression of CDK inhibitor proteins represent a potential therapeutic agent against the cancer.

The other putative proto-oncogene, cyclin E, has been isolated with the help of its ability to complement the triple CLN mutant in *S. cerevisiae*. So far, it is unclear as to whether the abnormal expression will perturb the normal cell cycle machinery; it has been reported that the cyclin E mRNA levels dramatically rise and reach its peak at the G1/S boundary. With the fact that very little information is available regarding cyclin E protein, it would be important to show that the protein follows a periodic expression pattern during cell cycle. The upregulation of Cyclin E expression has been demonstrated in leukemia, sarcoma, lymphoma, and breast, lung, gastrointestinal tract, cervix, and endometrial carcinoma. The nonfunctional pRb protein is also associated with the overexpression of Cyclin E and Cyclin D via the upregulation of E2F target genes. Till now, the exact mechanism of Cyclin E dysregulation is not known; however, mutation in the p16, Cyclin D, pRb, and E2F may be associated with the overexpression of Cyclin E.

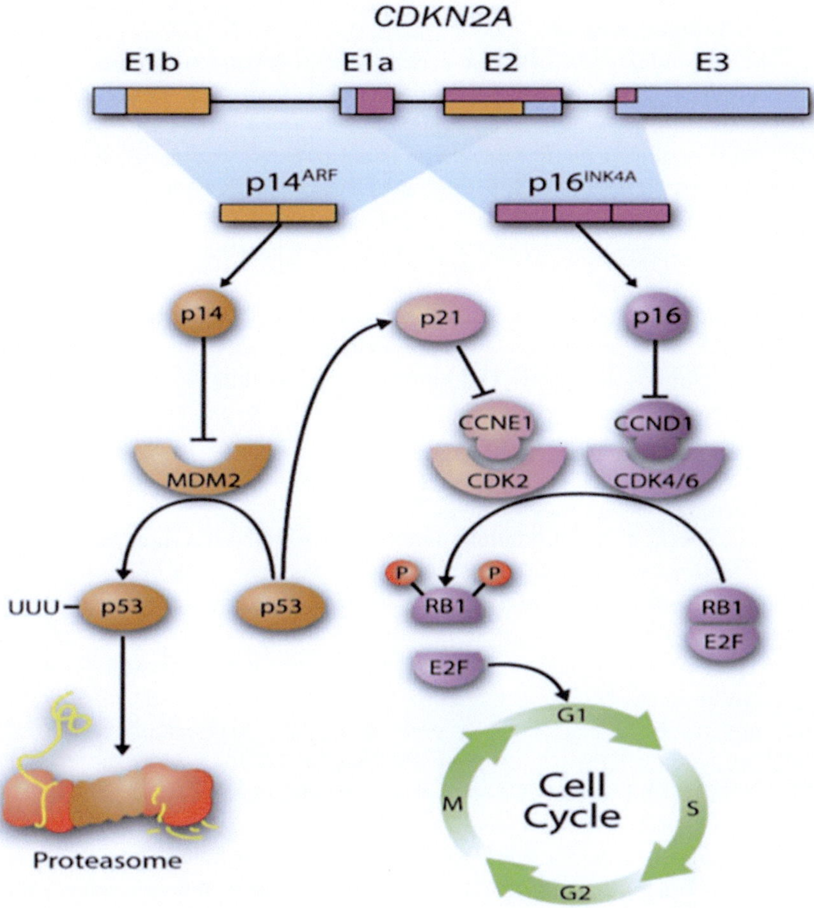

Fig. 18.30 Role of CDKN2A and CDK 4/6 in cell cycle progression (Adapted from Sekulic et al. 2008)

Box 18.1: Scientific Concept: Cell Cycle Checkpoint Pathway Alterations in Breast Cancer (Gargi Dan Basu et al.)

Breast cancer subtypes are molecularly determined based on immunohistochemical profile defined by estrogen receptor (ER), progesterone receptor (PR), and human epidermal growth factor receptor (HER2) status (Fig. 18.29). Dysregulation in the cell cycle is a major hallmark of cancer and has been reported in the various breast cancers as well.

(continued)

Box 18.1 (continued)

The communication between cyclins, cyclin-dependent kinases (CDKs), and cell cycle inhibitors is reported in cell cycle progression through various stages. During the progression, cyclins are involved in activating respective CDKs by binding to them, thereby causing phosphorylation of downstream target proteins. The cell cycle inhibitors such as p16INK4A and p14ARF, also known as tumor suppressor proteins, block cyclin-CDK complex activities, thereby suppressing the downstream processes. The p16INK4A and p14ARF proteins are implicated in the RB1 and p53 pathway, respectively, and the loss of any of these proteins can cause uncontrolled growth and proliferation inside a cell. RB1, a tumor suppressor gene, is acting as a gatekeeper of the cell cycle pathway. Unphosphorylated RB1 binds with and inhibits the E2F transcription factor, hence preventing the E2F from activating the target gene expression involved in G1/S progression. Cyclin D1 and CDK4 or CDK6 complex phosphorylates RB1 protein which in turn releases E2F and allows the induction of target gene expression (Fig. 18.30).

In the recent past, outstanding understanding in the management of breast cancer leads to expeditious detection and hence development of more effective treatments with remarkable death rate decline. Breast cancer is considered as multifaceted rather than a single disease consisting of a variety of biological subtypes with diverse natural histories, presenting a varied spectrum of molecular, pathological, and clinical features with different prognostic and therapeutic implications. According to preclinical data, palbociclib, a CDK4/6 inhibitor, induced growth arrest in RB-positive cell lines and xenograft, and higher RB1 and cyclin D1 expression levels as well as low levels of p16INK4A were also associated with sensitivity to palbociclib. Based on phase II trial results, the inhibitor was granted accelerated approval by the FDA in February 2015 as the first-line treatment in postmenopausal women with ER+/Her2- metastatic breast cancer. However, currently ER status is considered as the most reliable predictive biomarker for palbociclib response; fewer than half of the ER-positive breast cancer patients in the phase II PALOMA-1 trial responded to this drug, thereby highlighting the need for identification of additional biomarkers to help guide treatment decisions. Next-generation sequencing, mutation analysis, and somatic structural rearrangements involving genomic tumor DNA have identified dysregulation of the cell cycle pathway to occur in HR + HER2+ and HR + HER2- cohort of breast cancer patients, thus helping to identify potential targets for cell cycle-related treatment options. Genomic profiling of 48 breast cancer samples has revealed CDKN2A loss in 11% of the total population and was found to co-occur with CCND1 amplification or CDK4 amplification which was not observed in HR + HER2+ breast cancer.

(continued)

Box 18.1 (continued)

In conclusion, HER2-positive patients might be sensitive to CDK4/6 inhibitors due to the cell cycle pathway activation as evidenced by key gene alterations. CCND1 gain was found to be most frequent in HR + HER2+ cohort (40%), followed by HR + HER2- cohort (29%), and then by TNBC cohort (9.5%). Loss of RB1 was localized in the TNBC cohort which may indicate resistance from CDK4/6 inhibitor therapy.

18.15 Summary

- Cancer is considered to be a group of disorders in which the normal cell cycle regulation is lost. In fact, it is well established today that cancers usually arise from a single cell which has dramatically undergone a series of genetic mutations.
- Histological type of cancer as established by the International Classification of Diseases for Oncology includes hundreds of different cancer types which have been primarily grouped into six major categories, namely, carcinoma, leukemia, lymphoma, myeloma, sarcoma, and mixed types.
- Cancer treatment generally includes procedures such as surgery, chemical therapy, radiation therapy, or therapy including drug targeting.
- There exist a limited set of "mission-critical" events propelling the tumor and its progeny cells into uncontrolled division and invasion such as unregulated cell proliferation, which, along with suppression of apoptosis required for support, gives a minimal "stage" inevitable to support further neoplastic progression.
- Two initiation pathways, (a) mitochondrial (intrinsic) and (b) death receptor (extrinsic), eventually lead to a common pathway or execution phase of apoptosis. Perforin/granzyme pathway is a third pathway; however a lesser known fourth intrinsic endoplasmic reticulum pathway is also present for apoptosis.
- Telomeres are the protective structures present at the end of chromosomes and consist of repetitive nucleotide sequences (TTAGGG in case of vertebrates) along with linked proteins, commonly termed as shelterin.
- As regards overexpression of shelterin genes, such as TRF1, TRF2, TIN2, and, in some cases, POT1, many cancer types have been reported as compared to the non-cancerous tissues. Interestingly, telomere length is inversely correlated with the high expression of these shelterin genes and TERT or even telomerase activity.
- IARC has classified carcinogens under six groups: biological agents; arsenic, fibers, metals, and dust; pharmaceuticals; chemical agents and related occupation; radiation; and personal habits and indoor combustions.
- Viral oncogenes are homologues to the cellular oncogenes, for instance, *c-src* is the cellular homologue of the *v-src*. Proto-oncogenes are the representatives of

normal genes of cells, which show similarity with the nucleotide or protein sequences that are tumorigenic or have the potential of transforming.

- Cell cycle progression is a highly regulated process in normal cells to ensure that each step is completed before proceeding to the next. There are well-known three points: the G1/S, the G2/M, and M checkpoints in the whole cell cycle where equilibrium between external and internal signals is checked before a cell enters the next stage.

References

Aoki K, Taketo MM (2007) Adenomatous polyposis coli (APC): a multi-functional tumor suppressor gene. J Cell Sci 120(19):3327–3335

Baer R, Ludwig T (2002) The BRCA1/BARD1 heterodimer, a tumor suppressor complex with ubiquitin E3 ligase activity. Curr Opin Genet Dev 12(1):86–91

Bai L, Zhu WG (2006) p53: structure, function and therapeutic applications. J Cancer Mol 2(4): 141–153

Bates S, Peters G (1995) Cyclin D1 as a cellular proto-oncogene. Semin Cancer Biol 6(2):73–82

Bird RE, Glebov OK, Borellini F, Jacobson-Kram D, Ostrove JM (2003) U.S. Patent No. 6,593,084. U.S. Patent and Trademark Office, Washington, DC

Bishop JM (1985) Viral oncogenes. Cell 42(1):23–38

Bos JL (1989) Ras oncogenes in human cancer: a review. Cancer Res 49(17):4682–4689

Botezatu A, Iancu IV, Popa O, Plesa A, Manda D, Huica I et al (2016) Mechanisms of oncogene activation. In: Bulgin D (ed) New aspects in molecular and cellular mechanisms of human carcinogenesis. IntechOpen, Croatia, pp 1–52

Boxer LM, Dang CV (2001) Translocations involving c-myc and c-myc function. Oncogene 20(40):5595–5610

Butel JS (2000) Viral carcinogenesis: revelation of molecular mechanisms and etiology of human disease. Carcinogenesis 21(3):405–426

Cantley LC, Auger KR, Carpenter C, Duckworth B, Graziani A, Kapeller R, Soltoff S (1991) Oncogenes and signal transduction. Cell 64(2):281–302

Carnero A, Blanco-Aparicio C, Renner O, Link W, Leal JF (2008) The PTEN/PI3K/AKT signalling pathway in cancer, therapeutic implications. Curr Cancer Drug Targets 8(3):187–198

Cavenee WK, White RL (1995) The genetic basis of cancer. Sci Am 272(3):72–79

Chen C-Y, Chen J, He L, Stiles BL (2018) PTEN: tumor suppressor and metabolic regulator. Front Endocrinol 9:338

Chial H (2008) Tumor suppressor (TS) genes and the two-hit hypothesis. Nat Educ 1(1):177

Choisy-Rossi C, Reisdorf P, Yonish-Rouach E (1999) The p53 tumor suppressor gene: structure, function and mechanism of action. In: Apoptosis: biology and mechanisms. Springer, Berlin, Heidelberg, pp 145–172

Choudhuri, S, Chanderbhan R (2007) Carcinogenesis: mechanism and models. Veterinary Toxicology

Cichowski K, Jacks T (2001) NF1 tumor suppressor gene function: narrowing the GAP. Cell 104(4):593–604

DeCaprio JA (2009) How the Rb tumor suppressor structure and function was revealed by the study of Adenovirus and SV40. Virology 384(2):274–284

Delbridge AR, Valente LJ, Strasser A (2012) The role of the apoptotic machinery in tumor suppression. Cold Spring Harb Perspect Biol 4(11):a008789

Deng CX, Scott F (2000) Role of the tumor suppressor gene Brca1 in genetic stability and mammary gland tumor formation. Oncogene 19(8):1059–1064

Di Fiore R, D'Anneo A, Tesoriere G, Vento R (2013) RB1 in cancer: different mechanisms of RB1 inactivation and alterations of pRb pathway in tumorigenesis. J Cell Physiol 228(8):1676–1687
Diehl JA (2002) Cycling to cancer with cyclin D1. Cancer Biol Ther 1(3):226–231
Dimaras H, Corson TW, Cobrinik D, White A, Zhao J, Munier FL et al (2015) Retinoblastoma. Nat Rev Dis Primers 1(1):1–23
Dyson NJ (2016) RB1: a prototype tumor suppressor and an enigma. Genes Dev 30(13):1492–1502
Elenitoba-Johnson KS, Lim MS (2018) New insights into lymphoma pathogenesis. Ann Rev Pathol 13:193–217
Elmore S (2007) Apoptosis: a review of programmed cell death. Toxicol Pathol 35(4):495–516
Finlay CA, Hinds PW, Levine AJ (1989) The p53 proto-oncogene can act as a suppressor of transformation. Cell 57(7):1083–1093
Funk JO (2001) Cell cycle checkpoint genes and cancer. e LS
Garnett MJ, Marais R (2004) Guilty as charged: B-RAF is a human oncogene. Cancer Cell 6(4): 313–319
Harris CC (1996) Structure and function of the p53 tumor suppressor gene: clues for rational cancer therapeutic strategies. JNCI 88(20):1442–1455
Hassan M, Watari H, AbuAlmaaty A, Ohba Y, Sakuragi N (2014) Apoptosis and molecular targeting therapy in cancer. BioMed Res Int 2014:150845
Hay N (2005) The Akt-mTOR tango and its relevance to cancer. Cancer Cell 8(3):179–183
Helman LJ, Meltzer P (2003) Mechanisms of sarcoma development. Nat Rev Cancer 3(9):685–694
Hunter T, Pines J (1991) Cyclins and cancer. Cell 66(6):1071–1074
Hwang HC, Clurman BE (2005) Cyclin E in normal and neoplastic cell cycles. Oncogene 24(17): 2776–2786
Jhiang SM (2000) The RET proto-oncogene in human cancers. Oncogene 19(49):5590–5597
Joerger AC, Fersht AR (2010) The tumor suppressor p53: from structures to drug discovery. Cold Spring Harb Perspect Biol 2(6):a000919
Kamaraj B, Bogaerts A (2015) Structure and function of p53-DNA complexes with inactivation and rescue mutations: a molecular dynamics simulation study. PLoS One 10(8)
Karakas B, Bachman KE, Park BH (2006) Mutation of the PIK3CA oncogene in human cancers. Br J Cancer 94(4):455–459
Kastan MB, Bartek J (2004) Cell-cycle checkpoints and cancer. Nature 432(7015):316
Lee EY, Muller WJ (2010) Oncogenes and tumor suppressor genes. Cold Spring Harb Perspect Biol 2(10):a003236
Leslie NR, Downes CP (2004) PTEN function: how normal cells control it and tumour cells lose it. Biochem J 382(Pt 1):1–11
Ligresti G, Libra M, Militello L, Clementi S, Donia M, Imbesi R et al (2008) Breast cancer: molecular basis and therapeutic strategies. Mol Med Rep 1(4):451–458
Lin P, O'Brien JM (2009) Frontiers in the management of retinoblastoma. Am J Ophthalmol 148(2):192–198
Lowe SW, Lin AW (2000) Apoptosis in cancer. Carcinogenesis 21(3):485–495
Malumbres M, Barbacid M (2009) Cell cycle, CDKs and cancer: a changing paradigm. Nat Rev Cancer 9(3):153
Moschel RC (2001) Carcinogens. Encyclopedia of Genetics
Motokura T, Bloom T, Kim HG, Jüppner H, Ruderman JV, Kronenberg HM, Arnold A (1991) A novel cyclin encoded by a bcl1-linked candidate oncogene. Nature 350(6318):512–515
Nevins JR (2001) The Rb/E2F pathway and cancer. Hum Mol Genet 10(7):699–703
Nicholson RI, Gee JMW, Harper M (2001) EGFR and cancer prognosis. Eur J Cancer 37:9–15
Okamoto K, Seimiya H (2019) Revisiting telomere shortening in cancer. Cells 8(2):107
Osborne C, Wilson P, Tripathy D (2004) Oncogenes and tumor suppressor genes in breast cancer: potential diagnostic and therapeutic applications. Oncologist 9(4):361–377
Reesink-Peters N, Wisman GBA, Jéronimo C, Tokumaru CY, Cohen Y, Dong SM et al (2004) Detecting cervical cancer by quantitative promoter hypermethylation assay on cervical scrapings: a feasibility study. Molecular Cancer Research 2(5):289–295

Sachdeva UM, O'Brien JM (2012) Understanding pRb: toward the necessary development of targeted treatments for retinoblastoma. J Clin Invest 122(2):425–434

Sekulic A, Haluska Jr P, Miller AJ, De Lamo JG, Ejadi S, Pulido J S, ... & Melanoma Study Group of the Mayo Clinic Cancer Center (2008) Malignant melanoma in the 21st century: the emerging molecular landscape. Mayo Clin Proc 83(7):825–846, Elsevier

Senda T, Shimomura A, Iizuka-Kogo A (2005) Adenomatous polyposis coli (Apc) tumor suppressor gene as a multifunctional gene. Anat Sci Int 80(3):121–131

Shaikh Z, Niranjan KC (2015) Tumour biology: p53 gene mechanisms. J Clin Cell Immunol 6(344):2

Shay JW, Wright WE (2011) Role of telomeres and telomerase in cancer. Semin Cancer Biol 21(6): 349–353

Shih AH, Holland EC (2006) Platelet-derived growth factor (PDGF) and glial tumorigenesis. Cancer Lett 232(2):139–147

Siegel RL, Miller KD, Jemal A (2016) Cancer statistics, 2016. CA Cancer J Clin 66(1):7–30

Sigal A, Rotter V (2000) Oncogenic mutations of the p53 tumor suppressor: the demons of the guardian of the genome. Cancer Res 60(24):6788–6793

Singh CR, Kathiresan K (2014) Molecular understanding of lung cancers–A review. Asian Pac J Trop Biomed 4:S35–S41

Sugimura T (2000) Nutrition and dietary carcinogens. Carcinogenesis 21(3):387–395

Teixo R, Laranjo M, Abrantes AM, Brites G, Serra A, Proença R, Botelho MF (2015) Retinoblastoma: might photodynamic therapy be an option? Cancer Metastasis Rev 34(4):563–573

Visconti R, Della Monica R, Grieco D (2016) Cell cycle checkpoint in cancer: a therapeutically targetable double-edged sword. J Exp Clin Cancer Res 35(1):153

Vogt PK (2012) Retroviral oncogenes: a historical primer. Nat Rev Cancer 12(9):639–648

Wang SI, Parsons R, Ittmann M (1998) Homozygous deletion of the PTEN tumor suppressor gene in a subset of prostate adenocarcinomas. Clin Cancer Res 4(3):811–815

Wong RS (2011) Apoptosis in cancer: from pathogenesis to treatment. J Exp Clin Cancer Res 30(1): 87

Zheng ZM (2010) Viral oncogenes, noncoding RNAs, and RNA splicing in human tumor viruses. Int J Biol Sci 6(7):730

Zhu X, Han W, Xue W, Zou Y, Xie C, Du J, Jin G (2016) The association between telomere length and cancer risk in population studies. Sci Rep 6:22243

Part IV

Population Genetics

Developmental Genetics 19

Divya Vimal and Khadija Banu

19.1 Genetic Approach of Development

The foundation of the genetic approach to understand the development of an organism is that all the information is already encoded in an organism's DNA. These approaches help to understand how cells achieve different fates and what combinations of intercellular signaling and intracellular regulatory circuits generate spatially and temporally encoded patterns along the body axis. With the advent of enormous numbers of techniques in the field of genomics, molecular biology, and biochemistry, it is now possible to study the process of development at each and every step along with the role of individual gene, gene function, genetic pathways, and genome sequences. In order to unravel the mystery of developmental biology, integrative perspective is required where systems level approach combining computational and genomic approaches with cell and molecular biology techniques is required. Differential pattern of gene expression along with different transcription factors, which bind to promoters of genes thereby activating or repressing them, regulates the development of an organism. During development, there are control genes that regulate this cascade of genes and transcription factors.

Both authors have contributed equally.

D. Vimal (✉)
Columbia University Irving Medical Center, New York, NY, USA

K. Banu
Yale University, New Haven, CT, USA

D. Kar, S. Sarkar (eds.), *Genetics Fundamentals Notes*,
https://doi.org/10.1007/978-981-16-7041-1_19

19.1.1 Significance of Model Organism

Model organisms are non-human species that help scientists understand about a wide array of biological processes. The results obtained from these organisms are extrapolated to higher organisms like humans where experimentation is not possible due to ethical or practicality constraints. This extrapolation is possible due to the conservation of basic biological processes in all organisms during the course of evolution ranging from single-celled organisms to most complex humans. Researchers can control the variables that can affect the outcomes of the experiment by using varying animal models and by regulating the living conditions. Model organisms are generally selected based upon their unique characters like short life cycle, less breeding time, large litter size, and availability of mutant lines. These unique features of model organisms make them amenable to different types of manipulation under in vitro conditions (Müller 1997). Use of model organisms in research started with the realization that the data obtained from studies on these organisms can be used to understand the complex mechanism behind basic physiology and molecular mechanism of higher organisms (Fig. 19.1). Early uses of model organisms include discovery of germ theory by Louis Pasteur, natural selection by Charles Darwin, and genetics of heredity by Gregor Mendel. Model organisms can be categorized into genomic, experimental, and genetic model organisms. Genomic model organisms have a particular genomic size or the arrangement of genes in a specific manner which can be used for reference or manipulation during experiment. An experimental type of model organism has specific characteristics well suited for the particular type of experiment, while genetic model organisms are particularly useful in molecular manipulation as well as genetic studies where different mutants are generated by genetic crosses. The genome sequencing of these organisms has revealed the conservation of genes and cellular pathways which allows the manipulation of genes and studying their effect. In addition, it helps to apply reverse genetics to answer the functional roles of genes which can be extrapolated to the higher organism due to biological conservation. Discovery of many homologous genes of human diseases has allowed researchers to mimic the human disease pathological conditions and study them in simpler experimental systems. Knowledge of whole genome provides the opportunity to create genetic screen studies on a large scale that covers all genes of an organism. Libraries of gene knockouts, knockdowns, as well as overexpression are available which permit the study of function of each and every gene involved in a particular process.

Frequently used model organisms include *Escherichia coli*, yeast (*Saccharomyces cerevisiae*), nematode worm (*Caenorhabditis elegans*), fruit fly (*Drosophila melanogaster*), zebrafish (*Danio rerio*), Western clawed frog (*Xenopus tropicalis*), mouse (*Mus musculus*), etc. Libraries of different mutants along with a wide range of genetic tools are commercially available owing to the fact that all model organisms have been genetically sequenced. Model organism database (MODs) is a database dedicated to provide all the information available for a particular model organism like precise location of genes as well as regulatory regions present in the genome, gene expression patterns and phenotypes of individual genes, gene ontology annotations, pathway information, DNA/RNA/protein sequences, and stock centers

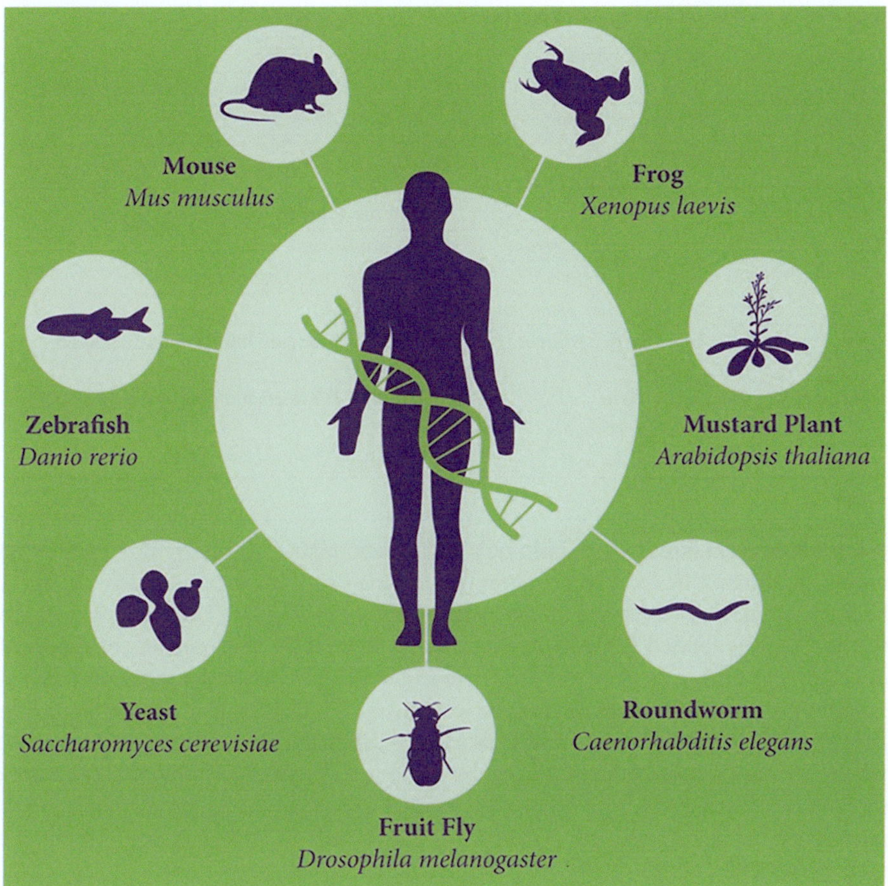

Fig. 19.1 Model organisms. Model organisms are the key players of biological research, and studies carried out on different organisms such as yeast, worm, fly, fish, and mouse have provided valuable insight into the biology of developmental as well as disease pathogenesis. (Adapted from https://biology.uiowa.edu/model-organisms)

(Table 19.1). These databases are also cross-referenced with many other useful tools and techniques which can be applied to carry out experiments feasibly. Some common keywords such as gene ontology are used to collect information regarding a specific gene like its function, genomic location, protein product, and biological process it is involved in.

For the study of development, nematode, fruit fly, frog, zebrafish, chick, and laboratory mouse are extensively investigated. In order to elucidate the biological mechanisms underlying any cellular process, different genetic approaches are employed. Studying the defects in single-gene mutants and comparing it with wild type allow to identify the new genes and their functions. This gene obtained in this way is then mapped to its genomic location, cloned using a specific vector, and thus identified at the molecular level. The protein product of this gene is then studied

Table 19.1 List of model organism databases

Common name	Scientific name	Database name
Baker's yeast	*Saccharomyces cerevisiae*	Saccharomyces Genome Database (SGD)
Fission yeast	*Schizosaccharomyces pombe*	PomBase
Clawed frog	*Xenopus*	Xenbase
Fruit fly	*Drosophila melanogaster*	FlyBase
Mouse	*Mus musculus*	Mouse Genome Informatics (MGI)
Nematode	*Caenorhabditis elegans*	WormBase
Rat	*Rattus norvegicus*	Rat Genome Database (RGD)
Social amoeba	*Dictyostelium discoideum*	dictyBase
Thale cress	*Arabidopsis thaliana*	The Arabidopsis Information Resource (TAIR)
Zebrafish	*Danio rerio*	Zebrafish Information Network (ZFIN)
–	*Candida albicans*	CGD
–	*Escherichia coli*	EcoCyc

using different methods of cell biology and biochemistry. Applying this method on model organisms has allowed mapping, cloning, and subsequent study of many heritable human diseases like breast cancer and cystic fibrosis. This genetic analysis method can be applied in two ways in order to dissect the mechanisms of action of developmental process: forward and reverse genetics. In forward genetics the investigation starts with a mutant phenotype (organism) which then leads to the gene. The first step in forward genetic is selecting a specific defective phenotype of interest which is easily recognizable. Next, using these mutagenized populations, saturation screens are carried out, and all the genes that are involved in producing a given phenotype in an organism or species are elucidated. In this way the screens are carried out until all the genes of a mechanism are determined and no new genes remain. This process was first used by Eric Wieschaus and Christiane Nüsslein-Volhard in *Drosophila* to study the genes involved in body plan patterning. Genetic mapping and complementation tests are then used to identify the genes responsible for these mutations. Effect of complete absence of gene function in null mutants either by deletion or abnormal function of gene is also determined. Further, in order to elucidate the sequence of function of these identified genes, double mutants are generated. In addition, modifier genes are identified that either reduce or worsen the phenotype of an existing mutant by employing screens for enhancer and suppressor sequences for a secondary mutation in a sensitized genetic background. Clones of individual gene in the form of complementary DNA (cDNA) libraries are generated for molecular analysis through genetic mapping. cDNAs corresponding to each gene are then sequenced, and respective sequence of each encoded protein is determined. This sequence is then used for similarity search in different databases to distinguish different domains and motifs which disclose the functional class of particular protein. To determine the developmental stages during which the transcripts and translated product of particular gene are expressed, nucleic acid probes, antibodies, and reporter constructs are used. The last step in the process is to express the candidate gene in the mutants to comprehend if the encoded protein product reverts the defective phenotype of mutant back to wild type.

In contrast to forward genetics, reverse genetic approaches are carried out from gene sequence to a phenotype. As the genomes of most organisms are fully sequenced, this vast information can be used to study biological processes using reverse genetics. In higher organisms such as humans where forward genetic approaches are not feasible, reverse genetics comes as rescue. The first step in reverse genetics process is cloning of gene of interest which is then used to generate mutant organisms with defective gene or abnormal expression studying its function. The gene of interest is inactivated or silenced by targeted inactivation or permanently by producing mutant organisms carrying null mutation called knocking out. This is achieved by inactivating the gene of interest by the process of RNA interference where a double-stranded antisense mRNA is injected that specifically prevents gene expression of that particular gene. The phenotypic consequences of this inactivation are then observed for function disruption. Different strategies can be followed in different model organisms for the inactivation of gene of interest; RNA interference using miRNA is widely used in *C. elegans*, whereas in *Drosophila* random transposition event or mutagen generates mutants which are then screened from a large population. In contrast, the most common and efficient method to generate mutants is by injecting the mutant DNA constructs into the germline cells which recombine with the host DNA. Mutations thus obtained are subjected to a wide array of genetic analyses either using traditional approaches or rapid large-scale approaches using microchips.

The genes identified in different model animals from both forward and reverse genetic approaches are then applied on higher organisms like humans where orthologous genes are isolated and characterized. Various tools and data from Genome Project Comparison are available for this task. For example, comparative analysis between genomic maps of human and mouse has revealed widespread conservation of linkage called synteny; arrangement of genes (orthologous) over a large region of DNA has been conserved similar to the last common ancestor. Much of the understanding and the key genes in human development and physiology have been studied in this way through the study of model organisms (Perlman 2016).

The most widely studied prokaryotic model organism is *Escherichia coli* owing to the ease of growing and culturing it inexpensively. *E. coli* is gram-negative bacteria with a rod shape, 2.0 μm in length and 0.5 μm in diameter inhabiting warm-blooded animals in the gastrointestinal tract. Different strains of *E. coli* like *E. coli* K12 are available that have been well improved and adjusted according to the lab environment. *E. coli* is called a molecular biologist tool box due to the availability of extensive molecular tools for different purposes. Discoveries made in *E. coli* have contributed a lot to the basic understanding or cellular biological processes and has also helped scientists to win well-deserved Nobel Prizes. Major keystone discoveries like DNA replication, genetic code, genetic regulation, gene organization, basis of mutation, evolution of organisms, and development of genetically modified organism have been carried out in *E. coli.* Manipulation of *E. coli* genome plays the most significant role in biotechnology.

Major advantages that *E. coli* provides as an ideal model organism include fast growth in relatively cheap chemically defined media; industrial scalability;

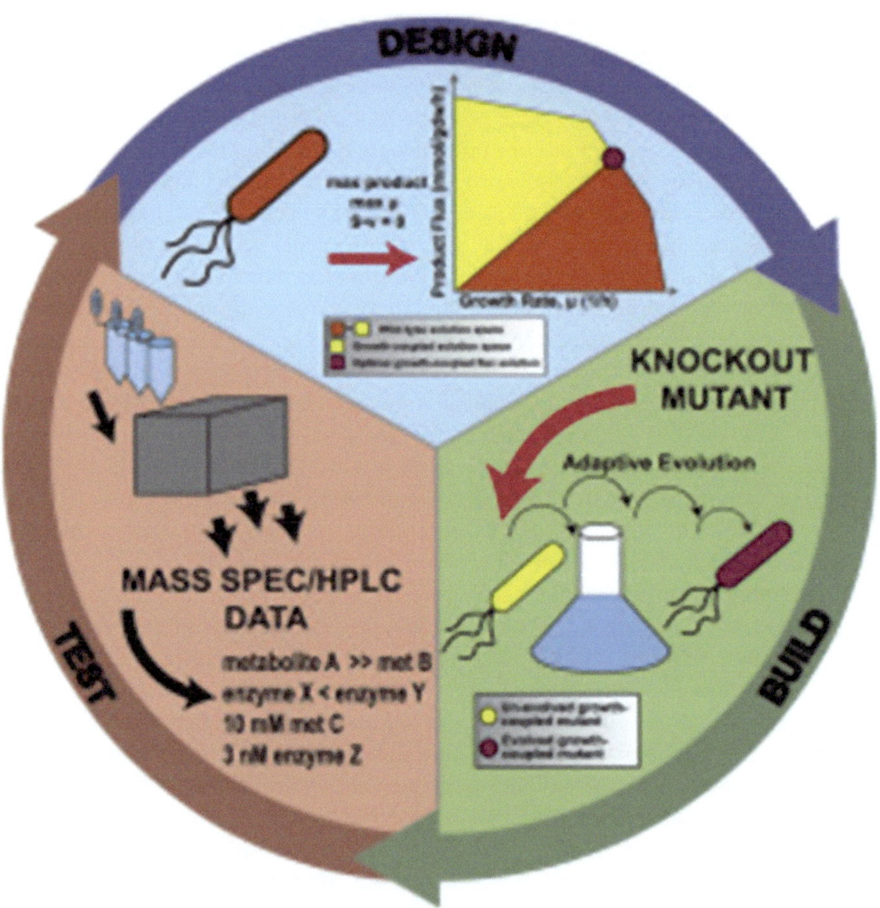

Fig. 19.2 *E. coli*. It is an ideal model organism for studies related to various aspects of molecular biology and biochemistry due to its simplicity. Any experiment utilizing *E. coli* as a model organism includes three stages, design, build, and test; firstly a design-specific strain is selected followed by suitable genomic modifications to produce a mutant strain which is then tested using various techniques. (Adapted from Adamczyk and Reed 2017)

enormous number of tools for genetic alterations; expansive knowledge of its genome; wide information of its transcriptomics, proteomics, as well as metabolomics; and lastly, availability of various strains that are nonpathogenic (biosafety-1). The most important feature of *E. coli* is that unlike multicellular model organism especially mammalian models, there are no ethical concerns (Fig. 19.2). Small size allows their usage in large numbers in a confined laboratory space. The doubling time of *E. coli* is about 20 min resulting in doubling of its population which helps researchers to study multiple generations within a short period of time. Another important advantage is that *E. coli* can endure a wide range of environmental conditions and survive under in vitro conditions. Culture media of different compositions depending upon the experimental conditions and consisting of various components and nutrients are sufficient for the growth of *E. coli*. The most common ingredients of any culture medium include carbon source generally glucose, nitrogen source like yeast extract, various salts, and different organic compounds. Some *E. coli* strains are well known for serious problems and diseases in humans like fever, diarrhea, and abdominal pain. *E coli* is a common inhabitant of the intestine constituting the gut flora and is essential for several functions like production of vitamin K2 and restricting the growth of harmful bacteria. In 1997, the first report of *E. coli* genome sequencing of a lab strain, K-12 derivative MG1655, was published. *E. coli* genome consists of approximately 4.6 million base pairs consisting of ~4000 genes with 7 rRNA and 86 tRNA genes, respectively.

Saccharomyces cerevisiae, also called as budding yeast, is the organism of choice for eukaryotic studies as it provides the basic eukaryotic features. Similar to any eukaryotic system, yeast cells have a nucleus enclosed by nuclear membrane containing its genome which is ~12.1 million base pairs organized in 16 chromosomes and consists of ~6500 genes. In addition, it also contains all the subcellular organelles suspended in cytoplasm. It is single-celled yeast commonly used in the bread-making industry with a genome size of 12,157,105 base pairs in length containing 6692 genes. It is a free-living organism residing primarily on substances having sugar like fruits and flowers. Yeast cells can survive a variety of environmental conditions ranging from freezing temperatures to ~55 °C with proliferative temperature of 15–35 °C. Yeast can grow in varying pH conditions ranging from pH 2.8 to 8.0, and after almost complete drying it can regenerate and thrive on any sugar-containing medium with osmotic pressure as high as 3 M sugar concentration and ferment it. The sequencing of *Saccharomyces cerevisiae* genome was completed and published in 1996 and was among the first eukaryotic organisms to be completely sequenced. Approximately 20% genes of human genome involved in various diseases have functional equivalents in yeast suggesting similar basic cellular machinery. *S. cerevisiae* offers advantages in various aspects such as small size, short generation time, easy accessibility, manipulation tools, known DNA sequences and regulatory regions, and conserved molecular mechanisms with higher organisms along with economic use. *S. cerevisiae* and *S. pombe* are the two most studied *Saccharomyces* species and provide excellent systems for studies related to DNA damage and its repair.

The life cycle of *S. cerevisiae* contains both haploid and diploid vegetative stages which allow generating mutant strains in haploid condition and studying its effect directly (Fig. 19.3). Furthermore, yeast offers an important tool in the form of complementation test used to assign a particular mutation to gene by studying the diploid organism with heterozygous condition for a mutation. In contrast, by studying the haploid progeny during meiosis, genetic relationship can be understood and used for gene mapping and to define the relationship between different genes on the basis of their function. The most valuable tool that yeast offer is the incredibly efficient systems for homologous recombination by which two DNA molecules having similar sequences or transformed DNA recombine. The recombination

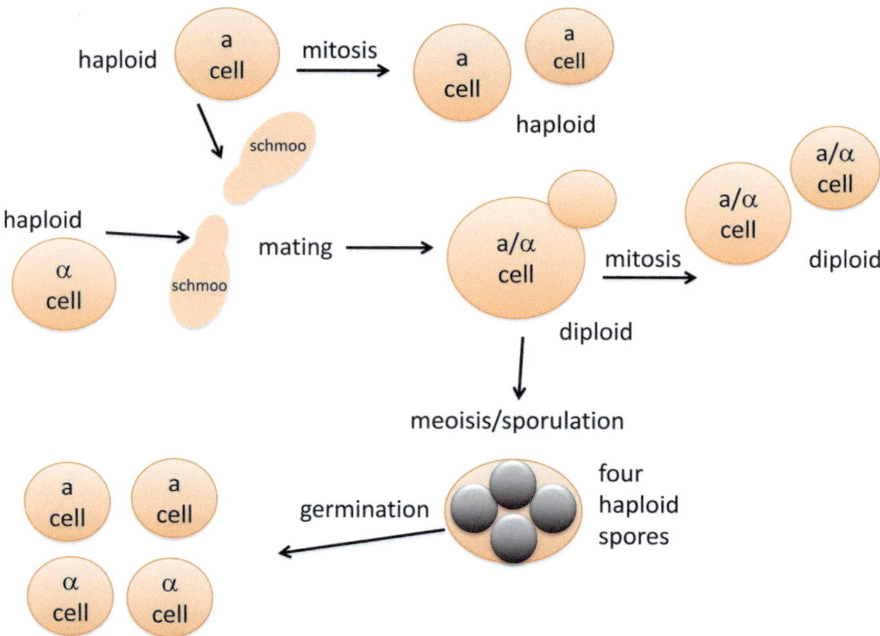

Fig. 19.3 Life cycle of *Saccharomyces cerevisiae.* Yeast is an extensively used eukaryotic model organisms, and studying homologs in yeast leads to the discovery of many vital proteins like those involved in cell cycle and signaling which are of utmost importance in human biology. Both diploid and haploid yeast cells undergo mitosis through budding producing daughter cells; however, diploid cells sometimes divide by meiosis into four haploid spores. (Adapted from Duina et al. 2014)

between DNA sequences in this way is highly precise and efficient and allows researchers to alter the genome with ease. The altered DNA thus generated is then used for transforming the yeast cells which then locate the precise location for incorporation based on the sequence similarity of few bases and bring out the predicted genetic change. *S. cerevisiae* has been extensively studied in order to understand the biology of aging and has led to the discovery of many genes involved. In corroboration, yeast cells provide excellent system for aging studies as it exhibits both chronological and replicative aging. In chronological aging the amount of time that a cell has survived is studied, while in replicative aging the number of progeny cells generated by a parent cell before senescence is generally studied. Yeast cells undergo replicative ageing where the cells divide finitely (30–40 division) by mitosis before the cell dies which is analogous to aging profile of human stem cells. *S. cerevisiae* is industrially manufactured and used in several ways, for example, as a probiotic, or used in numerous digestive tract-related problems. Another important advantage of using yeast is the microarray analysis which can be used to determine the expression of multiple genes at the same time. Microarray analysis combined with chromatin immunoprecipitation (ChIP sequencing) reveals the binding of transcription factors to specific sites. A complete set of more than 6000 deletion mutants (yeast deletion analysis) is available for research; the phenotypic analysis of these mutants can be done in high throughput to study the genetic networks. Yeast cells can be used to test the effect of new drugs as it has many genes common with humans. Mutated yeast cells carrying diseases of human gene can be used to study the effect of numerous drugs on their capability to rescue the normal function. Two mismatch repair system (MMR) genes of yeast that are almost identical in terms of sequence as well as function with humans are mutL homolog (Mlh1) and mutS homolog (Msh2). Both of these genes are well studied with respect to one of the most common types of cancer in humans, hereditary non-polyposis colorectal cancer (HNPCC). Mutations in either of the genes cause HPNCC, and their study in yeast has helped gain insight in their role in cancer.

The Saccharomyces Genome Database (SGD) is a database consisting of all the information on the biology of budding yeast *Saccharomyces cerevisiae*. In addition, it provides various searching and analyzing tools that can be used to carry out comparative studies with higher organisms at genomic as well as phenotypic level. *S. cerevisiae* has been used to study the role of α-synuclein during Parkinson's disease, dementia, and other neurodegenerative diseases. Different drugs or test compounds can be used to test their potential in reversing the adverse effects of α-synuclein in nervous system and thereby treating Parkinson's disease.

C. elegans is a nonparasitic, free-living soil nematode of ~1.3 mm in length belonging to the family of roundworms and can be found at various places. The use of *C. elegans* in research as a model system started since the nineteenth century with the completion of its genome sequencing published in 1998 (Marsh and May 2012). Earliest studies on *C. elegans* were carried out by Sydney Brenner due to various features like short duration of life cycle, large number of offspring, and ease of genetic manipulation. *C. elegans* has contributed a lot to the fundamental aspects of developmental and neuronal biology (Fig. 19.4). *C. elegans* development starts with an egg which then molts to L1 larvae followed by successive molts to L2, L3, and L4

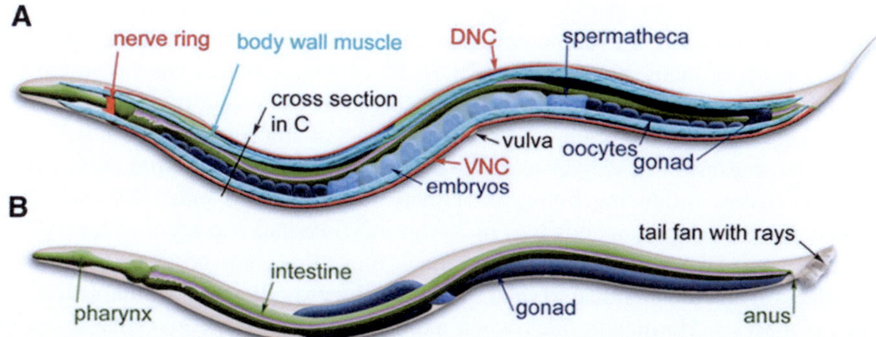

Fig. 19.4 Anatomy of *C. elegans*. Lateral view of a hermaphrodite (**a**) and male (**b**) showing nerve ring, vulva, gonads, intestine, and pharynx. (Adapted from Corsi et al. 2015)

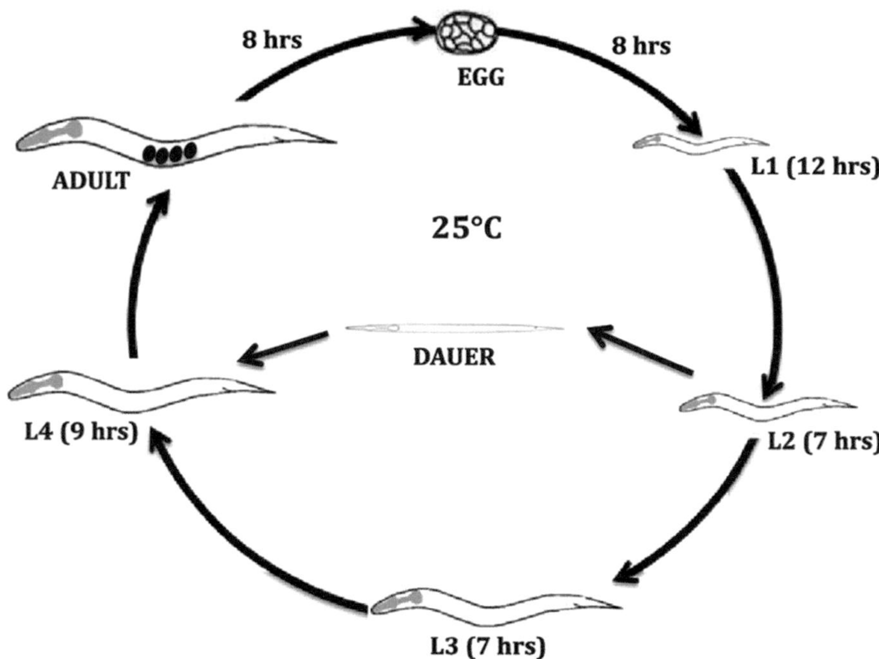

Fig. 19.5 Developmental stages of the nematode, *Caenorhabditis elegans*. Development of *C. elegans* starts from an egg which then develops through four larval stages, L1–L4, to the reproductive adult animal. (Adapted from Pandey 2014)

larvae and finally to an adult worm; this process takes about 3.5 days at 20 °C. *C. elegans* can be grown in large numbers inexpensively on nutrient plates which contain bacteria as food. *C. elegans* is a very small organism having high fecundity (~1000 eggs every day) but short life span (~2–3 weeks) which makes it very feasible for the developmental studies. Despite their short life span, under unfavorable conditions, worms adapt a unique developmental stage producing dauer larvae which resist and survive extreme unfavorable conditions like drying and absence of food even up to many months (Fig. 19.5). *C. elegans* cultures can be stored for indefinite time by freezing it and when needed can be defrosted, revived, and used.

C. elegans is a diploid organism having five pairs of autosomal (1–5) and one pair of sex chromosomes. *C. elegans* can exist in two sexes, hermaphrodite (XX) which can self-fertilize producing both sperms and oocytes, and a male (X0) which is formed due to the spontaneous loss of X chromosome and is present at a very low frequency (~0.2%). Hermaphrodites produce more than 300 progeny by asexual reproduction and are identical to each other, thereby providing an invaluable tool for genetic analysis. Hermaphrodites when exposed to heat shock give rise to males which can then mate with hermaphrodites producing cross-progeny. Another fascinating characteristic of nematode is that the complete development of fertilized egg to the adult worm along with cell lineage can be easily studied due to the simple body structure and limited number of somatic cells (1031 in male and 959 in hermaphrodite). The nervous system of *C. elegans* is relatively simple containing ~300 neurons in adult, thereby providing an excellent system for neurological studies as compared to much complex nervous system in other organisms. In *C. elegans*, most of the nerve cells are present in a large nerve ring, ventral and dorsal nerve cord, and a complex head sensory system. Additionally, well-known signaling components and neurotransmitters that function in mammalian nervous system can be seen in *C. elegans* nervous system. The genome size of *C. elegans* is ~100 million base pairs consisting of ~20,000 genes which make it feasible to both forward and reverse genetic approaches. In addition, *C. elegans* has ~43% of gene that have human homologs, including numerous disease genes. The transparent body makes studying the behavior of individual cells very easy throughout its development. In addition, *C. elegans* is an excellent system to study the expression patterns of genes under in vivo condition, and determining the localization of proteins within the cell is very feasible. The proteins can be tagged with various reporter enzymes like β-galactosidase or constructs like green fluorescence protein (GFP) either by injecting or bombarding the germ cells. The anatomy and development of *C. elegans* can be examined easily under a microscope, and each cell can be traced back to the embryo due to specific and specialized pattern of development. One of the most important tools that were discovered in *C. elegans* is RNA interference (RNAi) which can be used to generate thousands of mutants by gene silencing through double-stranded RNA (dsRNA) at transcriptional and post-transcriptional levels in a sequence-specific manner. There are various ways of delivering double-stranded RNA (dsRNA): injection of dsRNA into the worm, feeding the nematode bacterial food which produces dsRNA or directly soaking the worms into the dsRNA solution, as well as production of dsRNA using transgenic promoters in vivo.

C. elegans shares important molecular signaling pathways regulating its development with humans (Leung et al. 2008). Mutants with single-gene mutations can be easily produced to study the functions of specific genes. *C. elegans* genome has functional equivalents of many genes of humans that allow producing different mutants and disease models with single-gene mutations and studying their function. Different *C. elegans* mutant models are available to study various human diseases including cardiovascular diseases, neurological diseases, and renal disorders. Thousands of potential drugs for the treatment of several severe diseases can be screened using *C. elegans* mutants. By studying the mechanism behind apoptosis in *C. elegans*, the effects of aging can be offset in humans. In addition, studies in nematodes regarding the molecular mechanisms and hallmark gene that are generally found to be mutated in several diseases can provide important insights for curing these diseases.

Another important widely used model organism is zebrafish (*Danio rerio*) which is a freshwater fish of the Cyprinidae family. Zebrafish is one of the ideal model organisms for studying vertebrate development. Zebrafish derives its name due to the presence of zebra-like horizontal stripes on the side of the body which are five in number and of blue color. Zebrafish is used as an animal model due to various advantages including its regenerative abilities, small size and robustness, rich

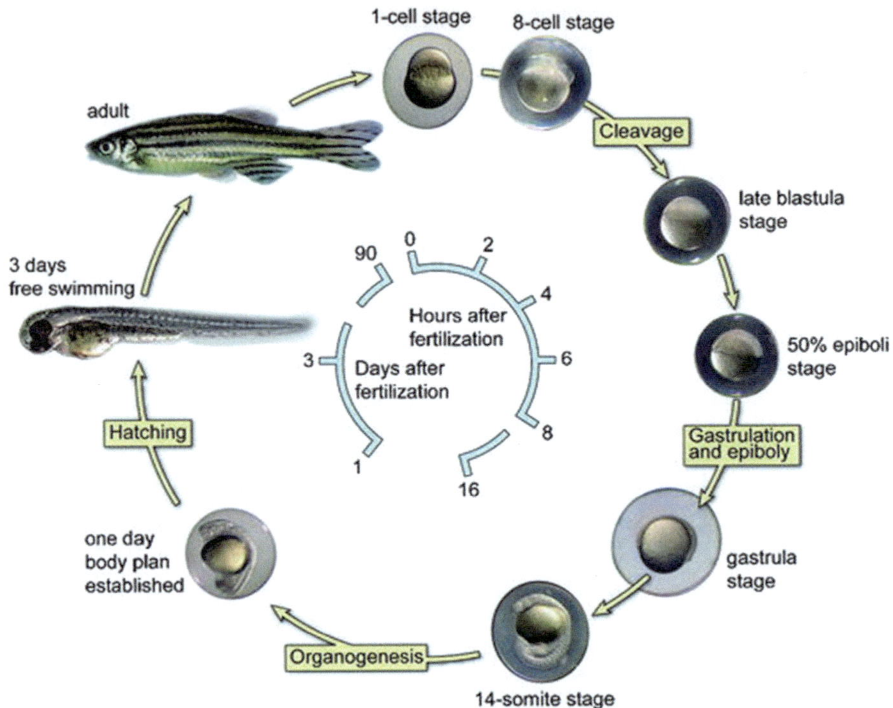

Fig. 19.6 Life cycle of zebrafish (*Danio rerio*). Zebrafish is one of the best studied vertebrate animal models that has contributed much of our understanding in the field of developmental biology, mis-regulation of molecular mechanisms in cancer development, toxicological studies, drug discovery, and screening. (Adapted from Willemsen et al. 2011)

genetics, and rapid and transparent development. The zebrafish is ~2.5 cm to 4 cm long and are cheaper to maintain in a laboratory (Fig. 19.6). Where most fishes lay eggs in the dark, zebrafish mates in daylight producing a large number of progeny every week providing excellent system for embryological studies. In addition, zebrafish embryo is an outstanding model system as it develops outside the adult worm and can be observed as well as altered at all stages of development. The embryos develop and grow at an exceptionally rapid rate and are almost transparent allowing the examination of internal structure development. Due to the transparent body, specific cells and their components can be tagged with fluorescent proteins, and the fate of the cells can be studied in the presence or absence of different genes. Zebrafish genome share 70 percent of genes with humans while containing 84 percent of genes that are known to be associated with various human diseases. Being a vertebrate organism, major organs and tissues similar to higher organisms like the kidney, eyes, muscle, and blood can be seen in zebrafish. One unique feature of zebrafish is its episodic memory associated with explicit memory systems where individuals can remember the context associated with respective locations, occasions, and objects. Another interesting feature of zebrafish is the capability to repair heart muscle in a matter of weeks which can be useful for patients with cardiovascular diseases. Completely sequenced zebrafish genome was published in 2001 and was carried out by the Sanger Institute, UK, which provided invaluable information for the generation of numerous mutant lines with single- or multiple-gene mutations.

Zebrafish is one of the most popular model systems for carrying out the behavioral studies and has helped researchers to understand the complex mechanism behind learning, sleep wake cycle, and depression. Recent advances in neuronal studies paired with behavioral analysis provide the understanding of the role of neural regulatory pathways in behavioral changes. Zebrafish is also being studied as a regenerative model as it can either restore or replace the damaged cells or tissues including major organs like the spinal cord, heart, as well as appendages. The repair can occur by either of the two mechanisms: dedifferentiation, proliferation of the neighboring cells, and replacement of the damaged cells or through the activity of stem cells. By unraveling the mechanisms and regulatory pathways behind the regeneration capacity of zebrafish, the knowledge might be applied to the mammals as well.

Being a fish, zebrafish can detect toxins present in the water and hence can be used as a predictive model for screening of toxic compounds to study the adverse effects of xenobiotics. In addition, zebrafish is widely used for studying a variety of diseases including cancer, kidney disease, diabetes, pigment cell disorders, aging-related disorders, nervous system-associated diseases, epilepsy, blood disorders, as well as addiction. Furthermore, zebrafish models of disease are used to carrying out pharmacological screening on a large scale by mixing the drug with water.

Fruit fly, *Drosophila melanogaster*, is one of the most extensively studied model organisms for development, behavior, neurobiology, genetics, diseases, and molecular-related studies. Flies have been used for basic research for more than a hundred years. Earliest uses of flies date back when Thomas Hunt Morgan carried out a series of experiments on the eye color and its gene providing the chromosomal theory of inheritance for which he won the Nobel Prize in 1933. This was followed by many more keystone discoveries in the field of mutations, genetic control of embryonic development, immunology, olfactory system, and molecular mechanisms of circadian rhythms which helped scientists bag five more Nobel Prizes.

Drosophila has 60% of the counterpart genes of the human diseases which allows the scientists to mutate, amplify, or delete diverse set of human disease-related genes and study their function in different scenarios. In addition, other attributes like short life cycle with high fecundity, ease of maintaining the cultures, manageable number of chromosomes, small genome size, and giant salivary gland chromosomes (polytene chromosomes) allow carrying out research with ease. The less number of chromosomes is an important feature of the fly that allows easy manipulation during genetic studies. Discoveries made using this feature of flies has helped to understand the mechanism behind transmission of genetic material from one generation to the next. Another major benefit of using *Drosophila* in research is that there are no ethical issues which are a major problem with mammalian models such as monkeys, rats, dogs, cats, and pigs. Behavioral studies such as eating, mating, and sleeping can also be done with ease in flies which allows observing the possible effect of genetic manipulation upon behavior and can be applied to humans also.

The fruit fly is ~3 mm in length; their small size allows raising them in a large number at once. Fruit flies are cultured in quarter-pint bottles using ripened banana or a maize and agar mixture as food. A vast range of mutant stocks (~80,000 *Drosophila* stock variants) are commercially available, and numerous experiments using only a few flies can be done in a limited lab space. The largest public collection of *Drosophila* lines is available at the Bloomington *Drosophila* Stock Center (BDSC), Indiana University, USA. Other major centers like RNAi library, Vienna *Drosophila* RNAi Center (VDRC), and the TRIP-RNAi Harvard collections from Japan, China, and Europe are all made available through BDSC. Another major stock center is at the *Drosophila* Genomics Resource Center (DGRC) which provides cDNAs and vectors.

A large number of embryos, larvae, or adults can be harvested at a time, and the material can be frozen in liquid nitrogen which can be later used to extract DNA, RNA, enzymes, or proteins. The completed sequence of *Drosophila melanogaster* genome was published in 2000, and a year later comparative studies with human genome were done which unraveled various aspects of similarity between the two organisms on both genetic and molecular levels establishing fruit fly as an excellent model organism. The fruit fly has four pairs of chromosomes: chromosome 1 is the sex chromosome (two X chromosomes in females while one X and one Y in males) and chromosomes 2–4 are autosomes (non-sex chromosomes). The smallest chromosome is the fourth and is called dot chromosome which represents only 2% of the total genome. Fly genome contains 132 million base pairs containing ~15,000 genes on 4 chromosomes as compared to the 3.2 billion base pairs containing ~22,000 genes on 23 chromosomes of humans. The *Drosophila* genome is 60% homologous to humans with ~75% of the homolog genes associated with various human diseases.

Drosophila is a holometabolous insect which undergoes many body plan changes throughout its development. Life cycle of a Drosophila completes in about 10–12 days at room temperature (25 °C) with the adult female fruit fly laying ~750–1500 eggs in her lifetime. After the fertilization of egg (~0.5 mm in length), embryo emerges in ~24 h (Fig. 19.7) and undergoes successive molts to become the first, second, and third instar larva which is a voracious eater consuming food in a large amount leading to rapid development which is then followed by a quiescent pupal stage. During the process of development through different stages, fly undergoes vivid reorganization of the body plan (metamorphosis) followed by the emergence of adult fly. Interaction between the two important hormones, prothoracicotropic and ecdysone, along with *Drosophila* insulin-like peptides (Dilps), ensures the proper development of the fly.

Arabidopsis thaliana belonging to Brassicaceae family is also called as rockcress or thale cress, is a small plant with white flowers, and is a very popular model organism for plant studies. It is a member of the mustard (which also includes cabbage and radish) (Fig. 19.8). *Arabidopsis* utilizes basic nutrients like water and few minerals to carry out the process of photosynthesis in the presence of light and complete its life cycle starting from germination to mature seed which takes

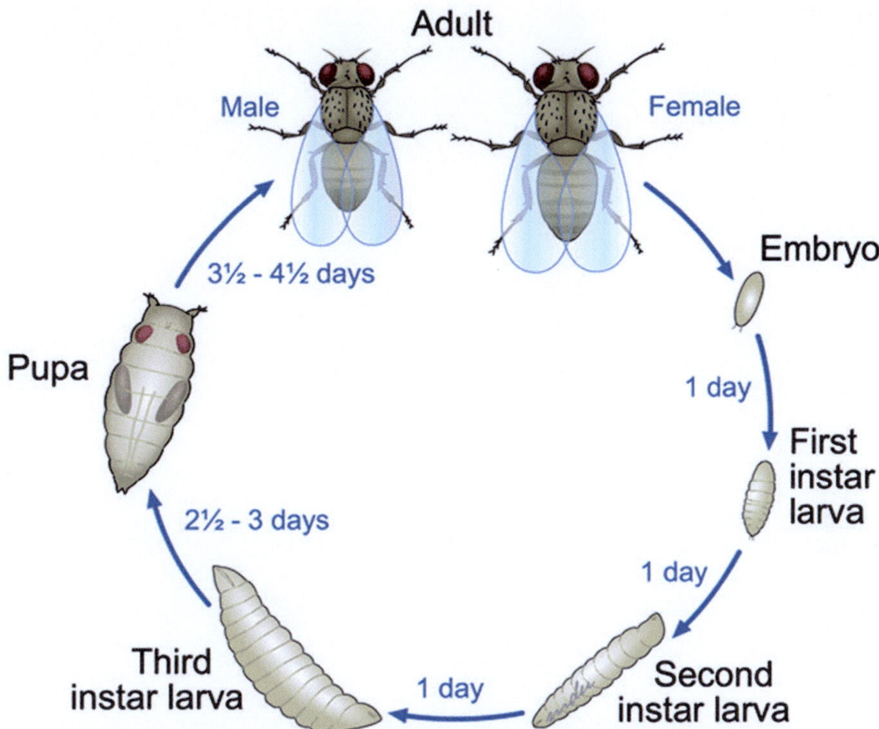

Fig. 19.7 Life cycle of fruit fly (*Drosophila melanogaster*). Being a holometabolous insect, *Drosophila* undergoes complete metamorphosis. The life cycle of fruit fly consists of an egg that hatches into first instar larva which successively molts into second and third instar larva which further molts into dormant pupa from which the adult fly eclosed. The development takes ~10 days and varies depending upon the temperature with rapid development at higher temperatures. (Adapted from Ong et al. 2014)

~6 weeks. Other important features of this model organism includes abundant seed production which can be easily cultivated in a small space, small genome size ~114.5 Mb which is well characterized, availability of extensive genetic and physical maps of all five chromosomes, easy transformation using *Agrobacterium tumefaciens* as vector, and availability of numerous mutants and wide range of genetic tools. The genome of *A. thaliana* is completely sequenced and was published in the year 2000. Diploid genome of the plant makes analysis of recessive mutations easy.

Flower development studies in *A. thaliana* have provided valuable insights on the mechanisms involved. Flower in *Arabidopsis* contains two whorls: outer whorl consisting of four sepals and inner whorl containing four petals and six stamens and a carpel in the center. Two scientists, E. Coen and E. Meyerowitz, studied homeotic genes in *A. thaliana* and found that mutations in these genes result in the change of one organ to another which formed the basis of formulation of the ABC

Fig. 19.8 *A. thaliana*. *Arabidopsis* is one of the most popular plant model organisms and is widely studied in the field of molecular biology and development. (Adapted from https://nl.wikipedia.org/wiki/Bestand:Zandraket.jpg)

model of flower development which is applicable to most flowering plants (Murai 2013). ABC model of flower development states that the genes responsible for defining floral development can be categorized into class A genes responsible for sepals and petals, class B genes for petals and stamens, and class C genes regulating stamen and carpel identity. The transcription factors encoded by these genes regulate the specification of individual region into the respective floral organ. Moreover *A. thaliana* have also been studied extensively to understand the genetics behind leaf morphogenesis. In addition *A. thaliana* is the system of choice to understand various aspects of development and flowering, genome organization, as well as gene regulation.

Although all the abovementioned organisms are excellent model organisms to gain insight into various cellular processes at molecular level, certain characteristics like placentation, intrauterine development, lactation, and aspects of immunology and carcinogenesis are restricted to mammals and hence demand a mammalian model. Being a mammal, intricate processes like aging and immunity can be studied at the molecular level. Their endocrine systems are very similar to humans with secreted hormones affecting a wide range of processes. The mice is among one of the first mammalian species to be completely sequenced. The mouse genome has a

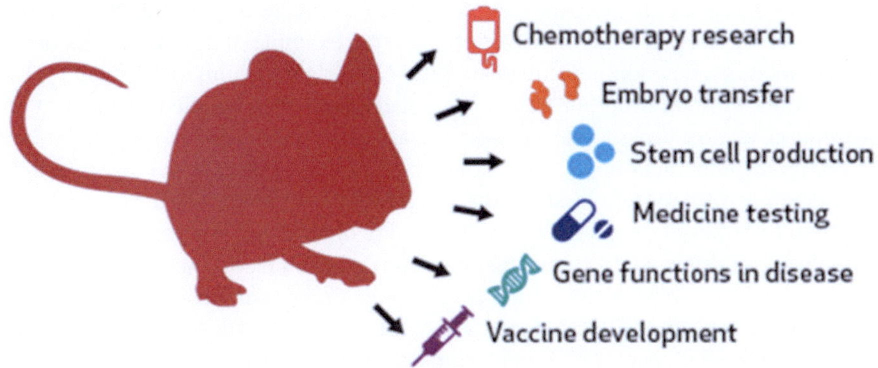

Fig. 19.9 Mouse as a mammalian model organism. Mouse is one of the most widely used mammalian model systems to understand the basic aspects of cellular and molecular machinery. Many groundbreaking discoveries have been done in this model system which unraveled mechanisms behind disease susceptibility and progression and has also helped to develop treatment for it. (Adapted from http://rotarynoidabloodbank.com/en/blog/2016-05/why-use-mice-medical-research.html)

similarity of ~85% with the human genome with a size of ~2.5 Gbp. In order to understand different aspects of biomedical research, various mammalian model organisms have been used, but among all of them, the mouse (*Mus musculus*) is the most flexible and extensively studied mammalian model organism (Fig. 19.9). Advantages of using mice include small size, low cost of maintenance, short generation time of around 10 weeks, prolific breeding, large litter size, and reproductive cycles that can be easily monitored especially during pregnancies (Monica et al. 2016). One of the most common problems with many model organisms is that in order to study a disease, it needs to be induced using artificial means; however, using mice model solves this problem as it develops many diseases like cancer, diabetes, and hypertension naturally. Studies from several years on different mouse models have culminated in our understanding of complex mechanisms underlying many grave diseases and the effectiveness of candidate drugs on these diseases as well as predicted the patient response against these drugs. Humanized mice models are being widely used in the field of biomedical research with the aim to minimize the risk in human therapeutics. Humanized mice models express human gene or contain human cells and tissues and are used to understand the involvement of that particular gene, cell, or tissue in disease development and biological response. Many breakthrough discoveries that have helped scientists win many Nobel Prizes have been done on mice models. Studies in mouse models have led to the discovery of vitamin K, vaccine development for various diseases, monoclonal antibody technology, and tuberculosis vaccine. Studies on cancer mouse models provide microscopic details about the process of metastasis and potential treatments. In the field of medical science, studies on various severe diseases like blood cancer have provided valuable insights which have helped to produce a treatment. In addition, studies on mouse model of cystic fibrosis have facilitated to create a gene transfer protocol which is being used to treat the condition. Another example of use of mice models in disease includes development of meningitis Hib (Haemophilus influenzae type b) vaccination. Further, drug tamoxifen is a widely used drug for the treatment of and prevention against breast cancer which was tested in mice to study its role in blocking hormone action. In strike contrast to various other model systems, mice provide an in vivo system to study the disease development as well as to study the response to different drugs. Additionally, a wide diversity of commercial strains are available with different exclusive features from which a researcher can select from, for example, the CBA mouse which is an inbred strain made from a cross between Bagg albino female and DBA male. CBA mouse is selected for its characteristic feature of low incidence of mammary tumors (breast cancer). Another specific mice strain is BALB/c nude mouse which lacks a thymus and is therefore immunodeficient. Mdx mice models lack mature dystrophin muscle protein and are used to study Duchenne muscular dystrophy. To develop new treatments for autoimmunity, non-obese diabetic (or NOD) mice models are generally used. Such function-specific mouse strains are produced by inbreeding of different mouse models and are used to study a specific disease. Furthermore, several genomic modification tools like CRISPR gene editing and Cre/lox system allow adding or removing a specific gene in a gene casket, thereby producing disease in the model system which helps to study its progression and developing new ways to treat it.

Apart from all the advantages of using mice as a model organism, there are some disadvantages associated with it, for instance, these organisms are complex animals with a genome as large as humans. Another point is that the embryos develop in utero hidden from view restricting access to study the developmental process. In addition, embryo culture is very difficult and limited. Moreover, the generation interval is long, ~3 months. It is difficult to search for the key genes involved in many important cellular processes and find genes through mutational screens. Despite all these difficulties, mouse is still an important model system as it allows researchers to study the development in a mammal unraveling many aspects of human development.

19.1.2 Analysis of Developmental Mechanism

Development is an intricate process leading to the remodeling of cellular differentiation where cells respond to different signaling pathways allowing specific localization of specialized cell subtypes in developing embryo and thus in adult. During the differentiation process, group of cells commits to a particular fate to differentiate into a given cell type; however the commitment is reversible. In the absence of retrogressive factors, inductive signals evoke commitment to convert into determination which is irreversible sealing the cell fate. Cells acquire specific behaviors according to distinct fates which induce growth and morphological changes in embryo. Studies on developmental biology are based upon the fact that most of the complex cellular processes along with the key molecular regulators are conserved throughout the animal kingdom. Completion of whole genome sequences and the advent of molecular tools over this time have allowed evolutionary developmental biologists to prove this by comparing the genes and their functions involved in the development of one organism to another. In order to dissect and gain an deep insight into the developmental systems, researchers utilize various tools like genetic maps, forward and reverse genetic screening, cell differentiation, etc.

During the development of an adult from zygote, the journey of any specific cell starting from origin to destination can be tracked, and a map can be made called the fate map. An extreme example of fate map comes from the completion of *C. elegans* cell lineage studies. This is achieved by asymmetric cell divisions in the daughter cells which obtain different fates from one another, either by segregation of some cytoplasmic determinants or through signaling pathways. Studies on the green algae, *Volvox*, has provided much of the information on evolution of multicellularity and asymmetric divisions. Further, signals from one cell or tissue induce another cell or tissue to affect its developmental fate. One model organism that has been extensively studied to understand this process is *Astyanax mexicanus*, which is a tropical freshwater fish. It is an inhabitant of dark caves which gave rise to completely different varieties or morphs that lack eyes and have several other unique physical, behavioral, and physiological changes. In the first few days of development, the expression of eye genes is inhibited through epigenesist resulting in loss of eye. The visual system of any organisms utilizes a big portion of energy; therefore the loss of eye adaptation to dark offers an energy advantage to these organisms. In the absence of visual system, these organisms depend on sucking instead in order to sense their environment. It was found that these cavefish have altered Pax6 expression and higher levels of a DNA methyltransferase called DNMT3B in their developing eyes. It was assumed that this occurred during the course of evolution of cavefish, leading to epigenetic suppression of eye development genes. Small genetic changes in a subset of genes may have a large impact in the evolution of organisms. A cell or tissue is also highly affected by extrinsic factors or signals often received from the neighboring cells or tissue which is called competence. Typical example of competence can be seen in nematodes where the position of the vulva is highly variable which results due to the change in size of equivalence group during evolution. Another feature of developmental process is genetic redundancy where multiple set of genes are responsible for a single function and silencing of any one of them has no or negligible effect on the respective phenotype. One such example is the function of bicoid, hunchback, and orthodenticle proteins of insects which are responsible for establishing the anterior-posterior axis. Moreover, during the development different mRNA and proteins make a blueprint that a cell can detect and use it for specification and pattern formation. This can also be observed during the development of insect is changes in the Hox gene, Ultrabithorax, Ubx, results in altered morphology of insect wings and in restricting the segments that can bear limbs. Determination is the capacity of a cell to acquire different fates, for example, a region of the embryo became committed to form a particular part of the body at a particular stage of development. Although the zygote is totipotent and has the capacity to make all the cells and tissues of a future organism, it is considered a highly polarized cell. In most of the organisms like insects, nematodes, and amphibians, the polarity of the zygote and cell fate potential of blastomeres is already highly restricted at the two-cell stage. In contrast, the mammalian embryo does not show polarity as the individual blastomeres retain totipotency through the four-cell stage such that an isolated blastomere is capable of forming a viable embryo. Additionally, blastomeres can be removed or added up to the eight-cell stage without affecting the viability of embryo. The last step in developmental process is lateral signaling by which neighboring cells inhibit each other from developing in a similar way. Excellent example of lateral inhibition is changes in bristle patterning in *Diptera* where nascent bristle precursors prevent neighboring cells from developing into bristles through achete-scute complex (AS-C) genes which regulate Notch lateral signaling. Moreover, in order to execute developmental processes, different genetic networks function as maps that represent interactions between discrete genes and modules.

19.1.3 Developmental Genetics: Overview

Developmental genetics explains how the genes control growth and development of an organism throughout its life cycle. A newly fertilized egg cell has all the necessary genes that carry information needed to transform it from a single cell into an embryo and then an adult. During the course of development, single cell transforms itself into an adult organism by developing complex structures. A vast variety of life forms and the intricate details of adult body plans arise from a unicellular stage through the process of embryonic development. This includes three key processes: cell division, cell differentiation, and morphogenesis where cells divide to produce more cells and change into different types of cell to do specific jobs in the body and groups of cells produce different structures of the organism, respectively. After the determination of cell into a specific type, different sets of genes are activated in the cells which are responsible for producing particular types of proteins which in turn carry out specific functions. There are some 350 different types of cells in an adult human being all express different sets of proteins. Some genes are activated (switched on), while others are inactivated (switched off) through DNA regulatory mechanisms. Similar cells express different sets of protein which differentiates them into different cell types and is regulated at the level of gene transcription, nuclear RNA processing, mRNA translation, and protein modification. For example, a nerve cell only produces the proteins required for the nervous system-related functions. There are master control genes or regulatory genes that produce proteins which in turn control the activity of other genes. The development of body plans in all animals is controlled by a remarkably small number of genes which are virtually identical in all animals. For example, homeotic or homeobox genes are responsible for the basic body plan of the embryo in most of the organisms by regulating different sets of genes, thereby differentiating specific body structures (Lawrence and Morata 1994). In addition to body plan, a common axial patterning system and other general architectural features in both vertebrates and invertebrates also appear to be controlled by common genetic mechanisms. These regulatory genes produce proteins called transcription factors that bind to specific DNA sequences called promoter and enhancer regions making a gene to switch on or off depending upon the requirement. In case of eukaryotes, RNA polymerase in addition with basal transcription factors binds to promoter sequences of the genes, thereby initiating transcription. Moreover these genes contain enhancer sequences that regulate their transcription in time and space. The transcription factor binds to the promoter of its own gene in order to maintain its activation. In order to inhibit the expression of nonspecific genes in any region of organism, enhancer sequences play an important role during transcription. Transcription factors act in different ways to regulate RNA synthesis. These transcription factors act in different ways; some stabilize the binding of RNA polymerase to DNA, some disrupt nucleosomes, while others increase the efficiency of transcription. One common mode of suppression of genes is through methylation of promoter and enhancer regions of genes. Difference in methylation pattern results in a process called genomic imprinting where the same gene transmitted through sperm and egg is expressed differentially. Different RNAs are selected which are transported to the cytoplasm from the nucleus, while others remain in the nucleus. Moreover, RNA splicing and combination of different exons and introns create a family of related proteins that function differently. During oogenesis, many mRNAs are localized at certain regions of the oocytes regulated by 3′ untranslated region of the mRNA. Translation of these mRNAs is timely regulated and is carried out only at a specific time during the development of oocyte either by inactivation of inhibitory proteins or by mRNA polyadenylation.

19.2 Genetic Control of Eye Development

The eye is one of the most fascinating organs of the living organisms and has been widely studied as well. The eye is an important organ for the functioning of the body but not essential for the survival of the organism, thereby allowing the study of lethal genotypes. Eye formation is a complex process which includes eye territory specification, polarity axis patterning, and regional specifications which are regulated by multiple genes. Further, morphogenetic movements, cell proliferation, and cell differentiation occur in the prospective eye, and lastly neural connections are established allowing visual function. Specification of eye organ primordia is a well-conserved developmental patterning pathway where mechanisms underlying the formation of eyes and photoreceptor cells in different animals exhibit striking similarity as shown by a series of experiments.

Studies on two mutations, aniridia defect in humans and small eye (Sey) mutation in mice and rats, played key role in unraveling the molecular pathways underlying the eye development. In aniridia defect in humans, the eyes become reduced in size and iris is absent in them, while in small eye mutation, there is complete absence of eyes and the mice dies in utero. Genetic and molecular analysis of both aniridia and the small eye mutants revealed problems in the same gene, Pax6, which belongs to the paired box/homeodomain family of transcriptional regulators. Pax6 protein is abundantly expressed in the eye cells from the early stages (optic sulcus) to later stages (eye vesicle, lens, retina, and finally cornea) of eye morphogenesis. Two genes, eyeless (ey) and twin of eyeless (toy), in *Drosophila* encode proteins that are homologs of Pax6. Both ey and toy are expressed at high levels in the eye primordial cells that form the photoreceptor cells of *Drosophila* eye. Heterozygous mutations in ey lead to the reduction or complete loss of compound eyes, whereas homozygous mutations are lethal. Ectopic expression of mouse Pax6 gene in various tissues led to the formation of small ectopic eyes on the wings, legs, and antennae of *Drosophila*. Pax6 and eyeless genes are considered as master control genes for eye morphogenesis as homologous genes are present in vertebrates, ascidians (sea squirts), insects, cephalopods (squids and octopus), and nematodes (worms), throughout the metazoan necessary for promoting eye development.

Visual system of flies has been studied extensively and represents an excellent model for studying the development, differentiation, and specification process of the retina and photoreceptors. *Drosophila* eyes develop from the posterior part of the monolayer epithelium called eye imaginal disc. During late larval and early pupal stages, neural as well as non-neural cell types in the retina are specified. *Drosophila* has compound eyes with each of them consisting of ~700 hexagonal unit eyes called ommatidia. One ommatidium comprises of light-sensing neural cells [photoreceptors (PRs)], 12 supporting non-neuronal cells (cone and pigment cells), and interommatidial cells (tertiary pigment cells and bristle complexes). The ommatidial structure is very precisely repetitive in normal individuals; therefore even the subtle abnormalities may be recognized. Rhabdomeres which are the microvillar structures extend toward the center of the ommatidium and bind to the six of the eight PRs (R1–R6) which are called outer PRs (Fig. 19.10). Rhabdomeres of each PR consist of the light-sensitive pigments called Rhodopsin (Rh). Outer PRs are arranged in a trapezoid shape and accompany the two other PRs, R7 and R8, also called as inner PRs and are positioned in the center of the ommatidium with R7 on top of R8. Outer PRs (R1–R6) express the broad-spectrum Rh1 and are thought to be functionally similar to vertebrate rod cells required for dim-light vision and motion detection. In contrast, inner PRs (R7 and R8) are considered similar to the vertebrate cone cells and function in mediating color vision and perception. Inner PR, R7, expresses the ultraviolet-sensitive Rh3 and Rh4, while R8 cells express either the blue-sensitive Rh5 or the green-sensitive Rh6. The expression of Rhs in both R7 and R8 cells is coupled in such a way that the expression of Rh3 in R7 is coupled to the expression of Rh5 in the R8, whereas Rh4 expression in the R7 cell is coupled to Rh6 expression in R8 cells. The ommatidia coupling of Rh3/Rh5 is called pale, while that of Rh4/Rh6 is called yellow. The distribution of pale to yellow is highly conserved among different fly species in a 30:70% ratio in the retina (Fig. 19.10).

Eye formation in *Drosophila* occurs at the posterior margin of the eye imaginal disc in response to a differentiation wave called morphogenetic furrow (MF) which proceeds from posterior to anterior end (Fig. 19.11). MF initiates and progresses with the help of secreted molecules and directs cell-cell signaling resulting in the sequential differentiation of PRs. MF initiation on the posterior margin of the eye imaginal disc occurs due to the Egfr and Notch signaling which further induces

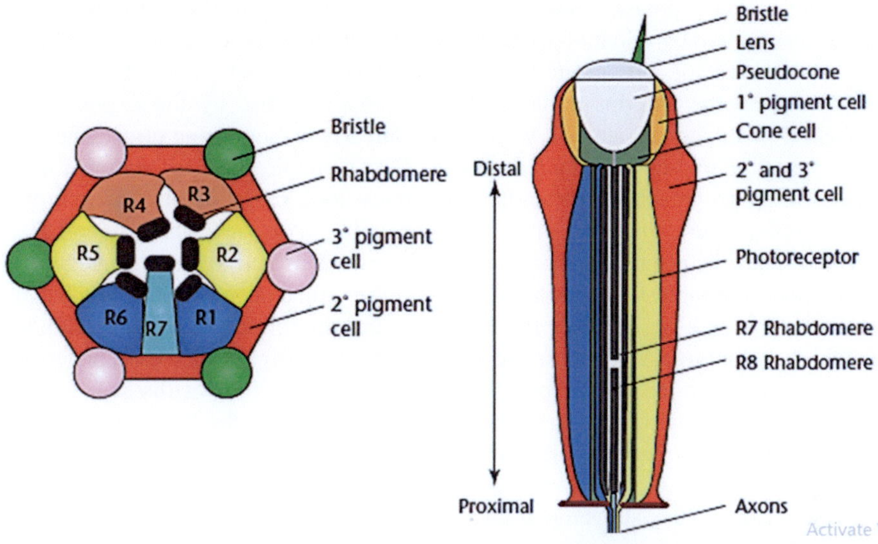

Fig. 19.10 Arrangement of photoreceptors during eye development. Transverse section of an adult ommatidium (left) showing six outer PRs and one of the inner PRs (R7). PRs specified simultaneously are represented in the same color. Rhabdomeres, light-sensing structures, are attached to PRs and are shown in black color. R3 rhabdomere is located further apart from the inner PR, giving rise to a trapezoid shape. Secondary and tertiary pigment cells surround the PRs with evenly localized bristle cells. Longitudinal section of an adult ommatidium (right), covered by lens and a pseudo-cone. (Adapted from Sahin and Celik 2013)

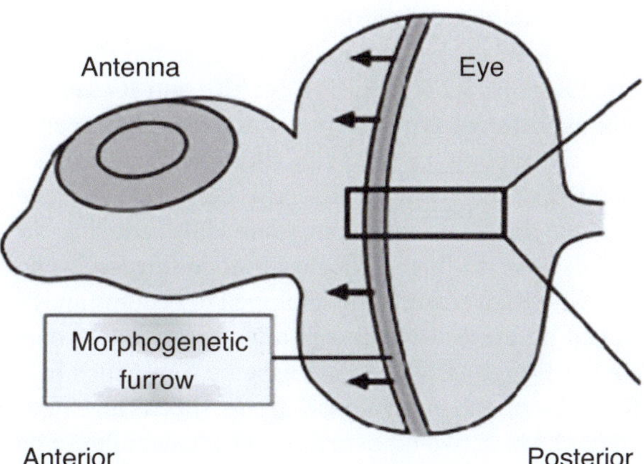

Fig. 19.11 Morphogenetic furrow (MF) in the eye-antennal imaginal disc. Eye imaginal disc is an epithelial tissue, where the eye is originated from the posterior part while the antenna and maxillary palps are formed from the anterior part. The morphogenetic furrow (MF) moves from the posterior to the anterior and is responsible for sequential differentiation of PRs. (Adapted from http://slideplayer.it/slide/999922)

Hedgehog (Hh) expression. Hh is one of the key players in the eye development process as evident by initiation of more than one MF in case of ectopic expression of Hh while arrest of MF progression in the absence of Hh. The MF is further progressed to the anterior parts of the eye imaginal disc by the expression of proneural gene, atonal (ato), as well as Hh. Hh has a short-range effect; therefore Hh induces the expression of a morphogen, Decapentaplegic (Dpp), member of the transforming growth factor-b family which has a long-range effect. Hh along with Dpp regulates the expression of Homothorax (Hth), Ato, and Notch ligand Delta (Dl) to induce the proneurogenesis in specific cells. Dpp activity is regulated by the expression of Wingless (Wg) which represses the retinal development. The balance between Dpp and Wg controls the MF wave in thin column of cells. In addition, Notch, receptor tyrosine kinases, and Ras-MAPK signaling define the PR patterning. Hh is in a positive feedback loop with Pointed (Pnt) and Sine oculis (So) while in negative feedback with Egfr ligand Spitz (Spi) which limits the MF to a thin line of cells. As the MF progresses, retinal differentiation occurs resulting in the generation of PRs.

In addition to the abovementioned signaling molecules, retinal determination genes which include eyeless (ey), eyes absent (eya), twin of eyeless (toy), teashirt (tsh), and dachshund (dac) play key roles in eye development. Ey, a homeodomain transcription factor, is the *Drosophila* ortholog of Pax6 and is considered the master regulatory gene for retinal development. Ey along with its transcription factor toy (an activator of ey) is expressed during embryogenesis which results in formation of the eye and antenna. Ey expression is restricted to only limited row of cells that become proneural at once; this is regulated by interplay between Hh and Dpp signaling which in turn control So and Ato. As the MF progresses, the epithelial cells in the imaginal disc proliferate asynchronously and form an evenly spaced cluster of ~20 cells called rosette. In order to complete the ommatidial assembly, first the posterior-most cells of the rosette are specified as the PR 8 (R8) which is then joined by pairs of PR cells, R2/R5, R3/R4, R1/R6, and R7 cells, followed by four cone cells as well as different types of pigment cells. This specification is mainly regulated by the transcription factor, Ato, through Wnt and Notch signaling in addition to Ey and So. The rosette cells now constitute the intermediate group, which starts to reduce due to apoptosis in some cells reducing the cluster to, first, five-cell cluster and, later, to three-cell cluster. The three-cell cluster is called the equivalence group with each cell equally equipped to become an R8 cell. R8 cell fate choice is determined by the Ato-Notch signaling through two transcription factors, Senseless (Sens) and Rough (Ro). Ato induces the Sen which in turn represses the Ro in one of the cells of the equivalence group specifying the cell as R8. Ro is expressed at a higher level in two other cells of the cluster repressing Sens and giving rise to R2 and R5 while repressing R3/R4 and R1/R6 fates by suppressing a nuclear receptor, Seven-up (Svp). After the specification of R8, Egfr signaling through its ligand Spi is responsible for the specification of all other PRs except R7. Moreover, Spalt complex, which consists of the two transcription factors, Spalt major and Spalt related, is responsible for inducing Svp resulting in the specification of R3/R4. Further, two actions in the cells surrounding the rosettes, first suppression of Spalt genes by Svp and second induction of another transcription factor, Lozenge (Lz), result in the generation of R1/R6 cells which are recruited to the cluster. Sev is expressed in other PR cells, R3/R4 and R1/R6; however, Svp represses its activity, thereby preventing them to differentiate into R7 cells. The last cell in the assembly of PRs to be specified is R7, which starts to express a receptor tyrosine kinase, Sevenless (Sev), which binds to the transmembrane ligand Bride of Sev (Boss) on the R8 cell. Activation of Lz as well as another transcription factor, Prospero (Pros), which is regulated by several zinc-finger transcription factors, Lz, So, Eya, and Glass, gives rise to R7. After the specification of all the PRs, support cells constituting of cone cells, pigment, and bristle complexes from the surrounding undifferentiated cell pool join the ommatidium. First, four cone cells join the ommatidium just above the PRs in order to minimize the surface area and secrete the pseudolens and the lens. Toward the basal membrane lies two primary pigment cells between the cone cells and PRs. Finally, in order to complete the visual cellular specification, secondary and tertiary pigment cells and bristle complexes join the ommatidium.

The eye consists of D-V polarity as the dorsal and ventral sides of the fly eye are different from each other in their cellular layout which is called planar cell polarity (PCP). Both the dorsal and ventral halves of the ommatidia are aligned as mirror images of each other around the line of symmetry called the equator, which is perpendicular to the MF along the A-P plane (Fig. 19.12). During early development, the entire disc has a ventral identity by default. During the first instar larval stage, several signaling pathways and transcription factors including Pannier (Pnr), Wg, Iro-C (araucan, caupolican, and mirror homeobox genes), Notch, and Janus kinase-signal transducer establishes the D-V polarity. Pnr functions at the dorsal side of the eye imaginal disc, whereas Wg functions at both sides. The dorsal identity is obtained via specific signals, such as Pnr, a zinc-finger transcription factor, Notch ligand Dl, and Iro-C, whereas the ventral identity is obtained with the help of glucosyltransferase Fringe (Fng) and Ser. Notch and its effector Eye gone (Eyg) play key roles to establish the dorsal-ventral polarity. Overexpression of Eyg results in the formation of additional eyes on the ventral side of the head, while Eyg mutants lose their eyes entirely. After the MF, symmetrical organization of ommatidia starts where the cell clusters start to rotate in opposite directions around the midline, which cuts the D-V axis into two halves. After the first 45^0 rotation, the clusters rotate another 45^0 in order to create a proper image. The PRs are arranged in an asymmetric trapezoidal arrangement in such a way that the distinct position of R3 and R4 gives rise to the chirality in the retina. Ommatidia exist in either of two chiral forms, depending on their position such that ommatidia in the dorsal half of the retina adopt one chiral form and the ventral half adopt the other. Establishment of chirality starts at the equator where one of the two anterior cells in the five-cell precluster, the cell with higher Frizzled expression, is destined to adopt the R3 fate, whereas the other cell adopts the R4 fate.

After the establishment of D-V polarity, the developing eye imaginal disc undergoes several morphological changes like establishment of cell junctions in the retina, attachment of PRs to each other through cell junctions, change in cell

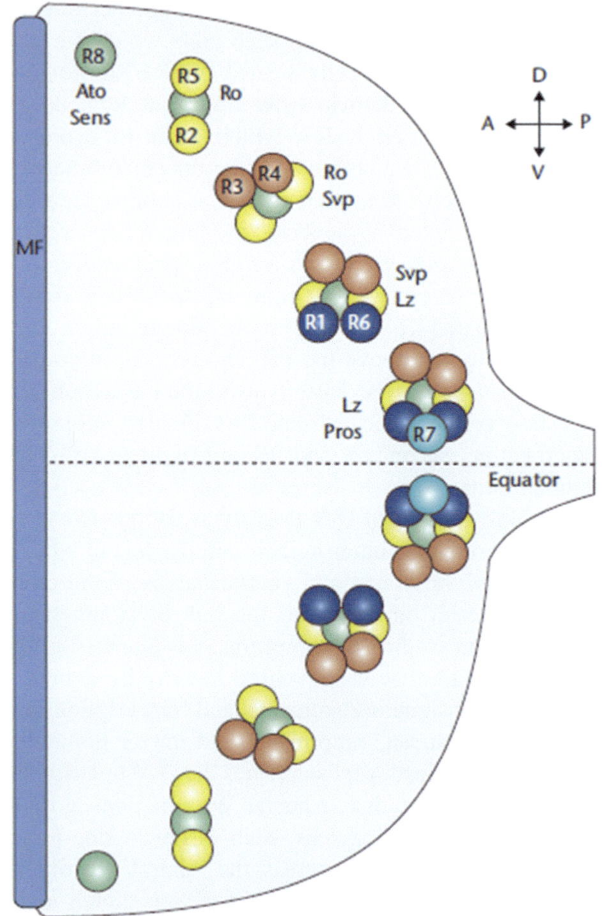

Fig. 19.12 Sequential assembly of ommatidia. Photoreceptor specification starts after the dorsal and ventral identities in the eye disc are obtained. The symmetrical organization of ommatidia starts right after the MF passes. The eye disc is symmetrical around the equator. This opposing alignment of the ommatidia should be precise in order to create a proper image. (Adapted from Sahin and Celik 2013)

shape of photoreceptors during early pupal stages, and extension of axons from PRs to different layers of the optic lobe to form proper connections with the brain. Among the cell junctions in the eye, the main homophilic adhesion molecules are the adherens junctions (AJs). In the fly, all three cadherin homologs function in the development of the retina. *Drosophila* E cadherin (DE-cad) is ubiquitously expressed, in all cone and pigment cells, whereas the expression of *Drosophila* N cadherin (DN-cad) is limited to different subsets of cells at each developmental stage. The asymmetric specification of R3/R4 in dorsal and ventral halves initiates polarization and ommatidia turn perpendicular to the equator in opposite directions in such a way that the R8 cell faces the equator. This motion requires cadherin function, DEcad promotes the rotation, whereas DN-cad has a specific function in R3/R4 rotation with both DN-cad and DN-cad2 having redundant functions. Further, the apical surfaces of PRs turn inside toward the center of the ommatidium detaching them from other cells. This leads to the formation of light-sensitive microvillar structures called rhabdomeres. Zonula adherens plays an important function of tissue integrity during this process. During the late pupal stages, PR cells start to elongate basally leading to the remodeling of zonula adherens junctions in the process to reach the brain. Before pupariation, larval photoreceptor neurons extend axons down the optic stalk toward the brain finding their target zones in the lamina and medulla, while glial cells migrate in the reverse direction, following photoreceptor axons from the brain back to the eye disc. R7 cells elongate sending their axon to the optic lobe, while R8 stays beneath the R7. Transmembrane protein, Crumbs, plays the most important role during this elongation event. In addition, cadherin-mediated adherens junctions play a key role in achieving the stereotypic arrangement as well as peculiar shape of the cone cells and pigment cells. Finally, PRs send their axons to the optic lobe of the brain in order to establish a functional connection with the brain. Axons of all the PRs send their axons through a fenestrated membrane which is formed by the cone cells and secondary and tertiary pigment cells at the basal part separating the retina from the brain. Thus, proper assembly and remodeling of AJs is absolutely required for different aspects of eye development to ensure the generation of a functional retina.

The last step in the eye morphogenesis is the specification of the photoreceptor subtypes which is dependent on the distinction between inner and outer PRs. This specification is dependent upon the transcription factors of Spalt family, for example, Sal is necessary for determining the inner cell fate. In addition, there is activity of additional transcription factors, Pros in R7 and Sens in R8, which then terminally differentiates these two inner cells into R7 and R8, respectively. Further, the next fate-determining step is the generation of the two major ommatidial subtypes pale and yellow. The yellow fate is determined by the transcription factor Spineless which is expressed in 70% of ommatidia in the R7 cell, thereby inducing rh4 expression in them which communicate with the underlying R8 cell including them to express rh6. All the remaining R7 cells that do not express spineless express rh3 by default and induce the underlying R8 cell to express rh5. Orthodenticle (otd) belongs to the paired class of homeobox genes and is expressed in all PRs which leads to the expression of pale Rhs, rh3 in R7 and rh5 in R8, as well as represses the expression of rh6 in outer PRs by activating the repressor defective proventriculus (dve). After the expression of correct subsets of Rh genes, two proteins, Melted, a Pleckstrin-homology (PH) domain protein, and lats, a serine/threonine kinase, function in a regulatory loop to ensure the maintenance of the correct Rh choice. The expression of both lats and melt is regulated by the Hippo signaling, through its members Mer and Kib. Melted is expressed in pale R8 cells promoting the expression of Rh5 and repressing lats, while lats is expressed in the yellow R8 cells promoting Rh6 in turn and represses melted expression. In addition to pale and yellow subtypes, the dorsal part of the eye contains two additional ommatidial subtypes called the dorsal rim area (DRA) and the dorsal y. The DRA is present at the two dorsal-most rows of the retina with R7 and R8 cells expressing Rh3 and is required for polarization vision. In contrast, dorsal y ommatidia are present in the dorsal third of the retina (with R7 cells expressing Rh3 and Rh4, whereas R8 cell expressing Rh6) and are required to discriminate between the solar and non-solar parts of the sky for proper navigation of insects. The expression of UV-sensitive Rh3 and Rh4 that are coexpressed in the R7 cells of dorsal ommatidia is through the action of iro-C. Once all the cells are specified, fly head eversion moves the eye and head tissues into their adult configuration before the end of pupation.

19.3 *Drosophila*: Embryonic Development

Drosophila is an excellent model organism for the study of developmental processes as the basic framework of early development between human and *Drosophila* closely resembles owing to the highly conserved network pathways and the key regulators in it. Numerous studies on *Drosophila* make the basis of our current understanding on developmental processes and pattern.

19.3.1 Overview

The transition of the oocyte to embryo marks the onset of development which employs complex and stringent regulations of the developmental signals, mRNA translation, as well as cell cycle. In almost all organisms, embryogenesis occurs in the absence of zygotic transcription, utilizing the maternal mRNAs, translational machinery components, and nutrients which are accumulated in the eggs during oogenesis well before embryogenesis. In drosophila, there are two meiotic arrests during oogenesis, first at the prophase I and a secondary meiotic arrest at metaphase I. This developmental strategy allows the maternal stores to be deposited into the oocyte during oogenesis. Further, several changes in protein levels, mRNA translation, polyadenylation, and egg activation lead to oocyte-to-embryo transition called maternal-to-zygotic transition (MZT) allowing activation of expression of the zygotic genome.

Accumulation of maternal mRNAs and proteins is a prerequisite for the normal production of functional oocyte embryo development. *Drosophila* oogenesis is extensively studied to understand the process of ovarian development, embryogenesis, as well as the underlying mechanism due to simplicity of oocyte development and the ease of studying it. *Drosophila* female contains a pair of ovary which in turn each has 20–30 ovarioles (Fig. 19.13). Each ovariole contains a germarium at the most anterior part and a mature egg at the most posterior end with 14 progressively more developed follicles or egg chambers in between. Germarium is the production house of the egg chambers, each containing 16 sister cells that share common cytoplasm attached through cytoplasmic channels called ring canals, out of which 15 cells become nurse cells, while 1 cell acquires the oocyte fate during oogenesis. Egg chamber is covered by follicle cells (FCs) which are polyploidy cells and are required for the patterning of early stages as well as for depositing the multilayered egg shell during late oogenesis (Fig. 19.13). An egg chamber provides a microenvironment for the development of the oocyte. Nurse cells are required for the synthesis of the DNA and RNA that are stored in the egg and are required for the early embryo

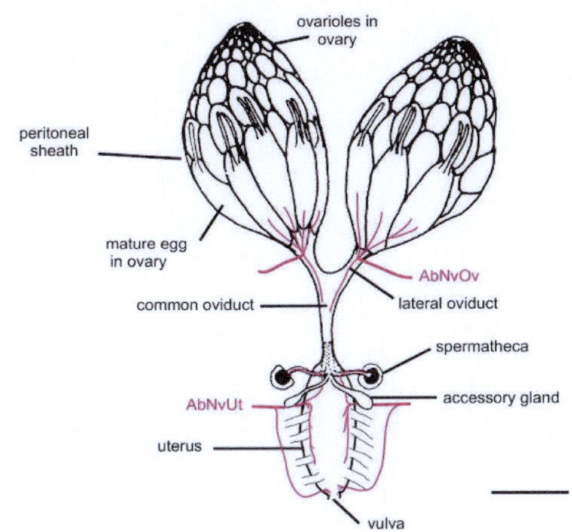

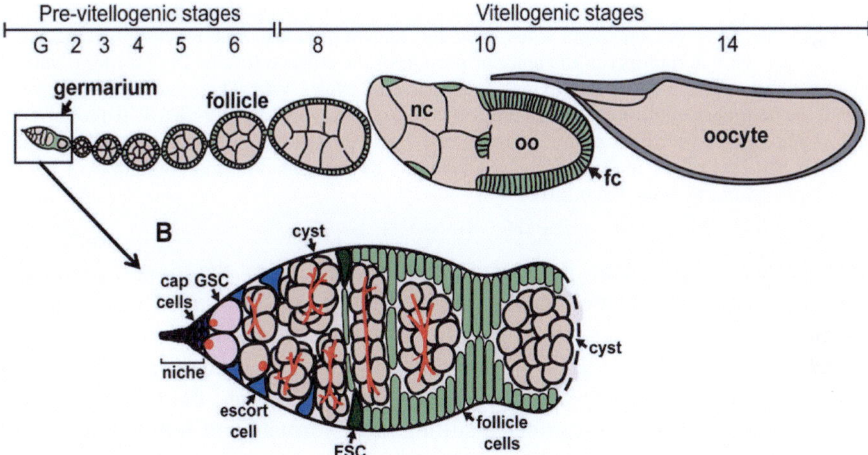

Fig. 19.13 Ovary morphology and oocyte development in *Drosophila melanogaster*. Each *Drosophila* female has a pair of ovaries covered in peritoneal sheath connected through lateral oviduct which descends into a common oviduct followed by the uterus and vulva. Spermatheca and seminal receptacles are the sperm storage organs, while accessory organs also called as parovaria are believed to have a secretory function. Each ovary is composed of 17 to 20 ovarioles each divided into 14 stages of development with stage 14 mature egg as the last stage. In each ovariole, the egg chambers are arranged in a developmental sequence with germarium at the most anterior end and most mature stage at the posterior end. Germline stem cells present in the germarium undergo karyokinesis simultaneously for four times with incomplete cytokinesis producing a 16-cell structure called cyst. In the cyst, 1 cell becomes the oocyte and the other 15 become the nurse cells. (Adapted from Middleton et al. 2006 and Ables et al. 2016)

development post-fertilization. The oocyte undergoes meiotic maturation at stage 13 of oogenesis where all the contents of the nurse cells are transferred to the oocyte in a process called dumping while they undergo apoptosis after that (Pritchett et al.

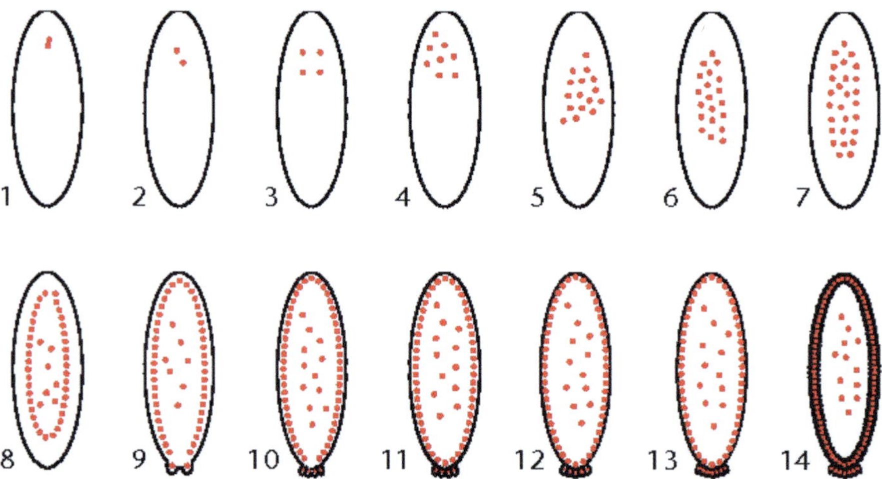

Fig. 19.14 Different stages of *Drosophila* embryogenesis. In *Drosophila*, embryonic development is divided into 17 stages and starts when male and female pronuclei fuse together (cycle 1). This nucleus then starts to multiply at the anterior portion of the embryo (cycles 2–3) forming multiple nuclei which spread along the A-P axis (axial expansion, cycles 4–6). These nuclei start migrating toward the periphery of the embryo in a process called cortical migration (cycles 8–10) followed by the formation of the pole cells on the posterior end (cycle 9). The embryo is enveloped by multiple layers by the end of all nuclear divisions followed by invagination and formation of cellular blastoderm. (Adapted from Tram 2011)

2009). Mature egg (stage 14 oocyte) is released from the ovary (ovulation) descending into the oviduct, where it undergoes a process called egg activation which prepares the oocytes for fertilization using the mechanical forces and hydration. The sperm then fertilizes the egg entering through a pore called micropyle present on the anterior side of the oocyte. After fertilization, the meiotic arrest is released, and the meiosis is resumed resulting in the formation of four female meiotic products. In the absence of fertilization, development is arrested and the four female meiotic products assemble to form a single polar body. After fertilization, the male and female pronuclei fuse and undergo 13 rounds of mitosis in a common shared cytoplasm (syncytia) which is regulated by maternal mRNAs and proteins.

The fused male and female pronuclei undergo 13 rounds of dynamic nuclear divisions in the shared cytoplasm resulting in the formation of a syncytial blastoderm containing ~6000 nuclei (Fig. 19.14).

Activation of a cascade of genes sets up the *Drosophila* body plan. The first in this sequence are maternal genes that are expressed in the ovaries and are accumulated at different but specific regions of the developing embryos. Translated products of these mRNAs act as morphogens and form a gradient along the embryo. Bicoid and Hunchback regulate the production of anterior structures, while Nanos and Caudal are responsible for the posterior parts of the embryo. Next in this sequence are zygotic genes that are, depending on the condition, either activated or suppressed by maternal genes. Zygotic genes include gap genes, pair-rule genes, and segment polarity genes in the same sequence of action. Gap genes are expressed in broad domains throughout the embryo, and mutations in them result in gaps between these segments. Next, pair-rule genes that are responsible for dividing the embryo into seven bands perpendicular to the A-P axis are activated by the gap genes. Further, segment polarity genes come in action that are regulated by the pair-rule genes and are responsible for dividing the embryo in 14 equal segments. Once the combined action of all three abovementioned set of genes has divided the embryo in periodic segments, homeotic selector genes are activated and regulate the fate of individual segments. In the next 24 h, the embryo converts to the larva that keeps growing for the next ~4 days (at 25 °C). The larva then molts two times to second instar larva in ~24 h and third instar larva in ~48 h. The larva then converts into pupa that is a dormant stage where it undergoes metamorphosis for ~4 days before the adult or imago ecloses. The adult fly can be divided into head, thorax, and abdomen. The head region contains the eyes, mouth, and antennae; the thoracic region is divided into three segments T1 (contains a pair of legs), T2 (contains a pair of leg and a pair of wings), and T3 (contains a pair of legs and a pair of halteres), while the abdomen region is divided into eight segments (A1 to A8).

19.3.2 Anterior-Posterior Body Axis

The anterior-posterior (A-P) as well as dorso-ventral (D-V) body axes are determined way before the embryonic development during the egg development. The anterior-posterior axis of the embryo is broadly specified by three sets of genes: the first set defines the anterior organizing center, the second set defines the posterior organizing center, and the third set defines the terminal boundary region. Different mRNAs are localized at the different regions of egg predestining the embryo development. The A-P patterning starts after the completion of gastrulation and is regulated by two sets of genes, maternal effect genes and zygotic genes (Fig. 19.15). Maternal effect genes are expressed when the egg is in the ovary of the fly and their transcripts are distributed in the egg. Maternal effect genes include Bicoid, Nanos, Hunchback, Caudal, Torso, and Toll. Among these genes, bicoid and hunchback mRNAs determine the head and thorax formation, while nanos and caudal mRNAs are required for the formation of abdominal segments. A-P patterning is established due to the intercellular communication between the oocyte and the somatic follicle cells. The nurse cells present in the egg chamber of ovary deposit the transcripts of the different maternal effect genes in the egg. These transcripts are localized due to the action of microtubules which are arranged in such a way that their positive and negative ends are oriented toward the posterior and anterior sides of the egg chambers, respectively. The localization of these maternal effect genes in the egg predetermines the A-P polarity of the embryo. After fertilization, the proteins from these maternal effect gene transcripts act as the transcriptional activators for the next set of genes called zygotic genes.

During egg development in the ovary, bicoid mRNA from maternal bicoid gene is accumulated in the anterior region of the egg cells by the nurse cells in the dormant

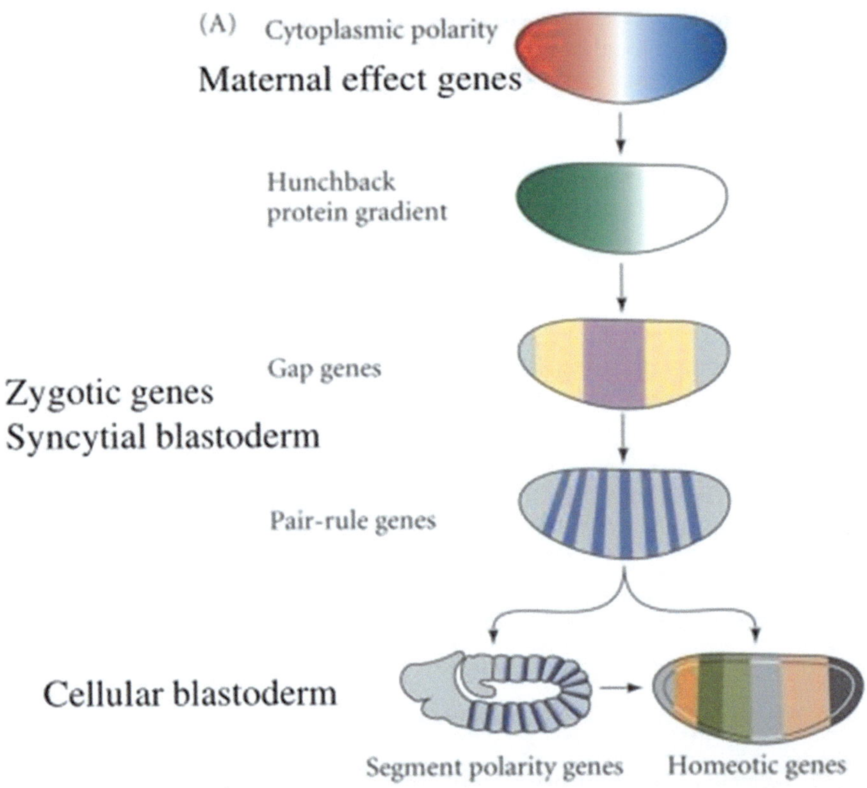

Fig. 19.15 Action of maternal and zygotic genes in *Drosophila* pattern formation. Maternal effect genes first establish the anterior-posterior (A-P) pattern with bicoid at the anterior tip and nanos at the posterior tip of the fertilized egg. Gap genes are regulated by maternal genes and divide the embryo into broad regions. Pair-rule genes are then activated by gap genes (hunchback and Krüppel) which are responsible for the segment formation in embryo. Two pair-rule genes, fushi tarazu and even stripped, are expressed in alternate strips, resembling the zebra stripes, along the A-P axis of the embryo. All these genes regulate the expression of homeotic genes that define the identity of each segment. Expression of segment polarity genes (engrailed) divides the embryo into a repeated series of segmental primordia along the anterior-posterior axis. (Adapted from Mundlos 2010)

condition (Chang et al. 2011). Cytoskeleton (microtubule motor protein dynein) at the anterior region of the egg anchors the bicoid mRNA through 3′ untranslated region (UTR). After fertilization, a polyadenylate (poly-A) tail is added to the bicoid mRNA resulting in its translation forming a gradient with highest concentration in the anterior of the egg and the lowest in the posterior third of the egg. Different mutation studies have confirmed the function of bicoid in the anterior region of the embryo. When bicoid is injected in the anterior region of the bicoid-deficient embryos, normal anterior-posterior polarity was established in it. When bicoid is added in the center of the embryo, it developed in the head, while both ends became thorax. In contrast, if bicoid is added in the posterior region of the wild-type embryo (having normal bicoid level at anterior end) two heads develop at either side. Another evidence for bicoid function was observed in the exuperantia and swallow mutant embryo; both these genes are required for restricting bicoid in the anterior region. In these mutants, bicoid diffuses toward the posterior end of the egg such that the gradient cannot be formed resulting in the absence of anterior structures and presence of extended mouth and thoracic region. Bicoid protein not only ensures the formation of anterior organs but also ensures that the proteins required for the posterior region are localized there only. One such protein is caudal which is required for the formation of posterior domains of the embryo. Bicoid protein binds to the 3′UTR of the caudal region inhibiting the translation of caudal mRNA allowing its translation only in posterior region. In addition, bicoid works as a transcriptional activator of the hunchback gene in the nucleus. The Bicoid and Hunchback proteins act synergistically as the enhancers to promote the transcription of the genes required for head formation.

Next in the array of maternal effect genes is nanos, which similar to bicoid is synthesized by nurse cells during egg development and accumulated in the posterior region of the egg. Similar to bicoid, nanos expression remains repressed by the binding of the Smaug protein to its 3′ UTR. Nanos protein remains bound to the cytoskeleton in the posterior region of the egg through its 3′ UTR. Nanos along with oskar, valois, vasa, staufen, and tudor ensures normal embryonic abdomen formation. In addition, nanos along with pumilio protein binds with hunchback mRNA and prevents its translation in the posterior region. As a result, early *Drosophila* embryo shows a gradient of four proteins, Bicoid and Hunchback proteins at anterior and Nanos and Caudal proteins at the posterior end. Bicoid, Hunchback, and Caudal proteins function as transcription factors which further activate or repress different zygotic genes.

Another maternal effect gene is torso (encodes a receptor tyrosine kinase) which is required for the formation of extreme ends in the embryo; the most anterior head segments are called acron and most posterior abdominal segments are called telson (tail). Similar to all maternal effect genes, torso mRNA is synthesized by the ovarian cells, deposited in the oocyte, and translated after fertilization. The ligand for the activation of Torso protein is trunk protein which is secreted in an inactive form. Trunk is activated by the proteolytic cleavage carried out by Torso-like protein which is activated by trunk-like protein present only at the two poles of the oocyte. Thus, Torso protein is expressed only at the extreme anterior and posterior regions of the oocyte membrane. Torso protein functions through the Receptor tyrosine kinase cascade which results in the activation of tailless and huckebein gap genes. These gap genes along with bicoid specify the termini of the embryo. In the presence of bicoid, these genes form acron, while in its absence the terminal regions differentiate into telson.

19.3.3 Embryogenesis

Embryonic development also called embryogenesis is the process from fertilization of the egg followed by embryo formation and its development. The study of *Drosophila* embryogenesis has contributed most of the understanding of the molecular basis of animal development. *Drosophila* and human development are homologous processes. *Drosophila* utilizes conserved set of genes and regulatory networks which makes the studies on this animal model to be utilized to understand the process in higher organisms. The embryonic development of *Drosophila* melanogaster has been subdivided into 17 stages (Fig. 19.16, Table 19.2). *Drosophila* egg is bilaterally symmetrical where only difference between the two sides is flat dorsal side and slightly convex ventral side. The average length of the egg is

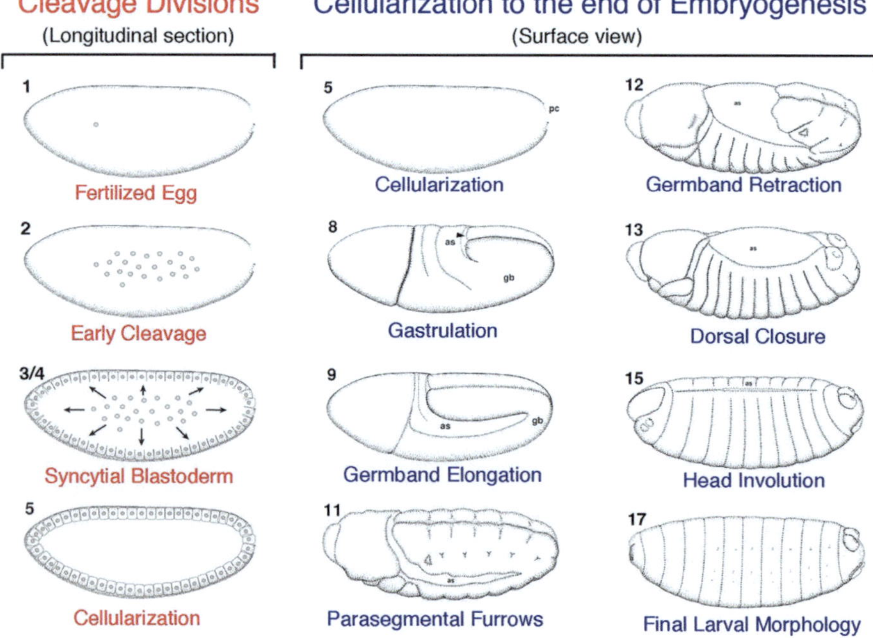

Fig. 19.16 Different stages of embryonic development. After the fertilization, embryonic development starts and comprises of 17 stages of development marked by specialized process occurring in that stage. Earlier stages are characterized by nuclear divisions without division of cytoplasm (cytokinesis), after ten rounds of synchronized nuclear divisions, nuclei start to migrate to the periphery of embryo where they become encapsulated by actin-based furrow canals (stages 3–4). Stage 5 is characterized by cellularization followed by gastrulation which determines the three germ layers (stage 8). Stage 9 is characterized by germband extension which remodels the body plan as cells from the posterior end of the embryo migrate toward the anterior end. Germband retraction in stage 12 is followed by the migration of epithelial cells toward the dorsal midline called dorsal closure (stage 13). Further, in stage 15, head involution occurs where head structures mature, and finally the larva reaches its mature state (stage 17) and hatches from the eggshell. (Adapted from Hales et al. 2015)

Table 19.2 List of *Drosophila* embryogenesis stages with key process

Stage	Developmental events
Stage 1	First two nuclear divisions
Stage 2	Syncytial divisions 3–8, cytoplasmic clearing
Stage 3	Syncytial division 9, polar bud formation
Stage 4	Syncytial divisions 10–13, pole cell formation
Stage 5	Cellularization
Stage 6	Onset of gastrulation, formation of ventral and cephalic furrow, dorsal shift of pole cells
Stage 7	Completion of gastrulation, pole cells in a pocket
Stage 8	Rapid phase of germ band extension, to 60% egg length
Stage 9	Germ band elongation to 70% egg length, early neuroblast delamination
Stage 10	Germ band elongates to 75% egg length, stomodeum invaginates
Stage 11	Segmentation, tracheal pits arise, posterior midgut invagination reaches the posterior pole
Stage 12	Onset of germ band retraction, fusion of anterior and posterior midgut
Stage 13	Completion of germ band shortening
Stage 14	Dorsal closure to 80% along the dorsoventral axis, head involution
Stage 15	Dorsal closure of epidermis and midgut
Stage 16	Gastric caeca are formed, somatic musculature becomes visible
Stage 17	Completion of organogenesis, movements of the embryo within the vitelline envelope

~500 μm; the diameter is about 180 μm. The mature egg is covered by a tough, opaque outermost layer called chorion. Below chorion is an additional transparent homogeneous membrane called vitelline membrane. Multiple hexagonal and pentagonal patterns can be seen on the chorion which is the impressions of the ovarian follicle cells. A pair of filament is present at the anterior dorsal surface as an extension of the chorion. There is an opening in the vitelline membrane called micropyle required for the entry of sperms.

Embryogenesis in *Drosophila* starts with the fertilization, followed by multiple rounds of nuclear division through mitosis without cytokinesis resulting in a multinucleated cell called syncytium or syncytial blastoderm with shared cytoplasm. Syncytial blastoderm allows different proteins to form gradient along the cytoplasm which regulates pattern formation in embryo. After eight rounds of nuclear division, 256 nuclei are produced in the central portion of the egg. A group of cells reach the surface of the posterior pole of the embryo and become enclosed by a cell membrane forming pole cells which in the future give rise to the gametes of the adult fly. After the tenth nuclear division, the nuclei in the center starts to migrate to the periphery of the embryo where the nuclear divisions continue. The shared cytoplasm is not uniform in nature, but each nucleus is surrounded by its cytoskeletal proteins (microtubule and microfilament). The nuclei and its associated cytoplasmic islands are called energids. At the 13th division, ~6000 nuclei, arranged at the periphery of the embryo, are partitioned into separate cells by the invagination of the oocyte cell membrane. This process produces an embryo with peripheral cells and yolk center and is called cellular blastoderm. At the 14th cycle of embryo development also called as midblastula transition, division is asynchronous producing different types of cells. Midblastula transition is followed by gastrulation where the presumptive mesoderm, endoderm, and ectoderm are formed. Approximately 1000 cells of future mesoderm forming ventral midline of the embryo start to fold inward to produce the ventral furrow. The ventral furrow pinches off from the embryo surface forming the ventral tube within the embryo. The endodermal cells form two pockets at the anterior and posterior ends of the ventral furrow followed by pole cell internalization. At this time, the embryo bends to form the cephalic furrow. The ectodermal and mesodermal cells migrate toward the ventral midline forming the germ band. The cells of germ band are destined to become the trunk of the embryo. The germ band extends posteriorly and covers the dorsal surface of the embryo. At this stage the cephalic furrow separates the future head region (procephalon) from the germ band that will form the thorax and abdomen, and body segments start to appear. While the germ band is in extended position, various important processes like organogenesis, segmentation, segregation of imaginal discs, and nervous system formation occur. All the developmental stages of *Drosophila*, embryo, larva, and adult have the segmented body plan with three thoracic and eight abdominal segments. The first thoracic segment contains legs, the second thoracic segment has legs and wings, and the third thoracic segment has legs and halteres (balancers).

19.4 *Drosophila*: Zygotic Gene

The first set of genes that are expressed in the embryo are zygotic genes. This includes segmentation genes (gap genes, pair-rule genes, and segment polarity genes) that are responsible for the transition of the syncytial embryo to segmented bodied fly. Maternal effect genes work as the transcriptional activators for the zygotic genes. There are two steps for a cell to commit to its fate in *Drosophila*: firstly a cell is specified and then later on its fate is determined. In *Drosophila* cell specification is based upon its environment where different maternal effect morphogens guide the cell. This process is reversible and can be altered by modifying the surrounding morphogens. The next step in cell commitment is cell determination, which is irreversible as well as cell intrinsic and occurs due to the expression of segmentation genes. Expression of segmentation genes divides the early embryo into a series of repeating segmental primordia along the A-P axis. Segmentation genes divide the embryo into major anatomical divisions and 14 parasegments. Each parasegment includes the posterior part of an anterior segment and the anterior portion of the posterior segment.

19.4.1 Gap Genes

First in the array of segmentation genes are gap genes that are either activated or repressed by the maternal effect genes. Gap genes were discovered in the mutant embryos which lacked groups of consecutive segments. All gap genes function as transcription factors and include hunchback, Krüppel, giant, knirps, tailless, huckebein (zygotic), orthodenticle, buttoned, and empty spiracles. Gap genes are responsible for the segment formation, and thereby they are expressed in the overlapping domains (Fig. 19.17). A significant amount of hunchback protein accumulation can be detected at the anterior portion of the embryo by the completion of 12th cycle. The transcription pattern of the next gap genes is regulated by the levels of the hunchback and bicoid. In the anterior region, high levels of hunchback promote the expression of giant while suppressing the expression of posterior gap genes such as knirps. Caudal protein at the posterior end activates the expression of abdominal gap genes knirps and giant. Giant produces two bands, one anterior expression band and another posterior expression band. In addition to maternal effect genes, different gap gene expression itself establishes their expression patterns. After gap gene expression, the early embryo has a broad anterior

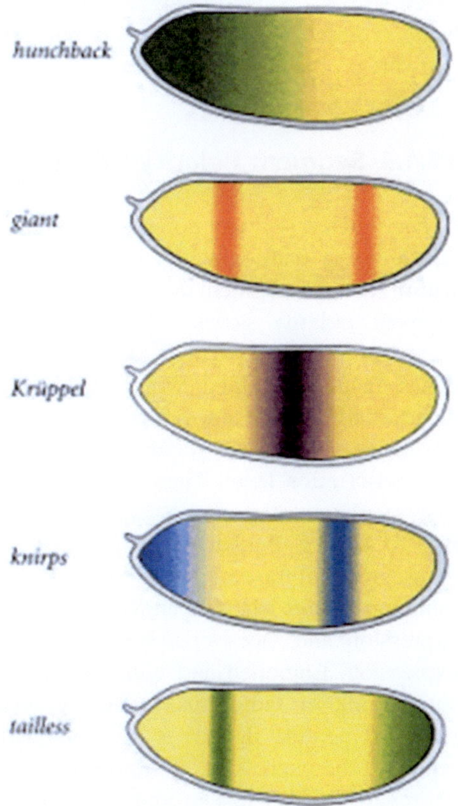

Fig. 19.17 Expression pattern of different gap genes. The anterior gap genes include Hunchback, Giant, and Krüppel, while posterior gap genes include Knirps and Giant. The terminal gap genes include Tailless and Huckebein. These genes are required for the formation of the unsegmented terminal regions of the larvae: acron at the anterior and telson at the posterior end. (Adapted from Gilbert 2000)

hunchback band, giant bands at middle anterior and middle posterior, Krüppel band in the middle, tailless bands at middle anterior and extreme posterior, and knirps bands at extreme anterior and middle posterior end.

19.4.2 Pair-Rule Genes

Next in the line of segmentation genes are pair-rule genes that are activated by gap genes during the 13th division cycle and divide the embryo in seven vertical bands perpendicular to the A-P axis. Pair-rule genes are expressed in the zebra stripe pattern along the A-P axis, dividing the embryo into 7 transverse bands and 15 subunits. Pair-rule genes are of two types, primary pair rule genes which includes hairy, even-skipped, and runt and secondary pair-rule genes which include fushi tarazu, odd-skipped, odd-paired, and paired. Gap genes activate the primary pair-rule genes which are essential for the formation of the periodic pattern in the embryo. Gap gene protein concentrations regulate the function of pair-rule genes through the enhancer sequences. Mutation in the enhancer of a particular pair-rule gene can delete its particular stripe. Product of primary pair-rule genes acts as a transcriptional activator for the secondary pair-rule genes. Expression of each pair-rule gene in seven stripes divides the embryo into 14 parasegments, with each pair-rule gene being expressed in alternate parasegments with particular and unique combination of pair-rule products which in turn activate the segment polarity genes.

19.4.3 Segment Polarity Genes

In *Drosophila*, segment polarity genes carry out two important functions: first, they reinforce the parasegmental periodicity, and second they establish the cell fates within each parasegment due to cell to cell signaling (Lee et al. 2016). In the embryo, segment polarity genes establish the A-P polarities within each embryonic parasegment through Wnt and Hedgehog signaling pathway. Major segment polarity genes include engrailed, wingless, hedgehog, fused, patched, cubitus interruptus, dishevelled, frizzled, gooseberry, pangolin, and armadillo. The pair-rule genes either positively or negatively regulate the expression patterns of segment polarity genes at the transcription level. There are two phases of segment polarity gene regulation: first, regulation by pair-rule genes, and second, cell to cell signaling. Pair-rule genes like ftz or eve regulate the expression of engrailed as well as wingless in each parasegment of the embryo. Evidence for cell to cell communication comes from the mutation studies where in engrailed mutant embryos there is no detectable wingless expression and vice versa suggesting both genes are required for each other's expression. Engrailed and wingless are expressed in different cells; therefore for such regulation there must be cell-cell communication. The receptor for wingless (secreted peptide ligand) is present on the posterior cells, which upon binding regulates the expression of engrailed at transcription level. Engrailed and wingless expression is also lost in another segment polarity gene mutant called hedgehog suggesting its involvement in the genetic circuit regulating wingless and engrailed expression. Hedgehog ligand binds to its receptor on the wingless-expressing cells which results in the maintenance of wingless transcription. The expression of segment polarity genes acts as morphogens by accumulating in different concentrations in individual segments where they regulate the cell differentiation fate. The expression of paired as well as even skipped is required for the regulation of odd engrailed stripes, whereas fushi tarazu along with the odd paired is required for even engrailed stripes. The expression of Engrailed is required for the determination of the A-P compartment boundaries. Wingless and hedgehog expression is responsible for the anterior and posterior compartment of each segment, respectively. Mutations in the segment polarity gene result in the segment errors like deletion, mirror image, duplication, and segment polarity reversal.

19.5 *Drosophila*: Homeotic Gene

Homeotic selector genes are a group of genes that control the pattern of body formation as well as for establishing the characteristic structures of each segment during early embryonic development of organisms (Mallo and Alonso 2013). Homeotic genes include multiple subsets of Hox and ParaHox genes that are the key regulators of segmentation in flies. These genes encode proteins which function as transcription factors directing different cells to form various parts of the body. Mutations or misexpression of homeotic genes causes displaced body parts or transformation of one organ into another which is called homeosis. For example, flies with antennapedia mutation (lethal mutation) have ectopic legs on the head in the place of antennas.

The elucidation of homeotic gene function in the embryonic development can be attributed to Edward B. Lewis, Eric F. Wieschaus, and Christiane Nüsslein-Volhard. Their discovery of the genetic control of early embryonic development helped them win the Nobel Prize in the field of Physiology or Medicine in the year 1995. During the early research, Edward B. Lewis at the California Institute of Technology in Los Angeles observed flies with occasional malformations. In one such case, a mutation leads to the transformation of halteres (balancing organ of the fly) into an extra pair of wings. In Greek, homeosis means malformations, and from there the homeotic genes acquired their name. It was later found that the mutation in segmentation gene of bithorax complex leads to the doubling of same body segment leading to the development of additional pair of wings. According to the colinearity principle, genes at the beginning of the complex controlled anterior body segments, while genes further down the complex regulated posterior body segments (Gummalla et al. 2014) (Fig. 19.18). It was also found that the regions controlled by the individual genes of this complex overlapped each other and a complex interplay between them specified the individual body segments during development. The inactivity of the first gene of the bithorax complex in the segment where halteres should be produced leads to the formation of extra pair of wings resulting in a fly with four wings. This inactivity caused other homeotic genes to re-specify this particular segment into one that forms wings.

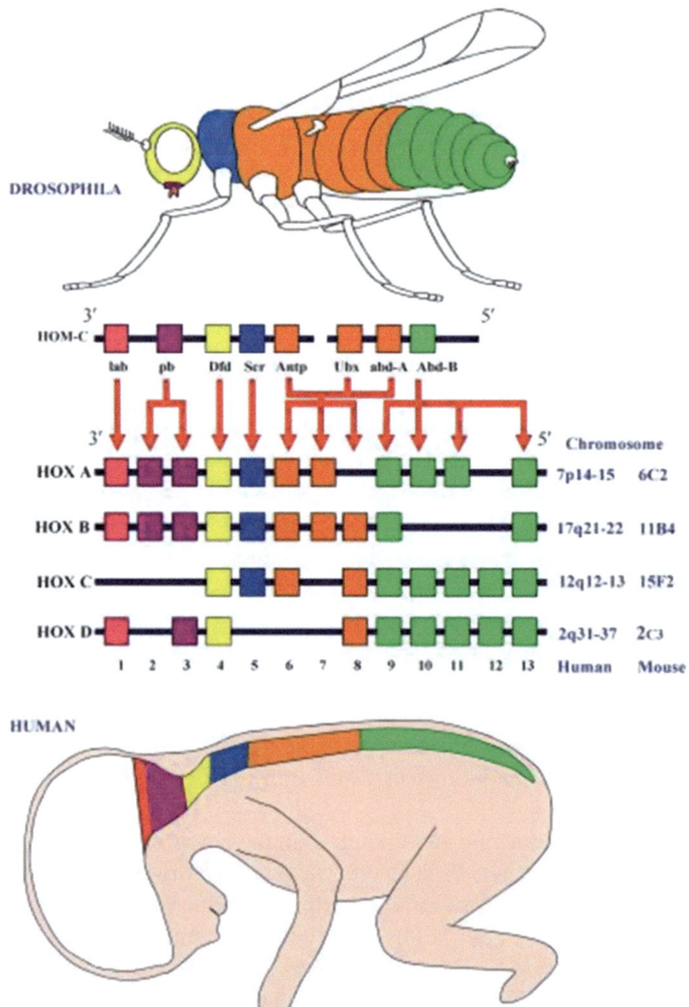

Fig. 19.18 Conservation between the HOM-C (*Drosophila*) and HOX (human) gene clusters. In terms of nucleotide sequence and relative position to each other (colinear expression), the two *Drosophila* Hom-C complex clusters can be seen distributed over four Hox gene clusters in mammals. (Adapted from Lappin et al. 2006)

Homeotic genes contain a unique DNA sequence of ~180 base pair length known as a homeobox, which encodes a segment of 60 amino acids within the homeotic transcription factor protein. Homeoboxes were first discovered in three *Drosophila* homeotic and segmentation genes: (1) Antennapedia (Antp), ultrabithorax (Ubx), and fushi tarazu (Ftz), mutations which caused homeotic transformations. The homeodomain protein is composed of three alpha helixes; helix 2 and 3 form a helix-turn-helix (HTH) structure, where the two alpha helices are connected by a short loop region. These two helices are present at the N-terminal in antiparallel position, while helix 1 is present at the C-terminal in perpendicular position. Helix 1 directly interacts with the DNA by a large number of hydrogen bonds, by hydrophobic interactions, as well as by indirect interactions between specific side chains and the exposed bases within the major groove of the DNA. Due to the DNA-recognition properties, homeodomain proteins induce cascade of coregulated targeted genes which direct the formation of many body structures during early embryonic development.

The homeodomain proteins bind to the DNA at a particular conserved nucleotide sequence, TAAT present at the 5′ terminal with the thymine being the most important for binding. All homeodomain proteins recognize this initial sequence; however the base pairs following this initial sequence are used to distinguish between different homeodomain proteins. For example, amino acid lysine present at the position 9 of homeodomain protein Bicoid recognizes the nucleotide guanine after the initial sequence. Similarly, glutamine present at the ninth position in Antennapedia recognizes and binds to adenine. If lysine in Bicoid is switched with glutamine of Antennapedia homeodomain, the resulting protein starts to recognize Antennapedia-binding enhancer sites. Moreover, Hox proteins bind to protein cofactors that provide DNA sequence specificity. Two such examples of Hox cofactors are Extradenticle (Exd) and Homothorax (Hth) which upon binding induce conformational changes in the Hox protein, thereby increasing the sequence specificity.

Homeotic genes homologous to those of *Drosophila* were later found in a wide range of organisms, including fungi, plants, and vertebrates. In vertebrates, these genes are commonly referred to as HOX genes. Humans possess ~39 HOX genes, which are divided into four different clusters, A, B, C, and D, located on different chromosomes, 7p15, 17q21.2, 12q13, and 2q31, respectively (Fig. 19.18). These genes have been assumed to have arisen due to the duplication and divergence from a primordial homeobox gene. On the basis of sequence similarity and relative position within the cluster, each cluster consists of 13 paralog groups with 9 to 11 members. A high degree of homology can be seen between the human HOX genes and the Hom-C genes of *Drosophila*. The human paralog groups 1–8 are more closely related to antennapedia (Antp), while groups 9–13 are more closely related to abdominal-B (abd-B).

19.5.1 Hox Gene in *Drosophila*

Hox genes are a subset of homeobox genes that specify regions of the body plan of an embryo along the A-P axis of organism (Pavlopoulos and Akam 2007). The products of Hox genes are Hox proteins that function as transcription factors by binding to a specific nucleotide sequences on DNA called enhancers through their homeodomain. The same Hox protein can act as a repressor for one gene and as activator for another. Hox genes are arranged in clusters, and their order on the chromosome is the same as the order in which they appear along the body such that

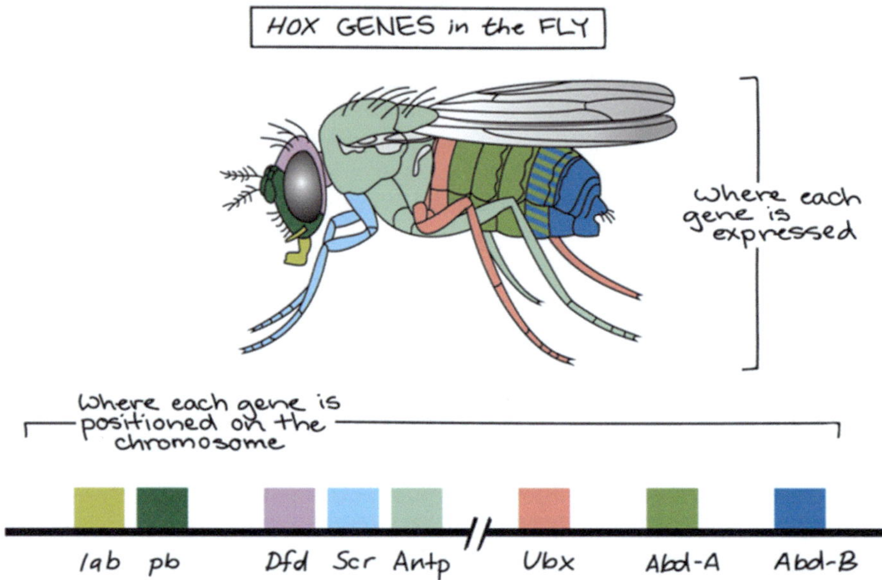

Fig. 19.19 Hox gene clusters in *Drosophila*. The two Hox gene clusters Antennapedia complex (left) and bithorax complex (right). The break mark (//) in the chromosome indicates that these two clusters of genes are separated by a long intervening region. (Adapted from Hox genes of fruit fly by PhiLiP, public domain)

the genes on the left control patterning of the head, while the genes on the right control patterning of tail. *Drosophila* has eight Hox genes that are clustered into two complexes, antennapedia complex (ANT-C) and bithorax complex (BX-C) collectively called homeotic complex (HOM-C), both of which are located on chromosome 3 (Fig. 19.19). The ANT-C is also called anterior homeotic gene complex as it controls the identity of parasegment present prior to five, while BX-C regulates the identity of fifth to 13th parasegment and is called posterior homeotic gene. The Antennapedia complex contains the homeotic genes labial (lab), Antennapedia (Antp), sex combs reduced (scr), deformed (dfd), and proboscipedia (pb). The labial and deformed genes of the antennapedia complex specify the head segments, while sex combs reduced and antennapedia define the thoracic segments. The bithorax complex contains ultrabithorax (ubx), abdominal A (abdA), and Abdominal B (AbdB) genes. Ubx is required for the identity of the third thoracic segment; abdA and AbdB genes are responsible for the segmental identities of the abdominal segments.

The lab gene is the most anteriorly expressed gene of the Antennapedia complex. It is expressed in the head, mainly in the intercalary segment between the antenna and mandible as well as in the midgut. The lab gene was initially named because it disrupted the labial appendage; however it was later found that it was due to the broad disorganization resulting from the failure of head involution. Mutation in lab results in defective head involution process where embryos fail to internalize the

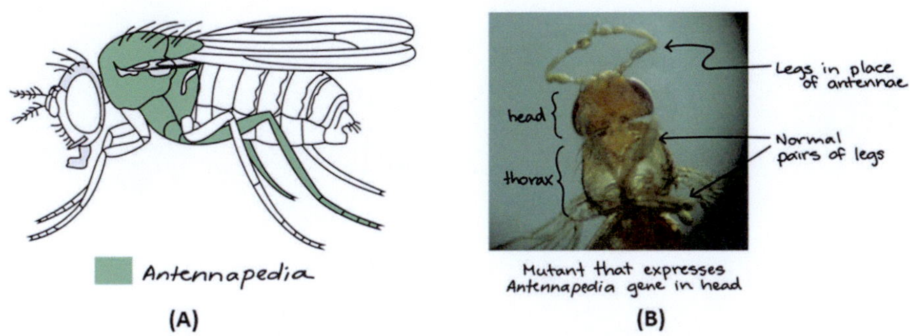

Fig. 19.20 Antennapedia expression and mutation. In *Drosophila*, antennapedia is expressed in the second segment of a fly's thorax (**a**). Ectopic expression of antennapedia gene into the fly's head results in the activation of normal, second-segment leg development program in the head producing a mutation where legs grow from the fly's head in place of antennae (**b**). (Adapted from Hox genes of fruit fly by PhiLiP, public domain)

mouth and head structures that initially develops on the outside of the body. Failure of head involution disrupts or deletes the salivary glands and pharynx. The pb gene of the ANT-C is responsible for the formation of the labial and maxillary palps. The Dfd gene is responsible for the formation of the maxillary and mandibular segments in the larval head. Similar to lab, mutation in Dfd results in a failure of head involution. The Scr gene is responsible for cephalic and thoracic development in *Drosophila* embryo and adult. The Antp gene specifies the development of a pair of legs and a pair of wings on the second thoracic segment, T2. The classical example of homeosis is due to dominant Antp mutation caused by a chromosomal inversion leading to the Antp expression in the antennal imaginal disc resulting in the formation of leg coming out of the fly's head in place of antenna (Fig. 19.20).

The first gene in the bithorax complex is the Ubx which is responsible for the determination of a pair of legs and a pair of halteres, highly reduced wings that function in balancing during flight, on the third thoracic segment, T3. Ubx functions mainly by repressing the genes involved in wing formation like blistered and spalt which play important role in wing development. Another classical example of homeosis is the four-winged flies due to the loss-of-function Ubx mutation (Fig. 19.21). In these mutants Ubx is no longer able to represses the wing development genes resulting in the transformation of halteres as a second pair of wings eventually resulting in four-winged flies. In contrast, upon Ubx misexpression in the second thoracic segment, it represses wing genes and the wings develop as halteres, resulting in a four-haltered fly also called as Cbx enhancer mutation. The next gene in the BX-C is abd-A which is expressed from abdominal segments A1 to A8 and is required for the specification of most of the abdominal segment identity. Moreover, it also affects the pattern of cuticle generation muscle generation in the ectoderm and mesoderm, respectively. One of the main functions of abd-A in insects is to repress limb formation. In abd-A loss-of-function mutants, abdominal segments A2–A8 are transformed into A1 similar segments. The last gene in the assembly of BX-C is

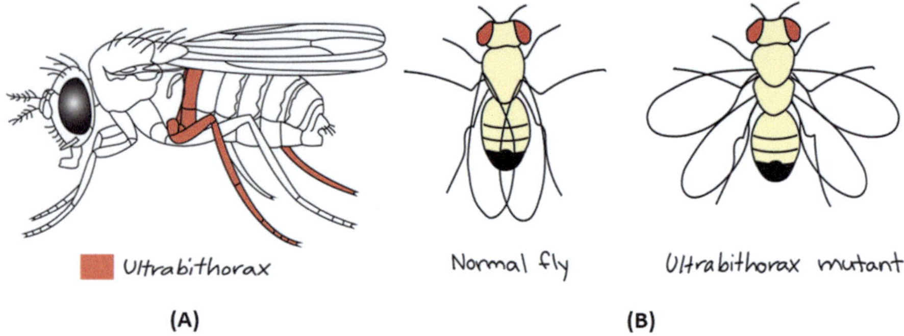

Fig. 19.21 Ultrabithorax gene expression and mutation. Ubx is strongly expressed in the third segment of the thorax. Inactivation of Ultrabithorax results in the conversion of halteres into a second set of wings behind the normal set of wings producing a four-winged fly. (Adapted from Hox genes of fruit fly by PhiLiP, public domain)

Table 19.3 List of Hox genes, target genes, their function

Hox genes	Target gene regulated by Hox genes	Function of target gene
ULTRABITHORAX (represses distal-less)	distal-less	Activates gene pathway for limb formation
ABDOMINAL-A (represses distal-less)	distal-less	activates gene pathway for limb formation
ULTRABITHORAX (activates decapentaplegic)	decapentaplegic	Triggers cell shape changes in the gut that are required for normal visceral morphology
DEFORMED (activates reaper)	reaper	Apoptosis: localized cell death creates the segmental, boundary between the maxilla and mandible of the head
ABDOMINAL-B (represses decapentaplegic)	decapentaplegic	Prevents the above cell changes in more posterior positions

abd-B which is transcribed into two different forms, a regulatory protein and a morphogenetic protein. Regulatory abd-B suppresses embryonic ventral epidermal structures in the eighth and ninth segments of the *Drosophila* abdomen. Both the regulatory protein and the morphogenetic protein are involved in the development of the tail segment.

Hox proteins with identical homeodomains are assumed to have identical DNA-binding properties as well as functions and are classified based on the phylogenetic inference, synteny, and sequence similarity. Hox genes regulate many genes that in turn regulate large developmental signaling networks. In addition, they also regulate realisator genes or effector genes which are directly responsible for forming the tissues, structures, and organs of each segment (Table 19.3). Hox genes are regulated by gap genes and pair-rule genes which themselves are regulated by maternally supplied mRNA. In this way, maternal factors activate gap or pair-rule genes which in turn activate Hox genes which further activate realisator genes that cause the segments in the developing embryo to differentiate.

Initial regulation of homeotic genes is carried out by the gap and pair-rule genes as they act as transcription factors for homeotic genes by cis-regulatory elements called initiator enhancer elements. For instance, Hunchback and Krüppel proteins repress the expression of abdA and AbdB genes from the head and thorax restricting their expression only in the abdomen. In contrast, Ultrabithorax gene is activated by the Hunchback protein expressing it in a broad band in the middle of the embryo, while Antennapedia is activated by Krüppel. The Fushi tarazu and Even-skipped proteins confine the expression of homeotic genes to the parasegments only. In addition, homeotic gene themselves act as the transcriptional factors, as ANT-C and BX-C homeotic gene complexes repress the expression of each other in their expression region. Once the expression pattern of the homeotic genes have become stabilized, chromatin conformation occurs which locks them in their respective positions. The repression of homeotic genes is regulated by polycomb family, while chromatin conformation is regulated by Trithorax proteins.

Mutation in homeotic genes leads to the abnormal development of the fly body parts. Normal fly body contains three thoracic segments, the first segment contains only a pair of legs, the second thoracic segment contains both a set of legs and a set of wings, and the third thoracic segment produces a set of wings and a set of balancers known as halteres (Fig. 19.22). Upon deletion of ultrabithorax gene, the third thoracic segment becomes transformed into another second thoracic segment

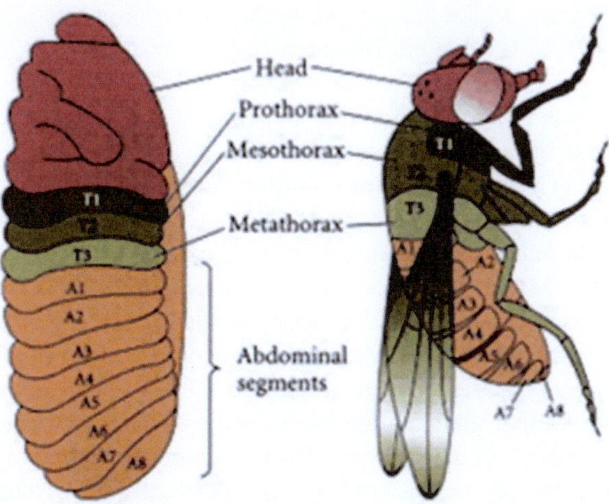

Fig. 19.22 Organization of the *Drosophila* body into segments. Drosophila adult as well as larvae is broadly divided into three regions called head, thorax, and abdomen which further contains segments. The bodies of both the larval and adult insect are divided into 14 segments along the A-P axis. In the adult fly, each thoracic segment (T1-T3) contains a pair of legs with the middle segment T2 having a pair of wings and the most posterior segment T3 containing a pair of halteres. (Adapted from Gilbert 2000)

resulting in a fly with four wings. Similarly, mutation in the Antennapedia gene results in the fly head, having legs in place of antennae in the head sockets. The ParaHox gene cluster is an array of homeobox genes from the Gsx, Xlox (Pdx), and Cdx gene families and is involved in morphogenesis and regulation of anatomical development patterns.

19.5.2 Genetic Disorder in Humans

The Human Genome Project which was completed in April 2003 revealed that it is composed of 46 chromosomes or 22 pairs of autosomal chromosomes and 2 sex chromosomes made up of ~3 billion base pairs of DNA and contains ~20,500 protein-coding genes with coding region of only ~5%. Most genetic diseases are the direct result of a single or multiple mutations in one or multiple genes.

Genetic disorder is any disease caused by the abnormalities in the genetic makeup of an individual ranging from single-base mutation to chromosomal abnormality like addition, subtraction, or inversion of an entire chromosome or set of chromosomes. These diseases can be hereditary or acquired due to some mutation exposure to some chemicals. There are four types of inherited genetic disorders: single-gene disorder, multifactorial inheritance, chromosome abnormalities, and mitochondrial inheritance.

Single-gene disorder is also called Mendelian or monogenetic disorders as there is mutation in the DNA sequence of a single gene. Currently there are ~4000 single gene disorders known which are caused by the mutation in one gene (Table 19.4). Common examples of single-gene disorders include cystic fibrosis, alpha- and beta-thalassemias, sickle cell anemia (sickle cell disease), Marfan syndrome, Fragile X syndrome, muscular dystrophy, familial hypercholesterolemia (FH), Huntington's disease, and hemochromatosis. Single-gene diseases affect at least 1 in 500 people around the globe. These diseases may follow autosomal dominant or recessive as well as X-linked dominant or recessive inheritance pattern. Pedigree analyses of large families with many affected members can be used for tracking the inheritance of many diseases. Many databases provide the accumulated and comprehensive information on diseases and related genes; one such database for genes following Mendelian inheritance is Online Mendelian Inheritance in Man (OMIM™). This database was initially started by Dr. Victor A. McKusick in the early 1960s for recording the Mendelian traits and disorders and was called Mendelian Inheritance in Man (MIM). OMIM till date report ~387 human genes with a known phenotype, ~2310 human phenotypes with a known molecular basis, ~1621 confirmed Mendelian phenotypes with unknown molecular basis, and ~ 2084 phenotypes with suspected Mendelian basis. OMIM was made online by the collaborative effort of National Library of Medicine and the William H. Welch Medical Library at Johns Hopkins in 1985.

In addition to gene mutations, environmental factors also contribute to many disease, and in combination they cause a specific set of disorder called multifactorial

Table 19.4 Examples of single-gene diseases in humans, their mode of inheritance, and associated genes

Disease	Type of inheritance	Gene responsible
Phenylketonuria (PKU)	Autosomal recessive	Phenylalanine hydroxylase (PAH)
Cystic fibrosis	Autosomal recessive	Cystic fibrosis conductance transmembrane regulator (CFTR)
Sickle-cell anemia	Autosomal recessive	Beta hemoglobin (HBB)
Albinism, oculocutaneous, type I	Autosomal recessive	Oculocutaneous albinism II (OCA2)
Huntington's disease	Autosomal dominant	Huntingtin (HTT)
Myotonic dystrophy type 1	Autosomal dominant	Dystrophia myotonica protein kinase (DMPK)
Hypercholesterolemia, autosomal dominant, type B	Autosomal dominant	Low-density lipoprotein receptor (LDLR), apolipoprotein B (APOB)
Neurofibromatosis, type 1	Autosomal dominant	Neurofibromin 1 (NF1)
Polycystic kidney disease 1 and 2	Autosomal dominant	Polycystic kidney disease 1 (PKD1) and polycystic kidney disease 2 (PKD2)
Hemophilia A	X-linked recessive	Coagulation factor VIII (F8)
Muscular dystrophy, Duchenne type	X-linked recessive	Dystrophin (DMD)
Hypophosphatemic rickets, X-linked dominant	X-linked dominant	Phosphate-regulating endopeptidase homolog, X-linked (PHEX)
Rett's syndrome	X-linked dominant	Methyl-CpG-binding protein 2 (MECP2)
Spermatogenic failure, non-obstructive, Y-linked	Y-linked	Ubiquitin-specific peptidase 9Y, Y-linked (USP9Y)

inheritance disorder which includes neurodegenerative diseases, heart-related diseases, various cancers, etc.

Any change at chromosomal level either in the structure or addition or subtraction of a part or whole chromosome can result in various severe genetic disorders as they are the genetic carriers for the next generation. Chromosomal abnormalities typically occur due to a problem during cell division. The most common examples of chromosomal abnormalities include Down's syndrome (trisomy 21), Prader-Willi syndrome, chronic myeloid leukemia (CML), Turner syndrome (45, X0), Klinefelter syndrome (47, XXY), and Cri du chat syndrome (cry of the cat syndrome; 46, XX or XY, 5p-). Numerical and structural abnormalities are the two types of chromosomal abnormalities. Addition or subtraction of one of the chromosomes from the pair causes numerical abnormalities, whereas structural errors are caused due to the change in the structure of a chromosome. Change in the structure of chromosomes can be of many types: deletion where a portion of the chromosome is missing or deleted or duplication which results in the copying of a segment. Translocation is the transfer of a segment from one chromosome to another and can be of two types, reciprocal and Robertsonian translocation. During inversion, a segment gets inverted before attaching again, whereas a ringed structure is formed when a segment is deleted or broken off.

Mutation in non-nuclear genetic material like in mitochondria and cytoplasm forms another class of genetic disorders. Each mitochondrion contains 5–10 circular DNA and is carried by egg cells from maternal side. Mitochondrial genetic disorders include some rare diseases like Leber hereditary optic neuropathy (LHON), myoclonic epilepsy with ragged red fibers (MERRF), lactic acidosis, and mitochondrial encephalopathy.

With the sequencing of human genome, maps of each chromosome were generated showing the precise location of every gene. This also revealed the parts of the genome that differ from one person to the next and called as polymorphic. The human genome contains various single nucleotide changes throughout the genetic material (one SNP per 1000 base pairs) and is called single nucleotide polymorphisms (SNPs). The International HapMap Project provides the human SNP map of all the chromosomes. High throughput screening can be done using SNP chip and can be used to map disease-associated genes. A small silicon glass chip is used on which short, single-stranded DNA molecules are adhered in a grid-like pattern that corresponds to known SNP variants; this chip is called a SNP chip. The isolated DNA from the individual is then converted into labeled probes (fluorescently labeled single-strand segments) and is incubated with the SNP chip. The probes then bind to their respective SNPs due to the complementarity between them; the fluorescence is then read with a scanner to identify the specific SNPs. Whole SNP map can be made accordingly for each individual person and can later be used for various types of comparison. This comparison can be used to determine which SNPs segregate with a particular disease. Because the SNP sequences have already been mapped to specific chromosomal locations, researchers can use the obtained information to tag the specific disease-related gene to individual chromosomes.

In addition to this, various bioinformatics tools are available that can compare the sequence of a gene with unknown function to the rest of the genome and find similar genes with known functions. Researchers can also try to predict the function of unknown proteins from different genes. Large-scale studies of genes and their expression levels in healthy and diseased individual can be done and are called genomics.

In order to study gene expression, gene chips are created that are bound to small single-stranded DNA segments. Next a sample from tissue of interest is produced that contains the fluorescently labeled single-stranded complementary DNA (cDNA). This sample is then incubated with the gene chip. Using the complementary binding property, a laser then scans for the fluorescent signal producing an expression profile for the target tissue. This expression data from healthy individuals and those with diseases can then be used for the comparison and identification of disrupted process. Proteomics utilizes the similar approach through large-scale gene chips to study thousands of proteins within a given cell population at the same time. Similar to gene chips in order to carry out the translational studies, protein chip grids are used. In this approach, fluorescently labeled protein sample from the target individual is prepared and is incubated with the antibody-coated chip so that the specific protein binds to their respective antibody which is read by a laser. Using proteomic approach, the protein profile of any individual can be made and compared with the diseased individual's profile in order to determine the altered proteins.

Owing to the conservation of key processes and their molecular components across the evolutionary tree, model organisms like mice, frogs, worms, flies, and yeast are studied to understand the basic cellular processes and their underlying molecular mechanisms. In a similar fashion, genes associated with single-gene diseases and the underlying molecular mechanisms can be elucidated using model organisms. One such model organism is mice in which most of the homologous human disease genes are found and much of our understanding of human disease comes from these studies. Various disease models in mice have been generated using mutation or deletion of disease-associated genes where detailed phenotypic as well as functional analyses can be carried out.

19.5.3 Control of Hox Gene Expression

Expression of Hox genes is regulated by three mechanisms, transcriptional regulation from earlier segmentation genes, a cellular memory system based on the action of Polycomb (PcG)/trithorax (trxG) group proteins, and cross-regulatory interactions among the Hox genes themselves. Hox genes regulate each other by posterior prevalence; posterior Hox genes are able to repress the expression (and suppress the function) of more anterior genes. In vertebrates, Hox genes are regulated by temporal collinearity.

where these genes are kept globally silent but are activated during initial stages of development through temporal sequence that correlates with the gene's position within the cluster in a 3′ to 5′ direction. After initial activation, PcG and trxG system of proteins regulates the Hox genes to produce specific expression patterns. The major determinant of Hox gene regulation is functionally autonomous transcriptional cis-regulation at spatial level. Several studies have revealed a role of retinoic acid in the regulation of Hox gene expression in higher animals including humans along the anterior-posterior axis. Retinoic acid induces the expression of genes at the 3′ ends of the Hox clusters, thereby regulating the body plan of anterior embryo while having no effect on 5′ end Hox genes. In addition, Hox clusters have also been known to harbor various noncoding RNAs (ncRNAs). One such example is HOTAIR which is negatively regulated by polycomb-group proteins (PRC2) in trans.

19.6 *Arabidopsis thaliana*: Homeotic Gene

Homeotic genes comprise a highly conserved gene set responsible for regulating the anatomical structure development throughout the evolution tree. Mutations in homeotic genes result in homeosis, ectopic placement of body parts which are usually lethal. Flower formation is initiated when vegetative meristem is converted to floral meristem due to the activity of heterochrony or flowering time genes. The flower

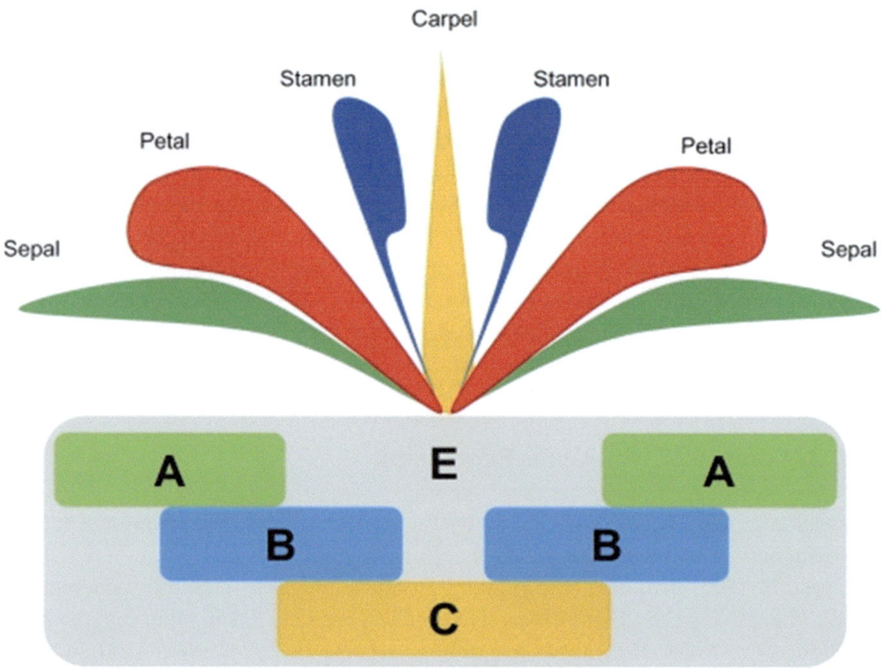

Fig. 19.23 The ABCE model of floral organ identity. According to this model, the flower meristem is divided into three regions, A, B, and C, with overlapping gene activity that altogether defines two adjacent whorls of a flower. Region A determines sepals, while region C specifies carpels. However, the joint activity of regions A and B determines petals, and activity of regions B and C determines stamens. Different genes are required for this specification process such as APETALA1(AP1) and APETALA2(AP2) are necessary for region A, while two genes, APETALA3(AP3) and PISTILLATA(PI), are required for region B and AGAMOUS(AG) gene is required for region C. (Adapted from Chanderbali et al. 2010)

meristem is then converted into a flower and is regulated by flower meristem identity genes. Once the flowering starts in plants, the whorl (sepal, petal, stamens, and carpel) formation, development, pattern, and structure are regulated by two sets of genes called cadastral genes and homeotic genes in the same sequence (Fig. 19.23). The floral development was studied in detail by George Haughn and Chris Somerville in 1988 in two plant species, *Arabidopsis thaliana* and *Antirrhinum majus*, and ABC model of floral development was given. According to ABC model, five homeotic genes, A, B, C, D, and E, regulate the formation and development of floral organs. All the homeotic genes have been studied in detail through single gene mutation studies, and it was found that the sepal whorl is regulated by the action of gene A, while both A and B genes are responsible for petal whorl. The stamen and carpel whorl are regulated by the co-expression of the B and C and C genes, respectively. Genes A and C work in an antagonistic manner. Gene D has been known to regulate the ovule formation and development, whereas gene E is responsible for the regulation of normal physiological functioning of whole flower.

19.7 *Caenorhabditis elegans*: Development

C. elegans is a free-living transparent soil nematode of ~1 mm in length and shows bilateral body symmetry. It derives its name from two Greek words: *caeno* which means recent and *rhabditis* which means rod-like. It is a sexually dimorphic organism with most of the population being hermaphrodites (XX) producing both eggs and sperm while only 0.001 percent population being males. Hermaphrodites may self-fertilize or cross-fertilize with the males. *C. elegans* genome consists of 5 autosomes and X chromosome comprising of ~97,000 kb per haploid genome which encodes for ~18,000 proteins. *C. elegans* genome was the first multicellular genome to be sequenced and served as the basis for human genome project framework. The basic anatomy of *C. elegans* includes a mouth, pharynx, intestine, gonad, and collagenous cuticle without any respiratory or circulatory system (Fig. 19.24). The outer body wall consists of an outer protective epidermis and underlying muscular layer. There are two tubes in the body of *C. elegans*: one that is made of endodermal cells forms the intestine, and the other present between the intestine and

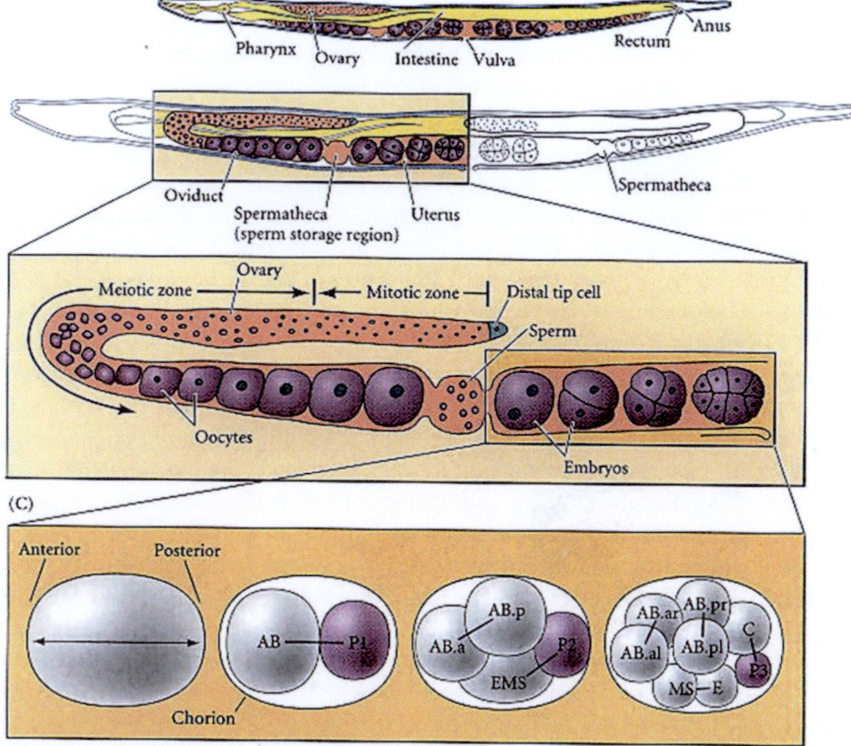

Fig. 19.24 Lateral view of adult hermaphrodite. In the hermaphrodite nematode, sperms are stored in the storage organ, spermatheca, such that eggs passing through them get fertilized before reaching the vulva. (Adapted from Gilbert 2000)

body wall made of somatic cells constitutes the gonad. The adult has 945 cells with 959 somatic nuclei and 302 neuronal cells derived from 407 neural precursor cells.

Sperm enters the oocyte in spermatheca followed by fertilization and is also responsible for the establishment of A-P axis. Fertilized zygote then enters the uterus where it undergoes rotational holoblastic cleavage. The microtubule organizing center of the sperm directs the movement of the sperm pronucleus to the future posterior pole of the embryo also inducing the movement of PAR proteins in the embryo which results in asymmetric and asynchronous first cleavage divisions. Each asymmetrical division produces one founder cell (denoted AB, MS, E, C, and D) which further produces different cell types and one stem cell (P1-P4 lineage). All the 558 cells of the small worm inside the egg shell are generated by the divisions in the descendants of each founder cell at specific times and are named according to their positions relative to their sister cells. The first cell division results in the formation of cleavage furrow which is located asymmetrically along the A-P axis of the egg resulting in the formation of a larger blastomere founder cell (AB) and a smaller blastomere stem cell (P1). At the four-cell stage, the second cell division results in the formation of ABp and ABa cells due to equatorial division of anterior founder cell (AB) and a posterior stem cell (P2) and EMS due to meridionally (transversely) dividing P1 cell. Before the egg is laid, mRNAs of the maternal genes are accumulated in the egg which interacts with one another directing the early development; these genes are called maternal effect genes.

The A-P axis is decided on the basis of the position of sperm pronucleus. When sperm enters the oocyte, sperm pronucleus is pushed to the nearest end of the oblong oocyte by the centriole due to cytoplasmic movements; this end becomes the posterior pole. Another determinant of the A-P axis is the migration of the P granules, ribonucleoprotein (RNP) complexes that function in specifying the germ cells. P granules are initially distributed uniformly in oocyte but associate at posterior pole just before the first cleavage so that they only enter the P1 blastomere. The P granules from the P1 cell are passed to the P2, P3, and eventually to P4 cell, descendants of which become the sperm and eggs of the adult. The partitioning of the P granules is thought to be regulated by the par (partition-defective) genes. Par (partitioning-defective) gene family is a maternal effect gene family containing six genes that are responsible for organizing cell asymmetry and polarization of the cytoskeleton in nematode egg. Par genes have homologs in vertebrates. Par proteins orient the mitotic spindle in such a way that P granules are segregated to only the posterior daughter cell and not to the anterior daughter cells. In this way, at the 16-cell stage, there is just one cell that contains the P granules which gives rise to the germline.

The dorsal-ventral axis of the nematode is defined through Wnt and Notch signaling, by the second division of the AB and P1 cells with their descendant cells. P2 cell expresses a homolog of the Notch ligand, Delta, while the ABa and ABp cells express the corresponding transmembrane receptor, homolog of Notch. Due to the elongated shape of the nematode egg, ABa acquires the anterior end, while ABp and EMS acquire leaving only ABp and EMS cells exposed to the signal from P_2. This signal acts on ABp, making it different from ABa and defining the

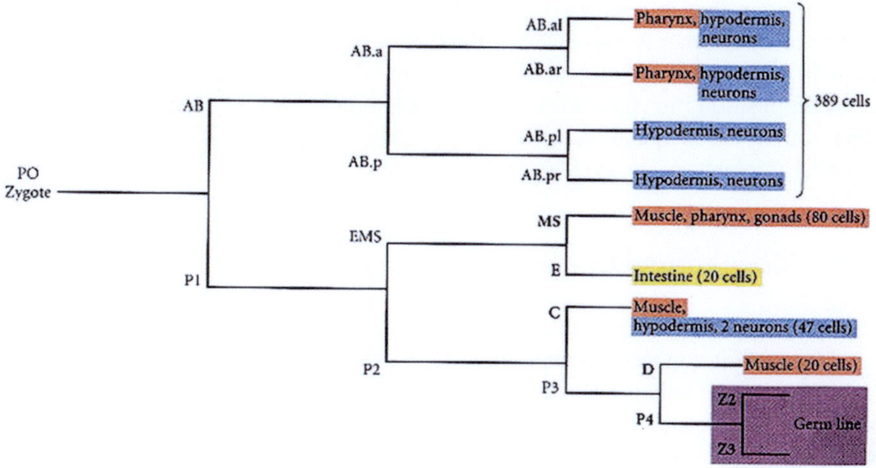

Fig. 19.25 Cell lineage chart. The germline segregates into the posterior portion of the most posterior (P) cell. Three cell lineages, AB, C, and MS, are produced from the primary cell divisions. The newly hatched larva contains a total of 558 cells, with different numbers of cells (in brackets) belonging to different tissue types, while more divisions further produce the 959 somatic cells of the adult. (Adapted from Gilbert 2000)

future dorsal-ventral axis of the worm (Fig. 19.25). The ABp cell defines the future dorsal side of the embryo, while the EMS cell, the precursor of the muscle and gut cells, marks to future ventral surface of the embryo. In addition, P_2 also expresses Wnt protein, which acts on the Frizzled protein (Wnt receptor) present on the membrane of the EMS cell resulting in the orientation of the mitotic spindle. As a result of the Wnt signal from P_2, EMS gives rise to two daughter cells, one MS cell, which give rise to muscles and various other body parts, and an E cell, which is the precursor for all the cells of the gut. The left-right axis is specified at the 12-cell stage where the descendant of EMS cell, MS blastomere, contacts the descendant cells of the ABa cell, distinguishing the right side of the body from the left side.

The molecular mechanism involved in the developmental potential of the individual cells in the embryo can be determined by various methods which includes laser microbeam microsurgery and genetic screens. Laser microbeam microsurgery can be used to alter the cell's environment by killing the neighboring cells or rearranging the cells inside the eggshell. Such studies revealed that if the relative positions of ABa and ABp are flipped at the four-cell stage of development, Aba acquires the fate of the ABp cell and vice versa showing that the two cells initially have the same developmental potential and depend on the signals from their neighbors to make them different. Genetic screens with different mutant worm strains are used to study the cell-cell interaction. Two mutants, one in which no gut cells are induced (mom mutants) and another in which extra gut cells are induced (pop mutants), were used in order to understand the P2-EMS cell interaction. Genetic screening of the *mom* genes revealed that these genes encode for the Wnt signal protein that is expressed in the P2 cell, as well as for the Frizzled protein (a Wnt receptor) that is expressed in the EMS cell. In the absence of this signaling between P2 and EMS in mutant worms, both the daughter cells of the EMS cell fail to induce resulting in more mesoderm. On the other hand, in the absence of pop-1 gene activity both daughter cells of the EMS receive the Wnt signal from P2 resulting in plenty of pharynx due to extra gut.

19.8 *Caenorhabditis elegans*: Vulva Formation

The vulva is a hermaphrodite female-specific organ of utmost importance in *C. elegans* as it connects the hermaphrodite uterus to the outer environment. It is required for egg laying after internal fertilization and for sperm injection during copulation with male. Vulval development is extensively studied in *C. elegans* as it provides an excellent method for studying the process of morphogenesis and tissue remodeling (Ririe et al. 2008). Another advantage in studying vulva formation is the ease of genetic analyses of signal transductions and regulatory pathways that work as the bridge between network pathways and organogenesis. Vulval development occurs during the L3-L4 larval development stages, but the opening can only be seen after L4-adult molting (Fig. 19.26). Vulval development includes some major steps:

1. Generation of vulval precursor cells (VPCs): After hatching, L1 larva of *C. elegans* consists of a set of 12 ventral epidermal blast cells called the Pn. p cells from P1.p to P12.p. Anterior Pn.p cells give rise to neuronal daughter cells, while posterior Pn.p cells acquire different fates, viz., primary, secondary, or tertiary. During larval development six epidermal blast cells named P3.p–P8.p on the ventral surface are induced by a somatic gonadal cell, anchor cell, to become the vulva. Through LIN-3/epidermal growth factor (EGF) and LIN-12/Notch pathways, these epidermal cells are induced to become the VPCs. P6.p acquires the primary fate to become the 1° vulval precursor cell (VPC), while P5.p and P7. p are induced to acquire secondary fates and become 2° VPCs. P1.p–P4.p and P8. p–P11.p acquire the tertiary fate where they fuse either directly or after a single cell division with the syncytial hypodermis hyp7. P12.p gives rise to two daughter cells of which the posterior progeny cell, P12.pp, undergoes apoptosis, while the anterior daughter cell, P12.pa, forms anal hypodermis called hyp12
2. In order to understand the signals that provoke different Pn.p cells to assume and acquire a particular developmental fate, different experimental methods like laser ablation and mutation studies are done. In the absence of induction by anchor cell both due to ablation of anchor cell or signal cascade, all the VPCs acquire the tertiary fate and fuse with hyp7, and no vulva formation takes place. The phenotype thus produced is called Vul for vulvaless. In case of removal or incompetence of any Pn.p cells, the remaining ones can acquire and assume its position and function depending upon the time of ablation. Thus all the Pn.p cells from P3.p to P8.p form the vulval equivalence group where all the cells are

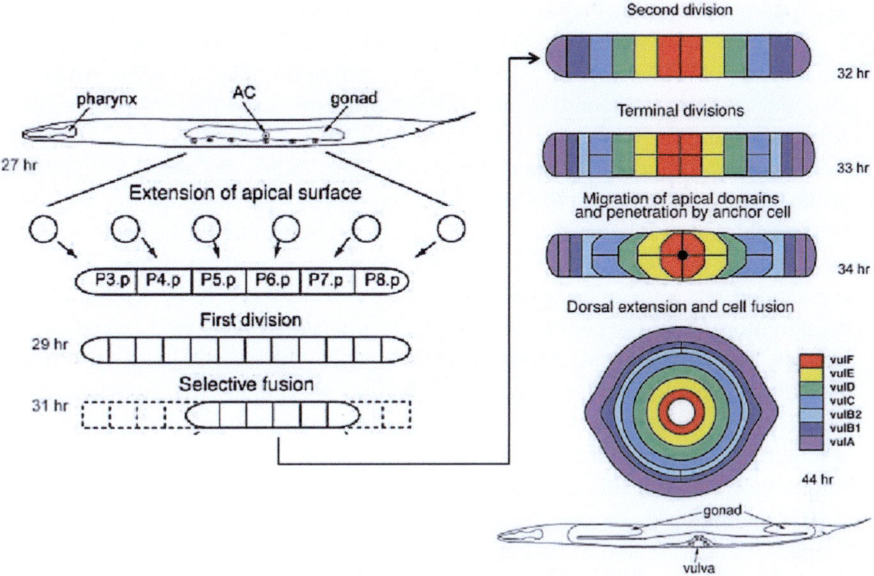

Fig. 19.26 Morphogenesis of the vulva in *C. elegans*. In the L3 stage larvae, a somatic gonadal cell, anchor cell (AC), induces three precursor cells, P5.p, P6.p, and P7.p, out of the total six (P3.p–P8.p) vulval precursor cells (VPCs) to adopt the vulval fates. The non-vulval cells, P3.p, P4.p, and P8.p, undergo a single proliferation step and fuse with the underlying syncytial cell. After the formation of vulva primordium, the cells from the outer regions carry out short-range migrations in a symmetrical fashion toward the central midline fusing with specific partners. A seven-toroid (vulA to vulF) stack is formed in the center where component cells from each ring fuse in a specific manner resulting in the formation of epithelial tube connecting the uterus to the outside. (Adapted from Sharma-Kishore et al. 1999)

competent enough to acquire each other's role ensuring normal vulva formation. For example, in case of P6.p ablation, the surrounding Pn.p cells, P5.p or P7.p, change their fates and adopt the primary fate, and P4.p or P8.p adopts the secondary fate resulting in the normal vulva formation. Ablation of gonads or anchor cells results in all VPCs acquiring the 3° fate. Multiple genes affect the VPC generation and induction, including let-23, lin-12, lin-39, and sem-4. In the absence of lin-39 expression, VPC generation is terminated due to the fusion of Pn.p cells with hyp7 epidermis. Sem-4 encodes a zinc finger protein and is required for the expression of the lin-39, which can also directly affect the VPC generation

3. Division and patterning of VPCs to produce different progeny cells: After the VPCs have acquired their sublineage fate, they start to divide and form their progeny cells. During L3 larval development stage, VPCs is induced by anchor cell and regulated through let-23 and lin-12 signaling pathways. Three cell divisions occur in both 1° (P6.p) and 2° VPCs (P5.p and P7.p) to produce eight cells and seven cells each respectively to form a vulval primordium consisting of

22 progeny cells. Both 1° and 2° VPCs undergo different morphogenesis regime: 1° VPC produces progeny cells vulE and vulF, while 2° VPCs produce progeny cells of type vulA, vulB1, vulB2, vulC, and vulD. The division pattern in secondary fate is asymmetric, and all the progeny cells thus formed result in vulval structure formation.

4. Anchor cell invasion: After the cell division, all the VPCs are arranged in a linear fashion with anchor cell in the middle. During the L3-L4 molt, the 1° VPC starts to move dorsally such that the homotypic cells on both sides of the anchor cell are migrated toward the center facing each other. As a result a lumen is formed surrounded by VPC's progeny cells in early L4 stage. Ring formation and invagination of VPCs continue resulting in a stack of cells in a dorsal-ventral manner. The structure thus formed is called toroid with a central hole surrounded by circular group of cells. Vulval invagination is regulated by eight sqv genes (sqv-1 to sqv-8). sqv genes encode enzyme for glycosaminoglycan synthesis that is required to maintain the shape of forming lumen. During toroid formation the similar cells on both sides of the midline starts to fuse together except vulB1 and vulB2 generating a multinucleate syncytial structure. In the uterine lumen that formed, the uterine and vulval epithelial are separated by a thin syncytium utse membrane (Fig. 19.27).

After the formation of lumen, a connection is established between lumen and uterus. Uterine anchor cell invades the basement membrane reaching the central 1° VPCs. On the dorsal side, vulF cells contact the uv1 cells on the ventral side of uterus. The vulF and uv1 cells are covered by a syncytial utse membrane formed by the fusion of AC cell with uterine cells. The utse membrane separates the uterine and vulval lumens. The vulE cell extends laterally and contacts with the hypodermal seam cells that stabilize the vulval structure. The opening and closing of the vulva is controlled by eight muscle cells, four vm1 and four vm2 that are in turn controlled by motor neurons. vm1 cells connects the vulva between vulC and vulD toroids, and vm2 connects vulF and the uterus. On the ventral side, vulA cells contact the hyp7 hypodermal syncytium. During hermaphrodite to adult molt in, finally, the vulva turns inside out resulting in an everted vulva with a slit for egg laying and copulation.

19.9 Cell Signaling Network in Development

During development, the ability of cells to respond to their microenvironment is dependent upon the communication process through a numerous number of signaling pathways. This communication occurs between different components within a cell, adjacent cell, as well as environment that governs basic activities of cells and coordinates multiple-cell actions to bring about development, tissue repair, and immunity, as well as normal tissue homeostasis (Fig. 19.28). The signaling network comprising of binding of a ligand to their specific receptors leading to the activation of intracellular machinery is a highly conserved mechanism. This general signaling

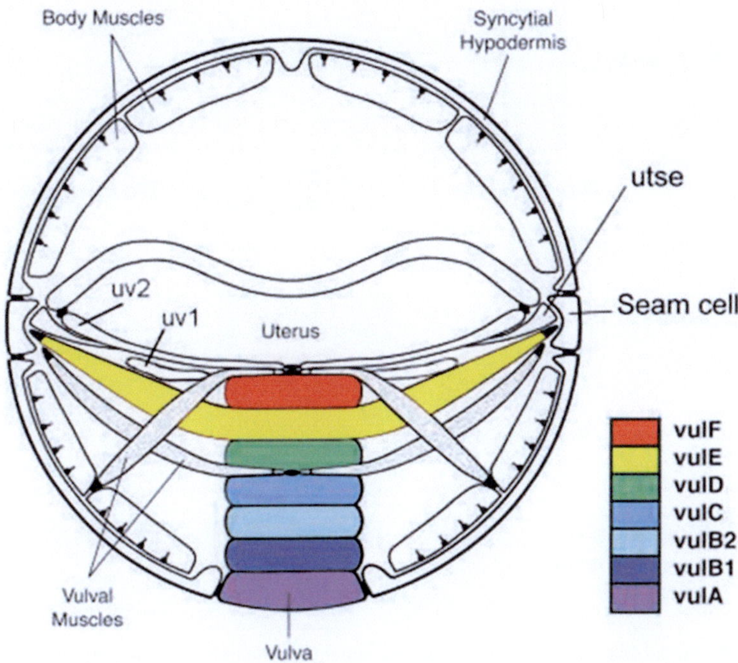

Fig. 19.27 Transverse section of late L4 stage worm showing vulval region and uterus. Vm1s and vm2s are the vulvar muscles present in two sets of four, where vm1s connects body wall and vulva while vm2 muscles connect ventral body wall and vulF. (Adapted from Sharma-Kishore et al. 1999)

network comprises of various interconnected signaling pathways as the same signaling component is capable of receiving signals from multiple inputs.

Cell signaling can be of two types depending upon the type of signal: mechanical where cell sense and response to the forces exerted on the cell or biochemical in which biomolecules like lipids, proteins, ions, and gases acts as signal. Signals can be categorized as intracrine, paracrine, or endocrine depending upon the distance between the secretory cell and its target. Intracrine signals are produced by the target cell and act inside a cell, regulating intracellular events. Examples of intracrine signaling include several protein or peptide hormones like renin-angiotensin system: angiotensin II and angiotensin, fibroblast growth factor 2, and parathyroid hormone-related protein. Similar to intracrine signaling, autocrine signals can function on the same cell, but they can also act on nearby cells if they are of same type. Autocrine signals, hormone or chemical messenger, are produced by the target cell, secreted, and bind to the same cell through receptors. Examples of autocrine signaling include immune cells like T-cell lymphocytes and cytokine interleukin-1 in monocytes. Juxtacrine signaling is also called contact-dependent signaling as it targets the adjacent cells. A juxtacrine signal can be the ligand from a membrane or a combination of two membrane proteins from neighboring cells. An intracellular protein or

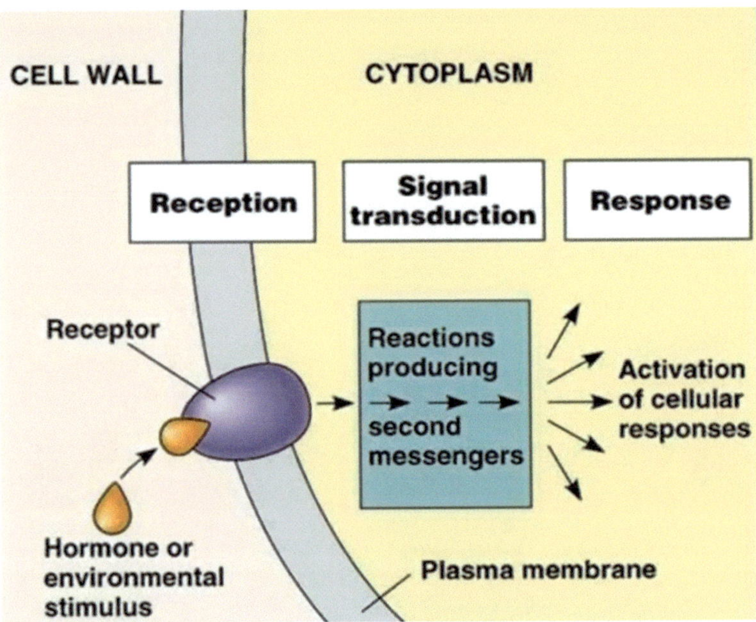

Fig. 19.28 Signal transduction. Mechanism of signal transduction includes three steps, reception, transduction, and response. Ligand binds on specific receptors that produce secondary messengers which carry the message to the nucleus or other organelles which in turn respond to this message by carrying out different metabolic reactions. (Adapted from https://biologydictionary.net/signal-transduction)

compartments of single cell or combination of two adjacent cells can also function as juxtacrine signal. These signals may function to carry out communication either between two cells or a cell and its extracellular matrix. In cell to cell signaling, a cell expresses a specific ligand on the surface of its membrane which binds to the appropriate cell surface receptor or cell adhesion molecule present on the adjacent cell. This type of communication can be seen in the Notch signaling pathway involved in neural development. In one type of juxtacrine signaling, two adjacent cells can construct communicating channel between their intracellular compartments, for example, gap junctions in animals and plasmodesmata in plants. Cell to extracellular matrix signaling can be seen during cell cycle and cellular differentiation where the cells interact with the glycoproteins secreted by extracellular matrix through a receptor integrin. Another type of cell-to-cell communication is paracrine signaling where cell produces a signal into the immediate extracellular environment that diffuses over a relatively short distance to induce changes in nearby cells altering its behavior. The highly conserved receptors and pathways of the paracrine signaling can be organized into four major families based on similar structures: fibroblast growth factor (FGF) family, Hedgehog family, Wnt family, and TGF-β superfamily. Endocrine signaling targets distant cells through hormones produced by endocrine cells which travel through the blood to reach all parts of the body. A number of endocrine glands signal each other in sequence and regulated by feedback loops are usually referred to as an axis, for example, the hypothalamic-pituitary-adrenal axis. In comparison, exocrine glands secrete hormones to the outside of the body utilizing ducts to distribute them throughout the body; common examples include sweat glands, gastrointestinal tract glands, and salivary glands. Major endocrine systems include TRH-TSH-T3/T4, GnRH-LH/FSH sex hormones, CRH-ACTH-cortisol, renin-angiotensin-aldosterone, and leptin-insulin systems. Hormones secreted by endocrine systems can be categorized into proteins, steroids, and eicosanoids. All the cellular, physiological, and behavioral function of any organism is dependent upon the intricate crosstalk between these hormones and different tissues and organs.

19.9.1 Wnt Pathway

Wnt pathway is one of the most conserved signaling pathways throughout the animal kingdom that regulates multiple cellular processes (Komiya and Habas 2008). Wnt signaling plays a critical role in embryonic development, axis patterning, cell fate specification, cell proliferation, cell migration, and insulin sensitivity (Cadigan and Liu 2006). The name Wnt is derived from *Drosophila* segment polarity gene wingless (Wg) and its vertebrate homolog integrated or int-1. Wnt signaling pathway was first identified in retroviruses for its role in carcinogenesis. Further, it was found that int-1 is actually a homologous gene of Wg in *Drosophila* extensively involved in embryonic development. Wnt pathway is known to function in at least two major ways: canonical or Wnt/β-catenin pathway, non-canonical planar cell polarity, or non-canonical Wnt/$Ca2^{+}$. Wnt-protein ligand binds to the receptors of Frizzled family, thereby transmitting the signal intracellularly to the Dishevelled protein.

Wnt ligands are the secreted glycoproteins of ~40 kDa containing several conserved cysteine residues and are extensively palmitoylated at a conserved serine residue. The lipid modification on Wnts is required for efficient signaling, binding of Wnt to its secretion carrier protein, Wntless (WLS), so it can be transported to the plasma membrane for secretion and for binding to its receptor Frizzled. Similar to many secretory proteins, Wnt ligands undergo glycosylation in endoplasmic reticulum and are then secreted into the extracellular matrix. The palmitoylation of the Wnt proteins is regulated by porcupine protein which in turn is regulated by wntless or evenness interrupted proteins and the retromer complex (Fig. 19.29).

In addition to the receptor, some molecules like low-density lipoprotein-related protein 5/6 (LRP5/6), receptor tyrosine kinase (RTK), and ROR2 work as the co-receptor for the signaling. In contrast, some molecules like Dickkopf (Dkk) proteins, Wnt Inhibitory Factor-1 (WIF-1), and secreted Frizzled-Related Proteins (sFRPs) work as antagonists to Wnt signaling by binding either to Wnt ligand or to its receptor.

A basic mechanism of Wnt signaling includes the binding of Wnt ligand to the extracellular N-terminal cysteine-rich domain of a Frizzled (Fz) family receptor. Frizzled (Fz) receptor is a transmembrane receptor spanning the plasma membrane

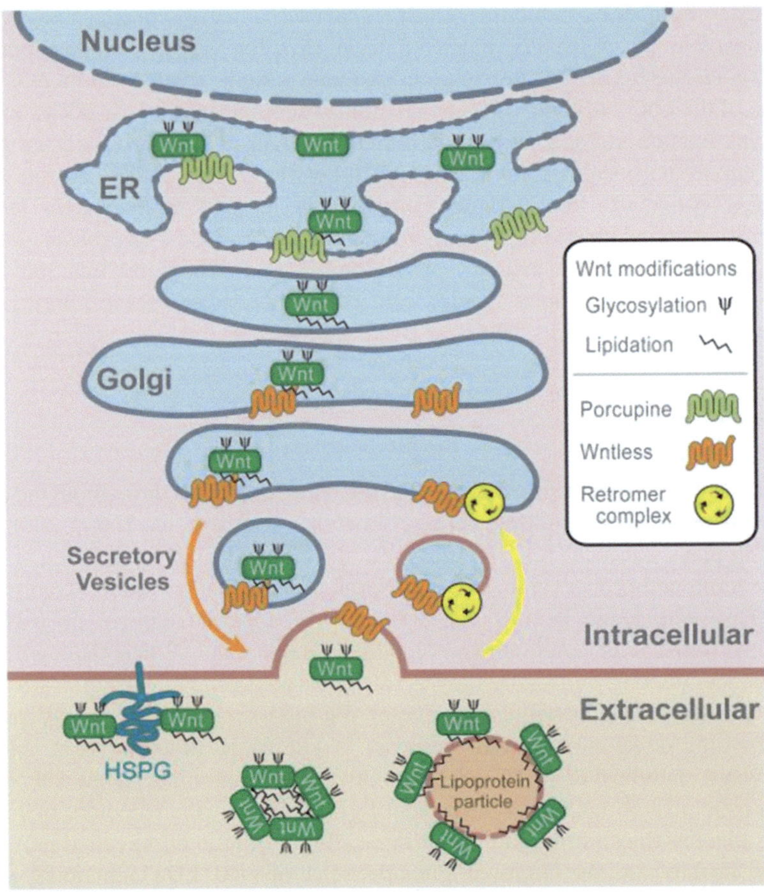

Fig. 19.29 Wnt biogenesis and secretion. Wnt proteins are highly modified before becoming mature as a ligand. Wnt proteins are first glycosylated and then lipid modified in the endoplasmic reticulum which is regulated by porcupine. After the maturation these proteins are transported from Golgi body to the plasma membrane for secretion by wntless. (Adapted from Macdonald et al. 2009)

seven times and constitutes a distinct family of G protein-coupled receptors (GPCRs). After activation Fz directly interacts with the intracellular phosphoprotein Dishevelled (Dsh) and transmits the signal. Dsh is a highly conserved Wnt signaling molecule present in all organisms and contains conserved domains: an amino-terminal DIX domain, a central PDZ domain, and a carboxy-terminal DEP domain. Dsh protein activation might then branch Wnt pathway into various directions (Fig. 19.30).

Canonical Wnt pathway: Canonical pathway is also called Wnt/β-catenin pathway as it is dependent on the protein β-catenin. The hallmark of the canonical pathway is the accumulation of the protein β-catenin into the cytoplasm which is

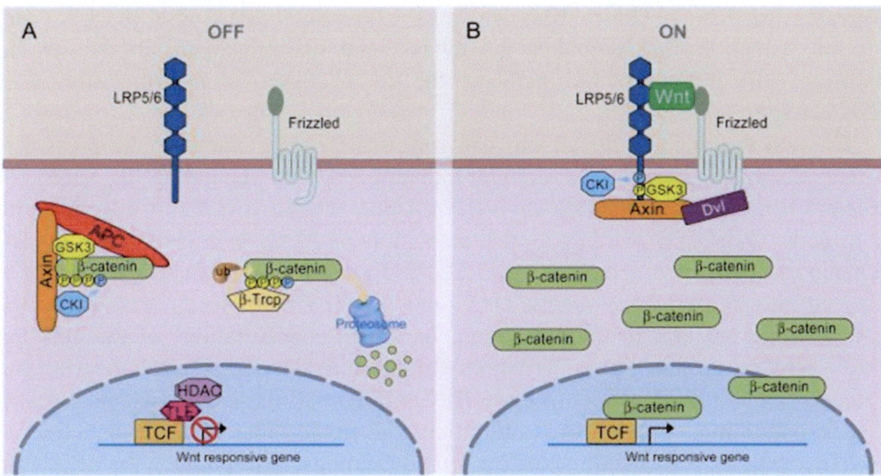

Fig. 19.30 Overview of Wnt/β-catenin signaling. In the absence of Wnt, cytoplasmic β-catenin levels are very low due to the proteasome-mediated degradation which is regulated through GSK-3/APC/Axin complex. However, in the presence of Wnt, this complex is inactivated leading to an increase in the levels of β-catenin in cytoplasm as well as nucleus which in turn interacts with transcription factors inducing different target genes. (Adapted from MacDonald et al. 2009)

subsequently translocated to the nucleus. In the absence of inducing Wnt signal, β-catenin in the cytoplasm is destined to degradation by a cytoplasmic destruction complex that comprises of Axin, adenomatosis polyposis coli (APC), protein phosphatase 2A (PP2A), glycogen synthase kinase 3 (GSK3), and casein kinase 1α (CK1α). β-catenin is targeted for its ubiquitination by proteasomal machinery after its phosphorylation. In contrast, when Wnt binds to the Fz and LRP5/6 receptor, it results in the translocation of the Axin to the plasma membrane inactivating the β-catenin destruction complex. CK1γ or GSK3 phosphorylates LRP5/6 which in turn dephosphorylates Axin resulting in the reduction of Axin stability and concentration. Further, Dsh becomes phosphorylated and activated resulting in inhibition of GSK3 of destruction complex by DIX and PDZ domains of Dsh protein. Upon inactivation of destruction, complex β-catenin starts to accumulate in the nucleus and induces target genes with TCF/LEF (T-cell factor/lymphoid enhancing factor) transcription factors.

Non-canonical Wnt pathway: Non-canonical pathway can further branch into two pathways, viz., planar cell polarity (PCP) pathway and Wnt/calcium pathway. Both of these pathways work in a β-catenin independent fashion. In non-canonical pathway, receptor Fz uses NRH1, Ryk, PTK7, or ROR2 as a co-receptor for signal transduction instead of LRP-5/6. Using one of these co-receptors, the signal is then transmitted to the Dsh protein activating its PDZ and DIX domains which in turn recruits Dishevelled-associated activator of morphogenesis 1 (DAAM1) protein to form a complex which in turn activates Rho and Rac protein pathways. In the first pathway, Dsh-Damm1 activates a Rho guanine exchange factor which in turn activates the Rho GTPase. Rho activates Rho-associated kinase (ROCK) and myosin, which further regulates the cytoskeletal rearrangement. In the second pathway, DEP domain of Dsh activates the Rac GTPase which further stimulates JNK activity.

The non-canonical Wnt/calcium pathway is also independent of the β-catenin. This pathway controls the intracellular calcium levels by regulating the release of calcium from endoplasmic reticulum (ER). Upon activation of Fz, it interacts with a trimeric G protein leading to the activation of PLC or cGMP-specific PDE domain of Dsh. Upon activation of PLC, PIP2 present in the plasma membrane is cleaved into its two components, DAG and IP3. IP3 binds to its receptor on the ER releasing calcium in the cytoplasm. Increased concentrations of calcium and DAG activate protein kinase C which further activates Cdc42, calcineurin, and calcium/calmodulin-dependent kinase II (CaMKII). All these components regulate ventral patterning, cell adhesion, migration, and tissue separation. Calcineurin can interfere with TCF/ß-catenin signaling in the canonical Wnt pathway by activating TGF-β-activated kinase (TAK1) and Nemo-like kinase (NLK). Activation of PDE domain of Dsh leads to the inhibition of PKG which in turn impedes the calcium release from the ER.

19.9.2 Hedgehog Pathway

The Hedgehog (Hh) signaling pathway is one of the key conserved pathways in a wide range of organisms regulating many developmental processes (Jia et al. 2015). The Hedgehog (Hh) signaling pathway was first identified in the *Drosophila* as one of the genes required for establishing anterior-posterior body axis of the fly. Christiane Nüsslein-Volhard and Eric Wieschaus performed genetic screens in *Drosophila* embryo in order to understand the body segmentation. The protein derived its name from the appearance of short and spiked phenotype of the cuticle in Hh mutant embryos which resembles the spikes of a hedgehog.

Drosophila contains a single Hh gene as compared to three homologs in vertebrate with different spatial and temporal distribution patterns: Sonic Hedgehog (Shh), Indian Hedgehog (Ihh), and Desert Hedgehog (Dhh). Hh undergoes several post-translational modifications in order to become fully functional. A precursor Hh protein consist of a signaling N-terminal called "Hedge" domain and a protease C-terminal called "Hog" domain. The Hog domain can further be divided into a Hint domain at N-terminal and sterol-recognition region (SRR) at the C-terminal. At the C-terminal hint domain, a cholesterol moiety is added, while at the N-terminus, palmitoyl acyltransferase adds a palmitoyl moiety. After autocatalytic cleavage of the precursor Hh molecule, a dually lipidated active signaling molecule called HhNp is released from the secreting cell by a transmembrane transporter protein called Dispatched (Disp). In addition, Scube2, cell surface protein LRP2, and the Glypican family of heparan sulfate proteoglycans (GPC1–6) also help in trafficking over multiple cells. Functional Hh proteins initiate signaling at the plasma membrane by binding to transmembrane receptor Patched (PTCH1) facilitated by co-receptors GAS1, CDON, and BOC.

Hh signaling is regulated by a Hh signaling complex, HSC, comprising of the transcription factor Cubitus interruptus (Ci), serine/threonine kinase Fused (Fu), kinesin-like molecule Costal 2 (Cos2), and suppressor of fused (Sufu). In the absence of Hh, this complex remains bonded to the microtubule in the cytoplasm. In the absence of Hh signaling, the Ci- Cos2 complex present in the cytoplasm is targeted by a degradation complex which contains protein Slimb in *Drosophila*. This degradation complex targets the Ci protein for proteasome-dependent cleavage resulting in the generation of protein fragment called CiR. CiR diffuses into the nucleus from the cytoplasm where it acts as a co-repressor for Hedgehog (Hh) target genes. In contrast, upon activation by Hh, the complex dissociates and the Cos2-Fu-Ci complex interacts with the C-tail of Smo domains regulated by Cos2 phosphorylation leading to formation of active Ci and consequent pathway activation. In addition, Hh binds and inhibits a cell surface transmembrane protein called Patched (PTCH) which in turn results in the accumulation of Smoothened (SMO) inhibiting the cleavage of Ci. Thus Ci and CiR levels increase and decrease respectively in the cell allowing Ci to act as transcription factor for many genes in nucleus.

19.9.3 TGF-β Pathway

It is an evolutionary conserved pathway regulating a wide number of cellular processes including cell growth, differentiation, and development in a wide range of organisms. TGF superfamily includes various ligands like growth and differentiation factors (GDFs), bone morphogenetic proteins (BMPs), anti-Müllerian hormone (AMH), Activin, Nodal, and TGF-βs. The result of TGF-β signaling pathway depends upon the control of the spatial and temporal expression of more than 500 genes which in turn is dependent on the functions of the SMAD protein and cofactors. There are three major classes of SMAD protein family, R-SMAD (SMAD, SMAD1–3, and SMAD5), co-Smad (SMAD 4), and I-SMAD (SMAD6–7). An SSXS motif is present at the C terminal of the receptor-regulated Smads (R-SMAD) which is phosphorylated by the type I receptors which results in the formation of phosphorylated R-Smad (p-R- SMAD) which in turn associates with SMAD4. TGF-β signaling regulates the transcription of more than 500 genes which contains one or more SMAD binding elements (SBEs) in their promoter region. In corroboration to SMADs, several SMAD interacting transcription factors modulate specific target gene expression depending on the cellular type in both physiological and pathological conditions.

TGF signaling starts when TGF-β ligand binds to the serine/threonine type II receptor kinase dimer which catalyzes the phosphorylation of the Type I receptor dimer resulting in the formation of hetero-tetrameric complex. Ligand binding also

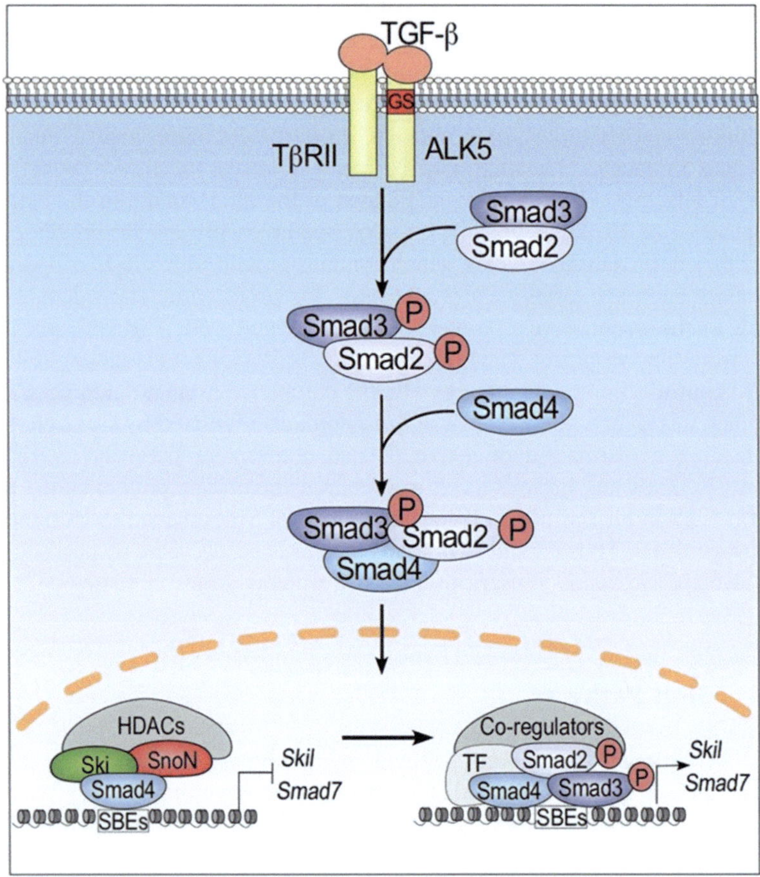

Fig. 19.31 TGF-β Pathway. TGF-β ligand binding results in the assembly of type I and type II receptors which further transmits the signal by phosphorylation of R-Smad proteins. The activated R-Smad proteins along with co-Smads translocate to the nucleus, thereby regulating the transcription of target genes. (Adapted from Tecalco-Cruz et al. 2018)

phosphorylates cytoplasmic signaling molecules, Mothers against decapentaplegic homolog (SMAD) family members, Smad1/2/3/5/9, called R-SMADs. The TGF-β/activin pathway is induced by Smad2 and Smad3, while BMP pathway is induced by Smad1/5/9. Phosphorylation at the C-terminal of the Smads allows the results in the activation of co-SMAD (Smad4) forming a RSMAD/coSMAD complex which translocate to the nucleus resulting in the transcription of target genes through different transcription factors (Fig. 19.31). Smad6 and Smad7 function in the antagonistic manner inhibiting the R-Smad activation; therefore both of these Smads are used in negative feedback loop in activin/TGF-β and BMP signaling pathway. In addition, two SMAD interacting proteins, Ski (Sloan-Kettering

Institute) and SnoN (Ski novel), disrupt the formation of R-Smad/Smad4 complex as well as inhibit the SMAD association with the p300/CBP coactivators resulting in the negative regulation of the TGF-β signaling pathway. Ski and SnoN indirectly bind to the consensus sequence called SBE (5′-GTCTAGAC-3′), thereby acting as SMAD corepressors.

19.9.4 Receptor Tyrosine Kinase Pathway

Receptor tyrosine kinases (RTKs) comprise of the family of receptors that in addition to working as receptors for ligand binding also function as enzymes leading to intracellular phosphorylation event. RTKs are single-pass, type I receptors embedded in the plasma membrane. Each receptor of RTK is comprised of two domains, an extracellular domain present at the N terminal region and an intracellular domain present at the C terminal region. Both the domains together comprise a single hydrophobic transmembrane-spanning domain. Binding of a ligand (growth factor or hormone) to the extracellular N terminal domain results in the oligomerization generally dimerization of the RTKs called cross-phosphorylation. In cross-phosphorylation, neighboring RTKs associate with each other resulting in autophosphorylation and thus activation of the tyrosine kinases present in the intracellular C domain of RTK (Fig. 19.32). The cross-phosphorylation of tyrosine kinases produces binding sites for a variety of intracellular protein that further propagate signal transduction. The proteins bind to the tyrosine kinases through special domains called Src homology 2 (SH2) domain- and phosphotyrosine-binding (PTB) domain (Hubbard and Miller 2007). The most common downstream signaling

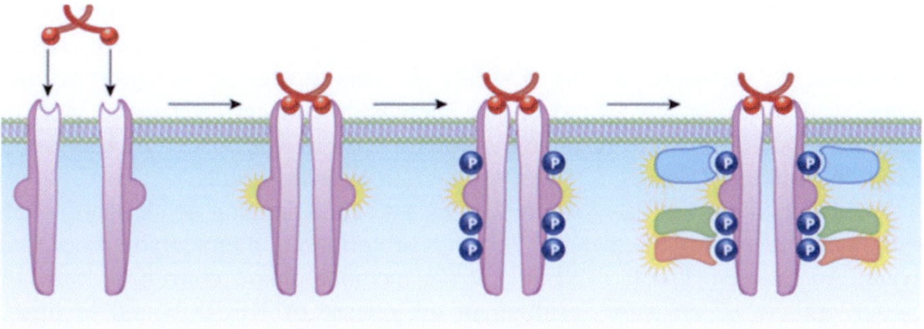

Fig. 19.32 Receptor tyrosine kinase activation. Upon binding of ligand to the inactivated RTK receptor (left), the receptors dimerize together and recruit various different proteins that in turn activate each other through phosphorylation. These phosphorylated proteins in turn alter the gene expression of target genes, thereby completing the signaling pathway. (Adapted from https://www.nature.com/scitable/topicpage/rtk-14050230/)

pathways activated by RTKs are proto-oncogene tyrosine-protein kinase, phosphoinositide 3-kinase, mitogen-activated protein kinase, janus kinases (JAKs), signal transducer and activator of transcription proteins, etc.

RTKs target various intracellular signaling pathways; however the commonly employed pathway is mitogen-activated protein (MAP) kinase cascade regulating cell proliferation, cell death, differentiation, migration, and angiogenesis. When SH2-containing proteins bind the activated RTKs, they cause Ras to bind GTP in place of GDP and become active which in turn activates MAP kinase leading to the amplification of original signal. The final step in this cascade is the phosphorylation of transcription regulators, leading to a change in gene transcription.

19.9.5 Notch Pathway

It is a highly conserved signaling pathway evolutionarily present in most animals regulating cell fate determination and tissue homeostasis during development in adult. Notch was discovered in the *Drosophila* due to the appearance of a notch in the wings of mutant flies. Most of the knowledge on the notch signaling pathway comes from the studies on *Drosophila* and *C. elegans* mutants. Notch receptor is composed of two domains, an extracellular (NECD) domain and an intracellular (NICD) domain, both of which together form a single-pass transmembrane protein. The NECD consists of multiple epidermal growth factor (EGF)-like repeats and interacts with the Delta/Serrate/LAG-2 (DSL) family ligands mediating juxtacrine signaling. The NICD of the Notch receptor consists of a transcriptional activation domain (TAD) which acts as a transcriptional activator in association with CBF1/Suppressor of Hairless/LAG-1 (CSL) family transcription factors. Notch receptors are further processed in the endoplasmic reticulum and Golgi body through S1 cleavage and glycosylation resulting in a heterodimer consisting of NECD non-covalently attached to the TM-NICD. This processed receptor is then transported by endosome to the plasma membrane and gets embedded in it. In mammals, Delta-like and Jagged family proteins serve as ligands for Notch signaling receptors, while in *Drosophila* Delta and Serrate work as ligands.

Notch signaling is initiated when an adjacent cell expressing notch ligand binds with notch receptor on another cell; this in result promotes a proteolytic processing event. An enzyme TACE (TNF-α ADAM metalloprotease converting enzyme) carries out the S2 cleavage resulting in the release of NECD bound to ligand which is then endocytosed/recycled by the ligand-expressing cell. The remaining receptor is then further cleaved by γ-secretase enzyme (S3 cleavage) releasing the NICD which then translocates to the nucleus resulting in the activation of the target genes: Myc, p21, and the HES-family members through CSL transcription factors (Fig. 19.33).

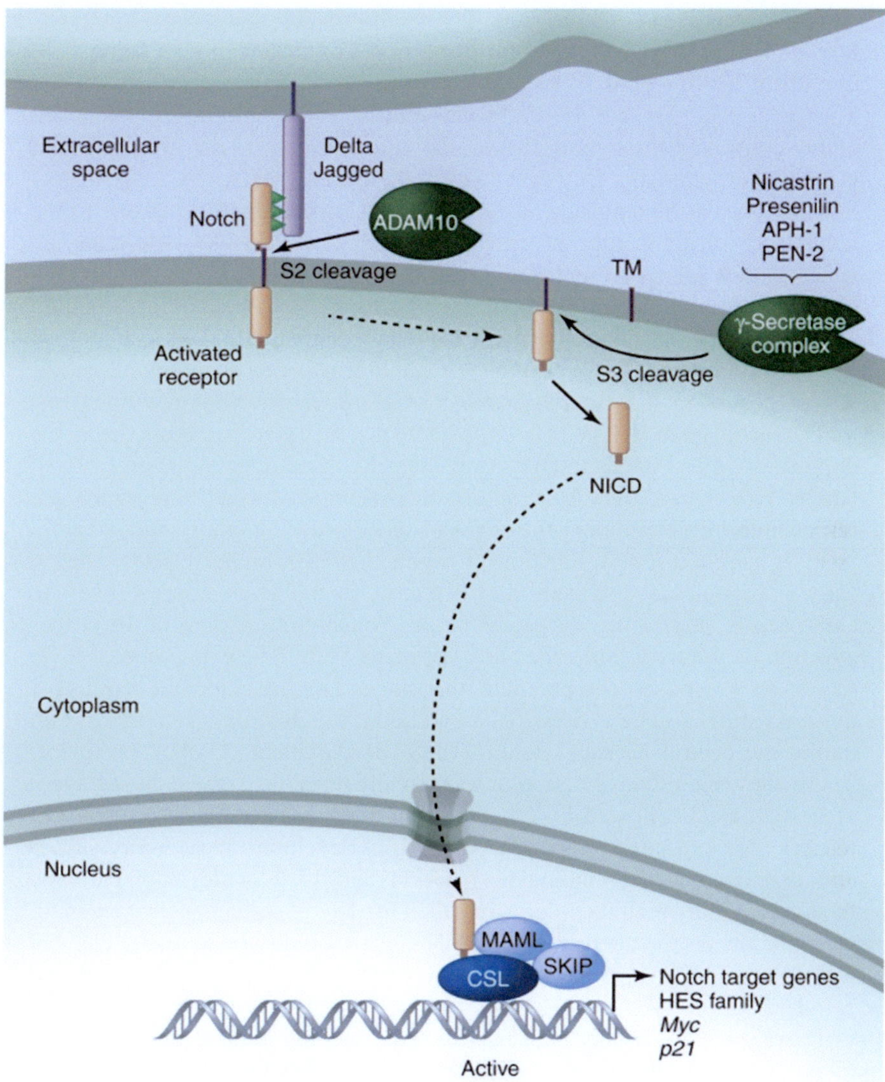

Fig. 19.33 Notch pathway. Notch ligand is a cell-surface receptor that communicates with other transmembrane ligands such as Delta (termed Delta-like in humans) and Serrate (termed Jagged in humans) on adjacent cells, thereby transmitting short-range signals. Upon ligand binding the Notch intracellular domain (NICD) gets cleaved and is released into the cytoplasm from where it translocates to the nucleus and regulates transcriptional activity of target genes. (Adapted from Kopan 2012)

Box 19.1: Scientific Concept: Variation and Constraint in Hox Gene Evolution (Heffer et al. 2013)

The genes that regulate the body plan during embryonic development are highly conserved throughout the animal kingdom despite the diversification in body plan symmetry. However some of these conserved genes acquired new function during the course of evolution which is thought to occur due to the redundancy or duplication of genes. One such class of genes is Hox family genes that are highly regulated and play important roles in body plan patterning during embryonic development. This opens up a new field of investigation as how these genes acquired new roles due to redundancy and why did not get lost during the course of evolution. In addition was this acquisition a single step process or took several stages to convert into a gene with new functions.

The fushi tarazu (ftz) gene in *Drosophila* is a Hox gene and arose due to the duplication of an Antennapedia (Antp)-like Hox gene. Due to the presence of nearby Hox genes like Antp and/or Sex-combs reduced (Scr) which are responsible for determining the segment identity during embryonic development, it acquired a new function of segmentation in higher insects. During early embryonic development in *Drosophila*, ftz and even-skipped (eve) are expressed in alternating 7 vertical stripes constituting a total of 14 stripes which in the future become the body segments of fly. The expression of ftz in the seven stripes as compared to the single Hox-like domain requires a dynamic shift in gene expression. In addition, ftz also plays important roles during the central nervous system (CNS) development. Midline precursor-2 cells show high ftz expression in segmentally repeating pattern during 5–6 h of development followed by expression in other neuronal precursors which reaches a second peak of abundance at 9 h of development. Moreover, ftz is also expressed in the hindgut between 11 and 15 h of development. The functioning of ftz requires a cofactor called Ftz-F1, an orphan nuclear receptor, which facilitates the binding of ftz to DNA. Ftz-F1 binds the sequence element 5′-YCYYGGYCR-3′ in the zebra element of ftz; this interaction is mediated by a LXXLL motif.

Despite of the dynamic changes in sequence and expression of ftz, it is found in all arthropod genomes examined to date. Tracking the ftz expression and Ftz protein domains in different arthropod phylogeny revealed that in myriapods and chelicerates, ftz retains the Hox-like expression unlike in crustaceans, basal insect *Thermobia*, and several holometabolous insects where striped expression can be seen. Ftz expression as a segmentation gene or Hox gene in addition with loss or gain of cofactor interaction motifs is highly variable throughout arthropods. Despite the variations in the ftz expression as segmentation or Hox gene, its expression is well conserved in the developing CNS. Interestingly, it was found that the Antp homeodomain can substitute for the Ftz homeodomain in the core function of ftz in CNS by

(continued)

Box 19.1 (continued)
acquiring a neurogenic ftz cis-regulatory elements (CRE) leading to its expression in a unique group of cells, thereby differentiating its role from neighboring Hox genes. It was found that homeodomain present in all Ftz proteins is indispensable for its function unlike YPWM or LXXLL motifs which were variably present in some organisms. Ftz acquired the neurogenic element (NE) in the cis-regulatory region which regulates the ftz expression in specific cells of the developing CNS. Since the function of Antp is also conserved in the CNS, the question arises that Antp could have taken up the function of ftz by acquiring the NE; however, this is not the case. This is evident by the inability of entire Antp coding region under the control of ftz NE to rescue the eve expression in neurons. Moreover, degeneration of CREs results in the further change in the ftz expression in the stripped fashion, and change in the sequences resulted in a switch in cofactor interaction. Ftz requires the homeodomain for its function in developing CNS, while LXXLL motif is indispensable for the segmentation function. Based upon all the above observation, it was proposed that functional loss and acquisition of new function occur in the stepwise manner during evolution.

In conclusion, the requirement of gene function along with its protein domain in a specific tissue applies a constraint on the gene, thereby leading to its retention during evolution as seen in ftz, where CNS activity requires the homeodomain. In comparison, other domains which are not essential for the core function underwent extensive diversification leading to variations in the gene. In this way the balance between constraint and variation allows functional diversification as well as long-term retention during animal evolution (Fig. 19.34).

19.10 Summary

- Developmental biology explains how different cellular processes interact with each other at the molecular level to produce an anatomically as well as morphologically complex organism starting from an embryo to adult.
- Different intrinsic as well as extrinsic signals direct the cell to acquire their fate during development. Intrinsic signals are also called lineage information and are inherited from the mother cell, while extrinsic signals, also called as positional information, are received from the cell's surroundings.
- During evolution the basic developmental pathways have been conserved ranging from unicellular to higher organisms. This conservation has allowed the researchers to dissect these processes and gain insights into it using various model organisms.

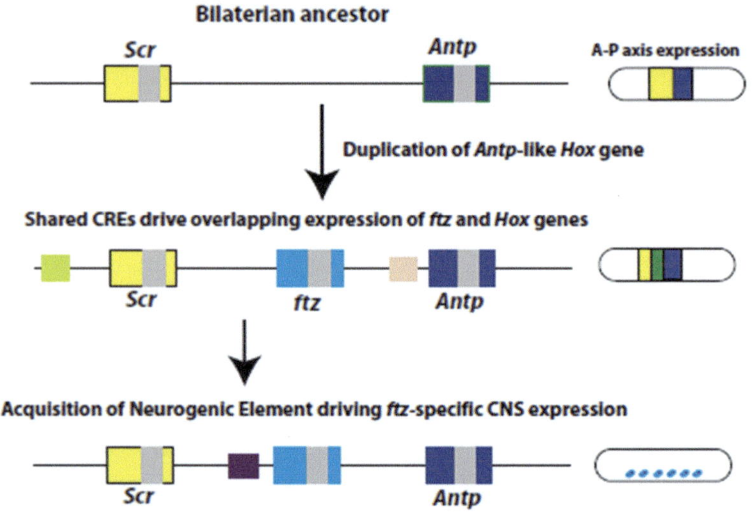

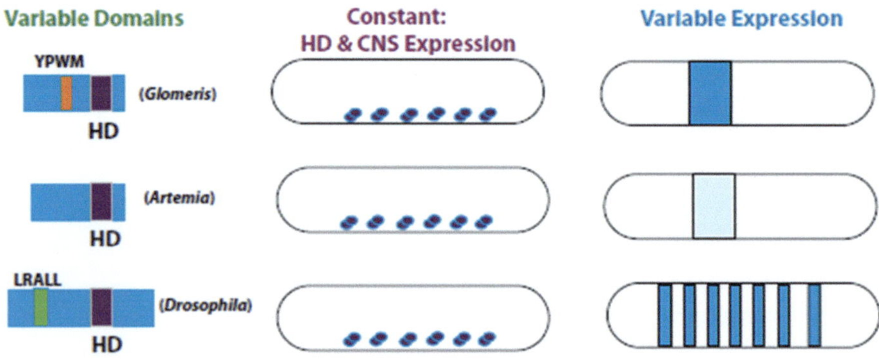

Fig. 19.34 Ftz neurogenic element (NE). Duplication of Hox gene produced ftz which acquired NE in the homeodomain which allowed it to be retained in almost all the organisms and is essential for the CNS development. While new domains like LXXLL were acquired, other domains like YPWM are degenerated. (Adapted from Heffer et al. 2013)

- Model organisms are animals that have certain characteristic features that allow researchers to study them with ease like large cultures in confined laboratory space, short generation time, large number of progeny, easy manipulation at the molecular level, etc. The most widely used model organisms include *Saccharomyces cerevisiae*, *Xenopus*, *Drosophila melanogaster*, *Mus musculus*, *Caenorhabditis elegans*, *Arabidopsis thaliana*, *Danio rerio*, and *Escherichia coli*.

- Various genetic approaches are extensively applied in order to study the developmental processes and understand the mechanisms behind it. These approaches can be divided into two classes, forward and reverse genetics; the goal of both processes is to associate a gene with its biological function.
- Forward genetics starts with the discovery of a mutant organism with distinct phenotype followed by the identification of gene responsible. In contrast, reverse genetics deals with the known candidate gene and then elucidating its function.
- With the identification of a sufficient number of genes and proteins that are involved, researchers can acquire insights on the underlying molecular processes. Despite the basic knowledge on these processes, they cannot be used to elucidate the exact mechanism as behavior of any organism is the culmination of the very intricate network of molecular processes.
- Therefore, various genetic tools are now being employed to gain insight on the underlying complex but basic cellular processes.
- There exists a functional homology in developmental genetics, for example, the Hox genes are conserved among animal kingdom and play important role in axial patterning. The mutation in Hox genes is known to cause various developmental defects in different organisms.
- The development of any organisms is based on the interplay and intercommunication between the underlying signaling networks. A great deal of research is focused on unraveling the signaling cascade with upstream and downstream interactions which regulate and relay the information carried on the specific binding of ligands to their receptors thereby by a ligand to cell surface receptors and finally to the cellular effectors such as metabolic enzymes, channels, or transcription factors.
- This cascade is not essentially linear in its function; rather it is highly branched leading to the interaction between the components of different pathways which help to regulate multiple functions in a context-dependent manner.
- Recent studies indicate that there is a conserved set of signaling components as well as pathways that receives signals from cell type-specific inputs and engages cell type-specific machinery.
- The interconnection between different pathways may be on two levels: first at a junction which functions as a signal integrator and second, nodes which split the signal and route them to multiple outputs; this signaling can be both positive and negative.
- Studies deciphering the signaling mechanisms involved in various human diseases help to understand the deregulations in human disorders by considering a broader range of molecular process types.
- In order to understand the larger scheme of cellular networks where different signaling and metabolic networks function in an integrated fashion, there is a need to find the common regulatory components of these pathways responsible for the interconnection.
- By integrating the knowledge obtained in this way, it is possible to elucidate the crosstalk between metabolic and signaling networks and to find the potential targets to cure the said disease.

- During evolution, major pathways and key genes involved are both preserved and modified to take on more specialized roles, for example, in vertebrates like humans and mice, Hox genes have been duplicated over evolutionary history and now exist as four similar gene clusters.

References

Ables ET, Hwang GH, Finger DS, Hinnant TD, Drummond-Barbosa D (2016) A genetic mosaic screen reveals ecdysone-responsive genes regulating Drosophila oogenesis. G3 (Bethesda) 6(8): 2629–2642. https://doi.org/10.1534/g3.116.028951

Adamczyk PA, Reed JL (2017) Escherichia coli as a model organism for systems metabolic engineering. Curr Opin Syst Biol 6:80–88. https://doi.org/10.1016/j.coisb.2017.11.001

Cadigan KM, Liu Y (2006) Wnt signaling: complexity at the surface. J Cell Sci 119(Pt 3):395–402

Chanderbali AS, Yoo MJ, Zahn LM, Brockington SF, Wall PK, Gitzendanner MA, Albert VA, Leebens-Mack J, Altman NS, Ma H, dePamphilis CW, Soltis DE, Soltis PS (2010) Conservation and canalization of gene expression during angiosperm diversification accompany the origin and evolution of the flower. Proc Natl Acad Sci U S A 107(52):22570–22575. https://doi.org/10.1073/pnas.1013395108

Chang CW et al (2011) Anterior–posterior axis specification in *Drosophila* oocytes: identification of novel bicoid and oskar mRNA localization factors. Genetics 188(4):883–896

Corsi AK, Wightman B, Chalfie M (2015) A transparent window into biology: a primer on Caenorhabditis elegans. Genetics 200(2):387–407. https://doi.org/10.1534/genetics.115.176099

Duina AA, Miller ME, Keeney JB (2014) Budding yeast for budding geneticists: a primer on the Saccharomyces cerevisiae model system. Genetics 197(1):33–48. https://doi.org/10.1534/genetics.114.163188

Gilbert SF (2000) Early development of the nematode Caenorhabditis elegans. Sinauer Associates, Sunderland, MA

Gummalla M, Galetti S, Maeda RK, Karch F (2014) Hox gene regulation in the central nervous system of Drosophila. Front Cell Neurosci 8:96. https://doi.org/10.3389/fncel.2014.00096

Hales KG, Korey CA, Larracuente AM, Roberts DM (2015) Genetics on the fly: a primer on the drosophila model system. Genetics 201(3):815–842. https://doi.org/10.1534/genetics.115.183392

Heffer A, Xiang J, Pick L (2013) Variation and constraint in Hox gene evolution. Proc Natl Acad Sci U S A 110(6):2211–2216

Hubbard SR, Miller WT (2007) Receptor tyrosine kinases: mechanisms of activation and signaling. Curr Opin Cell Biol 19(2):117–123. https://doi.org/10.1016/j.ceb.2007.02.010

Jia Y, Wang Y, Xie J (2015) The Hedgehog pathway: role in cell differentiation, polarity and proliferation. Arch Toxicol 89(2):179–191

Komiya Y, Habas R (2008) Wnt signal transduction pathways. Organogenesis 4(2):68–75

Kopan R (2012) Notch signaling. Cold Spring Harb Perspect Biol 4(10):a011213. https://doi.org/10.1101/cshperspect.a011213

Lappin TRJ, Grier DG, Thompson A, Halliday HL (2006) HOX Genes: Seductive science, mysterious mechanisms. Ulster Med J 75(1):23–31

Lawrence PA, Morata G (1994) Homeobox genes: their function in *Drosophila* segmentation and pattern formation. Cell 78(2):181–189

Lee RTH, Zhao Z, Ingham PW (2016) Development at a glance: Hedgehog signaling. Development 143:367–372. https://doi.org/10.1242/dev.120154

Leung MCK et al (2008) Caenorhabditis elegans: an emerging model in biomedical and environmental toxicology. Toxicol Sci 106(1):5–28

MacDonald BT, Tamai K, He X (2009) Wnt/β-catenin signaling: components, mechanisms, and diseases. Dev Cell 17(1):9–26. https://doi.org/10.1016/j.devcel.2009.06.016

Mallo M, Alonso CR (2013) The regulation of Hox gene expression during animal development. Development 140:3951–3963. https://doi.org/10.1242/dev.068346

Marsh EK, May RC (2012) Caenorhabditis elegans, a model organism for investigating immunity. Appl Environ Microbiol 78(7):2075–2081

Middleton CA, Nongthomba U, Parry K, Sweeney ST, Sparrow JC, Elliott CJ (2006) Neuromuscular organization and aminergic modulation of contractions in the Drosophila ovary. BMC Biol 4:17

Monica J, Justice MJ, Dhillon P (2016) Using the mouse to model human disease: increasing validity and reproducibility. Dis Model Mech 9:101–103. https://doi.org/10.1242/dmm.024547

Müller WA (1997) Model organisms in developmental biology. In: Developmental biology. Springer, New York, NY. https://doi.org/10.1007/978-1-4612-2248-4_3

Mundlos S (2010) Gene action: developmental genetics. Vogel and Motulsky's Human Genetics:417–450. https://doi.org/10.1007/978-3-540-37654-5_15

Murai K (2013) Homeotic genes and the ABCDE model for floral organ formation in wheat. Plan Theory 2:379–395. https://doi.org/10.3390/plants2030379

Ong C, Yung LY, Cai Y, Bay BH, Baeg GH (2014) *Drosophila melanogaster* as a model organism to study nanotoxicity. Nanotoxicology 9(3):396–403. https://doi.org/10.3109/17435390.2014.940405

Pandey A (2014) The UNC-53 mediated interactome

Pavlopoulos A, Michael Akam M (2007) Hox go omics: insights from *Drosophila* into Hox gene targets. Genome Biol 8(3):208. https://doi.org/10.1186/gb-2007-8-3-208

Perlman RL (2016) Mouse models of human disease an evolutionary perspective. Evol Med Public Health 2016(1):170–176

Pritchett TL, Tanner EA, McCall K (2009) Cracking open cell death in the *Drosophila* ovary. Apoptosis 14(8):969–979. https://doi.org/10.1007/s10495-009-0369-z

Ririe TO, Fernandes JS, Sternberg PW (2008) The Caenorhabditis elegans vulva: a post-embryonic gene regulatory network controlling organogenesis. Proc Natl Acad Sci U S A 105(51): 20095–20099

Sahin HB, Celik A (2013) *Drosophila* eye development and photoreceptor specification. In: ells. John Wiley & Sons, Ltd, Chichester. https://doi.org/10.1002/9780470015902.a0001147.pub2

Sharma-Kishore R, White JG, Southgate E, Podbilewicz B (1999) Formation of the vulva in Caenorhabditis elegans: a paradigm for organogenesis. Development 126:691–699

Tecalco-Cruz AC, Ríos-López DG, Vázquez-Victorio G, Rosales-Alvarez RE, Macías-Silva M (2018) Transcriptional cofactors Ski and SnoN are major regulators of the TGF-β/Smad signaling pathway in health and disease. Signal Transduct Target Ther 3:15. https://doi.org/10.1038/s41392-018-0015-8

Tram U (2011) Drosophila's unusual syncytial blastoderm: an overview

Willemsen R, Padje S, van Swieten JC, Oostra BA (2011) Zebrafish (Danio rerio) as a model organism for dementia. In: De Deyn P, Van Dam D (eds) Animal models of dementia, neuromethods, vol 48. Humana Press, New Jersey

Quantitative Genetics 20

Anindo Chatterjee

20.1 Introduction to Quantitative Genetics

An overwhelming amount of genetics that is taught in classrooms is derived from a common assumption, which is that a given phenotype is the output of a given gene. This one gene-one phenotype assumption actually does not hold true for the majority of phenotypic traits. Most of the traits that you would have studied till now can be classified as qualitative traits. They encompass different types or kinds of a trait. For example, curly hair, wavy hair and straight hair are different kinds of qualities of hair. Similarly, red or white eye colour in *Drosophila* refers to a qualitative trait, the quality in question being the colour. Other examples include human blood-group types (A, B, AB and O) and the shape of the fruit in squash (spherical, disc-shaped, elongated, etc.). However, careful observation of the world around us informs us that a given trait can be given a numerical value (or a unique phenotypic class), and these numerical values tend to be part of a range of possible degrees and intensity of the trait in question. For example, the protein content in a seed will have a numerical value, and these values are going to vary even for seeds of plants of the same species. Thus, protein content of a seed is a typical quantitative trait. Such traits are also called continuous traits to distinguish them from traits that can be categorized in a few phenotypic classes. The latter traits are known as discontinuous traits. The tall and dwarf pea plants of Mendel are considered to be discontinuous traits. However, we can also study the range of tallness or dwarfness of these plants and analyse it quantitatively. Therefore, a given qualitative trait can sometimes be further studied quantitatively. Human IQ scores are also considered to be a quantitative trait.

Quantitative traits are also called as *polygenic traits*. This is because the phenotypic expression of most quantitative traits is regulated by more than one gene. The polygenic control of a trait was demonstrated by James Crow in an experiment

A. Chatterjee (✉)
Jain University, Bangalore, India

D. Kar, S. Sarkar (eds.), *Genetics Fundamentals Notes*,
https://doi.org/10.1007/978-981-16-7041-1_20

which showed that the resistance to DDT in a population of DDT-resistant *Drosophila* was conferred by genes present on chromosomes 2 and 3 and the X-chromosome. Quantitative traits are also called *multifactorial traits* because the range of phenotypes exhibited for a given trait is influenced by more than one factor. The expression of quantitative traits is regulated by both genetic and environmental factors. This was convincingly proven by Wilhelm Johannsen through his work on *Phaseolus vulgaris* (broad bean) wherein he showed that both genetic and environmental factors contribute to the 150 mg–900 mg range of seed weight observed in this plant.

An important factor which is usually ignored in the study of classical Mendelian genetics is the environment the organism grows in. The same genotype has the potential to exhibit a range of overlapping phenotypes due to differing environmental factors such as temperature, pH, moisture content, etc. The phenotypic range (or variation) shown by a given genotype due to variation in environmental factors is called the *norm of the reaction*. Similarly, multiple forms of genetic interactions such as pleiotropy and epistasis also result in phenotypic ranges for a given trait.

The study of quantitative genetics is considered very significant both from the purview of basic sciences and its application potential. Selection pressures which have modified multiple traits over an evolutionary scale of time have invariably acted upon mutations in polygenic and/or multifactorial traits. An understanding of quantitative genetics also enables us to predict the possible phenotype of the offspring if the parental phenotypes are known. This is invaluable in artificial selection and breeding, done to improve the yields of commercially important plant and animal products.

20.1.1 Types of Quantitative Traits

Quantitative traits can be broadly classified into three types:

1. *Meristic traits* are traits that can be quantified and exhibit a phenotypic range, but the numerical value associated with each phenotypic class is a whole number. For example, the number of children borne by a female of any species, or the number of seeds in a pod, can only be whole numbers. The number of children borne cannot be 4.6 or 5.3. Similarly, the number of seeds in a pod cannot be a fraction. Other examples include the number of eggs laid by a hen and the number of bristles on the thorax of a fruit fly.
2. *Threshold traits* resemble qualitative traits superficially. The phenotype is either present or absent for such traits. However, the underlying mechanism determining whether a phenotype will be expressed or not is polygenic and/or multifactorial. For example, the symptoms of a given disease may only be expressed if a certain minimum number of mutant alleles are present in an individual (Fig. 20.1). The minimum number of mutant alleles required is the threshold; however, the susceptibility to the disease increases progressively with the increase in the number of mutant alleles present. Sometimes, the susceptibility

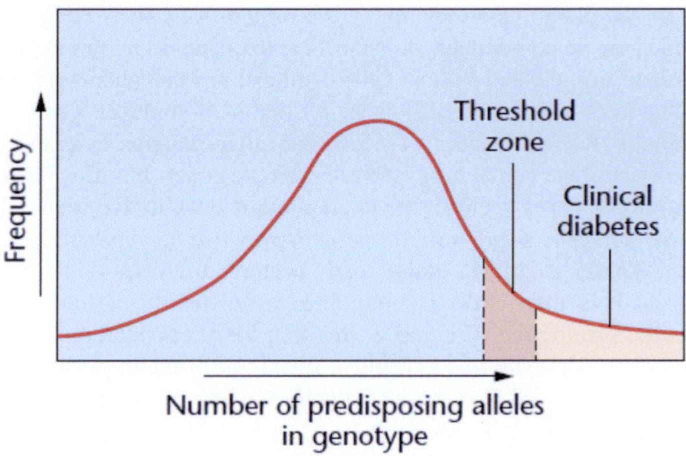

Fig. 20.1 Diabetes as a threshold trait: The phenotype of the disease is not expressed until a threshold number of predisposing alleles is not reached. The *Y*-axis plots the frequency or number of individuals who carry these predisposing alleles in the human population

to a disease might be the output of an interplay between genetic and environmental factors; however, till this output does not reach a threshold value, the disease phenotype will not be expressed. Such a mechanism is generally agreed to regulate the phenotypic expression of type II diabetes.

3. *Continuous variations* are characterized by a potentially infinite number of numerical values that can be assigned to the trait of interest. For example, the variation in height in a population of students can range from a lower extreme to an upper extreme value. Within this range of extreme values, one can observe an infinite number of variations in height. Two people having the same height up to 1 decimal point may vary in height if it is calculated up to 2 decimal points, hence the expression of infinite variation.

20.1.2 Multiple Gene Hypothesis (or the Polygene Hypothesis)

The study of quantitative traits is a pre-Mendelian effort. Famous statisticians F. Galton and K. Pearson had already shown that quantitative traits like height and intelligence are heritable. Due to the pre-Mendelian nature of that time, they were at a loss to explain how the underlying system of inheritance functioned. The rediscovery of Mendel's work resulted in a heated debate over the application of Mendelian principles to explain the inheritance of quantitative traits and the phenotypic variation seen in them. Biometricians used statistical techniques to study quantitative traits and staunchly opposed the idea that single genes (of the nature controlling Mendelian traits) can explain the inheritance of quantitative traits. On the other hand, the Mendelians strongly believed that genes ultimately regulate the inheritance and expression of quantitative traits. Experiments conducted by Wilhelm

Johannsen on the phenotypic variation in the weight of beans (*Phaseolus vulgaris*) indicated that genetic components do contribute to quantitative trait expression. The mathematician George Uduy Yule in 1906 demonstrated that phenotypic variation in a quantifiable trait can be derived from a system of multiple genes interacting together. Finally, Ronald Fisher in 1918 showed that variation in quantitative traits is an end product of the cumulative effect of multiple genes. It is now broadly agreed upon that multiple genes regulate the expression of quantitative traits. This is also known as the *polygene hypothesis* or *the multiple gene hypothesis* for quantitative inheritance. These multiple genes can occupy multiple loci on different chromosomes. It is their collective output in a population of individuals that is reflected as the phenotypic variation of that trait in that population. The collective output can be an aggregate of the additive effects of different alleles belonging to different genes. It can also be a result of dominance interactions between alleles of a given gene and epistatic interactions between genes. The additive effects of genes are of great value as it makes the trait manipulable and therefore amenable to artificial selection. We will study about these phenomena in the coming sections.

20.1.3 Variation in a Quantitative Trait and the Number of Genetic Loci Involved

Variation in the phenotype of a trait can be caused either due to genetic or environmental factors. One of the main genetic factors contributing to phenotypic variation is additive. It simply means that for a polygenic quantitative trait, the alleles present in each of the participating loci contribute to a certain definite amount to the final phenotype. In the case of a coloured trait, each allele might contribute to a certain degree of pigmentation. Different F2 phenotypic ratios can be obtained based on how many number of genetic loci regulate a trait and by the nature of the alleles in each of those genetic loci.

Trait controlled by a single genetic locus: To understand quantitative variation controlled by a single gene, let us consider the hypothetical example of the number of heads on a sea monster. Let us assume that there are two varieties of sea monsters: one having eight heads and the other having two heads. Let the eight-headed sea monster be assigned the genotype AA, wherein each allele "A" contributes to the formation of four heads. The two-headed monster has the genotype aa, wherein each allele "a" contributes to the formation of one head. All the F1 have the genotype Aa and hence have five heads (4 + 1). Selfing of the F1 generates a phenotypic ratio of 1: 2:1 of an eight-headed, five-headed and two-headed sea monster, respectively (Fig. 20.2a). Interestingly, the phenotypic class with the maximum representation in the F2 population (two sea monsters with five heads in this case) has the same phenotype as all the F1. However, unlike the F1, the F2 also displays a variation around this mean phenotype.

Trait controlled by two genetic loci: Let us take the hypothetical example of the number of dragon wings to understand phenotypic variation when a trait is controlled by two genes. Let us assume that the alleles "A" and "B" contribute to one

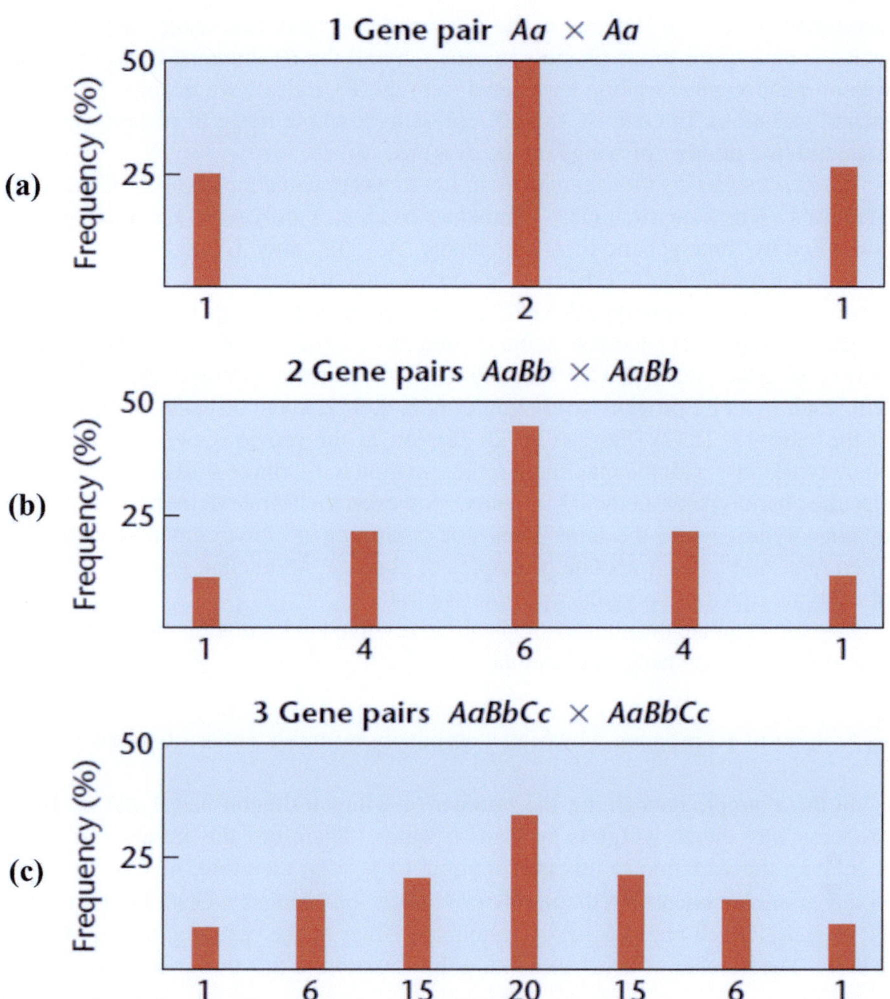

Fig. 20.2 The number of phenotypic classes in relation to the number of underlying genes governing that trait additively: (**a**) Three distinct phenotypic classes are observed when a quantitative trait is regulated additively by two alleles of one genetic locus. (**b**) Five distinct phenotypic classes are observed when a quantitative trait is regulated additively by four alleles of two genes. (**c**) Up to seven distinct phenotypic classes are observed when a quantitative trait is regulated additively by six alleles spread across three genetic loci

wing each, while the alleles "a" and "b" do not contribute to the development of a wing at all. One parent, having the genotype AABB, has four wings and the other parent having the genotype aabb is wingless. Their progeny (F1) will have the genotype AaBb, and hence all of them will have two wings. Selfing of F1 generates five distinct phenotypic classes in the F2 population, with number of wings ranging from 4 to 0. The F2 generation exhibits a phenotypic ratio of 1:4:6:4:1 (Fig. 20.2b).

Maximum number of dragons in the F2 generation have two wings (six dragons) which is equivalent to the phenotype shown by all the F1 dragons. However, there was no phenotypic variation associated with the F1 trait (if we disregard environmental variation). In contrast, the F2 represents a whole range of phenotypes with respect to the number of wings on the dragons.

Trait controlled by three genetic loci: Let us assume that the number of snakes on Medusa's (Greek mythological character) head is a polygenic quantitative trait controlled by three genetic loci. The alleles "A", "B", and "C" code for one snake each, while the alleles "a", "b", and "c" do not code for any snakes. The mating of a Medusa having the genotype AABBCC (having six snakes on its head) with another Medusa having the genotype aabbcc (snakeless head) results in an F1 progeny having the genotype AaBbCc (three snakes on their head). Selfing of the F1 progeny will result in an F2 progeny consisting of 6, 5, 4, 3, 2, 1 and 0 snakes on their heads in the ratio 1:6:15:20:15:6:1 (Fig. 20.2c). As in the previous two examples, the phenotypic class with the maximum representation in F2 (three snakes in the head) is also the phenotype of all the F1 progeny. However, the former demonstrates a range of phenotypes, while the latter does not (assuming no environmental variation). Therefore, one can extrapolate principles of classical Mendelian genetics to derive phenotypic ratios of polygenic quantitative traits.

One can estimate the number of genes (n) involved in regulating the variation of a phenotypic trait by using the formula:

$$\mathbf{\frac{1}{4^n}} = \text{Ratio of F2 progeny which phenotypically resemble either of the two parents}$$

In the example concerning the number of wings a dragon has, 1 out of 16 F2 progeny was either wingless or had 4 wings. Therefore the number of genes regulating the variation in this trait is equal to $\frac{1}{4^n} = \frac{1}{16}$; therefore, $n = 2$. Similarly, in the example concerning the number of snakes on Medusa's head, the number of F2 progeny which phenotypically resembled either of the two original parents was 1 in 64. The number of genes regulating the variation in this trait is equal to $\frac{1}{4^n} = \frac{1}{64}$; therefore, $n = 3$. It is important to bear in mind that this formula assumes that the parental phenotypes represent the opposite extremes of the phenotypic range for the trait under consideration.

The number of phenotypic classes that are generated in the F2 generation by a trait controlled by n number of genes is provided by the formula $2n + 1$. This means that a trait controlled by two genes will generate five [= (2 × 2) +1] distinctly observable phenotypic classes in the F2 generation. Similarly, a trait controlled by three genes will generate seven [= (2 × 3) +1] distinct phenotypic classes in the F2 generation. This is also what we observe in the examples given above.

20.1.4 Examples of Polygenic Inheritance

1. *Kernel colour in wheat*: In 1909, Nielsen and Ehle provided one of the earliest proofs of the polygene hypothesis for a quantitative trait. They studied the colour of kernels (seeds) in wheat. They selected a variety which produced purple (deep dark-red) kernels and crossed it with a variety of wheat which produced white kernels. The F1 produced kernels of an intermediate phenotype of red colour. Selfing of the F1 generation generated five distinct phenotypic classes of colours in F2 wheat kernels: purple (deep dark-red), dark red, red, light red and white in the phenotypic ratio 1:4:6:4:1. This phenotypic ratio can be explained if we assume that the trait is regulated by two genetic loci, each having two alleles. One allele at each locus, "A" and "B", contributes to red pigment formation, whereas the other two alleles, "a" and "b", do not synthesize any pigment. Applying the principle of additive genetic contribution, we can derive all the observed phenotypic classes in the experiment. It was later found that kernel colour in wheat is actually controlled by three genetic loci. The third loci had had the same set of alleles (CC or cc) in both the varieties in the previous experiment; therefore, its contribution went unnoticed. Crosses carried out between two wheat strains that differ in kernel colour due to differences in allele identity at all the three loci resulted in a F2 phenotypic ratio of 1:6:15:20:15:6:1 (Fig. 20.3). This represented seven phenotypic classes of kernel colour ranging from purple (deep dark-red) to successively lighter shades of red and finally white. This experiment also demonstrated how simple Mendelian nature of genetic characteristics can successfully explain polygenic quantitative inheritance.
2. *Ear length in maize*: In 1913, while working on the ear length in maize (*Zea mays*), Emerson and East provided an example of continuous variation. They generated two parental strains, each exhibiting an extreme phenotype of ear length. The Black Mexican variety of maize had a mean ear length of 16.8 cm, while the Tom Thumb popcorn variety had a mean length of only 6.6 cm. The two varieties were obtained after generations of inbreeding. In crossing the above two parental varieties, F1 plants were obtained exhibiting a relatively intermediate ear length of 12.1 cm. The F1 were considered to be all homogenously heterozygous at the gene loci controlling ear length as the parents used were purebreds and hence assumed to be homogeneously homozygous at all the genetic loci. The F2 plants produced as a result of selfing F1 exhibited a mean ear length of 12.9 cm and a large phenotypic variation. Even amongst 646 plants analysed, the extremely long or short ear length parental phenotype was not observed. This suggests that ear length in maize is controlled by five or more genes.
3. *Skin colour in humans*: Skin colour in humans is considered to be a polygenic trait exhibiting continuous variation. Skin colour depends on the deepness of pigmentation provided by the pigment melanin and is a function of both genetic components and environmental factors.

Studies have implicated two to six loci contributing to this trait; however, it is generally accepted that three to four loci controlling this trait adequately explains

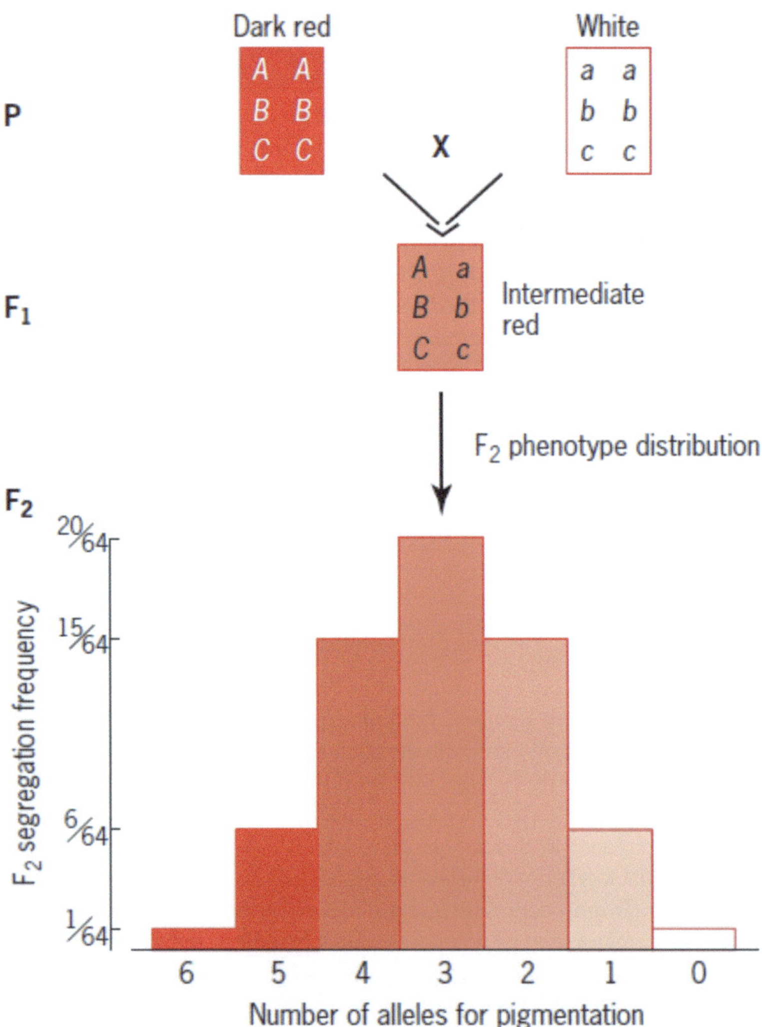

Fig. 20.3 Variation observed in kernel colour in wheat: Assuming three genes are involved in determination of kernel colour in wheat, each of the alleles contributing to the red pigmentation of the kernel has been denoted as an upper-case letter (A, B, C). The alleles not contributing to kernel pigmentation have been denoted in lower-case letters (a, b, c). A gradation in kernel colour is observed in the F2 generation in the ratio 1:6:15:20:15:6:1, as can be seen on the *Y*-axis

the observed phenotypic variation in skin colour. Assuming that three pairs of genes (A/a, B/b, C/c) control the degree of skin pigmentation, with the dominant alleles "A", "B", and "C" contributing to melanin synthesis and the alleles "a", "b", and "c" not coding for the pigment, we can arrive at a situation where we observe seven different F2 phenotypic classes in the ratio 1:6:15:20:15:6:1, wherein 1/64 of the F2 progeny will be black (or white) and the remaining phenotypes will exhibit skin colour darker than white but lighter than black. The parental genotypes for the above cross were AABBCC (black) and aabbcc (white), and the F1 were brown with the genotype AbBbCc (intermediate phenotype). The F2 generation will exhibit a fine gradation of skin tones, wherein the larger the number of dominant alleles the child inherits, the darker he/she will be. The F2 generation will show continuous variation, and it will be tough to demarcate distinct phenotypic classes of skin colour amongst them. From an evolutionary perspective, populations living close to the equator synthesize more melanin than those in the temperate regions in order to absorb the excess UV radiation in such areas and hence prevent its harmful effects (such as cancer and excess vitamin D synthesis which is toxic for the body).

20.1.5 Polygenic Traits and Oligogenic Traits

Polygenic traits are phenotypic traits that are regulated by multiple genes in the genome of the individual. Polygenic traits are quantitative traits that possess a magnitude and a range and display a number of overlapping phenotypic classes. Oligogenic traits are phenotypic traits that are controlled by one to a few genes and are usually qualitative in nature. Mendelian traits are considered to be oligogenic.

There are multiple features which differentiate polygenic and oligogenic traits:

1. Discontinuous variation is seen in case of qualitative traits, while continuous variation is observed in quantitative traits. Distinct phenotypic classes are often hard to discern in case of quantitative characteristics due to overlap of phenotypic expression.
2. Most quantitative traits are polygenic with each gene (and the alleles within those genes) having a small effect on the overall phenotype. Therefore, they are known as *minor genes*. Qualitative traits on the other hand are regulated by one or a few genes, each having a large and easily detectable contribution to the phenotypic trait. Therefore, they are known as *major genes*.
3. Quantitative characters are usually the output of additive gene action and in some cases are regulated by dominance and epistatic gene interactions. Qualitative traits are usually governed by non-additive gene interactions such as dominance-recessive relations and epistasis.
4. Environmental factors can contribute heavily to the phenotypic variation seen in quantitative traits. The effect of environmental factors on qualitative traits is considered to be minimal. Because quantitative traits are susceptible to environmental factors, they are relatively less stable than qualitative traits.
5. Numerical values can usually be assigned to the magnitude or intensity of quantitative traits. Qualitative traits are usually studied in regard to features such as overall colour and shape of the trait.
6. *Transgression* is the appearance of a phenotype which is higher or lower than either of the two parental phenotypes. This happens when the parental phenotypes to begin with were intermediates within a range of possible phenotypic classes. Transgression is not seen for qualitative traits.

7. As quantitative traits are susceptible to environmental influence, their heritability is much lesser than qualitative traits, which are predominantly dependent on the genotype of the individual.
8. Qualitative traits are mostly analysed using segregation ratios and applying chi-square tests to determine the statistical significance between observed and expected values. Analysis of quantitative values utilizes the statistical concepts of variance, standard deviation, correlation and regression. These concepts are used over and above the calculation of segregation ratios for quantitative traits in a population.

There are certain features which are common to oligogenic (qualitative) traits and polygenic (quantitative) traits:

1. Both these kinds of traits are regulated by genes in chromosomes.
2. Definite F2 generation segregation ratios are observed for both qualitative and quantitative traits.
3. Qualitative characters also show variation due to environmental factors, albeit to a much lesser degree than quantitative traits.
4. Dominance and epistatic genetic interactions contribute to both qualitative and quantitative trait manifestation.
5. Polygenes also exhibit linkage like oligogenes. Linkage can also be observed between a polygene and an oligogene.

20.2 Statistical Analysis of Quantitative Traits

Quantitative traits exhibit continuous variation. Analysis of these kinds of traits requires statistical descriptions of the phenotype in a population. This differs from the simple segregation ratio analysis of classical Mendelian traits.

20.2.1 Frequency Distributions, Samples and Population

Individuals in a group display a phenotypic range for a given trait. The magnitude or intensity of the phenotype in different individuals tends to vary. In order to summarize such a trend, geneticists often construct *frequency distribution* graphs. Such graphs depict the range of magnitudes in an increasing (or decreasing) order on the *X*-axis and the number of individuals exhibiting a given magnitude of the trait on the *Y*-axis. The latter is also known as the frequency of individuals exhibiting a particular trait. Connecting all the points on a frequency distribution graphs forms a curve. For most quant traits, this curve is bell-shaped and is called a normal distribution (Fig. 20.4). The curve might even have two peaks (instead of the characteristic single peak of a normal distribution), in which case the curve represents a bimodal distribution. Different kinds of curves can be generated depending on the nature of the acquired data; however, the normal distribution is the most common.

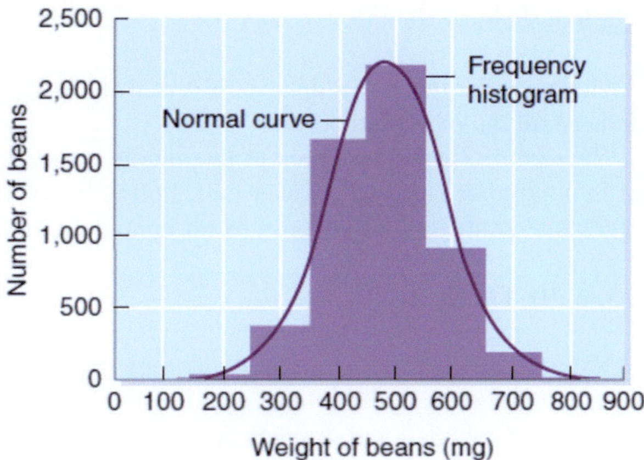

Fig. 20.4 A representative example of a normal distribution: The frequency distribution graph depicts the frequency (or number) of beans over a range of weights. A normal curve fits the frequency histogram

Data is acquired from the individuals exhibiting the phenotype of interest. The *population* consists of all the individuals who exhibit the phenotype. It is, however, impractical and tedious to acquire data from an entire population. For most purposes, data is acquired from a subsection of the population, called the *sample*. In order to faithfully depict the phenotypic range, a sample must be accurately representative of the entire population. To achieve this, a sample must be chosen at random from the population and must be large enough to incorporate the phenotypic range usually exhibited by the population. For example, if we wish to study the range of tusk lengths in Indian elephants, then collecting data only from the first five elephants that we come across might not reflect the true phenotypic range. Furthermore, consideration of elephants from only one region of the country might again not be representative of the elephant population of the entire country. Values of traits extracted from the population are called parameters, and values derived from representative samples are called statistics.

20.2.2 Mean, Median, Mode: Measures of Central Tendency

Once the frequency distribution graph for a given trait had been generated, a number of essential quantitative characteristics can be extracted from it. Most of the data points are distributed around a central value, which in a normal distribution tends to have the highest frequency. The computation of this value is known as measuring the central tendency of the data. Central tendencies can be summarized via the mean, median, or the mode of the distribution.

The *mode* is the value which occurs the maximum number of times in a data set. For example, the following is a data set of the number of lizards found in nine random quadrants of a forest:

4, 5, 6, 6, 6, 6, 7, 10, 25.

The mode for the above data set is 6 as it is the value with maximum frequency amongst all the other numbers.

The *median* can be described as the middle value. A median value divides a data set into a higher half and a lower half. Following is the data set of the number of students who failed their genetics exam in nine different colleges:

2, 5, 6, 10, 12, 15, 21, 21, 22.

12 is the median value in the above data set (it does not matter that 21 figures twice in the data set). There are equal number of data points above 12 and below 12, in this case 4 data points each. If there are an even number of total data points, then the average of the two middle data points constitutes the median.

The most useful statistic is the *mean* of the population. It refers to the arithmetic average of all the data points in the distribution. For a sample it is denoted by the symbol $\overline{x}$ and is computed as:

$$\overline{x} = \frac{\sum x}{n}$$

wherein

$\overline{x}$: the mean of the sample data set
$\sum x$: the summation of all the data values in the data set
n: the number of data points in the data set

In the above example,

$$\overline{x} = (2 + 5 + 6 + 10 + 12 + 15 + 21 + 21 + 22)/9 = 12.66$$

20.2.3 Variance and Standard Deviation: Measures of Dispersion

On generating a number of frequency distribution graphs, one soon realizes that two distributions might have the same mean, but the spread of the data points on either side of the mean is unique for each distribution (Fig. 20.5). This spread of data around the mean is referred to as dispersion and is computed using the *variance* (s^2) and *standard deviation* (s) of a given data set. The variance is calculated as:

$$s^2 = \frac{\sum (x_i - \overline{x})^2}{n - 1}$$

wherein

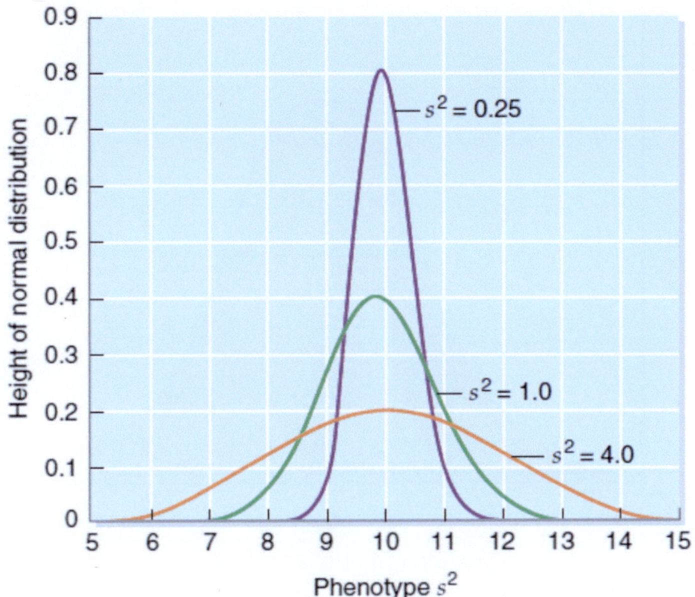

Fig. 20.5 Variance of a distribution: Three different frequency distributions having the same mean value but different spread of the values around the mean. The spread is represented by the variance s^2. The smaller is the value of the variance, the lesser is the spread of the data points around the mean. The larger is the value of the variance, the more is the spread of the data points around the mean

$\sum(x_i - \overline{x})$: the summation of the difference between the *i*th value (a given value) of x and the mean $\overline{x}$ of the distribution.

The reason we square the above value is because $(x_i - \overline{x})$ is always equal to 0. This happens due to the fact that the mean $\overline{x}$ represents the exact mathematical average and therefore the sum of deviations higher than $\overline{x}$ will equal the sum of deviations below $\overline{x}$. The denominator n-1 is used, instead of n. The value n-1 denotes the degree of freedom. This means that if the values of n-1 data points have been provided, then the last value can be derived even if it is unknown.

An important feature of variance is that it is additive. Multiple variance values can be added and/or subtracted normally. This helps in quantitating components of phenotypic variance as we shall soon see.

The problem with using variance is that its units are squared which can be hard to interpret. In order to measure the spread (or dispersion) in the units the original data values were recorded in, we can compute the *standard deviation* (s) of the sample.

$$s = \sqrt{s^2}$$

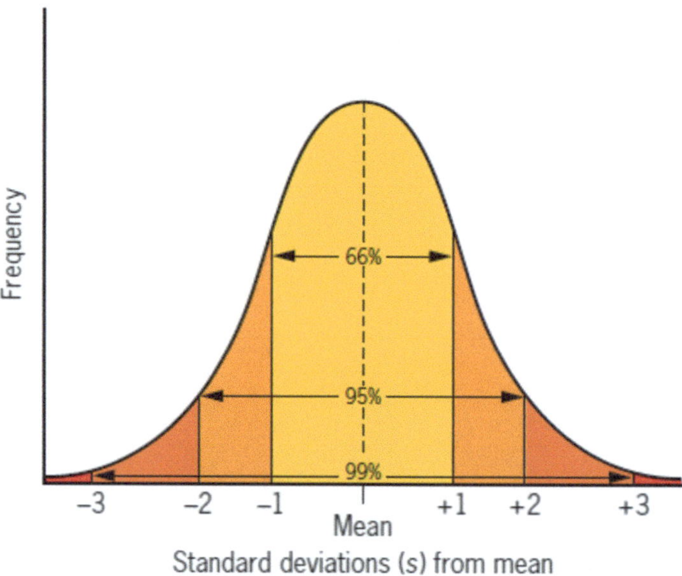

Fig. 20.6 Standard deviation and distribution of measurements: A normal frequency distribution curve showing that 66% of the values in a data set lie within 1 standard deviation around the mean, 95% of the values in a data set lie within 2 standard deviations around the mean, and 99% of the values lie within 3 standard deviations around the mean

A normal distribution can be completely defined by the ($\overline{x} \pm$ standard deviation), wherein 66% of the observed data lies within $\overline{x} \pm 1$ s; 95% of the observed data lies within $\overline{x} \pm 2$ s, and 99% of the observed data lies within $\overline{x} \pm 3$ s (Fig. 20.6). This also means that a data point selected at random from the distribution has a 66% probability of lying in the $\overline{x} \pm 1$ s range of measurements, 95% probability of lying within the $\overline{x} \pm 2$ s range of measurements and 99% probability of lying within the $\overline{x} \pm 3$ s range of measurements.

Sometimes data is gathered for the same phenotype from multiple samples. The mean value for each sample can be computed independently. The *standard error of means* (*SEM*) computes the standard deviation within a distribution of sample means.

$$\text{SEM} = \frac{s}{\sqrt{n}}$$

wherein

s: the standard deviation of the data set.
n: the number of data points within the data set.

A note regarding symbols: $\overline{x}$ is denoted as μ, standard deviation as σ, and variance as σ^2 if the data is collected from a population instead of a sample.

20.2.4 Correlation and Regression: Measures of Relation

Sometimes two or more phenotypic characteristics might vary together. In such cases we can say that the two characteristics are correlated. A change in the magnitude of one characteristic is associated with a change in magnitude of the other. For example, the number of eggs laid by hens is correlated to the size of the eggs produced. A larger clutch of eggs is associated with smaller size of each individual egg. *Correlation* between traits (cov_{xy}) is measured using the correlation coefficient (r). To compute the correlation coefficient, we need to calculate the covariance between the two traits of interest.

$$\mathrm{cov}_{xy} = \frac{\sum (x_i - \overline{x})(y_i - \overline{y})}{n - 1}$$

wherein

x_i: the ith value of x in the data set for trait 1.
$\overline{x}$: the mean value of x in the data set for trait 1.
y_i: the ith value of y in the data set for trait 2.
y: the mean value of y in the data set for trait 2.
n: the number of xy pairs.

Once the covariance is known, we can calculate the correlation coefficient using the following formula:

$$r = \frac{\mathrm{cov}_{xy}}{s_x s_y}$$

wherein

cov_{xy}: the covariance of x and y.
$s_x s_y$: the product of the standard deviation of the data sets for trait 1 and trait 2.

In the above example, the two traits are clutch size and individual egg size.

The correlation coefficient (r) can range from -1 to +1. A positive value of r denotes that an increase in the magnitude of one trait is associated with a concomitant increase in the value of the other. A negative value of r denotes that an increase in the magnitude of one trait is associated with a decrease in the magnitude of the other. The absolute value of r denotes the strength of the association. A correlation coefficient nearing either +1 or -1 means that a change in the magnitude of one trait is nearly always associated with a change in the magnitude of

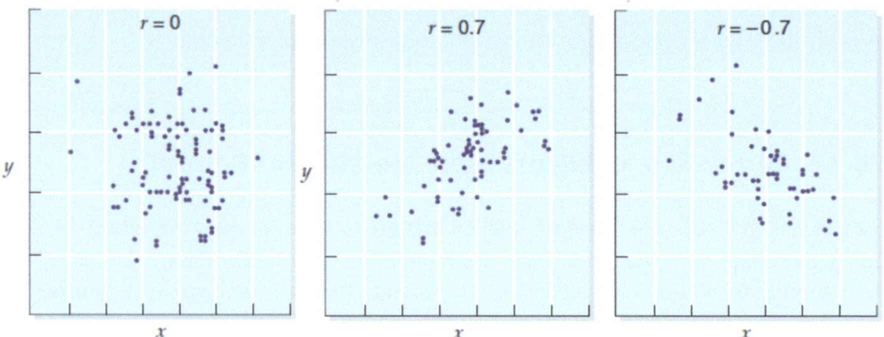

Fig. 20.7 Data points plotted to show correlation between x and y variables: The leftmost graph shows a random scattering of points with a $r = 0$; therefore, variations in magnitudes of the two variables are not associated with each other. A $r = 0.7$ is a strong positive correlation wherein an increase in magnitude of the x variable is associated with an increase of magnitude of the y variable. The opposite to the previous trend is a $r = -0.7$ wherein an increase in magnitude of an x variable is accompanied by a decrease in magnitude of the y variable

the other trait. A correlation coefficient near 0 means that either the association between the two traits is very weak or there is no association between the change in magnitudes of the two traits under consideration (Fig. 20.7). It is important to remember that an association does not automatically imply a cause-effect relation between the changes. It only means that a change in one trait is associated with a change in the other.

Regression is a type of statistical prediction wherein the value of the magnitude of one trait can be computed if the value of the magnitude of the other trait is provided. This plays an important role in breeding experiments as one can predict the offspring characteristics from the parental traits.

Regression can be calculated by plotting a graph between values on the *X*-axis (reflecting the different magnitudes of trait (1) and values on the *Y*-axis (reflecting the corresponding values of trait (2). For example, the values on the *X*-axis can represent the average wing length of parents in *Drosophila*, while values on the *Y*-axis represent the wing length of the corresponding offspring (Fig. 20.8). The line that best fits all the points on the graph is called the *regression line* and is represented by the equation:

$$y = a + bx$$

wherein

a: the value of y when $x = 0$ (the y-intercept of the line).

b: the *regression coefficient* or the slope of the line. It denotes the change in magnitude of y per unit change in magnitude of x.

Also,

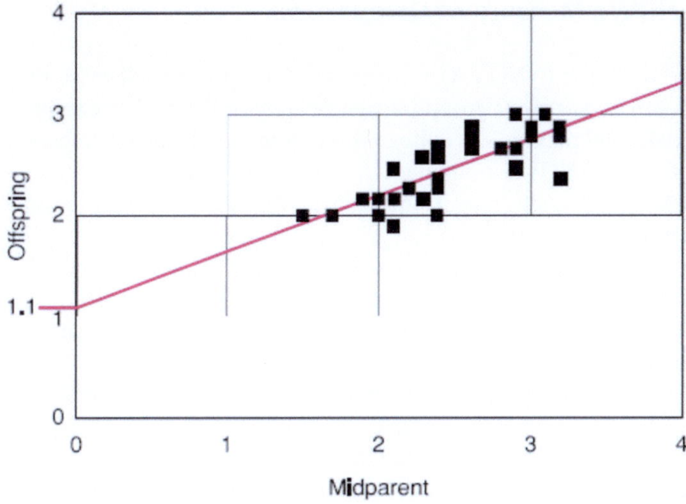

Fig. 20.8 Plotting of a regression graph: A regression line drawn as the best fit for a data set depicting the correlation between wing lengths (in mm.) of offspring *Drosophila* and the corresponding mid-parent wing length values. Mid-parent refers to the average wing length of both parents. 1.1 is the y intercept of the line

$$b = \frac{\text{cov}_{xy}}{s_x^2}$$

wherein

cov_{xy}: the covariance between values of trait 1 (x) and trait 2 (y).
s_x^2: the variance of values of trait 1.

Once the value of b has been calculated, the value of a can be derived using the following formula:

$$a = \overline{y} - b\overline{x}$$

wherein

$\overline{y}$: the mean value for the data set of trait 2.
$\overline{x}$: the mean value for the data set of trait 1.

Once the values of a and b are known, then for any given value of x, the value of y can be computed.

20.2.5 ANOVA: Measure of Comparison

Analysis of variance (ANOVA) is often used to confirm whether the computed means between two or more groups is significantly different. For example, the mean height of basketball players in a college is 6 ft. 4 in., while the mean height of cricket players is 6 feet. ANOVA computes the probability that the difference in means between the two samples arises from chance. The analysis might provide a probability value of less than 0.02 ($p < 0.02$), which means that there is less than 2% probability that the difference in mean heights between the basketball and cricket players results from random chance. One can then infer that the difference in heights between the players of the two sports is a consequence of definite factors, which can be both genetic and environmental.

ANOVA can also be used to calculate what proportion of a phenotypic variance is genetic and how much of the variance is environmental. In the above example, ANOVA has been used to determine whether the means of two groups are statistically significant; however, it can also be used to determine statistical significance between multiple means derived from multiple groups.

20.3 Components of Phenotypic Variance

Variation in any given phenotype exists in the parental generation, the F1 generation and the F2 generation. This variation is represented as the standard deviation from the mean value for the trait. Variance is calculated as the square of this standard deviation. Fisher in 1918 proceeded to dissect this phenotypic variance and deduce the various contributing factors to it. The *phenotypic variance* (V_p) can foremost be divided into variance (or variation) originating due to *genetic differences* between individuals (V_g) in the sample or population and variance (or variation) arising due to *differences in the environment* that different individuals in a sample or population are exposed to (V_e).

Therefore,

$$V_p = V_g + V_e$$

wherein

V_p: the total phenotypic variance.
V_g: the total genotypic variance.
V_e: the total environmental variance.

20.3.1 Genetic Contribution to Phenotypic Variance

The total genetic variance, V_g, can be further divided into three main sub-categories—*additive variance* (V_a), *dominance variance* (V_d) and *epistatic variance* (V_{epi}).

Additive genetic variance is observed when different alleles contribute a definite amount or quanta to the final magnitude of the phenotype. For example, for a given trait controlled by the alleles "A" and "a" belonging to the same gene, let us assume that the allele "A" contributes 10 units to the phenotype of interest, while the allele "a" contributes 2 units to the trait of interest. In such a scenario, the genotype AA will express 20 units of the trait, Aa will express 12 units of the trait, and aa will express only 4 units of the trait. The reader will recognize that the examples given so far in this chapter all follow the additive model of genetic contribution to the phenotype.

The dominance component of genetic variance is what most of the classical Mendelian traits display. In such cases the phenotype depends on the identity of the two alleles that make up the gene. Let us assume that a given hypothetical gene is composed of two alleles, "A" and "a". Different dominance relations can be at play between these two alleles. In a situation where the expression of the allele "A" completely masks the expression of the allele "a", the heterozygotes Aa will completely resemble the homozygotes AA. This is called *complete dominance*. It is to be appreciated that the phenotypic variation herein is not caused by the quantitative contributions of each individual allele. Other forms of dominance include incomplete dominance and codominance which have been explained in detail in the previous chapters. Sometimes the heterozygote Aa will have a phenotypic value that is larger than the dominant homozygote (AA) or lower than the recessive homozygote (aa). The former is called *overdominance*, and the latter is termed *underdominance*. Depending on the fitness value of the genotype, the heterozygous condition may have a higher or lower frequency of occurrence in a population. For example, in cases of sickle-cell anaemia, the heterozygous condition is known to confer resistance to malaria; therefore, in regions prone to malaria outbreaks, heterozygotes will be seen more in number in comparison to regions where malaria infections are uncommon. Overdominance or underdominance may also reduce fitness values. If we assume the wingspan in butterflies to be regulated by dominance relation between alleles, then a larger wingspan in heterozygotes (Aa) as compared to the dominant homozygotes (AA) would be exemplary of overdominance. The larger wingspan might be burdensome to the heterozygote butterfly and might impede flight speed, thereby increasing its chances of being predated upon. Similarly, a smaller wingspan in heterozygotes (Aa) as compared to the observed wingspan in recessive homozygotes (aa) would be exemplary of underdominance. This might again reduce flight efficiency due to the reduced size of the wings and increase its chances of being captured by a predator. In such populations the homozygous butterflies (AA and aa) will be higher in number than the heterozygote butterflies (Aa). *Partial dominance* is seen when the heterozygote has an intermediate phenotype, but the phenotype expressed resembles the dominant trait relatively more than the recessive trait.

Both additive and dominance genetic variance represents variations originating from within a single genetic locus. However, polygenic quantitative traits are controlled by multiple genes. This introduces the possibility of the final phenotypic variance being a result of genic interactions. This is called epistatic variance. Epistatic variance can be further partitioned into *additive x additive*, *dominance x dominance* and *additive x dominance*. The additive x additive interaction summarizes the contribution to phenotypic variance by two genetic loci, each regulating the trait of interest in an additive manner. These loci are of immense interest to plant and animal breeders because they allow for easy prediction of phenotype in the F1 and F2 generations if the parental phenotype is known. The dominance x dominance interactions summarize the interaction between two genetic loci which regulate phenotypic variance using dominance-recessive relations between the alleles. The additive x dominance genetic interactions summarize the contribution to phenotypic variance by two genetic loci wherein one of the genes is composed of alleles contributing additively to the trait of interest, while the other gene expresses itself on the basis of the principles of dominance relation between the alleles.

By incorporating all the above information into the formula for phenotypic variance, we can generate a more nuanced formula:

$$V_p = V_a + V_d + V_{epi} + V_e$$

wherein

$V_g = V_a + V_d + V_{epi}$.

20.3.2 Environmental Contribution to Phenotypic Variance

Ideally a given genotype should only express a given phenotype. However, general observation shows that a given fixed genotype can express a range of phenotypes, that is, it can display phenotypic variation. The source for such variation lies in the environment the organism is growing in. It is now generally agreed upon that while the genotype of an individual organism decides the range of possible traits that it can display, its environment determines where in that range the organism stands. This specially becomes important when considering the environmental conditions under which one wants to rear economically important domestic animals or grow food crops. Under favourable environmental conditions, a given genotype might double its productivity as compared to its conspecifics being raised in a relatively poor environment. The entire range of phenotypes that a given genotype can express when exposed to all possible environments is called the range of reaction or norm of reaction for that particular genotype.

Most phenotypes are a result of several gene products interacting with each other. Therefore one can imagine the humming of an entire genetic background engine at play to express a single phenotype. One can also assume that over the course of evolution, the optimum expression of a trait has been standardized by that organism. This means that the permutation and combination of interaction between genes and gene products in an organism have already been tuned towards the maximization of the probability of its survival and its reproductive success. The visible output of this interaction is the population mean for a given trait in a species. Any deviation from this mean, either due to environmental changes or genetic mutations, is resisted. This property of resistance is called *canalization of development* or *developmental homeostasis*.

Environmental effects are of two kinds—*external environmental effects* and *internal environmental effects*. The external environment encompasses all the sources of variation which originate from outside the body of the organism, while changes in internal environment originate from within the concerned organism. External environmental factors can either be abiotic (or non-living) or biotic (living). Temperature, water content and soil properties are some typical abiotic factors at play, while age and sex constitute an organism's internal environment.

20.3.2.1 External Environment

Temperature is an important environmental factor due to its ability to control rates of chemical reactions. Most phenotypes are dependent on chains of chemical reactions being catalysed by different enzymes coded by different genes. In a mutant of the Chinese primrose plant *Primula sinensis*, the colour of the flower depends on the temperature the plant was exposed to at an early critical time of flower development. Flowers blooming at a relatively higher temperature of 23 °C tend to be red in colour, while flowers blooming at colder temperatures of below 18 °C are usually white in colour. This is a clear example of a given genotype expressing two different phenotypes based on its environment. This is also seen in barley wherein colder temperatures usually cause albinism while warmer climates promote the development of the normal green colour in the plants. In virescent maize the seedlings tend to be yellowish green; however, they turn completely green if grown in higher temperatures. Certain thermosensitive crops such as wheat and maize depend on perceiving a certain degree of temperature before flowering.

Phenotypic variance due to temperature has also been observed in animals. Many genes in *Drosophila*, called heat shock protein genes (hsp genes), are transcriptionally activated when flies are exposed to higher temperatures. An interesting example is the determination of hair colour in the Himalayan strain of rabbits. While the normal coat colour tends to be white, portions of the body which are susceptible to heat loss like the tail, nose, paws and ears tend to be black. This is due to inactivation of an enzyme (tyrosinase) governing melanin content at higher temperatures, leading to the white coat colour. If one shaves off the fur from a region of the body naturally consisting of white fur and artificially reduces the temperature locally in that region, then one will observe the growth of black fur instead of white in that region. The determination of coat colour in Siamese cats also follows similar principles.

Sometimes the external temperature can be lethal for an organism. The water flea *Daphnia* dies if the temperature exceeds 28 °C. Interestingly, a mutant within the species requires higher temperatures for survival and dies at the temperatures normally required for survival by its conspecifics.

A certain mutation in *Drosophila* (tetraptera) causes the halteres to develop as wings. The probability of active expression of the mutant gene is a function of temperature with higher temperatures associated with a higher tendency for the expression of the mutant phenotype. Another mutation (Bar) in *Drosophila* reduces the number of facets in the eyes of the fly. The number of facets in the eyes of this mutant decreases as the temperature increases. In another eye mutant (infrabar), the opposite trend has been observed.

Light is also a crucial factor. The expression of chlorophyll is highly dependent on exposure to sunlight. Colourless kernels in maize can become bright red on exposure to sunlight. Certain photoperiod sensitive plants only flower if they receive light more, or less, than a specific number of hours. This determines the season of flowering for that plant. Some genes, for example, *rbcs* (encoding a small RuBISCO subunit) and *cab1* (encoding chlorophyll binding proteins), have upstream light response elements regulating their transcriptional activity. Freckling in humans is also controlled by the amount of exposure to light. Identical twins freckle to different degrees with the twin working outdoors developing more freckles than the twin staying indoors.

Nutrition affects the phenotype of multiple organisms. Certain mutants of *Drosophila* can grow to attain giant sizes; however, the final size attained is dependent on the amount of food resources available. Under conditions of scarce food supply, these mutants develop into wild-type-sized flies. On the other hand, a sufficient amount of food supply leads to the attainment of giant sizes amongst these flies. Yellow fat (y) mutants in rabbits store yellow-coloured subcutaneous fat if their diet includes xanthophyll-containing green vegetables. In the absence of green vegetables in their diet, these mutants store white-coloured fat.

Auxotrophic bacteria are nutritional mutants that are unable to synthesize a common compound required for its survival. However, they grow normally, like wild-type prototrophs, if the compound they cannot synthesize is added artificially to their nutrient substrate.

Soil acidity regulates the colour of the flowers in hydrangeas. An acid pH causes the flowers to be blue in colour, while a relatively basic pH causes the flowers to be pink or off-white.

Maternal environment provided to the progeny affects its phenotype as well. Consumption of nicotine, drugs, or alcohol can be deleterious to the developing foetus. Incompatibility in the blood Rh factor can cause Rh-negative mother to mount an immune response against their Rh-positive foetus.

Mice homozygous for the allele hair-loss (hl) tend to lose their hl+/hl- heterozygous progeny due to calcium loss. These progeny survive if the mother is hl+/hl + or hl+/hl-. Therefore the genotype of the mother can establish an external environment which can affect the phenotype of the progeny. Seed characteristics such as seed size and protein content in crop plants are also dependent on the genotype of the parent plant.

Moisture content or *humidity* affects the morphology of the abdomen in the abnormal abdomen *Drosophila* mutants. These mutants develop a distorted abdominal appearance due to irregular chitinous bands. The mutation expresses itself predominantly under moist culture conditions, but flies show normal abdominal banding under dry conditions. Also, disease susceptibility as well as the resistance to lodging in plants is often dictated by moisture content.

Other factors such as the presence of symbionts, immune response to parasites, as well as the surrounding population density of conspecifics can cause the same genotypes to be phenotypically different.

20.3.2.2 Internal Environment

Internal factors originate from within the body of the organisms. Many genes are actively transcribed only after the concerned organism attains a certain age. In plants the genes determining floral, fruit and seed characteristics become active once the plant reaches sexual maturity. Similarly, genes regulating milk production in cattle and egg production in poultry are expressed only after attainment of the appropriate age. While genes determining the blood group of a person are active even before his/her birth, genes regulating alkaptonuria, phenylketonuria and rickets are expressed postnatally. Genes responsible for baldness, diabetes and Huntington's disease are predominantly expressed in adults. Therefore *age* is an important internal factor.

Phenotypic expression of traits can vary based on the *sex of the individual.* Recessive mutant alleles for genes exclusively located on the X-chromosome are expressed more often in males than females. This is because even a single mutant copy of the allele is sufficient to express the mutant phenotype in the heterogametic sex, a phenomenon known as *pseudodominance*. Sometimes, though a gene is present in both the male and female of a species, only one of the sexes expresses the trait. For example, the genes controlling milk production and egg laying are active in the females of the species only. Such traits are called *sex-limited characters*. *Sex-influenced characters* are expressed in both the sexes of a given species, but the dominance relation between the alleles of the gene changes, that is, the dominant allele for a trait in males behaves as the recessive allele in females. Examples of this include presence of horns in sheep and crown baldness in humans.

The *presence of a particular substrate* can also decide whether a phenotype will be expressed. Inclusion of tyrosine and phenylalanine in the diet of patients suffering from alkaptonuria causes their urine to turn black on exposure to air. This is because of the build-up of alkapton or homogentisic acid in the urine as a result of the partial breakdown of tyrosine and phenylalanine. A tyrosine- and phenylalanine-deficient diet leads to a significant decrease in the symptoms of the disease. Similarly, the levels of sugar in patients suffering from diabetes can be controlled on consuming a diet restricted in both fats and carbohydrates. This helps in reducing the intense symptoms of the disease.

Several genes are only expressed in places where their products are physiologically required. The human liver, pancreas and lungs will have many different genes which are expressed exclusively in those organs. For example, the genes for insulin are expressed in the pancreas. Similarly, genes responsible for protein storage in seeds are going to be exclusively expressed in seeds. Therefore, the *identity of the tissue* is an important factor determining gene expression. Furthermore, a phenotype is the end result of a number of gene products interacting with each other, and thus the genetic background of the individual finally decides whether a phenotype will be expressed or not.

Therefore the formula for phenotypic variance can be further dissected as:

$$V_p = V_a + V_d + V_{epi} + V_{e-ext} + V_{e-int}$$

wherein

$V_e = V_{e\text{-}ext} + V_{e\text{-}int}$.
$V_{e\text{-}ext}$: the phenotypic variance caused due to external environmental factors.
$V_{e\text{-}int}$: the phenotypic variance caused due to internal environmental factors.

20.3.2.3 General and Special Environmental Variance

Environmental variance can also be described as *general environmental variance* and *special environmental variance*. Special environmental variance measures the intra-individual (or within-individual) variance arising from temporary or localized circumstances. For example, measuring the milk yield from the same cow during different months of the year will aid in estimating special environmental variance. Such effects result in transient changes in the phenotype. General environmental variance originates from permanent or non-localized environmental factors. Such factors may originate from differences in the amount of nutrition available during childhood of an organism which results in permanent effects on the phenotype. Such factors will remain constant between the months and seasons of a year and therefore do not contribute to special environmental variance.

Therefore, another equation for phenotypic variance can be written as:

$$V_p = V_a + V_d + V_{epi} + V_{e-gen} + V_{e-sp}$$

wherein

$V_e = V_{e\text{-}gen} + V_{e\text{-}sp}$.
$V_{e\text{-}gen}$: the phenotypic variance caused due to general environmental factors.
$V_{e\text{-}sp}$: the phenotypic variance caused due to special environmental factors.

20.3.3 Other Factors Contributing to Phenotypic Variance

Having considered the genetic and environmental contributions to the overall phenotypic variance, we must also include variance due to genotype and environmental interactions ($V_{g \ x \ e}$). This factor arises due to the reason that different genotypes might interact uniquely to a given range of environmental conditions. For example, in Fig. 20.9c, the height of both the plants was different to begin with. An increase in

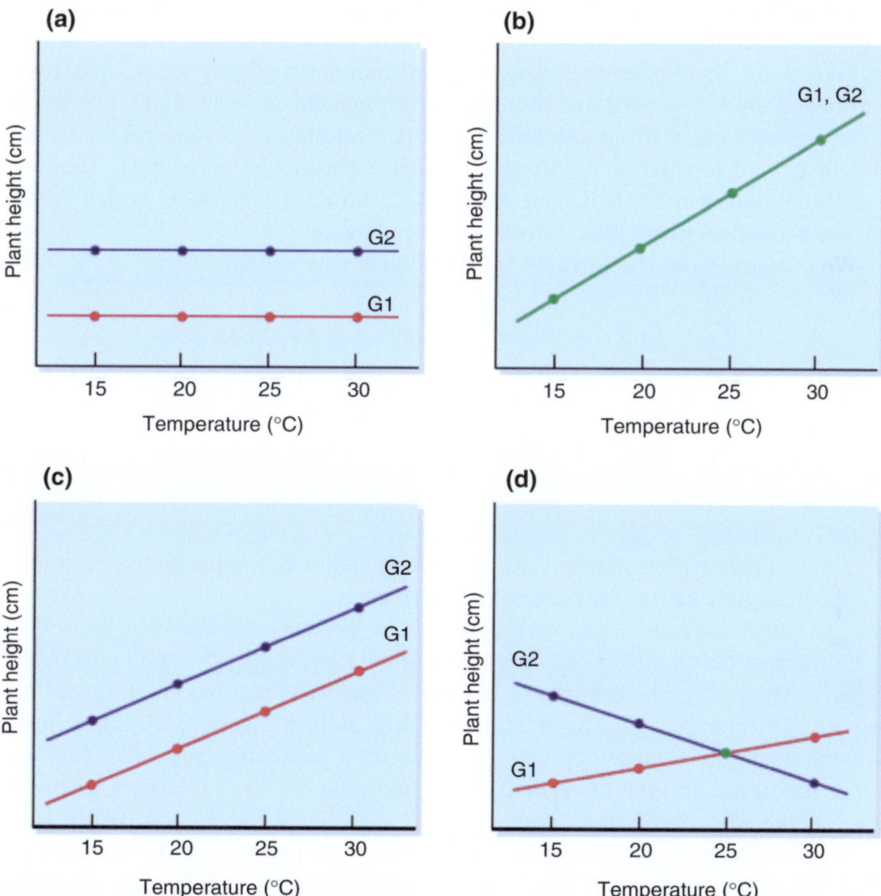

Fig. 20.9 Different sources of phenotypic variation: A hypothetical variation in plant height as a function of temperature has been shown in the graphs above. G1 (blue) and G2 (red) denote the genotypes of the two plants. Green represents an overlap of the phenotypic trend in both the plants. (**a**) Represents a situation where neither plant is affected by temperature and the height is determined genetically only. (**b**) Plant height herein is completely dependent on temperature, and there is no genetic source of height variance. (**c**) Temperature and the individual genotypes of the plants independently and additively determine plant height. (**d**) Each plant genotype interacts uniquely with a given temperature value and produces independent trends of plant height. This is a case of genetic x environment interaction

temperature is correlated with an increase in plant height; however, the difference in height of the two plants remains constant at any given temperature. In this case the genetic and environmental factors are not interacting, and each has an independent effect on the final phenotype. In contrast, in Fig. 20.9d, the height of one plant is negatively correlated with an increase in temperature, while that of the other is positively correlated to the same environmental factor. The genotype of each plant in this case interacts with temperature in a different way. Variances caused due to such interactions are labelled as *genetic-environmental interaction variance* ($V_{g\ x\ e}$). Figure 20.9a, b represents pure genetic and environmental contribution to phenotypic variation, respectively.

Sometimes the underlying reason for phenotypic variance cannot be easily partitioned into a genetic, environmental, or genetic-environmental interaction. Variances originating from unknown causes are referred to as *intangible variation* (V_I). It is hard to discern V_I through planned experimental paradigms. These are usually random and unpredictable fluctuations during development which end up having permanent phenotypic effects on the individual.

We can now write the equation for phenotypic variance as:

$$V_p = V_a + V_d + V_{epi} + V_{e-gen} + V_{e-sp} + V_{g\ x\ e} + V_I$$

Or,

$$V_p = V_a + V_d + V_{epi} + V_{e-ext} + V_{e-int} + V_{g\ x\ e} + V_I$$

wherein

$V_{g\ x\ e}$: the phenotypic variance caused due to genetic-environmental interactions.
V_I: the intangible causes for phenotypic variance.

Some geneticists include another factor while considering the source of phenotypic variance. It is the covariance between a genotype and the environment it is exposed to. For example, for a farmer selling milk, the cows yielding a higher volume of milk are more important than the ones producing less milk. This may cause the farmer to give more food (hence more nutrition) to the cows producing more milk. Also, the farmer might simultaneously give lesser food (hence lesser nutrition) to the cows producing less milk, especially if the amount of food is limited. This will result in an even higher yield of milk from cows which were producing more milk to start with and a decrease in milk production from cows which were yielding less milk to start with. In this situation the genotype and environment for an individual covary. This fraction of the contribution to phenotypic variance is called *genetic-environmental covariance*.

20.4 Heritability

The partitioning of the phenotypic variance results in the identification of the multiple contributors to the variation within a given phenotype. For commercial purposes such as artificial breeding, one is often specifically interested in that component of the phenotypic variance which is genetically determined. The proportion of the phenotypic variance in a population that is controlled by genetic factors is termed as *heritability*. The concept of heritability plays an important role in the process of artificial selection.

20.4.1 Types of Heritability

Broad-sense heritability (H^2): The value of H^2 estimates the fraction of phenotypic variance that is regulated by genetic factors. The superscript 2 is there to remind us that heritability is a type of variance.

$$H^2 = \frac{V_g}{V_p}$$

wherein

V_g: the genetic component of phenotypic variance.
V_p: the total phenotypic variance.

Let us try to calculate the H^2 of inter-moulting interval time in a hypothetical snake species. Let us assume that this species has two distinct varieties of snakes, one with an inter-moulting interval time of 7.6 days $\pm$ 1.2 days and the other 23.3 days $\pm$ 2.7 days. The F1 has an intermediate inter-moulting interval time of 15.7 days $\pm$ 3.1 days. The F2 generation exhibits an inter-moulting interval time of 15.4 days $\pm$5.5 days. As expected the standard deviation for the trait has increased considerably in the F2 generation. It is important to remember that the phenotypic variance in the parental and the F1 generation is due to environmental factors as the genotypes do not vary between the individuals in these generations (or at least we assume so). We can therefore average the variances observed in the parental and F1 generation to obtain a mean value of V_e in this species for this trait. Please keep in consideration the fact that the variance is the squared value of the standard deviation.

$V_p = V_g + V_e$
$(5.5)^2$ days^2 $= V_g$ + [{$(1.2)^2 + (2.7)^2 + (3.1)^2$}/3] days^2
30.25 days^2 $= V_g$ + [(1.44 + 7.67 + 9.61)/3] days^2
Therefore, V_g = [30.25 – 6.24] days^2 = 24.01 days^2
Hence, H^2 = 24.01 days^2 / 30.25 days^2 = 0.79

This means that 79% of the variation in inter-moulting interval time in the species of snakes can be attributed to genetic factors.

Table 20.1 Narrow-sense heritability values for some quantitative traits

Trait	h^2 value
Stature in human beings	0.65
Milk yield in dairy cattle	0.35
Litter size in pigs	0.05
Egg production in poultry	0.10
Tail length in mice	0.40
Body size in *drosophila*	0.40

Adapted from *Principles of Genetics* by Snustad and Simmons, 6th edn. (2012)

Genetic contribution to a trait can be due to additive, dominance, or epistatic relations. Of these the most manipulable and therefore of the most interest to plant and animal breeders is the additive component of genetic variance. One is usually not privy to the actual genotype of each individual in a population. Furthermore, for most polygenic traits, the exact genes regulating the trait of interest have not been deciphered; therefore, manipulating dominance and epistatic genetic interactions is very tough. Under such circumstances one relies on the observable phenotype of a trait in the parents and tries to predict the possible phenotypic range of the trait in the offspring. This can be done most faithfully by calculating the additive components of genetic variance for the trait of interest. The fraction of the total phenotypic variance that is regulated by additive components of genetic variance is called as the *narrow-sense heritability* (h^2).

$$h^2 = \frac{V_a}{V_p}$$

wherein

V_a: the additive component of the genetic contribution to phenotypic variance.
V_p: the total phenotypic variance.

The values of H^2 and h^2 range from 0 to 1. A value of H^2 close to 0 means that the genetic contribution to phenotypic variance is nearly non-existent. A H^2 value close to 1 means that the genetic contribution to the phenotypic variance for the given trait is very high. Similarly, a value of h^2 close to 0 means that the additive components of genetic variance do not regulate the phenotypic variance, while a value close to 1 means that the additive components of genetic variance can predominantly account for the variance in the phenotype. The h^2 values for some quantitative traits have been presented in Table 20.1.

Before we proceed further, it is important that we correctly understand how to interpret the values of heritability. The need for this arises from the fact genetic studies in this field have been historically politicized and used to unfairly target certain sections of the population. Please keep the following points in mind while interpreting heritability values:

1. The heritability value calculated does not hold true for an individual. It is an estimate of the genetic contribution to phenotypic variance for a population. For example, a H^2 value of 0.67 for plant height in a population (or sample) of sunflowers does not mean that for each individual sunflower, 67% of its height is regulated by genes. The correct interpretation would be that 67% of the variance in plant height amongst sunflowers can be attributed to genetic factors.
2. A high value of heritability does not mean that the phenotype is regulated by genes only. It simply means that in that particular sample or population, the cause of phenotypic variance was genetic in nature. This might be because all the individuals in the sample or population were reared in near identical environmental conditions, thereby eliminating variations due to environmental factors. Similarly, a very low heritability value does not mean that genes do not regulate that particular phenotype. For example, in a population of lizards that is genetically inbred (and therefore homogenously homozygous), the genotype will nearly be the same in all the individuals. This prevents the genotype from contributing to the phenotypic variance of a given trait like tail length. The variation in tail length in this population of lizards will be only due to environmental causes. This does not mean that genes do not play an important role in determining tail length.
3. Heritability values only allow for an estimation of the fraction of the genetic and environmental contribution to phenotypic variance. It does not tell us anything about the actual genes or environmental factors controlling the trait.
4. Similarity amongst relatives of a family is not to be confused with heritability. Members of a family might resemble each other solely due to spatially and temporally shared environmental factors. Shared resemblance of phenotypic traits amongst family members is called **familiality** and is not the same as heritability.
5. Heritability is not fixed for a trait. Heritability estimates for milk yield in an endemic population of cows in England cannot be extrapolated to breeds of cows in India. Moreover, this estimate will probably be untrue for other breeds within England.
6. If two distinct populations exhibit very high heritability values for the same trait, it does not automatically mean that the trait is predominantly regulated genetically. For example, the estimation of heritability of human height in a developed country and in a developing country might yield very high values. One can assume that individuals in the developed country have access to sufficient food sources and hence are uniformly nourished. In a poor region of a developing country, the access to food might be compromised and hence the population may be uniformly undernourished. In such a situation, the environmental factor of nutrition exerts an equal effect on individuals belonging to a given population and thus does not contribute to the phenotypic variance of height within a given population (one population being from the developed country and the other being from the developing country). In this case the calculated heritability for height will only reflect the genetic component to variance in height. It would be incorrect to deduce from these results that environmental factors like nutrition do not play an important role in determining human height.

20.4.2 Factors Affecting Heritability

Measures of heritability can be affected by the following factors:

1. Genetic diversity of the parental generation decides the degree of phenotypic variation that will be observed in the F2 generation. The more is the allelic variability at the loci amongst the parents, the larger is the number of novel permutation and combination of genotypes at the F2 level and hence greater is the phenotypic variability.
2. As it is impractical to study the entire population for a trait of interest, usually smaller samples are chosen for analysis. However, if the sample is too small, then the measured genetic variance might not faithfully represent the actual population-level value. Larger samples are preferred.
3. Random sampling of a population provides genetic variance values (heritability) closer to the actual population-level values. Biased or non-random sampling (repeatability) is unable to estimate the true value of the same.
4. Manual errors in measurement of values can lead to a deviation in variance values from the true value. Such errors can be minimized by increasing the number of samples chosen for evaluation, increasing the sample size itself, measuring the variance multiple times from a given sample, and conducting collection of data from multiple testing sites instead of just one.
5. The type of heritability being calculated matters. Broad-sense heritability (H^2) will usually have a higher value than narrow-sense heritability (h^2). This is because H^2 estimates also include dominance and epistatic contributions to genetic variance, apart from the additive component.
6. Genetic linkage between two loci controlling a quantitative trait can lead to an overestimation or underestimation of heritability values. Linkage in a cis-arrangement (AB/ab) leads to an overestimation of the dominance and additive components of genetic variance. Linkage in a trans-arrangement (Ab/aB) again results in an overestimation of the dominance component of genetic variance; however, it leads to an underestimation of the additive component.

20.4.3 Measures of Heritability

There are multiple ways of measuring heritability for a given trait. Here we discuss three such methods.

1. *Elimination of one variance component*: The phenotypic variance is a sum of the genetic and environmental components of variance. If we are successfully able to prevent one of the factors from affecting phenotypic variance, then any variation in the trait must have arisen from the other component. For studies in plants, one can reduce $V_e = 0$ by growing them in a greenhouse. Greenhouses can efficiently maintain a uniform temperature and moisture content, thereby precluding any variation in the phenotype due to different plants being exposed to different

intensities of a given environmental factor. Any variation observed for a given trait within a given species in such a greenhouse can be attributed to differences in the genotype between the individual plants. In such a situation, $V_p1 = V_g$. The same species can then be grown outside a greenhouse where the environmental factors are not controlled. The phenotypic variance for the same trait can be calculated; however, this time the variance in the phenotype (V_p2) originates from both the differing genotypes and the environment. We can utilize the Vg value estimated previously and divide it by V_p2 to compute the heritability value for the chosen trait. Needless to say that we are assuming that the genotypes of the plants grown inside and outside the greenhouse are the same. This is not always true and hence can be considered a limitation of the technique.

We can also adopt the alternate approach of reducing $V_g = 0$. The study of variation of a given trait in an inbred line will yield $V_p1 = V_e$ for the simple reason that individuals belonging to an inbred line are assumed to be composed of the same genotype. Therefore any genetic contribution to phenotypic variance is precluded, and all the variation observed in the trait of our choice is because of environmental causes. V_p2 can be calculated by estimating the variance for the same phenotype in a non-inbred line of the same species. Utilizing the previously calculated value of V_e, we can compute V_g (= $V_p2 - V_e$). Heritability can then be estimated by using the formula V_g/V_p2. We do assume that the effects of the environment on both the inbred and non-inbred line are going to be the same. This is however not always the case, especially because the inbred and non-inbred lines are genotypically dissimilar. This is a lacuna in the above technique.

2. *Computation of parent-offspring regression*: In the earlier sections of this chapter, we have examined how a change in the magnitude of one trait can be associated with a change in the magnitude of another trait. This is called correlation. Furthermore we learnt how to calculate the corresponding value on the *Y*-axis for a given value on the *X*-axis. This is called regression. Regression analysis can be used to estimate heritability values for a given trait if that trait is shared between the parents and the offspring. For example, we can plot a graph between the wingspan length of parental *Drosophila* (trait 1 on the *X*-axis) and the corresponding wingspan length of their offspring (trait 2 on the *Y*-axis). As there are two parents for any given offspring, geneticists usually plot the mid-parental value on the *X*-axis. This value reflects the average of the two parental values for the trait. This usage of the mid-parental value also introduces the element of treating the phenotypic variance of the trait as having an additive genetic component. This is because the mid-parent value is an intermediate value of the trait and using this value to compare the trait with the offspring means that we have assumed that there is an additive genetic component to the trait. Remember that only in cases of additive genetics did the F1 express a phenotype which was intermediate with regard to the parental phenotype (this is not true in cases of genetic components of variance derived from dominance and epistatic relations). Thus one can compute narrow-sense heritability values using these graphs. The mathematical derivations for the relation between the slope of a correlation graph

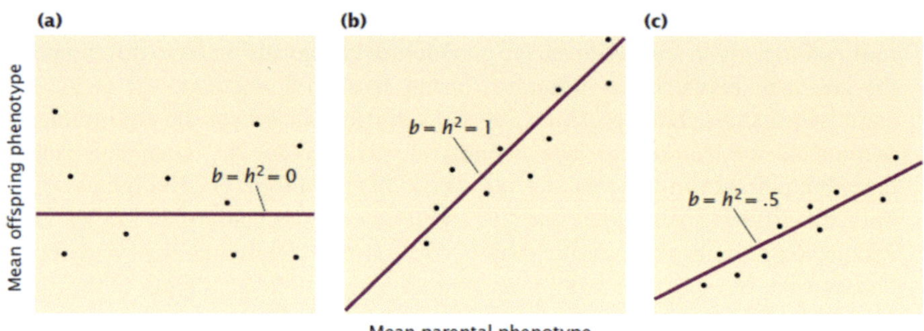

Fig. 20.10 Parent-offspring regression plots: The value of h^2 can be estimated from the value of the slope of the graph having the mean parental value on the X-axis and the mean offspring phenotype values on the Y-axis. (**a**) There is no relation between the parental and offspring values of the phenotype. (**b**) The phenotype values of the offspring are entirely dependent on the phenotype values of the parents. (**c**) The phenotypic range in the offspring is a product of additive genetic, non-additive genetic, and environmental influences

(b) and h^2 are beyond the scope of this chapter; however, it is important to know the following two results of the said derivations:

$$h^2 = b$$

wherein

h^2: the narrow-sense heritability value for the phenotypic trait.

b: the slope of the regression graph.

If the phenotypic values of only a single parent are available for analysis, then the mid-parent values cannot be plotted. In such a case, we modify the above equation.

$$h^2 = 2b$$

wherein

h^2: the narrow-sense heritability value for the phenotypic trait.

$2b$: twice the value of the slope of the regression graph. This is done to compensate for the absence of the phenotypic values for the other parent.

If the absolute value of $b = 1$, then all the genetic variance is derived entirely from additive components. If the absolute value of the slope is smaller than 1 but larger than 0, then the genetic contribution to phenotypic variance is a mixture of additive and non-additive components (Fig. 20.10). If the absolute value of $b = 0$, then there is no contribution of additive genetic components to phenotypic variance; however, this does not rule out dominance and epistatic genetic contribution to the variation observed in the phenotype

3. *Comparison of phenotypic variances for the same trait in individuals with varying degrees of relatedness*: Individuals related to each other are expected to bear a resemblance due to shared genes. The closer the relatedness, the higher is

the proportion of genes shared. That would mean that siblings are genetically more similar than first cousins, who in turn would be more genetically similar than second cousins. It is known that siblings have 50% of their genes in common on an average. Similarly, it is known that half-siblings, who share only one of the parents, have 25% of their genes in common on an average. The fraction of the observed correlation coefficient between two relatives to the expected correlation coefficient between the same relatives for a given phenotypic trait computes the narrow-sense heritability for that trait.

$$h^2 = \frac{r_{\text{obs}}}{r_{\text{exp}}}$$

wherein

h^2: the narrow-sense heritability value for the phenotypic trait.

r_{obs}: the observed value for correlation coefficient between the relatives for the phenotypic trait.

r_{exp}: the expected value for correlation coefficient between the relatives for the phenotypic trait.

The occurrence of monozygotic and dizygotic twins also allows for the calculation of heritability estimates for a given phenotype. Monozygotic twins, or identical twins, share a complete set of genes with each other, that is, they are genetically the same. Dizygotic twins, or fraternal twins, share only 50% of their genes on an average like any two siblings. An estimate of the broad-sense heritability value can be obtained on doubling the difference between the correlation coefficients for a given trait calculated for both monozygotic and dizygotic twins.

$$H^2 = 2\,(r_{MZ} - r_{DZ})$$

wherein

H^2: the broad-sense value for the phenotypic trait.

r_{MZ}: the correlation coefficient for the phenotypic trait amongst monozygotic twins.

r_{DZ}: the correlation coefficient for the phenotypic trait amongst dizygotic twins.

Such estimates, however, should be interpreted with caution as one assumes that the environment shared by monozygotic twins is no more different to the one shared by dizygotic twins. That is not always the case. Monozygotic twins might share a far more similar environment than dizygotic twins as they are often treated very similarly. The above assumption is a limitation of this technique.

A trait that is expressed by both individuals of a twin-pair is considered to be *concordant* if both the twins express it, or neither of them does. A trait is considered to be *discordant* if one of the individuals of the twin-pair expresses it and the other one does not. A comparison of the concordance values for a given trait between monozygotic and dizygotic twins helps in the estimation of the genetic contribution to a given phenotypic trait. If two individuals of a monozygotic pair, who have been reared apart, show very high concordance values for a given trait, then it is quite certain that there is a heavy genetic contribution to the trait. On the other hand, if a large concordance value is observed for a given trait in both monozygotic and dizygotic twins, then there might be environmental factors contributing to the phenotype as well. This is because dizygotic twins only share 50% of their genes, and a high concordance value might mean that the similar phenotype is a result of shared environment too.

20.5 Artificial Selection

Evolution works on the principle of selecting beneficial traits in a population and increasing the frequency of individuals exhibiting that beneficial trait. This kind of selection is called natural selection because it happens in nature as a default process. However, through the process of selective breeding, humans have over the last 10,000 years artificially changed the mean values of the phenotypic traits that are of interest to him. This includes milk yield from cattle, wool yield from sheep, egg production, etc. Most of these traits are of commercial value and artificial selection for these traits has greatly improved standards of living in recent human history. The parameter to be exploited for artificial selection is the phenotypic variance exhibited by the original (non-selected) population. Within the range of phenotypes exhibited by the original population, a few individuals will have relatively superior traits as compared to others within the same population. Artificial selection works if there is a large additive genetic component to the phenotypic variance. Dominance, epistatic, and environmental components of phenotypic variance are not useful for long-term, permanent changes in trait characteristics. Dominance and epistatic components of phenotypic variance are hard to artificially control and manipulate while the environmental components are not transferrable form one place to another.

To further understand artificial selection, let us assume the mean volume of milk produced by a cow in a local herd to be 50 L/week ($\bar{Y}$). In order to increase the milk yield, we select breeding pairs which have an average output of 80 L/week (Y_p). The difference between the milk yield in the general population and the selected parents for breeding is 80–50 = 30 L/week. This difference is known as the *selection differential* (S). After breeding, let us assume that the average milk yield of an offspring is 65 L/week. The difference between the output of the offspring and the original population is known as the *genetic gain*, or the *response to selection* (G). In this case $G = 65–50 = 15$ L/week (Y_o). Considering that the observed increase in average milk yield is due to additive components of phenotypic variance, we can calculate the narrow-sense heritability using the formula:

$$h^2 = (Y_o - \bar{Y})/\left(Y_p - \bar{Y}\right) = G/S$$

wherein

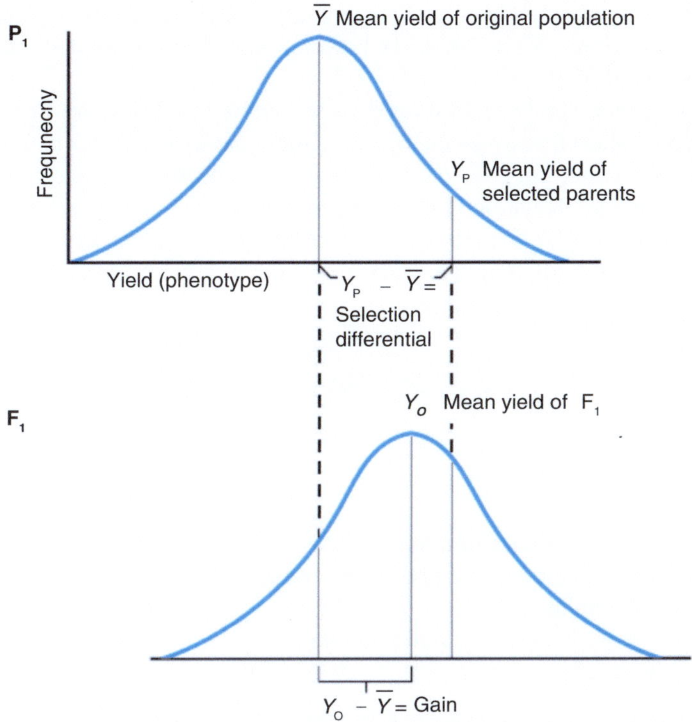

Fig. 20.11 Change in mean value of a trait due to artificial selection: Realized heritability is the genetic gain ($Y_o - \bar{Y}$) divided by the selection differential ($Y_p - \bar{Y}$). The mean of the phenotypic trait has shifted to the left in the F1 generation denoting a positive selection for the trait under consideration

h^2: the narrow-sense heritability value for the phenotypic trait.
$\bar{Y}$: the mean milk yield by a cow in the original population.
Y_p: the mean milk yield by a cow in the population selected for breeding.
Y_o: the mean milk yield by a cow in the offspring population.

In this case $h^2 = 15/30 = 0.5$.

The narrow-sense heritability in the above example is called realized heritability since the value is computed after the breeding is done (Fig. 20.11). It is an estimate of true heritability which encompasses both narrow-sense and broad-sense heritability.

Once the h^2 for a trait is known, then predictions can be made about the change of magnitude to be expected in the offspring generation. G. Clayton et al. studied the number of abdominal bristles in *Drosophila melanogaster*. Using multiple techniques, they calculated the h^2 for the number of abdominal bristles to be 0.52. They selected parents with mean bristle number of 40.6, while the population mean was 35.3. This means the selection differential (S) is 5.3 (40.6–35.3). Therefore the response to selection (or genetic gain, G) $= 0.52 \times 5.3 = 2.8$. This value can be interpreted as the expected increase in magnitude of abdominal bristle number in the offspring generation; 35.3 + 2.8 = 38.1, which turned out to be very close to the observed value of 37.9.

Every phenotypic trait has a variance associated with it. Therefore we see a range of phenotypic values for any given trait. It is important to also factor in this variance for artificial selection. The *selection intensity* (i) is defined as the selection differential per unit standard deviation of the phenotypic variance of the original population. *Genetic advance* (GA) is defined as the improvement in the mean genotypic value of a selected trait in the offspring generation over the original parental population.

$$i = S/\sigma_p;$$

$$GA = i \times {\sigma^2}_p \times H^2 = S \times \sigma_p \times H^2$$

wherein

i: selection intensity.
${\sigma^2}_p$: the phenotypic variance of a trait.
σ_p: the standard deviation of the phenotypic variance.
S: selection differential.
H^2: the broad-sense heritability value of a trait.
GA: genetic advance.

Different populations being reared in different environments can give different values for h^2 for the same trait under study. Therefore, it is prudent to conduct independent research on heritability of a trait for each individual population.

Artificial selection has been practiced for multiple commercially important traits such as milk yield, egg number, egg size, and wool yield. Milk yield can be measured in successive lactations, and several measurements from each individual can be obtained. Usually both the first and the second lactation are measured for milk yield. Variance per yield per lactation can be analysed within the same individual and between individuals. The within individual component is entirely environmental in origin, while the between individual component is partly environmental and partly genetic in origin. A positive correlation has been reported for milk yield in the first and second lactation. The h^2 for milk yield has been estimated to be around 0.35 which means that 35% of the variance in milk yield can be explained by additive genetic variance. Milk yield cannot be measured in males; hence, its breeding value can only be judged by assessing the performance of its female relatives. Milk yield has been reported to decrease over time as a result of inbreeding depression. An increase in milk yield has also been correlated to a decrease in protein and butterfat content in the milk. Milk yielding varieties of cattle include Dutch Holstein – Friesian, Finnish Ayrshire and Holstein Dairy. Indian milk yielding cattle include Sahiwal and Frieswal. Jakhrana is a milk-yielding goat variety in India. Genes regulating milk yield have been localized to chromosomes 6, 14, 17 and 20 (Arranz et al. 1998 and Zhang et al. 1998). The *DGAT1* gene on chromosome

14 has been implicated in regulating milk yield in cattle (Grisart et al. 2002). Genes implicated in milk yield and composition in dairy sheep have been localized to chromosomes 1, 2, 3, 20, 23 and 25 (Gutiérrez-Gil et al. 2009).

The number of eggs laid is known as the clutch size. Egg size and egg number are both considered to be quantitative traits. It has been studied both in poultry and *Drosophila*. The h^2 estimate for egg weight in poultry is around 0.50. The h^2 estimate for egg number in poultry and *Drosophila* is 0.10 and 0.20, respectively. Egg number and egg size variation cannot be studied in males; therefore, their selection potential is calculated by assessing their female relatives (as was noted previously for artificial selection concerning milk yield). Both these traits have been improved over the years using artificial selection. The level of circulating gonadotropin hormone has also been implicated in determining the number of eggs laid by poultry. A positive correlation has been observed between the body weight of the mother and the weight of the eggs she lays, while a negative correlation has been reported between the weight of the eggs and the number of eggs laid. Variation in egg numbers and egg size can be regulated by additive, dominance and epistatic genetic interactions. Genes on chromosomes 2, 4 and 5 have been implicated in regulating variation in egg weight and number in poultry (Wolc et al. 2012). Candidate genes responsible for regulating egg weight include *CECR2*, *MEIS1* and *SPRED2* (Liu et al. 2018). Genes implicated in regulating egg number include *GTF2A1* and *CLSPN*. An SNP mutation on chromosome 5 can cause a phenotypic difference of nearly seven eggs between two homozygous genotypes (Yuan et al. 2015). The breeds on which such studies have been done include the red junglefowl and white leghorn.

Approximately 90% of the world's sheep produce wool. One sheep can produce anywhere between 1 kg to 13 kg of wool annually. Wool yield is usually measured at the first and second shearing. The amount of wool a sheep produces is a function of its breed, genetics, nutrition and shearing interval. The wool length and diameter of wool fibres predominantly determine wool yield. The sex of the lamb can also affect wool yield. Wool yield is of immense economic interest and has been subjected to artificial selection. It is also considered to be a polygenic trait. Breeds of sheep that have been used in an attempt to increase wool yield include Merino, Romney-Marsh and Lincoln. Indian breeds of sheep used for wool production include Muzaffarnagari sheep and Garole sheep. Heritability estimates for wool yield are quite low – approximately 0.15 ± 0.07 for Muzaffarnagari sheep (Sinha and Singh 1997). Heritability estimates for wool yield range from 0.23 to 0.37 for breeds such as Rambouillet and Romnolet (Vesely et al. 1970). Genes regulating wool yield have been localized to chromosomes 3, 4 and 24 of the Merino sheep (Bidinost et al. 2008). Additional genes regulating wool yield have been found on ovine chromosomes 1 and 11 in later studies (Roldan et al. 2010).

Selection can continue till the time there is observable phenotypic variance in the trait of choice. But constant selection for a trait finally leads to the appearance of genotypes which are nearly homozygous for all the genes contributing to the trait. At this point $h^2 = 0$, and there is not much scope to introduce new variations, and the selection process comes to an end (Fig. 20.12). Remember that $h^2 = 0$ does not mean

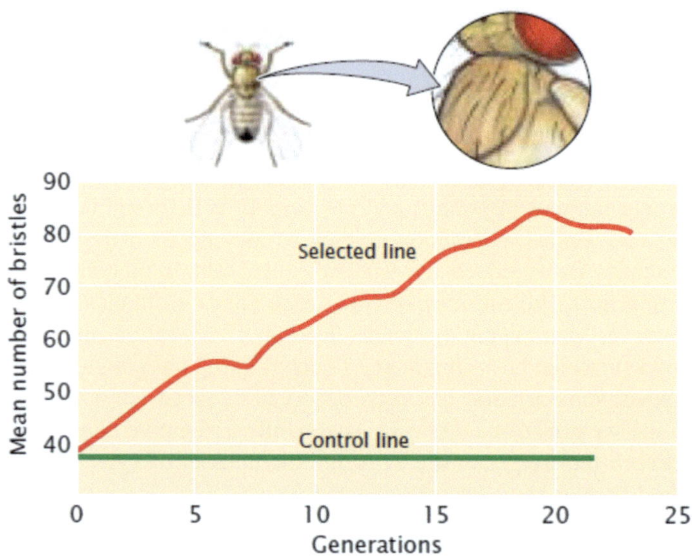

Fig. 20.12 Response to selection of a trait plateaus after a number of generations: In an experiment to increase the number of abdominal bristles in female fruit flies, the response to selection levelled off after 20 generations of selection. The trend for both the selected line (the line which was subjected to selection for the trait) and the control line has been shown

that the phenotypic variance does not have a genetic component. It means that variation due to genetic reasons cannot be introduced anymore. Also, as homozygosity increases in a selected line, detrimental mutations start expressing themselves which leads to a general decrease in yield and vigour. This is known as *in-breeding depression*. Limitations to selection are also due to phenotypic and genetic correlations, which is the following topic.

20.6 Phenotypic and Genotypic Correlations

A connection between two or more things is defined as a correlation. The variations of two phenotypic traits that are correlated influence each other. For example, tall people also tend to weigh more than shorter people. Herein the quantitative traits of height and weight are correlated to each other. Such correlations are called phenotypic correlations. Another example of phenotypic correlation is the observation that fair skin, blond hair and blue eyes are often found in the same individual. Phenotypic correlations can be either due to environmental correlations or genetic correlations. A non-zero correlation coefficient between two traits is seen if they are phenotypically correlated.

If two or more traits are affected by the same environmental factor, then the traits are considered to be *environmentally correlated*. For example, plants that are grown in climates affording enough moisture tend to grow taller and produce more number of seeds in comparison to plants that are made to grow in places of limited water supply. Similarly, plants growing in soil supplemented with fertilizers grow taller and bear more number of flowers in comparison to plants which are reared without any fertilizers. In both cases, a common environmental factor was the cause of correlation between two traits. In the first case, the height of the plant and the number of seeds are phenotypically correlated due to the moisture content in their environment. In the second case, the height of the plant and the number of flowers are correlated due to the use of fertilizers.

Phenotypic correlations can also be due to genetic factors. When two phenotypes are correlated due to underlying genetic causes, it is called a *genetic correlation*. This might happen due to pleiotropy or genetic linkage. *Pleiotropy* is the phenomenon wherein a single gene regulates multiple phenotypes. For example, people who are taller tend to have bigger hands and vice versa. This is due to the fact that in general the size of various body parts is dependent on growth hormones, and there are genes which regulate the amount of growth hormone being secreted by the pituitary. Therefore, a common group of genes is able to affect the size of multiple parts of the human body. Furthermore, genetic linkage is observed when two genes are physically located very close to each other because of which they have a very high tendency of being inherited together generation after generation. Two such genes tend to show genetic correlation if they are controlling two different phenotypic traits.

Genetic correlations can either be positive or negative. *Positive correlation* is seen when the genes that cause an increase in the measure of one quantitative trait also simultaneously increase the measure of another trait. This would also mean that a decrease in the magnitude of one trait is accompanied by a decrease in the other. For example, the genes that control thorax length and wing length in *Drosophila* are common; therefore, an increase in thorax length is accompanied by an increase in wing length. This is also true for the size of a chicken and the mean weight of the eggs it lays. *Negative correlation* is seen when the genes contributing to an increase in the measure of one trait cause a decrease in the measure of another trait. The amount of milk production in cattle is negatively correlated to the percentage of butterfat in the milk. Also, the size of the eggs and the number of eggs laid by a chicken are negatively correlated.

Genetic correlations are important from the standpoint of both natural and artificial selection. This is because a change in one trait can be accompanied by a change in magnitude of another trait (Table 20.2). This becomes a problem because many traits are optimized for the environments we have to survive in. Genetic correlations, especially negative ones, can lead to the decrease or increase in measure of a trait that results in the organism to be unfit for its own environment. For example, there is a negative correlation between body size and fertility in turkeys. Attempts at producing larger turkeys for commercial purposes led to the decrease in fertility of the population. This provided an upper limit to artificial selection for size in turkeys. Selection pressure in the natural world also fine-tunes the effects of genetic correlation. Garter snakes tend to prey on toxic newts which produce the neurotoxin tetrodotoxin. These snakes also have to be fast in order to escape their own predators.

Table 20.2 Genetic correlations between traits in multiple organisms

Organism	Traits	Genetic correlation
Humans	IgG, IgM	0.07
Cattle	Butterfat content, milk yield	−0.38
Pigs	Weight gain, back-fat thickness	0.13
Chickens	Body weight, egg production Body weight, egg weight	−0.17 0.42
Mice	Body weight, tail length	0.29
Jewelweed	Seed weight, germination time	−0.81
Milkweed bugs	Wing length, fecundity	−0.57
Wood frogs	Developmental rate, size at metamorphosis	−0.86
Drosophila	Early life fecundity, resistance to starvation	−0.91

Adapted from *iGenetics: A Molecular Approach* by Peter J. Russell, 3rd edn. (2014)

However, there is a negative correlation between speed and resistance to tetrodotoxin in these snakes. This means that neither trait can be allowed to increase indefinitely, and therefore only a mutually beneficial range of expression of the two traits are selected for.

20.7 QTL Mapping

Our analysis and understanding of quantitative traits have predominantly stemmed from statistical analysis of observable phenotypes in successive generations of artificially bred plants and animals. However, each of the traits studied must have an underlying group of genes and alleles which interact to produce the observable phenotype. In other words, the true physical basis of quantitative traits lies within segments of DNA which, via multiple modes of intercommunication, finally produce a phenotype. Unlike classical qualitative Mendelian traits, the search for multiple genes and alleles responsible for quantitative traits is quite a daunting task. Unlike simple recombination frequency calculations required to map out the former kind of genes, different strategies are employed to map the genes contributing to a quantitative trait. These strategies predominantly include the use of different observable or quantifiable markers and then checking whether any of these markers consistently get inherited along with the quantifiable trait under study. This kind of consistent inheritance of the marker with a trait of choice is called *co-segregation*. When co-segregation happens, one can assume that the chosen marker is physically close to a segment of DNA which is contributing to the trait under study. This is because co-segregation happens when two segments of DNA are very close to each other and therefore very rarely get separated due to crossing over. In case co-segregation is seen, then we can say that the marker is close to a *Quantitative Trait Loci* (QTL). A QTL is a segment of DNA which may contain a single gene or a group of genes which regulate a quantitative trait.

A number of markers have been used by multiple research teams. These include using visible phenotypes, like eye colour, which are distinct from the trait under study but can be checked for co-segregation. Other markers include using proteins having different electrophoretic mobilities, for example, isozymes. However, the most commonly used markers are DNA markers such as *single nucleotide polymorphisms* (SNPs), *variable number of tandem repeats* (VNTRs) and *restriction fragment length polymorphisms* (RFLPs). These markers are spaced out throughout the genome of all species, and their locations on different chromosomes for most model organisms are already known. One can check whether any of the above markers co-segregate with the trait under study, and this may tell us the possible location of a QTL.

The experimental paradigm employed to find QTLs includes using two artificially bred or naturally found inbred lines which exhibit two extremes of a phenotypically variable trait, for example, short longevity versus long longevity in *Drosophila*. Each of these lines is assumed to be homozygous for the alleles regulating the quantitative trait, as generations of inbreeding cause the formation of pure-bred lines. These two lines are used as the parental generation to breed the F1 generation. The F1 generation usually exhibits a phenotype which is the mean of the quantitative traits in the parental generation and is assumed to be heterozygous for all the alleles regulating the quantitative trait. The F1 progeny are selfed to produce the F2 generation. Individuals of the F2 generation exhibit a large phenotypic variance for the quantitative trait under study. The F2 generation is also known as the *QTL Mapping Population*. The large phenotypic variance in F2 is assumed to be the result of the mixing up of all the alleles contributing to the quantitative trait in various permutations and combinations in the gametes of the F1 generation. This gives rise to a large phenotypic variance. Before we execute the above paradigm, we must have a list of the DNA markers which differ between the parental lines. To begin with we do not know which of these DNA markers is close to a QTL. After breeding up to the F2 generation, we can now check which of the original DNA markers have co-segregated with the quantitative traits of interest (Fig. 20.13).

The above paradigm can be better understood using an example. A study was conducted by Steven Tanksley and colleagues to find the QTLs responsible for fruit weight in tomatoes. They used two varieties of tomatoes which differ drastically in their weight, *Lycopersicon esculentum* and *Lycopersicon pimpinellifolium*. While the tomatoes from the former plant weigh around 500 g, those from the latter weigh 1 g on an average. They positioned 88 RFLPs on the 12 chromosomes of *Lycopersicon* and used these RFLPs as DNA markers. The F1 tomatoes produced weighed on an average 10.5 g (not exactly the mean of the two parental weights but more than *L. pimpinellifolium* and less than *L. esculentum*). The F1 plants were self-fertilized and the subsequent.

F2 generation exhibited a large range of fruit weights, from less than 5 g up to 45 g. This is due to the segregation of all the genes (and alleles) affecting fruit weight. Extraction of DNA from each of the F2 plants helped in analysing whether any of the original parental RFLPs co-segregated with a given fruit weight. Most RFLPs were not seen to co-segregate with weight; however, there were a few RFLPs

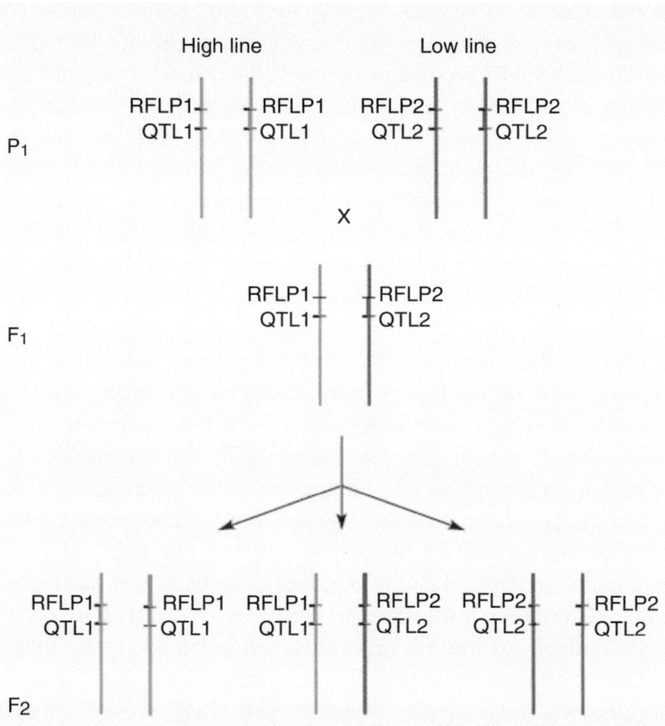

Fig. 20.13 QTL mapping using known RFLP markers in the genome: RFLP1 is a DNA marker for QTL1, and RFLP2 is a DNA marker for QTL2 in two chromosomes of a hypothetical organism. RFLP1 is found in a line which exhibits a high magnitude of a given quantitative trait (high line), and RFLP2 is found in a line which exhibits a low magnitude of the same trait (low line). In the F1 generation, all such loci are heterozygous for the DNA marker. In the F2 generation, co-segregation of the RFLP markers to individuals exhibiting a certain magnitude of the trait helps in locating the positions of possible QTLs. In the F2 generation, RFLP1 co-segregates with individuals exhibiting high magnitude of the trait, while RFLP2 co-segregates with individuals exhibiting lower magnitudes of the trait. This means that RFLPs 1and 2 are close to the genetic region regulating this trait

which were always associated with a given fruit weight. These RFLPs were then considered to mark the position of a possible QTL nearby. Cloning of different segments of the chromosomal segment housing the possible QTL can finally allow the identification of the gene responsible for the quantitative trait. Also, candidate genes in the QTL region having the desired functional annotations can be studied further. This means that if a known gene having a known function is known to be located near the co-segregating DNA marker, then that gene becomes a strong candidate gene for further study. The gene in the above QTL was identified as *ORFX*, and it was cloned. Different *ORFX* alleles have been correlated with plants producing tomatoes of varied sizes. The product of the *ORFX* gene seems to inhibit cell division and when artificially made to express in a plant causes a reduction in the size of tomatoes produced by the plant. However, the entire range of phenotypic variance observed in tomato size cannot be explained by the ORFX gene, and there are certainly multiple genes in other QTLs which contribute to this trait.

Sometimes the observed phenotypic variance is not due to genetic contribution of multiple genes from several QTLs but due to the existence of multiple alleles all belonging to the same genetic locus. This was found to be the case for phenotypic variations in haltere development in fruit flies, as well as acid phosphatase activity in humans. In both cases, variations in phenotypes were due to the presence of multiple alleles of a single gene in a given population; however, each of these alleles affected their phenotypes quantitatively to a definite degree. The cumulative effect of all the different combinations of these alleles produces a phenotypic variation for the trait in the population. Furthermore, not all QTLs contribute to protein levels (protein QTLs). Some are known to regulate phenotypes by altering the RNA transcript levels of genes (expression QTLs).

Identification of genes in QTLs has advantages. A knowledge of the genes regulating the quantitative trait of interest allows us to manipulate the genome more effectively. This allows for larger genetic gains in artificial selection and a better understanding of genetic predisposition to many medical conditions, especially those which can be categorized as threshold traits. It also aids in the development of more authentic theories to explain evolutionary processes. Recent studies on QTLs have disproven the assumption of polygenic models that each quantitative locus contributes nearly equally to a phenotype. It is now widely believed that a few of the loci, or even one, contribute to a major component of the phenotype; therefore, mutations in even one important loci can bring about evolutionarily significant phenotypic changes.

Often closely linked loci do not get separated. Linked QTLs might share a common DNA marker, and the presumed effect of one QTL might be due to multiple such loci in that local segment of the chromosome. Sometimes loci with relatively small effects on the phenotype go unnoticed due to the resolution limit of the experiment. Importantly, we must remember that the only loci we can detect are those that were different in the two parental lines to begin with. Genes which are homozygous in the QTLs of the two parental lines will remain undetected. The above points are a few reasons why the number of QTLs is usually underestimated.

Box 20.1: Scientific Concept: Genetic Architecture of Natural Variation in *Drosophila melanogaster* Aggressive Behaviour – John Shorter et al. (2015)

Many traits of interest in both plants and animals seem to be controlled by more than one gene. This kind of polygenic inheritance is often associated with phenotypes ranging from phenotypic variations in kernel colour in wheat to the pattern of inheritance of skin colour in humans. The study of polygenic inheritance influences our understanding of complex traits such as IQ variation

(continued)

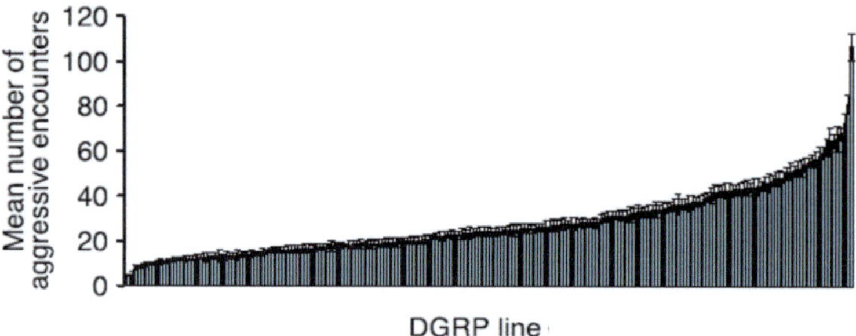

Fig. 20.14 Gradation in displayed aggression levels: Aggression levels were measured in 200 available inbred lines of *Drosophila* (from Drosophila Genetic Reference Panel, or DGRP). Low and high aggression parental lines were chosen from this population to generate the outbred lines

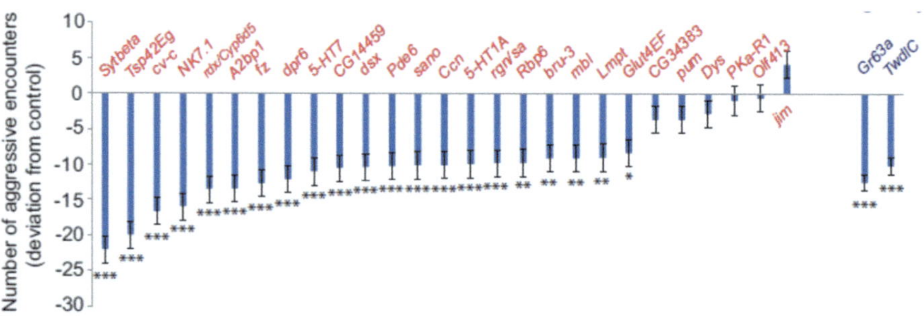

Fig. 20.15 Change in gene expression is associated with change in aggression levels: Insertional mutants of different candidate genes exhibited a decreased magnitude of aggression (except *jim*, which showed increased aggression) implicating their role in the polygenic control of the trait. The two genes on the right were downregulated using RNAi. Decreased aggression levels were observed in each case

Box 20.1 (continued)

and helps in making economically important decisions concerning the improvement in quality of domestic animals and cash crops. Though derived from Mendelian genetics, an understanding of the inheritance of polygenic traits deviates from the simpler analysis of classical Mendelian examples. It requires a unique set of concepts and definitions, the majority of which this chapter has been dedicated to. The sheer number of genes that can possibly participate in dictating the phenotypic variation of a complex trait makes it a daunting task to find the genetic correlates of any trait under investigation. One must factor in the added component of the possible permutation and combination of interactions that are possible between the genes regulating the expression of the trait. These can range from simple additive interactions to the more

(continued)

Box 20.1 (continued)
elusive dominant and epistatic genetic interactions. Furthermore, interactions between the genetic components and the environment of the organism further complicate the matter of genetically dissecting a complex trait. One such trait is aggression.

Aggression is an evolutionarily conserved, complex and universal behaviour that influences access to food resources, escape from predators, obtaining mates and establishing dominance hierarchies. It is considered to be a quantitative trait and is influenced by both genetic and environmental factors. Common genes and neural pathways have been implicated in aggressive behaviour in *Drosophila*, mice and humans. The bioamine signalling pathway plays a key role in regulating this behaviour. However, the genetic basis of the phenotypic variation in aggression remains unknown. Using *Drosophila* as a model system, Shorter et al. proceeded to find the genes responsible for genotypic (or genetic) variation in aggression. They performed genome-wide association (GWA) analysis in available inbred *Drosophila* lines after determining their aggression levels (Fig. 20.14). They also performed GWA analysis on outbred lines. These outbred lines were derived from mating inbred parental *Drosophila* lines which differed drastically in their levels of aggression (as ascertained from Fig. 20.14). One chosen parental line was a low aggression line, while the other was a high aggression line. SNPs were used as DNA markers to locate the position of possible QTLs. SNPs were scored for co-segregation with the magnitude of aggression exhibited by the fly harbouring that particular SNP. Genes significantly associated with aggression post-analysis included those that are involved in motor activity, reproduction, chemosensation and neurotransmitter reception. Interestingly, they calculated $H^2 = 0.69 \pm 0.07$, but, $h^2 = 0$ for this trait. This suggests that though genetic components of variance in aggression exist, they are predominantly non-additive in nature. Furthermore, analysis of the data did not find much overlap between the candidate genes identified in the inbred parental population and the outbred offspring population of *Drosophila*. This lack of overlap also hints at the near absence of additive components of genetic variance in aggression. In fact, the construction of a genetic interaction network showed that the genes identified in the original parental population and the offspring population were epistatically linked to each other (gene-gene interactions). In order to validate the biological relevance of the candidate genes, insertional mutants (of the candidate genes) were generated and RNAi knockdown experiments were conducted in *Drosophila*. In nearly all the cases, disruption of the expression of the candidate gene resulted in a reduction in aggression (Fig. 20.15). Additionally, it was shown that epistatic interactions between different candidate genes can also regulate the degree of expressed aggression. It is now believed that aggression has an underlying polygenic genetic architecture dominated by epistatic gene action. It is yet to be determined how these genetic interactions establish the neural underpinnings of aggression.

20.8 Summary

- Quantitative genetics is the study of phenotypic traits which express a range of values. Such traits are usually regulated by more than one gene and hence are also called polygenic traits. They are considered to be multifactorial in nature as both genetic and environmental factors can contribute to the range of values (variation) observed in the trait. Such traits can exhibit meristic, threshold, or continuous characteristics. They are distinct from oligogenic (or Mendelian) traits. Examples of such traits include kernel colour in wheat, ear length in maize and skin colour in humans.
- Quantitative or polygenic traits are studied at a population level. The collective variation exhibited by the individuals of a population (or sample) is described using various statistical terms such as mean, standard deviation and variance. The relationship between two different phenotypic traits is computed using correlation and regression analysis. ANOVA helps in the estimation of the degree by which measures of two traits are different from each other.
- Variation in a phenotypic trait can be caused due to variation originating from the genes as well as variation arising from environmental factors. Genetic variation can be due to additive and dominance factors operating within a locus, or due to epistatic interactions between loci. Environmental factors can be internal or external in nature. They can also be special or general in nature. Other factors contributing to phenotypic variation include gene-environment interaction, gene-environment covariance and intangible variation.
- Heritability measures the contribution of genetic factors to phenotypic variation (variance). It can be estimated as broad-sense heritability and narrow-sense heritability. The former takes into consideration all genetic sources of phenotypic variation, while the latter only considers the additive genetic component of phenotypic variance. There are multiple ways of calculating heritability. These include elimination of one variance component, parent-offspring regression and comparing phenotypic variance in individuals with varying degree of relatedness.
- Anthropomorphic manipulation of the mean value of phenotypic traits is artificial selection. Such selection is usually done for commercially important traits such as milk yield, egg production and wool yield. Quantities such as genetic gain, selection intensity and genetic advance help describe the selection process. There are limits to artificial selection as excessive change in the magnitude of one trait usually has an undesirable effect on the phenotype of another trait. This is due to the phenomenon of phenotypic correlations.
- Localizing the multiple genetic elements responsible for regulating the phenotypic expression of a trait is known as QTL mapping. The consistent co-segregation of molecular markers in the genome, such as RFLPs and SNPs, with a specific magnitude of the quantitative trait of interest helps in deciphering the possible location of genes controlling a polygenic trait.

References

Arranz JJ, Coppieters W, Berzi P, Cambisano N, Grisart B, Karim L, Marcq F, Moreau L, Mezer C, Riquet J, Simon P (1998) A QTL affecting milk yield and composition maps to bovine chromosome 20: a confirmation. Anim Genet 29:107–115

Bidinost F, Roldan DL, Dodero AM, Cano EM, Taddeo HR, Mueller JP, Poli MA (2008) Wool quantitative trait loci in merino sheep. Small Rumin Res 74:113–118

Grisart B, Coppieters W, Farnir F, Karim L, Ford C, Berzi P, Cambisano N, Mni M, Reid S, Simon P, Spelman R (2002) Positional candidate cloning of a QTL in dairy cattle: identification of a missense mutation in the bovine DGAT1 gene with major effect on milk yield and composition. Genome Res 12:222–231

Gutiérrez-Gil B, El-Zarei MF, Alvarez L, Bayón Y, De La Fuente LF, San Primitivo F, Arranz JJ (2009) Quantitative trait loci underlying milk production traits in sheep. Anim Genet 40:423–434

Liu Z, Sun C, Yan Y, Li G, Wu G, Liu A, Yang N (2018) Genome-wide association analysis of age-dependent egg weights in chickens. Front Genet 9

Roldan DL, Dodero AM, Bidinost F, Taddeo HR, Allain D, Poli MA, Elsen JM (2010) Merino sheep: a further look at quantitative trait loci for wool production. Animal 4:1330–1340

Russell PJ (2014) iGenetics, a molecular approach, 3rd edn. Pearson New International Edition, Harlow

Shorter J, Couch C, Huang W, Carbone MA, Peiffer J, Anholt RR, Mackay TF (2015) Genetic architecture of natural variation in Drosophila melanogaster aggressive behavior. Proc Natl Acad Sci 112:E3555–E3563

Sinha NK, Singh SK (1997) Genetic and phenotypic parameters of body weights, average daily gains and first shearing wool yield in Muzaffarnagri sheep. Small Rumin Res 26:21–29

Snustad DP, Simmons MJ (2012) Principles of genetics, 6th edn. John Wiley and Sons, Inc., Hoboken, NJ

Vesely JA, Peters HF, Slen SB, Robison OW (1970) Heritabilities and genetic correlations in growth and wool traits of Rambouillet and Romnelet sheep. J Anim Sci 30:174–181

Wolc A, Arango J, Settar P, Fulton JE, O'sullivan NP, Preisinger R, Habier D, Fernando R, Garrick DJ, Hill WG, Dekkers JCM (2012) Genome-wide association analysis and genetic architecture of egg weight and egg uniformity in layer chickens. Anim Genet 43:87–96

Yuan J, Sun C, Dou T, Yi G, Qu L, Qu L, Wang K, Yang N (2015) Identification of promising mutants associated with egg production traits revealed by genome-wide association study. PLoS One 10:e0140615

Zhang Q, Boichard D, Hoeschele I, Ernst C, Eggen A, Murkve B, Pfister-Genskow M, Witte LA, Grignola FE, Uimari P, Thaller G (1998) Mapping quantitative trait loci for milk production and health of dairy cattle in a large outbred pedigree. Genetics 149:1959–1973

Further Reading

Falconer DS, Mackay TFC (1996) Introduction to quantitative genetics, 4th edn. Addison Wesley Longman Limited, Harlow

Griffiths AJ, Wessler SR, Lewontin RC, Gelbart WM, Suzuki DT, Miller JH (2011) An introduction to genetic analysis, 10th edn. W H Freeman and Company, New York

Klug WS, Cummings MR, Spencer CA, Palladino MA (2012) Concepts of genetics, 10th edn. Pearson Education, Inc., San Francisco, CA

Lynch M, Walsh B (1998) Genetics and analysis of quantitative traits. Sinauer Associates, Inc., Sunderland, MA

Pierce BA (2010) Genetics: a conceptual approach. W H Freeman and Company, New York
Powar CB (2003) Genetics (volume 1), 1st edn. Himalaya Publishing House, Mumbai
Singh BD (2005) Genetics, 1st edn. Kalyani Publishers, New Delhi
Tamarin RH (2001) Principles of genetics, 7th edn. The McGraw-Hill Companies, New York

Population Genetics

21

Payal Gupta

21.1 Identification of Genetic Variation

With the exception of identical twins, all humans show variations in their looks, features and habits. The basis of these fundamental differences lies in the genetic make-up of all individuals. The gigantic genome and the multiple processes of recombination, mutation, random assortment, linkage, etc. provide each individual with his/her unique genetic make-up. Therefore, as humans we have the same basic genomic structure, but every human has a different genetic constitution.

Thus *genetic variation* defines the differences in DNA sequences between individuals of the same population or the gross sequence differences between two populations. It can also imply genetic differences between members of the same species or members of different species. Genetic variations are at the core of all the natural diversity that we observe within a population or between populations.

The variations can arise from differences in sequences within coding or non-coding segments of the genome. Since most of the genes that code for important peptides are relatively conserved, variations in non-coding sections are usually more informative.

When certain variations present themselves at specific sites along the chromosome and can be uniquely characterized using techniques such as polymerase chain reaction (PCR) and gel electrophoresis, they can serve as *molecular markers*. Also, the genetic distance between molecular markers that are linked can be estimated from the outcomes of crosses.

A single molecular marker can exist in more than two "forms," that is, the same site can have more than two types of sequence variations. This property is termed *polymorphism*, literally meaning existing in multiple forms. If a specific locus in the genome of species always has a fixed nucleotide/s, it is said to be monomorphic.

P. Gupta (✉)
University of Calcutta, Kolkata, India

D. Kar, S. Sarkar (eds.), *Genetics Fundamentals Notes*,
https://doi.org/10.1007/978-981-16-7041-1_21

However, if the genomes of different members of that species show sequence variation at that locus, then the locus is said to be polymorphic. A particular variation for a locus should be present in at least 1% of the population to qualify as a polymorphic marker. The study of distribution of polymorphic markers in a given population provides vital information about the population. Further, polymorphic markers serve as important connotations for specific genes and traits, allowing us to perform genetic association studies to predict genetic identity of the linked gene.

In population genetics, polymorphic markers generally display as observable variations in a trait within a particular population. Population geneticists have been interested in variations in traits that can be observed with naked eyes, like differences in colour, shapes and patterns. A classic example is the Hawaiian happy-face spider that displays an umbrella of colour and pattern polymorphism (Fig. 21.1). The abdominal colour and pattern of the Maui population range from plain yellow to varying red, white and/or black patterns on a yellow background, and the colour morphism is controlled by simple Mendelian alleles.

In this discussion two dimensions of polymorphism become important: phenotypic polymorphism and genetic polymorphism. Phenotypic polymorphism describes two or more morphs/variation of a single trait that is visibly noticeable in individuals of a population. Genetic polymorphism describes changes or variations at genetic level in a population. Genetic polymorphism may not always be evident as phenotypic variations.

21.1.1 Single Nucleotide Polymorphism (SNP)

Single nucleotide polymorphism (*SNP*, frequently pronounced as "snips") denotes genetic variation present at a single nucleotide position. When a single nucleotide position is polymorphic among individuals, it is recognized as SNP. SNPs are fairly common in all genomes. They are extremely useful in the mapping of disease-causing alleles and genes that contribute to quantitative traits. In humans alone, SNPs represent 90% of all the variation in sequences that are found in different individuals. A classic example of how polymorphism at a single base can result in stark phenotypic changes is the human β-globin gene (Hb). The example shown in Fig. 21.2 shows that the Hb gene can exist as two morphs Hb^A and Hb^S both of which differ from each other at single base position; however, the recessive Hb^S allele can cause sickle cell disease.

21.1.2 Microsatellite

Molecular markers can also present themselves as short tandem repeats (STRs), also known as microsatellites, simple sequence repeats (SSR) and simple sequence length polymorphisms (SSLP). These are di-, tri-, tetra- or penta-nucleotide sequences repeated multiple times in a stretch of a chromosome. Many such stretches of short tandem repeats are found across the genome, and their length varies

Fig. 21.1 Colour polymorphism displayed by the Hawaiian happy-face spider *Theridion grallator*. Maui spiders display an array of colour and pattern polymorphism with different colours and patterns appearing on a yellow background on the abdomen

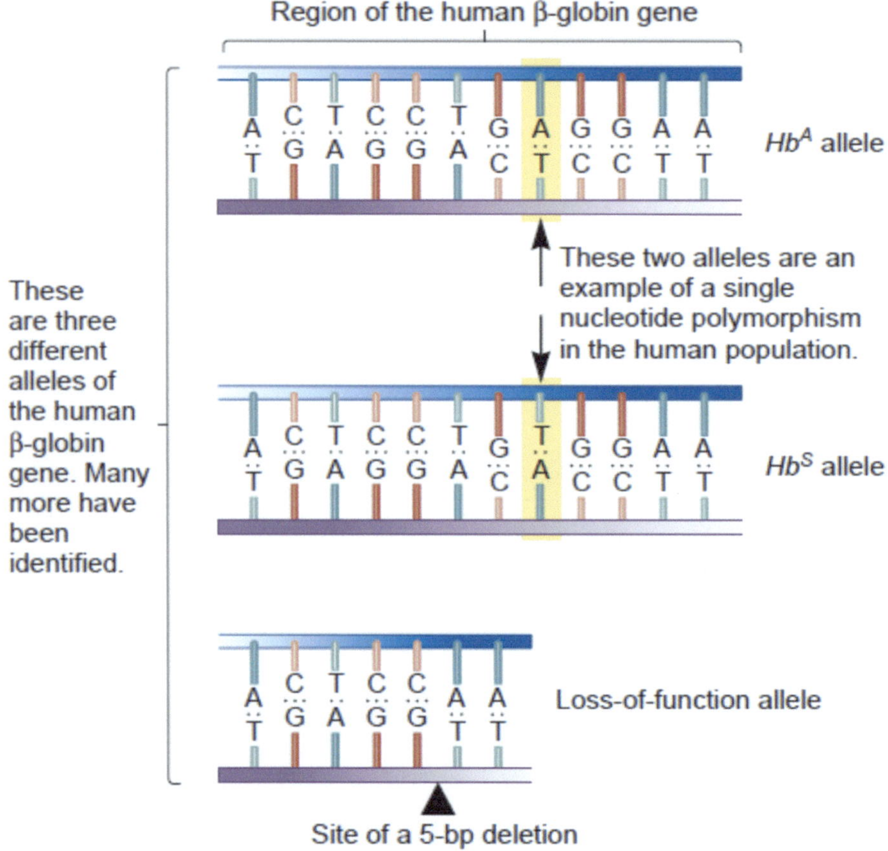

Fig. 21.2 The single nucleotide polymorphs of the human β-globin gene. The β-globin gene has two alleles, Hb^A (top sequence) and Hb^S (middle sequence), which differ from each other at single nucleotide position (highlighted base pair) and serve as an example of single nucleotide polymorphism. However, a 5-bp deletion (bottom sequence) can result in a different loss of function allele

considerably between individuals. For instance, a dinucleotide sequence—$(CA)_n$, where n is the number of times CA sequence is repeated and can be anywhere between 5 and 50—is commonly found in humans. In fact, thousands of different DNA segments contain the $(CA)_n$ microsatellite. During replication of a microsatellite region, the replication machinery often slips and causes copy number variation of the repeat region. For this reason, the microsatellite segment is regarded as *hypervariable*. This makes them a very informative molecular marker and one that rapidly evolves. They are in fact more polymorphic than SNPs and are more helpful in detailed genetic polymorphism studies. Microsatellites are inherited in Mendelian patterns, are often co-dominant and are easily characterized. Small amount of samples can be used to conduct PCR for microsatellites, and the information can help in pedigree analysis for the study of population structure.

21.1.3 Haplotype

Today innumerable genetic markers are available at one's disposal. However, it is more informative to analyse multiple markers simultaneously to make mapping and genetic analysis more robust. Therefore, the inheritance of multiple markers as a single linked haplotype becomes essential. *Haplotype* is a set of genes present on a chromosome that are transferred from parents to progeny en bloc or as a single block unit. This inheritance of a group of genes as a single cluster occurs due to tight genetic linkage. Similarly, a set of SNPs or other polymorphisms can also be inherited as a block due to their close genetic proximity. The study of haplotype structures can help in various ways. Haplotype mapping can help us to identify a disease-carrying allele by virtue of its proximity to a particular morph of a haplotype (Fig. 21.3). It can also help in the identification of genetic variants that are responsible for different traits in a population.

When markers and alleles show linkage frequencies considerably higher than one would expect out of a random cross, the phenomenon is known as *linkage disequilibrium (LD)*. A study of STR and Alu deletion polymorphism at the CD4 locus at 42 populations worldwide first outlined the importance of LD in tracing population history.

21.1.4 The International HapMap Project

The understanding of the usefulness of SNPs on a block gave rise to *The International HapMap Project*. Each haplotype block of SNPs may contain many SNPs, but a combination of a few variants itself can give rise to a pattern unique to an

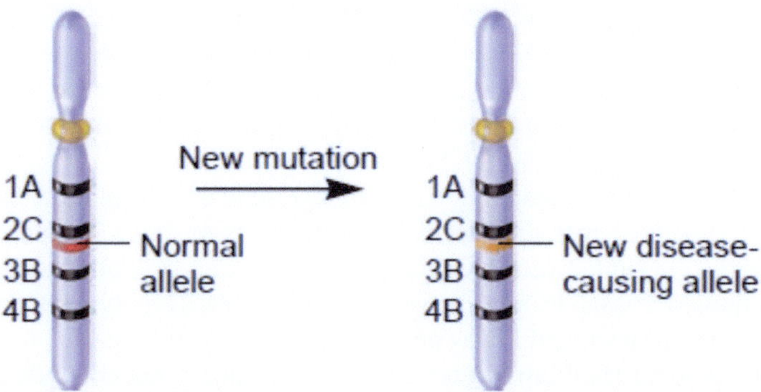

Fig. 21.3 Schematic representation of disease allele mapping using haplotype. Four sites (1–4, and) are linked and occur as a haplotype along a particular chromosome. These sites have SNP variants (A, B and C) that serve as their markers. If a founder disease-causing mutation arises near to the second locus which harbours the SNP variant "C", the disease allele becomes tightly linked to the "C" variant at the second locus. Thus the presence of the "C" variant at this locus would indicate a higher probability of disease allele

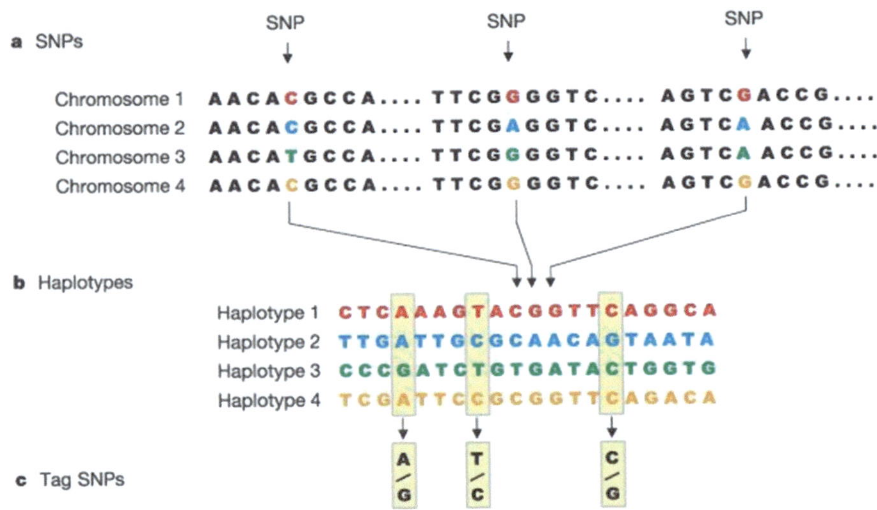

Fig. 21.4 Tag SNPs can help identify haplotype variants. (**a**) The same chromosomal segments from four different people show SNP variations. The DNA sequence is identical for maximum bases in these chromosomes, but three bases exhibit sequence variation. Each SNP has two possible alleles; the first SNP in panel (**a**) has the alleles C and T. (**b**) A haplotype comprises a specific set of alleles linked to proximal SNPs. The figure shows only the variable bases consisting of 20 SNPs that stretch of 6000 bases of DNA and mark a particular haplotype. This includes the three SNPs shown in panel (**a**). For this segment of DNA, the majority of population would have haplotypes 1–4. (**c**) The three highlighted SNPs serve as tag SNPs such that genotyping only these three tag SNPs out of the 20 SNPs is enough to uniquely identify these four haplotypes

individual and a haplotype. These few SNPs that can successfully and specifically identify a haplotype are called *tag SNPs*. The genomic map of these tag SNP haplotype blocks is known as a *HapMap*. The HapMap is crucial as it narrows the nearly ten million known SNPs to roughly 50,000 SNP tags that are required to examine the entire genome for association. This makes genome scan for identification of linked disease alleles both efficient and comprehensive as now less number of SNPs needs to be mapped (Fig. 21.4).

To execute this task of mapping haplotypes, researchers from academic centres, non-profit biomedical research groups and private companies in Canada, China, Japan, Nigeria, the United Kingdom and the United States collaborated to undertake the International HapMap Project. A meeting on October 27 to 29, 2002, marked the beginning of the project, and it was projected to take about 3 years.

The International HapMap Project aims at determining the common features in DNA sequence variation in the human genome. This involves characterization of sequence variants, their prevalence and associations between them. For this purpose DNA samples from populations with ancestral lineage from parts of Africa, Asia and Europe were analysed. The project makes the expense of whole genome sequencing dispensable by allowing the indirect association approach of marking the linkage of a trait to a particular haplotype. This can be applied to any functional candidate gene in the genome, to any DNA segment suggested by family-based linkage analysis, or finally to the entire genome for scans for disease risk factors.

Apart from the study of genetic associations with diseases, the HapMap can also be applied to the study of the genetic factors that contribute to variation in response to environmental factors and susceptibility to infection and in the physiological responses to drugs and vaccines in specific populations or individuals. These studies are based on the understanding that contributing genetic variants occur with a higher frequency in a group of people having a certain disease or exhibiting a particular response to environmental factors, drugs or pathogens than in a group of people that do not exhibit the disease or response phenotype. The read-outs from analysis of tag SNPs can indicate chromosomal segments displaying different haplotype distributions in the two groups of people, those with a disease or response and those without. Each segment can be further analysed in greater detail to identify the exact variants or genes in the genetic region contributing to the disease or response phenotype. This can eventually lead to more effective interventions and also allow the development of diagnostic markers that can predict which drugs or vaccines would be most effective in individuals with particular genotypes for genes affecting drug metabolism.

21.2 The Gene Pool Concept

Since population genetics involves the study of genetic variation in a group of individuals over time, it becomes mandatory to look at the genetic make-up of the entire population rather than the genetic composition of an individual. From this idea evolves the *Gene Pool Concept*. The gene pool essentially comprises all of the alleles of every single gene present in a given population. It is precisely a collection of all the genetic information that a population harbours, and this determines the set of all genetic variants from which the genetic make-up of forthcoming generation will be determined. The gene pool of a population determines the genetic characteristics of the next generation except if there is a change in the composition of the gene pool due to a founder mutation, migration or any of the forces of evolution. With each passing generation, the dynamics of the gene pool will change depending on allelic variations and evolutionary forces.

To completely comprehend the concept of gene pool, it becomes important to discuss the meaning of "population" from the perspective of a geneticist. For a sexually breeding species, a population signifies a group of interbreeding individuals of the same species that inhabit the same region. A species may occupy a wide geographic area wherein they are subdivided into smaller, discrete populations. Therefore smaller local populations or demes can make up a larger population, with the members of a deme having a higher probability of interbreeding.

The aim of a population geneticist is to understand the prevalence of different alleles of a polymorphic gene in a given population, study the changes that these alleles undergo from generation to generation and identify the underlying causes for such changes. In order to make an objective study, the population geneticist works with two fundamental concepts: *allele frequency* and *genotype frequency*. Allele frequency of a particular allele is the ratio of the number of copies of that allele in the population to the total number of all alleles for that gene in the said population. The genotype frequency is denoted as the ratio of the number of individuals with a particular genotype in a population to the total number of individuals in that said population.

This can be understood with an example of a particular population. Let us suppose that there is a population of 100 mice with the following genotype:

64 mice have black coat colour with the genotype *BB*.
32 mice have grey coat colour with the genotype *Bb*.
4 mice have white coat colour with the genotype *bb*.

Therefore, for homozygous individuals an allele frequency calculation would account for two copies of that allele, while for heterozygous individuals, it will account for one copy of that allele. Likewise, while calculating the frequency of the "*b*" allele, the heterozygotes will be counted once, but the homozygous "*bb*" individuals will be counted twice for that allele. Hence the frequency of allele "*b*" would be as follows:

$$b = \frac{32 + 2(4)}{2(64) + 32 + 2(4)}$$

$$\text{or}, b = \frac{40}{200} = 0.2, \text{or } 20\%$$

This tells us that the frequency of the "*b*" allele is 20% or that the "*b*" allele represents 20% of the total alleles of this gene in this particular population.

Let us now calculate the genotype frequency of "*bb*" in this population of mice.

$$bb = \frac{4}{64 + 32 + 4}$$

$$\text{or}, bb = \frac{4}{100} = 0.04, \text{or } 4\%$$

This indicates that 4% of the total mice in this population have "*bb*" genotype and therefore have white coat colour.

21.3 The Hardy-Weinberg Equilibrium

With the understanding that a population inherits the alleles for all genes from an ancestral gene pool, it becomes essential to address how this inheritance works in context to the frequency of these alleles and the genotypic proportions. It is also vital to derive a mathematical correlation between these concepts of change of genotype and allele frequency over generations. Godfrey Harold Hardy, a British mathematician, and Wilhelm Weinberg, a German Physician, independently proposed a mathematical expression that could help estimate how allele and genotype frequencies change over generations. This came to be known as the Hardy-Weinberg equilibrium.

The Hardy-Weinberg equilibrium postulates that given that there are no evolutionary forces acting on a geographically isolated interbreeding population, the genotype and allele frequencies remain constant over generations. Since there is no predicted change in the allele and genotype frequencies, hence the term "equilibrium" applies.

21.3.1 Assumption of the Law

The Hardy-Weinberg equilibrium provides a mathematical proof of concept that allele and genotype frequencies remain constant over generation for a population which is not undergoing any evolutionary changes. However, this derivation is based on an exclusive set of assumption, and the law of equilibrium applies only to a population which fulfils the following assumptions:

1. No new mutations: The genes of interest should not undergo any new mutation.
2. No genetic drift: Random sampling should not affect the allele frequency, and therefore only infinitely large populations are considered.
3. No migration: The population of interest is assumed to be geographically isolated, such that there is no immigration or emigration.
4. No natural selection: No genotype (dominant or recessive) should be favoured by the environment, and all genotypes should have equal chances of survival and mating.
5. Random mating: There is no mate selection or mating preference, and the members of the population mate randomly with regard to their genotype and phenotype such that the gene of interest is evenly mixed and distributed in the population.

21.3.2 Prediction of the Law

When all of the above assumptions hold, then the allele and genotype frequency of the gene of interest would be constant over generations. If we follow the example of the mice with black (*BB*), brown (*Bb*) or white (*bb*) coat colour, the Hardy-Weinberg Law would predict that the allele frequency of "*b*" would remain to be 0.2 or 20% and the frequency of the "*bb*" genotype would remain 0.4 or 40% in the next and forthcoming generations of the population in question. Therefore, Hardy-Weinberg Law provides a quantitative relationship between allele and genotype frequencies in a population.

21.3.3 Derivation of the Law

To further understand the Hardy-Weinberg equilibrium, let us reconsider the hypothetical example of mice population that possesses polymorphic gene for the coat colour trait with alleles "*B*" and "*b*". Let the allele frequency for "*B*" be denoted by "*p*" and the allele frequency for "*b*" be denoted by "*q*". According to the Hardy-Weinberg Law,

$$p + q = 1$$

Therefore, if the frequency of "*b*" (q) is calculated to be 0.2 or 20%, then the frequency of "*B*" (p) would be 0.8 or 80% so that their frequencies would add up to 1 or 100%. Since Hardy-Weinberg equilibrium assumes random mating, therefore every individual in this biallelic system would inherit two alleles, and these would be randomly and independently assorted. Therefore, we can apply the product rule and multiply individual probability sums, $p + q$ together. This would be applied as follows to the earlier equation:

$$(p + q)(p + q) = 1$$

$$\text{or}, p^2 + 2pq + q^2 = 1$$

In our example, p^2 = genotype frequency of "*BB*"; $2pq$ = genotype frequency of "*Bb*"; q^2 = genotype frequency of "*bb*". So if we say that $p = 0.2$ and $q = 0.8$, then, based on the Hardy-Weinberg equilibrium, we can calculate the allele frequencies as follows:

$BB = p^2 = (0.8)^2 = 0.64$
$Bb = 2pq = (2)\ (0.8)\ (0.2) = 0.32$
$bb = q^2 = (0.2)^2 = 0.04$

Therefore, the allele frequency for "*B*" is 0.8 or 80%, and the allele frequency for "*b*" is 0.2 or 20%. The genotype frequency for "*BB*" is 0.64 or 64%, for "*Bb*" is 0.32 or 32% and for "*bb*" is 0.04 or 4%.

The Hardy-Weinberg equilibrium can also be explained with the help of a Punnett's square used to calculate offspring genotype frequencies from randomly combining gametes. The frequency of an allele is equal to the frequency of gametes carrying it in the population. In our example, the frequency of the allele "*B*" is 0.8, and so the frequency of the gametes carrying allele "*B*" is also 0.8. We can now apply the product rule to calculate the probability of the genotype "*BB*" as $0.8 \times 0.8 = 0.64$ (Fig. 21.5). The frequency of "*Bb*" or heterozygote (given as $2pq$) would be (2) (0.8) (0.2) or (0.16) + (0.16) or 0.32.

The Hardy-Weinberg equilibrium forms the basis of our understanding of how the alleles of a gene mix and change in a population. It can help us predict changes in the frequency of one allele or genotype depending on the changes that the other

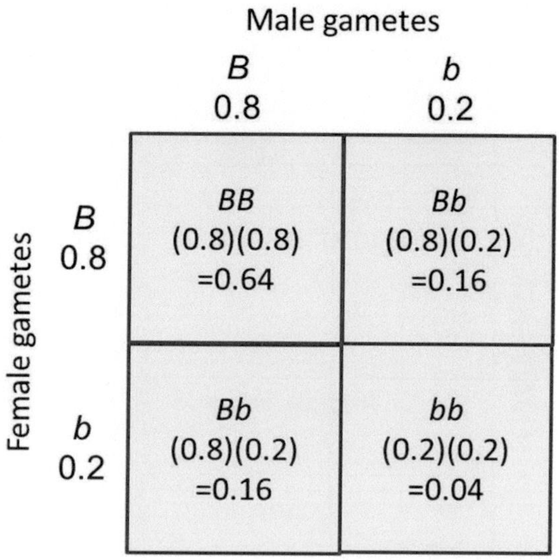

Fig. 21.5 Punnett's square can be used to estimate allele and genotype frequencies

alleles go through. For instance, when the frequency of the "*b*" allele is low, the genotype "*BB*" will be predominant. Conversely when the frequency of allele "*b*" is high, the genotype "*bb*" will have a higher probability of representation in a population.

21.3.4 Extension of the Law with More than Two Alleles

Given the stringent assumptions, no population can practically follow the Hardy-Weinberg equilibrium; however, the Law can be extended and modified to fit practical examples like multiple allelic systems. The basic understanding is that the Hardy-Weinberg equations predict the probability of finding a particular genotype combination if the frequency of alleles is known. Since the premise of independent assortment still stands, therefore the probability of finding a particular genotype is calculated by applying the product rule using individual allele frequencies. Let us understand this better with the help of the example of the ABO blood grouping with three alleles, I^A, I^B, I^O with their frequencies represented as p_A, p_B, p_O, respectively. Let us suppose that the values for the frequencies are as follows:

$p_A = 0.3$
$p_B = 0.1$
$p_O = 0.6$

So we see that the sum of allele frequencies, i.e. $p_A + p_B + p_O$ would be equal to 1 [$0.3 + 0.1 + 0.6 = 1$] according to the basic principle of Hardy-Weinberg equilibrium. Now we can draw a Punnett's square similar to the one used for a

FEMALE GAMETES	MALE GAMETES		
	I^A 0.3	I^B 0.1	I^O 0.6
I^A 0.3	I^AI^A (0.3) (0.3) = 0.09	I^AI^B (0.3) (0.1) = 0.03	I^AI^O (0.3) (0.6) = 0.18
I^B 0.1	I^AI^B (0.3) (0.1) = 0.03	I^BI^B (0.1) (0.1) = 0.01	I^BI^O (0.1) (0.6) = 0.06
I^O 0.6	I^AI^O (0.3) (0.6) = 0.18	I^BI^O (0.1) (0.6) = 0.06	I^OI^O (0.6) (0.6) = 0.36

Fig. 21.6 Punnett's square for calculation of allele frequencies in a tri-allelic system of ABO blood grouping

trihybrid cross for gamete combination. Again, it is implicit that the frequency of a particular allele in a population is equivalent to the frequency of that gamete in that population. As shown in Fig. 21.6, the frequency of genotypes can be calculated as follows:

Frequency of the homozygote = square of allele frequency;
Frequency of the heterozygote = (2) (frequency of the first allele) (frequency of the second allele).

Therefore, the frequency of the I^A homozygote can be calculated as:

$$(p_A)^2 = (0.3)^2 = 0.09$$

Also, the frequency of the I^AI^B heterozygote genotype can be calculated as:

$$(2)\ (p_A)\ (p_B) = (2)\ (0.3)\ (0.1) = 0.06$$

The above basis of product rule can be applied to any number of alleles for a gene in a population.

21.3.5 Extension of the Law to X-Linked Alleles

In our discussions so far, we have focussed on autosomal genes. However, applying the principles of Hardy-Weinberg equilibrium to alleles on X-chromosomes in organisms, such as *Drosophila* and humans, in which females are homozygous and males are hemizygous for the X-chromosome, requires a different approach. The fundamental principles remain the same. Since females carry two X-chromosomes, the frequency of a genotype would be calculated by simply applying the logistics of the Hardy-Weinberg equations as follows:

Frequency of homozygous females = (Frequency of allele)^2.

Frequency of heterozygotes females = (2) (frequency of first allele) (frequency of second allele).

However, since males are hemizygous for the X-chromosome, the frequency of the genotypes is equal to the individual allele frequency. Hence, X-linked traits are more common in males compared to females. It is now important to appreciate that in a randomly mating population, the males will receive the X-chromosomes from the female parent, while the females will receive one X-chromosome from each male and female parent. Hence, the X-linked allele frequency in females will be the average of both parents. Therefore, the frequency pattern of the alleles between two sexes will alternate in each generation.

21.3.6 Test for Hardy Weinberg Proportion

Since Hardy-Weinberg's equilibrium applies only to populations that follow specific assumptions and do not undergo any evolutionary changes, hence it might be expected that very few examples follow this in practice. However, a chi-square test can be performed to test if a particular population is following Hardy-Weinberg's equilibrium or not. If the equilibrium is not in place, then one can analyse which of the five assumptions have been violated.

Let us test the Hardy-Weinberg law on a hypothetical population using the chi-square (χ^2) test. In our hypothetical population, a gene with alleles "R" and "W" controls the colour of wings in a species of butterfly population. The allele frequency of "R" is denoted by "p", while the frequency of "W" is denoted by "q". Now let us assume that in the population of our interest, the number of "RR" butterflies is 12, the number of "RW" butterflies is 53 and the number of "WW" butterflies is 12. The allele frequency "p" can be calculated as follows:

$$p = \frac{(2 \times \text{number of ``RR'' homozygotes}) + \text{number of ``RW'' heterozygotes}}{2 \times \text{total number of individuals}}$$

Therefore,

$$p = \frac{(2 \times 12) + 53}{(2 \times 77)} = 0.5$$

And,

$$q = 1 - p = 1 - 0.5 = 0.5$$

Using the above values for "p" and "q", we can calculate the expected genotype frequencies as per Hardy-Weinberg law.

$p^2 = (0.5)^2 = 0.25$
$2pq = (2)\ (0.5)\ (0.5) = 0.5$
$q^2 = (0.5)^2 = 0.25$

To implement a chi-square test, we need to calculate the actual numbers for the individual genotype populations and compare them with the numbers one would expect if the Hardy-Weinberg law were true for the population. The same has been shown in Table 21.1. We can then compute the chi-square (χ^2) value by calculating the value of "d" or the deviation for each genotype by subtracting the expected values (e) from the observed values (o). Then d^2 is estimated followed by estimation of d^2/e values for each genotype. Since two degrees of freedoms are lost, the probability (p-value) for degree of freedom 1 and chi-square (χ^2) value, from Tables 21.1, 10.98 is estimated from the chi-square (χ^2) table. The chi-square (χ^2) value 10.98 for 1 degree of freedom indicates a p-value of 0.01 which implies that there is less than 1% probability that the difference between observed and expected values is due to chance alone. Therefore, we can conclude that our hypothetical population does not follow the Hardy-Weinberg law.

21.4 Mating System

As we have discussed, one of the basic assumptions of the Hardy-Weinberg equilibrium is random mating, which implies that individuals will choose their mating partners, regardless of their genotype and phenotype. This ensures that there is even mixing and shuffling of alleles in a population. However, in an actual scenario, this is often not the case. There are many different types of mating systems that exist in real-life populations. *Mating system* essentially describes the norms that define the sexual behaviour and mate selection patterns of a group under the given circumstances. One system that we have already discussed is random mating. This system entails that there will be no bias in the mate choice and that any given individual will choose a mate from a population by complete chance and in a random fashion. The opposite of this would be the concept of *non-random mating*. In this type of mating system, there are various factors that govern the mate choice of an individual in a population. Due to this phenomenon of mate preference, even

Table 21.1 The chi-square (χ^2) value table for testing of the Hardy-Weinberg equilibrium

Genotype	Genotype frequency	Expected number (e)	Observed number (o)	$d = (o\text{-}e)$	d^2	d^2/e	$\chi^2 = \sum(d^2/e)$
RR	$p^2 = 0.25$	$0.25 \times 77 = 19.3$	12	−7.3	53.29	2.76	10.98
RW	$2pq = 0.5$	$0.5 \times 77 = 38.5$	53	14.5	210.25	5.46	
WW	$q^2 = 0.25$	$0.25 \times 77 = 19.3$	12	−7.3	53.29	2.76	

shuffling of alleles does not take place. The non-random mating system can be divided into the following categories based on the criterion for mate selection:

- Mating according to index values.
- Mating according to relationship.
- Mating between breeds—crossbreeding.

In the following section, we will discuss different types of non-random mating at length and also understand their subcategories.

21.4.1 According to Index Value

In this type of non-random mating system, the mate is chosen based on the phenotypes. This is also known as *assortative mating*. Certain phenotypes are desirable in a prospective mate, while others are undesirable. This often happens in natural populations, but it is a tool which is extensively exploited by breeders for the purpose of creation and maintenance of a practically desirable population. Each desirable trait has certain index value, and the breeding mate is chosen based on the sum of the index values. Assortative mating ensures that certain phenotypes are preferred over others; thus, even mixing of alleles is not possible in this type of mating system. Based on the type of selection, assortative mating can be divided into two groups: positive assortative mating and negative assortative mating.

21.4.1.1 Positive Assortative Mating

Positive assortative mating means the mating of "like-to-like." In this system of mating, individuals with similar phenotypes are more likely to mate. This is often observed in human populations, where a tall man would prefer a tall woman as a mate and a short man and short woman would be more likely mating partners. This would result in more genetic and phenotypic variability than that in the case of random mating. Further, positive assortative mating would result in higher proportion of homozygotes.

21.4.1.2 Negative Assortative Mating

Negative assortative mating is the system where individuals with dissimilar phenotypes have higher probabilities of mating. This mating system would produce offspring that would represent the mean of the population. For example, the continuous mating of a tall and short individual for over generations would finally give rise to a population with a height which represents the mean of the tall and the short heights. Therefore, genetic and phenotypic variability is likely to be reduced. *Negative assortative mating (or disassortative mating)* is a system wherein the males with highest index value are likely to mate the lowest ranking females and vice versa. This would result in a higher proportion of heterozygotes.

21.4.2 According to Relationship

In this type of non-random mating, mates are chosen based on their genotype. In other words, genetically related individuals are chosen as mating partners. This type of non-random mating is also known as *inbreeding* or *consanguinity*. This is a common practice in some human societal structures and often takes place in nature when the population strength is low. Since inbreeding involves the mixing of similar types of alleles, it favours homozygosity. Highly inbred populations would have an exceptionally high proportion of homozygous individuals.

21.4.2.1 Line Breeding

In this type of inbreeding, the mating partners come from the same line of descent. *Line breeding* generally involves the mating of genetically related individuals apart from the individual parents or brothers and sisters. Typical line bred mates are grandchild/grandparent, great-granddaughter/great-grandson, uncle/niece, aunt/nephew and cousin. Line breeding is a less intense form of inbreeding. It is widely practiced by a number of animal breeders who want to concentrate a certain desirable trait and therefore an allele within the population without undertaking conventional inbreeding route.

21.4.2.2 Deliberate Inbreeding

As discussed before, populations with small number of individuals, certain societal structures and animal breeders favour inbreeding. In the latter two cases, the choice of an inbred mating partner is deliberate and is often made so that a certain trait or allele remains "fixed" in the prospective generation. However, in the process of trying to "fix" one or many traits, other allelic positions also start to become homozygous. So many unfavourable recessive traits that were previously not expressed owing to heterozygosity become homozygous and prevalent in a population. This results in the reduction of mean fitness of the population and is termed as *inbreeding depression*. In small populations, because there is a lack of choice in terms of mating partners, inbreeding is an unavoidable choice. In such populations, recessive diseases start becoming rampant and often lead to the extinction of these populations. Conservation biologists try to overcome this problem by introducing members from other populations into the inbreeding population.

21.4.2.3 Inbreeding Avoidance

To minimize the effects of inbreeding depression, conservation biologists and breeders tend to minimize the inbreeding coefficient by the process of outcrossing. *Outcrossing* is a term given to matings within a breed that are as unrelated as possible. The purpose is to avoid inbreeding but also to maximize heterozygosity of gene loci to capitalize on non-additive genetic effects.

21.4.3 Crossbreeding

Another type of non-random mating is *crossbreeding* which implies the mating of individuals from different breeds, varieties or populations within a species. The underlying idea is that each of the selected breeds has been line bred for several generations. By doing so, favourable genes have become "fixed" in that breed, and these are fundamentally different from those that have become fixed in partner breed. Thus, by mixing breeds, the favourable alleles of each breed are combined in the heterozygous offspring. Heterozygosity reaches its maximum in the offspring generation. A key feature of this type of mating is *heterosis*, H, which is defined as the superiority of crossbred offspring compared to the average of the two parental breeds. Heterosis is also known as *hybrid vigour*.

21.4.3.1 Single Cross

This system of crossbreeding involves two pure bred strains which are interbred. This is also known as *rotational crossbreeding system*. For example, suppose there are two breeds of cows "A" and breed "*B*" each having a set of desirable traits. In a single crossbreed system, cows of breed "A" will be mated with bulls of breed "*B*" and vice versa. The first generation offspring will be a true hybrid and possess maximum hybrid vigour. In subsequent crosses, only the crossbred females will be mated with pure bred males and never the reverse. However, the alleles of the pure bred will be concentrated, and an allelic equilibrium will be achieved by the seventh generation of mating.

21.4.3.2 Terminal Sire System

As we have seen that in single cross breed system, the heterosis is eventually lost, therefore to circumvent this problem, a *terminal crossbreed system* is often used. In this system, there are two rotational crossbreds. The female offspring resulting from the mating of the crossbreeding is then mated to the male of a third cross breed. The resulting offspring has highest heterosis and is not used for further breeding purpose. Therefore, this type of breeding is known as *terminal sire system* of breeding.

21.4.3.3 Composite Breed

When animals from two or more breeds with similar genetic composition are mated to each other and selection for further breeding mates is applied within this group, the crossbreed is known as *composite breeding* system. The crossbred animals become a composite breed. Composite breeds show more phenotypic variance compared to either purebreds or F offspring. This happens because of the random segregation of alleles. The more the number of breeds used to establish the composite breed, the more is the heterosis. A composite breed is created to have the "good" qualities of each breed that has gone into it.

21.5 Measurement of Genetic Variation

Now that we understand and appreciate how the frequencies of different alleles change in a population, we must move on to addressing a fundamental question of population genetics. Most geneticists are concerned with understanding how much genetic variation actually exists in a population. There are a number of reasons why this question is so central to the understanding of changes in population genetics dynamics. First, the amount of genetic variation determines the potential of a population to adapt to evolutionary changes. This adaptability plays a significant role in the survival or extinction of a particular population. Second, the variations give us an idea of the types of evolutionary forces acting on the population as some forces increase variations while others work to decrease it. The genetic variation can be measured at the protein or DNA levels as discussed below.

21.5.1 At Protein Level

Genetic codes translate into proteins which are finally responsible for functional execution. These are also perceived as phenotypic variations. In order to understand variation and polymorphism at protein levels, a population-wide analysis of differences in forms of proteins is conducted for a particular locus. This can be achieved by electrophoretically resolving the proteins for a particular gene. This will separate all different variants of the protein based on their charge and mass. Significant changes in amino acid will result in protein bands in different positions. However, the silent or same sense mutations at the DNA level will not translate into different proteins, and therefore these variations will not be noted. Understanding protein variation at population is basic and informative. However, it is not as robust as an analysis of DNA sequence variation.

21.5.2 At DNA Level

We have already discussed at some length about the various polymorphisms observed at DNA sequence levels. These polymorphisms at most loci are very informative as they can account for synonymous mutations (those that do not translate into a structurally or functionally different protein) and non-synonymous mutations (that which translate into functionally and/or structurally different protein). *Polymerase chain reaction* (PCR) is a uniquely useful tool in this direction. For example, *restriction fragment length polymorphism (RFLP)* is a process where changes in length of PCR products that can indicate sites of polymorphisms. Variations or polymorphism in this case results in the addition or deletion of a restriction enzyme cut site. Further, *DNA fingerprinting* studies are performed to analyse differences in DNA banding patterns. This will lead to identification of various alleles and variations in the genetic sequence. Another technique which is now being increasingly used to locate polymorphic loci is DNA sequencing. With the advent of *next-generation sequencing* techniques, obtaining whole genome sequence data is becoming feasible, and significant changes in the defining genetic pool of a population can be assessed.

21.6 Modulation of Genetic Variation

Population genetics involves both experimental and theoretical aspects. On the experimental side, the actual genetic variations are recorded, and an estimate of the mean heterozygosity in the population is made. The theoretical facet of population genetics deals with understanding how the current trends in genetic variation will change with respect to time and circumstances. It also involves making predictions about what the genetic composition should be and what changes are expected in response to evolutionary forces. Therefore, it is important to appreciate how different forces of evolution concert to modulate genetic variation in a population.

21.6.1 Mutation

One of the most important factors that can modulate genetic variation is *mutation*. Mutation is defined by a sudden, random and irreversible change in the genetic make-up of an individual. If the mutation is in the germ line, it will also become heritable. It is perhaps the strongest tool of evolution for the creation of new alleles and for the generation of population-wide genetic variation. Mutation creates new genetic variations; however, the fate of these variations is decided by the environment and the forces of evolution acting upon it. Whether a mutation is neutral or detrimental or beneficial to the organism is decided by the environment it is in and the functioning of natural selection. Therefore, mutation provides the raw material for the forces of evolution to work on. Let us take the example insecticide resistance in a pest population. If an insecticide "X" is widely used for the eradication of the insect "Y", the population will grossly suffer. However, if a mutation is able to confer resistance to the insecticide "X", then the insects with the mutation will have a survival advantage, and the frequency of this allele responsible for the resistant phenotype will start to rise. Therefore, mutation creates variation, and the process of natural selection decides which mutations would be retained or propagated and which ones would perish.

21.6.2 Genetic Drift

Any change in the relative frequencies of different genotypes in a small population occurring due to chance factors is termed as *genetic drift*. When such change is random in nature, it is known as random genetic drift. There are multiple reasons behind such observations. We must appreciate that a basic assumption of the Hardy-

Weinberg equilibrium is that the size of the population is infinitely large. An infinitely large population ensures that there is very little effect of chance factors, like a catastrophic event, or sampling errors on the observed frequencies of different genotypes. However when the population size is small, the expected and observed probabilities of genotypic distribution show a considerable change. For example, let us suppose that on an island there are ten individuals, five with black hair and five with blond hair colours. Also, let us assume that hair colour is controlled by a single gene inherited in Mendelian proportions. Now as an earthquake hits the island, and 50% of the population dies, there is a chance that all five individuals with black hair perish as the colour of hair cannot confer any survival advantage in this case. Thus the genetic variant resulting in black hair colour will be lost. Now, suppose that the population size were 1000, with 500 people having black and 500 having brown hair colour. In the event of an earthquake, it becomes very improbable that of the 50% of the population dying out, all 500 will have black hair colour.

This example also illustrates an important aspect of the factor that gives rise to genetic drifts. It is known as the *bottleneck effect*. A population bottleneck is an event which randomly wipes out a major segment of the population and considerably decreases its size. The event could be a natural disaster or a man-made disaster (like hunting). As discussed, in the event of a catastrophe, if the population size is small, the probability that the genotypic frequencies would reduce in proportion to the reduction in the population is extremely low. It may in fact so happen that certain variants become extinct and others become more concentrated in the surviving population.

The flip-side of this is the *founder's effect* which defined the loss in genetic variation that occurs due to the population arising from a small ancestral subset of a larger population. In other words, if few members of a large population migrated and established their own line of descent, the descendants would carry only the alleles that the few members had and the variants from the original population would be lost. A very compelling practical example of this phenomenon is the high prevalence of rare disease alleles in Ashkenazi Jews. The current estimates suggest that every one in three Eastern European descent Ashkenazi Jews is a carrier for certain rare genetic disorders, like Gaucher disease. A strong theory that explains this phenomenon is that Ashkenazi genetic diseases arise because of the common ancestry many Jews share. While people from any ethnic group can develop genetic diseases, Ashkenazi Jews are at higher risk for certain diseases because of specific gene mutations.

21.6.3 Balance Between Mutation and Drift

We have now discussed two important regulators of genetic variation: one being mutation and the other being genetic drift. However, these factors affect the variation in a population in reciprocal manner. While mutation is responsible for the creation of new alleles and thus in the increase in variation, genetic drift is responsible for sampling errors and chance factors resulting in decrease in genetic variation. It becomes interesting to understand how these forces together would influence genetic variation dynamics in a population. This is often understood by the *mutation-drift balance* concept.

Given that the rate of mutation and the effective population size are relatively stable, the amount of genetic variation will tend towards an equilibrium known as *mutation-drift balance* in which the rate at which variation is lost through drift is equal to the rate at which new variation is created by mutation.

21.6.4 Migration

An important assumption of the Hardy-Weinberg equilibrium is that there is not migration between populations of the same species. This ensures that the gene pool and relative frequency of alleles are stabilized in the given population. However, in practicality, migration is an important factor influencing genetic variation in a population. *Migration* implies the movement of organisms. For the purpose of this discussion, we will consider the movement of the organisms of the same species between two different populations. An immigration or addition of a new member from a different population would imply the addition of new genetic variants into the population of interest. Therefore, we concern ourselves with the movement of genes or alleles from donor to recipient population, and this can happen when a new allele migrates from one population to a recipient population. This is also known as *gene flow*.

Gene flow introduces a new allele into the recipient population. Given that mutation is a rare event, a mutation that arises in one population would remain limited to that population only if the process of gene flow does not happen. Therefore, gene flow helps in the spread of a new allele and the creation of genetic variation across receiving populations. Further, the gene frequencies of the migrating population and recipient populations are generally different. Through exchange of genes, gene flow helps in homogenizing the frequencies of alleles across populations. This ensures that two separate populations do not amass gross genetic differences between them.

21.6.5 Natural Selection

So far, we have focussed on factors that can affect changes in the genetic variations within a population, viz. mutation, drift and migration. However, these factors can create or change variations, but they alone cannot affect the adaptability of the change thus created. *Adaptation*, which implies the tendency of an organism to adjust to its habitat and environment, is controlled by the process of *natural selection*. Natural selection is a process by which organisms that can adapt better to their environment survive and reproduce. Natural selection is the driver of evolution. Migration, mutation and drift affect and modulate the pattern of adaptation, but adaptation arises from natural selection.

Let us go back to the example of the insecticide-resistant pests. Mutation gives rise to the allele for resistance. However, if there is no insecticide in the environment, the resistant insects will not gain any survival advantage, and given the low number of initial mutants, the frequency of such allele will be extremely low. However, if there is rampant use of the insecticide, the frequency of the resistant allele will rise considerably as the mutants will gain a survival advantage. This process which decides the acceptability and propagation of a variation is natural selection.

21.6.6 Balance Between Mutation and Selection

Now that we understand how variations are created and selected, let us look at how natural selection and mutation function to balance the frequency of alleles. This is often referred to as the *mutation-selection balance*. There is a pertinent mutation pressure on a population. Mutations arising naturally are often neutral or detrimental. Beneficial mutations are rare in nature. So, the process of mutation is continually adding detriment genes into the pool. On the other hand, the process of natural selection works to weed out unfavourable alleles by hindering their survival or reproducibility. Therefore, by the process of natural selection, the frequency of a detrimental allele will continue to decrease until it becomes rare. When the allele becomes extremely rare in a population, no significant change in its frequency is brought about by natural selection. So, when these opposing forces of mutation and selection work on a population, detrimental alleles are added by mutation and reduced by natural selection. Eventually, the population will achieve a mutation-selection balance, where the addition of a detrimental allele by mutation will be counter-balanced by the removal of alleles by selection. Consequently, a state of equilibrium will be established, and no effective change in allele frequency will be observed unless there is a new factor that is introduced into this dynamics.

21.7 Conservation Genetics

All the discussions so far have been involving genetic variability in a population and how the allele frequencies respond to evolutionary forces. One of the main areas where the application of such understanding is focussed is *conservation genetics* or the conservation of a gene pool of a species. This branch of population genetics aims at working out how the changes in allele frequencies and gene pool of a population can affect the viability of this population. For this purpose population viability techniques are employed, and researchers work out ways to prevent a population from going extinct.

The environment has suffered a great change. Part of it is natural, but human activities account for the majority of the detrimental changes to the environment. As such many species have lost their habitat and have become extinct. This process of extinction takes time and generally happens with gradual decrease in the size of the population in a particular habitat. The aim of a conservation geneticist is to determine the minimum number of members in a population required for the species to maintain its gene pool and a genotypic frequency. A population or a particular part of genetic variation can be maintained either in situ or ex situ.

21.7.1 Ex Situ Conservation: Captive Breeding

Ex situ *conservation* essentially means the conservation of the genetic diversity outside of its natural habitat and ecosystem. The genetic diversity is preserved in the form of a population of species, their germ line cells or somatic cell lines. These are preserved, maintained and kept available.

One method of doing so is *captive breeding*, where small populations of a species are captured and bred outside of their natural habitat and maintained in controlled environments. Some examples are wildlife reserves, zoos, botanic gardens, etc. This is done to prevent the species from extinction due to gross changes in their ecological environment. However, the fruitful preservation of a species in such environments requires a deep understanding of genetic, ecological, behavioural and ethical issues. One must consider the effect of genetic drift on a small population maintained outside of its habitat. The impact of founder's effect and bottleneck effect is higher in small population especially outside of its natural habitat.

21.7.2 Ex Situ Conservation: Gene Bank

Another way of ex situ conservation is the creation of a *gene bank* which acts as a repository of genetic material. This could be done in the form of cutting and freezing parts of plants, storing seeds in a seed bank and freezing and maintaining germ line of somatic cells of organisms.

Gene banks help in the preservation and rehabilitation of genetic diversity. For example, frozen plant material could be revived and propagated artificially. Further, frozen mammalian sperms can be used for artificial insemination, surrogacy and revival of the mammalian species.

21.7.3 In Situ Conservation

In situ *conservation* is the process of preserving populations of organisms as they occur in nature. The population is maintained in a natural state and in its natural habitat. This helps not only in the preservation of the organism itself but also in the preservation of the evolutionary processes that help maintain the adaptability of the population to its environment. For example, large ecosystems may be left intact as protected reserve areas with minimal intrusion or alteration by humans.

In situ conservation adds to the ex situ method of conservation. It is extremely advantageous as it maintains not only a population but an ecosystem as a whole. As such, it safeguards many genetic variations that we might not be aware of. It also offers a window through which a biologist can study the forces of evolution in action. However, maintaining in situ conservatories requires the combined efforts of the local and government level support.

Box 21.1: Scientific Concept: Analysis of Genetic Variation and Potential Applications in Genome-Scale Metabolic Modelling – Joao G. R. Cardoso et al.

An understanding of genetic variability in microorganisms is very crucial in the current scheme of things. The advent of next-generation sequencing (NGS) has reduced the cost of sequencing and made re-sequencing possible. Re-sequencing implied the sequencing of closely related strains in search of genetic variations. Re-sequencing has implication in biotechnology, understanding of pathology of microorganism-based diseases, evolution in environment and effect of human activity on microorganism diversity. It also helps in the making of in silico genome-scale metabolic models (GSMs) which consists of gene-protein reaction (GPR) association maps that help in determining how biochemical reactions relate to the genome and proteome of a cell.

In a very comprehensive review, Cardoso et al. talked about the various tools that are used for analysis of genetic variants. They also discussed techniques that could detect both synonymous and non-synonymous single nucleotide variants (SNVs) in coding regions or variants in promoter or other regulatory regions.

They focus on different in silico tools that can be used to determine how a particular variation in the genetic make-up of an organism can affect its phenotype. It is important to appreciate that the effect of a variation is not binary in nature for practical purposes, i.e. a single variation can lead to multiple types of effects. A variation can have no effect on the phenotype; it could lead to a loss in function of the coded protein; or it could improve the protein's activity. The software uses a different criterion to analyse the effect of a genotype on a particular phenotype. Figure 21.7 depicts the different softwares available and the criterion they use for analysing the effects of a variation.

The article also discusses deep mutational scanning, a technique that uses traditional methods like error-prone PCR to construct an exhaustive library of all possible variants. The variants are sequence using NGS methods and then assigned scores based on fitness measure (e.g. growth rate, ligand binding or product fluorescence) for each amino acid substitution discovered in the selected pool.

This article describes, in extensive details, the emergence of in silico, in vivo and in vitro technologies that help in the prediction of genetic variations and their possible phenotypic outcomes.

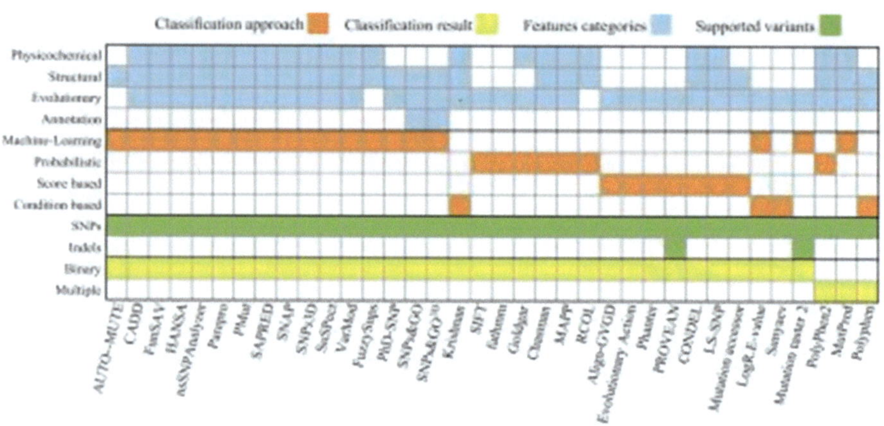

Fig. 21.7 Different criterion used in software to analyse the effect of a genotype

21.8 Chapter Summary

In this particular chapter, we have tried to explore the various facets of population genetics and genetic variation.

- Genetic variation is the key to the diversity present in nature and is often evident in the form of polymorphism.
- Polymorphism is alternating forms of a genetic sequence present in at least 1% of a population. If the variation is at a single nucleotide level, it is known as single nucleotide polymorphism (SNP).
- Microsatellites are short sequences that display copy number polymorphism and can be indicative of genetic variability.
- Haplotype is a segment of DNA that is inherited en bloc and thus displays very tight linkage. A map of haplotypes can indicate which polymorphisms are tightly associated with traits, and very few polymorphs can thus be informative of a considerable segment of the genome.
- All the alleles and polymorphs present in a population are known as its gene pool. Allele frequency and genotypic frequency are crucial parameters to understand the dynamics of inheritance of this gene pool in the population.
- The Hardy-Weinberg equilibrium explains the dynamics of genetic inheritance through a mathematical expression. Given that no evolutionary forces are acting on a population, the allele frequency for a particular gene inherited in a Mendelian pattern would remain constant over generations.
- Genetic drift, mutation, migration and natural selection act on a population and have distinctive effects on the genetic variability and genotypic frequency of the population.

- The random mating system forms a basic assumption of the Hardy-Weinberg equilibrium. However, non-random mate selection is prevalent in nature.
- Assortative mating is a form of non-random mating where the mates are selected on the basis of presence or absence of a certain panel of phenotypes.
- Inbreeding is the mating of individuals with common ancestry. It works to decrease heterozygosity in a population.
- Outcrossing is the opposite of inbreeding and requires mating of individuals from different ancestral lineages.
- Given that the gross changes in the environment are leading to the loss of genetic variation and extinction of certain populations, genetic conservation becomes necessary.
- Gene banks and captive breeding are forms of ex situ conservation where individuals are conserved outside of their natural habitat.
- In situ conservation is the conservation of a population in its natural habitat.

References

Abdul-Muneer PM (2014) Application of microsatellite markers in conservation genetics and fisheries management: recent advances in population structure analysis and conservation strategies. Gen Res Int 2014:691759

Brooker RJ Genetics: analysis and principles, 4th edn. McGraw-Hill, New York

Bubendorf L, Grote HJ, Syrjänen K (2008) Molecular techniques comprehensive cytopathology, 3rd edn. Saunders Elsevier, London, pp 1071–1090

Cardoso JG, Andersen MR, Herrgård MJ, Sonnenschein N (2015) Analysis of genetic variation and potential applications in genome-scale metabolic modeling. Front Bioeng Biotechnol 3:13

Jarne P, Lagoda PJ (1996) Microsatellites, from molecules to populations and back. Trends Ecol Evol 11:424–429

Leimar O (2005) The evolution of phenotypic polymorphism: randomized strategies versus evolutionary branching. Am Nat 165:669–681

Oxford GS, Gillespie RG (1996) Genetics of a colour polymorphism in Theridion grallator (Araneae: Theridiidae), the Hawaiian happy-face spider, from Greater Maui. Heredity 76: 238–248

Panoutsopoulou K, Wheeler E (2018) Key concepts in genetic epidemiology. Methods Mol Biol 1793:7–24

Russell PJ iGenetics: a molecular approach, 3rd edn. Pearson Education, Inc., Publishing as Pearson Benjamin Cummings (publisher), San Francisco, CA

Strachan T, Read AP (2011) Human molecular genetics, 4th edn. Garland Science/Taylor & Francis Group, New York

The International HapMap Consortium (2003) The international HapMap project. Nature 426:789–796

Zeldovich L (2017) Genetic drift: the ghost in the genome. Lab Anim (NY) 46:255–257

Zhao H, Pfeiffer R, Gail MH (2003) Haplotype analysis in population genetics and association studies. Pharmacogenomics 4:171–178

Evolutionary Genetics

22

Ankita Dua and Aeshna Nigam

22.1 Concept of Evolution

Nothing in biology makes sense except in light of evolution—Theodosius Dobzhansky

Evolution may be defined as the changes in gene pool which will lead to progressive adaptation of the population to the environment. The concept of natural selection proposed by Charles Darwin and Alfred Russel Wallace combined with Mendelian inheritance gives an insight into the mechanism of evolution. Evolution is basically a two-step process which can only occur when there are heritable variations in the gene pool.

In words of Darwin, evolution is 'descent with modification' that species change over time giving rise to new species sharing a common ancestor. Each species has its own set of heritable differences from the common ancestor, which accumulate over periods of time gradually. In the 'tree of life,' repeated branching events produce a multilevel tree that links all organisms. Evolution, with respect to extent of change and geological time scales, can be classified as:

- Microevolution is defined as a systematic change in allele frequencies and chromosomal segments in local populations.
- Macroevolution involves changes of greater magnitude. Changes of macro extent occur in the development of characteristics that distinguish groups such as genera, families, orders, classes and phyla. These take place over a long span of geological time scale.

A. Dua · A. Nigam (✉)
Shivaji College, University of Delhi, New Delhi, India
e-mail: aeshnanigam@shivaji.du.ac.in

D. Kar, S. Sarkar (eds.), *Genetics Fundamentals Notes*,
https://doi.org/10.1007/978-981-16-7041-1_22

Various evidences have supported the theory of evolution. Homologous structures are evidence of presence of a common ancestor, whereas analogous structures evolve independently without common descent and are similar due to inhabitance of similar environment as a result of convergent evolution. Species relatedness can be evaluated by similarities and differences at the molecular level, assessing the DNA and protein structure. Study of biogeography shows distribution of organisms all over the globe and gives clues about relationships between them. One of the most valuable evidences comes from fossil records that unfold the story of timelines of existence of species over various periods (Futuyma 1998; Hall and Hallgrimsson 2008; Ridley 2004).

22.1.1 Theories of Evolution: Lamarckism

Jean Baptiste de Lamarck is a French naturalist of the nineteenth century, known for his speculations on evolution, published in his book *The Philosophie Zoologique* in the year of 1809. 'Inheritance of Acquired characters' is his significant proposition albeit unaccepted widely.

Lamarck's strong doctrine is, species undergo modifications concerning the environment, contradicting its fixity. He claims domestication of plants and animals modifies their structure unrecognizable to the wild variety. For instance, domestic ducks and geese lost their ability to fly compared to the wild birds of their race due to prolonged captivity. If the captivity is extended, even more, there might not be a change only in their ability but also in their morphology, claims Lamarck. He further endorsed, *Ranunculus hederaceus* of terrestrial habitat grown in a damp soil has been found to have a smaller stem and devoid of small segmental leaves which are dissimilar to the same species, *Ranunculus aquatilis* of aquatic habitat. According to Lamarck, the impact of the environment on living species causes an imperceptible alteration in structure and organization. Unconcealed modifications in animals can be observed with substantial changes in the environment leading to novel requirements. The emergence of new habits due to these long-lasting changes in the habitat will lead to the development of a new pertinent organ which further develops stronger and larger with perpetual use. These modifications in animals curtail inefficient organs to disuse. The disappearance of inefficient organs is coupled with prolonged disuse. Lamarck also observed these permanent modifications become inherited giving rise to distinct species. In a nutshell, external stimulus causes a heritable beneficial genomic mutation in species for adaptation (Fig. 22.1). Based on his extensive studies and observation of species, he discerned two laws of nature:

First Law: 'In every animal which has not passed the limit of its development, more frequent and continuous use of any organ gradually strengthens, develops and enlarges that organ, and gives it a power of the proportional length of time it has been so used; while the permanent disuse of any organ imperceptibly weakens and deteriorates it, and progressively diminishes its functional capacity until it finally disappears.'

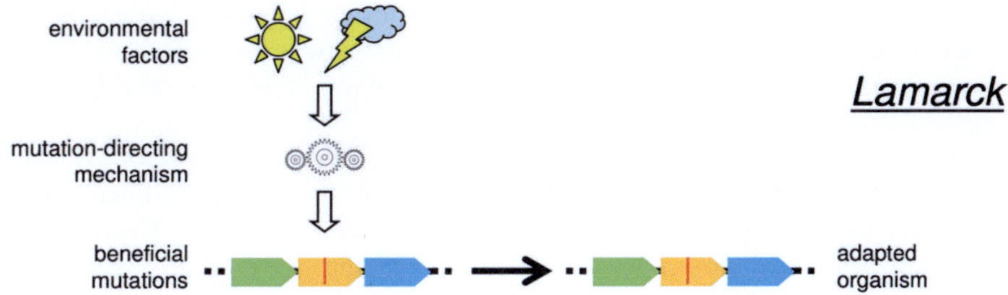

Fig. 22.1 Lamarck's theory of evolution. (Adapted from Koonin and Wolf 2009)

Lamarck has demonstrated the authenticity of the laws with various examples and believed these laws are certainly true and permanent. He gave definitive examples of use and disuse of organs rising due to contemperory habits. Certain changes in the habitat of animals have induced swallowing of the feed without primitive mastication, eventually leading to the absence of teeth in vertebrates (e.g. whale, anteater). Further, Lamarck articulates, disuse of eyes had constricted the organ in moles. Living beneath the soil, where sunlight is arduous to percolate, mole has tiny eyes based on its utility. In addition to that, *Spalax* which lives in the similar habitat of mole rats is blind with the vestiges of the organ, as a result of lack of utility. Snakes are the exceptional class of reptiles which do not have four limbs like crocodiles, frogs and turtles. Lamarck reasoned this unique feature of snakes with two facts: (1) Their peculiar adoption to crawl with the elongated body helped them to hide in the grass and move in confined places with ease (2). Their long legs have put to perpetual disuse, which eventually disappeared. It is unlikely for a snake to have short legs which makes them incompetent, and they cannot have more than four legs under the reptile criteria.

Lamarck illustrated his opinion on the development of a new organ or stronger and prominent development due to recurrent use of an existing organ supporting altered environment with exemplifications. Perpetual use of skin between the three-digit feet to capture aquatic life forms of prey has given rise to the palmate or webbed foot essential for swimming in ducks and geese. Lamarck explained, in a few cases, birds developed long stretchy legs with feathers above the thighs when they are reluctant to swim and depend on prey in the shore. In forests, prey-predator stress is inevitable and especially ruminants like deer need to protect themselves from predators as well as hunters. Lamarck claims, the '**inner feeling**' of the animals to safeguard them from dangers has allowed the secretion of a blend of horny-bony substance which gave rise to antlers and horns. He claims giraffe has developed long necks compared to its ancestors. The gradual transformation of African forest grasslands to arid areas coerced the animals to depend on trees for nourishment. This obligation has resulted in elongated necks and limbs in giraffe. Lamarck's other doctrine is, the adaptations in species that help in survival are inherited and are called use-inheritance. This gives rise to the second law of Lamarckism.

Second Law: 'All the acquisitions or losses wrought by nature on individuals, through the influence of the environment in which their race has long been placed, and hence through the influence of predominant use or permanent disuse of any organ; all these are preserved by reproduction to the new individuals which arise, provided that the acquired modifications are common to both sexes or at least the individuals which produce the young.'

This theory of use-inheritance is considered improbable due to lack of evidence by Lamarck. This theory was condemned by many people, and experiments were conducted to prove or disprove it. August Weismann, a German Evolutionary biologist, was the first person to propose the germplasm theory in animals (Weismann 1893). He proposed any metamorphosis in the somatoplasm does not affect the germplasm. Weismann argued Lamarck's proposition of 'Inheritance of acquired characters,' claiming germ cells give rise to somatic cells, therefore for a variation to occur, preliminary change must occur in the germplasm to be inherited. During one of his lectures delivered in 1888 on 'A supposed transmission of mutilations,' he presented results of his experimental investigation on mutilation inheritance in mice (Weismann 1891). In the first generation, he amputated tails of 12 mice comprising 7 females and 5 males. The offspring from the first generation were found with perfectly grown tail, and even subtle presence of the acquired inheritance was not found. Surprisingly, the fifth generation with 901 offspring developed from the mutilated parents did not show the trivial presence of rudimentary tail defects or tail-less condition. Weismann set forth a plausible assumption that the expression of mutilations in the progeny might take place after many generations. Unfortunately, use-inheritance can be widely accepted if there was at least one proof to support this theory.

Furthermore, McDougall in 1938 (McDougall 1938) conducted experiments on learning as an acquired inheritance. He designed a T-shaped tank, with two exits: one exit with the electric shock was illuminated, whereas the free exit was kept dim. Rats which chose the lighted pathway received an electric shock for 3 s, and animals which chose the dim exit were rewarded. He trained the rats six times daily to accustom to the experiments and halted the training only when the rats learnt to discriminate the exits and chose the dim exit successively. He bred these rats for the second generation. McDougall found mistakes reduced gradually from generation to generation and claimed learning is an acquired trait. Drew (1939) criticized McDougall experiments for biased learning in animals, and inheritance of avoidance behaviour interlinked with various factors is impossible. When repeated, contrast results were obtained by Crew and Agar (Agar et al. 1954; Crew 1936). Further, technical errors found in McDougall's experiment led to severe criticism.

22.1.2 Theories of Evolution: Darwinism and Neo-Darwinism

Charles Darwin and Alfred Russel Wallace independently developed theories of evolution based on natural selection which were communicated to a meeting of Linnaean Society of London on July 1, 1858. Both went on voyages of discovery before stating their ideas: Darwin went around the world on H.M.S. Beagle (Fig. 22.2), and Wallace travelled to Brazil, Malaysia and Indonesia.

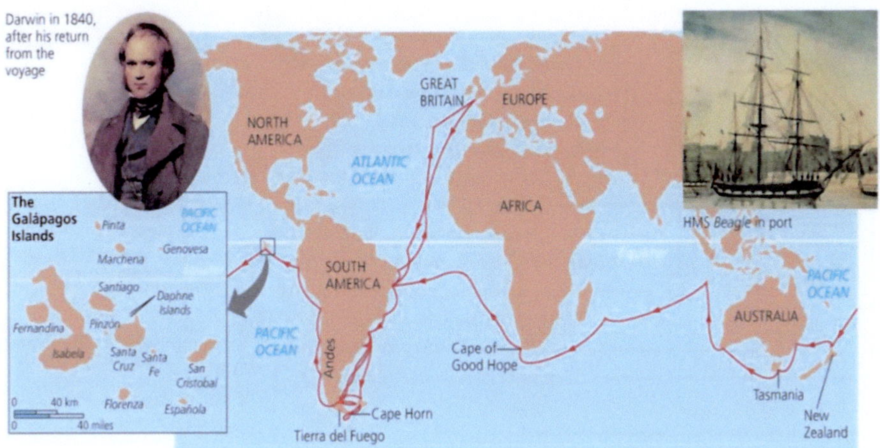

Fig. 22.2 The voyage of HMS Beagle. The path traced by HMS Beagle in 1831 in its 5-year journey that led to Darwin's postulates of natural selection and origin of species. (Adapted from Campbell et al. 2008)

Wallace is considered to have begun the study of biogeography, and both of them were posthumously awarded the 'Gold Medal' by Linnaean Society of London for the 50th anniversary of their publication. Darwin's voyage spanned a period of 5 years from 1826 to 1830 that enabled him to observe the wide range of species and geological forms around the globe. His breakthrough discovery was the exotic collection of flora and fauna evolving on the Galapagos Islands off the coast of Ecuador. These are located in the Pacific Ocean, approximately 960 km west of the South American coast, straddling the equator at the 90th meridian west. The archipelago was made of 13 major islands, 6 smaller islands, over 40 islets and many smaller unnamed islets and rocks, for a total of approximately 8000 km^2 of land spread over 45 000 km^2 of water.

He noted that different islands with similar habitats were not always occupied by identical species. He proposed that:

- Excess reproduction and limited resources lead to competition, and as a result of natural selection, only the organisms best adapted to the habitat could survive and pass their characters to the next generation.
- Changing environments and hereditary variations and natural selection together result in modification of existing characters or origin of new characters that become established throughout a species.

Darwin's work was documented in his book *The Origin of Species* in 1859 which is said to have revolutionized the foundation of evolutionary biology.

> *The most curious fact is the perfect gradation in the size of the beaks in the different species of Geospiza, from one as large as that of a hawfinch to that of a chaffinch, and . . . even to that of a warbler. . . . Seeing this gradation and diversity of structure in one small, intimately related group of birds, one might really fancy that from an original paucity of birds in this*

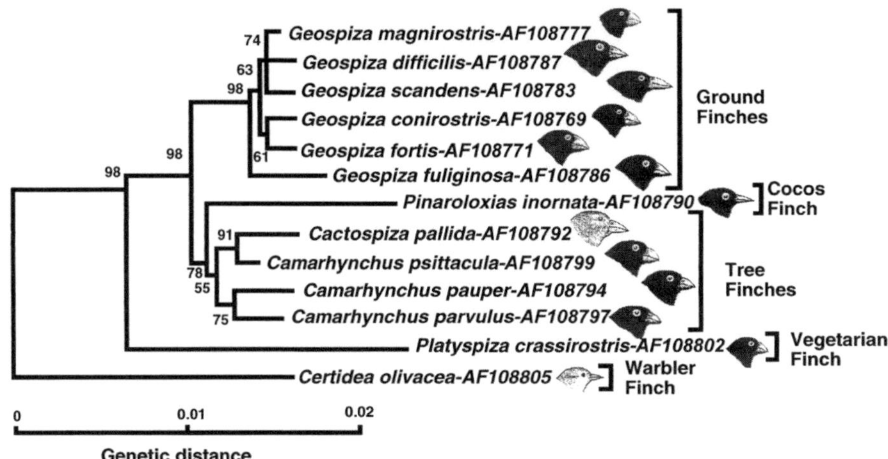

Fig. 22.3 Phylogenetic analysis of Darwin's finches. Combined analysis of the cytb and cr sequences of Darwin's finches done by neighbour-joining tree construction method. Shape of the beak is illustrated by the drawings made on the right side (Sato et al. 1999)

> *archipelago, one species had been taken and modified for different ends.—Darwin (1839)* (Abzhanov 2010)

He studied the concept of adaptive radiation (diversification of a founder population into a collection of species differentially adapted to diverse habitats) under natural selection of 14 closely related finch species (belonging to the Avian order Passeriformes). These were diverse in their beak shapes and sizes. Later on, he asked John Gould at the museum of Zoological Society, London, to catalogue these species. Gould realized that these diverse species actually reflected differences in diets that were closely related to each other and distantly to a South American mainland species. Developmental basis of this variation was analysed by a comparative analysis of expression patterns of the growth factor Bmp4 of the species of genus *Geospiza* (Abzhanov et al. 2004; Parent et al. 2008). The expression of Bmp4 in upper beak mesenchyme correlated with the deep and broad beak morphology. When this protein was misexpressed in chick embryo, it caused morphological transformations similar to beak morphology of the large ground finch (Abzhanov et al. 2004). In another analysis, two mitochondrial DNA segment sequences (cytb, cytochrome b, and cr, control region) have been used for evolutionary history of the group. The results reveal that Darwin's finches are a monophyletic group with the closest species to the founder being Warbler's finch, followed by vegetarian finch and two sister groups of ground and tree finches. The Cocos finch found on the Cocos Island of the Pacific Ocean is related to the tree finches (Fig. 22.3). The mtDNA and microsatellite data found was consistent with the theory that the finches originated from a single common ancestor from a possible founder population that reached the islands from South/Central America (Bluestone 2009; Bowman 1961; Parent et al. 2008).

Post Darwin, a number of workers such as J. Huxley, T. Dobzhansky, J.B.S. Haldane, S. Wright, T.H. Morgan, G.G. Simpson, G.L. Stebbins, E. Mayr and C. Darlington came about the theory of 'Modern Synthesis' or Neo-Darwinism.

> *Neo-Darwinism is an attempt to reconcile Mendelian genetics, which says that organisms do not change with time, with Darwinism, which claims they do.*—Lynn Margulis

The term 'Neo-Darwinism' was coined by physiologist George Romanes in 1883. It is the modern concept of evolution where a gene pool of a population is a unit of selection versus the individual. There is a synthesis between Darwin's theory of natural selection and gene mutations playing a central role in production of variation in a population. By this time, Mendel's work had also been rediscovered, and it was significant in adding to the theory. A connection was established between genes as a unit of evolution and natural selection as the mechanism. Evolution is a gradual process which is a result of variations accumulated at the genetic level in populations over a period of time (via mutations). The result of these variations is the phenotypic changes in the population and in allele frequencies as a result of natural selection (Denis 2011).

The pioneer workers who strengthened the 'Concept of Modern Synthesis' were as follows:

- Theodosius Dobzhansky worked on evolution of fruit fly populations (*Drosophila melanogaster*). His major publication was *Genetics and the Origin of Species* in 1937.
- E.B. Ford worked on 'Ecological Genetics,' and his book titled the same was published in 1964.
- H.B.D. Kettlewell was the pioneer worker on industrial melanism in peppered moth *Biston betularia*.
- Julian Huxley wrote the book 'Evolution: The Modern Synthesis' in 1942 that brought to light the concepts introduced by Fisher, Haldane and Wright.

22.2 Genetic Variation in Population

Variation is an interplay of heredity and environment and is the most impressive characteristic of any sexually breeding population. The capacity to undergo variation is due to a complex set of heritable traits. Variations are the reason present-day organisms have evolved from a few primordial forms of life and diversified into complex life systems; they are seen at phenotypic, chromosomal and molecular levels.

Phenotypic variation is seen in different varieties of species in animals as well as in different flowers in the flower kingdom. These visible morphological or phenotypic differences are called polymorphism, meaning 'many forms.' For example, land snails have different coloured bands on their shells, mammals have different coloured coats, and insects have patterned wings. The human species have also been understood to be polymorphic; the most interesting example being their blood groups where the antigen coded is present on the surface of the blood cell. The antigens are polymorphic. In the Duffy blood typing system, there are two antigens present on the surface of cells. Alleles coding these antigens called the 'Duffy alleles' encoded by a gene on chromosome 1 are often polymorphic. Various human ethnic groups have varied status of Duffy polymorphism (Anstee 2010). Variation in chromosomes is often an indication of polymorphism at the phenotypic level. Researchers found abundant comparative data on comparing polytene chromosomes from various species of *Drosophila* (Zykova et al. 2018). These chromosomes develop from diploid nuclei chromosomes by successive duplication of each chromatid without the segregation. The formed elements associate lengthwise and form a cable-like structure. In *Drosophila melanogaster*, they are >100 times longer than regular metaphase chromosomes. Here, level of variation in chromosomes can be studied at an unparalleled level. In every polytene chromosome, banding patterns are significant as there is alternation between compacted and decompacted regions of chromosomes known as bands and interbands. Dobzhansky and his team members identified various patterns of banding in *Drosophila* species. Variations are also seen at the nucleotide and at the protein level. Genetic variation in natural population was studied by R.C. Lewontin, J.L. Hubby and H. Harris by application of gel electrophoresis to study amino acid differences in proteins of various species. Amino acids are building blocks of proteins, and their differences in shape, molecular weight and charge can be studied while migrating in gels. This technique was applied to various other creatures as well, and different forms of proteins could be studied as the mobility of a protein was specific through the gel. The ultimate data on genetic variation is obtained on DNA sequencing. All sequences—exons or introns—can be sequenced and analysed. At present high end sequencing technologies have been successful in decoding even the billion base pair human genome.

Variations have been classified by various workers into a number of categories:

22.2.1 Continuous Vs. Discontinuous Variations (Table 22.1)

Evolutionary biologists also classify variations as follows.

Table 22.1 Comparing continuous and discontinuous mutations

Continuous variations	Discontinuous variations
• Numerous, small-scale variations occurring between individuals of a population with respect to characters such as height, weight, skin colour, texture, etc. • These are fluctuations on either side of the mean value of a character.	• Sudden, drastic changes in traits of individuals of a population • Usually referred to as mutations which are responsible for variations that lead to formation of a new species

22.2.2 Environmental Variations

These are acquired changes and may not be inherited at the gene level. Environmental influences act on nutrition, competition, disease and biotic and abiotic factors. Phenotypic plasticity is defined as the ability of a genotype to produce more than one phenotype when exposed to different environments. For example, in the semiaquatic plant *Ranunculus*, the leaves that are submerged in water have a dissected leaf lamina, whereas those above water have a single lamina (Cook and Johnson 1968).

22.2.3 Mutational Variations

According to T.H. Morgan, the ultimate source of heritable variation is mutation, and he stressed on the fact that the latter are changes in single genes with effects ranging from minute to severely pronounced. The term 'gene' was coined by Danish botanist Johannsen in 1909, and Wilhelm Waagen gave the term 'mutation.' However, it was Hugo de Vries and William Bateson who described mutations as major changes in genes at the molecular level leading to deviations from the parental types. Mutations occur in genes bringing about variation, and natural selection favours the advantageous variants and weeds out the disadvantageous ones (Marshall 2002). Mutations can be point mutations or chromosomal aberrations:

- Gene/point mutations (discovered by T.H. Morgan during his work on white-eyed trait in a laboratory population of *Drosophila*) are the elemental source of variations as they bring about changes in a pure line gene of a parent and lead to origin of a new gene that is inheritable. Genes are generally stable and mutate at a slow rate. These changes could be in composition—transition and transversion—or change in number by addition, deletion and frame shift. Mutation rate is defined as 'probability that an allele copy changes to another allelic form in one generation.' As mutation is a spontaneous process, the rate of change in gene frequency from mutation is very low. Mutation of a single gene usually has minimal effect on the population; it is amplified by interaction with other genes, for example, cases of multiple allele interactions, epistasis, pleiotropy and polygenic effects. Generally, mutations that are less severe occur at a higher frequency, and a mutant gene may not display its effects under normal environmental conditions but may be deleterious in a different environment.

Table 22.2 Change in number of chromosomes along with amount of DNA

Aneuploidy	Loss or gain of individual chromosome of a normal diploid set of an organism. It can be nullisomy ($2n - 2$), monosomic ($2n - 1$), trisomic ($2n + 1$), tetrasomic ($2n + 2$) and polysomic
Haploidy	Loss of an entire set of chromosome
Polyploidy	Gain of one or more sets of chromosomes. It can be triploidy ($3n$), tetraploidy ($4n$) and polyploidy

n refers to the haploid set of chromosomes

- Aberrations on the other hand refer to loss or gain of genes and change in placement or position within the chromosome or between different chromosomes. Structural changes involve deficiency (loss), duplication (repetition of a DNA segment) and polyteny (multiple copies of entire DNA strands) which bring about changes in the amount of total DNA (Table 22.2). Changes in location of genes (no change in DNA amount) are done by inversion (reversal of gene order within same chromosome)—paracentric/pericentric (depending on presence/absence of centromere in the inverted segment) and translocation. Loss or gain of genes with change in amount of DNA is done by change in chromosome number.

22.2.4 Recombination

Variations introduced as a result of permutations and combinations of existing genes are termed as recombination, and these are the raw materials for natural selection and evolutionary changes. Such events may add up with point mutations, and their combined effect makes a great impact. Recombination maybe a result of:

- Crossing over of two homozygous gametes to give rise to a heterozygous offspring.
- Random assortment of genetic material.

22.3 The Neutral Theory of Molecular Evolution

22.3.1 Neutral Theory

The neutral theory of evolution was given by Motoo Kimura in 1968 and it states:

> This neutral theory claims that the overwhelming majority of evolutionary changes at the molecular level are not caused by selection acting on advantageous mutants, but by random fixation of selectively neutral or very nearly neutral mutants through the cumulative effect of sampling drift (due to finite population number) under continued input of new mutations (Kimura 1991).

As sequencing technologies developed, more data was available to test this hypothesis. Kimura predicted that in protein molecules the substitutions of amino acids that occur more often are conservative. These replacements do not affect the function of the protein as the changed amino acid shares similar biochemical properties to the original one. Introns and pseudogenes also evolve at a high rate similar to these base substitutions.

There were several similarities of the neutral theory with Darwin's postulates:

- Natural selection is the driving evolutionary force and results in adaptation of organisms to their environment.
- Mutations in the functionally important regions are deleterious, and selection rapidly removes them from the population.

Kimura's theory challenged Darwin's work as:

- It was proposed that intraspecific nucleotide differences are a product of neutral mutations rather than genetic drift.
- Most intraspecific polymorphisms are also neutral.

Alterations or substitutions in protein-coding DNA sequences can be classified as synonymous (do not affect encoded amino acids) and non-synonymous (affect encoded amino acids). Synonymous substitutions are mostly neutral as they do not alter the protein sequence.

The neutral theory focuses on three main ingredients: mutation, genetic drift, and purifying selection:

- Mutations are the driving forces of evolution in proteins as well as DNA. Every generation, approximately 10^8–10^9 events of mutation occur. As discussed, mutations can be beneficial or detrimental to the fitness of an organism or may be selectively neutral.
- If the mutations are advantageous, they end up getting fixed in the population. The negative mutations are eliminated from the population by the action of purifying selection.
- Selectively neutral mutations have no effect on fitness, and their fate is dependent on random genetic drift. Most are lost from the population shortly after they appear.

Evolution rate according to the neutral theory depends on the neutral mutation rate which is constant in different lineages over time. Highest rates of evolution are found in molecules in which any mutational change may have least effect on the function. On the contrary, the lowest rate is found in the molecules where selection pressure is the highest (Duret 2008).

22.3.2 Rate of Neutral Substitution and Molecular Clock

Mutations in DNA can be of three types: deleterious (which may affect the fitness of the individual negatively), may increase efficiency of organism and can be neutral. When a mutation has no effect on survival and reproduction of an individual, it is termed as a neutral mutation.

Investigations about the molecular clock started with study of the proteins—haemoglobin, cytochrome c and fibrinopeptides in the early 1960s. E. Zuckerkandl, L.B. Pauling, E. Margoliash, R.F. Doolittle and B. Blomback concluded that

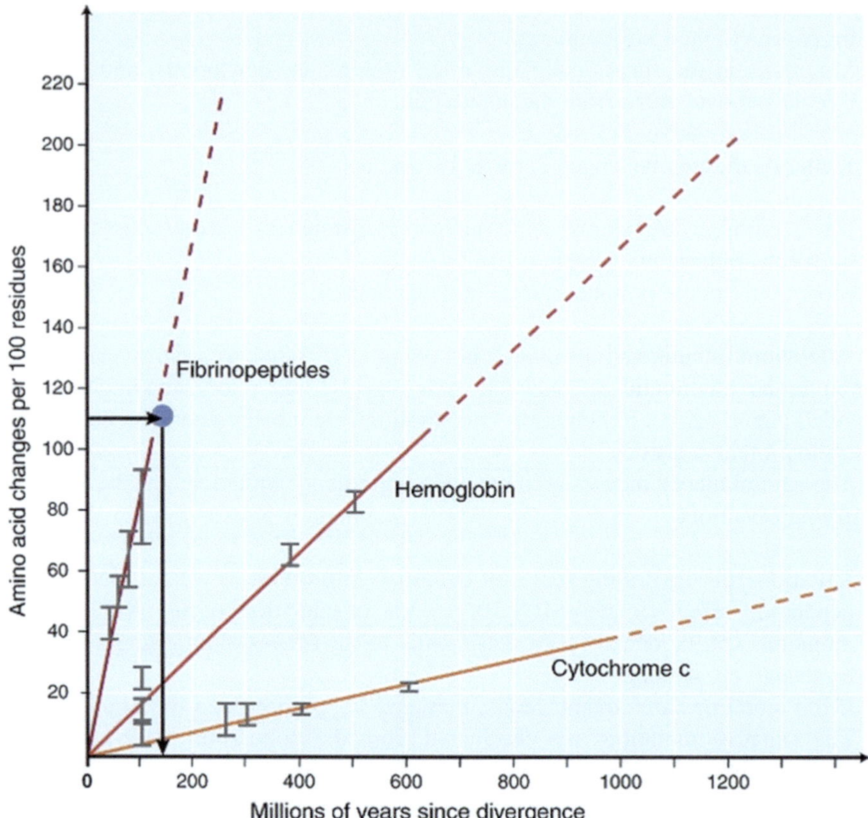

Fig. 22.4 Amino acid changes in cytochrome c, haemoglobin and fibrinopeptides. All three proteins display different rates of changes per unit time, but the rate is constant for each. (Adapted from Yi, S.: Neutrality and Molecular Clocks. Nature Education Knowledge. 4(2), 3 (2013))

differences in the protein sequences of different species were in accordance to their timeline of divergence (Fig. 22.4). Regular ticks of clocks were considered synonymous to accumulation of amino acid/nucleotide substitutions over time in a lineage and hence the name 'molecular evolutionary clock' coined by Zuckerkandl and Pauling in 1965. Evolutionary rate or rate of molecular clock is the number of substitutions per a defined unit of time.

The concept of molecular clock measures the absolute time of evolutionary change based on the observation that various genomic regions evolve at steady rates. It talks about orthologous and paralogous genes. Both are homologous, that is, arising from a common ancestor, where orthologous genes diverged post a speciation event and code for proteins with similar function in different species and paralogous genes diverge within the same species and encode for proteins with similar not identical function. Nucleotide substitutions in orthologous genes are proportional to the speciation event where the species diverged from the common ancestor and in paralogous genes to the time of duplication.

Initially it was hypothesized that the evolutionary force driving these substitutions was natural selection. However, Kimura said that the changes at the molecular level were neutral, that is, no consequence over fitness, and these occurred completely by random chances. Hence, it could not be predicted whether a specific neutral mutation will be or not fixed in a population.

Rate at which neutral substitutions occur in a population depends on the mutation rate and can be predicted as:

Total number of haploid individuals in a population = N.
Rate of neutral mutations = u/individual/generation.
Total number of mutations in one generation = Nu.

As all the mutations are neutral, their success rate is dependent on simply chance probability. Hence all mutations have an equal chance of getting fixed (equivalent to substitution).

$$\text{Probability that each mutation is fixed} = 1/\text{N}.$$

$$\begin{aligned}\text{Rate of substitution} &= \text{Number of new mutations in each generation (Nu)} \\ &\times \text{probability that each mutation gets fixed } (1/\text{N}) = \text{u}\end{aligned}$$

Therefore, for neutral mutations, rate of substitution = rate of mutation.

It can be predicted that according to the neutral theory, if mutation rates are constant over time, substitutions will occur constantly as well and clock-like regular rates of substitutions will occur constantly over time.

22.4 Natural Selection

Natural selection is defined as a directional, non-random and guiding force that leads to evolution of organisms to a better state of adaptiveness. Natural selection is the differential reproduction of genotypes; it is measured by the relative reproductive successes (fitness) of genotypes.

Salient features of natural selection:

- Key force in driving evolution.
- Shifting of allele frequencies in a large population.
- Fitness is an attribute of a genotype due to its genetic contribution to the future generation. High fitness is a measure of high rate of reproductive success and hence increased genetic contribution to the gene pool. Natural selection clearly favours individuals with higher fitness (Orr 2009).
- Intensity of natural selection acting on genotypes in a population is defined as selection coefficient.

- If a mutation is selected for or against in a population, it is termed as 'positive or negative' selection.

Modes of selection in nature vary at different levels of biological organization leading to varied evolutionary outcomes:

A. At the population level:
   - Directional selection.
   - Stabilizing selection.
   - Disruptive selection.
B. At the local population level or within species:
   - Kin selection.
   - Group selection.
C. At the level of individuals:
   - Sexual selection.

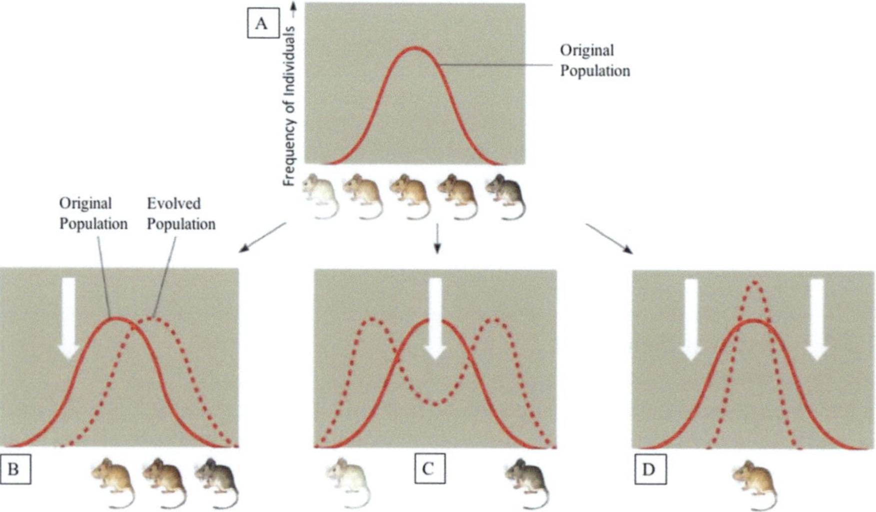

Fig. 22.5 Modes of selection. A hypothetical deer mouse population (with heritable variation in fur colour from light to dark) as an example of three types of natural selection. White arrows indicate patterns of evolution, selective pressures against certain phenotypes: *A*: Original population graph where frequency of individuals is plotted against the fur colour phenotypes. *B*: Directional selection favours the extreme phenotypes, in this case the dark fur individuals. The darker mice are saved from their predators as they take refuge under dark rocks in the environment. *C*: Disruptive selection favours variants at both ends of the spectrum. The mice inhabit patches of light and dark coloured rocks, and the mice of an intermediate colour are at a disadvantage. *D*: Stabilizing selection favours the intermediate/average phenotypes. If the environment consists of intermediate colour rocks, the light and dark mice will be eliminated. (Adapted from Campbell et al. 2008)

22.4.1 Directional, Stabilizing and Disruptive Selection

Effect of the genotype at various loci combined with environmental effects define the phenotype of an organism (Fig. 22.5).

22.4.1.1 Directional/Progressive Selection (Under Changing Environmental Conditions)

Under pressure of changing environmental conditions (biotic and abiotic factors), a species population is faced with the challenge of adjusting and keeping pace with change. Hence the genetic variables that were ideal in the older environment cannot continue to do so in an altered one. With respect to the change in the environment, a new adaptive peak is reached, an extreme being favoured and others eliminated. Also known as progressive selection, a unidirectional change in the genetic composition of the gene pool occurs where the peripheral variants with weak selective advantages originally may find themselves to be better adapted. Phenotypes at one end of the spectrum are selected causing the allele frequency to continuously shift in one direction.

The classical example of directional selection is the phenomenon of industrial melanism in England in the nineteenth century. The role of natural selection in evolution of melanic forms in peppered moths (*Biston betularia*) was studied in detail by H.B.D. Kettlewell and E.B. Ford. Peppered moths exist in two forms—melanic (black in colour; dominant allele C) and non-melanic (mottled grey; homozygous for recessive allele c). Prior to the industrial revolution, the melanic forms were a rare sight, and the non-melanic light coloured moths were abundant in the natural habitat. Predators could easily spot and prey upon the dark melanic forms that rested on the light coloured lichen-covered trees. The non-melanic moths easily blended with the environment and were not easily visible. With the start of the industrial revolution, these numbers underwent a dramatic change as did the environment. The sooty smoke ejected from the factories (heavy industrial areas around Manchester and Birmingham) resulted in attributing a black cover to the trees and prevented growth of lichens. This resulted in protection of the melanic forms and more visibility of the light coloured moths to the predators. Industrial melanism owes its name to the increase in proportion of the melanic forms as selection favoured the moths for the specific character—protective coloration. Kettlewell and Ford who studied this phenomenon in detail reasoned that a single dominant gene for the melanic colour was present in the population and its frequency increased from a mere 1% to 90% with the onset of the revolution in less than 50 generations of the insect. Therefore, changing environment led to change in the genetic constitution of the population as in the changed environment the alleles favoured by selection were different from the ones found in the earlier environment (Cook and Saccheri 2013).

The action of natural selection on a character such as body size which is continuously distributed is interesting to study. Based on the assumption that small-sized individuals have higher fitness (greater reproductive success) compared to larger individuals, natural selection is then directional and favours these, and if the character is inherited, a decrease in body size is seen over the generations. The same assumption can be vice versa, i.e., if large-sized individuals had higher fitness.

For example, the pink salmon (*Oncorhynchus gorbuscha*) in the Pacific Northwest known for performing extensive migrations has been decreasing in body size. In 1945, fisherman chose the salmon based on not their number but by their pound weight. And for such a screening, they increased the use of gill netting that selects for larger fish. This further led to the increased survival of the smaller fish, and as a result of such a selection, the average weight of salmon decreased by about one-third in the next 25 years.

22.4.1.2 Stabilizing/Normalizing/Centripetal Selection (Under Constant Environmental Conditions)

This selection takes place in a uniform environment, with the sole aim of increasing the adaptive peak by eliminating peripheral variants on both sides of the mean population. The mean population is well adapted to the given environment that remains stable, and the variants arise by mutation, gene flow, recombination and chromosomal segregation. The ill-adapted variants are weeded out, and the well-adapted are preserved and increase in number over time. Hence, this selection aims to keep the population genetically constant for hundreds or thousands of generations. In such conditions, individuals with extreme measurements of any trait are at a disadvantage in comparison to the ones with average measurements. It is unspectacular, ubiquitous and most common mode of selection in nature.

At the genetic level, variability is preserved by stabilizing selection. The intermediate phenotype having the average characteristics is usually heterozygous compared to the extremes that are homozygous and usually disadvantageous.

The most spectacular study of stabilizing selection was published by American scientist H.C. Bumpus at Rhode Island. Post a severe snow storm, he studied 136 immobilized sparrows in his lab of which 72 were revived but 64 died. He identified the sex and measured several morphological traits of the sparrows. His conclusion was that 'the birds which perished had either very long/ short wings as compared to those with average structural measurements.' This indicated that individuals on the extremities of the populations do not survive abnormal or catastrophic events.

Another example of stabilizing selection is the data regarding birth and survival rate of human babies. Karn and Penrose (1952) showed that mortality for infants is greater for those who are either underweight or overweight. Newborns born at average weight (7.5–8.5 lb) show minimum mortality. The underweight babies are born premature (not ready for independent existence), and the overweight babies often suffer physical damage during delivery.

22.4.1.3 Disruptive/Diversifying/Centrifugal Selection (Under Heterogeneous Environmental Conditions)

In disruptive selection, both the extreme phenotypes in a population are favoured, and the intermediates are selected against. Here, genetic uniformity becomes a drawback when a population encounters a range of ecological discontinuities. The population as a result splits into several groups with different sets of genotypes, each capable of successfully exploiting a different environment. This leads to adaptive polymorphism with respect to ecological opportunities. Diversifying selection hence facilitates a polymorphic population to adapt to different niches of a heterogeneous environment (consisting of different microhabitats). Individuals of the intermediate category have lower fitness compared to the extremes (homozygous for various alleles) and fail to survive. Hence, disruptive selection promotes genetic diversity as a previous homogenous population gets split into different adaptive forms as a result of being subjected to divergent selection pressures.

22.4.2 Kin and Group Selection

Kin selection and group (multilevel) selection are two evolutionary phenomena which form the framework explaining the social behaviour in animals.

22.4.2.1 Kin Selection

Charles Darwin while explaining his theory of natural selection proposed that individuals have a tendency towards increasing the chances of their own survival and reproduction. However, cases have been seen in nature where the close genetic relatives aid each other selectively. This social phenomenon known as altruistic behaviour was a Darwinian puzzle. Why should the individuals save their close relatives at the risk of their own fitness? It was several years later that this question was answered by J.B.S. Haldane, one of the founders of population genetics. This phenomenon was termed as 'indirect selection' by Jerry Brown. This is in contrast to direct selection where the individual increases its own fitness by self-reproduction. Another term 'kin selection' was introduced by John Maynard Smith which includes the parental aid given to the offspring (which are descendent kin) as well as altruistic behaviour for close relative (non-descendent kin). Kin selection has been mainly used to explain why the phenomenon of social and altruistic behaviour has evolved in animals.

The total contribution of an individual including direct and indirect fitness is termed as inclusive fitness. Direct and indirect fitness can be expressed in the same genetic unit. For example, an individual A has one offspring and adopts three nephews. The total genetic contribution of A to the next generation would be:

$$\begin{aligned}\text{Inclusive fitness} &= \text{Direct fitness} + \text{Indirect fitness}\\ &= 1 \times 0.5 + 3 \times 0.25\\ &= 0.5 + 0.75 = 1.25 \text{ genetic units}\end{aligned}$$

The mathematical concept of inclusive fitness by W.D. Hamilton introduced in 1954 can be used to calculate the adaptive value of altruistic behaviour. The altruistic behaviour would evolve only when the indirect fitness gained through the altruistic allele is greater than the direct fitness gained by self-reproduction. This is stated as Hamilton's rule and can be represented as:

$$r_b B > r_c C$$

where

'r' is the coefficient of relatedness.
'B' is the number of genetic relatives that have survived due to the altruistic act of the individual.
'C' is the number of offspring produced by the individual.

Using the equation for the previous example:

$$r_b B = 0.25 \times 3 = 0.75$$
$$r_c C = 0.5 \times 1 = 0.5$$

Through calculations we conclude $r_b B > r_c C$; thus, any allele associated with this altruistic act would increase in frequency.

Studies have been carried out to show the universality of kin selection. Łukasiewicz et al. (2017) worked on bulb mites (*Rhizoglyphus robini*) to show how the effect of relatedness promotes female productivity and cooperation in sex. They carried out experiment in two evolutionary groups: one of relatives and the other of non-relatives in the laboratory during the reproductive phase of the cycle. The result was in sync with the kin selection theory, where the evolution in the group of relatives resulted in increased reproductive output by the females (Kin slection, http://nectunt.bifi.es/to-learn-more-overview/kin-selection/; Kramer and Meunier 2016).

22.4.2.2 Group Selection

Group selection or multilevel selection, as against the Darwinian concept of selection, refers to the idea that selection does not take place at the individual level but can act on multiple levels of biological organizational level including a whole group of organisms. Group selection looks into the direction and strength selection acting on multiple hierarchical levels. There are several arguments which emerge and discard the theory of group selection.

22.4.3 Sexual Selection

Sexual selection is a subcategory of natural selection which provides advantage to certain individuals over the others to be able to successfully mate. It accounts for the evolutionary phenomenon of the existence of expensive secondary traits which make them much more vulnerable to the predators. Sexual selection is also the evolutionary phenomenon which has also resulted in sexual dimorphism. A common example

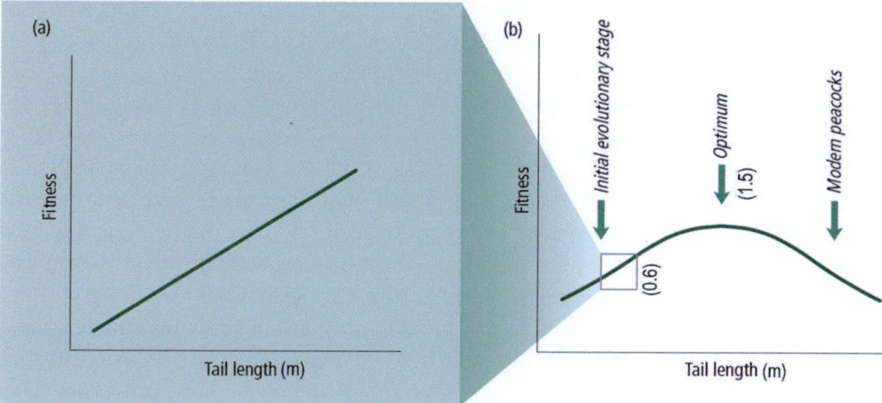

Fig. 22.6 Sexual selection in peacocks. (**a**) The relation between the tail length of the peacock and fitness; (**b**) the correlation in the exaggeration of character and fitness results in bell-shaped curve. The peak is the optimal value. The modern peacocks lie on the right of the optimal value (Adapted from Ridley 2004)

of this is the peacock and peahen. Peacock has elaborate tail feathers and is brightly coloured which is costly for its survival. Firstly, there is immense physiological investment in development of these tail feathers. The elaborate courtship display through these colourful plumages consumes time and energy. Lastly these feathers make it easy for the predators to spot them. In contrast the peahen is drab coloured with a short tail exhibiting sexual dimorphism between male and female. Why have these secondary sexual traits evolved in male? Darwin argued that even though these traits are expensive and may shorten the life, but it ensures that the individual is able to contribute its genes to the next generation, thus increasing their fitness. The peacock with the bigger, colourful and elaborate tail feathers would be able to perform an elaborate courtship display, thus ensuring successful reproduction. A peahen would choose such a peacock as the mate because the elaborate and colourful plumage and courtship display would indicate healthy genes, ensuring good genes are passed onto her offspring. The male ornamentation is basically targets of female choice.

There are two types of sexual selection: intersexual (female's choice) and intrasexual (male-male competition). Intersexual selection is when the female has a choice of the mate, and she chooses the mate on the basis of the sexual traits or the exhibition of male dominance (Byers and Waits 2006). The gain obtained could either be direct such as resources and safety, or in other case the gain could be indirect, i.e. creation of offspring with superior quality of genes. The above-stated example also exhibits mate choice of peahen of peacock with highest number of ocelli in the tail feathers, thus exhibiting superior genes. Fisher attempted to explain this evolution of costly characters with his 'runaway theory.' Earlier before the evolution of female choice, the peacock might not have been prevalent with long tails. Randomly a mutant female chose peacock with long tail which also had higher fitness associated with it. The peahen will now produce peacock with averagely long tail and higher fitness. Slowly, the population will be replaced by peacocks with long tail and peahen who choose the peacock with this attribute. The evolution of long tail of peacock and peahen (with mother's genes of choice and father's genes for long tail) with such preferential choice would reinforce each other resulting in the evolution of long tails (Fig. 22.6).

A study was carried out on the pronghorns of the National Bison Range in northwestern Montana. The sample population was ear tagged and genotyped to ten microsatellite loci. In the study which spanned for 4 years, it was seen that 59% of the fawns were fathered by a small group of males who were physically attractive. Male attractiveness was associated with the offspring survival to weaning with the help of general linear latent and mixed models program, a statistical model. Fawn deaths are basically caused by vulnerability to coyote predation. 50 days post their birth, the fawns are able to gain speed in their sprint and thus escape predation. Thus there would be a differential growth rate of the fawns depending on the rate of weaning. The study showed that the fawns of preferred attractive males had faster growth (hind foot length was measured) and thus greater chances of survival. Thus female investment in mate sampling results in higher fitness by producing offspring with superior genes.

Intrasexual selection mainly includes competition among male to be able to mate with the female. The male-male competition is usually observed in polygynous mating system. This competition often results in intense contests to prove the superiority of the individual. Sometimes the competition is extremely intense resulting in fights. This has resulted in evolution of large body size or modes of fighting (such as horns and antlers). The winner of these intense competitions usually gets access to the males as they display superior and good quality genes.

22.4.4 Concept of Fitness, Selection Coefficient, Genetic Load and Genetic Death

22.4.4.1 Fitness

Fitness in the simplest term may be defined as the ability of an organism, rarely a population or species, to survive and reproduce in its adapted environment. If the organism reproduces successfully, it consequently contributes its genes to the next generation.

Thus, in order to estimate fitness, it is thus important to understand the various components of fitness:

- Viability which defines the survival of the newly formed zygote up to the reproductive age.
- Fecundity which defines the number of offspring produced in the next generation.

Together these components decide the extent of contribution of a genotype or allele to the next generation. Each allele does not enjoy the same fitness through time but changes with the changing environment. The terms dominant and recessive allele, only define the expression of the phenotype at a particular locus. The changing frequency of allele however is influenced by the caring for young one. The fancy display of feather by male might seem to endanger the survival of the adult, as they become vulnerable to the predator. They are important in attracting the opposite sex and ensuring the survival of the young ones (Alcock 2005). This ensures the increased fitness of the individual and contribution of its genes in the next generation.

Fitness mathematically involves two terms: absolute and relative fitness.

Absolute fitness is the total fitness of a genotype which includes viability, successful reproduction, no. of viable offspring produced, etc.; it is represented as W and can be greater than or equal to 1.

The geneticists, however, more often use the term 'fitness' for relative fitness of an organism. It is represented by w and may be defined as the survival/reproductive rate of a genotype in comparison to the maximum survival/reproductive rate of other genotypes in the environment. It is also known as survival or adaptive value.

For example, $w_{AA} = 1$ represents the relative fitness of the genotype AA, $w_{Aa} = 0.8$ represents the relative fitness of genotype Aa and $w_{aa} = 0.7$ represents the relative fitness of genotype aa. This means all individuals of genotype AA, 80% of genotype Aa and 70% of the genotype aa would survive in the given environment. Of the three genotypes, AA is considered to be most fit.

22.4.4.2 Selection Coefficient

The measure of fitness and effect of selection is measured in terms of selection coefficient. It is the measure of the extent of natural selection against the contribution of a genotype to the next generation. It is symbolized as 's' and lies between 0 and 1. Let us look at a hypothetical case where the population comprises the genotype AA, Aa and aa to understand this concept. These genotypes have different chances of survival as seen in Table 22.3.

The selection acts against the recessive allele 'a', and thus there will be a smaller number of surviving adults of genotype 'aa'. If $s = 0.1$ then the chances of survival of genotype 'aa' are 90% as against 100% survival of genotypes AA and Aa. This is a mathematical model; however, in a real environment, the chances of survival, from birth to reproductive age, of an organism of even the best genotype are only 50%.

22.4.4.3 Genetic Load and Genetic Death

The concept of genetic load was introduced by Haldane where he observed the loss of population fitness due to selection-mutation balance. Therefore, in order to understand genetic load, we need to once again refer to our concept of fitness. As we have already discussed, a population comprises genotypes of varying fitness. The genotype with highest fitness is assigned a relative fitness value of 1. The rest of the genotype have relative fitness values less than 1. In spite of the fact that the selection

Table 22.3 Chances of survival of different genotypes

Genotype	Chances of survival
AA, Aa	1
aa	1-s

Table 22.4 Relative fitness of different genotypes

Genotype	Survival to reproduce	Relative fitness
AA	40%	0.8
Aa	50%	1
aa	30%	0.6

is continually acting, these alleles with suboptimal fitness remain in the gene pool either because the allelic variant resulting in reduction in fitness is being replenished in the gene pool due to mutation (known as mutational load) or because they remain in combination with advantageous alleles (known as segregational load). We can also measure the average fitness of a population. It is defined as mean fitness (w) and equals to the frequency multiplied by fitness of genotype. The genetic load basically measures the relative chance that an average individual will die before the reproductive age and thus have no contribution to the next generation because of the deleterious allele in it. It can also be defined as the sum of deleterious genes in the genome. It is symbolized as 'L' and lies between 0 and 1.

$$\mathrm{L} = 1 - \mathrm{w}$$

If all the individuals of a population have fitness 1, then there is no load on the population.

Let us look at an example to understand genetic load: The frequency of the two alleles A and a is 0.5.

Calculating from Table 22.4:

$$\text{Mean fitness} = 0.25(0.8) + 0.5(1) + 0.25(0.6) = 0.85$$

$$\text{The load of the population } \mathrm{L} = 1 - 0.85 = 0.15$$

That means 15% of the offspring will die before the reproductive age, i.e. undergo genetic death.

The failure of individuals to produce offspring and contribute to the next generation is called **genetic death**. High genetic load can put the population in the danger of extinction. The marine life has been observed to have maximum genetic load in comparison to freshwater or terrestrial species. The bivalves among the marine species show highest genetic load due to small population size and high mutation rate.

22.5 Population Genetics

A population is a group of organisms of the same species living in a defined geographical area and can reproductively interbreed. Population genetics is an important aspect of evolution and speaks about gene and allele frequencies. It is the study of genetic make-up of the population, i.e. gene pool and change in the gene pool over time.

Evolution and variation in gene pool are population phenomena so they are best understood as changes in allele frequencies. There are four characteristics which account for most of the changes in allele frequencies:

1. Mutation: produces genetic variation in gene pool and contributes to the first step of evolution.
2. Natural selection: adaptive, directional changes in allele frequencies.
3. Genetic drift: random, non-adaptive and non-directional changes in allele frequencies.
4. Migration: presence of gene flow.

The rates of mutation are very low in nature, so the primary contribution of mutation is only in production of genetic variation. It is migration, genetic drift or natural selection which acts on these genetic variations to produce change in the allele frequencies.

It might seem that a population which is well adapted to a geographic area would actually harbour high levels of homozygosity. But to the surprise of many evolutionary biologists, there is considerable genetic variation in the gene pool. In fact, for a population to succeed, it should have genetic variability.

22.5.1 Hardy-Weinberg Model

While studying single gene locus in a population, we observe the changes in allele and genotype frequencies. The determination of the change in allele and genotype frequencies of the population, from one generation to the other, forms the major study of the population geneticists. The relation between changing allele frequencies leading to change in genotype frequencies has been explained by Hardy-Weinberg law. G.H. Hardy and Wilhelm Weinberg in 1908 independently formulated a simple equation which can be used to trace the allele and genotype frequencies of the population in an ideal scenario. The Hardy-Weinberg law states that in an ideal, infinitely large population with random mating, on which no evolutionary force is acting, the allele frequency does not change and the genotype frequencies stabilize after one generation.

Assumptions made in Hardy-Weinberg law:

1. Infinitely large population.
2. Random mating.
3. Absence of evolutionary forces like:
   (a) Natural selection—thus all individuals have equal rates of survival and reproduction.
   (b) Migration—absence of gene flow.
   (c) Mutation—as a result no new alleles are created.

Fig. 22.7 Punnett square showing permutation and combinations of genotype produced by random matings. Each allele has equal probability of pairing with other alleles producing genotypes p^2, $2pq$, q^2

sperm / egg	f(A)=p	f(a)=q
f(A)=p	p x p= p^2	p x q= pq
f(a)=q	q x p= qp	q x q= q^2

A population which meets the above criteria are said to be in Hardy-Weinberg equilibrium. The Hardy-Weinberg law is a mathematical model and uses the Mendelian laws of segregation.

If two alleles A and a of a single locus are considered, then the genotype frequencies AA, Aa and aa if in Hardy-Weinberg equilibrium would be p^2, q^2 and $2pq$, where the frequency of $A = p$ and frequency of $a = q$.

The above genotype can only be obtained when there is random mating in the population resulting in the random combination of gametes. This can be understood by the study of Punnett square often used by the twentieth-century geneticists to predict probability of genotypes.

From the Punnett square (Fig. 22.7), it is clear that each allele has equal probability with pairing with the other producing the genotypes p^2, $2pq$ and q^2 if there is random mating.

The Hardy-Weinberg equilibrium equation is:

$$p^2 + 2pq + q^2 = 1$$

Since we are taking into consideration a single locus with two alleles (A and a), these alleles should account for the 100% frequency of the gene in the gene pool.

Or in other words, $p + q = 1$

To demonstrate how Hardy-Weinberg law of equilibrium can be used, let us consider a population with $T = 0.7$ and $t = 0.3$.

$$f(T) + f(t) = p + q = 0.7 + 0.3 = 1$$

If the population undergoes random mating, then we attain three genotypes TT, Tt and tt in the proportion as discussed above: p^2, $2pq$ and q^2:

$$f(TT) = p^2 = 0.7 \times 0.7 = 0.49$$

$$f(tt) = q^2 = 0.3 \times 0.3 = 0.09$$

$$f(Tt) = 2pq = 2 \times 0.3 \times 0.7 = 0.42$$

0.49 + 0.42 + 0.09 = 1 showing that we have accounted for all zygotes formed.

Assuming they follow Hardy-Weinberg law, these offspring have equal chances of survival, and they become adults. They also have equal probability to reproduce and mate randomly. Forty-nine percent of the gametes would be contributed by the genotype *TT*, 42% by *Tt* and only 9% by *tt*.

The gamete *T* would be contributed by both the genotypes *TT* and *Tt*. Thus

$$f(T) = 0.49 + ½\,0.42 = 0.7$$

The gamete *t* would be contributed by both the genotypes *TT* and *Tt*. Thus

$$f(t) = 0.09 + ½\,0.42 = 0.3$$

The initial gene pool that we started with had the frequency of alleles *T* and t in the proportion 0.7 and 0.3. After a generation of random mating, the alleles remain in the same proportion. Thus, we can say in absence of evolutionary force, this population has remained in Hardy-Weinberg equilibrium and has not exhibited any change in allele frequencies.

Thus, we can draw two inferences from this model:

1. Allele frequencies remain the same from one generation to another.
2. Genotype frequencies stabilize after one generation of random selection.

However, the real population is rarely in Hardy-Weinberg equilibrium. Then why is this model of great importance to the population geneticists? The Hardy-Weinberg model helps the population geneticists to understand what the evolutionary forces are acting on the real-world population and changing the allele frequencies (Jankowska et al. 2011; Klugs et al. 2012).

22.5.1.1 Application of Hardy-Weinberg Law on Human Population

Hardy-Weinberg law can be applied on human population to test if the population is undergoing evolution or not. It is used to calculate the percentage of population carrying autosomal recessive alleles causing inherited diseases and for biomedical research. One such study was done on population of the USA to study the occurrence of phenylketonuria (PKU), a metabolic disease due to homozygosity of recessive allele. It results in mental retardation, stunted growth and other symptoms. The frequency of occurrence of this disease is 1 in every 10,000 babies born in the USA. Considering that no new mutations for PKU are being added into the gene pool, we can use this data to calculate the percentage of carriers in the population.

Using the Hardy-Weinberg equation: $p^2 + 2pq + q^2 = 1$

$q^2 = 1/10000 = 0.0001$

$q = \sqrt{}0.0001 = 0.01$
$p + q = 1$
$p = 1 - q = 1 - 0.01 = 0.099$
The number of carrier (heterozygotes) $= 2pq = 2 \times 0.099 \times 0.01 = 0.198$.

Approximately 2% of the population are the carriers of the recessive allele. This estimate is not exact, but we can get an idea of an approximate idea of the carriers in the population.

Another such genetic disease is haemophilia, a genetic disease which impairs the body's ability to make blood clot. The occurrence of haemophilia A and B in the population is 1/12000. Using Hardy-Weinberg's equation, we can estimate the carriers in population.

$q^2 = 1/12000 = 0.00008333$
$q = \sqrt{}0.00008333 = 0.0091$
$p + q = 1$
$p = 1 - q = 1 - 0.0091 = 0.9909$
The number of carrier (heterozygotes) $= 2pq = 2 \times 0.9909 \times 0.0091 = 0.01803$.

Thus the frequency of occurrence of carriers in the population is 0.01803 which is very low (Klugs et al. 2012; Pierce 2012).

22.5.2 Genetic Drift

When the size of the population is small, then chance alone can result in the change in allele frequency. The smaller the size, the greater can be the degree of fluctuation of allele frequencies. Any random, non-adaptive, non-directional change in allele frequency occurring in a small population is called genetic drift. Once it begins, the phenomenon of genetic drift will continue in subsequent generations till an allele is either completely lost or fixed in a population. The concept of genetic drift was introduced by one of the founding fathers of population genetics, Sewall Wright in 1931, and is also known as Sewall Wright effect (Wade 2008).

This concept can be understood by random sampling in genetics. Let us consider a single locus gene with two alleles A and a, in a population of ten individuals. The allele frequency of A and a is 0.5 each in a gene pool of genotype AA, Aa and aa. Each of these genotypes has equal probability of survival and reproduction, i.e. in absence of natural selection, the fitness of all the genotypes is 1. In accordance with the Hardy-Weinberg principle, the allele frequency should remain the same in the next generation. However random sampling may change the scenario, and by chance the allele 'A' might get a better environment for reproduction resulting in increase in the frequency of allele 'A'. The allele A has no adaptive value for the environment, and this increase in frequency is just a matter of chance. Such random change in allele frequency is significant only in a small population and results in subsequent changes in genotype frequencies.

The genetic drift can act through two phenomena:

(a) Founder's effect.
(b) Bottleneck effect.

22.5.2.1 Founder's Effect

The establishment of new population formed by a few founding members from the parental population which are not representative of the original gene pool is known as founder's effect. This concept was introduced by Ernst Mayr in 1942.

Several evidences of founder's effect have been seen in humans such as prevalence of some recessive diseases in selective populations. These usually occur in small or isolated population as observed in Finns. There are 30 recessive diseases which are common in Finland than rest of the world. Diseases like cystic fibrosis and phenylketonuria are common in Caucasian population. Studies have shown that these diseases have arisen from one remote mutation in the immigrants of the founder's population, thus exhibiting the founder's effect (Peltonen 2001).

22.5.2.2 Bottleneck Effect

Bottleneck effect occurs usually because of a disaster or natural calamity which drastically reduces the population. The surviving population does not represent the genetic make-up of the parental population.

An example illustrating bottleneck effect is the loss of genetic variation in Northern elephant seals. Due to unchecked hunting, their number reduced to as low as 20 at the end of the nineteenth century. Conservation acts resulted in rebound of their population but with low genetic variation showing the mark of bottleneck effect.

Similar loss of genetic diversity has also been observed in cheetah (*Acinonyx jubatus*). Cheetahs are inbred with extremely low genetic diversity. Investigations have been carried out by Stephen J. O'Brien and his colleagues where they examined the genomes of seven cheetahs, four from Namibia and three from Tanzania. Genome sequencing confirmed the fact that there is very low genetic diversity in the gene pool of this animal. Low genetic diversity poses a threat to the extinction of the species.

The initial loss of diversity happened around 10,000 years ago, when they just managed to survive the ice age. Studies have been carried out on hypervariable minisatellite loci and mitochondrial DNA to date this bottleneck event to the late Pleistocene epoch. Modern times have increased the problems of habitat encroachment, hunting and poaching making them vulnerable to extinction. Conservation biologists are working hard to repopulate them, but their gene pool exhibits low genetic diversity due to these events (Raymond and O'Brien 1993; Emerling 2016).

22.5.3 Role of Mutation and Migration in Changing Allele Frequency

22.5.3.1 Mutation

Mutation is any change in DNA sequence of the gene within the chromosome which occurs due to error in DNA replication. Mutation in fact is the only phenomenon which can produce new alleles. All other evolutionary forces only reshuffle the gene pool to produce variable genotypes. Mutations are very important as they produce variation in the gene pool on which evolution acts. In other words, mutations are the raw material for evolution or are the engine of evolution. Mutations are random events without any adaptation value. They can either be selected and become abundant or completely lost from the gene pool.

Mutation, in itself, is a weak evolutionary force to change allele frequency but a strong force to create genetic variation. However, it is difficult to measure mutation in a diploid organism as most mutations are recessive in nature.

22.5.3.2 Migration

Migration is the movement of a subpopulation from one place to another. The migrating population carries its own ancestral genes and interbreeds with the native subpopulation resulting in sudden influx of alleles. This transfer of genes results in gene flow. Thus if the two populations have different set of genes and in absence of any selection phenomenon, migration alone can result in change in genotype frequency. In order to understand it, let us look at an example where a hypothetical species has two alleles A and a. The species also has two populations, one residing in the mainland and the other in the island. The frequency of A on mainland is represented by p_m and frequency of A on island by p_i. If migration takes place from the mainland to island, then the frequency of A in the next generation on island p_{i1} is represented by:

$$p_{i1=} (1 - m)\, p_i + mp_m$$

where m represents the migrants from mainland to island. Putting in the value of $p_m = 0.7$ and $p_i = 0.3$ and that at a time 10% of the migrants have moved from mainland to island,

$$\begin{aligned} p_{i1} &= (1 - 0.1)\, 0.3 + 0.1 \times 0.7 \\ &= 0.9 \times 0.3 + 0.07 \\ &= 0.27 + 0.07 \\ &= 0.34 \end{aligned}$$

These calculations show change in allele frequency due to migration from 0.3 to 0.34 in one generation itself.

There is considerable influence of human migration in distribution of ABO blood group (Mourant et al. 1976). Karl Landsteiner, an Austrian physician, has been credited with the discovery of ABO blood group. The ABO blood group is controlled by a single gene with three alleles, present on chromosome 9. The native Americans can be traced to a founder population of 10–20 individuals who had migrated to the American mainland. This has been illustrated through study of mtDNA and Y chromosome. The Americans thus have a high percentage of O blood group. The D_i blood group polymorphism tracks the migration of humans from East Asia to America. Another such example is the prevalence of D_i antigen in South East Poland which provides the measure of extent of invasion of Mongolians in recent times in Europe.

22.5.4 Impact of Positive Selection

Selection can be of two types: positive and negative or purifying selection. Purifying selection prohibits the spread of deleterious mutations in the gene pool.

Positive selection is also known as Darwinian selection. It is the phenomenon of natural selection by which advantageous mutation becomes fixed in a population, or in other words it promotes the spread of advantageous mutation in a gene pool. Positive selection thus promotes the emergence of new phenotype.

Charles Darwin in his explanation of selection had stated that those organisms which have the best attributes are the ones that survive in an environment. He was mainly concerned with phenotypic evolution. As we are looking at evolution in terms of genetics, let us redefine this outlook. The organisms that harbour mutations which increase their fitness in the environment, are the ones which survive and reproduce (Forsdyke 2007).

Whether the mutations that occur in the nucleic acid sequence result in positive or negative selection depends on which part of the gene product or protein they are affecting. If a mutation occurs in the active site of the enzyme such that it lowers the catalysing rate of the enzyme, it might result in lowering the fitness of the organism. In other case, mutation in the antigen might enhance the ability of the pathogen to invade the host, thus increasing its fitness and resulting in positive selection.

Extensive studies have been done on positive selection. One such study reviews large number of genes of the human population which have undergone positive selection (Wu and Zhang 2008). Darwinian selection has intensively acted in the modern human population resulting in high genetic diversity which has resulted in differences in appearances, metabolism of drugs and resistance to diseases. One such set of genes are those that are involved in the development of the brain. The brain size has increased in primates specially in the *Homo sapiens* and the species closely related to them. Some genes involved in development of brain have undergone rapid positive selection. Microcephalin is the key regulating gene of the brain of human and is still evolving. *FOXP2* is another gene present both in human and birds. It regulates the singing ability of the birds and speech expression in man. The copy of FOXP2 in human has high evolutionary rate under the influence of positive selection.

Another such example was a study carried out by Zhang et al. (2002), on the evolution of a duplicated pancreatic ribonuclease gene (RNASE1) of leaf-eating colobines. Like ruminants these old world monkeys extract nutrients by breaking down the symbiotic bacteria with a set of enzymes including RNASE1. Phylogenetic analyses of the RNASE1 gene of the non-colobine monkeys with the Asian colobine douc langur (*Pygathrix nemaeus*) show the substantial difference in the sequence. A closer examination shows that one copy of the gene RNASE1 has remained conserved but the other copy of the gene RNASE1B has accumulated many non-synonymous substitutions post recent duplication. These rapid substitutions have accumulated due to positive selection pressure for adaptation of enhanced ribonuclease activity at low pH of the colobine intestine (Zhang 2010).

22.6 Speciation

22.6.1 Definition of Species

Several workers have tried to define the concept of 'species,' and the most important fact to be noted is that these individuals belonging to a species co-exist in a particular span of time.

22.6.1.1 Typological Species Concept

This concept suggests that 'variety in nature can be reduced to a few distinct types.' Varied individuals often belong to a single type of species that is defined by a certain set of norms. The concept was proposed centuries ago by Plato and fails to recognize a species as a unit of evolution. Hence it was not accepted by modern evolutionists.

22.6.1.2 Biological Species Concept

The 'Biological Species Concept' proposed by Harvard evolutionist Ernst Mayr (1963) states 'Species are groups of interbreeding natural populations that are reproductively isolated from other such groups.' Species is an ecological unit sharing resources of the environment and interacting as a unit with other such groups. This concept focuses on species, in genetic terms as a gene pool, a unit within which the gene frequencies can change. The movement of genes through a species via mechanisms such as immigration, emigration and interbreeding is termed as gene flow. Natural selection favours inheritance of genes that are beneficial to the organisms as well as that help them in better co-existence with the environment. A gene that does not favour these aspects is selected against. An important limitation of the biological species concept is that it does not apply to the organisms that reproduce asexually.

22.6.1.3 Recognition Species Concept

The concept states 'species as a collection of individuals with shared specific mate recognition system (SMRS).' It was given by H. Paterson in 1993. The idea behind placing individuals together as a species is interbreeding and production of a viable, fertile offspring. And this concept highlights the method by which potential conspecific mates are recognized for breeding. The sensory systems may be visual, acoustic, chemical, etc. An example citing the same is a population of crickets within a single habitat in the USA. Their courtship begins with the males singing for attracting the females. The songs are species specific and hence confine interbreeding to a single species.

22.6.1.4 Morphological Species Concept

Species are characterized by body shape and other structural features and can be applied to sexual as well as asexual organisms.

22.6.1.5 Ecological Species Concept

The ecological niche, that is, the microenvironment of a species, defines a species population. The concept highlights the interaction of the members with the living and non-living parts of their environment.

In a population that is sexually reproducing, speciation leads to the division of a parental gene pool into two or more distinct gene pools. The concept of genetic changes accumulating to form a new species is a macroevolutionary event. The underlying cause of speciation is the adaptation of populations to discontinuous ecological niches, leading to reproductive isolation that allows them to evolve independently.

22.6.2 Modes of Speciation

Speciation is classified as allopatric, sympatric, peripatric and parapatric (Fitzpatrick et al. 2009).

- If a geographical barrier is responsible for divergence of a population from its ancestor, it is termed as **allopatric speciation** (Fig. 22.8). The physical barrier prevents mixing of the population, interbreeding and hence gene flow. The divided populations hence formed start developing independently of each other, accumulating genetic differences between them as a result of mutations, genetic drift and natural selection. The presence of related flora and fauna in Galapagos Islands off the west coast of South America is a key evidence of allopatric speciation. These have been discussed in detail by Charles Darwin in his book *Origin of Species*. Over 14 species of finches are distributed all over these islands that have accumulated differences from their ancestors with respect to their food preferences and beaks. The antelope squirrel *Ammospermophilus harrisii* inhabits the southern rim of the Grand Canyon and the northern rim by the white-tailed *Ammospermophilus leucurus*. Small rodents are unable to cross such barriers unlike birds and hence have diverged into two different species.
- **Sympatric speciation** occurs in geographically overlapping areas (Fig. 22.9). Divergence occurs due to reduced gene flow by polyploidy, sexual selection and habitat differentiation. An example is presence of cichlid species in East African Great Lakes that are known to be monophyletic (Schliewen et al. 1994). Analysis of their mitochondrial DNA revealed that these flocks have undergone sympatric diversification. Sympatric speciation is also said to be underway in pest of apples, North American maggot fly *Rhagoletis pomonella*. Its original habitat was the hawthorn tree, but now it colonizes apple trees. Apples mature sooner compared to hawthorn fruits; natural selection favoured the apple-feeding flies. The apple-feeding flies show temporal isolation from the hawthorn feeders providing a prezygotic isolating barrier to gene flow as well.
- Ernst Mayr described the founder effect speciation which was later termed as **peripatric speciation** by him. His idea elaborated that in isolated populations with restricted distributions, the peripherally located species to the parent species

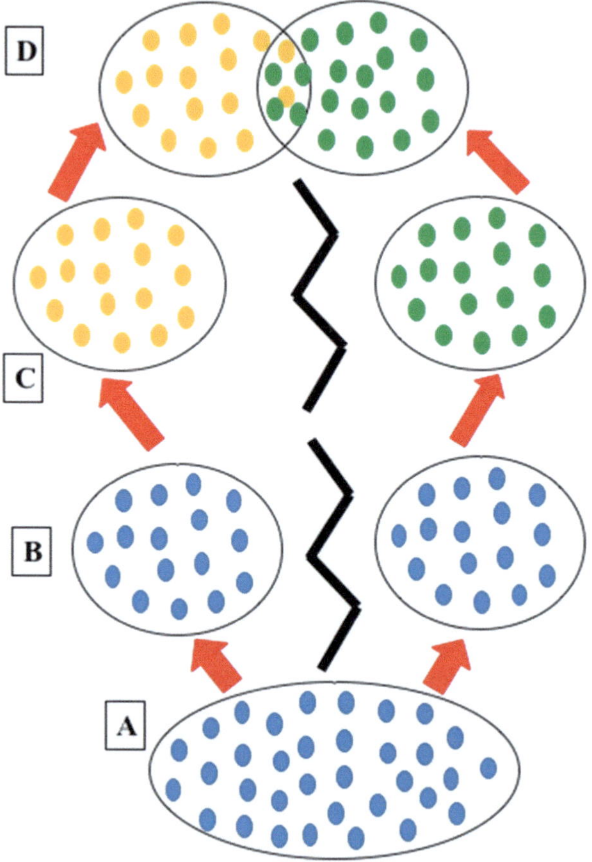

Fig. 22.8 Sequence of events in allopatric speciation. *A*: The original parent population. *B*: The population is split by a geographical barrier. *C*: Owing to the distance and isolating mechanisms that act, the populations become genetically different by means of drift and selection. *D*: As a result of the events, the populations have become independent species, have adapted to their unique niche and cannot interbreed anymore. Speciation is complete when no gene flow occurs even if they occupy the same geographical area

tend to diverge genetically and form new genera. He reasoned that the allele frequencies at various loci differed from the parent population due to genetic drift. A common example is of the paradise kingfishers *Tanysiptera* in New Guinea. The *T. galatea* is present throughout lowlands of New Guinea, whereas several distinct forms (*T. riedelii*, *T. carolinae*) are distributed on the islands along its coast.

- If gene flow is weak between populations residing in adjacent regions with varied selection pressures, it leads to **parapatric speciation**. The hybrids so formed may be weak, with lower fitness, inviable and sterile. Steady genetic divergence would lead to complete reproductive isolation. An example is of the three-spined sticklebacks (*Gasterosteus aculeatus*) in lakes (each with outlet streams) of Vancouver Island in western Canada. Though there was an absence of any physical barriers between streams and lake populations, the subpopulations have evolved with various different morphological features. Genomic analysis has shown that genetic differences between the populations were pronounced in the central chromosomal regions (Roesti et al. 2012).

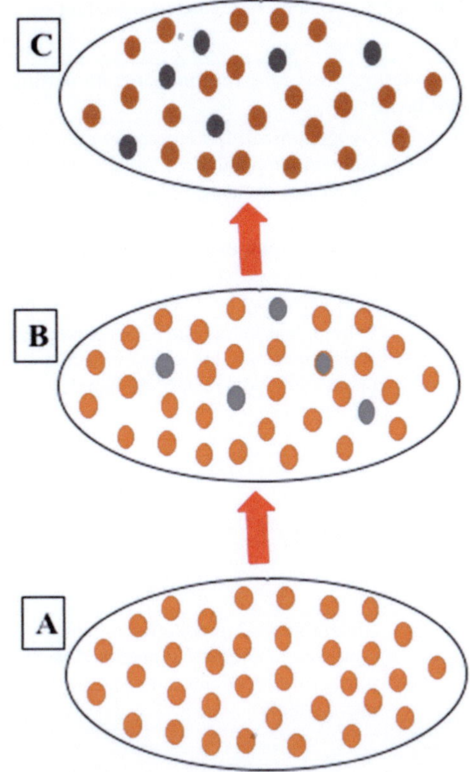

Fig. 22.9 Sequence of events in sympatric speciation. *A*: The original parent population. *B*: The subpopulations start getting formed in the absence of any geographical barrier and begin their genetic differentiation. *C*: The subspecies hence formed become completely reproductively isolated and cannot interbreed, thus forming independent species

22.6.3 Isolating Mechanisms

The barriers that lead to these events may be physical, geographical, physiological, temporal or ethological and are collectively termed as reproductive isolating mechanisms. These gene flow barriers evolve as a result of divergence between populations that accumulate genetic differences over time. Accumulation of these isolating mechanisms results in the process of speciation. The strength of isolation is assessed if the species merge back into a single lineage on coming back into contact with each other. Speciation is complete if no intermixing takes place and no hybrids are formed. However, if isolating mechanisms are weak and overcome by gene flow, they end up merging the species into a single lineage.

These are classified into two main categories with respect to their occurrence pre- or post-mating (Table 22.5):

(a) Prezygotic isolating mechanisms: These act before fertilization and hence no zygote formation takes place. As a result of these mechanisms, no mating can take place.
(b) Post-zygotic isolating mechanisms: These act post fertilization, the population members are willing to mate and the hybrid hence formed isn't fit to be either

Table 22.5 Isolating mechanisms operating in nature

S. no.	Mechanism	Example
Pre-mating/prezygotic mechanisms: no mating, no zygote formation (interspecific pairs do not form)		
1.	**Geographical**: Differences in preferred habitat/biotope hinder mating in animals and pollen exchange in plants	Two species of fern *primula*, *P. vulgaris* and *P. elatior*, have different soil, temperature and moisture preferences (Keller et al. 2016)
2.	**Seasonal/temporal**: Members of different populations differ in breeding time/time of attaining sexual maturity or flowering	Species of toads: *Bufo americanus* and *Bufo fowleri* have different breeding times (early and late in spring, respectively)
3.	**Ethological/behavioural/psychological**: Isolation occurs due to incompatible behaviour of the males and females of different species	Behaviour, courtship signals are species specific and enable mate recognition. For example, in birds and arthropods, songs, courtship dances, display and tactics are responsible for conspecific mating. Marine blue footed booby birds are recognized by their blue feet, a sexually selected trait. They mate post a courtship display that calls for the male to high step that enables the female to focus on his bright feet
4.	**Mechanical:** Reproductive organs (genitalia in animals, stamen and pistil in plants) of the species are physically incompatible	The 'Lock & key' theory has been proposed for genitalia of animals to correspond and mate with each other. For entomophilous flowers, floral parts become highly specialized to attract a single kind of pollinator and prevent interspecific pollination. Shells of two species of snail *Bradybaena* spiral in different directions (clockwise and counter-clockwise). As a result, their genital openings are not aligned, and mating is unsuccessful
Post-mating prezygotic mechanisms: Mating takes place, but the gametes do not get fertilized		
1.	**Gamete viability**: The sperms do not survive in the female reproductive tract; pollen fails to germinate	In some crosses of *drosophila*, an insemination reaction can result in swelling in the vagina of the female and hence unsuccessful fertilization (Turissini et al. 2018)
2.	**Gamete recognition**: Sperms and eggs of the two species are incompatible with each other; pollen and egg nuclei are also incompatible/fail to attract each other	Sea urchins of the genus *Echinometra* have external fertilization, where the sperms and eggs are laid into surrounding water and fuse to form zygotes. Hetero-specific gametes do not fuse with each other as the proteins on their surfaces cannot bind to each other
Post-zygotic mechanisms: mating takes place, and hybrids hence formed are sterile/non-viable		
1.	**Hybrid inviability**: Interspecific hybrids are weak, are non-viable with lower	Salamander subspecies *Ensatina* occasionally hybridize, and the resultant hybrids either do not complete their

(continued)

Table 22.5 (continued)

S. no.	Mechanism	Example
	fitness and hence get eliminated at any stage before attaining sexual maturity	development or, if they do, are frail in stature
2.	**Hybrid sterility**: Hybrids are sterile due to abnormal developments of their gonads or failure of gametes to develop	Genetic causes of such an event are at three levels: (a) Genic: Incompatible gene combinations create a disharmony and failure to form gametes (common in animal hybrids) (b) Chromosomal: Different gene arrangements and chromosomal aberrations (inversions, translocations, etc.) lead to sterility (common in plant hybrids) (c) Cytoplasmic: Unfavourable interactions between nuclear and cytoplasmic factors
3.	**F_2 breakdown**: Inviability or sterility of F2 hybrid even if F1 is vigorous	

viable or reproduce further. Hybrids are sterile, or even if they manage to reproduce, the progeny does not have high levels of fitness. Post-zygotic mechanisms are uneconomical as the hybrids hence produced have no genetic contributions to the gene pool and hence a waste of sexual energy. Mostly the parent populations have diverged so much that hybrids are inviable or sterile.

Reproductive isolation is not a product of a single gene mutation between two isolated populations; rather, it arises by gradual accumulation of differences in gene combinations, frequencies, arrangements and interactions between them.

22.6.4 Genetics of Speciation and Reproductive Isolation

Differences in various chromosomal segments and genes lead to strengthening of isolating mechanisms and finally result in speciation (Noor and Feder 2006). The hybrids formed due to a hetero-specific mating are inviable, weak and sterile as both paternal and maternal sets of chromosomes form disharmonious association. This may lead to a loss in function of several indispensable housekeeping genes due to epistasis or occurrence of homologous alleles on different chromosomes. Prezygotic mechanisms hence are more economical and do not allow investment of biological energy which may lead to formation of a wasteful end product (Wolf et al. 2010).

In prezygotic isolation, several differences in gene combinations and chromosomal segments have been established firmly by natural selection over a long span of time. When two such populations come in proximity, then do not seek potential mates within each other. Post-zygotic mechanisms reveal that the two populations under scrutiny haven't yet accumulated enough differences yet. Strongest genetic dissimilarities exist in the mechanism resulting in inviable hybrids compared to others. Hence, natural selection is yet to intensify its divergent action on them. Hence, it can be concluded that species do not arise at a single step, but through an accumulation of various genic and chromosomal differences. At the genetic level, four problems are known to cause hybrid inviability/sterility/breakdown: chromosomal rearrangements, incompatibility at the gene level, ploidy and interaction between endosymbionts and the nuclear genome. When a gene at a specific locus from one parent does not interact well with a gene at another locus from the other parent, such genic incompatibilities affect the hetero-specific hybrid viability.

Also proposed in the 1930s was the Dobzhansky-Muller theory for post-zygotic isolation. According to the theory, an ancestral population splits into different populations in which gene flow is absent. Each population is well adapted to its local environment as a result of genetic changes that have accumulated over time. If the populations encounter each other later and mate, the genetic changes at the various loci will now allow the hybrids to be successful, and hence they will be sterile. The main aspect to be impressed upon is that post-zygotic isolation is due to interaction of multiple loci. An interesting pattern in post-zygotic isolation was inferred by J.B.S. Haldane, called Haldane's rule, that 'the heterogametic hybrid of a population will have lower fitness compared to the homogametic one.'

22.7 Evolution of Man

Apart from fossil evidences, the origin and evolution of man have been widely studied with the help of DNA sequences to trace the history of modern man. The relationship among man and its nearest relatives 'The great apes' has long been studied. There are major morphological differences including bipedalism, presence of an apposable thumb and body proportions distinguishing chimpanzees, gorillas and man.

All early hominid fossils have been procured from Africa, but the first fossils found outside were of *Homo erectus* (China and Indonesia) (Fig. 22.10). It is postulated that *H. erectus* gave rise to archaic European, Asian and African populations. The best known of such hominids are the Neanderthals (*Homo neanderthalensis*) that lived in Europe and Western Asia about 3,00,000 years ago. These fossils were obtained from Feldhofer cave in Germany and Mezmaiskaya cave in Caucasus Mountains east of Black Sea. Mitochondrial DNA analysis from Neanderthal fossils with 2000 present-day human samples suggests that they did not contribute to the mitochondrial DNA of *Homo sapiens*. They may have competed with ancestors of modern humans and lost out in the competition and became extinct. Another approach was the whole genome sequencing of the Neanderthal species. Their genome is made up of 3.2 billion base pairs and is 99.7% identical to the modern human genome. Comparisons of the Neanderthal genome, five present-day humans and chimpanzee genome revealed that there are amino acid coding differences in 78 genes. Interestingly it was revealed that 1–4% of Neanderthal sequences were found in humans from Europe and Asia but not Africa. Phylogenetic

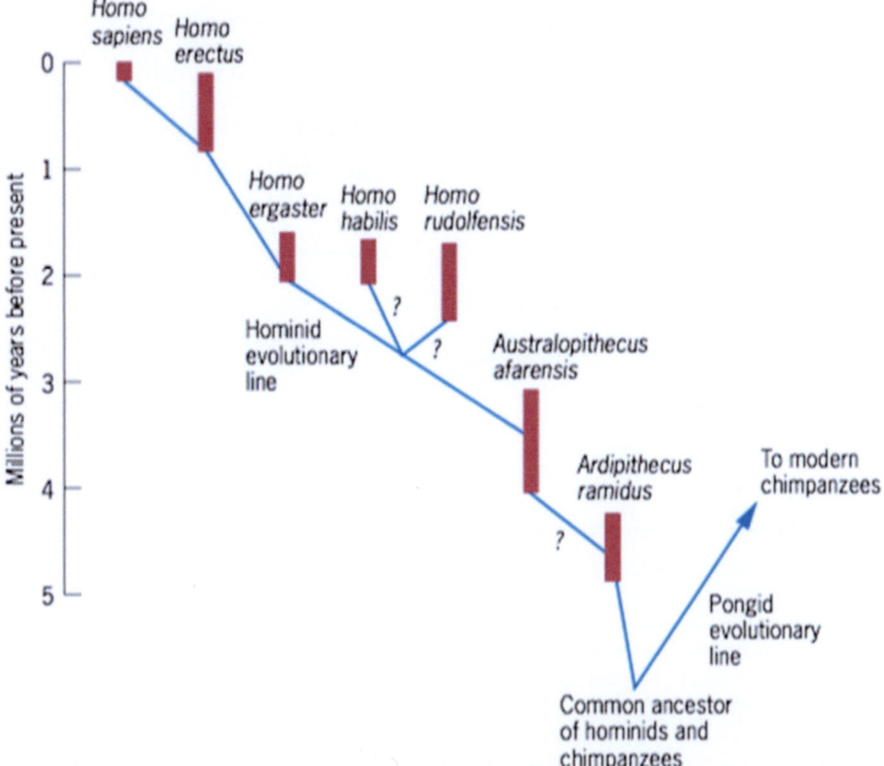

Fig. 22.10 Evolutionary history of the modern man. *Homo sapiens* evolved from a common ancestor in parallel with chimpanzees as traced by fossil evidences procured from over the globe. Uncertainties in the line are indicated by question marks. (Adapted from Snustad & Simmons; 6th Edition)

analysis done to compare and analyse the divergence of the species revealed that humans and Neanderthals last shared an ancestor 706,000 years ago (Fig. 22.11). The mitochondrial genome of 12 Neanderthal specimens has been completely sequenced, and these are quite different from the known human mtDNA. It is unlikely that their mtDNA made any significant contributions to modern human mtDNA. Hence, modern man and Neanderthal man are considered as clear biological species (Hartl and Jones 2009).

The initial publication of the chimpanzee (*Pan troglodytes*) genome and comparison with the human genome (The Chimpanzee Sequencing and Analysis Consortium, 2005) has shed light on the formation of the human species and the complex speciation between the two. In addition to this was the sequencing of the genome of rhesus macaque (*Macaca mulatta*) by the Rhesus Macaque Genome Sequencing and Analysis Consortium in 2007. It was possible to compare the three primate genomes and construct an ancestral primate genome. Comparative genomics could also

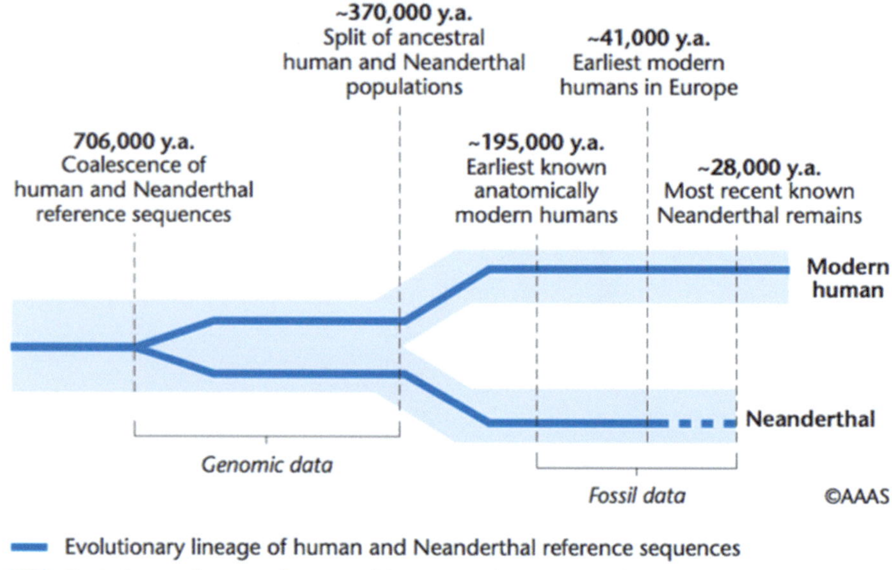

Fig. 22.11 Divergence between the human and the Neanderthal species. The separation events led to major evolutionary events in both populations. Data obtained on sequencing comparisons between the genomes of modern humans and the Neanderthal DNA. (Adapted from Concepts of Genetics by Klug – 10th Edition)

determine the regions of the ancestor that could have contributed to human evolution. Chimpanzees and humans have 98% nucleotide level similarity (Portin 2007). The main difference between the haploid chromosomal sets of the two in a karyotype is a big metacentric chromosome 2 in man vs acrocentric in chimpanzee. The sequencing of the Y chromosome of both the species revealed that there is an accelerated rate of evolution of the chromosome vs the entire genome. It is a huge challenge till date to explain the reasons of emergence of man post their separation less than 6.3 million years ago.

As of today, genetic data such as variations at the level of blood groups, restriction fragment length polymorphisms, lengths of repeat DNA sequences and DNA composition have been used to investigate relatedness amongst various populations, races and ethnic groups. Most analysis of human evolution has been done using mitochondrial DNA as it evolves faster than the nuclear DNA and is passed on only through the maternal parent. Hence, in evolutionary terms researchers can detect changes at a genic level over short period of time and trace it back to a common female ancestor.

22.8 Concept of Molecular Phylogeny

The concept of classification and systematic organization of biological hierarchy is an age-old concept introduced by the naturalist Linnaeus in the eighteenth century. He classified mainly to arrange the organisms in a hierarchical pattern, but unknowingly he also paved path for understanding phylogeny of organisms. Phylogeny is the study of evolutionary relationships. Earlier the direct phylogenetic studies were mainly based on the fossil evidence. Other studies included comparison of morphological traits. The main problem with these methods was that the fossil data was often incomplete and comparison of morphological traits was often biased. Moreover, study on microorganisms could not use any of these mentioned methods. The introduction of molecular phylogeny broadened the horizons where phylogenetic relationships could also be derived among evolutionary distant organisms. The molecular phylogeny uses the sequences of the biomolecules to look into the evolution of organisms. It is basically selection acting on the mutation in the biomolecules. The accumulated mutations over millions of years can act as molecular fossils. Though sequencing was first introduced only in the 1970s, but Nuttall had use immunological assays in the early 1900s to understand the evolutionary position of man in relation to other primates. Even though Nuttall had displayed the successful use of biomolecules in understanding phylogeny as early as 1904 yet it gained momentum only in the 1950s because of several technical challenges. There has now been gradual shift towards molecular phylogeny, which involves the use of DNA, RNA and protein as molecular markers to study modern phylogenetics. The major reason of this shift is to be able to obtain large datasets for the studies. The evolution of sequencing has also ensured easy availability of molecular data. The molecular data can be easily converted into mathematical and statistical data which makes it easier to analyse. Unlike the morphological traits, the molecular data (A, T/U, G, C for DNA/RNA and amino acids for proteins) are unambiguous (Brown 2002).

22.8.1 Phylogenetic Tree

The molecular data has been largely used to construct phylogenetic trees. A phylogenetic tree (Fig. 22.12) is a visual display of the evolutionary relationships among organisms. Even though the phylogenetic tree is being constructed now to display phylogenetic events, but tree-like illustrations (Fig. 22.13) were also observed in Darwin's book *Origin of Species*, where he used it to show that accumulation of slow modification can lead to speciation event.

22.8.1.1 Objectives of Molecular Phylogeny

1. To infer a tree displaying true phylogenetic relationships among organisms.
2. To study and recover the order of evolutionary events and represent it as phylogenetic tree, which is a graphical representation of phylogenetic relationship.

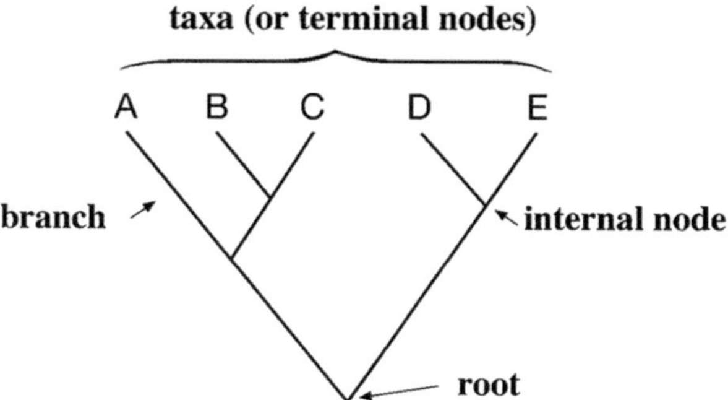

Fig. 22.12 A typical rooted phylogenetic tree. Diagram showing various parts of the tree—terminal nodes or taxa (operational taxonomic unit, OTU) are the extant species, internal nodes are the recent common ancestors, branching shows the event of divergence and root is the common ancestor. Often the details of common ancestor are not available, so to root a tree an outgroup species is used. Outgroup is the species which is distantly related to the group of organisms. The pattern of branching of the tree is known as tree topology. Ninety-nine percent of the species on earth have become extinct. A tree like this gives a visual display inferring what would have been the phylogenetic relationship of the extant species with the extinct species

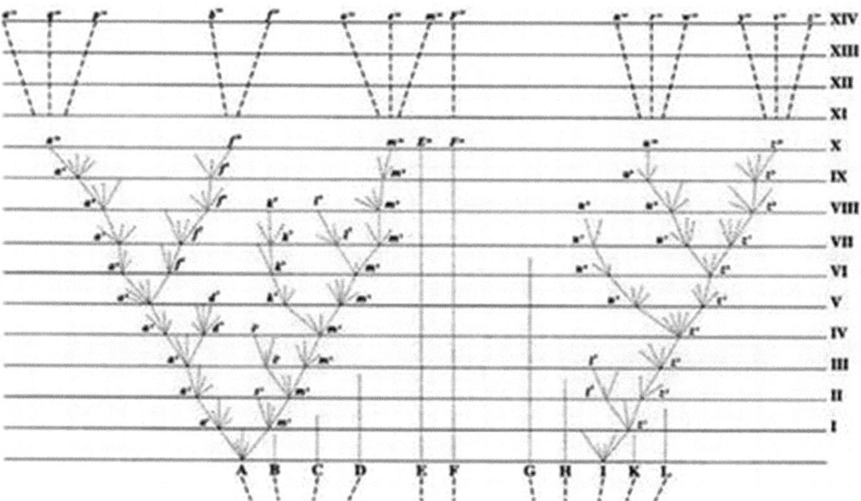

Fig. 22.13 Darwin's illustration. Darwin too in his book *Origin of species* had made a tree-like illustration which expressed evolution. (Adapted from Karen Dowell 2008)

3. To be able to estimate the evolutionary date of divergence of organisms from their common ancestor.

Since it is thought that all organisms have arisen from the Last Universal Common Ancestor (LUCA), objectively there should be a single tree of life. However, it is

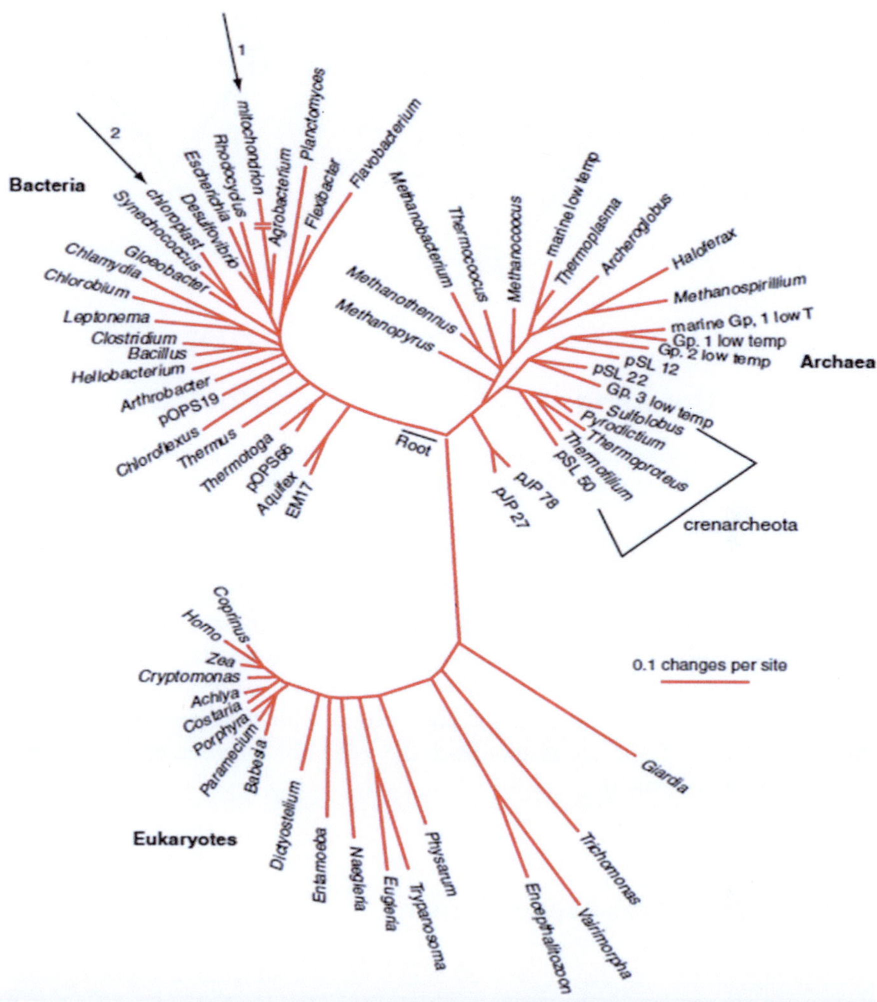

Fig. 22.14 16S rRNA tree of life. This is a rooted tree of life made by analyses of 16 s rRNA gene. It has three major branches—bacteria, archaea and eukaryotes. This is a phylogram (scale 0.1 changes per site) (explained in the next section) showing hypothetically how life originated 3.8 bya from primordial soup and diverged into various life forms. (Adapted from Pevsner 2009)

close to impossible to construct a true tree of life; rather, we construct 'inferred trees' which are based on the mutation in biomolecules or available data, showing hypothesized phylogenetic relationships.

The most popular tree of life has been constructed based on the phylogenetic analyses of 16 s rRNA gene (molecular marker). This tree has three branches showing the major divergence—bacteria, archaea and eukarya (Fig. 22.14).

A tree may be rooted showing the common or may be unrooted (Fig. 22.15) without the common ancestor. Often the data for the ancestors is not available that is

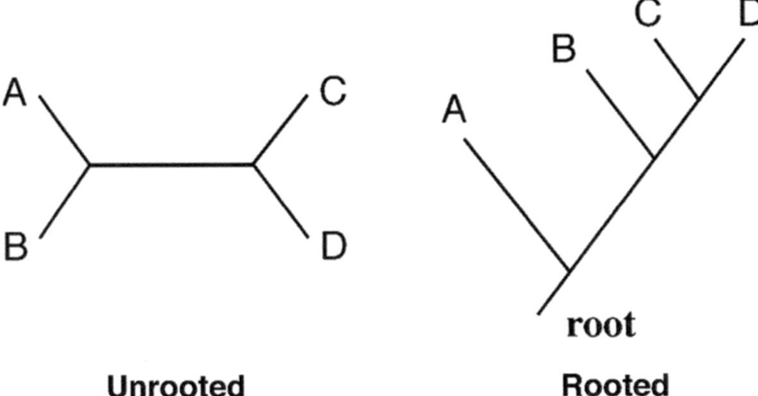

Fig. 22.15 A comparison between unrooted and rooted tree. The unrooted tree does not have any common ancestor. On the other hand, the rooted tree always shows the divergence from the common ancestor

when the unrooted tree is constructed which just shows the phylogenetic relationship among the organisms. A phylogenetic tree is not always constructed to observe the phylogeny of various species; it can also be constructed to chart the evolutionary path of the individual gene. Such a study is known as gene phylogeny. The evolutionary path of the gene might not overlap with the speciation events. A phylogenetic tree of species results from the evolution of genome (total genetic make-up of the organism).

22.8.2 Types of Tree Representation

The topology of the tree can be defined in two ways: cladogram and phylogram. A cladogram (Fig. 22.16) is a basic representative tree. It is a relative tree based on the order of phylogenetic events. The branch is unscaled and of the same length. Phylogram (Fig. 22.16) on the other hand has scaled branches. The branch represents the amount of evolution that has taken place since the time of divergence from the ancestor.

22.8.2.1 Clade

Clade is a group which includes the ancestor and the descendants. The tree also exhibits different type of branching pattern or the types of clade formed. There are three types of clade—monophyletic, paraphyletic and polyphyletic. Monophyletic clade includes the recent common and all its descendants. The paraphyletic clade excludes a few of the descendants, and polyphyletic clade includes distantly related species (OTU) (Fig. 22.17).

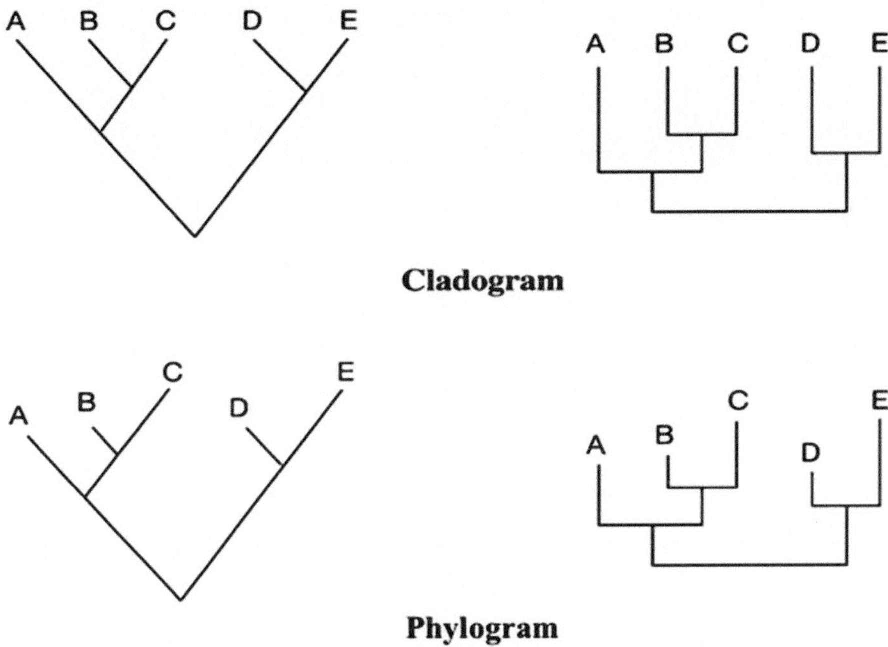

Fig. 22.16 Types of tree representation. Cladogram which has unscaled branches and phylogram which has scaled branches showing amount of evolution that has taken place. (Adapted from Jin Xiong 2006)

22.8.3 Procedure for Tree Construction

The phylogenetic tree construction involves the following steps (Fig. 22.18):

1. Choice of the molecular marker and collecting data.
2. Alignment of the data.
3. Choice of evolutionary model.
4. Constructing the phylogenetic tree.
5. Testing the reliability of the tree.

Let us look into the details of each of these steps:

1. Choice of the molecular marker and assembling data: Molecular marker is the biomolecule whose sequence would be taken into consideration to study the evolution. It may be nucleotide or protein sequence. The correct choice of molecular marker is an important step as it helps in construction of a true tree. If we are working with closely related organisms, then the rapidly evolving nucleotides should be the choice. For studying the evolution of slightly divergent organisms, the relatively conserved rRNA gene should be used. For more divergent organisms, protein sequences are used as they are relatively more conserved

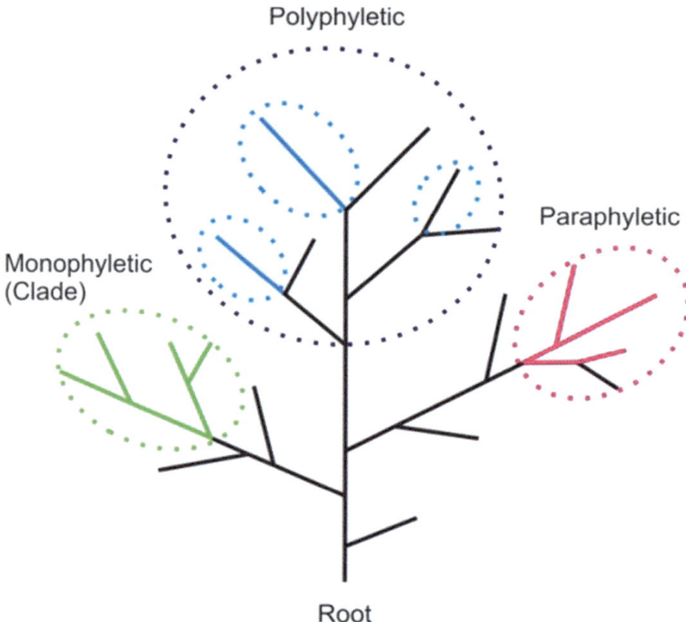

Fig. 22.17 The three types of clades. Monophyletic (green), polyphyletic (blue) and paraphyletic (pink). (Adapted from Karen Dowell 2008)

due to degeneracy of genetic code. DNA sequences are also biased than protein due to preferential usage of codon in some organisms. The protein has 20 amino acids as against only 4 bases of nucleotides and thus can be used for sensitive alignment. Globins are popularly used as molecular marker and were one of the first proteins to be sequenced. They are also used as molecular clocks (concept explained in earlier section).

Molecular marker can be used to spot positive and negative selection. For this it is important to distinguish between synonymous (results in no change in amino acid sequences) and non-synonymous substitution (results in change in amino acid). If non-synonymous substitution is higher, then it means a part of protein is undergoing evolution to bring about change in function of protein.

Once the molecular marker is chosen, the next step is to assemble the data of the organisms. For this there are several databases available from which the data can be extracted. For DNA the databases are DNA Data Bank of Japan (DDBJ), GenBank, etc. and for protein are SWISSPROT, etc. There are online tools like BLAST which can carry out the search and extract the data from databases.

2. Alignment of the data: Once the data is collected, the next step is to align these sequences according to homology in these sequences. The homology describes the phylogenetic relationships. There are two types of homologs—orthologs and paralogs. Orthologs are genes which have same ancestor but have diverged due to speciation event. Paralogs are duplicated genes of the same ancestor. Sequence

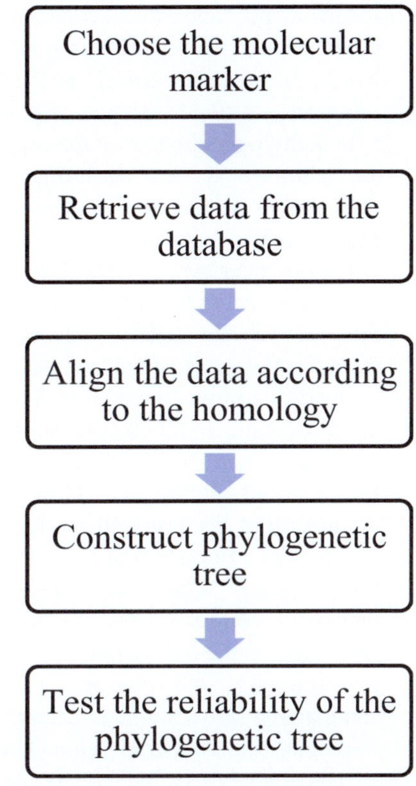

Fig. 22.18 Brief overview of the steps involved in phylogenetic tree construction. The flowchart enlists the various steps sequentially described in the text to construct a phylogenetic tree

alignment helps in the identification of homologous region and thus defines the evolutionary path. The multiple sequence alignment can be done by several tools—CLUSTAL, MSA, T-Coffee, etc.

3. Choice of evolutionary/substitution model: Substitution models are statistical methods to analyse the amount of evolution taking place. Several models are available for scoring the nucleotide substitution. One of the simplest models is Jukes-Cantor model which assumes that each nucleotide is replaced with equal probability. The other slightly more complex model is Kimura 2 parameter which differentiates between transition (mutation from purine to purine or pyrimidine to pyrimidine) and transversion (mutation from purine to pyrimidine and vice versa). In accordance with this model, transition occurs much more frequently than transversion, which is logical. For amino acid substitution, there are models like PAM. However these models assume that all positions in sequence have equal mutation rates. But this is not the case. For example, the wobble position of the codon mutates at a faster rate than others.
4. Construction of phylogenetic tree: There are two basic methods for tree construction: character based and distance based. The character-based method considers the molecular sequence as character, and after alignment each of these characters shares homology. It is also assumed that each of these characters evolves

independently and thus is considered separate evolutionary units. The character-based methods are maximum parsimony and maximum likelihood. On the other hand, the distance-based method calculates the dissimilarity between the sequences during sequence alignment and converts it into a matrix based on which phylogenetic tree of the organisms can be constructed. The branch lengths are additive, i.e. the evolutionary distance between two organisms can be obtained by adding the length of all the branches connecting them. The commonly used distance-based methods are unweighted pair group method using arithmetic average (UPGMA) and neighbour-joining method.

5. Testing the reliability of the tree: The last step of the phylogenetic tree construction involves statistically analysing the reliability of the inferred tree. This can be done through a statistical analysis model, bootstrapping. It tests for the sampling errors of the phylogenetic tree. This is done by repeated sampling of the datasets by introducing slight changes in them. If there is error in alignment, it will result in construction of biased tree. Random fluctuation in dataset will result in formation of altered tree. However if the alignment is correct, this random fluctuation in datasets will produce the same tree showing statistical confidence in the tree. Bootstrapping is of two types—parametric and non-parametric bootstrapping. When the changes introduced in the datasets are random, it is known as non-parametric bootstrapping. However when the new datasets are generated on the basis of particular sequential changes, it is known as parametric bootstrapping. All phylogenetic trees constructed after bootstrapping are summarized into a single consensus tree with each node of the branching pattern displaying the bootstrap value. This value evaluates the confidence level of the statistical analyses (Fig. 22.19) (Xiong 2006).

Box 22.1: Scientific Concept

Eweleit, L., Reinhold, K., Sauer, J.: Speciation Progress: A Case Study on the Bushcricket *Poecilimon veluchianus*. PLoS One 10(10): e0139494. https://doi.org/10.1371/journal.pone.0139494 (2015).

A speciation study was carried out on subspecies of the flightless bushcricket *Poecilimon veluchianus*, endemic to Central Greece. These are *P. v. veluchianus* and *P. v. minor* and are differentiated on body size, timing of male signalling and sperm transfer rate. The two subspecies are parapatrically distributed in a V-shaped zone in Central Greece. The Iti mountain could be a geographical barrier to gene flow. Also, as speciation is a long process that could span over centuries, it can be considered that fragmentations occurring previously could be barriers to gene flow. Laboratory experiments done earlier suggest that females do not differentiate between songs of the two subspecies and there is presence of partial post-zygotic isolation that diminishes fertility

(continued)

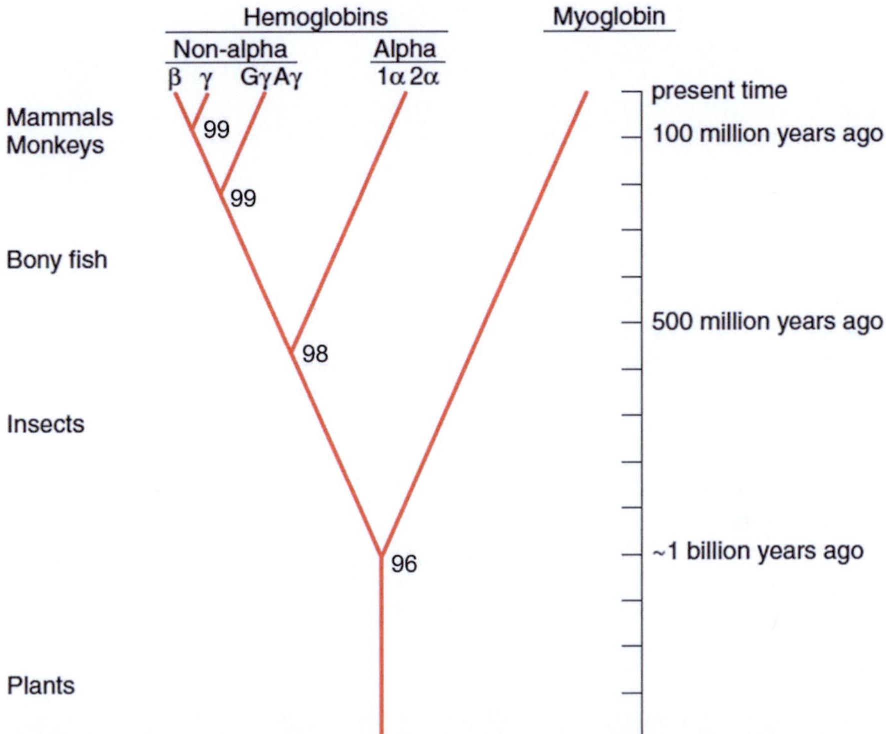

Fig. 22.19 A representative dendrogram. It is a representation of tree where branches have a scale showing evolutionary time showing evolution of globin family of genes (bootstrap values are shown at branching points)

Box 22.1 (continued)

of F1 by reduction in amount of sperm transfer. Also, F1 female hybrids were mostly fertile compared to males with lower sperm count. Hence, there is evidence for presence of 'premating barriers vs post-mating barriers' and it indicates that speciation is an ongoing process.

Genetic differences between the subspecies were evaluated in this study using the sequences of mitochondrial control region (CR marker) and internal transcribed spacers (ITS marker 1 and 2). As mentioned earlier, due to absence of premating isolating mechanisms, hybridization is possible in the area of contact zone that would encourage gene flow in comparison to the distantly located sites. The occurrence of shared haplotypes was investigated in the contact zone vs the distant sites to predict the site of hybridization. Single site substitution differences were found in between various haplotypes. Also to

(continued)

Box 22.1 (continued)
determine if geographic isolation has shaped the population structure, a distance-based redundancy analysis (dbDA) was done.

The results demonstrated that the aligned sequences of the CR dataset contained 794 bp and the ITS dataset consisted of 706 bp and both showed high levels of genetic variation especially due to low number of variable sites. A characteristic of *Poecilimon* species is a high number of exclusive haplotypes occurring at one of the different sampling sites in the mitochondrial CR marker. In contrast there was lesser diversity in the nuclear ITS marker that displayed a lower genetic variation.

The genetic analysis done based on the ITS marker revealed one main barrier to gene flow, hence indicating incomplete reproductive isolation. The contact zone has been proposed to extend from north-east of Central Greece to the south-west. The CR marker in contrast does not clearly support the speciation with formation of two subspecies, restricted gene flow and a clear contact zone.

No influence of sex was observed on the genetic pattern for *P. veluchianus* on investigation with dbRDA. But on testing for isolation by distance (IBD) for both sexes, 19% variability was found in females versus 10% in males. Hence, it can be concluded that IBD has a stronger impact on females. Males produce sounds prior to mating, waiting for the females, and this result indicates that females need to walk around to locate the correct partner.

Individuals of both subspecies are tougher to distinguish in the field as they are phenotypically similar with the exception of body size. This feature depends on the size of the mother, and hence the hybrids' body size depends on the identity of the mothers' subspecies. Laboratory experiments have revealed that hybrids with a *P. v. veluchianus* mother grow bigger than pure *P. v. minor* individuals and hybrids with a *P. v. minor* mother stay smaller than pure *P. v. veluchianus* individuals.

Speciation is in progress for these subspecies, as there is a lack of a strong prezygotic isolation barrier between these two parapatrically distributed subspecies. It could also be predicted that the species experienced a bottleneck and are now in a phase of range expansion. IBD and sexual selection have shown to have a great influence on the population structure, and it has been hypothesized that *P. veluchianus* may be a case of widely distributed ring species. To investigate this further, closely related species of *Poecilimon* are necessary. Both subspecies are also distributed in various altitudes with *P. v. veluchianus* occurring above 380 m and *P. v. minor* occurring also below this altitude, and the lack of a strong prezygotic barrier probably supports a secondary contact zone. It is suggested that a secondary contact after an allopatric phase is likely for the two subspecies. The missing premating barrier suggests a rather weak selection against hybrids and might also indicate speciation in progress. Further study using microsatellite data as well as AFLP could shed light on this ongoing speciation process.

22.9 Summary

- Evolutionary genetics is the modern field of study which integrates genetics with the Darwinian view of evolution. It attempts to account for any change in nature in terms of allele, genes and genotypes and how the variations at population level can bring permanent variations in the species leading from microevolutionary to macroevolutionary changes.
- Mutation, natural selection, genetic drift and migration are the microevolutionary changes. Mutation is the most important variation which acts as the raw material for the evolution of the gene pool. Most of the mutations are neutral, but some are positive which might improve the fitness of the organism in its environment resulting in adaptive evolution. The other variations may occur at the chromosomal level through recombination and aberrations.
- The Hardy-Weinberg law was introduced to understand how these variations affect the allele and genotype frequencies in a population. However, Hardy-Weinberg law functions under ideal conditions in an infinite population in absence of evolutionary forces. But in a real, finite population, the evolutionary forces like natural selection, genetic drift and migration act to affect the variation in gene pool every generation.
- Natural selection is the force which results in the adaptation of the fittest in the environment. It is directional in nature and acts on a large population. Selection can occur at individual level, population level or sexual level.
- On the contrary genetic drift is non-adaptive and results in random fixation of allele in a small population. Migration or gene flow also affects the variation in a gene pool. Magnification of these variations over a long period of time will lead to speciation.
- Earlier evolution was studied through fossil record. However fossil records were often incomplete leaving a number of question marks. With the onset of new technology and development of molecular biology techniques, the field of molecular phylogeny gained momentum. Using biomolecules and genes as markers, phylogenetic tree can be constructed which gives a bird-eye's view of the phylogenetic relationship among organisms.
- Human evolution studies have also been carried out using the mitochondrial DNA which has helped chart out the divergence of **Homo neanderthalensis** and **Homo sapiens**.

References

Abzhanov A (2010) Darwin's Galapagos finches in modern biology. Philo Trans R Soc Lond B Biol Sci 365(1543):1001–1007

Abzhanov A, Protas M, Grant BR, Grant PR, Tabin CJ (2004) Bmp4 and morphological variation of beaks in Darwin's finches. Science 305(5689):1462–1465

Agar W, Drummond F, Tiegs O, Gunson M (1954) Fourth (final) report on a test of McDougall's Lamarckian experiment on the training of rats. J Exp Biol 31(3):307–321

Alcock J (2005) Animal behaviour: an evolutionary approach, 8th edn. Sinaeur Associates, Sunderland, MA

Anstee DJ (2010) The relationship between blood groups and disease. Blood 115:4635–4643

Bluestone CD (2009) Galapagos: Darwin, evolution and ENT. Laryngoscope 119(10):1902–1905

Bowman RI (1961) Morphological differentiation & adaptation in Galapagos finches. Univ Calif Publ Zool 58:1–302

Brown TA (2002) Molecular phylogenetics. genomes, 2nd edn. Wiley-Liss, U.K, Oxford

Byers JA, Waits L (2006) Good genes sexual selection in nature. PNAS 103(44):16343–16345

Campbell NA, Reece JB, Urry LA, Cain ML, Wasserman SA, Minorsky PV, Jackson RB (2008) Biology, 8th edn. Pearson Education Inc., San Francisco, CA

Cook SA, Johnson MP (1968) Adaptation to heterogeneous environments. I. Variation in heterophylly in *Ranunculus flammula*. Evolution 22:496–516

Cook LM, Saccheri IJ (2013) The peppered moth and industrial melanism: evolution of a natural selection case study. Heredity 110(3):207–212

Crew F (1936) A repetition of McDougall's Lamarckian experiment. J Genet 33(1):61–102

Denis N (2011) Neo-Darwinism, the modern synthesis and selfish genes: are they of use in physiology? J Physiol 589(5):1007–1015

Dowell K (2008) Molecular Phylogenetics: an introduction to computational methods and tools for analysing evolutionary relationships. http://www.math.umaine.edu/~khalil/courses/MAT500/papers/MAT500_Paper_Phylogenetics.pdf

Drew G (1939) McDougall's experiment of the inheritance of acquired habits. Nature 143:188–191

Duret L (2008) Neutral theory: the null hypothesis of molecular evolution. Nat Educ 1(1):218

Emerling CA (2016) Will evolution doom the cheetah? Understanding Evolution https://evolutionberkeleyedu/evolibrary/news/160201_cheetahs

Fitzpatrick BM, Fordyce JA, Gavrilets S (2009) Pattern, process & geographic modes of speciation. J Evol Biol 22:2342–2347

Forsdyke DR (2007) Positive Darwinian selection: does the comparative method rule? JBS 15:95–108

Futuyma DJ (1998) Evolution, 3rd edn. Oxford University Press, Boston, MA

Hall BK, Hallgrimsson B (2008) Strickberger's evolution, 4th edn. Jones & Bartlett Publishers, LLC, Burlington, MA

Hartl DL, Jones EW (2009) Genetics: analysis of genes & genomes, 7th edn. Jones & Bartlett Publishers, LLC, Burlington, MA

Jankowska D, Milewski R, Górska U, Milewska AJ (2011) Application of Hardy-Weinberg law in biomedical research. Stud InLog Gramm Rhetor 25(38)

Karn MN, Penrose LS (1952) Birth rate and gestation time in relation to maternal age, parity and infant survival. Ann Eugenics 16:147–164

Keller B, Vos JM, Schmidt-Lebuhn AM, Thomson JD, Conti E (2016) Both morph- and species-dependent asymmetries affect reproductive barriers between heterostylous species. Ecol Evol 6(17):6223–6244

Kimura M (1991) The neutral theory of molecular evolution: a review of recent evidence. Jpn J Genet 66(4):367–386

Klugs WS, Cummings MR, Spencer CA, Palladino MA (2012) Concepts of genetics, 10th edn. Pearson Education Inc., San Francisco, CA

Koonin EV, Wolf YI (2009) Is evolution Darwinian or/and Lamarckian? Biol Direct 4(42):1–14

Kramer J, Meunier J (2016) Kin and multilevel selection in social evolution: a never-ending controversy? F1000 Res 5:Faculty Rev-776. https://doi.org/10.12688/f1000research.8018.1

Łukasiewicz A, Szubert-Kruszyńska A, Radwan J (2017) Kin selection promotes female productivity and cooperation between the sexes. Evol Biol 3(3):e1602262

Marshall JH (2002) On the changing means of mutation. Hum Mutat 19:76–78

McDougall W (1938) Fourth report on Lamarckian experiment. Br J Psychol 28(3):321–325

Mourant AE, Kopeć AC, Domaniewska-Sobczak K (1976) The distribution of the human blood groups and other polymorphisms, 2nd edn. Oxford University Press, London

Noor MAF, Feder JLF (2006) Speciation genetics: evolving approaches. Nat Rev Genet 7:851–861

Orr HA (2009) Fitness and its role in evolutionary genetics. Nat Rev Genet 10(8):531–539

Parent CE, Caccone A, Petren K (2008) Colonization and diversification of Galapagos terrestrial fauna: a phylogenetic & biogeographical synthesis. Philo Trans R Soc Lond B Biol Sci 363(1508):3347–3361

Peltonen L (2001) Founder effect. https://www.sciencedirect.com/sdfe/pdf/download/eid/3-s2.0-B0122270800004742/first-page-pdf

Pevsner J (2009) Bioinformatics and functional genomics, 2nd edn. John Wiley & Sons, Hoboken, NJ

Pierce BA (2012) Genetics: a conceptual approach, 4th edn. W.H. Freeman & Company, New York

Portin P (2007) Evolution of man in the light of molecular genetics: a review. Part I. our evolutionary history and genomics. Hereditas 144(3):80–95

Raymond MM, O'Brien SJ (1993) Dating the genetic bottleneck of the African cheetah. PNAS 90(8):3172–3176

Ridley M (2004) Evolution, 3rd edn. Blackwell Publishing, Malden, MA

Roesti M, Hendry AP, Salzburger W, Berner D (2012) Genome divergence during evolutionary diversification as revealed in replicate lake–stream stickleback population pairs. Mol Ecol 21: 2852–2862

Sato A, O'hUigin C, Figueroa F, Grant PR, Grant R, Tichy H, Klein J (1999) Phylogeny of Darwin's finches as revealed by mtDNA sequences. PNAS 96(9):5101–5106

Schliewen UK, Tautz D, Pääbo S (1994) Sympatric speciation suggested by monophyly of crater lake cichlids. Nature 368:629–632

Snustad DP, Simmons MJ (2012) Principles of genetics, 6th edn. John Wiley & Sons, Inc., Hoboken, NJ

Turissini DA, McGirr JA, Patel SS, David JR, Matute DR (2018) The rate of evolution of post mating-prezygotic reproductive isolation in drosophila. Mol Biol Evol 35(2):312–334

Wade M Evolutionary genetics. In: Zalta EN (ed) The Stanford encyclopedia of philosophy (Fall 2008 edition). https://plato.stanford.edu/archives/fall2008/entries/evolutionary-genetics/

Weismann A (1891) A supposed transmission of mutilations. In: Essays upon heredity and kindred biological problems. Oxford University Press, Oxford

Weismann A (1893) The germplasm: a theory of heredity. Charles Scribner's and Sons, New York

Wolf JBW, Lindell J, Backstrom N (2010) Introduction speciation genetics: current status and evolving approaches. Phil Trans R Soc B 365:1717–1733

Wu DD, Zhang Y (2008) Positive Darwinian selection in human population: a review. Chin Sci Bull 53(10):1457–1467

Xiong J (2006) Essential bioinformatics, 1st edn. Cambridge University Press, New York

Yi S (2013) Neutrality and molecular clocks. Nat Educ Knowl 4(2):3

Zhang J (2010) Positive darwinian selection in gene evolution. https://pdfs.semanticscholar.org/6c91/7d8d705af18a3d0e5e139076b5335f046f6c.pdf

Zhang J, Zhang YP, Rosenberg HF (2002) Adaptive evolution of a duplicated pancreatic ribonuclease gene in a leaf eating monkey. Nat Genet 30:411–415

Zykova TY, Levitsky VG, Belyaeva ES, Zhimulev IF (2018) Polytene chromosomes-a portrait of functional organization of the *drosophila* genome. Curr Genomics 19(3):179–191

MIX
Papier aus verantwortungsvollen Quellen
Paper from responsible sources
FSC® C105338

If you have any concerns about our products,
you can contact us on
ProductSafety@springernature.com

In case Publisher is established outside the EU,
the EU authorized representative is:
Springer Nature Customer Service Center GmbH
Europaplatz 3, 69115 Heidelberg, Germany

Printed by Libri Plureos GmbH
in Hamburg, Germany